CEREBRAL ASYMMETRIES

HANDBOOK OF CLINICAL NEUROLOGY

Series Editors

MICHAEL J. AMINOFF, FRANÇOIS BOLLER, AND DICK F. SWAAB

VOLUME 208

ELSEVIER

CEREBRAL ASYMMETRIES

Series Editors

MICHAEL J. AMINOFF, FRANÇOIS BOLLER, AND DICK F. SWAAB

Volume Editors

COSTANZA PAPAGNO AND PAUL CORBALLIS

VOLUME 208

3rd Series

ELSEVIER

ELSEVIER
Radarweg 29, PO Box 211, 1000 AE Amsterdam, Netherlands
125 London Wall, London EC2Y 5AS, United Kingdom
50 Hampshire Street, 5th Floor, Cambridge, MA 02139, United States

ISBN: 978-0-443-15646-5

Printed in Great Britain

Last digit is the print number: 10 9 8 7 6 5 4 3 2

Publisher: Stacy Masucci
Acquisitions editor: Megan McManus
Editorial Project Manager: Kristi Anderson
Production Project Manager: Omer Mukthar
Cover Designer: Christian Bilbow

Typeset by STRAIVE, India

In Memoriam

Michael Corballis

Emeritus Professor Michael Corballis passed away while this volume on cerebral asymmetries—for which he had agreed to serve as volume editor—was being developed and he helped in its planning. Michael was a preeminent scholar in human cognitive neuroscience and neuropsychology. In a career that spanned seven decades, he made significant contributions to various research areas, including perception, attention, memory and mental time travel, language origins, recursive thought, and mind wandering. Central to his endeavors were the questions of human uniqueness and the role of hemispheric asymmetry in enabling and shaping the human mind.

Michael was born on September 10, 1936. His university education coincided with the foundation of academic departments of psychology in his native New Zealand, and with the evolution and expansion of neuropsychology internationally. He was among the first students to study psychology at Victoria University (Wellington) and the University of Auckland before moving to Montreal for his doctoral training at McGill University. Montreal had emerged as a hub of neuropsychologic thought, and the notion of grounding psychology in the anatomy and physiology of the brain inspired Michael's early interest in how cerebral lateralization might underlie asymmetries in perception and action.

Cerebral asymmetry remained a key focus after Michael returned to New Zealand in 1977. He was particularly interested in how apparently unrelated asymmetries might have common origins or lead to unexpected consequences. In *The Lopsided Ape* (1991), he reflected on the strong lateralization of both language and fine motor control, and advanced the hypothesis that language may have its origin in gesture rather than speech—an idea that would motivate much of his thinking for the next three decades.

Michael authored or coauthored more than 300 scientific articles, 14 books, and many commentaries, published reviews, and public-interest pieces. His status as a scholar was recognized through numerous fellowships and awards, including the New Zealand Order of Merit in 2002 and the Rutherford Medal from the Royal Society of New Zealand in 2016.

Aside from his prolific scholarship, Michael will be remembered for his urbane wit, humility, and generosity of spirit. He passed away peacefully on November 13, 2021, following a brief battle with cancer. He was predeceased by Barbara, his wife of 58 years, and is survived by his sons Paul and Timothy, and three granddaughters, Simone, Natasha, and Lena.

Photo by Paul Corballis.

Michael Aminoff
Francois Boller
Paul Corballis
Dick F. Swaab
September 2024

Handbook of Clinical Neurology 3rd Series

Available titles

Vol. 79, The human hypothalamus: Basic and clinical aspects, Part I, D.F. Swaab, ed. ISBN 9780444513571
Vol. 80, The human hypothalamus: Basic and clinical aspects, Part II, D.F. Swaab, ed. ISBN 9780444514905
Vol. 81, Pain, F. Cervero and T.S. Jensen, eds. ISBN 9780444519016
Vol. 82, Motor neurone disorders and related diseases, A.A. Eisen and P.J. Shaw, eds. ISBN 9780444518941
Vol. 83, Parkinson's disease and related disorders, Part I, W.C. Koller and E. Melamed, eds. ISBN 9780444519009
Vol. 84, Parkinson's disease and related disorders, Part II, W.C. Koller and E. Melamed, eds. ISBN 9780444528933
Vol. 85, HIV/AIDS and the nervous system, P. Portegies and J. Berger, eds. ISBN 9780444520104
Vol. 86, Myopathies, F.L. Mastaglia and D. Hilton Jones, eds. ISBN 9780444518996
Vol. 87, Malformations of the nervous system, H.B. Sarnat and P. Curatolo, eds. ISBN 9780444518965
Vol. 88, Neuropsychology and behavioural neurology, G. Goldenberg and B.C. Miller, eds. ISBN 9780444518972
Vol. 89, Dementias, C. Duyckaerts and I. Litvan, eds. ISBN 9780444518989
Vol. 90, Disorders of consciousness, G.B. Young and E.F.M. Wijdicks, eds. ISBN 9780444518958
Vol. 91, Neuromuscular junction disorders, A.G. Engel, ed. ISBN 9780444520081
Vol. 92, Stroke – Part I: Basic and epidemiological aspects, M. Fisher, ed. ISBN 9780444520036
Vol. 93, Stroke – Part II: Clinical manifestations and pathogenesis, M. Fisher, ed. ISBN 9780444520043
Vol. 94, Stroke – Part III: Investigations and management, M. Fisher, ed. ISBN 9780444520050
Vol. 95, History of neurology, S. Finger, F. Boller and K.L. Tyler, eds. ISBN 9780444520081
Vol. 96, Bacterial infections of the central nervous system, K.L. Roos and A.R. Tunkel, eds. ISBN 9780444520159
Vol. 97, Headache, G. Nappi and M.A. Moskowitz, eds. ISBN 9780444521392
Vol. 98, Sleep disorders Part I, P. Montagna and S. Chokroverty, eds. ISBN 9780444520067
Vol. 99, Sleep disorders Part II, P. Montagna and S. Chokroverty, eds. ISBN 9780444520074
Vol. 100, Hyperkinetic movement disorders, W.J. Weiner and E. Tolosa, eds. ISBN 9780444520142
Vol. 101, Muscular dystrophies, A. Amato and R.C. Griggs, eds. ISBN 9780080450315
Vol. 102, Neuro-ophthalmology, C. Kennard and R.J. Leigh, eds. ISBN 9780444529039
Vol. 103, Ataxic disorders, S.H. Subramony and A. Durr, eds. ISBN 9780444518927
Vol. 104, Neuro-oncology Part I, W. Grisold and R. Sofietti, eds. ISBN 9780444521385
Vol. 105, Neuro-oncology Part II, W. Grisold and R. Sofietti, eds. ISBN 9780444535023
Vol. 106, Neurobiology of psychiatric disorders, T. Schlaepfer and C.B. Nemeroff, eds. ISBN 9780444520029
Vol. 107, Epilepsy Part I, H. Stefan and W.H. Theodore, eds. ISBN 9780444528988
Vol. 108, Epilepsy Part II, H. Stefan and W.H. Theodore, eds. ISBN 9780444528995
Vol. 109, Spinal cord injury, J. Verhaagen and J.W. McDonald III, eds. ISBN 9780444521378
Vol. 110, Neurological rehabilitation, M. Barnes and D.C. Good, eds. ISBN 9780444529015
Vol. 111, Pediatric neurology Part I, O. Dulac, M. Lassonde and H.B. Sarnat, eds. ISBN 9780444528919
Vol. 112, Pediatric neurology Part II, O. Dulac, M. Lassonde and H.B. Sarnat, eds. ISBN 9780444529107
Vol. 113, Pediatric neurology Part III, O. Dulac, M. Lassonde and H.B. Sarnat, eds. ISBN 9780444595652
Vol. 114, Neuroparasitology and tropical neurology, H.H. Garcia, H.B. Tanowitz and O.H. Del Brutto, eds. ISBN 9780444534903
Vol. 115, Peripheral nerve disorders, G. Said and C. Krarup, eds. ISBN 9780444529022
Vol. 116, Brain stimulation, A.M. Lozano and M. Hallett, eds. ISBN 9780444534972
Vol. 117, Autonomic nervous system, R.M. Buijs and D.F. Swaab, eds. ISBN 9780444534910
Vol. 118, Ethical and legal issues in neurology, J.L. Bernat and H.R. Beresford, eds. ISBN 9780444535016
Vol. 119, Neurologic aspects of systemic disease Part I, J. Biller and J.M. Ferro, eds. ISBN 9780702040863
Vol. 120, Neurologic aspects of systemic disease Part II, J. Biller and J.M. Ferro, eds. ISBN 9780702040870
Vol. 121, Neurologic aspects of systemic disease Part III, J. Biller and J.M. Ferro, eds. ISBN 9780702040887
Vol. 122, Multiple sclerosis and related disorders, D.S. Goodin, ed. ISBN 9780444520012
Vol. 123, Neurovirology, A.C. Tselis and J. Booss, eds. ISBN 9780444534880
Vol. 124, Clinical neuroendocrinology, E. Fliers, M. Korbonits and J.A. Romijn, eds. ISBN 9780444596024
Vol. 125, Alcohol and the nervous system, E.V. Sullivan and A. Pfefferbaum, eds. ISBN 9780444626196
Vol. 126, Diabetes and the nervous system, D.W. Zochodne and R.A. Malik, eds. ISBN 9780444534804
Vol. 127, Traumatic brain injury Part I, J.H. Grafman and A.M. Salazar, eds. ISBN 9780444528926
Vol. 128, Traumatic brain injury Part II, J.H. Grafman and A.M. Salazar, eds. ISBN 9780444635211
Vol. 129, The human auditory system: Fundamental organization and clinical disorders, G.G. Celesia and G. Hickok, eds. ISBN 9780444626301
Vol. 130, Neurology of sexual and bladder disorders, D.B. Vodušek and F. Boller, eds. ISBN 9780444632470

Vol. 131, Occupational neurology, M. Lotti and M.L. Bleecker, eds. ISBN 9780444626271
Vol. 132, Neurocutaneous syndromes, M.P. Islam and E.S. Roach, eds. ISBN 9780444627025
Vol. 133, Autoimmune neurology, S.J. Pittock and A. Vincent, eds. ISBN 9780444634320
Vol. 134, Gliomas, M.S. Berger and M. Weller, eds. ISBN 9780128029978
Vol. 135, Neuroimaging Part I, J.C. Masdeu and R.G. González, eds. ISBN 9780444534859
Vol. 136, Neuroimaging Part II, J.C. Masdeu and R.G. González, eds. ISBN 9780444534866
Vol. 137, Neuro-otology, J.M. Furman and T. Lempert, eds. ISBN 9780444634375
Vol. 138, Neuroepidemiology, C. Rosano, M.A. Ikram and M. Ganguli, eds. ISBN 9780128029732
Vol. 139, Functional neurologic disorders, M. Hallett, J. Stone and A. Carson, eds. ISBN 9780128017722
Vol. 140, Critical care neurology Part I, E.F.M. Wijdicks and A.H. Kramer, eds. ISBN 9780444636003
Vol. 141, Critical care neurology Part II, E.F.M. Wijdicks and A.H. Kramer, eds. ISBN 9780444635990
Vol. 142, Wilson disease, A. Członkowska and M.L. Schilsky, eds. ISBN 9780444636003
Vol. 143, Arteriovenous and cavernous malformations, R.F. Spetzler, K. Moon and R.O. Almefty, eds. ISBN 9780444636409
Vol. 144, Huntington disease, A.S. Feigin and K.E. Anderson, eds. ISBN 9780128018934
Vol. 145, Neuropathology, G.G. Kovacs and I. Alafuzoff, eds. ISBN 9780128023952
Vol. 146, Cerebrospinal fluid in neurologic disorders, F. Deisenhammer, C.E. Teunissen and H. Tumani, eds. ISBN 9780128042793
Vol. 147, Neurogenetics Part I, D.H. Geschwind, H.L. Paulson and C. Klein, eds. ISBN 9780444632333
Vol. 148, Neurogenetics Part II, D.H. Geschwind, H.L. Paulson and C. Klein, eds. ISBN 9780444640765
Vol. 149, Metastatic diseases of the nervous system, D. Schiff and M.J. van den Bent, eds. ISBN 9780128111611
Vol. 150, Brain banking in neurologic and psychiatric diseases, I. Huitinga and M.J. Webster, eds. ISBN 9780444636393
Vol. 151, The parietal lobe, G. Vallar and H.B. Coslett, eds. ISBN 9780444636225
Vol. 152, The neurology of HIV infection, B.J. Brew, ed. ISBN 9780444638496
Vol. 153, Human prion diseases, M. Pocchiari and J.C. Manson, eds. ISBN 9780444639455
Vol. 154, The cerebellum: From embryology to diagnostic investigations, M. Manto and T.A.G.M. Huisman, eds. ISBN 9780444639561
Vol. 155, The cerebellum: Disorders and treatment, M. Manto and T.A.G.M. Huisman, eds. ISBN 9780444641892
Vol. 156, Thermoregulation: From basic neuroscience to clinical neurology Part I, A.A. Romanovsky, ed. ISBN 9780444639127
Vol. 157, Thermoregulation: From basic neuroscience to clinical neurology Part II, A.A. Romanovsky, ed. ISBN 9780444640741
Vol. 158, Sports neurology, B. Hainline and R.A. Stern, eds. ISBN 9780444639547
Vol. 159, Balance, gait, and falls, B.L. Day and S.R. Lord, eds. ISBN 9780444639165
Vol. 160, Clinical neurophysiology: Basis and technical aspects, K.H. Levin and P. Chauvel, eds. ISBN 9780444640321
Vol. 161, Clinical neurophysiology: Diseases and disorders, K.H. Levin and P. Chauvel, eds. ISBN 9780444641427
Vol. 162, Neonatal neurology, L.S. De Vries and H.C. Glass, eds. ISBN 9780444640291
Vol. 163, The frontal lobes, M. D'Esposito and J.H. Grafman, eds. ISBN 9780128042816
Vol. 164, Smell and taste, Richard L. Doty, ed. ISBN 9780444638557
Vol. 165, Psychopharmacology of neurologic disease, V.I. Reus and D. Lindqvist, eds. ISBN 9780444640123
Vol. 166, Cingulate cortex, B.A. Vogt, ed. ISBN 9780444641960
Vol. 167, Geriatric neurology, S.T. DeKosky and S. Asthana, eds. ISBN 9780128047668
Vol. 168, Brain-computer interfaces, N.F. Ramsey and J. del R. Millán, eds. ISBN 9780444639349
Vol. 169, Meningiomas, Part I, M.W. McDermott, ed. ISBN 9780128042809
Vol. 170, Meningiomas, Part II, M.W. McDermott, ed. ISBN 9780128221983
Vol. 171, Neurology and pregnancy: Pathophysiology and patient care, E.A.P. Steegers, M.J. Cipolla and E.C. Miller, eds. ISBN 9780444642394
Vol. 172, Neurology and pregnancy: Neuro-obstetric disorders, E.A.P. Steegers, M.J. Cipolla and E.C. Miller, eds. ISBN 9780444642400
Vol. 173, Neurocognitive development: Normative development, A. Gallagher, C. Bulteau, D. Cohen and J.L. Michaud, eds. ISBN 9780444641502
Vol. 174, Neurocognitive development: Disorders and disabilities, A. Gallagher, C. Bulteau, D. Cohen and J.L. Michaud, eds. ISBN 9780444641489
Vol. 175, Sex differences in neurology and psychiatry, R. Lanzenberger, G.S. Kranz, and I. Savic, eds. ISBN 9780444641236
Vol. 176, Interventional neuroradiology, S.W. Hetts and D.L. Cooke, eds. ISBN 9780444640345
Vol. 177, Heart and neurologic disease, J. Biller, ed. ISBN 9780128198148
Vol. 178, Neurology of vision and visual disorders, J.J.S. Barton and A. Leff, eds. ISBN 9780128213773
Vol. 179, The human hypothalamus: Anterior region, D.F. Swaab, F. Kreier, P.J. Lucassen, A. Salehi and R.M. Buijs, eds. ISBN 9780128199756
Vol. 180, The human hypothalamus: Middle and posterior region, D.F. Swaab, F. Kreier, P.J. Lucassen, A. Salehi and R.M. Buijs, eds. ISBN 9780128201077
Vol. 181, The human hypothalamus: Neuroendocrine disorders, D.F. Swaab, R.M. Buijs, P.J. Lucassen, A. Salehi and F. Kreier, eds. ISBN 9780128206836
Vol. 182, The human hypothalamus: Neuropsychiatric disorders, D.F. Swaab, R.M. Buijs, F. Kreier, P.J. Lucassen, and A. Salehi, eds. ISBN 9780128199732
Vol. 183, Disorders of emotion in neurologic disease, K.M. Heilman and S.E. Nadeau, eds. ISBN 9780128222904

Vol. 184, Neuroplasticity: From bench to bedside, A. Quartarone, M.F. Ghilardi, and F. Boller, eds. ISBN 9780128194102
Vol. 185, Aphasia, A.E. Hillis and J. Fridriksson, eds. ISBN 9780128233849
Vol. 186, Intraoperative neuromonitoring, M.R. Nuwer and D.B. MacDonald, eds. ISBN 9780128198261
Vol. 187, The temporal lobe, G. Miceli, P. Bartolomeo, and V. Navarro, eds. ISBN 9780128234938
Vol. 188, Respiratory neurobiology: Physiology and clinical disorders, Part I, R. Chen and P.G. Guyenet, eds. ISBN 9780323915342
Vol. 189, Respiratory neurobiology: Physiology and clinical disorders, Part II, R. Chen and P.G. Guyenet, eds. ISBN 9780323915328
Vol. 190, Neuropalliative care: Part I, J.M. Miyasaki and B.M. Kluger, eds. ISBN 9780323850292
Vol. 191, Neuropalliative care: Part II, J.M. Miyasaki and B.M. Kluger, eds. ISBN 9780128245354
Vol. 192, Precision medicine in neurodegenerative disorders: Part I, A.J. Espay, ed. ISBN 9780323855389
Vol. 193, Precision medicine in neurodegenerative disorders: Part II, A.J. Espay, ed. ISBN 9780323855556
Vol. 194, Mitochondrial diseases, R. Horvath, M. Hirano, and P.F. Chinnery, eds. ISBN 9780128217511
Vol. 195, Motor system disorders, Part I: Normal physiology and function and neuromuscular disorders, D.S. Younger, ed. ISBN 9780323988186
Vol. 196, Motor system disorders, Part II: Spinal cord, neurodegenerative, and cerebral disorders and treatment, D.S. Younger, ed. ISBN 9780323988179
Vol. 197, Brain and crime, H. Swaab and G. Meynen, eds. ISBN 9780128213759
Vol. 198, Migraine biology, diagnosis, and co-morbidities, J.W. Swanson and M. Matharu, eds. ISBN 9780128233566
Vol. 199, Migraine management, J.W. Swanson and M. Matharu, eds. ISBN 9780128233573
Vol. 200, Paraneoplastic neurological disorders, B. Giometto and S.J. Pittock, eds. ISBN 9780128239124
Vol. 201, Hematopoietic stem cell transplantation for neurological diseases, M. Inglese and G.L. Mancardi, eds. ISBN 9780323902427
Vol. 202, Focal neuropathies, C. Chalk, ed. ISBN 9780323901086
Vol. 203, Neurological channelopathies, M.G. Hanna, ed. ISBN 9780323908207
Vol. 204, Inherited white matter disorders and their mimics, D.S. Lynch and H. Houlden, eds. ISBN 9780323992091
Vol. 205, Cell and gene therapies for neurologic diseases, H.J. Federoff and R.A. Barker, eds. ISBN 9780323901208
Vol. 206, Circadian rhythms, sleep, and sleep disorders, L. Ferini-Strambi and C. Cajochen, eds. ISBN 9780323909181
Vol. 207, Coma, stupor, and related disorders of consciousness, P.W. Kaplan and R.G. Geocadin, eds. ISBN 9780443134081

All volumes in the 3rd Series of the **Handbook of Clinical Neurology** are published electronically, on Science Direct: http://www.sciencedirect.com/science/handbooks/00729752.

Foreword

We have two hands, we have two feet, and we also have two cerebral hemispheres. Functional differences between the limbs (handedness) have been recognized since antiquity, but recognition of a functional asymmetry between the left and the right cerebral hemispheres took a long time in coming. It took even longer to recognize that there are anatomic differences as well. As recently as 1962, a review by the distinguished neuroanatomist Gerhardt von Bonin claimed that there were no significant anatomic differences between the two hemispheres. Actually, we now know that asymmetry extends not only to other animal species but also to the entire living world, and lateralization of the brain is found in both vertebrates and invertebrates, suggesting that it provides significant fitness for life. In his introductory chapter in the present volume, Professor Giovanni Berlucchi reminds us that Primo Levi, a monumental writer and chemist (1919–1987), since he was a student, had been attracted to chiral molecules like the amino amides, of which only the levorotatory enantiomers build all biologically active proteins. Levi argued that the emergence of chiral molecules may go back to the origins of life and may even constitute the primary basis of left–right asymmetries in living organisms. Levi concluded with a famous aphorism "Asymmetry is life."

It is now clear that functional and structural differences exist between the two parts of the brain. Each is dominant in processing specific cognitive tasks, and this has considerable clinical significance. On this basis, the series editors of the *Handbook of Clinical Neurology* decided to have a volume dedicated to Cerebral Asymmetries, and we approached one of the leaders in the field, Professor Michael Corballis, to serve as volume editor. It was agreed that he would be joined by a neurologist, Professor Costanza Papagno, former Director of the Department of Psychology at the University of Milano-Bicocca and now Head of a Neurocognitive Rehabilitation Center at the University of Trento (Italy). Then, just as the table of contents had been decided and authors were being invited, Michael Corballis tragically passed away. His life and achievements are presented in an *In Memoriam* obituary published hereafter. Professor Papagno took over as primary volume editor while Paul Corballis—Michael's son—provided assistance. Together, they have gathered a remarkable set of authors, thus ensuring the right mix of continuity and highly updated information on the topic. As series editors, we reviewed all the chapters and made suggestions for improvement, but we are delighted that the volume editors and chapter authors have produced such scholarly and comprehensive accounts of different aspects of the topic. We hope that the volume will appeal to clinicians as a state-of-the-art reference. We hope also that basic researchers will find within it the foundations for new constructive approaches to solving the complex issues involved.

In addition to the print version, the volume is available electronically on Elsevier's ScienceDirect website: https://www.sciencedirect.com/. ScienceDirect is popular with readers and will improve the volume's accessibility. Indeed, all the volumes in the present series of the *Handbook* are available electronically on this website. This should make them even more accessible to readers and facilitate searches for specific information.

As always, it is a pleasure to thank Elsevier, our publisher, and in particular Kristi Anderson in San Diego, Megan McManus in Salt Lake City, Punithavathy Govindaradjane and Omer Mukthar at Elsevier Global Book Production in Chennai, India, for their assistance in the development and production of the *Handbook of Clinical Neurology.*

Michael J. Aminoff
François Boller
Dick F. Swaab

Foreword

Preface

A general preference for the right hand has been observed throughout recorded history and is evident in all human cultures, so far without exception. The proportion of left-handers is typically around 10%–12%, with some variations probably due to cultural pressure to conform to the right-handed norm. Although not recognized as such until the 19th century, right-handedness can be considered the earliest manifestation of brain asymmetry. It then began to emerge that the left side of the brain was dominant for language. The left hemisphere was found to be responsible for both right-handedness and language in most people—although 5% of right-handers and around 30% of left-handers have a right lateralization of language—and came to be regarded as the *dominant* or major hemisphere, with the right hemisphere considered *nondominant* or minor. This terminology persisted well into the 20th century. Nevertheless, some observations started to suggest that the right hemisphere also had its own specializations, such as for perception or emotion. Actually, it is proposed that the two hemispheres are complementary rather than one dominating the other.

Further research and ideas on cerebral asymmetry occurred with the split-brain studies in the 1960s, leading to a belated Nobel Prize awarded to Roger W. Sperry in 1981 for "his discoveries concerning the functional specialization of the cerebral hemispheres." The sectioning of the forebrain commissures allowed each side of the brain to be tested for cognitive and perceptual functions independently of the other, and this research tended to support the idea of complementary function rather than cerebral dominance. This work also sparked what might be termed a cottage industry of studies on neurologically intact individuals, using techniques such as dichotic listening or tachistoscopic perception to measure perceptual asymmetries, which were interpreted as signs of cerebral asymmetries.

From around the mid-20th century, asymmetries in the intact brain could be more directly corroborated through the development of brain-imaging techniques, including electroencephalography, positron emission tomography, and magnetic resonance imaging. This advancement enabled the measurement of both structural and functional asymmetries and their correlations with various psychologic components. Brain imaging also allowed the noninvasive detection of brain anomalies in patients with neurologic disorders. Techniques such as transcranial magnetic stimulation were developed to stimulate the brain asymmetrically and observe the consequences. These imaging and technologic developments, along with the emergence of cognitive neuroscience in the late 1950s, enabled a more detailed study of how brain asymmetries are involved in various aspects of human mental processes, such as attention, face recognition, calculation, emotion, music, and art. Research on brain asymmetries also enhances our understanding of conditions such as psychoses, dyslexia, autism, and attention-deficit disorders.

The development of techniques for measuring brain function and structure has deepened the understanding of cerebral asymmetries, including their development and evolution. Systematic asymmetries can be detected in the fetus. Until recently, it was widely held that cerebral asymmetry was a uniquely human phenomenon; however, it is now known to be present in animals, with particularly striking examples observed in birds and nonhuman primates. There has long been interest in the genetics of handedness, and with the development of genomics and gene sequencing, this has now extended to cerebral asymmetries.

Part of the fascination with cerebral asymmetry is that the brain remains markedly symmetric from an anatomic perspective. Despite our asymmetries, we humans belong to the vast and nearly universal clade of species known as the *bilateria,* characterized by bilateral symmetry in body structure. From an evolutionary perspective, symmetry can provide certain advantages in biologic fitness, but specific asymmetries offer some benefits. The trade-off between symmetry and asymmetry is one of the many challenges in research.

Many books on cerebral asymmetry have been published following the split-brain work of the 1960s and beyond, but most of these are now outdated. Moreover, they focused on specific aspects of cerebral asymmetry. Therefore, for the present volume, experts from around the world have been gathered to provide up-to-date information and a comprehensive overview of both the anatomic and clinical aspects of brain asymmetries in humans throughout their lifespan, as well as the types of asymmetries observed in other species, including nonhuman primates and nonprimates.

In addition to neurologists, the book is intended to appeal to biologists, psychologists, and interested members of the public. This volume of the *Handbook* consists of 31 chapters, beginning with an anatomic description of brain asymmetries and their relation to visceral asymmetries. It then explores the clinical outcomes of these asymmetries with respect to pathologic processes involving attention, language, emotion, and more. Evolutionary as well as genetic aspects are also discussed.

We sincerely thank our colleagues and friends for their contributions to this volume. We greatly appreciate their time, commitment, and resilience. We also thank the Series Editors of the *Handbook of Clinical Neurology*, particularly François Boller, for his support during the most challenging moments of this journey, and the editorial office manager, Kristi Anderson, for her patience and help in monitoring every detail. Special thanks go to Stephanie Forkel who agreed to cover a last-minute crucial gap—without her help, this book would have been incomplete.

Costanza Papagno
Paul Corballis

Contributors

A. Albonico
Departments of Medicine (Neurology), Ophthalmology and Visual Sciences, University of British Columbia, Vancouver; Department of Psychology, University of the Fraser Valley, Abbotsford, BC, Canada

I. Babik
Department of Psychological Science, Boise State University, Boise, ID, United States

M. Baciu
Université Grenoble Alpes, Université Savoie Mont Blanc, CNRS, LPNC, Grenoble, France

S. Banjac
Université Grenoble Alpes, Université Savoie Mont Blanc, CNRS, LPNC, Grenoble, France

P. Bartolomeo
Sorbonne Université, Institut du Cerveau/Paris Brain Institute—ICM, Inserm, CNRS, APHP, Hôpital de la Pitié Salpêtrière, Paris, France

J.J.S. Barton
Departments of Medicine (Neurology), Ophthalmology and Visual Sciences, University of British Columbia, Vancouver, BC, Canada

M. Behrmann
Department of Ophthalmology, University of Pittsburgh School of Medicine, Pittsburgh, PA, United States

S. Benavides-Varela
Padova Neuroscience Center; Department of Developmental Psychology and Socialisation, University of Padova, Padova, Italy

G. Berlucchi
Department of Neuroscience, Biomedicine and Movement Sciences, University of Verona, Verona, Italy

C. Bortolami
Optical Approaches to Brain Function Laboratory, Istituto Italiano di Tecnologia; Università di Genova, Genova, Italy

E. Bruner
Departamento de Paleobiología, Museo Nacional de Ciencias Naturales, CSIC; Reina Sofia Alzheimer Center, CIEN Foundation, ISCIII, Madrid, Spain

J.M. Campbell
Department of Psychology, Illinois State University, Normal, IL, United States

S.F. Cappa
University School for Advanced Studies (IUSS-Pavia); Dementia Research Center, IRCCS Mondino Foundation, Pavia, Italy

G.A. Carlesimo
Laboratory of Neuropsychology of Memory, IRCSS Santa Lucia Foundation; Department of Systems Medicine, Tor Vergata University, Rome, Italy

R. Cavallaro
Department of Clinical Neurosciences, IRCCS San Raffaele Scientific Institute; School of Medicine, Vita-Salute San Raffaele University, Milan, Italy

J. Diedrichsen
Western Institute for Neuroscience; Department of Computer Science; Department of Statistical and Actuarial Sciences, Western University, London, ON, Canada

L. Dulyan
Donders Institute for Brain Cognition Behaviour, Radboud University, Nijmegen, The Netherlands; Brain Connectivity and Behaviour Laboratory, Sorbonne Universities, Paris, France

C.L. Ferre
Department of Occupational Therapy, College of Health & Rehabilitation Sciences, Boston University, Boston, MA, United States

S.B. Festini
University of Tampa, Tampa, FL, United States

S.J. Forkel
Donders Institute for Brain Cognition Behaviour, Radboud University; Max Planck Institute for Psycholinguistics, Nijmegen, The Netherlands; Brain Connectivity and Behaviour Laboratory, Sorbonne Universities, Paris, France; Centre for Neuroimaging Sciences, Department of Neuroimaging, Institute of Psychiatry, Psychology and Neuroscience, King's College London, London, United Kingdom

C. Francks
Language & Genetics Department, Max Planck Institute for Psycholinguistics; Department of Cognitive Neuroscience & Donders Community for Medical Neuroscience, Radboud University Medical Center, Nijmegen, The Netherlands

E. Frasnelli
CIMeC, Center for Mind/Brain Sciences, University of Trento, Rovereto, Italy

A.D. Friederici
Department of Neuropsychology, Max Planck Institute for Human Cognitive and Brain Sciences, Leipzig, Germany

G. Gainotti
Institute of Neurology, Università Cattolica del Sacro Cuore, Fondazione Policlinico A. Gemelli, IRCCS, Rome, Italy

M.S. Gazzaniga
Department of Psychological and Brain Sciences, University of California, Santa Barbara, CA, United States

O. Güntürkün
Institute of Cognitive Neuroscience; Research Center One Health Ruhr, Research Alliance Ruhr, Ruhr University Bochum, Bochum, Germany

M. Hausmann
Department of Psychology, Durham University, England, United Kingdom

S. Hodgetts
Department of Psychology, Durham University, England, United Kingdom

W.D. Hopkins
Department of Comparative Medicine, The University of Texas MD Anderson Cancer Center, Bastrop, TX, United States

G. Kegler
University of Tampa, Tampa, FL, United States

X.-Z. Kong
Department of Psychology and Behavioral Sciences & The State Key Lab of Brain-Machine Intelligence, Zhejiang University; Department of Psychiatry of Sir Run Shaw Hospital, Zhejiang University School of Medicine, Hangzhou, China

E.C. Marcinowski
School of Kinesiology, Louisiana State University, Baton Rouge, LA, United States

F. Martini
Department of Clinical Neurosciences, IRCCS San Raffaele Scientific Institute, Milan, Italy

C.A. Marzi
Department of Neuroscience, Biomedicine and Movement, University of Verona, Verona, Italy

A. Meguerditchian
Laboratoire de Psychologie Cognitive, CNRS, Aix-Marseille University, Institute of Language, Communication and the Brain, Marseille, France

G.F. Michel
Department of Psychology, University of North Carolina Greensboro, Greensboro, NC, United States

M.B. Miller
Department of Psychological and Brain Sciences, University of California, Santa Barbara, CA, United States

E.L. Nelson
Department of Psychology, Florida International University, Miami, FL, United States

C. Nettekoven
Western Institute for Neuroscience; Department of Computer Science, Western University, London, ON, Canada

S. Ocklenburg
ICAN Institute for Cognitive and Affective Neuroscience, MSH Medical School Hamburg, Hamburg; Institute of Cognitive Neuroscience, Ruhr University Bochum, Bochum, Germany

C. Papagno
CIMeC, Center for Mind/Brain Sciences, University of Trento, Rovereto, Italy

Y. Pu
Shanghai Key Laboratory of Brain Functional Genomics (Ministry of Education), School of Psychology and Cognitive Science, East China Normal University, Shanghai, China

P.A. Reuter-Lorenz
University of Michigan, Ann Arbor, MI, United States

E. Salillas
Department of Psychology and Sociology, University of Zaragoza, Zaragoza, Spain

J. Sapienza
Department of Clinical Neurosciences, IRCCS San Raffaele Scientific Institute, Milan; Department of Humanities and Life Sciences, University School for Advanced Studies IUSS, Pavia, Italy

D. Satoer
Department of Neurosurgery, Erasmus MC University Medical Center Rotterdam, The Netherlands

S. Schintu
CIMeC—Center for Mind/Brain Sciences, University of Trento, Trento, Italy; Department of Psychology, George Washington University, Washington, DC, United States

C. Semenza
Department of Neuroscience; Padova Neuroscience Center, University of Padova, Padova, Italy

Z. Sha
Language & Genetics Department, Max Planck Institute for Psycholinguistics, Nijmegen, The Netherlands

J.M. Simonson
Department of Psychological and Brain Sciences, University of California, Santa Barbara, CA, United States

M. Spangaro
Department of Clinical Neurosciences, IRCCS San Raffaele Scientific Institute, Milan, Italy

R. Starrfelt
Department of Psychology, Center for Visual Cognition, Copenhagen University, Copenhagen, Denmark

P.C. Trettenbrein
Department of Neuropsychology, Max Planck Institute for Human Cognitive and Brain Sciences; International Max Planck Research School on Neuroscience of Communication: Structure, Function, and Plasticity (IMPRS NeuroCom), Leipzig; Experimental Sign Language Laboratory (SignLab), Department of German Philology, University of Göttingen, Göttingen, Germany

L. Trojano
Department of Psychology, University of Campania 'Luigi Vanvitelli', Caserta, Italy

G. Vallar
Department of Psychology, University of Milano-Bicocca; MiBTec, University of Milano-Bicocca, Milan, Italy

G. Vallortigara
CIMeC, Center for Mind/Brain Sciences, University of Trento, Rovereto, Italy

L. Veronelli
Department of Psychology, University of Milano-Bicocca; MiBTec, University of Milano-Bicocca; Department of Neurorehabilitation Sciences, Casa di Cura Igea, Milan, Italy

G. Vingerhoets
Department of Experimental Clinical and Health Psychology, Ghent University, Ghent, Belgium

L.J. Volz
Department of Neurology, University of Cologne, Cologne, Germany

E. Zaccarella
Department of Neuropsychology, Max Planck Institute for Human Cognitive and Brain Sciences, Leipzig, Germany

D.W. Zaidel
Department of Psychology, University of California, Los Angeles (UCLA), Los Angeles, CA, United States

G.D. Zannino
Laboratory of Neuropsychology of Memory, IRCSS Santa Lucia Foundation, Rome, Italy

C. Zhang
Department of Psychology and Behavioral Sciences & The State Key Lab of Brain-Machine Intelligence, Zhejiang University, Hangzhou, China

Contents

Handbook of Clinical Neurology, Vol. 208 (3rd series)
Cerebral Asymmetries
C. Papagno and P. Corballis, Editors
https://doi.org/10.1016/B978-0-443-15646-5.00006-3

Chapter 1

Cerebral asymmetry: Historical introduction

GIOVANNI BERLUCCHI*

Department of Neuroscience, Biomedicine and Movement Sciences, University of Verona, Verona, Italy

Abstract

Cortical localization and hemispheric asymmetry of function are key concepts for analyzing the relationship between the brain, mind, and behavior in clinical neurology, neuropsychology, cognitive neuroscience, and modern philosophy. This chapter presents the history of asymmetries in the human central nervous system within the context of right–left asymmetries of bodily structure, as can be detected in individuals of virtually all living species. It has been known since antiquity that behavioral deficits on one side of the body could be caused by brain lesions on the opposite side. Yet, the idea that the two cerebral hemispheres could subserve different functions emerged only in the 19th century following the work of Franz Gall, Marc Dax, and especially Paul Broca. The concept of functional duality flourished during the golden age of cortical localization. Doubts occurred following the antilocalizationist tendencies in neurology and psychology, a movement mainly led by Karl Lashley. Later, a disciple of Lashley, Roger Sperry, and his coworkers unequivocally demonstrated the functional specialization of the cerebral hemispheres, thanks to their work on split brain. Finally, Norman Geschwind and the many scholars who were inspired by him have clarified, complemented, or changed the way of thinking of clinical neurologists about the great neuropsychologic sectors of aphasia, agnosia, and apraxia, as well as consciousness.

In the arts, symmetry is an important element of perfection and beauty. In biology, the bilateral symmetry of the body of most animals is important for function as much as for esthetics (Corballis, 2021). But life also requires asymmetry, as eloquently argued in the essay "L'asimmetria e la vita" (asymmetry and life) by the Italian writer Primo Levi, a survivor of Auschwitz and an accomplished chemist. As a student, Levi was attracted to chiral molecules like amino acids, of which only the levorotatory enantiomers build all biologically active proteins (Levi, 2016). The emergence of chiral molecules may go back to the origins of life and may even constitute the primary basis of left–right asymmetries in living organisms (McManus, 2002). Asymmetric gene expression and directional flow of extraembryonic fluid generated by motile embryonic cilia are also involved in the systematic asymmetric placement of some internal unpaired organs (Hamada, 2020). Fluctuating or persistent right–left asymmetries of bodily structure can be detected in individuals of virtually all living species (Vogt, 2015). Some of these asymmetries may be useless or even detrimental to the organism, while others are endowed with a fundamental functional significance. If we believe, as many of us do, that we are our brain, another fundamental asymmetry in our life arises due to the functional differences between our right and left cerebral hemispheres. Like hands, the cerebral hemispheres are regarded as chiral structures, as nonsuperimposable mirror images of one another, although a closer inspection shows that they are not morphologically identical. The history of functional hemispheric asymmetry is strictly intertwined with that of cortical localization because the total or prevalent assignment of a function to a single hemisphere is by itself a clear instance of localization (Stein, 2001). Cortical localization and hemispheric asymmetry of function have been and are key concepts for the analysis of the relations between brain, mind, and behavior in clinical neurology, neuropsychology, cognitive neuroscience, and modern philosophy. Since antiquity, it has been known that behavioral deficits on one side of the body could be caused by brain lesions on the opposite side, a fact explained by the principle of crossed innervation established by Giovanni Battista Morgagni in the 18th century. Yet the idea that the two cerebral hemispheres could subserve different functions emerged only in the 19th century (Benton and Joynt, 1960; Benton, 1976, 1977, 1984; Harris, 1999). Most professional historians of neuroscience, as well as clinical neurologists, attribute the birth of the theory of cortical localization to Franz Joseph Gall (Critchley, 1965, 1979b; Finger, 2009) and that of hemispheric differences to Marc Dax (Critchley, 1979a; Cubelli and Montagna, 1994). By contrast, for the great French historian of science Jules Soury, it was Paul Broca who single-handedly introduced into brain science the concepts of both cortical localization and hemispheric asymmetry of function (Soury, 1899). Between 1861 and 1865,

*Correspondence to: Giovanni Berlucchi, Department of Neuroscience, Biomedicine and Movement Sciences, University of Verona, Verona, Italy. Tel: +39-0458027141, E-mail: giovanni.berlucchi@univr.it

Broca forcefully claimed that articulate speech is a function of the sole left hemisphere, after he had found that clinical speech loss was associated with autopsy lesions of the foot of the third frontal convolution in the left but not the right hemisphere. Broca's claim of hemispheric functional difference was a memorable turning point in the neurologic way of thinking, which at that time was dominated by the fundamental biologic law of symmetry enunciated by Xavier Bichat at the beginning of the 19th century: bilateral symmetric organs, such as the eyes, the lungs, kidneys, etc., have the same function, and the two brain hemispheres are no exception to the law (Bichat, 1800). In this chapter, I describe some trends in the scientific thinking about symmetry and asymmetry of the cerebral hemispheres from the early 19th century to the revolutionary effects of Broca's discovery and to the developments and elaborations by modern and contemporary neurosciences. Since I am not a historian of neuroscience, this chapter belongs more to neuroscience storytelling than to science historiography, though the information provided has been drawn from primary sources. A history of brain asymmetry studies with an ample bibliography has been recently published by Christman (2024).

BEFORE BROCA

The functional equality of the two hemispheres, as stated by Bichat's law, was accepted by Franz Joseph Gall's organologic or phrenological theory (Gall, 1825). Gall is recognized by science historians as the man who introduced the study of mind into biology and medicine by identifying the brain as the organ of the mind and by searching for relations between mind features and brain features (Critchley, 1965, 1979b). Phrenology anticipated the localizationist approach to the study of the brain by dividing the mind into discrete faculties and by locating each of them in a specific and isolated brain organ, consisting of a compound of two mirror-symmetric isofunctional halves, one in the cortex of the right hemisphere and the other in the cortex of the left hemisphere. The functionality of each mental faculty was thought to express itself in the size of the corresponding cerebral organ, which in turn was thought to be indicated by the size of the overlying part of the cranium. Cranioscopy, the measurement of cranial bumps, was thus regarded as a technique that could provide objective information on the functionality of different mental faculties and their cortical organs in different persons. As a result, personality traits and intellectual abilities of specific individuals were inferred from their physiognomic and bodily characteristics, but above all from their cranial bumps. On these grounds, phrenology was to become first a cult and then to be discredited as a ridiculous, scientifically unfounded fabrication. Yet Gall was a masterful brain anatomist, and his important inspiration for the scientific theory of cortical localization has been acknowledged by Broca himself and other scholars of past and present times. After Gall's death, the Phrenological Journal published numerous case reports with careful correlations between symptoms and brain lesions. In 1836, a paper by the young English botanist and evolutionist Hewett Watson, published in that journal, bore the title "What is the Use of the Double Brain?" Watson rejected the idea that the two hemispheres always perform the same function, so that each of them may supply the other's activities in case of injury. He suggested instead the possibility of an interhemispheric division of labor. As an example, by referring to the chess player who must analyze all his possible future moves as well as those of his opponent, he argued that the two tasks are performed concurrently by dividing them between the hemispheres (Watson, 1836). Many years later, a similar mechanism of parallel and complementary processing was proposed by modern students of brain lateralization in chicks. In chicks, the visual pathways are completely crossed; the left hemisphere is better at visually detecting food stimuli, and the right hemisphere is better at detecting emotional stimuli such as predators. As a result, a chick can search for food with the right eye while watching for predators with the left eye (Rogers et al., 2004; Güntürkün et al., 2020). An efficient division of critical information between the hemispheres of the human brain is also known to increase the total information processing capacity of the whole brain (Levy, 1977; Banich and Belger, 1990). An extreme view of the independent working of the two hemispheres was proposed in 1844 by the eccentric English physician Arthur Wigan in a book titled "A new view of insanity: the duality of the mind proved by the structure, functions, and diseases of the brain, and by the phenomena of mental derangement and shown to be essential to moral responsibility" (Wigan, 1844). The author claimed that each hemisphere is a complete brain and as such a completely independent organ of thought and volition. Separate thinking processes can be carried out in each hemisphere, and each hemisphere can have volitions of its own, either in agreement or in opposition to those of the other hemisphere. Wigan assumed that in the healthy brain, one hemisphere (probably the left, based on the prevalent preference for the right hand and Morgagni's principle of crossed innervation) is almost always superior in power to the other hemisphere. It can thus hold in check unwanted volitions of the weaker hemisphere and disallow them to give rise to manifest actions. This was probably the earliest conception of hemispheric dominance combined with partial hemispheric independence, which was revived and exploited in the 1960s and 1970s by the so-called neowiganists. They sustained the personification of the hemispheres and the theory of hemisphericity, which asserts the division of the general population into right-brained and left-brained individuals (Bogen, 1969, 1986). Just a year before Broca's first announcement of the cerebral localization of speech loss, the physicist Gustav Fechner published his magnum opus Elemente der Psychophysik (Fechner, 1860). He had a keen interest in the relations between the mind and the brain, a relation which he felt Gall and the phrenologists had treated with arguments unsupportable by either experience or aprioristic considerations. He was impressed by clinical reports of patients who had suffered an extensive loss of cerebral substance, even of an entire hemisphere, and yet had remained conscious. To explain the relations between mental activities and the cooperation or interference between the two cerebral hemispheres, he used the metaphor of a wagon pulled by two horses. If one of the horses is unharnessed, the other horse can continue to pull the wagon with greater effort, but if one of the horses becomes wild or paralyzed, the progression of the other horse and the wagon is disturbed. Fechner also believed that the unity of consciousness requires a continuity of nervous tissue, such that a surgical separation of the two hemispheres, in his view an operation impossible to perform in a human being, would bring about the existence of two different minds within the same head (Fechner, 1860). In a similar vein, the philosopher Eduard von Hartmann proposed that if the brains of two men could be united by a bridge of conducting substance, the two men would share a single common consciousness (von Hartmann, 1869). This hypothetical situation is aptly captured by a 2003 New Yorker cartoon reproduced on this page.

According to the British psychologist Oliver Zangwill, Fechner was the father not only of psychophysics but also of neuropsychology because of his interest in the brain bases of consciousness (Zangwill, 1977). The psychologic problem of the unity of consciousness and its relations with cerebral hemispheric organization continues to be discussed in the present-day neuropsychologic literature (de Haan et al., 2020).

"That's amazing—I was just thinking the same thing."

BROCA AND THE CANCEL CULTURE

Paul Broca, a surgeon, anthropologist, and brain scientist, did the best work of his life between 1861 and 1869. He acknowledged that his discovery of unilateral brain representation of the faculty of articulate speech grew out of the previous bilateral localization of that function in the frontal lobes by Gall, Jean-Baptiste Bouillaud, and Ernest Auburtin. Gall had localized the organ of language in the frontal lobes based on the association between good verbal memory, bulging eyes, and big frontal bumps (Gall, 1825). Bouillaud used reports of speech loss cases from brain lesions checked at autopsy to build what today would be called a meta-analysis. He found an association between speech loss and frontal lesions but did not pay attention to unilateral lesions and their sides (Bouillaud, 1825). Auburtin limited himself to confirming the findings of Bouillaud, his father-in-law, and to reporting the probable first instance in history of speech arrest from a brain manipulation, i.e., temporary inhibition of speech from pressing a spatula on the exposed frontal lobes of a man who had blown off his forehead in a suicide attempt (Auburtin, 1861). By contrast, Broca never admitted that his discovery had been influenced by Marc Dax's 1836 report at the Montpellier Medical Society linking between speech losses and damage in the left hemisphere. A written version of this report was eventually produced by Dax's son Gustave, also a neurologist who gathered evidence of his own on language disorders and unilateral brain lesions (Critchley, 1979a; Cubelli and Montagna, 1994). There seems to be a consensus among present-day experts in neuroscience history that Broca's discovery should also be credited to Marc Dax. This appears acceptable from a chronologic viewpoint, but Broca's view of the relations between hand preference, hemispheric speech lateralization, and brain development was much broader and more profound than that of Dax. Broca linked hand preference and the cerebral side of the speech center with the rule that right-handers are left-brained and left-handers are right-brained. He hypothesized that the left hemisphere matures earlier than the right hemisphere, a concept developed and elaborated by Corballis and Morgan (1978) more than a century later, and that the right hemisphere can assume control of articulate speech in cases of very early lesions of the left hemisphere. Above all, Broca has left permanent marks in the history of neuroscience with his memorable writing style. Regarding cortical localization, Broca (1866) wrote "nous savons ... que toutes les parties du cerveau proprement dit n'ont pas les même memes attributions, que l'ensemble des circonvolutions ne constitue pas un seul organe, mais plusieurs organes ou plusieurs groupes d'organes, et qu'il y a dans le cerveau des régions distinctes correspondant aux grandes régions de l'esprit." (We know that all parts of the brain properly defined do not have the same attributions, that the assembly of the convolutions does not constitute a single organ, but many organs or groups of organs, and that there are in the brain great distinct regions which correspond to the great regions of the mind.) And about speech, right-handedness, and the left hemisphere, he (Broca, 1865) wrote "De même que nous dirigeons les mouvements de l'écriture, du dessin, de la broderie, etc., avec l'hémisphère gauche, de même nous parlons avec l'hémisphère gauche. C'est une habitude que nous prenons dès notre première enfance." (As we steer the movements of writing, drawing, embroidering, etc. with the left hemisphere, so we speak with the left hemisphere. It is a habit that we form since early infancy.) Broca maintained that the two hemispheres are morphologically identical except for minimal differences between the small convolutions (Broca, 1866). A historical mystery is the fact that Broca's interest in cortical localization and hemispheric lateralization, leading to the best work of his life, was limited to a period of a few years, almost incidental to his very productive scientific career (Broca, 1861a,b, d, 1863, 1865, 1868). In a book titled Fearful Asymmetry, Leblanc (2017) has recently argued that Bichat's theory of the unitary function of symmetric organs, including the brain hemispheres, had almost a religious character that scared Broca into abandoning the pursuit of hemispheric differences. But Broca knew that at least the minor convolutions of the two sides of the brain could be different in form (Broca, 1866), a fact that his contemporary Jules Bernard Luys, discoverer of the subthalamic nucleus and of the emotional role of the right hemisphere, documented objectively a few years later. He would take a sheet of tracing paper, apply it to a fresh vertical section of the brain, and mark with a brush of watercolor the contour of the cortical substance of one hemisphere. After folding the paper over, he consistently saw that the outline of the convolutions of one side did not match those of the other side

(Luys, 1876). The most likely reason why Broca left the investigation of aphasia is that he had many scientific interests, with a decided preference for physical anthropology and the search for relations between intelligence and brain size (Broca, 1861c). It is not unlikely that he regarded the latter kind of work as more attractive than cortical localization and lateralization. Unfortunately for him, this choice has seriously challenged his hero image in the Neuroscience Hall of Fame. In the 1970s, many decades after his death, two celebrities of science popularization, Sagan (1979) and Gould (1981), argued that Broca had tailored his data to demonstrate that whites are superior in intelligence to blacks and that males are superior in intelligence to females. Broca was a freethinker and a believer in the objectivity of science, but he could not rid himself of the social prejudices of his time. By the late 20th-century standards, he thus qualified as a racist, a colonialist, and a male supremacist. In 2016, at the University of Bordeaux, local neuroscientists voted to name the newly built neuroscience center after Paul Broca, who was born less than a hundred kilometers away. The ships operating in the transatlantic slave trade had regularly utilized Bordeaux's harbor between the 17th and early 19th century. In 2021, the Association Mémoires et Partages, seeking recognition of Bordeaux's slave-owning past, questioned the wisdom of giving a university building the name of a man who as a scientist had proclaimed that blacks are inferior to whites, even though he was personally against slavery (Boraud and Forkel, 2022). In retrospect, if Broca had continued to work on aphasia rather than on socially sensitive problems of physical anthropology, he would have protected himself from posthumous shame.

THE GOLDEN AGE OF CORTICAL LOCALIZATION

Soon after Broca's discovery, the notion of the functional duality of the brain was pursued with great vigor in clinical neurology in relation to hysteria and double personality, along with the attempts to treat these disorders with metallotherapy and hypnosis. In agreement with Wigan, the basic assumption was that the head houses two relatively independent conscious systems, one in each hemisphere, with different pathologic reactivities and different responsiveness to treatment attempts. Despite the initial involvement and interest of great men of science like Bernard, Binet, Charcot, and Brown-Sequard, the endeavor suffered from the quackery and sheer ignorance of brain anatomy of most of its adherents and fell rapidly into disrepute (Harrington, 1985, 1987). The field of inquiry implicitly or explicitly spurred by Broca's discovery was the search for the localization of different functions in the cerebral cortex. In 1870, Fritsch and Hitzig reported that punctate electrical stimulation of the frontal cortex in each hemisphere of dogs gave rise to discrete movements on the opposite side of the body, a clear indication that motor function was restricted to a specific cortical region (Fritsch and Hitzig, 1870). Soon thereafter, in 1874, a young Carl Wernicke burst onto the scene of aphasiology by asserting that speech comprehension was represented in the temporal lobe of the left hemisphere in the context of a general system for language reception and production. Broca believed that speech comprehension was localized in the foot of the third frontal convolution of both hemispheres and that his speechless patients with left frontal lesions understood spoken communications thanks to the intact corresponding convolutions in the right hemisphere (Hecaen, 1970). Wernicke (1874) used anatomic and pathologic information from his teacher Meynert to affirm that the hemispheres are asymmetric also with respect to speech comprehension. He proposed that Broca's area in the frontal lobe is the speech output station of an interconnected collection of centers subserving language generation and comprehension in the posterior temporal lobe and in the angular gyrus of the parietal lobe. Lesions in these regions with an intact Broca's area caused impairments in language comprehension with the preservation of a fluent though paraphrastic and partly garbled spoken output. This was the sensory or Wernicke aphasia, a counterpart to Broca's motor aphasia. Wernicke tried to match the various clinical symptoms of aphasia with largely hypothetical collections of language centers and their connections, represented as point-line diagrams superimposed on a badly drawn sketch of a supposedly human but simian-like cerebral hemisphere. For unexplained reasons, the sketch was always that of a right hemisphere, even though Wernicke's aphasia, like Broca's aphasia, was in most cases a disease of the left hemisphere. With appropriate qualifications, the classic Broca–Wernicke localization of language functions, with its tight coupling between function and structure, is still used in the clinical practice of neurologists and neurosurgeons (Rutten, 2022). In experiments on animals, Fritsch and Hitzig's (1870) discovery spurred the use of extirpation and stimulation of cortical sites for identifying circumscribed parts of the cortex as specific sensory, motor, or sensory-motor centers. The British neurologist David Ferrier maintained that most cortical centers are sensory motor and that in the last analysis, mental activities must be merely the subjective side of sensory-motor operations (Ferrier, 1876). For example, he localized the hearing center in the superior temporal gyrus in monkeys because (1) stimulation of that gyrus on one side produced ear and eye movements toward the opposite side, as in response to an acoustic hallucination generated by brain stimulation and (2) unilateral or bilateral lesions of the same gyrus caused hearing defects. Similar localization of centers for other sensory modalities, including sight, touch, smell, and taste, was proposed by Ferrier himself, Munk, Hitzig, Luciani, Beevor, Horsley, and others. In the early years of the 20th century, the cyto- and myeloarchitectural parcellations of the cortex of humans and other mammals by Brodmann, Campbell, and the Vogts promised to provide a histologic counterpart to functional localization inferred from clinical and pathologic observations. The combination of stimulation–extirpation experiments by Grünbaum and Sherrington on the pericentral cortex of apes, Campbell's histologic analyses of the same cortical region, and Cushing's finding of somatic sensations from postcentral stimulations in awake neurosurgical patients established the localization of the primary motor cortex in the precentral convolution and the primary somatic sensory cortex in the postcentral convolution of both apes and humans (Berlucchi and Vallar, 2018). Grünbaum and Sherrington (1903) stimulated the homolog of the human Broca's area in both hemispheres in chimpanzees, failing to obtain vocalizations. They admitted that electrical stimulation of the Broca speech cortex may be inefficient by itself to evoke vocalization but seemed more favorable to the conclusion that no Broca speech center, distantly foreshadowing the human, exists in these anthropoid brains. Sherrington was by no means a materialist, but even today some card-carrying materialists are inclined to draw an insurmountable boundary line between humans and other primates because of the uniqueness of human language. The evolutionary origin of human language is still mysterious and debated, but everybody agrees that no other animal species exhibits a capacity for communication even distantly comparable in kind and degree to that of humans. If human language is unique, its demonstrated asymmetric representation in the brain does not imply that functional differentiation between the hemispheres is limited to language and to humans. Hemispheric differentiation for various functions has now been shown to exist in many nonhuman species (Güntürkün et al., 2020), and the supposed uniqueness of a lopsided human brain has been regarded as an added myth to the rich and ancient mythology compounding the history of the concepts of right and left (Corballis, 1980; McManus, 2002). The epoch in which it was hoped that the function of each region of the brain, from a cortical convolution to an entire hemisphere, could be identified by the joint effort of histologists, experimental physiologists, and clinical neurologists has been defined as the golden age of cortical localization (Hécaen and Lanteri-Laura, 1977). It was an age full of research efforts and achievements, but also of excesses and misconceptions that shortened its duration.

ANTILOCALIZATIONIST TENDENCIES IN NEUROLOGY AND PSYCHOLOGY

Hughlings Jackson, the father of British neurology, was not opposed to cortical localization per se. He maintained that different cortical areas in both hemispheres participate in different aspects of language, with the left hemisphere taking care of voluntary speech and the right hemisphere taking care of automatic speech (Jackson, 1874). However, he always objected to the localization of a physiologic mechanism in a locus of brain destruction. He regarded this kind of inference as a typical example of continental loose thinking. His dictum "to locate the lesion which destroys speech is not to localize speech" has been an axiom for most British aphasiologists up to this day. Jackson was also one of the first to suggest that the right hemisphere may have its own functions, especially in relation to spatial perception. The effects of a local brain lesion depend on more than just the loss of the damaged tissue, as pointed out by Constantin von Monakow, director of the Brain Institute in Zurich, who introduced the concept of diaschisis (von Monakow, 1910, 1914). An acute lesion of a brain locus destroys not only the bodies of the neurons in that locus but also their short and long connections with anatomically intact regions in both near and far parts of the brain. The overall pattern of clinical symptomatology immediately following a brain lesion depends therefore on the loss of the destroyed neurons as well as on diaschisis, i.e., a malfunctioning of intact but disconnected brain regions. In the case of focal lateralized lesions, diaschisis may disrupt the functions of anatomically intact distant mechanisms not only in the damaged hemisphere but also in the opposite, supposedly healthy hemisphere. Von Monakow attributed functional recovery after brain damage to a gradual disappearance of diaschisis and to a normalization of the regions affected by it. In America, diaschisis (misspelled as diaschesis) was discussed in scientific journals by psychologist Shepherd Ivory Franz, a fierce opponent of cortical localization and hemispheric dominance (Franz, 1911, 1912). In 1912, in a Science magazine essay titled "New Phrenology," he wrote that we may believe that mental processes are due to cerebral activities, but we do not know with what anatomic elements the individual mental processes may be connected. "Notwithstanding our ignorance, it would appear best and most scientific that we should not adhere to any of the phrenological systems, however scientific they may appear to be on the surface," a statement in which phrenological stands for localizationist. The phrase "new phrenology" continues to be used as disparagingly in the neuroscientific literature, a recent example being its use by the critics of present-day techniques for brain imaging, especially functional magnetic resonance. Franz is also remembered for introducing Karl Lashley to the experimental study of learning and memory in rats with cerebral lesions (see later). In the field of aphasia, the English neurologist Henry Head recognized that the World War I had offered the opportunity for neurologists to study the effects of clean missile wounds on the brain, as opposed to the messy lesions resulting from vascular, tumoral, infectious, or degenerative processes. In his writings on aphasia and related disorders of speech (Head, 1926), he expressed his conviction that the mechanisms of language cannot be localized if not in the whole brain. He was uninterested in the question of possible hemispheric differences in language functions, to the point that he did not systematically record the hand preference of his aphasic patients (Critchley, 1961), and never mentioned the thoughts of his teacher Hughlings Jackson on the different roles of the two hemispheres in verbal communication (Harrington, 1987). He wrote well and critically, and the jibes of his sharp pen were mostly aimed at the so-called diagram-makers, especially the arrogant Teutonic Wernicke and Lichtheim and Bastian in England. The diagram-makers were trying to explain speech and language disorders in terms of dysfunctions of independent centers and their connections (Zangwill, 1971). Head's main thrust against this approach was that such centers and connections were not real but made up for fitting the symptoms of the various clinical varieties of aphasia into tidy anatomic schemes. Right or wrong, the reaction against cortical localizationism had a detrimental influence on the search for hemispheric asymmetries and specializations. In the early 20th century, there was a general acceptance by neurologists that at least in right-handers, i.e., the great majority of the human population, linguistic activities and the movements of the preferred hand were controlled by the left hemisphere. Further cognitive attributions to the left hemisphere had been implicated by the work of Liepmann on apraxia, Marie and Head on intellectual deficits in aphasics, and Gerstmann on disorders of the body schema. This led to the establishment of the concept of hemispheric dominance and the definition of the left hemisphere as major or leading and the right hemisphere as minor or subordinate (Benton, 1972, 1976, 1977, 1984). Hugo Liepmann, perhaps the chief diagram-maker in the field of apraxia, had used an elegant similitude to describe the subordination of the right hemisphere to the left in the production of communicative gestures and generally of volitional movements. He compared the hemispheres to two brothers, of whom the left one is stronger and more dexterous. The weaker and clumsier right hemisphere can use the left hand to perform threatening or beckoning gestures, but only with the essential aid of the left hemisphere, as it loses this ability after some left brain or callosal injuries (Liepmann, 1908). In the 1920s and 1930s, the doctrine of left hemispheric dominance gained almost universal acceptance among clinical neurologists and neurosurgeons. The dominant left hemisphere was seen as the essential cerebral mediator of almost all linguistic abilities and language-related intellectual functions, as well as the executive for directing volitional movements and the repository of the motor memories underlying learned purposeful actions. The minor right hemisphere was depicted as being mute, word-blind, word-deaf, agraphic, apraxic, and generally lacking in the more evolved intellectual functions. The scanty participation in intellectual functions attributed to the minor right hemisphere by workers in clinical neurology is made clear by the following trenchant assertion by the influential Johns Hopkins neurosurgeon Walter Dandy, who claimed that it is possible to remove all of the right cerebral hemispheres above the basal ganglia with no appreciable disturbance of mentality (Dandy, 1933). However, a reading of Dandy's case reports makes it clear that all his informal and simplistic tests of mentality after right hemisphere removal were strictly verbal in character. The general trend of clinical neurology toward a strict belief in left hemisphere dominance was countered only by a quiet and respectable group of clinical researchers who sought evidence for a more democratic view of the brain halves (e.g., Zangwill, 1960: see Harrington, 1987). Certain syndromes like simultaneous agnosia were shown to depend on bilateral hemispheric lesions, while others like anosognosia, face perception, and a series of disorders of spatial cognition including unilateral neglect were found, in a few cases, to be associated with right hemisphere lesions. Among the many clinicians and experimenters who established neuropsychology as an independent discipline (Walusinski et al., 2019), an experimental psychologist like Zangwill in England and clinical neurologists like Hecaen in France and De Renzi in Italy stand out as those in Europe who best succeeded in supporting a functional division of labor between the hemispheres with complementary specializations of the left and right hemispheres.

LASHLEY AND THE COALESCENCE BETWEEN NEUROLOGY AND PSYCHOLOGY

Karl Lashley was perhaps the most important psychologist in the world from the late 1920s to the late 1950s. He studied learning and memory in rats and monkeys and interpreted his results in the framework of a new concept of the organization of the nervous system, which aimed for a complete coalescence with psychology (Lashley, 1941). He argued that the classic organization of synaptically connected neurons in pathways running from sensory organs to motoneurons and muscles could explain simple reflexes like scratching, but not higher-order behavior. He did not believe that specific cortical neurons were involved in perception, nor did he accept that memory engrams can be formed by learning-dependent synaptic changes in specific pathways in some specific part of the brain. He thought that adjacent neurons interacted more by extrasynaptic than synaptic mechanisms in a nonspecialized dynamic function of the tissue as a whole and was somewhat enticed by the idea that long fiber connections such as the corpus callosum were skeletal structures keeping the brain from falling apart rather than functional links between distant neurons. Concerning cortical localization, Lashley declared his position by citing Head (1926) in his 1929 monograph on brain mechanisms and intelligence: "No function is localized strictly in any part of the cortex and no form of activity, somatic or psychical, is built up into a mosaic of elementary processes that become evident when it is disturbed by a lesion of the brain. On the other hand, local destruction of tissue prevents the normal fulfilment of some form of behavior and the reaction that follows expresses the response of the organism as a whole under these changed conditions." Lashley's (1929) monograph reported his quantitative studies on the effects of cortical lesions of different sizes on the learning and memory of rats trained to run through a maze or to discriminate visual stimuli. He established two general principles. According to the principle of mass action, learning ability is impaired in proportion to the extent of the association cortex lesion and not in relation to its locus. As for equipotentiality, the term designates the apparent capacity of any intact part of a functional area to carry out, with or without reduction in efficiency, the functions that are lost by destruction of the whole. For example, the ability to perform some visual tasks that are lost after a complete removal of the visual cortex is totally or partially preserved after incomplete visual cortex lesions regardless of the location of the spared tissue in that cortex. Lashley also supported radical learning theories that depicted the developing nervous system as a random, unstructured, and essentially equipotential network to be molded and shaped into a functionally adaptive system by use, practice, and conditioning. He never studied the effects of brain lesions in humans and regarded the evidence of clinical neurology as too vague and uncontrolled to deserve serious scientific attention, but manifested his sympathy for clinicians like Head, Goldstein, Gelb, and others who argued that the brain acted as a whole and brain damage location was unimportant. Between the 1920s and the 1950s, Lashley's ideas succeeded to some extent in outmoding orthodox connectionist models of the central nervous system based on the histology of Cajal and the physiology of Sherrington and Adrian (Lashley, 1930, 1931, 1950, 1958; Lashley et al., 1951). The notions of mass action and functional equipotentiality of the cortex had a decided influence on clinical neurology that was still present in the 1960s (Zangwill, 1961). For example, clinical investigators continued to look for, and often found, systematic relations between the degree of intellectual impairment and the amount of tissue removed from the association cortex for circumscribed neoplasms or arteriovenous anomalies, regardless of the locus of lesions. A study on numerous patients by Chapman and Wolff (1959) showed that the degree of impairment of the highest integrative functions was directly related to the estimated total number of malfunctioning cortical neurons regardless not only of site but also of whether the defective neurons were aggregated in one site of the cortex or diffusely distributed throughout the hemispheres. It is of theoretical interest that Lashley's principles of mass action and equipotentiality can apply only to functionally symmetric hemispheres. If two bilaterally symmetric regions function in an additive manner, a lesion of one of them would produce a graded deficit in accordance with Lashley's principle of mass action. If the function can be subserved completely by either one of the two bilaterally symmetric regions alone, then we have a special instance of Lashley's principle of equipotentiality (Piercy, 1964). As we will see, Lashley's ideas about the principles of brain functioning have been sharply criticized by some of his pupils, but one must recognize the cogency of his attempt to understand how local and general cerebral activities produce the integration evident in thought and behavior (Zangwill, 1961).

DONALD HEBB AND BRENDA MILNER

Lashley taught at the Universities of Minnesota and Chicago, at Harvard University, where he was the first in the world to hold the title of professor of neuropsychology, and at the Yerkes laboratory at Orange Grove in Florida. Among the many talented students and disciples of Lashley, two gained permanent world fame through their work that contradicted and disproved many of their teacher's favorite ideas. One of them was Donald Hebb, whose book "The Organization of Behavior" (Hebb, 1949) marked a turning point in the history of neuroscience by introducing the concepts of Hebbian synapses and cell assemblies, concepts that are still very much alive and influential in contemporary neuroscience and psychology. Connections, synapses, specialized neurons, and linear reflex pathways, all of which Lashley disliked, instead held a prominent position in Hebb's book. But Hebb also understood that the reciprocal connections between cortical neurons demonstrated in histology by Lorente de No′ could underlie recurrent and reverberatory activities in neuronal circuits, resulting in a self-sustained endogenous activity of the brain independent of any exogenous input (Lorente de No', 1938). Lashley (1958) liked the idea of spontaneous neuronal activity and appropriated it with his concept of trace systems, hypothetical neuronal associations akin to Hebb's cell assemblies (Nadel and Maurer, 2020). Although Hebb was known for being skeptical about women in science, his influence on theoretical and empiric developments in clinical and experimental neuroscience was largely through the work of his pupil Brenda Milner. Brenda Langford, born in Manchester, United Kingdom, was trained in experimental psychology as an undergraduate at Cambridge University by Bartlett and Oliver Zangwill. After moving to Canada in 1944 with her husband Peter Milner, who was engaged in a wartime research project, in 1949 Brenda Milner was eventually accepted as a PhD student by Hebb, then a professor of psychology at McGill University in Montreal. Hebb was interested in the possible visual functions of the temporal lobes, as suggested by the psychic blindness or visual object agnosia of monkeys affected by the Klüver and Bucy syndrome (Klüver and Bucy, 1939). The syndrome is produced by bilateral temporal lobe ablations extending into the hippocampal formation and the amygdala; in addition to the visual deficits, it includes memory impairment, changes in temperament in the direction of docility, hypersexuality, and stimulus-bound oral tendencies. Milner worked on her thesis at the Montreal Neurological Institute, headed by the famous neurosurgeon Wilder Penfield with whom Hebb had collaborated years before in studies on frontal lobe functions. She investigated visual and later auditory deficits in epileptic patients with unilateral temporal excisions (Milner, 1954, 1958), while another PhD student of Hebb, Mortimer Mishkin, worked under the supervision of Karl Pribram at Yale University on a thesis on the parcellation of the Klüver and Bucy syndrome in monkeys, localizing visual functions in the inferior temporal lobe and temperamental and memory functions in the amygdala and hippocampus (Mishkin, 1951). At the Montreal Neurological Institute, Milner had an advantage over the investigators who studied the effects of unihemispheric lesions from incidental spontaneous pathologies in a limited number of cases. She could compare scores of patients before and after planned, controlled, and comparable surgical excisions of epileptogenic lesions in the right or left temporal lobes, usually extending into the hippocampal formation and the amygdala. She looked first for perceptual deficits on complex visual tasks, finding mild but significant impairments after the right temporal lobectomy, but not after the left. These relatively minor perceptual visual deficits were present before the operation and persisted, or tended to be accentuated, following it. Much stronger results emerged when Milner switched to memory tasks, where she found clear differences related to the side of the lesion: lesions of the left temporal lobe were accompanied by difficulties in verbal acoustic stimulation, although recall of nonverbal material was normal; lesions of the right temporal lobe were accompanied by impaired comprehension of complex abstract drawings and impaired memory for faces, though verbal skills were intact. Milner also studied auditory discriminations in patients with right or left temporal lobe removals. She found that while left temporal lesions impaired comprehension of spoken material, right temporal lobes impaired auditory discriminations such as the comparison of tonal patterns or judgment of tonal quality, which were not affected by left lesions. At that time, there was no certain way to assess which hemisphere was dominant for speech in any non-aphasic individual since clinical and autopsy findings had demonstrated various exceptions to Broca's rule that speech is represented in the brain contralaterally to the preferred hand. A test developed in Japan was introduced to the MNI by neurologist Juhn Wada and applied since 1959 for determining the lateralization of speech in each patient undergoing neurosurgery. Using the test in collaboration with neurosurgeons Branch and Rasmussen, Milner was able to quantify the relations between dominance for speech and handedness: the left hemisphere was found to be dominant for language in 95% of right-handers and in 70% of left-handers; the right hemisphere was found to be dominant for language in 5% of right-handers and 15% of left-handers; in the remaining 15% of left-handers, language was represented bilaterally (Branch et al., 1964). Milner's PhD student Doreen Kimura developed Milner's findings by demonstrating complementary specializations of the two hemispheres in patients whose speech lateralization had previously been assessed with the Wada test. Kimura employed Broadbent's acoustic stimulation technique, which involves the simultaneous presentation of different auditory stimuli to the two ears and the selective channeling of signals from each ear to the opposite hemisphere. Kimura found that when the stimuli were verbal signals (digits), patients reported significantly more digits presented to the ear opposite the hemisphere dominant for speech compared to the other ear. When stimuli were tonal patterns constituting fragments of melodies, recognition of the melody was significantly better for signals delivered to the ear opposite the hemisphere subordinate for speech. This was true regardless of whether speech was represented in the right or left hemisphere as determined by the Wada test. Kimura was a pioneer in the search for functional hemispheric differences in healthy subjects using the dichotic technique as well as the tachistoscopic stimulation of one-half of the visual field for channeling visual signals selectively into the contralateral hemisphere (Kimura, 1961a,b, 1963, 1964, 1973). In 1971, massive evidence had been gathered for complementary hemispheric specializations in various clinical and experimental investigations of neurologic patients and neurologically intact subjects alike. In that year, Milner published a masterful review of the findings from all different approaches to functional cerebral asymmetry (Milner, 1971), including her personal contributions to the subject that have been somewhat overshadowed by her outstanding work on memory and bilateral hippocampus lesions in the famous patient Henry Molaison (Corkin, 2013). The complex demonstration of hemispheric differences in the intact brain reached its peak in the late 1970s and early 1980s (Bryden, 1982; Corballis, 1983).

ROGER SPERRY, THE RIDDLE OF THE CORPUS CALLOSUM AND THE SPLIT BRAIN

The other famous disciple of Lashley who did not agree with his teacher's ideas was Roger Sperry, an unusual scientific personality (Berlucchi and Marzi, 2024). He solved the riddle of the corpus callosum and won a Nobel Prize for Physiology or Medicine for his discoveries concerning the functional specialization of the cerebral hemispheres (Sperry, 1982). While Lashley did not believe in the specificity of neurons and neuronal connections, Sperry did ingenious experiments to show that early in development the billions of neurons in a mammalian brain acquire and retain individual chemical identification tags that allow the establishment of lasting functional connections only between neurons that are matched by inherent chemical affinities. Sperry believed that the brain's wiring diagram must be largely innate and formed through an enormously elaborate chemical guidance program under genetic control. He thought that learning does not require the formation of new neural paths but depends on rapid or instantaneous changes in the efficacy of specific synaptic connections which facilitate the behavioral output momentarily required (Sperry, 1992). In the early 1950s, when Sperry began to be interested in the corpus callosum, many investigations in animals failed to show specific symptoms after a complete callosal section uncomplicated by additional cortical damage (Glickstein and Berlucchi, 2008a). The only exception was a study by two pupils of Pavlov, Bykov (or Bykoff) and Speranski, who had seen in dogs that conditioned salivary and motor reflexes evoked by tactile stimuli of skin points on one side of the body no longer transferred to corresponding points on the other side after a callosal section. These results were published in extenso only in Russian and therefore had little impact on the international literature, also because at that time scarce or no functional importance was attributed to the specificity of cortical connections (Glickstein and Berlucchi, 2008b). In humans, the neurosurgeon Dandy, who had found almost no deficits after removal of the right hemisphere, also proclaimed that the entire body of the corpus callosum may be split in the midline without any appreciable disturbance of function (Dandy, 1930). Paradoxically, only a pathologic activity could be attributed with certainty to the corpus callosum in both experimental animals and humans; this was the propagation of epileptic discharges from one hemisphere to the other (review in Bremer et al., 1956). Between the late 1930s and the early 1940s, another American neurosurgeon, William Van Wagenen, decided to section the corpus callosum totally or partially to reduce epileptic seizures in patients who did not respond to pharmacologic treatment. About 30 cases were studied before and after the operation by the psychiatrist Andrew Akelaitis and the psychologists Smith and Parsons (for a synthesis of the results see Akelaitis, 1944). The therapeutic results were uncertain; those related to the functions of the corpus callosum were as surprising as they were easy to summarize. Suffice it to say that the operation did not produce any definite functional symptoms but for a slight slowing of reaction time, ipsilateral as well as contralateral. Akelaitis's report could only reinforce Lashley's belief that the main integrative brain functions are carried out efficiently without the main associative tracts. Sperry began tackling the riddle of the corpus callosum in the early 1950s at the University of Chicago with his student Ronald Myers. According to Lashley, memories are spread throughout the cerebral cortex and can be accessible through afferent pathways other than those employed in memory formation during learning (Lashley, 1950). For example, after learning visual discrimination with one eye covered, rats can perform the same discrimination when they are forced to utilize the eye that was never exposed to the stimuli to be discriminated during learning. This interocular transfer could depend on strengthening of synaptic connections in cortical or subcortical sites known to receive afferents from the two eyes via the optic chiasm or the corpus callosum. Lashley had shown that interocular transfer persists after midline sections of the optic chiasm or the corpus callosum but had never examined interocular transfer following a combined section of both structures. Myers showed in cats that such a combined callosal and chiasmatic section eliminates interocular transfer altogether and that the corpus callosum is totally responsible for visual interhemispheric transfer after splitting the optic chiasm (Myers, 1956). After moving to the California Institute of Technology, Sperry showed in cats and monkeys that the corpus callosum is the essential link for allowing interhemispheric sensory-sensory and sensory-motor integration as well as the sharing of memories in vision and other sensory modalities as well. In a review in *Science* magazine, he spelled out the potential of the split-brain preparation for exploring cerebral organization in experimental animals, hinting at the possibility that similar callosal mechanisms could be involved in the development of language and its unilateral dominance in the human brain (Sperry, 1961). The neurologist Norman Geschwind immediately acknowledged that the physiologic studies in experimental animals described in Sperry's (1961) review provided a new impetus to clinicians to reinvestigate callosal functions in humans. In a paper with Edith Kaplan on callosal disconnection in a patient (Geschwind and Kaplan, 1962), he stated: it was Sperry's work that alerted us to the possibility of disconnection syndromes in man. He tried to resurrect intra- and interhemispheric disconnection syndromes hypothesized by the old diagram-makers using modern knowledge of brain anatomy in animals (Geschwind, 1965; Mesulam, 2015). But the clinician most influenced by Sperry's findings was the Los Angeles neurosurgeon Joseph Bogen, who conducted research work at Caltech and knew that possibly van Wagenen had never sectioned the corpus callosum completely. Bogen and his chief Philip Vogel cut the corpus callosum and the other forebrain commissure in a war veteran with progressively worsening drug-resistant seizures, achieving a reduction in the number and severity of the patient's seizures (Bogen and Vogel, 1962). This split-brain operation was repeated over a few years in a small series of similar patients who were tested by Bogen, Sperry, and Gazzaniga and later by Sperry's graduate students and collaborators Trevarthen, Levy, and Eran and Dahlia Zaidel. A routine medical exam of split-brain patients does not reveal any abnormality; interhemispheric disconnection symptoms and differences in psychologic performances between the two hemispheres only appear in special testing conditions that restrict sensory input and motor output to one hemisphere in the absence of interfering right–left cross-communication. The experience with split-brain animals and humans convinced Sperry that a section of the forebrain commissure results in two separate conscious entities or minds running in parallel within the same cranium, each with its own mental events and cognitive operations (Sperry, 1982). He later admitted that after complete forebrain commissurotomy, the cerebral hemispheres may use brainstem channels for cross-communication of at least some conscious states and contents (Myers and Sperry, 1985). Recent results that challenge the disunity of consciousness in the split-brain syndrome (de Haan et al., 2020) can indeed be accounted for by subcortical interhemispheric interactions (Corballis et al., 2018). In the classic split-brain syndrome, the most conspicuous difference between the disconnected hemispheres is that the left hemisphere retains the ability to speak about its mental processes much as before the operation, whereas the right hemisphere can express itself through signs but not in speech or in writing. Nevertheless, the right hemisphere has proven to be a conscious system capable of perceiving, thinking, remembering, reasoning, willing, and emoting at a characteristically human level. It outperforms the left hemisphere in tasks involving the global apprehension of spatial relations and complex images, such as recognizing faces and deciphering facial emotions, perceiving wholes from a collection of parts, discriminating musical chords, and the intuitive discernment of geometric principles. We have already seen that the evidence for functional differentiation between hemispheres in split-brain syndrome has given rise to the theory of hemisphericity, which emphasized the division of the neurotypical population into right-brained and left-brained individuals and proposed far-fetched hypotheses about the relevance of this division to cultural domains outside the neurosciences, like philosophy, sociology, anthropology, political science, education, literature, and so on (Bogen, 1986). The limitations and excesses of these approaches have been analyzed and criticized by Corballis (2007). In this volume of the *Handbook of Clinical Neurology*, it is appropriate to explain how split-brain research has clarified, complemented, or changed the way of thinking of clinical neurologists regarding major neuropathologic conditions such as aphasia, agnosia, and apraxia. This will be done in the thematic chapters of this volume.

References

Akelaitis AJ (1944). A study of gnosis, praxis and language following section of the corpus callosum and anterior commissure. J Neurosurg 1: 94–102.

Auburtin E (1861). Reprise de la discussion sur la forme et le volume du cerveau. Bull Soc Anthrop Paris 2: 209–220.

Banich MT, Belger A (1990). Interhemispheric interaction: how do the hemispheres divide and conquer a task? Cortex 26: 77–94.

Benton AL (1972). The 'minor' hemisphere. J Hist Med Allied Sci 27: 5–14.

Benton A (1976). Historical development of the concept of hemispheric cerebral dominance. In: SF Spicker, H Tristram Engelhardt Jr (Eds.), Philosophical dimensions of the neuro-medical sciences. Springer, Dordrecht, pp. 35–57.

Benton AL (1977). Historical notes on hemispheric dominance. Arch Neurol 34: 127–129.

Benton A (1984). Hemispheric dominance before Broca. Neuropsychologia 22: 807–811.

Benton AL, Joynt RJ (1960). Early descriptions of aphasia. Arch Neurol 3: 205–222.

Berlucchi G, Marzi CA (2024). Roger Sperry, the maverick brain scientist who was haunted by psyche. Front Hum Neurosci 11 (18): 1392660.

Berlucchi G, Vallar G (2018). The history of the neurophysiology and neurology of the parietal lobe. Handb Clin Neurol 151: 3–30.

Bichat X (1800). Recherches physiologiques sur la vie et la mort, Brosson, Gab on et Cie, Paris.

Bogen JE (1969). The other side of the brain: II. An appositional mind. Bull Los Angeles Neurol Soc 34: 135–162.

Bogen JE (1986). Mental duality in the intact brain. Bull Clin Neurosci 51: 3–29.

Bogen JE, Vogel PJ (1962). Cerebral commissurotomy in man: preliminary case report. Bull L A Neurol Soc 27: 169–172.

Boraud T, Forkel SJ (2022). Paul Broca: from fame to shame? Brain 145: 801–804.

Bouillaud JB (1825). Recherches cliniques propres a demontrer, et a'confirmer l'opinion de M. Gall, que la perte de la parole correspond a la lesion des lobules anthérieurs du cerveau. Arch Gen Med 8: 25–45.

Branch C, Milner B, Rasmussen T (1964). Intracarotid sodium amytal for the lateralization of cerebral speech dominance; observations in 123 patients. J Neurosurg 21: 399–405.

Bremer F, Brihaye G, André-Balisaux G (1956). Physiologie et pathologie du corps calleux. Schweizer Arch Neurol Psychiatr 78: 51–87.

Broca PP (1861a). Perte de la parole, ramollissement chronique et destruction partielle du lobe anterieur gauche du cerveau. Bull Soc Anthropol 2: 235–238.

Broca P (1861b). Remarques sur le siège de la faculte du langage articule, suivies d'une observation d'aphemie (perte de la parole). Bull Soc Anat 6: 330–357.
Broca P (1861c). Sur le volume et la forme du cerveau suivant les individus et suivant les races, Hennuyer, Paris.
Broca P (1861d). Nouvelle observation d'aphemie produite par une lesion de la troisième circonvolution frontale. Bull Soc Anat VI: 398–407.
Broca P (1863). Localisation des fonctions cérébrales: Siége de langage articulé. Bull Soc Anthropol 4: 200–208.
Broca P (1865). Du siège de la faculté du langage articulé. Bull Soc Anthropol 6: 377–393.
Broca P (1866). Sur le crâne de Dante Alighieri. Bull Mem Soc Anthropol Paris 1: 206–210.
Broca P (1868). Sur la siège de la faculte du langage articule. Trib Med 4: 254–256 and 75: 263–269.
Bryden MP (1982). Laterality. Functional asymmetry in the intact brain, Academic Press, New York.
Chapman LF, Wolff HG (1959). The cerebral hemispheres and the highest integrative functions of man. Arch Neurol 1: 357–424.
Christman SD (2024). A history of brain asymmetry studies. In: LA Bieliauskas, WB Barr (Eds.), The Oxford handbook of the history of clinical neuropsychology. Oxford University Press177–213.
Corballis MC (1980). Laterality and myth. Am Psychol 35: 284–295.
Corballis MC (1983). Human laterality, Academic Press, New York.
Corballis MC (2007). The dual-brain myth. In: S Della Sala (Ed.), Tall Tales about the mind and brain: separating fact from fiction. Oxford Academic291–313.
Corballis MC (2021). How asymmetries evolved: hearts, brains, and molecules. Symmetry 13: 914.
Corballis MC, Morgan MJ (1978). On the biological basis of human laterality: I. Evidence for a maturational left–right gradient. Behav Brain Sci 1: 226–269.
Corballis MC, Corballis PM et al. (2018). Perceptual unity in the split brain: the role of subcortical connections. Brain 141: e46.
Corkin S (2013). Permanent present tense: the unforgettable life of the amnesic patient, H. M. Basic Books, New York.
Critchley M (1961). Head's contribution to aphasia. Brain 84: 551–560.
Critchley M (1965). Neurology's debt to F. J. Gall (1758-1828). Br Med J 2: 775–781.
Critchley M (1979a). The Broca–Dax controversy. In: M Critchley (Ed.), The divine banquet of the brain and other essays. Raven Press, New York, pp. 72–82.
Critchley M (1979b). God and the brain: medicine's debt to phrenology. In: M Critchley (Ed.), The divine banquet of the brain and other essays. Raven Press, New York, pp. 235–253.
Cubelli R, Montagna CG (1994). A reappraisal of the controversy of Dax and Broca. J Hist Neurosci 3: 1–12.
Dandy WE (1930). Changes in our conception of the localization of function in the brain. Am J Physiol 93: 643.
Dandy WE (1933). Physiological studies following extirpation of the right cerebral hemisphere in man. Bull Johns Hopkins Hosp 53: 31–51.
de Haan EHF, Corballis PM, Hylliard SA et al. (2020). Split-brain: what we know now and why this is important for understanding consciousness. Neuropsychol Rev 30: 224–233.
Fechner GT (1860). Elemente der Psychophysik, Breitkopf und Härtel, Leipzig.
Ferrier D (1876). The functions of the brain, Smith, Elder and Company, London.
Finger S (2009). The birth of localization theory. Handb Clin Neurol 95: 117–128.
Franz SI (1911). The functions of the cerebrum. Psychol Bull 8: 111–119.
Franz SI (1912). New phrenology. Science 35: 321–328.
Fritsch GT, Hitzig E (1870). Uber die elektrische Erregbarkeit des Grosshirns. Arch Anat Phys 37: 300–332.
Gall FJ (1825). Sur les fonctions du cerveau et sur celles de chacune de ses parties: avec des observations sur la possibilité de reconnaître les instincts, les penchans, les talens, ou les dispositions morales et intellectuelles des hommes et des animaux, par la configuration de leur cerveau et de leur tête, J.-B. Baillière, Paris.
Geschwind N (1965). Disconnection syndromes in animals and man. I. Brain 88: 237–294. https://doi.org/10.1093/brain/88.2.237.
Geschwind N, Kaplan E (1962). A human cerebral deconnection syndrome. Neurology 12: 675–685.
Glickstein M, Berlucchi G (2008a). Classical disconnection studies of the corpus callosum. Cortex 44: 914–927.
Glickstein M, Berlucchi G (2008b). K.M. Bykov and transfer between the hemispheres. Brain Res Bull 77: 117–123.
Gould SJ (1981). The mismeasure of man, Norton and Co., New York.
Grünbaum ASF, Sherrington CS (1903). Observations on the physiology of the cerebral cortex of the anthropoid apes. Proc R Soc Lond 72: 152–155.
Güntürkün O, Ströckens F, Ocklenburg S (2020). Brain lateralization: a comparative perspective. Physiol Rev 100: 1019–1063.
Hamada H (2020). Molecular and cellular basis of left–right asymmetry in vertebrates. Proc Jpn Acad Ser B Phys Biol Sci 96: 273–296.
Harrington A (1985). Nineteenth-century ideas on hemisphere differences and "duality of mind". Behav Brain Sci 8: 617–660.
Harrington A (1987). Medicine, mind, and the double brain. A study in nineteenth century thought, Princeton University Press, Princeton.
Harris LJ (1999). Early theory and research on hemispheric specialization. Schizophr Bull 25: 11–39.
Head H (1926). Aphasia and kindred disorders of speech, Cambridge University Press, London.
Hebb DO (1949). The organization of behavior, Wiley, New York.
Hecaen H (1970). La symetrie en neuropsychologie. Totus Homo 2: 8–16.
Hécaen H, Lanteri-Laura G (1977). Evolution des connaissances et de doctrine sur les localisations cerebrales, Desclee de Brouwer, Paris.
Jackson JH (1874). On the nature of the duality of the brain. Brain 38: 80–103. Medical Press and Circular. Reprinted in Brain.
Kimura D (1961a). Some effects of temporal-lobe damage on auditory perception. Can J Psychol 5: 156–165.
Kimura D (1961b). Cerebral dominance and the perception of verbal stimuli. Can J Psychol 15: 166–171.
Kimura D (1963). Right temporal lobe damage. Arch Neurol 8: 264–271.
Kimura D (1964). Left-right differences in the perception of melodies. Q J Exp Psychol 16: 355–358.
Kimura D (1973). The asymmetry of the human brain. Sci Am 228: 70–78.
Klüver H, Bucy PC (1939). Preliminary analysis of functions of the temporal lobes in monkeys. 1939. J Neuropsychiatry Clin Neurosci 9: 606–620.
Lashley KS (1929). Brain mechanisms and intelligence: a quantitative study of injuries to the brain, University of Chicago Press, Chicago.
Lashley KS (1930). Basic neural mechanisms in behavior. Psychol Rev 37: 1–24.

Lashley KS (1931). Mass action in cerebral function. Science 73: 245–254.
Lashley KS (1941). Coalescence of neurology and psychology. Proc Am Philos Soc 84: 461–470.
Lashley KS (1950). In search of the engram. In: Society for experimental biology, physiological mechanisms in animal behavior, Academic Press, New York, pp. 454–482.
Lashley KS (1958). Cerebral organization and behavior. Res Pub Ass Res Nerv Ment Dis 36: 1–4. discussion 14–18.
Lashley KS, Chow K, Semmes J (1951). An examination of the electrical field theory of cerebral integration. Psychol Rev 58: 123–136.
Leblanc R (2017). Fearful asymmetry: Bouillaud, Dax, Broca, and the localization of language, Paris, 1825–1879, McGill-Queen's University Press, Montreal & Kingston, London, Chicago.
Levi P (2016). L'asimmetria e la vita. In: M Belpoliti (Ed.), Articoli e Saggi 1955–1987. Einaudi, Torino.
Levy J (1977). The mammalian brain and the adaptive advantage of cerebral asymmetry. Ann N Y Acad Sci 299: 264–272.
Liepmann H (1908). Über die Funktionen des Balkens beim Handeln. In: H Liepmann (Ed.), Drei Aufsätze aus dem Apraxiegebiet. Karger, Berlin.
Lorente de No´ R (1938). Architectonics and structure of the cerebral cortex. In: JF Fulton (Ed.), Physiology of the nervous system. Oxford University Press, New York, pp. 291–330.
Luys J (1876). Le cerveau et ses fonctions. Librairie Germer Baillière et C, Paris.
McManus C (2002). Right hand, left hand. The origins of asymmetry in brains, bodies, atoms and cultures, Weidenfeld & Nicholson, London.
Mesulam M-M (2015). Fifty years of disconnexion syndromes and the Geschwind legacy. Brain 138: 2791–2799.
Milner B (1954). Intellectual function of the temporal lobes. Psychol Bull 51: 42–62.
Milner B (1958). Psychological defects produced by temporal lobe excision. Res Publ Assoc Res Nerv Ment Dis 36: 244–257.
Milner B (1971). Interhemispheric differences in the localization of psychological processes in man. Br Med Bull 27: 272–277.
Mishkin M (1951). Effects of selective ablations of the temporal lobes on the visually guided behavior of monkeys and baboons. Unpublished doctor's thesis. McGill University.
Myers RE (1956). Function of corpus callosum in interocular transfer. Brain 79: 358–363.
Myers JJ, Sperry RW (1985). Interhemispheric communication after section of the forebrain commissures. Cortex 21: 249–260.
Nadel L, Maurer AP (2020). Recalling Lashley and reconsolidating Hebb. Hippocampus 30: 776–794.
Piercy M (1964). The effects of cerebral lesions on intellectual function: a review of current research trends. Br J Psychiatry 110: 319–352.
Rogers LJ, Zucca P, Vallortigara G (2004). Advantages of having a lateralized brain. Proc R Soc Lond B 271: S420–S422.
Rutten G-J (2022). Broca-Wernicke theories: a historical perspective. Handb Clin Neurol 185: 26–34.
Sagan C (1979). Broca's brain, Random House, New York.
Soury J (1899). Le système nerveux central, structure et fonctions, histoire critique des theories et des doctrines, Georges Carre et C. Naud, Paris.
Sperry RW (1961). Cerebral organization and behavior. Science 133: 1749–1757.
Sperry R (1982). Some effects of disconnecting the cerebral hemispheres. Nobel Lecture Sci 217: 1223–1226.
Sperry RW (1992). In search of psyche. In: FG Worden, JP Swazey, Adelman (Eds.), The neurosciences: paths of discovery. Birkhäuser, Boston.
Stein J (2001). The concept of hemispheric lateralization. In: CF Rose (Ed.), Twentieth century neurology. The British contribution. Imperial College Press, London, pp. 47–58.
Vogt G (2015). Stochastic developmental variation, an epigenetic source of phenotypic diversity with far-reaching biological consequences. J Biosci 40: 159–204.
von Hartmann E (1869). Philosophie des Unbewussten, Versuch einer Weltanschauung, Carl Duncker, Berlin.
von Monakow C (1910). Über Lokalisation der Hirnfunktion, Bergmann, Wiesbaden.
von Monakow C (1914). Die Lokalisation im Grosshirn und der Abbau der Funktion durch kortikale Herde, Bergmann, Wiesbaden.
Walusinski O, Boller F, Henderson VW (2019). Shining a light on some of the most famous 19th and 20th Century's neuropsychologists. Front Neurol Neurosci 44: 192–229.
Watson H (1836). What is the use of the double brain? Phrenological J Miscellany 9: 608–611.
Wernicke C (1874). Der aphasische Symptomencomplex: eine psychologische Studien auf anatomischer Basis, Cohn and Weigert, Breslau.
Wigan AL (1844). A new view of insanity: the duality of the mind proved by the structure, functions, and diseases of the brain, and by the phenomena of mental derangement, and shewn to be essential to moral responsibility, Longman, Brown, Green, and Longmans, London.
Zangwill OL (1960). Cerebral dominance and its relation to psychological function, Oliver & Boyd, Edinburgh.
Zangwill OL (1961). Lashley's concept of cerebral mass action. In: WH Thorpe, OL Zangwill (Eds.), Current problems in animal behaviour. Cambridge University Press, Cambridge, pp. 59–86.
Zangwill OL (1971). Diagram makers old and new. Totus Homo 3: 53–58.
Zangwill OL (1977). Consciousness and the Brain. In: RE Butts, J Hintikka (Eds.), Foundational problems in the special sciences. The University of Western Ontario Series in philosophy of science. vol. 10. Springer, Dordrecht, pp. 153–165.

Section I

Anatomic asymmetries

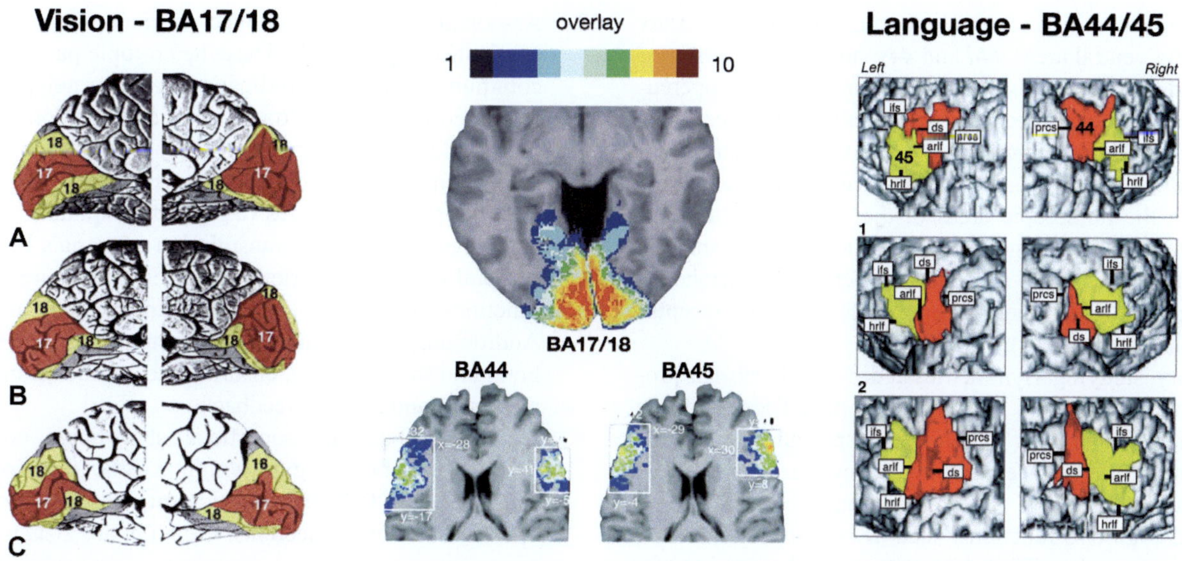

Fig. 2.2. Cytoarchitectonic probability maps with stereotaxic information about intersubject variability for the occipital (vision) and inferior frontal gyrus (language). Adapted from Amunts K, Weiss PH, Mohlberg H, et al. (2004). Analysis of neural mechanisms underlying verbal fluency in cytoarchitectonically defined stereotaxic space—the roles of Brodmann areas 44 and 45. Neuro Image 22: 42–56. https://doi.org/10.1016/j.neuroimage.2003.12.031 and Uylings HBM, Rajkowska G, Sanz-Arigita E, et al. (2005). Consequences of large interindividual variability for human brain atlases: converging macroscopical imaging and microscopical neuroanatomy. Anat Embryol 210: 423–431. https://doi.org/10.1007/s00429-005-0042-4.

pattern—such as reduced or absent asymmetry—are often observed in individuals with developmental language disorders like dyslexia (Vanderauwera et al., 2018).

To explore cortical asymmetries in detail, the ENIGMA (Enhancing NeuroImaging Genetics through Meta-Analysis; Thompson et al., 2020) Consortium conducted a large-scale study to map cortical brain asymmetries in 17,141 healthy young adults by examining cortical thickness (the measurement of the distance between the outer and inner layers of the brain's cortex) and surface area (the total exposed area of the brain's cortex; Kong et al., 2018). The cortical thickness analysis revealed an asymmetry pattern along the anterior–posterior axis, characterized by pronounced leftward asymmetry in the frontal regions and rightward asymmetry in the posterior cortex (temporal, parietal, occipital cortex). The study observed that the inferior frontal gyrus, including Broca's area, exhibited a bilateral distribution in cortical thickness, contrasting with previous literature findings (Koelkebeck et al., 2014; Plessen et al., 2014). However, the surface area analysis demonstrated leftward lateralization in two language-related regions: the pars opercularis of the inferior frontal gyrus (part of Broca's area) and the transverse gyrus (or Heschl's gyrus) in the temporal lobe, while the pars triangularis in the inferior frontal gyrus and the inferior parietal lobule showed asymmetry in the opposite direction (rightward). Further examples of surface and sectional asymmetry are detailed in the book "The Lateralized Brain" in Chapter 4 (Ocklenburg and Güntürkün, 2024).

Subcortical asymmetries are rarely reported in the literature on healthy adults and are minor compared to macroanatomical asymmetries at the cortical level. This observation aligns with a study by Croxson and colleagues, which shows that the subcortical areas that are phylogenetically older demonstrate less variability between individuals (Croxson et al., 2018). A study by Guadalupe et al. (2017) conducted the largest-ever analysis of subcortical asymmetries in healthy individuals (Guadalupe et al., 2017). Their findings revealed that the thalamus, caudate nucleus, amygdala, and hippocampus exhibit a rightward lateralization in volume. Additionally, they identified sex differences and age-related changes in the asymmetry of the putamen. The study also reported that the asymmetry of the globus pallidus ($h^2 = 0.148$), putamen ($h^2 = 0.270$), hippocampus ($h^2 = 0.180$), nucleus accumbens ($h^2 = 0.114$), and thalamus ($h^2 = 0.228$) is highly heritable. In this context, the heritability estimate (h^2) represents the proportion of variation in brain asymmetry attributed to genetic as opposed to environmental factors. This means that a heritability estimate of 0.270 for the putamen represents approximately 27% of the variability in the asymmetry of this structure due to genetic factors, making it the most heritable subcortical structure (for more details on heritability, check Section V in this book).

Cytoarchitectural asymmetries in the brain refer to differences in the cellular structure and organization between the two hemispheres (Fig. 2.2), which play a crucial role in functional lateralization. Much focus has been put on the cytoarchitectural asymmetry in language areas (Amunts et al., 1999) showing that pars the opercularis (BA44) is left-lateralized. In contrast, the pars triangularis (BA45) did not reveal hemispheric differences (for more details, see Broca's Region by Grodzinsky and Amunts, 2006). The subdivision of the inferior frontal gyrus into two cytoarchitectonic areas (44 and 45) has been further elucidated using a **multireceptor approach**. The regional distribution of transmitter receptors in the cerebral cortex has been proven to be a powerful indicator of functional diversity. In vitro receptor autoradiography identified a ventral precentral transitional cortex *6r1*, dorsal and ventral areas *44d* and *44v*, anterior and posterior areas *45a* and *45p*, and areas *op8* and *op9* in the frontal operculum. The cholinergic M_2 receptor was significantly lateralized, particularly in the pars opercularis (area 44v+d; Amunts et al., 1999). In another study, cluster analyses over 15 receptors were performed on the homotopic areas of the left and right hemispheres, including language-relevant regions. This analysis revealed distinct clustering patterns between the hemispheres, suggesting potential interhemispheric differences in receptor densities (Zilles et al., 2015).

In summary, studying asymmetries at different levels has provided valuable insights into how structural differences between the hemispheres contribute to functional specialization in the human brain. From surface anatomy to cytoarchitecture and receptor distribution, these asymmetries have been linked to behavior and cognitive functions, such as sensorimotor, visual, and language processing. The white matter, which connects cortical regions and facilitates communication between hemispheres, also exhibits significant asymmetry, which will be explored in detail in the remainder of this chapter.

CONNECTIONAL ANATOMY

Earlier reports have indicated selective asymmetry patterns for some white matter connections (e.g., Catani et al., 2010; Rojkova et al., 2016; Thiebaut de Schotten et al., 2011a). We offer an extensive update on the asymmetry of a wide range of white matter tracts using high-resolution data from the Human Connectome Project (7T HCP www.humanconnectome.org, preprocessed diffusion data openly available from www.BCBlab.com). We performed manual dissection of the commissural (connections between the two hemispheres), association (cortico-cortical connections within the same hemisphere), projection (cortico-subcortical) pathways, and short U-shaped fibers (connecting adjacent gyri). For each connection, we extracted the volume (referred to as voxel count—VC), the streamlines count (track count—TC), and hindrance-modulated orientational anisotropy (HMOA) to calculate the connectional asymmetry index.

Advancements in neuroimaging techniques, enhanced computational power, and algorithm refinements have ushered in an era of unprecedented anatomical resolution (Bates et al., 2023). This progress has enabled researchers to meticulously dissect and describe more minor connections and segments within larger fiber bundles. Notably, studies such as Rojkova et al. (2016) have leveraged these technological advancements to study the intricate details of neural pathways, mapping previously unexplored brain connections using advanced tractography. Within the intricate framework of the brain's white matter, an exquisite tapestry of connections emerges, classified into four distinct groups:

1. Interhemispheric connections, or commissural pathways, establish links between the brain's two hemispheres. These pathways facilitate communication and integration of information between corresponding (homotopic) and different (heterotopic) regions in each hemisphere.
2. Association pathways connect regions within the same hemisphere (ipsilateral). These heterotopic pathways enable communication and coordination between neighboring and distant cortical areas, allowing for the integration of complex information and cognitive processes across brain regions.
3. Projection pathways connect subcortical and cortical regions with peripheral structures (e.g., cortico-spinal tract). These pathways facilitate the transmission of signals between the cerebral cortex and various subcortical structures, enabling functions such as motor control and sensory processing. Additionally, they facilitate communication between the brain and the periphery, ensuring the coordination of motor responses and sensory feedback.
4. U-shaped fibers are superficial short-range connections linking neighboring gyri within a lobe (intralobar, e.g., Catani et al., 2017) or two adjacent lobes (interlobar, e.g., opercula-insular connections, Catani et al., 2012).

Such connections are evident in both hemispheres, leading to the formation of bilateral networks. This systematic organization was initially documented in the 19th century, with longer connections positioned deeper and shorter connections closer to the surface (Fig. 2.3). This arrangement facilitates a sequential progression during Klingler postmortem dissections, starting with the removal of the gray matter mantle to reveal the short U-shaped fibers, followed by the exposure of deeper association fibers, and ultimately reaching the corona radiata and thalamic radiations at the core (Ludwig and Klingler, 1956).

CONNECTIONAL ASYMMETRIES

Connectional asymmetry refers to differences in the organization, composition, or connectivity of white matter tracts between the brain's two hemispheres. The availability of diffusion-weighted imaging (DWI) in the 1980s ushered in a revolution in studying the brain's white matter in the living human brain (Basser et al., 2000; Le Bihan, 2003). This methodological advance facilitated the study of large cohorts, allowing for the systematic mapping of white matter connections and their hemispheric asymmetries, leading to the observation that some white matter tracts are asymmetric (e.g., arcuate fasciculus, Catani et al., 2007), while others are more symmetric (e.g., IFOF, Forkel et al., 2014b). These patterns have been related to cognitive functions (Thiebaut de Schotten et al., 2011b; Forkel et al., 2022) and motor skills (Howells et al., 2018), clinical presentation (Forkel et al., 2014a, 2020; Koshiyama et al., 2020; Chen et al., 2022), and evolution (e.g., Rilling et al., 2008; Croxson et al., 2018; Eichert et al., 2020).

As such, connectional asymmetries constitute a pivotal element in the brain's organization, playing a significant role in functional lateralization and hemispheric specialization. While the literature has investigated the lateralization of some tracts, it is essential to acknowledge that other tracts remain relatively understudied (see Forkel et al., 2022 for review) and have been investigated using small sample sizes or low-resolution data. Some divergent findings reported in the literature can also be attributed to methodological differences using different tracking algorithms

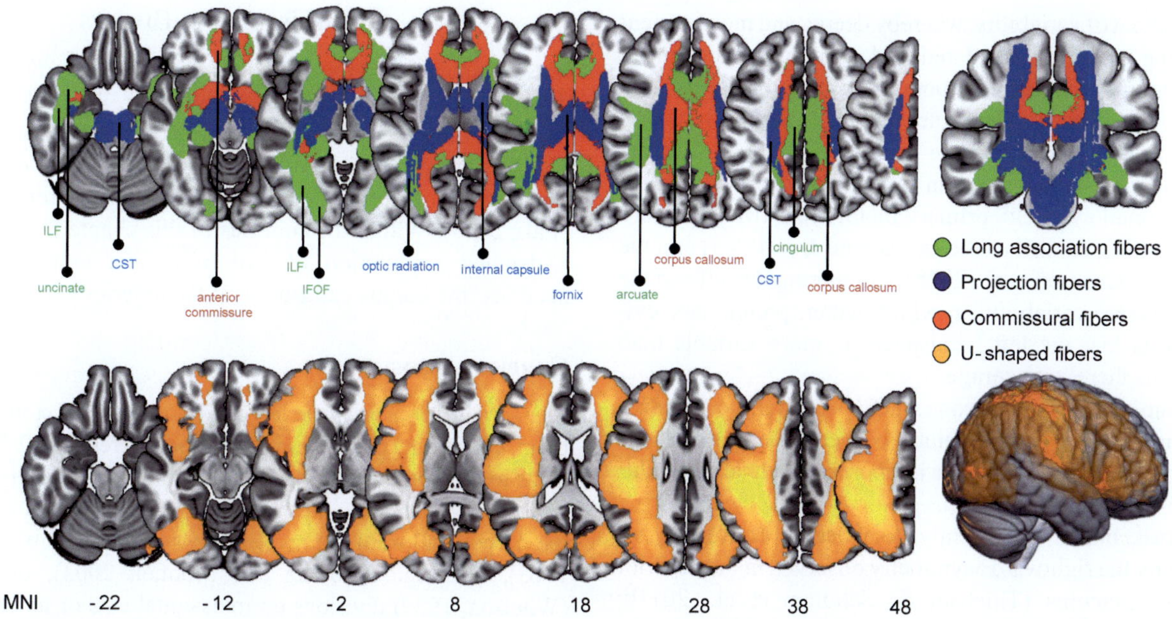

Fig. 2.3. The gradient of fiber arrangement in the brain. The brain's center contains projection fibers (*blue*) on the coronal section. The next layer is composed of interhemispheric commissural fibers (*red*). These central elements are encircled by intrahemispheric cortico-cortical association fibers (*green*, data from https://www.natbrainlab.co.uk/atlas-maps) and a superficial layer of short U-shaped fibers (*orange*, data from https://storage.googleapis.com/bcblabweb/open_data.html). All shown U-shaped fibers have been anatomically validated for the frontal (Catani et al., 2012; Rojkova et al., 2016), parietal (Catani et al., 2017), and occipital lobes (Vergani et al., 2014; Forkel et al., 2015). *ILF*, inferior longitudinal fasciculus; *CST*, cortico-spinal tract; *IFOF*, inferior fronto-occipital fasciculus.

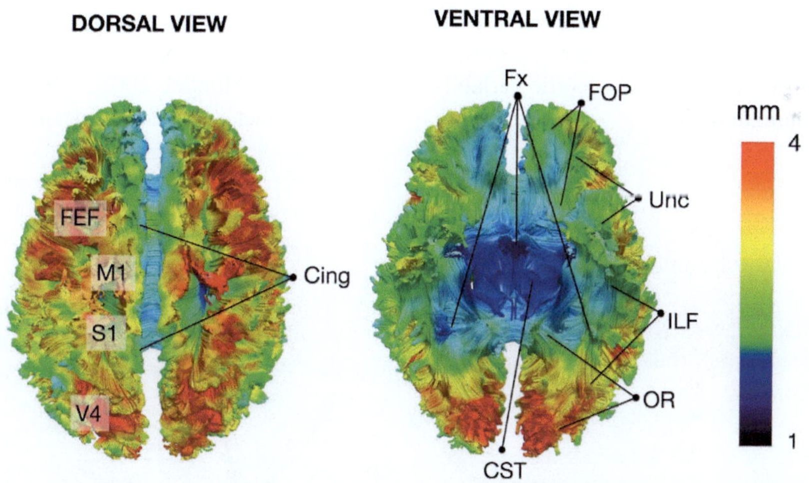

Fig. 2.4. Dorsal and ventral views of the white matter in the left and right hemispheres displaying a variability gradient. *FEF*, frontal eye field; *Cing*, cingulum; *Fx*, fornix; *FOP*, fronto-orbito-polar tract; *Unc*, uncinate fasciculus; *ILF*, inferior longitudinal fasciculus; *OR*, optic radiation; *CST*, cortico-spinal tract. With permission from the authors Croxson PL, Forkel SJ, Cerliani L, et al. (2018). Structural variability across the primate brain: a cross-species comparison. Cereb Cortex 28: 3829–3841. https://doi.org/10.1093/cercor/bhx244.

(e.g., tensor-based models vs spherical deconvolution, Dell'Acqua and Tournier, 2019; Maier-Hein et al., 2017) and dissection techniques (e.g., automatic vs manual; Beyh et al., 2022). For the sake of this chapter and to mitigate some of these limitations, we used an advanced tracking algorithm optimized on postmortem Klingler dissections (i.e., spherical deconvolution, Dell'Acqua et al., 2013; Dell'Acqua and Tournier, 2019). The high-resolution in vivo data was provided by the Human Connectome Project (www.humanconnectome.org) and we used a manual dissection approach (Forkel et al., 2024).

Connectional variability

A seminal study conducted in 2018 marked a significant milestone by unveiling, for the first time, the overarching pattern of interindividual variability within the white matter at the population level (Fig. 2.4, Croxson et al., 2018). This investigation shed light on a gradient of variability, whereby deeper and more ancient regions of the brain exhibited reduced variation, indicating a higher degree of similarity in structures such as the fornix (fx), cingulum (cing), cortico-spinal tract (CST), and optic radiation (OR). The variability increased in the superficial white matter, exhibiting distinct characteristics in connection to critical regions such as the frontal eye field, primary motor areas (M1), primary sensory areas (S1), as well as the extrastriate area 4 (V4) and the striate area (V1; Croxson et al., 2018). Comparing the left and the right hemispheres' variability in white matter, preliminary evidence suggests that the left hemisphere is more variable than the right hemisphere, on average.

When examining individual connections, however, a captivating pattern of intraindividual variability between the hemispheres unfolds. Some connections, like the inferior longitudinal fasciculus (ILF), exhibit a bilateral nature.

By contrast, others consistently demonstrate asymmetry, as exemplified by the rightward asymmetry observed in the superior longitudinal fasciculus (Thiebaut de Schotten et al., 2011b; Amemiya et al., 2021). Additionally, certain connections display interindividual variability, as exemplified by the arcuate fasciculus, which can manifest as bilateral, slightly asymmetric, or highly asymmetric across the population (Catani et al., 2007; Berthier et al., 2012). While this hemispheric difference in the arcuate is less prominent using advanced tracking algorithms, the general trend persists.

In the subsequent sections, we explore individual brain connections stratified by their classification into commissural, association and projections. We will specify their anatomical characteristics, asymmetric attributes, and tract-specific variability for 56 connections in the human brain.

Commissural pathways

Interhemispheric commissures serve as vital conduits, bridging the left and right hemispheres of the brain. While these connections are often perceived as symmetric, it should be noted that a significant proportion, potentially up to 75%, of commissural fibers may be heterotopic in healthy participants (Szczupak et al., 2023). Within the confines of this chapter, we embark on a detailed exploration of two prominent commissural connections: the corpus callosum and the anterior commissure.

Corpus callosum

The corpus callosum is the largest white matter tract in the human brain, connecting the left and right hemispheres (Fig. 2.5A). It connects to all lobes except the anterior temporal lobe, served by that anterior commissure, and the insular cortex. Several divisions of the corpus callosum are available based on anatomical (Crosby, 1963), histologic (Aboitiz and Montiel, 2003), or geometric (Witelson, 1989) divisions on midsagittal cuts or its connectivity profile (Catani and Thiebaut de Schotten, 2012).

Anatomical asymmetry

Most anatomical callosal studies examined the cross-section through the center of this large white matter tract on the midsagittal plane. Regional fiber composition analysis along the callosal midline indicates that midsagittal regions linking primary and secondary sensory areas have a more coarse diameter and extensively myelinated fibers than regions connecting higher-order areas (Aboitiz and Montiel, 2003). This leads to observations that the axonal diameter, or the connection strength, is correlated with

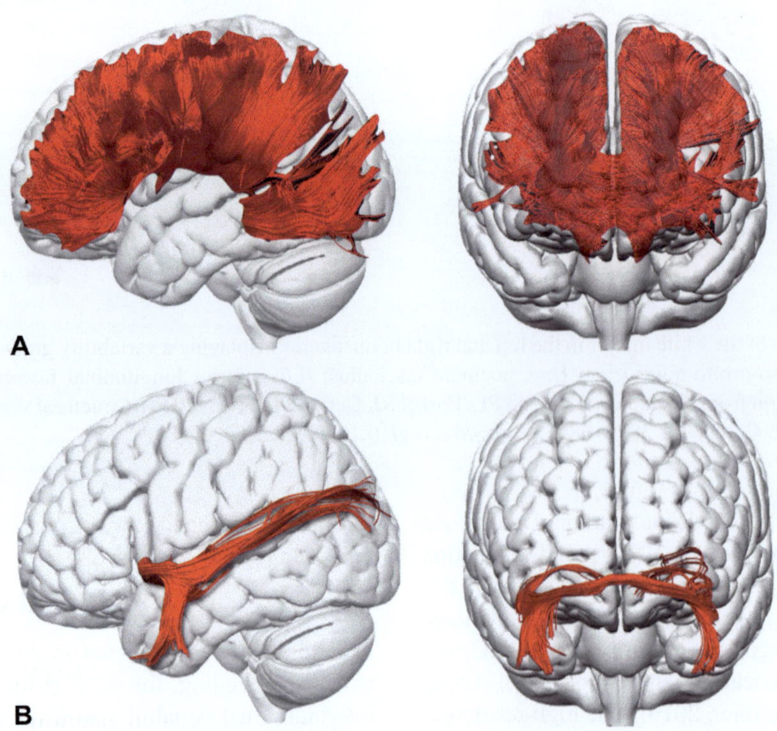

Fig. 2.5. Commissural pathways. (A) Corpus callosum and (B) anterior commissure.

the interhemispheric conduction speed for visual and tactile modalities (Horowitz et al., 2015). Hand preference, a remarkably prominent functional asymmetry observed in nearly 90% of individuals favoring the right hand, has been extensively associated with variations in the structure and function of the corpus callosum (for a comprehensive review, refer to Beaton, 1997; Budisavljevic et al., 2021; Westerhausen and Papadatou-Pastou, 2022). Hand preference is, however, also associated with fronto-parietal association pathways mediating attentional motor-visual integration (Howells et al., 2020).

Studying the cortical callosal projection beyond the midline has become possible using tractography. For example, Putnam et al. (2010) provide evidence for the anatomical asymmetries in the posterior part of the corpus callosum, the so-called splenium or forceps major. They demonstrated that more connections run from the splenium to the right hemisphere than from the splenium to the left hemisphere (Putnam et al., 2010). This anatomical asymmetry of the corpus callosum poses relevant implications for interhemispheric communication and coordination (Gazzaniga, 2000).

Interindividual variability

Throughout development, the posterior parts of the CC reach the peak of myelination earlier than the anterior regions of the CC (Lebel et al., 2008; Lebel and Deoni, 2018). In infants and toddlers, the most significant variability in terms of callosal thickness was observed in the posterior midbody (Tsuzuki et al., 2022). An interruption in the development of the corpus callosum can lead to a lack or partial development of the white matter tract, known as agenesis of the corpus callosum (ACC, see Paul, 2011), which alters the interhemispheric connectivity pattern (Forkel et al., 2014a). While these patients often present with subclinical scores on neuropsychologic and cognitive assessments, their anatomical features appear vastly different, often forming an ipsilateral connection of aberrant callosal fibers that did not cross the hemispheric midline (Aboitiz et al., 1992).

Healthy adults demonstrate higher variability in the anterior thirds of the corpus callosum and at the border between the splenium and isthmus, with the lowest variance in the body and the splenium and rostrum (Luders et al., 2006).

A recent tractography study identified another common variability feature within the corpus callosum. Up to 15% of people not only present with commissural (i.e., left/right) connections but also longitudinal (i.e., anterior/posterior) fibers in the middle of the corpus callosum (Van der Voorta et al., n.d.).

Another key observation is that lesions in the posterior part of the corpus callosum can result in a broad range of neurologic and neuropsychiatric disorders, such as left-hand tactile agnosia, apraxia, hemispatial neglect, dichotic listening suppression, bilateral crossed optic ataxia, optic aphasia, color anomia, alexia without agraphia, and hemialexia (Catani and Thiebaut de Schotten, 2012). In contrast, lesions in the anterior part of the corpus callosum tend to produce fewer symptoms, such as the alien hand syndrome or anarchic hand syndrome. These examples demonstrate the degree of variability in callosal development, differential presentation of callosal lesions, and interindividual variability in anatomical presentation.

Anterior commissure

The anterior commissure, an interhemispheric connection situated ventrally, is the primary connection between the anterior temporal lobes and the amygdalae (Peltier et al., 2011). It holds significant prominence in radiology and neurosurgery, serving as a classical landmark (i.e., the anterior–posterior commissure axis, AC-PC) due to its distinct appearance as a well-defined white dot on the midsagittal plane on structural imaging. As a white matter tract, it has a unique shape that resembles the handlebar of a racing bike when looked upon dorsally (Fig. 2.5B). Its midline is gently cushioned anteriorly and posteriorly by the columns of the fornix (Singh, 2014; Akeret et al., 2022), while the horns elegantly curve into the anterior temporal lobes on either side. Posteriorly, the anatomy of the anterior commissure is less well-defined, with some studies showing a fiber system extending posteriorly (Di Virgilio et al., 1999; Patel et al., 2010; Peltier et al., 2011; Barrett et al., 2020; Baudo et al., 2022), while others do not (Catani and Thiebaut de Schotten, 2008; Akeret et al., 2022). The functional significance of the anterior commissure is not yet well established beyond the olfactory system (Lavrador et al., 2019) and isolated reports of involvement in language functions (Northam et al., 2012).

Anatomical asymmetry

The anterior commissure is not a commonly studied white matter tract (Forkel et al., 2022) despite its prominence as an anatomical landmark in stereotactic and functional neurosurgery and general radiology to orient brain scans along the AC-PC line. One study used diffusion spectrum imaging (DSI) to study the size of the AC in the left and right hemispheres (Baudo et al., 2022). In this study, the anterior branch of the anterior commissure demonstrates slight left lateralization, while the posterior branches showed right dominance, although these results did not reach statistical significance. Using tractography, the anterior commissure often separates into the left and right branches due to crossing fibers with the fornix in the midline. This limitation can be partially overcome using optimized acquisitions (e.g., cardiac gating), appropriate preprocessing (e.g., alignment), and advanced tracking algorithms that can resolve crossings (e.g., spherical deconvolution).

Interindividual variability

Using postmortem histologic methods, the cross-section of the anterior commissure was studied, indicating a higher fiber count in women compared to men. At the same time, this pattern was inversed in patients with schizophrenia (Highley et al., 1999). Sex-related differences were also reported in the context of aging: the volume of the anterior commissure reduced in men within two decades (20–40 years), while there was no such statistically significant difference in women (Choi et al., 2011).

Association pathways

Long-range association pathways are interlobar white matter tracts (i.e., spanning across one or more lobes). Their pattern of

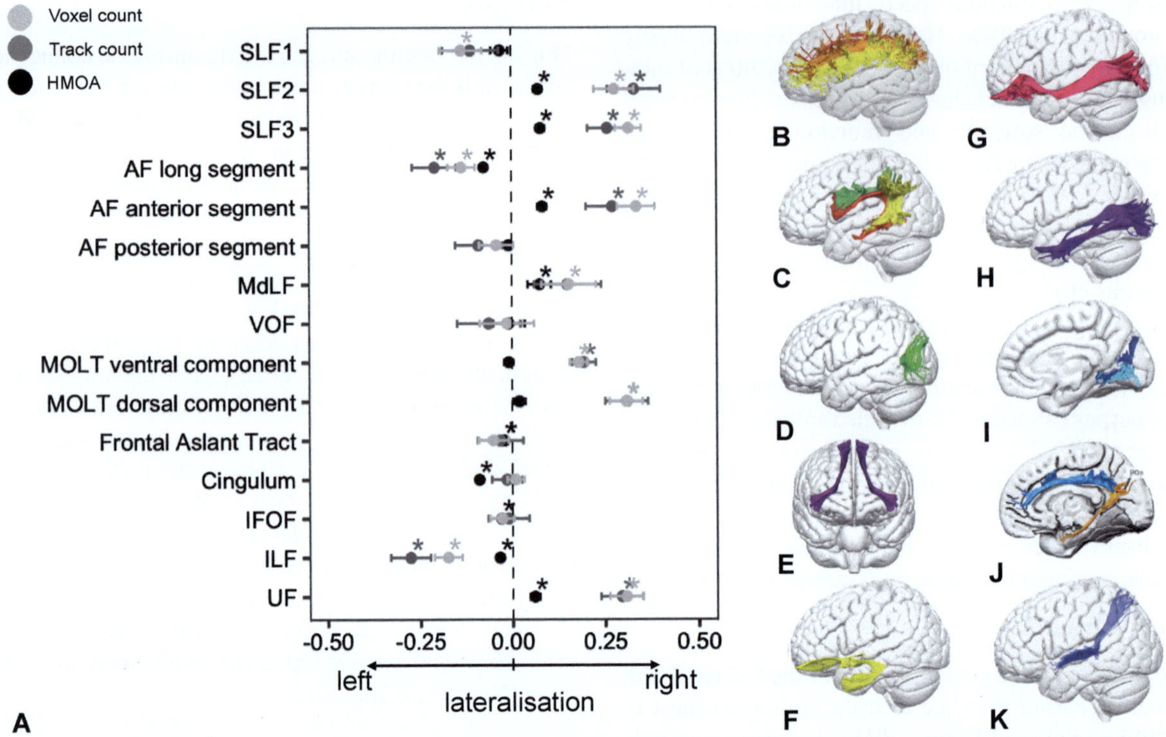

Fig. 2.6. Association pathways. (A) Lateralization index for the number of streamlines (tract count), volume (voxel count), and hindrance modulated orientational anisotropy (HMOA) (mean ± 95% confidence interval) for the association pathways. $^{*}P<0.05$ after Bonferroni correction for multiple comparisons. The lateralization index (LI) was calculated [LI = (right − left)/(right + left)] and compared using a one-sample *t*-test (comparison with zero). LI < 0 equals left lateralized or leftward asymmetry; LI > 0 equals right lateralized or rightward asymmetry, and LI = 0 means bilateral distribution or no asymmetry. (B) Superior longitudinal fasciculus [I (*red*), II (*orange*), III (*yellow*)], (C) arcuate fasciculus (long (*red*), anterior (*green*), and posterior (*yellow*) segments), (D) vertical occipital fasciculus, (E) frontal aslant tract, (F) uncinate fasciculus, (G) inferior fronto-occipital fasciculus, (H) inferior longitudinal fasciculus, (I) medial occipital longitudinal tract (ventral and dorsal components), (J) cingulum, (K) middle longitudinal fasciculus. *SLF*, superior longitudinal fasciculus; *AF*, arcuate fasciculus; *MdLF*, middle longitudinal fasciculus; *VOF*, vertical occipital fasciculus; *MOLT*, medial occipital longitudinal tract; *IFOF*, inferior fronto-occipital fasciculus; *ILF*, inferior longitudinal fasciculus; *UF*, uncinate fasciculus.

asymmetry can vary significantly, ranging from strongly asymmetric (e.g., arcuate fasciculus) to bilateral (e.g., inferior fronto-occipital fasciculus). Here, we will describe the primary long-range and vertical association fibers (Fig. 2.6).

Inferior fronto-occipital fasciculus

The history surrounding the fronto-occipital network is a multifaceted and intricate tale, as explored in studies by Forkel et al. (2014b) and Schmahmann and Pandya (2007). This story is still evolving and awaiting further experimental and empirical evidence. In monkeys, the brain exhibits a dorsal (or superior) fronto-occipital fasciculus (FOF), which can be visualized using axonal tracing techniques (Schmahmann et al., 2007). Conversely, initial reports suggested the existence of a superior FOF (sFOF) in the human brain, although it typically remained unseen due to its course amid the callosal fibers. This hypothesis arose from observations of patients with congenital absence of the corpus callosum (callosal agenesis, see section *Corpus callosum above*), who presented with a superior fronto-occipital fiber system (Onufrowicz, 1887). However, subsequent postmortem microsurgical studies (Türe et al., 1997) and diffusion-weighted tractography studies (e.g., Forkel et al., 2014b) were unable to reconstruct an sFOF in the human brain consistently. The observed longitudinal fibers were considered to be callosal fibers unable to cross the hemispheric midline and forming a longitudinal fiber system that resembles a fronto-occipital fasciculus (Forkel et al., 2014b). Early tractography atlases—still widely used today—include a mask for the sFOF (e.g., Mori et al., 2006), potentially leading to mislabeling anatomical findings in atlas-based studies.

In the case of the ventral connection, initial observations pointed to an interspecies disparity, where the inferior fronto-occipital fasciculus (IFOF) was visualized in the human brain using postmortem dissections (e.g., Curran, 1909; Kier et al., 2004) and tractography (e.g., Kier et al., 2004; Lawes et al., 2008; Forkel et al., 2014b), but not in the monkey brain using axonal tracing (Schmahmann et al., 2007). Consequently, the IFOF was regarded as a unique tract exclusive to humans and speculated to play a role in language processing (Duffau et al., 2005). However, comparative neuroimaging studies have since demonstrated that the IFOF can be visualized in monkeys using tractography (Mars et al., 2016), casting doubt on its uniqueness and exclusive involvement in language-related functions.

The inferior fronto-occipital fasciculus is a long-range association tract that connects the occipital lobe, enters the stem of the temporal lobe, narrows down at the level of the dorsal external capsule, and fans out, reaching the ventral frontal lobe (Forkel et al., 2014b; Wu et al., 2016). The triad of the IFOF, the uncincate fasciculus, and the inferior longitudinal fasciculus is often referred to as the ventral stream—especially in the neurobiology of language literature.

Anatomical asymmetry

Forkel et al. (2014b) and Wu et al. (2016) did not detect statistically significant differences in the volume between the left and right IFOF. While this is consistent with the volumetric results of (Thiebaut de Schotten et al., 2011a) they reported a right lateralization of the IFOF based on streamline count (Thiebaut de Schotten et al., 2011a).

Interindividual variability

The organization and characteristics of the inferior fronto-occipital fasciculus (IFOF) exhibit variability among individuals. Notably, studies conducted by Wu et al. (2016) involving a sample size of 10 participants and Forkel et al. in 2014 with a larger cohort of 30 participants reported a range of variation regarding the extent of overlap within the core of the tract. Specifically, the overlap between different individuals was observed to span from 40% to 90%. This variation underscores the diverse anatomical configurations this brain pathway can take across individuals.

Uncinate fasciculus

The uncinate fasciculus is a multicomponent white matter bundle connecting the orbitofrontal and anterior temporal lobes (Catani and Thiebaut de Schotten, 2008; Hau et al., 2017). It traverses the inferior floor of the external capsule to run ventral to the inferior fronto-occipital fasciculus (Catani and Thiebaut de Schotten, 2008). The uncinate was divided into five subcomponents: dorsal (subcomponent 1, SC1), ventrolateral (SC2), ventromedial (the hook-shaped bundle, SC3), short posterior (SC4), and anterior ventromedial (SC5) branches (Hau et al., 2017). The functional division of these branches remains to be established. Anatomically, the uncinate fasciculus is part of the limbic system and contributes to emotional regulation (e.g., Craig et al., 2009) and, more recently, has been included as an integral part of the language semantic system (Catani and Thiebaut de Schotten, 2008; D'Anna et al., 2016; Cocquyt et al., 2020; Forkel et al., 2022).

Anatomical asymmetry

The uncinate overall is not significantly lateralized (Thiebaut de Schotten et al., 2011a). Looking at the subcomponents, Hau et al. (2017) described the SC2 and SC4 as bilateral, while the dorsal SC1 segment was left-lateralized. In contrast, the ventromedial segment (SC3) was right-lateralized for streamlines and volume, while the anterior ventromedial branch (SC5) showed right lateralization only for volume (Hau et al., 2017). In an early dissection microscope study, Highley et al. (2002) reported that the right uncinate fasciculus is 27% larger and contains 33% more fibers in the right hemisphere than in the left in 80% of healthy participants.

Interindividual variability

Hau et al. (2017) reported interindividual variability where some uncinate branches are absent: the right ventromedial segment (SC3) was present in all participants ($n = 30$) but absent in four participants in the left hemisphere. The left anterior ventromedial branch (SC5) was present in all participants but absent in six participants in the right hemisphere (Hau et al., 2017). These variability patterns await replication or validation through cognitive correlations.

Inferior longitudinal fasciculus

The inferior longitudinal fasciculus (ILF) is a long-range multi-component white matter tract terminating in the occipital and ventral anterior temporal lobes (Catani et al., 2003). The ILF runs ventral and lateral to the inferior fronto-occipital fasciculus, and their terminations intermingle in posterior occipitotemporal regions (Herbet et al., 2018). The ILF's trajectory in the anterior temporal lobe is ventromedial to the uncinate fasciculus. The ILF is accompanied by a network of short U-shaped connections along its entire course (Catani et al., 2003).

Anatomical asymmetry

Based on the fractional anisotropy analysis, the inferior longitudinal fasciculus demonstrates a leftward asymmetry (Shu et al., 2015; Thiebaut de Schotten et al., 2011a). However, Latini et al. (2017) described a general right lateralization in volume and no statistically significant differences in fractional anisotropy and length measurements in 24 participants (Latini et al., 2017). However, the asymmetry (including volume) was absent when measurements of different subcomponents of the inferior longitudinal fasciculus (e.g., fusiform segment) were compared (Latini et al., 2017).

Superior longitudinal fasciculi (SLF I–III)

The superior longitudinal fasciculus (SLF) is a group of fronto-parietal connections first described in the monkey (Schmahmann and Pandya, 2007). With the availability of advanced tractography, it was possible to trace these connections in the human brain for the first time in 2011 (Thiebaut de Schotten et al., 2011a). It is classically divided into SLF I (connecting superior parietal and superior frontal regions), SLF II (connecting angular gyrus and caudal–lateral prefrontal region), and SLF III (connecting supramarginal gyrus, the ventral premotor, and prefrontal regions) (Thiebaut de Schotten et al., 2011a,b). The most ventral branch, the SLF III, is the most controversial as it shares its trajectory with the horizontal or anterior segment of the arcuate fasciculus (Catani, 2006; Frey et al., 2008; Catani and Thiebaut de Schotten, 2012; Forkel et al., 2022). Per definition, the SLF is considered a longitudinal frontoparietal system and, as such, should be distinct from the frontotemporal arching fibers of the arcuate fasciculus (see Forkel et al., 2022 for discussion).

Anatomical asymmetry

The most dorsal branch (SLF I) demonstrates a symmetric distribution between the two hemispheres (Thiebaut de Schotten et al., 2011b). Thiebaut de Schotten et al. (2011b) and Amemiya et al. (2021) did not detect statistically significant volumetric asymmetry in the second branch (SLF II). However, Amemiya et al. reported a rightward asymmetry for FA. The third branch consistently shows a right lateralization in volume (Thiebaut de Schotten et al., 2011b; Howells et al., 2018; Amemiya et al., 2021).

Interindividual variability

The asymmetry of SLFs varies across the lifespan (Manning et al., 2022). For example, fractional anisotropy in the SLF II follows a bilateral U-distribution in children and seniors but is significantly right-lateralized in adolescents and adults (Amemiya et al., 2021). Similarly, the volumetric asymmetry was only observed for the SLF III in the adult group (Amemiya et al., 2021). Janelle et al. (2022) stress that there is interindividual and intraindividual variability of the SLF, in line with the findings of Amemiya et al. (2021). An example of interindividual variability in the superior longitudinal fasciculus (SLF) was reported by Howells et al. (2018), where right-handed participants demonstrated left lateralization of SLF I, while the SLF I of left-handed participants was right lateralized. Both left- and right-handed participants showed rightward lateralization of SLF III, with left-handers exhibiting significant rightward asymmetry (Howells et al., 2018). They concluded that the effect of handedness is observed in the fronto-parietal tracts rather than in the projection tracts that are primarily involved in the execution of motor actions.

Medial occipital longitudinal tract

The medial occipital longitudinal tract (MOLT) is a recently defined occipital-temporal connection that links the cuneus and lingual gyrus with the medial temporal cortex, namely, the parahippocampal gyrus (Beyh et al., 2022). The MOLT's ventral (lingual) component is larger than the dorsal (cuneus) component, and the tract shows slight right lateralization. Functionally, the MOLT has been implicated in the visuospatial processing domain, particularly in encoding spatial configurations.

Anatomical asymmetry and interindividual variability

Beyh et al. (2022) described statistically significant right lateralization in volume and surface area for the ventral and dorsal branches. Looking at diffusion indices along the tracts (i.e., hindrance modulated orientational anisotropy), only the ventral branch reached significance. Due to its very recent description, the interindividual variability of the MOLT has not been described yet. However, based on the distribution of the lateralization indices, the dorsal MOLT seems to have greater interindividual variability than the ventral MOLT (see Figure 6 in Beyh et al., 2022).

Middle longitudinal fasciculus

The middle longitudinal fasciculus (MdLF) was first described in monkeys connecting the anterior temporal cortex to the inferior parietal lobe (Seltzer and Pandya, 1984). In the human brain, its trajectory was extended to include superior parietal and occipital cortices (Makris et al., 2009; Maldonado et al., 2013; Wang et al., 2013; Kalyvas et al., 2020). In a probabilistic tractography study (Bryant et al., 2020), additional temporal lobe connections with the lateral occipital cortex and superior parietal lobule have been reported in humans. A recent anatomical dissection (n = 12) and in vivo tractography study (n = 24 participants from the Human Connectome Project database; https://db.humanconnectome.org) provided evidence for two subcomponents of the MdLF in the human brain that were consistent across both samples (Latini et al., 2021). According to this work, a posterior branch (pMdLF) connects the posterior regions of the superior temporal gyrus, the anterior portion of the transverse gyrus, and the planum temporale with the superior parietal lobule and lateral occipital cortex. An anterior branch (aMdLF) was reported to connect the anterior portion of the superior temporal gyrus and the occipital lobe. The IFOF runs medial to the MdLF (Latini et al., 2021).

Anatomical asymmetry and interindividual variability

The MdLF is reported to be symmetric in the healthy population with no differences in FA, volume, length, mean diffusivity, axial diffusivity, and radio diffusivity (Makris et al., 2009, 2013; Wang et al., 2013). The analysis of the subcomponents, as described by Latini et al. (2021), also did not reveal any asymmetries.

Arcuate fasciculus (frontotemporal)

The arcuate fasciculus is a large white matter tract that connects the posterior inferior frontal gyrus with the posterior temporal cortex (Catani et al., 2005). Three main branches have been described using different terminology: anterior, posterior, and long segments (Catani and Ffytche, 2005) or horizontal and vertical branches (Frey et al., 2008). The frontotemporal connection, the classical arcuate fasciculus, runs medially, while the shorter segments connecting to the inferior parietal lobe run laterally.

In relation to adjacent white matter tracts, it is one of the most lateral fasciculi, with its two shorter segments (anterior and posterior) being the most lateral. It is inferior to the SLF II; sometimes, it is referred to as SLF III or SLF IV (Janelle et al., 2022), which is misleading terminology (see Forkel et al., 2022 for discussion). There is no consensus yet about the exact anatomical termination of this connection based on postmortem or in vivo dissections. For a review of the temporal terminations, see Giampiccolo and Duffau (2022). Functionally, the arcuate fasciculus is part of the language system but is also involved in other cognitive functions (Forkel et al., 2022). A recent study investigated the functional anatomy and topographic organization of the fronto-temporal arcuate fasciculus, suggesting that the arcuate fasciculus may not function as a uniform structure but instead could be subdivided into distinct segments that connect different cortical regions, each potentially serving specialized roles in cognitive processes such as language, auditory processing, or motor functions (Basile et al., 2024).

Anatomical asymmetry

Developmental changes in asymmetry have been reported and linked to genetic and environmental factors (Lebel and

Beaulieu, 2011; Budisavljevic et al., 2021). In the adult brain, the arcuate is left-lateralized. The individual segments show a bilateral pattern for the posterior segment and right lateralization for the anterior segment. The volume and the number of streamlines of the long segment demonstrated left lateralization (Thiebaut de Schotten et al., 2011a), which aligns with the group-level lateralization distribution shown in the supplementary materials of Catani et al. (2007).

Interindividual variability

Despite the group-level anatomical asymmetries described above, the lateralization in the different segments of the arcuate fasciculus varies (Bernard et al., 2019; Forkel et al., 2022). For example, the long segment is bilaterally distributed in 40% of participants, correlated with a higher performance in a verbal working memory task (Catani et al., 2007). The anatomical variability also predicts a language recovery from aphasia poststroke (Forkel et al., 2014a). The parietal–temporal branch (posterior segment) has also been demonstrated to change as a function of experience, in this case, literacy acquisition (Thiebaut de Schotten et al., 2014).

Cingulum

The cingulum is an encircling white matter tract that runs atop the corpus callosum and then arches around the splenium and downward into the depth of the temporal lobes. It comprises short cortico-cortical connections (U-shaped fibers) and longitudinal fibers (Gong et al., 2005). This structure is anatomically categorized as part of the limbic system (Catani et al., 2013a). Utilizing techniques involving postmortem dissection and tractography, investigations often indicate the presence of a dorsal and ventral segment within the cingulum, with their separation occurring at the level of the parieto-occipital fissure (Jones et al., 2013; Budisavljevic et al., 2016).

Anatomical asymmetry

The cingulum has a symmetric distribution across the two hemispheres regarding volume, fractional anisotropy, and the number of streamlines (Thiebaut de Schotten et al., 2011a).

Interindividual variability

In a data-centered investigation, an endeavor was made to uncover shared patterns and distinctions within the developmental cohort across 20 primary white matter tracts encompassing the cingulum (Bathelt et al., 2019). Clustering machine-learning algorithms were employed to accomplish this. The outcome revealed two major clusters, exhibiting divergence solely for the fractional anisotropy measurements of the cingulum. This divergence correlated with the outcomes of various cognitive assessments, including general intelligence, vocabulary, and different forms of memory. The positive correlation demonstrated that the lower the FA in the cingulum, the lower the task performance scores (Bathelt et al., 2019).

Vertical systems

Frontal aslant tract

First described as the fontal blade (Oishi et al., 2008), the tract was later described in detail and renamed to its current terminology as the frontal aslant tract (FAT; Catani and Thiebaut de Schotten, 2012). The FAT has recently been described alongside various frontal lobe short fibers. It connects pre- and supplementary motor areas with the posterior portion of the inferior frontal gyrus (Catani et al., 2012, 2013b).

Anatomical asymmetry and interindividual variability. In a seminal study, the volume of the FAT was left lateralized (Catani et al., 2012). Several studies have investigated the FAT over the last decade, especially in the neurosurgical and language literature. Many of these results are summarized in reviews (e.g., Burkhardt et al., 2021; Gallet et al., 2022). In recent studies, the FAT has been described as a symmetric white matter tract regarding volume (Briggs et al., 2020). The discrepancies in results likely stem from the high variability of the FAT across populations. Szmuda et al. (2017) reported that approximately 64% ($n = 54$) of neurosurgical patients exhibited a left-lateralized FAT (Szmuda et al., 2017).

The FAT has been implicated in motor-speech functions and speech initiation processes (see Dick et al., 2019 for the discussion; Kemerdere et al., 2016). It has recently been implicated in lexical decision-making tasks where greater leftward lateralization of the FAT in volume, number of streamlines, and FA was correlated with faster processing speed (Vallesi and Babcock, 2020). In contrast, verbal fluency was not correlated with the FAT in this study (Vallesi and Babcock, 2020).

Vertical occipital fasciculus

The anatomy of the vertical occipital fasciculus (VOF) has been verified using tractography (Yeatman et al., 2014; Keser et al., 2016; Briggs et al., 2018; Panesar et al., 2019; Schurr et al., 2019), postmortem dissections (Vergani et al., 2014; Güngör et al., 2017; Palejwala et al., 2020), and comparative anatomy (Takemura et al., 2017). These descriptions were scrutinized against historical postmortem descriptions from Wernicke (Yeatman et al., 2014), Sachs (Vergani et al., 2014; Forkel et al., 2015), and the Dejerines (Bugain et al., 2021). Anatomically, the vertical occipital fiber system is distinct from the parietal–temporal branch of the arcuate fasciculus (e.g., posterior segment) in its trajectories and cortical terminations (Bugain et al., 2021). The intralobar VOF projects between the ventral and dorsal occipital cortices while the interlobar posterior segment of the arcuate fasciculus connects the posterior temporal and inferior parietal regions (i.e., supramarginal and angular gyrus) (Catani et al., 2005; Weiner et al., 2017). The first comprehensive description of the occipital lobe white matter was provided in the atlas "*The white matter of the human cerebrum. Part 1. The Occipital Lobe*" by Heinrich Sachs (Forkel et al., 2015). In recent years, the occipital white matter was revisited with the help of tractography and Klingler dissections (Vergani et al., 2014;

Bugain et al., 2021). Yeatman et al. (2014) demonstrated in vivo that the VOF connects ventral and dorsal parts of the occipital, temporal, and parietal cortices. In relation to other white matter bundles, the VOF is posterior to the parietotemporal arcuate branch (posterior segment) and lateral to the ILF (Yeatman et al., 2014; Weiner et al., 2017).

Anatomical asymmetry. In a deterministic tractography study, Keser and colleagues (2016) demonstrated no statistically significant differences between the left and right VOF for FA, volume, and diffusivity (Keser et al., 2016). This is not in line with a recent study that used a quantitive T1 mapping technique to delineate the VOF and reported a right lateralization of the VOF (Schurr et al., 2019).

Interindividual variability. Weiner et al. (2017) showed several examples of VOF variability across participants. For some participants, the VOF had a sharp boundary with the posterior arcuate fasciculus; for others, the separation was hardly identifiable (Weiner et al., 2017). The VOF has been suggested as the link between the dorsal and ventral visual streams and to contribute to reading through the integration of eye movement with word recognition (anterior branch) and communication with visual processes (posterior branch) (Yeatman et al., 2014).

A recent review identified an association between the VOF and visual tasks, including object recognition and integrating the dorsal and ventral stream during hand actions (Budisavljevic et al., 2018). The VOF terminations have also been shown to predict reading-related responses in the human ventral temporal cortex (Grotheer et al., 2021) and early literacy skills in children (Broce et al., 2019). In the clinical setting, the VOF has recently been associated with visual scores in assessments of patients with multiple sclerosis (Abdolalizadeh et al., 2022).

Projection pathways

Fornix

Owing to its anatomy, automatic methods that rely on priors of commissural-projection-association connections might miss the connection in its entirety due to its commissural (midsagittal body), projection (fimbria), and longitudinal orientations (subcallosal body) along its course. The fornix connects the medial temporal lobe to the mammillary bodies and hypothalamus. The fornix is part of the limbic system and is associated with memory functions (Catani and Thiebaut de Schotten, 2008). Anteriorly, it comprises pre- and postcommissural fibers, known as the anterior and posterior columns, that hug the central part of the anterior commissure from above. The fornix fibers converge just above the anterior commissure and funnel into the body of the fornix, coursing underneath the corpus callosum as a single bundle. At the splenial level of the corpus callosum, the fornix diverges into two branches known as fimbriae, which traverse into the bilateral medial temporal lobes (Choi et al., 2021).

At this juncture, a structure emerges between the fimbriae, commonly called the commissure of the fornix or the hippocampal commissure. A multimodal approach to the forniceal anatomy recently revisited the existence of the forniceal commissure and concluded that it is a thin membrane rather than a commissure (Akeret et al., 2022).

Anatomical asymmetry

A statistically significant leftward FA asymmetry was identified along the fimbria, while the body of the fornix appears symmetric (Sboto-Frankenstein et al., 2014). In an earlier work, Supprian and Hofmann (1997) reported that the left column of the fornix is situated caudal to the right column for most of the participants. However, this pattern was not systematically observed and sometimes also inverse (Supprian and Hofmann, 1997).

Interindividual variability

Given its location in the depth of the brain, the fornix is expected to be less variable across participants (Croxson et al., 2018). The variability in the fornix has been related to accuracy in the performance of scene discrimination but not face discrimination tasks in healthy participants (Postans et al., 2014) and symptom severity with memory deficits (O'Sullivan et al., 2023).

Cortico-spinal tract

The cortico-spinal tract (CST) is a prominent pathway that originates bilaterally in the primary motor cortex with some representations in the premotor and somatosensory cortices (Dum and Strick, 1991; Karadimas et al., 2020). Descending, it passes through the posterior limb of the internal capsule and cerebral peduncles of the midbrain, following the ventral part of the pons before entering the spinal cord at the level of the medulla. Around 75%–90% of fibers of the cortico-spinal tract are decussating to the contralateral side, forming the lateral crossed cortico-spinal tract (Lemon, 2008; Nieuwenhuys et al., 2008). The remaining 10%–25% fibers form the uncrossed anterior cortico-spinal tract (Jang, 2014; Singh, 2014). The CST is part of an extensive projection system between the cortex, the internal capsule, and the brainstem. The CST is the most studied pathway in neurologic and neurosurgical patients (Forkel et al., 2022).

Anatomical asymmetry

Angstmann et al. (2016) described that the asymmetry of the CST is positively correlated with a right-hand advantage in a dexterity task (circle-drawing) in adolescents. However, in older adults (60 years +), the authors did not find a linear relationship between fluency in performing the task and the lateralization of the CST (Angstmann et al., 2016). Notably, the absence of the relationship in the latter group was not driven by the microstructure of the CST per se since there were no significant differences in lateralization between the two groups (adolescents and old adults). In other words, the deterioration of fluency in a dexterity task in older adults cannot be explained by the CST asymmetries alone. It might include regions beyond the areas of the CST (Demnitz et al., 2021). In general, a leftward asymmetry was described in right-handed participants (Hervé et al., 2006; Thiebaut de

Schotten et al., 2011a). However, Hervé et al. (2006) also observed leftward volumetric asymmetry for the left-handed participants albeit less evident than in the right-handed group (Hervé et al., 2006).

Interindividual variability

The lack of consistency in the results of the different studies that attempt to relate the laterality of the cortico-spinal tract with the hand preference might relate to structural brain variability of the CST across healthy populations. First, the number of fibers that decussate at the level of the medulla is not fixed across the population (Seizeur et al., 2014). Second, Dalamagkas et al. (2020) generated the individual variability map of the CST pathway on 37 healthy participants. They described significant individual differences in the CST closer to the cortex (Dalamagkas et al., 2020).

Optic radiation

The optic radiation connects the thalamus' lateral geniculate nucleus (LGN) with the calcarine fissure in the occipital lobe—hence, it is also known as the geniculocalcarine fasciculus. The OR loops from the LGN anteriorly, forming Meyer's loop, before it takes a posterior turn toward the occipital lobe and terminates in the striate cortex of the calcarine sulcus. Meyer's loop might be challenging to dissect with tractography using stringent angular thresholds and can benefit from advanced algorithms (Chamberland et al., 2018). The optic radiation is divided into three bundles: dorsal, central, and anterior, equivalent to Meyer's loop (Mandelstam, 2012). The dorsal and central bundles are the shortest connections to the visual cortex. In contrast, before arriving at the visual cortex, the anterior bundle curves around the tip of the temporal horn of the lateral ventricle (Dreessen de Gervai et al., 2014).

Anatomical asymmetry

The optic radiations are right lateralized (Thiebaut de Schotten et al., 2011a). However, a bundle-wise analysis demonstrated significant hemispheric asymmetry in Meyer's loop in relation to neighboring structures. Specifically, the distances between Meyer's loop and the temporal pole, temporal horn, and uncinate fasciculus in the left hemisphere were shorter than in the right (Rodrigues et al., 2022).

Interindividual variability

A study by Nilsson et al. (2007) revealed a 17-mm variability in the distance between the anterior extent of the Meyer's loop and the temporal pole in a sample of seven healthy participants (Nilsson et al., 2007). This result aligns with findings from other research groups showing a 15-mm variability in sample sizes ranging from 17 to 20 participants (Ebeling and Reulen, 1988; Bürgel et al., 1999; Peuskens et al., 2004).

Thalamic radiations

The thalamic radiations consist of four projection tracts that fan out from the thalamus to reach different parts of the cerebral cortex (Radwan et al., 2022) (Fig. 2.7). The anterior thalamic radiation connects the thalamus to the frontal lobe passing through the

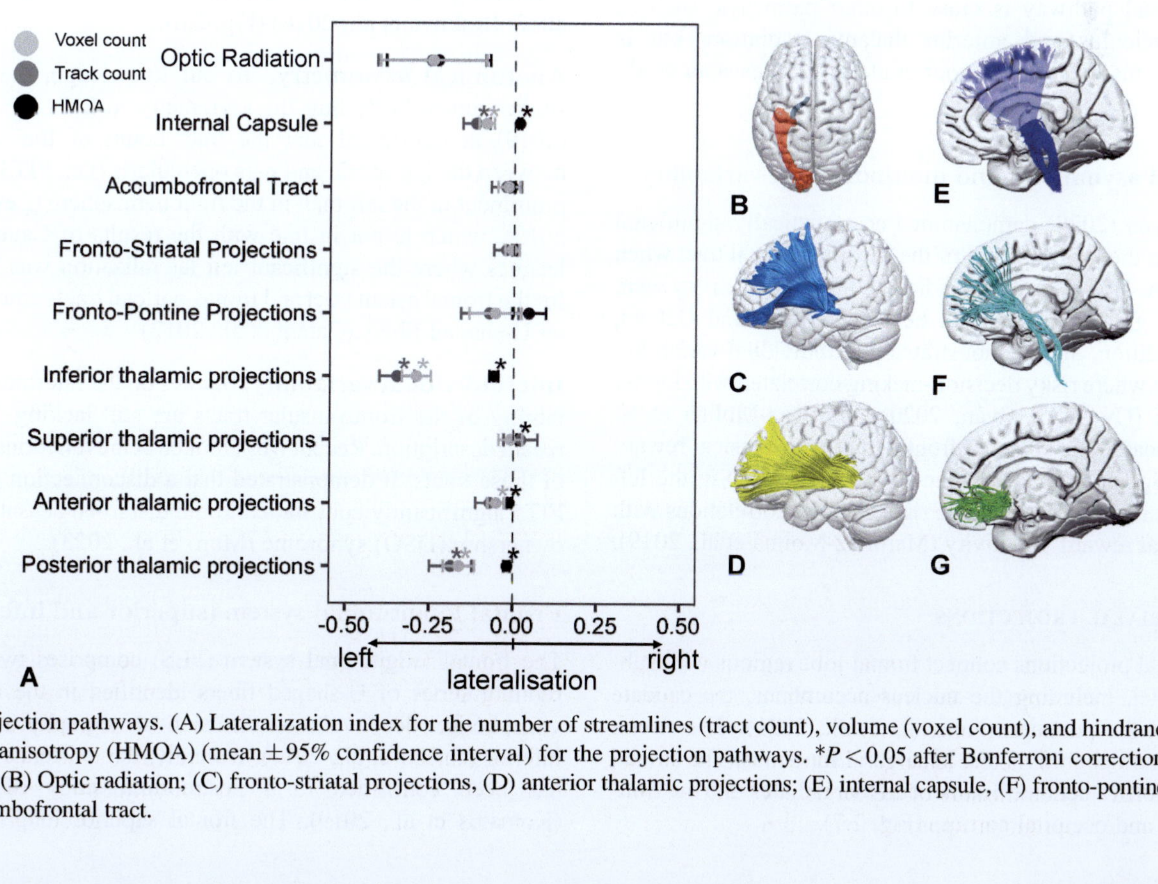

Fig. 2.7. Projection pathways. (A) Lateralization index for the number of streamlines (tract count), volume (voxel count), and hindrance modulated orientational anisotropy (HMOA) (mean ± 95% confidence interval) for the projection pathways. *$P < 0.05$ after Bonferroni correction for multiple comparisons. (B) Optic radiation; (C) fronto-striatal projections, (D) anterior thalamic projections; (E) internal capsule, (F) fronto-pontine projections, and (G) accumbofrontal tract.

anterior thalamic peduncle. The posterior thalamic radiation connects the thalamus to the occipital and parietal lobes through the superior thalamic peduncle. The superior thalamic radiation connects the thalamus to the motor and sensory regions through the superior thalamic peduncle. The inferior thalamic radiation connects the thalamus with the temporal lobe, insula, and frontal lobe (George and Das, 2024). Further small connections to the thalamus include the mammillo-thalamic tract (also known as the bundle of Vicq d'Azyr; (Vicq-d'Azyr, 1786) with fibers originating from the mammillary bodies. After a brief trajectory, they conclude their course by reaching the anterior and dorsal nuclei of the thalamus. A separate branch extending ventrally from the mammillary bodies projects toward the tegmental nuclei, forming what is referred to as the mammillo-tegmental tract (Nauta, 1958).

Anatomical asymmetry and interindividual variability

Wakana et al. (2007) analyzed the asymmetry of the anterior thalamic radiation and observed no statistically significant lateralization in FA or volume (Wakana et al., 2007).

Accumbofrontal tract

The accumbofrontal pathway is a white matter pathway that connects the nucleus accumbens, a brain region involved in reward and pleasure, with the orbital/prefrontal cortex, a brain region involved in decision-making and executive function (Karlsgodt et al., 2015). The accumbofrontal pathway is thought to play a role in regulating reward-related behavior and may be involved in the development of addiction (Feltenstein and See, 2008). The accumbofrontal pathway is close to other pathways, such as uncinate fasciculus and anterior thalamic radiation, but is located more medially (Karlsgodt et al., 2015; Pascalau et al., 2018) (Fig. 2.7).

Anatomical asymmetry and interindividual variability

Uy and Galván (2020) demonstrated no statistically significant differences in the lateralization of the accumbofrontal tract when looking at FA. However, axial, radial, and mean diffusivity were significantly greater in the right hemisphere (Uy and Galván, 2020). In addition, they demonstrated interindividual variability in the sample where risky decision-making correlated with higher tract indices (Uy and Galván, 2020). Martínez-Molina et al. (2019) implicated the accumbofrontal tract with musical reward sensitivity. Specifically, participants with higher AD in the left hemisphere and higher MD in the right showed correlations with lower musical reward sensitivity (Martínez-Molina et al., 2019).

Fronto-striatal projections

Fronto-striatal projections connect frontal lobe regions with subcortical nuclei, including the nucleus accumbens, the caudate nucleus, and the putamen (Komaitis et al., 2021). The circuit mediates motor, cognitive, and behavioral functions. The striatal projections form a delicate brain circuit that radiates into the frontal, parietal, and occipital cortices (Fig. 2.7).

Anatomical asymmetry and interindividual variability

Silk et al. (2016) showed right tract volume lateralization in healthy children in a diffusion-weighted imaging study. In contrast, a postmortem Klingler dissection study did not reveal a hemispheric asymmetry of the tract (Komaitis et al., 2021).

Fronto-pontine tract

The fronto-pontine tract is a white matter pathway that connects the frontal lobe with the pontine nuclei via the internal capsule (Fig. 2.7). Concerning other tracts, the fronto-pontine tract passes through the anterior limb of the internal capsule, while the CST occupies the posterior limb of the internal capsule (Nieuwenhuys et al., 2008; Catani and Thiebaut de Schotten, 2012).

To our knowledge, no quantitative study has been conducted so far focused on the lateralization and variability of parietal fronto-pontine tracts.

U-shaped tracts

U-shaped fibers are the chain-like short association white matter tracts that connect neighboring cortical areas. Here, we describe the U-shaped systems that have thus far been anatomically validated using Klingler postmortem dissections.

Short frontal tracts

Fronto-insular tracts (FIT 1–5)

Fronto-insular tracts (FITs) consist of five short U-shape fibers connecting the frontal operculum with the insula (Catani et al., 2012; Rojkova et al., 2016) (Fig. 2.8).

Anatomical asymmetry. To our knowledge, there are two studies aimed at finding the asymmetry in FITs. Levman et al. (2019) demonstrated that the tract count of the connections between the left insula and pars opercularis (i.e., FIT3) was more prominent in the left than in the right hemisphere (Levman et al., 2019), which is not in line with the results of Catani and colleagues where the significant left lateralization was found only for the frontal aslant tract and fronto-parietal tracts among all frontal U-shaped fibers (Catani et al., 2012).

Interindividual variability. Studies on the interindividual variability of the fronto-insular tracts are still lacking, given their recent description. Recent work looked at the functional relevance of these tracts. It demonstrated that a disconnection of the right FIT5 significantly contributed to the disturbed sensation of limb ownership (DSO) syndrome (Moro et al., 2023).

Frontal longitudinal system (superior and inferior)

The frontal longitudinal system (FLS) comprises two parallelly running series of U-shaped fibers identified in the frontal lobe via spherical deconvolution diffusion tractography: superior and inferior (Catani et al., 2012). Recently, the existence of the FLS has been confirmed in a postmortem study of 15 brains (Komaitis et al., 2019). The frontal superior longitudinal tract

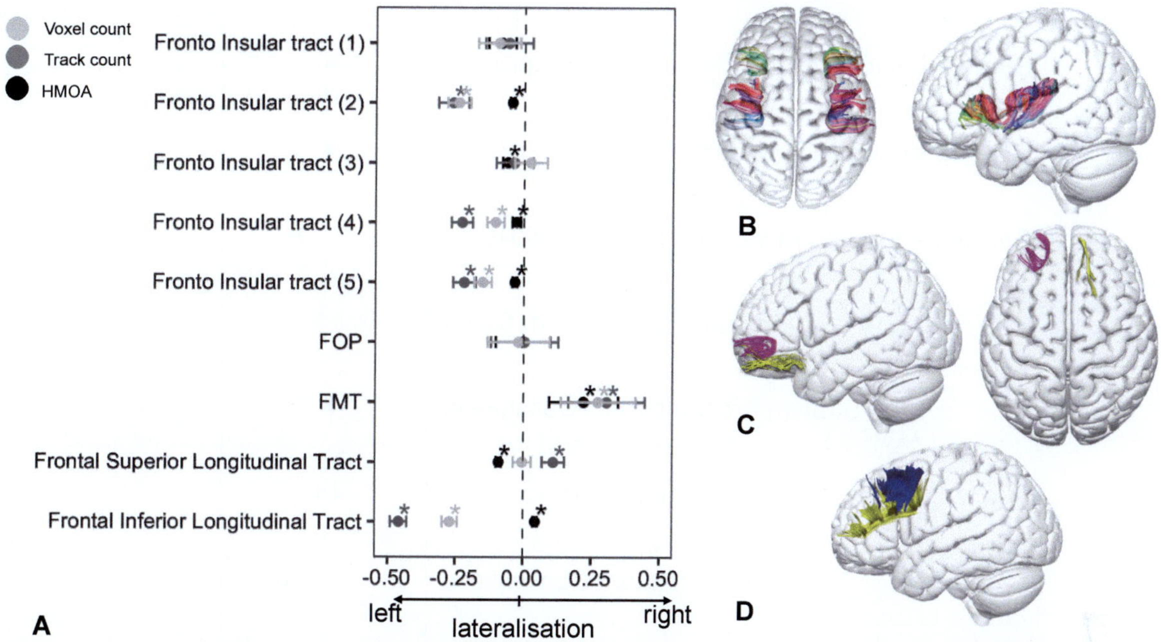

Fig. 2.8. Frontal U-shape tracts. (A) Lateralization index for the number of streamlines (tract count), volume (voxel count), and hindrance modulated orientational anisotropy (HMOA) (mean ± 95% confidence interval) for the frontal U-shape tracts. $*P < 0.05$ after Bonferroni correction for multiple comparisons. (B) Fronto-insular tracts (1–5); (C) fronto-orbito-polar tract (*yellow*) and fronto-marginal (*pink*), (D) inferior (*yellow*) and superior (*blue*) frontal longitudinal fasciculi. *FOP*, fronto-orbito-polar tract; *FMT*, fronto-marginal tract.

connects the precentral and middle frontal gyrus with the superior frontal gyrus. The frontal inferior longitudinal tract connects the precentral gyrus, pars triangularis, and opercularis with the middle frontal gyrus (Catani et al., 2012; Rojkova et al., 2016).

Anatomical asymmetry and interindividual variability. Bakhit et al. (2020) analyzed the frontal superior longitudinal tract. They observed that half of the participants presented with a symmetric frontal superior longitudinal tract, while the FA distribution was skewed toward the right side. In addition, there was no correlation between lateralization indices with sex and handedness (Bakhit et al., 2020).

Fronto-orbito-polar and marginal tract (FOP, FMT)

In the frontal pole, there are two distinct intralobar pathways. The Fronto-Orbito-Polar Tract (FOP), establishes a connection between the posterior orbital gyrus (known as Brodmann area 12) and the anterior orbital gyrus (Brodmann area 11), reaching toward the ventromedial area of the frontal pole. The fronto-marginal tract (FMT), travels below the fronto-marginal sulcus, linking the medial and lateral sections of the frontal pole at Brodmann area 10 (BA10; Rojkova et al., 2016).

To our knowledge, no quantitative study focused on the lateralization and variability of fronto-orbito-polar tract and marginal tract has been conducted so far.

Occipital U-shaped tracts

The occipital U-shaped tracts are a group of occipital short intralobar white matter tracts that have been ignored in the scientific community due to methodological challenges (Bugain et al., 2021). Only 100 years later, Bugain et al. (2021) revived the forgotten white matter tracts from a work of a pioneering neurologist Joseph Jules Dejerine (1849–1917) with a constrained spherical deconvolution method (Bugain et al., 2021). Dejerine and his wife described five short tracts in the occipital lobe: the stratum calcarinum, the VOF, the transverse bundle of the lingual lobule of Vialet, the transverse fasciculus of the cuneus, and the stratum proprium cunei (Dejerine, 1849-1917; Bugain et al., 2021). These tracts were found in the postmortem Klingler dissection study (Vergani et al., 2014; Forkel et al., 2015).

These tracts are awaiting further studies to determine their interindividual variability (Fig. 2.9).

Parietal U-shaped tracts

Intraperietal U-shaped white matter tracts have been recently described in a comprehensive study by Catani et al. (2017): parietal inferior-to-superior tract (PIST), parietal inferior-to-postcentral (PIP), parietal superior-to-postcentral (PSP), parietal angular-to-supramarginal (PAS), parietal intra-gyral (PIG) tracts of the supramarginal gyrus, of the precuneus, of the superior parietal lobule (Catani et al., 2017).

These tracts are awaiting further studies to determine their interindividual variability (Fig. 2.10 bottom).

Fronto-parietal U-shaped tracts

The fronto-parietal U-shaped tracts consist of short fibers connecting the precentral and the postcentral gyrus (Fig. 2.10 top). They consist of the paracentral lobule tract ("foot" area),

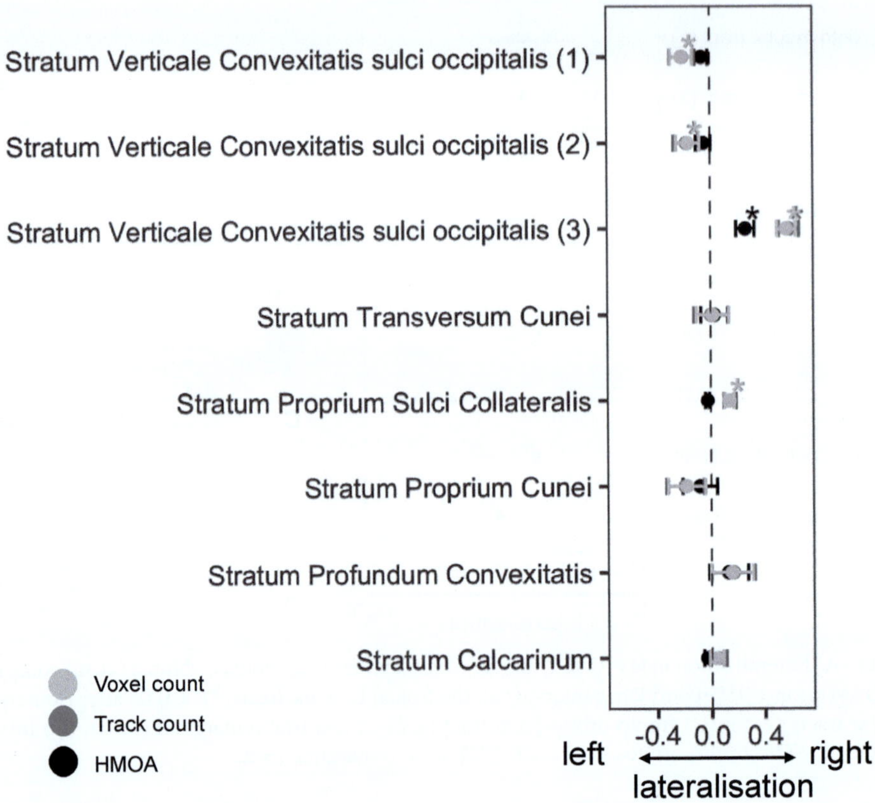

Fig. 2.9. Occipital U-shape tracts. Lateralization index for the number of streamlines (tract count), volume (voxel count), and hindrance modulated orientational anisotropy (HMOA) (mean ± 95% confidence interval) for the occipital U-shape tracts. *$P < 0.05$ after Bonferroni correction for multiple comparisons.

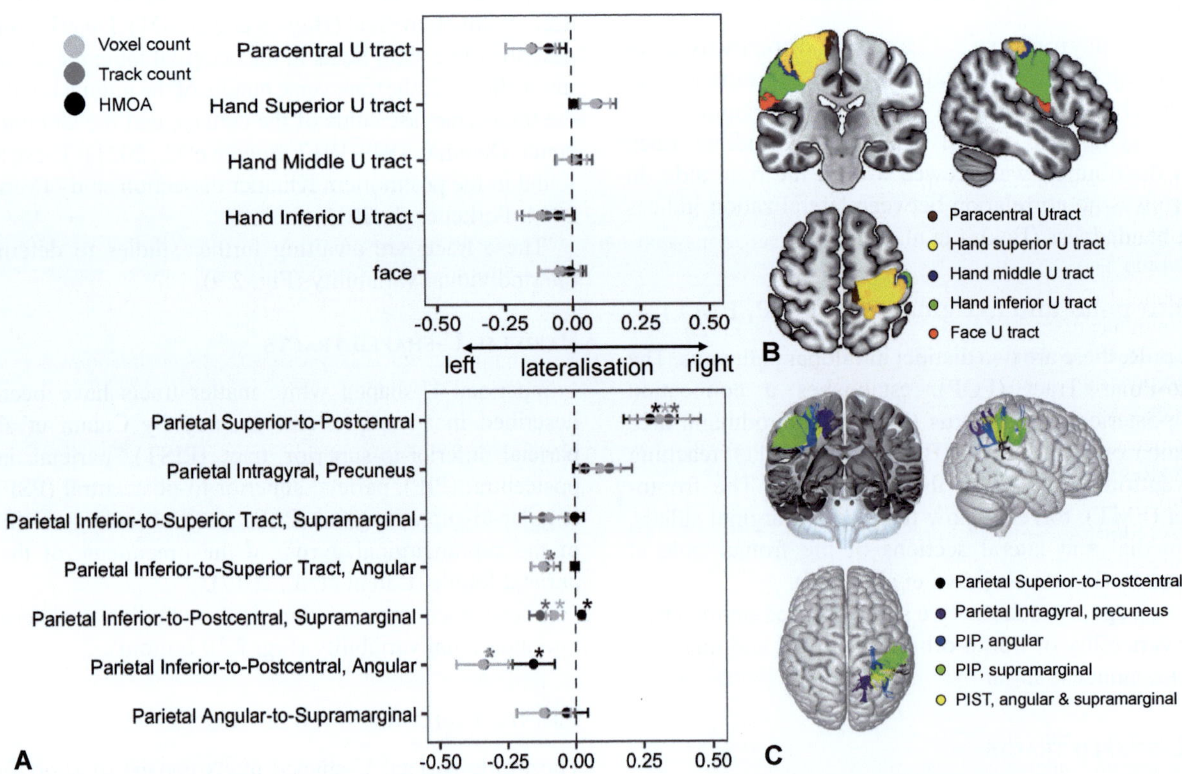

Fig. 2.10. Fronto-parietal and parietal U-shaped tracts. (A) Lateralization index for the number of streamlines (tract count), volume (voxel count), and hindrance modulated orientational anisotropy (HMOA) (mean ± 95% confidence interval) for the fronto-parietal (B), maps from Rojkova et al., 2016) and parietal (C) U-shaped tracts. *$P < 0.05$ after Bonferroni correction for multiple comparisons. *PIP*, parietal inferior-to-postcentral; *PIST*, parietal inferior-to-superior tract.

"hand" area (superior, inferior, and middle), and the ventral group of U-shaped fibers ("face, mouth, tongue" areas; Catani et al., 2012).

Anatomical asymmetry and interindividual variability

Catani el al.'s (2012) results demonstrated that the fronto-parietal U-shaped tracts corresponding to the hand area showed statistically significant left lateralization.

CONCLUSION

In this chapter, we offered a brief overview of cortical and subcortical asymmetries and conducted an exhaustive review of the asymmetry and interindividual variability across 56 distinct white matter tracts. New data from the Human Connectome Project supplement this review. Our analysis reinforces previous findings concerning the lateralization of white matter connections, highlighting that these patterns vary between the left and right hemispheres, though not consistently across all tracts. Notably, several recently identified pathways require further investigation to comprehensively map their variability and functional significance in larger populations. This ongoing research will enhance our understanding of the neurologic diversity underlying human functions.

ACKNOWLEDGMENTS

This project received funding from the Donders Mohrmann Fellowship No. 2401515 (S.J.F., NEUROVARIABILITY). We thank the BCBLab, particularly Michel Thiebaut de Schotten and Valentina Pacella, for constructive feedback.

REFERENCES

Abdolalizadeh A, Mohammadi S, Aarabi MH (2022). The forgotten tract of vision in multiple sclerosis: vertical occipital fasciculus, its fibre properties, and visuospatial memory. Brain Struct Funct 227: 1479–1490. https://doi.org/10.1007/s00429-022-02464-3.

Aboitiz F, Montiel J (2003). One hundred million years of interhemispheric communication: the history of the corpus callosum. Braz J Med Biol Res 36: 409–420. Scopus. https://doi.org/10.1590/S0100-879X2003000400002.

Aboitiz F, Scheibel AB, Fisher RS et al. (1992). Individual differences in brain asymmetries and fibre composition in the human corpus callosum. Brain Res 598: 154–161. https://doi.org/10.1016/0006-8993(92)90179-d.

Akeret K, Forkel SJ, Buzzi RM et al. (2022). Multimodal anatomy of the human forniceal commissure. Commun Biol 5: 1–11. https://doi.org/10.1038/s42003-022-03692-3.

Amemiya K, Naito E, Takemura H (2021). Age dependency and lateralization in the three branches of the human superior longitudinal fasciculus. Cortex 139: 116–133. https://doi.org/10.1016/j.cortex.2021.02.027.

Amunts K, Schleicher A, Bürgel U et al. (1999). Broca's region revisited: cytoarchitecture and intersubject variability. J Comp Neurol 412: 319–341. https://doi.org/10.1002/(sici)1096-9861(19990920)412:2<319::aid-cne10>3.0.co;2-7.

Amunts K, Weiss PH, Mohlberg H et al. (2004). Analysis of neural mechanisms underlying verbal fluency in cytoarchitectonically defined stereotaxic space—the roles of Brodmann areas 44 and 45. Neuro Image 22: 42–56. https://doi.org/10.1016/j.neuroimage.2003.12.031.

Angstmann S, Madsen KS, Skimminge A et al. (2016). Microstructural asymmetry of the corticospinal tracts predicts right–left differences in circle drawing skill in right-handed adolescents. Brain Struct Funct 221: 4475–4489. https://doi.org/10.1007/s00429-015-1178-5.

Bakhit M, Fujii M, Hiruta R et al. (2020). The superior frontal longitudinal tract: a connection between the dorsal premotor and the dorsolateral prefrontal cortices. Sci Rep 10: 15855. https://doi.org/10.1038/s41598-020-73001-7.

Barrett RLC, Dawson M, Dyrby TB et al. (2020). Differences in frontal network anatomy across primate species. J Neurosci 40: 2094–2107. https://doi.org/10.1523/JNEUROSCI.1650-18.2019.

Basile GA, Nozais V, Quartarone A et al. (2024). Functional anatomy and topographical organization of the frontotemporal arcuate fasciculus. Preprint avilable at Research Square https://doi.org/10.21203/rs.3.rs-4614103/v1.

Basser PJ, Pajevic S, Pierpaoli C et al. (2000). In vivo fibre tractography using DT-MRI data. Magn Reson Med 44: 625–632. https://doi.org/10.1002/1522-2594(200010)44:4<625::aid-mrm17>3.0.co;2-o.

Bates S, Dumoulin SO, Folkers PJM et al. (2023). A vision of 14 T MR for fundamental and clinical science. Magma (New York, Ny) 36: 211–225. https://doi.org/10.1007/s10334-023-01081-3.

Bathelt J, Johnson A, Zhang M et al. (2019). The cingulum as a marker of individual differences in neurocognitive development. Sci Rep 9: 1–16. https://doi.org/10.1038/s41598-019-38894-z.

Baudo MC, Colombo EE, Pérez MA et al. (2022). Three-dimensional anatomy of the anterior commissure: a tractography and anatomical study. World Neurosurg 159: e365–e374. https://doi.org/10.1016/j.wneu.2021.12.059.

Beaton AA (1997). The relation of planum temporale asymmetry and morphology of the corpus callosum to handedness, gender, and dyslexia: a review of the evidence. Brain Lang 60: 255–322. https://doi.org/10.1006/brln.1997.1825.

Bernard F, Zemmoura I, Ter Minassian A et al. (2019). Anatomical variability of the arcuate fasciculus: a systematical review. Surg Radiol Anat 41: 889–900. https://doi.org/10.1007/s00276-019-02244-5.

Berthier ML, Lambon Ralph MA, Pujol J et al. (2012). Arcuate fasciculus variability and repetition: the left sometimes can be right. Cortex 48: 133–143. https://doi.org/10.1016/j.cortex.2011.06.014.

Beyh A, Dell'Acqua F, Cancemi D et al. (2022). The medial occipital longitudinal tract supports early stage encoding of visuospatial information. Commun Biol 5: 318. https://doi.org/10.1038/s42003-022-03265-4.

Briggs RG, Conner AK, Sali G et al. (2018). A connectomic atlas of the human cerebrum-chapter 16: tractographic description of the vertical occipital fasciculus. Oper Neurosurg (Hagerstown, Md) 15: S456–S461. https://doi.org/10.1093/ons/opy270.

Briggs RG, Khan AB, Chakraborty AR et al. (2020). Anatomy and white matter connections of the superior frontal gyrus. Clin Anat (New York, NY) 33: 823–832. https://doi.org/10.1002/ca.23523.

Broca P (1861). Perte de la parole, ramouissement chronique et destruction partielle du lobe antérieur gauche du cerveau. Bull MémSoc Anthropol Paris 2: 235–238. https://doi.org/10.1212/WNL.0000000000002579.

Broce IJ, Bernal B, Altman N et al. (2019). Fibre pathways supporting early literacy development in 5-8-year-old children. Brain Cogn 134: 80–89. https://doi.org/10.1016/j.bandc.2018.12.004.

Bryant KL, Li L, Eichert N et al. (2020). A comprehensive atlas of white matter tracts in the chimpanzee. PLoS Biol 18: e3000971. https://doi.org/10.1371/journal.pbio.3000971.

Budisavljevic S, Kawadler JM, Dell'Acqua F et al. (2016). Heritability of the limbic networks. Soc Cogn Affect Neurosci 11: 746–757. https://doi.org/10.1093/scan/nsv156.

Budisavljevic S, Dell'Acqua F, Castiello U (2018). Cross-talk connections underlying dorsal and ventral stream integration during hand actions. Cortex 103: 224–239. https://doi.org/10.1016/j.cortex.2018.02.016.

Budisavljevic S, Castiello U, Begliomini C (2021). Handedness and white matter networks. Neuroscientist 27: 88–103. https://doi.org/10.1177/1073858420937657.

Bugain M, Dimech Y, Torzhenskaya N et al. (2021). Occipital intralobar fasciculi: a description, through tractography, of three forgotten tracts. Commun Biol 4: 1–11. https://doi.org/10.1038/s42003-021-01935-3.

Bürgel U, Schormann T, Schleicher A et al. (1999). Mapping of histologically identified long fibre tracts in human cerebral hemispheres to the MRI volume of a reference brain: position and spatial variability of the optic radiation. Neuroimage 10: 489–499. https://doi.org/10.1006/nimg.1999.0497.

Burkhardt E, Kinoshita M, Herbet G (2021). Functional anatomy of the frontal aslant tract and surgical perspectives. J Neurosurg Sci 65: 566–580. https://doi.org/10.23736/S0390-5616.21.05344-3.

Catani M (2006). Diffusion tensor magnetic resonance imaging tractography in cognitive disorders. Curr Opin Neurol. https://doi.org/10.1097/01.wco.0000247610.44106.3f.

Catani M, Ffytche DH (2005). The rises and falls of disconnection syndromes. Brain. https://doi.org/10.1093/brain/awh622.

Catani M, Thiebaut de Schotten M (2008). A diffusion tensor imaging tractography atlas for virtual in vivo dissections. Cortex 44: 1105–1132. https://doi.org/10.1016/j.cortex.2008.05.004.

Catani M, Thiebaut de Schotten M (2012). Atlas of human brain connections, OUP Oxford.

Catani M, Jones DK, Donato R et al. (2003). Occipito-temporal connections in the human brain. Brain 126: 2093–2107. https://doi.org/10.1093/brain/awg203.

Catani M, Jones DK, ffytche, D. H. (2005). Perisylvian language networks of the human brain. Ann Neurol 57: 8–16. https://doi.org/10.1002/ana.20319.

Catani M, Allin MPG, Husain M et al. (2007). Symmetries in human brain language pathways correlate with verbal recall. Proc Natl Acad Sci U S A 104: 17163–17168. https://doi.org/10.1073/pnas.0702116104.

Catani M, Forkel S, de Schotten MT (2010). Asymmetry of white matter pathways. In: K Hugdahl, R Westerhausen (Eds.), The two halves of the brain: information processing in the cerebral hemispheres. The MIT Press. https://doi.org/10.7551/mitpress/9780262014137.003.0136.

Catani M, Dell'Acqua F, Vergani F et al. (2012). Short frontal lobe connections of the human brain. Cortex 48: 273–291. https://doi.org/10.1016/j.cortex.2011.12.001.

Catani M, Dell'acqua F, Thiebaut de Schotten M (2013a). A revised limbic system model for memory, emotion and behaviour. Neurosci Biobehav Rev 37: 1724–1737. https://doi.org/10.1016/j.neubiorev.2013.07.001.

Catani M, Mesulam MM, Jakobsen E et al. (2013b). A novel frontal pathway underlies verbal fluency in primary progressive aphasia. Brain 136: 2619–2628. https://doi.org/10.1093/brain/awt163.

Catani M, Robertsson N, Beyh A et al. (2017). Short parietal lobe connections of the human and monkey brain. Cortex 97: 339–357. https://doi.org/10.1016/j.cortex.2017.10.022.

Chamberland M, Tax CMW, Jones DK (2018). Meyer's loop tractography for image-guided surgery depends on imaging protocol and hardware. NeuroImage Clin 20: 458–465. https://doi.org/10.1016/j.nicl.2018.08.021.

Chen A, Deng Y, Zuo X et al. (2022). Alteration in asymmetry of white matter network of Parkinson's disease. Contrast Media Mol Imaging 2022: 8493729. https://doi.org/10.1155/2022/8493729.

Choi MH, Kim JH, Yeon HW et al. (2011). Effects of gender and age on anterior commissure volume. Neurosci Lett 500: 92–94. https://doi.org/10.1016/j.neulet.2011.06.010.

Choi YJ, Lee EJ, Lee JE (2021). The fornix: functional anatomy, normal neuroimaging, and various pathological conditions. Investig Magn Reson Imaging 25: 59. https://doi.org/10.13104/imri.2021.25.2.59.

Cocquyt E-M, Lanckmans E, van Mierlo P et al. (2020). The white matter architecture underlying semantic processing: a systematic review. Neuropsychologia 136: 107182. https://doi.org/10.1016/j.neuropsychologia.2019.107182.

Corballis MC (2012). Lateralization of the human brain. Prog Brain Res 195: 103–121. https://doi.org/10.1016/B978-0-444-53860-4.00006-4.

Corballis MC (2014). Left brain, right brain: facts and fantasies. PLoS Biol 12. https://doi.org/10.1371/journal.pbio.1001767.

Craig MC, Catani M, Deeley Q et al. (2009). Altered connections on the road to psychopathy. Mol Psychiatry 14: 946–953907. https://doi.org/10.1038/mp.2009.40.

Crosby EC (1963). Correlative anatomy of the nervous system. Acad Med 38: 526.

Croxson PL, Forkel SJ, Cerliani L et al. (2018). Structural variability across the primate brain: a cross-species comparison. Cereb Cortex 28: 3829–3841. https://doi.org/10.1093/cercor/bhx244.

Curran EJ (1909). A new association fibre tract in the cerebrum with remarks on the fibre tract dissection method of studying the brain. J Comp Neurol Psychol 19: 645–656. https://doi.org/10.1002/cne.920190603.

D'Anna L, Mesulam MM, Thiebaut de Schotten M et al. (2016). Frontotemporal networks and behavioral symptoms in primary progressive aphasia. Neurology 86: 1393–1399. https://doi.org/10.1212/WNL.0000000000002579.

Dalamagkas K, Tsintou M, Rathi Y et al. (2020). Individual variations of the human corticospinal tract and its hand-related motor fibres using diffusion MRI tractography. Brain Imaging Behav 14: 696–714. https://doi.org/10.1007/s11682-018-0006-y.

Dax MG (1865). Lésions de la moitié gauche de l'encéphale coïncident avec l'oublie des signes de la pensée: Lu au Congrès méridional tenu à Montpellier en 1836, par le docteur Marc Dax. Gaz Hebd Méd et de Chirurgie 17: 259–260. https://doi.org/10.1007/s00429-020-02054-1.

Dejerine J (1849-1917). Anatomie des centres nerveux. Méthode générale d'études, embryogénie, histogénèse et histologie, anatomie du cerveau/par. A. du texte. (1895). *J. Dejerine,... ; Avec la collaboration de Mme Dejerine-Klumpke*. https://gallica.bnf.fr/ark:/12148/bpt6k76498h.

Dell'Acqua F, Tournier JD (2019). Modelling white matter with spherical deconvolution: how and why? NMR Biomed32: . https://doi.org/10.1002/nbm.3945.

Dell'Acqua F, Simmons A, Williams SCR et al. (2013). Can spherical deconvolution provide more information than fibre orientations?

Hindrance modulated orientational anisotropy, a true-tract specific index to characterize white matter diffusion. Hum Brain Mapp 34: 2464–2483. https://doi.org/10.1002/hbm.22080.
Demnitz N, Madsen KS, Johnsen LK et al. (2021). Right-left asymmetry in corticospinal tract microstructure and dexterity are uncoupled in late adulthood. Neuroimage 240. https://doi.org/10.1016/j.neuroimage.2021.118405.
Di Virgilio G, Clarke S, Pizzolato G et al. (1999). Cortical regions contributing to the anterior commissure in man. Exp Brain Res 124: 1–7. https://doi.org/10.1007/s002210050593.
Dick AS, Garic D, Graziano P et al. (2019). The frontal aslant tract (FAT) and its role in speech, language and executive function. Cortex 111: 148–163. https://doi.org/10.1016/j.cortex.2018.10.015.
Dreessen de Gervai P, Sboto-Frankenstein UN, Bolster RB et al. (2014). Tractography of meyer's loop asymmetries. Epilepsy Res 108: 872–882. https://doi.org/10.1016/j.eplepsyres.2014.03.006.
Duffau H, Gatignol P, Mandonnet E et al. (2005). New insights into the anatomo-functional connectivity of the semantic system: A study using cortico-subcortical electrostimulations. Brain 128: 797–810. https://doi.org/10.1093/brain/awh423.
Dum RP, Strick PL (1991). The origin of corticospinal projections from the premotor areas in the frontal lobe. J Neurosci 11: 667–689. https://doi.org/10.1523/jneurosci.11-03-00667.1991.
Ebeling U, Reulen HJ (1988). Neurosurgical topography of the optic radiation in the temporal lobe. Acta Neurochir 92: 29–36. https://doi.org/10.1007/BF01401969.
Eichert N, Robinson EC, Bryant KL et al. (2020). Cross-species cortical alignment identifies different types of anatomical reorganization in the primate temporal lobe. Elife 9: e53232. https://doi.org/10.7554/eLife.53232.
Feltenstein MW, See RE (2008). The neurocircuitry of addiction: an overview. Br J Pharmacol 154: 261–274. https://doi.org/10.1038/bjp.2008.51.
Finger S, Koehler PJ, Jagella C (2004). The Monakow concept of diaschisis. Arch Neurol. https://doi.org/10.1001/archneur.61.2.283.
Forkel SJ, Thiebaut de Schotten M, Dell'Acqua F et al. (2014a). Anatomical predictors of aphasia recovery: a tractography study of bilateral perisylvian language networks. Brain 137: 2027–2039. https://doi.org/10.1093/brain/awu113.
Forkel SJ, Thiebaut de Schotten M, Kawadler JM et al. (2014b). The anatomy of fronto-occipital connections from early blunt dissections to contemporary tractography. Cortex 56: 73–84. https://doi.org/10.1016/j.cortex.2012.09.005.
Forkel SJ, Mahmood S, Vergani F et al. (2015). The white matter of the human cerebrum: part I the occipital lobe by Heinrich Sachs. Cortex 62: 182–202. https://doi.org/10.1016/j.cortex.2014.10.023.
Forkel SJ, Rogalski E, Drossinos Sancho N et al. (2020). Anatomical evidence of an indirect pathway for word repetition. Neurology 94: e594–e606. https://doi.org/10.1212/WNL.0000000000008746.
Forkel SJ, Friedrich P, Thiebaut de Schotten M et al. (2022). White matter variability, cognition, and disorders: a systematic review. Brain Struct Funct 227: 529–544. https://doi.org/10.1007/s00429-021-02382-w.
Forkel S, Bortolami C, Dulyan L et al. (2024). Dissecting white matter pathways: a neuroanatomical approach. PsyArXiv. https://doi.org/10.31234/osf.io/9xwgm.
Frey S, Campbell JSW, Pike GB et al. (2008). Dissociating the human language pathways with high angular resolution diffusion fibre tractography. J Neurosci 28: 11435–11444. https://doi.org/10.1523/JNEUROSCI.2388-08.2008.
Gainotti G (2023). Some historical notes orienting towards brain mechanisms that could underlie hemispheric asymmetries. Cortex 163: 26–41. https://doi.org/10.1016/j.cortex.2023.03.001.
Gallet C, Clavreul A, Bernard F et al. (2022). Frontal aslant tract in the non-dominant hemisphere: a systematic review of anatomy, functions, and surgical applications. Front Neuroanat 16. https://doi.org/10.3389/fnana.2022.1025866.
Gazzaniga MS (2000). Cerebral specialization and interhemispheric communication: does the corpus callosum enable the human condition? Brain 123: 1293–1326. https://doi.org/10.1093/brain/123.7.1293.
George K, Das JM (2024). Neuroanatomy, thalamocortical radiations. In: StatPearls, StatPearls Publishing. http://www.ncbi.nlm.nih.gov/books/NBK546699/.
Geschwind N, Levitsky W (1968). Human brain: left-right asymmetries in temporal speech region. Science (New York, NY)161: 186–187. https://doi.org/10.1126/science.161.3837.186.
Giampiccolo D, Duffau H (2022). Controversy over the temporal cortical terminations of the left arcuate fasciculus: a reappraisal. Brain145: 1242–1256. https://doi.org/10.1093/brain/awac057.
Gong G, Jiang T, Zhu C et al. (2005). Asymmetry analysis of cingulum based on scale-invariant parameterization by diffusion tensor imaging. Hum Brain Mapp 24: 92–98. https://doi.org/10.1002/hbm.20072.
Grodzinsky Y, Amunts K (2006). Broca's region, Oxford University Press.
Grotheer M, Yeatman J, Grill-Spector K (2021). White matter fascicles and cortical microstructure predict reading-related responses in human ventral temporal cortex. Neuroimage 227: 117669. https://doi.org/10.1016/j.neuroimage.2020.
117669.
Guadalupe T, Mathias SR, vanErp TGM et al. (2017). Human subcortical brain asymmetries in 15,847 people worldwide reveal effects of age and sex. Brain Imaging Behav 11: 1497–1514.
Güngör A, Baydin S, Middlebrooks EH et al. (2017). The white matter tracts of the cerebrum in ventricular surgery and hydrocephalus. J Neurosurg 126: 945–971. https://doi.org/10.3171/2016.1.JNS152082.
Hau J, Sarubbo S, Houde JC et al. (2017). Revisiting the human uncinate fasciculus, its subcomponents and asymmetries with stem-based tractography and microdissection validation. Brain Struct Funct 222: 1645–1662. https://doi.org/10.1007/s00429-016-1298-6.
Herbet G, Zemmoura I, Duffau H (2018). Functional anatomy of the inferior longitudinal fasciculus: from historical reports to current hypotheses. Front Neuroanat 12. https://doi.org/10.3389/fnana.2018.00077.
Hervé PY, Crivello F, Perchey G et al. (2006). Handedness and cerebral anatomical asymmetries in young adult males. Neuroimage 29: 1066–1079. https://doi.org/10.1016/j.neuroimage.2005.08.031.
Highley JR, Esiri MM, McDonald B et al. (1999). The size and fibre composition of the corpus callosum with respect to gender and schizophrenia: a post-mortem study. Brain 122: 99–110. https://doi.org/10.1093/brain/122.1.99.
Highley JR, Walker MA, Esiri MM et al. (2002). Asymmetry of the uncinate fasciculus: a post-mortem study of normal subjects and patients with schizophrenia. Cereb Cortex (New York, NY: 1991) 12: 1218–1224. https://doi.org/10.1093/cercor/12.11.1218.
Horowitz A, Barazany D, Tavor I et al. (2015). In vivo correlation between axon diameter and conduction velocity in the human brain. Brain Struct Funct 220: 1777–1788. https://doi.org/10.1007/s00429-014-0871-0.

Howells J, Matamala JM, Park SB et al. (2018). In vivo evidence for reduced ion channel expression in motor axons of patients with amyotrophic lateral sclerosis. J Physiol 596: 5379–5396. https://doi.org/10.1113/JP276624.

Howells H, Puglisi G, Leonetti A et al. (2020). The role of left fronto-parietal tracts in hand selection: evidence from neurosurgery. Cortex 128: 297–311. https://doi.org/10.1016/j.cortex.2020.03.018.

Ide A, Dolezal C, Fernández M et al. (1999). Hemispheric differences in variability of fissural patterns in parasylvian and cingulate regions of human brains. J Comp Neurol 410: 235–242.

Janelle F, Iorio-Morin C, D'amour S et al. (2022). Superior longitudinal fasciculus: a review of the anatomical descriptions with functional correlates. Front Neurol 13. https://doi.org/10.3389/fneur.2022.794618.

Jang SH (2014). The corticospinal tract from the viewpoint of brain rehabilitation. J Rehabil Med 46: 193–199. https://doi.org/10.2340/16501977-1782.

Jones DK, Christiansen KF, Chapman RJ et al. (2013). Distinct subdivisions of the cingulum bundle revealed by diffusion MRI fibre tracking: implications for neuropsychological investigations. Neuropsychologia 51: 67–78. https://doi.org/10.1016/j.neuropsychologia.2012.11.018.

Kalyvas A, Koutsarnakis C, Komaitis S et al. (2020). Mapping the human middle longitudinal fasciculus through a focused anatomo-imaging study: shifting the paradigm of its segmentation and connectivity pattern. Brain Struct Funct 225. https://doi.org/10.1007/s00429-019-01987-6.

Karadimas SK, Satkunendrarajah K, Laliberte AM et al. (2020). Sensory cortical control of movement. Nat Neurosci 23: 75–84. https://doi.org/10.1038/s41593-019-0536-7.

Karlsgodt KH, John M, Ikuta T et al. (2015). The accumbofrontal tract: diffusion tensor imaging characterization and developmental change from childhood to adulthood. Hum Brain Mapp 36: 4954–4963. https://doi.org/10.1002/hbm.22989.

Kemerdere R, de Champfleur NM, Deverdun J et al. (2016). Role of the left frontal aslant tract in stuttering: a brain stimulation and tractographic study. J Neurol 263: 157–167. https://doi.org/10.1007/s00415-015-7949-3.

Keser Z, Ucisik-Keser FE, Hasan KM (2016). Quantitative mapping of human brain vertical-occipital fasciculus. J Neuroimaging 26: 188–193. https://doi.org/10.1111/jon.12268.

Kier EL, Staib LH, Davis LM et al. (2004). MR imaging of the temporal stem: anatomic dissection tractography of the uncinate fasciculus, inferior occipitofrontal fasciculus, and Meyer's loop of the optic radiation. AJNR Am J Neuroradiol 25: 677–691.

Koelkebeck K, Miyata J, Kubota M et al. (2014). The contribution of cortical thickness and surface area to gray matter asymmetries in the healthy human brain. Hum Brain Mapp 35: 6011–6022. https://doi.org/10.1002/hbm.22601.

Komaitis S, Kalyvas AV, Skandalakis GP et al. (2019). The frontal longitudinal system as revealed through the fibre microdissection technique: structural evidence underpinning the direct connectivity of the prefrontal-premotor circuitry. J Neurosurg 133: 1503–1515. https://doi.org/10.3171/2019.6.JNS191224.

Komaitis S, Koutsarnakis C, Lani E et al. (2021). Deciphering the frontostriatal circuitry through the fibre dissection technique: direct structural evidence on the morphology and axonal connectivity of the fronto-caudate tract. J Neurosurg 135: 815–827. https://doi.org/10.3171/2020.7.JNS201287.

Kong X-Z, Mathias SR, Guadalupe T et al. (2018). Mapping cortical brain asymmetry in 17,141 healthy individuals worldwide via the ENIGMA Consortium. Proc Natl Acad Sci U S A 115: E5154–E5163. https://doi.org/10.1073/pnas.1718418115.

Koshiyama D, Fukunaga M, Okada N et al. (2020). White matter microstructural alterations across four major psychiatric disorders: mega-analysis study in 2937 individuals. Mol Psychiatry 25: 883–895. https://doi.org/10.1038/s41380-019-0553-7.

Latini F, Mårtensson J, Larsson EM et al. (2017). Segmentation of the inferior longitudinal fasciculus in the human brain: a white matter dissection and diffusion tensor tractography study. Brain Res 1675: 102–115. https://doi.org/10.1016/j.brainres.2017.09.005.

Latini F, Trevisi G, Fahlström M et al. (2021). New insights into the anatomy, connectivity and clinical implications of the middle longitudinal fasciculus. Front Neuroanat 14. https://doi.org/10.3389/fnana.2020.610324.

Lavrador JP, Ferreira V, Lourenço M et al. (2019). White-matter commissures: a clinically focused anatomical review. Surg Radiol Anat 41: 613–624. https://doi.org/10.1007/s00276-019-02218-7.

Lawes INC, Barrick TR, Murugam V et al. (2008). Atlas-based segmentation of white matter tracts of the human brain using diffusion tensor tractography and comparison with classical dissection. Neuroimage 39: 62–79. https://doi.org/10.1016/j.neuroimage.2007.06.041.

Le Bihan D (2003). Looking into the functional architecture of the brain with diffusion MRI. Nat Rev Neurosci 4: 469–480. https://doi.org/10.1038/nrn1119.

Lebel C, Beaulieu C (2011). Longitudinal development of human brain wiring continues from childhood into adulthood. J Neurosci 31: 10937–10947. https://doi.org/10.1523/JNEUROSCI.5302-10.2011.

Lebel C, Deoni S (2018). The development of brain white matter microstructure. Neuroimage 182: 207–218. https://doi.org/10.1016/j.neuroimage.2017.12.097.

Lebel C, Walker L, Leemans A et al. (2008). Microstructural maturation of the human brain from childhood to adulthood. Neuroimage 40: 1044–1055. https://doi.org/10.1016/j.neuroimage.2007.12.053.

LeMay M, Geschwind N (1975). Hemispheric differences in the brains of great apes. Brain Behav Evol 11: 48–52. https://doi.org/10.1159/000123623.

Lemon RN (2008). Descending pathways in motor control. Annu Rev Neurosci 31: 195–218. https://doi.org/10.1146/annurev.neuro.31.060407.125547.

Levman J, Fang Z, Zumwalt K et al. (2019). Asymmetric insular connectomics revealed by diffusion magnetic resonance imaging analysis of healthy brain development. Brain Connect 9: 2–12. https://doi.org/10.1089/brain.
2018.0582.

Luders E, Narr KL, Zaidel E et al. (2006). Parasagittal asymmetries of the corpus callosum. Cereb Cortex. https://doi.org/10.1093/cercor/bhi112.

Ludwig E, Klingler J (1956). Atlas cerebri humani, Karger.

Maier-Hein KH, Neher PF, Houde J-C et al. (2017). The challenge of mapping the human connectome based on diffusion tractography. Nat Commun 8: 1349. https://doi.org/10.1038/s41467-017-01285-x.

Makris N, Papadimitriou GM, Kaiser JR et al. (2009). Delineation of the middle longitudinal fascicle in humans: a quantitative, in vivo, DT-MRI study. Cereb Cortex 19: 777–785. https://doi.org/10.1093/cercor/bhn124.

Makris N, Preti MG, Asami T et al. (2013). Human middle longitudinal fascicle: variations in patterns of anatomical connections. Brain Struct Funct 218: 951–968. https://doi.org/10.1007/s00429-012-0441-2.

Maldonado IL, de Champfleur NM, Velut S et al. (2013). Evidence of a middle longitudinal fasciculus in the human brain from fibre dissection. J Anat 223: 38. https://doi.org/10.1111/joa.12055.

Mandelstam SA (2012). Challenges of the anatomy and diffusion tensor tractography of the Meyer loop. Am J Neuroradiol 33: 1204–1210. https://doi.org/10.3174/ajnr.A2652.

Manning KY, Reynolds JE, Long X et al. (2022). Multimodal brain features at 3 years of age and their relationship with pre-reading measures 1 year later. Front Hum Neurosci 16. https://doi.org/10.3389/fnhum.2022.965602.

Mars RB, Foxley S, Verhagen L et al. (2016). The extreme capsule fibre complex in humans and macaque monkeys: a comparative diffusion MRI tractography study. Brain Struct Funct 221: 4059–4071. https://doi.org/10.1007/s00429-015-1146-0.

Martínez-Molina N, Mas-Herrero E, Rodríguez-Fornells A et al. (2019). White matter microstructure reflects individual differences in music reward sensitivity. J Neurosci 39: 5018–5027. https://doi.org/10.1523/JNEUROSCI.2020-18.2019.

Mori S, Wakana S, Nagae-Poetscher LM et al. (2006). MRI atlas of human white matter. AJNR Am J Neuroradiol 27: 1384–1385.

Moro V, Pacella V, Scandola M et al. (2023). A fronto-insular-parietal network for the sense of body ownership. Cereb Cortex (New York, NY: 1991) 33: 512–522. https://doi.org/10.1093/cercor/bhac081.

Nauta WJH (1958). Hippocampal projections and related neural pathways to the mid-brain in the cat. Brain 81: 319–340. https://doi.org/10.1093/brain/81.3.319.

Nieuwenhuys R, Voogd J, Van Huijzen C (2008). The human central nervous system, Springer. https://doi.org/10.1007/978-3-540-34686-9.

Nilsson D, Starck G, Ljungberg M et al. (2007). Intersubject variability in the anterior extent of the optic radiation assessed by tractography. Epilepsy Res 77: 11–16. https://doi.org/10.1016/j.eplepsyres.2007.07.012.

Northam GB, Liégeois F, Tournier J-D et al. (2012). Interhemispheric temporal lobe connectivity predicts language impairment in adolescents born preterm. Brain 135: 3781–3798. https://doi.org/10.1093/brain/aws276.

O'Sullivan MJ, Li X, Galligan D et al. (2023). Cognitive recovery after stroke: memory. Stroke 54: 44–54. https://doi.org/10.1161/STROKEAHA.122.041497.

Ocklenburg S, Güntürkün O (2024). The lateralized brain: the neuroscience and evolution of hemispheric asymmetries, Elsevier Science.

Oishi K, Zilles K, Amunts K et al. (2008). Human brain white matter atlas: identification and assignment of common anatomical structures in superficial white matter. Neuroimage 43: 447–457. https://doi.org/10.1016/j.neuroimage.2008.07.009.

Ono M, Kubik S, Abernathey CD (1990). Atlas of the cerebral sulci, G. Thieme Verlag; Thieme Medical Publishers. http://books.google.com/books?id=L7tqAAAAMAAJ.

Onufrowicz W (1887). Das balkenlose Mikrocephalengehirn Hofmann: Ein Beitrag zur pathologischen und normalen Anatomie des menschlichen Grehirnes. Arch Psychiatr Nervenkr18: 305–328. https://doi.org/10.1007/BF02054370.

Palejwala AH, O'Connor KP, Pelargos P et al. (2020). Anatomy and white matter connections of the lateral occipital cortex. Surg Radiol Anat 42: 315–328. https://doi.org/10.1007/s00276-019-02371-z.

Panesar SS, Belo JTA, Yeh F-C et al. (2019). Structure, asymmetry, and connectivity of the human temporo-parietal aslant and vertical occipital fasciculi. Brain Struct Funct 224: 907–923. https://doi.org/10.1007/s00429-018-1812-0.

Pascalau R, Popa Stănilă R, Sfrângeu S et al. (2018). Anatomy of the limbic white matter tracts as revealed by fibre dissection and tractography. World Neurosurg 113: e672–e689. https://doi.org/10.1016/j.wneu.2018.02.121.

Patel MD, Toussaint N, Charles-Edwards GD et al. (2010). Distribution and fibre field similarity mapping of the human anterior commissure fibres by diffusion tensor imaging. Magma (New York, NY) 23: 399–408. https://doi.org/10.1007/s10334-010-0201-3.

Paul LK (2011). Developmental malformation of the corpus callosum: a review of typical callosal development and examples of developmental disorders with callosal involvement. J Neurodev Disord 3: 3–27. https://doi.org/10.1007/s11689-010-9059-y.

Peltier J, Nicot B, Baroncini M et al. (2011). Anatomy of the periventricular white matter. Neurochirurgie 57: 151–155. https://doi.org/10.1016/j.neuchi.2011.09.017.

Peuskens D, Van Loon J, Van Calenbergh F et al. (2004). Anatomy of the anterior temporal lobe and the frontotemporal region demonstrated by fibre dissection. Neurosurgery 55: 1174–1183. https://doi.org/10.1227/01.NEU.0000140843.62311.24.

Plessen KJ, Hugdahl K, Bansal R et al. (2014). Sex, age, and cognitive correlates of asymmetries in thickness of the cortical mantle across the life span. J Neurosci 34: 6294–6302. https://doi.org/10.1523/JNEUROSCI.3692-13.2014.

Postans M, Hodgetts CJ, Mundy ME et al. (2014). Interindividual variation in fornix microstructure and macrostructure is related to visual discrimination accuracy for scenes but not faces. J Neurosci 34: 12121–12126. https://doi.org/10.1523/JNEUROSCI.0026-14.2014.

Putnam MC, Steven MS, Doron KW et al. (2010). Cortical projection topography of the human splenium: hemispheric asymmetry and individual differences. J Cogn Neurosci. https://doi.org/10.1162/jocn.2009.21290.

Radwan AM, Sunaert S, Schilling K et al. (2022). An atlas of white matter anatomy, its variability, and reproducibility based on constrained spherical deconvolution of diffusion MRI. Neuroimage 254: 119029. https://doi.org/10.1016/j.neuroimage.2022.119029.

Rilling JK, Glasser MF, Preuss TM et al. (2008). The evolution of the arcuate fasciculus revealed with comparative DTI. Nat Neurosci 11: 426–428. https://doi.org/10.1038/nn2072.

Rodrigues EM, Isolan GR, Becker LG et al. (2022). Anatomy of the optic radiations from the white matter fibre dissection perspective: a literature review applied to practical anatomical dissection. Surg Neurol Int 13: 1–6. https://doi.org/10.25259/SNI_1157_2021.

Roe JM, Vidal-Pineiro D, Amlien IK et al. (2023). Tracing the development and lifespan change of population-level structural asymmetry in the cerebral cortex. Elife 12: e84685. https://doi.org/10.7554/eLife.84685.

Rojkova K, Volle E, Urbanski M et al. (2016). Atlasing the frontal lobe connections and their variability due to age and education: A spherical deconvolution tractography study. Brain Struct Funct 221: 1751–1766. https://doi.org/10.1007/s00429-015-1001-3.

Sboto-Frankenstein UN, Lazar T, Bolster RB et al. (2014). Symmetry of the fornix using diffusion tensor imaging. J Magn Reson Imaging 40: 929–936. https://doi.org/10.1002/jmri.24424.

Schmahmann JD, Pandya DN (2007). The complex history of the fronto-occipital fasciculus. J Hist Neurosci 16: 362–377. https://doi.org/10.1080/09647040600620468.

Schmahmann JD, Pandya DN, Wang R et al. (2007). Association fibre pathways of the brain: parallel observations from diffusion spectrum imaging and autoradiography. Brain 130: 630–653. https://doi.org/10.1093/brain/awl359.

Schurr R, Filo S, Mezer AA (2019). Tractography delineation of the vertical occipital fasciculus using quantitative T1 mapping. Neuroimage 202: 116121. https://doi.org/10.1016/j.neuroimage.2019.116121.

Seizeur R, Magro E, Prima S et al. (2014). Corticospinal tract asymmetry and handedness in right- and left-handers by diffusion tensor tractography. Surg Radiol Anat 36: 111–124. https://doi.org/10.1007/s00276-013-1156-7.

Seltzer B, Pandya DN (1984). Further observations on parieto-temporal connections in the rhesus monkey. Exp Brain Res 55: 301–312. https://doi.org/10.1007/BF00237280.

Shapleske J, Rossell SL, Woodruff PW et al. (1999). The planum temporale: a systematic, quantitative review of its structural, functional and clinical significance. Brain Res Rev 29: 26–49. https://doi.org/10.1016/s0165-0173(98)00047-2.

Shu N, Liu Y, Duan Y et al. (2015). Hemispheric asymmetry of human brain anatomical network revealed by diffusion tensor tractography. Biomed Res Int 2015. https://doi.org/10.1155/2015/908917.

Silk T, Vilgis V, Adamson C et al. (2016). Abnormal asymmetry in frontostriatal white matter in children with attention deficit hyperactivity disorder. Brain Imaging Behav 10. https://doi.org/10.1007/s11682-015-9470-9.

Singh I (2014). Inderbir Singh's textbook of human neuroanatomy. In: Williams textbook of endocrinology. https://doi.org/10.1016/B978-0-323-29738-7.00032-0.

Steinmetz H, Rademacher J, Huang YX et al. (1989). Cerebral asymmetry: MR planimetry of the human planum temporale. J Comput Assist Tomogr 13: 996–1005.

Supprian T, Hofmann E (1997). The fornix of the human brain: evidence of left/right asymmetry on axial MRI scans. Surg Radiol Anat 19: 105–109. https://doi.org/10.1007/BF01628134.

Szczupak D, Lent R, Tovar-Moll F et al. (2023). Heterotopic connectivity of callosal dysgenesis in mice and humans. Front Neurosci 17. https://doi.org/10.3389/fnins.2023.1191859.

Szmuda T, Rogowska M, Słoniewski P et al. (2017). Frontal aslant tract projections to the inferior frontal gyrus. Folia Morphol 76: 574–581. https://doi.org/10.5603/FM.a2017.0039.

Takemura H, Pestilli F, Weiner KS et al. (2017). Occipital white matter tracts in human and macaque. Cereb Cortex 27: 3346–3359. https://doi.org/10.1093/cercor/bhx070.

Thiebaut de Schotten M, Ffytche DH, Bizzi A et al. (2011a). Atlasing location, asymmetry and inter-subject variability of white matter tracts in the human brain with MR diffusion tractography. Neuroimage. https://doi.org/10.1016/j.neuroimage.2010.07.055.

Thiebaut de Schotten MT, Dell'Acqua F, Forkel SJ et al. (2011b). A lateralized brain network for visuospatial attention. Nat Neurosci 14: 1245–1246. https://doi.org/10.1038/nn.2905.

Thiebaut de Schotten M, Cohen L, Amemiya E et al. (2014). Learning to read improves the structure of the arcuate fasciculus. Cereb. Cortex (New York, N.Y.: 1991) 24: 989–995. https://doi.org/10.1093/cercor/bhs383.

Thompson PM, Jahanshad N, Ching CRK et al. (2020). ENIGMA and global neuroscience: a decade of large-scale studies of the brain in health and disease across more than 40 countries. Transl Psychiatry 10: 1–28. https://doi.org/10.1038/s41398-020-0705-1.

Tsuzuki D, Taga G, Watanabe H et al. (2022). Individual variability in the nonlinear development of the corpus callosum during infancy and toddlerhood: a longitudinal MRI analysis. Brain Struct Funct 227: 1995–2013. https://doi.org/10.1007/s00429-022-02485-y.

Türe U, Yaşargil MG, Pait TG (1997). Is there a superior occipitofrontal fasciculus? A microsurgical anatomic study. Neurosurgery 40: 1226–1232. https://doi.org/10.1097/00006123-199706000-00022.

Uy JP, Galván A (2020). Individual differences in accumbofrontal tract integrity relate to risky decisions under stress in adolescents and adults. Dev Cogn Neurosci 45: 100859. https://doi.org/10.1016/j.dcn.2020.100859.

Uylings HBM, Rajkowska G, Sanz-Arigita E et al. (2005). Consequences of large interindividual variability for human brain atlases: converging macroscopical imaging and microscopical neuroanatomy. Anat Embryol 210: 423–431. https://doi.org/10.1007/s00429-005-0042-4.

Vallesi A, Babcock L (2020). Asymmetry of the frontal aslant tract is associated with lexical decision. Brain Struct Funct 225: 1009–1017. https://doi.org/10.1007/s00429-020-02054-1.

Van der Voorta EC, de Lucaa A, Westerhausen R, Thiebaut de Schotten M, Leemans A. (n.d.) More than just left and right: A diffusion MRI study of the corpus callosum fibre architecture. (in preparation)

Vanderauwera J, Altarelli I, Vandermosten M et al. (2018). Atypical structural asymmetry of the planum temporale is related to family history of dyslexia. Cereb Cortex 28: 63–72. https://doi.org/10.1093/cercor/bhw348.

Vergani F, Mahmood S, Morris CM et al. (2014). Intralobar fibres of the occipital lobe: a post mortem dissection study. Cortex 56: 145–156. https://doi.org/10.1016/j.cortex.2014.03.002.

Vicq-d'Azyr F (1786). Traité d'anatomie et de physiologie, avec des planches coloriées représentant au naturel les divers organes de l'homme et des animaux. Impr. de F.A. Didot l'ainé.

Wakana S, Caprihan A, Panzenboeck MM et al. (2007). Reproducibility of quantitative tractography methods applied to cerebral white matter. Neuroimage 36: 630–644. https://doi.org/10.1016/j.neuroimage.2007.02.049.

Wang Y, Fernández-Miranda JC, Verstynen T et al. (2013). Rethinking the role of the middle longitudinal fascicle in language and auditory pathways. Cereb Cortex 23: 2347–2356. https://doi.org/10.1093/cercor/bhs225.

Weiner KS, Yeatman JD, Wandell BA (2017). The posterior arcuate fasciculus and the vertical occipital fasciculus. Cortex 97: 274–276. https://doi.org/10.1016/j.cortex.2016.03.012.

Westerhausen R, Papadatou-Pastou M (2022). Handedness and midsagittal corpus callosum morphology: a meta-analytic evaluation. Brain Struct Funct 227: 545–559. https://doi.org/10.1007/s00429-021-02431-4.

Witelson SF (1989). Hand and sex differences in the isthmus and genu of the human CORPUS callosum: a postmortem morphological study. Brain 112: 799–835. https://doi.org/10.1093/brain/112.3.799.

Wu Y, Sun D, Wang Y et al. (2016). Subcomponents and connectivity of the inferior fronto-occipital fasciculus revealed by diffusion spectrum imaging fibre tracking. Front Neuroanat 10. https://doi.org/10.3389/fnana.2016.00088.

Yaşargil MG (1994). CNS tumors: surgical anatomy, neuropathology, neuroradiology, neurophysiology, clinical considerations, operability, *treatment options*, Thieme.

Yeatman JD, Weiner KS, Pestilli F et al. (2014). The vertical occipital fasciculus: a century of controversy resolved by in vivo measurements. Proc Natl Acad Sci U S A 111: E5214–E5223. https://doi.org/10.1073/pnas.1418503111.

Yousry TA, Schmid UD, Alkadhi H et al. (1997). Localization of the motor hand area to a knob on the precentral gyrus. A new landmark. Brain 120: 141–157. https://doi.org/10.1093/brain/120.1.141.

Zilles K, Bacha-Trams M, Palomero-Gallagher N et al. (2015). Common molecular basis of the sentence comprehension network revealed by neurotransmitter receptor fingerprints. Cortex 63: 79–89. https://doi.org/10.1016/j.cortex.2014.07.007.

Handbook of Clinical Neurology, Vol. 208 (3rd series)
Cerebral Asymmetries
C. Papagno and P. Corballis, Editors
https://doi.org/10.1016/B978-0-443-15646-5.00027-0

Chapter 3

Latent dimensions of brain asymmetry

CHENGHUI ZHANG[1,†], YI PU[2,†], AND XIANG-ZHEN KONG[1,3*]

[1]*Department of Psychology and Behavioral Sciences & The State Key Lab of Brain-Machine Intelligence, Zhejiang University, Hangzhou, China*

[2]*Shanghai Key Laboratory of Brain Functional Genomics (Ministry of Education), School of Psychology and Cognitive Science, East China Normal University, Shanghai, China*

[3]*Department of Psychiatry of Sir Run Shaw Hospital, Zhejiang University School of Medicine, Hangzhou, China*

Abstract

Functional lateralization represents a fundamental aspect of brain organization, where certain cognitive functions are specialized in one hemisphere over the other. Deviations from typical patterns of lateralization often manifest in various brain disorders, such as autism spectrum disorder, schizophrenia, and dyslexia. However, despite its importance, uncovering the intrinsic properties of brain lateralization and its underlying structural basis remains challenging. On the one hand, functional lateralization has long been oversimplified, often reduced to a unidimensional perspective. For instance, individuals are sometimes labeled as left-brained or right-brained based on specific behavioral measures like handedness and language lateralization. Such a perspective disregards the nuanced subtypes of lateralization, each potentially attributed to distinct factors and associated with unique functional correlates. On the other hand, traditional studies of brain structural asymmetry have typically focused on localized analyses of homologous regions in the two hemispheres. This perspective fails to capture the inherent interplay between brain regions, resulting in an overly complex depiction of structural asymmetry. Such conceptual and methodological discrepancies between studies of functional lateralization and structural asymmetry pose significant obstacles to establishing meaningful links between them. To address this gap, a shift toward uncovering the dimensional structure of brain asymmetry has been proposed. This chapter introduces the concept of latent dimensions of brain asymmetry and provides an up-to-date overview of studies regarding dimensions of functional lateralization and structural asymmetry in the human brain. By transcending the traditional analysis and employing multivariate pattern techniques, these studies offer valuable insights into our understanding of the intricate organizational principles governing the human brain's lateralized functions.

INTRODUCTION

The organization of the brain into left and right hemispheres stands as one of its most enduring features deeply ingrained in evolutionary history (Toga and Thompson, 2003; Corballis, 2019; Gunturkun et al., 2020). This division not only amplifies the brain's processing capacity by expanding its surface area (Calvin and Ojemann, 1994) but also lays the foundation for functional specialization within each hemisphere—a phenomenon known as lateralization (Bradshaw and Nettleton, 1981; Herve et al., 2013). Lateralization is considered crucial for optimal brain function, with deviations from typical patterns often linked to various cognitive and neuropsychiatric conditions, including dyslexia, autism, and mood disorders (Corballis, 2014; Kong et al., 2022b). Understanding these deviations is crucial for accurate diagnosis, effective treatment strategies, and gaining insights into the underlying neurobiologic mechanisms of these disorders. Despite its recognized significance, the correlation between functional/behavioral lateralization and hemispheric differences in brain structure (structural asymmetry) remains elusive, posing a persistent challenge in neuroscience research (Ocklenburg et al., 2021).

Functional lateralization has historically been oversimplified, often reduced to a unidimensional perspective (Corballis, 2021).

[†]These authors are equally contributed.

[*]Correspondence to: Xiang-Zhen Kong, PhD, Department of Psychology and Behavioral Sciences, Zhejiang University, Yuhangtang Rd 866, 310058 Hangzhou, Zhejiang, China. Tel: 0571-88273337, E-mail: xiangzhen.kong@zju.edu.cn

For instance, while individuals may prefer to use both left and right hands for different activities, they are usually categorized as either left-handed or right-handed, or assigned a single continuous handedness score (Oldfield, 1971), which often serves as a proxy measure of brain asymmetry (e.g., Huber and Marsolek, 2022). Similarly, various experiments have also been employed to determine one's language dominance, being either typical (leftward) or atypical (Wada and Rasmussen, 2007; Woodhead et al., 2021; Guadalupe et al., 2022). Such different lateralizations have often been considered to correlate with each other, and thus individuals are sometimes labeled as left-brained or right-brained simply using only one of these lateralization measures. Researchers may consider that these lateralizations potentially share common genetic and developmental factors with the asymmetry of internal organs (e.g., Kennedy et al., 1999; Vingerhoets et al., 2018; see also Chapter 5 in this volume). However, more studies have emerged challenging this simplistic view. These studies suggest that the nature of brain asymmetry is more nuanced than previously believed and that more than one dimension of lateralization exists, each likely with distinct factors and functional correlates (Corballis, 2021).

In terms of brain structural asymmetry, traditional studies have typically focused on localized analyses of homologous regions in the two hemispheres. This perspective fails to capture the inherent interplay between brain regions, resulting in an overly complex depiction of structural asymmetry (e.g., Kong et al., 2018). Yet linking these structural differences with functional lateralization has proven challenging. One possible reason for this limitation is the narrow focus on localized analyses of structural differences between hemispheres, which may provide limited insights into broader patterns of hemispheric differences. To address this gap, a shift toward a pattern recognition perspective has been proposed for brain asymmetry studies. The community has embraced multivariate approaches, such as principal component analysis (PCA) and diffusion mapping (Table 3.1), to map the latent dimensions of brain asymmetry (e.g., Karolis et al., 2019; Woodhead et al., 2019; Wan et al., 2023). Drawing on evidence such as brain torque and large-scale cortical thickness studies (e.g., Kong et al., 2018, 2021; Zhao et al., 2022), this paradigm shift underscores the importance of exploring structural asymmetry from a holistic viewpoint.

In this chapter, we present an up-to-date overview of these innovative studies, offering insights into the dimensional structure of both functional lateralization and structural asymmetry as well as their inherent relationship (Fig. 3.1). Additionally, we propose avenues for future research, aiming to deepen our understanding of the fundamental organizational principles of the human brain.

DIMENSIONS OF FUNCTIONAL LATERALIZATION

Regarding functional lateralization, handedness and language lateralization have been the most prominent areas of investigation. In this section, we reviewed studies regarding handedness and language lateralization, along with several other functional lateralities.

Handedness

Traditionally, handedness is treated as a binary concept (i.e., left- or right-handed) and only refers to one manual task (i.e., writing). Of note, many studies in the field of laterality have recognized this issue and satisfy an assessment of handedness using several items. For instance, many researchers use the Edinburgh Handedness Inventory (EHI) (Oldfield, 1971), a questionnaire that queries hand preference for various manual tasks (e.g., handling a knife, brushing teeth). Such an approach ultimately allows the calculation of a laterality quotient which satisfies handedness as a continuous variable (i.e., handedness index). Still, only 10 manual tasks are included in EHI, ignoring other manual activities that could result in a different handedness index if taken into consideration.

Early studies have also demonstrated subtypes of hand preference and proposed the multidimensional nature of hand preference. For instance, Healey et al. (1986) employed a 55-item questionnaire designed to study hand preference for a wide range of activities and revealed four distinct factors, which accounted for 80% of the variance in a PCA (Healey et al., 1986). Among these latent factors, Factor 1 included tasks such as writing, drawing, and brushing teeth, and was characterized as a "general handedness factor," similar to the handedness index commonly used. The other factors were independent of Factor 1, and each showed unique characteristics regarding involvement of body movement and existence of movement targets. While the four-dimensional categorization may not seem very intuitive, Gonzalez and Nelson proposed a two-dimensional scale of handedness: the Home Handedness Questionnaire measures skills of unimanual actions and role-differentiated bimanual manipulation, and its two-dimensional structure has been confirmed in both children and adults (Nelson et al., 2019; Gonzalez and Nelson, 2021).

Overall, while the handedness index has been widely used in the field, e.g., for investigating potential links with brain disorders (e.g., Wiberg et al., 2019) and scanning for candidate genetic variants (e.g., Cuellar-Partida et al., 2021), the latent dimensions would provide new insights for a better understanding of the development of handedness.

Language lateralization

Measurement of language lateralization is of interest both to researchers investigating brain organization and to clinicians, for example, needing to establish an individual's hemispheric dominance for language prior to surgery, as in patients with intractable epilepsy. Traditionally, language lateralization has been treated as a singular dimension, focusing on whether and to what extent language functions are predominantly localized in the left hemisphere. The Wada test is the clinical gold standard for the assessment of hemispheric language dominance, which involves unilateral intracarotid injection of a short-acting barbiturate (amobarbital) that immediately suspends several ipsilateral brain functions and allows for 5–10 min of language testing of the contralateral hemisphere (Wada and Rasmussen, 2007). As a noninvasive measurement approach, the relative blood flow in the two middle cerebral arteries can be measured using functional

Table 3.1

Methods for identifying latent dimensions of brain asymmetry.

Method name	Brief description
Principal component analysis (PCA)	A statistical technique used for dimensionality reduction in data analysis. PCA identifies patterns and relationships in high-dimensional data by transforming it into a lower-dimensional space while preserving most of its variability. PCA achieves this by finding orthogonal axes, called principal components, along which the data varies the most. These components allow for a simplified representation of the original data, aiding in the visualization and interpretation of complex datasets.
Independent component analysis (ICA)	A computational technique used to separate a multivariate signal into additive, independent components. ICA assumes that the observed data are a linear combination of independent sources, each with a different statistical distribution. By estimating the mixing matrix that relates the observed data to the sources, ICA can recover the original sources from their linear mixtures. ICA finds applications in various fields, such as signal processing, image processing, and neuroscience, where it is used for blind source separation and extracting meaningful information from mixed signals.
Singular value decomposition (SVD)	A method in linear algebra used to factorize a matrix into three smaller matrices. SVD decomposes a matrix into singular vectors and singular values, which capture the directions and magnitudes of the most significant variability in the data, respectively. SVD is widely used in various applications such as data compression, image processing, and recommendation systems. It is particularly useful in PCA, where it forms the mathematical basis for finding principal components.
Structural equation modeling (SEM)	A statistical technique used to test and validate complex relationships among variables. SEM combines factor analysis and path analysis to analyze both observed and latent variables simultaneously. SEM allows researchers to specify and test theoretical models that represent causal relationships between variables, including direct and indirect effects. It is particularly useful in social sciences, psychology, and economics for modeling complex systems of variables and understanding underlying mechanisms. SEM provides a comprehensive framework for hypothesis testing, model comparison, and theory development.
Laplacian eigenmaps (LE)	A dimensionality reduction technique used for nonlinear manifold learning. LE constructs a graph representation of the data, where each data point is a node and edges represent similarity. By computing the graph Laplacian matrix, it preserves local neighborhood relationships. The eigenvalue decomposition of this matrix yields low-dimensional embeddings, where each dimension captures the intrinsic structure of the data manifold. Laplacian eigenmaps are effective for tasks like clustering, visualization, and feature extraction from high-dimensional data.
Diffusion mapping (DM)	Another manifold learning technique that focuses on capturing the global structure of the data manifold. DM operates by modeling the diffusion process of information flow between data points. By computing the eigenvectors of the diffusion kernel matrix, it uncovers the underlying low-dimensional representation of the data. DM is robust to noise and can handle complex, nonlinear structures in high-dimensional data. It is often used for tasks such as dimensionality reduction, denoising, and clustering in various fields including bioinformatics and computer vision.
Louvain community detection algorithm	A popular method for detecting communities in complex networks. This algorithm works by optimizing a modularity measure that quantifies the quality of network partitioning. In each iteration, nodes are greedily moved between communities to maximize modularity. This process continues until no further improvement can be made, resulting in a partition that maximizes modularity. This algorithm can be used for spatial clustering of correlated variables.

transcranial Doppler sonography (fTCD) to give an index of lateralization as participants perform a specific task, e.g., a word generation task (Woodhead et al., 2021). Functional MRI (fMRI) is another noninvasive technique for the assessment of language lateralization. By analyzing the activation patterns observed in specific brain regions during language tasks, researchers can infer language lateralization. Typically, left-lateralization is indicated by greater activation in regions of the left hemisphere, such as Broca's area and Wernicke's area, while right-lateralization may be inferred if there is significant activation in corresponding regions of the right hemisphere (Rutten et al., 2002). Comparisons of activation patterns between hemispheres, e.g., via the laterality index, provide insights into individual differences in language lateralization. In addition, the right ear advantage in dichotic listening is a marker of left hemisphere language dominance in right-handed individuals (Van der Haegen et al., 2013; Guadalupe et al., 2022).

It is commonly assumed that language laterality is a unidimensional trait: if an individual is strongly lateralized (relative to other individuals) on one language task, they will be strongly lateralized on all other language tasks. As research methods have advanced, it has become increasingly evident that language lateralization is a more complex phenomenon than previously thought. In particular, there has been increasing awareness of the possibility that different language processes may develop hemispheric lateralization independently and to varying degrees (e.g., Bradshaw et al., 2017). Woodhead et al. (2019) used fTCD and Structural Equation Modeling to test the factorial structure of lateralization across six different language tasks, including list generation, phonologic decision, semantic decision, sentence generation, sentence comprehension, and syntactic decision. A bifactor model of language lateralization was proposed, where covariance in lateralization in the tasks varied along two independent dimensions (Woodhead et al., 2019, 2021). While list generation only

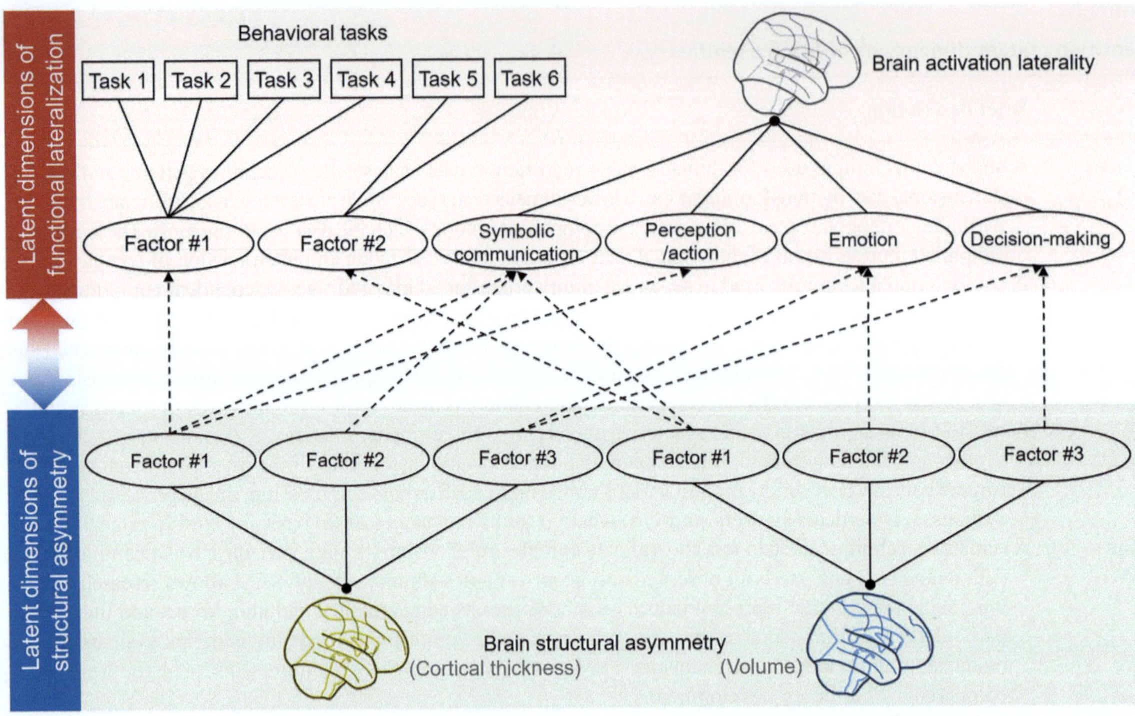

Fig. 3.1. Conceptual diagram of latent dimension analysis of brain functional lateralization and structural asymmetry. The four dimensions of brain activation laterality refer to the principal component analysis results from a previous study (Karolis et al., 2019). *Dashed lines* represent potential links between the latent dimensions of brain asymmetry.

showed a weak loading on the first factor and syntactic decision only showed a strong loading on the second factor, the other four have moderate loadings on both factors.

At one time it was thought that handedness was a reliable indicator of cerebral dominance for language, with right-handers being left-cerebrally dominant and left-handers right-cerebrally dominant—a relation once known as "Broca's rule" (Corballis, 2019). We now know that the correlation is weak at best (e.g., Van der Haegen et al., 2013; Packheiser et al., 2020). About 95% of right-handers are left-cerebrally dominant for language, but so are 70% or more of left-handers, contravening Broca's rule. Little is known about the relationship between handedness and language laterality when taking their latent dimensions.

Functional laterality beyond handedness and language

Besides handedness and language laterality, there are several other fascinating aspects of functional lateralization. One prominent area is emotional processing. Research suggests that the left hemisphere is generally more involved in positive emotions, such as happiness and excitement, while the right hemisphere tends to play a larger role in processing negative emotions like fear and sadness. Additionally, spatial abilities are often lateralized in the brain. The right hemisphere tends to excel in tasks involving spatial awareness, such as navigating through a new environment or mentally rotating objects (e.g., van der Ham et al., 2011; Kong et al., 2017). An early analysis of left-hemispheric asymmetry for language and right-hemispheric asymmetry for spatial tasks, based on evidence from unilateral brain injury, suggested that the two asymmetries were independent (Bryden et al., 1983). The relationship between different hemispheric specializations could be complementary or independent, which is still in debate (Badzakova-Trajkov et al., 2016).

Accordingly, researchers have turned to factor analysis to better understand the dimensional structure of functional lateralization. Badzakova-Trajkov et al. (2016) carried out a factor analysis of functional asymmetries while participants undertook language tasks, an attentional task, and a face-recognition task, which yielded three independent factors, a left-lateralized one corresponding to the language network and two right-lateralized networks corresponding to the face-processing network and the attentional network (Badzakova-Trajkov et al., 2016). The right-lateralized face-processing network was largely homologous to the language network, yet correlational independent of it. Haberling and Corballis (2016) undertook a factor analysis of laterality indices while participants performed various left-lateralized tasks and found three independent factors, representing a language circuit, a gesture-related circuit associated with handedness, and another gesture-related circuit independent of handedness (Haberling and Corballis, 2016). As the first attempt to map the dimensional structure of functional lateralities, Karolis et al. (2019) applied PCA to a large-scale set of neuroimaging maps of 590 terms related to specific cognitive processes, which revealed four functional axes: symbolic communication, perception/action, emotion and decision-making. While the former two axes are related to the most prominent lateralized functions/behaviors, that is, language and handedness, the latter two present the additional dimensions of functional laterality (Karolis et al., 2019). These new findings give rise to new hypotheses regarding the development of the lateralized organization of the human brain.

Overall, factor analysis studies have revealed distinct factors underlying functional laterality, such as motor skills, language abilities, and visuospatial processing. This approach underscores the complexity of brain organization and emphasizes the importance of considering multiple dimensions when investigating functional lateralization.

Network analysis of functional laterality

The connectome, a sophisticated network of neural connections within the brain, constitutes the foundational architecture supporting cognitive functions and behaviors (Sporns et al., 2005; Sporns, 2013). Leveraging advanced neuroimaging methodologies, such as resting-state fMRI, researchers have adopted a fresh perspective to elucidate functional laterality. In line with the rationale of preceding studies, investigations have sought to delineate brain laterality by directly assessing homologous connections. Employing graph theory techniques, researchers have scrutinized interhemispheric disparities in network-level metrics, encompassing measures like small-worldness, global efficiency, and modularity, across hemispheres. Intriguing findings have emerged from analyses conducted on both healthy brains and those afflicted with neurological disorders, which can be referred to a comprehensive review article (Wang et al., 2023). Here our focus centers on exploring the latent dimensions of the laterality index within functional networks.

Liu et al. (2009) analyzed the laterality indices of functional connectivity derived from intrinsic brain activity using resting-state fMRI and factor analysis and revealed four independent factors. Two were left-lateralized, one corresponding to the language network and the other the default mode network, and the other two were right-lateralized, corresponding to the visual and attentional networks (Liu et al., 2009). Wu et al. (2022) integrated dynamics network approaches with hemispheric laterality and proposed a framework for dynamic laterality analysis. With resting-state fMRI and a spatial clustering approach (i.e., the Louvain community detection algorithm, see Table 3.1), they identified four major clusters of brain regions which demonstrated distinct laterality dynamics (Wu et al., 2022). Three of the four identified clusters roughly correspond to the top 3 factors identified in Liu et al. (2009), which include the visual, attention, and the default mode network, respectively. In addition, Wan et al. (2023) employed the diffusion mapping approach on the functional connectome derived from resting-state fMRI and revealed the top gradients of different connectivity patterns. The asymmetry index scores along these organizational axes for intra-hemispheric functional connectivity patterns and interhemispheric functional connectivity patterns were calculated (Wan et al., 2023).

Network analysis of functional laterality captures associated organizational features of the brain networks and provides new insights into understanding the organization of the human brain from a system-level perspective.

DIMENSIONS OF STRUCTURAL ASYMMETRY

Within the domain of structural brain asymmetry, the predominant methodology involves the direct comparison between homologous regions within the left and right hemispheres. The left–right asymmetry of the planum temporale is one of the most prominent structural asymmetries in the human brain (Altarelli et al., 2014). Studies have consistently shown a larger planum temporale in the left hemisphere (e.g., Geschwind and Levitsky, 1968; Carrion-Castillo et al., 2020). Another example is the Sylvian fissure, a prominent fissure that separates the frontal and temporal lobes. The Sylvian fissure often exhibits asymmetry in its morphology and position between the left and right hemispheres (Rubens et al., 1976). With advanced image analysis techniques and higher-resolution MRI, studies have revealed various local asymmetries in the human brain. For instance, as an international effort for mapping structural brain asymmetries, Kong et al. (2018) collected data of multiple morphologic metrics, including regional cortical thickness, surface area, and volume, and presented asymmetries of different cortical regions across the whole brain (Kong et al., 2018). It is important to note that the majority of these asymmetries show considerable variability across individuals.

Brain torque

These local asymmetries showed complex patterns across the brain, suggesting existing global dimensions of hemispheric differences. For example, a global anterior–posterior asymmetry pattern in cortical thickness was revealed at the population level (Kong et al., 2018). The global twisting seems to be related to the Yakovlevian torque, which is manifested by several local features of hemispheric differences, including a more anteriorly protruding right frontal lobe (frontal petalia) and posteriorly protruding left occipital lobe (occipital petalia), a so-called "bending" of the right frontal and left occipital lobes across the midline, and relative increases in the dimensions (e.g., volume and width) of the right frontal and left occipital poles (Toga and Thompson, 2003). The global asymmetry pattern in cortical thickness has been well reproduced using a different approach for brain image analysis (i.e., vertex-wise analysis) (Sha et al., 2021). Note that these studies only provide a qualitative explanation of localized analysis results.

To provide a quantitative measure of this global pattern, researchers have proposed various approaches. For example, Lancefield et al. (2006) applied PCA of the asymmetry indices of the regional gray matter volume and uncovered the single dimension of the fronto-occipital torque in children (Lancefield et al., 2006). Kong et al. (2021) proposed a registration-based approach for measuring, such as global asymmetry. The derived brain skew metric can effectively capture the global asymmetric shape of the human brain and shows excellent test–retest reliability (Kong et al., 2021). In addition, a Typicality score was proposed which indexes how much a given participant deviated from the population mean cortical thickness asymmetry profile when considering simultaneously the regional asymmetries across the whole brain (Kong et al., 2020). Xiang et al. (2019)

proposed an image processing approach for calculating the three principal aspects of brain torque: (1) petalia, whereby one cerebral hemisphere protrudes anteriorly and the other posteriorly in the anteroposterior direction, (2) shift, whereby one cerebral hemisphere moves upward anteriorly and downward posteriorly relative to the other in the dorsoventral direction bending, and (3) bending, whereby the brain tissue in one hemisphere crosses the midline to displace tissue in the other hemisphere. It is shown that all three aspects of the torque are specific to the human brain (Xiang et al., 2019).

Global asymmetries beyond brain torque

Besides the brain torque, studies have attempted to reveal additional asymmetry patterns of the human brain in structure. Traditional studies typically focused on examining differences of homologous regions and mostly resulted in minimal and even inconsistent structural differences (e.g., Kong et al., 2022a). The basic idea of new studies is to decompose brain asymmetries across the whole brain into relatively independent spatial patterns.

For instance, a large-scale work investigating hemispheric differences in gray matter volume revealed multiple dissociable asymmetry patterns, one of which showed a similar pattern with the brain torque mentioned above (Saltoun et al., 2023). Specifically, the authors employed singular value decomposition on asymmetry index measures of 85 brain regional volume measures from over 37,000 UK Biobank participants, revealing dissociable asymmetry patterns with distinct phenotypic associations. Similarly, Roe and colleagues utilized a clustering algorithm on asymmetry age trajectories, identifying three clusters with unique asymmetric thinning properties across the adult lifespan (Roe et al., 2021). Both studies applied a multivariate approach to reveal the brain asymmetry patterns in the brain. The decomposition of the asymmetry patterns was based on the inter-regional similarity in either the individual differences of asymmetry measures or the age trajectories in asymmetry. An unsupervised approach guided both studies.

In contrast, later work also adopted a supervised approach for investigating interhemispheric differences. Specifically, multivariate prediction models were trained to explicitly classify the two hemispheres, enabling the identification of the left–right asymmetry patterns. In a yet unpublished work, we employed an ensemble machine learning approach and trained predictive models based on morphologic metrics including cortical thickness, surface area, and gray matter volume for classifying the two hemispheres. As expected, our multivariate modeling results of all the three morphologic features showed high classification performance (over 87%). These results are in line with an earlier study which applied a lasso-based linear regression approach but only focused on volumetric differences in a much smaller sample size ($N = \sim 200$) (Friedrich et al., 2022). These results suggest that the two hemispheres possess unique multivariate structural patterns, which might have the potential to explain significant lateralization in cognitive functions and behaviors like language and handedness. In addition, we can identify the brain regions with the highest contribution scores in these trained models which construct the global asymmetry patterns in the human brain.

Network analysis of structural asymmetry

Diffusion MRI and T1-weighted MRI stand out as pivotal tools in brain imaging for probing the structural architecture of the brain. Among these modalities, diffusion MRI is particularly adept at delineating white matter connections, providing invaluable insights into the intricate pathways that facilitate neural communication across different brain regions (Sotiropoulos and Zalesky, 2019). Meanwhile, T1-weighted MRI offers complementary perspectives, allowing researchers to construct covariance networks that capture the covariation patterns of various brain regions (Evans, 2013). Moreover, T1-weighted MRI enables the creation of morphologic similarity networks, which illuminate the structural similarities between different brain areas (Kong et al., 2014, 2015; Wang and He, 2024). With these modalities for brain structural network modeling along with graph theory techniques for network analysis, researchers have examined interhemispheric differences by directly comparing the homologous regions, connections, and networks. Many intriguing findings have been reported (refer to Wang et al., 2023). Here our focus centers on exploring the latent dimensions of the asymmetry index within structural networks.

Relatively few studies have examined the dimensional information of structural network asymmetry. Yoo et al. (2024) utilized diffusion MRI and dimensionality reduction techniques (i.e., diffusion mapping, see Table 3.1) to present cortex-wide structural connectivity with a set of macroscale eigenvectors. Then they calculated the asymmetry index for each eigenvector and found atypical structural connectome asymmetry in autism (Yoo et al., 2024).

DIMENSIONAL ASSOCIATIONS BETWEEN FUNCTIONAL LATERALITY AND STRUCTURAL ASYMMETRY

It has proven surprisingly difficult to link brain structural asymmetries to lateralized functions (Ocklenburg et al., 2021). Traditional studies usually investigated the structural basis of functional lateralization by correlating a structural asymmetry variable (e.g., planum temporale asymmetry) with a functional lateralization variable (e.g., language lateralization). The latent dimensions of brain asymmetry provide a new avenue for understanding such structure–function relationships (Fig. 3.1). To the best of our knowledge, only a few studies have explicitly examined the associations between dimensions of functional lateralization and dimensions of structural asymmetry.

In terms of handedness, studies focusing on the simple correlation between handedness and regional brain measures have shown mixed results (e.g., Kong et al., 2018). New studies employed multivariate approaches to examine the brain–behavioral association. For instance, Kong et al. (2021) made use of the registration-based brain torque metric and revealed replicable association with handedness (Kong et al., 2021). With a multivariate machine learning approach, Chormai et al. (2022) showed that the brain networks with the highest importance in the hand preference prediction model showed significant functional relevance to lateralized functions, including hand movement and several higher-level cognitive functions such as language (Chormai et al., 2022). In an independent study, Wiberg et al. (2019) reported that left-handers showed a stronger connectivity between the right and left homologous language networks and weaker connectivity between the right homologous language network and the default-mode network and salience network. These functional networks were computed using independent component analysis beforehand (Wiberg et al., 2019). Together, these large-scale studies using multivariate machine learning have shed new light on the neural correlates of human handedness.

In terms of language lateralization, morphologic asymmetries favoring the left hemisphere in the planum temporale and Heschl's gyrus have both been presumed to be related to the typical left-hemisphere dominance for language functions (Dorsaint-Pierre et al., 2006). Theories also suggest that atypical language organization would entail a diffuse functional network, rather than single local regions (Brown and Hecaen, 1976). Bidula et al. (2017) used a threshold-free classification method of language lateralization pattern similarity and divided their participants into two groups: typical and atypical (Bidula et al., 2017). By comparing brain structural measures between the two groups, they found no significant differences related to the structure of the brain. In addition, they found that the left-hemisphere cerebellar seed exhibited significantly stronger connectivity with the right occipital pole in the typically lateralized group. A large neuroimaging study proposed a more refined classification of the atypical subpopulation (Berl et al., 2014) using hierarchical clustering analysis based on the language lateralization index. Such a refined categorization of atypical language lateralization would provide new insights into understanding the mechanisms underlying the (a)typical language lateralization in healthy individuals.

The multivariate and dimensional approaches of brain asymmetry have been applied in studies of brain disorders. Altered brain torque properties have been suggested to be linked to schizophrenia (Guerguerian and Lewine, 1998) and depression (Maller et al., 2014). The registration-based brain torque measure showed a significant genetic correlation with genetic risks of autism (Kong et al., 2021). Kong et al. (2020) calculated the typicality score regarding the brain structural asymmetry patterns in cortical thickness, surface area, and volume, but did not find significant differences in patients with obsessive–compulsive disorder (Kong et al., 2020). Chen et al. (2024) used canonical correlation analysis to examine the associations between shape asymmetries and psychosis symptoms. A single canonical model linked the asymmetry measures to symptoms: higher cortical asymmetry was correlated with more severe excitement symptoms and less severe emotional distress (Chen et al., 2024).

CONCLUDING REMARKS AND FUTURE DIRECTIONS

In this chapter, we present an up-to-date overview of studies on dimensions of brain asymmetry using multivariate pattern approaches. The first set of studies underscores the intricate interplay between individual localized asymmetries in both structure and function.

A review paper on network analysis of brain asymmetry provided a summary of existing studies on the hemispheric differences in structural and functional connectivity in both health and diseases (Wang et al., 2023). The authors noted the limited existing literature. Our current review focused on the latent dimensional analysis of brain asymmetry. We found that the field is still emerging. Many questions remain unresolved in the field. One of the fundamental questions regards the associations between structural asymmetry and lateralized functions, such as handedness and language lateralization. While traditional univariate approaches prove difficult to link brain structure and functional laterality, the latent dimensions of brain asymmetries would provide a new avenue for understanding their inherent relationship. Another critical question is about the nature of the latent dimensions. Dimensionality reduction algorithms, such as PCA and diffusion mapping, provide a convenient mathematical way to identify asymmetry components that are independent of each other and provide a useful means of determining how many dimensions of brain asymmetry there are in the brain. Current studies usually focused on a small number of specific tasks or imaging modalities with small sample sizes. Little is known about how reproducible these findings are and what they indicate for understanding brain organization. As the first try to map the dimensional structure of functional lateralization, Karolis et al. (2019) integrated a large scale of neuroimaging maps and revealed four functional dimensions: symbolic communication, perception/action, and emotion-related and decision-making functions (Karolis et al., 2019). These results provide new data on the lateralized organization of the human brain. Compared to the dimensions of functional lateralization, the dimensional structure of structural asymmetry could be much more complicated given the various morphologic features (e.g., cortical thickness, surface area, and volume).

To the best of our knowledge, no studies have addressed this critical question. Future work can use large cohorts of brain imaging data such as the UK Biobank and international consortia, such as ENIGMA working groups, to achieve a more definitive picture of the nature of the dimensional structure of brain asymmetry. Moreover, it remains to determine the associations between the functional dimensions and structural dimensions and how such dimensional structure contributes to the understanding of brain organization, its evolutional and developmental trajectories, and relevant brain disorders.

Overall, brain asymmetry is not a single trait but is better conceptualized as a multifactorial feature of brain organization that involves a mixture of latent components. By leveraging advanced computational techniques, such as dimensionality reduction and graph theory algorithms, researchers have attempted to unravel the inherent dimensional structure of brain asymmetry. Such latent dimensions of brain asymmetry provide new perspectives for understanding the underlying mechanisms driving (a)typical lateralized cognitive functions in health and diseases.

References

Altarelli I, Leroy F, Monzalvo K et al. (2014). Planum temporale asymmetry in developmental dyslexia: revisiting an old question. Hum Brain Mapp 35: 5717–5735.

Badzakova-Trajkov G, Corballis MC, Haberling IS (2016). Complementarity or independence of hemispheric specializations? A brief review. Neuropsychologia 93: 386–393.

Berl MM, Zimmaro LA, Khan OI et al. (2014). Characterization of atypical language activation patterns in focal epilepsy. Ann Neurol 75: 33–42.

Bidula SP, Przybylski L, Pawlak MA et al. (2017). Unique neural characteristics of atypical lateralization of language in healthy individuals. Front Neurosci 11: 525.

Bradshaw JL, Nettleton NC (1981). The nature of hemispheric specialization in man. Behav Brain Sci 4: 51–63.

Bradshaw AR, Thompson PA, Wilson AC et al. (2017). Measuring language lateralisation with different language tasks: a systematic review. PeerJ 5: e3929.

Brown JW, Hecaen H (1976). Lateralization and language representation. Neurology 26: 183–189.

Bryden MP, Hecaen H, DeAgostini M (1983). Patterns of cerebral organization. Brain Lang 20: 249–262.

Calvin WH, Ojemann GA (1994). Conversations with Neil's brain: the neural nature of thought and language, Addison-Wesley Publishing Company, Reading, MA.

Carrion-Castillo A, Pepe A, Kong X-Z et al. (2020). Genetic effects on planum temporale asymmetry and their limited relevance to neurodevelopmental disorders, intelligence or educational attainment. Cortex 124: 137–153.

Chen YC, Tiego J, Segal A et al. (2024). A multiscale characterization of cortical shape asymmetries in early psychosis. Brain Commun 6: fcae015.

Chormai P, Pu Y, Hu H et al. (2022). Machine learning of large-scale multimodal brain imaging data reveals neural correlates of hand preference. Neuroimage 262: 119534.

Corballis MC (2014). Left brain, right brain: facts and fantasies. PLoS Biol 12: e1001767.

Corballis MC (2019). Evolution of cerebral asymmetry. Prog Brain Res 250: 153–178.

Corballis MC (2021). How many lateralities? Laterality 26: 307–309.

Cuellar-Partida G, Tung JY, Eriksson N et al. (2021). Genome-wide association study identifies 48 common genetic variants associated with handedness. Nat Hum Behav 5: 59–70. https://doi.org/10.1038/s41562-020-00956-y.

Dorsaint-Pierre R, Penhune VB, Watkins KE et al. (2006). Asymmetries of the planum temporale and Heschl's gyrus: relationship to language lateralization. Brain 129: 1164–1176.

Evans AC (2013). Networks of anatomical covariance. Neuroimage 80: 489–504.

Friedrich P, Patil KR, Mochalski LN et al. (2022). Is it left or is it right? A classification approach for investigating hemispheric differences in low and high dimensionality. Brain Struct Funct 227: 425–440.

Geschwind N, Levitsky W (1968). Human brain: left-right asymmetries in temporal speech region. Science 161: 186–187.

Gonzalez SL, Nelson EL (2021). Factor analysis of the home handedness questionnaire: unimanual and role differentiated bimanual manipulation as separate dimensions of handedness. Appl Neuropsychol Adult 28: 173–184.

Guadalupe T, Kong XZ, Akkermans SEA et al. (2022). Relations between hemispheric asymmetries of grey matter and auditory processing of spoken syllables in 281 healthy adults. Brain Struct Funct 227: 561–572.

Guerguerian R, Lewine RR (1998). Brain torque and sex differences in schizophrenia. Schizophr Res 30: 175–181.

Gunturkun O, Strockens F, Ocklenburg S (2020). Brain lateralization: a comparative perspective. Physiol Rev 100: 1019–1063.

Haberling IS, Corballis MC (2016). Cerebellar asymmetry, cortical asymmetry and handedness: two independent networks. Laterality 21: 397–414.

Healey JM, Liederman J, Geschwind N (1986). Handedness is not a unidimensional trait. Cortex 22: 33–53.

Herve PY, Zago L, Petit L et al. (2013). Revisiting human hemispheric specialization with neuroimaging. Trends Cogn Sci 17: 69–80.

Huber KB, Marsolek CJ (2022). Do cerebral motivational asymmetries mediate the relationship between handedness and personality? Laterality 27: 21–56.

Karolis VR, Corbetta M, Thiebaut de Schotten M (2019). The architecture of functional lateralisation and its relationship to callosal connectivity in the human brain. Nat Commun 10: 1417.

Kennedy DN, O'Craven KM, Ticho BS et al. (1999). Structural and functional brain asymmetries in human situs inversus totalis. Neurology 53: 1260–1265.

Kong XZ, Wang X, Huang L et al. (2014). Measuring individual morphological relationship of cortical regions. J Neurosci Methods 237: 103–107.

Kong X-z, Liu Z, Huang L et al. (2015). Mapping individual brain networks using statistical similarity in regional morphology from MRI. PLoS ONE 10: e0141840. https://doi.org/10.1371/journal.pone.0141840.

Kong XZ, Wang X, Pu Y et al. (2017). Human navigation network: the intrinsic functional organization and behavioral relevance. Brain Struct Funct 222: 749–764.

Kong XZ, Mathias SR, Guadalupe T et al. (2018). Mapping cortical brain asymmetry in 17,141 healthy individuals worldwide via the ENIGMA consortium. Proc Natl Acad Sci U S A 115: E5154–E5163.

Kong XZ, Boedhoe PSW, Abe Y et al. (2020). Mapping cortical and subcortical asymmetry in obsessive-compulsive disorder: findings from the ENIGMA consortium. Biol Psychiatry 87: 1022–1034.

Kong XZ, Postema M, Schijven D et al. (2021). Large-scale phenomic and genomic analysis of brain asymmetrical skew. Cereb Cortex 31: 4151–4168.

Kong XZ, ENIGMA Laterality Working Group, Francks C (2022a). Reproducibility in the absence of selective reporting: an illustration from large-scale brain asymmetry research. Hum Brain Mapp 43: 244–254.

Kong XZ, Postema MC, Guadalupe T et al. (2022b). Mapping brain asymmetry in health and disease through the ENIGMA consortium. Hum Brain Mapp 43: 167–181.

Lancefield K, Nosarti C, Rifkin L et al. (2006). Cerebral asymmetry in 14 year olds born very preterm. Brain Res 1093: 33–40.

Liu HS, Stufflebeam SM, Sepulcre J et al. (2009). Evidence from intrinsic activity that asymmetry of the human brain is controlled by multiple factors. Proc Natl Acad Sci U S A 106: 20499–20503.

Maller JJ, Thomson RHS, Rosenfeld JV et al. (2014). Occipital bending in depression. Brain 137: 1830–1837.

Nelson EL, Gonzalez SL, El-Asmar JM et al. (2019). The home handedness questionnaire: pilot data from preschoolers. Laterality 24: 482–503.

Ocklenburg S, Berretz G, Packheiser J et al. (2021). Laterality 2020: entering the next decade. Laterality 26: 265–297.

Oldfield RC (1971). The assessment and analysis of handedness: the Edinburgh inventory. Neuropsychologia 9: 97–113.

Packheiser J, Schmitz J, Arning L et al. (2020). A large-scale estimate on the relationship between language and motor lateralization. Sci Rep 10.

Roe JM, Vidal-Pineiro D, Sorensen O et al. (2021). Asymmetric thinning of the cerebral cortex across the adult lifespan is accelerated in Alzheimer's disease. Nat Commun 12: 721.

Rubens AB, Mahowald MW, Hutton JT (1976). Asymmetry of the lateral (sylvian) fissures in man. Neurology 26: 620–624.

Rutten GJ, Ramsey NF, van Rijen PC et al. (2002). Reproducibility of fMRI-determined language lateralization in individual subjects. Brain Lang 80: 421–437.

Saltoun K, Adolphs R, Paul LK et al. (2023). Dissociable brain structural asymmetry patterns reveal unique phenome-wide profiles. Nat Hum Behav 7: 251–268.

Sha Z, Schijven D, Carrion-Castillo A et al. (2021). The genetic architecture of structural left–right asymmetry of the human brain. Nat Hum Behav 5: 1226–1239. https://doi.org/10.1038/s41562-021-01069-w.

Sotiropoulos SN, Zalesky A (2019). Building connectomes using diffusion MRI: why, how and but. NMR Biomed 32: e3752.

Sporns O (2013). Structure and function of complex brain networks. Dialogues Clin Neurosci 15: 247–262.

Sporns O, Tononi G, Kotter R (2005). The human connectome: a structural description of the human brain. PLoS Comput Biol 1: e42.

Toga AW, Thompson PM (2003). Mapping brain asymmetry. Nat Rev Neurosci 4: 37–48.

Van der Haegen L, Westerhausen R, Hugdahl K et al. (2013). Speech dominance is a better predictor of functional brain asymmetry than handedness: a combined fMRI word generation and behavioral dichotic listening study. Neuropsychologia 51: 91–97.

van der Ham IJM, van Zandvoort MJE, Frijns CJM et al. (2011). Hemispheric differences in spatial relation processing in a scene perception task: a neuropsychological study. Neuropsychologia 49: 999–1005.

Vingerhoets G, Li X, Hou L et al. (2018). Brain structural and functional asymmetry in human situs inversus totalis. Brain Struct Funct 223: 1937–1952.

Wada J, Rasmussen T (2007). Intracarotid injection of sodium amytal for the lateralization of cerebral speech dominance. 1960. J Neurosurg 106: 1117–1133.

Wan B, Hong SJ, Bethlehem RAI et al. (2023). Diverging asymmetry of intrinsic functional organization in autism. Mol Psychiatry 28: 4331–4341.

Wang J, He Y (2024). Toward individualized connectomes of brain morphology. Trends Neurosci 47: 106–119.

Wang B, Yang L, Yan W et al. (2023). Brain asymmetry: a novel perspective on hemispheric network. Brain Sci Adv 9: 56–77.

Wiberg A, Ng M, Al Omran Y et al. (2019). Handedness, language areas and neuropsychiatric diseases: insights from brain imaging and genetics. Brain 142: 2938–2947.

Woodhead ZVJ, Bradshaw AR, Wilson AC et al. (2019). Testing the unitary theory of language lateralization using functional transcranial Doppler sonography in adults. R Soc Open Sci 6: 181801.

Woodhead ZVJ, Thompson PA, Karlsson EM et al. (2021). An updated investigation of the multidimensional structure of language lateralization in left- and right-handed adults: a test-retest functional transcranial Doppler sonography study with six language tasks. R Soc Open Sci 8: 200696.

Wu X, Kong X, Vatansever D et al. (2022). Dynamic changes in brain lateralization correlate with human cognitive performance. PLoS Biol 20: e3001560.

Xiang L, Crow T, Roberts N (2019). Cerebral torque is human specific and unrelated to brain size. Brain Struct Funct 224: 1141–1150.

Yoo S, Jang Y, Hong SJ et al. (2024). Whole-brain structural connectome asymmetry in autism. Neuroimage 288: 120534.

Zhao L, Matloff W, Shi Y et al. (2022). Mapping complex brain torque components and their genetic architecture and phenomic associations in 24,112 individuals. Biol Psychiatry 91: 753–768.

Handbook of Clinical Neurology, Vol. 208 (3rd series)
Cerebral Asymmetries
C. Papagno and P. Corballis, Editors
https://doi.org/10.1016/B978-0-443-15646-5.00022-1

Chapter 4

The relationship between brain and visceral asymmetry: Evidence from situs inversus in humans

GUY VINGERHOETS*

Department of Experimental Clinical and Health Psychology, Ghent University, Ghent, Belgium

Abstract

This review examines the relationship between visceral and brain asymmetry and explores whether their alignment observed in some vertebrate species also exists in humans. While the development of visceral and brain asymmetry may have occurred for different reasons, it is possible that the basic mechanisms for left–right differentiation of the visceral system were duplicated in the brain. We describe the main phenotypical anomalies and the general mechanism of left–right differentiation in vertebrates, followed by a systematic review of available human studies on behavioral and brain asymmetry in individuals with reversed visceral organization. The available evidence shows no direct link between human visceral and brain laterality. Most individuals with situs inversus totalis (SIT) show typical population biases for handedness and brain functional asymmetry, although an increased prevalence of atypical hemispheric segregation may be present. Perisylvian brain structural asymmetries also reveal the expected population bias in participants with SIT. However, several independent studies indicate that SIT is associated with a general reversal of the gross morphologic asymmetry of brain torque. Potential differences in brain structural and functional asymmetries between subtypes of situs inversus with ciliary and nonciliary causes remain to be determined.

INTRODUCTION

In vertebrates, the visceral and central nervous systems display a strong population bias toward typical left–right asymmetry (Hamada et al., 2002; Vingerhoets, 2019). Since both systems serve biologically different functions, it is plausible to assume that the reasons for their asymmetry may be entirely different. Nevertheless, basic mechanisms for establishing left–right differentiation of the viscera may have been reused for the establishment of central nervous system laterality. Animal studies have shown that some vertebrate species align their brain and visceral asymmetry, suggesting a link mediated by nodal-related events (Wehrmaker, 1969; Concha et al., 2000; Long et al., 2003). The frequent-situs-inversus (fsi) line of zebrafish, which exhibits visceral reversal, also shows neuroanatomic reversals in the diencephalon, particularly in epithalamic nuclei believed to be involved in the functional lateralization of the vertebrate central nervous system (Concha and Wilson, 2001). Furthermore, diencephalic reversals of fsi zebrafish correlate with the reversal of some lateralized behavioral responses (Barth et al., 2005).

To investigate the possible association between human visceral and neural asymmetry, we will describe the phenotypes and development of left–right asymmetry in both systems and discuss the potential links between their mechanisms of differentiation. The strong population bias in visceral and brain asymmetry, however, makes it challenging to determine whether they develop independently or are related. Ultimately, a valid test for this hypothesis is to investigate the prevalence of atypical brain asymmetry in participants with visceral situs anomalies. In a systematic review, we will examine the studies that provide empiric evidence on behavioral, brain functional, and structural asymmetries in participants with situs anomalies.

VISCERAL ASYMMETRY

Phenotypes of visceral asymmetry

During embryogenesis, humans establish left–right asymmetry in the position and shape of the thoracic and abdominal organs (Hamada et al., 2002; Nakamura and Hamada, 2012). The internal organs in the human body exhibit a strong population bias toward an asymmetric organization, with the heart's apex and aorta, bilobed lung, stomach, and spleen on the left side of the body midline, and the heart's vena cava, most of the liver, and the trilobed lung on the right side (Aylsworth, 2001). This typical configuration, called situs

*Correspondence to: Guy Vingerhoets, Department of Experimental Clinical and Health Psychology, Ghent University, Ghent, Belgium. E-mail: guy.vingerhoets@ugent.be.

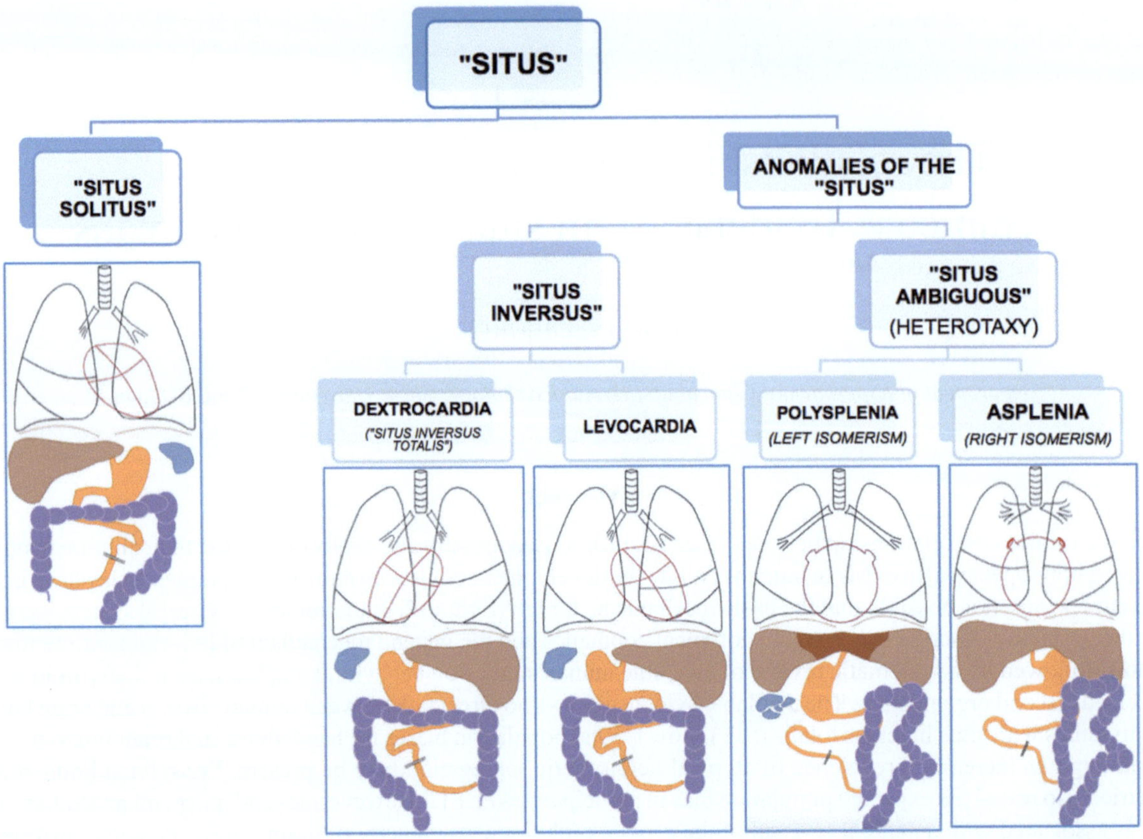

Fig. 4.1. Anatomic dispositions of the viscera in the different types of "situs" (radiologic convention). Reprinted with kind permission by Dr. Francisco Barqueros Escuer, https://dx.doi.org/10.26044/ecr2019/C-2735

solitus (from Latin, *situs*: position and *solitus*: habitual), is taken to reflect optimal packing and transfer of body fluids and is present in about 99.99% of the human population (Aylsworth, 2001).

Anomalies in the arrangement of internal organs result in a range of laterality defects, but the classification of these anomalies lacks consensus, hindering pathologic, genetic, and epidemiologic research (Houyel et al., 2011; Lin et al., 2014). However, two main phenotypic subgroups of situs anomalies are generally distinguished: situs inversus (from Latin, meaning inverted), which involves the complete or partial reversal of the typical condition, and heterotaxy (from Greek *heteros*: other, different, and *taxis*: arrangement), which involves the mirroring of either the typical left or right visceral configuration (Fig. 4.1). Situs inversus and heterotaxy usually occur in different families, but are occasionally present in the same (often consanguineous) family (Casey, 1998; Kosaki and Casey, 1998). The prevalence of human visceral laterality defects is estimated to be between 1/5000 and 1/11,000 live births (Lin et al., 2000; Lin et al., 2014; Evans et al., 2015).

Situs inversus

Situs inversus totalis, a complete reversal of the visceral arrangement with the heart now on the right side of the body, is the most common form of situs inversus, and estimates of its prevalence range from 1/6000 to 1/33,000 live births (Peeters and Devriendt, 2006; Lin et al., 2014; Evans et al., 2015). Situs inversus totalis is generally not associated with adverse medical complications and may be underdiagnosed due to its limited clinical repercussions (Aylsworth, 2001; Sutherland and Ware, 2009). In rare cases, situs inversus may not be complete, and the heart is in its usual position while the other organs are reversed. This isolated levocardia is associated with severe cardiovascular malformations due to the unusual position of the heart compared to the other organs and their connections (Schmutzer and Linde, 1958). Situs inversus may occur as part of a congenital syndrome in about a quarter of cases, where medical complications are more prominent. Primary ciliary dyskinesia is one such syndrome that highlights the importance of tiny hair-like organelles called cilia in the development of visceral asymmetry.

Heterotaxy

Heterotaxy, also known as situs ambiguus, is a type of situs anomaly that is fundamentally different from situs inversus. Unlike situs inversus, heterotaxy describes a complete loss of left–right laterality in the arrangement of the visceral organs along the superior–inferior axis. This disorder affects the structure of visceral organs, particularly the heart and its large blood vessels, with major morbidity and mortality resulting from complex cardiovascular malformations (Lin et al., 2000; Kennedy et al., 2007; Houyel et al., 2011). Prevalence figures for heterotaxy are estimated to be between 1/8000 and 1/12,000 live births (Lin et al., 2000, 2014; Evans et al., 2015). Although classic heterotaxy accounts for only 3% of all congenital heart defects, gene mutations causing heterotaxy are also known to result in isolated cardiovascular malformations with no other visceral abnormalities, indicating that the actual prevalence of genetic heterotaxy is likely higher (Sutherland and Ware, 2009; Gabriel and Lo, 2020).

There are two general types of heterotaxy, known as isomerisms, but the exact morphology and its resulting abnormalities vary from patient to patient (Aylsworth, 2001; Sutherland and Ware, 2009). In left isomerism, the morphologically left structures are present on both sides of the body in the same individual. Atrial cavities are morphologically left, both lungs will be bilobar with long main bronchial branches, and the spleen is present but consists of multiple small and poorly functioning parts (polysplenia). In right isomerism, the right-sided visceral configuration is copy-mirrored to the left resulting in morphologically right atrial cavities, two trilobar lungs with short main bronchi, and an absent spleen (asplenia). In both conditions, the morphologically altered liver lies across the midline of the body, and intestinal malrotation is a typical feature, as well as cardiac malformations, which are more severe and sometimes life-threatening in right isomerism.

Cause of visceral situs anomalies

Visceral situs anomalies are congenital conditions that are caused by genetic mutations that affect the left–right patterning during early embryonic development (Sutherland and Ware, 2009). The development of the left–right axis involves a complex genetic cascade that emerges from animal studies (Casey, 1998). Most cases of situs anomalies occur due to sporadic mutations, and different genetic factors or genes cause the condition in different people or families (Deng et al., 2015). Environmental and stochastic factors may also play a role, as no clear monogenetic basis for the condition can be found in a significant number of cases (Postema et al., 2020).

In some families, situs viscerum anomalies present with an autosomal dominant, autosomal recessive, or even X-linked pattern of inheritance (Aylsworth, 2001; Sutherland and Ware, 2009). Situs anomalies may also be a variable manifestation of a syndrome encompassing a broader spectrum of defects (Aylsworth, 2001). For instance, situs inversus sometimes occurs in renal cystic disease, Bardet–Biedl syndrome, and retinitis pigmentosa (Deng et al., 2015). The best-known example of syndromal situs inversus is when it arises as a symptom of primary ciliary dyskinesia, accounting for about 20%–25% of its cases (Sutherland and Ware, 2009; Lucas et al., 2014; Chen et al., 2020).

Primary ciliary dyskinesia (PCD) is a group of autosomal recessive disorders that are causally heterogeneous and are characterized by a defect in the motility of small hair-like organelles called cilia. Cilia protrude from the cell surface into the extracellular space and perform various transport-related functions in the human body. Ciliary motility is crucial for moving fluids and particles over epithelial surfaces, and cilia play central roles in various signal transduction pathways. The genesis of motile cilia requires a complex genetic program, and mutations of the genes involved (e.g., *DNAH5*, *DNAH11*, *DNAI1*) have been associated with ciliopathies, including PCD (Bartoloni et al., 2002; Leigh et al., 2009; Zariwala et al., 2011; Lucas et al., 2014). The diagnosis of PCD is commonly based on electron microscopy showing abnormalities in the structure and function of dynein arms or outright absence of cilia (Lucas et al., 2014). Its prevalence is estimated between 1/10,000 and 1/20,000 live births (Afzelius and Stenram, 2006; Zariwala et al., 2011).

Affected individuals with PCD have chronic upper respiratory tract infections (sinusitis), lower respiratory tract infections (bronchiectasis), and chronic ear infections (otitis media) due to defective mucociliary clearance. They may also experience reduced male fertility caused by decreased sperm motility, variable female infertility, and decreased sense of smell. About half of the patients with PCD and associated sinusitis and bronchiectasis also have situs inversus, a triad of symptoms known as Kartagener syndrome, while the other half have situs solitus (Leigh et al., 2009). Given the specificity of the ciliary mutation causing visceral inversion vs those causing respiratory problems, most (but not all) subgroups of the PCD syndrome will affect the genetic cascade induced by ciliary motion at the embryonic node that determines organ situs. Therefore, the incidence of situs inversus in PCD is estimated to be slightly less than the often-reported 50%, and the Kartagener triad is expected in 1/22,000 live births (Afzelius and Stenram, 2006).

Cardiac malformations suggestive of heterotaxy are found in 6%–12% of individuals with primary ciliary dyskinesia (Engesaeth et al., 1993; Kennedy et al., 2007). It is generally believed that the condition is associated with a near-randomization of left–right directionality rather than a loss of left–right specification (Casey, 1998). The occurrence of a monozygotic twin pair with primary ciliary dyskinesia and discordant visceral situs underlines the arbitrary nature of situs directionality in this condition.

Ontogenesis of visceral asymmetry

The vertebrate left–right axis is established after the development of the dorsal–ventral and anterior–posterior axes and involves the correct positioning and morphogenesis of internal organs (Hamada et al., 2002; Nakamura and Hamada, 2012). The formation of the left–right axis involves several steps that have been studied in different model organisms like frogs, zebrafish, chicks, pigs, and mice (Bisgrove and Yost, 2006; Hirokawa et al., 2009; Komatsu and Mishina, 2013; Vandenberg and Levin, 2013).

Motile cilia at the primitive node

Symmetry breaking is established in most model organisms at the primitive node, which is a cavity that forms at the anterior tip of the primitive streak during gastrulation. The primitive streak is a line of cells that establishes bilateral symmetry in the embryo and marks its future posterior side. The formation of the node coincides with the development of motile cilia that produce a coordinated and unidirectional flow of extracellular fluid. This flow induces symmetry breaking during gastrulation, which is an important period in embryogenesis involving the differentiation of cells into ectoderm, mesoderm, and endoderm layers.

Some genetic mechanisms, such as the expression of *nodal*, *lefty1*, *lefty2*, and *pitx2*, are shared between vertebrates in the establishment of the left–right axis. However, other steps of the process seem to have diverged in evolution. For example, earlier asymmetries in the localization of some molecules have been established in some species, and some species (e.g., pig, chick) do not have fluid-producing nodal cilia yet show similar asymmetries in the visceral population (Vandenberg and Levin, 2010, 2013).

In species with nodal cilia (e.g., mouse, fish, frog), experimental disruption of cilia functioning results in situs anomalies (Nonaka et al., 2002; Basu and Bruedner, 2008). Reversal of the flow in wildtype embryos results in left–right inversion, and introducing a leftward flow in mutants with ciliopathy restores typical left–right asymmetry (Nonaka et al., 2002; Watanabe et al., 2003). The Kartagener syndrome, where ciliary malfunction impacts visceral asymmetry, confirms the importance of nodal cilia in establishing the left–right axis in humans.

Propagation of the signal to the lateral plate mesoderm and organ primordia

The cilia that arise from nodal cells at the center of the nodal pit are responsible for creating a leftward "nodal flow" toward the left periphery of the node. This is due to the tilt and chiral nature of the cilia, which produce a clockwise rotational motion (from tip to base) (Smith et al., 2019). The flow is detected by mechanosensory and/or chemosensory cilia in peripherally located crown cells at the lateral ends of the pit (Nonaka et al., 2002). These events cause an increase in intracellular Ca^{2+} levels on the left side of the node, resulting in asymmetries in gene expression and the establishment of a left–right (L-R) axis (Bisgrove and Yost, 2006). The resulting asymmetric gene expression is propagated to the lateral plate mesoderm, where a cascade of asymmetric left-sided gene expression is established. This includes the *nodal*, *lefty2*, and *pitx2* genes. Several mechanisms have been proposed to explain the propagation of signaling from the node to the lateral plate mesoderm, including either direct diffusion of *nodal* or a cascade of signaling events via sonic hedgehog (*shh*) or bone morphogenetic protein (*bmp*) that asymmetrically affect nodal expression (Hamada et al., 2002).

The expression of *nodal* and *lefty* genes is transient and exclusively on the left side. Finally, this asymmetric signaling is propagated from the lateral plate mesoderm to organ primordia for proper morphogenesis of the viscera to occur (*pitx2*). It is proposed that nodal acts as a determinant for leftness because cells that receive nodal signals will adopt left-side morphology, and those that lack nodal signals will adopt right-side morphology. In mutations where nodal is bilaterally expressed in the lateral plate mesoderm, embryos will develop left isomerism, and in those that lack nodal signal on either side, embryos will develop right isomerism (Hamada et al., 2002).

Deficits in any of the above steps can result in heterotaxy, but they more often occur at one of the later stages. Situs inversus, on the other hand, is believed to originate from a more initial deficit in nodal flow caused by defectively operating cilia when the total direction of left–right asymmetry is determined. Over 100 genes involved in left–right patterning have been identified in animal models, and more are expected to be discovered (Deng et al., 2015). Mutations in these genes, combined with reduced penetrance and variable expressivity, predict vast differences in the phenotypical presentation of situs anomalies.

BRAIN ASYMMETRY

Despite its overall appearance, the human brain shows many structural asymmetries at the macroscopic, microscopic, and molecular levels. Several of these asymmetries have been associated with functional asymmetries such as handedness and language lateralization as they show a similar population bias and are located in or near regions relevant to motor or language functions (Ocklenburg and Gunturkun, 2018).

Asymmetric hallmarks around the Sylvian fissure are suggested to be associated with language including a larger left hemispheric planum temporale (Geschwind and Levitsky, 1968), larger left pars opercularis of Broca's area and insula (Keller et al., 2009; Greve et al., 2013; Bidula and Kroliczak, 2015), shallower left superior temporal asymmetric pit (Leroy et al., 2015), and larger left arcuate fasciculus (Powell et al., 2006). Unfortunately, there is little evidence that these structural asymmetries predict individual language dominance as people with left and right language dominance reveal similar (leftward) structural differences (Vernooij et al., 2007; Tzourio-Mazoyer et al., 2018). There is no data on the relation between the different structural asymmetries as most studies focused on one, or only a few, key hallmarks. Moreover, given they all show strong leftward population biases correlations are likely to be high even if they would lateralize independently of one another.

Asymmetries associated with handedness include the "hand knob" area in the central sulcus (Sun et al., 2012) and the brain torque. The brain torque is a gross morphologic cerebral asymmetry characterized by a counterclockwise twist (if seen from above) that shifts the right frontal part of the brain to the left and front, and the left parieto-occipital region to the right and back of the cranial cavity (Toga and Thompson, 2003) and was proposed to result from a lateralizing gradient in the neuroembryologic development of the cerebral hemispheres (Best, 1988). Large-scale studies revealed its interindividual variations to be associated with age, sex, and handedness although effect sizes were small (Kong et al., 2021; Zhao et al., 2022).

Despite the many structural brain asymmetries described, the focal approach of most studies does not allow to disentangle the principal components underlying clusters of structural asymmetries that, perhaps, could offer a better link with functional asymmetries (Ocklenburg and Gunturkun, 2018). Let us therefore turn to brain functional asymmetries, where more data are available on phenotypes of hemispheric functional segregation.

Left–right asymmetry of the neurocognitive system

Just like our visceral organs, our mental organs, the cognitive functions, are also represented asymmetrically in the brain. The functional lateralization of the brain offers advantages such as improved parallel processing, saved neural space, and avoiding competition between redundant control centers (Vallortigara et al., 1999; Rogers et al., 2004). In humans, there is a strong bias toward a prototypical asymmetric configuration of cognitive functions in most individuals, with the left hemisphere being responsible for language, manual dexterity, and praxis, and the right hemisphere controlling spatial attention, face recognition, and prosody of speech (Vingerhoets, 2019). While the population-wide bias of this functional segregation over both hemispheres suggests a potential biologic advantage of this particular configuration, the underlying mechanisms for this advantage remain unclear. One way to investigate this is to explore alternative configurations of brain organization and their relationship with behavior.

Phenotypes of brain functional organization

I recently proposed three major categories to describe the phenotypes of brain functional segregation: typical, reversed typical, and atypical (Vingerhoets, 2019). This classification was based on evidence from studies investigating the asymmetry of more than one function in the same individuals. Studies investigating a *random sample* of the population most often used a language task as a typically left hemispheric function and a spatial task to investigate right hemisphere dominance. Results showed that the majority of people had typical lateralization of the investigated functions, while around 30% did not conform to this pattern. Since many studies oversampled left-handers, this may have boosted the estimate of atypical segregation as left-handers are known to have increased variability in functional lateralization (Carey and Johnstone, 2014). That being said, in about two-thirds of participants not showing typical functional segregation, usually segregated functions were lateralized in the same hemisphere, a condition called crowding. In the remaining one-third of individuals not showing typical segregation, participants had all investigated functions lateralized in the atypical hemisphere, resulting in a mirrored image of the prototypical functional segregation (Vingerhoets, 2019).

A second type of study investigated the lateralization of other functions in participants recruited based on the atypical lateralization of one function, usually language. These *selective sample* studies were performed on left-handers as they are known to have a higher prevalence of atypical language dominance. Concomitant reversal of the other investigated function was often observed in these participants (Cai et al., 2013; Vingerhoets et al., 2013; Gerrits et al., 2019). In a recent study, five different lateralized functions were tested, and about 80% of participants who had atypical language lateralization demonstrated complete or near-complete reversal of all other functions as well (Gerrits et al., 2020). In the remaining 20%, typical (or reversed typical) functional segregation was compromised more substantially, with two functions showing atypical lateralization, while the other three functions had conventional lateralization (Gerrits et al., 2020).

Reversed typical functional segregation

The data show that typical functional segregation is common in most people, but alternative arrangements are not rare (Vingerhoets, 2019). One such alternative arrangement is a mirrored version of typical functional segregation, which thus far has only been observed in left-handed individuals (Gerrits et al., 2020). The reversed typical segregation phenotype is similar to the visceral anomaly of situs inversus totalis, but it is about 100 times more prevalent. A complete reversal of up to five asymmetric cognitive functions in the same individual is difficult to explain if one assumes independent lateralization of functions. And yet, most studies have found no correlation between the laterality of different functions, suggesting that functions indeed lateralize independently of one another.

One way to reconcile independent lateralization and the observation of reversed typical functional segregation is to assume the existence of a generic blueprint of functional brain organization. This blueprint provides a directional building plan that, in rare cases, can be flipped, resulting in the mirrored mind phenotype (Vingerhoets, 2019). This assumption explains the phenotype of the mirrored mind and allows for the independence of functional laterality indices. It also predicts that the frequency of deviation from the standard pattern is similar in both typical and reversed typical conditions because both mechanisms (independency of lateralization degree and reversal of the directional blueprint) are likely unrelated (Gerrits, 2024).

Atypical functional segregation

A second alternative phenotype is a more chaotic pattern of lateralization, where some functions show typical and others show atypical asymmetry. In these cases, the habitual functional segregation is lost (Vingerhoets, 2019). This phenotype is called atypical functional segregation and could be interpreted as the brain functional homolog of heterotaxy, where a loss of left–right asymmetry in the arrangement of visceral organs is assumed (Francks, 2015). The individual presentation of heterotaxy is very diverse, and similarly, atypical functional segregation can result from one or multiple functions deviating from the prototypical constellation. Heterotaxy is associated with more frequent and severe medical problems than situs inversus, and some findings likewise suggest that healthy participants with increased deviation from standard brain functional segregation perform significantly worse on neuropsychologic tests compared to participants with typical or reversed typical segregation, suggesting that atypical functional segregation may be cognitively disadvantageous (Powell et al., 2012; Vingerhoets et al., 2018a; Gerrits et al., 2020). The available evidence is based on small samples and a more convincing answer on a possible association between atypical brain organization and cognition awaits more research, particularly large-scale epidemiologic work.

The brain functional homolog of the two isomerisms in heterotaxy might be a bilateral functional representation. Some right- and left-handers do not show clear-cut lateralization for language, and this group is said to have mixed or bilateral representation for language (Mazoyer et al., 2014). Using a machine learning approach to analyze participants' left and right hemispheric activation patterns, researchers were able to differentiate those with bilaterally dominant language representation from those with a nondominant pattern (Zago et al., 2017).

In conclusion, alternative organizations of hemispheric functional segregation can be divided into two phenotypical categories with some commonalities with the main phenotypical subgroups of visceral anomalies. Whether these similarities reflect fundamental principles that share a biologic mechanism remains to be determined.

Ontogenesis of brain asymmetry

Neurulation

The mechanism behind the breaking of brain symmetry is not yet fully understood, but it is believed that the uneven distribution of molecules initiates left–right patterning (Sun and Walsh, 2006). The process starts during gastrulation when the ectodermic tissue thickens and flattens to become the neural plate. The notochord appears below where the primitive streak and node used to be in the mesodermic tissue, inducing the start of neurulation, which is the process where the ectodermal neural plate folds into a neural tube, which later becomes the central nervous system. Primary cilia are involved in this process by regulating Sonic Hedgehog signaling, neural stem cell pool regulation, neural differentiation, and migration (Thomas et al., 2019).

The ventral part of the developing neural tube, adjacent to the notochord, becomes the floor plate, and its dorsal part becomes the roof plate. These plates project ventralizing (*nodal*, *lefty*, *shh*) and dorsalizing (bone morphogenic protein (*bmp*)) inductive signals to the developing neural tube, of which its most rostral part will develop into the forebrain. Asymmetric secretion of morphogens from the floor and roof plates to the left and right sides of the neural tube is believed to break the symmetry of neural patterning and induce asymmetric expression of downstream genes (Sun and Walsh, 2006; Medina, 2009).

In addition to the floor and roof plates, the most rostral part of the neural tube has a third patterning center, the anterior neural ridge. The anterior neural ridge emits rostralizing signals essential for the development of the secondary prosencephalon that will form the telencephalon, thalamus, hypothalamus, and epithalamus (Sun and Walsh, 2006; Medina, 2009). It has been suggested that the asymmetric expression of morphogens secreted from this region could reflect asymmetric topographic mapping of functional regions in the cortex (Fukuchi-Shimogori and Grove, 2001; Sun et al., 2005).

Asymmetric development of the central nervous system

Research has shown that there are differences in gene expression between the left and right forebrains and midbrains of human embryos, beginning at 5 weeks postconception (de Kovel et al., 2018). The data were collected by pooling information from voluntary medical abortions of healthy pregnancies and the Human Developmental Biology Resource in the United States. The researchers found evidence of age-dependent laterality of transcriptomic profiles for most structures, indicating subtle differences in maturation rates between left and right CNS structures (de Kovel et al., 2017, 2018).

The asymmetry in gene expression appears to be initiated and amplified at multiple locations, with different functional destinations showing varying maturation rates between left and right structures. For example, the left spinal cord shows faster maturation rates at 5–5.5 weeks postconception, while the opposite pattern is observed for the midbrain and hindbrain (de Kovel et al., 2017). By 7.5 weeks postconception, the left choroid plexus, basal ganglia, diencephalon, and temporal cortex show faster maturation rates, while the rest of the cerebral cortex matures faster on the right side (de Kovel et al., 2018).

The researchers suggest that these early and differentiated patterns in brain asymmetry may play a role in the laterality of certain functions, such as handedness, but not necessarily influence other functions or regions. An earlier maturation of the left spinal cord, for instance, as observed by predominant right arm movements before the innervation of the corticospinal tracts, may lead to cortical laterality of handedness but is unlikely to affect the laterality of other regions or functions (de Kovel et al., 2018). This is consistent with the weak correlations between the adult laterality of different brain functions like handedness and language (Mazoyer et al., 2014), and with the results of gene ontology analysis that suggest handedness and language lateralization are ontogenetically independent phenotypes (Schmitz et al., 2017).

While subtle brain asymmetries in gene expression are already measurable at 5 weeks postconception, structural human fetal brain asymmetries become visible with current methods by the 11th week of gestational age for the choroid plexus (Abu-Rustum et al., 2013), by the 16th week for fetal cortex volume (Vasung et al., 2020), by the 18th week for temporal lobe morphology (Kasprian et al., 2011), by the 20th week for sulcal folding (Habas et al., 2012; Yun et al., 2020), and by the 26th week for perisylvian hallmarks associated with language (Dubois et al., 2010). However, the gap between genetic and morphologic or functional brain asymmetries remains to be detailed.

Are asymmetries of visceral and brain development related?

During the third and fourth weeks of human gestation, visceral and neural patterning begins in close temporal proximity, but it is not clear if the mechanisms that regulate visceral asymmetry also impact brain asymmetry. Asymmetric gene expression and the role of cilia seem to be potential factors for a link between visceral and brain manifestations of asymmetry.

Studies have shown that *nodal* and *shh* pathways are expressed during neurulation, but none of the reported genes found to be differentially expressed in the left and right hemispheres of 12- to 14-week-old human fetal brains have known essential roles in visceral organ asymmetry (Sun et al., 2005). Similarly, lateralized gene expression in postmortem temporal cortex from embryo to old age has been reported, but none of the reported genes have been associated with visceral anomalies (Karlebach and Francks, 2015). However, relative hand skill in a cohort of individuals with reading disability has been associated with a variant in the gene *pcsk6*, which cleaves *nodal* into an active form (Brandler et al., 2013). *Pcsk6* knock-out mice display heterotaxy, and human variants of this gene are associated with heterotaxy and situs inversus, suggesting that handedness is at least in part controlled by genes that contribute to the determination of visceral asymmetry (Brandler et al., 2013). Genes like *GPC3*, associated with relative hand skill in the general population, cause situs anomalies when their orthologs are knocked out in mice (Brandler et al., 2013; Brandler and Paracchini, 2014).

Clinical evidence shows that cilia play a significant role in human neurulation. Ciliopathy-associated hereditary cerebral anomalies include neural tube defects, corpus callosum malformations, cerebellar hypoplasia, and hydrocephaly. Less severe neurologic features such as cognitive deficits, autism spectrum disorders, and seizures are also frequently observed in individuals with ciliopathies and suggest the possibility of more subtle cortical deficiencies (Thomas et al., 2019). In terms of laterality, genes most strongly associated with relative hand skill in a dyslexia cohort are involved in ciliogenesis, and their disruption in mice causes situs inversus (Brandler et al., 2013; Brandler and Paracchini, 2014). Additionally, cilia-related gene sets are more highly expressed in the right choroid plexus in the 7.5–13 post-conception age range (de Kovel et al., 2018). The choroid plexus is the first brain structure to show morphologic asymmetry and is associated with the circulation of cerebrospinal fluid in the ventricles. However, there is no clear evidence that cilia play a role in the initiation or propagation of central nervous system asymmetry (de Kovel et al., 2018).

ATYPICAL BRAIN ASYMMETRY IN HUMAN VISCERAL SITUS ANOMALIES

Although the link between visceral and brain asymmetry is not yet fully understood, one strategy to investigate this relationship is to compare people with atypical organ situs to those with typical situs to see if there are any differences in brain structure or function. However, this approach is limited by the rarity of situs anomalies, which makes it difficult to recruit a large enough sample size, particularly when using more intensive research methods like neuroimaging. In addition, the heterogeneity of genetic and other factors that contribute to brain and visceral asymmetries makes it challenging to draw conclusions from such studies.

Different manifestations of situs anomalies have been associated with different genetic mutations, suggesting that genetic screening or a detailed description of the situs condition and family history should be used to categorize participants. Furthermore, many gene mutations are known to affect different steps and mechanisms of the complex process of visceral left–right patterning, making it likely that some gene mutations bear no relation to brain asymmetrization while others do.

For example, primary ciliary dyskinesia, a ciliopathy resulting in randomization of organ situs, is caused by genetic mutations that affect ciliary function. Although cilia play a role in neurulation, it is unclear whether this includes lateralization of morphogens that induce brain asymmetry. Therefore, a ciliopathy like PCD may have no effect on the development of brain asymmetry at all.

In people with situs inversus who do not have PCD, the origin of their situs anomaly may be due to a temporary or local malfunction of nodal cilia or caused by a different mechanism altogether. Within the subgroup of situs inversus, etiologic heterogeneity is substantial and extends beyond the role of genes. In a recent genome sequencing study of 15 cases with situs inversus totalis (Postema et al., 2020), six participants with PCD-associated mutations presented with likely recessive mutations. Similar mutations were also detected in two non-PCD SIT participants, and two other non-PCD SIT participants had recessive mutations in genes linked to SIT outside the context of PCD. In five of the nine non-PCD cases, no monogenic basis for their situs anomaly was found, leading the authors to consider early environmental or stochastic effects as possible causative factors.

Systematic review

In June 2023, a systematic literature search was conducted to investigate whether visceral situs anomalies have a different prevalence of brain and behavioral asymmetry (Moher et al., 2009). The search for manuscripts in the English language was performed on three platforms: Web of Science (Indexes: SCI-EXPANDED, SSCI, AHCI, CPCI-S, CPCI-SSH, ESCI; range: 1955–2023), PubMed, and Scopus. The search strategy used was: (1) Topic: situs inversus OR heterotaxy; (2) Topic: brain asymmetry OR brain functional asymmetry OR brain structural asymmetry OR behavio(*u*)ral asymmetry OR hemispheric dominance OR brain laterality; and (3) #1 AND #2.

We obtained a total of 112 records (WoS $n = 94$; PubMed $n = 11$; Scopus $n = 5$). Of these, 94 records were screened after the removal of duplicates. Records on animal research ($n = 25$), genetics ($n = 21$), review/theoretical ($n = 19$), and papers on medical comorbidities of visceral laterality defects ($n = 17$) were excluded. A total of 12 full-text articles were addressed for eligibility. In the references of these articles, six further (mostly older) studies were identified that reported empiric data on the research question. A preprint study on brain functional segregation in SIT (10.31234/osf.io/erpx8) was also added to the list. A total of 19 studies were included in the qualitative synthesis for this systematic review.

However, many studies did not describe the situs condition of their participants in detail or differentiate their small samples of participants into separate categories or explore their genetic background. The majority of brain- and behavior-related research in this field has been performed within the subgroup of participants with situs inversus, and there is a lack of research on other situs anomalies. Therefore, the following section will summarize the behavioral, brain structural, and brain functional data on atypical asymmetry in participants with situs anomalies, specifically in participants with situs inversus.

Handedness in situs inversus

An overview of the findings on handedness in situs inversus is presented in Table 4.1. Early studies have reported a prevalence of left-handedness in situs inversus of around 6%–7% (Cockayne, 1938; Torgersen, 1950; Everett, 1963; Matsumoto et al., 1997), which is lower than contemporary estimates of 10% for left-handedness in the general population (Papadatou-Pastou et al., 2020), suggesting that cultural pressure against left-handedness may have underestimated natural left-hand

Table 4.1

Overview of handedness studies in situs inversus.

Authors	Date	Sample size	Situs anomaly	Etiology	Source	Percent left handedness	Handedness determination
Cockayne	1938	115	Situs inversus	Not provided	From literature and author's own series; most included were examined on account of illness or a congenital malformation	6.1% LH and 1.7% ambidexter	Unknown
Torgersen	1950	Convenience sample of 160 from a total sample of 270	Situs inversus	Not provided	Mass X-ray photographs of the Norwegian population (200) +additional cases through hospitals and municipal health departments (70)	6.9% in SI compared to 3.5% in (715) siblings and 5.3% in (320) parents	Self-report obtained from the individual with SI
Everett	1963	Convenience sample of 6 from a total sample of 10	Situs inversus totalis	Not provided	Hospital records search	0%	Self-report or hospital records
Matsumoto et al.	1997	16	Situs inversus	Not provided	Not reported	6.3%	Writing, drawing, throwing
McManus et al.	2004	46 PCD-SI & 42 PCD-SS, 334 controls	PCD-related situs inversus	PCD-only	Family PCD-support group	15.2% in PCD-SI and 14.3 in PCD-SS, 8.1 in controls	Writing hand and laterality questionnaires
Afzelius & Stenram	2006	Convenience sample of 112 from a total sample of 239 PCD patients (105 PCD-SI & 134 PCD-SS)	Situs inversus	PCD-only	Obtained from contacting a large number of Swedish clinicians	12.5% LH and 3.6 ambidexter in PCD, 14.3% LH in subgroup with PCD-SI	Self-report

Adapted from Vingerhoets G, Gerrits R, Verhelst H (2021). Atypical brain asymmetry in human situs inversus: gut feeling or real evidence? Symmetry-Basel 13.

preference in these cohorts. In addition, these early studies suffer from poor behavioral assessment and poor etiologic description of the situs anomaly. Later studies that focused on handedness in primary ciliary dyskinesia (PCD)-related situs inversus totalis (SIT) found no difference in the prevalence of left-handedness compared to the general population (McManus et al., 2004; Afzelius and Stenram, 2006). Typical handedness in PCD-related SIT may seem surprising given the reports of a possible genetic association between relative hand skill and ciliogenesis. Compensatory mechanisms that allow the typical development of handedness to overrule the influence of ciliopathy have been proposed in explanation (Dubois et al., 2010; Karlebach and Francks, 2015).

For nonsyndromal SIT, the issue of handedness is less clear due to the paucity of available data. Some information can be gathered from studies that investigated brain functional asymmetry in SIT (and in addition registered handedness), which predominantly featured nonsyndromal cases. These studies reported on 22 sporadic cases that were explicitly reported to be free of PCD symptoms (Woods, 1986; Kennedy et al., 1999; Tanaka et al., 1999; Schuler et al., 2017; Vingerhoets et al., 2018b). Of these, 17 participants were right-handed and 5 were left-handed, representing a 29% prevalence of left-handedness. One study reported that five out of nine non-PCD-related SIT cases had left-hand preference, representing an increased prevalence of left-handedness (55%) in this cohort (Vingerhoets et al., 2018b). Interestingly, this study also recruited six PCD-related SIT cases, of which only one was left-handed, consistent with previous findings on handedness in PCD-related SIT. The seemingly random hand preference in the non-PCD-related SIT participants of the Ghent cohort may be an accidental finding, as none of the other studies even remotely suggested anything of the kind. Recently, however, this group presented the results of a replication study with 23 SIT cases (two with PCD, 9%). Of the SIT participants, 26% (6/23) were classified as adextral (five left-handed: EHI <0; 1 ambidextrous: EHI $=0$). The proportion of adextrals in this sample was significantly higher ($P=0.029$) than the prevalence of adextrality in the general population which was set at 10.6% based on a meta-analysis of over two million individuals (Papadatou-Pastou et al., 2020). The replication study thus confirmed their earlier report of increased prevalence of adextrality in SIT (Gerrits and Vingerhoets, 2023). Future research, preferably in a larger cohort of SIT, is necessary to confirm this finding as well as a potential differential effect of SIT etiology on handedness.

Brain functional asymmetry in situs inversus

The first observations date from the late 1980s with reports of two right-handed stroke patients with visceral anomalies (see Table 4.2 for an overview of the relevant studies). One of them became aphasic following a left hemisphere cerebrovascular lesion while the other, a patient with left isomerism heterotaxy, showed crossed-aphasia after a right hemisphere stroke (Woods, 1986; Cohen et al., 1993). Later, nine SIT participants (only one with PCD-related SIT) underwent a dichotic listening paradigm, showing typical right ear advantage in all but one case (Tanaka et al., 1999). With the advent of MRI research, the first fMRI study showed typical left hemisphere lateralization for language in three nonsyndromal SIT participants (Kennedy et al., 1999). Still, a decade later, a second fMRI study reported atypical right hemisphere lateralization for language in two out of three SIT cases (Ihara et al., 2010). Until then, all studies, including a longitudinal case study that used fMRI (Schuler et al., 2017) had focused on language. Recently, research has broadened to other lateralized functions such as praxis, spatial attention, and face recognition in an fMRI study of 15 SIT participants, of which 6 had PCD-related SIT and 9 had non-PCD-related SIT (Vingerhoets et al., 2018a). While 80% of this cohort had left hemisphere language dominance, suggesting generally typical language lateralization, a control group matched for handedness showed 93% leftward lateralization. The same trend was found for the other three tested functions that showed more typical asymmetry in the matched controls compared to the SIT participants. The authors concluded that atypical functional segregation is more frequent in participants with SIT, regardless of the PCD-relatedness. However, the small sample size limits proper statistical comparison. As atypical lateralization occurred equally frequently in the right-handed SIT participants, the increased prevalence of atypical lateralization is unlikely to be caused by the high number of left-handers in this sample (Vingerhoets et al., 2018a). The recent replication of their MRI study in 23 SIT participants (of which two have PCD-related SIT) confirms the exact same findings. SIT participants in general show the expected population bias for language, praxis, spatial attention, and face recognition, albeit less pronounced—but not significantly so—compared to controls. The authors concluded that complete situs inversus in humans does not co-occur with an obligatory transposition of the brain's functional architecture and that the determinants of visceral and neural asymmetries appear largely independent from one another. The authors further suggested that the increased atypical hemispheric segregation and adextrality observed in SIT, across two separate cohorts, could be attributed to partial genetic pleiotropy between visceral reversal and atypical brain laterality (Gerrits and Vingerhoets, 2023).

Brain structural asymmetry in situs inversus

An overview of studies on structural asymmetry in situs inversus is shown in Table 4.3. A consistent finding in individuals with situs inversus (SI) is that their cerebral torque is generally reversed in comparison to the typical human population bias. The cerebral or "Yakovlevian" torque refers to an anticlockwise twist of the brain about the ventral-dorsal axis and is most often described in terms of its petalia, where the right frontal pole protrudes anteriorly relative to the right frontal pole, and the left occipital pole protrudes posteriorly relative to the right occipital pole (Xiang et al., 2019). Typical petalia asymmetry is observed in 44% of modern human brains (Balzeau et al., 2012) and appears to be absent in nonhuman primates (Xiang et al., 2018). In 15 out of 23 SIT participants (65%), most of which were sporadic cases, the petalia were reported to be reversed. Interestingly, a recent study reported typical petalia in all six PCD-related SIT

Table 4.2

Overview of brain functional asymmetry studies in situs inversus.

Authors	Date	Sample size	Hand preference[a]	Situs anomaly	Etiology	Source	Method[b]	Function	Laterality[c]
Woods et al.	1986	1 SI	RH	Situs inversus totalis	Sporadic case, no PCD-symptoms	Stroke patient	Lesion	Language	Aphasia following left stroke
Cohen et al.	1993	1 SI	RH	Heterotaxy (polysplenia)	Sporadic case	Stroke patient	Lesion	Language	Aphasia following right stroke (crossed aphasia)
Tanaka et al.	1999	9 SI and 24 controls	9 RH	Situs inversus totalis	8 sporadic cases, 1 PCD-SIT	Hospital search	DLT	Language	88.9% of SIT shows REA, 79.1% of controls shows REA
Kennedy et al.	1999	3 SI and 15 controls	3 RH (Annett Handedness Battery)	Situs inversus totalis	3 sporadic cases, normal general health	Hospital records search	fMRI	Language	100% of SIT show left lateralized activation on each of two language tasks (word stem completion and semantics)
Ihara et al.	2010	3 SI and 11 controls	1 weak RH (EHI = 38), 1 weak LH (EHI = −20), 1 strong RH (EHI = 100)	Situs inversus totalis	Unknown, 1 SIT case has left temporal epilepsy	Unknown	MEG	Language	33% of SIT show left lateralized activation during reading task, 66% show right lateralized activation
Schuler et al.	2017	1 SIT and 1 control	RH (EHI = 100)	Situs inversus totalis	Sporadic case, no PCD-symptoms	Fetal screening	fMRI	Language	Typical left lateralization (LI = 0.48)
Vingerhoets et al.	2018	15 SIT and 15 controls	9 RH, 6 LH; 20% LH in PCD-SIT, 55% LH in non-PCD SIT, 40% LH overall (EHI)	Situs inversus totalis	9 sporadic cases, 6 PCD-related	Hospital records search	fMRI	Language	80% of SIT left lateralized (93% in controls)
								Praxis	73% of SIT left lateralized (87% in controls)
								Spatial attention	73% of SIT right lateralized (93% in controls)
								Face recognition	80% of SIT right lateralized (87% in controls)
Gerrits & Vingerhoets	2023	23 SIT and 23 controls	17 RH, 5 LH, 1 ambidexter 24% adextral in non-PCD SIT, 50% adextral in PCD SIT	Situs inversus totalis	21 sporadic cases, 2 PCD-related	Hospital records search	fMRI	Language	83% of SIT left lateralized (100% in controls)
								Praxis	83% of SIT left lateralized (87% in controls)
								Spatial attention	83% of SIT left lateralized (96% in controls)
								Face recognition	74% of SIT left lateralized (74% in controls)

Adapted from Vingerhoets G, Gerrits R, Verhelst H (2021). Atypical brain asymmetry in human situs inversus: gut feeling or real evidence? Symmetry-Basel 13.

[a]EHI, Edinburgh Handedness Inventory.

[b]DLT, dichotic listening test; fMRI, functional magnetic resonance imaging; MEG, magneto-encephalography.

[c]REA, right ear advantage.

Table 4.3

Overview of studies on brain structural asymmetry in situs inversus.

Authors	Date	Sample size	Hand preference	Situs anomaly	Etiology	Source	Method[a]	Cerebral torque (petalia)	Planum temporale	Sylvian fissure	Other[b]
Kennedy et al.	1999	3 SIT and 15 controls	RH (AHB =1.24)	Situs inversus totalis	Sporadic cases, normal general health	Hospital records search	MRI	Reversed in all SIT	Volume L > R in 2 SIT and R > L in one SIT	Higher R > L in 2 SIT	
Tubbs et al.	2003	1 SIT	Unknown	Situs inversus totalis	Unknown, died from "natural causes"	Unknown	Auto-psy	Reversed		Higher L > R	Reversed intracranial vasculature
Ihara et al.	2010	3 SIT and 11 controls	2 RH, 1 LH	Situs inversus totalis	Unknown, 1 SIT has temporal lobe epilepsy	Unknown	MRI	Reversed in all SIT	Volume L > R in 3 SIT		IFG volume L > R in 2 SIT, R > L in1 SIT
Leroy et al.	2015	6 SIT and 95 controls (48 RH)	5 RH, 1 LH	Situs inversus totalis	Cases from the Kennedy et al. and Ihara et al. studies	Cases from the Kennedy et al. And Ihara et al. studies	MRI				STS deeper on right (typical)
Schuler et al.	2017	1 SIT and 1 control	RH (EHI =100)	Situs inversus totalis	Sporadic case, no PCD-symptoms	Fetal screening	MRI	Reversed in SIT	Absent asymmetry in SIT		STS deeper on right (typical)
Vingerhoets et al. & Mannaert et al.	2018 & 2019	15 SIT and 15 controls	9 RH, 6 LH	Situs inversus totalis	9 sporadic cases, 6 PCD-related	Hospital records search	MRI	Reversed in 78% of non-PCD SIT and in 0% of PCD SIT	Not different from controls	Same as controls	IFG volume same as controls Heschl's gyrus and anterior insula volume same as controls Arcuate fasciculus volume same as controls Uncinate fasciculus lower volume in SIT than controls
Kienast et al.	2021	5 fetuses with ciliopathies and 21 fetuses with laterality defects; 26 age-matched controls	Not applicable	4 SIT; 8 dextrocardia; 4 situs ambiguus; 5 situs inversus abdominalis	Prenatal diagnosis reports	Hospital records search	MRI				Perisylvian fetal patterns and asymmetry indices do not differ between cases with laterality defects, ciliopathies, and normal controls

Adapted from Vingerhoets G, Gerrits R, Verhelst H (2021). Atypical brain asymmetry in human situs inversus: gut feeling or real evidence? Symmetry-Basel 13.

[a]MRI, magnetic resonance imaging.

[b]IFG, inferior frontal gyrus; STS: superior temporal sulcus.

participants, whereas complete reversal of the petalia was observed in seven out of nine non-PCD-related SIT participants (78%), suggesting a possible distinction between syndromal and nonsyndromal SIT. Recently, we confirmed the reversal of intracranial posterior venous asymmetry in SIT as previously suggested by one postmortem study (Tubbs et al., 2003). In addition to a reversal of the brain torque and typical transverse sinus asymmetry in 38 participants with SIT compared to 38 matched controls, the extent of transverse sinus asymmetry was significantly associated with the magnitude of occipital petalia and occipital and frontal bending, suggesting that a reversed intracranial venous configuration in SIT might contribute to its reversed torque and indeed should be taken into account in a more encompassing explanation of the enigmatic brain torque (Vingerhoets et al. 2024).

There have been no systematic reversals in other structural brain asymmetries reported in individuals with SI. Alleged language-related markers like the planum temporale, Sylvian fissure, inferior frontal gyrus, depth of the superior temporal sulcus, and arcuate fasciculus show the same variability and directional bias as the general population. However, the available data are scarce, and further research is needed to discover more subtle effects or differences between SIT types.

DISCUSSION

Human visceral laterality anomalies and atypical brain organization are two complex phenomena that have similarities in their overall appearance of main phenotype subgroups. However, it is important to keep in mind that the visceral and neural systems serve fundamentally different biologic functions, and the reasons for the development of asymmetry in each system are likely to be dissimilar. Two broad categories of visceral laterality anomalies, situs inversus and heterotaxy, are generally distinguished, while two main categories of unconventional brain organization, reversed functional segregation and atypical functional segregation, are advanced. Although the developmental cascades of visceral and brain laterality remain to be elucidated, the possibility of selfsame basic mechanisms for left–right differentiation being employed by both systems to generate and/or propel asymmetry is uncertain.

The prevalence of atypical visceral and brain organization varies significantly, which raises the concept of evolutionary canalization. This refers to a trait's increased resistance to genetic and environmental perturbations over evolutionary time. While left-sided heart anatomy is preserved in all living vertebrates, the incidence of spontaneous reversal decreases with evolution. It is 5% in fish, 1%–2% in amphibians, 0.1% in mammals, and 0.01% in humans (Palmer, 2004). The evolutionary increase in canalization may be due to the increased predictability of symmetry breaking by cilia-controlled nodal flow or the more stable conditions of the placental environment (Palmer, 2004). There are no cladistic estimates of reversals in brain organization, but the concept of evolutionary canalization can be used to explore the origin and timing of brain structural and functional asymmetries in humans. This can be done by comparing prevalence measures of atypical laterality, which could provide an interesting avenue for research.

In addition to comparing the developmental cascades and scrutinizing the molecular genetics underlying visceral and brain asymmetry is to directly compare phenotypes by investigating brain and behavioral asymmetries in individuals with situs anomalies. However, the low prevalence and substantial phenotypical variability of human visceral laterality anomalies have postponed consensus on clear classification criteria for subgroup determination, and the assessment of brain asymmetries is hampered by the scarcity of data and small sample sizes. Most studies on behavioral and brain asymmetry in situs anomalies focused on situs inversus, which is not inseparably associated with a reversal of brain and behavioral asymmetries in humans as seen in some other species. On the contrary, most people with situs inversus seem to present with typical patterns of hemispheric specialization, although a higher prevalence of functional crowding in this group remains a possibility.

The findings of some studies hint at more subtle effects that distinguish between types of situs inversus with different etiologies. In PCD-related syndromal situs inversus, handedness and perhaps also brain torque reveal the same laterality bias as the general population, suggesting that nodal ciliopathy and the eventually reversed subsequent molecular cascade that gives rise to visceral laterality has only a little effect on hand preference and gross brain morphology. By contrast, situs inversus caused by nonciliary factors does seem to be accompanied by a reversal of the brain torque, suggesting a possible relation between human visceral and venous asymmetry and the asymmetric shape of the brain organ.

To corroborate and extend these findings, future research should provide detailed phenotypical information of participants supplemented by genetic data if possible. A consensus should be reached on core information to be reported that will allow open science and meta-analytic initiatives to gather larger samples of participants with situs anomalies and further the understanding of possible interactions between human visceral and brain asymmetry.

References

Abu-Rustum RS, Ziade MF, Abu-Rustum SE (2013). Reference values for the right and left fetal choroid plexus at 11 to 13 weeks an early sign of "developmental" laterality? J Ultrasound Med 32: 1623–1629.

Afzelius BA, Stenram U (2006). Prevalence and genetics of immotile-cilia syndrome and left-handedness. Int J Dev Biol 50: 571–573.

Aylsworth AS (2001). Clinical aspects of defects in the determination of laterality. Am J Med Genet 101: 345–355.

Balzeau A, Gilissen E, Grimaud-Hervé D (2012). Shared pattern of endocranial shape asymmetries among great apes, anatomically modern humans, and fossil hominins. PLoS One 7: e29581. https://doi.org/10.1371/journal.pone.0029581.

Barth KA, Miklosi A, Watkins J et al. (2005). Fsi zebrafish show concordant reversal of laterality of viscera, neuroanatomy, and a subset of behavioral responses. Curr Biol 15: 844–850.

Bartoloni L, Blouin JL, Pan YZ et al. (2002). Mutations in the DNAH11 (axonemal heavy chain dynein type 11) gene cause one form of situs inversus totalis and most likely primary ciliary dyskinesia. Proc Natl Acad Sci U S A 99: 10282–10286.

Basu B, Bruedner M (2008). Cilia: multifunctional organelles at the Center of Vertebrate Left-Right Asymmetry. Ciliary Function in Mammalian Development 85: 151–174.

Best CT (1988). The emergence of cerebral asymmetries in early human development: A literature review and a neuroembryological model. In: DL Molfese, SJ Segalowitz (Eds.), Brain lateralization in children: developmental implications. Guilford Press, New York.

Bidula SP, Kroliczak G (2015). Structural asymmetry of the insula is linked to the lateralization of gesture and language. Eur J Neurosci 41: 1438–1447.

Bisgrove BW, Yost HJ (2006). The roles of cilia in developmental disorders and disease. Development 133: 4131–4143.

Brandler WM, Paracchini S (2014). The genetic relationship between handedness and neurodevelopmental disorders. Trends Mol Med 20: 83–90.

Brandler WM, Morris AP, Evans DM et al. (2013). Common variants in left/right asymmetry genes and pathways are associated with relative hand skill. PLoS Genet 9: e1003751. https://doi.org/10.1371/journal.pgen.1003751.

Cai Q, Van der Haegen L, Brysbaert M (2013). Complementary hemispheric specialization for language production and visuospatial attention. Proc Natl Acad Sci U S A 110: E322–E330.

Carey DP, Johnstone LT (2014). Quantifying cerebral asymmetries for language in dextrals and adextrals with random-effects meta analysis. Front Psychol 5. https://doi.org/10.3389/fpsyg.2014.01128. https://www.frontiersin.org/journals/psychology/articles/10.3389/fpsyg.2014.01128.

Casey B (1998). Two rights make a wrong: human left-right malformations. Hum Mol Genet 7: 1565–1571.

Chen W, Guo Z, Qian L et al. (2020). Comorbidities in situs inversus totalis: a hospital-based study. Birth Defects Res 112: 418–426.

Cockayne EA (1938). The genetics of transposition of the viscera. Q J Med New Ser 27: 479–493.

Cohen L, Geny C, Hermine O et al. (1993). Crossed aphasia with visceral situs-inversus. Ann Neurol 33: 215–218.

Concha ML, Wilson SW (2001). Asymmetry in the epithalamus of vertebrates. J Anat 199: 63–84.

Concha ML, Burdine RD, Russell C et al. (2000). A nodal signaling pathway regulates the laterality of neuroanatomical asymmetries in the zebrafish forebrain. Neuron 28: 399–409.

de Kovel CGF, Lisgo S, Karlebach G et al. (2017). Left-right asymmetry of maturation rates in human embryonic neural development. Biol Psychiatry 82: 204–212.

de Kovel CGF, Lisgo SN, Fisher SE et al. (2018). Subtle left-right asymmetry of gene expression profiles in embryonic and foetal human brains. Sci Rep 8: 12606. https://doi.org/10.1038/s41598-018-29496-2.

Deng H, Xia H, Deng S (2015). Genetic basis of human left-right asymmetry disorders. Expert Rev Mol Med 16: e19.

Dubois J, Benders M, Lazeyras F et al. (2010). Structural asymmetries of perisylvian regions in the preterm newborn. Neuroimage 52: 32–42.

Engesaeth VG, Warner JO, Bush A (1993). New associations of primary ciliary dyskinesia syndrome. Pediatr Pulmonol 16: 9–12.

Evans WN, Acherman RJ, Restrepo H (2015). Heterotaxy in southern Nevada: prenatal detection and epidemiology. Pediatr Cardiol 36: 930–934.

Everett HC (1963). Situs inversus totalis—a survey of laterality and some observations on frequency of mental disorder. Am J Psychiatry 119: 884.

Francks C (2015). Exploring human brain lateralization with molecular genetics and genomics. Ann N Y Acad Sci 1359: 1–13.

Fukuchi-Shimogori T, Grove EA (2001). Neocortex patterning by the secreted signaling molecule FGF8. Science 294: 1071–1074.

Gabriel GC, Lo CW (2020). Left-right patterning in congenital heart disease beyond heterotaxy. Am J Med Genet C Semin Med Genet 184: 90–96.

Gerrits R (2024). Variability in hemispheric functional segregation phenotypes: a review and general mechanistic model. Neuropsychol Rev 34: 27–40. https://doi.org/10.1007/s11065-022-09575-y.

Gerrits R, Vingerhoets G (2023). Brain functional segregation, handedness and cognition in situs inversus totalis: a replication study. PsyArXiv Preprints.

Gerrits R, Van der Haegen L, Brysbaert M et al. (2019). Laterality for recognizing written words and faces in the fusiform gyrus covaries with language dominance. Cortex 117: 196–204.

Gerrits R, Verhelst H, Vingerhoets G (2020). Mirrored brain organization: statistical anomaly or reversal of hemispheric functional segregation bias? Proc Natl Acad Sci U S A 117: 14057–14065.

Geschwind N, Levitsky W (1968). Human brain—left-right asymmetries in temporal speech region. Science 161: 186.

Greve DN, Van der Haegen L, Cai Q et al. (2013). A surface-based analysis of language lateralization and cortical asymmetry. J Cogn Neurosci 25: 1477–1492.

Habas PA, Scott JA, Roosta A et al. (2012). Early folding patterns and asymmetries of the Normal human brain detected from in utero MRI. Cereb Cortex 22: 13–25.

Hamada H, Meno C, Watanabe D et al. (2002). Establishment of vertebrate left-right asymmetry. Nat Rev Genet 3: 103–113.

Hirokawa N, Tanaka Y, Okada Y (2009). Left-right determination: involvement of molecular motor KIF3, cilia, and nodal flow. Cold Spring Harb Perspect Biol 1: a000802. https://doi.org/10.1101/cshperspect.a000802. PMID: 20066075; PMCID: PMC2742083.

Houyel L, Khoshnood B, Anderson RH et al. (2011). Population-based evaluation of a suggested anatomic and clinical classification of congenital heart defects based on the International Paediatric and Congenital Cardiac Code. Orphanet J Rare Dis 6: 64. https://doi.org/10.1186/1750-1172-6-64. PMID: 21968022; PMCID: PMC3198675.

Ihara A, Hirata M, Fujimaki N et al. (2010). Neuroimaging study on brain asymmetries in situs inversus totalis. J Neurol Sci 288: 72–78.

Karlebach G, Francks C (2015). Lateralization of gene expression in human language cortex. Cortex 67: 30–36.

Kasprian G, Langs G, Brugger PC et al. (2011). The prenatal origin of hemispheric asymmetry: an in utero neuroimaging study. Cereb Cortex 21: 1076–1083.

Keller SS, Crow T, Foundas A et al. (2009). Broca's area: nomenclature, anatomy, typology and asymmetry. Brain Lang 109: 29–48.

Kennedy DN, O'Craven KM, Ticho BS et al. (1999). Structural and functional brain asymmetries in human situs inversus totalis. Neurology 53: 1260–1265.

Kennedy MP, Omran H, Leigh MW et al. (2007). Congenital heart disease and other heterotaxic defects in a large cohort of patients with primary ciliary dyskinesia. Circulation 115: 2814–2821.

Komatsu Y, Mishina Y (2013). Establishment of left-right asymmetry in vertebrate development: the node in mouse embryos. Cell Mol Life Sci 70: 4659–4666.

Kong XZ, Postema M, Schijven D et al. (2021). Large-scale phenomic and genomic analysis of brain asymmetrical skew. Cereb Cortex 31: 4151–4168.

Kosaki K, Casey B (1998). Genetics of human left-right axis malformations. Semin Cell Dev Biol 9: 89–99.

Leigh MW, Pittman JE, Carson JL et al. (2009). Clinical and genetic aspects of primary ciliary dyskinesia/Kartagener syndrome. Genet Med 11: 473–487.

Leroy F, Cai Q, Bogart SL et al. (2015). New human-specific brain landmark: the depth asymmetry of superior temporal sulcus. Proc Natl Acad Sci U S A 112: 1208–1213.

Lin AE, Ticho BS, Houde K et al. (2000). Heterotaxy: associated conditions and hospital-based prevalence in newborns. Genet Med 2: 157–172.

Lin AE, Krikov S, Riehle-Colarusso T et al. (2014). Laterality defects in the National Birth Defects Prevention Study (1998-2007): birth prevalence and descriptive epidemiology. Am J Med Genet A 164: 2581–2591.

Long S, Ahmad N, Rebagliati M (2003). The zebrafish nodal-related gene southpaw is required for visceral and diencephalic left-right asymmetry. Development 130: 2303–2316.

Lucas JS, Burgess A, Mitchison HM et al. (2014). Diagnosis and management of primary ciliary dyskinesia. Arch Dis Child 99: 850–856.

Matsumoto T, Kuriya N, Akagi T et al. (1997). Handedness and laterality of the viscera. Neurology 49: 1751. https://doi.org/10.1212/wnl.49.6.1751. PMID: 9409390.

Mazoyer B, Zago L, Jobard G et al. (2014). Gaussian mixture modeling of hemispheric lateralization for language in a large sample of healthy individuals balanced for handedness. PLoS One 9: e101165. https://doi.org/10.1371/journal.pone.0101165. PMID: 24977417; PMCID: PMC4076312.

McManus IC, Martin N, Stubbings GF et al. (2004). Handedness and situs inversus in primary ciliary dyskinesia. Proc R Soc B-Biol Sci 271: 2579–2582.

Medina L (2009). Evolution and Embryological Development of Forebrain. In: MD Binder, N Hirokawa, U Windhorst (Eds.), Encyclopedia of neuroscience. Springer, Berlin, Heidelberg.

Moher D, Liberati A, Tetzlaff J et al. (2009). Preferred reporting items for systematic reviews and Meta-analyses: the PRISMA statement. PLoS Med 6: e1000097. https://doi.org/10.1371/journal.pmed.1000097. Epub 2009 Jul 21. PMID: 19621072; PMCID: PMC2707599.

Nakamura T, Hamada H (2012). Left-right patterning: conserved and divergent mechanisms. Development 139: 3257–3262.

Nonaka S, Shiratori H, Saijoh Y et al. (2002). Determination of left-right patterning of the mouse embryo by artificial nodal flow. Nature 418: 96–99.

Ocklenburg S, Gunturkun O (2018). The lateralized brain: the neuroscience and evolution of hemispheric asymmetries, Academic Press Imprint Elsevier Science & Technology Books, San Diego.

Palmer AR (2004). Symmetry breaking and the evolution of development. Science 306: 828–833.

Papadatou-Pastou M, Ntolka E, Schmitz J et al. (2020). Human handedness: a meta-analysis. Psychol Bull 146: 481–524.

Peeters H, Devriendt K (2006). Human laterality disorders. Eur J Med Genet 49: 349–362.

Postema MC, Carrion-Castillo A, Fisher SE et al. (2020). The genetics of situs inversus without primary ciliary dyskinesia. Sci Rep 10: 3677. https://doi.org/10.1038/s41598-020-60589-z. PMID: 32111882; PMCID: PMC7048929.

Powell HWR, Parker GJM, Alexander DC et al. (2006). Hemispheric asymmetries in language-related pathways: a combined functional MPI and tractography study. Neuroimage 32: 388–399.

Powell JL, Kemp GJ, Garcia-Finana M (2012). Association between language and spatial laterality and cognitive ability: an fMRI study. Neuroimage 59: 1818–1829.

Rogers LJ, Zucca P, Vallortigara G (2004). Advantages of having a lateralized brain. Proc R Soc B-Biol Sci 271: S420–S422.

Schmitz J, Lor S, Klose R et al. (2017). The functional genetics of handedness and language lateralization: insights from gene ontology, pathway and disease association analyses. Front Psychol 8: 1144. https://doi.org/10.3389/fpsyg.2017.01144. PMID: 28729848; PMCID: PMC5498560.

Schmutzer KJ, Linde LM (1958). Situs Inversus Totalis associated with complex cardiovascular anomalies. Am Heart J 56: 761–768.

Schuler AL, Kasprian G, Schwartz E et al. (2017). Mens inversus in corpore inverso? Language lateralization in a boy with situs inversus totalis. Brain Lang 174: 9–15.

Smith DJ, Montenegro-Johnson TD, Lopes SS (2019). Symmetry-breaking cilia-driven flow in embryogenesis. Annu Rev Fluid Mech 51: 105–128.

Sun T, Walsh CA (2006). Molecular approaches to brain asymmetry and handedness. Nat Rev Neurosci 7: 655–662.

Sun T, Patoine C, Abu-Khalil A et al. (2005). Early asymmetry of gene transcription in embryonic human left and right cerebral cortex. Science 308: 1794–1798.

Sun ZY, Kloppel S, Riviere D et al. (2012). The effect of handedness on the shape of the central sulcus. Neuroimage 60: 332–339.

Sutherland MJ, Ware SM (2009). Disorders of left-right asymmetry: heterotaxy and situs inversus. Am J Med Genet C Semin Med Genet 151c: 307–317.

Tanaka S, Kanzaki R, Yoshibayashi M et al. (1999). Dichotic listening in patients with situs inversus: brain asymmetry and situs asymmetry. Neuropsychologia 37: 869–874.

Thomas S, Boutaud L, Reilly ML et al. (2019). Cilia in hereditary cerebral anomalies. Biol Cell 111: 217–231.

Toga AW, Thompson PM (2003). Mapping brain asymmetry. Nat Rev Neurosci 4: 37–48.

Torgersen J (1950). Situs inversus, asymmetry and twinning. Am J Hum Genet 2: 361–370.

Tubbs RS, Wellons JC, Salter G et al. (2003). Intracranial anatomic asymmetry in situs inversus totalis. Anat Embryol 206: 199–202.

Tzourio-Mazoyer N, Crivello F, Mazoyer B (2018). Is the planum temporale surface area a marker of hemispheric or regional language lateralization? Brain Struct Funct 223: 1217–1228.

Vallortigara G, Rogers LJ, Bisazza A (1999). Possible evolutionary origins of cognitive brain lateralization. Brain Res Rev 30: 164–175.

Vandenberg LN, Levin M (2010). Far from solved: a perspective on what we know about early mechanisms of left-right asymmetry. Dev Dyn 239: 3131–3146.

Vandenberg LN, Levin M (2013). A unified model for left-right asymmetry? Comparison and synthesis of molecular models of embryonic laterality. Dev Biol 379: 1–15.

Vasung L, Rollins CK, Yun HJ et al. (2020). Quantitative in vivo MRI assessment of structural asymmetries and sexual dimorphism of transient fetal compartments in the human brain. Cereb Cortex 30: 1752–1767.

Vernooij MW, Smits M, Wielopolski PA et al. (2007). Fiber density asymmetry of the arcuate fasciculus in relation to functional hemispheric language lateralization in both right- and left-handed healthy subjects: a combined fMRI and DTI study. Neuroimage 35: 1064–1076.

Vingerhoets G (2019). Phenotypes in hemispheric functional segregation? Perspectives and challenges. Phys Life Rev 30: 1–18.
Vingerhoets G, Alderweireldt AS, Vandemaele P et al. (2013). Praxis and language are linked: evidence from co-lateralization in individuals, with atypical language dominance. Cortex 49: 172–183.
Vingerhoets G, Gerrits R, Bogaert S (2018a). Atypical brain functional segregation is more frequent in situs inversus totalis. Cortex 106: 12–25.
Vingerhoets G, Li X, Hou L et al. (2018b). Brain structural and functional asymmetry in human situs inversus totalis. Brain Struct Funct 223: 1937–1952.
Vingerhoets G, Gerrits R, Karlsson EM (2024). Brain (Yakovlevian) torque direction is associated with volume asymmetry of the intracranial transverse sinuses: evidence from situs inversus totalis. Brain Struct Funct 229: 1461–1470. https://doi.org/10.1007/s00429-024-02810-7.
Watanabe D, Saijoh Y, Nonaka S et al. (2003). The left-right determinant Inversin is a component of node monocilia and other 9+0 cilia. Development 130: 1725–1734.
Wehrmaker A (1969). Right-left asymmetry and situs Iversus in Triturus alpestris. Wilhelm Roux Arch Entwickl Mech Org 163: 1–32.
Woods RP (1986). Brain asymmetries in situs-Inversus - a case-report and review of the literature. Arch Neurol 43: 1083–1084.
Xiang L, Crow TJ, Hopkins WD et al. (2018). Human torque is not present in chimpanzee brain. Neuroimage 165: 285–293.
Xiang L, Crow T, Roberts N (2019). Cerebral torque is human specific and unrelated to brain size. Brain Struct Funct 224: 1141–1150.
Yun HJ, Vasung L, Tarui T et al. (2020). Temporal patterns of emergence and spatial distribution of Sulcal pits during fetal life. Cereb Cortex 30: 4257–4268.
Zago L, Herve PY, Genuer R et al. (2017). Predicting hemispheric dominance for language production in healthy individuals using support vector machine. Hum Brain Mapp 38: 5871–5889.
Zariwala MA, Omran H, Ferkol TW (2011). The emerging genetics of primary ciliary dyskinesia. Proc Am Thorac Soc 8: 430–433.
Zhao L, Matloff W, Shi YG et al. (2022). Mapping complex brain torque components and their genetic architecture and Phenomic associations in 24,112 individuals. Biol Psychiatry 91: 753–768.

Section II

Clinical

Handbook of Clinical Neurology, Vol. 208 (3rd series)
Cerebral Asymmetries
C. Papagno and P. Corballis, Editors
https://doi.org/10.1016/B978-0-443-15646-5.00041-5

Chapter 5

Oncology: Brain asymmetries in language-relevant brain tumors

DJAINA SATOER[1*], LILIT DULYAN[2,3], AND STEPHANIE FORKEL[2,3,4]

[1]*Department of Neurosurgery, Erasmus MC University Medical Center Rotterdam, The Netherlands*

[2]*Donders Institute for Brain Cognition Behaviour, Radboud University, Nijmegen, The Netherlands*

[3]*Brain Connectivity and Behaviour Laboratory, Sorbonne Universities, Paris, France*

[4]*Centre for Neuroimaging Sciences, Department of Neuroimaging, Institute of Psychiatry, Psychology and Neuroscience, King's College London, London, United Kingdom*

Abstract

Brain tumors are classified as rare diseases, with an annual occurrence of 300,000 cases and account for an annual loss of 241,000 lives, highlighting their devastating nature. Recent advancements in diagnosis and treatment have significantly improved the management and care of brain tumors. This chapter provides an overview of the common types of primary brain tumors affecting language functions—gliomas and meningiomas. Techniques for identifying and mapping critical language areas, including the white matter language system, such as awake brain tumor surgery and diffusion-weighted tractography, are pivotal for understanding language localization and informing personalized treatment approaches. Numerous studies have demonstrated that gliomas in the dominant hemisphere can lead to (often subtle) impairments across various cognitive domains, with a particular emphasis on language. Recently, increased attention has been directed toward (nonverbal) cognitive deficits in patients with gliomas in the nondominant hemisphere, as well as cognitive outcomes in patients with meningiomas, a group historically overlooked. A patient-tailored approach to language and cognitive functions across the pre-, intra-, and post-operative phases is mandatory for brain tumor patients to preserve quality of life. Continued follow-up studies, in conjunction with advanced imaging techniques, are crucial for understanding the brain's potential for neuroplasticity and optimizing patient outcomes.

INCIDENCE, PREVALENCE, AND PROGNOSIS OF BRAIN TUMORS

Types of brain tumors

A brain tumor is characterized by an accumulation of abnormal cells that proliferate, resulting in a growing mass of tissue infiltrating into or pressing on healthy brain tissue. This chapter focuses on primary (as opposed to metastatic) brain tumors, which originate in the brain (gliomas) or its surrounding tissues (meninges).

Gliomas and meningiomas are the prevailing types of brain tumors in humans. Among them, malignant gliomas are the most aggressive and with fatal variation. Conversely, meningiomas are typically benign but often reappear following surgical intervention. These tumors differ concerning their origin and their effects on surrounding brain areas. However, both tumor types can affect brain areas that are involved in sensorimotor language [often referred to as "eloquent areas," although this is dependent on the neurosurgeon's definition of "functional" tissue; see Rammeloo et al., 2023 for a systematic review on the assessment of eloquence in neurosurgery], and other higher cognitive functions. These tumors can be further dichotomized into low-grade, grade I or II, and high-grade, grade III or IV, according to the World Health Organization (WHO) (Louis et al., 2016, 2021). This distinction is based on the cell type a tumor originates from and is determined by histology. Low-grade brain tumors have relatively slow growth rates, whereas high-grade brain tumors have a more aggressive evolution (see later the molecular features). Early diagnosis is linked to longer survival times (NCIN, 2013).

Gliomas are intra-axial tumors within the brain tissue (brain parenchyma). These infiltrating tumors originate from glial

*Correspondence to: Djaina Satoer, Department of Neurosurgery, Erasmus MC University Medical Center Rotterdam, The Netherlands. Tel: +31-650008179, E-mail: d.satoer@erasmusmc.nl

cells. Gliomas are the most frequently occurring primary brain tumors with approximately 1000 new diagnoses (incidence) of adult gliomas every year in the Netherlands alone, of which 20% are low grade (Ho et al., 2014). Worldwide, more than 250,000 new cases of primary brain tumors are diagnosed every year, of which 77% are gliomas (Walsh and Hiroko Ohgaki, 2016). Furthermore, low-grade gliomas (LGG) affect relatively young individuals resulting in a mean age at diagnosis of 42 years (Ho et al., 2014). LGGs are usually discovered rather late when a patient develops overt symptoms, such as epileptic seizures, which is the most frequent symptom (58% according to a Danish Neuro-Oncology Registry (Rasmussen et al., 2017), mild language, and/or other cognitive impairments (43%), or headaches (35%). However, it may take many years before symptoms of an LGG become apparent because of its slow growth rate with a diameter expansion of approximately 4 mm/year (Mandonnet et al., 2003). This slow growth pattern offers the brain time to slowly adjust and compensate by enhancing neuroplasticity leading to functional and/or structural reorganization. LGGs, which tend to be WHO grade II tumors in adults, eventually progress to higher grades of malignancy (WHO grades III and IV; high-grade glioma—HGG) within approximately 5–10 years (Hervey-Jumper and Berger, 2016). For LGGs residing in language relevant areas, the gold standard treatment is considered to be a maximal resection with the help of direct electrical stimulation (DES) during awake surgery. Using a personalized and direct approach to mapping the language-relevant network in each patient minimizes the damaging of functional tissue, which would otherwise have resulted in irreversible neurologic and cognitive damage (Duffau et al., 2002; Thiebaut de Schotten et al., 2005; De Witt Hamer et al., 2012) (see Gliomas (awake surgery) section for more details). HGGs (otherwise referred to as anaplastic astrocytoma or oligodendroglioma, glioblastoma) are very aggressive tumors with an unfavorable prognosis. Among them, glioblastoma has the most devastating prognosis with 12–15 months survival time after diagnosis (Stupp et al., 2005), whereas the anaplastic and mixed type are less aggressive (WHO grade III). Due to their faster growth rate, neurologic deficits are common in this patient group and if they affect language systems, patients present with (moderate to severe) aphasia. Although HGGs may develop at all ages, they typically occur between 50 and 60 years of age (Preusser et al., 2011).

More recently, several molecular markers have been identified to aid prognosis in gliomas (Cohen and Colman, 2015; Louis et al., 2021; Rudà et al., 2022). Some of the key markers are the following: (1) Isocitrate dehydrogenase 1 and 2 and wild-type (IDH1/2) mutations: IDH mutations are commonly observed in gliomas with IDH1 and 2 mutations and tend to have a more favorable outcome compared to IDH-wild type and to those without these mutations; (2) 1p/19q co-deletion: co-deletion of chromosome arms 1p and 19q is commonly observed in oligodendrogliomas, a subtype of glioma. This genetic alteration is associated with a better response to chemotherapy and longer overall survival; (3) O6-methylguanine-DNA methyltransferase (MGMT) promoter methylation: MGMT is a DNA repair enzyme that can affect the response of gliomas to alkylating chemotherapy agents, such as temozolomide. Promoter methylation of the MGMT gene is associated with improved response to chemotherapy and a better prognosis; (4) Telomerase reverse transcriptase (TERT) promoter mutations: TERT promoter mutations are frequently found in gliomas and have been associated with a worse prognosis. They are more common in HGGs and are associated with increased tumor aggressiveness. The assessment of these markers, along with other clinical and histopathologic factors, can be of help in the prognosis and guidance for treatment decisions. Traditionally, surgical removal of glioblastomas is performed under general anesthesia with the goal of achieving a maximal safe resection (see Glioma (awake procedure) section for more discussion).

Meningiomas are extra-axial tumors, which means that they lie outside the brain parenchyma. These tumors arise from the meninges, which are the protective membranes surrounding the brain (i.e., dura mater, dura, and arachnoid). Even though meningiomas generally do not infiltrate brain tissue, they can compress adjacent brain areas and displace white matter (Piper et al., 2016). Meningiomas represent approximately one-third of all primary brain tumors, of which 90% are classified as WHO low-grade I, which is considered benign (Whittle et al., 2004; Wiemels et al., 2010). Approximately 500 adults with a symptomatic meningioma are diagnosed every year in the Netherlands (the Netherlands Cancer Registry, https://iknl.nl/en/ncr). However, many meningiomas are asymptomatic and remain undiagnosed for many years if not decades (Wiemels et al., 2010). Therefore, the overall meningioma incidence is presumably at least twice as high (Larjavaara et al., 2008). Symptomatic meningiomas are usually diagnosed between the ages of 40–70 years (Radhakrishnan et al., 1995). They can manifest through a wide range of symptoms, depending on size and location of the tumor. Possible symptoms include epileptic seizures (typically located in the temporal lobe), disturbed vision (occipital location), sensorimotor dysfunction (fronto-parietal location), cognitive decline, headaches, and other signs of increased intracranial pressure (Whittle et al., 2004). A significant number of meningiomas are discovered incidentally during brain scanning for other purposes (Whittle et al., 2004). Due to its extra-axial location, meningioma surgery is performed under general anesthesia without waking the patients up for language or sensorimotor testing. About 90% of meningiomas are benign (WHO grade I), 5%–7% are atypical (WHO grade II), and 1%–3% are considered anaplastic or malignant (WHO grade III) (Louis et al., 2007). Grade I meningioma patients have a relatively long life expectancy: more than half of the patients survive the first 20 years after surgery (van Alkemade et al., 2012). For grade II and III meningiomas, the 5-year and 10-year overall survival rates were 93.5% and 83.4%, respectively (Wang et al., 2016).

BRAIN TUMORS AND DIFFERENCES

Sex differences

Sex differences are observed between men and women with regard to glioblastomas (GBMs), primarily confined to incidence rates and outcomes. Our understanding of sex differences in GBM at the level of disease phenotype and genetic/molecular factors remains limited. Between 2008 and 2012 in the United States, approximately 55% of malignant brain tumors were diagnosed in men, while women accounted for 45%. In contrast, during the same period, around 36% of nonmalignant brain tumors were observed in men, with women accounting for 64% (Gould, 2018) (see Fig. 5.1). Not only do the rates of occurrence of tumors vary but also the side effects differ between the sexes. For example, women tend to have more inflammation while men tend to have more edema as sex is linked to macrophage content and inflammatory gene expression in glioblastomas (Ochocka et al., 2023). These differences have partly been attributed to hormones affecting the microenvironment of a tumor whereby estrogens seemed to have a protective impact, while increased androgen receptor expression and elevated testosterone levels had a negative impact on GBM (Carrano et al., 2021). In LGG, sex-based differences have not been investigated thoroughly. However, a recent retrospective cohort study demonstrated that male sex was an independent risk factor for unfavorable outcomes as opposed to female sex. In addition, 1p19q co-deletion status is beneficial for overall survival and "next intervention free survival," which was defined as "any treatment subsequent to adjuvant or salvage therapy or death" (Tewari et al., 2022). According to the Central Brain Tumor Registry of the United States (CBTRUS-CBTRUS, 2013–2017) and a population-based study in California, meningioma is more commonly observed in women and the risk was elevated between the ages of 27 and 37 years (Cote et al., 2022).

Cognitive difference with left- and right-hemispheric tumors

Neuroscience studies have recently identified that cognitive functions are much better understood for the left hemisphere (LH) than the right hemisphere (RH) (Thiebaut de Schotten et al., 2020; Talozzi et al., 2023). This discrepancy is mainly due to language processing being obviously impaired with left hemisphere lesions and most neuropsychologic assessments being language-based. Looking at the impact of LH and RH tumors, evidence suggests that patients with left-hemispheric GBMs have a worse prognosis compared to right GBMs. This observation was related to the shorter progression time with a faster functional decline possibly due to a more conservative surgical approach leaving more tumor tissue behind (Coluccia et al., 2018). Left-hemisphere lesions that impact

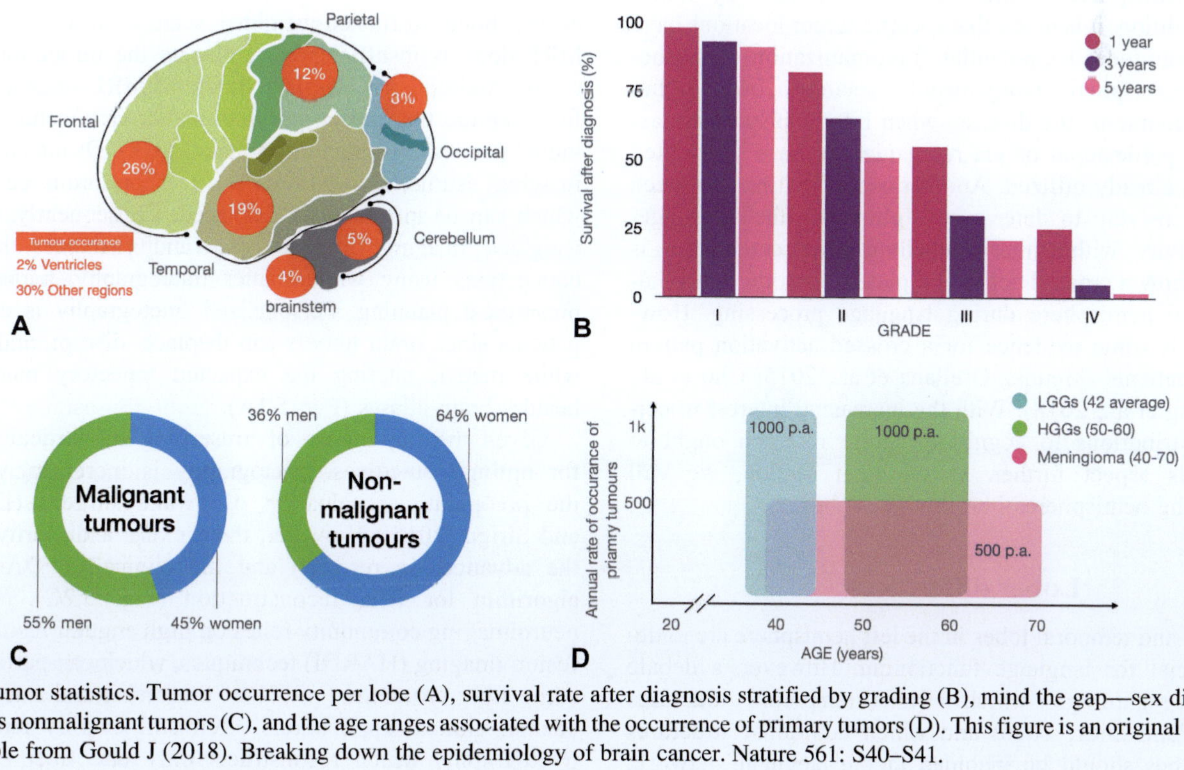

Fig. 5.1. Tumor statistics. Tumor occurrence per lobe (A), survival rate after diagnosis stratified by grading (B), mind the gap—sex differences in malignant vs nonmalignant tumors (C), and the age ranges associated with the occurrence of primary tumors (D). This figure is an original based on the data available from Gould J (2018). Breaking down the epidemiology of brain cancer. Nature 561: S40–S41.

language processing likely interfere with articulation, comprehension, syntactic sentence formation, and semantic access while right-hemisphere lesions tend to affect prosody, intonation, and emotional affect (Durfee et al., 2021; Sheppard et al., 2022). Impairments in the suprasegmental language functions of the right hemisphere have psychologic impacts on patients (e.g., monotonous speech) but interfere less with their daily life as patients are understood by others and can understand language. As such, cognitive testing has focused on left-hemisphere language functions. Consequently, tumors considered "eloquent" are typically left-hemisphere tumors, which can be closely linked to the definition of language and language in the brain. In a study of 489 tumor patients, no significant differences were observed between the hemispheres for the major tumor types. That said, LGGs and meningiomas had a tendency to occur slightly more frequently on the left side, albeit not significantly. Conversely, HGG exhibited a nonsignificant tendency to occur more often on the right side (Inskip et al., 2003).

The linguistic competency of each hemisphere is also relevant when looking at recovery after surgery. Different patterns of recovery in language and other cognitive functions have been associated with several neuronal adaptations. For example, Nieberlein et al. (2023) performed an extensive systematic review and described four different types of neural (language) reorganization in gliomas: (1) persisting (language) functions within the tumor, (2) perilesional reorganization, (3) reorganization in a distributed network of the affected hemisphere, (4) contralesional reorganization (i.e., RH in the case of LH dominant hemisphere), which is also associated by tumor grade. In addition, it is noted that specific tumor locations have different magnitudes for potential of reorganization. Nevertheless, right-hemispheric reorganization seems to occur rather late in the course of the disease, when interhemispheric plasticity (e.g., perilesional or via reorganization in a distributed network) is already utilized. Another aspect that has not been studied extensively to date is how gliomas affect language-related activity within the cerebellum. The cerebellum is known to show a crossed activation pattern with the contralateral cerebral hemisphere during language processing. However, there is some evidence for a crossed activation pattern in glioma patients (Méndez Orellana et al., 2015; Cho et al., 2018; Zhang et al., 2018). With the increasing interest in cerebellar contributions to cognition, future research ought to explore this aspect further. In the next section, we will parcellate the hemispheres into separate lobes.

Lobar differences

The frontal and temporal lobes in the left hemisphere are undeniably crucial for language functioning. However, a debate continues regarding, which frontal/temporal structures (Tremblay and Dick, 2016) and which additional structures in other lobes should be included in the language network. While the parietal and anterior temporal lobe also play roles in language processes, they are not always incorporated into models that focus exclusively on syntactic processing. Similarly, as discussed earlier, this uncertainty extends to the role of the entire right hemisphere. A statistical review of primary brain tumors diagnosed in the United States in 2008–2012 revealed that brain tumors are inhomogeneously distributed across the brain (Fig. 5.1A). Most tumors (26%) are in the frontal lobe, followed by temporal lobe tumors (19%), parietal tumors (12%), and cerebellar tumors (5%) while the brainstem (4%), occipital lobes (2%), and ventricular tumors (2%) are less common. The remaining 30% are in other regions (Gould, 2018) (see Fig. 5.1).

TECHNIQUES TO IDENTIFY TUMORS AFFECTING THE LANGUAGE SYSTEM

Noninvasive structural and functional techniques

Noninvasive techniques offer a valuable means to map the altered brain anatomy resulting from the presence of a tumor. These techniques can assess structural changes, functional alterations, and plasticity changes. The noninvasive nature of these techniques means they are applied externally, typically without medication administration, except in cases where patients have a low tolerance for confined spaces (e.g., claustrophobia).

Structural techniques primarily rely on magnetic resonance imaging (MRI) methods to investigate the cortical and subcortical structures (T1-weighted scans), to identify pathologic changes (T2-weighted scans) and to explore the connectivity of the brain (diffusion-weighted scans). However, structural MRI alone is insufficient to delineate the tumor margin precisely. Additionally, classical structural MRI sequences do not fully capture the impact on white matter architecture and individual pathways (Forkel and Catani, 2018). Diffusion-weighted imaging estimates the tumor's effect on brain connections, which can be infiltrated or displaced. Consequently, diffusion-weighted imaging often assesses and visualizes the brain's connections using white matter tractography, particularly in presurgical planning. Personalized tractography is crucial for patients since brain tumors can displace, disrupt, and infiltrate white matter, altering the expected trajectory compared to healthy brain atlases (Fig. 5.1A).

Given the importance of preserving subcortical pathways for optimal outcomes, tractography is increasingly used in the preoperative evaluation of awake surgery (Desmurget and Sirigu, 2015). However, there exists a disparity between the advances in research and the clinically FDA-approved algorithm for fiber reconstruction (Fig. 5.2C). While the neuroimaging community relies on high angular resolution diffusion imaging (HARDI) techniques, which can partially overcome fiber crossings (~90% of the brain) and track through edema, clinical applications often still employ tensor-based tractography, which reconstructs only one fiber orientation

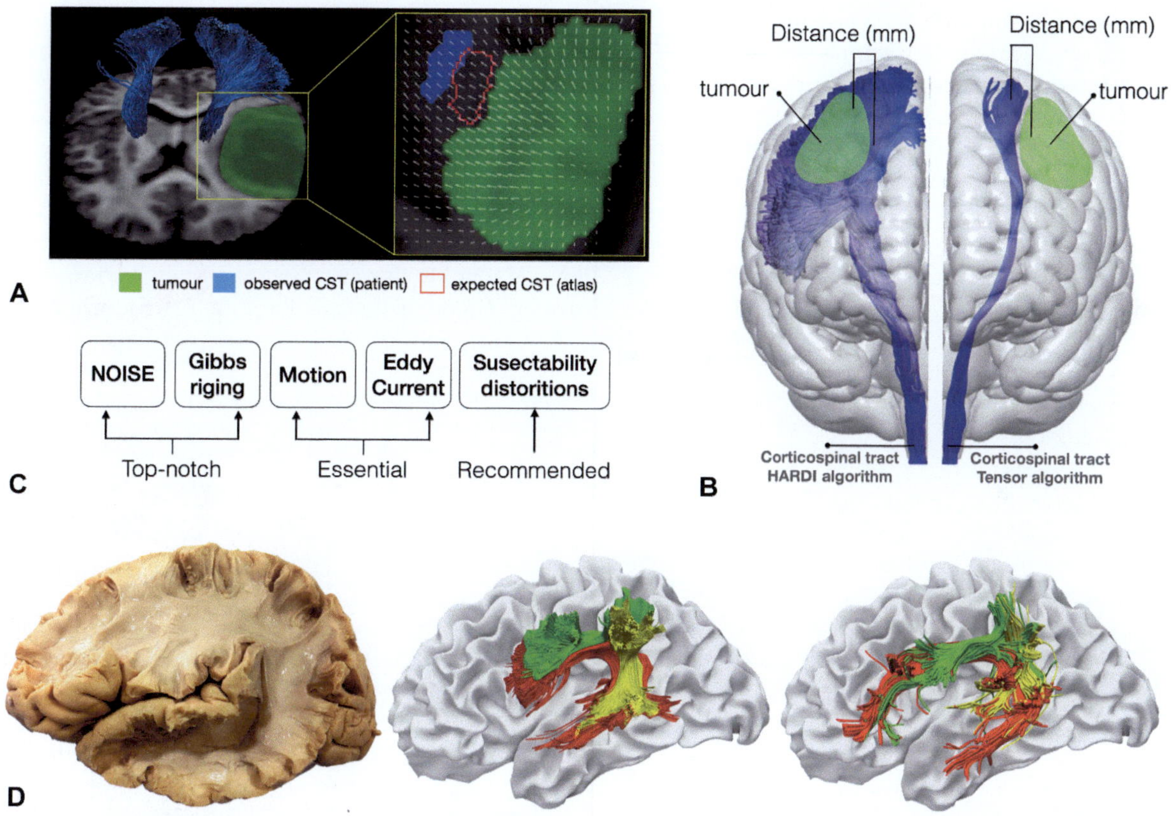

Fig. 5.2. Technical considerations in tractography reconstructions for brain tumors. Panel (A) emphasizes the limitations of atlas-based approaches due to white matter displacement. Panel (B) underscores the impact of algorithm selection on reconstruction reliability. Panel (C) presents the recommended preprocessing pipeline for tractography. Finally, panel (D) showcases the comparison between different techniques for visualizing the arcuate fasciculus using Klingler dissection (*left*, diffusion tensor imaging (DTI, *middle*) and spherical deconvolution (SD, *right*). Panels (A) and (B) are modified from Forkel SJ, Bortolami C, Dulyan L, et al. (in press). Dissecting white matter pathways: a neuroanatomical approach. In: Leemands A and Dell'Acqua F. (Eds). Handbook of Diffusion MRI Tractography. Reused and modified with permission from my coauthors. (DWI methods) Panels (C) and (D) are original.

per voxel, thereby underrepresenting the complexity of the connectional anatomy [for a review see Beyh et al., 2024]. Due to the alterations caused by the lesion, manual dissections of the white matter are recommended over atlas-based tools (Fig. 5.2A).

The steps for preprocessing diffusion-weighted imaging data for tractography depend on the acquisition quality but typically involve motion and eddy current corrections (Fig. 5.2B). It is highly recommended that susceptibility distortion corrections be performed, and research data should also undergo denoising and correction for Gibbs ringing where the data quality is sufficient (Perrone et al., 2015; Kellner et al., 2016). The choice of processing parameters and tracking algorithms directly influences the reconstruction of white matter. Traditional tensor-based algorithms, such as DTI, cannot resolve crossing fibers, leading to false positive and false negative reconstructions of the anatomy (Fig. 5.2C). Advanced algorithms based on high angular resolution diffusion imaging (HARDI) offer an improved resolution of fiber crossings and can visualize lateral projections that have been demonstrated using postmortem dissections (Dell'Acqua et al., 2013; Dell'Acqua and Tournier, 2019). However, these algorithms are more susceptible to false positive reconstructions and require comprehensive anatomic knowledge. In the case of tumors affecting language functions, tractography can selectively visualize the language network using these techniques [for detailed information, see Beyh et al., 2024, Ohlerth et al., in press]. A recent review has highlighted the variability observed in the cortical terminations of the arcuate fasciculus in the temporal lobe (Giampiccolo and Duffau, 2022) (see Fig. 5.2D). However, another recent review identified no one tract-one function association and suggests that language is a spatiotemporally dynamic process in the brain beyond the arcuate fasciculus (Forkel et al., 2022). The current anatomic understanding of the connectional language system includes seven pathways dichotomized into a dorsal and ventral stream depending on whether they arch above or below the lateral fissure. However, an intriguing

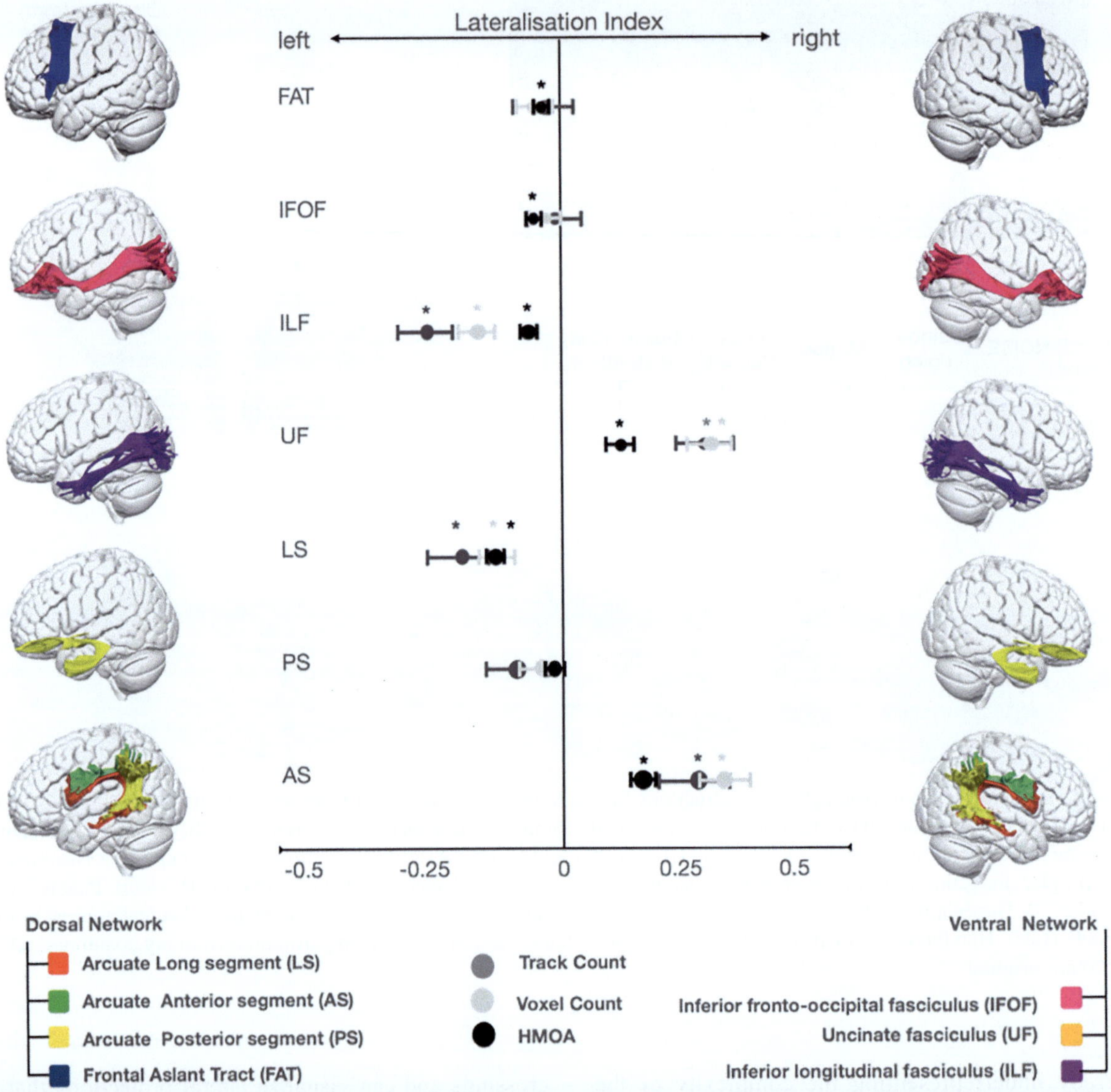

Fig. 5.3. White matter network asymmetry in the dorsal and ventral networks associated with language. Data shown are from the Human Connectome Project 7T data set (www.humanconnectome.com) and are available preprocessed from www.bcblab.com; HMOA: Hinderance Modulated Orientational Anisotropy. See Chapter 2 in this volume (Dulyan et al., 2025) for the asymmetry pattern of all white matter connections. This is an original figure.

question arises regarding the bilateral nature of these networks, considering that language is typically described as a unilateral function (Fig. 5.3).

On the functional side, two sequences can be acquired using MRI scanning: task-based functional MRI (fMRI) and non-task-based resting state functional imaging (rs-fMRI). fMRI is the most used noninvasive imaging technique to map brain functions. The word-verb association task and finger tapping are often used to localize linguistic and motor functions. In the neurosurgical setting, fMRI contributes to determining dominance for these functions. In addition, fMRI can be easily repeated in the postoperative period to study brain plasticity (Young et al., 2010). However, fMRI has several shortcomings. First, fMRI cannot discriminate between the activated "essential" and "compensable" regions. Second, motion-related artifacts decrease the reliability of fMRI findings. Third, the activation zones may vary depending on the paradigm chosen, the statistical method, and the underlying neurologic disorder. Fourth, the sensitivity of fMRI is limited. While fMRI has mapped the cortical language network of brain areas involved in language processing (e.g., Price, 2010), in clinical populations, its predictive power for the language-relevant cortex is limited (Kullmann, 2020) as revealed by comparison with DES (see Invasive techniques section) during awake surgery (Giussani et al., 2010). fMRI-DES match for motor mapping was estimated at approximately 71%, while the correlation for language mapping was only 66% (Bartos et al., 2009).

A technique that has proven reliable in the presurgical mapping of language functions is navigating transcranial magnetic stimulation (nTMS). nTMS is a noninvasive neurophysiologic technique that stimulates brain regions using an electromagnetic coil. Single-pulse nTMS is used as an activation tool to map the motor cortex, while repetitive nTMS has an inhibitory effect. Linguistic processes can be interrupted, leading to temporary speech errors (paraphasia) or even "speech arrest" (no speech output). This technique best approximates the effects of DES so that the patient also experiences preoperatively what a disruption of speech, such as a "speech arrest," feels like. A disadvantage is that TMS can be painful in the temporal regions due to stimulation of the facial nerves. Good correlations between nTMS and DES were reported for mapping the motor functions (Krieg et al., 2012). Localizing the language functions with nTMS, on the other hand, is complex; however, there is growing evidence with promising results for clinical neuro-oncologic settings (Picht et al., 2013; Tarapore et al., 2013; Sollmann et al., 2020) and the design of new linguistics test batteries for nTMS interventions is also in development (Ohlerth et al., 2020). Compared to fMRI, nTMS has a significantly higher sensitivity for identifying language functions. For example, nTMS correlates significantly higher with intraoperative DES and demonstrates better sensitivity and specificity (exceeding 99% and 83%, respectively) (Muir et al., 2022). This indicates that TMS can consistently identify critical language sites that match DES more reliably than fMRI-positive sites. Given the limited spatial resolution and low specificity of nTMS in posterior language regions, further research is needed before nTMS can fully replace fMRI for preoperative functional screening.

Brain activity can also be registered using electroencephalography (EEG). This noninvasive technique directly measures brain signals in real-time by electrodes affixed to the scalp. These signals are amplified, digitalized, and plotted as changes in voltage over time. EEG registration has a high temporal resolution, but its spatial resolution is low compared to structural MRI imaging. Alternatively, brain activity can be registered using magnetoencephalography (MEG), combining high temporal and relatively high spatial resolutions. However, EEG has the advantage of being less expensive and more widely available than MEG. Moreover, EEG is frequently used in clinical settings when brain dysfunction or epileptic seizures are suspected.

Finally, a recent ground-breaking and upcoming noninvasive technique concerns ultrasound stimulation, which can open the blood–brain barrier to optimize deliverance of an increased concentration of medication to a targeted brain area; see, for instance, GBM patients and patients with Alzheimer disease (Arrieta et al., 2024; Rezai et al., 2024).

Invasive techniques

Wilder Penfield (1891–1976) was the first to systematically examine language with DES in 190 patients with a brain lesion. He initially used the technique to detect motor functions and epileptic foci but also described speech and language areas in the brain by eliciting verbal paraphasias when stimulating a widely distributed surface of the brain. Based on this data, he proposed that dynamic connections exist within a cortical network that can compensate when "classic" linguistic areas "Broca's and Wernicke's area" in the left hemisphere are damaged (Penfield and Roberts, 1959) (for a critical evaluation of the terminology see (Dulyan and Forkel, 2024)). These findings were confirmed by George Ojemann in a study of 117 patients (Ojemann et al., 1989), where even more variation of critical language areas in the brain was observed in the language-dominant hemisphere.

The pioneers of white matter neurosurgery were the scholars of the Zurich school (Dziedzic et al., 2021). They paid close attention to the connectional anatomy of the brain when planning their entry routes to a tumor, aiming to minimize damage to the connective tissue. Neurosurgeon Hugues Duffau combined these methods and developed an entire "modern neurosurgical research" field. He discovered, among others, that apart from cortical language areas, it is extremely important to localize language functions at the subcortical level with DES (Duffau et al., 2002; Duffau, 2014). Not only the arcuate fasciculus (a dorsal tract) but also the frontal aslant tract (FAT, dorsal tract) and ventral tracts such as the inferior fronto-occipital fasciculus (IFOF), the inferior longitudinal fasciculus (ILF), and the uncinate fasciculus (UF) were described to play a role in language (Duffau et al., 2005; Duffau, 2014; Chang et al., 2015; Kemerdere et al., 2016). An exciting and recent development is digitally combining postmortem white matter dissections with in vivo virtual tractography-based dissections using brain dissection photogrammetry (https://bradipho.eu/). Merging diffusion tractography with photogrammetric dissection of the human brain offers a user-friendly framework to explore microdissection and tractography data within the same environment and can significantly improve neurosurgical training.

Nowadays, awake surgery with DES is the gold standard treatment for resecting LGGs in eloquent brain regions (De Witt Hamer et al., 2012). The motor mapping may optionally be performed under local anesthesia (using evoked potentials), while linguistic, cognitive, visual, and spatial mapping generally use an asleep-awake-asleep procedure. Cortical stimulation is usually performed with a bipolar stimulator generating biphasic and rectangular pulses of 1 ms each time. Frequencies from 50 to 60 Hz are used by default, and the intensity progressively increases from 1 to 10 mA in 0.5 mA increments. A stimulus time of 1 or 2 s is sufficient to elicit a motor response, while language inhibition requires 3–4 s. Subcortical stimulation is done in a similar way (stimulus time may be extended to 6 s) to identify functional subcortical pathways (Ojemann et al., 1989; Berger and Ojemann, 1992; Talacchi et al., 2013a,b).

Based on the location of the tumor and the patient's personal history (preoperative cognitive linguistic level, line of work, hobbies), a test battery is tailored to the patient's needs in the preoperative phase by a clinical linguist/speech-language therapist/neuropsychologist. The tests a patient can perform well preoperatively are selected for use during cortico-subcortical stimulation and throughout the resection. For example, an intraoperative test battery may consist of motor tasks (e.g., precentral gyrus, inferior frontal gyrus) and language tasks (e.g., perisylvian region), and also arithmetic tasks (e.g., angular gyrus), visual tasks (e.g., occipital lobe), visuospatial tasks (e.g., parietal regions), and mnestic tasks (e.g., hippocampal regions) or according to latest developments also (nonlinguistic) cognitive tasks for right-hemispheric functions, such as prosody, visuospatial and social cognitive functions to prevent unilateral neglect or deficits in facial emotion recognition, prosody, empathy, and theory of mind (Lemée et al., 2018) (see Gliomas (awake surgery) section for an elaborate discussion).

In the event of a motor response or inhibition of linguistic/cognitive/visual functions following DES, these critical areas are marked on the brain using numbered sterile tags reporting a particular type of response (see Fig. 5.4).

One of the main advantages of DES is its high sensitivity. The individual critical language zones are accurately identified, and differentiation between "essential" and "compensable" language zones is possible. Its high sensitivity explains why DES is used to validate noninvasive mapping techniques such as fMRI and TMS (Mandonnet et al., 2010). Furthermore, DES can be used to examine subcortical structures. As a result, essential subcortical connections can be detected during surgery. Since disconnection of the subcortical pathways can lead to severe neurologic deficits, cortical mapping is increasingly combined with subcortical mapping (Trinh et al., 2013). Hence, DES makes it possible to remove a tumor after identification of the functional (sub)cortical boundaries and can reduce the risk of postoperative disturbances. For this reason, DES is often associated with a good postoperative outcome and quality of life (De Witt Hamer et al., 2012). Intraoperative findings were found to have prognostic value for the level or presence of postoperative aphasia. For instance, intraoperative performance on The Pyramids and Palm Tree Test (PPTT) appeared to predict postoperative language outcomes (Chang et al., 2018), underlining the application of a semantic task intraoperatively. A recent systematic review on intraoperative language errors and outcome measures reported that the occurrence of anomia and speech initiation problems were significant predictors of postoperative acute language impairments in the context of production and spontaneous speech deficits (Collée et al., 2022), stressing the need for the administration of at least an object-naming task and the monitoring of spontaneous speech during awake surgery (see "Supplementary motor area syndrome and spontaneous speech" section for more details). Besides monitoring sensorimotor, language, and cognitive functions, awake surgery teams are exploring the potential of other skills relevant to the patient's daily life, such as mapping musical skills. A recent systematic review revealed the feasibility of mapping musical skills by asking patients to play in theater (reported so far: piano, saxophone, violin, guitar, and clarinet, e.g., violin: https://www.youtube.com/watch?v=o-oGpD_Nx5U, opera singer: https://www.youtube.com/watch?v=obiARnsKUAo&t=3s, guitar: https://youtu.be/dnzjZqgfZSc), and sing during awake surgery (Kappen et al., 2022a). They identified studies describing brain regions relevant for music production suggesting the added

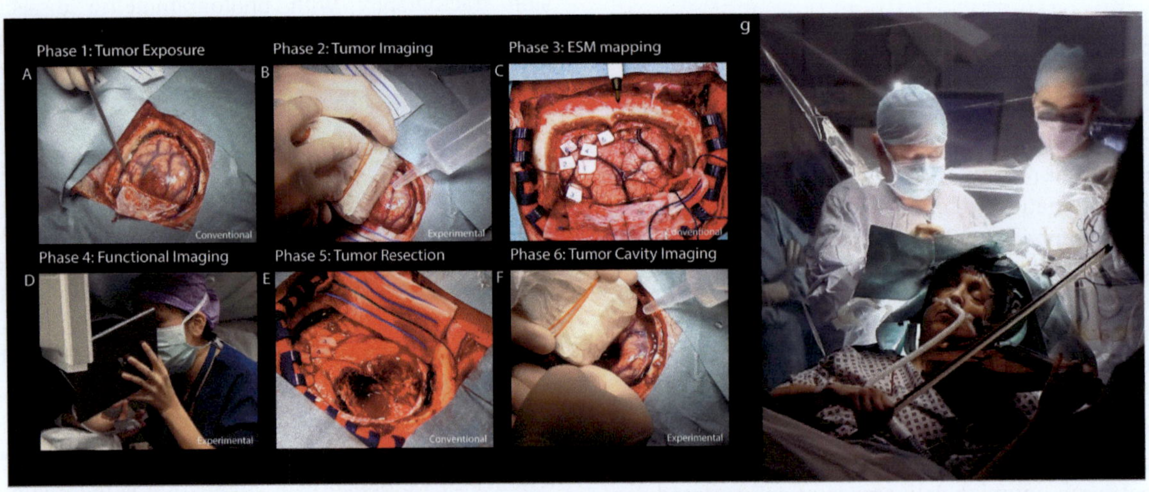

Fig. 5.4. Step-by-step principle of the direct electrical stimulation procedure in an awake patient. Panel (A–F) figure taken from Soloukey et al. (2019) during mapping in theater G) music mapping—in this case playing the violin—courtesy of Professor Ashkan and team at King's College London, video available online: https://www.youtube.com/watch?v=o-oGpD_Nx5U) and we obtained permission from the surgeons in the photograph to use it.

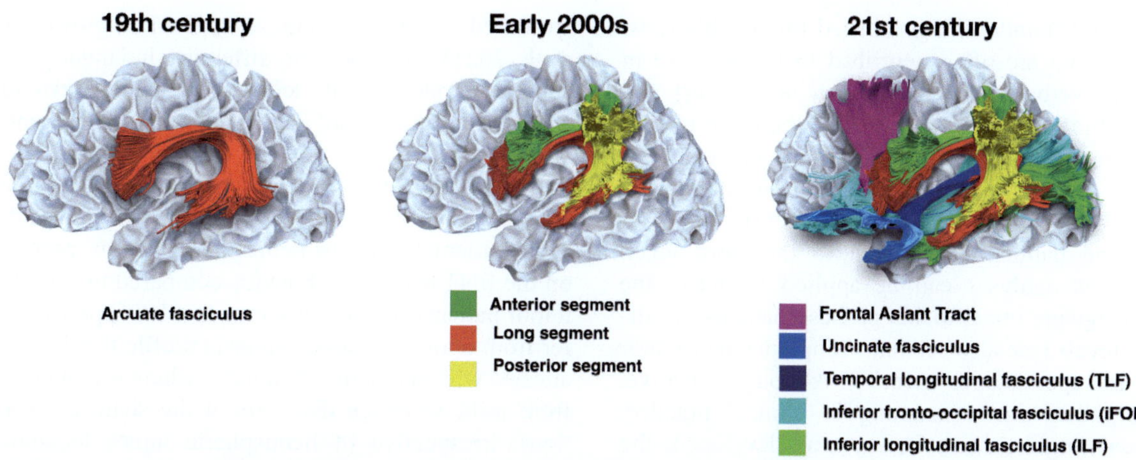

Fig. 5.5. Language network in transition. Since the discovery of the first language areas in the brain by the physician Paul Broca (1824–1880) and psychiatrist Carl Wernicke (1848–1905), knowledge about the neuronal language networks has changed considerably. Whereas in the 19th century, we assumed that there was only one fiber tract—the arcuate fasciculus, which connects the inferior frontal gyrus (i.e., "Broca" area) and the posterior temporal cortex (i.e., "Wernicke" area). In the early 2000s, there were already three connecting tracts that also rely on the inferior parietal lobe (i.e., "Geschwind's" territory, Catani et al., 2005). Today, we consider a minimum of eight pathways relevant for speech and language (three segments of the arcuate fasciculus, frontal aslant tract, uncinate fasciculus, inferior fronto-occipital fasciculus, inferior longitudinal fasciculus, and temporal longitudinal fasciculus), most of which have already been clinically validated through neurosurgical approaches, lesion studies, and DES. Modified from Catani M, Forkel SJ (2019). Diffusion imaging methods in language sciences. In: de Zubicaray GI, Schiller, NO (eds). Oxford Handbook of Neurolinguistics. Oxford University Press. S.F. is the author of the original figure.

relevance of mapping music for left- and right-hemispheric tumors. Another small patient study by the same research group found that musicality contributes to language recovery after awake surgery, underlining its relevance (Kappen et al., 2022b). They also found a trend toward a larger size of the corpus callosum in musical patients, which may have contributed to functional reorganization toward the contralateral side of the lesion. In combination with advances in tractography, a new understanding of anatomy of language has emerged over the past three decades (Fig. 5.5).

Apart from DES, other direct techniques for detecting functional boundaries are under development. A promising method is functional ultrasound (fUS) (Imbault et al., 2017; Deffieux et al., 2018; Soloukey et al., 2019). Functional ultrasound is a mobile neuroimaging technique with unprecedented spatiotemporal resolution. This technique allows the detection of small changes in blood dynamics reflecting changes in the metabolic activity of activated neurons through neurovascular coupling. Soloukey et al. (2019) showed the clinical potential of applying this technique during conventional awake surgery with motor, language, and visual tasks in 10 glioma patients. fUS could detect functional areas defined with DES and deep inside the brain. They were able to image (real-time) tumor-specific and healthy vascular characteristics (see Video 3 in Supplementary Materials of Soloukey et al. (2019) for an illustration of fUS during an awake surgery with a verbal and silent word repetition: https://www.frontiersin.org/articles/10.3389/fnins.2019.01384/full#supplementary-material).

LANGUAGE AND COGNITIVE PROFILES IN BRAIN TUMOR PATIENTS

Gliomas

Regardless of the origin of the exact neural disturbance, it is without doubt that glioma patients suffer from language and other cognitive impairments before receiving any type of treatment [for systematic reviews see: Satoer et al., 2016; van Kessel et al., 2017). Performance on tests for word retrieval was reported to be most frequently impaired before surgery (Satoer et al., 2016). However, many studies investigated a mixed group of LGG and HGG patients; language impairments in LGG patients seem less severe than those in HGG patients (Campanella et al., 2015; Noll et al., 2015). This difference is presumably due to the slow growth rate of an LGG (i.e., 5 mm p/year), which allows for neuroplasticity and reorganization of language functions (Mandonnet et al., 2003; Desmurget et al., 2007).

Most studies used "classical" aphasia tests in glioma patients that were developed for stroke patients, who typically suffer from moderate to severe language impairment, such as the Boston Diagnostic Aphasia Examination (BDAE) (Goodglass et al., 2001), the Aachener Aphasia Test (AAT), (Huber et al., 1984) or the Comprehensive Aphasia Test (Swinburn et al., 2005). These tests are not always sensitive to mild language deficits (Davie et al., 2009; Satoer et al., 2012; Brownsett et al., 2019). In addition, despite the often-absent objectified aphasia, LGG patients do report language complaints (Satoer et al., 2012; Brownsett et al., 2019;

Mooijman et al., 2022) underlining the need for suitable tests. Standardized tests that are often described to be sensitive in LGG patients are verbal fluency (category and letter) and (rapid) naming (objects and actions) and are therefore recommended in a minimal test battery to tap into the language domain [Papagno et al., 2012; see also Rofes et al. (2017) for a European survey on cognitive assessment]. However, verbal fluency and naming are limited to the word level. Spontaneous speech analyses can be applied to detect the generally mild language impairments in these patients on different linguistic levels (see also section "Supplementary motor area syndrome and spontaneous speech"). Although sensitive, this method is rather time-consuming in clinical practice. Therefore, a new linguistic test battery was developed, the Diagnostic Instrument for Mild Aphasia (DIMA; Satoer et al., 2022), with complex production and comprehension tests at the levels of phonology, semantics, and syntax derived from the Dutch Linguistic Intraoperative Protocol (DuLIP, De Witte et al., 2015). Most complex tasks and items were selected from DuLIP and new subtasks and items were added to DIMA. It consists of four subtests: (1) repetition (words, nonwords, and sentences), (2) semantic odd-picture-out (under time constraints), (3) sentence completion, and (4) sentence judgment. In the latter subtest, correct and incorrect sentences (at all linguistic levels) are randomly presented on a laptop screen, and both accuracy and reaction time in milliseconds are recorded (see, e.g., Table 5.1).

Case illustrations showed the sensitivity and clinical application of DIMA in a glioma and a meningioma patient before surgery. A preliminary analysis at the group level showed that preoperatively, patients deviated on DIMA sentence repetition and DIMA sentence completion. As expected, HGG patients performed worse than LGG patients, on word, nonword, and sentence repetition tasks, which is probably attributed to more aggressive tumor growth (Mooijman et al., 2021). There was no difference in language scores when comparing patients with left- to right-hemispheric tumor localization. As for standard language tests, LGG patients were preoperatively impaired on the Boston Naming Test (BNT), Category and Letter Fluency. In contrast, HGG patients performed worse than LGG patients on the BNT and the Token Test. Patients with left-hemispheric tumors performed worse on the BNT and Letter Fluency compared to right-hemispheric tumor patients. In sum, it seems that HGG patients tend to suffer from a more "classic" aphasia profile that is left-lateralized attested with standard instruments whereas LGG patients suffer from mild language disorders at the sentence level (production), irrespective of hemispheric tumor location (see also Davie et al., 2009; Brownsett et al., 2019), which is supported by other (neuroimaging) studies as well [see extensive review from Nieberlein et al., 2023]. In addition, as DIMA has subtests at all linguistic levels and was derived from DuLIP, it can be used to select DuLIP tasks for the intraoperative procedure [see Satoer et al., 2022 for case illustrations]. DIMA is currently digitized to an app format but also adapted and standardized to other languages (Clément et al., 2022), aiming for a more uniform international language assessment in glioma patients.

As for other cognitive functions, there is also no clear consensus about the selection of diagnostic instruments. van Loon et al. (2015) demonstrated in a systematic review that 46 different instruments were used for the evaluation of cognition in glioma patients. Similarly, as with language testing, nonlinguistic cognitive tests should also be sensitive to detect the more subtle deviations. Correa et al. (2007) suggested the use of a test battery with at least the Digit Span, Trail Making Test A/B, Brief Test of Attention, Hopkins Verbal Learning Test, the Grooved Pegboard test, and the Barona

Table 5.1

Diagnostic Instrument for mild aphasia.

Linguistic level and modality	Subtest	Example
Phonology—production	Repetition words Repetition compounds Repetition sentences	*Gorilla (gorilla)* *Ontdekkingsreiziger (explorer)* *De Griek ontdekte vier nietjes in de band van zijn fiets* *(The Greek discovered four staples in the tire of his bike)*
Phonology—comprehension	Sentence judgment	*De wikker gaat naar Parot* *(The wikker goes to Parot)*
Semantics—production	Semantic odd-picture-out	*Snake—dog—cat*
Semantics—comprehension	Sentence judgment	*De sigaar verveelde zich* *(The cigar was bored)*
Syntax—production	Sentence completion	*Hij viel van … (He fell off …)* *Om 5 uur … (At 5 o'clock …)*
Syntax—comprehension	Sentence judgment	*Hij gaat geschilderd op de muur* *(He goes painted on the wall)*

Index tapping into attention, executive functioning, motor function, verbal memory, and premorbid IQ estimation and quality of life. Papagno et al. (2012) suggested The Milano-Bicocca Battery (MIBIB) with tests for the domains of (language) memory, executive functions, apraxia, and spatial cognition. Other tests are to be selected according to tumor location. It should also be noted that the ideal testing time should consist of no more than 45 min (Klein and De Witt Hamer, 2011). Indeed, it appeared that LGG patients often deviate in the domains of verbal/working memory as assessed with a verbal learning test or digit span and attention/executive functions as assessed with Trail Making Test A, B, and B/A (Papagno et al., 2012; van Loon et al., 2015; Satoer et al., 2016). Sensitive cognitive screeners (via tablet/computer) with limited administration time duration have been proposed by several research groups (e.g., Rijnen et al., 2020; de Sain et al., 2023). van Kessel et al. (2021) indicated that preoperative memory deficits, which were revealed with their neuropsychologic examination, were of additional prognostic value in HGG to other well-known predictors of survival, underlining the relevance of including a memory test. Another study also demonstrated a significant predictive value of preoperative cognitive functioning with regard to functional independence in patients with glioma, providing evidence that cognitive functioning, assessed with neuropsychologic tests, can be translated into "real-world" functions and activities (Noll et al., 2018). Deficits in cognition are associated with gliomas in both the left and right hemisphere, although tasks with a verbal component are more frequently affected in patients with a glioma in the left hemisphere, in particular the left fronto-parietal network (Noll et al., 2015; Habets et al., 2019).

Meningiomas

As opposed to gliomas, less research has been done related to cognitive functioning in meningioma patients. It has only been for about the last decade that most nonlinguistic cognitive functions have been investigated in this patient group. As these tumors do not infiltrate eloquent brain areas, they were considered "harmless" to cognitive functions. However, a systematic review by Meskal et al. (2016) demonstrated that most meningioma patients suffer from one or more impaired cognitive functions in (working) memory, attention, and executive functioning before surgery. Recent studies have replicated these findings (van Nieuwenhuizen et al., 2018).

Previous studies on cognition in meningioma patients included a small number of language measures. Before surgery, performance on letter fluency (phonology) and category fluency (semantics) tests, which at least partly assess language processing (Whiteside et al., 2016), was reported to be impaired at the group and the individual level (Tucha et al., 2003; Bommakanti et al., 2016; Liouta et al., 2016; Hendrix et al., 2017). Some studies, in contrast, reported intact performance (van Nieuwenhuizen et al., 2013; Butts et al., 2017). Similarly, contradictory findings are reported for object naming tests (assessing word retrieval) and the Token Test [assessing language comprehension, presence and severity of aphasia; Campanella et al., 2015; Butts et al., 2017; Di Cristofori et al., 2018]. Word retrieval deficits were also revealed in a study including elderly patients (>70 years) with meningioma (Di Cristofori et al., 2018). However, other linguistic functions were not considered.

The first extensive neurolinguistic (longitudinal) study in a small number of patients with a meningioma in the language-dominant hemisphere (N = 10) was done by Wolthuis et al. (2021, 2022), who investigated the linguistic levels of phonology, semantics, and syntax. They revealed deficits in word retrieval, grammar, and writing compared to healthy controls. In a reading test, patients scored higher than the healthy population, as all but one patient made no errors on this test. At the individual level, four patients (40%) had a preoperative language impairment based on performance measures. No effects of preoperative tumor and treatment characteristics (localization, tumor volume, use of anti-epileptic drugs, and extent of resection) were found on the presence of a language impairment before surgery (nor on worsening of language performance). A limitation of this study was that they did not include right-sided meningiomas as a control group.

In addition to objective testing, van der Linden et al. (2020) showed the relevance of adding a subjective measure, studying patient-proxy agreement on executive functioning and the potential of performance-based measures on the level of agreement. It was remarkable that patients rated their executive functioning lower compared to what was objectively tested. They also rated themselves lower in comparison to their proxies (partner/caretaker). The subjective-objective relation and patient-proxy agreement should be investigated further in other domains of cognitive functioning. For language at the subjective level, Wolthuis (2022) administered structured interviews, and the study revealed that six meningioma patients (60%) self-reported one or more speech and/or language problems in daily life before surgery. They mentioned difficulties with word finding ($N = 6$), speaking (producing words, formulating sentences, expressing thoughts; $N = 5$), articulation ($N = 1$), reading ($N = 5$), and writing/typing ($N = 2$). No patient-proxy comparison was made. As meningiomas do not infiltrate into eloquent brain areas, the question arises as to why these patients suffer from cognitive and/or language deficits. One of the reasons could be a (long-term) disturbed brain activity pattern, such as abnormal slow-wave activity or disturbed functional networks. Increased slow-wave activity in low-grade glioma patients was associated with poorer working memory, information processing, and executive functioning (Bosma et al., 2009). Wolthuis et al. (2022) described that meningioma patients did not have preoperative abnormal slow-wave activity at the group level, as was seen in the study conducted by

Oshino et al. (2007). In their study, meningiomas were associated with less delta activity and less consistent delta source locations than intra-axial tumors (gliomas and metastatic tumors), while correcting for tumor size and surrounding edema. The difference between tumor types is presumably because meningiomas are not infiltrative but "only" cause a mass effect on surrounding brain tissue, especially subcortical tissue. The white matter pathways "only" tend to be displaced but are intact. Therefore, meningiomas probably cause minimal or no white-matter injury; hence, only minimal or no delta activity. However, a functional connectivity analysis showed that a greater small worldness (entire network configuration) was related to worse language performance (Wolthuis et al., 2021).

Supplementary motor area syndrome and spontaneous speech

The supplementary motor area (SMA) is a dorsomedial prefrontal region where gliomas often reside. Surgery in this area can result in the SMA syndrome, a well-known clinical phenomenon, which is associated with loss of initiation of motor or speech function (Laplane et al., 1977; Palmisciano et al., 2022). The SMA syndrome can occur after resection in either a left- or a right-sided glioma. However, the loss of speech initiation is more frequently observed after surgery in the left hemisphere (Palmisciano et al., 2022). Typically, the decline is at its worst in the acute phase, followed by spontaneous recovery within weeks to months after surgery. This "non-fluent" type of aphasia is classically known as transcortical aphasia. There are two forms of transcortical aphasia: the sensory form and the motor form. In sensory transcortical aphasia, language comprehension is impaired, production is fluent and contains paraphasias and/or neologisms, while the repetition of words is intact (unlike Wernicke's aphasia). In motor transcortical aphasia (spontaneous), language production is not fluent, often with intact language comprehension. Repetition of words and sentences is often intact even though a patient with transcortical motor aphasia can barely produce spontaneous spoken and written language. It is sometimes possible to elicit an appropriate target word via a carrier phrase (for example, "I cut bread with a … [knife]." The latter type is typically observed in lesions located in the supplementary motor area (SMA) (Naeser and Hayward, 1978; Krainik et al., 2003) and also known as "dynamic aphasia" (Luria and Tsvetkova, 1968). Dynamic aphasia resembles transcortical motor aphasia, apart from the patient not initiating (Ardila and Lopez, 1984). Luria described this pattern as a failure of an early stage of verbal planning, translating the "internal speech" into the linear scheme of the sentence [see also Costello and Warrington, 1989]. An alternative explanation viewed this syndrome as an inability to select a unique response in "open" situations, in which one out of many possible verbal responses or competitors must be selected (Robinson et al., 1998). The underlying mechanism of dynamic aphasia has been associated with a language-specific impairment or a more general impairment of executive functions (Esmonde et al., 1996; Robinson et al., 2006). Apart from the "classical" left-hemispheric involvement in language production, there are also studies in which spontaneous speech was affected after right-hemispheric damage (Thomson et al., 1998; Jehna et al., 2017). Intraoperative results reporting on (semi-)spontaneous speech production deficits during DES or resection have shed light on the location of this dynamic function in addition to the SMA. Dissociations were found between sentence production/generation, sentence completion, and spontaneous speech vs other isolated language tasks (Chernoff et al., 2019; Chivukula et al., 2018; Dragoy et al., 2020; Ojemann and Mateer, 1979; De Witte et al., 2015; Satoer et al., 2014a, 2018, 2021).

The frequent occurrence of this transient syndrome underlines the relevance of monitoring spontaneous speech (in context), as this can be missed when only applying isolated language tasks in the perioperative phase (e.g., object naming). In addition, the added value of a spontaneous speech analysis has been demonstrated by Satoer et al. (2013, 2018) in addition to verbal fluency and object naming. One of the most sensitive parameters in their perioperative spontaneous speech was the occurrence of incomplete sentences, which deteriorated in the long run after surgery. A subsequent study by Rofes et al. (2018) partially confirmed these findings. It argued that a spontaneous speech analysis can be of added value when there is a lack of time to administer a comprehensive neurolinguistic protocol or when the patient is too tired. To the best of our knowledge, no thorough spontaneous speech analyses have been conducted in meningioma patients, apart from one case study that described postoperative dynamic aphasia in a patient with a left frontal meningioma involving Brodmann area 45 followed by full recovery within 2 weeks (Robinson et al., 1998). Finally, spontaneous speech is the most ecologically valid way of daily conversation, hence, crucial for quality of life and should therefore be adopted as a standard measure in clinical care for LGGs.

ANATOMIC AND FUNCTIONAL ALTERATIONS

White matter dysfunctions and disconnections

Deprivation of connections in a brain region triggers dendritic and synaptic pruning, leading to the degeneration or death of neurons (Capurso et al., 1997). Consequently, both functional disconnection, characterized by impaired communication (Stephan et al., 2006), and structural disconnection occur within the affected region. This phenomenon, known as diaschisis (Carrera and Tononi, 2014), highlights the critical role of connections in maintaining the integrity and functionality of distant brain regions.

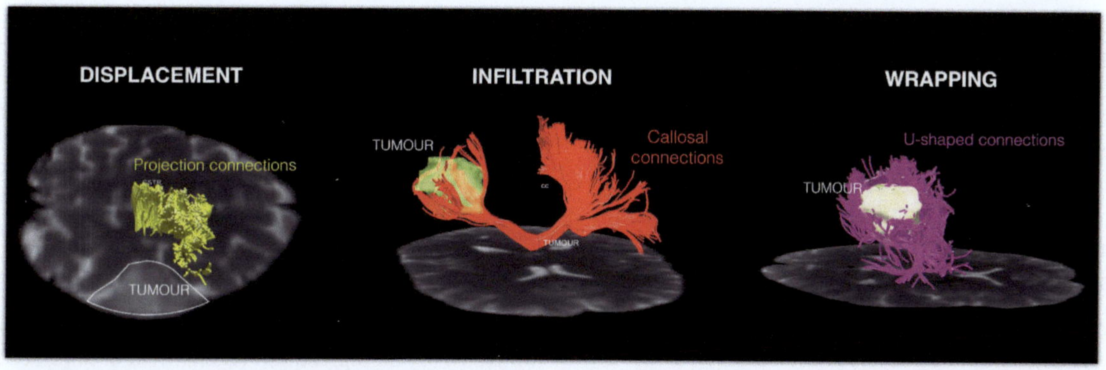

Fig. 5.6. Various impacts of tumor lesions on white matter. In the *left panel*, low-grade gliomas (LGGs) are depicted as gradually exerting pressure on the surrounding brain tissue, including the white matter, leading to displacement. The *middle panel* represents tumors infiltrating the white matter and migrating along the connective tissue. Finally, in the *right panel*, a phenomenon is observed where the white matter appears to wrap around the tumor, which could either be indicative of a pathologic process or a methodological artifact. This figure is previously unpublished data and was modified with permission from the creator, Dr. Henrietta Howells.

In the case of brain injuries caused by masses like gliomas, the mass can exert pressure on or displace neighboring white matter fibers, causing deviations from the expected anatomy in terms of shape and location. For example, a lateral frontal lobe tumor can compress and displace a corticospinal tract (CST) segment toward the medial wall. However, this alteration only affects the immediate vicinity of the tumor, resulting in a change in the overall trajectory of that specific portion of the CST (Fig. 5.6). It is important to note that these deviations are unique to each patient, and tractography is necessary to visualize the precise shape of the CST in individual cases. Additionally, tumors can indirectly impact white matter through their pathologic cascade, such as edema around a tumor, which temporarily compromises the white matter and may cause transient clinical symptoms. For example, a tumor in the motor cortex surrounded by edema can affect the arcuate fasciculus, leading to language deficits resembling Broca-like aphasia (i.e., effortful articulation, telegraphic speech).

Tumors can impact brain regions and their connections through various mechanisms. Minor displacement of white matter typically has minimal or no significant effect on structural or functional integrity (Fig. 5.6, *left*). However, if the strain on the white matter becomes too severe, it can cause structural damage and disconnection. Quantifying the extent of disconnection relies on analyzing the white matter of the healthy hemisphere, as atlas-based approaches fail to account for interindividual variability. Tractography emerges as a valuable tool in personalized diagnostics, enabling the assessment of a lesion's impact on white matter, aiding in presurgical planning, and determining postsurgical consequences (Thiebaut de Schotten et al., 2005; Duffau et al., 2008; Leclercq et al., 2010; Kemerdere et al., 2016; Cochereau et al., 2020; Dragoy et al., 2020; Mirchandani et al., 2020; Sollmann et al., 2020).

Functional connectivity alterations due to tumor lesions

Intracranial tumors can also lead to disruptions of functional connectivity, that is the extent to which brain areas can functionally interact, which can in turn cause brain dysfunctions. Increased slow-wave activity in adults is considered pathologic and may also indicate brain dysfunction in wakeful restful conditions. In brain tumor patients, increased slow-wave activity and deviant functional connectivity are observed compared to healthy participants (Baayen et al., 2003; Bartolomei et al., 2006a,b).

Characteristics of slow-wave activity and functional connectivity network may also be a language or cognitive dysfunction marker. In stroke patient groups, brain activity in the delta and theta frequency band (i.e., slow-wave activity) and functional connectivity networks were predictive of language outcome (Szelies et al., 2002; Nicolo et al., 2015). Bosma et al. (2009) demonstrated a relationship between functional connectivity networks and cognitive impairments in brain tumor patients. Wolthuis et al. (2022) also showed with EEG that LGG patients had increased slow-wave activity compared to healthy individuals before (and 1 year after surgery). Increased preoperative slow-wave activity was associated with language impairment and poor language performance in specific language domains. Slow-wave activity before surgery was also related to word retrieval after surgery.

As for functional connectivity, Wolthuis et al. (2022) did not observe differences between LGG patients and healthy participants, although hub presence (i.e., nodes with many connections and a central position in a network) and higher focal and global functional connectivity were associated with poorer language functioning (in particular with production and comprehension at the sentence level). For meningioma patients, a greater small worldness (entire network configuration) was related to worse preoperative language performance (word retrieval, semantics, and writing) and hub presence. Preoperative better average clustering (local connectivity) and global integration predicted worse outcomes on language function 1 year after surgery. However, these results must be interpreted cautiously as small numbers of patients were included, and patients with and without language impairments collapsed. Investigation into FC in specific brain areas (in larger, specified groups) may provide further insights into language impairments in brain tumor patients.

CORTICAL AND SUBCORTICAL AREAS: FUNCTIONS AND TASKS

Gliomas (awake surgery)

For a long time, only counting and/or naming pictures were used as intraoperative tasks during awake glioma surgery. However, counting is automatized language and confrontational object naming does not capture complex linguistic functions; hence, they are insufficiently sensitive to investigate critical language functions (Petrovich Brennan et al., 2007; De Witte and Mariën, 2013). Although object naming does not tap into complex linguistic functions, the relevance of this test was underlined by several studies; preoperative (subnormal) performance on an object naming task appeared to be predictive for postoperative language outcome (Ilmberger et al., 2008) and object naming appeared to be involved in large cortico-subcortical areas important for preservation (Papagno et al., 2011; Collée et al., 2023). However, by solely applying an object naming test, other linguistic functions can be missed, such as auditory language comprehension (De Witte et al., 2015).

Hence, a decade ago, for the first time an extensive standardized language test battery was developed in Dutch (DuLIP), consisting of phonologic, semantic, and syntactic language tasks [De Witte et al., 2015 and see Collée et al., 2023 for an update of DuLIP]. Ever since, adaptations of DuLIP to other languages has been applied (e.g., in Portuguese, Alves et al., 2021) and other linguistic tasks or test batteries have been developed [see Martín-Monzón et al., 2022 for a review]. There is no uniform consensus about the application of specific language/cognitive tasks in neurooncologic patients. However, based on the literature and clinical studies, several schemas and proposals have been made (Fernández Coello et al., 2013; De Witte et al., 2015; Collée et al., 2023) (see a selection later in Tables 5.2 and 5.3).

A thorough systematic review on intraoperative errors in awake surgery with 102 studies made an attempt to sketch an overview of eloquent stimulation sites. The most frequently reported errors concerned: speech arrest, anomia, dysarthria/anarthria, and semantic and phonologic errors in both the left and right hemisphere [for a review see Collée et al., 2023].

Meningiomas

As opposed to gliomas, in meningioma patients it is not so clear yet whether there is a relation between tumor location and cognitive performance. However, due to greater diagnostic imaging, an increasing number of meningiomas are found, albeit (more or less) asymptomatic. Clinical (and research-related) radiologic follow-up allows us to gain more knowledge about the effects of a meningioma and its location on cognition.

Mixed findings of localization were found on cognition ranging from: (1) no signs at all (Steinvorth et al., 2003; Tucha et al., 2003; Meskal et al., 2016), (2) better performance in patients with right-sided meningiomas in (verbal) cognitive tests (Yoshii et al., 2008; Dijkstra et al., 2009; Sleurs et al., 2022) than left-sided meningioma, and (3) more positive findings in patients with falx meningioma in comparison to convexity and fronto-basal location (Tucha et al., 2003). A recent large lesion-symptom mapping study showed that patients with tumor locations in the left middle and superior frontal gyrus were more at risk for lower performance on cognitive flexibility and complex attention, respectively (De Baene et al., 2019). As for language, a nonsignificant observation showed that the majority of patients with an anterior (frontal) meningioma did not have a language impairment. In contrast, the majority of patients with a posterior (parietal/parieto-occipital) meningioma did have a language impairment before surgery. Cipolotti et al. (2015) compared deviating cognitive scores in low-grade glioma and meningioma patients with frontal lesion location. In a 3-year follow-up study, Liouta et al. (2016) found that patients with anterior convexity meningiomas performed worse in executive functioning tests than patients with posterior tumor location. Left-hemispheric meningioma patients performed worse in verbal tests than right-hemispheric patients. In sum, due to this great variety, it seems that even in non-infiltrating tumors, functional neuroplasticity takes place.

EFFECT OF SURGERY ON LANGUAGE/ COGNITIVE FUNCTIONS

Glioma (awake procedure)

The goal of awake surgery is to preserve neurologic outcomes and obtain a larger extent of resection. Risk factors for postoperative language impairment are preoperative aphasia, non-frontal tumor location, intraoperative complications, and language-positive stimulation sites within the tumor (Ilmberger et al., 2008). Over the last two decades, the research field tapping into the effects of awake surgery on language mapping cognitive functions intraoperatively has grown exponentially.

Previously, it was believed that language and cognitive abilities declined immediately after surgery but fully recovered within 3 months (e.g., Duffau et al., 2002; Whittle et al., 2003;

Table 5.2

Overview of cortical brain regions with sensorimotor, language/cognitive functions and tasks.

Cortical level—lobe	Function	Task (example)	Hemisphere
Frontal lobe			
Precentral gyrus	Motor	Movement upper limb	L+R
Inferior frontal gyrus	Speech articulation	Counting, repetition, verbal diadochokinesis	L+R
	Word-finding	Picture naming	L
	Syntax	Action naming, verb generation	L
	Semantics	Semantic association and judgment, odd-picture/word-out	L
(+ middle frontal frontal gyrus)	Writing	Writing	L
Prefrontal areas	Executive functions, inhibition	Stroop	L+R
	Working memory	Double task	L+R
Supplementary Motor Area	Language initiation	Sentence completion, spontaneous speech	L
Parietal lobe			
Postcentral gyrus	Sensory-motor, articulatory processing, motor speech	Movements, counting, verbal diadochokinesis, repetition	L+R
	Articulation/motor processing	Verbal diadochokinesis	
Supramarginal gyrus	Word-finding	Picture naming	L
	Reading	Reading, odd-word-out	L
	Semantics	Odd-picture/word-out, semantic association, PPTT	L
	Working memory	Double task	L+R
	Visuo-spatial	Line bisection	R
	Set-shifting, inhibition	Trail Making Test	R
Angular gyrus	Arithmetics	Calculation (see also Gerstmann syndrome, e.g., Ardila, 2020))	L
	Finger recognition	Finger naming (see also Gerstman syndrome, e.g., Ardila, 2020)	L
	Reading	Reading, odd-word-out	L
	Writing	Writing	L
Superior parietal gyrus	Writing	Writing	L
Insular lobe			
	Word-finding	Picture naming	L
	Sensory-motor	Movements	L+R
	Working memory	Double task	L+R
Temporal lobe			
Superior temporal gyrus	Word-finding	Picture naming	L
	Phonologic network	Phonologic judgment	L
Inferior temporal gyrus	Semantics (verbal)	Semantic association and judgment, odd-picture-out	L
	Semantic (nonverbal)	Semantic association and judgment, odd-picture-out	R
Middle temporal gyrus	Face naming/recognition	Famous face naming	L
	Reading	Reading, odd-word-out	L
Posterior temporal areas	Visual field/recognition	Visual tasks	L+R
	Reading	Reading	L
Occipital lobe			
	Visual field/recognition	Visual tasks	L+R

Table 5.3

Overview of subcortical brain regions with sensorimotor, language/cognitive functions and tasks.

Subcortical level—tracts	Function	Task	Hemisphere
Frontal Aslant Tract	Initiation of speech, motor speech	Verbal fluency, sentence completion, spontaneous speech	L+R
Corticospinal tract	Motor speech	Verbal diadochokinesis, word repetition	L+R
Arcuate fasciculus	Phonology	(Non) word repetition	L
Superior longitudinal fasciculus	Visuospatial	Line bisection	R
Inferior frontal occipital fasciculus	Word retrieval Semantics, reading	Naming	L
		Semantic judgment and association	L+R
	Visuospatial	Line bisection	R
Inferior Longitudinal Fasciculus	Reading, phonology, semantics	Reading, (non) word repetition, semantic judgment, association, object naming	L
Uncinate fasciculus	Word/proper name retrieval, semantics, phonology, face naming/recognition	Object naming/proper noun naming, semantic judgment, famous face naming, (non) word repetition	L

Sanai et al., 2008). Further comprehensive neuropsychologic examinations revealed that while there is a temporary decline in language functions shortly after surgery (between 3 and 6 months), language performance tends to improve over time. Nevertheless, certain tests, such as category fluency, showed impairment even 1 year after the surgery (Satoer et al., 2014b, 2016). Additionally, Satoer et al. (2013, 2018) found that characteristics of spontaneous speech, particularly the occurrence of incomplete sentences, worsened in the year following the surgery (see also Rofes et al., 2018). A preliminary study from the same group showed that subtests from DIMA even appeared more sensitive to detecting surgical effects than standard tests: all phonologic DIMA subtests captured short-term decline (pre- and 3 months postoperatively), in line with earlier evidence for the value of (non)word repetition (Sierpowska et al., 2017; Mooijman et al., 2021). A long-term decline was detected with DIMA sentence completion in the year after surgery, reflecting an earlier spontaneous speech analysis with incomplete sentences as a sensitive parameter. The short-term postoperative decline affected verbal fluency as well. Only the localization of left-hemispheric tumors had an impact on performance in standard language tests. HGG patients showed more significant impairments compared to LGG patients in DIMA repetition and standard tests such as the BNT and the Token Test. Therefore, it is recommended to include DIMA in the standard language evaluation of glioma patients to enable more precise counseling regarding language outcomes.

As for higher cognitive functions, Rijnen et al. (2019) investigated a large group of LGG patients, who underwent surgery (with or without DES). There was no hemispheric effect on performance compared to normative data, and cognitive impairments were identified in both groups. Verbal memory and sustained attention were more prone to postoperative deterioration in patients with LGG in the left hemisphere. Patients with DES surgery scored worse than patients without DES on tests for executive functions and verbal fluency (over time). The authors conclude that the addition of tests for higher cognitive functions to the intraoperative procedure may enhance cognitive outcome. Noll et al. (2015) showed that in patients with tumors in the left temporal lobe, a decline was most frequent in verbal learning, memory, executive functioning, and object naming. Postoperative worsening was relatively less often observed in patients with right-sided lesions, but still present on measures of processing speed, executive functioning, and verbal learning and memory. It is notable that both left- and right-hemispheric patients exhibited a postoperative decline in verbal learning and memory, yet more pronounced in the left-hemispheric patient group.

Although awake surgery is the gold standard treatment in LGGs, it is not widely accepted to be a standard treatment for HGGs, in particular glioblastoma (GBM) patients. The main reason is that this patient group suffers from more severe neurologic and cognitive deficits (Noll et al., 2015; van Kessel et al., 2017, 2021). Furthermore, severe aphasia is regarded as an exclusion criterion (Dziedzic and Bernstein, 2014; Bonifazi et al., 2020; Zigiotto et al., 2020) as it challenges intraoperative monitoring and defining a functional boundary. However, there seems to be a trend toward a higher percentage of total resection, a lower complication rate, and a longer median survival in awake versus surgery with general anesthesia (Gerritsen et al., 2019, 2022; Li et al., 2021). A recent case series showed that resection of glioblastoma patients with severe aphasia undergoing awake surgery was feasible without perioperative complications. By using a patient-tailored peri- and intraoperative language monitoring, intraoperative language deterioration was distinguished from preoperative aphasia. Postoperatively, language measures were stable or even improved (Donders-Kamphuis et al., 2022). A prospective multicenter randomized clinical trial [SAFE-trial NL7589 Gerritsen et al., 2020] is ongoing to assess the added value of awake surgery in glioblastoma patients. High expertise of a multidisciplinary awake team is mandatory including a specialized speech and language pathologist/clinical linguist. More standardized intraoperative tasks for severe aphasia are needed for this population. An extended version of DuLIP (De Witte et al., 2015) with standardized tasks designed especially for patients with severe aphasia is in preparation.

Meningioma (asleep procedure)

A systematic review by Meskal et al. (2016) reported that meningioma patients tend to improve after surgery, but impairments are still present in 44% of the patients. A large observational study by Rijnen et al. (2019) investigated meningioma (WHO grade I and II) patients preoperatively and at 3 and 12 months after surgery with a neuropsychologic assessment. Seven cognitive domains were tested: Verbal Memory, Visual Memory, Processing Speed, Psychomotor Speed, Reaction Time, Complex Attention, and Cognitive Flexibility. At the group level, impairments were found in all cognitive domains across all time points. On the individual patient level, impairments were significantly more common in meningioma patients compared with normative controls on six out of seven domains at 3 months after surgery and on three out of seven domains 1 year after surgery. Performance on Psychomotor Speed, Reaction Time, and Complex Attention was most frequently, as well as most severely, impaired. The authors postulated that patients in their study may have difficulties with the agility and adequacy of movements (related to Psychomotor Speed), may respond slowly to stimuli (related to Reaction Time), and may struggle to adapt behaviors and thoughts to new, changing, or unexpected events (related to Complex Attentions, but also to higher order executive functions) potentially harming daily activities, such as driving a car or cooking dinner. They claim that they underline the necessity to implement routine cognitive screening in meningioma patients to adequately monitor cognitive deficits in order to maximize quality of life and functional independence in home, work, and social settings. According to another study, no long-term decline was observed at the group level in executive functioning tests; instead, some of the tests showed improvement over time (Liouta et al., 2016).

Studies that included language tests in their protocol, such as object naming, verbal fluency or the Token Test found generally no effect of surgery on patients' performance (Tucha et al., 2003; Campanella et al., 2015; Bommakanti et al., 2016; Hendrix et al., 2017). It is to be debated whether these tests are sensitive enough to detect more subtle (as in LGGs) language deficits. Wolthuis et al. (2022) reported both postoperative improvement in the domains of word retrieval and syntax, whereas writing remained impaired compared to healthy population, which suggest that further research with larger groups must be conducted.

CONCLUSION

This chapter discussed the incidence, prevalence and prognosis of gliomas and meningiomas, considering their impact on language and cognitive functions while discussing the available techniques to map language in brain tumor patients. It is without doubt that these tumors can have a profound impact on language and cognitive processes, which highlights the significance for early detection and treatment to preserve patients' quality of life. The presence and severity of functional impairment can be influenced by several factors such as tumor type (e.g., intra- or axial location, slow or fast tumor growth) and lesion location (e.g., hemisphere, lobe, and white matter connections), which in turn may affect the potential of neuroplasticity. Up to now, evidence shows that patients with left-hemispheric tumors still run the highest risk for neurocognitive impairments with emphasis on language/verbal components. The added value of more sensitive (and nonlinguistic) measurements should be investigated further in patients with right-sided tumors.

The emergence of noninvasive neuroimaging techniques, such as fMRI and tractography, and invasive awake brain surgery have revolutionized our ability to map and monitor brain tumors. These methods provided valuable insights in understanding the neurobiology of language and revealed high interindividual variability. In the last decade, a patient-tailored approach for language and cognitive tests considers demographic and clinical variables and is inevitable in neuro-oncologic care.

Despite significant progress, challenges remain. Longitudinal studies with large patient cohorts are necessary to explore the temporal and dynamic evolution of linguistic and cognitive changes caused by a brain tumor. Addressing this intricate labyrinth of factors contributing to language and cognitive impairments in brain tumor patients requires a multidisciplinary and open approach. Neurosurgeons, neurologists, speech language pathologists, neuropsychologists, neuroimaging specialists, and industry partners must work together to facilitate the mapping of cognitive functions in the surgical setting to improve patients' outcomes and quality of life.

References

Alves J, Cardoso M, Morgado M et al. (2021). Language assessment in awake brain surgery: the Portuguese adaptation of the Dutch linguistic intraoperative protocol (DuLIP). Clin Linguist Phon 35: 1113–1131.

Ardila A (2020). Gerstmann syndrome. Curr Neurol Neurosci Rep 20: 48.

Ardila A, Lopez MV (1984). Transcortical motor aphasia: one or two aphasias? Brain Lang 22: 350–353.

Arrieta VA, Gould A, Kim KS et al. (2024). Ultrasound-mediated delivery of doxorubicin to the brain results in immune modulation and improved responses to PD-1 blockade in gliomas. Nat Commun 15: 4698.

Baayen JC, de Jongh A, Stam CJ et al. (2003). Localization of slow wave activity in patients with tumor-associated epilepsy. Brain Topogr 16: 85–93.

Bartolomei F, Bosma I, Klein M et al. (2006a). Disturbed functional connectivity in brain tumor patients: evaluation by graph analysis of synchronization matrices. Clin Neurophysiol 117: 2039–2049.

Bartolomei F, Bosma I, Klein M et al. (2006b). How do brain tumors alter functional connectivity? A magnetoencephalography study. Ann Neurol 59: 128–138.

Bartos R, Jech R, Vymazal J et al. (2009). Validity of primary motor area localization with fMRI versus electric cortical stimulation: a comparative study. Acta Neurochir 151: 1071–1080.

Berger MS, Ojemann GA (1992). Intraoperative brain mapping techniques in neuro-oncology. Stereotact Funct Neurosurg 58: 153–161.

Beyh A, Ohlerth A-k, Forkel S (2024). Harnessing advanced tractography in neurosurgical practice. In: K Picht (Ed.), Navigated transcranial magnetic stimulation in neurosurgery. https://osf.io/preprints/psyarxiv/emhrn.

Bommakanti K, Somayajula S, Suvarna A et al. (2016). Preoperative and post-operative cognitive deficits in patients with supratentorial meningiomas. Clin Neurol Neurosurg 143: 150–158.

Bonifazi S, Passamonti C, Vecchioni S et al. (2020). Cognitive and linguistic outcomes after awake craniotomy in patients with high-grade gliomas. Clin Neurol Neurosurg 198: 106089.

Bosma I, Reijneveld JC, Klein M et al. (2009). Disturbed functional brain networks and neurocognitive function in low-grade glioma patients: a graph theoretical analysis of resting-state MEG. Nonlinear Biomed Phys 3: 9.

Brownsett SLE, Ramajoo K, Copland D et al. (2019). Language deficits following dominant hemisphere tumor resection are significantly underestimated by syndrome-based aphasia assessments. Aphasiology 33: 1163–1181.

Butts AM, Weigand S, Brown PD et al. (2017). Neurocognition in individuals with incidentally-identified meningioma. J Neurooncol 134: 125–132.

Campanella F, Fabbro F, Ius T et al. (2015). Acute effects of surgery on emotion and personality of brain tumor patients: surgery impact, histological aspects, and recovery. Neuro Oncol 17: 1121–1131.

Capurso SA, Calhoun ME, Sukhov RR et al. (1997). Deafferentation causes apoptosis in cortical sensory neurons in the adult rat. J Neurosci 17: 7372–7384.

Carrano A, Juarez JJ, Incontri D et al. (2021). Sex-specific differences in glioblastoma. Cells 10.

Carrera E, Tononi G (2014). Diaschisis: past, present, future. Brain 137: 2408–2422.

Catani M, Jones DK, Ffytche DH (2005). Perisylvian language networks of the human brain. Ann Neurol 57: 8–16.

Chang EF, Raygor KP, Berger MS (2015). Contemporary model of language organization: an overview for neurosurgeons. J Neurosurg 122: 250–261.

Chang WH, Pei YC, Wei KC et al. (2018). Intraoperative linguistic performance during awake brain surgery predicts postoperative linguistic deficits. J Neurooncol 139: 215–223.

Chernoff BL, Sims MH, Smith SO et al. (2019). Direct electrical stimulation of the left frontal aslant tract disrupts sentence planning without affecting articulation. Cogn Neuropsychol 36: 178–192.

Chivukula S, Pikul BK, Black KL et al. (2018). Contralateral functional reorganization of the speech supplementary motor area following neurosurgical tumor resection. Brain Lang 183: 41–46.

Cho NS, Peck KK, Zhang Z et al. (2018). Paradoxical activation in the cerebellum during language fMRI in patients with brain tumors: possible explanations based on neurovascular uncoupling and functional reorganization. Cerebellum 17: 286–293.

Cipolotti L, Healy C, Chan E et al. (2015). The impact of different aetiologies on the cognitive performance of frontal patients. Neuropsychologia 68: 21–30.

Clément A, Perez A, Mandonnet E et al. (2022). DIMA-fr: a French adaptation and standardization of the Dutch diagnostic instrument for mild aphasia (DIMA-nl). Clin Linguist Phon 36: 954–967.

Cochereau J, Lemaitre AL, Wager M et al. (2020). Network-behavior mapping of lasting executive impairments after low-grade glioma surgery. Brain Struct Funct 225: 2415–2429.

Cohen AL, Colman H (2015). Glioma biology and molecular markers. Cancer Treat Res 163: 15–30.

Collée E, Vincent A, Dirven C et al. (2022). Speech and language errors during awake brain surgery and postoperative language outcome in glioma patients: a systematic review. Cancers (Basel) 14.

Collée E, Vincent A, Visch-Brink E et al. (2023). Localization patterns of speech and language errors during awake brain surgery: a systematic review. Neurosurg Rev 46: 38.

Coluccia D, Roth T, Marbacher S et al. (2018). Impact of laterality on surgical outcome of glioblastoma patients: a retrospective single-center study. World Neurosurg 114: e121–e128.

Correa DD, Maron L, Harder H et al. (2007). Cognitive functions in primary central nervous system lymphoma: literature review and assessment guidelines. Ann Oncol 18: 1145–1151.

Costello AL, Warrington EK (1989). Dynamic aphasia: the selective impairment of verbal planning. Cortex 25: 103–114.

Cote DJ, Wang R, Morimoto LM et al. (2022). Birth characteristics and risk of meningioma in a population-based study in California. Neurooncol Adv 4: vdac173.

Davie GL, Hutcheson KA, Barringer DA et al. (2009). Aphasia in patients after brain tumor resection. Aphasiology 23: 1196–1206.

De Baene W, Rijnen SJM, Gehring K et al. (2019). Lesion symptom mapping at the regional level in patients with a meningioma. Neuropsychology 33: 103–110.

de Sain AM, Mantione MHM, Wajer I et al. (2023). A timeline of cognitive functioning in glioma patients who undergo awake brain tumor surgery. Acta Neurochir 165: 1645–1653.

De Witt Hamer PC, Robles SG, Zwinderman AH et al. (2012). Impact of intraoperative stimulation brain mapping on glioma surgery outcome: a meta-analysis. J Clin Oncol 30: 2559–2565.

De Witte E, Mariën P (2013). The neurolinguistic approach to awake surgery reviewed. Clin Neurol Neurosurg 115: 127–145.

De Witte E, Satoer D, Robert E et al. (2015). The Dutch linguistic intraoperative protocol: a valid linguistic approach to awake brain surgery. Brain Lang 140: 35–48.

Deffieux T, Demene C, Pernot M et al. (2018). Functional ultrasound neuroimaging: a review of the preclinical and clinical state of the art. Curr Opin Neurobiol 50: 128–135.

Dell'Acqua F, Tournier JD (2019). Modelling white matter with spherical deconvolution: how and why? NMR Biomed 32: e3945.

Dell'Acqua F, Simmons A, Williams SC et al. (2013). Can spherical deconvolution provide more information than fiber orientations? Hindrance modulated orientational anisotropy, a true-tract specific index to characterize white matter diffusion. Hum Brain Mapp 34: 2464–2483.

Desmurget M, Sirigu A (2015). Revealing humans' sensorimotor functions with electrical cortical stimulation. Philos Trans R Soc Lond B Biol Sci 370: 20140207.

Desmurget M, Bonnetblanc F, Duffau H (2007). Contrasting acute and slow-growing lesions: a new door to brain plasticity. Brain 130: 898–914.

Di Cristofori A, Zarino B, Bertani G et al. (2018). Surgery in elderly patients with intracranial meningioma: neuropsychological functioning during a long term follow-up. J Neurooncol 137: 611–619.

Dijkstra M, van Nieuwenhuizen D, Stalpers LJ et al. (2009). Late neurocognitive sequelae in patients with WHO grade I meningioma. J Neurol Neurosurg Psychiatry 80: 910–915.

Donders-Kamphuis M, Vincent A, Schouten J et al. (2022). Feasibility of awake brain surgery in glioblastoma patients with severe aphasia: five case illustrations. Aphasiology 37: 1944–1963.

Dragoy O, Zyryanov A, Bronov O et al. (2020). Functional linguistic specificity of the left frontal aslant tract for spontaneous speech fluency: evidence from intraoperative language mapping. Brain Lang 208: 104836.

Duffau H (2014). The huge plastic potential of adult brain and the role of connectomics: new insights provided by serial mappings in glioma surgery. Cortex 58: 325–337.

Duffau H, Capelle L, Sichez N et al. (2002). Intraoperative mapping of the subcortical language pathways using direct stimulations. An anatomo-functional study. Brain 125: 199–214.

Duffau H, Gatignol P, Mandonnet E et al. (2005). New insights into the anatomo-functional connectivity of the semantic system: a study using cortico-subcortical electrostimulations. Brain 128: 797–810.

Duffau H, Thiebaut de Schotten M, Mandonnet E (2008). White matter functional connectivity as an additional landmark for dominant temporal lobectomy. J Neurol Neurosurg Psychiatry 79: 492–495.

Dulyan L, Forkel S (2024). The complexity of anatomical language: unravelling terminological discrepancies in the neuroanatomy of language. In: J Grafmann (Ed.), Encyclopedia of the human brain. Elsevier.

Dulyan, L., Bortolami, C. & Forkel, S.J. (2025) Asymmetries in the human brain. In: C Papagno, P Corballis (Eds.), Handbook of clinical neurology, pp. 15–36. Elsevier

Durfee AZ, Sheppard SM, Blake ML et al. (2021). Lesion loci of impaired affective prosody: a systematic review of evidence from stroke. Brain Cogn 152: 105759.

Dziedzic T, Bernstein M (2014). Awake craniotomy for brain tumor: indications, technique and benefits. Expert Rev Neurother 14: 1405–1415.

Dziedzic TA, Balasa A, Jeżewski MP et al. (2021). White matter dissection with the Klingler technique: a literature review. Brain Struct Funct 226: 13–47.

Esmonde T, Giles E, Xuereb J et al. (1996). Progressive supranuclear palsy presenting with dynamic aphasia. J Neurol Neurosurg Psychiatry 60: 403–410.

Fernández Coello A, Moritz-Gasser S, Martino J et al. (2013). Selection of intraoperative tasks for awake mapping based on relationships between tumor location and functional networks. J Neurosurg 119: 1380–1394.

Forkel SJ, Catani M (2018). Lesion mapping in acute stroke aphasia and its implications for recovery. Neuropsychologia 115: 88–100.

Forkel SJ, Labache L, Nachev P et al. (2022). Stroke disconnectome decodes reading networks. Brain Struct Funct 227: 2897–2908.

Gerritsen JKW, Viëtor CL, Rizopoulos D et al. (2019). Awake craniotomy versus craniotomy under general anesthesia without surgery adjuncts for supratentorial glioblastoma in eloquent areas: a retrospective matched case-control study. Acta Neurochir 161: 307–315.

Gerritsen JKW, Klimek M, Dirven CMF et al. (2020). The SAFE-trial: safe surgery for glioblastoma multiforme: awake craniotomy versus surgery under general anesthesia. Study protocol for a multicenter prospective randomized controlled trial. Contemp Clin Trials 88: 105876.

Gerritsen JKW, Zwarthoed RH, Kilgallon JL et al. (2022). Effect of awake craniotomy in glioblastoma in eloquent areas (GLIOMAP): a propensity score-matched analysis of an international, multicentre, cohort study. Lancet Oncol 23: 802–817.

Giampiccolo D, Duffau H (2022). Controversy over the temporal cortical terminations of the left arcuate fasciculus: a reappraisal. Brain 145: 1242–1256.

Giussani C, Roux FE, Ojemann J et al. (2010). Is preoperative functional magnetic resonance imaging reliable for language areas mapping in brain tumor surgery? Review of language functional magnetic resonance imaging and direct cortical stimulation correlation studies. Neurosurgery 66: 113–120.

Goodglass H, Kaplan E, Barresi B (2001). Boston diagnostic aphasia examination, third edn. Lippincott Williams & Wilkins, Philadelphia, PA.

Gould J (2018). Breaking down the epidemiology of brain cancer. Nature 561: S40–S41.

Habets EJJ, Hendriks EJ, Taphoorn MJB et al. (2019). Association between tumor location and neurocognitive functioning using tumor localization maps. J Neurooncol 144: 573–582. https://doi.org/10.1007/s11060-019-03259-z. Epub 2019 Aug 13. PMID: 31410731.

Hendrix P, Hans E, Griessenauer CJ et al. (2017). Neurocognitive function surrounding the resection of frontal WHO grade I meningiomas: a prospective matched-control study. World Neurosurg 98: 203–210.

Hervey-Jumper SL, Berger MS (2016). Maximizing safe resection of low- and high-grade glioma. J Neurooncol 130: 269–282.

Ho VK, Reijneveld JC, Enting RH et al. (2014). Changing incidence and improved survival of gliomas. Eur J Cancer 50: 2309–2318.

Huber W, Poeck K, Willmes K (1984). The Aachen aphasia test. Adv Neurol 42: 291–303.

Ilmberger J, Ruge M, Kreth FW et al. (2008). Intraoperative mapping of language functions: a longitudinal neurolinguistic analysis. J Neurosurg 109: 583–592.

Imbault M, Chauvet D, Gennisson JL et al. (2017). Intraoperative functional ultrasound imaging of human brain activity. Sci Rep 7: 7304.

Inskip PD, Tarone RE, Hatch EE et al. (2003). Laterality of brain tumors. Neuroepidemiology 22: 130–138.

Jehna M, Becker J, Zaar K et al. (2017). Symmetry of the arcuate fasciculus and its impact on language performance of patients with brain tumors in the language-dominant hemisphere. J Neurosurg 127: 1407–1416.

Kappen PR, Beshay T, Vincent A et al. (2022a). The feasibility and added value of mapping music during awake craniotomy: a systematic review. Eur J Neurosci 55: 388–404.

Kappen PR, van den Brink J, Jeekel J et al. (2022b). The effect of musicality on language recovery after awake glioma surgery. Front Hum Neurosci 16: 1028897.

Kellner E, Dhital B, Kiselev VG et al. (2016). Gibbs-ringing artifact removal based on local subvoxel-shifts. Magn Reson Med 76: 1574–1581.

Kemerdere R, de Champfleur NM, Deverdun J et al. (2016). Role of the left frontal aslant tract in stuttering: a brain stimulation and tractographic study. J Neurol 263: 157–167.

Klein M, De Witt Hamer PC (2011). Neurocognitive outcome and resective brain tumor surgery in adults. In: H Duffau (Ed.), Brain mapping. From neural basis of cognition to surgical applications. Springer, Vienna.

Krainik A, Lehéricy S, Duffau H et al. (2003). Postoperative speech disorder after medial frontal surgery: role of the supplementary motor area. Neurology 60: 587–594.

Krieg SM, Shiban E, Buchmann N et al. (2012). Utility of presurgical navigated transcranial magnetic brain stimulation for the resection of tumors in eloquent motor areas. J Neurosurg 116: 994–1001.

Kullmann DM (2020). Editorial. Brain 143: 1045. https://doi.org/10.1093/brain/awaa082.

Laplane D, Talairach J, Meininger V et al. (1977). Clinical consequences of corticectomies involving the supplementary motor area in man. J Neurol Sci 34: 301–314.

Larjavaara S, Haapasalo H, Sankila R et al. (2008). Is the incidence of meningiomas underestimated? A regional survey. Br J Cancer 99: 182–184.

Leclercq D, Duffau H, Delmaire C et al. (2010). Comparison of diffusion tensor imaging tractography of language tracts and intraoperative subcortical stimulations. J Neurosurg 112: 503–511.

Lemée JM, Bernard F, Ter Minassian A et al. (2018). Right hemisphere cognitive functions: from clinical and anatomical bases to brain mapping during awake craniotomy. Part II: neuropsychological tasks and brain mapping. World Neurosurg 118: 360–367.

Li YC, Chiu HY, Lin YJ et al. (2021). The merits of awake craniotomy for glioblastoma in the left hemispheric eloquent area: one institution experience. Clin Neurol Neurosurg 200: 106343.

Liouta E, Koutsarnakis C, Liakos F et al. (2016). Effects of intracranial meningioma location, size, and surgery on neurocognitive functions: a 3-year prospective study. J Neurosurg 124: 1578–1584.

Louis DN, Ohgaki H, Wiestler OD et al. (2007). The 2007 WHO classification of tumors of the central nervous system. Acta Neuropathol 114: 97–109.

Louis DN, Perry A, Reifenberger G et al. (2016). The 2016 World Health Organization classification of tumors of the central nervous system: a summary. Acta Neuropathol 131: 803–820.

Louis DN, Perry A, Wesseling P et al. (2021). The 2021 WHO classification of tumors of the central nervous system: a summary. Neuro Oncol 23: 1231–1251.

Luria AR, Tsvetkova LS (1968). The mechanism of dynamic aphasia. Found Lang 4: 296–307.

Mandonnet E, Delattre JY, Tanguy ML et al. (2003). Continuous growth of mean tumor diameter in a subset of grade II gliomas. Ann Neurol 53: 524–528.

Mandonnet E, Winkler PA, Duffau H (2010). Direct electrical stimulation as an input gate into brain functional networks: principles, advantages and limitations. Acta Neurochir 152: 185–193.

Martín-Monzón I, Rivero Ballagas Y, Arias-Sánchez S (2022). Language mapping: a systematic review of protocols that evaluate linguistic functions in awake surgery. Appl Neuropsychol Adult 29: 845–854.

Méndez Orellana C, Visch-Brink E, Vernooij M et al. (2015). Crossed cerebrocerebellar language lateralization: an additional diagnostic feature for assessing atypical language representation in presurgical functional MR imaging. AJNR Am J Neuroradiol 36: 518–524.

Meskal I, Gehring K, Rutten GJ et al. (2016). Cognitive functioning in meningioma patients: a systematic review. J Neurooncol 128: 195–205.

Mirchandani AS, Beyh A, Lavrador JP et al. (2020). Altered corticospinal microstructure and motor cortex excitability in gliomas: an advanced tractography and transcranial magnetic stimulation study. J Neurosurg 134: 1368–1376.

Mooijman S, Vincent A, De Witte E et al. (2021). OS07.5A diagnostic instrument for mild aphasia (DIMA): sensitive and valuable addition to standard language assessment in glioma patients. Neuro Oncol 23: ii10–ii11.

Mooijman S, Bos LS, De Witte E et al. (2022). Language processing in glioma patients: speed or accuracy as a sensitive measure? Aphasiology 36: 1467–1491.

Muir M, Patel R, Traylor J et al. (2022). Validation of non-invasive language mapping modalities for eloquent tumor resection: a pilot study. Front Neurosci 16: 833073.

Naeser MA, Hayward RW (1978). Lesion localization in aphasia with cranial computed tomography and the Boston diagnostic aphasia exam. Neurology 28: 545–551.

NCIN (2013). NCIN data briefing astrocytic brain tumors: survival rates in England, UNCIN.

Nicolo P, Rizk S, Magnin C et al. (2015). Coherent neural oscillations predict future motor and language improvement after stroke. Brain 138: 3048–3060.

Nieberlein L, Rampp S, Gussew A et al. (2023). Reorganization and plasticity of the language network in patients with cerebral gliomas. Neuroimage Clin 37: 103326.

Noll KR, Sullaway C, Ziu M et al. (2015). Relationships between tumor grade and neurocognitive functioning in patients with glioma of the left temporal lobe prior to surgical resection. Neuro Oncol 17: 580–587.

Noll KR, Bradshaw ME, Weinberg JS et al. (2018). Neurocognitive functioning is associated with functional independence in newly diagnosed patients with temporal lobe glioma. Neurooncol Pract 5: 184–193.

Ochocka N, Segit P, Wojnicki K et al. (2023). Specialized functions and sexual dimorphism explain the functional diversity of the myeloid populations during glioma progression. Cell Rep 42: 111971.

Ohlerth AK, Valentin A, Vergani F et al. (2020). The verb and noun test for peri-operative testing (VAN-POP): standardized language tests for navigated transcranial magnetic stimulation and direct electrical stimulation. Acta Neurochir 162: 397–406.

Ohlerth AK, Lavrador JP, Vergani F et al. (in press). Combining nTMS and tractography for language mapping: an integrated paradigm for neurosurgical planning. In: E Picht (Ed.), Navigated transcranial magnetic stimulation in neurosurgery. Springer.

Ojemann G, Mateer C (1979). Human language cortex: localization of memory, syntax, and sequential motor-phoneme identification systems. Science 205: 1401–1403.

Ojemann G, Ojemann J, Lettich E et al. (1989). Cortical language localization in left, dominant hemisphere. An electrical stimulation mapping investigation in 117 patients. J Neurosurg 71: 316–326.

Oshino S, Kato A, Wakayama A et al. (2007). Magnetoencephalographic analysis of cortical oscillatory activity in patients with brain tumors: synthetic aperture magnetometry (SAM) functional imaging of delta band activity. Neuroimage 34: 957–964.

Palmisciano P, Haider AS, Balasubramanian K et al. (2022). Supplementary motor area syndrome after brain tumor surgery: a systematic review. World Neurosurg 165: 160–171.e162.

Papagno C, Gallucci M, Casarotti A et al. (2011). Connectivity constraints on cortical reorganization of neural circuits involved in object naming. Neuroimage 55: 1306–1313.

Papagno C, Casarotti A, Comi A et al. (2012). Measuring clinical outcomes in neuro-oncology. A battery to evaluate low-grade gliomas (LGG). J Neurooncol 108: 269–275.

Penfield W, Roberts L (1959). Speech and brain-mechanisms, Princeton University Press, Princeton, New Jersey.

Perrone D, Aelterman J, Pižurica A et al. (2015). The effect of Gibbs ringing artifacts on measures derived from diffusion MRI. Neuroimage 120: 441–455.

Petrovich Brennan NM, Whalen S, de Morales BD et al. (2007). Object naming is a more sensitive measure of speech localization than number counting: converging evidence from direct cortical stimulation and fMRI. Neuroimage 37: S100–S108.

Picht T, Krieg SM, Sollmann N et al. (2013). A comparison of language mapping by preoperative navigated transcranial magnetic stimulation and direct cortical stimulation during awake surgery. Neurosurgery 72: 808–819.

Piper RJ, Mikhael S, Wardlaw JM et al. (2016). Imaging signatures of meningioma and low-grade glioma: a diffusion tensor, magnetization transfer and quantitative longitudinal relaxation time MRI study. Magn Reson Imaging 34: 596–602.

Preusser M, de Ribaupierre S, Wöhrer A et al. (2011). Current concepts and management of glioblastoma. Ann Neurol 70: 9–21.

Price CJ (2010). The anatomy of language: a review of 100 fMRI studies published in 2009. Ann N Y Acad Sci 1191: 62–88.

Radhakrishnan K, Mokri B, Parisi JE et al. (1995). The trends in incidence of primary brain tumors in the population of Rochester, Minnesota. Ann Neurol 37: 67–73.

Rammeloo E, Schouten JW, Krikour K et al. (2023). Preoperative assessment of eloquence in neurosurgery: a systematic review. J Neurooncol 165: 413–430.

Rasmussen BK, Hansen S, Laursen RJ et al. (2017). Epidemiology of glioma: clinical characteristics, symptoms, and predictors of glioma patients grade I-IV in the Danish neuro-oncology registry. J Neurooncol 135: 571–579.

Rezai AR, D'Haese PF, Finomore V et al. (2024). Ultrasound blood-brain barrier opening and Aducanumab in Alzheimer's disease. N Engl J Med 390: 55–62.

Rijnen SJM, Meskal I, Bakker M et al. (2019). Cognitive outcomes in meningioma patients undergoing surgery: individual changes over time and predictors of late cognitive functioning. Neuro Oncol 21: 911–922.

Rijnen SJM, Meskal I, Emons WHM et al. (2020). Evaluation of normative data of a widely used computerized neuropsychological battery: applicability and effects of sociodemographic variables in a Dutch sample. Assessment 27: 373–383.

Robinson G, Blair J, Cipolotti L (1998). Dynamic aphasia: an inability to select between competing verbal responses? Brain 121: 77–89.

Robinson G, Shallice T, Cipolotti L (2006). Dynamic aphasia in progressive supranuclear palsy: a deficit in generating a fluent sequence of novel thought. Neuropsychologia 44: 1344–1360.

Rofes A, Mandonnet E, Godden J et al. (2017). Survey on current cognitive practices within the European low-grade glioma network: towards a European assessment protocol. Acta Neurochir 159: 1167–1178.

Rofes A, Talacchi A, Santini B et al. (2018). Language in individuals with left hemisphere tumors: is spontaneous speech analysis comparable to formal testing? J Clin Exp Neuropsychol 40: 722–732.

Rudà R, Capper D, Waldman AD et al. (2022). EANO - EURACAN - SNO guidelines on circumscribed astrocytic gliomas, glioneuronal, and neuronal tumors. Neuro Oncol 24: 2015–2034.

Sanai N, Mirzadeh Z, Berger MS (2008). Functional outcome after language mapping for glioma resection. N Engl J Med 358: 18–27.

Satoer D, Vork J, Visch-Brink E et al. (2012). Cognitive functioning early after surgery of gliomas in eloquent areas. J Neurosurg 117: 831–838.

Satoer D, Vincent A, Smits M et al. (2013). Spontaneous speech of patients with gliomas in eloquent areas before and early after surgery. Acta Neurochir 155: 685–692.

Satoer D, Kloet A, Vincent A et al. (2014a). Dynamic aphasia following low-grade glioma surgery near the supplementary motor area: a selective spontaneous speech deficit. Neurocase 20: 704–716.

Satoer D, Visch-Brink E, Smits M et al. (2014b). Long-term evaluation of cognition after glioma surgery in eloquent areas. J Neurooncol 116: 153–160.

Satoer D, Visch-Brink E, Dirven C et al. (2016). Glioma surgery in eloquent areas: can we preserve cognition? Acta Neurochir 158: 35–50.

Satoer D, Vincent A, Ruhaak L et al. (2018). Spontaneous speech in patients with gliomas in eloquent areas: evaluation until 1 year after surgery. Clin Neurol Neurosurg 167: 112–116.

Satoer D, De Witte E, Dragoy O (2021). Spontaneous speech. In: EHG Mandonnet (Ed.), Intraoperative mapping of cognitive networks. Which tasks for which locations. Springer, Cham.

Satoer D, De Witte E, Bulté B et al. (2022). Dutch diagnostic instrument for mild aphasia (DIMA): standardisation and a first clinical application in two brain tumor patients. Clin Linguist Phon 36: 929–953.

Sheppard SM, Stockbridge MD, Keator LM et al. (2022). The company prosodic deficits keep following right hemisphere stroke: a systematic review. J Int Neuropsychol Soc 28: 1075–1090.

Sierpowska J, Gabarrós A, Fernandez-Coello A et al. (2017). Words are not enough: nonword repetition as an indicator of arcuate fasciculus integrity during brain tumor resection. J Neurosurg 126: 435–445.

Sleurs C, Zegers CML, Compter I et al. (2022). Neurocognition in adults with intracranial tumors: does location really matter? J Neurooncol 160: 619–629.

Sollmann N, Zhang H, Fratini A et al. (2020). Risk assessment by presurgical tractography using navigated TMS maps in patients with highly motor- or language-eloquent brain tumors. Cancers (Basel) 12: 1264.

Soloukey S, Vincent A, Satoer DD et al. (2019). Functional ultrasound (fUS) during awake brain surgery: the clinical potential of intra-operative functional and vascular brain mapping. Front Neurosci 13: 1384.

Steinvorth S, Welzel G, Fuss M et al. (2003). Neuropsychological outcome after fractionated stereotactic radiotherapy (FSRT) for base of skull meningiomas: a prospective 1-year follow-up. Radiother Oncol 69: 177–182.

Stephan KE, Baldeweg T, Friston KJ (2006). Synaptic plasticity and dysconnection in schizophrenia. Biol Psychiatry 59: 929–939.

Stupp R, Mason WP, van den Bent MJ et al. (2005). Radiotherapy plus concomitant and adjuvant temozolomide for glioblastoma. N Engl J Med 352: 987–996.

Swinburn K, Porter G, Howard D (2005). The Comprehensive Aphasia Test (CAT), Psychology Press, Hove, UK.

Szelies B, Mielke R, Kessler J et al. (2002). Prognostic relevance of quantitative topographical EEG in patients with poststroke aphasia. Brain Lang 82: 87–94.

Talacchi A, Santini B, Casagrande F et al. (2013a). Awake surgery between art and science. Part I: clinical and operative settings. Funct Neurol 28: 205–221.

Talacchi A, Santini B, Casartelli M et al. (2013b). Awake surgery between art and science. Part II: language and cognitive mapping. Funct Neurol 28: 223–239.

Talozzi L, Forkel SJ, Pacella V et al. (2023). Latent disconnectome prediction of long-term cognitive-behavioural symptoms in stroke. Brain 146: 1963–1978.

Tarapore PE, Findlay AM, Honma SM et al. (2013). Language mapping with navigated repetitive TMS: proof of technique and validation. Neuroimage 82: 260–272.

Tewari S, Tom MC, Park DYJ et al. (2022). Sex-specific differences in low-grade glioma presentation and outcome. Int J Radiat Oncol Biol Phys 114: 283–292.

Thiebaut de Schotten M, Urbanski M, Duffau H et al. (2005). Direct evidence for a parietal-frontal pathway subserving spatial awareness in humans. Science 309: 2226–2228.

Thiebaut de Schotten M, Foulon C, Nachev P (2020). Brain disconnections link structural connectivity with function and behaviour. Nat Commun 11: 5094.

Thomson AM, Taylor R, Whittle IR (1998). Assessment of communication impairment and the effects of resective surgery in solitary, right-sided supratentorial intracranial tumors: a prospective study. Br J Neurosurg 12: 423–429.

Tremblay P, Dick AS (2016). Broca and Wernicke are dead, or moving past the classic model of language neurobiology. Brain Lang 162: 60–71.

Trinh VT, Fahim DK, Shah K et al. (2013). Subcortical injury is an independent predictor of worsening neurological deficits following awake craniotomy procedures. Neurosurgery 72: 160–169.

Tucha O, Smely C, Preier M et al. (2003). Preoperative and postoperative cognitive functioning in patients with frontal meningiomas. J Neurosurg 98: 21–31.

van Alkemade H, de Leau M, Dieleman EM et al. (2012). Impaired survival and long-term neurological problems in benign meningioma. Neuro Oncol 14: 658–666.

van der Linden SD, Gehring K, De Baene W et al. (2020). Assessment of executive functioning in patients with meningioma and low-grade glioma: a comparison of self-report, proxy-report, and test performance. J Int Neuropsychol Soc 26: 187–196.

van Kessel E, Baumfalk AE, van Zandvoort MJE et al. (2017). Tumor-related neurocognitive dysfunction in patients with diffuse glioma: a systematic review of neurocognitive functioning prior to anti-tumor treatment. J Neurooncol 134: 9–18.

van Kessel E, Schuit E, Huenges Wajer IMC et al. (2021). Added value of cognition in the prediction of survival in low and high grade glioma. Front Neurol 12: 773908.

van Loon EM, Heijenbrok-Kal MH, van Loon WS et al. (2015). Assessment methods and prevalence of cognitive dysfunction in patients with low-grade glioma: a systematic review. J Rehabil Med 47: 481–488.

van Nieuwenhuizen D, Ambachtsheer N, Heimans JJ et al. (2013). Neurocognitive functioning and health-related quality of life in patients with radiologically suspected meningiomas. J Neurooncol 113: 433–440.

van Nieuwenhuizen D, Douw L, Klein M et al. (2018). Cognitive functioning and functional brain networks in post-operative WHO grade I meningioma patients. J Neurooncol 140: 605–613.

Walsh KM, Hiroko Ohgaki MRW (2016). Epidemiology. In: S Mitchel, MW Berger (Eds.), Handbook of clinical neurology. Elsevier.

Wang YC, Chuang CC, Wei KC et al. (2016). Long term surgical outcome and prognostic factors of atypical and malignant meningiomas. Sci Rep 6: 35743.

Whiteside DM, Kealey T, Semla M et al. (2016). Verbal fluency: language or executive function measure? Appl Neuropsychol Adult 23: 29–34.

Whittle IR, Borthwick S, Haq N (2003). Brain dysfunction following 'awake' craniotomy, brain mapping and resection of glioma. Br J Neurosurg 17: 130–137.

Whittle IR, Smith C, Navoo P et al. (2004). Meningiomas. Lancet 363: 1535–1543.

Wiemels J, Wrensch M, Claus EB (2010). Epidemiology and etiology of meningioma. J Neurooncol 99: 307–314.

Wolthuis N (2022). Language impairments and resting-state EEG in brain tumor patients: revealing connections, Doctor of Philosophy. University of Groningen, Groningen, The Netherlands.

Wolthuis N, Satoer D, Veenstra W et al. (2021). Resting-state electro-encephalography functional connectivity networks relate to pre- and postoperative language functioning in low-grade glioma and meningioma patients. Front Neurosci 15: 785969.

Wolthuis N, Bosma I, Bastiaanse R et al. (2022). Distinct slow-wave activity patterns in resting-state electroencephalography and their relation to language functioning in low-grade glioma and meningioma patients. Front Hum Neurosci 16: 748128.

Yoshii Y, Tominaga D, Sugimoto K et al. (2008). Cognitive function of patients with brain tumor in pre- and postoperative stage. Surg Neurol 69: 51–61discussion 61.

Young RJ, Brennan N, Fraser JF et al. (2010). Advanced imaging in brain tumor surgery. Neuroimaging Clin N Am 20: 311–335.

Zhang N, Xia M, Qiu T et al. (2018). Reorganization of cerebro-cerebellar circuit in patients with left hemispheric gliomas involving language network: a combined structural and resting-state functional MRI study. Hum Brain Mapp 39: 4802–4819.

Zigiotto L, Annicchiarico L, Corsini F et al. (2020). Effects of supra-total resection in neurocognitive and oncological outcome of high-grade gliomas comparing asleep and awake surgery. J Neurooncol 148: 97–108.

Web References

https://iknl.nl/en/ncr - Netherlands Cancer Registry (NCR).

https://cbtrus.org/ - CBTRUS.

https://bradipho.eu/.

https://www.youtube.com/watch?v=o-oGpD_Nx5U.

https://www.youtube.com/watch?v=obiARnsKUAo&t=3s.

https://youtu.be/dnzjZqgfZSc.

https://www.frontiersin.org/articles/10.3389/fnins.2019.01384/full#supplementary-material.

Handbook of Clinical Neurology, Vol. 208 (3rd series)
Cerebral Asymmetries
C. Papagno and P. Corballis, Editors
https://doi.org/10.1016/B978-0-443-15646-5.00018-X

Chapter 6

Cerebral asymmetries in schizophrenia

FRANCESCA MARTINI[1], MARCO SPANGARO[1], JACOPO SAPIENZA[1,2]*, AND ROBERTO CAVALLARO[1,3]

[1]*Department of Clinical Neurosciences, IRCCS San Raffaele Scientific Institute, Milan, Italy*

[2]*Department of Humanities and Life Sciences, University School for Advanced Studies IUSS, Pavia, Italy*

[3]*School of Medicine, Vita-Salute San Raffaele University, Milan, Italy*

Abstract

Historically, the first observations of a lower prevalence of right-handed patients among subjects with schizophrenia led to the hypothesis that brain asymmetry could play a significant role in the etiopathogenesis of the disease. Over the last decades, a growing number of findings obtained through many different techniques such as EEG, MEG, MRI, and fMRI, consistently reported reduction/loss of brain asymmetries as a core feature of schizophrenia, further suggesting such alterations to play a cardinal role in the pathogenesis of the disease.

Moreover, several cognitive and psychopathologic dimensions have shown significant correlations with the reduced degree of asymmetry. In particular, the absence or even reversal of structural asymmetries has been documented in language-related brain such as the Sylvian fissure and planum temporale.

These findings have been reprocessed within an evolutionary and psychopathologic framework pointing at the loss of asymmetry and the consequent language impairment as primum moves in the pathogenesis of schizophrenia. Overall, despite growing evidence demonstrating a heterogeneous and multifaced etiopathogenesis in schizophrenia, the "old concept" of brain asymmetry still offers intriguing hints and thought-provoking elements for clinicians and researchers who deal with schizophrenia.

INTRODUCTION

Left–right asymmetry is a crucial feature of human brain organization, which can result from various genetic, epigenetic, or neural mechanisms (Corballis, 2014).

As early as the second trimester of pregnancy, the developing brain exhibits asymmetries, with cortical fissures appearing earlier in the right hemisphere. Moreover, the most pronounced brain asymmetry, that of the planum temporale (PT), is evident already in fetal and infant brains (Petty, 1999).

The human brain exhibits asymmetries at several levels, from morphology (volume and fissurization), to cytoarchitecture and dendritic arborization, to neurotransmitter sensitivity (Toga and Thompson, 2003).

In healthy subjects, an extensive amount of postmortem studies and, later, brain imaging studies consistently reported cerebral asymmetries. In particular, the frontal right hemisphere protrudes more anteriorly and is generally wider than the left, while the left occipital lobe is often wider than the right, a phenomenon referred to as the Yakovlevian or cerebral torque (Mitchell, 2005). Asymmetries have been described also in the posterior ends of the Sylvian fissures, with the left reaching further back and being more horizontal than the right. Finally, the PT, a triangular area lying on the superior surface of the superior temporal gyrus, between Heschl's gyrus (the auditory cortex) and the most posterior portions of the Sylvian fissure, is generally more developed in the left hemisphere (Petty, 1999).

Over the last decades, the association between structural and functional asymmetries has been increasingly studied and demonstrated. Several cognitive and motor functions, such as handedness, language perception and production, linear reasoning, and numeric and arithmetic skills are often lateralized to one hemisphere in most of the general population (Oertel-Knöchel and Linden, 2011). In particular, the left hemisphere has been identified as dominant for language and handedness (Corballis, 2014), with the normal right-frontal and left-occipital asymmetries being more prevalent in right-handed than left-handed individuals. Conversely, the right hemisphere appears dominant for some nonverbal functions, such as spatial attention and face processing (Luchins et al., 1982; Petty, 1999; Oertel-Knöchel and Linden, 2011; Schijven et al., 2023).

*Correspondence to: Jacopo Sapienza, Department of Humanities and Life Sciences, University School for Advanced Studies IUSS, Pavia, Italy. Tel: 00390226433299, E-mail: sapienza.jacopo@hsr.it.

Disturbances of structural and functional brain asymmetry have been associated with several psychiatric disorders with extensive evidence (Ocklenburg et al., 2015). In particular, atypical asymmetry has been long studied and proposed as a potential biomarker of schizophrenia.

Evidence shows that several asymmetric brain structures are affected in schizophrenia, including the Sylvian fissure (Falkai et al., 1992), the frontal and the occipital lobe (Bilder et al., 1994), and the PT (Petty, 1999).

Not surprisingly, two lateralized functions are of particular interest in relation to schizophrenia: language and handedness (Oertel-Knochel et al., 2012).

Theoretical work has focused on altered laterality for language in relation to disorganized speech perception and production, frequently observed in subjects with schizophrenia. Decreased left-lateralized language dominance, together with an absence or even reversal of structural asymmetries of language-related regions around the Sylvian fissure, have been reported in affected individuals (Schijven et al., 2023). Moreover, reversal/symmetry of PT, containing a portion of the heteromodal association cortex responsible for the comprehension and generation of language (Petty, 1999), has been described in several studies as characteristic of schizophrenia (Delisi et al., 1994; Falkai, 1995; Petty et al., 1995; Sakuma et al., 1996).

Furthermore, while the vast majority of the general population is right-handed, the prevalence of left-handedness and mixed-handedness in schizophrenia is higher than in healthy subjects (Oertel-Knochel et al., 2012). Interestingly, some genomic loci influencing structural brain asymmetry or handedness overlap with those associated with schizophrenia, suggesting a possible etiologic link between brain asymmetry abnormalities and schizophrenia (Schijven et al., 2023).

Drawing upon extensive research in structural and functional brain imaging, in this chapter we will explore how alterations in cerebral asymmetry may contribute to the etiology and clinical features of schizophrenia.

SCHIZOPHRENIA

Schizophrenia is a severe and chronic mental illness affecting approximately 1% of the general population, ranking among the most frequent causes of disability in developed countries, due to the pervasiveness of related deficits and life-long course. The disorder is characterized by a great heterogeneity of symptoms, encompassing positive, negative, and cognitive symptoms whose severity varies along the course of illness and across individuals. Positive symptoms refer to phenomena that are not normally experienced by healthy subjects, such as hallucinations, delusions, disorganization of thoughts, speech, and behavior. Negative symptoms are core features of the disorder and identify different declinations of the same psychopathologic element, disinterest/detachment from the outside world. They include alogia, blunted affect, avolition, anhedonia, and asociality. Positive and negative symptoms are often accompanied by a range of neurocognitive deficits, mainly in attention, working memory, and executive functions, together with alterations in social cognitive abilities (Oertel-Knöchel and Linden, 2011; Schijven et al., 2023). Cognitive impairment is often detectable before the onset of illness, remains stable or worsens through the course of illness, and is almost completely resistant to antipsychotic treatment. Moreover, it represents a significant determinant of a patient's global functional outcome and quality of life (Bosia et al., 2022).

A significant amount of evidence supports the presence of structural and functional brain alterations in schizophrenia, such as whole brain and gray matter volume reductions, especially in frontal, temporal, and postcentral cortical regions, and neurotransmitter abnormalities in multiple pathways, mainly dopaminergic and glutamatergic systems (Bosia et al., 2022).

The diagnosis of schizophrenia is based entirely on clinical manifestations and guided by the criteria given in current psychiatric classification systems: the International Statistical Classification of Diseases and Related Health Problems 10th Edition (ICD-10) and the Diagnostic and Statistical Manual of Mental Disorders 5th Edition (DSM-5). However, no clinical symptoms are pathognomonic, and associated biochemical, neuroimaging, and neuropsychologic correlates lack sensitivity and specificity.

Cerebral asymmetry in schizophrenia: An historical overview

Among the vast range of functional, structural, neurochemical, and molecular brain alterations associated with schizophrenia (Harrison, 1999; Shenton et al., 2001; Goghari et al., 2010; Van Den Heuvel and Fornito, 2014), the finding of atypical laterality patterns (Gur and Chin, 1999; Li et al., 2009; Oertel-Knöchel and Linden, 2011; Crow et al., 2013) is one of the soundest and most reliable. Consequently, it has been proposed that laterality and schizophrenia may have a shared genetic basis and that schizophrenia might even result from a failure in brain lateralization (Crow, 1989, 1990; Ocklenburg et al., 2015).

The possibility of a relationship between disturbances in cerebral asymmetry and schizophrenia dates back to the 17th century. The idea that disturbances within the corpus callosum resulting in aberrant intrahemispheric communication might underlie psychotic illnesses was first proposed by Wigan in 1841 (Clarke, 1987).

A few decades later, in 1879, Crichton-Browne postulated an evolutionary theory in the context of mental illness hypothesizing an association between psychosis and left hemisphere dysfunction and suggesting that the most highly evolved and last organized cortical centers, then supposed to be located on the left side of the brain, "might suffer first in insanity" (Crichton-Browne, 1879).

In 1915, in his detailed neuropathology studies, Southard described a tendency of focal atrophies and aplasias to occur in the left cerebral hemisphere (Southard, 1915). In addition, he reported that the pathologic lesion would preferentially affect the "association centers of Flechsig" (Flechsig, 1908, 1920) which is the interlinked system of heteromodal association cortex, which has subsequently been identified as implicated in the pathology of schizophrenia (Schlaepfer et al., 1994).

A development of the idea that asymmetry of the hemispheres may play some role in schizophrenia was then proposed by Jaynes (1977). The author hypothesizes that some of the symptoms characteristic of schizophrenia could be ascribed to a relative overactivity in the right hemisphere of the brain. He suggested that the experience of auditory hallucination could represent the reception by the left hemisphere of verbal activity produced by the right (Petty, 1999).

From then on, abnormalities in the left hemisphere of subjects with schizophrenia have been reported in several studies using different techniques, including neuropsychologic testing, air encephalography, CT, EEG, positron emission tomography, single-photon emission tomography (SPECT), magnetic resonance imaging (MRI), and biochemistry (Crow, 1990; Petty, 1999).

Structural and functional brain asymmetry in schizophrenia

Air encephalographic studies

Air encephalographic studies allowed the first observation of ventricular enlargement in schizophrenia, which appeared to be asymmetric in a part of the subjects (Jacobi and Winkler, 1927; Hunter et al., 1968). In their original study, Jacobi and Winkler reported a greater enlargement in the left ventricle in 9 of 19 patients affected by schizophrenia, and in the right ventricle just in one case. Hunter et al. (1968) subsequently confirmed that the enlargement was sometimes unilateral, with the left ventricle being larger than the right in most cases. Moreover, in 1982, Haug described a greater enlargement in patients with chronic schizophrenia as compared to subjects in an acute phase of the disease (Crow, 1990).

CT scan studies

With the advent of CT, several studies confirmed the presence of left hemisphere dysfunction in schizophrenia. Consistently with the finding of a left ventricle enlargement, some authors described a left hemisphere reduction in scan density in patients with schizophrenia (Golden, 1981; Largen et al., 1983; Coffman et al., 1984) and in monozygotic twins discordant for schizophrenia (Reveley, 1987).

Moreover, several authors reported alterations of the "typical" pattern of anatomic brain asymmetries in subjects affected.

In 1982, Luchins and colleagues described that the "reversal" of normal frontal and occipital asymmetries (i.e., brain width wider on the right in the occipital region and wider on the left in the frontal region) was more frequent in patients with schizophrenia compared to healthy subjects. In particular, the author reported an increase in reversals in schizophrenia subjects without generalized brain atrophy and an increase in normal asymmetries in subjects with brain atrophy. Interestingly, the authors found that subjects with reversal asymmetry had a higher performance IQ than verbal IQ. This finding led the authors to hypothesize that abnormal laterality could interfere with normal language functions (Luchins, 1982).

In subsequent studies, although with some inconsistencies, reductions or loss, more than reversals, of brain asymmetries were described.

In 1989, Crow and colleagues reported significant reductions in the width of the left hemisphere in the occipital and temporal regions of the brain in chronic patients with an early disease onset (Crow, 1989), suggesting that an early onset could have interfered with the development of normal asymmetries in the posterior part of the brain.

Similarly, Daniel et al. (1989) reported a loss of asymmetry in the occipital region in schizophrenia and reduced asymmetry of occipital skull measures in a group of patients with schizophrenia (with right-hand and right-eye dominance) compared to healthy controls.

Although other studies confirmed altered brain asymmetry in the occipital but not the frontal lobes (Tsai, 1983; Lee et al., 1985), others failed to find the same results. In particular, Andreasen and colleagues measured the frontal and occipital width of each hemisphere on the CT scans of 43 right-handed subjects with schizophrenia and 40 right-handed healthy controls, finding no significant differences in structural asymmetry between the groups. However, the authors found greater ventricular enlargement in left-handed patients (Andreasen et al., 1982). This finding was confirmed in a subsequent study, which reported that left-handed schizophrenic patients showed lateral ventricular enlargement and generally poorer cognitive performance compared with right-handed schizophrenic patients (Katsanis and Iacono, 1989). It was thus suggested that left-handedness might be associated with altered brain symmetry in schizophrenia (Crow, 1990).

Postmortem studies

Postmortem studies confirmed left ventricle enlargement, which was described as more marked in the temporal horns of subjects with schizophrenia (Brown, 1986; Crow, 1989, 1990; DeLisi et al., 1992).

In detail, in 1986, Brown and colleagues found a brain weight reduction, a modest but not significant increase in the lateral ventricular area, a significant increase in the temporal horn area, and a reduction in the width of the parahippocampal gyrus in patients with schizophrenia compared to subjects with affective disorders. The authors reported that the differences between groups were significantly greater in the left hemisphere (Brown, 1986).

In another postmortem study, Crow and colleagues compared the brains of patients with schizophrenia and subjects with Alzheimer-type dementia with the brains of age-matched controls. The authors reported that the posterior area and the temporal horn of the lateral ventricle were enlarged in subjects with schizophrenia. In subjects with Alzheimer, ventricular enlargement was more general, affecting the anterior horns and body as well as the temporal and posterior horns. Interestingly, the alteration in schizophrenia was described as selective to the left side of the brain, while in subjects with dementia it was not lateralized. These findings supported the assumption that the disease process in schizophrenia is somehow associated with the mechanisms that determine brain asymmetries (Crow, 1989).

In addition to the ventricular system and the frontal and occipital regions, some postmortem studies have described alterations in brain asymmetry in temporal lobe size itself. In two postmortem studies, the normal asymmetry of the lateral (Sylvian) fissure (left longer and flatter than the right) was reported to be lost in subjects with schizophrenia (Falkai et al., 1992; Lam et al., 2012). Moreover, in a subsequent postmortem study, Falkai and colleagues (Falkai, 1995) reported a 20% cortical volume reduction in the left PT of subjects with schizophrenia compared with age- and sex-matched controls.

Magnetic resonance imaging (MRI) studies

Cortical areas

Abnormalities of cerebral asymmetry in schizophrenia have been confirmed in the following decades also by multiple neuroimaging studies.

Several MRI studies have documented a reduction of cerebral torque in individuals with schizophrenia compared to their neurotypical counterparts (Crow et al., 2013). Moreover, both structural and functional data converge to indicate a diminished lateralization of language to the left hemisphere in schizophrenia, with certain studies reporting inverse lateralization to the right hemisphere (Gruzelier et al., 1999; Gur and Chin, 1999; Kwon et al., 1999; Aydin et al., 2001; Sommer et al., 2001; Mathiak et al., 2002; Sumich et al., 2002). In this view, Haijma et al. (2013) reported notable asymmetry exclusively within the PT, where a more conspicuous reduction was observed on the left side. The PT is both anatomically and functionally lateralized, with left-sided structures, in particular, playing a key role in language processes and constituting a component of Wernicke's region. Interestingly, diminished asymmetry in the PT has recently been linked to increased severity of positive symptoms in individuals diagnosed with schizophrenia. Moreover, the loss of left-hemisphere dominance and structural irregularities in cerebral asymmetry correlate with the progression of the disorder (Lam et al., 2012).

Another morphologic feature of schizophrenia pertains to abnormal cortical folding. Investigations have revealed elevated or more intricate cortical folding in the right hemisphere compared to the left hemisphere in both patients with schizophrenia and high-risk subjects (Harris et al., 2004; Narr et al., 2004; Stanfield et al., 2008). Conversely, regions exhibiting substantial reductions in gyrification (hypogyria) are predominantly observed in the left hemisphere (Palaniyappan and Liddle, 2012), supporting the hypothesis of aberrant brain lateralization in the disease. Of note, schizophrenia patients with resistant auditory hallucinations exhibit a pronounced decrease in the local sulcal index in the superior temporal sulcus bilaterally, and notably, in the left middle frontal sulcus and the diagonal branch of the left Sylvian fissure (Cachia et al., 2008). A close association between gyrification and disruptions in neural connectivity in schizophrenia has been postulated (White and Hilgetag, 2011). For instance, increased prefrontal gyrification in schizophrenia has been linked to increased local short-range prefrontal connectivity and diminished long-range connectivity (Dauvermann et al., 2012), although without investigation into the asymmetry of these connections. Thus, morphologic evidence of both the loss of cerebral torque and gyrification abnormalities may represent an anatomic basis for the disrupted connectivity observed in schizophrenia.

Subcortical structures

Relevant findings concerning altered brain asymmetry in schizophrenia also derive from studies focused on subcortical areas. Subcortical structures, encompassing the basal ganglia and components of the limbic system, play crucial roles in learning, memory, and fundamental functions such as motor control, attention, and emotion. Additionally, they contribute significantly to higher-order executive functions, including inhibitory control and working memory, through their structural and functional connectivity with prefrontal cortices. The majority of these functions are known to be altered among patients with schizophrenia, and previous research has identified associated volumetric alterations in subcortical areas, such as hippocampal and thalamic volume reductions.

Several differences in hemispheric asymmetries of subcortical volumes have also been reported, such as greater rightward asymmetry of the thalamus and the hippocampus, and reduced rightward asymmetry of the amygdala. Okada et al. (2016), in a large-scale multisite study (ENIGMA Consortium) including 1680 healthy individuals and 884 patients with schizophrenia, showed that healthy controls exhibited leftward asymmetry in the lateral ventricles, thalamus, caudate, and putamen, while rightward asymmetry was observed in the hippocampus and amygdala. Notably, volumetric asymmetries in subcortical regions did not significantly differ between controls and schizophrenia patients, except for the globus pallidus. In this region, healthy subjects showed no asymmetry, while schizophrenia patients demonstrated leftward asymmetry. This finding suggests potential implications for altered structural and functional connectivity, especially ipsilateral connectivity, between basal ganglia and frontal cortices. Indeed, among patients with schizophrenia altered lateralization may contribute to dysfunctions in the dorsolateral prefrontal cortex, which are associated with cognitive impairments in executive functions.

More recently, Gutman et al. (2022) conducted a comprehensive analysis of data derived from 2833 individuals diagnosed with schizophrenia and 3929 healthy controls (ENIGMA Schizophrenia Working Group). Their study unveiled distinct patterns that transcended individual structures, revealing a predominantly greater asymmetry pattern across the hippocampus, amygdala, and thalamus in individuals with schizophrenia, contrasting with a pattern of diminished asymmetry encompassing the ventral striatum, as well as both ventral and dorsal thalamus. In contrast to prior meta-analyses, such as that conducted by Haijma et al. (2013), which included over 18,000 subjects and found no statistically significant differences in global left- vs right subcortical volume effect sizes between individuals with schizophrenia and healthy controls, Gutman et al. (2022) observed distinct left-sided decreases in hippocampus and amygdala volumes that surpassed the right-sided decreases.

The observed exaggerated asymmetry across multiple subcortical structures in schizophrenia may be partially attributed to genetic mechanisms influencing cerebral asymmetry development, as suggested by previous research (Crow, 1989; Thompson et al., 2001; Brun et al., 2008; Satizabal et al., 2019; van der Meulen et al., 2020). Conversely, diminished asymmetry, as evidenced in the ventral striatum and ventral and dorsal thalamus, might result from epigenetic dysregulation (Abdolmaleky et al., 2019). Environmental insults during neurodevelopment are also posited as potential contributors to the observed disturbances in brain asymmetry later in life. Additionally, abnormal neuroanatomic asymmetry developments could underlie the disrupted asymmetry patterns observed in the brain network connectivity of individuals with schizophrenia (Gutman et al., 2022).

White matter

Diffusion tensor imaging (DTI) studies have described widespread white matter alterations in schizophrenia, which appear even before illness onset (Kubicki and Shenton, 2014). Of note, distinct patterns of disrupted connectivity persist across various stages of the disorder, encompassing medication-naïve first-episode patients (Zhang et al., 2015), high-risk subjects (Carletti et al., 2012), and schizotypal subjects (Smallman et al., 2014), suggesting a potential neurobiologic trait marker for schizophrenia spectrum disorders. Indeed, reduced fractional anisotropy (FA) values in left temporal white matter are observed in high-risk subjects who later develop schizophrenia, both at baseline and after illness onset (Carletti et al., 2012). Concerning hemispheric asymmetry of white matter integrity in schizophrenia, overall FA asymmetries display patterns similar to healthy controls (Miyata et al., 2012). However, individuals with schizophrenia exhibit a significant reduction of leftward asymmetry in specific white matter tracts, including the uncinate fasciculus, inferior occipitofrontal fasciculi, superior occipitofrontal fasciculus, and the anterior cingulum bundle. Notably, the rightward shift of FA in the uncinate fasciculus was reported to correlate with the severity of negative symptoms (Miyata et al., 2012). The integrity of white-matter tracts in the left hemisphere appears more disrupted compared to the right hemisphere, as evidenced by a meta-analysis including 407 patients with schizophrenia or related diagnoses. Specifically, altered white-matter integrity primarily was mainly reported in the cerebello-thalamo-cortical circuit and in the temporal network connecting the frontal lobe, insula, hippocampus/amygdala, and occipital lobe (Ellison-Wright and Bullmore, 2009).

Although DTI findings consistently support the hypothesis of reduced leftward asymmetry in schizophrenia, some studies reported increased FA in specific white-matter tracts, indicative of hyperconnectivity or deficient axonal pruning. Interestingly, positive correlations between FA and the severity of negative symptoms have been observed (Bijanki et al., 2015). Additionally, altered FA values in certain tracts, such as the left arcuate fasciculus, are associated with auditory hallucinations, further suggesting reduced leftward asymmetry in schizophrenia as a potential neurobiologic pathway of the disease.

More recently, different studies have focused on functional connectivity through fMRI analyses. Consistently with data concerning gray and white matter, investigations of asymmetry patterns of resting-state networks revealed an overall reduction of functional connectivity in chronic schizophrenia compared to healthy controls (Rotarska-Jagiela et al., 2010). The degree of abnormal asymmetry increased with illness duration and negative symptoms. Notably, positive symptoms resulted to correlate with increased leftward asymmetry of functional connectivity, while more pronounced negative symptoms were associated with increased rightward asymmetry (Ribolsi et al., 2014).

Electroencephalography (EEG) studies

Evidence of altered brain asymmetry in schizophrenia has been provided also by electroencephalography.

In the healthy brain cortex, the electrophysiologic activity tends to be symmetric and synchronous. On the contrary, patients with schizophrenia show intra and interhemispheric asymmetries during electroencephalography (EEG) recordings (Manchanda et al., 2008). The most replicated findings in the resting state EEG activity are a typical triad: (1) delta (1–4 Hz) and/or theta (4–8 Hz) excesses in frontal areas; (2) reduced mean frequency and lower power in the alpha band (8–12 Hz); and (3) increased beta power (12–20 Hz) (Hughes and John, 1999). A fast frequency beta activity (28–31.5 Hz) was found to be left-sided in the parietal region in patients not treated with antipsychotics, whereas in medicated patients was right-sided at the level of central and temporal electrodes (Morihisa, 1983). Other findings on the resting state activity also showed an increased delta and theta range frequency activity over the fronto-temporal-occipital cortex and a reduced activity of anterior-inferior areas in the theta frequency associated with reduced left anterior-inferior alpha frequencies. Moreover, increased activity of beta-band frequencies was found in the right posterior-superior regions and posterior limbic structures (Pascual-Marqui et al., 1999; Mientus et al., 2002; Koles et al., 2004). Interestingly, the degree of asymmetry (mainly right-sided) in the distribution of the spectral power probably depends on the severity of the disease, as patients in the acute phase of illness show greater functional asymmetry compared to chronic patients (Strelets et al., 2006). In healthy subjects "left-less-than-right" mismatch negativity asymmetry was widely demonstrated in EEG studies (Hirayasu et al., 1998; Mathiak et al., 2002; Kircher et al., 2004). Interestingly, such difference is reduced in patients with schizophrenia (Javitt et al., 1993; Hirayasu et al., 1998; Kasai et al., 1999; Winterer et al., 2001). Concerning event-related potentials (ERP) studies, auditory P300 occurs at about 300 ms poststimulus, and it is thought to be involved in the mechanisms of orientation and maintenance of attention (Polich and Criado, 2006). Patients with schizophrenia, their relatives, and subjects at high risk for schizophrenia, showed a reduction of P300 amplitude over midline electrodes (Morstyn, 1983; Ford et al., 1999; van der Stelt et al., 2005). A left-greater-than-right amplitude reduction of P300 was also described in patients with schizophrenia over the temporal region (Özgürdal et al., 2008). Such left temporal P300 abnormality has been shown to be related to the severity of positive symptoms (Shenton et al., 1989; Debruille et al., 2005), particularly thought disorder (Meisenzahl et al., 2004; Kirihara et al., 2005).

Asymmetry of language structures and implications for symptoms

Historical framework

The classification of symptoms has been a matter of debate since the first clinical observations and conceptualizations of the illness nowadays called schizophrenia. Specifically, there were disagreements among the most eminent theorists, particularly on the core features of the disease (Insel, 2010). Starting from the concept of dementia praecox and the central role played by cognitive impairment, the focus shifted to Bleuler's concept of autism and the withdrawal from reality first proposed in 1911, symptoms nowadays ascribed to the negative symptoms dimension. Then, the attention was put predominantly on positive symptoms (in particular the so-called Schneider's first-rank symptoms), resembling the current diagnostic criteria (Hoenig, 1983). More recently, a growing body of literature has been focusing on language as a core feature of schizophrenia and a phenotypic trait of the disease.

Language disturbances have been considered a key symptom since the first descriptions of schizophrenia as a mental disorder (Bleuler, 1911; Kraepelin, 1919). They are evident at different levels, encompassing disrupted speech production (such as tangentiality, and derailment), reduced verbal output (the negative symptom of alogia), and aberrant speech perceptions (such as auditory hallucinations). Language disorders seem to manifest as an early distinctive feature of schizophrenia, preceding the disease onset, and occurring not only in patients but also in individuals at high risk as well as in patients' relatives. Moreover, language dysfunction, such as "incoherent and disorganized speech," may adversely impact the patient's capacity to communicate effectively in social contexts (Bambini et al., 2022; Parola et al., 2023).

Concerning the structural correlates of the different symptom dimensions of schizophrenia, the loss of cerebral asymmetry and the failure to establish hemisphere dominance was proposed as a pathogenic noxa, particularly concerning the language dimension (Ceccherini-Nelli and Crow, 2003). Indeed, as previously mentioned, the human brain is structurally and functionally asymmetric. The left hemisphere is typically associated with language ability, while the right hemisphere is typically associated with nonverbal functions (Galaburda et al., 1978; Toga and Thompson, 2003; Hervé et al., 2013). Although the cerebral hemispheres are similar in weight and volume, several differences exist in brain tissue distribution. The right hemisphere is significantly more interconnected than the left, whereas the left hemisphere shows more clustered regions that are more central/crucial for the whole-brain structural network function (Iturria-Medina et al., 2011). Overall, such considerations reflect functional principles: the left hemisphere demands specialized networks for language and motor actions, whereas the right hemisphere is deputed to more general processes such as the integration of information (Iturria-Medina et al., 2011). During the last decades of the 20th century, Timothy J. Crow formulated his famous and influential theory on the failure of the lateralization of the language process in schizophrenia, postulating schizophrenia as a disorder of the genetic mechanisms that control the development of cerebral asymmetry (Crow, 1989). In this view, the nuclear symptoms of schizophrenia can be understood as a lack of lateralization of brain structures and functions due to the polygenic susceptibility to the disease (Ceccherini-Nelli and Crow, 2003; Crow, 2004). The main supporting evidence at the time was the prevalence of non-righthandedness, reported to be significantly higher in patients with schizophrenia than in healthy individuals, which suggested a failure to establish cerebral asymmetry (Sommer et al., 2001).

Recent supporting findings: Asymmetry and the genetic load

Nowadays, it is well known that left–right asymmetry is an important organizing feature of the healthy brain that may be altered in schizophrenia as demonstrated by a large-scale analysis of structural brain asymmetries in 5080 patients with schizophrenia and 6015 controls (Schijven et al., 2023). Differences were observed for thickness asymmetries of the rostral anterior cingulate and the middle temporal gyrus, both driven by thinner left-hemispheric cortices in schizophrenia. Notably, reduced left middle temporal cortical thickness is consistent with the altered left-hemisphere language network organization in schizophrenia and, interestingly, the 7% of the variance across all structural asymmetries was explained by case–control status, which in turn may reflect differences at the molecular, cytoarchitectonic, or circuit levels that have functional relevance for the disorder (Zhu et al., 2021; Schijven et al., 2023). This finding supports the hypothesis of the genetic load in determining the attenuated cerebral asymmetry in schizophrenia. Similarly, other studies reported an association between polygenic risk and altered asymmetry, highlighting that schizophrenia-related risk genes may influence aberrant alterations in hemispheric asymmetry. Specifically, it was found that such genes were involved in signal transduction, neural development, neuron structure, and calcium signaling pathways (Zhu et al., 2021). More specifically, findings from genome-wide association studies (GWAS) showed the correlation between the polygenic risk for schizophrenia and autism and asymmetries of regions important for language and executive functions, two cardinal dimensions impaired in schizophrenia (Sha et al., 2021). Finally, even subjects at genetically high risk for psychosis showed altered folding patterns of the cingulate sulcus/paracingulate in the anterior cingulate cortex (ACC) (loss of leftward sulcal asymmetry) and language-related regions, which endorse the hypothesis of a developmental alteration dependent on the polygenic genetic load for schizophrenia (Francis et al., 2012; Park et al., 2013). All these findings corroborate the hypothesis of Crow as, in this view, the genetic load for schizophrenia contrasts the radical change in brain function underlying the development of human language: the process of lateralization and the resulting asymmetry.

The importance of language: An evolutionary perspective

Maynard-Smith and Szathmary indicated the origin of language as the seventh of the major transitions in evolution and the pivotal step in the speciation process (Szathmáry and Smith, 1995). According to this hypothesis, the symptoms of the disease can arise from the dysfunctional lateralization of language structures, as mentioned above (Crow, 1989). As a matter of fact, the ability to communicate is necessary for survival in a socially organized species, and it is closely related to intelligence, which is difficult to define (Howard, 1993; Bates et al., 2001). Distinguishing features of human language are the arbitrariness of the association between the sign (word) and what it signifies (object) (de Saussure, 1916), and its generativity (Chomsky, 1985) defined as the ability to generate an infinite number of sentences by the syntactical combination of words. Indeed, "Universal grammar" has been postulated as the unique feature of the human brain as generativity distinguishes the human brain from computational systems such as chat-GPT (Friederici et al., 2017). Interestingly, syntactic and semantic capacities rely on the multitude of connections and circuits that take place in our brain and are mostly independent from the phonologic and acoustic modalities; they relate to the brain as a whole and patients with schizophrenia do not show patterns of right-more-than-left efficient global integration that are observable in healthy control participants (Iturria-Medina et al., 2011).

Disrupted asymmetry and symptoms

The dysfunctional process of lateralization may account for the disruption of the mechanism of "indexicality" that allows the speaker to distinguish his/her thoughts from the speech output that he/she generates and the speech input that he/she receives and decodes from others (Crow, 2004). Considering positive symptoms, we can find unusual and incomprehensible experiences concerning one's thoughts as they are not conceived as one's own. Typically, patients think that these thoughts are inserted into or removed from one's mind by an outside agency, or that thoughts are broadcast to others or to the surrounding environments. Similarly, auditory hallucinations are hearing one's thoughts spoken aloud, running commentaries on one's thoughts or actions, or other kinds of voices (Ceccherini-Nelli and Crow, 2003; Moscarelli, 2020). Interestingly, auditory hallucinations or these unusual phenomena occur in all populations independently of the specific language spoken, being intrinsic to the human capacity for language. Overall, all these symptoms could arise from an anomalous system of reference of language to the self (consciousness of the self). Indeed, it emphasized the need for a frame of reference (coordinates) that relates the phonologic output, and what has led up to it, to the "I" of the speaker, to the present moment in time "now" and to the location of the speaker in space "here," and the failure of this reference system could underly the nuclear symptoms of schizophrenia (Crow, 1989, 2004). Neural activity with the characteristic of thought but that lacks autonomy could acquire the characteristic of thought insertion, thought broadcast or thought withdrawal, becoming dissociated from the "I." The physiologic transition between thought and speech that normally occurs between nondominant and dominant frontal association areas might be disrupted in patients with schizophrenia due to reduced asymmetry of these regions (Crow, 1989). Similarly, auditory hallucinations may arise from neural activity with the characteristic of speech input interpreted in the nondominant occipito-parieto-temporal regions (Altamura et al., 2020). Overall, the distinction between the "I" of the speaker and his/her interface with the rest of the linguistic community, is lost. To summarize, Crow suggested that the nuclear symptoms of schizophrenia represent "language at the end of its tether" (Crow, 1997), providing a window on the transition between speech and thought (Crow, 1998). In addition, he said that schizophrenia is the price that the homo sapiens species must pay to have language (Crow, 2000). In other terms, given the human genetics that encloses the "speciation event" of the neurodevelopmental process of lateralization of language structures, the 1% of the population affected by schizophrenia worldwide is burdened by polygenic alterations that hamper such process, which allowed the two hemispheres to develop with a degree of independence, thus ending in the loss of "indexicality" in the transition from language to thoughts.

Recent supporting findings: Asymmetry and symptoms

Considering recent findings on the association between positive and negative symptoms and asymmetry, patients with prominent positive symptoms showed increased leftward asymmetry (left-greater-than-right) in terms of functional connectivity, while patients displaying more negative symptoms exhibited increased rightward asymmetry (right-greater-than-left) of functional connectivity. Specifically, connectivity asymmetries in the frontal lobe, temporal lobe, cingulate gyrus, and precuneus were significantly associated with the severity of positive symptoms pointing to a pivotal role of frontal and temporal areas (Ke et al., 2010). Similarly, it was reported that a left-sided asymmetry in the activation of the temporal lobe, and particularly of the superior temporal gyrus, was significantly related to the severity of hallucinations (Zhang et al., 2008; Oertel-Knochel et al., 2012) and that fluctuating asymmetry was significantly associated to the severity of hallucinations and disorganization (Stephan-Otto et al., 2020). Moreover, Zhang et al. reported that the auditory-related asymmetry is attenuated in schizophrenia and the severity of auditory hallucinations was associated with such attenuation (Zhang et al., 2008). In addition, different cortical sulcation and left–right sulcal asymmetry were found to correlate with the severity of hallucinations (Cachia et al., 2015) and overall, differences in the cortical folding patterns between patients with schizophrenia and healthy controls are widely reported sustaining the neurodevelopmental hypothesis of Crow (Rollins et al., 2020; Ćurčić-Blake et al., 2023). As mentioned above, there is evidence that altered brain asymmetry is responsible for negative symptoms, particularly in rightward structural asymmetry in amygdala volumes and superior temporal regions (Huang et al., 2022). Concerning cognitive symptoms, Sakuma and colleagues found that healthy subjects perform significantly better compared to patients on several measures of functional laterality. The patients exhibited a significantly smaller difference (i.e., less laterality) between left and right ears in the dichotic listening tests, and between left and right hands in the dexterity tasks, than controls did, highlighting that a decreased or reversed pattern of nonverbal functional laterality might be related to a poorer cognitive performance (Sakuma et al., 1996). Studies performed in subjects at high genetic risk found that reversed structural asymmetry of the pars triangularis was associated with verbal fluency deficits, while a more prominent left paracingulate sulcus and leftward paracingulate sulcus asymmetry were both associated with higher scores on working memory tasks (Bhojraj et al., 2009; Park et al., 2013). Concerning EEG findings, deviations from the normal brain activity along an anterior-left/posterior-right axis were found during cognitive tasks with different patterns of EEG alpha asymmetry during verbal and spatial cognitive tasks, that highlight the predominance of left- vs right-sided cortical activation (Merrin and Floyd, 1997). Moreover, several studies reported a correlation between MMN of the left temporal lobe and symptoms of schizophrenia (Umbricht and Krljes, 2005), particularly hallucinations (Hirayasu et al., 1998). Differently, the severity of negative symptoms is positively related to the amplitude of MMN in the left frontal regions and negatively related to MMN scalp activity in the right hemisphere (Hirayasu et al., 1998; Kasai et al., 2001).

CONCLUSIONS

This chapter examined the extensive evidence indicating altered laterality and hemispheric asymmetry in schizophrenia. Despite some inconsistencies, numerous studies employing various techniques, including CT, MRI, and EEG, report asymmetry alterations in both brain cortical and subcortical structures, as well as in the structural and functional connectivity of affected individuals. The most robust and replicable finding points toward a reduction or loss of the cerebral torque resulting in a failure of the left-hemisphere dominance. In particular, both structural and functional data converge to suggest a diminished lateralization of language to the left hemisphere in schizophrenia, with certain studies noting an inverse lateralization to the right. An absence or even reversal of structural asymmetries of language-related regions around the Sylvian fissure has also consistently been observed. Moreover, asymmetry abnormalities have been extensively associated with different cognitive and psychopathologic dimensions, encompassing positive symptoms, negative symptoms, and language disturbances.

As some genomic loci influencing structural brain asymmetry overlap with those associated with schizophrenia, a possible etiologic link between brain asymmetry abnormalities and schizophrenia has long been proposed. Altered asymmetry in schizophrenia may thus emerge during development through the disruption of a genetically regulated program of asymmetric brain development, but also through different trajectories of lifespan-related changes in brain asymmetries.

Although some of the behavioral and biologic indicators of altered laterality in schizophrenia exhibit reasonable stability across studies, their lack of sensitivity and specificity precludes their qualification as reliable biologic markers. Consequently, the use of laterality parameters for diagnostic and therapeutic purposes remains distant. Nonetheless, laterality research has proven to be an intriguing field that bridges clinical psychiatry with evolutionary psychology and biology. It continues to offer avenues for future research on the molecular and cellular basis of altered lateralized cognitive functions in schizophrenia, potentially contributing to the identification of the pathophysiologic mechanisms underlying the disease.

REFERENCES

Abdolmaleky HM, Gower AC, Wong C et al. (2019). Aberrant transcriptomes and DNA methylomes define pathways that drive pathogenesis and loss of brain laterality/asymmetry in schizophrenia and bipolar disorder. Am J Med Genet B Neuropsychiatr Genet 180: 138–149.

Altamura M, Prete G, Elia A et al. (2020). Do patients with hallucinations imagine speech right? Neuropsychologia 146: 107567.

Andreasen NC, Olsen SA, Dennert JW et al. (1982). Ventricular enlargement in schizophrenia: relationship to positive and negative symptoms. Am J Psychiatry 139: 297–302.

Aydin N, Dane Ş, Öztürk I et al. (2001). Left ear (right temporal hemisphere) advantage and left temporal hemispheric dysfunction in schizophrenia. Percept Mot Skills 93: 230–238.

Bambini V, Frau F, Bischetti L et al. (2022). Deconstructing heterogeneity in schizophrenia through language: a semi-automated linguistic analysis and data-driven clustering approach. Schizophrenia 8: 102.

Bates E, Tager-Flusberg H, Vicari S et al. (2001). Debate over language's link with intelligence. Nature 413: 565–566.

Bhojraj TS, Francis AN, Rajarethinam R et al. (2009). Verbal fluency deficits and altered lateralization of language brain areas in individuals genetically predisposed to schizophrenia. Schizophr Res 115: 202–208.

Bijanki KR, Hodis B, Magnotta VA et al. (2015). Effects of age on white matter integrity and negative symptoms in schizophrenia. Schizophr Res 161: 29–35.

Bilder RM, Wu H, Bogerts B et al. (1994). Absence of regional hemispheric volume asymmetries in first-episode schizophrenia. Am J Psychiatry 151: 1437–1447.

Bleuler E (1911). Dementia praecox oder Gruppe der Schizophrenien, Verlag Franz Deuticke.

Bosia M, Spangaro M, Cocchi F et al. (2022). Psychotic disorders. In: Fundamentals of psychiatry for health care professionals, Springer International Publishing, Cham, pp. 85–119.

Brown R (1986). Postmortem evidence of structural brain changes in schizophrenia. Arch Gen Psychiatry 43: 36.

Brun C, Leporé N, Pennec X et al. (2008). A tensor-based morphometry study of genetic influences on brain structure using a new fluid registration method. Med Image Comput Comput Assist Interv 11: 914–921.

Cachia A, Paillère-Martinot M-L, Galinowski A et al. (2008). Cortical folding abnormalities in schizophrenia patients with resistant auditory hallucinations. Neuroimage 39: 927–935.

Cachia A, Amad A, Brunelin J et al. (2015). Deviations in cortex sulcation associated with visual hallucinations in schizophrenia. Mol Psychiatry 20: 1101–1107.

Carletti F, Woolley JB, Bhattacharyya S et al. (2012). Alterations in white matter evident before the onset of psychosis. Schizophr Bull 38: 1170–1179.

Ceccherini-Nelli A, Crow TJ (2003). Disintegration of the components of language as the path to a revision of Bleuler's and Schneider's concepts of schizophrenia. Br J Psychiatry 182: 233–240.

Chomsky N (1985). Cambios de perspectiva sobre el conocimiento y uso del lenguaje. Rev Int Filos 15: 11–71.

Clarke B (1987). Arthur Wigan and *the duality of the mind*. Psychol Med Monogr Suppl 11: 1–52.

Coffman JA, Andreasen NC, Nasrallah HA (1984). Left hemispheric density deficits in chronic schizophrenia. Biol Psychiatry 19: 1237–1247.

Corballis MC (2014). Left brain, right brain: facts and fantasies. PLoS Biol 12: e1001767. https://doi.org/10.1371/journal.pbio.1001767.

Crichton-Browne J (1879). On the weight of the brain and its component parts in the insane. Brain 2: 42–67.

Crow TJ (1989). Schizophrenia as an anomaly of development of cerebral asymmetry. Arch Gen Psychiatry 46: 1145.

Crow TJ (1990). Temporal lobe asymmetries as the key to the etiology of schizophrenia. Schizophr Bull 16: 433–443.

Crow TJ (1997). Is schizophrenia the price that Homo sapiens pays for language? Schizophr Res 28: 127–141.

Crow TJ (1998). Nuclear schizophrenic symptoms as a window on the relationship between thought and speech. Br J Psychiatry 173: 303–309.

Crow T (2000). Schizophrenia as the price that Homo sapiens pays for language: a resolution of the central paradox in the origin of the species. Brain Res Rev 31: 118–129.

Crow T (2004). Auditory hallucinations as primary disorders of syntax: an evolutionary theory of the origins of language. Cogn Neuropsychiatry 9: 125–145.

Crow TJ, Chance SA, Priddle TH et al. (2013). Laterality interacts with sex across the schizophrenia/bipolarity continuum: an interpretation of meta-analyses of structural MRI. Psychiatry Res 210: 1232–1244.

Ćurčić-Blake B, de Vries A, Renken RJ et al. (2023). Paracingulate sulcus length and cortical thickness in schizophrenia patients with and without a lifetime history of auditory hallucinations. Schizophr Bull 49: S48–S57.

Daniel DG, Myslobodsky MS, Ingraham LJ et al. (1989). The relationship of occipital skull asymmetry to brain parenchymal measures in schizophrenia. Schizophr Res 2: 465–472.

Dauvermann MR, Mukherjee P, Moorhead WT et al. (2012). Relationship between gyrification and functional connectivity of the prefrontal cortex in subjects at high genetic risk of schizophrenia. Curr Pharm Des 18: 434–442.

de Saussure F (1916). Cours de linguistique générale.

Debruille JB, Schneider-Schmid A, Dann P et al. (2005). The correlation between positive symptoms and left temporal event-related potentials in the P300 time window is auditory specific and training sensitive. Schizophr Res 78: 117–125.

DeLisi LE, Hoff AL, Kushner M et al. (1992). Left ventricular enlargement associated with diagnostic outcome of schizophreniform disorder. Biol Psychiatry 32: 199–201.

Delisi LE, Hoff AL, Neale C et al. (1994). Asymmetries in the superior temporal lobe in male and female first-episode schizophrenic patients: measures of the planum temporale and superior temporal gyrus by MRI. Schizophr Res 12: 19–28.

Ellison-Wright I, Bullmore E (2009). Meta-analysis of diffusion tensor imaging studies in schizophrenia. Schizophr Res 108: 3–10.

Falkai P (1995). Disturbed planum temporale asymmetry in schizophrenia. A quantitative post-mortem study. Schizophr Res 14: 161–176.

Falkai P, Bogerts B, Greve B et al. (1992). Loss of sylvian fissure asymmetry in schizophrenia. Schizophr Res 7: 23–32.

Flechsig P (1908). Bermerkungen tiber die Horsphare dochen gehirns. Neurologische Zentralblatt 27: 2–7.

Flechsig P (1920). Anatomie des menschlichen Gehirns und Riickenmarks auf myelogenetischer Grundlage, G. Thieme, Leipzig, Germany.

Ford JM, Mathalon DH, Marsh L et al. (1999). P300 amplitude is related to clinical state in severely and moderately ill patients with schizophrenia. Biol Psychiatry 46: 94–101.

Francis AN, Seidman LJ, Jabbar GA et al. (2012). Alterations in brain structures underlying language function in young adults at high familial risk for schizophrenia. Schizophr Res 141: 65–71.

Friederici AD, Chomsky N, Berwick RC et al. (2017). Language, mind and brain. Nat Hum Behav 1: 713–722.

Galaburda AM, LeMay M, Kemper TL et al. (1978). Right-left asymmetries in the brain. Science 199: 852–856.

Goghari VM, Sponheim SR, MacDonald AW (2010). The functional neuroanatomy of symptom dimensions in schizophrenia: a qualitative and quantitative review of a persistent question. Neurosci Biobehav Rev 34: 468–486.

Golden CJ (1981). Structural brain deficits in schizophrenia. Arch Gen Psychiatry 38: 1014.

Gruzelier J, Wilson L, Richardson A (1999). Cognitive asymmetry patterns in schizophrenia: retest reliability and modification with recovery. Int J Psychophysiol 34: 323–331.

Gur RE, Chin S (1999). Laterality in functional brain imaging studies of schizophrenia. Schizophr Bull 25: 141–156.

Gutman BA, van Erp TGM, Alpert K et al. (2022). A meta-analysis of deep brain structural shape and asymmetry abnormalities in 2,833 individuals with schizophrenia compared with 3,929 healthy volunteers via the ENIGMA Consortium. Hum Brain Mapp 43: 352–372.

Haijma SV, Van Haren N, Cahn W et al. (2013). Brain volumes in schizophrenia: a meta-analysis in over 18 000 subjects. Schizophr Bull 39: 1129–1138.

Harris JM, Whalley H, Yates S et al. (2004). Abnormal cortical folding in high-risk individuals: a predictor of the development of schizophrenia? Biol Psychiatry 56: 182–189.

Harrison PJ (1999). The neuropathology of schizophrenia. A critical review of the data and their interpretation. Brain 122: 593–624.

Hervé P-Y, Zago L, Petit L et al. (2013). Revisiting human hemispheric specialization with neuroimaging. Trends Cogn Sci 17: 69–80.

Hirayasu Y, Potts GF, O'Donnell BF et al. (1998). Auditory mismatch negativity in schizophrenia: topographic evaluation with a high-density recording montage. Am J Psychiatry 155: 1281–1284.

Hoenig J (1983). The concept of schizophrenia Kraepelin–Bleuler–Schneider. Br J Psychiatry 142: 547–556.

Howard RW (1993). On what intelligence is. Br J Psychol 84: 27–37.

Huang Z, Ruan D, Huang B et al. (2022). Negative symptoms correlate with altered brain structural asymmetry in amygdala and superior temporal region in schizophrenia patients. Front Psych 13: 1000560.

Hughes JR, John ER (1999). Conventional and quantitative electroencephalography in psychiatry. J Neuropsychiatry Clin Neurosci 11: 190–208.

Hunter R, Jones M, Cooper F (1968). Modified lumbar air encephalography in the investigation of long-stay psychiatric patients. J Neurol Sci 6: 593–596.

Insel TR (2010). Rethinking schizophrenia. Nature 468: 187–193.

Iturria-Medina Y, Pérez Fernández A, Morris DM et al. (2011). Brain hemispheric structural efficiency and interconnectivity rightward asymmetry in human and nonhuman Primates. Cereb Cortex 21: 56–67.

Jacobi W, Winkler H (1927). Encephalographische Studien an chronisch Schizophrenen. Arch Psychiatr Nervenkr 81: 299–332.

Javitt DC, Doneshka P, Zylberman I et al. (1993). Impairment of early cortical processing in schizophrenia: an event-related potential confirmation study. Biol Psychiatry 33: 513–519.

Jaynes J (1977). The origin of consciousness in the breakdown of the bicameral mind, Houghton Mifflin, Boston, MA.

Kasai K, Okazawa K, Nakagome K et al. (1999). Mismatch negativity and N2b attenuation as an indicator for dysfunction of the preattentive and controlled processing for deviance detection in schizophrenia: a topographic event-related potential study. Schizophr Res 35: 141–156.

Kasai K, Yamada H, Kamio S et al. (2001). Brain lateralization for mismatch response to across- and within-category change of vowels. Neuroreport 12: 2467–2471.

Katsanis J, Iacono WG (1989). Association of left-handedness with ventricle size and neuropsychological performance in schizophrenia. Am J Psychiatry 146: 1056–1058.

Ke M, Zou R, Shen H et al. (2010). Bilateral functional asymmetry disparity in positive and negative schizophrenia revealed by resting-state fMRI. Psychiatry Res Neuroimaging 182: 30–39.

Kircher TTJ, Rapp A, Grodd W et al. (2004). Mismatch negativity responses in schizophrenia: a combined fMRI and whole-head MEG study. Am J Psychiatry 161: 294–304.

Kirihara K, Araki T, Kasai K et al. (2005). Confirmation of a relationship between reduced auditory P300 amplitude and thought disorder in schizophrenia. Schizophr Res 80: 197–201.

Koles ZJ, Lind JC, Flor-Henry P (2004). A source-imaging (low-resolution electromagnetic tomography) study of the EEGs from unmedicated men with schizophrenia. Psychiatry Res Neuroimaging 130: 171–190.

Kraepelin E (1919). Dementia praecox and paraphrenia.

Kubicki M, Shenton ME (2014). Diffusion tensor imaging findings and their implications in schizophrenia. Curr Opin Psychiatry 27: 179–184.

Kwon JS, McCarley RW, Hirayasu Y et al. (1999). Left planum temporale volume reduction in schizophrenia. Arch Gen Psychiatry 56: 142.

Lam M, Collinson SL, Sim K et al. (2012). Asymmetry of lexico-semantic processing in schizophrenia changes with disease progression. Schizophr Res 134: 125–130.

Largen Jr JW, Calderon M, Smith RC (1983). Asymmetries in the densities of white and gray matter in the brains of schizophrenic patients. Am J Psychiatry 140: 1060–1062.

Lee JH, Lee YJ, Oh SW (1985). Schizophrenia and cerebral laterality. Ann Acad Med Singapore 14: 91–94.

Li X, Branch CA, DeLisi LE (2009). Language pathway abnormalities in schizophrenia: a review of fMRI and other imaging studies. Curr Opin Psychiatry 22: 131–139.

Luchins DJ (1982). Computed tomography in schizophrenia. Arch Gen Psychiatry 39: 859.

Luchins DJ, Weinberger DR, Jed Wyatt R (1982). Schizophrenia and cerebral asymmetry detected by computed tomography. Am J Psychiatry 139: 753–757.

Manchanda R, Norman R, Malla A et al. (2008). EEG abnormalities and 3-year outcome in first episode psychosis. Acta Psychiatr Scand 117: 277–282.

Mathiak K, Rapp A, Kircher TTJ et al. (2002). Mismatch responses to randomized gradient switching noise as reflected by fMRI and whole-head magnetoencephalography. Hum Brain Mapp 16: 190–195.

Meisenzahl EM, Frodl T, Müller D et al. (2004). Superior temporal gyrus and P300 in schizophrenia: a combined ERP/structural magnetic resonance imaging investigation. J Psychiatr Res 38: 153–162.

Merrin EL, Floyd TC (1997). Clinical symptoms of schizophrenia affect reference-independent measures of task-induced EEG alpha asymmetry. Psychiatry Res Neuroimaging 74: 47–62.

Mientus S, Gallinat J, Wuebben Y et al. (2002). Cortical hypoactivation during resting EEG in schizophrenics but not in depressives and schizotypal subjects as revealed by low resolution electromagnetic tomography (LORETA). Psychiatry Res Neuroimaging 116: 95–111.

Mitchell RLC (2005). Right hemisphere language functions and schizophrenia: the forgotten hemisphere? Brain 128: 963–978.

Miyata J, Sasamoto A, Koelkebeck K et al. (2012). Abnormal asymmetry of white matter integrity in schizophrenia revealed by voxelwise diffusion tensor imaging. Hum Brain Mapp 33: 1741–1749.

Morihisa JM (1983). Brain electrical activity mapping (BEAM) in schizophrenic patients. Arch Gen Psychiatry 40: 719.

Morstyn R (1983). Altered P300 topography in schizophrenia. Arch Gen Psychiatry 40: 729.

Moscarelli M (2020). A major flaw in the diagnosis of schizophrenia: what happened to the Schneider's first rank symptoms. Psychol Med 50: 1409–1417.

Narr KL, Bilder RM, Kim S et al. (2004). Abnormal gyral complexity in first-episode schizophrenia. Biol Psychiatry 55: 859–867.

Ocklenburg S, Güntürkün O, Hugdahl K et al. (2015). Laterality and mental disorders in the postgenomic age—a closer look at schizophrenia and language lateralization. Neurosci Biobehav Rev 59: 100–110.

Oertel-Knöchel V, Linden DEJ (2011). Cerebral asymmetry in schizophrenia. Neuroscientist 17: 456–467.

Oertel-Knochel V, Knochel C, Stablein M et al. (2012). Abnormal functional and structural asymmetry as biomarker for schizophrenia. Curr Top Med Chem 12: 2434–2451.

Okada N, Fukunaga M, Yamashita F et al. (2016). Abnormal asymmetries in subcortical brain volume in schizophrenia. Mol Psychiatry 21: 1460–1466.

Özgürdal S, Gudlowski Y, Witthaus H et al. (2008). Reduction of auditory event-related P300 amplitude in subjects with at-risk mental state for schizophrenia. Schizophr Res 105: 272–278.

Palaniyappan L, Liddle PF (2012). Does the salience network play a cardinal role in psychosis? An emerging hypothesis of insular dysfunction. J Psychiatry Neurosci 37: 17–27.

Park HY, Hwang JY, Jung WH et al. (2013). Altered asymmetry of the anterior cingulate cortex in subjects at genetic high risk for psychosis. Schizophr Res 150: 512–518.

Parola A, Lin JM, Simonsen A et al. (2023). Speech disturbances in schizophrenia: assessing cross-linguistic generalizability of NLP automated measures of coherence. Schizophr Res 259: 59–70.

Pascual-Marqui RD, Lehmann D, Koenig T et al. (1999). Low resolution brain electromagnetic tomography (LORETA) functional imaging in acute, neuroleptic-naive, first-episode, productive schizophrenia. Psychiatry Res Neuroimaging 90: 169–179.

Petty RG (1999). Structural asymmetries of the human brain and their disturbance in schizophrenia. Schizophr Bull 25: 121–140.

Petty RG, Barta PE, Pearlson GD et al. (1995). Reversal of asymmetry of the planum temporale in schizophrenia. Am J Psychiatry 152: 715–721.

Polich J, Criado JR (2006). Neuropsychology and neuropharmacology of P3a and P3b. Int J Psychophysiol 60: 172–185.

Reveley MA (1987). Left cerebral hemisphere hypodensity in discordant schizophrenic twins. Arch Gen Psychiatry 44: 625.

Ribolsi M, Daskalakis ZJ, Siracusano A et al. (2014). Abnormal asymmetry of brain connectivity in schizophrenia. Front Hum Neurosci 8: 1010.

Rollins CPE, Garrison JR, Arribas M et al. (2020). Evidence in cortical folding patterns for prenatal predispositions to hallucinations in schizophrenia. Transl Psychiatry 10: 387.

Rotarska-Jagiela A, van de Ven V, Oertel-Knöchel V et al. (2010). Resting-state functional network correlates of psychotic symptoms in schizophrenia. Schizophr Res 117: 21–30.

Sakuma M, Hoff AL, DeLisi LE (1996). Functional asymmetries in schizophrenia and their relationship to cognitive performance. Psychiatry Res 65: 1–13.

Satizabal CL, Adams HHH, Hibar DP et al. (2019). Genetic architecture of subcortical brain structures in 38,851 individuals. Nat Genet 51: 1624–1636.

Schijven D, Postema MC, Fukunaga M et al. (2023). Large-scale analysis of structural brain asymmetries in schizophrenia via the ENIGMA consortium. Proc Natl Acad Sci U S A 120: e2213880120.

Schlaepfer TE, Harris GJ, Tien AY et al. (1994). Decreased regional cortical gray matter volume in schizophrenia. Am J Psychiatry 151: 842–848.

Sha Z, Schijven D, Francks C (2021). Patterns of brain asymmetry associated with polygenic risks for autism and schizophrenia implicate language and executive functions but not brain masculinization. Mol Psychiatry 26: 7652–7660.

Shenton ME, Faux SF, McCarley RW et al. (1989). Correlations between abnormal auditory P300 topography and positive symptoms in schizophrenia: a preliminary report. Biol Psychiatry 25: 710–716.

Shenton ME, Dickey CC, Frumin M et al. (2001). A review of MRI findings in schizophrenia. Schizophr Res 49: 1–52.

Smallman RP, Barkus E, Azadbakht H et al. (2014). MRI diffusion tractography study in individuals with schizotypal features: a pilot study. Psychiatry Res 221: 49–57.

Sommer I, Aleman A, Ramsey N et al. (2001). Handedness, language lateralisation and anatomical asymmetry in schizophrenia. Br J Psychiatry 178: 344–351.

Southard EE (1915). General reviews and summaries: general psychopathology. Psychol Bull 12: 245–273.

Stanfield AC, Moorhead TWJ, Harris JM et al. (2008). Increased right prefrontal cortical folding in adolescents at risk of schizophrenia for cognitive reasons. Biol Psychiatry 63: 80–85.

Stephan-Otto C, Lombardini F, Núñez C et al. (2020). Fluctuating asymmetry in patients with schizophrenia is related to hallucinations and thought disorganisation. Psychiatry Res 285: 112816.

Strelets VB, Garakh ZV, Novototskii-Vlasov VY et al. (2006). Relationship between EEG power and rhythm synchronization in health and cognitive pathology. Neurosci Behav Physiol 36: 655–662.

Sumich A, Chitnis XA, Fannon DG et al. (2002). Temporal lobe abnormalities in first-episode psychosis. Am J Psychiatry 159: 1232–1235.

Szathmáry E, Smith JM (1995). The major evolutionary transitions. Nature 374: 227–232.

Thompson PM, Cannon TD, Narr KL et al. (2001). Genetic influences on brain structure. Nat Neurosci 4: 1253–1258.

Toga AW, Thompson PM (2003). Mapping brain asymmetry. Nat Rev Neurosci 4: 37–48.

Tsai LY (1983). Hemispheric asymmetries on computed tomographic scans in schizophrenia and mania. Arch Gen Psychiatry 40: 1286.

Umbricht D, Krljes S (2005). Mismatch negativity in schizophrenia: a meta-analysis. Schizophr Res 76: 1–23.

Van Den Heuvel MP, Fornito A (2014). Brain networks in schizophrenia. Neuropsychol Rev 24: 32–48.

van der Meulen M, Wierenga LM, Achterberg M et al. (2020). Genetic and environmental influences on structure of the social brain in childhood. Dev Cogn Neurosci 44: 100782.

van der Stelt O, Lieberman JA, Belger A (2005). Auditory P300 in high-risk, recent-onset and chronic schizophrenia. Schizophr Res 77: 309–320.

White T, Hilgetag CC (2011). Gyrification and neural connectivity in schizophrenia. Dev Psychopathol 23: 339–352.

Winterer G, Mulert C, Mientus S et al. (2001). P300 and LORETA: comparison of normal subjects and schizophrenic patients. Brain Topogr 13: 299–313.

Zhang Z, Shi J, Yuan Y et al. (2008). Relationship of auditory verbal hallucinations with cerebral asymmetry in patients with schizophrenia: an event-related fMRI study. J Psychiatr Res 42: 477–486.

Zhang R, Wei Q, Kang Z et al. (2015). Disrupted brain anatomical connectivity in medication-naïve patients with first-episode schizophrenia. Brain Struct Funct 220: 1145–1159.

Zhu Y, Wang S, Gong X et al. (2021). Associations between hemispheric asymmetry and schizophrenia-related risk genes in people with schizophrenia and people at a genetic high risk of schizophrenia. Br J Psychiatry 219: 392–400.

Handbook of Clinical Neurology, Vol. 208 (3rd series)
Cerebral Asymmetries
C. Papagno and P. Corballis, Editors
https://doi.org/10.1016/B978-0-443-15646-5.00009-9

Chapter 7

Hemispheric asymmetry in neurodegenerative diseases

STEFANO F. CAPPA*

University School for Advanced Studies (IUSS-Pavia), Pavia, Italy

Dementia Research Center, IRCCS Mondino Foundation, Pavia, Italy

Abstract

Hemispheric asymmetry in pathologic involvement is frequently observed in neurodegenerative disorders (NDD) and is responsible for differences in cognitive and motor clinical manifestations in individual patients. While asymmetry is modest in typical Alzheimer disease (AD), atypical AD presentations with prominent language impairment [logopenic/phonologic variant of primary progressive aphasia (L/Phv-PPA)] are associated with prevalent involvement of the language-dominant hemisphere. Similarly, in the frontotemporal dementia-amyotrophic lateral sclerosis (FTD-ALS) spectrum, the semantic (Sv) and nonfluent/agrammatic (Nf/Av) variants of PPA are due to asymmetric pathology involving the language-dominant hemisphere. A reversed (typically right-sided) pattern of asymmetry is often found in conditions associated with prominent disorders of behavior and social cognition (i.e., behavioral variant of frontotemporal degeneration—Bv FTD). Asymmetry is generally modest and less consistent in NDD with prevalent motor manifestations, such as Parkinson disease (PD). Overall, the pattern of hemispheric involvement reflects the network-specific selectivity of NDD and is compatible with the spreading of pathology along connection pathways.

Neurodegenerative disorders can affect the cerebral hemispheres asymmetrically, resulting in asymmetric clinical presentations of motor and/or cognitive manifestations [see Lubben et al., 2021 for a recent review]. The mechanisms responsible for the asymmetric presentation and clinical course of many neurologic disorders associated with different pathologies sharing the common feature of being "incurable and debilitating conditions that result in progressive degeneration and/or death of nerve cells" (JPND Research, 2023) are unknown, and there is the possibility that an improved comprehension may provide useful insights into the origin and functional role of hemispheric lateralization.

This chapter reviews the available evidence on hemispheric asymmetries in pathologic involvement and their clinical consequences in three major areas of neurodegeneration: the Alzheimer disease (AD) spectrum, the frontotemporal degeneration-amyotrophic lateral sclerosis (FTD-ALS) spectrum, and Parkinson disease (PD).

ASYMMETRY IN THE AD SPECTRUM[1]

In the words of the classic paper on the progression of neuropathology in AD (Braak and Braak, 1991) "The neuropathological hallmark of AD is the progressive accumulation of insoluble fibrous material which is normally not found in the human central nervous system. This material consists of extracellular amyloid (amyloid plaques, AP) and intraneuronal neurofibrillary (NF) changes. Furthermore, it is not distributed at random." The paper is based on the neuropathologic examination of 83 brains obtained at autopsy. Limited clinical information was available about the brain donors. Eight brains came from individuals who had been diagnosed with dementia, but "at neuropathological examination these brains did not meet the conventional criteria for diagnosis of fully developed AD." On the other hand, 21 brains were "from demented old-age individuals" and "displayed sufficient densities of isocortical NF changes … to confirm the clinical diagnosis of AD." Only one hemisphere was examined for each subject. The results of the neuropathologic study allowed the identification of three "stages" of amyloid deposition (A, B, and C) and six "stages" of NF changes. The practice of submitting only one hemisphere to neuropathologic examination was probably justified by the prevalence of cases in the advanced stages of disease at the time of death, where the extent of neurodegeneration is most likely to be bilateral. A study published in 1968 (Jamada and

[1]Abbreviations used in the chapter are listed at the end of the chapter before References section.

*Correspondence to: Stefano F. Cappa, University School for Advanced Studies (IUSS-Pavia), Pavia, Italy. Tel: +39-0382-375811, Fax: +39-0382-375899, E-mail: stefano.cappa@iusspavia.it.

Mehraein, 1968) was dedicated to a careful quantitative analysis of "senile modifications" in a sample of 48 demented patients, including 22 with AD pathologic hallmarks. Symmetric blocks of tissue were taken from several cortical and subcortical regions, with a special focus on the limbic system, and the number of senile plaques and neurofibrillary tangles were counted and compared. While clear intrahemispheric differences were found (with the highest counts in the amygdala), the differences between the right and left hemispheres were not significant in any brain area. A different finding was published in a study of the quantitative relationship between cell loss in subpopulations of the nucleus basalis of Meynert (Nb) and SP in the connected cortical areas (Arendt et al., 1985). Out of five brains of patients affected by AD, in which tissue was examined from both hemispheres, three showed a marked asymmetry of cell loss, both in the Nb and in cortical areas, with a more severe left hemispheric involvement. The handedness of the patients was not known, and there was no significant relationship with age or brain weight. Four out of five brains showed the same left-sided predominance also for SP counts. None of the differences, however, was statistically significant. Two years later, a significant hemispheric asymmetry (three out of five, but in one case with higher counts in the right hemisphere) was reported for SP and NF by Wilcock and Esiri (1987). Janota and Mountjoy (1988) reported a highly significant asymmetry for SP, NFT, and granulovacuolar degeneration in the temporal lobes of a large brain sample (56 subjects) for which no clinical information was available. In the latter case, however, no prevalent direction was observed. Similar findings were reported by Moossy et al. (1989): asymmetries were observed for SP, NF tangles, choline acetyltransferase, and acetylcholinesterase in 21 brains, without, however, a systematic difference between the two hemispheres. The asymmetry was reduced with increasing severity only in the case of SP. Hemispheric asymmetry in Braak staging has also been observed (Stefanits et al., 2012) in six patients (right more advanced than left in four). The authors of the latter report concluded by recommending a bilateral neuropathologic examination, underlining the necessity to evaluate the clinical correlates of the asymmetry and the influence of handedness. This information is actually minimal or absent in most neuropathologic studies. At the clinical level, however, reports of "extraordinary single cases or small series of patients with distinctive clinical features and focal degenerative change in the corresponding neuroanatomical areas" (Förstl and Fischer, 1994), in which pathologic evidence of focal hemispheric AD is associated with "atypical" clinical presentations, in particular aphasia with left frontotemporal neurodegeneration, are available from the 1980s (Morris et al., 1984; Pogacar, 1984). Four patients with clearly asymmetric clinical presentation (language impairment in three, visuospatial disturbances in one) showed postmortem asymmetric atrophy (respectively a worse on the left and on the right). Four additional cases of asymmetric pathology were described by Giannakopoulos et al. (1994): in this small series, however, while the three patients with more severe involvement of the left hemisphere had an aphasic presentation, the right-sided case had a global cognitive impairment. The authors make two important observations in this study. First, they report that severe neocortical involvement is associated with a relative sparing of the hippocampal region by AD pathology. Second, they suggest that the asymmetries may reflect the disruption of specific associative cortico-cortical circuits in "atypical" AD subtypes. A pathologic study on a large series of cases of aphasic presentation of AD has confirmed these findings. In contrast to typical AD, in this subgroup there was a male predominance, no association with the APOE e4 genotype, the neurofibrillary pathology was more severe in the language-dominant hemisphere, and the Braak pattern of hippocampo-entorhinal prominence was inverted in favor of the neocortex (Fig. 7.1) (Mesulam et al., 2014).

The progress in neuroimaging methods has had a huge impact on the number of observations of asymmetries in AD clinical presentation and their brain correlates and has opened the possibility to perform longitudinal studies. In particular, functional imaging methods, such as positron emission tomography with fluorodeoxyglucose (PET-FDG) (Foster et al., 1983; Haxby et al., 1985) and single-photon emission tomography (SPET) with perfusion tracers (Perani et al., 1988) has provided evidence of asymmetric involvement correlating with clinical phenotype from the early disease phase, and remaining stable during clinical progression (Jagust et al., 1988).

More recently, molecular imaging is providing additional information about the in vivo neuropathology of AD by means of tracers of beta-amyloid and tau deposition (Perani et al., 2021). Asymmetry in beta amyloid deposition has been frequently found in the very early phases of the AD continuum (i.e., asymptomatic, biomarker-positive, and prodromal). There was a nonsignificant trend toward left-sided prevalence and a progressive reduction of asymmetry with disease progression (Kjeldsen et al., 2022). Studies with tau tracers suggest that an asymmetric deposition is associated with an early age of onset and a malignant progression. There were no significant differences in the direction of the asymmetry, which was also not associated with specific cognitive profiles (Lu et al., 2023).

Taken together, the findings summarized above support the concept of a distinction between typical and atypical AD (Dubois et al., 2021), with an important role of hemispheric asymmetry in the latter condition.

An "atypical" AD form is the logopenic/phonologic variant of primary progressive aphasia (LPhv-PPA) (Gorno-Tempini et al., 2011). The primary progressive aphasias (PPA) are clinical syndromes characterized by progressive language dysfunction, reflecting neurodegeneration, which, at least in the early stages, is selective for the brain language networks, and can be due to a variety of neuropathologic mechanisms. While cases of PPA had been described in the classic neurologic literature (Luzzatti and Poeck, 1991), this syndromic group became a recognized clinical entity only after the reports of Wechsler (1977) and Mesulam (1982). Two main clinical presentations were originally described: semantic dementia (Hodges et al., 1992) and progressive nonfluent aphasia (Grossman et al., 1996). The clinical and linguistic profiles of several patients, however, did not fit into this binary subdivision. The third variety of PPA, logopenic aphasia, was described by Gorno-Tempini et al. (2004). The clinical

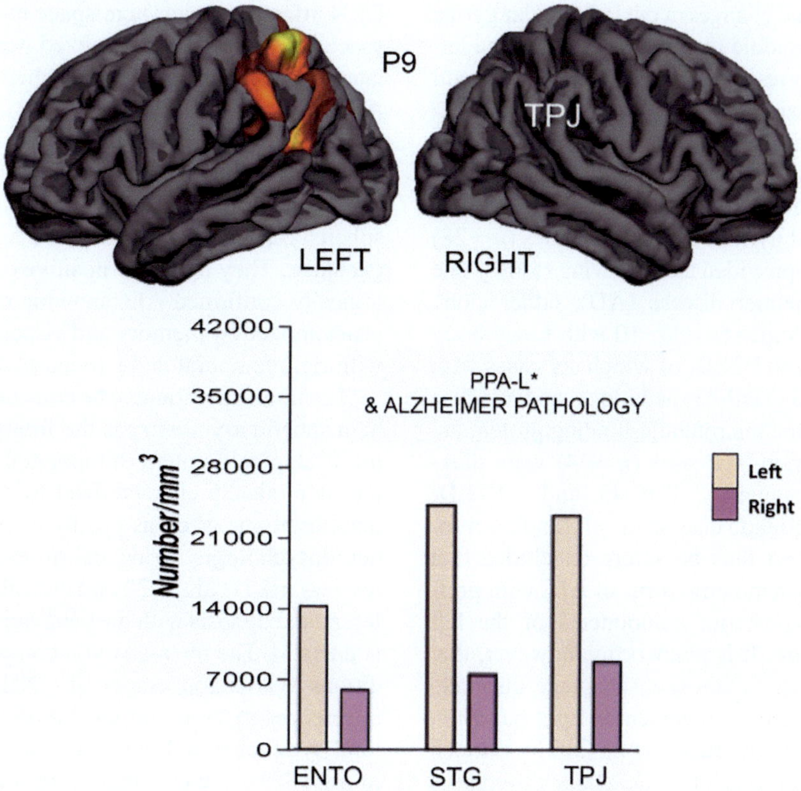

Fig. 7.1. Atypical distribution of pathology in a patient with Lv-PPA (PPA-L). Top: quantitative MRI imaging shows the peak of pathology in the left temporoparietal junction (TPJ). Bottom: the number of neurofibrillary tangles per cubic millimeter is greater in language-related neocortical areas [TPJ, superior temporal gyrus (STG)] than in the entorhinal cortex (ENTO) and in the left than in the right hemisphere. From Mesulam MM, Weintraub S, Rogalski EJ et al. (2014). Asymmetry and heterogeneity of Alzheimer's and frontotemporal pathology in primary progressive aphasia. Brain 137: 1176–1192; data taken from Gefen T, Gasho K, Rademaker A et al. (2012). Clinically concordant variations of Alzheimer pathology in aphasic versus amnestic dementia. Brain 135: 1554–1565. Used with permission from Oxford University Press.

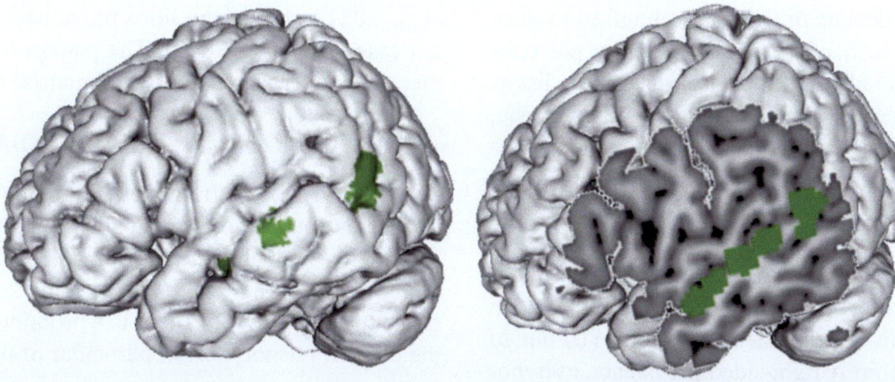

Fig. 7.2. Anatomic likelihood estimation meta-analysis of 201 foci and 175 subjects with a diagnosis of LPv-PPA, showing two left-lateralized clusters of atrophy, encompassing middle and superior temporal gyri. Volumetric rendering of the cluster-level FWE-corrected results, with a sagittal axial cut at $x = -58$ (right). From Conca F, Esposito V, Giusto G et al. (2022). Characterization of the logopenic variant of primary progressive aphasia: a systematic review and meta-analysis. Ageing Res Rev 82: 101760. Elsevier.

features were slow speech output with frequent pauses and simple, syntactically correct sentences, impaired confrontation naming, and defective repetition. Word comprehension and semantic association were preserved, while syntactic comprehension and single-word reading were impaired. While language impairment was the main clinical manifestation, some nonlinguistic deficits could also be present (acalculia and reduced free recall). A possible underlying cognitive dysfunction is a phonologic loop disorder, undermining the capability to maintain online verbal information, which is responsible for deficits in sentence repetition and comprehension (Gorno-Tempini et al., 2008). A comprehensive review and meta-analysis of the studies on this syndrome can be found in Conca et al. (2022) (Fig. 7.2). The typical neural correlates of L/Phv PPA, as shown with structural and functional imaging, are atrophic changes in the left angular gyrus, the posterior third of the left middle temporal gyrus, and the left superior temporal sulcus. The results of an activation likelihood estimation (ALE) meta-analysis (Conca et al., 2022) including 201 foci and 175 subjects yielded two left-lateralized clusters of atrophy, encompassing middle and superior temporal gyri. An exploratory analysis further identified three clusters in the left fusiform and inferior temporal gyri. Postmortem studies ($n=28$) including a total of 163 brains provided the following results. The majority ($n=129$) had Alzheimer disease (AD), either alone ($n=112$) or with other pathologies ($n=17$, 10 with Lewy body dementia (LBD) and seven with FTLD, of which six cases with TAR DNA-binding protein-43 (TDP-43) pathology, and one case with inclusions formed from the microtubule-binding protein tau (FTLD-tau). The remaining non-AD cases ($n=34$) were diagnosed as FTLD ($n=29$, of which 22 TDP-43, and 7 FTLD-tau), DLB ($n=3$), Creutzfeldt-Jakob disease ($n=1$), and cerebrovascular disease ($n=1$). It can thus be safely concluded that L/Phv-PPA is most often an asymmetric form of AD, with preferential involvement of the posterior components of the left language-dominant hemisphere. It is noteworthy, however, that the number of "crossed aphasia" cases (i.e., language disorders with prominent right-hemispheric involvement in right-handers), 7.8% in the large series reported by Buciuc et al. (2021), is much higher in PPA than in stroke aphasia (1–3%). Minor differences were observed in comparison to standard left hemispheric PPA (less severe, less phonologic impairment). The meaning of this difference is a matter of speculation.

The available evidence about the other classic "atypical" AD variant, posterior cortical atrophy (PCA) (Benson et al., 1988) does not directly support the notion of a selective "right-hemispheric" asymmetric AD. This clinical syndrome is characterized by the progressive decline of visuo-perceptual and visuospatial abilities, together with other functions of the posterior neocortical areas (Crutch et al., 2012). Consensus criteria list as most frequent features space perception deficits, simultanagnosia, object perception impairment, environmental disorientation, constructional apraxia, and Balint syndrome (Crutch et al., 2017). As in the case of PPA, episodic memory and insight are relatively preserved in the initial stages of disease. The neuroimaging features are focal structural or functional abnormalities in the occipital, parietal, and/or occipitoparietal-temporoparietal cortices, which are often asymmetric, with a right-sided prevalence (10 out of 15 cases reported by the clini A right-sided prevalence of hypometabolism has also been reported in late-onset cases) (Isella et al., 2022). Overall, these findings indicate that the PCA syndrome, characterized by a wide array of manifestations reflecting the involvement of multiple systems associated with high-order vision, is usually the consequence of bihemispheric involvement, with a possible right-sided prevalence.

It is noteworthy that unilateral spatial neglect (ULN) is not infrequently reported in PCA but without clear hemispace prevalence. In the relatively large series of patients reported by Andrade et al. (2013), 19/27 patients showed pathologic spatial biases in classic tasks to assess ULN, a disorder that is usually more severe and long-lasting after right-hemispheric damage. ULN affected the left hemispace in 10 cases and the right hemispace in 9 cases. The lateralized performance impairment correlated with the side of hemispheric hypoperfusion on SPET. A different approach was used in an interesting review by Abbate et al. (2019), which screened the AD literature for cases with well-documented right-sided asymmetry at neuroimaging and predominant cognitive deficits localizable on the right hemisphere; cases with an explicit diagnosis of atypical AD were excluded. They found 13 positive cases, including 3 with pathologically confirmed AD, showing clinical-radiologic syndrome, characterized by memory and visuospatial impairment associated with right temporal and parietal involvement.

The final AD variant to be considered from the point of view of hemispheric asymmetry is the frontal variant AD (Fv-AD). This third "atypical" pattern of topographic distribution of AD pathology is clinically characterized by the early onset of behavioral manifestations, such as apathy or disinhibition, associated with neuropsychologic evidence of executive dysfunction [recent reviews in Li et al., 2020 and Ossenkoppele et al., 2022]. The differential diagnosis with the behavioral variant of FTD (see below) is difficult. The meta-analytic comparison between the two conditions (Ossenkoppele et al., 2022) indicated minor clinical differences at the group level (with trends for less severe behavioral symptoms and worse memory dysfunction in fv-AD than in bv-FTD). Neuroimaging results were compatible with a "continuum" of cortical involvement, from an AD-like pattern of prominent temporoparietal atrophy or metabolic dysfunction to a bv-FTD pattern of predominant frontal and anterior temporal involvement. The differential diagnosis is thus based on the results of biomarker studies for beta-amyloid deposition (Rabinovici et al., 2011). PET tau tracers reflect the topographic distribution of AD pathology in the frontal variant (Therriault et al., 2021), and a right-sided prevalence is observed in individual cases (Li et al., 2020). The pattern of hemispheric involvement, however, is generally symmetric.

ASYMMETRY IN THE FTLD-ALS SPECTRUM

The FTLD-ALS spectrum includes multiple phenotypic presentations due to several different pathologies, associated according to a complex probabilistic mapping (Mackenzie and Neumann, 2016; Kovacs et al., 2022; Forrest and Kovacs, 2023). Asymmetric hemispheric involvement is a prominent feature of several of the clinical phenotypes, in particular of the aphasic variants.

The semantic variants

Two "semantic variants" have been reported, which are the consequences of mirroring patterns of hemispheric asymmetry in the early to mid-stages of the same condition. The semantic variant of primary progressive aphasia is the most frequently observed clinical presentation, and is due to prevalent left-sided hemispheric involvement (Gorno-Tempini et al., 2004), while the semantic-behavioral variant is the clinical expression of predominantly right-sided pathology (Younes et al., 2022). With the progression of disease resulting in bihemispheric involvement, the distinction progressively fades, and the resulting clinical picture is the severe clinical condition of semantic dementia, characterized by multiple cognitive impairments and loss of functional autonomy (Warrington, 1975; Snowden et al., 1989; Hodges et al., 1992).

The underlying pathologic process is in most instances a TDP43 proteinopathy, accounting for 70%–80% of cases (Spinelli et al., 2017). The pathology (most frequently TDP type C) has a striking predilection for the anterior part of the temporal lobes, with extension to the adjacent insula and orbitofrontal cortex (Mesulam et al., 2022). The same pathologic process is responsible for both left-sided and right-sided predominant presentations and is associated with a similar pattern of progression, medial to lateral and anterior to posterior in the temporal lobe, and toward a bihemispheric pattern of involvement (Borghesani et al., 2020). It is noteworthy that a similar topographic pattern can be observed also in patients affected by different pathologies, such as Pick disease (Giannini et al., 2019).

The clinical expression of the neuropathologic pattern of involvement is remarkable in the early stages for its selectivity for semantic knowledge. The left-sided sv-PPA is manifested initially by impairment in word finding, followed by deficits in word comprehension and nonverbal semantics (Gorno-Tempini et al., 2011). Anomias and semantic paraphasias are prominent in connected speech, while articulatory and morphosyntactic skills are spared, with the possible presence of paragrammatic phenomena and the production of complex sentences with little information content (Wilson et al., 2010). The semantic deficit is also reflected in reading impairment, leading in English-speaking patients to the classic picture of surface dyslexia (Wilson et al., 2012).

The recently proposed label for the less frequently reported, predominant right-sided presentation is the semantic-behavioral variant (Sbv-FTD) (Younes et al., 2022). Here, the clinical picture is characterized by behavioral changes and disorders of person recognition (Chan et al., 2009). In depth neuropsychologic assessment of social cognition abilities suggests that a progressive decline in semantic memory for concepts of socioemotional relevance may be considered as the basic disorder resulting in the clinically manifest loss of empathy and person-specific knowledge (Younes et al., 2022).

The two variants thus reflect the different contributions of the right and left anterior temporal lobes and their different connectivity patterns to the representation of semantic knowledge, characterized by a (relative) specialization pattern [see Woollams and Patterson, 2018 for a discussion].

Nonfluent/agrammatic variant of primary progressive aphasia

This is the second form of aphasic presentation of frontotemporal lobar degeneration. The main features of the language disorder are motor speech impairment and agrammatism (Gorno-Tempini et al., 2011). Both aspects are included in the classic concept of "nonfluency," i.e., quantitatively reduced, labored speech, interrupted by pauses, and with an altered melodic profile (dysprosodia). The agrammatic-type manifestations are detectable through analysis of extended production in narrative or figure description tasks (Ash et al., 2013; Boschi et al., 2017). The differential diagnosis with L/Ph-PPA is not always easy and crucially depends on the distinction between the features of speech production, which in L/Phv-PPA is characterized by hesitations, pauses, and phonologic errors rather than by phonetic distortions and prosodic disturbances (speech apraxia) typical of the Nf/Av-PPA. Often the disorder is associated with bucco-facial apraxia (Botha et al., 2014). Longitudinal studies have shown that some patients who present with progressive apraxia of speech (AS) progress to mutism, in the absence of other language deficits, while in others, apraxia remains predominant and is associated with mild language impairment (Josephs et al., 2013). In classic vNF/Av, marked length and complexity effects are present, whereas pure forms or those with predominant AS are characterized by syllabic segmentation with lengthening of intersegmental tenses (Josephs et al., 2013). Agrammatism is manifested by the omission of grammatical morphemes (articles, auxiliaries, verbal inflections, etc.), and simplified or absent syntactic structures (isolated words), resulting in a reduction in sentence length and decreased production of verbs. Language comprehension is preserved for single words but may be impaired for sentences, in particular of complex syntactic structure.

The Nf/Av is most frequently due to a tauopathy, i.e., Pick disease (3R), or, most frequently, with the neuropathologic features of the corticobasal disease (CBD) or progressive supranuclear palsy (PSP) (both 4R taupathies) (Mesulam et al., 2022). AD or TDP43 pathology has been infrequently reported (Spinelli et al., 2017). As in the case of Sv, the topographic involvement is predominantly left hemispheric, affecting the anterior part of the language network, with prominent involvement of left prefrontal areas and of the long dorsal tracts of the left hemisphere, i.-e., the superior longitudinal fasciculus and the aslant tract (Catani et al., 2013; Mandelli et al., 2014). Combining resting state functional connectivity and MRI tractography with the assessment of longitudinal progression of atrophy in patients with vNF/A (Mandelli et al., 2016) showed that the region of earliest involvement is the superior opercular portion of the inferior frontal gyrus. This region was used as a seed for assessment of the structural and functional connectivity pattern in control subjects, which included the left supplementary motor area and the left prefrontal, inferior parietal, and temporal regions. The aslant tract connected the inferior frontal gyrus with the medial frontal regions (anterior cingulate, supplementary motor region). The strength of functional connectivity in controls correlated with the longitudinal volumetric progression of atrophy in patients, supporting the hypothesis that the spread of neurodegeneration is following specific anatomic and functional neuronal network architectures. A similar, network-specific spreading model has been reported also in the case of Sv PPA (Agosta et al., 2023). It is noteworthy that in cases of progressive selective speech apraxia without language impairment there is no left hemisphere predominance of pathologic involvement (Robinson et al., 2023).

To summarize, both aphasic variants of FTD are characterized by a consistent pattern of asymmetric pathology, with a predominant involvement of the language-dominant hemisphere. There is evidence supporting the longitudinal spreading of pathology following the architecture of specific subnetworks (Battistella et al., 2020): the lexical-semantic memory network in the case of Sv-PPA, the speech production network for Nf/Av-PPA.

The behavioral variant

This variant is the most common clinical presentation of the FTLD spectrum (Logroscino et al., 2023) and a leading cause of early-onset dementia (Ratnavalli et al., 2002). The main clinical manifestations are behavioral changes, reflecting a progressive disorder of social cognition (Rascovsky et al., 2011). These include disinhibition, apathy and inertia, loss of empathy, perseverative behavior, hyperorality, and dietary changes. In the early stages, the behavioral manifestations dominate the clinical picture, in the context of relative preservation of other aspects of cognition. With progression, executive dysfunction and language impairment lead to further decline in functional abilities.

The Bv of FTLD is most often due to TDP 43 or tau proteinopathy, associated with a specific topographic pattern of brain involvement. A "staging system" has been proposed for pTDP43 pathologic progression. In a postmortem study, cases with the lowest burden of pathology (pattern I) showed pTDP-43 lesions in the orbital gyri, gyrus rectus, and amygdala. With increasing burden of pathology (pattern II) pTDP-43 lesions could be observed in the middle frontal and anterior cingulate gyrus, in anteromedial temporal lobe areas, the superior and medial temporal gyri, striatum, red nucleus, thalamus, and precerebellar nuclei. More advanced cases showed a third pattern (III) characterized by the involvement of the motor cortex, bulbar somatomotor neurons, and the spinal cord anterior horn. Cases with the highest burden of pathology (pattern IV) had pTDP-43 lesions extending to the visual cortex. The data are consistent with the hypothesis of spreading of pTDP-43 pathology along axonal pathways (Brettschneider et al., 2014).

A right hemispheric prevalence of involvement has been frequently reported, supporting, together with the Sbv described above, the concept of a network-selective pathologic process involving a system characterized by right-hemispheric asymmetry. It must however be underlined that, while asymmetry of pathologic involvement is frequently observed in imaging studies of the early to intermediate stage of disease, the direction of asymmetry is not consistently right-sided. For example, while a right-sided asymmetry is typically found in cases with predominant temporal atrophy (which can probably be reclassified as Sbv-FTD) (Whitwell et al., 2009), in a large series (85 cases) the majority of subjects had symmetric involvement (65%), and, in the asymmetric sample, 20% has more severe involvement of the left hemisphere, 15% of the right. The latter group was characterized by more severe behavioral manifestations, while the presence of frontal asymmetry in general was associated with progranulin mutations, suggesting a link with genetic differences (Whitwell et al., 2013). Neuropathologic evidence of hemispheric asymmetry, with a prevalence of TDP-43 inclusion density and microglia, and lower neuronal density in the left hemisphere than in the right hemisphere, has also been reported in PPA due to progranulin mutations (Kim et al., 2016). A neuropathologic study (Irwin et al., 2018) is specifically dedicated to the issue of asymmetry and provides evidence for a complex asymmetry pattern, i.e., more severe pathology in the right orbitofrontal gray matter compared to the left and higher pathology in ventrolateral temporal lobe gray matter of the left hemisphere compared to the right. Correlation with clinical data indicated an increased severity of behavioral disturbances with asymmetric right orbitofrontal gray matter pathology. In contrast, reduced semantic fluency was associated with asymmetric white matter pathology in the left ventrolateral temporal lobe.

Overall, the pathologic asymmetry data in a disorder characterized by social cognition impairment appear congruent with the involvement of a complex cognitive system associated with the function of both hemispheres, with differential asymmetry patterns. While the right temporal dominance appears to be confirmed for emotional function [see Gainotti (2019) for a review], other aspects, such as the theory of mind, have been associated with the participation of both hemispheres, although with a right-ward prevalence, by studies in stroke patients (meta-analysis in Adams et al. (2019) and by functional imaging (Boccadoro et al., 2019)).

Progressive supranuclear palsy and cortico-basal syndrome

These two neuropathologic forms of 4R tauopathy (Irwin et al., 2015) are associated with distinct but frequently overlapping clinical features, resulting in multiple clinical syndromes (Armstrong et al., 2013; Höglinger et al., 2017). To further complicate the clinicopathologic correlation, the syndromes are characterized, respectively, by a symmetric, akinetic-rigid syndrome with postural instability and vertical supranuclear gaze palsy (Richardson syndrome) and by a typically asymmetric apraxic-dystonic upper limb involvement with asymmetric cortical signs and distal myoclonus [typical cortico-basal syndrome (CBS)] can also be the expression of other neurodegenerative disorders, such as AD or Lewy body disease. In CBD, the pattern of asymmetry is most frequently left hemispheric (Niethammer et al., 2014). The presence of left-sided hypometabolism in patients clinically diagnosed as either PRS or CBS is associated with progressive aphasia, usually of the Nf/Av (Dodich et al., 2019) (Fig. 7.3). Right hemispheric-predominant pathologic involvement is less frequent, and is associated with right ULN (Silveri et al., 2011).

Amyotrophic lateral sclerosis

Amyotrophic lateral sclerosis (ALS), or motor neuron disease, belongs to the FTLD spectrum, as these two disease groups are due to an overlapping neuropathology, characterized by TDP-43-positive cytoplasmic neuronal inclusions in most ALS, and a large part of FTD brains (Neumann et al., 2006). Originally considered a pure motor disorder, infrequently associated with dementia sharing clinical features with FTD, ALS is now conceived as a disorder of movement and cognition, affecting motor and extra-motor brain systems (Strong et al., 2017).

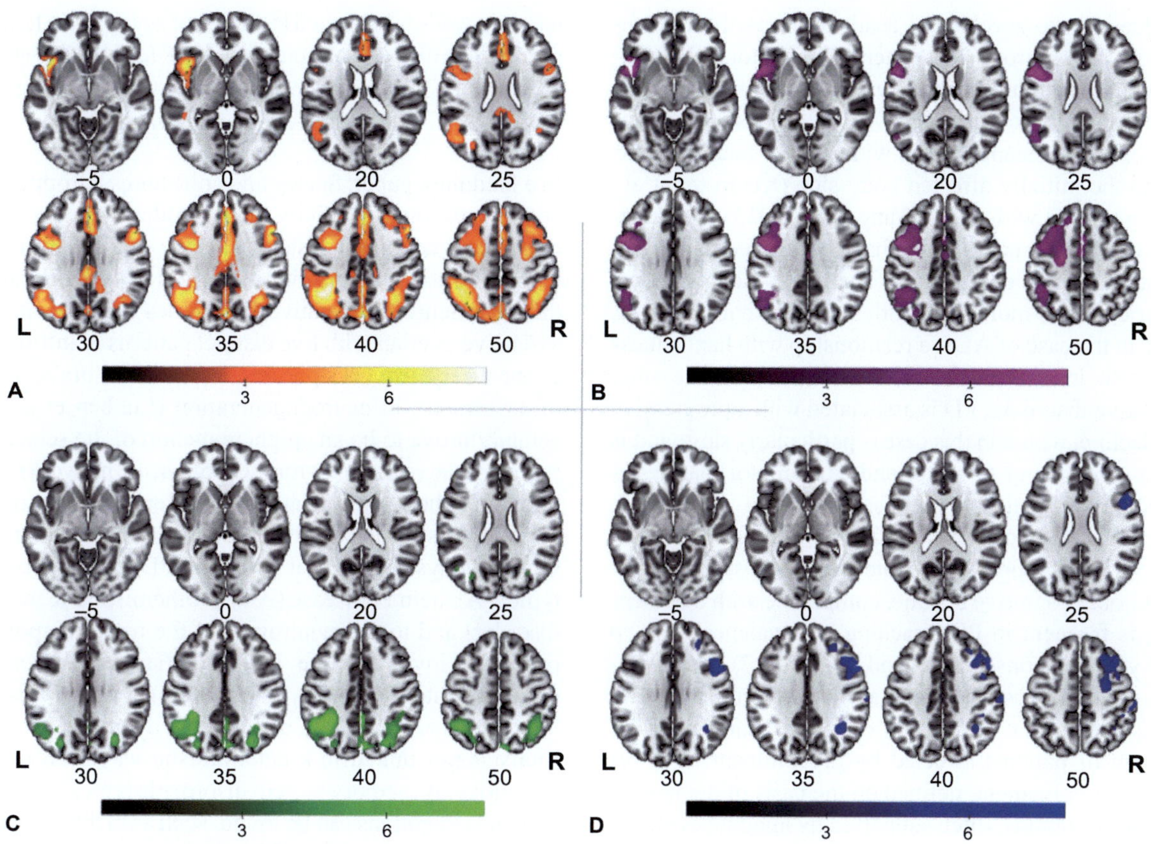

Fig. 7.3. Statistical parametric imaging FDG PET group analysis in 12 patients with a clinical diagnosis of Nf/Av-PPA. (A) In the whole PSPs sample ($n = 37$); and in patient subgroups according to language profiles: (B) nfv-PPA/PSPs ($n = 9$), (C) LANG (subtle language dysfunction not fulfilling PPA criteria)-PSPs ($n = 8$), (D) NOL (no language deficits)-PSPs ($n = 20$). From Dodich A, Cerami C, Inguscio E et al. (2019). The clinico-metabolic correlates of language impairment in corticobasal syndrome and progressive supranuclear palsy. NeuroImage Clin 24: 102009. Open access Elsevier.

ALS is a heterogeneous disorder, characterized by high phenotypic variability (Swinnen and Robberecht, 2014). At the motor level, the onset of weakness in ALS is often asymmetric. There has been some evidence supporting the hypothesis that the amount of physical use may play a role in the site of onset. Concordance between the side of onset and handedness, but only in the case of the upper limb, has been reported by Turner et al. (2011), and Zhang et al. (2014) found a general prevalence of left motor area atrophy in MRI studies of ALS patients. This pattern is however not consistent, as indicated by a study of 30 right-handers, where an equal number of subjects had a right or left-sided onset (Devine et al., 2015). It is noteworthy that, irrespective of the side of first limb weakness, cortical density was significantly reduced in the dominant (left) motor cortex hand area, as indicated by a left–right comparison of voxel-based morphometry data. Additional extension of atrophy in the left somatosensory and superior temporal cortex was found in the sample with right-sided atrophy. Cognitive and behavioral manifestations are found in about 50% of patients [see reviews in Beeldman et al. (2016) and Raaphorst et al. (2012). The features of cognitive and behavioral impairment are largely overlapping with those of Bv-FTD, and are associated with a similar patterns of predominantly frontotemporal cortical involvement (Consonni et al., 2018). Behavioral manifestations, in particular disinhibition, have a high correlation with right-sided atrophy in frontotemporal and cingulate areas (Consonni et al., 2019). While subtle language impairment can be present, clinically overt primary progressive aphasia is less frequent (Saxon et al., 2017; Gazulla et al., 2019) and appears to be associated with the rare TBK1 mutation (Caroppo et al., 2015).

Overall, also in the case of ALS, the presence of extra-motor manifestations is compatible with the spread of the pathologic process of inclusion and dissemination of phosphorylated TDP43 pathology from a disease epicenter in the motor cortical areas, following a trajectory based on proximity and connectivity (Brettschneider et al., 2015; Consonni et al., 2020).

ASYMMETRY IN PD AND LEWY BODY DISEASE (LBD)

Parkinson disease (PD) is a common neurodegenerative disorder, affecting millions of people worldwide (Bloem et al., 2021). The clinical manifestations (bradykinesia, rigidity, rest tremor) reflect the involvement by neurodegeneration of the motor system, but nonmotor manifestations, including cognitive and behavioral impairments, are also present. The core pathogenetic mechanism is pathologic aggregation of alpha-synuclein, leading through multiple mechanisms to accelerated death of primarily dopaminergic neurons of the nigrostriatal system. The motor manifestations are frequently asymmetric in the early stages of the disease, reflecting neuropathologic evidence of significant asymmetry of degeneration of the substantia nigra with greater neuronal loss contralateral to the initially affected body side (Kempster et al., 1989). Imaging studies with MRI (Wang et al., 2015) and dopaminergic ligands (Tatsch et al., 1997) have confirmed the neuropathologic data (Djaldetti et al., 2006). There is evidence that the left hemispheric side is more frequently affected than the right, suggesting, as in the case of ALS, a relationship with handedness [see meta-analysis in van der Hoorn et al. (2012)]. As the other neurodegenerative disorders, PD is associated with a progressive longitudinal decline, which in this case is particularly slow, and is associated with worsening of motor and extra-motor manifestations. The alpha-synuclein protein is capable of cell-to-cell, prion-like propagation (Goedert et al., 2017), leading to progressive neocortical involvement in the late stage of disease (Braak et al., 2003). Global cognitive decline, compatible with a dementia diagnosis, is frequent in PD, reaching a prevalence close to 50% after 10 years postonset (Aarsland et al., 2017). Some evidence is suggesting a possible role of asymmetry in modulating the characteristics and the severity of cognitive manifestations. The comparison of patients affected by predominant right- vs left-hemispheric involvement, defined on the basis of the laterality of motor manifestations, often supported by imaging evidence of asymmetric involvement of the dopaminergic system and its cortical projections, has indicated that language and verbal memory performance are typically more affected in the cases with prevalence of right-hemispheric involvement, while the reverse asymmetry is associated with more severe impairment in tasks of visuospatial cognition and nonverbal memory [see review in Verreyt et al., 2011]. A prevalence of left-hemispheric cortical atrophy in the early stages of the disease has been reported (Claassen et al., 2016), and is associated with a higher risk of cognitive decline than the right asymmetric presentation (Voruz et al., 2022). It is also noteworthy that patients with right-sided symptoms have an increased risk of both short- and long-term cognitive and behavioral side effects after deep-brain stimulation of the subthalamic nucleus for symptom control (Voruz et al., 2023).

Lewy body disease (LBD) is another form of alpha-synucleinopathy, accounting for 5%–7% of dementias in the elderly population (Walker et al., 2015). The main clinical features are progressive cognitive decline, characterized by fluctuation and changes in alertness, visual hallucinations, parkinsonian manifestations, and REM-behavior sleep disorder (McKeith et al., 2017). Motor manifestations appear to be more symmetric from onset than in the case of PD (Gnanalingham et al., 1997). A recent imaging study with [18F]-FE-PE2I PET reports a more symmetric striatal dopaminergic dysfunction in DLB than in PD patients (Fedorova et al., 2023). This finding may support the "body first" proposal for spreading alpha-synuclein pathology proposed for DLB (Borghammer, 2021), according to which the original enteric epicenter results in a bilateral propagation of pathology via the autonomic nervous system (Fedorova et al., 2023). It must be underlined, however, that there are well-documented instances of asymmetric LBD pathology, resulting in the clinical phenotype of L/Phv-PPA (Buciuc et al., 2021).

CONCLUSIONS

In a landmark paper, Seeley and colleagues introduced the crucial concept of the specificity of neurodegenerative disorders for large-scale cognitive systems (Seeley et al., 2009). An analysis of anatomic and connectivity changes in the early stages of five different neurodegenerative syndromes was shown to exhibit an extensive overlap with five distinct patterns of intrinsic functional connectivity networks present in the human brain. The "enigma" of asymmetry in neurodegeneration (Lubben et al., 2021) may actually prove to be an epiphenomenon of the selectivity of neurodegenerative diseases for the large-scale networks of cognition, reflecting their typical laterality pattern (Mesulam, 1990). The selectivity is evident in the case of disorders primarily involving the motor system, such as ALS and PD, but may be extended to typical Alzheimer disease (AD) (a bihemispheric memory system disorder) and to the syndromes of the frontotemporal spectrum, primarily involving the language (strongly asymmetric) and social cognition (partially asymmetric) systems. It is noteworthy, however, that in the case of cognitive disorders, the system specificity is not linked in a deterministic way with the underlying pathology. A memory system disorder is typically associated with AD, but exceptions can be found, such as in the case of the limbic-predominant age-related TDP-43 encephalopathy (LATE) syndrome (Nelson et al., 2019); in the case of language and social cognition, multiple pathologies can be associated with the system specificity (tauopathy, TDP43), but also AD and other less frequent conditions (Mackenzie and Neumann, 2016). The relationship between language selectivity and asymmetry in hemispheric involvement is well exemplified by the progressive aphasias, in which subcomponents of the language system are specific targets of the neurodegenerative process (Battistella et al., 2020). In the case of visuospatial and visuo-perceptual functions, the selectivity of the underlying large-scale networks has a more complex relationship with hemispheric asymmetries (Hervé et al., 2013). An exception is the ventral attentional network, responsible for bottom-up, sensory-driven attentional orienting, which is mainly lateralized to the right side of the brain and includes the temporoparietal junction and the ventral frontal cortex (Corbetta and Shulman, 2011). As reviewed above, a few cases of ULM have been reported in the case of predominant right hemispheric involvement by neurodegeneration due to focal AD [see also Venneri et al., 1998 and Lorusso et al., 2020 for a case of pure allocentric left-sided neglect in a PCA patient] or CBD (Silveri et al., 2011) pathology.

The same complexity applies to the social cognition brain systems, which have been the focus of intense investigations in recent years (Blakemore, 2008; Rankin, 2020). While the social cognition system involves multiple structures in both hemispheres, a right-sided, mostly temporal prevalence for specific aspects, such as emotion (Gainotti, 2019), social conceptual processing (Pobric et al., 2016), and human motion detection (Van Overwalle, 2009) has been suggested by clinical and imaging studies.

While the concept of "prion-like" spreading of misfolded proteins (Brettschneider et al., 2015) is suitable to explain the dissemination of pathology starting from an "epicenter" along the pathways of system-specific functional connectivity (Agosta et al., 2023), the mechanisms responsible for the different directions of dissemination of the same proteinopathy remain unknown. Most of the mechanisms proposed to explain asymmetry in NDD are based on the presence of predisposing factors resulting in a hemispheric-specific susceptibility to disease (Lubben et al., 2021). One hypothesis formulated for PPA, which has received some support from epidemiologic studies (Rogalski et al., 2008) is that a history of learning disabilities (LD) may represent a risk factor, as a persistent vulnerability, which can be compensated but becomes manifest in the case of onset of a ND condition, orienting the distribution toward the "weak" system (according to the traditional medical concept of "diathesis," i.e., "constitutional predisposition"). The increased prevalence of LD has however been confirmed only for the L/Phv of PPA (Miller et al., 2013), and the existence of a causal association between dyslexia, developmental speech disorders, and left-handedness with PPA has not been supported by the results of Mendelian randomization analysis (Nassan et al., 2023). The preferred pathways associated with neurodegeneration may have a name and a function, but who is deciding the choice of the route, and why, remains a deep mystery.

ABBREVIATIONS

AD, Alzheimer disease; ALE, activation likelihood estimation; ALS, amyotrophic lateral sclerosis; AP, amyloid plaques; Bv, behavioral variant; CBD, corticobasal degeneration; CBDS, corticobasal degeneration syndrome; FDG, fluoro deoxyglucose; FTD, frontotemporal dementia; FTLD, frontotemporal lobar degeneration; L/Phv-PPA, logopenic/phonologic variant; LBD, Lewy body dementia; Nb, nucleus basalis; NDD, neurodegenerative disorders; NF, neurofibrillary; Nf/Av, nonfluent/agrammatic variant; PCA, posterior cortical atrophy; PD, Parkinson disease; PET, positron emission tomography; PPA, primary progressive aphasia; PSP, progressive supranuclear palsy; SPET, single-photon emission tomography; Sv, semantic variant; TDP 43, TAR DNA-binding protein-43.

References

Aarsland D, Creese B, Politis M et al. (2017). Cognitive decline in Parkinson disease. Nat Rev Neurol 13: 217–231.

Abbate C, Trimarchi PD, Inglese S et al. (2019). Does the right focal variant of Alzheimer's disease really exist? A literature analysis. J Alzheimers Dis 71: 405–420.

Adams AG, Schweitzer D, Molenberghs P et al. (2019). A meta-analytic review of social cognitive function following stroke. Neurosci Biobehav Rev 102: 400–416.

Agosta F, Spinelli EG, Basaia S et al. (2023). Functional connectivity from disease epicenters in frontotemporal dementia. Neurology 100: e2290–e2303.

Andrade K, Kas A, Samri D et al. (2013). Visuospatial deficits and hemispheric perfusion asymmetries in posterior cortical atrophy. Cortex 49: 940–947.

Arendt T, Bigl V, Tennstedt A et al. (1985). Neuronal loss in different parts of the nucleus basalis is related to neuritic plaque formation in cortical target areas in Alzheimer's disease. Neuroscience 14: 1–14.

Armstrong MJ, Litvan I, Lang AE et al. (2013). Criteria for the diagnosis of corticobasal degeneration. Neurology 80: 496–503.

Ash S, Evans E, O'shea J et al. (2013). Differentiating primary progressive aphasias in a brief sample of connected speech. Neurology 81: 329–336.

Battistella G, Borghesani V, Henry M et al. (2020). Task-free functional language networks: reproducibility and clinical application. J Neurosci 40: 1311–1320.

Beeldman E, Raaphorst J, Twennaar MK et al. (2016). The cognitive profile of ALS: a systematic review and meta-analysis update. J Neurol Neurosurg Psychiatry 87: 611–619.

Benson DF, Davis RJ, Snyder BD (1988). Posterior cortical atrophy. Arch Neurol 45: 789–793.

Blakemore S-J (2008). The social brain in adolescence. Nat Rev Neurosci 9: 267–277.

Bloem BR, Okun MS, Klein C (2021). Parkinson's disease. Lancet 397: 2284–2303.

Boccadoro S, Cracco E, Hudson AR et al. (2019). Defining the neural correlates of spontaneous theory of mind (ToM): an fMRI multistudy investigation. Neuroimage 203: 116193.

Borghammer P (2021). The α-synuclein origin and connectome model (SOC model) of Parkinson's disease: explaining motor asymmetry, non-motor phenotypes, and cognitive decline. J Parkinsons Dis 11: 455–474.

Borghesani V, Battistella G, Mandelli M et al. (2020). Regional and hemispheric susceptibility of the temporal lobe to FTLD-TDP type C pathology. NeuroImage Clin 28: 102369.

Boschi V, Catricala E, Consonni M et al. (2017). Connected speech in neurodegenerative language disorders: a review. Front Psychol 8: 269.

Botha H, Duffy JR, Strand EA et al. (2014). Nonverbal oral apraxia in primary progressive aphasia and apraxia of speech. Neurology 82: 1729–1735.

Braak H, Braak E (1991). Neuropathological stageing of Alzheimer-related changes. Acta Neuropathol 82: 239–259.

Braak H, Del Tredici K, Rüb U et al. (2003). Staging of brain pathology related to sporadic Parkinson's disease. Neurobiol Aging 24: 197–211.

Brettschneider J, Del Tredici K, Irwin DJ et al. (2014). Sequential distribution of pTDP-43 pathology in behavioral variant frontotemporal dementia (bvFTD). Acta Neuropathol 127: 423–439.

Brettschneider J, Tredici KD, Lee VM-Y et al. (2015). Spreading of pathology in neurodegenerative diseases: a focus on human studies. Nat Rev Neurosci 16: 109–120.

Buciuc M, Duffy JR, Machulda MM et al. (2021). Clinical, imaging, and pathologic characteristics of patients with right vs left hemisphere–predominant logopenic progressive aphasia. Neurology 97: e523–e534.

Caroppo P, Camuzat A, De Septenville A et al. (2015). Semantic and nonfluent aphasic variants, secondarily associated with amyotrophic lateral sclerosis, are predominant frontotemporal lobar degeneration phenotypes in TBK1 carriers. Alzheimer's Dement 1: 481–486.

Catani M, Mesulam MM, Jakobsen E et al. (2013). A novel frontal pathway underlies verbal fluency in primary progressive aphasia. Brain 136: 2619–2628.

Chan D, Anderson V, Pijnenburg Y et al. (2009). The clinical profile of right temporal lobe atrophy. Brain 132: 1287–1298.

Claassen DO, McDonell KE, Donahue M et al. (2016). Cortical asymmetry in Parkinson's disease: early susceptibility of the left hemisphere. Brain Behav 6: e00573.

Conca F, Esposito V, Giusto G et al. (2022). Characterization of the logopenic variant of primary progressive aphasia: a systematic review and meta-analysis. Ageing Res Rev 82: 101760.

Consonni M, Contarino VE, Catricalà E et al. (2018). Cortical markers of cognitive syndromes in amyotrophic lateral sclerosis. NeuroImage Clin 19: 675–682.

Consonni M, Cappa SF, Dalla Bella E et al. (2019). Cortical correlates of behavioural change in amyotrophic lateral sclerosis. J Neurol Neurosurg Psychiatry 90: 380–386.

Consonni M, Dalla Bella E, Contarino VE et al. (2020). Cortical thinning trajectories across disease stages and cognitive impairment in amyotrophic lateral sclerosis. Cortex 131: 284–294.

Corbetta M, Shulman GL (2011). Spatial neglect and attention networks. Annu Rev Neurosci 34: 569–599.

Crutch SJ, Lehmann M, Schott JM et al. (2012). Posterior cortical atrophy. Lancet Neurol 11: 170–178.

Crutch SJ, Schott JM, Rabinovici GD et al. (2017). Consensus classification of posterior cortical atrophy. Alzheimers Dement 13: 870–884.

Devine MS, Pannek K, Coulthard A et al. (2015). Exposing asymmetric gray matter vulnerability in amyotrophic lateral sclerosis. NeuroImage Clin 7: 782–787.

Djaldetti R, Ziv I, Melamed E (2006). The mystery of motor asymmetry in Parkinson's disease. Lancet Neurol 5: 796–802.

Dodich A, Cerami C, Inguscio E et al. (2019). The clinico-metabolic correlates of language impairment in corticobasal syndrome and progressive supranuclear palsy. NeuroImage Clin 24: 102009.

Dubois B, Villain N, Frisoni GB et al. (2021). Clinical diagnosis of Alzheimer's disease: recommendations of the International Working Group. Lancet Neurol 20: 484–496.

Fedorova TD, Knudsen K, Horsager J et al. (2023). Dopaminergic dysfunction is more symmetric in dementia with Lewy bodies compared to Parkinson's disease. J Parkinsons Dis 13: 515–523.

Forrest SL, Kovacs GG (2023). Current concepts of mixed pathologies in neurodegenerative diseases. Can J Neurol Sci 50: 329–345.

Förstl H, Fischer P (1994). Diagnostic confirmation, severity, and subtypes of Alzheimer's disease: a short review on clinicopathological correlations. Eur Arch Psychiatry Clin Neurosci 244: 252–260.

Foster NL, Chase TN, Fedio P et al. (1983). Alzheimer's disease: focal cortical changes shown by positron emission tomography. Neurology 33: 961–965.

Gainotti G (2019). The role of the right hemisphere in emotional and behavioral disorders of patients with frontotemporal lobar degeneration: an updated review. Front Aging Neurosci 11: 55.

Gazulla J, Ferrer I, Izquierdo-Alvarez S et al. (2019). Hereditary primary lateral sclerosis and progressive nonfluent aphasia. J Neurol 266: 1079–1090.

Giannakopoulos P, Bouras C, Hof PR (1994). Alzheimer's disease with asymmetric atrophy of the cerebral hemispheres: morphometric analysis of four cases. Acta Neuropathol 88: 440–447.

Giannini LA, Xie SX, McMillan CT et al. (2019). Divergent patterns of TDP-43 and tau pathologies in primary progressive aphasia. Ann Neurol 85: 630–643.

Gnanalingham KK, Byrne EJ, Thornton A et al. (1997). Motor and cognitive function in Lewy body dementia: comparison with Alzheimer's and Parkinson's diseases. J Neurol Neurosurg Psychiatry 62: 243–252.

Goedert M, Masuda-Suzukake M, Falcon B (2017). Like prions: the propagation of aggregated tau and α-synuclein in neurodegeneration. Brain 140: 266–278.

Gorno-Tempini ML, Dronkers NF, Rankin KP et al. (2004). Cognition and anatomy in three variants of primary progressive aphasia. Ann Neurol 55: 335–346.

Gorno-Tempini ML, Brambati SM, Ginex V et al. (2008). The logopenic/phonological variant of primary progressive aphasia. Neurology 71: 1227–1234.

Gorno-Tempini ML, Hillis AE, Weintraub S et al. (2011). Classification of primary progressive aphasia and its variants. Neurology 76: 1006–1014.

Grossman M, Mickanin J, Onishi K et al. (1996). Progressive nonfluent aphasia: language, cognitive, and PET measures contrasted with probable Alzheimer's disease. J Cogn Neurosci 8: 135–154.

Haxby JV, Duara R, Grady CL et al. (1985). Relations between neuropsychological and cerebral metabolic asymmetries in early Alzheimer's disease. J Cereb Blood Flow Metab 5: 193–200.

Hervé P-Y, Zago L, Petit L et al. (2013). Revisiting human hemispheric specialization with neuroimaging. Trends Cogn Sci 17: 69–80.

Hodges JR, Patterson K, Oxbury S et al. (1992). Semantic dementia: progressive fluent aphasia with temporal lobe atrophy. Brain 115: 1783–1806.

Höglinger GU, Respondek G, Stamelou M et al. (2017). Clinical diagnosis of progressive supranuclear palsy: the movement disorder society criteria. Mov Disord 32: 853–864.

Irwin DJ, Cairns NJ, Grossman M et al. (2015). Frontotemporal lobar degeneration: defining phenotypic diversity through personalized medicine. Acta Neuropathol 129: 469–491.

Irwin DJ, McMillan CT, Xie SX et al. (2018). Asymmetry of post-mortem neuropathology in behavioural-variant frontotemporal dementia. Brain 141: 288–301.

Isella V, Licciardo D, Nastasi G et al. (2022). Clinical and metabolic imaging features of late-onset and early-onset posterior cortical atrophy. Eur J Neurol 29: 3147–3157.

Jagust W, Friedland R, Budinger T et al. (1988). Longitudinal studies of regional cerebral metabolism in Alzheimer's disease. Neurology 38: 909–912.

Jamada M, Mehraein P (1968). Verteilungsmuster der senilen Veranderungen im Gehirn. Die Beteiligung des limbischen Systems bei hirnatrophischen Prozessen des Seniums und bei Morbus Alzheimer. Arch Psychiatr Nervenkr 211: 308–324.

Janota I, Mountjoy CQ (1988). Asymmetry of pathology in Alzheimer's disease. J Neurol Neurosurg Psychiatry 51: 1011.

Josephs KA, Duffy JR, Strand EA et al. (2013). Syndromes dominated by apraxia of speech show distinct characteristics from agrammatic PPA. Neurology 81: 337–345.

JPND Research (2023). Available: https://neurodegenerationresearch.eu/what/.

Kempster P, Gibb W, Stern G et al. (1989). Asymmetry of substantia nigra neuronal loss in Parkinson's disease and its relevance to the mechanism of levodopa related motor fluctuations. J Neurol Neurosurg Psychiatry 52: 72–76.

Kim G, Ahmadian SS, Peterson M et al. (2016). Asymmetric pathology in primary progressive aphasia with progranulin mutations and TDP inclusions. Neurology 86: 627–636.

Kjeldsen PL, Parbo P, Hansen KV et al. (2022). Asymmetric amyloid deposition in preclinical Alzheimer's disease: a PET study. Aging Brain 2: 100048.

Kovacs GG, Ghetti B, Goedert M (2022). Classification of diseases with accumulation of Tau protein. Neuropathol Appl Neurobiol 48: e12792.

Li CH, Fan SP, Chen TF et al. (2020). Frontal variant of Alzheimer's disease with asymmetric presentation mimicking frontotemporal dementia: case report and literature review. Brain Behav 10: e01548.

Logroscino G, Piccininni M, Graff C et al. (2023). Incidence of syndromes associated with frontotemporal lobar degeneration in 9 European countries. JAMA Neurol 80: 279–286.

Lorusso S, Maioli C, Barvas E (2020). Pure allocentric unilateral spatial neglect in a case of Alzheimer's disease presenting with posterior cortical atrophy syndrome. Neurol Sci 41: 1619–1621.

Lu J, Zhang Z, Wu P et al. (2023). The heterogeneity of asymmetric tau distribution is associated with an early age at onset and poor prognosis in Alzheimer's disease. NeuroImage Clin 38: 103416.

Lubben N, Ensink E, Coetzee GA et al. (2021). The enigma and implications of brain hemispheric asymmetry in neurodegenerative diseases. Brain Commun Theory 3: fcab211.

Luzzatti C, Poeck K (1991). An early description of slowly progressive aphasia. Arch Neurol 48: 228–229.

Mackenzie IR, Neumann M (2016). Molecular neuropathology of frontotemporal dementia: insights into disease mechanisms from postmortem studies. J Neurochem 138: 54–70.

Mandelli ML, Caverzasi E, Binney RJ et al. (2014). Frontal white matter tracts sustaining speech production in primary progressive aphasia. J Neurosci 34: 9754–9767.

Mandelli ML, Vilaplana E, Brown JA et al. (2016). Healthy brain connectivity predicts atrophy progression in non-fluent variant of primary progressive aphasia. Brain 139: 2778–2791.

McKeith IG, Boeve BF, Dickson DW et al. (2017). Diagnosis and management of dementia with Lewy bodies: fourth consensus report of the DLB Consortium. Neurology 89: 88–100.

Mesulam MM (1982). Slowly progressive aphasia without generalized dementia. Ann Neurol 11: 592–598.

Mesulam MM (1990). Large-scale neurocognitive networks and distributed processing for attention, language, and memory. Ann Neurol 28: 597–613.

Mesulam M-M, Weintraub S, Rogalski EJ et al. (2014). Asymmetry and heterogeneity of Alzheimer's and frontotemporal pathology in primary progressive aphasia. Brain 137: 1176–1192.

Mesulam MM, Coventry CA, Bigio EH et al. (2022). Neuropathological fingerprints of survival, atrophy and language in primary progressive aphasia. Brain 145: 2133–2148.

Miller ZA, Mandelli ML, Rankin KP et al. (2013). Handedness and language learning disability differentially distribute in progressive aphasia variants. Brain 136: 3461–3473.

Moossy J, Zubenko GS, Martinez AJ et al. (1989). Lateralization of brain morphologic and cholinergic abnormalities in Alzheimer's disease. Arch Neurol 46: 639–642.

Morris JC, Cole M, Banker BQ et al. (1984). Hereditary dysphasic dementia and the pick-Alzheimer spectrum. Ann Neurol 16: 455–466.

Nassan M, Piras IS, Rogalski E et al. (2023). Evaluating the association between genetically proxied neurodevelopmental language phenotypes and the risk of primary progressive aphasia. Neurology 100: e1922–e1929.

Nelson PT, Dickson DW, Trojanowski JQ et al. (2019). Limbic-predominant age-related TDP-43 encephalopathy (LATE): consensus working group report. Brain 142: 1503–1527.

Neumann M, Sampathu DM, Kwong LK et al. (2006). Ubiquitinated TDP-43 in frontotemporal lobar degeneration and amyotrophic lateral sclerosis. Science 314: 130–133.

Niethammer M, Tang CC, Feigin A et al. (2014). A disease-specific metabolic brain network associated with corticobasal degeneration. Brain 137: 3036–3046.

Ossenkoppele R, Singleton EH, Groot C et al. (2022). Research criteria for the behavioral variant of Alzheimer disease: a systematic review and meta-analysis. JAMA Neurol 79: 48–60.

Perani D, Di Piero V, Vallar G et al. (1988). Technetium-99m HM-PAO-SPECT study of regional cerebral perfusion in early Alzheimer's disease. J Nucl Med 29: 1507–1514.

Perani D, Caminiti SP, Carli G et al. (2021). PET neuroimaging in dementia conditions. In: PET and SPECT in neurology, 211–282.

Pobric G, Lambon Ralph MA, Zahn R (2016). Hemispheric specialization within the superior anterior temporal cortex for social and nonsocial concepts. J Cogn Neurosci 28: 351–360.

Pogacar S (1984). Alzheimer's disease presenting as slowly progressive aphasia. This may be the first reported case of this sequence. RI Med J 67: 181–185.

Raaphorst J, Beeldman E, De Visser M et al. (2012). A systematic review of behavioural changes in motor neuron disease. Amyotroph Lateral Scler 13: 493–501.

Rabinovici G, Rosen H, Alkalay A et al. (2011). Amyloid vs FDG-PET in the differential diagnosis of AD and FTLD. Neurology 77: 2034–2042.

Rankin KP (2020). Brain networks supporting social cognition in dementia. Curr Behav Neurosci Rep 7: 203–211.

Rascovsky K, Hodges JR, Knopman D et al. (2011). Sensitivity of revised diagnostic criteria for the behavioural variant of frontotemporal dementia. Brain 134: 2456–2477.

Ratnavalli E, Brayne C, Dawson K et al. (2002). The prevalence of frontotemporal dementia. Neurology 58: 1615–1621.

Robinson CG, Duffy JR, Clark HA et al. (2023). Clinicopathological associations of hemispheric dominance in primary progressive apraxia of speech. Eur J Neurol 30: 1209–1219.

Rogalski E, Johnson N, Weintraub S et al. (2008). Increased frequency of learning disability in patients with primary progressive aphasia and their first-degree relatives. Arch Neurol 65: 244–248.

Saxon JA, Harris JM, Thompson JC et al. (2017). Semantic dementia, progressive non-fluent aphasia and their association with amyotrophic lateral sclerosis. J Neurol Neurosurg Psychiatry 88: 711–712.

Seeley WW, Crawford RK, Zhou J et al. (2009). Neurodegenerative diseases target large-scale human brain networks. Neuron 62: 42–52.

Silveri MC, Ciccarelli N, Cappa A (2011). Unilateral spatial neglect in degenerative brain pathology. Neuropsychology 25: 554.

Snowden J, Goulding PJ, Neary D (1989). Semantic dementia: a form of circumscribed cerebral atrophy. Behav Neurol 2: 167–182.

Spinelli EG, Mandelli ML, Miller ZA et al. (2017). Typical and atypical pathology in primary progressive aphasia variants. Ann Neurol 81: 430–443.

Stefanits H, Budka H, Kovacs GG (2012). Asymmetry of neurodegenerative disease-related pathologies: a cautionary note. Acta Neuropathol 123: 449–452.

Strong MJ, Abrahams S, Goldstein LH et al. (2017). Amyotrophic lateral sclerosis-frontotemporal spectrum disorder (ALS-FTSD): revised diagnostic criteria. Amyotroph Lateral Scler Frontotemporal Degener 18: 153–174.

Swinnen B, Robberecht W (2014). The phenotypic variability of amyotrophic lateral sclerosis. Nat Rev Neurol 10: 661–670.

Tatsch K, Schwarz J, Mozley PD et al. (1997). Relationship between clinical features of Parkinson's disease and presynaptic dopamine transporter binding assessed with [123 I] IPT and single-photon emission tomography. Eur J Nucl Med 24: 415–421.

Therriault J, Pascoal TA, Savard M et al. (2021). Topographic distribution of Amyloid-β, Tau, and Atrophy in patients with behavioral/dysexecutive Alzheimer disease. Neurology 96: e81–e92.

Turner M, Wicks P, Brownstein C et al. (2011). Concordance between site of onset and limb dominance in amyotrophic lateral sclerosis. J Neurol Neurosurg Psychiatry 82: 853–854.

van der Hoorn A, Burger H, Leenders KL et al. (2012). Handedness correlates with the dominant Parkinson side: a systematic review and meta-analysis. Mov Disord 27: 206–210.

Van Overwalle F (2009). Social cognition and the brain: a meta-analysis. Hum Brain Mapp 30: 829–858.

Venneri A, Pentore R, Cotticelli B et al. (1998). Unilateral spatial neglect in the late stage of Alzheimer's disease. Cortex 34: 743–752.

Verreyt N, Nys GM, Santens P et al. (2011). Cognitive differences between patients with left-sided and right-sided Parkinson's disease. A review. Neuropsychol Rev 21: 405–424.

Voruz P, Constantin IM, Péron JA (2022). Biomarkers and non-motor symptoms as a function of motor symptom asymmetry in early Parkinson's disease. Neuropsychologia 177: 108419.

Voruz P, Haegelen C, Assal F et al. (2023). Motor symptom asymmetry predicts cognitive and neuropsychiatric profile following deep brain stimulation of the subthalamic nucleus in Parkinson's disease: a 5-year longitudinal study. Arch Clin Neuropsychol 38: 904–912.

Walker Z, Possin KL, Boeve BF et al. (2015). Lewy body dementias. Lancet 386: 1683–1697.

Wang J, Yang QX, Sun X et al. (2015). MRI evaluation of asymmetry of nigrostriatal damage in the early stage of early-onset Parkinson's disease. Parkinsonism Relat Disord 21: 590–596.

Warrington EK (1975). The selective impairment of semantic memory. Q J Exp Psychol 27: 635–657.

Wechsler AF (1977). Presenile dementia presenting as aphasia. J Neurol Neurosurg Psychiatry 40: 303.

Whitwell JL, Przybelski SA, Weigand SD et al. (2009). Distinct anatomical subtypes of the behavioural variant of frontotemporal dementia: a cluster analysis study. Brain 132: 2932–2946.

Whitwell JL, Xu J, Mandrekar J et al. (2013). Frontal asymmetry in behavioral variant frontotemporal dementia: clinicoimaging and pathogenetic correlates. Neurobiol Aging 34: 636–639.

Wilcock G, Esiri M (1987). Asymmetry of pathology in Alzheimer's disease. J Neurol Neurosurg Psychiatry 50: 1384.

Wilson SM, Henry ML, Besbris M et al. (2010). Connected speech production in three variants of primary progressive aphasia. Brain 133: 2069–2088.

Wilson MA, Joubert S, Ferré P et al. (2012). The role of the left anterior temporal lobe in exception word reading: reconciling patient and neuroimaging findings. Neuroimage 60: 2000–2007.

Woollams AM, Patterson K (2018). Cognitive consequences of the left-right asymmetry of atrophy in semantic dementia. Cortex 107: 64–77.

Younes K, Borghesani V, Montembeault M et al. (2022). Right temporal degeneration and socioemotional semantics: semantic behavioural variant frontotemporal dementia. Brain 145: 4080–4096.

Zhang Q, Mao C, Jin J et al. (2014). Side of limb-onset predicts laterality of gray matter loss in amyotrophic lateral sclerosis. Biomed Res Int 2014: 1–11.

Section III

Consciousness/attention

Handbook of Clinical Neurology, Vol. 208 (3rd series)
Cerebral Asymmetries
C. Papagno and P. Corballis, Editors
https://doi.org/10.1016/B978-0-443-15646-5.00028-2

Chapter 8

Interhemispheric differences in visual attention

CARLO ALBERTO MARZI*

Department of Neuroscience, Biomedicine and Movement, University of Verona, Verona, Italy

Abstract

The chapter explores the difference between the cerebral hemispheres in the three categories of attention described in the fundamental classification of Posner and Petersen: Alerting, Orienting, and Executive Functions. The first section is concerned with the brain localization of visuospatial attention as studied in brain-damaged patients, mainly hemineglect and callosum-sectioned patients. Other important results have been provided more recently by means of brain imaging studies of cortical and subcortical attention networks. Most of these studies have shown a clear dominance of the right hemisphere (RH) in visuospatial attention, but there are some exceptions. Accordingly, the second section concerns the role of the left hemisphere (LH) in visuospatial attention. A third section describes the contribution of attention to interhemispheric communication. A fourth section is focused on a discussion of the existence of hemispheric asymmetries not only in conscious but also in unconscious attention. Further, a fifth section concerns the effects of emotion on hemispheric differences in visuospatial attention. Finally, the last section briefly discusses the controversial evidence concerning laterality in executive functions.

INTRODUCTION

This chapter deals mainly with the neural bases of visual attention and refers to one of the most appreciated neurally oriented theories of attention proposed in a seminal paper by Posner and Petersen (1990) and later updated by Petersen and Posner (2012). The model specified three different kinds of attentional processes: Alerting, Orienting and Executive Control. **Alerting** includes *arousal* whose neural basis has been discovered in the classic study by Moruzzi and Magoun (1949) on the crucial role of the brain stem reticular formation. Alerting defines processes that enable catching all or most of the information in the environment without operating a selection while maintaining vigilance (see Posner, 2008). **Orienting** is the ability to select modality, location, and other features of sensory input. **Executive control** is a supramodal function that concerns performance and control of behavioral operations including inhibition.

ALERTING

A large body of research demonstrates a prominent role of the locus coeruleus (LC) and other noradrenergic systems in the regulation of alerting, waking, and arousal (see Neal et al., 2023). The LC is bilaterally located in the upper part of the pons and, despite its small size, is the largest site of the release of norepinephrine (NE) in the brain. It is critically related not only to alerting and arousal effects but also to the modulation of various physiologic and psychologic processes including sensory information. Recent work has found that the LC is not a unitary structure but contains several anatomo-functional sites (Chandler et al., 2019) associated to multiple aspects of attention. Moreover, in addition to the release of NE, the LC provides specific signals, such as those related to the activation and deactivation of given functions, for example, prioritizing memories strength (Clewett et al., 2018). As to laterality, an overactivity of the right LC has been found to be related to the pathophysiology of attention deficit hyperactivity disorders (ADHD) and may be also related to inattention. LC activity is synchronous to pupil diameter, which reflects neural activity related to functions such as attention and arousal. Therefore, pupil study may represent a marker of ADHD and other disorders (Kumano et al., 2022). In a behavioral study in healthy participants, by using an attention task with lateralized stimuli (Lateralized Attention Network Test—LANT), Asanowicz et al. (2012) have shown the existence of hemispheric asymmetry for orienting and executive functions but no reliable hemispheric differences in alerting. In contrast, it has been found that lesions of the right parietal areas result in an impairment in vigilance tasks. Moreover, neuroimaging data have confirmed

*Correspondence to: Carlo Alberto Marzi, Department of Neuroscience, Biomedicine and Movement, University of Verona, Verona, Italy. Tel: +0039 055 2638491, E-mail: carloalberto.marzi@univr.it

that tonic alertness is lateralized to the right cerebral cortex (Posner, 2008). Thus, a function like alerting that was considered as less lateralized than other forms of attention turns out to show important hemispheric differences.

ORIENTING

Most of the neurally based studies of attention have been focused on orienting that concerns selective visuospatial attention. There are at least five kinds of spatial attention (i) Automatic (stimulus triggered) or Voluntary (goal triggered), also defined as Exogenous or Endogenous, respectively. (ii) Space- or Object-based. (iii) Explicit or Implicit, that is, with or without eye, head, or limb movements. (iv) Internal or External, that is, from working memory to behavior or from sensory stimuli to working memory (see van Ede and Nobre, 2023). Finally, (v) an important distinction, critically related to models of consciousness is: Conscious vs Unconscious selective attention (see Pedersini et al., 2020; Sanchez-Lopez et al., 2020).

A fundamental method of studying the neural substrate of selective attention is to use, together with neuroimaging tools, behavioral tasks that tap specific kinds of attention in neurotypical or brain-damaged individuals. Important information on attention asymmetries can also be obtained by studying healthy participants with sensory stimuli presented to one or the other hemisphere by using the so-called Divided Visual Field Paradigm (see Bourne, 2006). With this technique, brief visual stimuli are presented either to the left visual hemifield (right hemisphere) or to the right hemifield (left hemisphere). The chapter focuses on the most important conclusions gathered so far by using, during the performance of behavioral tasks, various neuroimaging devices such as EEG, event-related potentials (ERPs), transcranial magnetic or electrical simulation (TMS-TES), functional magnetic resonance imaging (fMRI), and other structural and dynamic techniques. Importantly, in this chapter data from the effects of brain lesions will be discussed in the context of laterality of attention while more comprehensive information will be provided by other chapters of this issue.

Clearly, there is no unique attention center orchestrating all attentional activities, however, it could be argued that there is an "attentional" hemisphere, i.e., the right hemisphere (RH) that plays a dominant role for visuospatial attention. However, it should be considered that also the left hemisphere (LH) contributes to various kinds of selective spatial attention, especially when related to language, short-term memory, and other cognitive processes including executive control.

Studies on brain-damaged patients

Hemispatial neglect

The first historical compelling evidence of hemispheric differences in attention comes from the hemineglect syndrome following a unilateral lesion of the RH. Briefly, these patients are impaired in attending to the left side of their body and may not detect visual, somatoesthesic, auditory, and other sensory events despite absence of sensory impairments. In contrast, they hyperattend to the right hemispace also by turning head and eyes to the right (see Natale et al., 2007) and tend to walk turning to the right. However, it should be pointed out that hemineglect is a complex neuropsychologic syndrome involving not only impairment of visuospatial attention but also of mental representations and other cognitive processes (see Bisiach, 1993; Rode et al., 2017a,b). On the basis of the hemineglect symptoms, Kinsbourne, 1977; Heilman and Van Den Abell, 1980; Mesulam, 1981, formulated theories of brain mechanism of visuospatial attention that are all focused, albeit with somewhat different neural mechanisms, on hemispheric laterality, and specifically, on a superiority of the RH in spatial attention. Briefly, the Heilman and Van Den Abell *hemispatial theory* hypothesizes that centers in the RH subserve shifts of attention in both contralateral and ipsilateral hemifields. Therefore, a patient with unilateral LH lesion, either does not have hemineglect, or may have short-lasting symptoms, because the intact RH can master attention in both hemifields. In contrast, lesion in the RH will yield full neglect of the left hemifield while attention to the right hemifield is preserved and often exaggerated. Kinsbourne's *opposite processors* theory hypothesizes that each hemisphere bias attention to the contralateral hemifield and the two processes are antagonistic by inhibiting each other interhemispherically through the commissural fibers. However, this mutual interhemispheric inhibition is asymmetric, being stronger in the RH, thus enabling the extension of spatial attention of this hemisphere to the ipsilateral left visual field. This asymmetry accounts well for the large effect on spatial hemineglect of RH, vs the lack of effect of the LH brain lesion. Interestingly, Koch et al. (2011) have provided experimental evidence of the neural basis of the asymmetry by showing that the posterior parietal cortex of the RH sends callosal inhibitory projections to the LH, while this effect is not reciprocal. For other theories (see, e.g., Rode et al., 2017a,b).

One important issue is *what kind of visuospatial attention* is impaired in hemineglect, namely, exogenous or endogenous. This question has been studied several times with different psychologic paradigms. Below an experiment is described to test exogenous and endogenous attention in hemineglect patients by using a simple, manual, visuomotor reaction time (RT) task with visual stimuli presented at various eccentricities in the left and right hemifield (Smania et al., 1998) (see Fig. 8.1). As is well known, in healthy individuals, simple visual RT progressively increases from the central to the peripheral position in the visual field. With randomized stimulus presentations, as a test of exogenous attention (since the location of the impending stimulus cannot be guessed) in the contralesional hemifield of neglect patients RT increases dramatically from the center to periphery, much more than in neurotypical participants. In contrast, in the ipsilesional hemifield, RT decreases paradoxically from the center to 5°–10° in contrast to what happens in extinction patients and healthy or brain-damaged controls. With blocked presentations where the site of stimulus presentation is known, (thus representing a test of endogenous attention), there is an overall shortening of RT but, in neglect patients, the abnormal difference between the contra- and ipsilesional hemifield remained, although reduced. Therefore, it is reasonable to conclude that both kinds of visuospatial attention are impaired but with exogenous to a reliably greater extent than endogenous attention (see Natale et al., 2005).

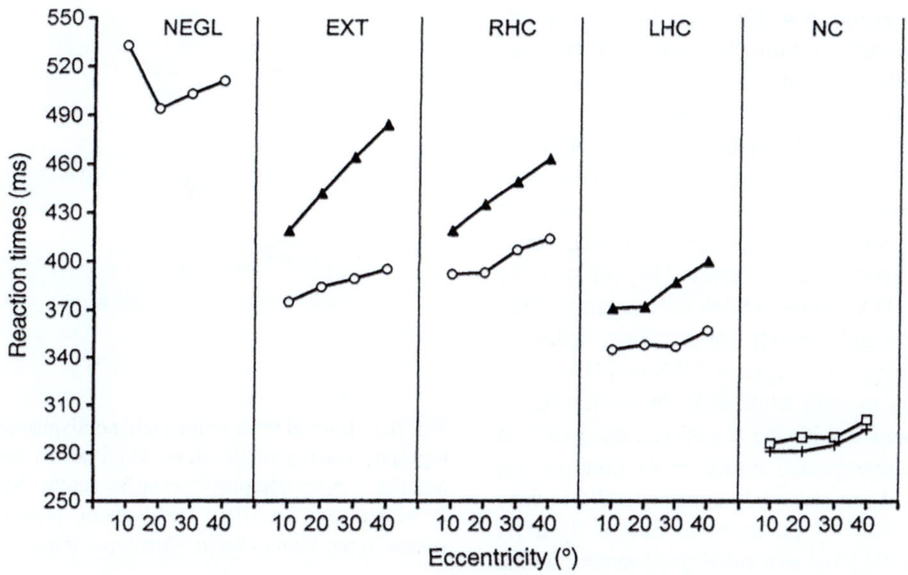

Fig. 8.1. Manual reaction time to brief visual stimuli presented at 10°, 20°, 30°, and 40° in the ipsilesional (circles) or contralesional (triangles) hemifield. NEGL: neglect patients (with RH damage), EXT: extinction patients (with RH damage), RHC: controls with RH damage, LHC: controls with LH damage, NC: neurotypical controls (right hemifield: squares; left hemifield vertical bars). Contralesional data of neglect patients are not included. From Smania N, Martini MC, Gambina G et al. (1998). The spatial distribution of visual attention in hemineglect and extinction patients. Brain 121: 1759–1770.

The typical ipsilesional increased speed of RT (obtained with patients fixating a central screen point) is in keeping with the characteristic postural right turning of head and body (the so-called hyper-attention) of neglect patients in spontaneous posture. Finally, as a consequence of hyper-attention, Natale et al. (2007) found abnormally speeded latency of saccadic eye movements in response to visual stimuli presented between 5° and 10° in the right ipsilesional hemifield of neglect patients.

Unilateral extinction

This term defines the missed detection of one target when presented simultaneously with another one in the opposite half of the visual field. With unilateral stimuli, hemineglect patients without hemianopia can detect the stimuli in either hemifield. While in RH parietal or frontal lesioned patients the left target is extinguished while in LH-damaged patients the missed target is on the right although extinction is usually milder and of shorter duration (e.g., Stone et al., 1993). Importantly, the right space hyper-attention is markedly reduced in unilateral extinction with respect to hemineglect and that could be considered as a reliable behavioral marker for distinguishing neglect from extinction.

There are several theories of the mechanisms of unilateral extinction whose thorough discussion is outside the scope of this chapter. In short, a reasonable one is based on competition between contralesional and ipsilesional input for accessing a decision center located in the LH. Following RH lesion, it is very likely that the contralesional input (left visual hemifield) would be weaker and/or slower than the ipsilesional one (right hemifield) and, therefore, it is bound to lose in the race for the perceptual decision that occurs in the LH prefrontal cortex (Marzi et al., 1997, 2000, 2001; Heekeren et al., 2006). Finally, it is important to note that a mild unilateral extinction effect has been found in healthy participants who, when asked to detect two stimuli presented simultaneously in both hemispaces, show a right hemispace bias. As to the neural locus of the extinction, Fink et al. (2000) in healthy adults found that this effect is detectable with fMRI in the striate and extrastriate cortex. A similar result has been obtained with event-related potentials (ERPs) by Marzi et al. (2000) in the contralesional visual cortex of RH-lesioned patients. These results indicate an important role of posterior visual areas in early attention processes.

Another neuropsychologic source of evidence on the role of the RH in attention comes from the study of ***split-brain patients*** (see Mangun et al., 1994), in whom the corpus callosum has been transected to avoid epileptic seizures spreading from one hemisphere to the other. The callosal disconnection enables studying cortical functions in one hemisphere isolated from the other. As to attention, a rightward bias similar to that observed in neglect patients has been found also in split-brain patients (Proverbio et al., 1994; Corballis, 1995; Berlucchi et al., 1997). This effect may be strong enough to cause extinction of the left hemifield (Reuter-Lorenz et al., 1995). This is likely due to the strong tendency of the LH to shift attention to the right hemifield that cannot be opposed by the RH as a consequence of the callosal section. This confirms the crucial role played by the corpus callosum in the distribution of attention resources between the cerebral hemispheres (for a more comprehensive description of this mechanism, see Banich, 1998).

Brain-imaging studies of attention networks

Cortical centers

Neurophysiologic and neuroimaging studies have provided important evidence on the brain structures involved in various aspects of brain asymmetries in visuospatial attention. Corbetta and Shulman (2002, 2011) inspired by the seminal work of Posner and Petersen (Posner and Petersen, 1990; Petersen and Posner,

2012), have provided a thorough description of the cortical networks subserving visuospatial attention. By means of fMRI cortical mapping, they described two distinct temporoparietal and frontal networks that mediate two basic attentional processes (see Fig. 8.2; Aboitiz et al., 2014): the Ventral Attention Network (VAN), which is responsible for stimulus-driven (exogenous) and the Dorsal Attention Network (DAN) for goal-driven attention (endogenous). VAN centers are located in the temporoparietal junction (TPJ) and the inferior ventral frontal cortex. They are strongly lateralized to the RH while DAN is more bilaterally organized and is located in the intraparietal and superior frontal cortex, including the frontal eye fields. In the right hemisphere, VAN and DAN subserve visuospatial attention in both hemifields. In the left hemisphere, VAN does not operate, with the occasional exception of TPJ that might respond to unexpected stimuli in the contralateral hemifield. In contrast, DAN responds only contralaterally in both hemispheres. Subsequent work by the same as well as by other authors have partially modified and expanded this account, which nonetheless remains valid. As to the TPJ, Doricchi et al. (2022) have formulated an interesting theory about the many functions of this area, which has been found to be parceled in several subareas. They propose that the left TPJ codes both matches and mismatches between various types of events, while the right TPJ codes mismatches. This hypothesis may explain the role of the TPJ in several functions, both cognitive and emotive (see Doricchi, 2022).

Superior longitudinal fasciculus

One important addition to the earlier picture concerns the intercommunication between the attention networks. The prefrontal and parietal components of VAN and DAN are innervated by three branches of the superior longitudinal fasciculus (SLF)

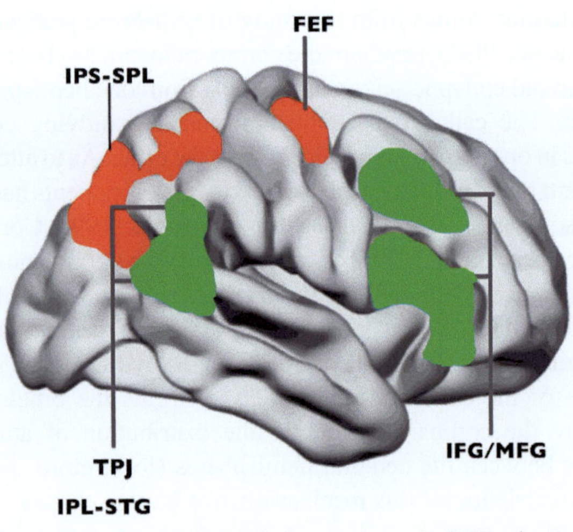

Fig. 8.2. Lateral view of the right hemisphere. In RED, the DAN network; in GREEN, the VAN network. FEF: frontal eye fields; IPS-SPL: intraparietal sulcus and superior parietal lobule; TPJ: temporoparietal junction; IPL-STG: intraparietal lobule and superior temporal gyrus; IFG-MFG: inferior frontal gyrus and middle frontal gyrus. Modified from Aboitiz F, Ossandón T, Zamorano F (2014). Irrelevant stimulus processing in ADHD: catecholamine dynamics and attentional networks. Front Psychol 5: 183. https://doi.org/10.3389/fpsyg.2014.00183.

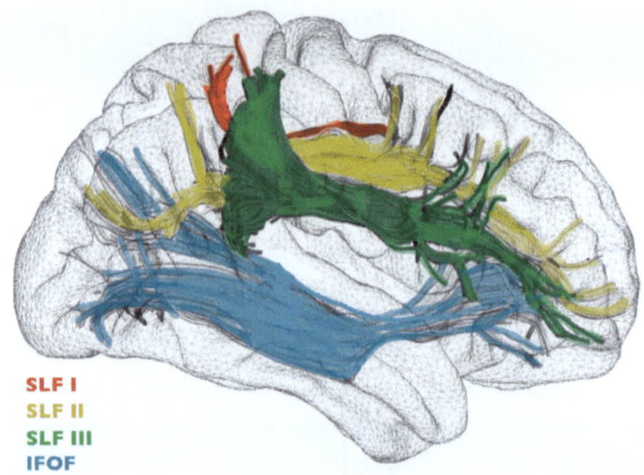

Fig. 8.3. Lateral view of the right hemisphere showing a diffusion tractography tracing of the three longitudinal frontal-occipital fasciculus and the inferior occipito-frontal fasciculus. Modified from Bartolomeo P, Malkinson, S (2019). Hemispheric lateralization of attention processes in the human brain. Curr Opin Psychol 29: 90–96.

(Bartolomeo and Malkinson, 2019) (see Fig. 8.3). In spatial congruence with their respective anatomic position, SLF I connects the DAN and SLF III connects the VAN while SLF II interconnects both networks with prevalence for the VAN. As to lateralization, SLF I is in prevalence bilaterally organized, while SLF III has a larger volume in the RH and this is in keeping with the strong lateralization of the VAN. By the same token, SFL II is larger in RH but to a lesser extent than SFL III (Amemiya et al., 2021). Interestingly, the lateralization pattern (assessed with fractional anisotropy) varies with age as found in a wide range of right-handed groups (6–81 years), with the right SLF III larger than the left in adults only, while SLF II is lateralized in the RH in adolescents as well as adults. In contrast, SLF I is lateralized to the LH in adults as found by measuring myelin levels (Amemiya et al., 2021). The above neuroimaging evidence has been substantiated by clinical neuropsychologic findings: accordingly, damage to the right SLF II predicts left neglect symptoms and this is true for SLF III lesion as well (Thiebaut de Schotten et al., 2005, 2011, 2014).

Subcortical attention centers

Most interestingly, recently, Alves et al. (2022) carried out a resting state functional connectivity analysis on VAN and DAN networks and revealed an involvement of subcortical structures such as pulvinar, superior colliculus (SC), caudate nucleus, as well as various brainstem nuclei related to acetylcholine nicotinic receptors and dopamine and serotonin transporters. These findings amplify considerably the overall picture of the neural bases of attention, extending it to subcortical centers. Actually, the role of subcortical extrageniculate centers in visual attention is not a recent issue (for a review see Esposito et al., 2024). It has been studied by experimental psychologists (Mizzi and Michael, 2016) as well as by neuroscientists (Sprague, 1991; Basso et al., 2021). One experimental psychologic approach that has yielded important evidence is that based on studying the retinal naso-temporal asymmetries in visual perception. The rationale being that neuroimaging studies in humans have demonstrated that the nasal hemiretina (i.e., temporal hemifield) is overrepresented in the contralateral SC while no asymmetry is present in the geniculate pathway (Sylvester et al., 2007). Therefore, if perceptual responses or eye movements are faster or more accurate following nasal vs temporal retinal input they are likely to be subserved by the extrageniculate pathways, mainly by the SC. Moreover, it has been found that nasal-temporal asymmetries in humans are abolished by damage to the SC (Sapir et al., 1999) or the pulvinar (Sapir et al., 2002). As to lateralization, studies of nasal-temporal asymmetries (e.g., Michael and Gálvez-García, 2011) show that attention to salient stimuli involves mainly extrageniculate processing in the RH. Thus, a general conclusion is that RH dominance in visuospatial attention is not limited to the geniculate-striate pathway but extends to extrageniculate routes.

Role of the left hemisphere in visuospatial attention

The role of the LH in visuospatial attention, i.e., a function that is clearly dominated by the RH, has been somewhat overlooked. Indeed, there are several conditions in which the LH is involved in selective visuospatial attention, especially for those which require the collaboration of both sides of the brain (Spagna et al., 2016). For example, in the acute phase of a LH lesion, *spatial hemineglect* is rather frequent (Stone et al., 1993) while is uncommon in the post-acute phase. After 3 months, only about 5% of patients are affected (see Weintraub and Mesulam, 1987; Ringman et al., 2004). However, although the overall proportion of hemineglect following LH lesions is much smaller with respect to RH lesions, several studies have described a broad range of patients with unilateral LH lesion, who show signs of right-side neglect (e.g., see a review by Bowen et al., 1999; Beis et al., 2004). Recently, Beume et al. (2017) studied a large number of acute patients with LH lesions and found that about 17% of them showed right-side neglect. Interestingly, the critical areas for visuospatial symptoms corresponded to those of the VAN network partly mirroring those involved in neglect after RH lesions. In another large study, Malherbe et al. (2018) tested 128 LH acute stroke patients in the line bisection and the Bells test and found an important role of the left SFL and of the arcuate fasciculus (AF) for spatial attention. Note that the AF also plays an important role in language, and this suggests that the LH contribution to visuospatial attention is likely to be related to language processing. Confirming Malherbe et al. (2018), Toba et al. (2022) have provided evidence of disconnection of the left SLF in right neglect, which is in keeping with what found for left neglect after an RH lesion. These data suggest that the role of LH in visuospatial functions is larger than is commonly conceived, perhaps, because language symptoms might have masked the presence of spatial impairment.

One interesting demonstration of the role of the LH in visuospatial attention comes from recent data obtained by using tasks testing the so-called ***"pseudoneglect"*** effect in healthy humans. This denomination describes the tendency of healthy humans, asked to centrally bisect a horizontal line, to slightly shift their bisection to the left (see Fig. 8.4). This effect has been found in visual and other modalities and represents an opposite shift with respect to what observed in RH-lesioned patients, who bisect the line at the right of the center. The percentage of healthy individuals that show pseudoneglect is high and the extent of the shift is somewhat variable, e.g., especially depending on the length of the line to bisect (the shorter the line the lesser the shift) but clear. Pseudoneglect can be reasonably interpreted as related to the dominant RH yielding a prevalent attentional allocation to the left visual field (Heilman and Van Den Abell, 1980). This enhances the perceived size of the left side of the horizontal line as well as inducing a size underestimation of the unattended right side with a consequent shift of the perceived center to the left (for a review, see Jewell and McCourt, 2000). However, an RH dominance in visuospatial attention may not be the unique explanation of pseudoneglect. Gerrits et al. (2020) have tested nondextrals with spatial attention centers lateralized either to the RH or LH, as assessed by fMRI. They found that both groups show pseudoneglect that is significantly smaller in LH individuals, but present. Therefore, there are probably other causes playing a role in pseudoneglect like, for example, scanning habits. Another factor that is important to consider is that the direction of pseudoneglect depends on the object to bisect. In right-handers, if instead of a simple horizontal line, an object to be semantically processed is to be bisected, there is a right rather than a left shift of the estimated midline of the object (Gray et al., 2021).

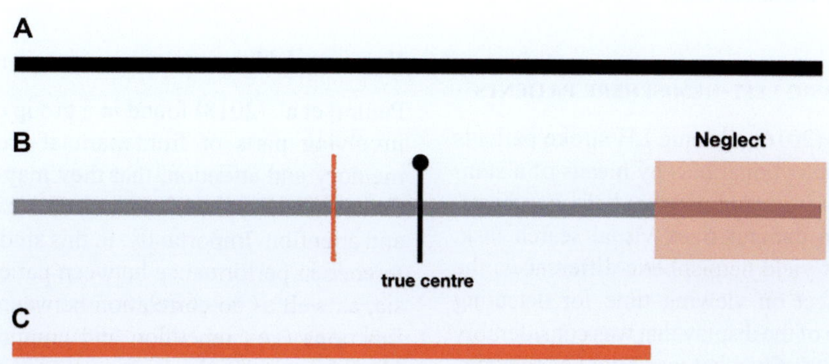

Fig. 8.4. Schematic description of the pseudoneglect effect. (A) Original line; (B) red: bisection line; black: real line center; pink area: neglected zone. (C) Estimated length of line to be bisected.

This effect can be interpreted on the basis of a rightward attention bias as a product of semantic processing lateralized to the LH. This result invites using visuospatial attention paradigms in more complex environments in order to investigate the hemispheric interplay between different processing modes. For example, in a study on pseudoneglect during object search in naturalistic scenes, Nuthmann and Clark (2023) used eye-movement recording and found that there was an initial leftward bias of the gaze followed by a significant reduction in search performance for right-side targets. This shows that pseudoneglect is present also with naturalistic material with a small but reliable advantage for searching left- than right-sided targets. Another avenue of promising research on pseudoneglect is that of studying patients with complex neurodevelopment disorders like, for example, autism. Given that, in right-handers the leftward bias in pseudoneglect is reasonably explained by the visuospatial dominance of the RH, it is of great interest to assess the performance of autistic patients. This is discussed in a recent review by English et al. (2023), who found in a wide number of studies a reduced left bias associated to autism. This might be considered as evidence of reduced hemispheric asymmetry in visuospatial attention and has important implications for a fuller understanding of the neural basis of autism. Interestingly, as to sex differences, a meta-analysis involving more than 2000 children has found that there are more leftward bisections in studies with a higher percentage of boys relative to girls (Kaul et al., 2023). In adults, Beste et al. (2006) had previously shown in a large group of participants that both females and males show a leftward bisection bias but that, in the former group, there is a decrease of the bias in the range of 50–69 years of age.

Effect of multitasking

It has been repeatedly shown that use of concomitant tasks increases the likelihood of finding signs of neglect following unilateral LH lesion. This effect is probably determined by an interaction between spatial and nonspatial components of attention and by attention load in a system with resource limitations. Blini et al. (2016), in a study devised to study multitasking in LH patients, found that a substantial percentage of patients, who did not present with neglect symptoms in traditional standard tests, showed large neglect effects with multitasking (see also a review by Bonato, 2012).

Visuospatial bias in chronic left-hemisphere patients

In a study by Schendel et al. (2016), chronic LH stroke patients and controls were tested for attentional bias by means of a standardized neuropsychologic assessment known as behavioral inattention test (BITC) and by a computerized visual search task. While the former test did not yield hemispheric differences, the latter revealed a reliable effect on viewing time for detecting visual stimuli on the right side of the display that was considerably slower than on the left, in contrast to what was found in controls. This result confirms the notion that the LH is not devoid of visuospatial processing functions and clinically invites to test patients with LH damage for visuospatial attention abilities.

The role of left and right angular gyrus in visuospatial attention and the Simon effect

The Angular Gyrus (AG) is situated in the Inferior Intraparietal Lobule of the Posterior Parietal Cortex. Usually, studies of visuospatial attention have been focused on the right AG with the left AG considered as a control to check for laterality effect. For example, Petitet et al. (2015) found that in healthy participants offline inhibitory TMS of the right AG, as expected, eliminated the leftward attentional bias, i.e., pseudoneglect, of healthy participants at baseline. However, fMRI showed that TMS consistently shifted from RH to LH the activity of the network of parietal and frontal cortex. Clearly, this result confirms the role of RH parietal areas in covert visuospatial attention and pseudoneglect; and the fMRI evidence highlights the role of the parietal cortex on the interhemispheric balance of visual attention.

Furthermore, a recent comprehensive review (Wagner and Rusconi, 2023) including TMS studies on the functions of right and left AG in various tasks of visuospatial attention, while confirming the dominant role of RH in spatial orientation, reported that stimulation of the left AG suppresses the Simon Effect (although with a longer latency than for the right AG). This is evidence for a role of the left AG in an attentional task that involves an automatic conflict between visuospatial information and lateralized manual response. As is well known, the Simon Effect defines a choice reaction time (RT) and accuracy task in which auditory (Simon and Rudell, 1967) or visual stimuli (Simon, 1969) are responded to more slowly or less accurately when they are presented contralateral to the responding hand. For example, when participants are to respond to a green stimulus with the right and to a red stimulus with the left hand, the relative position of stimulus and hand being irrelevant, they are slower and less accurate in incongruent than in congruent trials (i.e., stimulus and hand on the same side). Thus, the Simon Effect is characterized by the fact that spatial location, even if irrelevant, affects the response in a choice discrimination task (with the participant aware of that). Again, these results invite widening the scope of task paradigms used to study lateralization and interhemispheric interactions of selective attention by including material that requires the participation of the two hemispheres with their differential processing modes.

Role of LH in visuospatial memory

Paulraj et al. (2018) found in a group of LH patients with damage involving parts of frontoparietal areas that subserve working memory and attention, that they may suffer from an unexpected deficit in RH-typical functions such as visuospatial memory span and attention. Importantly, in this study, there was no reliable difference in performance between patients with and without aphasia, as well as no correlation between spatial span and language functions (i.e., repetition and comprehension). Thus, language played no role in the above symptoms. Clinically, it is important to note that this spatial impairment may negatively influence rehabilitation strategies, if it goes unnoticed.

Effects of unilateral cerebellar lesions

Finally, it is worth citing an effect that has been recently reported by Verbitsky et al. (2024) in a study on patients with selective unilateral stroke lesion of the right or left cerebellum. Patients were screened with a typical neglect test, i.e., line bisection, cancelation, and figure copying. It turned out that left cerebellar patients were significantly impaired in the detection of targets on the left side of cancelation tasks, as in hemineglect. In contrast, there were no significant effects for right cerebellar patients. Obviously, since the cerebellum cortex has contralateral connections with the cerebral cortex, this result is in keeping with the dominance of the RH in visuospatial operations. However, this finding has important clinical implications for the screening of neglect symptoms in cerebellar patients, especially those with LH damage.

Effects of attention on interhemispheric transfer

It is widely known that interhemispheric transfer time (ITT) can be measured either behaviorally, thanks to the pioneering study of Poffenberger, 1913 (see Marzi, 1999, for a description of the procedure), or with ERP (e.g., see Ipata et al., 1997). Behaviorally, by using a simple manual visuomotor RT paradigm, ITT is estimated by subtracting the RT in uncrossed conditions (hand and hemifield of stimulus presentation on the same side) from the crossed conditions (hand and stimulus on opposite sides). The latter condition typically yields a longer RT because it requires an interhemispheric callosal transfer. The difference between the two conditions, the so-called crossed-uncrossed difference (CUD), is about 3–5 ms in healthy individuals but is largely increased in split-brain patients (see Marzi et al., 1991). A similar rationale is employed in ERP studies where ITT is obtained by comparing the latency of early visual components of the evoked response (P1, N1) for direct responses (stimuli in one hemifield and recording in the contralateral hemisphere) vs indirect responses (stimuli presented to the ipsilateral hemisphere, thus requiring callosal transmission). Typically, ERPs yield ITT values longer than the behavioral. In a study by Ipata et al. (1997), it ranged between 5.77 and 12.54 ms, depending on the component analyzed (P1 or N1) and the location of the electrodes. An important question is whether attention might influence the interhemispheric transfer as assessed by the CUD. To cut a long story short: behaviorally, it has been found that varying the probability of stimulus occurrence in one or the other hemifield may affect the CUD (Weber et al., 2005; Wolski and Asanowicz, 2023). As a result of focusing of visuospatial attention, when the stimulus presentation is more probable on one hemifield, RT decreases and the CUD is shorter. This represents an attentional effect on interhemispheric transfer, but it is interesting to mention that even in conditions of equal probability of the side of stimulus presentation there is a small but significant speed advantage for responses to stimuli presented to the left visual hemifield. This effect is likely related to a tonic allocation of implicit visuospatial attention to the above hemifield (that is, an RH effect) (see Marzi et al., 1991; Bisiacchi et al., 1994). Incidentally, measuring the CUD is relatively simple and that has encouraged many behavioral studies. However, this measure is not reliable enough for individual clinical estimates and gives stable results only by averaging many individuals or with a high number of trials in a single participant and that is not always possible in clinical populations. With this caveat, the CUD is satisfactorily reliable for group studies. The same holds for electrophysiologic studies, as an example, Patston et al. (2007) in a ERP study found that nonmusicians showed significantly faster ITT from RH to LH than in the opposite direction and a shorter N1 latency in the LH than in the RH. In contrast, musicians showed neither hemispheric differences in ITT or in ERP latency. Thus, they seem to have a bilateral organization of visual attention in contrast to nonmusicians. This is probably related to the bimanual training necessary to learn an instrument that promotes a process of myelination of the cerebral commissures resulting in more bilateral communication between hemispheres with respect to individuals without musical training.

Finally, it is important to note that Banich (1998) has proposed the influential theory that interhemispheric interactions mediated by the corpus callosum play an important role in attentional processing by modulating the efficiency of interhemispheric communication, thus enabling a subdivision of processing labor according to the difficulty of the task. Under attention-demanding conditions, a high level of interhemispheric communication would be useful, while a smaller or even absent communication would be more convenient in less demanding conditions. Experimental evidence for that has been obtained in neurotypical participants, split-brain patients, and individuals with brain genetic disorders (see Banich, 1998).

Attention and category-specific cortical areas

The anatomo-functional organization of the visual cortical areas is such that at early processing level they receive input almost completely from the contralateral visual hemifield. However, at higher processing level, i.e., at the parietal–temporal level, this contralateral bias decreases, thus enabling a larger communication between hemispheres by having more access to the ipsilateral side. This trend has been found in nonhuman primates and other mammals. Recently, Herald et al. (2023) have used fMRI to study category-specific bias in healthy individuals and found evidence for a general stronger contralateral bias in the LH. This feature recalls the cortical organization of visuospatial attention where one hemisphere (the RH) has access to both hemifields while the other (LH) has a strong contralateral bias. By the same token, a complementary effect concerns face perception with the RH face form area being less contralaterally biased than the LH. Accordingly, unilateral lesion of the RH is more likely to yield severe prosopagnosia than the lesion of the LH. Moreover, it should be noted that Herald et al. (2023) also found that the general similarity between the organization of visuospatial attention and object categorization is limited, given that, in their study when participants were distracted by a concurrent attentional task in central vision, the strong LH contralateral bias found in their basic task did not diminish. This shows that attention is not the sole determinant of contralateral bias in category perception, although it can modulate it.

Importantly, there is ample evidence of a sex bias in visual perception of some object categories (Handa and McGivern, 2015). For example, women are biased to attending to color, and men to the spatial position of objects. McGivern et al. (2019) have shown that this difference probably relies on bottom-up processing because, when a top-down strategy is introduced in the perceptual task, the sex differences are greatly attenuated.

Hemispheric differences in unconscious vs conscious attention

The question of whether an unconscious form of attention does exist is crucial for theories of consciousness. So far, there is evidence for the presence of orienting of visuospatial attention to unseen stimuli and that supports the existence of unconscious attention. This has been found in healthy individuals by using psychophysically masked visual stimuli (see De Brigard and Prinz, 2010) and in hemianopic patients with "blindsight," that is, visually guided unconscious behavior for the presentation of visual stimuli to the blind hemifield (for reviews on this phenomenon see Weiskrantz, 2004; Trevethan and Sahraie, 2010; Danckert et al., 2019). The attention-consciousness dissociation in blindsight patients was found by Kentridge et al. (1999, 2004), Norman et al. (2015) and, more recently, by Pedersini et al. (2020), as well as by Sanchez-Lopez et al. (2020), who also found a prevalence of the RH over the LH in unconscious attention. In these two studies, the task used was the Posner endogenous attention paradigm in which a preliminary cue signals the side of occurrence of the successive imperative stimulus to be responded to. In that task, in most of the trials, the information provided by the cue is correct but in another condition the information is wrong, and the stimulus appears in the opposite side of that signaled by the cue with a consequent delay of RT (see Shenal et al., 2012 for further information on the cost of missed cues). In Pedersini et al. (2020), in an fMRI study as well as in Sanchez-Lopez et al. (2020), an EEG study, behaviorally, not only in the sighted, but also in the blind hemifield, there was a faster speed of response for valid vs invalid trials despite lack of stimulus perceptual awareness. Furthermore, clear neurophysiologic responses were found for stimulus presentations to the blind hemifield, and this pattern of activity was different according to the side of the brain lesion. LH-damaged patients showed attention-related ERP and oscillatory activity grossly similar to that of healthy participants, while RH-damaged patients showed a markedly different abnormal pattern. Thus, these data extend to neurophysiologic mechanisms the presence of unconscious visual attention and show an RH dominance for both unconscious and conscious visuospatial attention. Interestingly, Chokron et al. (2019) have found that RH patients with left hemianopia (or left hemineglect) might show an ipsilesional specific attentional deficit in the right visual field. This is not the case in patients with LH lesion and right hemianopia and, therefore, the ipsilesional RH impairment of hemianopics is in keeping with an attention role of RH visual primary cortical areas in addition to the frontal, parietal, and temporal areas. Finally, the overall results of the relationships between attention and consciousness gathered so far show that exogenous is more relevant than endogenous attention for the emergence of consciousness. This conclusion has been confirmed and extended by a recent study (Liu et al., 2023) with intracranial EEG recording from epileptic patients that has thoroughly analyzed the pattern of neural interactions associated to consciousness and attention. They found attention-related emergence of consciousness in RH frontal and temporal areas interconnected by SFL II–III as well later bilateral activity in the dorsofrontal and RH orbitofrontal cortex. Moreover, in RH there was an early, sustained, neural activity in the SFL III network when spatial reorienting affected consciousness. Finally, conscious report was accompanied by LH dorsolateral prefrontal activity. In conclusion, specific asymmetric cortical clusters in the frontal and parietal cortex subserve attentional effects on the emergence of consciousness.

Effects of emotion on hemispheric differences in visuospatial attention

The lateralization of emotion has been classically represented by two contrasting theories, namely, the RH dominance and the valence lateralization hypothesis (see Gainotti, 2019) and his chapter in this volume. Briefly, according to the former theory, all emotional processes are subserved by the RH while the latter assumes that the RH processes negative and the LH positive emotions. More recently, there is agreement that lateralization of emotion is more refined and specific with some functions subserved by the RH and others by the LH (Palomero-Gallagher and Amunts, 2022). LH would subserve positive stimuli and those inducing approach behaviors, while negative stimuli, and those inducing withdrawal behavior, would be processed by the RH. In particular, an important question is whether the so-called "emotion-induced blindness" that is, the disruption of visuospatial attention by emotional, especially negative, threatening stimuli, is larger for the RH or LH. Recently, an answer to that question has been provided by Moeck et al. (2023), who have shown, by presenting visual stimuli to one or the other hemifield-hemisphere in healthy participants, that emotional stimuli disrupted visuospatial attention for either presentation side. This represents convincing evidence of a bilateral hemispheric representation of emotion at least as far as emotion-induced blindness is concerned.

EXECUTIVE FUNCTIONS

For executive functions, namely, a series of higher level cognitive processes that control lower level operations, laterality effects have not been much studied so far. In contrast to RH dominance for Orienting and partly for Alerting, the evidence about an RH dominance for the Executive Control of attention is somewhat controversial (Spagna et al., 2020; Vallesi, 2021). One of the first behavioral hints of laterality effects has been provided by the well-known Stroop task (Stroop, 1935) in a modified paradigm with stimuli presented to one or the other hemifield/hemisphere. The Stroop task is an interference effect belonging to the category of attention executive control and one of its most used versions is the color-naming task in which participants are to name the ink color of a printed colored word. Notably, trials may be congruent, when the word and the ink color are the same (e.g., RED-RED), or incongruent, when word and ink are different (e.g., RED-GREEN).

In the latter condition, responses are less accurate and slower as a result of an interference effect. The results reported in the literature on Stroop laterality are controversial. In accord with that, recent neuroimaging work has shown that the neural bases of the Stroop effect are complex (see a review by Banich, 2019) with various cerebral areas involved in both hemispheres. However, recently, strong evidence for an RH superiority in contrasting the language interference in speed of response has been found by Spagna et al. (2020) by increasing alerting and spatial orienting information, that is, by stimulating functional attention interactions. Furthermore, evidence stressing the importance of interactions between the hemispheres in the Stroop effect has been reported by Okayasu et al. (2023) and by Vanderhasselt et al. (2009), both involving the left prefrontal cortex interacting either with the right cerebellum (the former study) or the right prefrontal cortex (the latter). Finally, lateralization may be also present at higher level of executive control and, according to Vallesi (2012), "criterion setting," i.e., the capacity to flexibly set up and select task rules, is more left-lateralized while "monitoring," i.e., the process of continuously evaluating, is more right-lateralized. Clearly, for hemispheric asymmetries in executive functions, there is ample space for future important research.

CONCLUSIVE REMARKS

The thrust of this chapter relies on the evidence provided by various research sources of the differences between the cerebral hemispheres in several aspects of attention. Certainly, attention can be considered, together with language, as the most lateralized cerebral cognitive function. This chapter is mainly focused on Attention Orienting studies in both brain-damaged- and healthy individuals. A reason for that is related to the wealth of available experimental data with respect to the other great subdivisions of attention, namely Alerting and Executive Functions. Hopefully, in the near future, the number of laterality studies on all aspects of attention as well as on association between attention and other cognitive functions will increase. The dynamic interplay of the exchange of function between the two sides of the brain is a fascinating subject whose understanding is being greatly increased by the continuous development of neuroimaging and neuropsychologic studies. Given for granted the dominance of the right hemisphere in many aspects of visuospatial attention, it will be important in clinical studies to also focus on the contribution of the LH on these aspects, especially for the sake of rehabilitation programs.

References

Aboitiz F, Ossandón T, Zamorano F (2014). Irrelevant stimulus processing in ADHD: catecholamine dynamics and attentional networks. Front Psychol 5: 183.

Alves PN, Forkel SJ, Corbetta M (2022). The subcortical and neurochemical organization of the ventral and dorsal attention networks. Commun Biol 5: 1343.

Amemiya K, Naito E, Takemura H (2021). Age dependency and lateralization in the three branches of the human superior longitudinal fasciculus. Cortex 139: 116–133.

Asanowicz D, Marzecová A, Jaśkowski PP et al. (2012). Hemispheric asymmetry in the efficiency of attentional networks. Brain Cogn 79: 117–128.

Banich MT (1998). The missing link: the role of interhemispheric interaction in attentional processing. Brain Cogn 36: 128–157.

Banich MT (2019). The Stroop effect occurs at multiple points along a Cascade of control: evidence from cognitive neuroscience approaches. Front Psychol 10: 2164.

Bartolomeo P, Malkinson TS (2019). Hemispheric lateralization of attention processes in the human brain. Curr Opin Psychol 29: 90–96.

Basso MA, Bickford ME, Cang J (2021). Unraveling circuits of visual perception and cognition through the superior colliculus. Neuron 109: 918–937.

Beis JM, Keller C, Morin N et al. (2004). Right spatial neglect after left hemisphere stroke: qualitative and quantitative study. Neurology 63: 1600–1605.

Berlucchi G, Aglioti S, Tassinari G (1997). Rightward attentional bias and left hemisphere dominance in a cue-target light detection task in a callosotomy patient. Neuropsychologia 35: 941–952.

Beste C, Hamm JP, Hausmann M (2006). Developmental changes in visual line bisection in women throughout adulthood. Dev Neuropsychol 30: 753–767.

Beume LA, Martin M, Kaller CP et al. (2017). Visual neglect after left-hemispheric lesions: a voxel-based lesion-symptom mapping study in 121 acute stroke patients. Exp Brain Res 235: 83–95.

Bisiacchi P, Marzi CA, Nicoletti R et al. (1994). Left-right asymmetry of callosal transfer in normal human subjects. Behav Brain Res 64: 173–178.

Bisiach E (1993). Mental representation in unilateral neglect and related disorders: the twentieth Bartlett Memorial Lecture. Q J Exp Psychol A 46: 435–461.

Blini E, Romeo Z, Spironelli C et al. (2016). Multi-tasking uncovers right spatial neglect and extinction in chronic left-hemisphere stroke patients. Neuropsychologia 92: 147–157.

Donato M (2012). Neglect and extinction depend greatly on task demands: a review. Front Hum Neurosci 6: 195.

Bourne VJ (2006). The divided visual field paradigm: methodological considerations. Laterality 373–393.

Bowen A, McKenna K, Tallis RC (1999). Reasons for variability in the reported rate of occurrence of unilateral spatial neglect after stroke. Stroke 30: 1196–1202.

Chandler DJ, Jensen P, McCall JG et al. (2019). Redefining noradrenergic neuromodulation of behavior: impacts of a modular locus Coeruleus architecture. J Neurosci 39: 8239–8249.

Chokron S, Peyrin C, Perez C (2019). Ipsilesional deficit of selective attention in left homonymous hemianopia and left unilateral spatial neglect. Neuropsychologia 128: 305–314.

Clewett DV, Huang R, Velasco R et al. (2018). Locus Coeruleus activity strengthens prioritized memories under arousal. J Neurosci 38: 1558–1574.

Corballis MC (1995). Visual integration in the split brain. Neuropsychologia 33: 937–959.

Corbetta M, Shulman GL (2002). Control of goal-directed and stimulus-driven attention in the brain. Nat Rev Neurosci 3: 201–215.

Corbetta M, Shulman GL (2011). Spatial neglect and attention networks. Annu Rev Neurosci 34: 569–599.

Danckert J, Tamietto M, Rossetti Y (2019). Definition: blindsight. Cortex 119: 569–570.

De Brigard F, Prinz J (2010). Attention and consciousness. Wiley Interdiscip Rev Cogn Sci 1: 51–59.

Doricchi F (2022). The functions of the temporal-parietal junction. Handb Clin Neurol 187: 161–177.

Doricchi F, Lasaponara S, Pazzaglia M et al. (2022). Left and right temporal-parietal junctions (TPJs) as "match/mismatch" hedonic machines: a unifying account of TPJ function. Phys Life Rev 42: 56–92.

English MCW, Maybery MT, Visser TAW (2023). A review of behavioral evidence for hemispheric asymmetry of visuospatial attention in autism. Autism Res 16: 1086–1100.

Esposito M, Palermo S, Nahi YC et al. (2024). Implicit selective attention: the role of the mesencephalic-basal ganglia system. Curr Neuropharmacol 22: 1497–1512.

Fink GR, Driver J, Rorden C (2000). Neural consequences of competing stimuli in both visual hemifields: a physiological basis for visual extinction. Ann Neurol 47: 440–446.

Gainotti G (2019). A historical review of investigations on laterality of emotions in the human brain. J Hist Neurosci 28: 23–41.

Gerrits R, Verhelst H, Vingerhoets G (2020). Hemispheric dominance for visuospatial attention does not predict the direction of pseudoneglect. Neuropsychologia 148: 107634.

Gray OJ, McFarquhar M, Montaldi D (2021). A reassessment of the pseudoneglect effect: attention allocation systems are selectively engaged by semantic and spatial processing. J Exp Psychol Hum Percept Perform 47: 223–237.

Handa RJ, McGivern RF (2015). Steroid hormones, receptors, and perceptual and cognitive sex differences in the visual system. Curr Eye Res 40: 110–127.

Heekeren HR, Marrett S, Ruff DA et al. (2006). Involvement of human left dorsolateral prefrontal cortex in perceptual decision making is independent of response modality. Proc Natl Acad Sci U S A 103: 10023–10028.

Heilman KM, Van Den Abell T (1980). Right hemisphere dominance for attention: the mechanism underlying hemispheric asymmetries of inattention (neglect). Neurology 30: 327–330.

Herald SB, Yang H, Duchaine B (2023). Contralateral biases in category-selective areas are stronger in the left hemisphere than the right hemisphere. J Cogn Neurosci 35: 1154–1168.

Ipata A, Girelli M, Miniussi C et al. (1997). Interhemispheric transfer of visual information in humans: the role of different callosal channels. Arch Ital Biol 135: 169–182.

Jewell G, McCourt ME (2000). Pseudoneglect: a review and meta-analysis of performance factors in line bisection tasks. Neuropsychologia 38: 93–110.

Kaul D, Papadatou-Pastou M, Learmonth G (2023). A meta-analysis of the line bisection task in children. Laterality 28: 48–71.

Kentridge RW, Heywood CA, Weiskrantz L (1999). Attention without awareness in blindsight. Proc Biol Sci 1430: 1805–1811.

Kentridge RW, Heywood CA, Weiskrantz L (2004). Spatial attention speeds discrimination without awareness in blindsight. Neuropsychologia 42: 831–835.

Kinsbourne M (1977). Hemi-neglect and hemisphere rivalry. Adv Neurol 18: 41–49.

Koch G, Cercignani M, Bonnì S et al. (2011). Asymmetry of parietal interhemispheric connections in humans. J Neurosci 31: 8967–8975.

Kumano H, Nobukawa S, Shirama A et al. (2022). Asymmetric complexity in a pupil control model with laterally imbalanced neural activity in the locus coeruleus: a potential biomarker for attention-deficit/hyperactivity disorder. Neural Comput 34: 2388–2400.

Liu J, Bayle DJ, Spagna A et al. (2023). Fronto-parietal networks shape human conscious report through attention gain and reorienting. Commun Biol 6: 730.

Malherbe C, Umarova RM, Zavaglia M et al. (2018). Neural correlates of visuospatial bias in patients with left hemisphere stroke: a causal functional contribution analysis based on game theory. Neuropsychologia 115: 142–153.

Mangun GR, Luck SJ, Plager R et al. (1994). Monitoring the visual world: hemispheric asymmetries and subcortical processes in attention. J Cogn Neurosci 6: 267–275.

Marzi CA (1999). The Poffenberger paradigm: a first, simple, behavioural tool to study interhemispheric transmission in humans. Brain Res Bull 50: 421–422.

Marzi CA, Bisiacchi P, Nicoletti R (1991). Is interhemispheric transfer of visuomotor information asymmetric? Evidence from a meta-analysis. Neuropsychologia 29: 1163–1177.

Marzi CA, Fanini A, Girelli M et al. (1997). Is extinction following parietal damage an interhemispheric disconnection phenomenon? In: P Thier, H-O Karnath (Eds.), Parietal lobe contributions to orientation in 3D space. Springer-Verlag, Heidelberg, pp. 431–445.

Marzi CA, Girelli M, Miniussi C et al. (2000). Electrophysiological correlates of conscious vision: evidence from unilateral extinction. J Cogn Neurosci 12: 869–877.

Marzi CA, Girelli M, Natale E et al. (2001). What exactly is extinguished in unilateral visual extinction? Neurophysiological evidence. Neuropsychologia 39: 1354–1366.

McGivern RF, Mosso M, Freudenberg A et al. (2019). Sex related biases for attending to object color versus object position are reflected in reaction time and accuracy. PLoS One 14: e0210272.

Mesulam MM (1981). A cortical network for directed attention and unilateral neglect. Ann Neurol 10: 309–325.

Michael GA, Gálvez-García G (2011). Salience-based progression of visual attention. Behav Brain Res 224: 87–99.

Mizzi R, Michael GA (2016). Exploring visual attention functions of the human extrageniculate pathways through behavioral cues. Psychol Rev 123: 740–757.

Moeck EK, Zhao JL, Most SB et al. (2023). Emotional stimuli similarly disrupt attention in both visual fields. Cognit Emot 37: 633–649.

Moruzzi G, Magoun HW (1949). Brain stem reticular formation and activation of the EEG. Electroencephalogr Clin Neurophysiol 1: 455–473.

Natale E, Posteraro L, Prior M et al. (2005). What kind of visual spatial attention is impaired in neglect? Neuropsychologia 43: 1072–1085.

Natale E, Marzi CA, Bricolo E et al. (2007). Abnormally speeded saccades to ipsilesional targets in patients with spatial neglect. Neuropsychologia 45: 263–272.

Neal J, Song I, Katz B et al. (2023). Association of intrinsic functional connectivity between the locus coeruleus and salience network with attentional ability. J Cogn Neurosci 35: 1557–1569.

Norman LJ, Heywood CA, Kentridge RW (2015). Exogenous attention to unseen objects? Conscious Cogn 35: 319–329.

Nuthmann A, Clark CNL (2023). Pseudoneglect during object search in naturalistic scenes. Exp Brain Res 241: 2345–2360.

Okayasu M, Inukai T, Tanaka D et al. (2023). The Stroop effect involves an excitatory–inhibitory fronto-cerebellar loop. Nat Commun14: 27. https://doi.org/10.1038/s41467-022-35397-w.

Palomero-Gallagher N, Amunts K (2022). A short review on emotion processing: a lateralized network of neuronal networks. Brain Struct Funct 227: 673–684.

Patston LL, Hogg SL, Tippett LJ (2007). Attention in musicians is more bilateral than in non-musicians. Laterality 12: 262–272.

Paulraj SR, Schendel K, Curran B et al. (2018). Role of the left hemisphere in visuospatial working memory. J Neurolinguistics 48: 133–141.

Pedersini CA, Lingnau A, Sanchez-Lopez J et al. (2020). Visuo-spatial attention to the blind hemifield of hemianopic patients: can it survive the impairment of visual awareness? Neuropsychologia 149: 107673.

Petersen SE, Posner MI (2012). The attention system of the human brain: 20 years after. Annu Rev Neurosci 35: 73–89.

Petitet P, Noonan MP, Bridge H et al. (2015). Testing the interhemispheric competition account of visual extinction with combined TMS/fMRI. Neuropsychologia 74: 63–73.

Posner MI (2008). Measuring alertness. Ann N Y Acad Sci 1129: 193–199.

Posner MI, Petersen SE (1990). The attention system of the human brain. Annu Rev Neurosci 13: 25–42.

Proverbio AM, Zani A, Gazzaniga MS (1994). ERP and RT signs of a rightward bias for spatial orienting in a split-brain patient. Neuroreport 5: 2457–2461.

Reuter-Lorenz PA, Nozawa G, Gazzaniga MS et al. (1995). Fate of neglected targets: a chronometric analysis of redundant target effects in the bisected brain. J Exp Psychol Hum Percept Perform 21: 211–230.

Ringman JM, Saver JL, Woolson RF et al. (2004). Frequency, risk factors, anatomy, and course of unilateral neglect in an acute stroke cohort. Neurology 63: 468–474.

Rode G, Fourtassi M, Pagliari C et al. (2017a). Complexity vs. unity in unilateral spatial neglect. Rev Neurol (Paris) 173: 440–450.

Rode G, Pagliari C, Huchon L et al. (2017b). Semiology of neglect: an update. Ann Phys Rehabil Med 60: 177–185.

Sanchez-Lopez J, Savazzi S, Pedersini CA et al. (2020). Neural bases of unconscious orienting of attention in hemianopic patients: hemispheric differences. Cortex 127: 269–289.

Sapir A, Soroker N, Berger A et al. (1999). Inhibition of return in spatial attention: direct evidence for collicular generation. Nat Neurosci 2: 1053–1054.

Sapir A, Rafal R, Henik A (2002). Attending to the thalamus: inhibition of return and nasal-temporal asymmetry in the pulvinar. Neuroreport 13: 693–697.

Schendel K, Dronkers NF, Turken AU (2016). Not just language: persisting lateralized visuospatial impairment after left hemisphere stroke. J Int Neuropsychol Soc 22: 695–704.

Shenal BV, Hinze S, Heilman KM (2012). The cost of action miscues: hemispheric asymmetries. Brain Cogn 79: 45–48.

Simon JR (1969). Reactions toward the source of stimulation. J Exp Psychol 81: 174–176.

Simon JR, Rudell AP (1967). Auditory S-R compatibility: the effect of an irrelevant cue on information processing. J Appl Psychol 51: 300–304.

Smania N, Martini MC, Gambina G et al. (1998). The spatial distribution of visual attention in hemineglect and extinction patients. Brain 121: 1759–1770.

Spagna A, Martella D, Fuentes LJ et al. (2016). Hemispheric modulations of the attentional networks. Brain Cogn 108: 73–80.

Spagna A, Kim TH, Wu T et al. (2020). Right hemisphere superiority for executive control of attention. Cortex 122: 263–276.

Sprague JM (1991). The role of the superior colliculus in facilitating visual attention and form perception. Proc Natl Acad Sci U S A 88: 1286–1290.

Stone SP, Halligan PW, Greenwood RJ (1993). The incidence of neglect phenomena and related disorders in patients with an acute right or left hemisphere stroke. Age Ageing 22: 46–52.

Stroop JR (1935). Studies of interference in serial verbal reactions. J Exp Psychol 18: 643.

Sylvester R, Josephs O, Driver J et al. (2007). Visual FMRI responses in human superior colliculus show a temporal-nasal asymmetry that is absent in lateral geniculate and visual cortex. J Neurophysiol 97: 1495–1502.

Thiebaut de Schotten M, Urbanski M et al. (2005). Direct evidence for a parietal-frontal pathway subserving spatial awareness in humans. Science 309: 2226–2228.

Thiebaut de Schotten M, Dell'Acqua F, Forkel SJ et al. (2011). A lateralized brain network for visuospatial attention. Nat Neurosci 14: 12451246. https://doi.org/10.1038/nn.2905. Erratum in: Nat Neurosci (2011) 14: 1617. PMID: 21926985.

Thiebaut de Schotten M, Tomaiuolo F, Aiello M et al. (2014). Damage to white matter pathways in subacute and chronic spatial neglect: a group study and 2 single-case studies with complete virtual "in vivo" tractography dissection. Cereb Cortex 24: 691–706.

Toba MN, Migliaccio R, Potet A et al. (2022). Right-side spatial neglect and white matter disconnection after left-hemisphere strokes. Brain Struct Funct 227: 2991–3000.

Trevethan CT, Sahraie A (2010). Blindsight. In: Encyclopedia of consciousness, Academic Press, pp. 107–122.

Vallesi A (2012). Organisation of executive functions: hemispheric asymmetries. J Cogn Psychol 24: 367–386.

Vallesi A (2021). The quest for hemispheric asymmetries supporting and predicting executive functioning. J Cogn Neurosci 33: 1679–1697.

van Ede F, Nobre AC (2023). Turning attention inside out: how working memory serves behavior. Annu Rev Psychol 74: 137–165.

Vanderhasselt MA, De Raedt R, Baeken C (2009). Dorsolateral prefrontal cortex and Stroop performance: tackling the lateralization. Psychon Bull Rev 16: 609–612.

Verbitsky R, Anderson B, Danckert J et al. (2024). Left cerebellar lesions may be associated with an increase in spatial neglect-like symptoms. Cerebellum 23: 431–443.

Wagner J, Rusconi E (2023). Causal involvement of the left angular gyrus in higher functions as revealed by transcranial magnetic stimulation: a systematic review. Brain Struct Funct 228: 169–196.

Weber B, Treyer V, Oberholzer N et al. (2005). Attention and interhemispheric transfer: a behavioral and fMRI study. J Cogn Neurosci 17: 113–123.

Weintraub S, Mesulam MM (1987). Right cerebral dominance in spatial attention. Further evidence based on ipsilateral neglect. Arch Neurol 44: 621–625.

Weiskrantz L (2004). Roots of blindsight. Prog Brain Res 144: 229–241.

Wolski P, Asanowicz D (2023). Does CUD measure interhemispheric transfer time? The allocation of attention influences the Poffenberger effect. Neuropsychologia 185: 108581.

Handbook of Clinical Neurology, Vol. 208 (3rd series)
Cerebral Asymmetries
C. Papagno and P. Corballis, Editors
https://doi.org/10.1016/B978-0-443-15646-5.00025-7

Chapter 9

Left- and right-side unilateral spatial neglect: Hemispheric differences

LAURA VERONELLI[1,2,3*] AND GIUSEPPE VALLAR[1,2*]

[1]*Department of Psychology, University of Milano-Bicocca, Milan, Italy*

[2]*MiBTec, University of Milano-Bicocca, Milan, Italy*

[3]*Department of Neurorehabilitation Sciences, Casa di Cura Igea, Milan, Italy*

Abstract

Neglect of one side of space, typically contralateral to a lesion of one cerebral hemisphere, is a multicomponent neurologic syndrome. In humans, left neglect after right brain damage is more frequent, severe, or both, than right neglect after left brain damage. Right neglect is behaviorally like left neglect. In the monkey, such a functional asymmetry is not present. In humans, left hemisphere-based spatial systems are weaker, likely due to the coexistence of language and spatial processes. This may account for the lateral asymmetry of neglect, which is present at birth. Except in a few patients, there is no global functional reversal of language and spatial cognition. Left brain-damaged patients often show both aphasia and right neglect, as many right brain-damaged patients with crossed aphasia show left neglect. Lateralized sensory stimulations temporarily improve both left and right neglect. Damage to the posterior parietal lobe (inferior parietal lobule), the temporo-parietal junction, the superior and middle temporal, and to the premotor and prefrontal cortices is associated with contralateral neglect; also, lesions in white matter fiber tracts and subcortical nuclei bring about neglect, with no definite left–right asymmetries.

INTRODUCTION

Unilateral spatial neglect in humans

Patients with unilateral spatial neglect (henceforth, neglect) fail to orient toward, consciously perceive, report, and reach for stimuli on one side of space. While patients are unaware of neglected events, a considerable amount of processing without awareness occurs, so that neglect has been considered a spatial-specific disorder of consciousness (Bisiach, 1992).

Neglect has been termed unilateral spatial agnosia (or inattention, Battersby et al., 1956; Hécaen, 1972) and hemiinattention (Friedland and Weinstein, 1977). However, the term unilateral spatial neglect is more appropriate (Williams et al., 2021): it does not make a confusing reference to disorders of object identification, as unilateral spatial agnosia does; it does not imply that the deficit precisely involves one half of space and is confined to the orientation of spatial attention, as hemiinattention does. Despite the diverse terminologies that have been used over the years to describe the phenomenon, the term unilateral spatial neglect is currently widely accepted.

During evolution, the human brain has developed a lateralization of functions, with differences between the left and the right hemisphere. In the 19th century, evidence from brain-damaged patients indicated a left hemispheric lateralization of language, with an association of aphasia with left brain damage, first with frontal Broca's area (Phillips et al., 2015; Tzourio-Mazoyer et al., 2020). The right hemispheric counterpart of aphasia is neglect, which, in humans, typically concerns the left, contralateral, side of space, and predominantly occurs after right brain damage, indicating a lateralization of several spatial abilities to the right hemisphere (De Renzi, 1982). These hemispheric differences, more conspicuous for disorders of language, are likely to be a matter of degree. In right-handed persons, damage to the right hemisphere infrequently brings about aphasia (crossed aphasia, Mariën et al., 2004). Much more frequently, damage to the left hemisphere causes right spatial neglect (Esposito et al., 2021). This may be taken as an indication of a more bilateral representation of space, with respect to language, in the two hemispheres.

Neglect cannot be traced back to elementary, lower-level sensorimotor deficits (Heilman and Valenstein, 1979), as

*Correspondence to: Laura Veronelli, PhD and Giuseppe Vallar, MD, Department of Psychology, University of Milano-Bicocca, Piazza dell'Ateneo Nuovo 1, 20126, Milan, Italy. Tel: +39-0264483704, E-mail: laura.veronelli@unimib.it, giuseppe.vallar@unimib.it

hemianopia, hemiplegia, and hemianesthesia. Right brain-damaged patients with left neglect may show no somatosensory, visual, and motor deficits (e.g., Bolognini et al., 2016). Furthermore, right brain-damaged patients with neglect and hemianopia and hemianesthesia may show preserved visual and somatosensory evoked potentials to stimuli delivered to the left half-field or to the left hand. These findings suggest that these putatively visual and somatosensory deficits are not due to impairments of elementary sensory processes, being instead a manifestation of neglect (Vallar et al., 1991b; Di Russo et al., 2008). Accordingly, they may be considered as pseudohemianopia (Kooistra and Heilman, 1989), and pseudohemianesthesia (Pitteri et al., 2013). Conversely, neglect may exacerbate these sensorimotor contralateral disorders (Halligan and Marshall, 2002), although the two types of deficits may occur independent of each other.

The history of neglect

The first descriptions of neglect can be traced back to the second half of the 19th century. In an early report by Jackson (1876), the term imperception denoted a patient, with a lesion affecting the right posterior temporal lobe, who occasionally failed to read letters on the left side of words. When asked to read the Snellen visual acuity chart, she "began at the right lower corner and tried to read backwards" (p. 8).

The Hungarian physician Balint (1909) described a patient with two nearly symmetric softenings, centered on the posterior parietal areas of the two hemispheres, and encroaching upon the surrounding posterior temporal and anterior occipital lobe. In addition to a disorder of reaching visual objects (optic ataxia), the patient exhibited two disorders of visual attention: an extreme restriction, which prevented him to focus on more than one stimulus at a time; and a rightward shift, approximately 35–40 degrees from the point he was fixating on. The latter disorder is a manifestation of left spatial neglect (De Renzi, 1996). Quoting Bálint (Bálint and Harvey, 1995), "he did not read all the letters in the right hemifield, but only the last letter on the right …. I managed to approach his left side silently from behind and sit next to him on the bench; he did not notice anything. When I performed the same movement from the right, he noticed immediately. A few times I extended my hand into his left visual field; he noticed nothing. … From these examinations it follows that the patient's attention was constantly drawn to the right side of space, so that he saw the far-right object only."

The British neurologist Brain (1941, p. 246), who first suggested an association between right posterior parietal damage and "inattention to, or neglect of, the left half of external space," provided a vivid description: "… In going from his bedroom, on the first floor downstairs to the sitting-room he tended to turn to the right instead of to the left and in consequence would arrive in the dining-room. … He complained that he did not know where the left side of his body was." Another patient (Paterson and Zangwill, 1944) "… often collided with objects on his left which he had clearly perceived a few moments before. He was liable at table to knock over dishes on his left-hand side and occasionally missed food on the left of his plate. He commonly failed to attend to the left-hand page in turning the pages of a book and in reading lines of disconnected words commonly omitted the first word or two" (historical review in Halligan and Marshall, 1993). These examples show that neglect may disrupt activities of daily life (see also Della Sala et al., 2018). In line with these observations, neglect leads to negative functional outcomes (Jehkonen et al., 2006), and impacts on activities of daily living, as self-care (Mark, 2003; Chen et al., 2012), balance (Nijboer et al., 2014), and walking (Tromp et al., 1995).

Etiology of neglect

Neglect has been most frequently described after stroke (Stone et al., 1992; Esposito et al., 2021). In the acute stage, neglect may be apparent at a clinical observation at the patient's bedside, as well as emerging from a simple, formal, quantitative assessment. Patients may show a deviation of the head and eyes toward the ipsilateral side and a defective orienting toward the contralateral side of space (Fruhmann Berger et al., 2008). In stroke patients with unilateral hemispheric damage, conjugate paresis of gaze, with defective ocular movements toward the contralateral side of space, and an eye deviation toward the ipsilateral side, is more frequent, severe, and long-lasting after right than left brain damage (De Renzi et al., 1982). Brain tumors (Battersby et al., 1956; Shallice et al., 2010; Sallard et al., 2012) and traumatic brain injuries (Maeshima et al., 1995; Chen et al., 2016) may cause neglect. Also neurodegenerative diseases may bring about neglect. The signs of posterior cortical atrophy, in the biparietal variant (Crutch et al., 2017), and of Alzheimer disease (Scheltens et al., 2021) include neglect. Neglect-like visuospatial asymmetries have been found in Parkinson Disease (Norton et al., 2015), and a transient left neglect has been described as the unique clinical expression of non-convulsive status epilepticus (Veronelli et al., 2016).

Rare types of neglect

As noted above, neglect typically concerns the side of space contralateral to the side of the hemispheric lesion, along a left/right dimension, with reference to the mid-sagittal plane of the body. Altitudinal neglect is a rare phenomenon, with a few reported patients showing a deficit involving the lower, or, less frequently, the upper portions of space (Glazer et al., 2017). Ipsilateral neglect is also infrequent, having been reported mainly after right (Harciarek et al., 2012), but also after left, hemispheric damage (Cubelli et al., 2011).

THE LATERALIZATION OF NEGLECT IN BRAIN-DAMAGED ANIMALS

Neglect has been reported in animals, but without the hemispheric asymmetry found in humans. In rats, unilateral lesions of the prefrontal medial agranular cortex, of the posterior parietal cortex, and of the dorso-central striatum (a subcortical convergence zone of the prefrontal medial agranular and of the posterior parietal cortices) produce manifestations of neglect broadly like those found in human patients, although symmetrically allocated. Neglect is multimodal, involves visual, tactile, and auditory stimuli, presented in the side of space contralateral to the side of the lesion, and is not associated with primary sensorimotor deficits (Reep et al., 2004). After a unilateral frontal lesion, monkeys and dogs make rotatory movement toward the ipsilateral side (Bianchi, 1895). Monkeys, from which one entire frontal lobe has been removed, do not appear to see objects in the contralateral field of vision; food placed in the cages is ignored on the contralateral side, but quickly reached when placed in the ipsilateral visual field; the blink response to threat is also absent on the contralateral side (Kennard, 1939). Similarly, monkeys show spontaneous cycling activity toward the side ipsilateral to the frontal lesion, and a multimodal bias toward ipsilateral visual, auditory, and tactile stimuli (Welch and Stuteville, 1958). A unilateral lesion to the white matter (parietal leucotomy) brings about a tendency to make erroneous choices of ipsilateral, non-target stimuli (Gaffan and Hornak, 1997). Reversible unilateral lesions into specific parietal circuits (the lateral intraparietal area, the medial bank of the intraparietal sulcus, and the parietal reach region) bias choices toward the ipsilateral side of space (Kubanek et al., 2015).

THE LATERALIZATION OF NEGLECT IN BRAIN-DAMAGED CHILDREN

Some insight into the innateness and development of spatial abilities (Buckley et al., 2018), and of language (van der Hulst, 2023), as well as into the hemispheric specialization of their neural underpinnings, comes from brain-damaged children. Several single case studies are on record. Hécaen (1976) briefly mentions one right-handed, 13-years-old, stroke right brain-damaged child with unilateral spatial agnosia and spatial dysgraphia. Ferro et al. (1984) reported three right brain-damaged children (6-, 9-, and 5-years-old, one left-handed, two strokes, one tumor), who showed left neglect in a line cancellation task, that recovered within 1 month. One right-handed 8-year-old child, with left hemiparesis and hemianopia, due to a wide right parieto-occipital lesion, showed neglect only in walking, colliding with obstacles on the left side (Grossi et al., 2001). Another 9-year-old child, with extensive right fronto-parieto-temporal and left pulvinar stroke lesions, exhibited left neglect in line cancellation and drawing, by copy and from memory, that recovered in 1 month (Kleinman et al., 2010). In a retrospective study (Laurent-Vannier et al., 2003), the most demonstrative 12 cases (mean age at lesion onset about 7 years, range 0–14) were selected: nine childen with right or mainly right brain damage, and three with left brain damage, due to stroke, traumatic brain injury, and infection (one child). Behavior indicative of neglect was observed in all children, according to their age: for example, bumping into contralateral obstacles, not playing with toys located in the contralateral side of space, missing the left-side part of the food in the dish, not looking to the left, drawing only in the right side of the sheet. According to their age, children performed drawing, writing, and reading tasks, and a test specifically devised for children (the teddy bear cancellation test).

Further information is provided by perinatal stroke, which refers to a focal vascular brain injury sustained between the 20th week of gestation and the 28th day of postnatal life, whose categories of morbidity include neglect (Kirton et al., 2021). A 14-year-old right-handed child suffered a left hemispheric damage at 4 h of age. The child, since approximately 3½ years of age, showed right neglect in activities of daily life, as running into the right edge of the kitchen counter, and bumping into peers on the right during social situations. Right neglect was found also in drawing by copy, and as a leftward bias in lateralized response selection; conversely, in line bisection the child showed a rightward bias, interpreted as overcompensation of right neglect (Johnston and Shapiro, 1986). In a group study including mostly right-handed children (19 with left and 19 with right brain damage, due to perinatal stroke), results from visual and tactile exploratory tasks seem to suggest a more severe contralateral deficit (Thareja et al., 2012; see also Kriksciun et al., 2021). In the study by Yousefian et al. (2015), 38 children, 6–14 years of age, with either right or left brain damage due to perinatal stroke, showed a bilateral deficit in a clock drawing test from memory, as compared with control participants. The impairment was more severe after right brain damage. At 6–8 years of age, right brain-damaged children exhibited a more severe contralateral (left-side) impairment, that largely disappeared by adolescence ($\geq$15 years). In a task requiring removal of objects on a board in front of the child, both left and right brain-damaged children preferentially removed objects from the ipsilateral side of the board first, while control children did not show a removal order preference. The deficit was mild, since accuracy (i.e., the number of removed objects) did not differ from that of control participants (Trauner, 2003). A recent review (Purpura et al., 2023) concludes that right brain-damaged children may show visual neglect, mainly assessed by target cancellation/removal and drawing tasks. The disorder mainly involves the contralateral side of space, compatible with the diagnosis of unilateral spatial neglect, as made in adult persons. In left brain-damaged children, conversely, a more bilateral deficit of spatial attention may be found. Left brain-damaged children with right-side neglect are, however, on record (Johnston and Shapiro, 1986; Laurent-Vannier et al., 2003). Similar to adults, paper-and-pencil tests are less able to identify neglect than functional scales (Hart et al., 2021). Finally, two 7- and 13-year-old right-handed right brain-damaged children showed ipsilateral, right, neglect in the drawing of a clock face from memory (Lewis and Geller, 1994). In sum, these findings suggest that the hemispheric specialization supporting spatial attention and representation is likely to be present already at birth.

NEGLECT AS A MULTICOMPONENT SYNDROME

In the 1970s, it became clear that neglect is a multicomponent syndrome, including defective (featuring the absence of behaviors appropriate in a setting) and productive (featuring inappropriate behaviors) manifestations. Deficits have been categorized according to several dimensions, originally based on studies in right brain-damaged patients with left neglect (Vallar, 1998; Halligan et al., 2003; Vallar and Ronchi, 2022).

Defective manifestations

Sectors of space

External space

Brain (1941) distinguished discrete patterns of extra-personal neglect concerning grasping (using hand and arm) and walking (using legs) distance. A division may be drawn between neglect of near, peri-personal space, within arm's reach, and neglect of far space, which can be reached through locomotion, or a tool as a rake (Vallar and Maravita, 2009). Neglect may also concern the contralateral side of the body (personal neglect).

Internal space

Neglect may involve the left side (from a given vantage point) of familiar images retrieved from long-term memory, featuring a horizontal dimension, with left-sided and right-sided elements. In this paradigm, the response is verbal. Participants may be required to report objects (e.g., buildings, shops) in a place, such as Milan's Piazza del Duomo (Bisiach et al., 1981), cities in the map of France (Rode and Perenin, 1994), rooms in a house, houses in a street, and a plaza (Meador et al., 1987). Neglect patients fail to report objects from the contralateral side of the image. In the mental number line task, participants hear two numbers that define a number interval (putatively represented in a continuous, quantity-based analogic format) and are asked to state the midpoint number without making calculations. Right brain-damaged patients displace the midpoint number rightward, making an error directionally similar to the one made in the bisection of physical lines (Zorzi et al., 2002). In the O'clock test (Grossi et al., 1993), participants are asked to communicate which out of the two hands on two clock faces forms the larger angle. The task assesses imaginal neglect, since the hours are proposed verbally by the examiner (imagine a clock that says nine "O'clock" and then one that says "seven thirty").

Spatial reference systems

In egocentric neglect (body-centered or viewer-centered), sensory events are encoded and neglected with reference to egocentric frames, as the body's mid-sagittal plane, the head, and the line of sight in retinotopic coordinates (Bisiach et al., 1985), or body parts, as the hand (Moscovitch and Behrmann, 1994). In allocentric/object-based neglect, the deficit occurs with reference to the axes of the object, independent of its position with respect to egocentric frames (Chatterjee, 1994; Ota et al., 2003). In reading, a further distinction is drawn between allocentric/object-centered (referred to the intrinsic coordinates of the physical stimulus) and allocentric word/grapheme-centered neglect, which refers to the abstract (left-to-right in languages as English) representation of the letter string (Hillis and Caramazza, 1995; review in Vallar et al., 2010).

Input vs output neural systems

Perceptual neglect (Bisiach et al., 1990), also termed sensory-attentional (Na et al., 1998), is a deficit of processing of inputs in different sensory modalities. Premotor neglect (Bisiach et al., 1990), also termed motor-intentional, and directional hypokinesia (Na et al., 1998) is a deficit of planning, initiating, and executing movements toward the contralateral side. A further distinction (Mattingley et al., 1992) has been drawn between the slowing in movement initiation (directional hypokinesia) and the slowing of movement time (directional bradykinesia). Motor neglect is a reduction or absence of movement of the contralateral limbs, mimicking hemiparesis or hemiplegia (Laplane and Degos, 1983; Toba et al., 2021).

Extinction to double simultaneous stimulation

When two stimuli are presented simultaneously, after a unilateral hemispheric lesion, the contralateral stimulus is not reported, but extinguished (Vallar and Bolognini, 2014). Extinction may occur in the auditory (Heilman et al., 1971), visual, tactile (Vallar et al., 1994), and olfactory (Bellas et al., 1988) modalities, and across modalities (cross-modal extinction, Brozzoli et al., 2006).

Sensory domains: Visual, tactile, auditory, and olfactory neglect

Neglect has been first and most frequently studied in the visual domain. However, right brain-damaged patients may also be impaired in searching for left-side stimuli in a tactile maze, with only somatosensory information being available (De Renzi et al., 1970; Haeske-Dewick et al., 1996). In the auditory domain, patients may show a rightward ipsilateral deviation in the localization of binaural stimuli (Clarke and Thiran, 2004). In the olfactory domain, with double simultaneous trigeminal and olfactory stimulation of the nostrils, patients may displace rightward the stimulus delivered to the left nostril (Bellas et al., 1988).

Domain specificity

Neglect dyslexia affects reading in the contralateral left side of letter strings (words and nonwords), sentences and passages of prose, with omission and substitution errors (Vallar et al., 2010). One right-brain-damaged patient with a defective recognition of the left side of faces (facial neglect), without other signs of neglect, has been reported (Young et al., 1990).

Anosognosia

Patients may exhibit anosognosia (Babinski, 1914; Papagno and Vallar, 2003), that features unawareness of contralateral neurologic deficits: anosognosia for hemiplegia (Orfei et al., 2007), and for visual half-field deficits (Bisiach et al., 1986b).

Productive manifestations

In extra-personal space, productive behaviors include: perseveration and gratuitous productions (Na et al., 1999; Rusconi et al., 2002), inappropriate for the task and the environmental setting; hyper-attention toward ipsilateral events (Làdavas et al., 1993); active avoidance (Denny-Brown and Chambers, 1958), up to repulsion, of contralateral events in the neglected side of space; hallucinations appearing in the ipsilateral, not neglected, side (Silberpfennig, 1941; Perani et al., 1987). In personal space, productive phenomena include a sense of unfamiliarity, separation from the patient's body, disownership (somatoparaphrenia, Vallar and Ronchi, 2009), morbid dislike or hatred of the left limbs (misoplegia, Pearce, 2007). All these defective and productive manifestations may be associated, but may also occur independent of one another, creating dissociations (Barbieri and Renzi, 1989; Halligan et al., 2003).

THE ASSESSMENT OF SPATIAL NEGLECT

Peri-personal/near extra-personal space

Within hand reach, tasks requiring the crossing out (cancellation) of targets as lines (Albert, 1973) or letters (Diller and Weinberg, 1977) are most used. Cancellation tasks may have distracter non-target stimuli (e.g., Diller and Weinberg, 1977), or not (e.g., Albert, 1973). The distinction between egocentric vs allocentric types of neglect can be made using complete stimuli, as circles and triangles (Ota et al., 2003), and the same stimuli incomplete on one side (i.e., with a left-sided or right-sided gap). The patients' task is to cross out complete stimuli. Egocentric neglect (with reference to the mid-sagittal plane of the body) is indexed by omissions in the contralateral side of the display, independent of stimulus type (complete or incomplete). Allocentric neglect is revealed by the crossing out of stimuli with a contralateral gap, judged as complete, independent of their position in the display.

Another tool is line bisection (Schenkenberg et al., 1980), whereby participants set manually the center of a horizontal line. Patients with left neglect typically make a rightward error that increases as line length increases. A variant is the landmark task (Milner et al., 1993) in which participants are asked to point to the end of a pre-bisected, horizontal line closer to the point of bisection. Most patients with left-side neglect point leftward, toward the neglected side, then perceptually underestimating the left lateral extent of the line; this would account for the rightward bias in manual line bisection. Some patients point rightward, showing then a premotor deficit (directional hypokinesia). Other tools are drawing by copy (Gainotti et al., 1972), and from memory to a verbal request (Bisiach and Vallar, 2000), and reading tasks (Vallar et al., 2010). For far extra-personal space, wheelchair navigation and walking may be used, whenever possible (Turton et al., 2009).

Figs. 9.1–9.3 show examples of left neglect in line bisection, target cancellation, and drawing by copy and from memory. Painters, after unilateral brain damage, offer the unique opportunity to

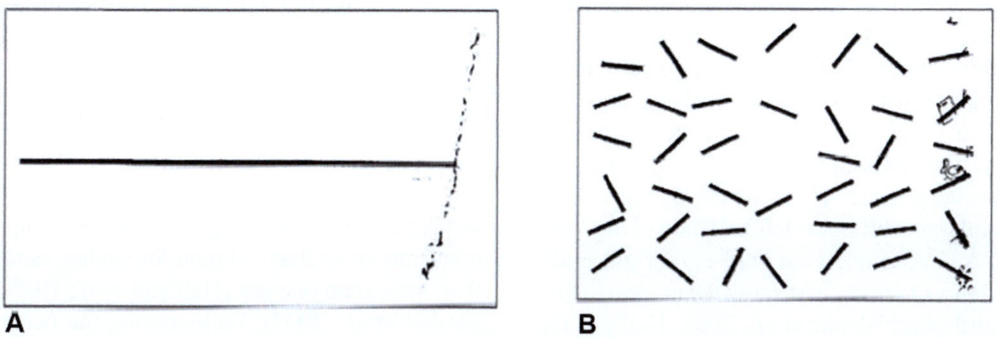

Fig. 9.1. Patient E.B. (A) Line bisection: the midline is set at the right end of the line, with repeated vertical marks (perseverations). (B) Line cancellation: only lines printed at the right end of the sheet are crossed out, with simple and complex perseverations: repeated marks, a square, and a goose. Reproduced with permission from Bottini G, Bisiach E, Sterzi R, et al. (2002). Feeling touches in someone else's hand. Neuroreport 13: 249–252. https://doi.org/10.1097/00001756-200202110-00015, fig. 2.

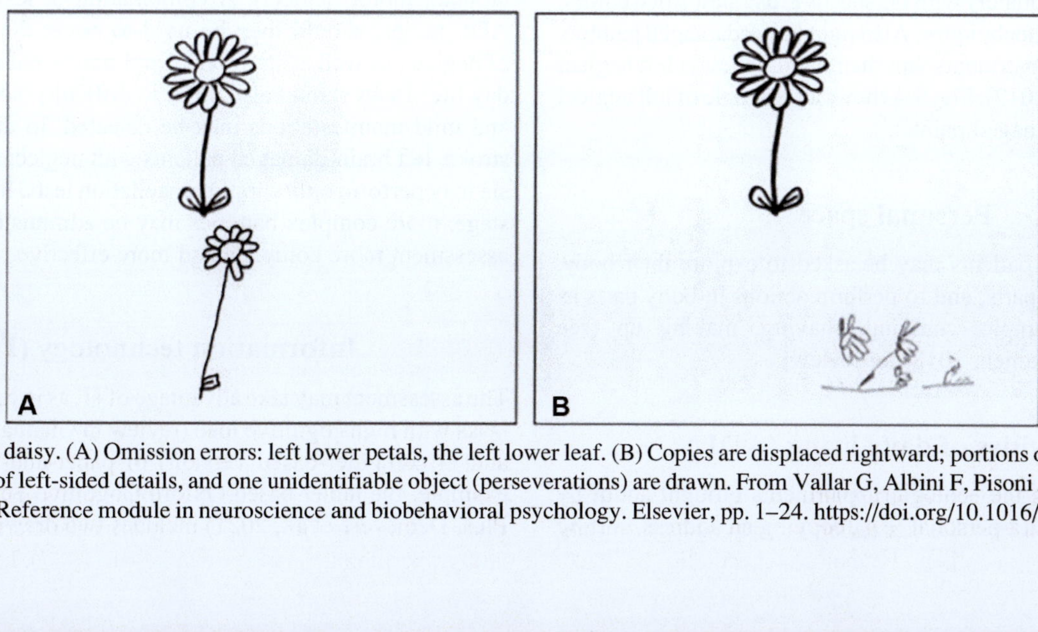

Fig. 9.2. Copy of a daisy. (A) Omission errors: left lower petals, the left lower leaf. (B) Copies are displaced rightward; portions of three daisies, each one with omission of left-sided details, and one unidentifiable object (perseverations) are drawn. From Vallar G, Albini F, Pisoni A (2024). Unilateral spatial neglect, In: Reference module in neuroscience and biobehavioral psychology. Elsevier, pp. 1–24. https://doi.org/10.1016/B978-0-12-820480-1.00053-X.

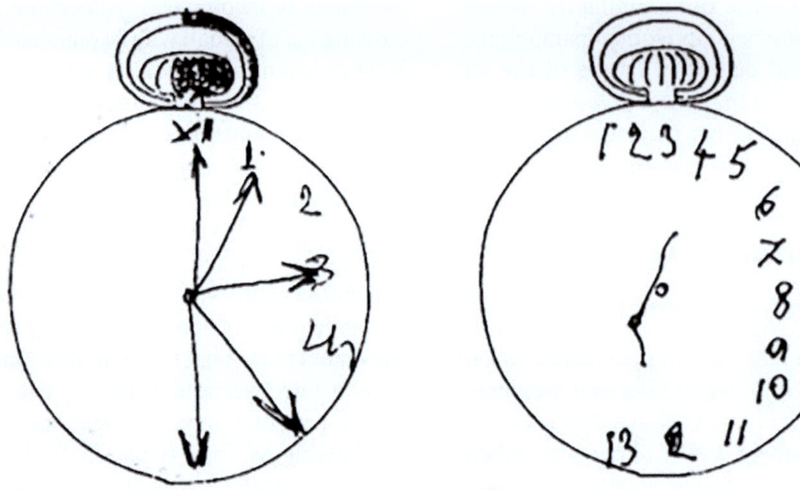

A B

Fig. 9.3. Different behaviors of two right brain-damaged patients with left neglect in setting the hours on a blank clock dial. (A) Only some hours are drawn in the right-hand side, and perseverative drawing of the clock hands is made; the right-hand side of upper components of the clock is colored black. (B) All hours are drawn, but the left-hand side ones are displaced rightward. Reproduced with permission from Bisiach E, Vallar G (2000). Unilateral neglect in humans, In: Boller F, Grafman J, Rizzolatti G. (Eds.), Handbook of neuropsychology. Elsevier Science, B.V., pp. 459–502, fig. 1.

gain further insight into deficits of visuospatial processes, including neglect, in patients with special skills in the visuo-spatial domain (see Zaidel, this volume). In the evaluation of such paintings, as well as in drawing from memory and by copy, the additional roles of two other disorders should be considered: constructional apraxia (Gainotti and Trojano, 2018), and, after right-brain damage, contralateral left hyperschematia, namely the disproportionate enlargement of the left half-side of drawings (Rode et al., 2018). A few, single case studies of right brain-damaged amateur painters or artists, who show left neglect in their paintings, are on record (e.g., Blanke et al., 2003; Halligan and Marshall, 1997, also in sculpting). Neglect in paintings may also be object-based (Halligan and Marshall, 2001). After damage to the left hemisphere, paintings are more simplistic with little, if any, left–right differences (Annoni et al., 2005). As assessed through a qualitative evaluation, paintings of left brain-damaged patients deal inadequately with perspective, use simplified colors, and simpler, naïve, techniques. Also right brain-damaged painters may show such impairments, but their main deficit is left neglect (Mazzucchi et al., 2012). Fig. 9.4 shows an example of left neglect in a right brain-damaged painter.

Personal space

In personal space, patients may be asked to explore their body, reaching for body parts, and to perform actions in body parts as the face, for example, combing, shaving, making up (see Caggiano and Jehkonen, 2018, for review).

Activities of daily living (ADL)

ADL scales assess the ecologic impairments brought about by neglect, both in extra-personal (e.g., copying an address, sorting coins, moving around by walking or in a wheelchair), and in personal space, for instance, dressing (Azouvi, 2016, for review).

Test batteries

The occurrence of neglect varies dependent on the use of a single test, typically target cancellation, vs a comprehensive multiple test battery, also including ADL. When compared, a lower occurrence rate of neglect is found for studies using a single test rather than more than one test (Halligan et al., 1989; Agrell et al., 1997; Lindell et al., 2007). Furthermore, the occurrence of neglect is much higher when assessed through ecologic, compared to non-ecologic, tests (53% vs 24%, review in Esposito et al., 2021). However, some ecologic tests (e.g., the baking tray test, Tham and Tegnér, 1996), do not permit disentangling between different manifestations of neglect (e.g., visual, tactile, representational, motor neglect). A combination of tests, together with ADL scales, should then be used to cover the various aspects of neglect, as well as its severity and functional impact on everyday life. Tasks should also differ in difficulty, so that both severe and mild manifestations may be detected. In the acute stage of stroke, left brain-damaged patients with neglect and severe aphasia may perform only simple cancellation tests. In a chronic stable stage, more complex batteries may be administered, making the assessment more complete and more effective.

Information technology (IT)

The assessment may take advantage of IT, as in computerized dual tasks with high cognitive load (review in Giannakou et al., 2022), and in computer-based versions of paper-and-pencil tests. For example, the tablet-based Oxford Cognitive Screen-Plus (OCS-Plus, Demeyere et al., 2021) includes two drawing tasks capable

Fig. 9.4. Two paintings by the German painter Anton Räderscheidt before (A) and after (B) a right hemisphere stroke. The picture made before the stroke shows the influence of Italian metaphysical art, a style of painting developed by the Italian artists Giorgio de Chirico and Carlo Carrà. By courtesy of Professor François Boller, see Boller F, Caputi N (2023). Effets du vieillissement et des lésions cérébrales chez les artistes [effects of ageing and brain damage in artists], in: Histoire des sciences Médicales, Séance Du Vendredi 15 Décembre.

of detecting neglect: figure copy/recall and target cancellation. Neurit-Space (Massetti et al., 2023), in which participants perform the tasks on a touch-screen monitor, using a digital pen, includes three tasks: target cancellation, line bisection, and drawing by copy five objects. In sum, computer-, graphics tablets-, virtual reality (VR)-based assessments, and immersive VR systems are promising IT tools (reviews in Kaiser et al., 2022; Terruzzi et al., 2023).

LEFT VS RIGHT SPATIAL NEGLECT

Overall frequency and severity

Since early observations in group studies, neglect has been found to be more frequent after right than after left hemispheric lesions. In 1963, in a series of 413 brain-damaged patients with cortical lesions, Hécaen and Angelergues found neglect in 59 patients: 51 out of 153 (33%) right brain-damaged patients, 4 out of 205 (1.95%) left brain-damaged patients, and 4 out of 55 (7.27%) patients with bilateral lesions (see Hécaen, 1972, pp. 211–12). This difference in the incidence of contralateral neglect was confirmed by studies using several behavioral tasks, such as response position preference in Raven's Progressive Matrices (Costa et al., 1969), copy of a drawing (Gainotti et al., 1972), copying crosses (Denes et al., 1982), a copy task and a measure of position preference on a multiple-choice test (Sunderland et al., 1987), and visual and tactile spatial exploratory tasks (Vallar et al., 1991a). In these studies, the proportion of right neglect after left brain damage was not as low as that reported by Hécaen and Angelergues. These findings suggest a right hemispheric dominance for the contralateral orientation of spatial attention. A systematic review of 17 studies found that the median occurrence of neglect is twice as high for left (median 43%, range 13%–82%) than for right neglect (median 21%, range 0%–76%, Bowen et al., 1999).

Several studies found that neglect is not only more frequent, but also more severe after right hemispheric damage (Gainotti et al., 1972; Vallar et al., 1991a; Agrell et al., 1997; Beis et al., 2004; Ringman et al., 2004; Ten Brink et al., 2017). In patients with acute stroke, a higher frequency of contralateral neglect (assessed by target cancellation) after right than left hemispheric lesions, but a comparable severity, have been reported; furthermore, right neglect resolved more completely (Stone et al., 1991). Chen et al. (2015) found left neglect to be more frequent and persistent, but not more severe (see also Costa et al., 1969, for a similar finding). Denes et al. (1982) found that left neglect is more frequent and persistent. Including patients with severe aphasia in their acute sample (48 patients with focal left-brain damage), Suchan et al. (2012) found no differences in the severity of neglect between left- and right-sided lesions, also supporting the clinical observation that severe aphasia and dense right neglect may occur together, potentially due to extensive lesions in the left perisylvian areas. The impact of neglect on rehabilitation outcomes does not differ between left and right brain-damaged stroke patients (Tarvonen-Schröder et al., 2020; Moore et al., 2021b; Yoshida et al., 2022), suggesting that neglect, as aphasia, and motor impairment affecting the dominant hand after left hemispheric damage, could be per se a negative prognostic factor for rehabilitation outcomes (Gialanella et al., 2011; Ginex et al., 2017).

However, the available data are heterogeneous, and a few reports show no hemispheric differences in the frequency of the disorder (Battersby et al., 1956; Albert, 1973). Patients' selection, locus of lesion, timing and type of assessment may account for such a large variability among studies (Bowen et al., 1999).

Furthermore, aphasia and right neglect after a left hemispheric lesion often coexist in the same patient. This might have limited patients' inclusion, due to defective understanding of the tests' instructions. However, the number of patients excluded because of aphasia cannot entirely explain the differences in the occurrence of contralateral neglect after left vs right hemispheric lesions. In the acute phase poststroke, Stone et al. (1991) excluded 5 out of 26 left brain-damaged patients (19.23%) and 2 out of 18 right brain-damaged patients (11.11%). Neglect was equally frequent in left- and right-brain-damaged patients (62% vs 72%) but was more severe and resolved less frequently after right brain damage. In the subacute phase, Beis et al. (2004) excluded 11 out of 89 (12.3%) left brain-damaged patients, mainly for disorders of language comprehension. Neglect was found in 34 out of 78 patients (43.5%). In the study of Suchan and Karnath (2011) aphasia was present in over 85% of 17 left brain-damaged patients with right neglect. In a recent review, including 6324 participants from 41 studies (3411, 54%, with right brain damage and 2913, 46%, with left brain damage) overall occurrence rates of neglect were 29%, 38% after right and 18% after left hemispheric damage (Esposito et al., 2021).

Concerning time after stroke, neglect occurrence decreases from the acute (34%) to the chronic (17%) stage, although the number of studies focused in later stages is limited (Esposito et al., 2021). In 44 patients with hemispheric stroke (Stone et al., 1991), in the acute stage (3 days after stroke onset) neglect was only slightly more frequent after damage to the right (72%) than to the left (62%) hemisphere, but more severe in right brain-damaged patients; 3 months later, frequency was greater in the right brain-damaged group (75%), than in the left one (33%). In 400 stroke survivors (Moore et al., 2021b), no significant differences have been found between left and right hemispheric lesions in the probability of neglect recovery at 6 months. Similarly, no differences in the severity of left and right neglect, assessed by a cancellation task, have been found 4 years after stroke (Halligan et al., 1992). In line with these findings, a study assessing chronic stroke survivors with aphasia found 5 out 40 patients (12.5%) of left brain-damaged patients with symptoms of mild spatial neglect (Hreha et al., 2017), suggesting the importance of routinely assessing neglect also after a left hemispheric lesion. Overall, these results indicate behavioral improvements during neglect recovery in the first months post stroke, sustained by a trend toward normalization of the neural network functional connectivity, altered by the brain lesion (Ramsey et al., 2016).

A possible confounding variable, as to the rate of occurrence, is the wide range of methods used to detect neglect and assess its severity (Colombo et al., 1976). In the previously revised studies on the occurrence of left and right neglect, the diagnostics varied from one single test, usually a cancellation task (line cancellation, Albert, 1973; bell cancellation, Becker and Karnath, 2007) to a comprehensive battery test, investigating several manifestations of neglect, and including ecologic tasks (Stone et al., 1991; Azouvi et al., 2006).

In general, a combination of tests, both domain-specific and ecologic, varying in difficulty, and of ADL scales, is recommended to cover the frequency and severity of neglect, its several components, possible dissociations (Rorden and Karnath, 2010; Ten Brink et al., 2017), and its functional impact on daily living (Azouvi, 2017). Patients with dense neglect in the acute stage of stroke (with associated severe aphasia after left hemispheric brain damage) typically have access only to very simple cancellation tests (Albert, 1973). In sum, when, as in most studies, hemispheric asymmetries are found, contralateral neglect is more frequent, severe, or both, and persistent after right brain damage.

Left neglect has been found to be more frequent than right neglect also in neurodegenerative disorders. Patients with Alzheimer disease show left neglect in line bisection (Ishiai et al., 1996, one patient), in visual search and target cancellation (Mendez et al., 1997), and in copy drawing (Oda et al., 2008, one patient with left allocentric neglect). In a group study, Ishiai et al. (2000) found left neglect in line bisection in 8 (25%) out of 32 patients. Venneri et al. (1998) reported left neglect in two patients, in a late stage of the disease, with cerebral atrophy more prounced in the right hemisphere. Negative findings are also on record (Kasai et al., 2007, who reported peripheral inattention and simplification). In posterior cortical atrophy (the visual variant of Alzheimer disease, Whitwell et al., 2007), in a series of 24 patients, neglect was found in 16 (66%), with left being more frequent and severe than right neglect (Andrade et al., 2010). Similarly, Silveri et al. (2011) reported left neglect in three patients, with atrophy more pronounced in the right hemisphere. Two patients showed vertical neglect (upper, Glazer et al., 2017; lower, Tariq et al., 2020). This left/right asymmetry of neglect is in line with findings that, in posterior cortical atrophy, hypometabolism (Nestor et al., 2003) and atrophy (Whitwell et al., 2007) are worse in the right than in the left hemisphere. Patients with right-side neglect have also been found. In Mendez et al.'s study (Mendez et al., 1997), some patients showed right neglect. Silveri et al. (2011) reported one such patient with corticobasal degeneration and left parietooccipital atrophy. One patient with probable Alzheimer disease, and cortical atrophy and hypoperfusion predominant in the left hemisphere, showed right neglect, worsened after 1 year (Bartolomeo et al., 1998b). In conclusion, patients with neurodegenerative disorders may show neglect, which, as in patients with stroke and tumors, involves more frequently and severely the right side of space.

Behavioral similarities and differences between left-side and right-side neglect

Extra-personal neglect

Early studies showed differences in copying drawings, with the deficit being more severe in right brain-damaged patients, who tend to leave unfinished the left half of drawings, while left brain-damaged patients tend to omit some figures in the right half of the models (Gainotti et al., 1972). In a later study (Gainotti et al., 1986) both right- and left-brain-damaged patients showed an asymmetric exploration of space in a visual search task, making more omissions in the contralateral than in the ipsilateral side; conversely, in an overlapping figures test, only right brain-damaged patients showed a clear tendency to omit contralateral figures, suggesting an inability to extract visual information from the left side of complex stimuli (but see Egelko et al., 1988, who reported a contralateral fine-grained deficit of perceptual discrimination also in left brain-damaged patients). In a study in 190 stroke patients with unilateral hemispheric damage (Halligan et al., 1992), the incidence of visual neglect in star cancellation was comparable after damage to either hemisphere. However, while right brain-damaged patients showed a linear relationship, whereby omission errors increased from right to left, in left brain-damaged patients two subgroups were discerned: one showing, as expected, more contralateral than ipsilateral errors, but the other being comparable to the sample with right hemisphere damage, making more left-side omissions (Halligan et al., 1992). Somewhat similarly, in a retrospective study using target cancellation (Weintraub et al., 1996), left contralateral neglect (scored as omission errors) was found in nearly all (18 out of 19) right brain-damaged stroke patients, while left brain-damaged patients showed a different, and more balanced, pattern, with seven patients committing more contralateral, right-side, errors, and six more ipsilateral, left-side, errors. Interestingly, all eight patients with bilateral damage (mixed etiology: cerebrovascular, tumoral, and neurodegenerative diseases) made more right-side omission errors. Overall severity was greater in right brain-damaged patients and in patients with bilateral damage, than in left brain-damaged patients.

Representational neglect

In right brain-damaged patients, left neglect of internally generated visuospatial images has been repeatedly reported, after Bisiach and Luzzatti's (1978) seminal observation in two patients. Different paradigms have been used in group studies, such as the description of familiar places (Bisiach et al., 1981; Meador et al., 1987; Rode et al., 1995), the reconstruction of mental images (Bisiach et al., 1979), the mental number line (Zorzi et al., 2002), and the O'clock test (Grossi et al., 1989). Right representational neglect after left brain damage is infrequent. van Dijck et al. (2013) reported one left brain-damaged patient with right representational neglect, associated with mild visuo-spatial neglect in near extra-personal space. The patient's performance was defective for the right side of internally generated visuospatial images in the map of Flanders test (a Dutch version of the map of France task, Rode et al., 1995), in the O'clock test (Grossi et al., 1989), in a task requiring same/different judgments, based on differences in the left and right sides of mentally reconstructed visual images (Bisiach et al., 1979). Another left brain-damaged patient, with right neglect, showed a leftward deviation in setting the mid-point of the mental number line (Pia et al., 2009).

As for group studies comparing left and right brain-damaged patients, contralateral imaginal neglect was found in both patients' groups by Ogden (1985a). Bartolomeo et al. (1994) assessed representational neglect using familiar mental visual images retrieved from long-term memory: details of three Roman squares (Piazza Venezia, Piazza del Popolo, and Piazza S. Pietro; countries in a map of Europe; towns on the western coast of Italy from the vantage point of the eastern coast of Sardinia). Visuospatial neglect was assessed by the overlapping figures test (Gainotti et al., 1986) and line cancellation and line bisection tasks. A group analysis revealed a greater ipsilateral bias in right-brain-damaged patients. The analysis of the performance of individual patients showed contralateral representational neglect in five out of 30 right-brain-damaged patients (17%), all with visuospatial neglect, which was detected in 17 patients (57%). As for left-brain-damaged patients, none showed representational, and two out of 30 (7%) had visuospatial neglect.

Input vs output neural systems

According to one model of neglect (Corbetta and Shulman, 2011), the right hemisphere has a dominant role in both input/perceptual and output/action attentional components of neglect. In one study (Mattingley et al., 1992), most of the 18 right-brain-damaged patients with left neglect showed leftward directional hypokinesia, particularly when the posterior cortex was damaged; patients with left neglect and anterior or subcortical lesions, or both, also showed leftward directional bradykinesia, namely: slowing in the execution phase of movements toward the contralateral side. One left-brain-damaged patient with right neglect showed rightward directional hypokinesia, but not bradykinesia. Contralateral directional hypokinesia may then occur after damage to either hemisphere, with an association between directional hypokinesia and neglect, which is more frequent after right (18 out of 24 patients, 75%), than after left (1 out of 5, 20%) brain damage. In another study, left brain-damaged patients with right neglect were slow, when producing arm movements toward the right (neglected) side, as compared to left-directed movements (Bartolomeo et al., 2001). This pattern differs from the finding that only a minority of right-brain-damaged patients produce slower left-directed movements (see Mattingley et al., 1992), suggesting that perceptual, rather than promotor, impairments may play a main role in producing left neglect. Conversely, output, premotor deficits may play a greater role for right neglect (Bartolomeo et al., 1998a). Directional hypokinesia and motor impersistence (impairment in sustaining a goal-directed action) have been found to be more frequent after right than after left brain damage, without, however, any association with neglect, assessed by bisection, cancellation, and figure copying tasks (Kim et al., 2013).

Reference frames

One of the most studied aspects of neglect, comparing left- and right-brain-damaged patients, concerns different spatial reference frames (egocentric vs allocentric). Although the distinction remains controversial (allocentric codes, see Humphreys et al., 2013; spatial reference frames may be all egocentric, see Filimon, 2015), egocentric and allocentric neglect may co-occur, but double dissociations have been found (Chatterjee, 1994; Ota et al., 2001; Kleinman et al., 2007; Marsh and Hillis, 2008; Demeyere and Gillebert, 2019). This suggests the independence of the two types of spatial representation. In a group study of 47 left-brain-damaged acute stroke patients, Kleinman et al. (2007) found neglect in nine patients (19.1%), using a cancellation task with complete and incomplete targets, and a task of copying multiple objects: right allocentric neglect in six out of nine patients (67%), right egocentric neglect in one patient (11%), and both types of neglect in two patients (22%). This pattern differs from the one found by the same research group in 50 right-brain-damaged patients: 16 patients (32%) showed left neglect, 11 (69%) egocentric, 4 (25%) allocentric, and 1 (6%) both types. These results have been interpreted as suggesting the involvement of the right hemisphere in processing global features of hierarchical stimuli, that may include the panoramic orientation of attention to the left and right sides of space, and then to the whole display. Conversely, the left hemisphere would play a role in the processing of local features, across the two sides of individual objects/stimuli (Robertson and Lamb, 1991; Fink et al., 1996; Volberg et al., 2009; Boksem et al., 2012). However, this hypothesis has not been confirmed by studies including larger samples of left- and right-brain-damaged neglect patients. Bickerton et al. (2011) found that egocentric and allocentric pure forms of each type of neglect are equally prevalent after right and left hemispheric lesions, while the presence of both deficits is associated with right hemisphere damage. In a consecutive sample of 366 acute stroke patients, Demeyere and Gillebert (2019) found that left egocentric neglect in right-brain-damaged patients is more severe and prevalent than the right deficit after left brain damage; conversely, left allocentric neglect is more severe, but not more prevalent, than right allocentric neglect. At 6 months, in a subsample of 160 patients, recovery rates were over 70% for both types of neglect. In sum, while contralateral egocentric and allocentric neglect may be present in both right- and left-brain-damaged patients, the interhemispheric differences are unclear.

Contralateral extinction

Patients may show contralateral extinction without contralateral neglect, neglect without extinction, and neglect and extinction in association, both after right (visual and tactile extinction, Vallar et al., 1994) and after left (visual extinction, Beume et al., 2020) brain damage. However, contralateral extinction is more frequently associated with right brain damage (tactile extinction, Gainotti et al., 1989; visual extinction, Becker and Karnath, 2007). Auditory extinction is more persistent after right brain damage (De Renzi et al., 1984). These asymmetries suggest that at least some mechanisms of extinction may be related to neglect, with the same lateral asymmetry.

Neglect dyslexia, dysgraphia, and dyscalculia

Spatial disorders of reading, writing and calculation, including unilateral neglect dyslexia (Vallar et al., 2010; Siéroff, 2017) and dysgraphia (Baxter and Warrington, 1983, one left-handed patient), have been described, most frequently after right brain damage, and involve the left side of meaningful and meaningless letter strings (words and nonwords, Ellis et al., 1987; Kinsbourne and Warrington, 1962) and sentences (Hécaen, 1972, pp. 78–81; Pizzamiglio et al., 1992). Reports of right neglect dyslexia and dysgraphia are sporadic. In one left-handed, left-brain-damaged patient with right extra-personal visual neglect, Caramazza and Hillis (1990) reported reading and spelling errors only on the right side of words, irrespective of the topographic arrangement of stimuli in reading (horizontal, vertical, or mirror-reversed words) and of the type of response in spelling (written, oral, and backward oral spelling). This pattern of performance has been interpreted as evidence that sequential information in orthographic representations is coded spatially in a "word-centered coordinate system," namely: a spatial frame whose center corresponds to the mid-point of a canonical, orientation-invariant representation of the word. Hashimoto et al. (2009) reported a right-handed, left brain-damaged patient with right neglect dysgraphia in the Japanese language, in which most letters are composed of two elements: "hen" (left) and "tsukuri" (right). The patient wrote recognizable, well-formed, but paragraphic, Kanji letters in both spontaneous and dictated writing. Neographism in tsukuri, the right-side component of Kanji letters (morphograms), was the more frequent error.

Anosognosia and personal neglect

Anosognosia for hemiplegia and personal neglect for the left side of the body after right brain damage have been reported in group studies, and are more frequent than for the right side after left brain damage (Appelros et al., 2002; Maguire and Ogden, 2002; Azouvi et al., 2006; Committeri et al., 2007; Di Vita et al., 2017; Caggiano and Jehkonen, 2018). Hécaen (1972, pp. 273) reported a large hemispheric difference in a series of 308 right-handed patients with retrorolandic lesions: 39 out of 136 right-brain-damaged patients (29%), and seven out of 172 left-brain-damaged patients (4%) showed anosognosia for contralateral hemiplegia. Four right-brain-damaged patients (3%), and 34 left-brain-damaged patients (20%) showed bilateral somatognosic disorders, namely: two symptoms of Gerstmann's syndrome (Gerstmann, 1924; Vallar, 2007; Rusconi and Cubelli, 2019), finger agnosia (the inability to recognize, name and select individual fingers of both hands), and left–right disorientation (the patients' inability to distinguish the left from the right side on their own body or on that of the examiner). The third symptom was autotopagnosia (a selective deficit in pointing to various parts of the body on verbal command, although they are correctly recognized and named when the patient touches them). The assessment was clinical, and no information was provided as to the etiology of the disease, and its duration. In line with Hécaen's results, Bisiach et al. (1986b), using a one-item interview in right-brain-damaged patients, reported medium-to-severe anosognosia for left hemiplegia in 12 out of 36 (33%) patients, and anosognosia for left hemianopia in 28 out of 32 (87%) patients. Berti et al. (1996) found anosognosia for left-sided motor impairment in nine out of 34 (26.5%) right-brain-damaged stroke patients. In another study by Berti et al. (2005), 18 out of 30 (60%) right-brain-damaged patients showed anosognosia for contralateral hemiplegia. Also following unilateral temporary suppression of activity of the left or the right hemisphere by intracarotid injection of sodium amytal (Wada Test, Baxendale, 2009), a hemispheric asymmetry has been found. According to a review of seven studies (Cocchini et al.,

2009), including 340 patients, the average percentage of patients with anosognosia was 88% (range 66%–100%) for left motor weakness, and 55% (range 0%–86%) for right motor weakness. In only one study, the percentage of patients with left and right weakness was identical (66%, Dywan et al., 1995). In the study of Adair et al. (1995), 30 out of 31 patients denied left (97%), and 15 out of 31 (48%) right hemiparesis. Since anosognosia for right motor weakness is typically assessed by a verbal interview (Bisiach et al., 1986b; Anderson and Tranel, 1989), in left brain-damaged patients aphasia could bring about an underestimation of unawareness (Nathanson et al., 1952). In line with this view, the frequency of anosognosia for right hemiplegia, assessed by tasks not requiring the use of language, is higher than when a verbal interview is used (Cocchini et al., 2009). However, temporary inactivation of the left hemisphere may bring about anosognosia for right hemiplegia, but not for aphasia and vice versa (Breier et al., 1995), suggesting that selective damage to one of two independent (modular) monitoring systems may cause anosognosia for hemiplegia or for aphasia. This implies that the presence of aphasia does not entirely account for the higher frequency of anosognosia for left vs right motor weakness. In a series of 66 stroke patients with acute left hemispheric brain lesions (Baier et al., 2014), anosognosia for right hemiplegia was assessed by a verbal interview (Bisiach et al., 1986b) and 22 patients (33, 3%) were excluded due to severe aphasia. Out of the 44 assessed patients, only one (2%) showed anosognosia for right hemiplegia. In this patient, dominance for language was reversed, namely right hemisphere-based, as assessed by functional magnetic resonance imaging (fMRI). This result is in line with Breier et al.'s (1995) conclusion that co-occurrent aphasia does not entirely account for the higher frequency of anosognosia for left than for right hemiplegia.

Similar to anosognosia for hemiplegia, left personal neglect is overall more frequent than the homologous right deficit (according to the review of Caggiano and Jehkonen, 2018, 24.5% vs 14%). However, single right-handed left-brain-damaged patients with a right-side deficit have been reported (Facchin et al., 2017, one patient with anosognosia for right hemiplegia and personal neglect; Peru and Pinna, 1997, one patient with right personal, and no extra-personal neglect, anognosia, and somatoparaphrenia).

Somatoparaphrenia

The occurrence of somatoparaphrenia for the right hand after left brain damage is a rare phenomenon, reported in individual case studies. In the review by Vallar and Ronchi (2009), 51 out of 56 patients (91%) showed somatoparaphrenia for the left side of the body, associated with a right-sided lesion. Two such patients (Ives and Nielsen, 1937, case #1; Roth, 1949, case #1) had also a lesion in the left hemisphere. Conversely, somatoparaphrenia for the right side of the body was observed in only five out of 56 (9%) patients (Schilder, 1935; Nielsen, 1938; one left-handed, non-aphasic patient, Halloran, 1946; Miura et al., 1996; Schiff and Pulver, 1999). Other patients were subsequently reported: a patient with chronic right somatoparaphrenia and extra-personal, but not personal, neglect, and aphasia (Cogliano et al., 2012); a patient with right somatoparaphrenia, anosognosia, and neglect, without aphasia (D'Imperio et al., 2017).

ANATOMO-FUNCTIONAL CORRELATES OF RIGHT AND LEFT SPATIAL NEGLECT

Extra-personal neglect

Damage to the right cerebral hemisphere: Left neglect

In the 1940s, anatomo-clinical correlation studies pointed to lesions of the right posterior parietal lobe (Brain, 1941; Critchley, 1953). Hécaen et al. (1956), in a neurosurgical series of 17 patients (15 right-handed, with right hemispheric excisions, 2 left-handed with left excisions) found an association of the syndrome of apractognosia, including extra-personal spatial neglect, and, particularly, personal neglect, with damage to the right inferior parietal lobule, including the supramarginal (BA 39) and the angular (BA 40) gyri, to the right temporo-parietal-occipital junction, and to the posterior part of the superior temporal gyrus (BA 22). Interestingly, the two left-handed patients of that series did not show apractognosia. Subsequent studies confirmed these findings, suggesting the involvement of the right inferior parietal lobule, specifically the supramarginal (Heilman and Valenstein, 1972; Vallar and Perani, 1986) and angular gyri (Mort et al., 2003), as well as the right temporo-parietal junction. Subsequently, damage to other cortical regions was associated with neglect: the right frontal cortex (dorso-lateral and medial, Heilman and Valenstein, 1972; dorso-lateral, Vallar and Perani, 1986; dorso-lateral prefrontal, Verdon et al., 2010), the right dorsal inferior frontal gyrus, BA 44, homologous to Broca's area in the left hemisphere (a single case study, Husain and Kennard, 1996). The right superior temporal gyrus has also been proposed to be a locus of damage (Karnath et al., 2001; Marshall et al., 2002, for discussion; see also Mort et al., 2003, who suggest the angular gyrus). Other possibly involved cortical regions are the right anterior cingulate cortex (Leibovitch et al., 1998, 1999) and the para-hippocampal region, on the medial surface of the temporal lobe (Mort et al., 2003). Damage to right subcortical regions may also bring about neglect: the thalamus (Watson and Heilman, 1979; Watson et al., 1981; Vallar and Perani, 1986; Karnath et al., 2002) and the basal ganglia (Damasio et al., 1980; Vallar and Perani, 1986; Karnath et al., 2002). Injuries to the gray subcortical nuclei are associated with neglect when a major cortical dysfunction is present, indicated by hypoperfusion (Perani et al., 1987; Hublet et al., 1995; Hillis et al., 2002; Sebastian et al., 2014), which may be taken as a pathogenetic mechanism of cortical diaschisis (von Monakow, 1914; Carrera and Tononi, 2014).

Also, damage to the white matter in the right hemisphere is associated with left neglect, particularly to the fronto-parietal (superior longitudinal fasciculus, SLF) and the occipito-frontal (inferior longitudinal fasciculus, ILF) connections (Leibovitch et al., 1998). In the SLF (see also Thiebaut de Schotten et al., 2011), lesions of the II branch play a relevant role (Thiebaut de Schotten et al., 2014). The direct stimulation of the right superior fronto-occipital fasciculus (Catani et al., 2002; but see Meola et al., 2015), which provides parieto-frontal connections, brings about contralateral neglect (Thiebaut de Schotten et al., 2005). Similarly, direct stimulation of the parieto-frontal connections (SLF-II) and the posterior superior parietal lobule (Brodmann's areas 5 and 7) is associated with left visual neglect (Vallar et al., 2014). The presence and severity of neglect in the subacute phase of stroke (i.e., <3 months after stroke onset) is associated with lesions of the II (connecting the anterior intraparietal sulcus and the angular gyrus with the posterior portions of the superior and middle frontal gyri), and of the III (connecting the temporo-parietal junction with the inferior frontal gyrus) branches of the right SLF. Also, a lesion of the right inferior fronto-occipital fasciculus (IFOF), in the presence of damage to the inferior parietal lobule of the right posterior parietal cortex, is associated with extra-personal neglect (Urbanski et al., 2008).

In patients with left visual half-field deficits and left neglect, due to an infarction in the vascular territory of the right posterior cerebral artery, a lesion involving the white matter in the occipital lobe brought about a disconnection between the para-hippocampal and the angular gyri (Bird et al., 2003). Another study in patients with a stroke in the vascular territory of the right posterior cerebral artery points instead to a major role of a perisylvian fronto-temporo-parietal network (Sperber et al., 2020), whose lesion is typically associated with neglect (Vallar, 2001; Vallar and Calzolari, 2018). Patients showing neglect without hemianopia are most relevant since they rule out the possibility of a contribution of primary sensory visual half-field deficits to the pathogenesis of neglect. In such patients, lesions involve the SLF and the supramarginal gyrus, spared in patients without neglect, suggesting a role of a parieto-frontal disconnection (Doricchi and Tomaiuolo, 2003).

Interhemispheric connections, through the corpus callosum (Glickstein and Berlucchi, 2008), also play a role. The persistence of neglect in the chronic phase is associated with a posterior interhemispheric disconnection, due to lesions in the posterior part (splenium) of the corpus callosum (Lunven et al., 2015). In line with these observations, microlesions of the posterior corpus callosum correlate with neglect severity (Bozzali et al., 2012). The left (unimpaired) hemisphere could support in compensating left neglect after right brain damage, provided interhemispheric communications are preserved (Lunven and Bartolomeo, 2017). Overall, these findings have been interpreted in terms of neglect as a disconnection syndrome (Bartolomeo et al., 2007). This view, however, should not be meant in the classical sense of the term (Lichtheim, 1885; Déjérine, 1914), but rather as the disruption, by the focal damage, of cortico-subcortical, also inter-hemispheric, neural networks (Corbetta and Shulman, 2011).

Damage to the left cerebral hemisphere: Right neglect

One early study (Damasio et al., 1980) reported three right-handed patients with left frontal (medial, dorso-lateral) lesions and right neglect, assessed by a visual target cancellation task. Language disturbances were also present: Broca's and then transcortical motor aphasia (patient #1), and akinetic mutism (patients #2 and #3). Patient #5, right-handed, with damage to the left basal ganglia (caudate nucleus, anterior putamen, the surrounding white matter) showed the infrequent, as mentioned above, phenomenon of left ipsilateral neglect. In a later study, in patients, mainly with cerebral tumors (Ogden, 1985b), contralateral neglect, assessed by cancellation and drawing tasks, was found to be associated with anterior lesions in left and with posterior lesions in right brain-damaged patients. In 89 left brain-damaged patients with subacute stroke, an association was found between right neglect and damage to the posterior cortex (Beis et al., 2004). Some studies investigated right neglect in left brain-damaged patients in the acute stage of stroke. Patients, assessed by target cancellation and copy drawing tests, had lesions of the superior and the middle temporal gyri, the inferior parietal lobule, and the insula (Suchan and Karnath, 2011). In another study in a large sample of patients, visual neglect, assessed by target cancellation and bisection tasks, was associated with damage to the left superior and middle temporal gyri, the temporal pole, the frontal operculum, and the insula; neglect severity, measured by the center of cancellation score (Rorden and Karnath, 2010) of the Bells test (Gauthier et al., 1989), was related with lesions centered in the left anterior temporal lobe and the left frontal operculum (Beume et al., 2017). Conversely, right visual extinction occurred after left parietal damage (anterior bank of the intraparietal sulcus and the supramarginal gyrus, Beume et al., 2020). In one study in left brain-damaged stroke patients (Malherbe et al., 2018), in target cancellation tasks, contralateral right (with a leftward bias) and ipsilateral left (with a rightward bias) deficits have been, respectively, associated with damage to the superior temporal and the inferior-posterior parietal regions, with damage to the arcuate fascicle being associated with both biases. In line bisection, leftward and rightward deviations were related with damage to the SLF and the inferior parietal lobule, correspondingly. However, in most patients these biases were mild, and hardly comply with current clinical definitions of neglect. Moore and Demeyere (2022) reported that right neglect was not associated with significant neural correlates, as compared with the relationship between left neglect and damage to the right temporal and posterior parietal cortices and the underlying white matter, specifically: parietal operculum, Heschl's gyrus, planum temporale, insular cortex, and white matter tracts, including the SLF. Damage to the left SLF, causing disconnection between fronto-parietal networks, has been found in three patients with right neglect (assessed by target cancellation, line bisection, drawing, and reading tasks) and aphasia; this complements analogous evidence from patients with left neglect (Toba et al., 2022). In the study by Moore et al. (2021a), right egocentric and allocentric neglect have been found to be respectively associated with damage to the left occipital cortex and the internal capsule, not homologs with the right hemispheric regions whose damage is associated with left neglect: egocentric (right temporo-parietal cortex and underlying white matter), allocentric (more posterior and ventral lesion sites). A recent review concludes that egocentric and allocentric neglect are anatomically dissociable conditions, whose pathologic anatomy may not be entirely homologous across hemispheres (Moore et al., 2023). Two patients, one with ischemic damage to the left and the other to the right intraparietal sulci, sparing the inferior parietal lobule, showed a deficit in selecting contralateral targets, suggesting a role of the superior parietal lobule in spatial attention; a leftward bias in line bisection was present in the left brain-damaged patient, indicating some relevance of this study to spatial neglect (Gillebert et al., 2011).

Personal neglect

Damage to the right cerebral hemisphere: Left neglect

For the more frequent left-side deficit, correlations have been found with damage to the right supramarginal, the postcentral parietal gyrus (Committeri et al., 2007), the temporo-parietal junction (Baas et al., 2011), and the underlying white matter in both studies. Recently, damage to a right hemispheric medial network (the temporal cortex, Heschl's gyrus, the ventrolateral nuclei of the thalamus and the fornix) has been reported (Bertagnoli et al., 2022). The review by Committeri et al. (2018) indicates damage to the right anterior parietal cortex and the underlying frontoparietal white matter as a neural correlate of personal neglect, suggesting a relevant role of somatosensory signals in the building up of an internal representation of the body.

Damage to the left cerebral hemisphere: Right neglect

Studies on the neural underpinnings of right personal neglect are scarce. Similar to the case of right extra-personal neglect, an association with damage to posterior areas has been found (Beis et al., 2004). A few single cases with left hemispheric damage and right personal neglect have been reported (Peru and Pinna, 1997, a small left insular hematoma; Maguire and Ogden, 2002, left frontotemporal, thalamic and basal ganglia damage; Facchin et al., 2017, an ischemic lesion in the left temporo-parietal cortex, the Rolandic operculum, the insula, the putamen, the internal capsule, and the SLF; review in Caggiano and Jehkonen, 2018).

Interim conclusion

Overall, left- and right-side neglect share, as main anatomic correlates, perisylvian regions, although not entirely homologous, which, in left brain-damaged right-handed patients, overlap with language areas (Suchan and Karnath, 2011; Vallar and Ronchi, 2022). One study (Moore and Demeyere, 2022) found no significant neuropathologic correlate of right neglect in the left hemisphere. Furthermore, the few reported behavioral differences between left vs right contralateral neglect (different impairments of allocentric vs egocentric reference frames, and of output vs input systems) are not fully consistent. Figs. 9.5 and 9.6 schematically show the main cortical regions and the main white matter fiber tracts involved in spatial representation and attention.

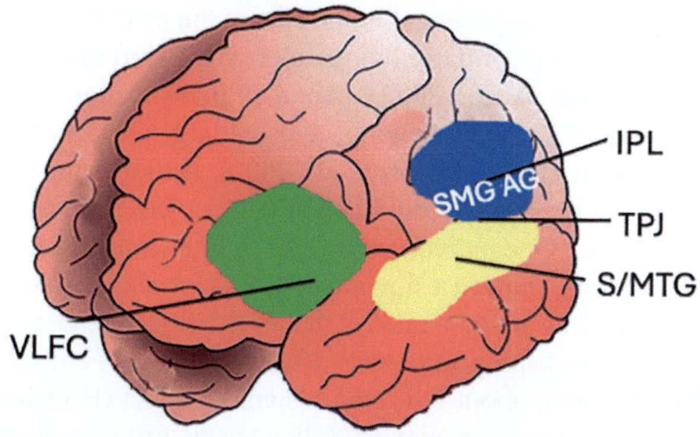

Fig. 9.5. Schematic main cortical left hemispheric regions whose damage may cause right neglect, illustrating the view that these regions are largely homologous with the right hemispheric lesions associated with left neglect: inferior parietal lobule (IPL, blue), comprising the supramarginal (SMG, BA 40), and the angular (AG, BA 39) gyri; temporo-parietal junction (TPJ); superior/middle temporal gyri (S/MTG, yellow); ventro-lateral premotor and prefrontal cortices (VLFC, green). Modified from Vallar G, Albini F, Pisoni A (2024). Unilateral spatial neglect, In: Reference module in neuroscience and biobehavioral psychology. Elsevier, pp. 1–24. https://doi.org/10.1016/B978-0-12-820480-1.00053-X.

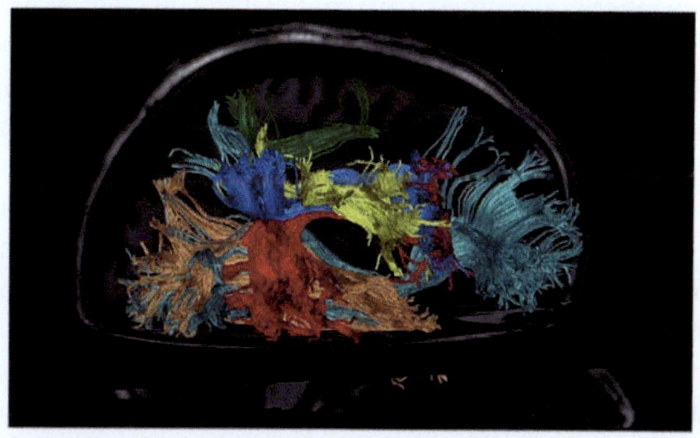

Fig. 9.6. Magnetic resonance tractography reconstructions of the main white-matter pathways relevant for left spatial neglect in the right hemisphere of a healthy person. The superior longitudinal fascicle (SLF) is divided into its four components, in lateral and axial view: SLF I (green), SLF II (blue), SLF III (yellow), and arcuate fascicle (red). The inferior fronto-occipital fasciculus (IFOF, cyan) and the inferior longitudinal fascicle (ILF, orange) are also shown. Left/right white matter asymmetries have been found (Honnedevasthana Arun et al., 2021; Verhelst et al., 2021), and possibly related to neglect (Büchel et al., 2004). Diffusion data acquired with high angular resolution diffusion-weighted imaging (HARDI: 60 gradient directions, b-value ¼ 3000 s/mm^2); tractography performed with probabilistic q-ball algorithm (Courtesy Dr. Antonella Castellano and Prof. Andrea Falini, Neuroradiology Department, Vita-Salute San Raffaele University and San Raffaele Scientific Institute, Milan, Italy). From Vallar G, Albini F, Pisoni A (2024). Unilateral spatial neglect, In: Reference module in neuroscience and biobehavioral psychology. Elsevier, pp. 1–24. https://doi.org/10.1016/B978-0-12-820480-1.00053-X.

Damage to the cerebellum and the brainstem

Overall, neglect associated with unilateral cerebellar damage is infrequent, as suggested by reports of single cases, and by a few group studies. One patient following left cerebellar stroke showed a mild left (ipsilateral to the side of the cerebellar lesion) neglect in activities of daily living, and a rightward bias in gaze position (Geiser et al., 2022). Another patient showed right neglect in cancellation, drawing, and reading tasks, following right cerebellar stroke (Silveri et al., 2001). As for damage also involving the brainstem, contralateral neglect has been reported in two patients. One patient, with right cerebellar and brainstem vascular lesions, showed a slowing of latencies to left visual stimuli, interpreted as an imbalance at the level of the ascending activating systems, disproportionately affecting the right hemisphere, and then bringing about left (contralateral) neglect (Hildebrandt et al., 2002). In another patient, following a stroke in the right pons (Cazzoli et al., 2023), a mild left visual neglect was detected, as revealed by a rightward shift in the spatial distribution of visual fixations. The deficit was interpreted as due to the interruption of descending pathways connecting the right cerebral cortex with the left cerebellum (cortico-pontocerebellar tracts), assessed by fiber tracking; conversely, ascending pathways from the cerebellum to the left cortex (dentato-rubro-thalamic tract) were spared.

Neglect in patients with cerebellar lesions extends the neural underpinnings of spatial cognition beyond the two cerebral hemispheres (cortex, white matter, and subcortical nuclei). Furthermore, these patients provide information as to whether the hemispheric asymmetry of contralateral neglect extends to the cerebellum. Each cerebellar hemisphere is mainly connected with the contralateral cortex, through pathways involving the thalamus and the pons as intermediate steps: the ascending cerebello-thalamo-cortical pathway and the descending cortico-ponto-cerebellar pathway. The fiber crossing takes place in the pontocerebellar descending projection and in the cerebellar-thalamic ascending projection. Through these pathways, cortical association areas (the prefrontal and the posterior parietal cortices) are reciprocally connected with each cerebellar hemisphere (Middleton and Strick, 1994; Schmahmann, 1996; Palesi et al., 2017). As to cerebellar functions, traditional views hold that the cerebellum is engaged exclusively in the control of action, with a specific role in the acquisition of motor skills. However, suggestions have been made for a cerebellar contribution to executive, visuospatial, language, and working memory functions (Schmahmann, 1996; Ramnani, 2006; Mariën and Borgatti, 2018). These effects on cognition may occur through diaschisis, a phenomenon originally hypothesized by von Monakow (1914), a condition in which lesions not only impair information processing at the site of the damage, but also have adverse, remote effects on connected regions of the brain. Diaschisis is indexed by the reduction of cerebral blood flow and metabolism in the structurally undamaged brain regions connected with the damaged one (Meyer et al., 1993; Carrera and Tononi, 2014). Crossed cerebellar diaschisis is a contralateral depression of cerebellar blood flow and metabolism following a supratentorial infarction. Conversely, in cerebello-cerebral diaschisis, there is a depression in the cerebral hemisphere contralateral to the side of a cerebellar stroke (Meyer et al., 1993). Therefore, the behavioral impairment associated with cerebellar lesions might be due, at least in part, to the impaired physiology of cortical areas, rather than to the direct effects of the structural cerebellar lesion, as brought about by a stroke. As for neglect, the crossed arrangement of the connections between each of the two cerebral and cerebellar hemispheres suggests that unilateral cerebellar lesions may bring about ipsilateral neglect, left following left cerebellar and right following right cerebellar damage. When the lesion involves the brainstem, in two patients contralateral neglect has been reported, and related to the site of the lesion (Hildebrandt et al., 2002; Cazzoli et al., 2023).

As for group studies, in a series of 28 patients with unilateral cerebellar stroke (Kim et al., 2008), neglect was found in 8 patients (28.6%, 4 with a left and 4 with a right cerebellar lesion). Three patients with left and one patient with right cerebellar damage showed contralateral neglect; four other patients, three with right and one with left cerebellar stroke, showed ipsilateral neglect. These results show that, in addition to cerebello-cortical diaschisis, responsible for neglect ipsilateral to the cerebellar lesion, other pathologic mechanisms at the cerebellar pontine-level, possibly involving the vestibular system (Shuren et al., 1998), might cause contralateral neglect. Finally, no definite left/right hemispheric asymmetries emerge from this study. Neglect following cerebellar lesions is however infrequent (Baier et al., 2010, one out of 16 patients, 6%, with a right-sided stroke and contralateral neglect; Frank et al., 2010, no neglect, assessed by letter cancellation and line bisection, in 22 patients, 12 with left, 10 with right acute cerebellar stroke). Eight children and adolescents with a cerebellar tumor, tested within days after surgery, showed no left–right asymmetries of omissions in letter cancellation, and no lateral biases in line bisection (Frank et al., 2008). In a recent group study of 20 patients (Verbitsky et al., 2024), left cerebellar stroke is associated with more left-side than right-side omissions in target cancellation; this asymmetry in the lateral distribution of omission errors persists up to more than 1 month after stroke onset. Conversely, right lesions do not bring about this asymmetry in error distribution. As for line bisection, no differences between the effects of left vs right cerebellar lesions were found.

Overall, these studies suggest that neglect after unilateral cerebellar damage, mainly stroke, is infrequent, as compared with the effects of unilateral hemispheric lesions. Reports of ipsilateral neglect after unilateral cerebellar damage, due to the crossed cerebellar-cerebral connections, support the view that neglect may be, also or mainly, due to contralateral cerebral diaschisis (Silveri et al., 2001; Geiser et al., 2022). These findings are in line with the view that each cerebral hemisphere directs spatial attention primarily toward the contralateral side of space, as discussed in the following section. Furthermore, the association between left ipsilateral neglect and left cerebellar damage supports the hypothesis of a major role of the right hemisphere, as after cortical damage (Verbitsky et al., 2024). Fine-grained anatomic lesion analysis may reveal that differences in the precise site of the lesions, such as in the brainstem, may bring about contralateral vs ipsilateral neglect, and are overall interpreted as a cerebello-cortical disconnection (Hildebrandt et al., 2002; Cazzoli et al., 2023).

HEMISPHERIC DIFFERENCES IN NEGLECT: ANATOMO-FUNCTIONAL MODELS

Two main models have been put forward to account for the hemispheric asymmetry of neglect. Both posit that neglect is a deficit of higher order spatial processes (i.e., spatial attention, spatial representation) rather than of elementary sensory and motor systems. Neglect may occur in the absence of sensorimotor (Bisiach et al., 1986a; Ofir-Geva et al., 2021) and visual half-field (Doricchi and Tomaiuolo, 2003; Saj et al., 2012) deficits, ruling out, together with its hemispheric asymmetry, these peripheral disorders as an explanation. Furthermore, any such interpretation cannot account for neglect for one side of visuospatial images (Bisiach and Luzzatti, 1978). Also, the finding of left neglect in right brain-damaged blind-folded patients, who explore a tactile maze using the right unaffected hand, runs counter to an explanation in terms of defective sensory processing or of oculomotor exploration (De Renzi et al., 1970).

Hemispheric rivalry (Kinsbourne, 1970, 1993)

Each hemisphere directs spatial attention toward the contralateral side of space, with mutual inhibition across the transverse commissures. Physiologically, in right-handers, the rightward (left hemisphere-based) contralateral orientation vector prevails slightly over the leftward (right hemisphere-based) one, also through transcallosal left-to-right inhibition, since the left hemisphere is overall more active than the right one, due to its role in language (Fig. 9.7A). Damage to one hemisphere impairs contralateral orientation, releasing the competitor from inhibition. The physiologic prevalence of the left hemisphere-based rightward-directed orientation system accounts for the hemispheric asymmetry of neglect (Fig. 9.7B). Contralateral neglect is also predicted to be worst on verbal tasks, which further engage and activate the unimpaired left hemisphere, therefore further inhibiting the right hemisphere. The empiric evidence concerning this aspect is, however, contradictory (Heilman and Watson, 1978; Caplan, 1985). This model also predicts a slight rightward bias in healthy persons, due to the physiologic prevalence of the rightward, left hemisphere-based, orientation vector. However, an opposite leftward bias is found in line bisection (pseudo-neglect, Jewell and McCourt, 2000; Learmonth and Papadatou-Pastou, 2022). For left neglect, the hypothesis of a hemispheric rivalry, with a disproportionately higher neural activity of the undamaged left hemisphere, released from the inhibition by the damaged right hemisphere, is strongly corroborated by studies using noninvasive brain stimulation (NiBS) techniques (Rossini et al., 2015). The rivalry hypothesis predicts that neglect symptoms are due not solely to the reduced activity of the lesioned area, but also to the pathologically and disproportionately increased activity of homologous regions of the opposite (undamaged) hemisphere. Accordingly, inhibitory stimulations, such as low-frequency repetitive transcranial magnetic stimulation (rTMS) over the posterior parietal cortex of the left (undamaged) hemisphere in right brain-damaged patients with left neglect, reduces the ipsilateral rightward bias (assessed by tests such as line bisection, target cancellation, visual search, landmark, clock drawing) and improves the deficit. There is also evidence that excitatory stimulations of the right (damaged) hemisphere, such as high frequency rTMS and anodal (a) tDCS, may improve left neglect (Koch et al., 2008, 2012; Sparing et al., 2009). Converging evidence for these directional effects comes from studies in healthy participants (see, for reviews about the effects of NiBS on spatial orientation biases, Cazzoli et al., 2010; Zebhauser et al., 2019; Gammeri et al., 2020; Veldema et al., 2020; Houben et al., 2021). For right neglect after left brain damage, laterally reversed NiBS are expected to bring about similar effects (Fig. 9.7C). In line with this prediction, Oliveri et al. (2001) found in six brain-damaged stroke patients with contralateral neglect (four with left neglect and right brain damage, and two with right neglect and left brain damage) that inhibitory rTMS to the unaffected hemisphere temporarily improved neglect, reducing the ipsilateral bias in a perceptual variant of the line bisection task (Fierro et al., 2000).

Hemispheric differences in orienting spatial attention toward the contralateral and ipsilateral sides of space (Heilman et al., 1987; Mesulam, 2000)

According to a second model, shown in Fig. 9.8A, each hemisphere subserves contralateral orientation of spatial attention. However, the right hemisphere-based networks, physiologically slightly stronger than the left hemisphere-based ones, also support ipsilateral rightward orientation. Conversely, the left hemisphere-based networks have little, if any, capability of leftward ipsilateral orientation. These different patterns account for the hemispheric asymmetry of neglect: after damage to the left hemisphere, the contralateral rightward spatial orientation system is impaired, but its function is compensated, at least in part, by the right hemisphere-based networks subserving ipsilateral rightward spatial orientation. This compensation does not occur after right brain damage, since the left, undamaged, hemisphere does not have the capability to orient leftward (ipsilaterally) spatial attention (Fig. 9.8B and C). The model also accounts for the phenomenon of pseudoneglect in healthy participants, due to the physiologic slight prevalence of the right hemisphere-based leftward orienting vector. However, to account also for the abovementioned effects of NiBS, the model should include a rivalry organization of inhibitory callosal connections.

RIGHT AND LEFT-SIDE NEGLECT: APHASIA AND CROSSED APHASIA

The several manifestations of neglect after left vs right brain damage are overall similar, but the neglected side is reversed. A question then arises, are all functions of the two hemispheres reversed, namely: does right neglect after left brain damage reflect a functional "reversal" between hemispheres, with the neural correlates of language becoming right-sided and those of spatial attention and representation left-sided? This reversal would make complete the linguistic skills of the right hemisphere, which are typically confined to some semantics and nonlinguistic

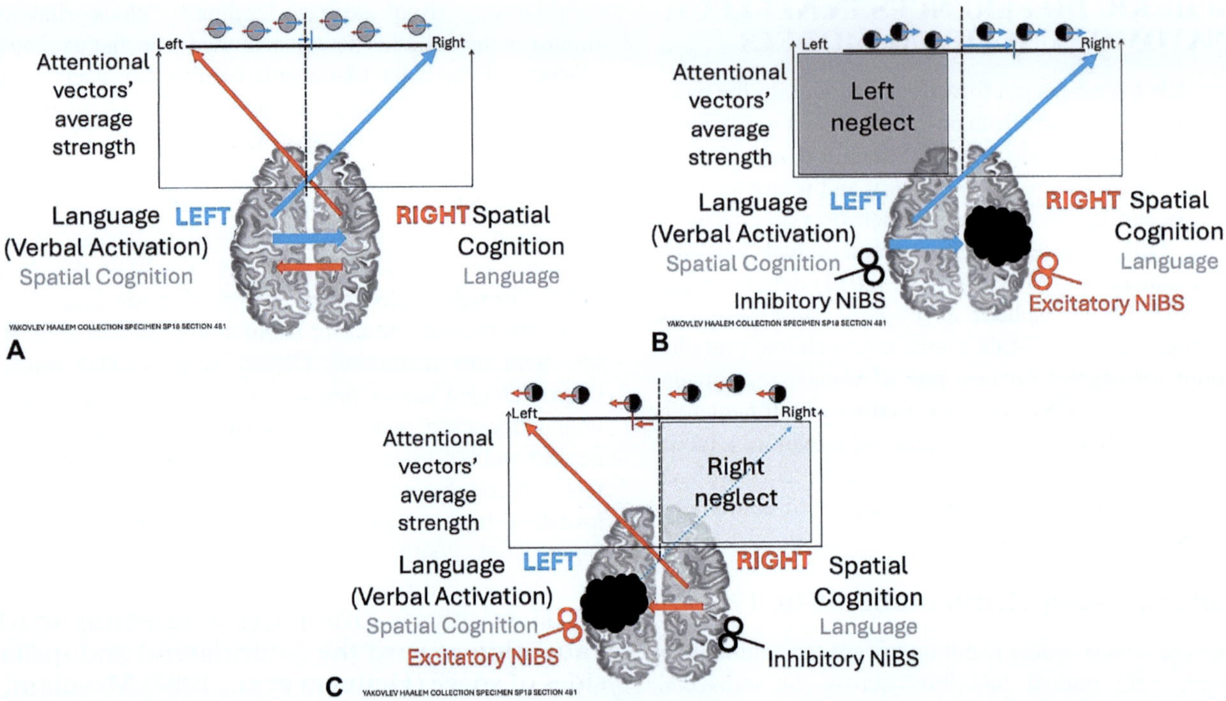

Fig. 9.7. The rivalry model of spatial attention in the two cerebral hemispheres (Kinsbourne, 1970, 1993). Axial brain slice, modified from image from The Yakovlev Haleem collection, Specimen SP18 Section 481. (A) Two systems in the left and in the right hemisphere for the contralateral spatial orientation of attention: the rivalry model. Blue continuous arrow: left hemisphere vector of rightward orientation of spatial attention. Red continuous arrow: right hemisphere vector of leftward orientation. Horizontal blue arrow: left-to-right vector of transcallosal modulation/inhibition of the right hemisphere by the left one. Horizontal red arrow: right-to-left vector of transcallosal modulation/inhibition of the left hemisphere by the right one. The left hemisphere-based attentional and transcallosal vectors (thicker rightward blue arrows) are stronger than the right hemisphere-based ones (thinner leftward red arrows), due to the linguistic activation of the left hemisphere. This asymmetry predicts a rightward bias in healthy persons (blue vertical line and horizontal arrow, above the black horizontal line). The gray circles above the black horizontal line represent allocentric, object-based, frames of reference, with similar attentional vectors. The linguistic and spatial competences of each hemisphere are marked black (major) and gray (minor). (B) Left neglect (left dark gray square). A unilateral right hemispheric lesion (black cloud) brings about contralateral left neglect, due to damage to the leftward orienting right hemisphere-based attentional vector. Neglect is also due to the increase of transcallosal inhibition of the right, damaged, hemisphere by the left, undamaged, one (thick left-to-right horizontal blue arrow). This increase is released, in turn, by the absence/decrease of the right-to-left inhibitory modulation by the right damaged hemisphere. After a right hemispheric lesion, contralateral neglect is more frequent/severe than after a left hemispheric lesion, since the rightward orienting vector is physiologically stronger and the left-to-right transcallosal inhibition greater (see panel A). The rightward pointing blue arrow above the black horizontal line indicates the rightward bias, for instance in line bisection. The left black portion of the gray circles indicates left allocentric neglect. In line with the rivalry hypothesis, inhibitory NiBS (black coil) of the posterior parietal cortex of the hyperactive left hemisphere, downregulating its activity, improves left neglect, releasing the right, damaged, hemisphere from inhibition by the left, undamaged, one. Conversely, it is excitatory NiBS (red coil) of the posterior parietal cortex of the hypoactive, damaged, right hemisphere, that, upregulating its activity, ameliorates left neglect. (C) Right neglect (right light gray square). A unilateral left hemispheric lesion (black cloud) brings about contralateral right neglect, due to damage to the rightward orienting attentional vector, which may be, however, at least in part, preserved (dotted rightward blue arrow), being physiologically stronger (see panel A). Accordingly, right neglect is less frequent/severe than left neglect. After left brain damage, inhibitory and excitatory NiBS are delivered, respectively, to the right undamaged and to the left damaged hemisphere. From Vallar G, Albini F, Pisoni A (2024). Unilateral spatial neglect, In: Reference module in neuroscience and biobehavioral psychology. Elsevier, pp. 1–24. https://doi.org/10.1016/B978-0-12-820480-1.00053-X

communicative properties of speech, with a limited lexicon and little, if any, phonology (Vigneau et al., 2011; Cattaneo, 2013; Gainotti, 2013; Bonandrini et al., 2020). The reversal hypothesis is supported by reports of right-handed left-brain-damaged patients with right-side neglect without aphasia, as the one described by Posteraro and Maravita (1996), who showed right extra-personal neglect (in line bisection and exploratory and drawing tasks), spatial (neglect) dysgraphia (Baxter and Warrington, 1983), and dyslexia (Vallar et al., 2010). Another right-handed (but with two left-handed children) left brain-damaged patient (Fujimori et al., 1994) showed right extra-personal neglect, constructional apraxia with neglect-related errors (Gainotti and Trojano, 2018), and anosodiaphoria (lack of concern over the existence of hemiparesis, Vallar and Ronchi, 2009), without aphasia and ideomotor apraxia (Wheaton and Hallett, 2007; Foundas, 2013). In a left-handed, left brain-damaged stroke patient, who showed, in the acute stage, right neglect without aphasia, the observed clinical pattern of impairment was similar (Dronkers and Knight, 1989).

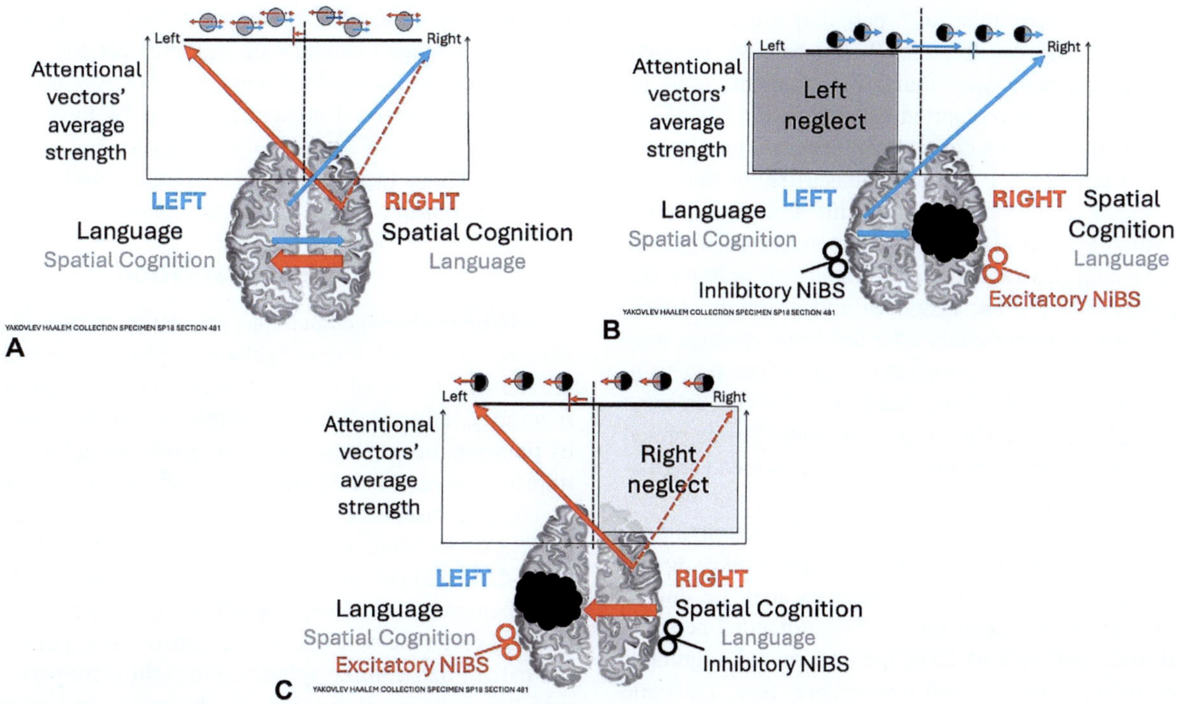

Fig. 9.8. A model of spatial attention in the two cerebral hemispheres (Heilman et al., 1987; Mesulam, 2000). Axial brain slice, modified from image from The Yakovlev Haleem Collection, Specimen SP18 Section 481. (A) Hemispheric differences in the contralateral and ipsilateral orientation of spatial attention. Each cerebral hemisphere orients spatial attention contralaterally: the left hemisphere rightward (blue continuous arrow), the right hemisphere leftward (red continuous arrow). The right hemispheric leftward orienting attentional spatial system is physiologically stronger than the left hemispheric rightward orienting one (thicker leftward pointing red continuous arrow). At variance with the rivalry model, the right hemisphere also supports ipsilateral rightward orientation (rightward red dashed arrow). The left hemisphere provides little, if any, ipsilateral leftward orientation. The horizontal rightward (blue) and leftward (red) arrows indicate reciprocal modulatory/inhibitory activity across the cerebral commissures. Since the right hemisphere-based leftward orientation system is stronger than the rightward one, the transcallosal right-to-left modulation/inhibition is assumed to be greater (thicker red arrow) than the left-to-right one (thinner blue arrow). The gray circles above the horizontal line represent allocentric, object-based, frames of reference, with similar spatial attentional vectors. As predicted by the model, in healthy persons a leftward bias is found in line bisection (pseudoneglect, Bowers and Heilman, 1980; Jewell and McCourt, 2000; Learmonth and Papadatou-Pastou, 2022): horizontal leftward red arrow, above the black horizontal line, and vertical red line. (B) Left neglect (left dark gray square). Right hemispheric lesions (black cloud) bring about left neglect, due to the damage to the right hemisphere-based system for leftward orientation of spatial attention. This asymmetry is not compensated by an ipsilateral leftward, left hemisphere-based, orienting system. Also, the increase of left-to-right transcallosal inhibition, due to the decrease/absence of the right-to-left reciprocal modulation, may contribute to the pathogenesis of left neglect. After right brain damage, inhibitory and excitatory NiBS are delivered, respectively, to the left undamaged and to the right damaged hemisphere. The left-sided black portion of the gray circles indicates left allocentric neglect. (C) Right neglect (right light gray square). Left hemispheric lesions (black cloud) bring about right neglect, due to the disruption of the left hemisphere-based neural systems supporting rightward orientation of attention. Right neglect is less frequent/severe, being compensated, at least in part, by the spared ipsilateral rightward orienting system of the right hemisphere. After left brain damage, inhibitory and excitatory NiBS are delivered, respectively, to the right undamaged and to the left damaged hemisphere. The right-sided black portion of the gray circles indicates right allocentric neglect. From Vallar G, Albini F, Pisoni A (2024). Unilateral spatial neglect, In: Reference module in neuroscience and biobehavioral psychology. Elsevier, pp. 1–24. https://doi.org/10.1016/B978-0-12-820480-1.00053-X

These findings do not imply, however, that the functional specialization of the two hemispheres is entirely reversed, but is in a few patients. In right-handed persons, damage to the right hemisphere may bring about crossed aphasia (Mariën et al., 2004), an infrequent disorder, found in approximately 1%–3% among stroke survivors (Coppens, 2011). The crossed patient reported by Paghera et al. (2003) showed global aphasia and left extrapersonal neglect in target cancellation, bisection, and drawing; in the 4-years follow-up, both neglect and aphasia improved substantially. The patient of Marangolo et al. (2003) showed a severe, non-fluent aphasia and left personal, but not extra-personal, neglect. According to the review by Coppens et al. (2002, see also Laiacona et al., 1999; De Witte et al., 2008), in right-handed right brain-damaged patients with crossed aphasia left neglect was present in 68 patients (40.7%), absent in 50 (29.9%), and unreported in 49 (29.3%). These findings suggest a coexistence of linguistic and spatial functions, in the right hemisphere of several crossed aphasic patients.

Furthermore, right neglect after left brain damage may occur independent of aphasia. In a series of 20 left brain-damaged patients with right neglect, Maeshima et al. (1992) found, in 11 right-handed patients, associated signs of damage to the left hemisphere, mainly severe aphasia (6 patients) and Gerstmann syndrome (4 patients); as for the 9 left-handed patients, moderate-to-severe aphasia was found in 6 patients, and no aphasia in 3. Reviews of the reported single case studies of left brain-damaged patients with a crossed nondominant hemisphere syndrome (i.e., right neglect without aphasia) show that they can be left-handed, right-handed, and with a mixed manual dominance, which also would be then independent of linguistic and visuospatial abilities (and deficits), including aphasia and neglect (Fujimori et al., 1994; Posteraro and Maravita, 1996). Similarly, right somatoparaphrenia after left brain damage may occur both with and without associated aphasia (see review in Perren et al., 2015). Overall, these findings show that, in most patients, right neglect after left brain damage cannot be interpreted as a functional hemispheric reversal between left-sided (linguistic) and right-sided (spatial) processes. Rather, right neglect may be due to damage of spatial attentional and representational left hemisphere-based networks, similar, though weaker, to the right hemisphere-based ones. This asymmetry may be due to the coexistence, in the left hemisphere of right-handed persons, of the neural underpinnings of language and spatial cognition. This "overcrowding" in the left-hemisphere (see Laiacona et al., 1999, for a similar line of reasoning, applied to crossed aphasia) may account for the lower frequency, severity, or both, of right neglect after left brain damage.

THE EFFECTS OF ASYMMETRIC SENSORY STIMULATIONS ON LEFT AND RIGHT-SIDE NEGLECT

The conclusion that the right and the left hemisphere-based spatial system are functionally largely homogeneous to the right hemisphere-based ones is also supported by the few studies that have attempted to temporarily ameliorate right spatial neglect in left brain-damaged patients, through the lateralized sensory stimulations that are known to temporarily improve left neglect in right brain-damaged patients, but delivered to obtain directionally opposite lateral effects (Vallar et al., 1997; Kerkhoff, 2003; Redding and Wallace, 2006; Jacquin-Courtois et al., 2013).

Caloric vestibular stimulation

The instillation of ice water into the left external ear canal, or of warm water into the right one, temporarily (up to 5 min) improves left neglect. Conversely, ice water into the right external ear canal, or warm water in the left one, worsen the disorder (e.g., Rubens, 1985, in target cancellation, pointing to people and reading tasks). In one left brain-damaged patient, stimulation of the right external ear canal with ice water improved right visual extra-personal and personal neglect, as observed clinically; however, no control caloric stimulation was delivered (Schiff and Pulver, 1999). The selectivity of caloric vestibular stimulation was shown in a right-handed, left brain-damaged patient with a severe fluent aphasia and right neglect (Vallar et al., 1995). Caloric vestibular stimulation (i.e., ice water into the right external ear canal) temporarily improved neglect, assessed by a line cancellation task (Albert, 1973), but not aphasia. These findings have been subsequently confirmed by the case study of an ambidextrous left-brain damaged patient with right neglect and fluent aphasia. Cold water in the right ear canal temporarily improved extra-personal and personal neglect (assessed by target cancellation, line bisection and body exploration tasks), and anosognosia for right hemiplegia (assessed by a verbal interview). No effects were found on aphasia, assessed clinically (Ronchi et al., 2013).

Prism adaptation

In right brain-damaged patients, adaptation to optical prisms displacing the visual scene rightward, through repeated pointing movements to visual targets, brings about leftward after-effects (namely, a leftward shift of the subjective straight ahead, assessed by pointing movements with participants being blindfolded), and improves several manifestations of left neglect, in target cancellation, line bisection, and drawing tasks (Rossetti et al., 1998; review in Redding and Wallace, 2006). One left-handed, left brain-damaged patient suffered a hemorrhagic infarction, with a large hematoma involving the left fronto-temporo-posterior parietal and cingulate cortices. At the time of the experimental study, the patient had anomic aphasia, and right extra-personal neglect was shown by copy drawing, line bisection, and target cancellation tasks. Adaptation to optical prisms displacing the visual scene leftward, obtained through pointing with the left hand to visual targets, brought about rightward aftereffects in pointing. Similar directional effects were found in control participants. In line bisection, the patient's leftward shift was reduced, measured up to approximately 10 days after adaptation. Conversely, the control condition of adaptation to sham prisms was ineffective (Bultitude and Rafal, 2010). Healthy participants also showed the expected minor leftward error in line bisection (pseudoneglect, Bowers and Heilman, 1980; reviews in Jewell and McCourt, 2000; Learmonth and Papadatou-Pastou, 2022). As for the long-term effects, sessions of 2 weeks of adaptation to prisms displacing the visual scene rightward are effective in the rehabilitation of left neglect in right brain-damaged patients (Frassinetti et al., 2002; Fortis et al., 2010, 2020; Zigiotto et al., 2021). In a retrospective study (Rich et al., 2023), 118 patients with left and right neglect, matched 1:1, completed at least three sessions of adaptation, with lenses displacing the visual fields toward the ipsilateral side (rightward in right brain-damaged patients with left neglect, leftward in left brain-damaged patients with right neglect). Results showed a global improvement, assessed by the Catherine Bergego Scale (Azouvi, 1996), larger for right than for left neglect, suggesting that prism adaptation is a viable treatment for both left and right neglect. Negative evidence is, however, also on record. Contralateral neglect after both left and right brain damage is not improved by prisms, as compared with sham adaptation for 2 weeks (Brink et al., 2017).

CONCLUSION

Over the years, the reported frequency of right neglect after left brain damage has increased, making it a sizeable phenomenon, rather than a rare observation, albeit left neglect after right brain damage is still reported as more frequent, severe, or both. Left and right neglect are overall behaviorally similar, both at a clinical observation, and as assessed by tests and ADL scales. A functional similarity is also suggested by the few studies that have investigated, in left brain-damaged patients with right neglect, the effects of the asymmetric sensory stimulations (vestibular, prism adaptation) effective on left neglect after right brain damage. To be effective, these stimulations are to be delivered reversed in side, with respect to the ones that ameliorate left neglect.

Right neglect does not reflect a complete functional reversal of left- and right-based hemispheric functions (i.e., language and spatial cognition), since right neglect and aphasia often coexist in the same left brain-damaged patient, and also in right brain-damaged patients with crossed aphasia. Rather, the overall minor frequency and severity of right neglect after left brain damage may be due to the anatomo-functional over-crowding of the left hemisphere of humans, by the neural network underpinnings of language and spatial cognition, reducing in turn the contribution of the left hemisphere to the ipsilateral orientation of attention. This may also account for the so far putative differences both in the behavioral manifestations, and in the localization of the responsible lesions, in patients with contralateral neglect after left or right brain damage. Future research, using more sensitive and theory-based paradigms, might further elucidate these issues.

ACKNOWLEDGMENTS

The authors are grateful to Professors Giovanni Berlucchi, Erminio Capitani, Monika Harvey, and Massimiliano Oliveri, as well as to Dr. Marcella Laiacona for their helpful suggestions.

References

Adair JC, Gilmore RL, Fennell EB et al. (1995). Anosognosia during intracarotid barbiturate anesthesia: unawareness or amnesia for weakness. Neurology 45: 241–243. https://doi.org/10.1212/wnl.45.2.241.

Agrell BM, Dehlin OI, Dahlgren CJ (1997). Neglect in elderly stroke patients: a comparison of five tests. Psychiatry Clin Neurosci 51: 295–300. https://doi.org/10.1111/j.1440-1819.1997.tb03201.x.

Albert ML (1973). A simple test of visual neglect. Neurology 23: 658–664. https://doi.org/10.1212/WNL.23.6.658.

Anderson SW, Tranel D (1989). Awareness of disease states following cerebral infarction, dementia, and head trauma: standardized assessment. Clin Neuropsychol 3: 327–339. https://doi.org/10.1080/13854048908401482.

Andrade K, Samri D, Sarazin M et al. (2010). Visual neglect in posterior cortical atrophy. BMC Neurol 10: 68. https://doi.org/10.1186/1471-2377-10-68.

Annoni JM, Devuyst G, Carota A et al. (2005). Changes in artistic style after minor posterior stroke. J Neurol Neurosurg Psychiatry 76: 797–803. https://doi.org/10.1136/jnnp.2004.045492.

Appelros P, Karlsson GM, Seiger A et al. (2002). Neglect and anosognosia after first-ever stroke: incidence and relationship to disability. J Rehabil Med 34: 215–220. https://doi.org/10.1080/165019702760279206.

Azouvi P (1996). Functional consequences and awareness of unilateral neglect: study of an evaluation scale. Neuropsychol Rehabil 6: 133–150. https://doi.org/10.1080/713755501.

Azouvi P (2016). The ecological assessment of unilateral neglect. Ann Phys Rehabil Med 60: 186–190. https://doi.org/10.1016/j.rehab.2015.12.005.

Azouvi P, Bartolomeo P, Beis J-M et al. (2006). A battery of tests for the quantitative assessment of unilateral neglect. Restor Neurol Neurosci 24: 273–285.

Baas U, de Haan B, Grässli T et al. (2011). Personal neglect—a disorder of body representation? Neuropsychologia 49: 898–905. https://doi.org/10.1016/j.neuropsychologia.2011.01.043.

Babinski J (1914). Contribution à l'étude des troubles mentaux dans l'hémiplégie organique cérébrale (anosognosie). Rev Neurol (Paris) 27: 845–848.

Baier B, Karnath H-O, Thömke F et al. (2010). Is there a link between spatial neglect and vestibular function at the cerebellar level? J Neurol 257: 1579–1581. https://doi.org/10.1007/s00415-010-5567-7.

Baier B, Vucurevic G, Müller-Forell W et al. (2014). Anosognosia for hemiparesis after left-sided stroke. Cortex 61: 120–126. https://doi.org/10.1016/j.cortex.2014.07.017.

Balint R (1909). Seelenlähmung des "Schauens", optische Ataxie, räumliche Störung der Aufmerksamkeit. Monatschr Psychiatr Neurol 25: 51–81.

Bálint R, Harvey M (1995). Psychic paralysis of gaze, optic ataxia, and spatial disorder of attention. Cogn Neuropsychol 12: 265–281. https://doi.org/10.1080/02643299508251999.

Barbieri C, Renzi ED (1989). Patterns of neglect dissociation. Behav Neurol 2: 13–24. https://doi.org/10.1155/1989/728487.

Bartolomeo P, D'Erme P, Gainotti G (1994). The relationship between visuospatial and representational neglect. Neurology 44: 1710–1714. https://doi.org/10.1212/wnl.44.9.1710.

Bartolomeo P, D'Erme P, Perri R et al. (1998a). Perception and action in hemispatial neglect. Neuropsychologia 36: 227–237. https://doi.org/10.1016/s0028-3932(97)00104-8.

Bartolomeo P, Dalla Barba G, Boissé MF et al. (1998b). Right-side neglect in Alzheimer's disease. Neurology 51: 1207–1209. https://doi.org/10.1212/wnl.51.4.1207.

Bartolomeo P, Chokron S, Gainotti G (2001). Laterally directed arm movements and right unilateral neglect after left hemisphere damage. Neuropsychologia 39: 1013–1021. https://doi.org/10.1016/s0028-3932(01)00042-2.

Bartolomeo P, Thiebaut de Schotten M, Doricchi F (2007). Left unilateral neglect as a disconnection syndrome. Cereb Cortex 17: 2479–2490. https://doi.org/10.1093/cercor/bhl181.

Battersby WS, Bender MB, Pollack M et al. (1956). Unilateral "spatial agnosia" ("inattention"). Brain 79: 68–92. https://doi.org/10.1093/brain/79.1.68.

Baxendale S (2009). The Wada test. Curr Opin Neurol 22: 185. https://doi.org/10.1097/WCO.0b013e328328f32e.

Baxter DM, Warrington EK (1983). Neglect dysgraphia. J Neurol Neurosurg Psychiatry 46: 1073–1078. https://doi.org/10.1136/jnnp.46.12.1073.

Becker E, Karnath H (2007). Incidence of visual extinction after left versus right hemisphere stroke. Stroke. https://doi.org/10.1161/STROKEAHA.107.489096.

Beis JM, Keller C, Morin N et al. (2004). Right spatial neglect after left hemisphere stroke: qualitative and quantitative study. Neurology 63: 1600–1605. https://doi.org/10.1212/01.wnl.0000142967.60579.32.

Bellas DN, Novelly RA, Eskenazi B et al. (1988). Unilateral displacement in the olfactory sense: a manifestation of the unilateral neglect syndrome. Cortex 24: 267–275. https://doi.org/10.1016/s0010-9452(88)80035-2.

Bertagnoli S, Pacella V, Rossato E et al. (2022). Disconnections in personal neglect. Brain Struct Funct 227: 3161–3171. https://doi.org/10.1007/s00429-022-02511-z.

Berti A, Làdavas E, Corte MD (1996). Anosognosia for hemiplegia, neglect dyslexia, and drawing neglect: clinical findings and theoretical considerations. J Int Neuropsychol Soc 2: 426–440. https://doi.org/10.1017/S135561770000151X.

Berti AM, Bottini G, Gandola M et al. (2005). Shared cortical anatomy for motor awareness and motor control. Science 309: 488–491. https://doi.org/10.1126/science.1110625.

Beume L-A, Martin M, Kaller CP et al. (2017). Visual neglect after left-hemispheric lesions: a voxel-based lesion–symptom mapping study in 121 acute stroke patients. Exp Brain Res 235: 83–95. https://doi.org/10.1007/s00221-016-4771-9.

Beume L-A, Rijntjes M, Dressing A et al. (2020). Dissociation of visual extinction and neglect in the left hemisphere. Cortex 129: 211–222. https://doi.org/10.1016/j.cortex.2020.04.010.

Bianchi L (1895). The functions of the frontal lobes. Brain 18: 497–522. https://doi.org/10.1093/brain/18.4.497.

Bickerton WL, Samson D, Williamson J et al. (2011). Separating forms of neglect using the apples test: validation and functional prediction in chronic and acute stroke. Neuropsychology 25: 567–580. https://doi.org/10.1037/a0023501.

Bird H, Howard D, Franklin S (2003). Verbs and nouns: the importance of being imageable. J Neurolinguistics 16: 113–149. https://doi.org/10.1016/S0911-6044(02)00016-7.

Bisiach E (1992). Understanding consciousness: clues from unilateral neglect and related disorders. In: AD Milner, MD Rugg (Eds.), The neuropsychology of consciousness. Academic Press113–137. https://doi.org/10.1016/B978-0-12-498045-7.50011-6.

Bisiach E, Luzzatti C (1978). Unilateral neglect of representational space. Cortex 14: 129–133. https://doi.org/10.1016/S0010-9452(78)80016-1.

Bisiach E, Vallar G (2000). Unilateral neglect in humans. In: F Boller, J Grafman, G Rizzolatti (Eds.), Handbook of neuropsychology. Elsevier Science, B.V, pp. 459–502.

Bisiach E, Luzzatti C, Perani D (1979). Unilateral neglect, representational schema and consciousness. Brain 102: 609–618. https://doi.org/10.1093/brain/102.3.609.

Bisiach E, Capitani E, Luzzatti C et al. (1981). Brain and conscious representation of outside reality. Neuropsychologia 19: 543–551. https://doi.org/10.1016/0028-3932(81)90020-8.

Bisiach E, Capitani E, Porta E (1985). Two basic properties of space representation in the brain: evidence from unilateral neglect. J Neurol Neurosurg Psychiatry 48: 141–144. https://doi.org/10.1136/jnnp.48.2.141.

Bisiach E, Perani D, Vallar G et al. (1986a). Unilateral neglect: personal and extra-personal. Neuropsychologia 24: 759–767. https://doi.org/10.1016/0028-3932(86)90075-8.

Bisiach E, Vallar G, Perani D et al. (1986b). Unawareness of disease following lesions of the right hemisphere: anosognosia for hemiplegia and anosognosia for hemianopia. Neuropsychologia 24: 471–482. https://doi.org/10.1016/0028-3932(86)90092-8.

Bisiach E, Geminiani G, Berti A et al. (1990). Perceptual and premotor factors of unilateral neglect. Neurology 40: 1278–1281. https://doi.org/10.1212/WNL.40.8.1278.

Blanke O, Ortigue S, Landis T (2003). Colour neglect in an artist. Lancet 361: 264. https://doi.org/10.1016/S0140-6736(03)12304-5.

Boksem MAS, Kostermans E, Tops M et al. (2012). Individual differences in asymmetric resting-state frontal cortical activity modulate ERPs and performance in a global-local attention task. J Psychophysiol 26: 51–62. https://doi.org/10.1027/0269-8803/a000067.

Bolognini N, Convento S, Casati C et al. (2016). Multisensory integration in hemianopia and unilateral spatial neglect: evidence from the sound induced flash illusion. Neuropsychologia 87: 134–143. https://doi.org/10.1016/j.neuropsychologia.2016.05.015.

Bonandrini R, Veronelli L, Licciardo D et al. (2020). Can the right hemisphere read? A behavioral and disconnectome study on implicit reading in a patient with pure alexia. Neurocase 26: 321–327. https://doi.org/10.1080/13554794.2020.1830118.

Bowen A, McKenna K, Tallis RC (1999). Reasons for variability in the reported rate of occurrence of unilateral spatial neglect after stroke. Stroke 30: 1196–1202. https://doi.org/10.1161/01.STR.30.6.1196.

Bowers D, Heilman KM (1980). Pseudoneglect: effects of hemispace on a tactile line bisection task. Neuropsychologia 18: 491–498.

Bozzali M, Mastropasqua C, Cercignani M et al. (2012). Microstructural damage of the posterior corpus callosum contributes to the clinical severity of neglect. PLoS One 7: e48079. https://doi.org/10.1371/journal.pone.0048079.

Brain WR (1941). Visual disorientation with special reference to lesions of the right cerebral hemisphere. Brain 64: 244–272. https://doi.org/10.1093/brain/64.4.244.

Breier JI, Adair JC, Gold M et al. (1995). Dissociation of anosognosia for hemiplegia and aphasia during left-hemisphere anesthesia. Neurology 45: 65–67. https://doi.org/10.1212/WNL.45.1.65.

Brink AFT, Visser-Meily JMA, Schut MJ et al. (2017). Prism adaptation in rehabilitation? No additional effects of prism adaptation on neglect recovery in the subacute phase poststroke: a randomized controlled trial. Neurorehabil Neural Repair 31: 1017–1028. https://doi.org/10.1177/1545968317744277.

Brozzoli C, Demattè ML, Pavani F et al. (2006). Neglect and extinction: within and between sensory modalities. Restor Neurol Neurosci 24: 217–232.

Büchel C, Raedler T, Sommer M et al. (2004). White matter asymmetry in the human brain: a diffusion tensor MRI study. Cereb Cortex 14: 945–951. https://doi.org/10.1093/cercor/bhh055.

Buckley J, Seery N, Canty D (2018). A heuristic framework of spatial ability: a review and synthesis of spatial factor literature to support its translation into STEM education. Educ Psychol Rev 30: 947–972. https://doi.org/10.1007/s10648-018-9432-z.

Bultitude JH, Rafal RD (2010). Amelioration of right spatial neglect after visuo-motor adaptation to leftward-shifting prisms. Cortex 46: 404–406. https://doi.org/10.1016/j.cortex.2009.06.002.

Caggiano P, Jehkonen M (2018). The 'neglected' personal neglect. Neuropsychol Rev 28: 417–435. https://doi.org/10.1007/s11065-018-9394-4.

Caplan B (1985). Stimulus effects in unilateral neglect. Cortex 21: 69–80. https://doi.org/10.1016/s0010-9452(85)80016-2.

Caramazza A, Hillis AE (1990). Spatial representation of words in the brain implied by studies of a unilateral neglect patient. Nature 346: 267–269. https://doi.org/10.1038/346267a0.

Carrera E, Tononi G (2014). Diaschisis: past, present, future. Brain 137: 2408–2422. https://doi.org/10.1093/brain/awu101.

Catani M, Howard RJ, Pajevic S et al. (2002). Virtual *in vivo* interactive dissection of white matter fasciculi in the human brain. Neuroimage 17: 77–94. https://doi.org/10.1006/nimg.2002.1136.

Cattaneo L (2013). In: AM Lozano, M Hallett (Eds.), Language. Handbook of Clinical Neurology. Brain Stimulation681–691.

Cazzoli D, Müri RM, Hess CW et al. (2010). Treatment of hemispatial neglect by means of rTMS—a review. Restor Neurol Neurosci 28: 499–510. https://doi.org/10.3233/RNN-2010-0560.

Cazzoli D, Kaufmann BC, Rühe H et al. (2023). A case study of left visual neglect after right pontine lesion: pathophysiological evidence for the infratentorial involvement in human visual attention. J Neurophysiol 129: 1534–1539. https://doi.org/10.1152/jn.00123.2023.

Chatterjee A (1994). Picturing unilateral spatial neglect: viewer versus object centred reference frames. J Neurol Neurosurg Psychiatry 57: 1236–1240. https://doi.org/10.1136/jnnp.57.10.1236.

Chen P, Hreha K, Fortis P et al. (2012). Functional assessment of spatial neglect: a review of the Catherine Bergego scale and an introduction of the Kessler foundation neglect assessment process. Top Stroke Rehabil 19: 423–435. https://doi.org/10.1310/tsr1905-423.

Chen P, Chen CC, Hreha K et al. (2015). Kessler Foundation neglect assessment process uniquely measures spatial neglect during activities of daily living. Arch Phys Med Rehabil 96: 869–876e1,. https://doi.org/10.1016/j.apmr.2014.10.023.

Chen P, Ward I, Khan U et al. (2016). Spatial neglect hinders success of inpatient rehabilitation in individuals with traumatic brain injury: a retrospective study. Neurorehabil Neural Repair 30: 451–460. https://doi.org/10.1177/1545968315604397.

Clarke S, Thiran AB (2004). Auditory neglect: what and where in auditory space. Cortex 40: 291–300. https://doi.org/10.1016/s0010-9452(08)70124-2.

Cocchini G, Beschin N, Cameron A et al. (2009). Anosognosia for motor impairment following left brain damage. Neuropsychology 23: 223–230. https://doi.org/10.1037/a0014266.

Cogliano R, Crisci C, Conson M et al. (2012). Chronic somatoparaphrenia: a follow-up study on two clinical cases. Cortex 48: 758–767. https://doi.org/10.1016/j.cortex.2011.08.008.

Colombo A, de Renzi E, Faglioni P (1976). The occurrence of visual neglect in patients with unilateral cerebral disease. Cortex 12: 221–231. https://doi.org/10.1016/s0010-9452(76)80003-2.

Committeri G, Pitzalis S, Galati G et al. (2007). Neural bases of personal and extrapersonal neglect in humans. Brain 130: 431–441. https://doi.org/10.1093/brain/awl265.

Committeri G, Piervincenzi C, Pizzamiglio L (2018). Personal neglect: a comprehensive theoretical and anatomo-clinical review. Neuropsychology 32: 269–279. https://doi.org/10.1037/neu0000409.

Coppens P (2011). Crossed aphasia. In: JS Kreutzer, J DeLuca, B Caplan (Eds.), Encyclopedia of clinical neuropsychology. Springer, New York, NY p. 747. https://doi.org/10.1007/978-0-387-79948-3_877.

Coppens P, Hungerford S, Yamaguchi S et al. (2002). Crossed aphasia: an analysis of the symptoms, their frequency, and a comparison with left-hemisphere aphasia symptomatology. Brain Lang 83: 425–463. https://doi.org/10.1016/S0093-934X(02)00510-2.

Corbetta M, Shulman GL (2011). Spatial neglect and attention networks. Annu Rev Neurosci 34: 569–599. https://doi.org/10.1146/annurev-neuro-061010-113731.

Costa LD, Vaughan HG, Horwitz M et al. (1969). Patterns of behavioral deficit associated with visual spatial neglect. Cortex 5: 242–263. https://doi.org/10.1016/S0010-9452(69)80033-X.

Critchley M (1953). The parietal lobes, Williams and Wilkins, Oxford, England.

Crutch SJ, Schott JM, Rabinovici GD et al. (2017). Consensus classification of posterior cortical atrophy. Alzheimers Dement 13: 870–884. https://doi.org/10.1016/j.jalz.2017.01.014.

Cubelli R, Beschin N, Della Sala S (2011). Ipsilateral neglect for nonverbal stimuli following left brain damage. Cortex 47: 899–901. https://doi.org/10.1016/j.cortex.2011.01.006.

D'Imperio D, Tomelleri G, Moretto G et al. (2017). Modulation of somatoparaphrenia following left-hemisphere damage. Neurocase 23: 162–170. https://doi.org/10.1080/13554794.2017.1329444.

Damasio AR, Damasio H, Chang Chui H (1980). Neglect following damage to frontal lobe or basal ganglia. Neuropsychologia 18: 123–132. https://doi.org/10.1016/0028-3932(80)90058-5.

De Renzi E (1982). Disorders of space exploration and cognition, Chichester, John Wiley.

De Renzi E (1996). Bálint-Holmes' syndrome. In: C Code, C-W Wallesch, Y Joanette et al. (Eds.), Classic cases in neuropsychology. Psychology Press, Hove, East Sussex, pp. 123–143.

De Renzi E, Faglioni P, Scotti G (1970). Hemispheric contribution to exploration of space through the visual and tactile modality. Cortex 6: 191–203. https://doi.org/10.1016/S0010-9452(70)80027-2.

De Renzi E, Colombo A, Faglioni P et al. (1982). Conjugate gaze paresis in stroke patients with unilateral damage: an unexpected instance of hemispheric asymmetry. Arch Neurol 39: 482–486. https://doi.org/10.1001/archneur.1982.00510200024004.

De Renzi E, Gentilini M, Pattacini F (1984). Auditory extinction following hemisphere damage. Neuropsychologia 22: 733–744. https://doi.org/10.1016/0028-3932(84)90099-x.

De Witte L, Verhoeven J, Engelborghs S et al. (2008). Crossed aphasia and visuo-spatial neglect following a right thalamic stroke: a case study and review of the literature. Behav Neurol 19: 177–194. https://doi.org/10.1155/2008/905187.

Déjérine J (1914). Sémiologie des affections du système nerveux, Masson, Paris.

Della Sala S, Beschin N, Cubelli R (2018). Persistent neglect in everyday life. Cortex 103: 382–384. https://doi.org/10.1016/j.cortex.2017.09.010.

Demeyere N, Gillebert CR (2019). Ego- and allocentric visuospatial neglect: dissociations, prevalence, and laterality in acute stroke. Neuropsychology 33: 490–498. https://doi.org/10.1037/neu0000527.

Demeyere N, Haupt M, Webb SS et al. (2021). Introducing the tablet-based Oxford cognitive screen-plus (OCS-plus) as an assessment tool for subtle cognitive impairments. Sci Rep 11: 8000. https://doi.org/10.1038/s41598-021-87287-8.

Denes G, Semenza C, Stoppa E et al. (1982). Unilateral spatial neglect and recovery from hemiplegia. Brain 105: 543–552. https://doi.org/10.1093/brain/105.3.543.

Denny-Brown D, Chambers RA (1958). The parietal lobe and behavior. Res Publ Assoc Res Nerv Ment Dis 36: 35–117.

Di Russo F, Aprile T, Spitoni G et al. (2008). Impaired visual processing of contralesional stimuli in neglect patients: a visual-evoked potential study. Brain 131: 842–854. https://doi.org/10.1093/brain/awm281.

Di Vita A, Palermo L, Piccardi L et al. (2017). Body representation alterations in personal but not in extrapersonal neglect patients. Appl Neuropsychol Adult 24: 308–317. https://doi.org/10.1080/23279095.2016.1174866.

Diller L, Weinberg J (1977). Hemiinattention in rehabilitation. The evolution of a rational remediation program. In: EA Weinstein, RP Friedland (Eds.), Hemi-inattention and hemisphere specialization. Raven Press, New York, pp. 62–82.

Doricchi F, Tomaiuolo F (2003). The anatomy of neglect without hemianopia: a key role for parietal-frontal disconnection? Neuroreport 14: 2239–2243. https://doi.org/10.1097/00001756-200312020-00021.

Dronkers NF, Knight RT (1989). Right-sided neglect in a left-hander: evidence for reversed hemispheric specialization of attention capacity. Neuropsychologia 27: 729–735. https://doi.org/10.1016/0028-3932(89)90118-8.

Dywan CA, McGlone J, Fox A (1995). Do intracarotid barbiturate injections offer a way to investigate hemispheric models of anosognosia? J Clin Exp Neuropsychol 17: 431–438. https://doi.org/10.1080/01688639508405134.

Egelko S, Riley E, Simon D et al. (1988). Unilateral spatial neglect: biases in contralateral search and fine spatial attention. Arch Clin Neuropsychol 3: 213–225.

Ellis AW, Flude BM, Young AW (1987). "Neglect dyslexia" and the early visual processing of letters in words and nonwords. Cogn Neuropsychol 4: 439–464. https://doi.org/10.1080/02643298708252047.

Esposito E, Shekhtman G, Chen P (2021). Prevalence of spatial neglect post-stroke: a systematic review. Ann Phys Rehabil Med 64: 101459. https://doi.org/10.1016/j.rehab.2020.10.010.

Facchin A, Beschin N, Daini R (2017). Rehabilitation of right (personal) neglect by prism adaptation: a case report. Ann Phys Rehabil Med 60: 220–222. https://doi.org/10.1016/j.rehab.2016.09.004.

Ferro JM, Martins IP, Tavora L (1984). Neglect in children. Ann Neurol 15: 281–284. https://doi.org/10.1002/ana.410150314.

Fierro B, Brighina F, Oliveri M et al. (2000). Contralateral neglect induced by right posterior parietal rTMS in healthy subjects. Neuroreport 11: 1519–1521.

Filimon F (2015). Are all spatial reference frames egocentric? Reinterpreting evidence for allocentric, object-centered, or world-centered reference frames. Front Hum Neurosci 9: 1–21. https://doi.org/10.3389/fnhum.2015.00648.

Fink GR, Halligan PW, Marshall JC et al. (1996). Where in the brain does visual attention select the forest and the trees? Nature 382: 626–628. https://doi.org/10.1038/382626a0.

Fortis P, Maravita A, Gallucci M et al. (2010). Rehabilitating patients with left spatial neglect by prism exposure during a visuomotor activity. Neuropsychology 24: 681–697. https://doi.org/10.1037/a0019476.

Fortis P, Ronchi R, Velardo V et al. (2020). A home-based prism adaptation training for neglect patients. Cortex 122: 61–80. https://doi.org/10.1016/j.cortex.2018.09.001.

Foundas AL (2013). Apraxia: neural mechanisms and functional recovery. Handb Clin Neurol 110: 335–345. https://doi.org/10.1016/B978-0-444-52901-5.00028-9.

Frank B, Schoch B, Hein-Kropp C et al. (2008). Aphasia, neglect and extinction are no prominent clinical signs in children and adolescents with acute surgical cerebellar lesions. Exp Brain Res 184: 511–519. https://doi.org/10.1007/s00221-007-1116-8.

Frank B, Maschke M, Groetschel H et al. (2010). Aphasia and neglect are uncommon in cerebellar disease: negative findings in a prospective study in acute cerebellar stroke. Cerebellum 9: 556–566. https://doi.org/10.1007/s12311-010-0197-2.

Frassinetti F, Angeli V, Meneghello F et al. (2002). Long-lasting amelioration of visuospatial neglect by prism adaptation. Brain 125: 608–623. https://doi.org/10.1093/brain/awf056.

Friedland RP, Weinstein EA (1977). Hemi-inattention and hemisphere specialization: introduction and historical review. Adv Neurol 18: 1–31.

Fruhmann Berger M, Johannsen L, Karnath HO (2008). Time course of eye and head deviation in spatial neglect. Neuropsychology 22: 697–702. https://doi.org/10.1037/a0013351.

Fujimori M, Wakisaka K, Yamadori A et al. (1994). Crossed non-dominant hemisphere syndrome in a right-hander. Behav Neurol 7: 123–126. https://doi.org/10.3233/BEN-1994-73-403.

Gaffan D, Hornak J (1997). Visual neglect in the monkey. Representation and disconnection. Brain 120: 1647–1657. https://doi.org/10.1093/brain/120.9.1647.

Gainotti G (2013). The contribution of language to the right-hemisphere conceptual representations: a selective survey. J Clin Exp Neuropsychol 35: 563–572. https://doi.org/10.1080/13803395.2013.798399.

Gainotti G, Trojano L (2018). Constructional apraxia. In: G Vallar, HB Coslett (Eds.), Handbook of clinical neurology. Elsevier, Amsterdam, pp. 331–348. https://doi.org/10.1016/B978-0-444-63622-5.00016-4.

Gainotti G, Messerli P, Tissot R (1972). Qualitative analysis of unilateral spatial neglect in relation to laterality of cerebral lesions. J Neurol Neurosurg Psychiatry 35: 545–550. https://doi.org/10.1136/jnnp.35.4.545.

Gainotti G, D'Erme P, Monteleone D et al. (1986). Mechanisms of unilateral spatial neglect in relation to laterality of cerebral lesions. Brain 109: 599–612. https://doi.org/10.1093/brain/109.4.599.

Gainotti G, De Bonis C, Daniele A et al. (1989). Contralateral and ipsilateral tactile extinction in patients with right and left focal brain damage. Int J Neurosci 45: 81–89. https://doi.org/10.3109/00207458908986219.

Gammeri R, Iacono C, Ricci R et al. (2020). Unilateral spatial neglect after stroke: current insights. Neuropsychiatr Dis Treat 16: 131–152. https://doi.org/10.2147/NDT.S171461.

Gauthier L, Dehaut F, Joanette Y (1989). The Bells test: a quantitative and qualitative test for visual neglect. Int J Clin Neuropsychol 11: 49–54.

Geiser N, Kaufmann BC, Rühe H et al. (2022). Visual neglect after PICA stroke—a case study. Brain Sci 12: 290. https://doi.org/10.3390/brainsci12020290.

Gerstmann J (1924). Fingeragnosie: eine umschriebene Störung der Orientierung am eigenen Körper. Wien Klin Wochenschr 37: 1010–1012.

Gialanella B, Bertolinelli M, Lissi M et al. (2011). Predicting outcome after stroke: the role of aphasia. Disabil Rehabil 33: 122–129. https://doi.org/10.3109/09638288.2010.488712.

Giannakou I, Lin D, Punt D (2022). Computer-based assessment of unilateral spatial neglect: a systematic review. Front Neurosci 16: 912626. https://doi.org/10.3389/fnins.2022.912626.

Gillebert CR, Mantini D, Thijs V et al. (2011). Lesion evidence for the critical role of the intraparietal sulcus in spatial attention. Brain 134: 1694–1709. https://doi.org/10.1093/brain/awr085.

Ginex V, Veronelli L, Vanacore N et al. (2017). Motor recovery in post-stroke patients with aphasia: the role of specific linguistic abilities. Top Stroke Rehabil 24: 428–434. https://doi.org/10.1080/10749357.2017.1305654.

Glazer H, Saadatpour L, Doty L et al. (2017). A case of posterior cortical atrophy with vertical neglect. Neurocase 23: 114–119. https://doi.org/10.1080/13554794.2017.1312692.

Glickstein M, Berlucchi G (2008). Classical disconnection studies of the corpus callosum. Cortex 44: 914–927. https://doi.org/10.1016/j.cortex.2008.04.001.

Grossi D, Modafferi A, Pelosi L et al. (1989). On the different roles of the cerebral hemispheres in mental imagery: the "o'Clock Test" in two clinical cases. Brain Cogn 10: 18–27. https://doi.org/10.1016/0278-2626(89)90072-9.

Grossi D, Angelini R, Pecchinenda A et al. (1993). Left imaginal neglect in heminattention: experimental study with the O'clock test. Behav Neurol 6: 155–158. https://doi.org/10.3233/BEN-1993-6307.

Grossi D, Lepore M, Napolitano A et al. (2001). On selective left neglect during walking in a child. Brain Cogn 47: 539–544. https://doi.org/10.1006/brcg.2001.1460.

Haeske-Dewick HC, Canavan AG, Hömberg V (1996). Directional hyperattention in tactile neglect within grasping space. J Clin Exp Neuropsychol 18: 724–732. https://doi.org/10.1080/01688639608408295.

Halligan PW, Marshall JC (1993). The history and clinical presentation of neglect. In: Unilateral neglect: clinical and experimental studies, Brain damage, behaviour & cognition series, Lawrence Erlbaum Associates, Inc, Hillsdale, NJ, US, pp. 3–25.

Halligan PW, Marshall JC (1997). The art of visual neglect. Lancet 350: 139–140. https://doi.org/10.1016/S0140-6736(97)03165-6.

Halligan PW, Marshall JC (2001). Graphic neglect—more than the sum of the parts. Neuroimage 14: S91–S97. https://doi.org/10.1006/nimg.2001.0821.

Halligan PW, Marshall JC (2002). Primary sensory deficits after right brain damage–an attentional disorder by any other name? In: H-O Karnath, AD Milner, G Vallar (Eds.), The cognitive and neural bases of spatial neglect. Oxford University Press, Oxford, pp. 326–340.

Halligan PW, Marshall JC, Wade DT (1989). Visuo-spatial neglect: underlying factors and test sensitivity. Lancet 334: 908–911. https://doi.org/10.1016/s0140-6736(89)91561-4.

Halligan PW, Burn JP, Marshall JC et al. (1992). Visuo-spatial neglect: qualitative differences and laterality of cerebral lesion. J Neurol Neurosurg Psychiatry 55: 1060–1068. https://doi.org/10.1136/jnnp.55.11.1060.

Halligan PW, Fink GR, Marshall JC et al. (2003). Spatial cognition: evidence from visual neglect. Trends Cogn Sci 7. https://doi.org/10.1016/S1364-6613(03)00032-9.

Halloran PM (1946). Delusion of body scheme due to subdural hematoma. Bull Los Angel Neuro Soc 11: 88–89.

Harciarek M, Williamson JB, Burtis B et al. (2012). Ipsilateral neglect from a subcortical lesion. Cogn Behav Neurol 25: 42–49. https://doi.org/10.1097/WNN.0b013e3182491465.

Hart E, Grattan E, Woodbury M et al. (2021). Pediatric unilateral spatial neglect: a systematic review. J Pediatr Rehabil Med 14: 345–359. https://doi.org/10.3233/PRM-200779.

Hashimoto M, Morii S, Uesaka Y et al. (2009). Right-sided neglect influences the writing of Kanji: a case study. Clin Neurol Neurosurg 111: 886–888. https://doi.org/10.1016/j.clineuro.2009.08.001.

Hécaen H (1972). Introduction à la neuropsychologie, Larousse, Paris.

Hécaen H (1976). Acquired aphasia in children and the ontogenesis of hemispheric functional specialization. Brain Lang 3: 114–134. https://doi.org/10.1016/0093-934x(76)90009-2.

Hécaen H, Penfield W, Bertrand C et al. (1956). The syndrome of apractognosia due to lesions of the minor cerebral hemisphere. Arch Neurol Psychiatry 75: 400–434. https://doi.org/10.1001/archneurpsyc.1956.02330220064007.

Heilman KM, Valenstein E (1972). Frontal lobe neglect in man. Neurology 22: 660–664. https://doi.org/10.1212/WNL.22.6.660.

Heilman KM, Valenstein E (1979). Mechanisms underlying hemispatial neglect. Ann Neurol 5: 166–170. https://doi.org/10.1002/ana.410050210.

Heilman KM, Watson RT (1978). Changes in the symptoms of neglect induced by changing task strategy. Arch Neurol 35: 37–49. https://doi.org/10.1001/archneur.1978.00500250051013.

Heilman KM, Pandya DN, Karol EA et al. (1971). Auditory inattention. Arch Neurol 24: 323. https://doi.org/10.1001/archneur.1971.0048 0340055005.

Heilman KM, Bowers D, Valenstein E et al. (1987). Hemispace and hemispatial neglect. In: M Jeannerod (Ed.), Neurophysiological and neuropsychological aspects of spatial neglect. 115–150. North-Holland, Amsterdam.

Hildebrandt H, Spang K, Ebke M (2002). Visuospatial hemi-inattention following cerebellar/brain stem bleeding. Neurocase 8: 323–329. https://doi.org/10.1076/neur.8.3.323.16197.

Hillis AE, Caramazza A (1995). Spatially specific deficits in processing graphemic representations in reading and writing. Brain Lang 48: 263–308. https://doi.org/10.1006/brln.1995.1012.

Hillis AE, Wityk RJ, Barker PB et al. (2002). Subcortical aphasia and neglect in acute stroke: the role of cortical hypoperfusion. Brain 125: 1094–1104. https://doi.org/10.1093/brain/awf113.

Honnedevasthana Arun A, Connelly A, Smith RE et al. (2021). Characterisation of white matter asymmetries in the healthy human brain using diffusion MRI fixel-based analysis. Neuroimage 225: 117505. https://doi.org/10.1016/j.neuroimage.2020.117505.

Houben M, Chettouf S, Van Der Werf YD et al. (2021). Theta-burst transcranial magnetic stimulation for the treatment of unilateral neglect in stroke patients: a systematic review and best evidence synthesis. Restor Neurol Neurosci 39: 447–465. https://doi.org/10.3233/RNN-211228.

Hreha K, Mulry C, Gross M et al. (2017). Assessing chronic stroke survivors with aphasia sheds light on prevalence of spatial neglect. Top Stroke Rehabil 24: 91–98. https://doi.org/10.1080/10749357.2016.1196906.

Hublet C, Demeurisse G, Paternot J et al. (1995). Remote cortical dysfunction as a possible cause of subcortical neglect? A regional cerebral blood flow study. Eur Neurol 35: 254–258. https://doi.org/10.1159/000117144.

Humphreys GW, Gillebert CR, Chechlacz M et al. (2013). Reference frames in visual selection. Ann N Y Acad Sci 1296: 75–87. https://doi.org/10.1111/nyas.12256.

Husain M, Kennard C (1996). Visual neglect associated with frontal lobe infarction. J Neurol 243: 652–657. https://doi.org/10.1007/BF00878662.

Ishiai S, Okiyama R, Koyama Y et al. (1996). Unilateral spatial neglect in Alzheimer's disease. A line bisection study. Acta Neurol Scand 93: 219–224. https://doi.org/10.1111/j.1600-0404.1996.tb00204.x.

Ishiai S, Koyama Y, Seki K et al. (2000). Unilateral spatial neglect in AD: significance of line bisection performance. Neurology 55: 364–370. https://doi.org/10.1212/wnl.55.3.364.

Ives ER, Nielsen JM (1937). Disturbance of body scheme. Delusion of absence of part of body in two cases with autopsy verification of the lesions. Bull Los Angel Neuro Soc 2: 120–125.

Jackson JH (1876). Case of large cerebral tumour without optic neuritis and left hemiplegia and imperception. In: Royal London Ophthalmic Hospital reports, No. 8, pp. 1–13.

Jacquin-Courtois S, O'Shea J, Luauté J et al. (2013). Rehabilitation of spatial neglect by prism adaptation: a peculiar expansion of sensorimotor after-effects to spatial cognition. Neurosci Biobehav Rev 37: 594–609. https://doi.org/10.1016/j.neubiorev.2013.02.007.

Jehkonen M, Laihosalo M, Kettunen J (2006). Anosognosia after stroke: assessment, occurrence, subtypes and impact on functional outcome reviewed. Acta Neurol Scand 114: 293–306. https://doi.org/10.1111/j.1600-0404.2006.00723.x.

Jewell G, McCourt ME (2000). Pseudoneglect: a review and meta-analysis of performance factors in line bisection tasks. Neuropsychologia 38: 93–110. https://doi.org/10.1016/S0028-3932(99)00045-7.

Johnston CW, Shapiro E (1986). Hemi-inattention resulting from left hemisphere brain damage during infancy. Cortex 22: 279–287. https://doi.org/10.1016/S0010-9452(86)80052-1.

Kaiser AP, Villadsen KW, Samani A et al. (2022). Virtual reality and eye-tracking assessment, and treatment of unilateral spatial neglect: systematic review and future prospects. Front Psychol 13: 787382. https://doi.org/10.3389/fpsyg.2022.787382.

Karnath H-O, Ferber S, Himmelbach M (2001). Spatial awareness is a function of the temporal not the posterior parietal lobe. Nature 411: 950–953. https://doi.org/10.1038/35082075.

Karnath H-O, Himmelbach M, Rorden C (2002). The subcortical anatomy of human spatial neglect: putamen, caudate nucleus and pulvinar. Brain 125: 350–360. https://doi.org/10.1093/brain/awf032.

Kasai M, Ishizaki J, Meguro K (2007). Alzheimer's patients do not show left unilateral spatial neglect but exhibit peripheral inattention and simplification. Dement Neuropsychol 1: 374–380. https://doi.org/10.1590/S1980-57642008DN10400008.

Kennard MA (1939). Alterations in response to visual stimuli following lesions of frontal lobe in monkeys. Arch Neurol Psychiatry 41: 1153–1165.

Kerkhoff G (2003). Modulation and rehabilitation of spatial neglect by sensory stimulation. Prog Brain Res 142: 257–271. https://doi.org/10.1016/S0079-6123(03)42018-9.

Kim EJ, Choi KD, Han MK et al. (2008). Hemispatial neglect in cerebellar stroke. J Neurol Sci 275: 133–138. https://doi.org/10.1016/j.jns.2008.08.012.

Kim E-J, Lee B, Jo M-K et al. (2013). Directional and spatial motor intentional disorders in patients with right versus left hemisphere strokes. Neuropsychology 27: 428–437. https://doi.org/10.1037/a0032824.

Kinsbourne M (1970). A model for the mechanism of unilateral neglect of space. Trans Am Neurol Assoc 95: 143–146.

Kinsbourne M (1993). Orientational bias model of unilateral neglect: evidence from attentional gradients within hemispace. In: IH Robertson, JC Marshall (Eds.), Unilateral neglect: clinical and experimental studies. Erlbaum, Hove, pp. 63–86.

Kinsbourne M, Warrington EK (1962). A variety of reading disability associated with right hemisphere lesions. J Neurol Neurosurg Psychiatry 25: 339–344. https://doi.org/10.1136/jnnp.25.4.339.

Kirton A, Metzler MJ, Craig BT et al. (2021). Perinatal stroke: mapping and modulating developmental plasticity. Nat Rev Neurol 17: 415–432. https://doi.org/10.1038/s41582-021-00503-x.

Kleinman JT, Newhart M, Davis C et al. (2007). Right hemispatial neglect: frequency and characterization following acute left hemisphere stroke. Brain Cogn 64: 50–59. https://doi.org/10.1016/j.bandc.2006.10.005.

Kleinman JT, Gailloud P, Jordan LC (2010). Recovery from spatial neglect and hemiplegia in a child despite a large anterior circulation stroke and Wallerian degeneration. J Child Neurol 25: 500–503. https://doi.org/10.1177/0883073809339060.

Koch G, Oliveri M, Cheeran B et al. (2008). Hyperexcitability of parietal-motor functional connections in the intact left-hemisphere of patients with neglect. Brain 131: 3147–3155. https://doi.org/10.1093/brain/awn273.

Koch G, Bonnì S, Giacobbe V et al. (2012). Theta-burst stimulation of the left hemisphere accelerates recovery of hemispatial neglect. Neurology 78: 24–30. https://doi.org/10.1212/WNL.0b013e31823ed08f.

Kooistra CA, Heilman KM (1989). Hemispatial visual inattention masquerading as hemianopia. Neurology 39: 1125–1127. https://doi.org/10.1212/wnl.39.8.1125.

Kriksciun J, Knight C, Trauner D (2021). Longitudinal stability of spatial inattention in children with perinatal stroke. Pediatr Neurol 118: 26–31. https://doi.org/10.1016/j.pediatrneurol.2021.01.005.

Kubanek J, Li JM, Snyder LH (2015). Motor role of parietal cortex in a monkey model of hemispatial neglect. Proc Natl Acad Sci U S A 112: E2057–E2072. https://doi.org/10.1073/pnas.1418324112.

Làdavas E, Umiltà C, Ziani P et al. (1993). The role of right side objects in left side neglect: a dissociation between perceptual and directional motor neglect. Neuropsychologia 31: 761–773. https://doi.org/10.1016/0028-3932(93)90127-L.

Laiacona M, Capitani E, Stangalino C et al. (1999). Crossed aphasia. II: why are deep lesions overrepresented with respect to standard aphasia? Behav Neurol 9: 163–170. https://doi.org/10.3233/BEN-1996-93-408.

Laplane D, Degos JD (1983). Motor neglect. J Neurol Neurosurg Psychiatry 46: 152–158. https://doi.org/10.1136/jnnp.46.2.152.

Laurent-Vannier A, Pradat-Diehl P, Chevignard M et al. (2003). Spatial and motor neglect in children. Neurology 60: 202–207. https://doi.org/10.1212/01.wnl.0000048201.26494.0b.

Learmonth G, Papadatou-Pastou M (2022). A meta-analysis of line bisection and landmark task performance in older adults. Neuropsychol Rev 32: 438–457. https://doi.org/10.1007/s11065-021-09505-4.

Leibovitch FS, Black SE, Caldwell CB et al. (1998). Brain-behavior correlations in hemispatial neglect using CT and SPECT: the Sunnybrook Stroke Study. Neurology 50: 901–908. https://doi.org/10.1212/wnl.50.4.901.

Leibovitch FS, Black SE, Caldwell CB et al. (1999). Brain SPECT imaging and left hemispatial neglect covaried using partial least squares: the Sunnybrook Stroke Study. Hum Brain Mapp 7: 244–253. https://doi.org/10.1002/(sici)1097-0193(1999)7:4<244::aid-hbm3>3.0.co;2-k.

Lewis DW, Geller T (1994). Ipsilateral constructional apraxia. J Child Neurol 9: 41–44. https://doi.org/10.1177/088307389400900109.

Lichtheim L (1885). On aphasia. Brain 7: 433–484.

Lindell AB, Jalas MJ, Tenovuo O et al. (2007). Clinical assessment of hemispatial neglect: evaluation of different measures and dimensions. Clin Neuropsychol 21: 479–497. https://doi.org/10.1080/13854040600630061.

Lunven M, Bartolomeo P (2017). Attention and spatial cognition: neural and anatomical substrates of visual neglect. Ann Phys Rehabil Med 60: 124–129. https://doi.org/10.1016/j.rehab.2016.01.004.

Lunven M, De Schotten MT, Bourlon CC et al. (2015). White matter lesional predictors of chronic visual neglect: a longitudinal study. Brain 138: 746–760. https://doi.org/10.1093/brain/awu389.

Maeshima S, Shigeno K, Dohi N et al. (1992). A study of right unilateral spatial neglect in left hemispheric lesions: the difference between right-handed and non-right-handed post-stroke patients. Acta Neurol Scand 85: 418–424. https://doi.org/10.1111/j.1600-0404.1992.tb06040.x.

Maeshima S, Terada T, Nakai K et al. (1995). Unilateral spatial neglect due to a haemorrhagic contusion in the right frontal lobe. J Neurol 242: 613–617. https://doi.org/10.1007/BF00866909.

Maguire AM, Ogden JA (2002). MRI brain scan analyses and neuropsychological profiles of nine patients with persisting unilateral neglect. Neuropsychologia 40: 879–887. https://doi.org/10.1016/S0028-3932(01)00169-5.

Malherbe C, Umarova RM, Zavaglia M et al. (2018). Neural correlates of visuospatial bias in patients with left hemisphere stroke: a causal functional contribution analysis based on game theory. Neuropsychologia 115: 142–153. https://doi.org/10.1016/j.neuropsychologia.2017.10.013.

Marangolo P, Piccardi L, Rinaldi MC (2003). Dissociation between personal and extrapersonal neglect in a crossed aphasia study. Neurocase 9: 414–420. https://doi.org/10.1076/neur.9.5.414.16554.

Mariën P, Borgatti R (2018). Language and the cerebellum. Handb Clin Neurol 154: 181–202. https://doi.org/10.1016/B978-0-444-63956-1.00011-4.

Mariën P, Paghera B, De Deyn PP et al. (2004). Adult crossed aphasia in dextrals revisited. Cortex 40: 41–74. https://doi.org/10.1016/s0010-9452(08)70920-1.

Mark VW (2003). Acute versus chronic functional aspects of unilateral spatial neglect. Front Biosci 8: e172–e189. https://doi.org/10.2741/973.

Marsh EB, Hillis AE (2008). Dissociation between egocentric and allocentric visuospatial and tactile neglect in acute stroke. Cortex 44: 1215–1220. https://doi.org/10.1016/j.cortex.2006.02.002.

Marshall JC, Fink GR, Halligan PW et al. (2002). Spatial awareness: a function of the posterior parietal lobe? Cortex 38: 253–257. https://doi.org/10.1016/s0010-9452(08)70654-3.

Massetti G, Albini F, Casati C et al. (2023). Validation of "Neurit. Space": three digital tests for the neuropsychological evaluation of unilateral spatial neglect. J Clin Med 12: 3042. https://doi.org/10.3390/jcm12083042.

Mattingley JB, Bradshaw JL, Phillips JG (1992). Impairments of movement initiation and execution in unilateral neglect. Directional hypokinesia and bradykinesia. Brain 115: 1849–1874. https://doi.org/10.1093/brain/115.6.1849.

Mazzucchi A, Sinforiani E, Boller F (2012). Focal cerebral lesions and painting abilities. Prog Brain Res 204: 71–98. https://doi.org/10.1016/B978-0-444-63287-6.00004-X.

Meador KJ, Loring DW, Bowers D et al. (1987). Remote memory and neglect syndrome. Neurology 37: 522–526. https://doi.org/10.1212/wnl.37.3.522.

Mendez MF, Cherrier MM, Cymerman JS (1997). Hemispatial neglect on visual search tasks in Alzheimer's disease. Neuropsychiatry Neuropsychol Behav Neurol 10: 203–208.

Meola A, Comert A, Yeh F-C et al. (2015). The controversial existence of the human superior fronto-occipital fasciculus: connectome-based tractographic study with microdissection validation. Hum Brain Mapp 36: 4964–4971. https://doi.org/10.1002/hbm.22990.

Mesulam M-M (Ed.), (2000). Principles of behavioral and cognitive neurology, second edn. Oxford University Press, New York.

Meyer JS, Obara K, Muramatsu K (1993). Diaschisis. Neurol Res 15: 362–366. https://doi.org/10.1080/01616412.1993.11740164.

Middleton FA, Strick PL (1994). Anatomical evidence for cerebellar and basal ganglia involvement in higher cognitive function. Science 266: 458–461. https://doi.org/10.1126/science.7939688.

Milner AD, Harvey M, Roberts RC et al. (1993). Line bisection errors in visual neglect: misguided action or size distortion? Neuropsychologia 31: 39–49. https://doi.org/10.1016/0028-3932(93)90079-F.

Miura N, Takeda A, Terao S et al. (1996). Somatoparaphrenia caused by the lesion in the dominant cerebral hemisphere—a case report. No To Shinkei 48: 275–279.

Moore MJ, Demeyere N (2022). Lesion symptom mapping of domain-specific cognitive impairments using routine imaging in stroke. Neuropsychologia 167: 108159. https://doi.org/10.1016/j.neuropsychologia.2022.108159.

Moore MJ, Gillebert CR, Demeyere N (2021a). Right and left neglect are not anatomically homologous: a voxel-lesion symptom mapping study. Neuropsychologia 162: 108024. https://doi.org/10.1016/j.neuropsychologia.2021.108024.

Moore MJ, Vancleef K, Riddoch MJ et al. (2021b). Recovery of visuospatial neglect subtypes and relationship to functional outcome six months after stroke. Neurorehabil Neural Repair 35: 823–835. https://doi.org/10.1177/15459683211032977.

Moore MJ, Milosevich E, Mattingley JB et al. (2023). The neuroanatomy of visuospatial neglect: a systematic review and analysis of lesion-mapping methodology. Neuropsychologia 180: 108470. https://doi.org/10.1016/j.neuropsychologia.2023.108470.

Mort DJ, Malhotra P, Mannan SK et al. (2003). The anatomy of visual neglect. Brain 126: 1986–1997. https://doi.org/10.1093/brain/awg200.

Moscovitch M, Behrmann M (1994). Coding of spatial information in the somatosensory system: evidence from patients with neglect following parietal lobe damage. J Cogn Neurosci 6: 151–155. https://doi.org/10.1162/jocn.1994.6.2.151.

Na DL, Adair JC, Williamson DJG et al. (1998). Dissociation of sensory-attentional from motor-intentional neglect. J Neurol Neurosurg Psychiatry 64: 331–338. https://doi.org/10.1136/jnnp.64.3.331.

Na DL, Adair JC, Kang Y et al. (1999). Motor perseverative behavior on a line cancellation task. Neurology 52: 1569–1576. https://doi.org/10.1212/wnl.52.8.1569.

Nathanson M, Bergman PS, Gordon GG (1952). Denial of illness. Arch Neurol Psychiatry 68: 380–387.

Nestor PJ, Caine D, Fryer TD et al. (2003). The topography of metabolic deficits in posterior cortical atrophy (the visual variant of Alzheimer's disease) with FDG-PET. J Neurol Neurosurg Psychiatry 74: 1521–1529. https://doi.org/10.1136/jnnp.74.11.1521.

Nielsen JM (1938). Gerstmann syndrome: finger agnosia, agraphia, confusion of right and left and acalculia. Arch Neurol Psychiatry 39: 536–559.

Nijboer TCW, Olthoff L, Van der Stigchel S et al. (2014). Prism adaptation improves postural imbalance in neglect patients. Neuroreport 25: 307–311. https://doi.org/10.1097/WNR.0000000000000088.

Norton DJ, Jaywant A, Gallart-Palau X et al. (2015). Normal discrimination of spatial frequency and contrast across visual hemifields in left-onset Parkinson's disease: evidence against perceptual hemifield biases. Vision Res 107: 94–100. https://doi.org/10.1016/j.visres.2014.12.003.

Oda H, Ohkawa S, Maeda K (2008). Hemispatial visual defect in Alzheimer's disease. Neurocase 14: 141–146. https://doi.org/10.1080/13554790802032226.

Ofir-Geva S, Serfaty C, Sacher Y et al. (2021). Unilateral spatial neglect without hemiplegia: the output-mode effect revisited. J Stroke Cerebrovasc Dis 30: 105777. https://doi.org/10.1016/j.jstrokecerebrovasdis.2021.105777.

Ogden JA (1985a). Contralesional neglect of constructed visual images in right and left brain-damaged patients. Brain Cogn 4: 273–277. https://doi.org/10.1016/0278-2626(85)90054-5.

Ogden JA (1985b). Anterior-posterior interhemispheric differences in the loci of lesions producing visual hemineglect. Brain Cogn 4: 59–75. https://doi.org/10.1016/0278-2626(85)90054-5.

Oliveri M, Bisiach E, Brighina F et al. (2001). rTMS of the unaffected hemisphere transiently reduces contralesional visuospatial hemineglect. Neurology 57: 1338–1340. https://doi.org/10.1212/WNL.57.7.1338.

Orfei MD, Robinson RG, Prigatano GP et al. (2007). Anosognosia for hemiplegia after stroke is a multifaceted phenomenon: a systematic review of the literature. Brain 130: 3075–3090. https://doi.org/10.1093/brain/awm106.

Ota H, Fujii T, Suzuki K et al. (2001). Dissociation of body-centered and stimulus-centered representations in unilateral neglect. Neurology 57: 2064–2069. https://doi.org/10.1212/WNL.57.11.2064.

Ota H, Fujii T, Tabuchi M et al. (2003). Different spatial processing for stimulus-centered and body-centered representations. Neurology 60: 1846–1848. https://doi.org/10.1212/01.wnl.0000065899.22903.07.

Paghera B, Mariën P, Vignolo LA (2003). Crossed aphasia with left spatial neglect and visual imperception: a case report. Neurol Sci 23: 317–322. https://doi.org/10.1007/s100720300008.

Palesi F, De Rinaldis A, Castellazzi G et al. (2017). Contralateral cortico-ponto-cerebellar pathways reconstruction in humans in vivo: implications for reciprocal cerebro-cerebellar structural connectivity in motor and non-motor areas. Sci Rep 7: 12841. https://doi.org/10.1038/s41598-017-13079-8.

Papagno C, Vallar G (2003). Anosognosia for left hemiplegia: Babinski's (1914) cases. In: C Code, C-W Wallesch, Y Joanette et al. (Eds.), Classic cases in neuropsychology. Psychology Press, pp. 171–189.

Paterson A, Zangwill OL (1944). Disorders of visual space perception associated with lesions of the right cerebral hemisphere. Brain 67: 331–358.

Pearce JM (2007). Misoplegia. Eur Neurol 57: 62–64. https://doi.org/10.1159/000097122.

Perani D, Vallar G, Cappa S et al. (1987). Aphasia and neglect after subcortical stroke. A clinical/cerebral perfusion correlation study. Brain 110: 1211–1229. https://doi.org/10.1093/brain/110.5.1211.

Perren F, Heydrich L, Blanke O et al. (2015). "Crossed" somatoparaphrenia: an unusual new case and a review of the literature. Exp Brain Res 233: 175–179. https://doi.org/10.1007/s00221-014-4101-z.

Peru A, Pinna G (1997). Right personal neglect following a left hemisphere stroke. A case report. Cortex 33: 585–590. https://doi.org/10.1016/S0010-9452(08)70240-5.

Phillips KA, Stimpson CD, Smaers JB et al. (2015). The corpus callosum in primates: processing speed of axons and the evolution of hemispheric asymmetry. Proc Biol Sci 282: 20151535. https://doi.org/10.1098/rspb.2015.1535.

Pia L, Corazzini LL, Folegatti A et al. (2009). Mental number line disruption in a right-neglect patient after a left-hemisphere stroke. Brain Cogn 69: 81–88. https://doi.org/10.1016/j.bandc.2008.05.007.

Pitteri M, Venneri A, Meneghello F et al. (2013). How to differentiate hemianesthesia from left tactile neglect: a preliminary case report. Behav Neurol 26: 151–155. https://doi.org/10.3233/BEN-2012-110225.

Pizzamiglio L, Antonucci G, Judica A et al. (1992). Cognitive rehabilitation of the hemineglect disorder in chronic patients with unilateral right brain damage. J Clin Exp Neuropsychol 14: 901–923. https://doi.org/10.1080/01688639208402543.

Posteraro L, Maravita A (1996). A new case of atypical cerebral dominance. Ital J Neurol Sci 17: 237–240. https://doi.org/10.1007/BF01995690.

Purpura G, Guzzetta A, Tinelli F (2023). Visual neglect: does it exist in children with unilateral brain lesion? A systematic review. Neuropsychol Rehabil 33: 703–717. https://doi.org/10.1080/09602011.2022.2032762.

Ramnani N (2006). The primate cortico-cerebellar system: anatomy and function. Nat Rev Neurosci 7: 511–522. https://doi.org/10.1038/nrn1953.

Ramsey LE, Siegel JS, Baldassarre A et al. (2016). Normalization of network connectivity in hemispatial neglect recovery. Ann Neurol 80: 127–141. https://doi.org/10.1002/ana.24690.

Redding GM, Wallace B (2006). Prism adaptation and unilateral neglect: review and analysis. Neuropsychologia 44: 1–20. https://doi.org/10.1016/j.neuropsychologia.2005.04.009.

Reep RL, Corwin JV, Cheatwood JL et al. (2004). A rodent model for investigating the neurobiology of contralateral neglect. Cogn Behav Neurol 17: 191–194.

Rich TJ, Pylarinos M, Parrott D et al. (2023). Prism adaptation treatment for right-sided and left-sided spatial neglect: a retrospective case-matched study. Arch Rehabil Res Clin Transl 5: 100263. https://doi.org/10.1016/j.arrct.2023.100263.

Ringman JM, Saver JL, Woolson RF et al. (2004). Frequency, risk factors, anatomy, and course of unilateral neglect in an acute stroke cohort. Neurology 63: 468–474. https://doi.org/10.1212/01.WNL.0000133011.10689.CE.

Robertson LC, Lamb MR (1991). Neuropsychological contributions to theories of part/whole organization. Cogn Psychol 23: 299–330. https://doi.org/10.1016/0010-0285(91)90012-D.

Rode G, Perenin M-T (1994). Temporary remission of representational hemineglect through vestibular stimulation. Neuroreport 5: 869–872. https://doi.org/10.1097/00001756-199404000-00004.

Rode G, Perenin MT, Boisson D (1995). Neglect of the representational space: demonstration by mental evocation of the map of France. Revue Neurologique (Paris) 151: 161–164.

Rode G, Vallar G, Chabanat E et al. (2018). What do spatial distortions in patients' drawing after right brain damage teach us about space representation in art? Front Psychol 9. https://doi.org/10.3389/fpsyg.2018.01058.

Ronchi R, Rode G, Cotton F et al. (2013). Remission of anosognosia for right hemiplegia and neglect after caloric vestibular stimulation. Restor Neurol Neurosci 31: 19–24. https://doi.org/10.3233/RNN-120236.

Rorden C, Karnath H-O (2010). A simple measure of neglect severity. Neuropsychologia 48: 2758–2763. https://doi.org/10.1016/j.neuropsychologia.2010.04.018.

Rossetti Y, Rode G, Pisella L et al. (1998). Prism adaptation to a rightward optical deviation rehabilitates left hemispatial neglect. Nature 395: 166–169. https://doi.org/10.1038/25988.

Rossini PM, Burke D, Chen R et al. (2015). Non-invasive electrical and magnetic stimulation of the brain, spinal cord, roots and peripheral nerves: basic principles and procedures for routine clinical and research application. An updated report from an I.F.C.N. Committee. Clin Neurophysiol 126: 1071–1107. https://doi.org/10.1016/j.clinph.2015.02.001.

Roth M (1949). Disorders of the body image caused by lesions of the right parietal lobe. Brain 72: 89–111. https://doi.org/10.1093/brain/72.1.89.

Rubens AB (1985). Caloric stimulation and unilateral visual neglect. Neurology 35: 1019–1024. https://doi.org/10.1212/WNL.35.7.1019.

Rusconi E, Cubelli R (2019). The making of a syndrome: the English translation of Gerstmann's first report. Cortex 117: 277–283. https://doi.org/10.1016/j.cortex.2019.03.021.

Rusconi ML, Maravita A, Bottini G et al. (2002). Is the intact side really intact? Perseverative responses in patients with unilateral neglect: a productive manifestation. Neuropsychologia 40: 594–604. https://doi.org/10.1016/S0028-3932(01)00160-9.

Saj A, Honoré J, Braem B et al. (2012). Time since stroke influences the impact of hemianopia and spatial neglect on visual-spatial tasks. Neuropsychology 26: 37–44. https://doi.org/10.1037/a0025733.

Sallard E, Duffau H, Bonnetblanc F (2012). Ultra-fast recovery from right neglect after 'awake surgery' for slow-growing tumor invading the left parietal area. Neurocase 18: 80–90. https://doi.org/10.1080/13554794.2011.556127.

Scheltens P, De Strooper B, Kivipelto M et al. (2021). Alzheimer's disease. Lancet 397: 1577–1590. https://doi.org/10.1016/S0140-6736(20)32205-4.

Schenkenberg T, Bradford DC, Ajax ET (1980). Line bisection and unilateral visual neglect in patients with neurologic impairment. Neurology 30: 509–517. https://doi.org/10.1212/wnl.30.5.509.

Schiff ND, Pulver M (1999). Does vestibular stimulation activate thalamocortical mechanisms that reintegrate impaired cortical regions? Proc Biol Sci 266: 421–423. https://doi.org/10.1098/rspb.1999.0654.

Schilder P (1935). The image and appearance of the human body, International Universities Press, Inc., New York.

Schmahmann JD (1996). From movement to thought: anatomic substrates of the cerebellar contribution to cognitive processing. Hum Brain Mapp 4: 174–198. https://doi.org/10.1002/(SICI)1097-0193(1996)4:3<174::AID-HBM3>3.0.CO;2-0.

Sebastian R, Schein MG, Davis C et al. (2014). Aphasia or neglect after thalamic stroke: the various ways they may be related to cortical hypoperfusion. Front Neurol 5: 231. https://doi.org/10.3389/fneur.2014.00231.

Shallice T, Mussoni A, D'Agostino S et al. (2010). Right posterior cortical functions in a tumour patient series. Cortex 46: 1178–1188. https://doi.org/10.1016/j.cortex.2010.04.005.

Shuren J, Hartley T, Heilman KM (1998). The effects of rotation on spatial attention. Neuropsychiatry Neuropsychol Behav Neurol 11: 72–75.

Siéroff E (2017). Acquired spatial dyslexia. Ann Phys Rehabil Med 60: 155–159. https://doi.org/10.1016/j.rehab.2015.07.004.

Silberpfennig J (1941). Contribution to the problem of eye movements. Disturbances of ocular movements with pseudo hemianopsia in frontal tumors. Confin Neurol 4: 1–13. https://doi.org/10.1159/000106147.

Silveri MC, Misciagna S, Terrezza G (2001). Right side neglect in right cerebellar lesion. J Neurol Neurosurg Psychiatry 71: 114–117. https://doi.org/10.1136/jnnp.71.1.114.

Silveri MC, Ciccarelli N, Cappa A (2011). Unilateral spatial neglect in degenerative brain pathology. Neuropsychology 25: 554–566. https://doi.org/10.1037/a0023957.

Sparing R, Thimm M, Hesse MD et al. (2009). Bidirectional alterations of interhemispheric parietal balance by non-invasive cortical stimulation. Brain 132: 3011–3020. https://doi.org/10.1093/brain/awp154.

Sperber C, Clausen J, Benke T et al. (2020). The anatomy of spatial neglect after posterior cerebral artery stroke. Brain Commun 2: fcaa163. https://doi.org/10.1093/braincomms/fcaa163.

Stone SP, Wilson B, Wroot A et al. (1991). The assessment of visuo-spatial neglect after acute stroke. J Neurol Neurosurg Psychiatry 54: 345–350. https://doi.org/10.1136/jnnp.54.4.345.

Stone SP, Patel P, Greenwood RJ et al. (1992). Measuring visual neglect in acute stroke and predicting its recovery: the visual neglect recovery index. J Neurol Neurosurg Psychiatry 55: 431–436. https://doi.org/10.1136/jnnp.55.6.431.

Suchan J, Karnath H-O (2011). Spatial orienting by left hemisphere language areas: a relict from the past? Brain 134: 3059–3070. https://doi.org/10.1093/brain/awr120.

Suchan J, Rorden C, Karnath H-O (2012). Neglect severity after left and right brain damage. Neuropsychologia 50: 1136–1141. https://doi.org/10.1016/j.neuropsychologia.2011.12.018.

Sunderland A, Wade DT, Langton Hewer R (1987). The natural history of visual neglect after stroke. Indications from two methods of assessment. Int Disabil Stud 9: 55–59. https://doi.org/10.3109/03790798709166235.

Tariq U, Parker A, Saadatpour L et al. (2020). Posterior cortical atrophy with right lower egocentric quadrantic neglect and lower vertical allocentric neglect. Arch Clin Neuropsychol 35: 448–457. https://doi.org/10.1093/arclin/acz078.

Tarvonen-Schröder S, Niemi T, Koivisto M (2020). Clinical and functional differences between right and left stroke with and without contralateral spatial neglect. J Rehabil Med 52: jrm00072. https://doi.org/10.2340/16501977-2699.

Ten Brink AF, Verwer JH, Biesbroek JM et al. (2017). Differences between left- and right-sided neglect revisited: a large cohort study across multiple domains. J Clin Exp Neuropsychol 39: 707–723. https://doi.org/10.1080/13803395.2016.1262333.

Terruzzi S, Albini F, Massetti G et al. (2023). The neuropsychological assessment of unilateral spatial neglect through computerized and virtual reality tools: a scoping review. Neuropsychol Rev. https://doi.org/10.1007/s11065-023-09586-3.

Tham K, Tegnér R (1996). The baking tray task: a test of spatial neglect. Neuropsychol Rehabil 6: 19–25. https://doi.org/10.1080/713755496.

Thareja T, Ballantyne AO, Trauner DA (2012). Spatial analysis after perinatal stroke: patterns of neglect and exploration in extra-personal space. Brain Cogn 79: 107–116. https://doi.org/10.1016/j.bandc.2012.02.009.

Thiebaut de Schotten M, Urbanski M, Duffau H et al. (2005). Direct evidence for a parietal-frontal pathway subserving spatial awareness in humans. Science 309: 2226–2228. https://doi.org/10.1126/science.1116251.

Thiebaut de Schotten M, Dell'Acqua F, Forkel SJ et al. (2011). A lateralized brain network for visuospatial attention. Nat Neurosci 14: 1245–1246. https://doi.org/10.1038/nn.2905.

Thiebaut de Schotten M, Tomaiuolo F, Aiello M et al. (2014). Damage to white matter pathways in subacute and chronic spatial neglect: a group study and 2 single-case studies with complete virtual "in vivo" tractography dissection. Cereb Cortex 24: 691–706. https://doi.org/10.1093/cercor/bhs351.

Toba MN, Pagliari C, Rabuffetti M et al. (2021). Quantitative assessment of motor neglect. Stroke 52: 1618–1627. https://doi.org/10.1161/STROKEAHA.120.031949.

Toba MN, Migliaccio R, Potet A et al. (2022). Right-side spatial neglect and white matter disconnection after left-hemisphere strokes. Brain Struct Funct 227: 2991–3000. https://doi.org/10.1007/s00429-022-02541-7.

Trauner DA (2003). Hemispatial neglect in young children with early unilateral brain damage. Dev Med Child Neurol 45: 160–166. https://doi.org/10.1111/j.1469-8749.2003.tb00924.x.

Tromp E, Dinkla A, Mulder T (1995). Walking through doorways: an analysis of navigation skills in patients with neglect. Neuropsychol Rehabil 5: 319–331. https://doi.org/10.1080/09602019508401475.

Turton AJ, Dewar SJ, Lievesley A et al. (2009). Walking and wheelchair navigation in patients with left visual neglect. Neuropsychol Rehabil 19: 274–290. https://doi.org/10.1080/09602010802106478.

Tzourio-Mazoyer N, Zago L, Cochet H et al. (2020). Development of handedness, anatomical and functional brain lateralization. Handb Clin Neurol 173: 99–105. https://doi.org/10.1016/B978-0-444-64150-2.00011-3.

Urbanski M, Thiebaut de Schotten M, Rodrigo S et al. (2008). Brain networks of spatial awareness: evidence from diffusion tensor imaging tractography. J Neurol Neurosurg Psychiatry 79: 598–601. https://doi.org/10.1136/jnnp.2007.126276.

Vallar G (1998). Spatial hemineglect in humans. Trends Cogn Sci 2: 87–97. https://doi.org/10.1016/S1364-6613(98)01145-0.

Vallar G (2001). Extrapersonal visual unilateral spatial neglect and its neuroanatomy. Neuroimage 14. https://doi.org/10.1006/nimg.2001.0822.

Vallar G (2007). Spatial neglect Balint-homes' and Gerstmann's syndrome, and other spatial disorders. CNS Spectr 12. https://doi.org/10.1017/S1092852900021271.

Vallar G, Bolognini N (2014). Unilateral spatial neglect. In: AC Nobre, S Kastner (Eds.), Oxford handbook of attention. Oxford University Press, Oxford, pp. 972–1027.

Vallar G, Calzolari E (2018). Unilateral spatial neglect after posterior parietal damage. Handb Clin Neurol. https://doi.org/10.1016/B978-0-444-63622-5.00014-0.

Vallar G, Maravita A (2009). Personal and extra-personal spatial perception. In: J Berntson, J., Cacioppo, J.T. (Eds.), Handbook of neuroscience for the behavioral sciences. Wiley, Hoboken, NJ, pp. 322–336.

Vallar G, Perani D (1986). The anatomy of unilateral neglect after right-hemisphere stroke lesions. A clinical/CT-scan correlation study in man. Neuropsychologia 24: 609–622. https://doi.org/10.1016/0028-3932(86)90001-1.

Vallar G, Ronchi R (2009). Somatoparaphrenia: a body delusion. A review of the neuropsychological literature. Exp Brain Res 192: 533–551. https://doi.org/10.1007/s00221-008-1562-y.

Vallar G, Ronchi R (2022). Unilateral spatial neglect. In: S Della Sala (Ed.), Encyclopedia of behavioral neuroscience, second edn. Elsevier, Oxford, pp. 605–618. https://doi.org/10.1016/B978-0-12-819641-0.00144-4.

Vallar G, Rusconi ML, Geminiani G et al. (1991a). Visual and nonvisual neglect after unilateral brain lesions: modulation by visual input. Int J Neurosci 61: 229–239. https://doi.org/10.3109/00207459108990740.

Vallar G, Sandroni P, Rusconi ML et al. (1991b). Hemianopia, hemianesthesia and spatial neglect. A study with evoked potentials. Neurology 41: 1918–1922. https://doi.org/10.1212/WNL.41.12.1918.

Vallar G, Rusconi ML, Bignamini L et al. (1994). Anatomical correlates of visual and tactile extinction in humans: a clinical CT scan study. J Neurol Neurosurg Psychiatry 57: 464–470. https://doi.org/10.1136/jnnp.57.4.464.

Vallar G, Papagno C, Rusconi ML et al. (1995). Vestibular stimulation, spatial hemineglect and dysphasia. Selective effects? Cortex 31: 589–593. https://doi.org/10.1016/S0010-9452(13)80070-6.

Vallar G, Guariglia C, Rusconi ML (1997). Modulation of the neglect syndrome by sensory stimulation. In: P Their, HO Karnath (Eds.), Parietal lobe contributions to orientation in 3D space. Springer-Verlag, Heidelberg, pp. 555–578.

Vallar G, Burani C, Arduino LS (2010). Neglect dyslexia: a review of the neuropsychological literature. Exp Brain Res 206: 219–235. https://doi.org/10.1007/s00221-010-2386-0.

Vallar G, Bello L, Bricolo E et al. (2014). Cerebral correlates of visuospatial neglect: a direct cerebral stimulation study. Hum Brain Mapp 35: 1334–1350. https://doi.org/10.1002/hbm.22257.

van der Hulst H (2023). A mind for language: an introduction to the innateness debate, Cambridge University Press, Cambridge. https://doi.org/10.1017/9781108647182.

van Dijck J-P, Gevers W, Lafosse C et al. (2013). Right-sided representational neglect after left brain damage in a case without visuospatial working memory deficits. Cortex 49: 2283–2293. https://doi.org/10.1016/j.cortex.2013.03.004.

Veldema J, Bösl K, Neumann G et al. (2020). Noninvasive brain stimulation in rehabilitation of hemispatial neglect after stroke. CNS Spectr 25: 38–49. https://doi.org/10.1017/S1092852918001748.

Venneri A, Pentore R, Cotticelli B et al. (1998). Unilateral spatial neglect in the late stage of Alzheimer's disease. Cortex 34: 743–752. https://doi.org/10.1016/s0010-9452(08)70777-9.

Verbitsky R, Anderson B, Danckert J et al. (2024). Left cerebellar lesions may be associated with an increase in spatial neglect-like symptoms. Cerebellum 23: 431–443. https://doi.org/10.1007/s12311-023-01542-4.

Verdon V, Schwartz S, Lovblad KO et al. (2010). Neuroanatomy of hemispatial neglect and its functional components: a study using voxel-based lesion-symptom mapping. Brain 133: 880–894. https://doi.org/10.1093/brain/awp305.

Verhelst H, Dhollander T, Gerrits R et al. (2021). Fibre-specific laterality of white matter in left and right language dominant people. Neuroimage 230: 117812. https://doi.org/10.1016/j.neuroimage.2021.117812.

Veronelli L, Bovo S, De Giampaulis P et al. (2016). Unilateral spatial neglect as unique clinical expression of non-convulsive status epilepticus. Cortex 76: 128–130. https://doi.org/10.1016/j.cortex.2015.11.010.

Vigneau M, Beaucousin V, Hervé P-Y et al. (2011). What is right-hemisphere contribution to phonological, lexico-semantic, and sentence processing? Insights from a meta-analysis. Neuroimage 54: 577–593. https://doi.org/10.1016/j.neuroimage.2010.07.036.

Volberg G, Kliegl K, Hanslmayr S et al. (2009). EEG alpha oscillations in the preparation for global and local processing predict behavioral performance. Hum Brain Mapp 30: 2173–2183. https://doi.org/10.1002/hbm.20659.

von Monakow C (1914). Die Lokalisation im Grosshirn und der Abbau der Funktion durch kortikale Herde, J.F. Bergmann, Wiesbaden.

Watson RT, Heilman KM (1979). Thalamic neglect. Neurology 29: 690–694. https://doi.org/10.1212/wnl.29.5.690.

Watson RT, Valenstein E, Heilman KM (1981). Thalamic neglect. Possible role of the medial thalamus and nucleus reticularis in behavior. Arch Neurol 38: 501–506. https://doi.org/10.1001/archneur.1981.00510080063009.

Weintraub S, Daffner KR, Ahern GL et al. (1996). Right sided hemispatial neglect and bilateral cerebral lesions. J Neurol Neurosurg Psychiatry 60: 342–344. https://doi.org/10.1136/jnnp.60.3.342.

Welch K, Stuteville P (1958). Experimental production of unilateral neglect in monkeys. Brain 81: 341–347. https://doi.org/10.1093/brain/81.3.341.

Wheaton LA, Hallett M (2007). Ideomotor apraxia: a review. J Neurol Sci 260: 1–10. https://doi.org/10.1016/j.jns.2007.04.014.

Whitwell JL, Jack CR, Kantarci K et al. (2007). Imaging correlates of posterior cortical atrophy. Neurobiol Aging 28: 1051–1061. https://doi.org/10.1016/j.neurobiolaging.2006.05.026.

Williams LJ, Kernot J, Hillier SL et al. (2021). Spatial neglect subtypes, definitions and assessment tools: a scoping review. Front Neurol 12: 742365. https://doi.org/10.3389/fneur.2021.742365.

Yoshida T, Mizuno K, Miyamoto A et al. (2022). Influence of right versus left unilateral spatial neglect on the functional recovery after rehabilitation in sub-acute stroke patients. Neuropsychol Rehabil 32: 640–661. https://doi.org/10.1080/09602011.2020.1798255.

Young AW, de Haan EHF, Newcombe F et al. (1990). Facial neglect. Neuropsychologia 28: 391–415. https://doi.org/10.1016/0028-3932(90)90068-Y.

Yousefian O, Ballantyne AO, Doo A et al. (2015). Clock drawing in children with perinatal stroke. Pediatr Neurol 52: 592–598. https://doi.org/10.1016/j.pediatrneurol.2014.07.036.

Zebhauser PT, Vernet M, Unterburger E et al. (2019). Visuospatial neglect—a theory-informed overview of current and emerging strategies and a systematic review on the therapeutic use of non-invasive brain stimulation. Neuropsychol Rev 29: 397–420. https://doi.org/10.1007/s11065-019-09417-4.

Zigiotto L, Damora A, Albini F et al. (2021). Multisensory stimulation for the rehabilitation of unilateral spatial neglect. Neuropsychol Rehabil 31: 1410–1443. https://doi.org/10.1080/09602011.2020.1779754.

Zorzi M, Priftis K, Umiltà C (2002). Neglect disrupts the mental number line. Nature 417: 138–139. https://doi.org/10.1038/417138a.

Handbook of Clinical Neurology, Vol. 208 (3rd series)
Cerebral Asymmetries
C. Papagno and P. Corballis, Editors
https://doi.org/10.1016/B978-0-443-15646-5.00015-4

Chapter 10

Split-brain patients: A clinical vs experimental perspective

MICHAEL B. MILLER[1], LUKAS J. VOLZ[2], JESSICA M. SIMONSON[1], AND MICHAEL S. GAZZANIGA[1*]

[1]Department of Psychological and Brain Sciences, University of California, Santa Barbara, CA, United States
[2]Department of Neurology, University of Cologne, Cologne, Germany

Abstract

For more than 80 years, the clinical perspective on the behavioral outcome of split-brain surgery suggested that it "may be carried out without any untoward effect on the patient," despite some initial and transient disconnection effects. The last 60 years of experimental studies, however, have shown quite the opposite. Using lateralized testing procedures, split-brain patients have demonstrated profound disconnection effects that can last a lifetime. This experimental work has transformed our understanding of hemispheric specializations and cerebral asymmetries. It paints a picture of patients with two distinct neural systems processing and operating independently with, nevertheless, a seemingly unified conscious experience. This chapter tracks how these two divergent perspectives have coexisted for so long and offers some explanations for why these patients appear normal from a clinical and social perspective.

INTRODUCTION

Comparative research across various species indicates that a hallmark of mammalian brains is the presence of two largely symmetric, laterally organized, and mirrored hemispheres. These hemispheres are interconnected by the corpus callosum, an evolutionary feature thought to enhance motor coordination, and bihemispheric representations (Aboitiz and Montiel, 2003). Recent evolutionary shifts have brought about cerebral asymmetries and specialized functions within these hemispheres, especially in humans.

In a series of papers, Michael Corballis posited that the recursive nature of certain evolved functions, such as fine motor coordination, speech, and complex social thinking, demand quick and precise control. He theorized that such precision is better managed within dedicated, localized circuits in a single hemisphere. This could be a critical underlying factor for the development of handedness, language, and sophisticated social behaviors (Corballis, 2009; Corballis et al., 2012). Indeed, the last six decades of research on split-brain patients has greatly enriched our knowledge of these hemispheric specializations (Gazzaniga, 2000, 2005), and the phenomenon that results from the surgical disconnection of the two hemispheres—primarily observed in laboratory settings—is among the most astonishing in neuropsychology. That is, two distinct minds operating independently within an individual who, nevertheless, appears to maintain a unified conscious experience.

Throughout these many years of both experiment and observation, one pressing question remains: *Why can a split-brain patient with such a profound disconnection effect seem perfectly normal in most clinical and social settings?*

EARLY HISTORY OF THE CALLOSOTOMY

The corpus callosum is the largest fiber system in the human brain, with 200–250 million axonal projections communicating information from the neocortex of one hemisphere to the neocortex of the other (Aboitiz and Montiel, 2003; Luders et al., 2010; De Benedictis et al., 2016; De León Reyes et al., 2020). Its size and reach far exceed all other commissures, including the anterior, hippocampal, posterior, and habenular commissures (of which, only the anterior commissure also projects axons to neocortical regions; LaMantia and Rakic, 1990a). The corpus callosum undergoes exuberant pruning shortly after birth, eliminating more than 70% of its fibers (LaMantia and Rakic, 1990b) and 21% of its total cross-sectional area (De León Reyes et al., 2020). One of the unique characteristics of the corpus callosum relative to other large fiber tracts in the association cortex (like the arcuate fasciculus and the superior longitudinal fasciculus) is the high proportion of *homotopic connections*: fibers that connect mirror regions of the two hemispheres. Histologic studies in nonhuman primates and high-resolution, diffusion-weighted MRI in humans estimate that there are more than twice as many homotopic connections as

*Correspondence to: Michael S. Gazzaniga, University of California, Santa Barbara, CA, United States. E-mail: mgazzaniga@ucsb.edu

heterotopic connections (Jarbo et al., 2012; De Benedictis et al., 2016; although see Szczupak et al., 2023). Severing the corpus callosum thus creates a nearly complete disconnection of communication between the hemispheres in the human brain (Gazzaniga, 2000, 2005; Zaidel et al., 2003). Left intact, this fiber connection enables many things, like the ability of the speaking left hemisphere to name an object being held in the left hand, even with the eyes closed.

Callosotomy effectively leads to an absolute isolation of the divided hemispheres—a seemingly inconspicuous yet profoundly impactful state. In this altered state, the speaking left hemisphere loses that ability to verbalize the contents held within the left hand. However, this kind of observation was not appreciated until three decades after the first callosotomies were performed. The first known surgical resection of the corpus callosum was conducted by William Van Wagenen in the late 1930s after observing that tumors that crossed the corpus callosum seemed to abate generalized seizures (Van Wagenen and Herren, 1940). Van Wagenen initially performed callosotomy on 24 patients, with some completed in two stages after determining that a partial callosotomy did not relieve those patients of generalized seizures. As he noted at the time, the surgery "may be carried out without any untoward effect on the patient." Van Wagenen did not offer specific explanations for why the surgery worked, other than to say that it seemed to isolate the seizure to one hemisphere. Later, the neurosurgeon Wilson and colleagues (Wilson et al., 1982) suggested that the surgery disconnected the interhemispheric ictal epileptic discharge along the CC. Clinical reports since then have determined that the surgery is most effective in suppressing generalized seizures associated with drop attacks (Kim et al., 2004).

Following these surgeries, the neurologist Andrew Akelaitis conducted a series of systematic studies in the 1940s assessing the behavioral impact of callosotomy (Akelaitis, 1941, 1943, 1945; Akelaitis et al., 1942; Smith and Akelaitis, 1942). One of the prevailing theories at the time was the concept of lateral cerebral dominance, or sidedness (Travis, 1931; Orton, 1937). That is, the idea that one hemisphere was generally dominant in its functions while the other was submissive. The evidence for this was based on handedness and general preferential use of one side of the body, as well as speech control and language understanding, which in most individuals were lateralized in the left hemisphere. At the time, it was believed that the corpus callosum was responsible for integrating the two sides of the body, mainly through inhibitory control, and that the lack of cerebral dominance could lead to motor incoordination, visual disorientation, speech defects, and more (Liepmann, 1908). Ambilaterality was even associated with mental deficiencies, psychoses, reading disabilities, and psychopathic conditions (Smith and Akelaitis, 1942). It was through this theoretical lens that Akelaitis studied a number of these patients on tests of motor skills, language, sensory processing, and general cognitive abilities, with the hypothesis that callosotomy should lead to severe impairments of the submissive or nondominant hemisphere. Remarkably, other than some initial fine motor discoordination between the two hands that recovered soon after, he found no behavioral effects. As Akelaitis wrote in 1942, "At the outset, we must distinguish the general concept of the lateral cerebral control of motor function from the hypothesis that the cerebral dominance (dictatorship of one hemisphere over the other) is responsible for the laterality of function. The latter belief is clearly disproved by the present results" (Smith and Akelaitis, 1942, p. 542).

In the 1960s, Roger Sperry collaborated with neurosurgeon Joseph Bogen and graduate student Michael Gazzaniga to study a new cohort of split-brain patients. Unlike the studies carried out by Akelaitis in the 1940s, Gazzaniga introduced innovative lateralized procedures that isolated the input and output to individual hemispheres. This pioneering research led to a new understanding of cerebral dominance and major discoveries in hemispheric asymmetries and specializations. As a result of these discoveries, Roger Sperry was awarded the Nobel Prize in 1981.

Their studies revealed that certain functional specializations, such as language production, predominantly resided in the left hemisphere (generally speaking). Conversely, other specialized functions, like spatial constructions, were more prevalent in the right hemisphere (Sperry et al., 1969). Their findings also indicated, for the first time, that surgically severing the corpus callosum led to a profound "functional disengagement of the right and left hemispheres with respect to nearly all cognitive and other psychic activities" (p. 273). Indeed, they did not find a "dictatorial" hemisphere and a more submissive counterpart, but two hemispheres that continued to function normally and independently in areas like learning, memory, sensation, perception, ideation, and other mental activities.

Yet, Sperry and colleagues also noted that "the discrepancy between the enormous size and strategic position of the corpus callosum on the one hand and the observed lack of any important functional disturbance following its complete surgical section on the other remained during this period one of the more puzzling enigmas of neurology" (Sperry et al., 1969, p. 273). That is, despite this profound "disengagement" between the hemispheres produced by callosotomy, these patients appear quite normal. This enigma, first reported by Sperry and colleagues in 1969 in the *Handbook of Clinical Neurology (Volume 4)*, continues to this day, almost 55 years later, as we report in this chapter, part of the third series of the *Handbook of Clinical Neurology*.

THE FIRST REPORTS OF A DISCONNECTION EFFECT

The first report of the disconnection effect as a result of callosotomy was published by Gazzaniga, Bogen, and Sperry in 1962. Their patient was a right-handed, 48-year-old male, W.J., with an I.Q. of 113 and no significant disturbances in sensory, motor, or associative processes. His surgical resection included the entire corpus callosum as well as the anterior and hippocampal commissures. The surgery successfully reduced his seizures and the behavioral testing was conducted 6–20 weeks after surgery either at his home or in the lab (Gazzaniga et al., 1962). Up until this report, the absence of any observable disconnection effects in callosotomy patients could be attributed to the lack of lateralized testing procedures.

Some of the procedures Gazzaniga and colleagues introduced are simple bedside tests that can be conducted in any clinical setting. For example, a stereoagnosia test revealed that W.J., blindfolded, could not name common objects like a pencil, cigarette, ring, pistol, hat, glasses, etc. in his left hand (sensed by his right hemisphere only), but he could manipulate and use those objects correctly. When those same objects were placed in his right hand (sensed by his speaking left hemisphere only), he could name them easily.

In a topognosia test, the experimenters touched a point on a finger of one of W.J.'s hands, again blindfolded, and W.J. could easily locate that site with the thumb of the same hand. He was completely unable, however, to cross-locate the correct finger with the thumb of his other hand, suggesting that the relevant sensory input could not be transferred across hemispheres.

Gazzaniga also introduced a new device that tachistoscopically presented visual stimuli laterally to one visual field or the other for a very short duration of 110 ms. They also carefully watched the gaze of W.J. to ensure that he was fixated at the center of the screen during the presentations. This effectively isolated the visual input to one hemisphere or the other. First, they found that W.J. had profound agnosia to anything presented to his left visual field (corresponding to the right hemisphere). Next, they found, using simple geometric designs, that either hand could accurately and reliably retrieve a corresponding figure to the visual stimuli presented in the same visual field (i.e., the left hand responding to the left visual field, both controlled and sensed by the right hemisphere, and vice versa), but neither hand could correctly retrieve the correct corresponding figure from the visual field opposite of the hand.

They also reported on the motor coordination of W.J. and noted that the left hand was quite "adept and dexterous" in common activities such as smoking a cigarette, lifting a cup, putting on glasses, etc. and that he could also coordinate the activity of both hands normally to do habitual tasks such as tying a knot on the belt of his robe, folding clothes, and putting on and removing clothing. Typically, callosotomy patients have difficulty with these two-handed tasks immediately after their surgeries, but they recover quite quickly. Gazzaniga found, however, that if W.J. was interrupted during these tasks and was asked to mimic the actions using his left hand only, his hand would often fail to do it. Moreover, intriguingly, his left hand and right hand were sometimes in conflict with each other. The report noted that W.J.'s wife described him reaching out with his right hand to read a newspaper but his left hand would put it back down, only to have his right hand try to pick it up again. As Gazzaniga and colleagues reported, "there were further indications that the separated hemispheres were each unaware of activity going on in the other in the case of those functions that are highly lateralized" (Gazzaniga et al., 1962, p. 1767).

Around the same time, a similar report came from Geschwind and Kaplan (1962) on a different patient, P.J.K., a 41-year-old male. The critical difference between the patients, however, was the impact on the corpus callosum. Instead of a surgical resection due to epilepsy, this patient's corpus callosum was impacted by a tumor from the left hemisphere. An undetermined amount of the corpus callosum was removed along with a portion of the left medial frontal lobe. Geschwind and Kaplan reported some similar disconnection effects: for example, when they placed an object or letter in a hand behind P.J.K.'s back, he was unable to select it correctly from a group of objects with his other hand. P.J.K., however, also had a number of other deficits including dense right hemiplegia and aphasia, likely resulting from damage to his left hemisphere.

These reports were the first to clearly document the role of the corpus callosum in integrating information across the corpus callosum in adult human patients.

CLINICAL PERSPECTIVES OF THE BEHAVIORAL OUTCOME OF CALLOSOTOMY

From this point on, there is a clear divergence in the reporting of the disconnection effect between experimental neuropsychologic studies, which normally utilize lateralized testing procedures (see the next section), and clinical reports that mainly focus on the long-term outcomes of seizure activity. As mentioned earlier, Akelaitis in the 1940s helped to alleviate the initial concern that callosotomy may lead to severe clinical deficits in sensory, motor, and cognitive processing. Instead, negligible differences were observed between pre- and postsurgical clinical examinations (Akelaitis, 1941), and soon after surgery, patients were able to walk and run while avoiding obstacles, swim, dance, and even play the piano. In fact, no neurologic deficits were routinely noted on clinical neurologic examinations (Bogen and Vogel, 1975). This somewhat surprising absence of a clear pattern of distinct clinical deficits even leads to speculations that the corpus callosum may hardly support any "psychologic function" (Tomasch, 1954, 1957) or "mental performances" (Fessard, 1954).

Conversely, various reports of clinical deficits observed immediately or shortly after callosotomy challenge this view. Following these accounts, patients exhibit neurologic symptoms ranging from weakness and apraxia of the left arm to bilateral Babinski responses (indicating damage to the pyramidal tracts) and mutism (cf. Zaidel et al., 2003). Even before the advent of callosotomy as a therapeutic approach for epilepsy patients, specific neurologic deficits had been described to result from callosal damage resulting from tumors or stroke. Liepmann and Maas (1907) described *callosal apraxia*, which comprises left-limb apraxia and left-hand agraphia (in the absence of aphasia) occurring after callosal damage in right-handed patients. An even more intriguing disturbance of hand motor control called *alien-hand syndrome* was first described in patients suffering from tumors of the corpus callosum (Brion and Jedynak, 1972), which has also been reported to occur after callosotomy. Ferguson et al. (1985) reported the case of patient P.O.V, who experienced trouble controlling her hands in various everyday situations. For example, when trying to pick out clothes from the closet, P.O.V described that while her right hand was fetching what she intended to wear, her left hand would grab another item she did not want to wear. Even worse, once her left hand was holding this competing item, she could not get her left hand to put it down and regularly had to call her daughter to help her resolve the situation (Ferguson et al., 1985). Such *alien-hand* or *anarchic-hand* syndromes may sometimes be hard to distinguish from *intermanual conflict* in which both hands act at cross purposes. This was strikingly described for patient P.Y., who tried buttoning up his shirt with his right hand while his left hand was simultaneously undoing the work of the right (Bogen and Vogel, 1975). Of note, however, such deficits are usually described as present immediately after surgery and often resolve within the ensuing weeks and months.

In summary, clinical neurologic deficits resulting from callosotomy are typically evident shortly after surgery and subside rather quickly. Hence, examining a split-brain patient months or years after surgery is likely to produce few to no abnormal findings when adhering to standard neurologic examination techniques. In other words, callosotomy patients may be hard to identify clinically, and "a pattern of sophisticated tests specifically designed to identify a commissurotomy patient would be needed for anyone to know the operation had occurred" (Deutsch and Springer, 1989). In particular, specific deficits such as callosal apraxia or the alien-hand syndrome are uncommon and usually subside after several weeks at most. This clinical impression is in stark contrast to the disconnection syndrome which becomes apparent during specified lateralized neuropsychologic testing. Given that deficits resulting from callosotomy can be extremely subtle in everyday social contexts, dedicated testing of split-brain patients is paramount to avoid overlooking the behavioral implications of sectioned interhemispheric fibers.

IMPORTANT QUALIFICATIONS REGARDING THE COMPLETE CORPUS CALLOSOTOMY

Before we delve into the experimental perspective on the behavioral outcome of full split-brain patients, it is important to note some important qualifications. Namely, the type of disconnection effect we report here is the almost complete disengagement of neocortically driven processing from one hemisphere to the other, which can easily be observed using lateralized testing procedures.

Additionally, unlike patient W.J., most full corpus callosotomies do not include severing the anterior commissure. As aforementioned, the anterior commissure is the only other commissure that projects axons to neocortical regions (LaMantia and Rakic, 1990a). In Rhesus monkeys, the anterior commissure contains almost 18 times fewer fibers than the corpus callosum (LaMantia and Rakic, 1990b), and in humans, it is about 1% the size of the corpus callosum (Foxman et al., 1986). Nevertheless, while diminished in capacity, the anterior commissure is still a significant crossing of neocortical fibers (LaMantia and Rakic, 1994), primarily extending to the anterior portion of the temporal lobe and toward the orbitofrontal cortex (Patel et al., 2010), and it has been shown to contribute to interhemispheric functional connectivity between specific brain regions (Hung et al., 2019). Despite this, severing the corpus callosum alone (sparing the anterior commissure) still creates an almost complete disconnection of communication between the hemispheres in the human brain (Gazzaniga, 2000; Zaidel et al., 2003; Gazzaniga, 2005). The functionality of the anterior commissure in humans is therefore still poorly understood; however, it is, from an evolutionary and comparative perspective, the most ancient of the forebrain commissures and has been linked to olfaction in rats and other species (Fenlon et al., 2021), as well as conveying visual information in monkeys (Jouandet and Gazzaniga, 1979)—although this is not the case in humans (Gazzaniga et al., 1965; Funnell et al., 2000b).

It has also been found over the years that certain low-level information can be shared and integrated across the hemispheres without the corpus callosum, most likely through subcortical projections. Corballis described two visual systems with different fates after sectioning of the complete corpus callosum (Corballis, 1995). One is high-resolution, dedicated to the analysis of form and to the exploration of scenes. The integration of this visual system is thought to rely on the corpus callosum and hence is completely disconnected by a callosotomy. The other visual system is of low resolution, is dedicated to the processing of movement and location. This system integrates information subcortically and is relatively unaffected by callosotomy. Along those lines, Holtzman and Gazzaniga (1982) found that there is a central repository of attentional resources that are shared between the hemispheres that do not depend on the corpus callosum (see also Reuter-Lorenz et al., 1996). Emotional valence and pain perception are also known to travel from one hemisphere to the other subcortically (Stein et al., 1989), as does the temporal coordination of bimanual movements. Spatial coordination, however, is disrupted by callosotomy (Franz et al., 1996).

Corpus callosotomies are often conducted in two phases. In the first phase, the surgeon will sever the anterior or posterior portion of the callosum and then determine if that was enough to reduce the seizures. If not, they will perform a second phase of the surgery in which they will complete the resection. The disconnection effect on these partial callosotomies will greatly depend on their location. The corpus callosum has a topographic organization that closely follows the rostral-caudal distribution of the functional areas of the hemispheres (De León Reyes et al., 2020). The early interpretations of the effects of a partially intact callosum were based on lesions that encroached on portions of the callosum. However, these interpretations were often confounded by extracallosal damage to either hemisphere (Gordon et al., 1971). Later, Gazzaniga and Freedman (1973) discovered that a patient with a relatively small section of the splenium (posterior callosum) left intact could transfer visual information but not tactile information, while a patient with a posterior section through the splenium (leaving a region just anterior to the splenium intact) could transfer tactile information but not visual. More systematic studies of patients that went through stages of callosotomy came later: while a small sectioning of the splenium disrupts the naming of visual stimuli lateralized to the right hemisphere, a larger sectioning of the posterior callosum will also disrupts sensorimotor integration and tactile naming. At the other end, an anterior section of the callosum will not disrupt tasks that require interhemispheric sensory integration, but it may interrupt verbally directed sensorimotor tasks (Volpe et al., 1982) and the transfer of semantic information (Sidtis et al., 1981) (see Fig. 10.1).

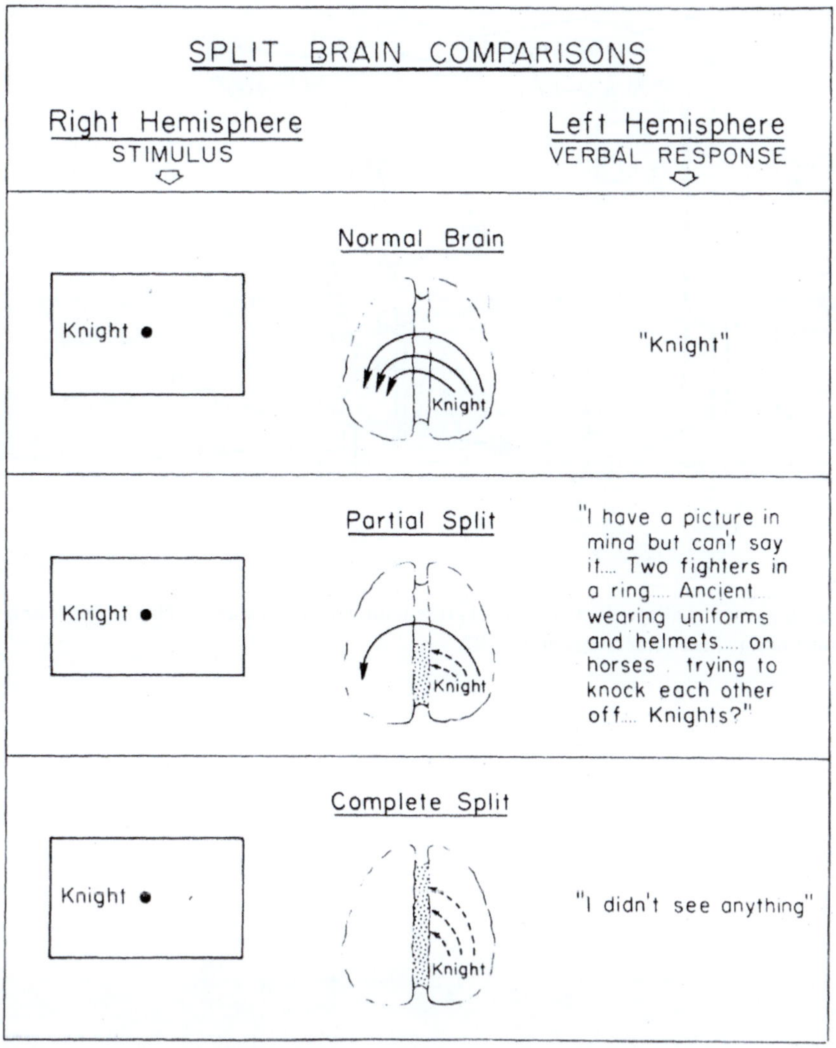

Fig. 10.1. A depiction of Patient J.W.'s ability to verbally report stimuli briefly presented to the right hemisphere before any surgery (top), after a partial surgical resection of the posterior callosum (middle), and after a second resection that completely severed the corpus callosum (bottom) (Sidtis et al., 1981).

Interestingly, some patients—thought to have had complete callosotomies by clinical assessment—were found later using neuroimaging techniques to have spared callosal fibers that may account for some limited visual integration between the hemispheres (Gazzaniga et al., 1989; Funnell et al., 2000a,b).

EXPERIMENTAL PERSPECTIVES ON THE BEHAVIORAL OUTCOME OF CALLOSOTOMY (OR, VIEWS FROM A VAN)

Going back to the divergent perspectives of the disconnection effect, we now focus on the findings of split-brain studies from experimental researchers. Unlike some of the earlier 20th-century thinking about cerebral dominance and hemispheric submissiveness, laboratory experiments in the last 60 years have demonstrated that in split-brain patients, the hemisphere traditionally considered "nondominant" for language abilities typically functions comparably close to the "dominant" hemisphere in most tasks (Gazzaniga, 2000, 2005). Even in the extreme case of hemispheric lateralization of language, the nondominant hemisphere still has remarkable language abilities (Sperry et al., 1969). Previous testing resulted in the inaccurate assumption that the right hemisphere has no capacity for language, but careful experimentation by Gazzaniga, Bogen, and Sperry revealed that both hemispheres typically have language abilities, while only *speech* is exclusively left-lateralized (Gazzaniga et al., 1965). Additionally, some patients have demonstrated semantic abilities in the right hemisphere (Baynes, 1990) (Fig. 10.2).

When communicating with a split-brain patient, it is important to keep in mind that you are essentially engaging exclusively with the "speaking" left hemisphere. A prime example comes from the pioneering simultaneous concept task led by Gazzaniga (Gazzaniga and LeDoux, 1978). In this task (see Fig. 10.3), two different images of objects were presented simultaneously to the left and right hemispheres using a tachistoscope. Split-brain patient P.S. was asked to point to the most related image with their left and right hands. When his "speaking" left hemisphere was shown a chicken claw and his "silent" right hemisphere was

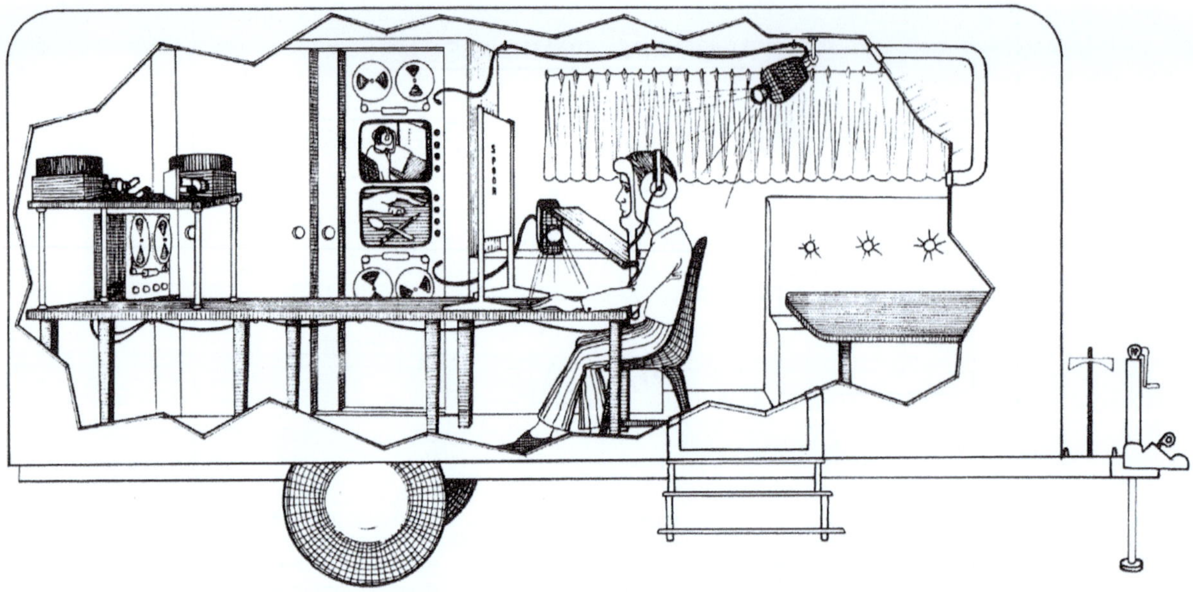

Fig. 10.2. A neuropsychologic testing unit that allowed for lateralized presentation of visual stimuli. This depicts a unit built inside of a van that could be driven to the clinic or the home of a patient (Gazzaniga et al., 1975).

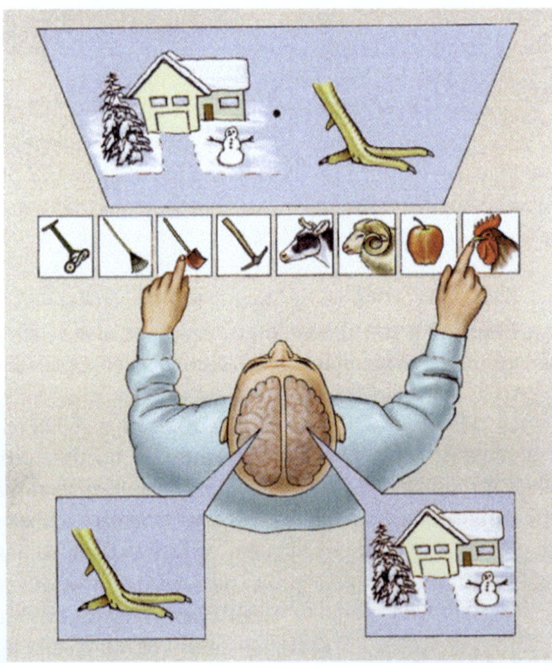

Fig. 10.3. Clinicians may misdiagnose split-brain patients as having a form of hemineglect since they seem unaware of the things presented exclusively to the left visual field. In this study, however, patient P.S. demonstrates that the silent right hemisphere is aware of the snow scene because it points to the snow shovel with his left hand. The speaking left hemisphere then quickly rationalizes the choice of the left hand with an interpretation based on what the left hemisphere saw: a chicken claw.

shown a snow scene, P.S. used his left hand to point to a shovel and his right hand to point to a chicken. His speaking left hemisphere had no conscious access to the snow scene image, yet because he had to explain why the left hand had pointed to the shovel, it justified the selection of the chicken and the shovel by stating, "The chicken claw goes with the chicken, and you need a shovel to clean out the chicken shed." This led to Gazzaniga's theory of the left-brain interpreter—that the left hemisphere is capable of creating remarkable rationalizations for behavior performed by the right hemisphere. When not considering this, clinicians may misdiagnose split-brain patients as having a form of neglect.

As noted previously, the right hemisphere has some unique specializations, including spatial orientation and face processing. One way of assessing spatial abilities uses the Wechsler Adult Intelligence Scale block design test, in which individuals must arrange red and white blocks to match a given pattern. The right hemisphere (which is functionally isolated by asking the patient to only use their left hand) performs this task without any problems, while the left hemisphere (controlling the right hand) is often unable to replicate the target pattern (Bogen and Gazzaniga, 1965). The right hemisphere also demonstrates a propensity for face processing, but this does not imply that the left hemisphere is prosopagnosic. As demonstrated by Kingstone's work, the right hemisphere is not only superior to the left in identifying upright faces, but it also shifts attention automatically in response to somebody else's gaze directions while the left does not (Kingstone et al., 2000).

Various correlational techniques, including fMRI, have also reported on hemispheric asymmetries, but when tested with split-brain subjects it would seem these results are more apparent than real (Miller et al., 2002). For example, one proposed asymmetry involving memory, the hemispheric encoding/retrieval asymmetry (HERA) model, proposes that the left hemisphere encodes episodic information while the right hemisphere is responsible for retrieving it. If this were the case, however, split-brain patients would be effectively amnesic, as their silent right hemisphere would be unable to verbally retrieve encoded memories. As described in Miller et al. (2002), split-brain patients do not demonstrate significant impairments in most memory domains. This suggests that both hemispheres are capable of encoding and retrieval, and therefore memory is not an exclusively lateralized process. Using a task designed to capture the encoding capacity for faces and words, Miller and colleagues found that hemispheric asymmetries in encoding are domain-dependent and correspond with hemispheric preferences for stimulus type. Furthermore, the left hemisphere is superior at encoding semantic information, while the right hemisphere excels at encoding faces.

Hemispheric differences have also been proposed in reasoning—though this may be due to processing preferences as well. Typically, the left hemisphere has a propensity for reducing uncertainty and forming explanations, whereas the right hemisphere is specialized for resolving inconsistencies when provided with new information (Marinsek et al., 2014). To this end, the left hemisphere tends to perseverate incorrect hypotheses and is unable to update, as seen in studies of split-brain patients performing a probability-matching task (Wolford et al., 2000). However, a meta-analysis of fMRI studies on reasoning failed to corroborate this hemispheric specialization model (Turner et al., 2015). Lateralized activations did not match the proposed pattern of left-hemisphere explanation and right-hemisphere monitoring. This highlights current methodological issues when aiming to assess these hemispheric specializations using neuroimaging.

While interpretative reasoning appears to be a critical property of the left hemisphere, the right hemisphere is capable of it as well. This is the case for probability matching—while it was originally believed that only the left hemisphere can successfully probability match, it was later shown that the right hemisphere will also match under certain conditions. In typical probability matching tasks, subjects try to guess which of two random events will happen next, each event having a different probability of occurrence (e.g., a blue stimulus might appear 75% of the time and a yellow one 25% of the time). The two possible strategies for responding to the task are either *matching* or *maximizing*. In the blue-yellow example, frequency matching would involve guessing blue 75% of the time and guessing yellow 25% of the time. Because the order of occurrence was random, this strategy potentially would result in a great number of errors. The second strategy, maximizing, involves simply guessing blue every time. That approach ensures an accuracy rate of 75% because blue will appear 75% of the time. In split-brain patients, the left hemisphere has been shown to match stimulus frequencies, while the right hemisphere tends to maximize (Wolford et al., 2000). A study by Miller and Valsangkar-Smyth, however, found that the right hemisphere of a split-brain patient was able to match the frequency of previous "face-type" occurrences, while the left hemisphere was not (Miller and Valsangkar-Smyth, 2005). This clarifies both hemispheres' abilities to probability match, although importantly, it is dependent on the type of stimulus used.

Various aspects of the perceptual system also demonstrate hemispheric asymmetries. Although illusory contours, defined as perceptually extrapolated lines or shapes not physically present, are equally perceivable by both hemispheres, it is the right hemisphere that predominantly completes amodal boundaries (Corballis et al., 1999). This was determined by having split-brain patients complete an established perceptual task, in which they were flashed "Kanisza rectangles" and were asked to judge whether the rectangle was "fat" or "thin." Kanisza rectangles are a common visual illusion in which four black Pac-Man-like shapes are arranged to create the illusion of a superimposed white square or rectangle (Kanizsa, 1976). These stimuli were nearly identical for the illusory contour and amodal completion tasks, only differing in the boundary completion of the Pac-Man shapes. While the illusory contour stimuli are composed of only four Pac-Man shapes arranged in a particular configuration, the amodal completion stimuli feature four outlined circles with a part of one quadrant missing, contributing to the illusion of a complete square or rectangle formed by the occluded regions. For the illusory contours task, split-brain patients performed equally well between hemifields, whereas they performed significantly better on the amodal boundaries task when stimuli were presented to the left visual field (right hemisphere) relative to the right visual field (left hemisphere). The authors concluded that illusory contours may be completed by low-level visual processes that equally exist in both hemispheres, while amodal completion may require higher-level processing that is much more lateralized (Corballis et al., 1999).

Over the years, it has been argued that higher-order information can be transferred subcortically in split-brain patients (Johnson, 1984; Cronin-Golomb, 1986; Sergent, 1990). As we discussed in the last section, however, there are many ways in which these studies can be confounded hindering straightforward conclusions (see Seymour et al., 1994). An example of this comes from a study conducted by Kingstone and Gazzaniga, in which patient J.W. had one word flashed to his left hemisphere and a different word flashed to his right, and was then asked to draw what he saw. When the left hemisphere was shown the word "ten" and the right hemisphere "clock," J.W. drew a clock and set the time to 10:00. They initially hypothesized that some representation of the word information had to have been transferred subcortically, resulting in the unified response. Further experimentation, however, revealed that each hemisphere can actually control the ipsilateral drawing hand, and subcortical transfer of higher-order information is not likely. This was elucidated through further divided visual field testing, where J.W. was again asked to draw what was presented to him, though the two words could now be shown as a literal pair or combined to create an emergent object. For example, the words "hot" and "dog" could be combined to depict a dog in warm weather, or a frankfurter. Patient J.W. never depicted the emergent object (frankfurter), and instead only drew a literal combination of the two objects, regardless of the hand used to draw. When the two words were both shown to the same hemisphere, J.W. did successfully depict the emergent objects, but only with his right hand (left hemisphere, right visual field presentation). These findings dispute the idea of subcortical transfer and instead support peripheral integration theories and dual-hemisphere control of the hand used to draw the objects (Kingstone and Gazzaniga, 1995; Miller and Kingstone, 2005).

Careful experimentation with split-brain patients can also reveal differences from normal, healthy individuals without using lateralized procedures—the kinds of tests that may not be conducted in standard clinical settings. For example, split-brain patients perform better than controls when asked to execute different motor tasks simultaneously with each hand, such as rubbing their stomach while patting their head, or copying two different shapes concurrently (Franz et al., 1996). Significant impairments, however, are revealed in split-brain patients when asked to perform *novel* bimanual actions. When asked to pantomime actions with varying degrees of familiarity with their eyes closed, they executed unfamiliar actions much less successfully than controls (Franz et al., 2000).

Although careful testing has shown that most information cannot be transmitted between the disconnected hemispheres in split-brain patients, crude spatial location information can (Trevarthen, 1968). Further exploration established that spatial attention can be directed by either hemisphere to either visual field in splits, but spatial attention cannot be divided (Reuter-Lorenz and Fendrich, 1990). Instead, the hemispheres appear to maintain a single resource of attention, regardless of which hemisphere is cued (Holtzman et al., 1981). Subsequent research conducted by Holtzman and Gazzaniga examined whether there is attentional sharing between the hemispheres of split-brain patients. They used a task in which patients viewed three sets of geometric shapes presented to each hemisphere simultaneously, including a difficult condition (same three shapes presented in both visual fields) and an easy condition (three shapes in one visual field, and one shape presented in three times in the other visual field), and were then asked whether a probe shape was presented. They found that split-brain patients were quicker to make their decision when the nonprobed hemisphere had the "easy" task compared to the "difficult" task—again suggesting a shared central attentional resource between the hemispheres (Holtzman and Gazzaniga, 1982).

More recent studies with split-brain patients also reveal interesting differences from normal subjects in moral reasoning. Previous research has implicated the involvement of the right temporal parietal junction (rTPJ) in moral and social reasoning (Young and Saxe, 2009). To test whether this implies hemispheric differences in moral reasoning in callosotomy patients, Miller and colleagues provided split-brain patients and controls with various fictional scenarios and asked them to decide whether the action performed was "permissible" or "forbidden" (Miller et al., 2010). For example, in one scenario, Grace fixes a cup of coffee for her friend. Believing it's sugar, she adds a white powder from a sugar bowl to her friend's coffee. The white powder, however, was not sugar but a toxic substance. Her friend drinks the coffee and dies. Each scenario had three factors: whether the action caused harm or not, whether the agent did or did not believe that the action would cause harm, and whether the agent's belief was revealed before or after the outcome. Furthermore, incongruent outcomes and beliefs constitute two main conditions: accidental harm, and failed attempts (for example, a waitress served a customer something she believed he was allergic to, but he wasn't and turned out to be fine). The subjects were then asked: Was it morally acceptable for Grace to give the coffee to her friend? Control subjects tended to base their judgments on the agent's belief in incongruent conditions (Grace believed the toxic substance was sugar), considering accidental harm as permissible and failed attempts as forbidden. In contrast, split-brain patients based their judgments largely on the outcome, ignoring the agent's belief for incongruent trials and saying that Grace's action was not permissible, while the waitress' was. Given that participants in this task responded verbally, this is likely a result of the speaking left hemisphere's inability to access the rTPJ for moral reasoning. Nonlinguistic morality scenarios were later used to test a split-brain patient in acknowledgment of the right hemisphere's limited language ability. Again, the right hemisphere was clearly focused on intent, indicating that it is "not only necessary but also sufficient for intent-based moral judgement" (Steckler et al., 2017).

COMMUNICATION BETWEEN THE HEMISPHERES OUTSIDE THE CORPUS CALLOSUM

Most signs of the disconnection syndrome readily observable in everyday life, such as *intermanual conflicting* or the *alien-hand syndrome*, have been reported to subside within weeks to months after surgery. Over time, patients typically exhibit increasingly integrated behavior allowing them to control complex actions such as swimming, dancing or even playing the piano (Akelaitis, 1941). However, we also have the aforementioned findings from the last 60 years that clearly demonstrate the existence of two distinct neural systems that can operate independently of each other, with no direct awareness of the processing in the opposite hemisphere. The mechanisms, neural or otherwise, that enable such increasing levels of integrated behavior despite the complete disconnection remain a topic of ongoing scientific debate (de Haan et al., 2020). At least three potential mechanisms mediating interhemispheric interactions between cortical areas seem plausible in the absence of the callosal fibers.

A first potential mechanism that does not rely on direct neural connections between hemispheres lies in *cross-cueing*. Cross-cueing strategies involve the targeted utilization of subtle behavioral cues given by one hemisphere to the other, allowing for complex behaviors that may seem impossible in the absence of direct neural connections (Gazzaniga and Hillyard, 1971; Seymour et al., 1994; Volz and Gazzaniga, 2017). For example, eye movements or movements of bilateral facial muscles can be initiated by either hemisphere to the direct attention of the other or to coordinate action. For cross-cueing to work, each hemisphere must learn to perceive potential cues sent by its contralateral counterpart, which can be assumed to involve considerable adaptation of perceptual systems and hence may gradually improve over time postsurgery. Thus, spatial attention and perceptual processing within the disconnected hemispheres may be assumed to gradually change and adapt after callosotomy. Mechanistically, such a reorganization of functional networks within hemispheres may resemble functional reorganization observed after a stroke. After stroke-induced tissue damage and disconnections of brain regions, numerous changes in functional connectivity have been reported to compensate for the damage to specialized neural tissue and allow the remaining circuitry to recover impaired function (Grefkes and Fink, 2014; Hallett et al., 2020; Hartwigsen and Volz, 2021). In other words, specialized neural networks adapt their functional architectures to contribute to the control of behavior previously performed by circuitry that was damaged by a lesion. Of note, such processes may help to enable cross-cueing between hemispheres via a reorganization of circuitry enabling action, attention, and perception of such cues.

An alternative transfer of information between hemispheres in the absence of callosal fibers may arise from subcortical interhemispheric connections left intact after callosotomy. In seminal early studies, Trevarthen and Sperry (Trevarthen, 1968; Trevarthen and Sperry, 1973) discussed that the integrated behavior observed in split-brain patients may arise from interhemispheric commissural fibers, such as anterior or posterior commissure or intrathalamic fiber bundles (Szczupak et al., 2021) readily transferring information between hemispheres. Yet another alternative subcortical route is offered by fibers projecting from subcortical nuclei to cortical areas in both hemispheres carrying perceptual information, as discussed for the *ambient visual system* (Trevarthen and Sperry, 1973), which includes the superior colliculus that is known to connect between hemispheres (Meredith and Stein, 1986). While the hypothesized information exchange via such subcortical routes certainly exceeds physiologic interhemispheric integration observed in the healthy brain, functional reorganization occurring after surgery may allow cortical signals to be transferred along spared interhemispheric connections. Support for this notion derives from recent studies in stroke patients which highlight that interhemispheric fibers may be used beyond their physiologic role to transfer information between hemispheres, thereby enabling motor recovery (Paul et al., 2023a,b).

A third aspect potentially contributing to unified behavior lies in noncrossing motor fibers innervating the ipsilateral side of the body. Most descending fibers of the main motor output pathway (the corticospinal tract; CST) cross over to innervate and control the contralateral side of the body (Yeo and Jang, 2012), explaining why the contralateral cerebral hemisphere mainly exerts motor control of one side of the body. However, a minority of CST fibers do not cross sides but rather project through the anterior part of the CST ipsilaterally to ultimately innervate ipsilateral proximal muscles (Lemon, 2008; Kwon et al., 2011). Specifically, viral tracing studies have shown that crossing and noncrossing neurons exhibit differential patterns of synaptic connections to spinal motor neurons: while crossing axons feature monosynaptic connections, noncrossing neurons offer a polysynaptic influence over spinal motoneurons (Porter and Lemon, 1995). While under physiologic circumstances, motor control is mainly attributed to crossing fibers, noncrossing fibers may allow each hemisphere to influence motor control of both sides of the body. Similar compensatory properties of these noncrossing descending motor fibers have been reported and discussed after unilateral brain damage, for example, induced by stroke. In particular, after damage to the CST, noncrossing CST fibers in the contralesional (undamaged) hemisphere may help to compensate for the loss of descending signals (Jankowska and Edgley, 2006; Raffin and Hummel, 2018; Paul et al., 2023a,b). Thus, the combination of crossing and non-crossing descending motor fibers originating in each hemisphere may allow each hemisphere to modulate motor control of both sides of the body. While the control of complex hand movements seems far beyond the capacity of noncrossing fiber systems, the modulation of proximal muscle activation may be more than sufficient to help facilitate integrated behavior.

There are, of course, many other potential avenues through which increasing degrees of integrated behavior may emerge. Including, for example, a combination of neuroplastic changes *within* hemispheres, facilitating behavioral cross-cueing and neuroplastic changes allowing for interhemispheric reorganization—which may still rely upon the use of spared subcortical pathways to exchange information between hemispheres (Gazzaniga et al., 1979; Baynes et al., 1995). Additionally, it is well-established that those actions which are *initiated* by the nonspeaking right hemisphere (and *not coordinated* with the left) are quickly rationalized by the speaking left hemisphere (Gazzaniga and LeDoux, 1978). To wit, we once flashed the command "Stand Up" to the right hemisphere of patient J.W., and he promptly stood up: when we asked him, "Why are you standing up," he was quite non-fazed and simply replied, "Oh, I felt like getting a Coke." His left hemisphere instantly came up with an explanation that made sense of an action without knowing what prompted him to stand up in the first place. This interpreter system in the left hemisphere thus enables a superficial appearance of a singular, unitary stream of processing despite the inability of each hemisphere to access the contents of its contralateral counterpart (Gazzaniga et al., 1996).

CONCLUSION

To this day, the clinical view of callosotomy is that it is a treatment of last resort, which is quite effective for the reduction of drop-attack seizures while simultaneously yielding transient and/or minimal impacts on cognition and behavior (despite some rare longer-term symptoms such as alien hand syndrome). From the perspective of the split-brain patient and their family and friends, they appear to be not much different of a person from before the surgery—as put by Zaidel et al. (2003), they are "socially normal." The patients continue to present as having a seemingly unified conscious experience, thanks in large part to a left-hemisphere interpreter that continuously strives to make sense of the world. Yet, remarkably, experimental studies from the last 60 years have convincingly demonstrated something very profound about split-brain patients. The surgical resection of the complete corpus callosum produces two neural systems, essentially two *minds*, that operate normally and independently, without direct awareness or knowledge of the processing occurring in the other hemisphere. And, despite this, within weeks or months of the surgery, the two hemispheres find a way to synchronize and coordinate their actions, through cross-cueing and perhaps via neuroplastic reorganization as well. The study of split-brain patients not only offers entirely unique insights into hemispheric specializations and cerebral asymmetries, but it also offers a window into the nature of consciousness and what it means to be human.

ACKNOWLEDGMENTS

The authors would like to thank Tyler Santander, and Becky Gazzaniga for their references, edits, and comments. The research was sponsored by the US Army Research Office and accomplished under contract W911NF-19-D-0001 for the Institute for Collaborative Biotechnologies.

References

Aboitiz F, Montiel J (2003). One hundred million years of interhemispheric communication: the history of the corpus callosum. Braz J Med Biol Res 36: 409–420.

Akelaitis AJ (1941). Studies on the corpus callosum: II. The higher visual functions in each homonymous field following complete section of the corpus callosum. Arch Neurol Psychiatry 45: 788–796.

Akelaitis AJ (1943). Studies on the corpus callosum: VII. Study of language functions (tactile and visual lexia and graphia) unilaterally following section of the corpus callosum. J Neuropathol Exp Neurol 2: 226–262.

Akelaitis AJ (1945). Studies on the corpus callosum: IV. Diagonistic dyspraxia in epileptics following partial and complete section of the corpus callosum. Am J Psychiatry 101: 594–599.

Akelaitis AJ, Risteen WA, Herren RY et al. (1942). Studies on the corpus callosum: III. A contribution to the study of dyspraxia and apraxia following partial and complete section of the corpus callosum. Arch Neurol Psychiatry 47: 971–1008.

Baynes K (1990). Language and reading in the right hemisphere: highways or byways of the brain? J Cogn Neurosci 2: 159–179.

Baynes K, Wessinger CM, Fendrich R et al. (1995). The emergence of the capacity to name left visual field stimuli in a callosotomy patient: implications for functional plasticity. Neuropsychologia 33: 1225–1242.

Bogen JE, Gazzaniga MS (1965). Cerebral commissurotomy in man: minor hemisphere dominance for certain visuospatial functions. J Neurosurg 23: 394–399.

Bogen JE, Vogel PJ (1975). Neurologic status in the long term following cerebral commissurotomy. In: F Michel, B Schott (Eds.), Les syndromes de disconnexion calleuse chez l'homme. Hopital Neurologique, Lyon, pp. 227–251.

Brion S, Jedynak CP (1972). Disorders of interhemispheric transfer (callosal disconnection). 3 cases of tumor of the corpus callosum. The strange hand sign. Rev Neurol 126: 257–266.

Corballis MC (1995). Visual integration in the split brain. Neuropsychologia 33: 937–959.

Corballis MC (2009). The evolution and genetics of cerebral asymmetry. Philos Trans R Soc Lond B Biol Sci 364: 867–879.

Corballis PM, Fendrich R, Shapley R et al. (1999). Illusory contours and amodal completion: evidence for a functional dissociation in callosotomy patients. J Cogn Neurosci 11: 459–466.

Corballis MC, Badzakova-Trajkov G, Haberling IS (2012). Right hand, left brain: genetic and evolutionary bases of cerebral asymmetries for language and manual action. Wiley Interdiscip Rev Cogn Sci 3: 1–17.

Cronin-Golomb A (1986). Subcortical transfer of cognitive information in subjects with complete forebrain commissurotomy. Cortex 22: 499–519.

De Benedictis A et al. (2016). New insights in the homotopic and heterotopic connectivity of the frontal portion of the human corpus callosum revealed by microdissection and diffusion tractography. Hum Brain Mapp 37: 4718–4735.

de Haan EH, Corballis PM, Hillyard SA et al. (2020). Split-brain: what we know now and why this is important for understanding consciousness. Neuropsychol Rev 30: 224–233.

De León Reyes NS, Bragg-Gonzalo L, Nieto M (2020). Development and plasticity of the corpus callosum. Development 147: dev189738.

Deutsch G, Springer SP (1989). Left brain, right brain, WH Freeman.

Fenlon LR, Suarez R, Lynton Z et al. (2021). The evolution, formation and connectivity of the anterior commissure. In: Seminars in cell & developmental biology, vol. 118. Academic Press50–59.

Ferguson SM, Rayport M, Corrie WS (1985). Neuropsychiatric observations on behavioral consequences of corpus callosum section for seizure control. In: Epilepsy and the corpus callosum, 501–514.

Fessard A (1954). Mechanisms of nervous integration and conscious experience. In: JF Delafresnaye (Ed.), Brain mechanisms and consciousness. CC Thomas, Springfield, IL, pp. 200–236.

Foxman BT et al. (1986). Proportional anterior commissure area in humans and monkeys. Neurology 36: 1513–1517.

Franz EA, Eliassen JC, Ivry RB et al. (1996). Dissociation of spatial and temporal coupling in the bimanual movements of callosotomy patients. Psychol Sci 7: 306–310.

Franz EA, Waldie KE, Smith MJ (2000). The effect of callosotomy on novel versus familiar bimanual actions: a neural dissociation between controlled and automatic processes? Psychol Sci 11: 82–85.

Funnell MG, Corballis PM, Gazzaniga MS (2000a). Cortical and subcortical interhemispheric interactions following partial and complete callosotomy. Arch Neurol 57: 185–189.

Funnell MG, Corballis PM, Gazzaniga MS (2000b). Insights into the functional specificity of the human corpus callosum. Brain 123: 920–926.

Gazzaniga MS (2000). Cerebral specialization and interhemispheric communication: does the corpus callosum enable the human condition? Brain 123: 1293–1326.

Gazzaniga MS (2005). Forty-five years of split-brain research and still going strong. Nat Rev Neurosci 6: 653–659.

Gazzaniga MS, Freedman H (1973). Observations on visual processes after posterior callosal section. Neurology 23: 1126–1130.

Gazzaniga MS, Hillyard SA (1971). Language and speech capacity of the right hemisphere. Neuropsychologia 9: 273–280.

Gazzaniga MS, LeDoux JE (1978). The integrated mind, Plenum Press.

Gazzaniga MS, Bogen JE, Sperry RW (1962). Some functional effects of sectioning the cerebral commissures in man. Proc Natl Acad Sci U S A 48: 1765–1769.

Gazzaniga MS, Bogen JE, Sperry RW (1965). Observations on visual perception after disconnexion of the cerebral hemispheres in man. Brain 88: 221–236.

Gazzaniga MS, Risse GL, Springer SP et al. (1975). Psychologic and neurologic consequences of partial and complete cerebral commissurotomy. Neurology 25: 10–15.

Gazzaniga MS, Volpe BT, Smylie CS et al. (1979). Plasticity in speech organization following commissurotomy. Brain 102: 805–815.

Gazzaniga MS, Kutas M, Van Petten C et al. (1989). Human callosal function: MRI-verified neuropsychological functions. Neurology 39: 942–946.

Gazzaniga MS, Eliassen JC, Nisenson L et al. (1996). Collaboration between the hemispheres of a callosotomy patient: emerging right hemisphere speech and the left hemisphere interpreter. Brain 119: 1255–1262.

Geschwind N, Kaplan E (1962). A human cerebral deconnection syndrome. Neurology 12: 675–685.

Gordon HW, Bogen JE, Sperry RW (1971). Absence of deconnexion syndrome in two patients with partial section of the neocommissures. Brain 94: 327–336.

Grefkes C, Fink GR (2014). Connectivity-based approaches in stroke and recovery of function. Lancet Neurol 13: 206–216.

Hallett M, de Haan W, Deco G et al. (2020). Human brain connectivity: clinical applications for clinical neurophysiology. Clin Neurophysiol 131: 1621–1651.

Hartwigsen G, Volz LJ (2021). Probing rapid network reorganization of motor and language functions via neuromodulation and neuroimaging. Neuroimage 224: 117449.

Holtzman JD, Gazzaniga MS (1982). Dual task interactions due exclusively to limits in processing resources. Science 218: 1325–1327.

Holtzman JD, Sidtis JJ, Volpe BT et al. (1981). Dissociation of spatial information for stimulus localization and the control of attention. Brain 104: 861–872.

Hung SC et al. (2019). Early recovery of interhemispheric functional connectivity after corpus callosotomy. Epilepsia 60: 1126–1136.

Jankowska E, Edgley SA (2006). How can corticospinal tract neurons contribute to ipsilateral movements? A question with implications for recovery of motor functions. Neuroscientist 12: 67–79.

Jarbo K, Verstynen T, Schneider W (2012). In vivo quantification of global connectivity in the human corpus callosum. Neuroimage 59: 1988–1996.

Johnson LE (1984). Bilateral visual cross-integration by human forebrain commissurotomy subjects. Neuropsychologia 22: 167–175.

Jouandet ML, Gazzaniga MS (1979). Cortical field of origin of the anterior commissure of the rhesus monkey. Exp Neurol 66: 381–397.

Kanizsa G (1976). Subjective contours. Sci Am 234: 48–53.

Kim DS et al. (2004). The surgical effect of callosotomy in the treatment of intractable seizure. Yonsei Med J 45: 233–240.

Kingstone A, Gazzaniga MS (1995). Subcortical transfer of higher order information: more illusory than real? Neuropsychology 9: 321.

Kingstone A, Friesen CK, Gazzaniga MS (2000). Reflexive joint attention depends on lateralized cortical connections. Psychol Sci 11: 159–166.

Kwon YH, Son SM, Lee J et al. (2011). Combined study of transcranial magnetic stimulation and diffusion tensor tractography for prediction of motor outcome in patients with corona radiata infarct. J Rehabil Med 43: 430–434.

LaMantia AS, Rakic P (1990a). Cytological and quantitative characteristics of four cerebral commissures in the rhesus monkey. J Comp Neurol 291: 520–537.

LaMantia AS, Rakic P (1990b). Axon overproduction and elimination in the Corpus callosum of the developing Rhesus monkey. J Neurosci 10: 2156–2175.

LaMantia AS, Rakic P (1994). Axon overproduction and elimination in the anterior commissure of the developing rhesus monkey. J Comp Neurol 340: 328–336.

Lemon RN (2008). Descending pathways in motor control. Annu Rev Neurosci 31: 195–218.

Liepmann H (1908). Die linke Hemisphäre und das Handeln. 1. In: Drei Aufsätze aus dem Apraxiegebiet, Karger Publishers17–50.

Liepmann H, Maas O (1907). Case of left-sided agraphia and apraxia with right-sided paralysis. J Psychol Neurol 10: 214–227.

Luders E, Thompson PM, Toga AW (2010). The development of the corpus callosum in the healthy human brain. J Neurosci 30: 10985–10990.

Marinsek N, Turner BO, Gazzaniga M et al. (2014). Divergent hemispheric reasoning strategies: reducing uncertainty versus resolving inconsistency. Front Hum Neurosci 8: 839. https://doi.org/10.3389/fnhum.2014.00839.

Meredith MA, Stein BE (1986). Visual, auditory, and somatosensory convergence on cells in superior colliculus results in multisensory integration. J Neurophysiol 56: 640–662.

Miller MB, Kingstone A (2005). Taking the high road on subcortical transfer. Brain Cogn 57: 162–164.

Miller MB, Valsangkar-Smyth M (2005). Probability matching in the right hemisphere. Brain Cogn 57: 165–167.

Miller MB, Kingstone A, Gazzaniga MS (2002). Hemispheric encoding asymmetry is more apparent than real. J Cogn Neurosci 14: 702–708.

Miller MB, Sinnott-Armstrong WA, Young L et al. (2010). Abnormal moral reasoning in complete and partial callosotomy patients. Neuropsychologia 48: 2215–2220.

Orton ST (1937). Reading, writing and speech problems in children, W. W. Norton & Company, Inc, New York.

Patel MD et al. (2010). Distribution and fibre field similarity mapping of the human anterior commissure fibres by diffusion tensor imaging. MAGMA 23: 399–408.

Paul T, Cieslak M, Hensel L et al. (2023a). The role of corticospinal and extrapyramidal pathways in motor impairment after stroke. *Brain*. Communications 5: fcac301.

Paul T, Wiemer VM, Hensel L et al. (2023b). Interhemispheric structural connectivity underlies motor recovery after stroke. Ann Neurol 94: 785–797.

Porter R, Lemon R (1995). Synthesis: corticospinal function and voluntary movement. In: Corticospinal function and voluntary movement, Oxford University Press348–366.

Raffin E, Hummel FC (2018). Restoring motor functions after stroke: multiple approaches and opportunities. Neuroscientist 24: 400–416.

Reuter-Lorenz PA, Fendrich R (1990). Orienting attention across the vertical meridian: evidence from callosotomy patients. J Cogn Neurosci 2: 232–238.

Reuter-Lorenz PA, Jha AP, Rosenquist JN (1996). What is inhibited in inhibition of return. J Exp Psychol Hum Percept Perform 22: 367.

Sergent J (1990). Furtive incursions into bicameral minds: integrative and coordinating role of subcortical structures. Brain 113: 537–568.

Seymour SE, Reuter-Lorenz PA, Gazzaniga MS (1994). The disconnection syndrome: basic findings reaffirmed. Brain 117: 105–115.

Sidtis JJ, Volpe BT, Holtzman JD et al. (1981). Cognitive interaction after staged callosal section: evidence for transfer of semantic activation. Science 212: 344–346.

Smith KU, Akelaitis AJ (1942). Studies on the corpus callosum: I. Laterality in behavior and bilateral motor organization in man before and after section of the corpus callosum. Arch Neurol Psychiatry 47: 519–543.

Sperry RW, Gazzaniga MS, Bogen JE (1969). Interhemispheric relationships: the neocortical commissures; syndromes of hemisphere disconnection. In: PJ Vinken, GW Bruyn (Eds.), Handbook of clinical neurology. North-Holland Publishing Company/John Wiley and Sons, Inc, Amsterdam/New York, pp. 273–290.

Steckler CM, Hamlin K, Miller MB et al. (2017). Moral judgment by the disconnected left and right hemispheres: a split-brain investigation. R Soc Open Sci 4: 170172.

Stein BE, Price DD, Gazzaniga MS (1989). Pain perception in a man with total corpus callosum transection. Pain 38: 51–56.

Szczupak D, Iack PM, Liu C et al. (2021). Direct interhemispheric cortical communication via thalamic commissures: a new white-matter pathway in the rodent brain. Cereb Cortex 31: 4642–4651.

Szczupak D, Iack PM, Rayêe D et al. (2023). The relevance of heterotopic callosal fibers to interhemispheric connectivity of the mammalian brain. Cereb Cortex 33: 4752–4760.

Tomasch J (1954). Size, distribution, and number of fibres in the human corpus callosum. Anat Rec 119: 119–135.

Tomasch J (1957). A quantitative analysis of the human anterior commissure. Acta Anat 30: 902–906.

Travis LE (1931). Speech pathology; a dynamic neurological treatment of normal speech and speech deviations, D. Appleton and Company, New York.

Trevarthen CB (1968). Two mechanisms of vision in primates. Psychol Forsch 31: 299–337.

Trevarthen C, Sperry RW (1973). Perceptual unity of the ambient visual system in human commissurotomy patients. Brain 96: 547–570.

Turner BO, Marinsek N, Ryhal E et al. (2015). Hemispheric lateralization in reasoning. Ann N Y Acad Sci 1359: 47–64.

Van Wagenen WP, Herren RY (1940). Surgical division of commissural pathways in the corpus callosum: relation to spread of an epileptic attack. Arch Neurol Psychiatry 44: 740–759.

Volpe BT, Sidtis JJ, Holtzman JD et al. (1982). Cortical mechanisms involved in praxis: observations following partial and complete section of the corpus callosum in man. Neurology 32: 645–650.

Volz LJ, Gazzaniga MS (2017). Interaction in isolation: 50 years of insights from split-brain research. Brain 140: 2051–2060.

Wilson DH, Reeves AG, Gazzaniga MS (1982). "central" commissurotomy for intractable generalized epilepsy: series two. Neurology 32: 687–697.

Wolford GL, Miller MB, Gazzaniga MS (2000). The left hemisphere's role in hypothesis formation. J Neurosci 20: 1–4.

Yeo SS, Jang SH (2012). Differences between the somatotopic corticospinal tract for the fingers and toes in the human brain. NeuroRehabilitation 31: 395–399.

Young L, Saxe R (2009). Innocent intentions: a correlation between forgiveness for accidental harm and neural activity. Neuropsychologia 47: 2065–2072.

Zaidel E, Iacoboni M, Zaidel DW et al. (2003). The callosal syndromes. Clin Neuropsychol 4: 347–403.

Section IV

Development and lifespan

Handbook of Clinical Neurology, Vol. 208 (3rd series)
Cerebral Asymmetries
C. Papagno and P. Corballis, Editors
https://doi.org/10.1016/B978-0-443-15646-5.00004-X

Chapter 11

Hemispheric organization of the brain and its prevailing impact on the neuropsychology of aging

SARA B. FESTINI[1*], GRACE KEGLER[1], AND PATRICIA A. REUTER-LORENZ[2]

[1]*University of Tampa, Tampa, FL, United States*
[2]*University of Michigan, Ann Arbor, MI, United States*

Abstract

Age differences in brain hemispheric asymmetry have figured prominently in the neuropsychology of aging. Here, a broad overview of these empirical and theoretical approaches is provided that dates back to the 1970s and continues to the present day. Methodological advances often brought new evidence to bear on older ideas and promoted the development of new ones. The deficit-focused hypothesis of accelerated right-hemisphere aging is reviewed first, followed by subsequent accounts pertaining to compensation, reserve, and their potential hemispheric underpinnings. Structural and functional neuroimaging reveal important and consistent age-related patterns, including indications of reduced brain asymmetry in older relative to younger adults. While not mutually exclusive, different neuropsychologic theories of aging offer divergent interpretations of such patterns, including age-related reductions in neural specificity (dedifferentiation) and age-related compensatory bilateral recruitment [e.g., Hemispheric Asymmetry Reduction in Older Adults (HAROLD); Compensation-Related Utilization of Neural Circuits Hypothesis (CRUNCH)]. Further, recent neurobehavioral evidence suggests that the right hemisphere plays a unique role in resisting the neurocognitive effects of aging via brain reserve. Future advances in human cognitive neuroscience, including neurostimulation methods for targeted interventions, along with analytic techniques informed by machine learning promise new insights into the neuropsychology of aging and the role of hemispheric processes in resilience and decline.

INTRODUCTION

In many respects, the brain is highly symmetrical. A longitudinal fissure separates the right and left cerebral hemispheres that are approximately equal in size (see Renteria, 2012). The brain receives input from symmetrically configured sensory structures (e.g., eyes, ears; Corballis, 2017). Additionally, our brain's hemispheres control somatosensation and movement for the opposite or contralateral side of the body (Roux et al., 2018, 2020), supporting the performance of many movements that require the symmetrical coordination of the whole body in action (e.g., walking, crawling, climbing; Corballis, 2017).

Nevertheless, decades of research document that rather than being represented symmetrically, the left and right hemispheres contribute asymmetrically to many cognitive abilities (e.g., Herve et al., 2013; Karolis et al., 2019), some of which are lateralized predominantly to one hemisphere of the brain or the other. While the foundations of functional lateralization are present at birth, the asymmetrical organization that characterizes the adult brain develops throughout childhood and adolescence (e.g., Hiscock and Kinsbourne, 2008; Olulade et al., 2020). Later adulthood and older age bring a myriad of changes in brain structure and function, including alterations in functional lateralization (e.g., Reuter-Lorenz, 2000; Dolcos et al., 2002; Cabeza et al., 2018). Indeed, the hemispheric organization of the brain and its potential alterations in later life have figured prominently in neuropsychologic theories of aging for nearly 50 years. Here, we review key evidence pertaining to age differences in functional brain asymmetry and offer a broad history of theoretical perspectives related to healthy aging and cerebral hemispheric organization. All in all, younger adults tend to show more hemispheric specialization than their older counterparts. Age-related differences in brain asymmetry are associated with greater behavioral symmetry in older adults in some performance domains (e.g., dichotic listening, motor tasks), whereas in other domains, the implications for age differences in performance are varied, difficult to link to neural underpinnings, and dependent on various factors, such as individual differences and task demands.

*Correspondence to: Sara B. Festini, University of Tampa, Tampa, FL, United States. Tel: +1 (813) 257-3677, E-mail: sfestini@ut.edu.

Lateralization of brain function

Although the entire brain contributes to most daily tasks that require multiple concurrent cognitive and motoric processes, evidence indicates that each cerebral hemisphere plays a dominant role in different types of cognitive abilities. In particular, the left cerebral hemisphere is typically dominant for fine motor control and language processing (e.g., Corballis, 1989; Corballis and Haberling, 2017), whereas the right cerebral hemisphere is typically dominant for many nonverbal abilities, including visuospatial tasks, face processing, spatial attention, and aspects of emotion processing (e.g., Corballis and Haberling, 2017; Corballis, 2017, 2021). In addition to functional specialization, there are neuroanatomic asymmetries in the structure of the left and right hemispheres, including enlargement of left-hemisphere regions known to be involved in language processing (see Thiebaut de Schotten and Beckmann, 2022 for a review). We briefly review evidence for hemispheric specialization to help frame how hemispheric organization may differ with advancing age.

Left-hemisphere dominance for right-handedness and language

Right-hand dominance, indicating left hemispheric specialization for fine motor control, is evident in evolutionary evidence from up to 2–3 million years ago (Corballis, 1989). Approximately 90% of humans have a right-hand preference (Corballis and Haberling, 2017). Left-hemisphere motor dominance often accompanies left-hemisphere language specialization. Estimates indicate that 95%–99% of right-handed people show left-hemisphere language dominance, and 70%–80% of left-handed people also show left-hemisphere language specialization (Corballis et al., 2012; Corballis, 2017).

Neural evidence from studies of patients with lateralized brain damage, as well as those who have undergone commissurotomy—surgical separation of the cortical hemispheres to treat intractable epilepsy (Gazzaniga, 1970)—also supports left-hemisphere language specialization. Speech deficits often arise from left-hemisphere damage to Broca's area or Wernicke's area (Broca, 1863; Wernicke, 1874; Corballis and Haberling, 2017). Studies of task-based brain activity measured with electrophysiology and neuroimaging further support a left-lateralized language network (Kutas and Van Petten, 1988; Posner and Raichle, 1994; Binder et al., 1997). Laterality indices, which compute a preference for asymmetrical vs symmetrical hemispheric processing, also indicate a left-hemisphere laterality bias for language tasks, such as word reading (e.g., see Damasio and Damasio, 1983), generating words or evaluating synonyms (Haberling and Corballis, 2016).

Right-hemisphere dominance for face and visuospatial processing

The idea that the right hemisphere plays a special role in face processing dates back to case studies of focal cortical lesions associated with the rare deficit of prosopagnosia—the inability to recognize faces of familiar individuals as well as those that are famous and well-known (e.g., Bodamer, 1947; see Rossion, 2018 for a review). In prosopagnosia, faces can no longer be recognized visually, yet other categories of visual information, such as letters, words, and numbers can still be recognized, indicating the selective nature of this face processing deficit. Although the location of brain damage that is necessary and sufficient to cause prosopagnosia has been debated (Papagno et al., 2021), decades of evidence converge to indicate that the right hemisphere is generally superior to the left in the domain of face processing (Corballis, 1991; see e.g., Behrmann et al., 2016 for a review; Blauch et al., 2022).

Right hemispheric specialization for visuospatial processing has also been well documented by investigating the effects of focal brain damage. Early studies comparing the consequences of stroke to the left and right hemispheres demonstrate that right-hemisphere damage leads to greater impairment in a variety of visual–spatial tasks, including copying or matching complex figures, non-namable designs, and faces, drawing such figures from memory, clock drawing, and other types of nonverbal problem solving or visuo-constructive tasks, as well as measures of visual attention (see also Bryden, 1982; Banich and Compton, 2018). Right-hemisphere dominance for visual–spatial processing is also well documented in studies of patients who had undergone commissurotomy (e.g., Gazzaniga, 1970; Corballis, 2003), and in neurotypical populations using brain imaging methods (e.g., Posner and Raichle, 1994; Corbetta and Shulman, 2002). When measured in young adults, greater right-hemisphere activity emerged for tasks requiring memory for spatial locations over brief delays, in contrast with predominant left-hemisphere activity when the memoranda were letters or words (Jonides et al., 1993; Smith and Jonides, 1998).

Benefits of hemispheric asymmetry and bilateral cooperation

Corballis (1989) hypothesized that numerous benefits could accrue from an organizational scheme that entails hemispheric specialization. First, it may be more efficient for complex cognitive functions to be performed preferentially by one hemisphere, rather than requiring the involvement of both hemispheres. Complete redundancy of complex mental functions in both the right and left hemispheres may waste neural resources, as the brain is constrained to fit in the skull. Furthermore, duplication of "expertise" in both hemispheres could promote between-hemisphere interference or conflict, resulting in diminished task performance, whereas specialization could minimize cross-hemisphere interference (see also Corballis et al., 2000).

Nevertheless, Corballis and Haberling (2017) are careful to point out that extreme notions of individuals or functions being fully "left-brained" or "right-brained" are exaggerated, as cerebral specializations do not indicate the use of only one hemisphere. For instance, even for language, which shows strong left-hemisphere dominance, the right hemisphere is capable of some basic language processing (e.g., Levy and Trevarthen, 1977) and can show right-hemisphere language dominance for certain subregions, including Wernicke's area, in approximately 15%–50% of epilepsy patients (Tailby et al., 2017). Furthermore, individuals who have undergone hemispherectomy in early childhood have remarkably spared capabilities even for such specialized domains as word reading and face recognition, regardless of which hemisphere is preserved (Reuter-Lorenz et al., 2011; Granovetter et al., 2022). Additionally, later in life, hypothesized mechanisms of recovery from brain damage due to stroke include the reorganization of intact homotopic regions in the nondamaged hemisphere that may serve to compensate for impaired functions (e.g., Raboyeau et al., 2008; Wilson and Schneck, 2021). Thus, there is evidence for considerable plasticity in hemispheric organization, and lateral specialization may best be characterized as graded, rather than absolute, with both sides of the brain possessing varying degrees of functionality in most domains (e.g., Behrmann et al., 2016).

There are also documented benefits from engaging both sides of the brain. That is, bilateral recruitment mediated predominantly by the corpus callosum—the major network of interhemispheric fibers that connect cortical regions of the left and right sides of the brain—can be beneficial to performance. For instance, greater creativity has been observed in individuals who exhibit mixed-handedness, indicative of less laterality over fine motor control (Shobe et al., 2009; see also Lindell, 2011). A recent study that compared individuals with highly creative achievements to age- and education-matched control participants also demonstrated greater functional connectivity, especially between the left and right sides of the brain (Chrysikou et al., 2024).

More generally, a range of evidence indicates that regional engagement becomes increasingly bilateral as tasks become more complex and demanding, suggesting that bilateral recruitment provides greater computational resources. Banich and her colleagues were among the first to document this phenomenon using behavioral laterality methods in neurotypical young adults (Banich and Belger, 1990; Belger and Banich, 1992). In their task, critical information can be projected within one visual field, and thus primarily to one hemisphere, or, the information can be divided between the two visual fields (i.e., on both sides of the screen). For simple matching tasks, performance was better for within-field than between-field presentations, however, as task complexity increased, the between-field (i.e., bihemispheric) processing became advantageous (Banich, 1998; see also Reuter-Lorenz et al., 1999). Brain imaging has also documented increasing bilateral recruitment in response to parametric increases in task demand (e.g., Braver et al., 1997). As we turn to theories of neurocognitive aging, bihemispheric recruitment will also figure prominently as a potential mechanism to benefit the aging mind and brain.

HEMISPHERIC ORGANIZATION AND THEORIES OF NEUROCOGNITIVE AGING

Accelerated right-hemisphere aging

The right-hemisphere aging hypothesis, one of the first neuropsychologic theories of cognitive aging, is rooted in the 1970s when work on hemispheric asymmetry was burgeoning, inspired by research on so-called "commissurotomy" patients, as well as by laterality studies of neurotypical adults. By this time, the general characterization of brain asymmetry had been established, with left-hemisphere dominance in the verbal domain and right-hemisphere specialization for nonverbal processing, including visuospatial tasks, face processing, and some aspects of emotion.

Against this backdrop, cross-sectional psychometric studies of lifespan samples, ranging from younger adults to older adults aged 60 and above, have begun to document the relative preservation of "crystallized intelligence" with age. This includes language abilities, such as vocabulary, semantic knowledge, and general world knowledge. In contrast, "fluid processing" abilities, whose measurement emphasized speeded tasks and nonverbal problem-solving showed marked age-related declines (Horn and Cattell, 1967; see Botwinick, 1970 for a review). Finding that aging affected some cognitive abilities more than others raised the possibility that patterns of age-related declines might resemble neuropsychologic profiles of clinical populations with documented brain damage (Goldstein and Shelly, 1975; Klisz, 1978)—a "deficit model" of aging cognition (Reuter-Lorenz, 2002). In particular, the relative preservation of language abilities coupled with impairments in nonverbal performance prompted the hypothesis that the hemispheres aged asymmetrically, and that accelerated right-hemisphere aging could be the source of differential cognitive decline.

Initial confirmatory tests of this idea emerged from Goldstein and Shelly (1981), who reported that the neuropsychologic profiles of older adults more closely resembled those of clinical populations with documented brain damage to the right hemisphere than those with left-hemisphere damage. Thus, initial support for more rapid right-hemisphere aging in otherwise healthy older adults came from behavioral performance on neuropsychologic tests. A related approach to test this hypothesis was to use experimental laboratory methods that compared the magnitude of left vs right performance asymmetries on cognitive tasks in healthy younger and older adults. This was done by measuring the speed and accuracy of performance under stimulus presentation conditions designed to maximize the unilateral engagement of the left vs right hemisphere. These methods included presenting lateralized visual input via brief stimulus exposure to the left or right visual hemifield, dichotic listening, or unimanual tactile/motor tasks. Early results from these approaches were inconsistent, with some studies reporting right-hemisphere performance deficiencies in older adults (see also Clark and Knowles, 1973; Riege et al., 1980), and others finding no evidence for selective effects of age on right-hemisphere performance (Schear and Nebes, 1980; Nebes et al., 1983). In a relatively comprehensive assessment of the right hemi-aging hypothesis using the visual half-field method and stimuli from verbal and nonverbal domains, Cherry et al. (1995) found no reliable hemisphere differences in younger and older adults leading them to conclude: "it seems unlikely changes in hemispheric asymmetry contribute significantly to age-related changes in cognitive functioning (p. 191)."

Several decades of research have continued to document the adverse effects of age on some visuospatial tasks, especially mental rotation, spatial memory, and spatial navigation (Colombo et al., 2017; Li and King, 2019; Learmonth and Papadatou-Pastou, 2022), although direct links to accelerated right-hemisphere aging are generally lacking. The selectivity of these deficits may be due to more rapid aging of the hippocampus and altered hippocampal–cortical interactions (see e.g., Klencklen et al., 2012; Colombo et al., 2017), rather than accelerated aging of right cortical regions. Furthermore, the multifaceted nature of way-finding ability and visual–spatial skills makes a direct and dominant link to right-hemisphere functioning unlikely (e.g., Richmond et al., 2018).

Nevertheless, if the right hemisphere ages faster than the left, anatomic measures should provide more direct evidence of differential decline. Advances in neuroimaging methods and the increasing availability of data sets from large lifespan samples have made such assessments possible. In particular, two recent large-scale studies of brain structure offer some enticing clues that continue to reinforce the importance of hemispheric brain organization in neurocognitive aging. One study used structural MRI to assess longitudinal changes in cortical thickness across homotopic regions in the left and right hemispheres in over 1000 individuals (Roe et al., 2021). Cortical thinning is likely due primarily to the loss of synapses and is a major contributor to age-related brain shrinkage or atrophy. Roe et al. (2021) found that regions characterized by the largest asymmetries in cortical thickness in younger adults showed the greatest *losses* in volume with advancing age. That is, thicker regions of the cortex showed the most cortical thinning over time. This pattern, which was evident for both left- and right-hemisphere brain regions, was replicated in four additional large-scale cohort samples (Roe et al., 2021). Thus, according to this work and in contrast with the right-hemisphere aging hypothesis, neuroanatomic declines are widely evident in *both* cortical hemispheres, such that neuroanatomic asymmetries become *less* pronounced with age.

Taking a very different approach to measuring brain aging, Yin et al. (2023) relied on "deep learning," a type of machine learning, to analyze MRI measures of brain structure. Typically, we think of age chronologically—the number of years a person has lived thus far. However, people vary in how rapidly they age biologically, with some being more youthful than others. This fact has given rise to the idea of biological age, and its neuroanatomic version, "brain age." Based on a sample of 650 cognitively unimpaired adults, Yin et al. (2023) used deep learning to estimate brain age and to identify the brain regions that most saliently distinguished brain age from chronological age. Salience regions differed based on sex, and for males, these regions were in the temporal and frontal regions of the right hemisphere. This pattern suggests that age-related structural properties of the right hemisphere are particularly relevant to the aging of the male brain, however, as the authors propose, right-hemisphere aging may be relatively slower in female brains. Across both sexes, measures of brain age were found to correlate with neurocognitive measures just as strongly as chronological age, indicating that any direct links between hemispheric differences in brain aging and cognition are lacking.

All in all, the past 50 years have failed to yield strong and consistent evidence that accelerated right-hemisphere aging is a general characteristic of the aging brain. Nevertheless, continued neuroimaging advances, and new analytic methods hold great promise for identifying unique regional, network, and hemispheric specific properties that may play a role in cognitive aging, along with compensatory or protective processes that preserve cognition in older age.

Asymmetry reduction

Theoretical advances in the cognitive neuroscience of aging accelerated with the advent of new neuroimaging methods. To measure brain function, positron emission tomography (PET) studies first emerged, followed by functional magnetic resonance imaging (fMRI), along with high-resolution imaging to measure brain structure. With functional neuroimaging, the laterality of brain activation and potential age differences could be assessed during the performance of various cognitive and motor tasks. Surprisingly, early studies revealed that older adults did not solely show reductions in brain activity in comparison to their younger counterparts—an observation that was contrary to the existing notions of widespread decline and aging (see also Festini et al., 2018). Instead, in many instances, older adults exhibited overactivations, including bilateral overactivation, and other neural differences (Grady et al., 1994; Cabeza et al., 1997, 2004; Reuter-Lorenz et al., 2000). Based on these novel neuroimaging findings, multiple influential theories emerged in an effort to explain functional brain differences with advancing age. One important early observation was that older adults exhibited decreased lateralization, or hemispheric asymmetry reduction, in that they tended to recruit both hemispheres for verbal and spatial tasks, while younger adults predominantly relied on the corresponding specialized hemisphere (Reuter-Lorenz et al., 2000).

Cognitive processing

The Hemispheric Asymmetry Reduction in Older Adults (HAROLD) model was proposed to describe the observed decline in prefrontal functional asymmetry with advancing age (Cabeza, 2002; Cabeza and Dennis, 2012). Older adults exhibited more bilateral prefrontal activation (i.e., reduced asymmetry), whereas younger adults exhibited primarily unilateral prefrontal activation (i.e., asymmetry). Evidence for greater bilateral activation with advancing age was originally documented in the memory domain, including neuroimaging assessments of paired-associate learning and recall (Cabeza et al., 1997) and verbal/spatial working memory (Reuter-Lorenz et al., 2000). Neuroimaging evidence suggesting greater bilateral engagement with age has been found in other cognitive domains as well (see Reuter-Lorenz and Lustig, 2005), including naming and semantic processing tasks (e.g., Obler et al., 2010; Berlingeri et al., 2013; Hoyau et al., 2017; Hoffman and Morcom, 2018), theory of mind (Castelli et al., 2010), and motor control (e.g., Heuninckx et al., 2008).

More recent behavioral and electrophysiological assessments of laterality also indicate age-related reductions in hemispheric specialization. For instance, during a rod-and-frame spatial orienting task, older adults displayed evidence of less right-hemisphere processing of left visual contexts (Abdul Razzak and Bagust, 2022). Moreover, during a flanker task with lateralized presentation, younger adults showed lateralized alerting and conflict effects, whereas older adults did not (Casagrande et al., 2021). Examining the time course of asymmetry reduction (bilateral activation) with EEG, Tagliabue et al. (2022) observed age-related differences in bilateral hemispheric activity during a visuospatial delayed match-to-sample task, a measure of working memory. Older adults exhibited bilateral frontal activation when visuospatial stimuli were being held within working memory (item retention phase). Further, compared to younger adults, older adults showed increased bilateral hemispheric activity that was modulated by memory load (Tagliabue et al., 2022).

Hand dominance and motor performance

Several distinct patterns emerge from the literature on hand dominance and brain laterality in aging, including evidence of hemispheric asymmetry reduction for motor control. When comparing younger adults and older adults, researchers have found a general decrease in hand dominance in older adults, indicating a shift toward ambidexterity with advancing age (Kalisch et al., 2006). At the same time, older adults are relatively unaware of this reduction in hand dominance, based on self-report data (Kalisch et al., 2006; Teixeira, 2008; Marcori et al., 2019). Older adults often self-report strong right motor preferences, yet these preferences do not match their true actions as well as younger adults (Marcori et al., 2019).

Studies investigating hand dominance and aging have used a variety of tasks to assess how each hand performs in comparison to the other. Common tasks include drawing and handwriting, grip strength, sequential finger movements, reaction time, circle/square tapping, and motor performance of hands, arms, wrists, and fingers (e.g., see Kalisch et al., 2006; Teixeira, 2008). Notably manual asymmetry (i.e., superiority of one hand vs the other) was present for some tasks but not others in older adults. For drawing and handwriting tasks, there was an increase in manual asymmetry for older adults in comparison to younger adults, in which the right hand was superior to the left (Teixeira, 2008). However, for grip strength, sequential finger movement, and reaction time tasks, no clear manual asymmetry was present, indicating a reduction in manual asymmetry in older adults compared to younger adults (Teixeira, 2008). Reduced manual asymmetry was also observed in older adults for a motor task that required a high degree of visuospatial processing (Flink and Matos, 2021). These results suggest that the preferred hand may no longer be behaviorally dominant in certain situations in older adults. That is, right-handed older adults still perceive their right side as being dominant (Kalisch et al., 2006; Teixeira, 2008; Marcori et al., 2019), despite this side performing at the same level as their left for certain tasks. In comparison, right-handed younger adults tended to use their right hand more frequently than older adults while also performing better at certain tasks with their right hand, in contrast to older adults who performed similarly with both hands (Kalisch et al., 2006).

This decrease in hand dominance with age may be related to neural underpinnings of motor control in older adults. When examining brain activity during motor tasks, there is evidence for some consistency across age. For simple motor movements (e.g., fast repetitive tapping), both younger and older adults rely more on the contralateral hemisphere, whereas for more complex motor movements (e.g., sequential finger movements), younger and older adults rely on both cerebral hemispheres (Teixeira, 2008). However, evidence also indicates increased bihemispheric control of movement in older adults. For instance, when performing one-handed tasks, older adults recruited both hemispheres compared to younger adults, who showed unilateral motor control (Bernard and Seidler, 2012b). Interestingly, younger adults who lacked a clear dominant hand also showed similar recruitment of both hemispheres (Bernard and Seidler, 2012b). Moreover, Naccarato et al. (2006) observed that older adults exhibited less lateralized (more bilateral) neural activity during a motor task that required tapping the index finger and thumb together at a given tempo, set with an auditory cue. Thus, these findings (Berger et al., 2020; see also Knights et al., 2021) suggest a link between reduced hand dominance and reduced hemispheric laterality over motor control in the brain in older adults. Nevertheless, importantly, a direct assessment of reduced brain asymmetry and reduced motor asymmetry has yet to be evaluated (see Hill et al., 2020). Furthermore, additional research is needed to assess situations when reduced motor hemispheric asymmetry may be absent in older adults, such as during motor preparation stages and bimanual tasks (see Woytowicz et al., 2020). Indeed, age differences between cortical motor maps are not always found (see Bernard and Seidler, 2012a; Hehl et al., 2020, 2022).

Dedifferentiation and decline

This observation of hemispheric asymmetry reduction could reflect less efficient neural processing with age (a deficit view), or it could index a means to bolster performance (a compensatory view). These views are not mutually exclusive, as it is likely that more widespread, bilateral brain activity could be detrimental or beneficial in different contexts (e.g., in different task domains, with different task demands, or at different points in the lifespan). First, we consider the deficit account of dedifferentiation of neural activity.

In some task domains, younger adults display focal activity in brain regions known to be specialized for particular stimulus categories or cognitive functions, whereas brain activity in older adults is more widespread and diffuse. This pattern was termed dedifferentiation, to reflect that older adults did not show differential processing of specialized brain regions as much as their younger counterparts (see Park et al., 2004; Festini et al., 2018; Rieck et al., 2021). This neural pattern resembled the behavioral findings of larger correlations among varied sensory and cognitive tasks with advancing age, where the term dedifferentiation was also used (Baltes and Lindenberger, 1997; Lindenberger and Baltes, 1997).

Neural data indicated that older adults exhibited dedifferentiation of the ventral visual pathway, including less specialized brain activity in response to domain-specific visual categories like faces and houses (Park et al., 2012; Burianova et al., 2013). Further, older adults exhibited dedifferentiation during nonsensory tasks, such as associative memory encoding (Saverino et al., 2016) and motor tasks (e.g., Knights et al., 2021). Interestingly, older adults who showed more neural specificity (i.e., less dedifferentiation) tended to exhibit better behavioral fluid processing (Park et al., 2010), indicating that, in certain instances, more specialized neural activity was associated with better performance in older adults.

Bihemispheric recruitment and compensatory overactivation

A rich literature on the cognitive neuroscience of healthy aging also includes evidence for superior task performance with bilateral brain activity, in certain instances. Older adults who recruit both hemispheres of the brain tend to outperform older adults who show unilateral brain activity (e.g., Reuter-Lorenz et al., 2001; Cabeza et al., 2002; Cappell et al., 2010; Chen et al., 2022). For example, Chen et al. (2022) observed that successful agers, as classified based on baseline cognition and longitudinal cognitive change, exhibited additional activation of bilateral prefrontal regions during a subsequent memory task.

Further, directly evaluating the question of whether the bilateral prefrontal activity is compensatory or not, Hennessee et al. (2022) compared the degree of left-lateralized prefrontal activity during semantic judgment performance in different age groups. In older adults, *bilateral* prefrontal activity was beneficial, whereas, in middle-aged adults, youth-like *left-lateralized* activity was associated with better performance. The authors interpreted these findings to indicate that earlier reliance on both cerebral hemispheres within middle age was an indicator of poorer cognition, but bilateral prefrontal activation could be beneficial in older adults. To the extent that bilateral hemispheric engagement supports preserved cognition in older age, we might also expect that greater functional connectivity characterizes successful aging. A recent study in very old adults without dementia, including centenarians, reported greater bilateral functional connectivity, especially involving frontoparietal networks in the left and right hemispheres compared to younger older adults (Jiang et al., 2020; see also Deng et al., 2021; cf. Knights et al., 2021).

However, there are important caveats to this general compensatory bilateral overactivation account, including evidence that more youth-like (often unilateral) brain activity can be beneficial in certain instances (e.g., Duzel et al., 2011; Pudas et al., 2013), and the importance of task demands as proposed in the Compensation-Related Utilization of Neural Circuits Hypothesis (CRUNCH) model (Reuter-Lorenz and Lustig, 2005; Reuter-Lorenz and Mikels, 2006; Reuter-Lorenz and Cappell, 2008). The CRUNCH theory notably posits that bilateral brain activity can be compensatory when task demands are sufficiently low, but there is a "crunch point," after which additional bilateral brain activation is no longer capable of assisting performance, such as when the task demands exceed the neural resources (see Mattay et al., 2006; Cappell et al., 2010; Festini et al., 2018). Evidence in support of the CRUNCH model was initially demonstrated using working memory tasks with varying set sizes. At low loads, older adults showed bilateral overactivation to include the right dorsolateral prefrontal cortex (DLPFC) and achieved equivalent task performance to younger adults, but, at high loads, older adults showed lower DLPFC activity than younger adults and worse task performance (see also Mattay et al., 2006; Cappell et al., 2010; Schneider-Garces et al., 2010; Crowell et al., 2020). More recently, it was shown that older adults who trained on a verbal working memory task showed greater activity at higher memory loads after training compared to pretraining baseline, suggesting that the "crunch" point can be shifted when task performance improves through practice thereby enabling greater responsivity to higher task demands (Iordan et al., 2020).

On a population level, many variables have been proposed to contribute to different trajectories of healthy aging, including lifestyle factors, genetics, and general health (Reuter-Lorenz and Park, 2014; Festini et al., 2016; Nyberg et al., 2020; de Godoy et al., 2021; Festini, 2022), and these factors likely contribute to differing patterns of hemispheric activation in older adults. For instance, the Scaffolding Theory of Aging and Cognition (STAC; Park and Reuter-Lorenz, 2009), now broadened and revised to incorporate longitudinal aging (STAC-r; Reuter-Lorenz and Park, 2014, 2024), considers how neural enrichment and depletion factors can influence brain structure, brain function, and the potential for "scaffolding" across the lifespan. These models propose that older adults engage neural scaffolding, often in both hemispheres, as a compensatory response to structural and physiological brain decline. Lifelong factors such as education, exercise, stress, and depression influence the potential for adaptive scaffolding, and associated variations in functional brain activity, including age-related changes in hemispheric recruitment. Indeed, the concept of neuroprotective factors is a prominent player in more recent theories of cognitive aging, including the proposal that the right hemisphere plays an important role in cognitive reserve (e.g., Robertson, 2014), which we turn to next.

The right hemisphere and cognitive reserve

The shift from a perspective on aging that emphasized deficits and loss, toward a consideration of resilience and compensatory processes, included the idea of *reserve,* which refers to neurocognitive resources that can counteract negative neurobiologic effects (e.g., Katzman et al., 1988; Bäckman and Dixon, 1992; Stern et al., 1992; Reuter-Lorenz et al., 1999; Grady, 2012). Persuasive evidence for the operation of reserve and its compensatory potential comes from evidence that higher education and greater engagement in cognitively stimulating activities through occupation, leisure, and social activities can be protective against the presence of brain pathology (Stern, 2009). For example, due to reserve, people who are more highly educated, have had cognitively demanding occupations, and frequently engage in enriching activities are less likely to show cognitive and behavioral signs of Alzheimer disease despite having considerable indications of brain pathology (see Stern, 2009; Robertson, 2014 for reviews; Mather, 2021; however see Nyberg et al., 2021).

Robertson (2014) proposed that the right hemisphere plays a special role in cognitive reserve (e.g., Stern, 2009), thereby prompting renewed interest in asymmetrical brain organization and its role in neurocognitive aging. The cornerstone of this account is the neurotransmitter noradrenaline, also known as norepinephrine, which is proposed to be neuroprotective and shield against the deleterious effects of age-related brain pathologies. The locus coeruleus (LC), a brainstem nucleus, is the primary source of noradrenergic innervation throughout the brain and plays a major role in arousal, alertness, and other attentional processes (Samuels and Szabadi, 2008). Robertson (2014) proposes that neurocognitive correlates of noradrenergic activity, including a range of attentional processes and responses to novelty, are primarily mediated by frontoparietal circuitry in the right hemisphere. This means that greater preservation or maintenance of right frontoparietal structure and function should be associated with more reserve and reduced age-related impairment in measures of attention.

Indeed, several recent studies provide support for this hypothesis in neurotypical older adults with evidence that greater reserve, as measured with questionnaires that ask about education, occupational history, and enriching activities among other things, is more strongly associated with right than left-hemisphere structure and function. For example, Brosnan et al. (2018) reported that higher levels of reserve were associated with a greater left visual field advantage and greater right-hemisphere responsivity to transcranial direct current stimulation (tDCS), as measured by more pronounced improvements in processing speed following right- than left-sided stimulation. In another study, select indices of reserve were associated with better performance on some measures of attention, which in turn were positively related to higher gray matter volumes in various regions of the right hemisphere (Shalev et al., 2020). Most recently, this same team reported a unique association between a cognitive reserve index and the integrity of a major white matter pathway in the right hemisphere—the superior longitudinal fasciculus. Moreover, using a causal mediation model, this relationship was found to positively influence measures of alertness obtained from the Attention Network Task (Brosnan et al., 2022).

One further study provides both cross-sectional and longitudinal evidence from the Harvard Aging Brain Study that links noradrenergic function, as indexed by LC activity, with attentional responsivity and more beneficial aging outcomes (Prokopiou et al., 2022). Using fMRI to compare responses to novel vs repeated face-name pairs in neurotypical older adults, Prokopiou et al. (2022) showed greater LC activity and greater connectivity with medial temporal regions, including the hippocampus, in response to novelty. However, individuals who performed more poorly on neuropsychologic measures of cognition showed weaker LC responses to novelty. Furthermore, lower novelty-induced LC activity was also associated with greater cognitive decline, an effect that was greater in individuals who had higher levels of beta-amyloid, an index of Alzheimer disease (AD)-related pathology. Finally, selective links to the right hemisphere were also evident in this study, in that lower right LC activity was related to steeper cognitive decline in individuals who were positive for AD pathology.

Despite these promising reports about the neuroprotective effects of noradrenergic activity, the reality is likely to be much more complex than the right-hemisphere reserve hypothesis can explain. For example, evidence reviewed by Mather (2020, 2021) points to potentially deleterious effects of noradrenergic activity in older age, as well as the questionable role in cognitive reserve of education and enriching environments after early childhood. Reconciling the right-hemisphere reserve hypotheses with these and other findings, as well as evidence pertaining to bilateral cortical thinning (Roe et al., 2021) and gender differences in right-hemisphere brain aging revealed with machine learning (Yin et al., 2023) will be important frontiers for future research.

ADDITIONAL CONSIDERATIONS

Bihemispheric neurostimulation

Some interventions aimed at boosting performance in older adults have also revealed laterality effects (e.g., Fiori et al., 2017; Ljubisavljevic et al., 2019; but see King et al., 2020 for an exception). For example, Fiori et al. (2017) observed that, for older adults only, language learning of novel pseudowords was differentially improved with bilateral tDCS compared to unilateral tDCS. Similarly, bilateral, as opposed to unilateral, tDCS reduced the dual-task cost in older adults (Ljubisavljevic et al., 2019). The benefit of tDCS has been proposed to be due to modulating interhemispheric communication of complex brain networks (Lindenberg et al., 2013) and heightening left lateralization in older adults (Martin et al., 2017). It is important to note, however, that some studies found comparable benefits of unilateral and bilateral stimulation (Martin et al., 2017), and at least one systematic review concluded that left prefrontal stimulation may be the most promising approach to improve age-related cognitive decline (Siegert et al., 2021).

Socioeconomic status and laterality

The development of laterality begins early in life and can be influenced by environmental factors. Socioeconomic status (SES) is one such factor reported to affect hemispheric asymmetry. Children in low SES groups tend to show patterns of reduced cerebral asymmetry, whereas children in higher SES groups reveal patterns of more pronounced asymmetry (Boles, 2011). These observations stem from assessments of dichotic listening and lateralized word processing/emotion recognition. Three potential mechanisms have been proposed to account for the variation between high and low SES groups: (1) individual differences in lateralization strength, (2) the maturity hypothesis, and (3) the specialization hypothesis (Boles, 2011). The individual differences view proposes that people who show strong lateralization do so across many different assessments of hemispheric asymmetry, contributing to the consistency of evidence between tasks. Second, lower maturity of the corpus callosum at the time of early cerebral lateralization (i.e., maturity hypothesis) may encourage greater hemispheric specialization. That is, children who develop more complex cognition earlier in life are predicted to show more asymmetry. Third, variations in the environmental context associated with SES (e.g., education, stress) may impact specialization and lateralization of certain brain areas (Boles, 2011).

Furthermore, SES has also been shown to impact structural brain asymmetry. In a study of 40- to 70-year-olds, higher-SES individuals had larger brain volumes in the *left* hemisphere and smaller brain volumes in the *right* hemisphere, whereas lower-SES participants exhibited the opposite pattern (Poeppl et al., 2022). One potential explanation for these structural hemispheric differences is the way education shapes the brain. For instance, higher education was associated with a larger left temporal pole, an area that plays a key role in semantic memory and language (Poeppl et al., 2022). Individuals with lower SES tend to have reduced access to quality education, potentially contributing to reduced left hemispheric specialization.

These observations call attention to the importance of considering SES and other overlooked demographic factors that could influence lateralization across the lifespan, and in later life (Park and Gutchess, 2002; see Brewster et al., 2019). Future research can aim to include more diverse samples of older adults to evaluate the impact of SES and other individual differences on changes in hemispheric lateralization with age.

CONCLUSION

For nearly 50 years, the brain's hemispheric organization has figured prominently in efforts to understand the causes of, and potential remedies for, age-related cognitive decline. Despite this unifying theme, there has yet to emerge a tidy picture that pulls the pieces together into a cohesive neuropsychologic account. On the one hand, the right cerebral hemisphere has been considered more vulnerable to the adverse effects of aging on cognition (e.g., Goldstein and Shelly, 1981; but see Roe et al., 2021). On the other hand, the right hemisphere is posited to be the source of neural processes supporting greater resilience to age-related decline (e.g., Brosnan et al., 2022). Both accounts may bear some veracity, as maintenance of right-hemisphere structure/function may afford more cognitive hardiness. Similarly, behavioral and neuroimaging evidence for greater reliance on the integrated engagement of both cerebral hemispheres by older compared to younger adults is interpreted as compensatory on the one hand (e.g., Reuter-Lorenz and Cappell, 2008; Cappell et al., 2010; Crowell et al., 2020), or as an indication of dedifferentiation and decline, on the other (e.g., Park et al., 2004; see Rieck et al., 2021). How can these divergent patterns be reconciled and discrepancies be resolved to provide a more complete understanding of the aging mind and brain?

The challenges are indeed numerous given the multifaceted nature of human cognition and performance, the wide variation in tasks and neuroscience methodologies used for assessment, and the tremendous differences between individuals with respect to biological, environmental, and lifestyle factors that influence the developmental course of later life (e.g., Stern, 2009; Reuter-Lorenz and Park, 2024). Progress will continue to be made as meta-analytic methods aggregate results from studies with smaller samples in order to identify regularities and the specific conditions or subgroups to which they do and do not apply (e.g., Thompson et al., 2022). At the same time, the field is moving away from exclusive reliance on smaller, cross-sectional experiments conducted primarily with relatively homogeneous participant samples, toward better-powered longitudinal studies of more diverse populations that can characterize changes in cognitive status and brain health within the same individuals over time (Falk et al., 2013; Zahodne et al., 2017). Together with computational advances and new human neuroscience techniques that image brain structure and function at multiple scales, these methodological developments will ultimately inform fuller, more nuanced theoretical accounts that will likely situate the brain's hemispheric organization within a complex, adaptive network of dynamic interacting circuits that underlie age-related cognitive change.

In the meantime, we view the overall accumulation of neuropsychological evidence to date to indicate differences between younger and older adults in hemispheric specialization, with younger adults characterized by more robust asymmetries and functional dominance compared with older adults. Key variations in brain hemispheric asymmetry and aging include: (1) heightened bilateral activation in older relative to younger adults (i.e., asymmetry reduction), (2) less specific, more widespread (including bilateral) neural activation in older adults (i.e., dedifferentiation), particularly for sensory and motor domains, and (3) bilateral recruitment serving a compensatory role, especially involving prefrontal regions for more demanding tasks. In sum, given the brain's enduring capacity for plasticity in the form of reorganization, repair, and compensation, it is surely true that two hemispheres are better than one, especially as we age.

References

Abdul Razzak R, Bagust J (2022). Perceptual lateralization on the rod-and-frame test in young and older adults. Appl Neuropsychol Adult 31: 405–411. https://doi.org/10.1080/23279095.2022.2030741.

Bäckman L, Dixon RA (1992). Psychological compensation: a theoretical framework. Psychol Bull 112: 259–283.

Baltes PB, Lindenberger U (1997). Emergence of a powerful connection between sensory and cognitive functions across the adult life span: a new window to the study of cognitive aging? Psychol Aging 12: 12–21.

Banich MT (1998). Integration of information between the cerebral hemispheres. Curr Dir Psychol Sci 7: 32–37.

Banich MT, Belger A (1990). Interhemispheric interaction: how do the hemispheres divide and conquer a task? Cortex 26: 77–94.

Banich MT, Compton RJ (2018). Cognitive neuroscience, Cambridge University Press, Cambridge.

Behrmann M, Scherf KS, Avidan G (2016). Neural mechanisms of face perception, their emergence over development, and their breakdown. Wiley Interdiscip Rev Cogn Sci 7: 247–263.

Belger A, Banich MT (1992). Interhemispheric interaction affected by computational complexity. Neuropsychologia 30: 923–929.

Berger A, Steinberg F, Thomas F et al. (2020). Neural correlates of age-related changes in precise grip force regulation: a combined EEG-fNIRS study. Front Aging Neurosci 12: 594810.

Berlingeri M, Danelli L, Bottini G et al. (2013). Reassessing the HAROLD model: is the hemispheric asymmetry reduction in older adults a special case of compensatory-related utilisation of neural circuits? Exp Brain Res 224: 393–410.

Bernard JA, Seidler RD (2012a). Evidence for motor cortex dedifferentiation in older adults. Neurobiol Aging 33: 1890–1899.

Bernard JA, Seidler RD (2012b). Hand dominance and age have interactive effects on motor cortical representations. PLoS One 7: e45443.

Binder JR, Frost JA, Hammeke TA et al. (1997). Human brain language areas identified by functional magnetic resonance imaging. J Neurosci 17: 353–362.

Blauch N, Behrmann M, Plaut D (2022). Connectivity constraints, viewing biases, and task demands within a bi-hemispheric interactive topographic network account for the layout of human ventral temporal cortex. Vision Sciences Society Annual Meeting. .

Bodamer J (1947). Die prosopagnosia [Prosopagnosia]. Arch Psychiatr Zeitschrift Neurol 179: 6–54.

Boles DB (2011). Socioeconomic status, a forgotten variable in lateralization development. Brain Cogn 76: 52–57.

Botwinick J (1970). Geropsychology. Annu Rev Psychol 21: 239–272.

Braver TS, Cohen JD, Nystrom LE et al. (1997). A parametric study of prefrontal cortex involvement in human working memory. Neuroimage 5: 49–62.

Brewster P, Barnes L, Haan M et al. (2019). Progress and future challenges in aging and diversity research in the United States. Alzheimers Dement 15: 995–1003.

Broca P (1863). Localisation des fonctions cérébrales: siége de langage articulé. Bull Mem Soc Anthropol Paris 4: 200–208.

Brosnan MB, Demaria G, Petersen A et al. (2018). Plasticity of the right-lateralized cognitive reserve network in ageing. Cereb Cortex 28: 1749–1759.

Brosnan MB, Shalev N, Ramduny J et al. (2022). Right fronto-parietal networks mediate the neurocognitive benefits of enriched environments. Brain Commun 4: fcac080.

Bryden MP (1982). Laterality: functional asymmetry in the intact brain, Academic Press, New York.

Burianova H, Lee Y, Grady CL et al. (2013). Age-related dedifferentiation and compensatory changes in the functional network underlying face processing. Neurobiol Aging 34: 2759–2767.

Cabeza R (2002). Hemispheric asymmetry reduction in older adults: the HAROLD model. Psychol Aging 17: 85–100.

Cabeza R, Dennis NA (2012). Frontal lobes and aging: deterioration and compensation. In: DT Stuss, RT Knight (Eds.), Principles of frontal lobe function, 2nd ed.Oxford University Press, New York.

Cabeza R, Grady CL, Nyberg L et al. (1997). Age-related differences in neural activity during memory encoding and retrieval: a positron emission tomography study. J Neurosci 17: 391–400.

Cabeza R, Anderson ND, Locantore JK et al. (2002). Aging gracefully: compensatory brain activity in high-performing older adults. Neuroimage 17: 1394–1402.

Cabeza R, Daselaar SM, Dolcos F et al. (2004). Task-independent and task-specific age effects on brain activity during working memory, visual attention and episodic retrieval. Cereb Cortex 14: 364–375.

Cabeza R, Albert M, Belleville S et al. (2018). Maintenance, reserve and compensation: the cognitive neuroscience of healthy ageing. Nat Rev Neurosci 19: 701–710.

Cappell KA, Gmeindl L, Reuter-Lorenz PA (2010). Age differences in prefontal recruitment during verbal working memory maintenance depend on memory load. Cortex 46: 462–473.

Casagrande M, Agostini F, Favieri F et al. (2021). Age-related changes in hemispherical specialization for attentional networks. Brain Sci 11: 1115.

Castelli I, Baglio F, Blasi V et al. (2010). Effects of aging on mindreading ability through the eyes: an fMRI study. Neuropsychologia 48: 2586–2594.

Chen X, Rundle MM, Kennedy KM et al. (2022). Functional activation features of memory in successful agers across the adult lifespan. Neuroimage 257: 119276.

Cherry BJ, Hellige JB, McDowd JM (1995). Age differences and similarities in patterns of cerebral hemispheric asymmetry. Psychol Aging 10: 191–203.

Chrysikou EG, Wintering NA, Hriso C et al. (2024). Resting-state functional connectivity reveals differences in large-scale network interactions between eminent and non-eminent thinkers. Creat Res J 36: 413–423. https://doi.org/10.1080/10400419.2023.2200617.

Clark LE, Knowles JB (1973). Age differences in dichotic listening performance. J Gerontol 28: 173–178.

Colombo D, Serino S, Tuena C et al. (2017). Egocentric and allocentric spatial reference frames in aging: a systematic review. Neurosci Biobehav Rev 80: 605–621.

Corballis MC (1989). Laterality and human evolution. Psychol Rev 96: 492–505.

Corballis MC (1991). The lopsided ape: evolution of the generative mind, Oxford University Press on Demand.

Corballis PM (2003). Visuospatial processing and the right-hemisphere interpreter. Brain Cogn 53: 171–176.

Corballis MC (2017). The evolution of lateralized brain circuits. Front Psychol 8: 1021.

Corballis MC (2021). How many lateralities? Laterality 26: 307–309.

Corballis MC, Haberling IS (2017). The many sides of hemispheric asymmetry: a selective review and outlook. J Int Neuropsychol Soc 23: 710–718.

Corballis PM, Funnell MG, Gazzaniga MS (2000). An evolutionary perspective on hemispheric asymmetries. Brain Cogn 43: 112–117.

Corballis MC, Badzakova-Trajkov G, Haberling IS (2012). Right hand, left brain: genetic and evolutionary bases of cerebral asymmetries for language and manual action. Wiley Interdiscip Rev Cogn Sci 3: 1–17.

Corbetta M, Shulman GL (2002). Control of goal-directed and stimulus-driven attention in the brain. Nat Rev Neurosci 3: 201–215.

Crowell CA, Davis SW, Beynel L et al. (2020). Older adults benefit from more widespread brain network integration during working memory. Neuroimage 218: 116959.

Damasio AR, Damasio H (1983). The anatomic basis of pure alexia. Neurology 33: 1573.

de Godoy LL, Alves C, Saavedra JSM et al. (2021). Understanding brain resilience in superagers: a systematic review. Neuroradiology 63: 663–683.

Deng L, Stanley ML, Monge ZA et al. (2021). Age-related compensatory reconfiguration of PFC connections during episodic memory retrieval. Cereb Cortex 31: 717–730.

Dolcos F, Rice HJ, Cabeza R (2002). Hemispheric asymmetry and aging: right hemisphere decline or asymmetry reduction. Neurosci Biobehav Rev 26: 819–825.

Duzel E, Schutze H, Yonelinas AP et al. (2011). Functional phenotyping of successful aging in long-term memory: preserved performance in the absence of neural compensation. Hippocampus 21: 803–814.

Falk EB, Hyde LW, Mitchell C et al. (2013). What is a representative brain? Neuroscience meets population science. Proc Natl Acad Sci U S A 110: 17615–17622.

Festini SB (2022). Busyness, mental engagement, and stress: relationships to neurocognitive aging and behavior. Front Aging Neurosci 14: 1–9.

Festini SB, McDonough IM, Park DC (2016). The busier the better: greater busyness is associated with better cognition. Front Aging Neurosci 8: 1–10.

Festini SB, Zahodne L, Reuter-Lorenz PA (2018). Theoretical perspectives on age differences in brain activation: HAROLD, PASA, CRUNCH—how do they STAC up?, Oxford University Press, Oxford Research Encyclopedia of Psychology.

Fiori V, Nitsche M, Iasevoli L et al. (2017). Differential effects of bihemispheric and unihemispheric transcranial direct current stimulation in young and elderly adults in verbal learning. Behav Brain Res 321: 170–175.

Flink TS, Matos JA (2021). Age-related reductions in manual asymmetry with increased visuospatial demands. Int J Motor Control Learn 3: 1–10.

Gazzaniga MS (1970). The bisected brain, Appleton-Century-Crofts.

Goldstein G, Shelly C (1975). Similarities and differences between psychological deficit in aging and brain damage. J Gerontol 30: 448–455.

Goldstein G, Shelly C (1981). Does the right hemisphere age more rapidly than the left? J Clin Neuropsychol 3: 65–78.

Grady C (2012). The cognitive neuroscience of ageing. Nat Rev Neurosci 13: 491–505.

Grady CL, Maisog JM, Horwitz B et al. (1994). Age-related changes in cortical blood flow activation during visual processing of faces and location. J Neurosci 14: 1450–1462.

Granovetter MC, Robert S, Ettensohn L et al. (2022). With childhood hemispherectomy, one hemisphere can support-but is suboptimal for-word and face recognition. Proc Natl Acad Sci U S A 119: e2212936119.

Haberling IS, Corballis MC (2016). Cerebellar asymmetry, cortical asymmetry and handedness: two independent networks. Laterality 21: 397–414.

Hehl M, Swinnen SP, Cuypers K (2020). Alterations of hand sensorimotor function and cortical motor representations over the adult lifespan. Aging 12: 4617–4640.

Hehl M, Swinnen SP, Van Malderen S et al. (2022). No evidence for a difference in lateralization and distinctiveness level of transcranial magnetic stimulation-derived cortical motor representations over the adult lifespan. Front Aging Neurosci 14: 971858.

Hennessee JP, Webb CE, Chen X et al. (2022). Relationship of prefrontal brain lateralization to optimal cognitive function differs with age. Neuroimage 264: 119736.

Herve PY, Zago L, Petit L et al. (2013). Revisiting human hemispheric specialization with neuroimaging. Trends Cogn Sci 17: 69–80.

Heuninckx S, Wenderoth N, Swinnen SP (2008). Systems neuroplasticity in the aging brain: recruiting additional neural resources for successful motor performance in elderly persons. J Neurosci 28: 91–99.

Hill C, Van Gemmert AWA, Fang Q et al. (2020). Asymmetry in the aging brain: a narrative review of cortical activation patterns and implications for motor function. Laterality 25: 413–429.

Hiscock M, Kinsbourne M (2008). Lateralization of language across the life span. In: B Stemmer, HA Whitaker (Eds.), Handbook of the neuroscience of language. Academic Press, New York.

Hoffman P, Morcom AM (2018). Age-related changes in the neural networks supporting semantic cognition: a meta-analysis of 47 functional neuroimaging studies. Neurosci Biobehav Rev 84: 134–150.

Horn J, Cattell RB (1967). Age differences in fluid and crystallized intelligence. Acta Psychol (Amst) 26: 107–129.

Hoyau E, Boudiaf N, Cousin E et al. (2017). Aging modulates the hemispheric specialization during word production. Front Aging Neurosci 9: 125.

Iordan AD, Cooke KA, Moored KD et al. (2020). Neural correlates of working memory training: evidence for plasticity in older adults. Neuroimage 217: 116887.

Jiang J, Liu T, Crawford JD et al. (2020). Stronger bilateral functional connectivity of the frontoparietal control network in near-centenarians and centenarians without dementia. Neuroimage 215: 116855.

Jonides J, Smith EE, Koeppe RA et al. (1993). Spatial working memory in humans as revealed by PET. Nature 363: 623–625.

Kalisch T, Wilimzig C, Kleibel N et al. (2006). Age-related attenuation of dominant hand superiority. PLoS One 1: e90.

Karolis VR, Corbetta M, Thiebaut de Schotten M (2019). The architecture of functional lateralisation and its relationship to callosal connectivity in the human brain. Nat Commun 10: 1417.

Katzman R, Terry R, DeTeresa R et al. (1988). Clinical, pathological, and neurochemical changes in dementia: a subgroup with preserved mental status and numerous neocortical plaques. Ann Neurol 23: 138–144.

King BR, Rumpf JJ, Heise KF et al. (2020). Lateralized effects of post-learning transcranial direct current stimulation on motor memory consolidation in older adults: an fMRI investigation. Neuroimage 223: 117323.

Klencklen G, Despres O, Dufour A (2012). What do we know about aging and spatial cognition? Reviews and perspectives. Ageing Res Rev 11: 123–135.

Klisz D (1978). Neuropsychological evaluation in older persons. In: M Storandt, IC Siegler, MF Elias (Eds.), The clinical psychology of aging. Springer, New York.

Knights E, Morcom AM, Henson RN (2021). Does hemispheric asymmetry reduction in older adults in motor cortex reflect compensation? J Neurosci 41: 9361–9373.

Kutas M, Van Petten C (1988). Event-related brain potential studies of language. In: P Ackles, JR Jennings, MGH Coles (Eds.), Advances in psychophysiology. JAI Press, Greenwich, CT.

Learmonth G, Papadatou-Pastou M (2022). A meta-analysis of line bisection and landmark task performance in older adults. Neuropsychol Rev 32: 438–457.

Levy J, Trevarthen C (1977). Perceptual, semantic and phonetic aspects of elementary language processes in split-brain patients. Brain 100: 105–118.

Li AWY, King J (2019). Spatial memory and navigation in ageing: a systematic review of MRI and fMRI studies in healthy participants. Neurosci Biobehav Rev 103: 33–49.

Lindell AK (2011). Lateral thinkers are not so laterally minded: hemispheric asymmetry, interaction, and creativity. Laterality 16: 479–498.

Lindenberg R, Nachtigall L, Meinzer M et al. (2013). Differential effects of dual and unihemispheric motor cortex stimulation in older adults. J Neurosci 33: 9176–9183.

Lindenberger U, Baltes PB (1997). Intellectual functioning in old and very old age: cross-sectional results from the Berlin aging study. Psychol Aging 12: 410–432.

Ljubisavljevic MR, Oommen J, Filipovic S et al. (2019). Effects of tDCS of dorsolateral prefrontal cortex on dual-task performance involving manual dexterity and cognitive task in healthy older adults. Front Aging Neurosci 11: 144.

Marcori AJ, Grosso NDS, Porto AB et al. (2019). Beyond handedness: assessing younger adults and older people lateral preference in six laterality dimensions. Laterality 24: 163–175.

Martin AK, Meinzer M, Lindenberg R et al. (2017). Effects of transcranial direct current stimulation on neural networks in young and older adults. J Cogn Neurosci 29: 1817–1828.

Mather M (2020). How do cognitively stimulating activities affect cognition and the brain throughout life? Psychol Sci Public Interest 21: 1–5.

Mather M (2021). Noradrenaline in the aging brain: promoting cognitive reserve or accelerating Alzheimer's disease? Semin Cell Dev Biol 116: 108–124.

Mattay VS, Fera F, Tessitore A et al. (2006). Neurophysiological correlates of age-related changes in working memory capacity. Neurosci Lett 392: 32–37.

Naccarato M, Calautti C, Jones PS et al. (2006). Does healthy aging affect the hemispheric activation balance during paced index-to-thumb opposition task? An fMRI study. Neuroimage 32: 1250–1256.

Nebes RD, Madden DJ, Berg WD (1983). The effect of age on hemispheric asymmetry in visual and auditory identification. Exp Aging Res 9: 87–91.

Nyberg L, Boraxbekk CJ, Sorman DE et al. (2020). Biological and environmental predictors of heterogeneity in neurocognitive ageing: evidence from Betula and other longitudinal studies. Ageing Res Rev 64: 101184.

Nyberg L, Magnussen F, Lundquist A et al. (2021). Educational attainment does not influence brain aging. Proc Natl Acad Sci U S A 118. https://doi.org/10.1073/pnas.2101644118.

Obler LK, Rykhlevskaia E, Schnyer D et al. (2010). Bilateral brain regions associated with naming in older adults. Brain Lang 113: 113–123.

Olulade OA, Seydell-Greenwald A, Chambers CE et al. (2020). The neural basis of language development: changes in lateralization over age. Proc Natl Acad Sci U S A 117: 23477–23483.

Papagno C, Barvas E, Tettamanti M et al. (2021). Selective defects of face familiarity associated to a left temporo-occipital lesion. Neurol Sci 42: 613–623.

Park DC, Gutchess AH (2002). Aging, cognition, and culture: a neuroscientific perspective. Neurosci Biobehav Rev 26: 859–867.

Park DC, Reuter-Lorenz PA (2009). The adaptive brain: aging and neurocognitive scaffolding. Annu Rev Psychol 60: 173–196.

Park DC, Polk TA, Park R et al. (2004). Aging reduces neural specialization in ventral visual cortex. Proc Natl Acad Sci U S A 101: 13091–13095.

Park J, Carp J, Hebrank A et al. (2010). Neural specificity predicts fluid processing ability in older adults. J Neurosci 30: 9253–9259.

Park J, Carp J, Kennedy KM et al. (2012). Neural broadening or neural attenuation? Investigating age-related dedifferentiation in the face network in a large lifespan sample. J Neurosci 32: 2154–2158.

Poeppl TB, Dimas E, Sakreida K et al. (2022). Pattern learning reveals brain asymmetry to be linked to socioeconomic status. Cereb Cortex Commun 3: tgac020.

Posner MI, Raichle ME (1994). Images of mind, Scientific American Library/Scientific American Books.

Prokopiou PC, Engels-Dominguez N, Papp KV et al. (2022). Lower novelty-related locus coeruleus function is associated with Aβ-related cognitive decline in clinically healthy individuals. Nat Commun 13: 1571.

Pudas S, Persson J, Josefsson M et al. (2013). Brain characteristics of individuals resisting age-related cognitive decline over two decades. J Neurosci 33: 8668–8677.

Raboyeau G, De Boissezon X, Marie N et al. (2008). Right hemisphere activation in recovery from aphasia: lesion effect or function recruitment? Neurology 70: 290–298.

Renteria ME (2012). Cerebral asymmetry: a quantitative, multifactorial, and plastic brain phenotype. Twin Res Hum Genet 15: 401–413.

Reuter-Lorenz PA (2000). Cognitive neuropsychology of the aging brain. In: DC Park, N Schwarz (Eds.), Cognitive aging: a primer. Psychology Press.

Reuter-Lorenz PA (2002). New visions of the aging mind and brain. Trends Cogn Sci 6: 394–400.

Reuter-Lorenz PA, Cappell KA (2008). Neurocognitive aging and the compensation hypothesis. Curr Dir Psychol Sci 17: 177–182.

Reuter-Lorenz PA, Lustig C (2005). Brain aging: reorganizing discoveries about the aging mind. Curr Opin Neurobiol 15: 245–251.

Reuter-Lorenz PA, Mikels JA (2006). The aging mind and brain: implications of enduring plasticity for behavioral and cultural change. In: PB Baltes, PA Reuter-Lorenz, F Roesler (Eds.), Lifespan development and the brain: the perspective of biocultural co-constructivism. Cambridge University Press, Cambridge, UK.

Reuter-Lorenz PA, Park DC (2014). How does it STAC up? Revisiting the scaffolding theory of aging and cognition. Neuropsychol Rev 24: 355–370.

Reuter-Lorenz PA, Park DC (2024). Cognitive aging and the life course: a new look at the scaffolding theory. Curr Opin Psychol 56: 101781.

Reuter-Lorenz PA, Stanczak L, Miller AC (1999). Neural recruitment and cognitive aging: two hemispheres are better than one, especially as you age. Psychol Sci 10: 494–500.

Reuter-Lorenz PA, Jonides J, Smith EE et al. (2000). Age differences in the frontal lateralization of verbal and spatial working memory revealed by PET. J Cogn Neurosci 12: 174–187.

Reuter-Lorenz PA, Marshuetz C, Jonides J et al. (2001). Neurocognitive ageing of storage and executive processes. Eur J Cogn Psychol 13: 257–278.

Reuter-Lorenz PA, Herter TM, Guitton D (2011). Control of reflexive saccades following hemispherectomy. J Cogn Neurosci 23: 1368–1378.

Richmond LL, Sargent JQ, Flores S et al. (2018). Age differences in spatial memory for mediated environments. Psychol Aging 33: 892–903.

Rieck JR, Baracchini G, Nichol D et al. (2021). Reconfiguration and dedifferentiation of functional networks during cognitive control across the adult lifespan. Neurobiol Aging 106: 80–94.

Riege WH, Metter EJ, Williams MW (1980). Age and hemispheric asymmetry in nonverbal tactual memory. Neuropsychologia 18: 707–710.

Robertson IH (2014). Right hemisphere role in cognitive reserve. Neurobiol Aging 35: 1375–1385.

Roe JM, Vidal-Pineiro D, Sorensen O et al. (2021). Asymmetric thinning of the cerebral cortex across the adult lifespan is accelerated in Alzheimer's disease. Nat Commun 12: 721.

Rossion B (2018). Damasio's error—prosopagnosia with intact within-category object recognition. J Neuropsychol 12: 357–388.

Roux FE, Djidjeli I, Durand JB (2018). Functional architecture of the somatosensory homunculus detected by electrostimulation. J Physiol 596: 941–956.

Roux FE, Niare M, Charni S et al. (2020). Functional architecture of the motor homunculus detected by electrostimulation. J Physiol 598: 5487–5504.

Samuels ER, Szabadi E (2008). Functional neuroanatomy of the noradrenergic locus coeruleus: its roles in the regulation of arousal and autonomic function part I: principles of functional organisation. Curr Neuropharmacol 6: 235–253.

Saverino C, Fatima Z, Sarraf S et al. (2016). The associative memory deficit in aging is related to reduced selectivity of brain activity during encoding. J Cogn Neurosci 28: 1331–1344.

Schear JM, Nebes RD (1980). Memory for verbal and spatial information as a function of age. Exp Aging Res 6: 271–281.

Schneider-Garces NJ, Gordon BA, Brumback-Peltz CR et al. (2010). Span, CRUNCH, and beyond: working memory capacity and the aging brain. J Cogn Neurosci 22: 655–669.

Shalev N, Brosnan MB, Chechlacz M (2020). Right lateralized brain reserve offsets age-related deficits in ignoring distraction. Cereb Cortex Commun 1: tgaa049.

Shobe ER, Ross NM, Fleck JI (2009). Influence of handedness and bilateral eye movements on creativity. Brain Cogn 71: 204–214.

Siegert A, Diedrich L, Antal A (2021). New methods, old brains—a systematic review on the effects of tDCS on the cognition of elderly people. Front Hum Neurosci 15: 730134.

Smith EE, Jonides J (1998). Neuroimaging analyses of human working memory. Proc Natl Acad Sci U S A 95: 12061–12068.

Stern Y (2009). Cognitive reserve. Neuropsychologia 47: 2015–2028.

Stern Y, Alexander GE, Prohovnik I et al. (1992). Inverse relationship between education and parietotemporal perfusion deficit in Alzheimer's disease. Ann Neurol 32: 371–375.

Tagliabue CF, Varesio G, Mazza V (2022). Inter- and intra-hemispheric age-related remodeling in visuo-spatial working memory. Front Aging Neurosci 13: 807907.

Tailby C, Abbott DF, Jackson GD (2017). The diminishing dominance of the dominant hemisphere: language fMRI in focal epilepsy. Neuroimage Clin 14: 141–150.

Teixeira LA (2008). Categories of manual asymmetry and their variation with advancing age. Cortex 44: 707–716.

Thiebaut de Schotten M, Beckmann CF (2022). Asymmetry of brain structure and function: 40 years after Sperry's Nobel prize. Brain Struct Funct 227: 421–424.

Thompson PM, Jahanshad N, Schmaal L et al. (2022). The enhancing neuroimaging genetics through meta-analysis consortium: 10 years of global collaborations in human brain mapping. Hum Brain Mapp 43: 15–22.

Wernicke C (1874). Der aphasische Symptomencomplex: eine psychologische Studie auf anatomischer Basis, Cohn & Weigert.

Wilson SM, Schneck SM (2021). Neuroplasticity in post-stroke aphasia: a systematic review and meta-analysis of functional imaging studies of reorganization of language processing. Neurobiol Lang (Camb) 2: 22–82.

Woytowicz EJ, Sainburg RL, Westlake KP et al. (2020). Competition for limited neural resources in older adults leads to greater asymmetry of bilateral movements than in young adults. J Neurophysiol 123: 1295–1304.

Yin C, Imms P, Cheng M et al. (2023). Anatomically interpretable deep learning of brain age captures domain-specific cognitive impairment. Proc Natl Acad Sci U S A 120: e2214634120.

Zahodne LB, Manly JJ, Smith J et al. (2017). Socioeconomic, health, and psychosocial mediators of racial disparities in cognition in early, middle, and late adulthood. Psychol Aging 32: 118–130.

Handbook of Clinical Neurology, Vol. 208 (3rd series)
Cerebral Asymmetries
C. Papagno and P. Corballis, Editors
https://doi.org/10.1016/B978-0-443-15646-5.00003-8

Chapter 12

Development of handedness and other lateralized functions during infancy and early childhood

GEORGE F. MICHEL[1*], IRYNA BABIK[2], ELIZA L. NELSON[3], CLAUDIO L. FERRE[4], JULIE M. CAMPBELL[5], AND EMILY C. MARCINOWSKI[6]

[1]*Department of Psychology, University of North Carolina Greensboro, Greensboro, NC, United States*

[2]*Department of Psychological Science, Boise State University, Boise, ID, United States*

[3]*Department of Psychology, Florida International University, Miami, FL, United States*

[4]*Department of Occupational Therapy, College of Health & Rehabilitation Sciences, Boston University, Boston, MA, United States*

[5]*Department of Psychology, Illinois State University, Normal, IL, United States*

[6]*School of Kinesiology, Louisiana State University, Baton Rouge, LA, United States*

Abstract

Using a historical or "development from" approach to study the development of hand-use preferences in infants and children, we show how various sensorimotor experiential events shape the cascade from initial to subsequent hand-use preferences. That cascade represents, creates, and shapes the lateralized asymmetry of neural circuits in the cerebral hemispheres. The control of the preferred hand requires neural circuits in the contralateral hemisphere that are capable of processing the organization of finely timed, sequentially organized movements and detecting haptic information derived from high-frequency transitions in the stimulus. We propose that the lateralized differences in these neural circuits underlie processes contributing to the development of other forms of hemispheric specialization of function. We show how the development of hand-use preferences contributes to the development of language skills, tool use, spatial skills, and other cognitive abilities during infancy and early childhood. Such evidence supports the proposal of Michael Corballis that the phylogeny of human language emerged during the evolution of hominins from the co-option of those neural circuits employed in the expression of manual skills involved in tool use, tool manufacture, and communication. Finally, we summarize evidence from children with cerebral palsy, which shows that their difficulties with sensorimotor processing, visuomotor coordination, anticipatory motor planning, and other cognitive abilities may stem from disturbances in the development of their hand-use preferences and hence the functional specialization of their hemispheres.

INTRODUCTION

It is a great honor to contribute to a volume celebrating the life of Michael Corballis. We all were impressed by his kindness, generosity, and thoughtfulness, although only two of us (GFM, ELN) had the pleasure of meeting Professor Corballis in person. His publications are an inspiring source of clever insights and meticulous experimental execution. He provided unconventional analyses of the developmental origins of hemispheric specialization and made theoretical and empirical contributions to understanding the phylogenetic history underlying human lateralization of function, especially the relation among various forms of functional lateralization (e.g., language and handedness). His range of interest and expertise spanned from perception/sensory abilities to motor/neural mechanisms. There was little in the domain of hemispheric specialization and interhemispheric communication and coordination that his intellect and curiosity did not illuminate. We dedicate this brief chapter to the inspiring research and conceptual acumen of Michael Corballis.

Recently, one of us (GFM) published a comprehensive but personal perspective on the issues of hemispheric specialization and interhemispheric coordination (Michel, 2021). Therefore, this chapter will reflect more the influence of the other authors as they continue their examination of the development of hemispheric

*Correspondence to: George F. Michel, Professor Emeritus, Psychology Department, University of North Carolina Greensboro, Greensboro, NC, United States. Tel: +1-336-686-4047, E-mail: gfmichel@uncg.edu

specialization of function (particularly the development of handedness) in relation to other forms of cognitive development in humans. We propose that knowledge of some of the ways handedness develops can improve the design of rehabilitative and therapeutic techniques for infants and children with neural dysfunction.

CONCEPTUALIZING THE DEVELOPMENT OF HEMISPHERIC LATERALIZATION

Corballis (2002, 2003) and Michel (2021) seemed to agree that the phylogeny of human language emerged during the evolution of hominins, most likely from the co-option of those neural circuits that were employed in the expression of the manual skills involved in communication, tool use, and manufacture. In a recent test of this notion, Corballis and colleagues examined the functional relation among the neural circuits underlying tool use, communication, language, handedness, etc., using pubertal and postpubertal subjects (e.g., Badzakova-Trajkov et al., 2010; Häberling et al., 2016). The results revealed *no relation among the activation of these circuits*. This lack of any activation relation led Corballis to propose that during the evolution of primates, "… the primate mirror neuron system became increasingly lateralized, and later fissioned onto subsystems with one mediating language and the other mediating the execution and observation of manual actions. The second network is further subdivided into one dependent on hand preference and one that is not…." (Häberling et al., 2016, p. 72). Although all of these functional circuits are lateralized in humans, their independent activation could be interpreted to mean that the neural circuits for handedness could not have been co-opted during evolution for employment in the control of language functions (Häberling et al., 2016).

In contrast, Michel took a more "historical" developmental approach (Moore, 2003; Michel and Tyler, 2007) to the question whether handedness is involved in the evolution of language skills that focused on the early development of handedness and its relation to the development of language skills (Michel et al., 2013a,b, 2018). Consistent with Corballis (2002, 2003, 2017), Michel proposed that the neural circuits involved in the neuromotor skills employed in the development of tool use and manufacture could be co-opted (cf., exaptation in Gould and Vrba, 1982) for employment in the neuromotor control of speech production and comprehension (cf., Michel, 2021). However, distinct from Corballis (Corballis, 2002, 2003, 2017), Michel et al. (2013a) proposed that the development of handedness provided the foundation for the development of the laterally asymmetric control of speech in the cerebral hemispheres. In part, Michel et al. (2013a) derived their thesis from embodiment theory (Thelen et al., 2001; Smith, 2005) which proposes that aspects of infant sensorimotor development contribute toward the development of symbol formation, language skills, concept formation, and reasoning. Michel et al. (2013a) argued that those early experiences, which stem from prenatal and infant asymmetrically lateralized manual actions, could shape the functional lateralization of the hemispheres.

THE HISTORICAL APPROACH VS THE INTENTIONAL STANCE APPROACH TO DEVELOPMENT

Since much of this chapter reflects the use of a "*historical*" approach to development, we need to specify what this approach is and how it differs from the typical "*intentional stance*" in developmental research. The historical or "development from" approach (Michel and Tyler, 2007) conceptualizes the development of traits as emerging from temporal antecedents rather than as proceeding toward specific outcomes (Lehrman, 1953, 1970; Moore, 2003). The construct "intentional stance" derives from Dennett's (1988) recognition that humans (whether engaged in science or not) typically make sense of reality and the behavior of others by using teleological explanations. This is especially true for behavior and other psychologic phenomena. There is nothing intrinsically wrong with using the intentional stance in developmental research. However, the historical approach is more likely to discover nonintuitive and counterintuitive influences (often environmental or experiential, Gottlieb, 1991; Michel, 2010) in the development of psychologic traits.

The way investigators seek the developmental factors influencing a trait reveals the difference between the intentional stance and the historical approach. For the intentional stance, an influence in the development of lateralization of functions would have to begin with some genetic program for lateralization of functions; i.e., the developmental outcome (in this case, lateralization of functions in the brain) is included in the account of its causal influences. Since the intentional stance exhibits a kind of "development to" the trait of interest (Michel and Tyler, 2007), it is reasonable to seek evidence of handedness, for example, at the very earliest of phases in development. Hence, if fetuses exhibit an asymmetry of hand activity, that is *ipso facto* an early manifestation of the familiar trait of handedness. From the intentional stance approach, fetal asymmetries of hand activity are neither prerequisite for, nor precursor to, the familiar handedness of humans; they are that handedness (see Michel, 2013 for a critique of such reasoning). The connection tightens if the asymmetry exhibits any sort of right-hand bias. Additionally, development before the manifestation of hand asymmetries of activity is explained as the consequence of genetic programs creating asymmetries in the neural circuits controlling the hands. Again, the trait (hand asymmetry of use) is in the explanation (genetic program for hand asymmetry).

Therefore, from the intentional stance, it is reasonable to seek evidence of the role of manual skills in the development of speech by examining the relations among the underlying neural circuits involved in the expression of each of these abilities. Since the explanation is reduced to neural circuits, the age of the participants for such examination is not relevant, if the development of the circuits is likely to have been completed (e.g., postpubertal) at the time of testing. However, ontogenetic research from the historical approach considers psychologic traits as emerging from early precursors and prerequisites (cf., Blumberg and Adolph, 2023).

Using the historical approach to identify how species-typical behavioral traits develop has been quite successful (Alberts and Cramer, 1988; Gottlieb, 1991; Michel and Moore, 1995). This is, in part, because many of the very same cellular processes used in the development of the earliest stages of the nervous system also are employed in the structural changes that underlie perceptual and motor development and the plasticity that occurs after the onset of behavioral function. We propose that adopting a historical approach (as described in detail below) when examining the development of hemispheric specialization of function and interhemispheric communication/coordination can provide valuable insights into the causal character of development that the intentional stance often misses. Moreover, the historical approach is capable of accounting for the development of traits that are specific to the individual's culture, as well as those that are common to human behavior without shifting explanatory constructs and frameworks (e.g., genetics vs culture/learning). In the following, we will provide an account of the development of handedness and its relation to other cognitive functions that derives from an historical approach.

DEVELOPMENT OF HAND-USE PREFERENCES DURING INFANCY AS AN ASPECT OF HANDEDNESS DEVELOPMENT

Typically, developmental studies employing the historical approach require large sample sizes that are measured longitudinally at least six to nine times within a developmental period. Thus, Michel et al. (2013b, 2014) observed that hand-use preferences for acquiring objects develop differently for at least four statistically justified groups of infants. The four groups were described as: early right-hand preferers, early left-hand preferers, later developing right-hand preferers, and those with no hand preference during their first 14 months of infancy (Michel et al., 2013b, 2014). The early and later right-hand preferring infants predominate in the population (~54% of infants) with left-hand preferring infants representing only about 12% of infants. Subsequent follow-up (18–24 months of age) revealed that many those infants without a hand-use preference do develop a hand-use preference beginning at 18 months, with most exhibiting a right-hand preference (Michel et al., 2014). The remaining no-preference infants continue to manifest no hand-use preference through their first 24 months after birth.

It is important to note that the early manifestation of a hand preference is not a derivative of the infant's postural development (Babik et al., 2014); nor are the developmental transitions in infant hand preferences derivatives of general neuromotor development (Campbell et al., 2018). Rather, these are distinct behavioral traits of the infant and toddler. Therefore, according to embodiment theory, these group differences in hand preference development ought to create differences among the groups in the development of their cognitive skills (e.g., tool use, artifact construction, and language), which in turn would contribute to the functional organization of their developing nervous system, including the development of hemispheric specialization of function (Michel, 2021).

Recently, Latta (2020) found that both the hand-use preferences for acquiring objects (identified during infancy as three groups—right-preferring, left-preferring, and no preference) and those preferences identified for the role-differentiated bimanual manipulation (RDBM) during toddlerhood (same three groups) successfully predicted the hand preferences of 5-year-old children for RDBM and for the differences between their hands in their speed on the Annett peg-moving task. This demonstrates the long-term consequences of the infant's hand preference. We propose that it is unlikely that the infant's hand preference development is just contributing to the further development of their hand preference; other functional differences must be occurring.

To understand how such developmental consequences are possible, Michel (2021) first had to show how the various measures of handedness (from kinematic to self-identity) were not mapping onto the same neural system and hence were not appropriate substitutions for one another when trying to use handedness as a factor for understanding hemispheric structure–function relations. Thus, Michel (2013, 2021) argued that handedness (as measured by analyses of differential use) at different phases of development was not a homologous representation of the same underlying trait but rather represented different aspects of hand-use (reflecting different neural circuits; cf., Blumberg and Adolph, 2023) that cascade into one another during development (e.g., Michel et al., 2013b; Campbell et al., 2015a). This cascade eventually contributes to a handedness construct that is employed in self-identity which involves more complex neural circuits than a hand preference. Therefore, when examining the development of handedness in relation to the development of other psychologic functions, research should compare actual sensorimotor differences in the actions of the limbs to sensorimotor measures of the other psychologic functions (e.g., speech and problem-solving skills). Comparing the development of the self-identified handedness construct to language functions is similar to comparing the individual's self-identified favorite sport team to language functions—any observed relation would be fortuitous at best and not likely meaningful.

WHAT IS HEMISPHERIC SPECIALIZATION?

Michel (2021) argued that we know too little about the neural processing underlying the presumed functional differences between the hemispheres to rely on intuitive notions of what makes the functions different. It is likely that hemispheric differences in function reflect different forms of information processing (Albouy et al., 2020) that depend on differences in neural network architecture which affects both pattern perception and motor organization (Deco et al., 2015). Therefore, descriptions of hemispheric specialization of function ought *not* to reflect intuitive notions about functional categories. Rather, descriptions should approximate the information processing likely manifested in the neural circuitry thought to underlie the linguistic categories we use to identify functional types. Such information processing descriptions ought to resolve some confusing observations like a right-hemisphere advantage for discriminating a face from non-face stimulus (Levine et al., 1988), but a left-hemisphere advantage for discriminating two faces that differ in just one feature (e.g., Patterson and Bradshaw, 1975).

To demonstrate how such an information processing approach could clarify some of the contradictory evidence about the functional differences between hemispheres, Michel (2021) revisited Sergent's (1982a,b,c, 1985) *frequency hypothesis* of hemispheric neural processing. Sergent (1982a,b,c, 1985) hypothesized that the hemispheres differed in their preferred processing of the frequency of stimulus transitions across spatial and/or temporal patterns of sensory experience. The left hemisphere seemed better able to process high-frequency transitions in spatial and temporal sensory input which permitted more efficient detection of speech-sound differences, syntactic transitions, facial expressions and body language, complex visual arrays, differences between two faces that differed in one feature, etc. By contrast, the right hemisphere seemed more efficient in processing low-frequency transitions in spatial and temporal stimuli which permitted more efficient detection of melody, emotional tone, prosody, relative position in visual space, simpler visual arrays (e.g., a face from a non-face pattern), etc.

Sergent (1982a,b,c, 1985) suggested that the two hemispheres are capable of processing all frequencies of spatial and temporal transitions but that they exhibited differences in their efficiency and "preference" for high- vs low-frequency processing. Michel (2021) argued that such hemispheric differences in efficiency and preference could reflect a developmental process whereby differences in sensorimotor experiences between the hands, as created by hand-use preferences, could concatenate into other forms of hemispheric processing differences. To clarify how hand-use preferences could contribute to the development of hemispheric differences in stimulus processing preferences, Michel (2021) introduced the feedback loop notions that Bizzi (e.g., Bizzi and Ajemian, 2020) used for describing the control of the arms and hands in primates, particularly humans.

Bizzi and Ajemian (2020) proposed that the precise control of arm and hand movements derived from the developmental establishment of *sensorimotor feedback loops* involving the spinal cord and cortex (especially with cortical primary motor neurons and dorsal premotor neurons, but also secondary motor area neurons). The postnatal development (cf., Tiriac et al., 2014) of these corticospinal loops would create synergies of neural action derived from the feedback experience of arm and hand movements. Therefore, via supraspinal variations in the timing and sequence of the activity of a relatively small set of spinal neuromuscular synergies, a wide diversity of actions can be created that likely are constrained only by the biomechanics of the skeletal-muscle system, the conditions of the task, and the environmental context (Kutch and Valero-Cuevas, 2012). Moreover, the same spinal neuromuscular synergy may be used in a variety of actions, and different behavioral actions may be constructed from the same set of synergies by altering their relative timing and scaling factors.

It is quite likely that the synergies established in the prenatal spinal cord are influenced, during their development, by feedback from the spontaneous myogenic and neuromotor activity of the fetus and newborn (cf., Brumley et al., 2015; Robinson, 2015). Any lateralized asymmetries in such feedback would impose asymmetries in the spinal synergies. With continued developmental experience, frontal–parietal, thalamic–cortical, corticobulbar, cortical–basal ganglia, cortical-cerebellar, and interhemispheric loops would be established (cf., Blumberg and Adolph, 2023) that would exhibit similar lateral asymmetries. Michel (2021) proposed that the developmental formation of these feedback loops could serve as the mechanism whereby feedback experiences provided by manual actions during prenatal and early postnatal development could shape the development of the neural circuits underlying hemispheric specialization of function.

HOW INFANT HAND PREFERENCES COULD AFFECT HEMISPHERIC SPECIALIZATION OF FUNCTION

Successful accomplishment of any manual action requires precisely timed, sequentially organized activation of the spinal synergies controlling the muscles moving the arm, wrist, and fingers within an approximately specified "neural map" of the body that is brought into register with an approximately specified "neural map" of the environment surrounding the body. As the neuromotor feedback loops become better organized during development (via the spontaneous activity of the infant's arms and hands and their engagement with objects), any hand preferences in such actions will establish differences between the hemispheres in the circuitry of these loops/synergies. The hemisphere controlling the preferred hand will be constructing neural synergies (feedback loops) capable of controlling high-frequency transitions by the fingers and wrist needed for the precisely timed, sequentially organized actions responsible for the acquisition and manipulation of objects. The hemisphere controlling the non-preferred hand will be establishing synergies relevant for controlling low-frequency spatial and temporal transitions in neuromotor control needed to support the manipulatory/exploratory actions of the preferred hand.

It would not be surprising if, during early development when so many neural circuits are highly overlapping and interactive, preferred patterns of sensorimotor control were co-opted for creating preferred patterns for stimulus processing. In this manner, as hypothesized by Sergent (1982a,b,c, 1985), differences in hemispheric information processing are established. The development of interhemispheric communication and coordination (primarily involving the corpus callosum but also involving other neural mechanisms) would become important for refining the differences between the hemispheres in their functional efficiency and for maintaining their organization. Moreover, as development continues into puberty and beyond, it would not be surprising if the neural architecture would become more refined by creating separations of the control systems for handedness, speech, tool use, etc.

Thus, Michel (2021) provided a detailed account of the neural organization and development of manual actions that would permit handedness development to be an important contributor to the sensorimotor developmental origins of hemispheric specialization of language skills, spatial comprehension, problem-solving skills, emotional expression, reasoning, and sociability. Via feedback mechanisms, the development of handedness would affect the development of the neural architecture underlying hemispheric differences in processing spatial/temporal frequencies. This account is in sharp contrast to the universally expressed and generally accepted account that functional lateralization is a consequence of some unspecified complex of interactions among genetic, environmental, and epigenetic factors (e.g., Schmitz et al., 2017; Güntürkün et al., 2020). Although trivially correct, the latter account remains vacuous because it seeks to place the origin of lateralization of function within the realm of an unknown (and likely unknowable) pattern of interaction among some presumably independently operating causal factors (genes, environment, and epigenetics) when evidence clearly demonstrates that these factors are not independent causal operators (Michel, 2010; Michel et al., 2018).

Of course, many studies have reported that measures reflecting each of these three factors are related to the manifestation of various lateralized functions. However, presuming an unspecified interaction among these three causal factors does not reveal how these functions develop within individuals. Perhaps, the failure of this simple causal account led some (e.g., Batheja and McManus, 1985) to propose a fourth causal factor—"chance"—which makes the discovery of the development of hemispheric specialization even more dicey.

There is no denying that multiple semi-independent factors become choreographed during development into a web of causation that exhibits sufficient feedback and reciprocal influences to ensure the developmental manifestation of functions appropriate for the individual's typical socioecological conditions (Moore and Lickliter, 2023). Nevertheless, although genes are intimately involved in the functioning of cells and in the physiologic functioning of systems of cells, they are not independent causal actors in development (Michel, 2010). The cell's DNA is recruited and inhibited in its contribution to protein expression according to the secretions of other cells, which in turn may reflect the translation of environmental events and conditions into molecular processes that affect cell function.

Moreover, the construct of "environment" should be clarified as to whether it refers to factors that operate as direct influences on cellular activities (e.g., radiation, toxins, heavy metals, free radicals) or indirectly, via an influence on the development of specialized receptors that transduce physical stimuli from the environment into physiologic processes that in turn can affect the functioning of physiologic systems (e.g., hormonal secretions, neural activity, immune reactions). These two types of environmental influences in development also must be distinguished from the influences of common learning conditions (Pavlovian and operant conditioning, habit formation, sensory adaptation, imitation, parental and cultural constraints and biases, etc.) that play an important role in plasticity observed during recovery of function after neural damage (Michel, 2012). Each of these three types of environmental factors plays different (albeit sometimes overlapping) roles during the development of functional traits and characteristics (see Michel, 2010) with epigenetic processes embedded in both the genetic and environmental constructs.

THE HISTORICAL APPROACH TO THE DEVELOPMENT OF HANDEDNESS

The *historical* or "*development from*" approach proposes that traits emerge from a complex transaction among the individual's intrinsic physiologic and biomechanical characteristics, the environmental context, and the experiences provided by the environment (Michel and Tyler, 2007; Michel, 2010). Thus, the development of handedness can emerge as a cascade of contingencies and developmental processes in which early asymmetries (e.g., on a cellular level and then in utero) concatenate into later hemispheric lateralization and handedness (Michel et al., 2013b; Ferre et al., 2020). Thus, initial zygotic asymmetries (e.g., Morgan, 1977) would couple with the asymmetries of the intrauterine environment (Dunn, 1976; Fong et al., 2005) to produce the prenatal postural asymmetries (about 85% of fetuses present in the left occiput anterior or transverse uterine position, with the fetus's head being oriented down and the left ear being pressed against the mother's pelvis; Previc, 1991) that further concatenate into the manifestation of brain stem-dependent neonatal postural preferences (Schulte, 1974; Turkewitz, 1977; Michel and Goodwin, 1979; Michel, 1981). The resulting neonate's (predominantly rightward) supine head orientation preference creates an asymmetry in the sensorimotor (visual, proprioceptive, and kinesthetic) feedback from the hands, with the right hand being more active and more often in the visual field of the neonate. This feedback creates early and strong sensorimotor feedback loops (circuits) that further promote the development of motor control and visuomanual coordination (Turkewitz, 1977; Michel and Goodwin, 1979; Michel and Harkins, 1986; McFarland, 2009). These early postural asymmetries concatenate into hand-use preferences for grasping (Turkewitz, 1977), which in turn would influence the development of handedness for unimanual manipulation (Hinojosa et al., 2003; Campbell et al., 2015a), RDBM (Babik and Michel, 2016a,b), artifact construction (Marcinowski et al., 2023), and tool use (Michel et al., 2013b).

Thus, to study the historical development of any trait requires: (1) detailed descriptions of the trait; (2) description of the course of its manifestation during the lifespan (or at least the period of focus for the research); (3) specification of the physiologic processes related to its manifestation; (4) specification of the social/physical environmental characteristics that are the context of development; and (5) identification of the ways those physiologic processes and contextual characteristics create the experiences involved in the individual's developmental expression of that trait (Michel, 2021). Moreover, developmental studies require research designs involving large numbers of participants, assessed at least 9–12 times during the developmental period of interest. These design requirements permit discovery of variation in the trajectory of development that may reveal latent group differences (in timing and/or duration of events or intensity of expression of the trait) that can be compared to trajectory variations in the development of other traits (Ferre et al., 2010; Michel et al., 2014; Gilbert and Epel, 2015).

For example, in the development of infant hand-use preferences, a hand preference for the acquisition of objects via reaching and grasping precedes a preference for unimanual manipulation of objects (Campbell et al., 2015a). A hand preference for unimanual manipulation is often revealed when the nearest hand acquires an object but then transfers it to the other hand for manual, visual, and oral exploration. A preference for unimanual manipulation in turn precedes a preference for RDBM in which the non-preferred hand supports the preferred hand's exploration of a grasped object (Michel et al., 2013b). To be sure, acts of acquisition, unimanual manipulation, and RDBM are observed in neonates; however, they exhibit no evidence of the "program-like" serial organization of actions leading up to their manifestation (which can be observed in older infants and children; Kotwica et al., 2008). Thus, the earlier manifestations of these actions are likely the fortuitous consequence of the combination of the biomechanics of the arm/hand, spinal "reflex-like" neuromotor synergies, the affordances of the objects, and their availability (e.g., eyeglasses on mom's face).

Except for acquisition, both unimanual manipulation and RDBM seem to occur with some "program-like" manifestation before children start exhibiting a hand-use preference in these actions and those eventual hand preferences seem to reflect the transfer from preferences manifested in acquisition (Campbell et al., 2015a). By examining the variability among the developmental trajectories of these three aspects of infant hand preference (i.e., object acquisition, unimanual manipulation, and RDBM), we can describe typical (and, by way of contrast, atypical) patterns of development. Moreover, it would be possible to apply the same type of investigation to the development of other psychologic functions to identify the relations among their trajectories. In this manner, typical patterns of developmental trajectories may be revealed across multiple psychologic functions.

For example, Marcinowski et al. (2023) assessed the developmental trajectory of object stacking skill monthly from 18 to 24 months of age for 61 children and compared that trajectory to the trajectories revealed in development of their hand-use preferences from 6 to 24 months of age. Object stacking skill is considered to be an important marker of developmental progress during the first 3 years postpartum (Folio and Fewell, 2000). Children exhibiting both infant and toddler consistency in their hand preferences were more successful at stacking objects and they exhibited a more consistent developmental trajectory in the development of their stacking skill. Children with inconsistent developmental trajectories for their hand-use preference from infancy through toddlerhood exhibited greater variability in the trajectory of their stacking skill and generally poorer performance. It is likely that the relation of hand preference development to the development of stacking skill represents only two potential differences in the development of other forms of sensorimotor development and hemispheric specialization of function.

In what follows, we would like to show some applications of the historical approach to development when investigating the role of handedness development in the development of such cognitive skills as language and tool use (Michel et al., 2016). We also will propose that understanding the development of handedness from the historical approach can improve the rehabilitation and therapeutic techniques employed for children born with various atypical neural conditions (cf., Michel, 2012).

THE ROLE OF HANDEDNESS IN THE DEVELOPMENT OF COGNITIVE FUNCTIONS

Language

Ideally, an investigation of the relation between handedness development and language development would involve longitudinal investigation of developing hand-use preferences during infancy and early childhood, along with the development of various aspects of language development (e.g., speech sound discrimination, production, word formation, and syntactic organization) during the same period. This would permit comparison among these developmental trajectories. Note that in what follows, we can only describe how handedness development is related to a common assessment of language development (the Preschool Language Scales) that is used to assess the child's auditory comprehension (PLSAC) and expressive communication (PLSEC) at 2, 3, 4, and 5 years of age (Nelson et al., 2014, 2017; Gonzalez et al., 2020; Contino et al., 2024). Nonetheless, we consider that relation indicative.

Our methodological approach is built on three key strengths: (1) a large sample size; (2) sufficient data to calculate hand preferences statistically at each time point; and (3) a rich longitudinal design spanning from 6 months to 5 years that tracked hand use across cascading manual skills from object acquisition to RDBM with language as an outcome variable. Briefly, we assessed 90 children at monthly intervals nine times from 6 to 14 months for infant acquisition hand-use preference and seven times from 18 to 24 months for toddler RDBM hand-use preference (Nelson et al., 2014). Annual assessments for language occurred three times: at 2, 3, and 5 years. The sample was predominantly white, educated, middle class, and English was the home language.

Initial analyses revealed that children with a consistent hand preference trajectory from infancy had higher language scores, but not higher general motor or cognitive scores at 2 years of age (Nelson et al., 2014). Subsequently, Nelson et al. (2017) reported that membership in one of three toddler hand preference trajectories predicted 3-year language outcomes. Children in the consistent right trajectory for RDBM as toddlers had higher receptive and expressive language scores than children in the inconsistent right trajectory and higher expressive language scores than children in the inconsistent left trajectory (in this sample, no children had a consistent left trajectory for RDBM).

Gonzalez et al. (2020) extended these findings by showing that toddler hand preference trajectory membership also predicted 5-year language outcomes. Children in the consistent right-hand preference toddler trajectory had higher receptive and expressive language scores than children in the inconsistent left-hand preference toddler trajectory, and children in the consistent right-hand preference toddler trajectory had higher receptive language scores than children in the inconsistent right-hand preference toddler trajectory. Finally, our most recent analysis revealed that those children who during infancy exhibited different types of developmental trajectories in their hand-use preferences for acquiring objects (not RDBM) had different language skills 4 years later (Contino et al., 2024). Those infants with a consistent hand preference trajectory had higher receptive and expressive language scores at age 5 when compared to those infants with an inconsistent hand preference trajectory. Moreover, infant hand preference trajectories accounted for the variance in language outcomes beyond the typical and well-established predictors of language development of mother's education and family income (e.g., Fernald et al., 2013).

Collectively, these findings support a robust link between the development of handedness and the development of language—a link only revealed via our trajectory analysis. Findings from others have been mixed, with handedness and language loosely or not at all linked (Ramsay, 1980, 1984, 1985; Bates et al., 1986; Vauclair and Imbault, 2009; Cochet et al., 2011; Esseily et al., 2011; Vauclair and Cochet, 2013; Nicoladis and Barbosa, 2021). We have attributed these differences in results to small sample sizes, too few data points to accurately assess hand preference at any particular assessment period, and use of a single assessment time point or too few time points for trajectory analysis (Campbell et al., 2015b). We want to caution that as many as 45% of infants may be misclassified when hand preference is based on a single assessment vs a trajectory analysis (Michel et al., 2014). Although there are many logistical obstacles when undertaking any longitudinal study of this magnitude, we hope that our work inspires confidence that such a project can be accomplished and that the results of such designs will be fruitful in uncovering the mechanism for the way that handedness development could contribute to the development of language (Michel et al., 2013a).

Tool use

While many (i.e., Connolly and Dalgleish, 1989; Willatts, 1990; Lockman, 2000; O'Regan et al., 2011) have studied the ontogeny of tool use during infancy, few have considered the role of hand preference in acquiring tool-using skills. One of the few studies (McCarty et al., 1999) observed an influence of hand preference when infants were presented with a spoon containing food in the bowl. Three groups of infants (9-, 14-, and 19-month-olds) were presented with the handle end of the spoon alternately pointing to the infant's right or left side. The 9-month-old infants were more likely to grasp the handle with their preferred hand, regardless of the direction of the handle. Older infants more often used the hand ipsilateral to the handle end of the spoon, resulting in a more efficient grasp for moving food toward the mouth. The authors inferred that older infants anticipate the goal of the action, based on their use of the nonpreferred hand to grasp the spoon because using the preferred hand would be more awkward (McCarty et al., 1999). The younger infants used their preferred hand to grasp the spoon, indicating that hand preference affects the outcome of the task. Unfortunately, neither the infant's hand-preference status nor the tool-use skill was derived from a trajectory analysis; therefore, it is difficult to generalize from these results.

Another study reported that infants would use their preferred hand to grasp the handle of a tool regardless of whether the handle is ipsilateral or contralateral to their preferred hand (McCarty et al., 2001). Four age groups of infants (9, 14, 19, and 24 months) were challenged on a task like that used by McCarty et al. (1999), with a spoon containing food in the bowl. Other tasks included a hairbrush, a hammer and peg apparatus, and a magnetic handle with a separate metal object. Again, the 9-month-old infants most often used the preferred hand to grasp the tools, which often resulted in a grasp that proved inefficient for accomplishing the tool-use aspect of the task. The authors concluded that the 9-month-olds engaged in object exploration rather than tool use (McCarty et al., 2001). The young infants did not appear to regard the objects as tools, and therefore hand preference influenced how they engaged with the objects.

Rat-Fischer et al. (2012) tested infants ranging from 14 to 22 months of age on a tool task in which a rake had to be used to obtain an object. The infant's hand preference was assessed by observing their unimanual object manipulation. The results show that right-hand preferring infants used their right hand to obtain the rake, which did not change across age groups. However, infants, increasingly with age, used their non-preferred, left hand to grasp the rake when the rake was presented on the left side. The authors concluded that hand preference rather than action planning more strongly influences the manual actions of younger infants (Rat-Fischer et al., 2012).

Thus, it is likely that hand preference plays a role in the ontogeny of tool use. For young infants (9 months, McCarty et al., 1999; 9 months, McCarty et al., 2001; 14 months, Rat-Fischer et al., 2012), the preferred hand was used to obtain a tool, even when that results in an inefficient grasp which has to be corrected in order to properly use the tool. Older infants readily used their non-preferred hand when a tool was presented ipsilaterally to the non-preferred hand. Note, however, that what is missing in these studies is information about the trajectory of hand preference development and its relation to the trajectory of tool-use development (cf., Marcinowski et al., 2023, which shows how the trajectory of hand preference development can be related to the trajectory of the development of artifact construction).

For example, we (Fraz et al., 2014) tested 60 infants longitudinally from 10 to 14 months for their hand preference for object acquisition and tool use. Results showed that consistent right- or left-hand preference for object acquisition was associated with more advanced tool-use skills (i.e., higher number of correctly completed tool-use actions) across ages 10–12 months in comparison to infants without a distinct hand-use preference. This association suggests that early established hand preference for the less complicated manual skill of object acquisition may facilitate the development of tool-use skills (Michel et al., 2016).

PERINATAL BRAIN INSULT AND THE DEVELOPMENT OF HANDEDNESS

Children who experience a perinatal brain insult provide a unique window into how early injury may constrain developmental trajectories of handedness and in turn affect the development of hemispheric specialization of function. In general, the side of unilateral brain injury, or the severity of damage in the bilateral brain injury, determines the more affected hand (typically the one contralateral to the site of the more extensive neural damage), with the less affected hand becoming the preferred one for daily manual activities (Yang et al., 2005). As there is an equal chance of either hemisphere to be injured during a prenatal or perinatal brain insult, it is unsurprising that there is a nearly equal distribution of left- vs right-handedness among children with cerebral palsy (CP). In contrast to the typical right-hand bias in adults, roughly half (45%–51%) of individuals with CP demonstrate left-handedness in their daily activities (Dubey et al., 2011; Lin et al., 2012). Thus, early damage to one of the hemispheres (e.g., hypoxic ischemic encephalopathy) likely results in a subsequent atypical organization of descending motor pathways (for a review, see Friel et al., 2014), migration of the functional hand control to the less damaged or undamaged contralateral hemisphere, and the establishment of a "pathologic" handedness (Dubey et al., 2011) or, more appropriately from a *historical* approach, an "atypical" organization of hand-use control.

In typical development, activity-dependent competition between the two hemispheres likely leads to the typical pattern of organization of the descending motor neural pathways. Thus, during the first 6–18 postnatal months, infants progress from an initially more bilateral pattern of motor connectivity to a predominantly contralateral pattern, with connections originating from M1 controlling the upper limb on the opposite side of the body (Friel et al., 2014). By contrast, early brain insults in children with unilateral CP may disrupt the development of contralateral connections and result in strengthening of the ipsilateral pathways (Eyre, 2007; Gordon et al., 2013; Reid et al., 2015). Thus, the control of the affected hand resides in the ipsilateral hemisphere in about 30% of children with hemiplegic CP (Farmer et al., 1991; Staudt et al., 2004; Guzzetta et al., 2007). In addition, the smaller corpus callosum and its suboptimal white matter integrity observed in children with CP (Davatzikos et al., 2003; Kułak et al., 2007; Weinstein et al., 2014) may reduce transcallosal inhibition, thus stimulating an atypical bilateral activation in primary sensorimotor cortices and encouraging expression of mirror movements (You et al., 2005; Weinstein et al., 2014; Qin et al., 2019; Kuo et al., 2023).

Behaviorally, atypical brain lateralization in children with CP may lead to the poor control of the contra-lesion limb resulting in impaired manual performance, disrupted bimanual coordination, spasticity, and uninhibited mirror movements (Farmer et al., 1991; Hung et al., 2004; Staudt et al., 2004; Guzzetta et al., 2007; Muetzel et al., 2008; Reid et al., 2015). In addition, the continuing competition between contralateral and ipsilateral corticospinal pathways may cause disconnection between sensorimotor inputs and outputs, further exacerbating deficits in sensorimotor information processing (Qin et al., 2019). Children with hemiplegic CP often "disregard" their contra-lesion hand, preferring to use a less affected hand for all daily activities (Charles and Gordon, 2005; Houwink et al., 2011). This tendency reduces regular practice of the affected hand and disrupts the development of the feedback loops (Bizzi and Ajemian, 2020) relevant for effective motor control of the arm/hand/fingers that is essential for sophisticated object exploration and use, bimanual coordination, and self-care (Charles and Gordon, 2005; Steenbergen and Gordon, 2006; Green et al., 2013). As a result, children with CP struggle with manual activities, such as grasping and object manipulation, due to the deficits in their active range of motion, muscle strength, selective motor control, and hyperreflexia (Golubović and Slavković, 2014; Sukal-Moulton et al., 2014, 2018; Xu et al., 2020).

The functional impairments seen in older children with CP are likely a result of atypical sensorimotor experiences during early infancy that canalize trajectories and result in the atypical organization of their nervous system. Furthermore, the *spontaneous movements*, which reflect mostly subcortical and spinal activity and which predominate in the early months of infancy, also contribute to motor and sensorimotor development (Babik, 2022). For infants at risk for motor impairment, the quality of early spontaneous movements may serve as a reliable marker of children's future neurobehavioral development (Prechtl et al., 1997; Hadders-Algra et al., 2004; Groen et al., 2005; Lüchinger et al., 2008; Babik et al., 2017, 2022b). In typically developing infants, variable and complex spontaneous movements contribute to the range of motor and sensorimotor feedback loops that enable the refinement of descending motor and sensorimotor corticospinal pathways and ascending sensory pathways and facilitate the optimal development of motor control and coordination (van der Meer et al., 1995; Piek, 2002; Einspieler and Prechtl, 2005; Babik et al., 2017). Infants at risk for CP often exhibit a reduction in the variability, frequency, and quality of their spontaneous movements or display atypical movement patterns (Hadders-Algra et al., 1997; Prechtl et al., 1997; Ferrari et al., 2002; Hadders-Algra, 2007). Also, atypical expression of spontaneous movements considerably reduces children's opportunities to experience the typical sensorimotor feedback involved in the refinement of motor and sensorimotor pathways, which likely disrupts their ability to predict the consequences of their own actions and engage in adequate anticipatory motor planning (Haggard, 2017; Ritterband-Rosenbaum et al., 2019).

In children with CP, the negative effects of atypical spontaneous activity are exacerbated by the overexpression of muscle group synergies and *mirror movements* (Cahill-Rowley and Rose, 2014; Sukal-Moulton et al., 2020). These mirror movements occur when voluntary, goal-directed motor activity in one hand results in nonvoluntary, often symmetric, movements in the other hand. Typically, mirror movements diminish during the prepubertal age period and are less prevalent after puberty (Wolff et al., 1983) due to pruning of ipsilateral corticospinal connections and the improved functioning of the corpus callosum (Eyre et al., 2001; Martin, 2005; D'Souza et al., 2017; Ritterband-

Rosenbaum et al., 2017). By contrast, children with CP continue exhibiting strong mirror movements even up to 18 years of age (Kuhtz-Buschbeck et al., 2000). Muscle synergies and mirror movements disrupt independent muscle control of the fingers, hands, and arms, thus impeding the execution of functional, goal-directed movements (Kuhtz-Buschbeck et al., 2000; Cahill-Rowley and Rose, 2014). Since sophisticated RDBM of objects requires independent control of the two hands that can only be achieved via inhibition of involuntary movements, overexpression of mirror movements often disrupt the development of bimanual coordination (Connolly and Stratton, 1968; Kuhtz-Buschbeck et al., 2000).

Thus, research on children with CP provides a lot of evidence for the role of feedback loops derived from movement in the establishment of the lateralized organization of the nervous system. Yet the *intentional* stance in developmental investigations of CP becomes apparent when investigators rely on a presumed lateralized brain to explain their results. For example, the pronounced spatial deficits (Stiles and Nass, 1991; Fontes et al., 2017; Nuara et al., 2019) and difficulties with drawing, writing, and copying designs (Abercrombie and Tyson, 1966; Stiles et al., 1997; Bumin and Kavak, 2008) in children with right hemisphere insults and the slower language acquisition in children with left hemisphere insults (Bates et al., 1997; Vicari et al., 2000; Chilosi et al., 2005) are interpreted as the consequence of the *reorganization* required when the relatively unaffected hemisphere must undertake atypical functions. However, the presumed reorganization of these functions is derived only from an *intentional* stance. From the historical approach, we need to identify how the developing nervous system is dealing with the very nontypical sensorimotor experiences created by neural insults such as those from CP. When we do so, we will be able to specify the ways that the individual's development is affecting the organization of their neural control processes. Then we can create and test rehabilitation therapies that are focused on providing the experiences that have been disrupted by the insult.

COGNITIVE DEVELOPMENTAL CONSEQUENCES OF CEREBRAL PALSY

As noted above, the deficits in motor and sensorimotor processing and integration in children born preterm or with CP are associated with disruptions in their cognitive development (Babik, 2022; Babik et al., 2022a,b). For example, a cognitive aspect of motor control—*anticipatory motor planning*—is disrupted in children with CP and shows no significant improvement with age (Crajé et al., 2010). When presented with end-state comfort tasks (which assess anticipatory motor planning), children with CP sacrifice the comfort of the end-state hand posture in favor of the initial-grasp comfort (Mutsaarts et al., 2007; Crajé et al., 2010). They do not anticipate that they need to implement an awkward hand posture while grasping a handle of the turned upside-down cup to achieve a biomechanically comfortable end state with a cup ready for drinking. Such motor planning deficits seem to be more pronounced in children with left hemisphere insults (Steenbergen et al., 2004; Mutsaarts et al., 2007; Crajé et al., 2010). Importantly, interventions (i.e., constraint-induced movement therapy in combination with bimanual training) focusing on the affected hand improved motor planning assessed in the less affected hand (Crajé et al., 2010), suggesting a possibility of an intermanual transfer (Gordon et al., 2006).

The impaired anticipatory motor control is not considered a function of purely motor or sensory deficits, but rather the individual's inability to appropriately integrate sensory and motor information coming from the manipulating hand (Gordon et al., 2006). Previous research (Jeannerod and Frak, 1999; Mutsaarts et al., 2007) revealed that deficits in *motor imagery* may impede anticipatory motor planning in children with CP: their delayed and/or reduced object manipulation did not provide enough opportunities for observing the perceptual-motor consequences of their own actions and impeded their ability to perform a mental simulation that would illuminate the appropriate initial hand posture in order to achieve a biomechanically appropriate end posture. Thus, training of mental imagery skills may serve as effective intervention for improving anticipatory motor control in children with CP (Crajé et al., 2010).

Again, researchers have adopted the *intentional* stance when assessing the consequences of early neural damage on hemisphere specialization of cognitive functioning. Consider that damage to either hemisphere disrupts the typical language lateralization in children with CP. The intra- vs interhemispheric organization of language after congenital brain damage would depend on the lesion site and the age (preterm or full term) at which the brain insult occurred (Brizzolara et al., 2002; Chilosi et al., 2005). The intentional stance becomes clear when it is proposed that "only when the left lesions involved cortical-subcortical regions encroaching the temporal lobe and occurred at term age, did language *reorganize* in the right hemisphere; when lesions (whether left or right) involved only the periventricular white matter and occurred at preterm age, language was *typically lateralized* in the left hemisphere" (Brizzolara et al., 2002, p. 620, emphasis added). Although left hemisphere damage was associated with higher atypical lateralization coefficients (see Bryden and Sprott, 1981) and correspondingly more significant lexical and grammatical delays (e.g., expressive vocabulary, receptive-expressive grammar, Thal et al., 1991; Vicari et al., 2000; Chilosi et al., 2005), "cortical-subcortical-periventricular lesions rather than solely periventricular damage, and larger lesions rather than small, were associated with the *most atypical lateralization coefficients*, irrespective of lesion side" (Chilosi et al., 2005, p. 725, emphasis added).

By adopting a more *historical* (or development from) approach, we propose that the reading difficulties that are common for children with CP likely stem from their deficits in visuospatial perception and phonologic processing (Peeters et al., 2009; Critten et al., 2019). Also, deficits in visuospatial processing may negatively affect not only reading but also math skills in children with CP. Previous research demonstrated that the skill of subitizing (ability to evaluate the number of items rapidly and accurately in a presented set without counting them) is an important prerequisite for the development of arithmetic skills

(Arp et al., 2006; Desoete et al., 2009; Kroesbergen et al., 2009; Jordan et al., 2010). Since subitizing requires automatic recognition of visual patterns, such as a triangle of three dots or a rectangle of four dots (Wender and Rothkegel, 2000; Ashkenazi et al., 2013), deficits in visuospatial processing in children with CP may significantly impede their subitizing (Arp and Fagard, 2005) and subsequently their math skills.

Indeed, children with CP reportedly have a much lower subitizing limit than their typically developing peers and a sharper decline in accuracy with an increase in the number of elements in the presented set. These subitizing difficulties have been linked to the deficits in visuospatial pattern recognition and the impaired visuospatial short-term memory (Arp and Fagard, 2005). The authors adopt an intentional stance to support this notion; i.e., when they note that difficulties with subitizing seem to predominate in children with right hemisphere lesions (Arp and Fagard, 2005; Arp et al., 2006). However, the development of more sophisticated math skills heavily relies on basic counting skills, which in turn depend on sensorimotor processing and visuomanual coordination (e.g., finger recognition and discrimination, finger and hand motor control during pointing that facilitates counting; Noël, 2005; Jordan et al., 2008). Therefore, a *historical* approach might reveal more about how the nervous system achieves subitizing and the development of arithmetic skills when typical development is disrupted by brain damage.

Thus, research on children with CP reveals much about the importance of early sensorimotor experience on the development of cognitive skills (Babik, 2022; Babik et al., 2023). Children with CP may struggle with tool use, problem-solving, language, reading, and math because of their deficits in motor, sensorimotor, visuomanual, and visuospatial skills. We propose that the role of infant hand-use preferences in the developmental organization of hemispheric specialization of function and interhemispheric communication and coordination is an important component for understanding what is disrupted by CP and for identifying how that disruption might be alleviated or even prevented.

CONCLUSIONS

Using our investigation of the early development of hand-use preferences in infants and children as an example, we hope to have demonstrated the value of adopting a *historical* (development from) approach to studying the development of hemispheric specialization and interhemispheric coordination/communication. The historical approach conceptualizes the development of traits as emerging from a complex transaction among the individual's intrinsic physiologic and biomechanical characteristics, the environmental context, and the experiences provided by the environment. Thus, the development of handedness represents a cascade of contingencies and developmental processes in which early neurobehavioral asymmetries concatenate into later hemispheric lateralization and handedness (Michel, 2021). This account contrasts with the more common *intentional* stance that considers development as proceeding toward the manifestation of traits such as handedness from genetic programs designed by evolution to govern the process. As a result, the historical approach is more likely to discover nonintuitive and counterintuitive influences in the development of a trait than will the intentional stance approach.

We showed that adoption of the historical approach in the development of hand-use preferences (combined with large sample sizes and frequent longitudinal assessments that permit trajectory analyses) can reveal the exact ways sensorimotor development contributes to the development of hemispheric specialization. In this manner, we provided support for the Corballis argument (Corballis, 2017) that the evolution of language and hemispheric specialization emerged from those neural circuits that were employed in the expression of manual skills involved in communication, tool use, and manufacture. We have proposed also that this historical approach may provide insights into more effective ways of treating children with CP or other forms of neural insult. We hope that this chapter will serve as encouragement for other researchers to undertake a historical approach for their developmental investigations.

ACKNOWLEDGMENTS

Except for the first author, the authorship order represents a pseudorandom order. All authors contributed equally to the construction and writing of this chapter. GFM takes full responsibility for any errors.

CONFLICT OF INTEREST

The authors have no conflict of interests to declare.

STATEMENTS OF ETHICS

No humans were challenged in this project and all of the data were collected following IRB guidelines.

References

Abercrombie MLJ, Tyson MC (1966). Body image and draw-a-man test in cerebral palsy. D Med Child Neurol 8: 9–15.

Alberts JR, Cramer CP (1988). Ecology and experience: sources of means and meaning of developmental change. In: EM Blass (Ed.), Handbook of behavioral neurobiology. Developmental psychobiology and behavioral ecology. vol. 9. pp. 1–39.

Albouy P, Benjamin L, Morillon B et al. (2020). Distinct sensitivity to spectrotemporal modulation supports brain asymmetry for speech and melody. Science 367: 1043–1047.

Arp S, Fagard J (2005). What impairs subitizing in cerebral palsied children? Dev Psychobiol 47: 89–102.

Arp S, Taranne P, Fagard J (2006). Global perception of small numerosities (subitizing) in cerebral-palsied children. J Clin Exp Neuropsychol 28: 405–419.

Ashkenazi S, Mark-Zigdon N, Henik A (2013). Do subitizing deficits in developmental dyscalculia involve pattern recognition weakness? Dev Sci 16: 35–46.

Babik I (2022). From hemispheric asymmetry through sensorimotor experiences to cognitive outcomes in children with cerebral palsy. Symmetry 14: 345.

Babik I, Michel GF (2016a). Development of role-differentiated bimanual manipulation in infancy: part 1. The emergence of the skill. Dev Psychobiol 58: 243–256.

Babik I, Michel GF (2016b). Development of role-differentiated bimanual manipulation in infancy: part 2. Hand preferences for object acquisition and RDBM—continuity or discontinuity? Dev Psychobiol 58: 257–267.

Babik I, Campbell JM, Michel GF (2014). Postural influences on the development of infant lateralized and symmetrical hand-use. Child Dev 85: 294–307.

Babik I, Galloway JC, Lobo MA (2017). Infants born preterm demonstrate impaired exploration of their bodies and surfaces throughout the first 2 years of life. Phys Ther 97: 915–925.

Babik I, Cunha AB, Lobo MA (2022a). A model for using developmental science to create effective early intervention programs and technologies to improve children's developmental outcomes. In: R Gilmore, J Lockman (Eds.), Advances in child development. vol. 62. Elsevier Inc, pp. 231–268.

Babik I, Galloway JC, Lobo MA (2022b). Early exploration of one's own body, exploration of objects, and motor, language, and cognitive development relate dynamically across the first two years of life. Dev Psychol 58: 222–235.

Babik I, Cunha AB, Srinivasan S (2023). Biological and environmental factors may affect children's executive function through motor and sensorimotor processing. Preterm birth and cerebral palsy. Infant Behav Dev 73: 101881.

Badzakova-Trajkov G, Häberling IS, Roberts RP et al. (2010). Cerebral asymmetries: complementary and independent processes. PLoS One 5: e9682.

Bates E, O'Connell B, Vaid J et al. (1986). Language and hand preference in early development. Dev Neuropsychol 2: 1–15.

Bates E, Thal D, Trauner D et al. (1997). From first words to grammar in children with focal brain injury. Dev Neuropsychol 13: 275–343.

Batheja M, McManus IC (1985). Handedness in the mentally handicapped. Dev Med Child Neurol 27: 63–68.

Bizzi E, Ajemian R (2020). From motor planning to execution: a sensorimotor loop perspective. J Neurophysiol 124: 1815–1823.

Blumberg MS, Adolph KE (2023). Protracted development of motor cortex constrains rich interpretations of infant cognition. Trends Cogn Sci 27: 233–245.

Brizzolara D, Pecini C, Brovedani P et al. (2002). Timing and type of congenital brain lesion determine different patterns of language lateralization in hemiplegic children. Neuropsychologia 40: 620–632.

Brumley MR, Kauer SD, Swann HE (2015). Developmental plasticity of coordinated action patterns in the perinatal rat. Dev Psychobiol 57: 409–420.

Bryden MP, Sprott DA (1981). Statistical determination of degree of laterality. Neuropsychologia 19: 571–581.

Bumin G, Kavak ST (2008). An investigation of the factors affecting handwriting performance in children with hemiplegic cerebral palsy. Disabil Rehabil 30: 1374–1385.

Cahill-Rowley K, Rose J (2014). Etiology of impaired selective motor control: emerging evidence and its implications for research and treatment in cerebral palsy. Dev Med Child Neurol 56: 522–528.

Campbell JM, Marcinowski EC, Babik I et al. (2015a). The influence of a hand preference for acquiring objects on the development of a hand preference for unimanual manipulation from 6 to 14 months. Infant Behav Dev 39: 107–117.

Campbell JM, Marcinowski EC, Latta J et al. (2015b). Different assessment tasks produce different estimates of handedness stability during the eight to 14-month age period. Infant Behav Dev 39: 67–80.

Campbell JM, Marcinowski EC, Michel GF (2018). The development of neuromotor skills and hand preference during infancy. Dev Psychobiol 60: 165–175.

Charles J, Gordon AM (2005). A critical review of constraint-induced movement therapy and forced use in children with hemiplegia. Neural Plast 12: 245–261.

Chilosi AM, Pecini C, Cipriani P et al. (2005). Atypical language lateralization and early linguistic development in children with focal brain lesions. Dev Med Child Neurol 47: 725–730.

Cochet H, Jover M, Vauclair J (2011). Hand preference for pointing gestures and bimanual manipulation around the vocabulary spurt period. J Exp Child Psychol 110: 393–407.

Connolly K, Dalgleish M (1989). The emergence of a tool-using skill in infancy. Dev Psychol 25: 894.

Connolly K, Stratton P (1968). Developmental changes in associated movements. Dev Med Child Neurol 10: 49–56.

Contino K, Campbell JM, Marcinowski EC et al. (2024). Hand preference trajectories as predictors of language outcomes above and beyond SES: infant patterns explain more variance than toddler patterns at 5 years of age. Infant Child Dev 33: e2468. https://doi.org/10.1002/icd.2468.

Corballis MC (2002). Hemisphere interactions in simple reaction time. Neuropsychologia 40: 423–434.

Corballis MC (2003). From mouth to hand: gesture speech, and the evolution of right handedness. Behav Brain Sci 26: 199–208.

Corballis MC (2017). Language evolution: a changing perspective. Trends Cogn Sci 21: 229–236.

Crajé C, Aarts P, Nijhuis-van der Sanden M et al. (2010). Action planning in typically and atypically developing children (unilateral cerebral palsy). Res Dev Disabil 31: 1039–1046.

Critten V, Messer D, Sheehy K (2019). Delays in the reading and spelling of children with cerebral palsy: associations with phonological and visual processes. Res Dev Disabil 85: 131–142.

Davatzikos C, Barzi A, Lawrie T et al. (2003). Correlation of corpus callosal morphometry with cognitive and motor function in periventricular leukomalacia. Neuropediatrics 34: 247–252.

Deco G, Tononi G, Boly M et al. (2015). Rethinking segregation and integration: contributions of whole-brain modelling. Nat Rev Neurosci 16: 430–439.

Dennett DC (1988). Précis of the intentional stance. Behav Brain Sci 11: 495–505.

Desoete A, Ceulemans A, Roeyers H et al. (2009). Subitizing or counting as possible screening variables for learning disabilities in mathematics education or learning? Educ Res Rev 4: 55–66.

D'Souza H, Cowie D, Karmiloff-Smith A et al. (2017). Specialization of the motor system in infancy: from broad tuning to selectively specialized purposeful actions. Dev Sci 20: e12409.

Dubey D, Amritphale A, Sawhney A et al. (2011). Impact of cerebral palsy on hand preference with special reference to premature birth. Pediatr Oncall 8: 97–99.

Dunn PM (1976). Congenital postural deformities. Br Med Bull 32: 71–76.

Einspieler C, Prechtl HF (2005). Prechtl's assessment of general movements: a diagnostic tool for the functional assessment of the young nervous system. Ment Retard Dev Disabil Res Rev 11: 61–67.

Esseily R, Jacquet A-Y, Fagard J (2011). Handedness for grasping objects and pointing and the development of language in 14-month-old infants. Laterality 16: 565–585.

Eyre JA (2007). Corticospinal tract development and its plasticity after perinatal injury. Neurosci Biobehav Rev 31: 1136–1149.

Eyre JA, Taylor JP, Villagra F et al. (2001). Evidence of activity-dependent withdrawal of corticospinal projections during human development. Neurology 57: 1543–1554.

Farmer SF, Harrison LM, Ingram DA et al. (1991). Plasticity of central motor pathways in children with hemiplegic cerebral palsy. Neurology 41: 1505.

Fernald A, Marchman VA, Weisleder A (2013). SES differences in language processing skill and vocabulary are evident at 18 months. Dev Sci 16: 234–248.

Ferrari F, Cioni G, Einspieler C et al. (2002). Cramped synchronized general movements in preterm infants as an early marker for cerebral palsy. Arch Pediatr Adolesc Med 156: 460–467.

Ferre CL, Babik I, Michel GF (2010). Development of infant prehension handedness: a longitudinal analysis during the 6- to 14-month age period. Infant Behav Dev 33: 492–502.

Ferre CL, Babik I, Michel GF (2020). A perspective on the development of hemispheric specialization, infant handedness, and cerebral palsy. Cortex 127: 208–220.

Folio M, Fewell R (2000). Peabody developmental motor scales – second edition: examiner's manual, Pro-Ed, Austin, TX.

Fong BF, Savelsbergh GJP, van Geijn HP et al. (2005). Does intrauterine environment influence fetal head-position preference? A comparison between breech and cephalic presentation. Early Hum Dev 81: 507–517.

Fontes PLB, Cruz TKF, Souto DO et al. (2017). Body representation in children with hemiplegic cerebral palsy. Child Neuropsychol 23: 838–863.

Fraz F, Babik I, Varholick J et al. (2014). Development of hand-use preference for tool-use in infancy. Dev Psychobiol 57: S13.

Friel KM, Williams PT, Serradj N et al. (2014). Activity-based therapies for repair of the corticospinal system injured during development. Front Neurol 5: 229.

Gilbert SF, Epel D (2015). Ecological developmental biology: the environmental regulation of development, health, and evolution, second edn. Sinauer.

Golubović Š, Slavković S (2014). Manual ability and manual dexterity in children with cerebral palsy. Hippokratia 18: 310.

Gonzalez SL, Campbell JM, Marcinowski EC et al. (2020). Preschool language ability is predicted by toddler hand preference trajectories. Dev Psychol 56: 699–709.

Gordon AM, Charles J, Steenbergen B (2006). Fingertip force planning during grasp is disrupted by impaired sensorimotor integration in children with hemiplegic cerebral palsy. Pediatr Res 60: 587–591.

Gordon AM, Bleyenheuft Y, Steenbergen B (2013). Pathophysiology of impaired hand function in children with unilateral cerebral palsy. Dev Med Child Neurol 55: 32–37.

Gottlieb G (1991). Experiential canalization of behavioral development: theory. Dev Psychol 27: 4–13.

Gould SJ, Vrba ES (1982). Exaptation—a missing term in the science of form. Paleobiology 8: 4–15.

Green D, Schertz M, Gordon AM et al. (2013). A multi-site study of functional outcomes following a themed approach to hand-arm bimanual intensive therapy for children with hemiplegia. Dev Med Child Neurol 55: 527–533.

Groen SE, de Blécourt ACE, Postema K et al. (2005). General movements in early infancy predict neuromotor development at 9 to 12 years of age. Dev Med Child Neurol 47: 731–738.

Güntürkün O, Ströckens F, Ocklenburg S (2020). Brain lateralization: a comparative perspective. Physiol Rev 100: 1019–1063.

Guzzetta A, Bonanni P, Biagi L et al. (2007). Reorganisation of the somatosensory system after early brain damage. Clin Neurophysiol 118: 1110–1121.

Häberling IS, Corballis PM, Corballis MC (2016). Language, gesture, and handedness: evidence for independent lateralized networks. Cortex 82: 72–85.

Hadders-Algra M (2007). Putative neural substrate of normal and abnormal general movements. Neurosci Biobehav Rev 31: 1181–1190.

Hadders-Algra M, Van den Nieuwendijk AWK, Maitijn A et al. (1997). Assessment of general movements: towards a better understanding of a sensitive method to evaluate brain function in young infants. Dev Med Child Neurol 39: 88–98.

Hadders-Algra M, Mavinkurve-Groothuis AM, Groen SE et al. (2004). Quality of general movements and the development of minor neurological dysfunction at toddler and school age. Clin Rehabil 18: 287–299.

Haggard P (2017). Sense of agency in the human brain. Nat Rev Neurosci 18: 196–207.

Hinojosa T, Sheu CF, Michel GF (2003). Infant hand-use preferences for grasping objects contributes to the development of a hand-use preference for manipulating objects. Dev Psychobiol 43: 328–334.

Houwink A, Aarts PB, Geurts AC et al. (2011). A neurocognitive perspective on developmental disregard in children with hemiplegic cerebral palsy. Res Dev Disabil 32: 2157–2163.

Hung YC, Charles J, Gordon AM (2004). Bimanual coordination during a goal-directed task in children with hemiplegic cerebral palsy. Dev Med Child Neurol 46: 746–753.

Jeannerod M, Frak V (1999). Mental imaging of motor activity in humans. Curr Opin Neurobiol 9: 735–739.

Jordan NC, Kaplan D, Ramineni C et al. (2008). Development of number combination skill in the early school years: when do fingers help? Dev Sci 11: 662–668.

Jordan NC, Glutting J, Ramineni C (2010). The importance of number sense to mathematics achievement in first and third grades. Learn Individ Differ 20: 82–88.

Kotwica KA, Ferre CL, Michel GF (2008). Relation of stable hand-use preferences to the development of skill for managing multiple objects from 7 to 13 months of age. Dev Psychobiol 50: 519–529.

Kroesbergen EH, Van Luit JEH, Van Lieshout ECDM et al. (2009). Individual differences in early numeracy: the role of executive functions and subitizing. J Psychoeduc Assess 27: 226–236.

Kuhtz-Buschbeck JP, Sundholm LK, Eliasson AC et al. (2000). Quantitative assessment of mirror movements in children and adolescents with hemiplegic cerebral palsy. Dev Med Child Neurol 42: 728–736.

Kułak W, Sobaniec W, Kubas B et al. (2007). Corpus callosum size in children with spastic cerebral palsy: relationship to clinical outcome. J Child Neurol 22: 371–374.

Kuo HC, Ferre CL, Chin KY et al. (2023). Mirror movements and brain pathology in children with unilateral cerebral palsy. Dev Med Child Neurol 65: 264–273.

Kutch JJ, Valero-Cuevas FJ (2012). Challenges and new approaches to proving the existence of muscle synergies of neural origin. PLoS Comput Biol 8: e1002434.

Latta J. (2020). Predicting child handedness from measures of infant and toddler handedness. (Publication No. 27739908). Master's thesis, University of North Carolina Greensboro, ProQuest dissertations and theses global.

Lehrman DS (1953). A critique of Konrad Lorenz's theory of instinctive behavior. Q Rev Biol 28: 337–363.

Lehrman DS (1970). Semantic and conceptual issues in the nature-nurture problem. In: LR Aronson et al. (Eds.), Development and evolution of behavior. Essays in memory of TC Schneirla. WH Freeman and Company, pp. 17–52.

Levine SC, Banich MT, Koch-Weser MP (1988). Face recognition: a general or specific right hemisphere capacity? Brain Cogn 8: 303–325.

Lin KR, Prabhu V, Shah H et al. (2012). Handedness in diplegic cerebral palsy. Dev Neurorehabil 15: 386–389.

Lockman JJ (2000). A perception–action perspective on tool use development. Child Dev 71: 137–144.

Lüchinger AB, Hadders-Algra M, van Kan CM et al. (2008). Fetal onset of general movements. Pediatr Res 63: 191–195.

Marcinowski EC, Nelson EL, Campbell JM et al. (2023). Early, concurrent, and consistent hand preferences predict stacking in toddlerhood. Dev Psychobiol 65: e22397.

Martin JH (2005). The corticospinal system: from development to motor control. Neuroscientist 11: 161–173.

McCarty ME, Clifton RK, Collard RR (1999). Problem solving in infancy: the emergence of an action plan. Dev Psychol 35: 1091.

McCarty ME, Clifton RK, Collard RR (2001). The beginnings of tool use by infants and toddlers. Inf Dent 2: 233–256.

McFarland K (2009). Motor control: pyramidal, extrapyramidal, and limbic motor control. In: GG Berntson, JT Cacioppo (Eds.), Handbook of neuroscience for the behavioral sciences. Wiley, pp. 431–453.

Michel GF (1981). Right-handedness: a consequence of infant supine head-orientation preference? Science 212: 685–687.

Michel GF (2010). The roles of environment, experience, and learning in behavioral development. In: K Hood, C Halpern, G Greenberg, R Lerner (Eds.), Handbook of developmental science, behavior and genetics. Wiley, pp. 123–165.

Michel GF (2012). Using knowledge of development to promote recovery of function after brain damage. Dev Psychobiol 54: 350–356.

Michel GF (2013). The concept of homology in the development of handedness. Dev Psychobiol 55: 84–91.

Michel GF (2021). Handedness development: a model for investigating the development of hemispheric specialization and interhemispheric coordination. Symmetry 13: 992.

Michel GF, Goodwin R (1979). Intrauterine birth position predicts newborn supine head position preferences. Infant Behav Dev 2: 29–38.

Michel GF, Harkins DA (1986). Postural and lateral asymmetries in the ontogeny of handedness during infancy. Dev Psychobiol 19: 247–258.

Michel GF, Moore CL (1995). Developmental psychobiology: an interdisciplinary science, MIT press.

Michel GF, Tyler AN (2007). Developing human nature: "development to" versus "development from?". Dev Psychobiol 49: 788–799.

Michel GF, Babik I, Nelson EL et al. (2013a). How the development of handedness could contribute to the development of language. Dev Psychobiol 55: 608–620.

Michel GF, Nelson EL, Babik I et al. (2013b). Multiple trajectories in the developmental psychobiology of human handedness. Adv Child Dev Behav 45: 227–260.

Michel GF, Babik I, Sheu CF et al. (2014). Latent classes in the developmental trajectories of infant handedness. Dev Psychol 50: 349–359.

Michel GF, Campbell JM, Marcinowski EC et al. (2016). Infant hand preference and the development of cognitive abilities. Front Psychol 7: 410.

Michel GF, Babik I, Nelson EL et al. (2018). Evolution and development of handedness: an Evo-Devo approach. Prog Brain Res 238: 347–374.

Moore CL (2003). Differences between the organism-environment systems conceived by Lehrman and Gibson: what's in the nest of reciprocities matters. Dev Psychobiol 45: 349–356.

Moore DS, Lickliter R (2023). Development as explanation: understanding phenotypic stability and variability after the failure of genetic determinism. Prog Biophys Mol Biol 178: 72–77.

Morgan M (1977). Embryology and inheritance of asymmetry. In: S Harnad, RW Doty, L Goldstein, J Jaynes, G Krauthamer (Eds.), Lateralization in the nervous system. Academic Press, pp. 173–194.

Muetzel RL, Collins PF, Mueller BA et al. (2008). The development of corpus callosum microstructure and associations with bimanual task performance in healthy adolescents. Neuroimage 39: 1918–1925.

Mutsaarts M, Steenbergen B, Bekkering H (2007). Impaired motor imagery in right hemiparetic cerebral palsy. Neuropsychologia 45: 853–859.

Nelson EL, Campbell JM, Michel GF (2014). Early handedness in infancy predicts language ability in toddlers. Dev Psychol 50: 809–814.

Nelson EL, Gonzalez SL, Coxe S et al. (2017). Toddler hand preference trajectories predict 3-year language outcome. Dev Psychobiol 59: 876–887.

Nicoladis E, Barbosa P (2021). Nine-month old infants' hand preference shows no link to concurrent or later vocabulary. Dev Psychobiol 63: e22160.

Noël MP (2005). Finger gnosia: a predictor of numerical abilities in children? Child Neuropsychol 11: 413–430.

Nuara A, Papangelo P, Avanzini P et al. (2019). Body representation in children with unilateral cerebral palsy. Front Psychol 354: 1–6.

O'Regan JK, Rat-Fischer L, Fagard J (2011). Mechanisms leading to tool use: a longitudinal study in human infants. In: Paper presented at the *IEEE ICDL-EPIROB 2011*, Frankfurt.

Patterson K, Bradshaw JL (1975). Differential hemispheric mediation of non-verbal visual stimuli. J Exp Psychol Hum Percept Perform 1: 246–252.

Peeters M, Verhoeven L, de Moor J et al. (2009). Importance of speech production for phonological awareness and word decoding: the case of children with cerebral palsy. Res Dev Disabil 30: 712–726.

Piek JP (2002). The role of variability in early motor development. Infant Behav Dev 25: 452–465.

Prechtl HF, Einspieler C, Cioni G et al. (1997). An early marker for neurological deficits after perinatal brain lesions. Lancet 349: 1361–1363.

Previc FH (1991). A general theory concerning the prenatal origins of cerebral lateralization in humans. Psychol Rev 98: 299.

Qin Y, Sun B, Zhang H et al. (2019). Aberrant interhemispheric functional organization in children with dyskinetic cerebral palsy. Biomed Res Int 2019: 4362539.

Ramsay DS (1980). Onset of unimanual handedness in infants. Infant Behav Dev 3: 377–385.

Ramsay DS (1984). Onset of duplicated syllable babbling and unimanual handedness in infancy: evidence for developmental change in hemispheric specialization? Dev Psychol 20: 64–71.

Ramsay DS (1985). Fluctuations in unimanual hand preference in infants following the onset of duplicated syllable babbling. Dev Psychol 21: 318–324.

Rat-Fischer L, O'Regan JK, Fagard J (2012). The emergence of tool use during the second year of life. J Exp Child Psychol 113: 440–446.

Reid LB, Rose SE, Boyd RN (2015). Rehabilitation and neuroplasticity in children with unilateral cerebral palsy. Nat Rev Neurol 11: 390–400.

Ritterband-Rosenbaum A, Herskind A, Li X et al. (2017). A critical period of corticomuscular and EMG–EMG coherence detection in healthy infants aged 9-25 weeks. J Physiol 595: 2699–2713.

Ritterband-Rosenbaum A, Justiniano MD, Nielsen JB et al. (2019). Are sensorimotor experiences the key for successful early intervention in infants with congenital brain lesion? Infant Behav Dev 54: 133–139.

Robinson SR (2015). Spinal mediation of motor learning and memory in the rat fetus. Dev Psychobiol 57: 421–434.

Schmitz J, Lor S, Klose R et al. (2017). The functional genetics of handedness and language lateralization: insights from gene ontology, pathway and disease association analyses. Front Psychol 8: 1144.

Schulte FJ (1974). The neurological development of the neonate. In: JA Davis, J Debbing (Eds.), Scientific foundations of pediatrics. Saunders, pp. 587–615.

Sergent J (1982a). Basic determinants in visual-field effects with special reference to the Hannay et al. (1981) study. Brain Lang 16: 158–164.

Sergent J (1982b). The cerebral balance of power: confrontation or cooperation? J Exp Psychol Hum Percept Perform 8: 253–272.

Sergent J (1982c). Theoretical and methodological consequences of variations in exposure duration in visual laterality studies. Percept Psychophys 31: 451–461.

Sergent J (1985). Influence of task and input factors on hemispheric involvement in face processing. J Exp Psychol Hum Percept Perform 11: 846–861.

Smith LB (2005). Cognition as a dynamic system: principles from embodiment. Dev Rev 25: 278–298.

Staudt M, Gerloff C, Grodd W et al. (2004). Reorganization in congenital hemiparesis acquired at different gestational ages. Ann Neurol 56: 854–863.

Steenbergen B, Gordon AM (2006). Activity limitation in hemiplegic cerebral palsy: evidence for disorders in motor planning. Dev Med Child Neurol 48: 780–783.

Steenbergen B, Meulenbroek RG, Rosenbaum DA (2004). Constraints on grip selection in hemiparetic cerebral palsy: effects of lesional side, end-point accuracy, and context. Cogn Brain Res 19: 145–159.

Stiles J, Nass R (1991). Spatial grouping activity in young children with congenital right or left hemisphere brain injury. Brain Cogn 15: 201–222.

Stiles J, Trauner D, Engel M et al. (1997). The development of drawing in children with congenital focal brain injury: evidence for limited functional recovery. Neuropsychologia 35: 299–312.

Sukal-Moulton T, Krosschell KJ, Gaebler-Spira DJ et al. (2014). Motor impairment factors related to brain injury timing in early hemiparesis, part I: expression of upper-extremity weakness. Neurorehabil Neural Repair 28: 13–23.

Sukal-Moulton T, Gaebler-Spira D, Krosschell KJ (2018). The validity and reliability of the Test of Arm Selective Control for children with cerebral palsy: a prospective cross-sectional study. Dev Med Child Neurol 60: 374–381.

Sukal-Moulton T, De Campos AC, Alter KE et al. (2020). Functional near-infrared spectroscopy to assess sensorimotor cortical activity during hand squeezing and ankle dorsiflexion in individuals with and without bilateral and unilateral cerebral palsy. Neurophotonics 7: 045001.

Thal DJ, Marchman V, Stiles J et al. (1991). Early lexical development in children with focal brain injury. Brain Lang 40: 491–527.

Thelen E, Schöner G, Scheier C et al. (2001). The dynamics of embodiment: a field theory of infant perseverative reaching. Behav Brain Sci 24: 1–34.

Tiriac A, Del Rio-Bermudez C, Blumberg MS (2014). Self-generated movements with 'unexpected' sensory consequences. Curr Biol 24: 2136–2141.

Turkewitz G (1977). The development of lateral differences in the human infant. In: S Harnad, RW Doty, I Goldstein, J Jaynes, G Krauthammer (Eds.), Lateralization in the nervous system. Academic Press, pp. 251–260.

van der Meer ALH, van der Weel FR, Lee DN (1995). The functional significance of arm movements in neonates. Science 267: 693–695.

Vauclair J, Cochet H (2013). Hand preference for pointing and language development in toddlers. Dev Psychobiol 55: 757–765.

Vauclair J, Imbault J (2009). Relationship between manual preferences for object manipulation and pointing gestures in infants and toddlers. Dev Sci 12: 1060–1069.

Vicari S, Albertoni A, Chilosi AM et al. (2000). Plasticity and reorganization during language development in children with early brain injury. Cortex 36: 31–46.

Weinstein M, Green D, Geva R et al. (2014). Interhemispheric and intrahemispheric connectivity and manual skills in children with unilateral cerebral palsy. Brain Struct Funct 219: 1025–1040.

Wender KF, Rothkegel R (2000). Subitizing and its subprocesses. Psychol Res 64: 81–92.

Willatts P (1990). Development of problem solving strategies in infancy. In: DF Bjorklund (Ed.), Children's strategies: contemporary views of cognitive development. Erlbaum, pp. 23–66.

Wolff PH, Gunnoe CE, Cohen C (1983). Associated movements as a measure of developmental age. Dev Med Child Neurol 25: 417–427.

Xu D, Wu YN, Gaebler-Spira D et al. (2020). Neural and non-neural contributions to ankle spasticity in children with cerebral palsy. Dev Med Child Neurol 62: 1040–1046.

Yang LJS, Anand P, Birch R (2005). Limb preference in children with obstetric brachial plexus palsy. Pediatr Neurol 33: 46–49.

You SH, Jang SH, Kim YH et al. (2005). Cortical reorganization induced by virtual reality therapy in a child with hemiparetic cerebral palsy. Dev Med Child Neurol 47: 628–635.

Section V

Genetics, hormones, and evolution

Handbook of Clinical Neurology, Vol. 208 (3rd series)
Cerebral Asymmetries
C. Papagno and P. Corballis, Editors
https://doi.org/10.1016/B978-0-443-15646-5.00007-5

Chapter 13

Handedness and brain asymmetries in nonhuman primates

WILLIAM D. HOPKINS[1]* AND ADRIEN MEGUERDITCHIAN[2]

[1]*Department of Comparative Medicine, The University of Texas MD Anderson Cancer Center, Bastrop, TX, United States*

[2]*Laboratoire de Psychologie Cognitive, CNRS, Aix-Marseille University, Institute of Language, Communication and the Brain, Marseille, France*

Abstract

A majority of humans are right-handed and exhibit left hemisphere specialization for the comprehension and production of language. To what extent population-level behavioral and brain asymmetries are unique to humans remains a topic of interest across a wide range of scientific disciplines. In this chapter, we present current findings on the expression of population-level behavioral and brain asymmetries in nonhuman primates. We further present data on the association between communication functions, and especially gestures and individual variation in neuroanatomic asymmetries in nonhuman primates, with an emphasis on data from chimpanzees and baboons. The collective data are interpreted within the context of different theories on the evolution of language lateralization.

INTRODUCTION

A fundamental feature of the human nervous system is hemispheric specialization. Hemispheric specialization refers to sensory, motor, and cognitive processes that are differentially performed by neural structures and connected regions within and between the left and right hemispheres. Hemispheric specialization is also manifest in morphologic features of the brain including left–right differences in size or area in specific regions of interest or, more recently, in white matter pathways that connect different brain regions. The most well-documented functional and neuroanatomic asymmetries in humans are associated with language and praxic functions, which are predominantly lateralized to the left hemisphere in the majority of human subjects (Knecht et al., 2000; Packheiser et al., 2020). Because language and speech are considered uniquely human, early theories postulated that lateralization in neural structure and function evolved in response to increasing selection for the motor and cognitive demands of praxic, communicative, and cognitive functions that underlie manual and linguistic functions (Warren, 1980; Ettlinger, 1988; Corballis, 1992; Bradshaw and Rogers, 1993; Ocklenburg and Gunturkun, 2018). Thus, until recently, many hypothesized that population-level behavioral, functional, and neuroanatomic asymmetries were unique to humans.

In the past 30 years, there has been a plethora of data demonstrating population-level behavioral, functional, and neuroanatomic asymmetries in nonhuman species (Ocklenburg et al., 2013b; Rogers et al., 2013; Strockens et al., 2013), thereby challenging previous views on their uniqueness to humans. Nonetheless, there remain fundamental questions regarding similarities and differences in behavioral and brain asymmetries between species, particularly primates, and how they may reflect adaptation to different ecologic and social factors (Ocklenburg et al., 2013a; Schmitz et al., 2019). To this end, the aim of this chapter is to provide a brief summary of existing data on handedness and neuroanatomic asymmetries and their associations in nonhuman primates.

In this chapter, we will also summarize and discuss current data on the theory of the gestural origins of language as it relates to lateralization in structure and function. As revisited and popularized by Corballis (2002), the gestural origin theory postulates that human speech evolved through the process of exaptation of neural systems that initially supported gestural forms of communication as well as related praxic functions, for example, tool manufacture and use. There have been a number of recent studies that have tested aspects of this theory in nonhuman primates, notably lateralization in manual gestures and their association with brain regions considered to be the homologues to language regions in the human brain including the inferior frontal gyrus (IFG) (i.e., Broca's area) and the planum temporale (PT) (i.e., Wernicke's area).

HANDEDNESS IN NONHUMAN PRIMATES

Since the seminal review paper by MacNeilage et al. (1987) in which they proposed the postural origins theory of handedness, there have been numerous studies on hand preferences for a variety of different tasks in nonhuman primates. Rather than

*Correspondence to: William D. Hopkins, Department of Comparative Medicine, Keeling Center for Comparative Medicine and Research, Bastrop, TX, 78602, United States. Tel: +1-512-332-7543, Fax: +1-512-332-5208, E-mail: wdhopkins@mdanderson.org

summarize data from all species and tasks, we will focus our overview on two recent meta-analyses of hand preference data in nonhuman primates for the so-called "tube" task. The tube task is a rather simple measure of handedness that requires coordinated bimanual hand use to extract foods placed inside a poly-vinyl-chloride (PVC) pipe (see Fig. 13.1). Specifically, a food that has some adhesive properties, such as peanut butter, yogurt, or frosting, is smeared on the inside the walls of a PVC pipe. To extract the food, the animals must hold the PVC pipe with one hand while dipping a single digit/hand with the opposite into the inside of the tube to get the food. Thus, the tube task requires that the two hands perform complementary roles to successfully extract the food (i.e., one hand must hold the tube while the opposite extracts the food). The tube task has some advantages over other measures of hand use traditionally used with nonhuman primates including (1) it tends to elicit significant hand preferences in a majority of subjects tested, (2) it is easy to use and can be administered to many species, (3) there is good test–retest reliability in hand use within individuals and across settings (Hopkins et al., 2003, 2013), and (4) positional or situational factors tend not impact hand use which can be problematic for other measures (i.e., simple reaching). Further, small but significant population-level biases have been found in several independent samples of apes and monkeys (Hopkins et al., 2004; Llorente et al., 2009, 2010; Meguerditchian et al., 2013).

As noted above, there are two recently published meta-analyses of hand use for the tube task and related measures requiring coordinated bimanual feeding. Specifically, Caspar et al. (2022) summarized all published data on the tube task in nonhuman primates. This analysis included data from 1786 subjects across 38 species of nonhuman primates. In addition to comparing hand preferences between species, Caspar et al. (2022) examined a number of "biologic" predictors of variation including ecologic factors, brain size, and whether the species is known to use tools. Caspar et al. (2022) reported that the majority of species tested on the TUBE task failed to exhibit population-level handedness with some exceptions including humans (right-handed), chimpanzees (right-handed), gorillas (right-handed), olive baboons (right-handed), orangutans (left-handed), and snub-nosed monkeys (left-handed). When considering the impact of different biologic factors, there was some evidence that ecologic factors, notably terrestrial versus arboreal habitats, had a significant effect with more terrestrial species showing greater rightward asymmetries in hand use, albeit the effect was not terribly robust.

Soto et al. (2022) reviewed the published data on handedness for coordinated bimanual tasks in nonhuman primates including the tube task, as well as related measures that required coordinated hand use. Thirty-three species were represented in the tube task data and 13 for other coordinated bimanual actions. Twenty-six studies were included in their sample of handedness for coordinated bimanual actions and included a total of 769 individuals. When considered as an entire sample, no significant population bias was found for the tube task nor for studies using other measures of coordinated bimanual actions; however, species had a significant moderating effect for both measures and, like the data from Caspar et al. (2022), some species exhibited population-level biases. For instance, for tasks requiring bimanual actions, gorillas and chimpanzees showed a significant right-hand preference and this pattern was consistent across different samples. For instance, shown in Fig. 13.2 are the mean handedness index values (± SEM) for five different cohorts of chimpanzees tested on the tube task. As can be seen, small but consistent significant right-hand preferences are found within each cohort.

Gestural communication

Because of the long-standing view that lateralization in hand preference is linked to hemispheric specialization in language functions, there have been some attempts to measure handedness for communicative functions, notably manual gestures. Far fewer studies on hand use for gestural communication have been

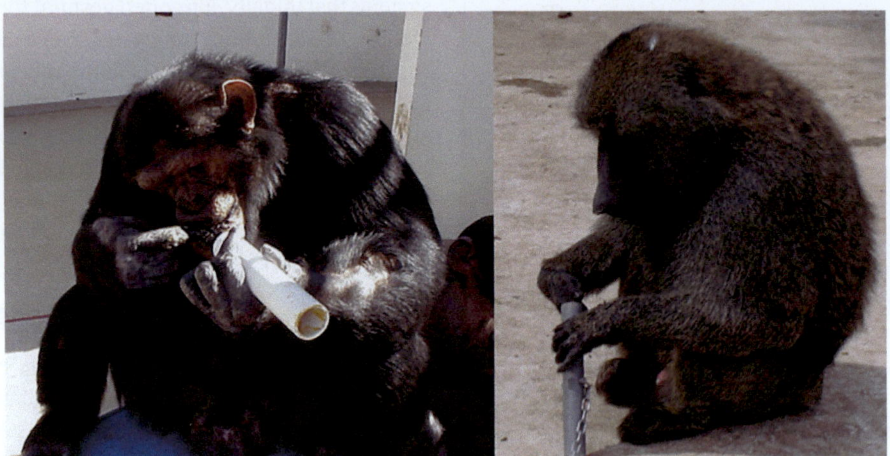

A B

Fig. 13.1. Handedness assessment in a chimpanzee (A) and baboon (B) using the tube task.

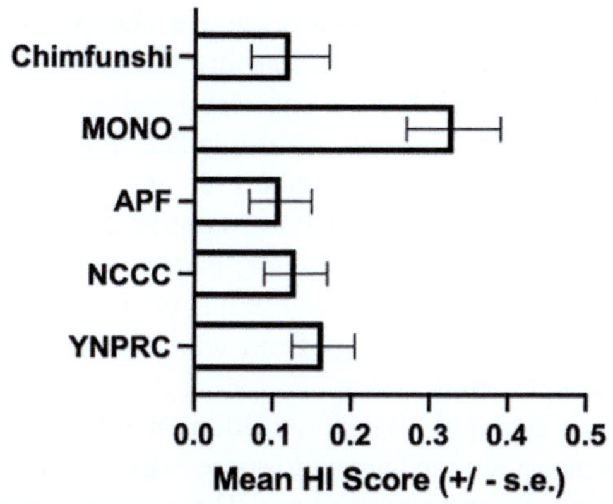

Fig. 13.2. Mean HI scores (±s.e.) for the tube task in different chimpanzee cohorts. Positive HI values reflect greater right-hand use and negative values indicate greater left-hand use. *APF,* Alamogordo Primate Facility; *Chimfunshi,* sanctuary in Africa; *MONA,* sanctuary in Spain; *NCCC,* National Center for Chimpanzee Care; *YNPRC,* Yerkes National Primate Research Center.

performed likely due to the fact that not all nonhuman primate species use manual gestures as a part of their communicative repertoire. Notwithstanding, both experimental and observational methods of hand use for gestural communication have produced some interesting results. In experimental tests, the available data clearly show that human infants, bonobos, chimpanzees, gorillas, and baboons all exhibit a significant population-level right-hand bias. Further, preferential use of the right hand was higher for gestures than for manual actions that are not communicative in functions (i.e., the tube task) in all five species (see Fig. 13.3).

In captive great apes, observational methods have been used in both chimpanzees (Meguerditchian et al., 2010; Prieur et al., 2016b) and gorillas (Prieur et al., 2016a, 2018), to record hand

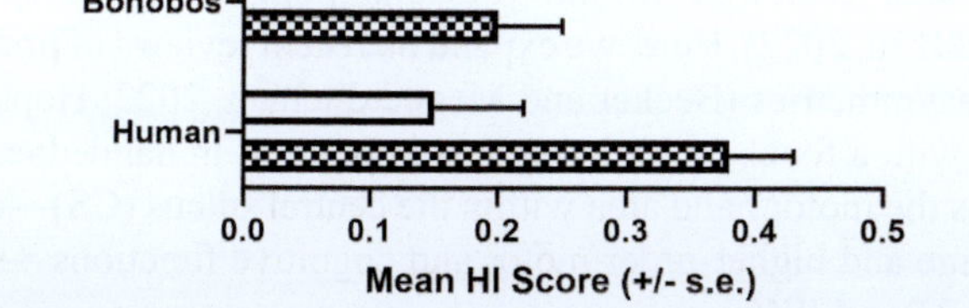

Fig. 13.3. Mean HI scores (±s.e.) for species-typical communicative gestures in 162 baboons, 18 gorillas, 70 chimpanzees, 51 bonobos, and whole-hand pointing in 37 human infants. Positive HI values reflect greater right-hand use and negative values indicate greater left-hand use. Adapted from Meguerditchian A, Vauclair J, Hopkins WD (2013). On the origins of human handedness and language: a comparative review of hand preferences for bimanual coordinated actions and gestural communication in nonhuman primates. Dev Psychobiol 55: 637–650.

use during intraspecific gestural communication events. Though some contextual factors influenced hand use, for both species, when considering all intraspecific gestures, a significant majority of apes preferred their right hand. In wild chimpanzees, Hobaiter and Byrne (2013) reported that a significant majority of individuals prefer their right hand for intraspecific gestures and this pattern increases during development, a pattern also reported in captive bonobos (Hopkins and de Waal, 1995). Similar observations of preferred right-hand use for different types of gestures in chimpanzees have also been reported by Roberts et al. (2019).

In more distantly related species, the best evidence of population-level handedness for gesture comes from studies in baboons (Meguerditchian and Vauclair, 2006). These researchers observed a group of living baboons and noted that they engage in a form of threat gesture that is referred to as a "hand slap" which is directed toward conspecifics and humans. They showed that the baboons are significantly right-handed for the slap gesture and that the directional bias exceeds those of noncommunicative tasks. In a subsequent study, Meunier et al. (2012) tested baboons and children on lateralization in manual gestures and grasping food items using an experimental paradigm. In this study, food items were placed within a distance where the subjects could grasp or at a further distance in which they had to gesture toward the food item. Importantly, the food items were placed at five positions that ranged from the far left to the right relative to the midline of the subject. Handedness scores were recorded for each task and food position in both species. For both human children and baboons, right-hand preferences for gesture were more resistant to contralateral left hemifield positions (positions 1, 2) of the food item than they were for reaching. This suggests that gesture induces a stronger right-hand bias than simple reaching. These findings are similar to data reported in chimpanzees (Hopkins and Wesley, 2002).

We note here that "gesture" is broadly defined in this summary and several studies have reported that lateralization in hand use can be influenced by the type of gesture and here we provide examples (Hopkins et al., 2012). In gorillas and chimpanzees, hand clapping is an auditory gesture that has been described as a behavior used to capture the attention of an otherwise inattentive audience. Thus, it is clearly communicative in function and, as can be seen in Table 13.1, is a strongly lateralized behavior in both chimpanzees and gorillas. With respect to the hand slap gesture described in baboons, there is also evidence of lateralization in hand slapping in other species, notably chimpanzees and gorillas. In both species, though it is not a common gesture, significant right-hand preferences are found, suggesting lateralization in this form of communication may have a common neurologic basis as simple manual gestures often used to request food.

Tool use

Many have suggested that selection for tool manufacture and use had a significant impact on brain evolution, including lateralization in structure and function (Johnson-Frey, 2004; Frey, 2008; Quallo et al., 2009; Stout and Chaminade, 2012). Chimpanzees are well known for their tool-using abilities across a wide

Table 13.1

Comparative data on handedness for specific manual gestures species.

	#L	#A	#R	Mean HI
Hand clapping				
Gorilla	1	5	11	0.508
Chimpanzees	28	4	42	0.262
Chimpanzees	8	1	8	0.040
Chimpanzees	4	1	3	−0.151
Hand slap				
Gorilla	0	6	16	0.530
Gorilla	1	12	5	NDA
Chimpanzees	0	16	17	0.391
Baboon	10	15	35	0.320

Data from Shafer DD (1993). Patterns of handedness: comparative study of nursery school children and captive gorillas. In: JP Ward JP, WD Hopkins (Eds.), Primate laterality: current behavioral evidence of primate asymmetries. Springer-Verlag, New York; Fletcher AW (2006). Clapping in chimpanzees: evidence of exclusive hand preference in a spontaneous, bimanual gesture. Am J Primatol 68: 1081–1088; Meguerditchian A, Gardner MJ, Schapiro SJ et al. (2012). The sound of one-hand clapping: handedness and perisylvian neural correlates of a communicative gesture in chimpanzees. Proc Biol Sci 279: 1959–1966; Prieur J, Pika S, Barbu S et al. (2016a). Gorillas are right-handed for their most frequent intraspecific gestures. Anim Behav 118: 165–170; Prieur J, Pika S, Barbu S et al. (2016b). A multifactorial investigation of captvie chimpanzees' intraspecific gestural laterality. Anim Behav 116: 31–43.

geographic range of regions in Africa, as well as in captive settings (Whiten et al., 1999, 2001; Shumaker et al., 2011); therefore, it is not surprising that there has been a number of studies on handedness for tool use in both wild and captive apes (mostly chimpanzees) and these were reviewed in Sanz et al. (2016) (see Fig. 13.4). In wild chimpanzees, significant population-level bias was found for three different types of tool use including termite fishing (left-handed), algae dipping (right-handed), and leaf sponging (right-handed). There was also a significant trend toward preferred right-hand use for nut cracking (see also Westergaard and Suomi, 1994; Neufuss et al., 2017). By contrast, evidence

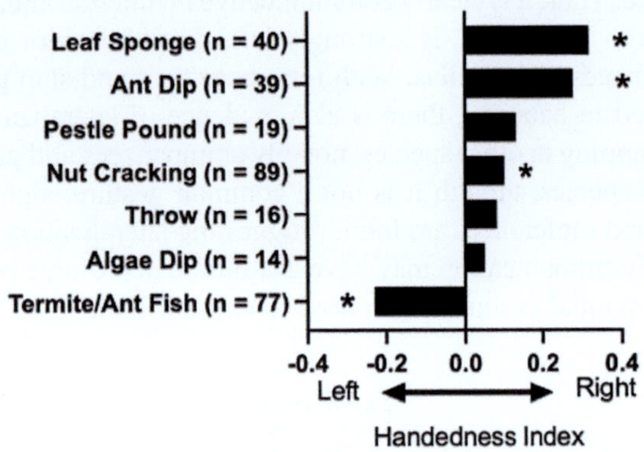

Fig. 13.4. HI scores for tool use in wild chimpanzees for different measures. Adapted from Sanz CM, Morgan DB, Hopkins WD (2016). Lateralization and performance asymmetries in the termite fishing of wild chimpanzees in the Goualougo Triangle, Republic of Congo. Am J Primatol 78: 1190–1200.

of population-level handedness in captive chimpanzees is less clear. For instance, on devices that are designed to simulate termite fishing in wild chimpanzees, hand preferences are equally distributed between right-handed, left-handed, and ambiguously handed apes (Hopkins et al., 2015b). Though not a frequently observed behavior, captive chimpanzees are strongly right-handed for throwing, particularly when doing so from an upright posture (Hopkins et al., 2005), which somewhat contradicts findings from wild chimpanzees, albeit in much smaller sample sizes.

In the Sanz et al. (2016) paper, in addition to hand preference, the authors measured the latency to successfully perform the tool use actions as a function of hand use. This allowed for an assessment of asymmetries in performance (as opposed to preference), which is uncommon in research with nonhuman primates (Hopkins, 2018). Sanz et al. (2016) reported that right-handed chimpanzees had significantly shorter latencies compared to left-handed individuals (see Fig. 13.5A). Further, Sanz et al. (2016) further found that the latencies of the dominant hand were significantly shorter than the nondominant within both right- and left-handed chimpanzees (see Fig. 13.5B). More recently, Hopkins et al. (2019) examined asymmetries in tool use performance in 206 captive chimpanzees on a simulated termite fishing task. In this study, response latencies were recorded for 30 dipping responses for both the left and right hand in 71 right-handed, 67 left-handed, and 68 ambiguously handed individuals. A significant hand preference by performance asymmetry was found. Right-handed chimpanzees performed better with their right—compared to left-hand, whereas no performance differences were found between the left and ambiguously handed apes (see Fig. 13.5C).

NEUROANATOMIC ASYMMETRIES AND THEIR BEHAVIORAL CORRELATES

Another prominent feature of the human brain is neuroanatomic asymmetries. As with handedness, there have also been historic (LeMay, 1985) and contemporary interests in whether nonhuman primates exhibit population-level neuroanatomic or neurofunctional asymmetries. Early studies focused on the measurement of sulci length measured from endocasts or directly from postmortem brains, as well as petalia asymmetries, and these results have been reviewed in the previous publications (Hopkins et al., 2015a, 2022). Here, we expand on recent reviews of primate brain asymmetries (Becker and Meguerditchian, 2022; Hopkins, 2022) with a focus on brain regions implicated in handedness—such as the motor hand area within the central sulcus (CS)—or in language and higher-order motor and cognitive functions—such as the PT and IFG.

In this section, where possible, we also report results of the association between hand use and structural asymmetries in the brain. We believe this is an important consideration when comparing directional asymmetries between species because of the difference in the prevalence of right-handedness among humans compared to other species. Specifically, the ratio of right-to-left handedness individuals is much higher in humans (~8:1 or 9:1)

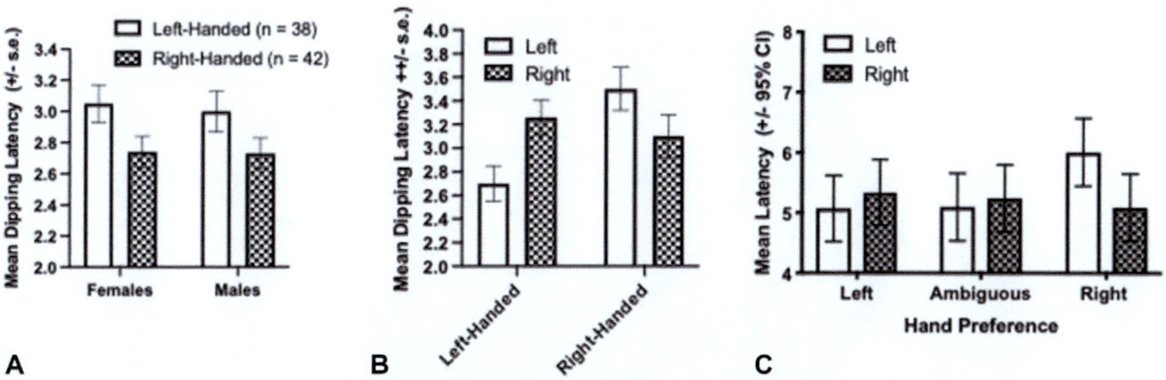

Fig. 13.5. 7 (A) Mean latency (±s.e.) for right- and left-handed chimpanzees on a termite fishing task. (B) Mean latency (±s.e.) for the left and right hands for right- and left-handed chimpanzees. (C) Mean latency (±95% CI) for the left and right hands for chimpanzees classified as right-handed, left-handed, or ambiguously handed (i.e., no preference). Data reproduced from Sanz CM, Morgan DB, Hopkins WD (2016). Lateralization and performance asymmetries in the termite fishing of wild chimpanzees in the Goualougo Triangle, Republic of Congo. Am J Primatol 78: 1190–1200; Hopkins WD, Mareno MC, Schapiro SJ (2019). Further evidence of a left hemisphere specialization and genetic basis for tool use skill in chimpanzees (Pan troglodytes): reproducibility in two genetically isolated populations of apes. J Comp Psychol 133: 512–519.

compared to nonhuman primates when population biases are revealed (~2:1–3:1). If left–right differences in the brain map onto handedness, one can assume that in any random sample of human subjects, ~90% would be right-handed, which would not be the case for nonhuman primate species.

Central sulcus

The CS divides the primary motor and somatosensory cortex, where the topographic motor and sensory regions of the body parts are represented (Penfield and Boldrey, 1936). Across the dorsal–ventral plane of the CS, the hand and finger regions, known as KNOB or motor hand area, are represented by the presence of omega or epsilon shape (Caulo et al., 2007; Yousry et al., 1997). In humans, asymmetries in the depth of the CS corresponding to the motor hand area have been found to be associated with hand dominance (e.g., Amunts et al., 1996, 2000) but not with lateralization in language functions (Ocklenburg et al., 2014; Haberling et al., 2016).

In nonhuman primates, there are reports in chimpanzees, baboons, capuchin monkeys, and squirrel monkeys that direction of handedness (left- vs right-handedness) for object reaching or for the coordinated bimanual tube task is associated with contralateral interhemispheric asymmetries of the motor hand area within the CS (Nudo et al., 1992; Hopkins and Cantalupo, 2004; Phillips and Sherwood, 2005; Dadda et al., 2006; Margiotoudi et al., 2019). For instance, right-handed chimpanzees for the tube task show deeper left-than-right asymmetries in the motor-hand area within the CS (Hopkins et al., 2010) and similar dissociations are found for hand preferences for tool use in this same cohort of chimpanzees (Hopkins et al., 2017).

Interestingly, the tube task handedness-CS link has also been found in baboons by comparing the depth of the CS fold along the dorsal–ventral plane (see Fig. 13.6). Though baboons lack a morphologically definable motor-hand area (Hopkins et al., 2014), in this study, differences between right- and nonright-handed baboons in CS depth corresponded to the central region which approximates the location of the motor-hand area. These findings were replicated in 30 juvenile baboons at the critical age when they are about to be weaned from their mothers (~9 months old), suggesting that early manifestation of handedness across development for the bimanual tube task is already represented within the CS at the exact same sulcus location as observed in adults (Margiotoudi et al., 2019).

Planum temporale

One of the most well-established neuroanatomic asymmetries in the human brain is the PT, which is the flat surface of the cortex that lies posterior to Heschl's gyrus. The PT overlaps with Wernicke's area (BA22) and therefore has been a region of interest for comparative studies of lateralization in nonhuman primates because of this brain region's role in language and related higher cognitive functions. As a proxy for measurement of the PT, early comparative studies examined the length of the Sylvian fissure in different primate species. For instance, Yeni-Komshian and Benson (1976) compared SF length between the left and right hemispheres in humans, chimpanzees, and macaques and found a significant leftward bias in humans and chimpanzees but not in macaques, a finding that has largely been replicated using alternative methods with some exceptions (Falk et al., 1986; Heilbronner and Holloway, 1989; Hopkins et al., 2000, 2015a; Cantalupo et al., 2003; Gannon et al., 2008; Gomez-Robles et al., 2016). Recent studies that have more concisely quantified the PT surface and/or its gray matter volume from magnetic resonance images have reported significant leftward asymmetries in chimpanzees and baboons but not in rhesus monkeys, bonnet monkeys, or vervet monkeys (Hopkins and Nir, 2010; Lyn et al., 2011; Marie et al., 2018; Spocter et al., 2020; Becker et al., 2022b) (see Fig. 13.7).

Further, cytoarchitectonic analysis of Brodmann's area 22 (BA22) has reported significant leftward asymmetries in volume and neuron number in a sample of 12 chimpanzees (Spocter et al., 2010). A similar pattern of results has been reported in

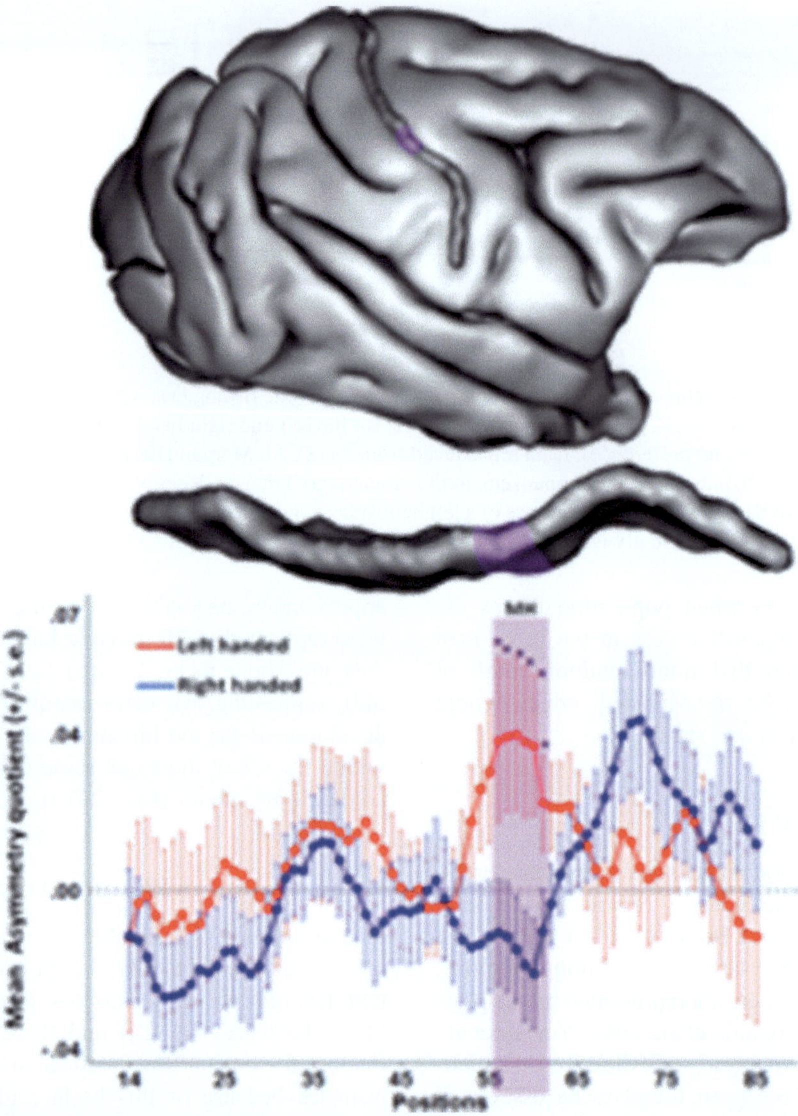

Fig. 13.6. Mean AQ values (±s.e.) in CS depth for right- and left-handed baboons on the tube task. Reprinted from Margiotoudi K, Marie D, Claidiere N et al. (2019). Handedness in monkeys reflects hemispheric specialization within the central sulcus. An in vivo MRI study in right- and left-handed olive baboons. Cortex 118: 203–211.

rhesus monkeys, albeit from a sample of only six individuals (Gannon et al., 2008).

Recently, in vivo longitudinal MRI studies conducted in 30 newborn baboons reported—in a proportion almost identical to the one found in human babies—early leftward PT structural asymmetry (i.e., surface and gray matter volume) which also increased with age (Becker et al., 2021, 2022b). While this brain asymmetry was considered an anatomic marker of a predisposition of the infant brain to rapidly acquire language, this finding suggests that this feature of the brain is neither human-specific nor language-specific, but rather extends the phenomena to at least one catarrhine primate species.

Interestingly, the PT structural asymmetry link with language in humans was found with gestural communication in chimpanzees and baboons. In fact, only subjects showing a right-hand

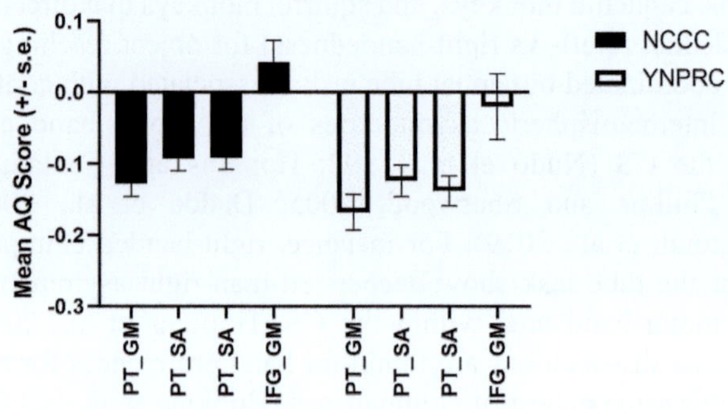

Fig. 13.7. Mean AQ scores (±s.e.) for PT surface area (*left panel*) and gray matter volume (*right*) in different primate species. Reprinted from Hopkins WD (2022). Neuroanatomical asymmetries in nonhuman primates in the homologs to Broca's and Wernicke's areas: a mini-review. Emerg Top Life Sci 6: 271–284.

preference for communicative gesturing (e.g., pointing and attention-getting clapping in chimpanzees; hand threat slapping on the ground in baboons) exhibited a leftward PT asymmetry, whereas subjects communicating preferentially with their left hand showed no PT asymmetry (Hopkins and Nir, 2010; Meguerditchian et al., 2012; Becker et al., 2022b). Early left PT asymmetry detected in newborn baboons is actually predictive of the further development of right-hand preference for communicative gestures. Moreover, the left PT asymmetry was also found related to handedness for tool use in chimpanzees but not to handedness in manipulative actions (Gilissen and Hopkins, 2013).

Inferior frontal gyrus

The IFG, sometimes referred to as Broca's area, is subdivided into two morphologic regions including the pars opercularis (Op) and pars triangularis (Tri) and is often considered synonymously to Brodmann's cytoarchitectonic area 44 and area 45, respectively. The seminal clinical research by Broca led to him describing this region as left-lateralized for speech production (Broca, 1861). Whether this region is also left-lateralized at the morphologic level of analysis in the human brain is less clear than the evidence for the PT (Kurth et al., 2020; Sprung-Much et al., 2022), likely due to the greater interindividual variability in the sulci that define its borders (Keller et al., 2009a). By contrast, cytoarchitectonic studies in human postmortem brains have reported a leftward asymmetry in the volume and neuron density of BA44 and BA45 (Uylings et al., 2006) which increases during development from 5 years old onward (Simonds and Scheibel, 1989; Amunts et al., 2003).

Comparative studies on the morphology of IFG are also challenging due to within and between species variation in sulci that define this region (Palomero Gallagher and Zilles, 2019). For instance, while the same sulcal landmarks can be used to define the Op in humans and chimpanzees, Tri cannot be defined in chimpanzees because of the absence of the horizontal ramus in chimpanzees (Keller et al., 2009b; Hopkins et al., 2022). In nonhuman primates, an early study reported a leftward asymmetry in the volume of the IFG in a sample of 20 great apes (including mostly chimpanzees but also gorillas and bonobos). In these studies, the IFG was defined using the precentral inferior sulcus, the inferior frontal sulcus, and the fronto-orbital sulcus as the posterior, superior, and anterior borders (Cantalupo and Hopkins, 2001; Keller et al., 2009b). The variability in sulci is by no means trivial, as they can impact the calculation of the volumes of the regions and each sulcus can have different directional asymmetries within the same cohorts of individuals (Keller et al., 2012; Hopkins et al., 2022). Indeed, a more recent study quantifying the volume of the IFG within a much larger sample of chimpanzees ($N=294$ individuals) showed a rightward asymmetry in the volume of the IFG but this finding was dependent on the landmarks used to define the region. Notably, ~50% of the chimpanzee sample had a bifurcated or an extra precentral inferior sulcus which defines the posterior border of the IFG. The directional asymmetry of the IFG within chimpanzees is influenced by which of the two PCI are used as the posterior border (see Fig. 13.8A). When using a more conservative criterion for defining the IFG, utilizing the most anterior limb of the PCI as the posterior border, chimpanzees showed a significant rightward asymmetry. By contrast, when the IFG was quantified using a more liberal criterion in which the posterior limb of the PCI was used as the posterior border, no significant population-level bias was found. Indeed, directional asymmetries within the IFG were strongly tied to the asymmetric presence of an extra or bifurcated PCI. Not surprisingly, chimpanzees with a unilateral extra or bifurcated PCI in the left showed significant leftward directional asymmetries, whereas rightward asymmetries were found in chimpanzees with a bifurcation in the PCI in the right hemisphere (see Fig. 13.8D).

At the microstructural level of analysis, the two cytoarchitectonic studies available so far in chimpanzees reported no population-level asymmetry within BA44 or BA45 (Schenker et al., 2010; Graic et al., 2020). In Old World monkeys, given

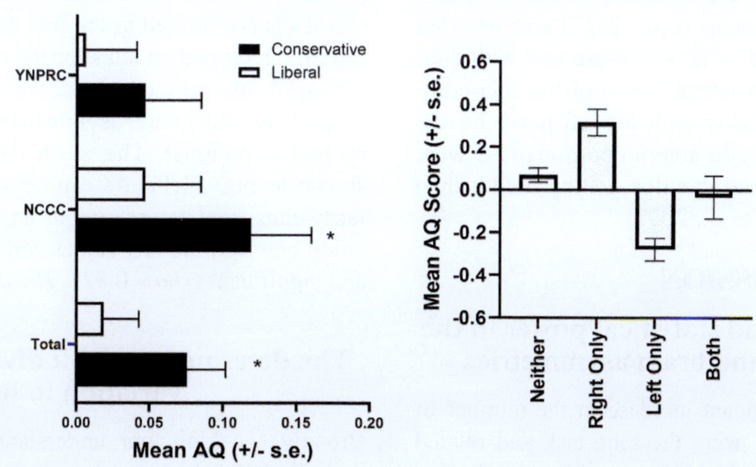

Fig. 13.8. Mean AQ values (±s.e.) for the IFG volume using either conservative or liberal criteria for defining the posterior border. *NCCC,* National Center for Chimpanzee Care; *Total,* total sample; *YNPRC,* Yerkes National Primate Research Center.

the absence of the three aforementioned sulci used in apes to delineate the IFG, the structural asymmetry of this region has been quantified by measuring the depth of the ventral portion of the inferior frontal sulcus (IA sulcus) in each hemisphere. Some suggest that this sulcus constitutes the homologue to the ascending branch of the inferior precentral sulcus (PCI) which, as noted above, delimits the Op posteriorly in humans and great apes (Amiez and Petrides, 2009; Amiez et al., 2023). In rhesus monkeys, while BA45 is situated in the anterior bank of the ventral *IA sulcus*, BA44 has been hypothesized to be located in its posterior bank (Petrides et al., 2005). In a recent study of 50 baboons, whereas no population-level depth asymmetry has been found in the IA sulcus, its ventral part showed a contralateral depth asymmetry relative to the direction of hand preference for communicative gestures—but not for noncommunicative manipulative actions (discussed below) (Becker et al., 2022a). In other words, baboons that prefer to communicate with their right hand have a deeper left-than-right IA sulcus, than those preferring to communicate with their left hand and vice versa. A contralateral association between IFG whole and gray matter volume and communicative gestures was also previously reported in chimpanzees (Taglialatela et al., 2006; Meguerditchian et al., 2012) and seems to extend to tool-use processing.

That stated, there are limitations to comparing the depth of the IA sulcus along the dorsal–ventral plane to the volume or surface area of the IFG in chimpanzees. If the IA sulcus is a homologue to the inferior limb of the PCI in chimpanzees, a more relevant comparison is to evaluate the asymmetry in the depth of the inferior limb of the PCI fold in apes that prefer to gesture with their left or right hand. When this analysis is done in a chimpanzee sample comprising 271 individuals, no significant population-level asymmetries in either depth or surface area are found; moreover, significant differences in asymmetry values are not found for the inferior limb of the PCI as a function of handedness for manual gestures in this cohort. By contrast, leftward asymmetries in the surface area and depth of the anterior border sulcus (fronto-orbital sulcus, FO) of the IFG were also found to be associated with right-handedness for tool use in chimpanzees (Hopkins et al., 2007). More recently, Hopkins et al. (2017) also reported that chimpanzees that performed better a tool-use task with their right hand showed a greater left lateralization of the IFG compared to those that performed better with the left hand. Lastly, chimpanzees with a bifurcation in the anterior portion of FO were found to be more likely to produce attention-getting sounds than individuals without it (Hopkins et al., 2022).

DISCUSSION

Rigor, reproducibility, and statistical power in the study of handedness and brain asymmetries

Though there has been a significant increase in the number of species studied for handedness using the tube task and related measures (Regaiolli et al., 2016a,b) in the past 50 years, limitations to the existing data remain in sample size which has been a continuous challenge in these studies (Hopkins, 2013). In some species, it is reasonable to expect fewer subjects given their prevalence in captive environments and the challenge of collecting experimental data in field settings. But, for some species, such as rhesus or marmosets, that are commonly used laboratory primates, it is curious why there are not more studies with larger numbers of subjects.

It is worth noting that in the meta-analyses on the tube task, the majority of species that showed population-level handedness also tended to be those with the largest sample sizes. That stated the ratio of right-to-left handedness (or vice versa) is ~2:1 or 3:1 at best in those species that were found to exhibit a population-level bias. This is in contrast to human handedness, where the ratio is more like 8:1 or 9:1 and these differences have implications for sample size and statistical power. Thus, population-level handedness effects in nonhuman primates are small (see Fig. 13.2, for example) and it has been estimated that a minimum of 52 subjects is needed to detect a significant population-level bias (Hopkins, 2006). Of the 38 species tested on the tube task and reviewed by Caspar et al. (2022), only five had sample sizes >52 individuals. Thus, in our view, it is difficult to make inferences about the existence of population-level handedness within and between species when most studies are statistically underpowered.

In some species, notably chimpanzees, there are consistencies in handedness for the tube task (see Fig. 13.2). This is not a trivial point because it speaks to the rigor and reproducibility of handedness effects, something often disregarded but of increasing interest to the scientific community and funding agencies. Reproducibility is not limited to measures of the tube task. Similar consistencies in handedness between chimpanzees from at least two different cohorts are found in measures of hand preference for simple reaching, manual gestures, and tool use (see Fig. 13.9) and tool use skills (Hopkins et al., 2019).

Among measures of brain asymmetry, shown in Fig. 13.10 are the mean AQ values for two genetically isolated populations of chimpanzees on measures of the PT and IFG. As can be seen, these results are consistent across the cohorts and therefore also appear to be reproducible at least with respect to the PT and volume of the IFG. Interestingly, consistency in morphologic asymmetries is not limited to the IFG and PT. Recently, Vickery et al. (2020) developed an atlas for the chimpanzee's brain and used a region-of-interest method (as well as voxel-based morphometry) to quantify gray matter asymmetries across 55 regions (excluding cerebellar regions). The mean AQ values for each cohort are shown in Fig. 13.11. As can be seen, the directional biases are fairly consistent across regions and a Spearman rank-order correlation between the AQ values across the 55 regions was positive and significant ($rho = 0.821$, $P < 0.001$).

The determinants of individual and phylogenetic variation in handedness

Broadly speaking, our understanding of the mechanisms that underlie individual variation in handedness and brain asymmetries in humans and nonhumans remains unresolved. Genetic models of human handedness have held a particular fascination in the scientific literature but, until recently, these models have

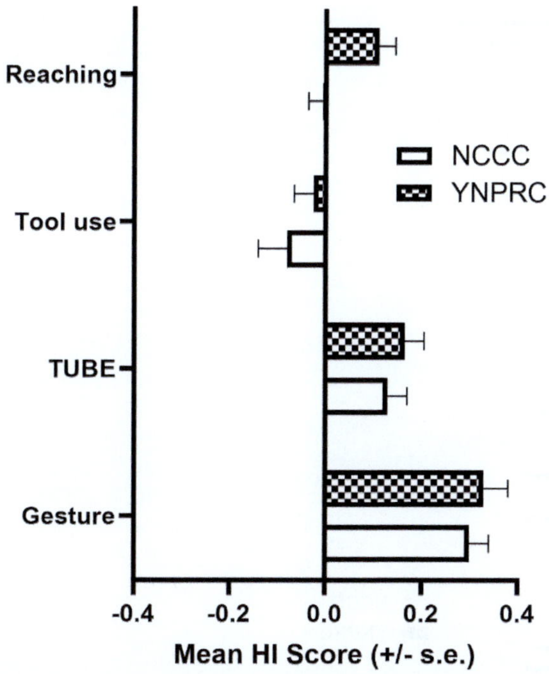

Fig. 13.9. Mean HI scores (±s.e.) for different handedness measures in two cohorts of chimpanzees. Positive HI values reflect greater right-hand use and negative values indicate greater left-hand use.

largely been heuristic and derived by fitting the genetic models to published family data (e.g., Annett, 2002). The advent of molecular biologic studies has provided an opportunity to identify specific genes that might be implicated in handedness and associated brain asymmetrics (Sha et al., 2021a,b). Indeed, in a recent genome-wide association study, only 54 genes were identified as distinguishing left- from right-handers and ambidextrous individuals (measured by self-report of which hand they prefer or write with) and they collectively explained only a small portion of variance (Cuellar-Partida et al., 2021). Interestingly, none of the genes identified in this study were the same as those previously reported to be linked to handedness in the previous genome studies. Moreover, some of the genes linked to handedness were implicated in neuropsychiatric disorders such as schizophrenia. Thus, it is difficult to determine whether handedness is a secondary trait or a byproduct of having schizophrenia, rather than being due to the genes themselves.

Arguably more importantly, hundreds if not thousands of genes could be identified as linked to handedness (or brain asymmetries) in humans but, even with this outcome, this would not be proof that they code for left–right differences in behavior or brain because in humans (and many primates), it is impossible to isolate genetic from nongenetic factors in determining directional asymmetries. The fundamental problem is that human infants do not prenatally or postnatally develop in a symmetric environment. Indeed, humans develop in a decidedly asymmetric world at multiple levels of input. For instance, at a societal or cultural level, there are religious taboos and cultural practices that promote specific expressions of laterality in hand use (Perelle and Ehrman, 1994). Prenatally, there are positional asymmetries of the developing fetus and intrinsic factors that induce asymmetric behaviors by the fetus, such as thumb sucking (Previc, 1991; Hepper et al., 2005). Postnatally, infants at birth show head orientation asymmetries (Michel, 1981), nursing preferences, and parents exhibit cradling biases (Vauclair, 2022).

If one truly sought to test for a purely genetic basis for population-level asymmetry in the most rigorous manner, they would need to study organisms that develop in a symmetric world which is certainly not the case in humans and indeed many nonhuman primates. Further, some of the early experiential factors that might shape behavioral and brain asymmetries are not randomly distributed which may or may not explain both individual and

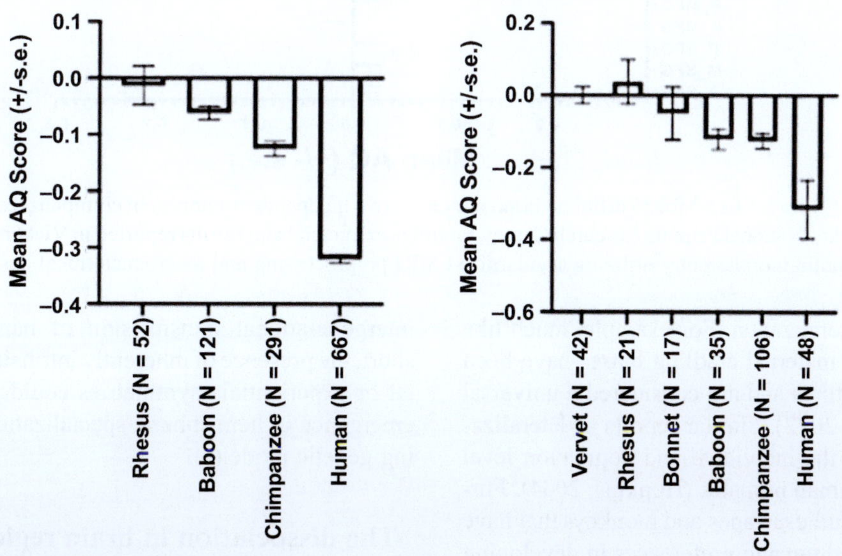

Fig. 13.10. Mean AQ scores (±s.e.) for different measures of brain asymmetry in two independent samples of chimpanzees. *IFG_GM*, inferior formal gyrus gray matter volume; *NCCC*, National Center for Chimpanzee Care; *PT_GM*, planum temporale gray matter volume; *PT_SA*, planum temporale surface area; *YNPRC*, Yerkes National Primate Research Center. Reprinted from Hopkins WD (2022). Neuroanatomical asymmetries in nonhuman primates in the homologs to Broca's and Wernicke's areas: a mini-review. Emerg Top Life Sci 6: 271–284.

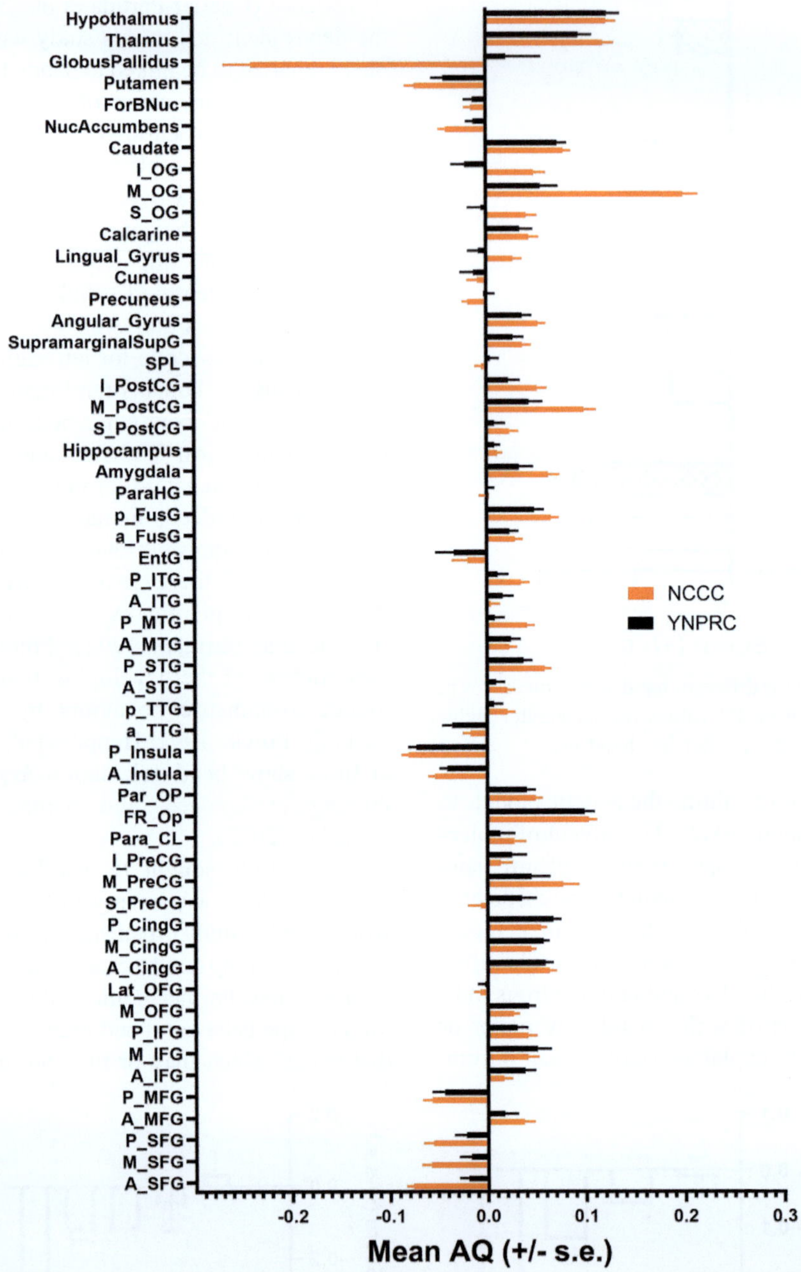

Fig. 13.11. Mean gray matter AQ scores (±s.e.) for 55 different brain regions in two independent samples of chimpanzees. *NCCC*, National Center for Chimpanzee Care; *YNPRC*, Yerkes National Primate Research Center. Findings extracted from results reported in Vickery S, Hopkins WD, Sherwood CC et al. (2020). Chimpanzee brain morphometry utilizing standardized MRI preprocessing and macroanatomical annotations. eLife 9.

phylogenetic variation in lateralization. For example, much like right-handedness, left-sided maternal cradling biases have been reported across multiple cultures and are considered a universal trait (Jones, 2014; Vauclair, 2022). Similar reports of lateralization in maternal cradling at the individual and population level have been reported in nonhuman primates (Hopkins, 2004). Further, there are a number of studies in apes and monkeys that have reported head orientation and nursing preferences in developing infants (Nishida, 1993; Tomaszycki et al., 1998; Damerose and Hopkins, 2002; Hopkins and De Lathouwers, 2006; Jaffe et al., 2006; Zhao et al., 2008; Regaiolli et al., 2018). Any of these early asymmetric inputs could impact individual, phylogenetic, and intergenerational transmission of handedness in primates. In short, the presence of inherently intrinsic or extrinsic environmental or experiential asymmetries could just as easily explain the emergence of hemispheric specialization compared to any existing genetic model(s).

The dissociation in brain regions associated with handedness in Primates

One emerging pattern of results from nonhuman primates is that asymmetries in different brain regions correlate with different measures of hand use. Specifically, when the data on handedness are combined with the neuroanatomic measures of asymmetry, the results converge toward the two following hypotheses: (1) the double dissociation between handedness for manipulation vs handedness for gestural communication/tool use and between their respective neuroanatomic correlates and (2) the specific signature of communicative gesture's lateralization in nonhuman primates—as well as tool use in apes—in the language-homologue hemispheric organization.

As can be seen in Fig. 13.3, in all primate species that have been investigated (children, chimpanzees, baboons, bonobos, and gorillas), communicative gestures elicited a more pronounced degree of right-handedness in comparison to biases for noncommunicative manipulation as assessed by the tube task (Meguerditchian et al., 2013). In addition, at least in chimpanzees and baboons, AQ scores for gestural communication are only weakly correlated with AQ values for the tube task. These findings suggest a greater dominance of the left hemisphere for communicative gestures which might involve a different lateralized system from the one related to handedness for manipulative functions.

This latter hypothesis is supported by their respective neuroanatomic correlates in both species. The direction of handedness for bimanual coordination in chimpanzees and baboons is associated with contralateral depth asymmetry largely within the central sulcus at the motor hand area section—but not to asymmetries in premotor areas including the language homologue areas. In contrast, right-handedness for gestural communication (broadly defined) in those two species—and for tool use in chimpanzees—predicts leftward neuroanatomic asymmetry of the two major language brain regions including the PT and IFG (Taglialatela et al., 2006; Meguerditchian et al., 2012; Becker et al., 2022a).

These collective findings in nonhuman primates are consistent with data from humans showing that typical object-manipulation handedness measures turn out to be a rather poor predictor of language lateralization (Fagard, 2013; Potdevin et al., 2023). This should not be that surprising in light of the fact that the significant majority of left-handers (78%) also show left-hemisphere dominance for language (Knecht et al., 2000; Mazoyer et al., 2014), just like right-handed people. In fact, it has recently been demonstrated that the neural substrates of typical handedness measures and language brain organization might not be related but rather independent from each other (Groen et al., 2013; Ocklenburg et al., 2014; Haberling et al., 2016; Labache et al., 2020).

CONCLUSION

It now seems clear that historical claims that population-level behavioral and brain asymmetries are uniquely human were incorrect (Warren, 1980; Ettlinger, 1988). There is now a plethora of data demonstrating population-level behavioral and brain asymmetries in vertebrates and invertebrates (Rogers et al., 2013; Ocklenburg and Gunturkun, 2018). In our view, previous evolutionary monolithic views postulating the emergence of population-level asymmetries to single human adaptive events such as bipedalism, tool manufacture and use, or language are no longer tenable (Bradshaw and Rogers, 1993). Indeed, based on the available data, it appears that lateralization for different brain regions or networks may have emerged in response to selection for a variety of different motor, cognitive, and sensory processes, some of which may be shared across species but others that may be unique to some species—including humans.

ACKNOWLEDGMENTS

This research was supported in part by NIH grants AG-067419, NS-73134, HD-60563, and NS-092988 (support for the National Chimpanzee Brain Resource), by the European Research Council under the European Union's Horizon 2020 research and innovation program grant agreement no. 716931—GESTIMAGE—ERC-2016-STG, as well as from grants ANR-16-CONV-0002 (ILCB). Chimpanzee maintenance at the National Center for Chimpanzee Care was previously funded by NIH/NCRR U42-OD-011197. American Psychological Association guidelines for the ethical treatment of animals were adhered to during all aspects of this study. We are grateful to the helpful assistance of the entire veterinary and caregiving staffs at the NCCC and YNPRC for their assistance in the collection of the MRI scans.

REFERENCES

Amiez C, Petrides M (2009). Anatomical organization of the eye fields in the human and non-human primate frontal cortex. Prog Neurobiol 89: 220–230.

Amiez C, Sallet J, Giacometti C et al. (2023). A revised perspective on the evolution of the lateral frontal cortex in primates. Sci Adv 9: eadf9445.

Amunts K, Schlaug G, Schleicher A et al. (1996). Asymmetry in the human motor cortex and handedness. Neuroimage 4: 216–222.

Amunts K, Jancke L, Mohlberg H et al. (2000). Interhemispheric asymmetry of the human motor cortex related to handedness and gender. Neuropsychologia 38: 304–312.

Amunts K, Schleicher A, Ditterich A et al. (2003). Broca's region: cytoarchitectonic asymmetry and developmental changes. J Comp Neurol 465: 72–89.

Annett M (2002). Handedness and brain asymmetry: the right shift theory, Psychology Press, Hove.

Becker Y, Meguerditchian A (2022). Structural brain asymmetries for language: a comparative approach across primates. Symmetry 14: 876.

Becker Y, Sein J, Velly L et al. (2021). Early left-planum temporale asymmetry in newborn monkeys (Papio anubis): a longitudinal structural MRI study at two stages of development. Neuroimage 227: 117575.

Becker Y, Claidiere N, Margiotoudi K et al. (2022a). Broca's cerebral asymmetry reflects gestural communication's lateralisation in monkeys (Papio anubis). eLife 11.

Becker Y, Phelipon R, Sein J et al. (2022b). Planum temporale grey matter volume asymmetries in newborn monkeys (Papio anubis). Brain Struct Funct 227: 463–468.

Bradshaw B, Rogers L (1993). The evolution of lateral asymmetries, language, tool-use and intellect, Academic Press, San Diego.

Broca P (1861). Remarques sur le siège de la faculté du langage articulé, suivies d'une observation d'aphémie (perte de la parole). Bull Soc Anat Paris 6: 330–357.

Cantalupo C, Hopkins WD (2001). Asymmetric Broca's area in great apes. Nature 414: 505.

Cantalupo C, Pilcher D, Hopkins WD (2003). Are planum temporale and sylvian fissure asymmetries directly related? A MRI study in great apes. Neuropsychologia 41: 1975–1981.

Caspar KR, Pallasdies F, Mader L et al. (2022). The evolution and biological correlates of hand preferences in anthropoid primates. eLife 11: e77875.

Caulo M, Briganti C, Mattei PA et al. (2007). New morphological variants of the hand motor cortex as seen with MR imaging in a large study population. Am J Neuroradiol 28: 1480–1485.

Corballis MC (1992). The lopsided brain: evolution of the generative mind, Oxford University Press, New York.

Corballis MC (2002). From hand to mouth: the origins of language, Princeton University Press, Princeton, NJ.

Cuellar-Partida G, Tung JY, Eriksson N et al. (2021). Genome-wide association study identifies 48 common genetic variants associated with handedness. Nat Hum Behav 5: 59–70.

Dadda M, Cantalupo C, Hopkins WD (2006). Further evidence of an association between handedness and neuroanatomical asymmetries in the primary cortex of chimpanzees (*Pan troglodytes*). Neuropsychologia 44: 2482–2486.

Damerose E, Hopkins WD (2002). A comparison of scan and focal sampling procedures in the assessment of laterality for maternal cradling and infant nipple preferences in olive baboons (*Papio anubis*). Anim Behav 63: 511–518.

Ettlinger GF (1988). Hand preference, ability and hemispheric specialization. How far are these factors related in the monkey? Cortex 24: 389–398.

Fagard J (2013). Early development of hand preference and language lateralization: are they linked, and if so, how? Dev Psychobiol 55: 596–607.

Falk D, Cheverud J, Vannier MW et al. (1986). Advanced computer graphics technology reveals cortical asymmetry in endocasts of rhesus monkeys. Folia Primatol (Basel) 46: 98–103.

Frey SH (2008). Tool use, communicative gesture and cerebral asymmetries in the modern human brain. Philos Trans R Soc Lond B Biol Sci 363: 1951–1957.

Gannon PJ, Kheck N, Hof PR (2008). Leftward interhemispheric asymmetry of macaque monkey temporal lobe language area homolog is evident at the cytoarchitectural, but not gross anatomic level. Brain Res 1199: 62–73.

Gilissen E, Hopkins WD (2013). Asymmetries in the parietal operculum in chimpanzees (*Pan troglodytes*) in relation to handedness for tool use. Cereb Cortex 23: 411–422.

Gomez-Robles A, Hopkins WD, Schapiro SJ et al. (2016). The heritability of chimpanzee and human brain asymmetry. Proc Biol Sci 283.

Graic JM, Peruffo A, Corain L et al. (2020). Asymmetry in the cytoarchitecture of the area 44 homolog of the brain of the chimpanzee Pan troglodytes. Front Neuroanat 14: 55.

Groen MA, Whitehouse AJ, Badcock NA et al. (2013). Associations between handedness and cerebral lateralisation for language: a comparison of three measures in children. PLoS One 8: e64876.

Haberling IS, Corballis PM, Corballis MC (2016). Language, gesture, and handedness: evidence for independent lateralized networks. Cortex 82: 72–85.

Heilbronner PL, Holloway RL (1989). Anatomical brain asymmetry in monkeys: frontal, temporoparietal, and limbic cortex in Macaca. Am J Phys Anthropol 80: 203–211.

Hepper PG, Wells DL, Lynch C (2005). Prenatal thumb sucking is related to postnatal handedness. Neuropsychologia 43: 313–315.

Hobaiter C, Byrne RW (2013). Laterality in gestural communication of wild chimpanzees. Ann N Y Acad Sci 1288: 9–16.

Hopkins WD (2004). Laterality in maternal cradling and infant positional biases: implications for the development and evolution of hand preferences in nonhuman primates. Int J Primatol 25: 1243–1265.

Hopkins WD (2006). Comparative and familial analysis of handedness in great apes. Psychol Bull 132: 538–559.

Hopkins WD (2013). Comparing human and nonhuman primate handedness: challenges and a modest proposal for consensus. Dev Psychobiol 55: 621–636.

Hopkins WD (2018). A review of performance asymmetries in hand skill in nonhuman primates with a special emphasis on chimpanzees. Prog Brain Res 238: 57–89.

Hopkins WD (2022). Neuroanatomical asymmetries in nonhuman primates in the homologs to Broca's and Wernicke's areas: a mini-review. Emerg Top Life Sci 6: 271–284.

Hopkins WD, Cantalupo C (2004). Handedness in chimpanzees is associated with asymmetries in the primary motor but not with homologous language areas. Behav Neurosci 118: 1176–1183.

Hopkins WD, De Lathouwers M (2006). Left nipple preferences in infant Pan paniscus and P. troglodytes. Int J Primatol 27: 1653–1662.

Hopkins WD, de Waal FBM (1995). Behavioral laterality in captive bonobos *(Pan paniscus):* replication and extension. Int J Primatol 16: 261–276.

Hopkins WD, Nir T (2010). Planum temporale surface area and grey matter asymmetries in chimpanzees (*Pan troglodytes*): the effect of handedness and comparison within findings in humans. Behav Brain Res 208: 436–443.

Hopkins WD, Wesley MJ (2002). Gestural communication in chimpanzees (Pan troglodytes): the influence of experimenter position on gesture type and hand preference. Laterality 7: 19–30.

Hopkins WD, Pilcher DL, MacGregor L (2000). Sylvian fissure length asymmetries in primates revisited: a comparative MRI study. Brain Behav Evol 56: 293–299.

Hopkins WD, Hook M, Braccini S et al. (2003). Population-level right handedness for a coordinated bimanual task in chimpanzees (*Pan troglodytes*): replication and extension in a second colony of apes. Int J Primatol 24: 677–689.

Hopkins WD, Wesley MJ, Izard MK et al. (2004). Chimpanzees are predominantly right-handed: replication in three colonies of apes. Behav Neurosci 118: 659–663.

Hopkins WD, Russell JL, Cantalupo C et al. (2005). Factors influencing the prevalence and handedness for throwing in captive chimpanzees (*Pan troglodytes*). J Comp Psychol 119: 363–370.

Hopkins WD, Russell JL, Lambeth S et al. (2007). Handedness and neuroanatomical asymmetries in captive chimpanzees: a summary of 15 years of research. In: WD Hopkins (Ed.), Evolution of hemispheric specialization in primates. Academic Press, London.

Hopkins WD, Coulon O, Mangin JF (2010). Observer-independent characterization of sulcal landmarks and depth asymmetry in the central sulcus of the chimpanzee brain. Neuroscience 171: 544–551.

Hopkins WD, Pika S, Liebal K et al. (2012). Handedness for manual gestures in great apes: a meta-analysis. In: S Pika, K Liebal (Eds.), Current developments in non-human primate gesture research. John Benjamins Publishing, Amsterdam.

Hopkins WD, Gardner M, Mingle M et al. (2013). Within- and between-task consistency in hand use as a means of characterizing hand preferences in captive chimpanzees (Pan troglodytes). J Comp Psychol 127: 380–391.

Hopkins WD, Meguerditchian A, Coulon O et al. (2014). Evolution of the central sulcus morphology in primates. Brain Behav Evol 84: 1930.

Hopkins WD, Misiura M, Pope SM et al. (2015a). Behavioral and brain asymmetries in primates: a preliminary evaluation of two evolutionary hypotheses. Ann N Y Acad Sci 1359: 65–83.

Hopkins WD, Reamer L, Mareno MC et al. (2015b). Genetic basis for motor skill and hand preference for tool use in chimpanzees (Pan troglodytes). Proc Biol Sci 282: 20141223.

Hopkins WD, Meguerditchian A, Coulon O et al. (2017). Motor skill for tool-use is associated with asymmetries in Broca's area and the motor hand area of the precentral gyrus in chimpanzees (Pan troglodytes). Behav Brain Res 318: 71–81.

Hopkins WD, Mareno MC, Schapiro SJ (2019). Further evidence of a left hemisphere specialization and genetic basis for tool use skill in chimpanzees (Pan troglodytes): reproducibility in two genetically isolated populations of apes. J Comp Psychol 133: 512–519.

Hopkins WD, Sprung-Much T, Amiez C et al. (2022). A comprehensive analysis of variability in the sulci that define the inferior frontal gyrus in the chimpanzee (Pan troglodytes) brain. Am J Biol Anthropol 179: 31–47.

Jaffe BD, Evans TA, Howell S et al. (2006). Left versus right nipple preference in free-ranging infant rhesus macaques (Macaca mulatta). Dev Psychobiol 48: 266–272.

Johnson-Frey SH (2004). The neural basis of complex tool use in humans. Trends Cogn Sci 8: 71–78.

Jones S (2014). Maternal cradling bias and early communicative interactions: implications for early identification of children at risk. Infant Behav Dev 37: 722–728.

Keller SS, Crow T, Foundas A et al. (2009a). Broca's area: nomenclature, anatomy, typology and asymmetry. Brain Lang 109: 29–48.

Keller SS, Roberts N, Hopkins WD (2009b). A comparative magnetic resonance imaging study of the anatomy, variability, and asymmetry of Broca's area in the human and chimpanzee brain. J Neurosci 29: 14607–14616.

Keller SS, Deppe M, Herbin M et al. (2012). Variability and asymmetry of the sulcal contours defining Broca's area homologue in the chimpanzee brain. J Comp Neurol 520: 1165–1180.

Knecht S, Drager B, Deppe M et al. (2000). Handedness and hemispheric language dominance in healthy humans. Brain 123 (Pt 12): 2512–2518.

Kurth F, Cherbuin N, Luders E (2020). Speaking of aging: changes in gray matter asymmetry in Broca's area in later adulthood. Cortex 129: 133–140.

Labache L, Mazoyer B, Joliot M et al. (2020). Typical and atypical language brain organization based on intrinsic connectivity and multitask functional asymmetries. Elife 9.

LeMay M (1985). Asymmetries of the brains and skulls of nonhuman primates. In: SD Glick (Ed.), Cerebral lateralization in nonhuman species. Academic Press, New York.

Llorente M, Mosquera M, Fabre M (2009). Manual laterality for simple reaching and bimanual coordinated task in naturalistic housed *Pan troglodytes*. Int J Primatol 30: 183–197.

Llorente M, Palou L, Carrasco L et al. (2010). Population-level right handedness for a coordinated bimanual task in naturalistic housed chimpanzees: replication and extension in 114 animals from Zambia and Spain. Am J Primatol 71: 1–10.

Lyn H, Pierre P, Bennett AJ et al. (2011). Planum temporale grey matter asymmetries in chimpanzees (Pan troglodytes), vervet (Chlorocebus aethiops sabaeus), rhesus (Macaca mulatta) and bonnet (Macaca radiata) monkeys. Neuropsychologia 49: 2004–2012.

MacNeilage PF, Studdertkennedy MG, Lindblom B (1987). Primate handedness reconsidered. Behav Brain Sci 10: 247–263.

Margiotoudi K, Marie D, Claidiere N et al. (2019). Handedness in monkeys reflects hemispheric specialization within the central sulcus. An in vivo MRI study in right- and left-handed olive baboons. Cortex 118: 203–211. https://doi.org/10.1016/j.cortex.2019.01.001. Epub 2019 Jan 14; PMID: 30738569.

Marie D, Roth M, Lacoste R et al. (2018). Left brain asymmetry of the planum temporale in a nonhominid primate: redefining the origin of brain specialization for language. Cereb Cortex 28: 1808–1815.

Mazoyer B, Zago L, Jobard G et al. (2014). Gaussian mixture modeling of hemispheric lateralization for language in a large sample of healthy individuals balanced for handedness. PLoS One 9: e101165.

Meguerditchian A, Vauclair J (2006). Baboons communicate with their right hand. Behav Brain Res 171: 170–174.

Meguerditchian A, Vauclair J, Hopkins WD (2010). Captive chimpanzees use their right hand to communicate with each other: implications for the origins of hemispheric specialization for language. Cortex 46: 40–48.

Meguerditchian A, Gardner MJ, Schapiro SJ et al. (2012). The sound of one-hand clapping: handedness and perisylvian neural correlates of a communicative gesture in chimpanzees. Proc Biol Sci 279: 1959–1966.

Meguerditchian A, Vauclair J, Hopkins WD (2013). On the origins of human handedness and language: a comparative review of hand preferences for bimanual coordinated actions and gestural communication in nonhuman primates. Dev Psychobiol 55: 637–650.

Meunier H, Vauclair J, Fagard J (2012). Human infants and baboons show the same pattern of handedness for a communicative gesture. PLoS One 7: e33959.

Michel GF (1981). Right-handedness: a consequence of infant supine head-orientation preference? Science 212: 685–687.

Neufuss J, Humle T, Cremaschi A et al. (2017). Nut-cracking behaviour in wild-born, rehabilitated bonobos (Pan paniscus): a comprehensive study of hand-preference, hand grips and efficiency. Am J Primatol 79: 1–16.

Nishida T (1993). Left nipple suckling preference in wild chimpanzees. Ethol Sociobiol 14: 45–51.

Nudo RJ, Jenkins WM, Merzenich MM et al. (1992). Neurophysiological correlates of hand preference in primary motor cortex of adult squirrel monkeys. J Neurosci 12: 2918–2947.

Ocklenburg S, Gunturkun O (2018). The lateralized brain: the neuroscience and evolution of hemispheric asymmetries, Academic Press, London.

Ocklenburg S, Beste C, Gunturkun O (2013a). Handedness: a neurogenetic shift of perspective. Neurosci Biobehav Rev 37: 2788–2793.

Ocklenburg S, Strockens F, Gunturkun O (2013b). Lateralisation of conspecific vocalisation in non-human vertebrates. Laterality 18: 1–31.

Ocklenburg S, Beste C, Arning L et al. (2014). The ontogenesis of language lateralization and its relation to handedness. Neurosci Biobehav Rev 43: 191–198.

Packheiser J, Schmitz J, Arning L et al. (2020). A large-scale estimate on the relationship between language and motor lateralization. Sci Rep 10: 13027.

Palomero-Gallagher N, Zilles K (2019). Differences in cytoarchitecture of Broca's region between human, ape and macaque brains. Cortex 118: 132–153.

Penfield W, Boldrey E (1936). Somatic motor and sensory representation in the cerebral cortex of man as studied by electrical stimulation. Brain 60: 389–443.

Perelle IB, Ehrman L (1994). An international study of human handedness: the data. Behav Genet 24: 217–227.

Petrides M, Cadoret G, Mackey S (2005). Orofacial somatomotor responses in the macaque monkey homologue of Broca's area. Nature 435: 1235–1238.

Phillips KA, Sherwood CC (2005). Primary motor cortex asymmetry is correlated with handedness in capuchin monkeys (Cebus apella). Behav Neurosci 119: 1701–1704.

Potdevin D, Adipour P, Garric C et al. (2023). Brain lateralization for language, vocabulary development and handedness at 18 months. Symmetry 15: 989.

Previc FH (1991). A general theory concerning the prenatal origins of cerebral lateralization in humans. Psychol Rev 98: 299–334.

Prieur J, Pika S, Barbu S et al. (2016a). Gorillas are right-handed for their most frequent intraspecific gestures. Anim Behav 118: 165–170.

Prieur J, Pika S, Barbu S et al. (2016b). A multifactorial investigation of captive chimpanzees' intraspecific gestural laterality. Anim Behav 116: 31–43.

Prieur J, Pika S, Barbu S et al. (2018). A multifactorial investigation of captive gorillas' intraspecific gestural laterality. Laterality 23: 538–575.

Quallo MM, Price CJ, Ueno K et al. (2009). Gray and white matter changes associated with tool-use learning in macaque monkeys. Proc Natl Acad Sci U S A 106: 18379–18384.

Regaiolli B, Spiezio C, Hopkins WD (2016a). Hand preference on unimanual and bimanual tasks in strepsirrhines: the case of the ring-tailed lemur (Lemur catta). Am J Primatol 78: 851–860.

Regaiolli B, Spiezio C, Hopkins WD (2016b). Three actions, two groups: looking for the origin of primate manual lateralization. J Comp Psychol 130: 259–268.

Regaiolli B, Spiezio C, Hopkins WD (2018). Asymmetries in mother-infant behaviour in Barbary macaques (Macaca sylvanus). PeerJ 6: e4736.

Roberts AI, Murray L, Roberts SGB (2019). Complex sociality of wild chimpanzees can emerge from laterality of manual gestures. Hum Nat 30: 299–325.

Rogers LJ, Vallortigara G, Andrew RJ (2013). Divided brains: the biology and behaviour of brain asymmetries, Cambridge University Press, New York.

Sanz CM, Morgan DB, Hopkins WD (2016). Lateralization and performance asymmetries in the termite fishing of wild chimpanzees in the Goualougo Triangle, Republic of Congo. Am J Primatol 78: 1190–1200.

Schenker NM, Hopkins WD, Spocter MA et al. (2010). Broca's area homologue in chimpanzees (Pan troglodytes): probabilistic mapping, asymmetry, and comparison to humans. Cereb Cortex 20: 730–742.

Schmitz J, Gunturkun O, Ocklenburg S (2019). Building an asymmetrical brain: the molecular perspective. Front Psychol 10: 982.

Sha Z, Pepe A, Schijven D et al. (2021a). Handedness and its genetic influences are associated with structural asymmetries of the cerebral cortex in 31,864 individuals. Proc Natl Acad Sci U S A 118: e2113095118.

Sha Z, Schijven D, Carrion-Castillo A et al. (2021b). The genetic architecture of structural left-right asymmetry of the human brain. Nat Hum Behav 5: 1226–1239.

Shumaker RW, Wallup KR, Beck BB (2011). Animal tool behavior: the use and manufacture of tools by animals, Johns Hopkins University Press, Baltimore, MD.

Simonds RJ, Scheibel AB (1989). The postnatal development of the motor speech area: a preliminary study. Brain Lang 37: 42–58.

Soto C, Gazquez JMM, Llorente M (2022). Hand preferences in coordinated bimanual tasks in non-human primates: a systematic review and meta-analysis. Neurosci Biobehav Rev 141: 104822.

Spocter MA, Hopkins WD, Garrison AR et al. (2010). Wernicke's area homolog in chimpanzees (Pan troglodytes) and its relation to the appearance of modern human language. Proc Biol Sci 277: 2165–2174.

Spocter MA, Sherwood CC, Schapiro SJ et al. (2020). Reproducibility of leftward planum temporale asymmetries in two genetically isolated populations of chimpanzees (Pan troglodytes). Proc Biol Sci 287: 20201320.

Sprung-Much T, Eichert N, Nolan E et al. (2022). Broca's area and the search for anatomical asymmetry: commentary and perspectives. Brain Struct Funct 227: 441–449.

Stout D, Chaminade T (2012). Stone tools, language and the brain in human evolution. Philos Trans R Soc Lond B Biol Sci 367: 75–87.

Strockens F, Gunturkun O, Ocklenburg S (2013). Limb preferences in non-human vertebrates. Laterality 18: 536–575.

Taglialatela JP, Cantalupo C, Hopkins WD (2006). Gesture handedness predicts asymmetry in the chimpanzee inferior frontal gyrus. Neuroreport 17: 923–927.

Tomaszycki M, Cline C, Griffin B et al. (1998). Maternal cradling and infant nipple preferences in rhesus monkeys (*Macaca mulatta*). Dev Psychobiol 32: 305–312.

Uylings H, Jacobsen A, Zilles K et al. (2006). Left-right asymmetry in volume and number of neurons in adult Broca's area. Cortex 42: 652–658.

Vauclair J (2022). Maternal cradling bias: a marker of the nature of the mother-infant relationship. Infant Behav Dev 66: 101680.

Vickery S, Hopkins WD, Sherwood CC et al. (2020). Chimpanzee brain morphometry utilizing standardized MRI preprocessing and macroanatomical annotations. Elife 9: e60136.

Warren JM (1980). Handedness and laterality in humans and other animals. Physiol Psychol 8: 351–359.

Westergaard GC, Suomi SJ (1994). Hand preference in the use of nut-cracking tools by tufted capuchin monkeys (*Cebus apella*). Folia Primatol 61: 38–42.

Whiten A, Goodall J, McGrew WC et al. (1999). Cultures in chimpanzees. Nature 399: 682–685.

Whiten A, Goodall J, McGrew W et al. (2001). Charting cultural variation in chimpanzees. Behaviour 138: 1489–1525.

Yeni-Komshian GH, Benson DA (1976). Anatomical study of cerebral asymmetry in the temporal lobe of humans, chimpanzees, and rhesus monkeys. Science 192: 387–389.

Yousry TA, Schmid UD, Alkadhi H et al. (1997). Localization of the motor hand area to a knob on the precentral gyrus. A new landmark. Brain 120: 141–157.

Zhao D, Gao X, Li B et al. (2008). First wild evidence of neonate nipple preference and maternal cradling laterality in Old World monkeys: a preliminary study from Rhinopithecus roxellana. Behav Processes 77: 364–368.

Handbook of Clinical Neurology, Vol. 208 (3rd series)
Cerebral Asymmetries
C. Papagno and P. Corballis, Editors
https://doi.org/10.1016/B978-0-443-15646-5.00011-7

Chapter 14

Brain and behavioral asymmetries in nonprimate species

ELISA FRASNELLI* AND GIORGIO VALLORTIGARA

CIMeC, Center for Mind/Brain Sciences, University of Trento, Rovereto, Italy

Abstract

Brain and behavioral asymmetries are widespread across the animal kingdom, suggesting that even simpler nervous systems benefit from such features. In the last 30 years, research conducted on several vertebrate (but also invertebrate) animal models has massively contributed to our understanding of the causation, development, evolution, and function of lateralization. Here, we review some of this research, highlighting the importance of studying this topic in nonprimate species for a deeper understanding of the mechanisms behind cerebral asymmetries. We report evidence of handedness and motor asymmetries as well as the results of research on perceptual and cognitive asymmetries in nonprimate animals, analyzing the contribution of such studies in the research field of cerebral asymmetries.

INTRODUCTION

In the last three decades, brain and behavioral asymmetries have been documented in a variety of vertebrate and invertebrate species. This suggests that having an asymmetric brain may confer advantages not only to complex brains like the human brain but also to simple nervous systems of a few hundred neurons. The presence of lateralization in species that are evolutionarily distant from our own species helps shed light on the evolution of this fundamental brain feature. Importantly, despite differences between the nervous systems of humans and other animals, there is good evidence that lateralization is implemented by similar mechanisms across different species and taxa (e.g., functional specialization of the right and left hemispheres seems to have been conserved throughout vertebrate evolution; see Rogers et al., 2012). Therefore, animal models help us uncover the mechanisms underlying asymmetries by virtue of their similarity with our species. Some species (e.g., the domestic chick *Gallus gallus*, the zebrafish *Danio rerio*, the domestic dog *Canis familiaris*, the nematode *Caenorhabditis elegans*, and the honeybee *Apis mellifera*) have been extremely valuable models for investigating causation, development, and function of brain and behavioral asymmetries. This chapter aims to demonstrate how nonprimate animals have contributed to our understanding of lateralization, but it is not intended to be a comprehensive review of the whole literature available. Specifically, we will present and discuss some of the existing evidence of lateralization manifesting as (1) limb preference and motor biases and (2) perceptual and cognitive asymmetries.

Before that, we would like to clarify a few concepts. First, as we are going to discuss studies conducted on both vertebrates and invertebrates, it is important to clarify that in vertebrate species there are *cross-lateral* projections from the brain to the rest of the body (apart from the sense of smell where the projections are ipsilateral), whereas in invertebrate species such projections are *ipsilateral*. This implies that in vertebrates, the right side of the brain controls the left part of the body, and the left side of the brain controls the right part of the body, but in invertebrates, the control happens on the same side, i.e., the right part of the brain controls the right part of the body and the same occurs for the left part (for details on the hypotheses behind decussation, see de Lussanet and Osse, 2015). A clear example to which we dedicate the first section of the chapter is the use of the limb: the use of the right limb in a vertebrate species indicates the activation of the left side of the brain, but in an invertebrate species it indicates the activation of the right side.

Second, we would like to introduce the concepts of *strength* and *direction* of asymmetry. The strength refers to how much an individual is lateralized. For example, an animal that consistently uses its right limb in all trials is considered strongly right lateralized compared to an individual that uses its right limb most of the time (but not all the time). The direction of the bias can be either right or left. Thus, referring to the example above an individual can be either right-, left-, or not lateralized, when it respectively uses mainly its right limb, its left limb, or it uses both the right and the left limb, showing no preference for one or the other.

Lastly, it is important to refer to the level at which lateralization expresses itself: in single individuals, regardless of a common

*Correspondence to: Elisa Frasnelli, CIMeC, Center for Mind/Brain Sciences, University of Trento, Piazza della Manifattura 1, 38068 Rovereto, (TN), Italy. Tel: +39-0464-808744, Fax: +39-0464-808654, E-mail: elisa.frasnelli@unitn.it

direction (bias) in the population (aka *individual-level* lateralization), or in single individuals and in the same direction in most of them so that the population is biased (aka *population-level* lateralization). Indeed, lateralization often occurs at the population level, with 60%–90% of individuals showing the same direction (right or left) of bias depending on species and behavior considered (e.g., right-handedness in humans varies between 70% and 90%—Annett, 2002). This is a very interesting aspect of lateralization and so far, not a definitive explanation has been provided. In fact, although the hypothesis that lateralization increases the brain's efficiency may explain the presence of individual-level lateralization (Vallortigara and Rogers, 2005), it cannot explain population bias, since individual brain efficiency is unrelated to the direction of the asymmetry in other individuals. The first hypothesis of the role of population-level lateralization was advanced by Rogers and Workman (1989), who suggested that population biases of asymmetry might be advantageous in social situations (see also Vallortigara and Rogers, 2005). This idea was developed by Ghirlanda and Vallortigara (2004) in a theoretical model showing that, in the context of prey–predator interactions, population-level lateralization can arise as an evolutionarily stable strategy (ESS) when individually asymmetric organisms must coordinate their behavior with that of other asymmetric organisms. If intraspecific interactions instead of interspecific prey–predator interactions are considered, the model suggests that asymmetry aligns in most individuals in a population (i.e., direction) when individually asymmetric organisms must interact with conspecifics and coordinate their activities (Ghirlanda et al., 2009). Differently than the original hypothesis of Rogers and Workman, the model based on ESS (Vallortigara, 2006) does not predict that population lateralization would necessarily emerge in social species but rather, depending on the kind of interaction between individuals, an ESS would emerge as an equilibrium condition either as an individual- or a population-level strategy (see for a discussion Frasnelli and Vallortigara, 2018). We will see that this prediction has found some empiric support in both vertebrate and invertebrate species.

LIMB PREFERENCE AND MOTOR BIASES

Handedness has been used as the major example of lateralization as it is very prominent in the asymmetric use of hands in humans. Asymmetry in the use of limbs, however, is not a uniquely human attribute as once believed but a well-known phenomenon across the animal kingdom. Several studies have shown that many groups of vertebrates and invertebrates exhibit lateralized motor behavior (Rogers et al., 2012; Stancher et al., 2018). The preferential use of one side of the body over the other as a proxy for the specialized control of one side of the brain has been extensively investigated in several animal taxa, from fish to mammals. In this section, we review some of the empiric evidence for limb asymmetry across different animal groups, excluding primates. Importantly, we refer to limb preference and motor biases when no other sensory cues are presented. In fact, some motor biases can be visually driven making it sometimes difficult to disentangle whether the bias is induced by an eye preference or it is a motor preference. Nevertheless, in both cases, the bias will indicate a functional specialized control by one side of the brain.

Fish

The only evidence of asymmetric use of limbs in fish, i.e., fins in this case, was observed in the blue gournami (*Trichogaster trichopterus*), a species where the ventral fins have gustatory and tactile functions, which preferentially use the left fin to initiate contact with novel objects (Bisazza et al., 2001a).

Lateralized motor behavior has instead been shown in several fish species both when visual stimuli are provided or not (hereafter referred to as *spontaneous*). Starting from the cases when no visual stimuli were provided, an example is the spontaneous rightward turning bias observed in four out of five species of minnows (Osteichthyes: Cyprinidae) when placed in a T-shaped arena (Stennett and Strauss, 2010). A further example is the rotational bias that some teleost species (Bisazza and Vallortigara, 1996; Izvekov and Nepomnyashchikh, 2010) and chondrostean fish (Izvekov et al., 2014) present when placed in a circular tank and swim along its wall. The direction of spontaneous swimming is affected by sex and time of the day in mosquitofish (*Gambusia holbrooki*) (Bisazza and Vallortigara, 1996; Bisazza et al., 1997b): while females have a population-level bias to swim clockwise in the morning and counterclockwise at night, males only present individual rotational preferences. The asymmetry observed in the females seems due to a sun-compass orientation mechanism as they do not present the bias anymore if tested under diffuse lighting conditions.

Exposure to CO_2 influences turning biases in fish when tested in a maze. Marine damselfish (*Pomacentrus wardi*) exposed to elevated CO_2 concentrations show a left-turning bias in a detour test, whereas the control group presented a population-level right-turning bias (Domenici et al., 2012, 2014). Exposure to elevated CO_2 concentrations decreases individual-level biases in turning behavior in temperate sand-smelt *Atherina presbyter* larvae, but a GABA-A receptor antagonist (gabazine) reverses this effect (Lopes et al., 2016). Instead, in small-spotted catsharks (*Scyliorhinus canicula*), exposure to elevated CO_2 concentrations produces an increased bias in turning behavior (Green and Jutfelt, 2014).

A well-studied motor bias induced by a stimulus—a threatening stimulus in this case—is the so-called C-start response, a fast escape response consisting of a C-shape body bending toward one side of the body that enables the animal to flee from a potentially threatening stimulus (Zottoli, 1977). This motor bias is either rightward or leftward and is present at the individual or at the population level depending on the species considered, but regardless of that, it allows very fast responses (Domenici and Blake, 1997). Indeed, zebrafish and goldfish show a population-level right bias (Heuts, 1999), Australian lungfish (*Neoceratodus forsteri*) a population-level left bias (Lippolis et al., 2009), whereas killifish (*Jenynsia multidentate*) and roach (*Rutilus rutilus*) present an individual-level bias with a similar number of individuals bending toward the right and the left (Bisazza et al., 1997a; Izvekov and Nepomnyashchikh, 2010). Interestingly, regardless of the direction and the level of this bias present in the species, what seems to matter is the strength of the bias as strongly lateralized shiner perch (*Cymatogaster aggregata*) have a higher escape reactivity and are thus better at escaping from predators than nonlateralized conspecifics (Dadda et al., 2010). The C-start response is caused by a unilateral muscle contraction triggered by two big reticulospinal neurons, called Mauthner cells, one of which elicits the muscle contraction, whereas the simultaneous activity of the opposite neuron is suppressed (Eaton et al., 2001; Korn and Faber, 2005). The size of the right and left Mauthner neurons has been shown to predict the direction of the C-start response in goldfish (*Carassius auratus*) as animals with the left larger neuron preferentially bend rightward and vice versa (Moulton and Barron, 1967; Mikhailova et al., 2006).

Not only threatening stimuli induce a lateralized escape response, but predatory pressure can modulate the strength of the bias. Individuals show a stronger turning bias when exposed to high risk of predation compared to individuals at low risk both in wild-caught coral reefs shooling fish (*Caesio teres*) (Chivers et al., 2016) and in captive breeding juvenile damselfish (*Pomacentrus* spp.), where the bias is also stronger in animals exposed to cortisol (Ferrari et al., 2015b, 2017). Furthermore, strongly lateralized shiner perch (*C. aggregata*) respond with shorter latencies in their escape response to mechanical stimulation (Dadda et al., 2010).

Animal models such as zebrafish have been extremely helpful in understanding the neural correlates and developmental signaling pathways for motor asymmetries. In fact, zebrafish larvae show individual-level motor biases across multiple behavioral assays to execute movements in a preferred direction on ~70% of the trials when tested in the absence of visual stimuli (Horstick et al., 2020) and maintain such preference over at least several days, despite handling and changes in the environment. This individual motor bias is mediated by a cluster of ~30 genetically identified neurons per hemisphere in the rostral lobe of the posterior tuberculum (PT) of the forebrain that project to the habenula nuclei as chemogenetic ablation of this cluster of neurons degrades the bias. However, it remains unclear how the individual preference is encoded in the PT-habenula unit. Moreover, heterozygous mutations in genes that regulate Notch signaling disrupt the lateralized behavior, suggesting that haploinsufficiency for genes in a core developmental pathway destabilizes left/right identity (see for a review Miletto Petrazzini et al., 2020).

Amphibians

Studies on lateralized behavior in amphibians are scarce but some evidence for limb asymmetry has been reported in frogs and toads. Green tree frogs (*Litoria caerulea*) preferentially use their right forelimb during climbing (Robins and Rogers, 2006a). The common European toad (*Bufo bufo*) (Bisazza and Vallortigara, 1996; Bisazza et al., 1996) and the cane toad (*Bufo marinus* or *Rhinella marina*) (Robins and Rogers, 2002; Bisazza et al., 1996), but not the Oriental fire-bellied toad (*Bombina orientalis*) (Goree and Wassersug, 2001) preferentially use their right forepaw to remove an object (i.e., a paper strip on the snout in the so-called "snout-wiping test" and a plastic balloon from the head); whereas the European green toads (*Bufotes viridis*) show a preferential use of the left forelimb in the same task (Malashichev and Nikitina, 2002) and also while bringing living larvae to the mouth (Sovrano, 2007; see also Bisazza et al., 1997). When overturned, toads also show a preferential use of the right or left hindlimb to push against a surface in a righting response with *B. bufo* and *B. marinus* using the right hindlimb and *B. viridis* the left (Robins et al., 1998). The motor biases reported above seem to be related to the species-specific type of locomotion. Species that alternate limbs during locomotion (*B. bufo, B. marinus, B. viridis, L. caerulea, Litoria latopalmata, Ceratophrys ornata, Pelobates fuscus*) show a bias at the population level, whereas those which jump or swim (*B. orientalis, Rana temporaria, Rana lessonae, Bombina*) do not show any bias (Bisazza and Vallortigara, 1996; Robins et al., 1998; Goree and Wassersug, 2001; Malashichev and Nikitina, 2002; Rogers, 2002; Malashichev, 2006), suggesting that quadruped locomotion may involve lateralization in the neural motor control (Malashichev, 2006). However, these motor biases are unlikely directly caused by a morphologic asymmetry (Goree and Wassersug, 2001; Malashichev, 2002; Robins and Rogers, 2002), with the exception of the position of epicoracoids (i.e., the asymmetric structure of the shoulder girdle) in *B. bombina* where individuals having the right epicoracoid in dorsal position use preferentially the right forelimb in the "snout-wiping test" and those having the left epicoracoid in dorsal position use the left forelimb (Malashichev and Nikitina, 2002).

Tadpoles of American bullfrog (*Lithobates catesbeianus*) (Wassersug et al., 1999) and the ornate narrow-mouthed frog (*Microhyla ornate*) (Yamashita et al., 2000) preferentially turn to the left when descending after surfacing to breathe air, whereas tadpoles of the clawed frog (*Xenopus laevis*) do not show any turning bias (Wassersug et al., 1999). Tadpoles also show a preferential direction in the startle response, i.e., the defensive response to sudden or threatening stimuli: to the left in *Agalychnis callidryas* and to the right in *Leptodactylus melanonotus* (Briggs-Gonzalez and Gonzalez, 2016). Interestingly, in free-moving hatchling *X. laevis* tadpole the direction of swimming seems to be controlled by the midbrain as different midbrain lesions alter the ratio between ipsilateral and contralateral swim initiations (Larbi et al., 2022). Furthermore, when raised with high doses of Roundup, the most used glyphosate-based herbicide, tadpoles show weaker rotational preferences compared to the control group (Bolis et al., 2020).

Motor asymmetries have been documented also in mating behavior in newt males (*Lissotriton vulgaris*) with a population-level bias to turn toward the left side after spermatophore deposition (Green, 1997) but not in the alpine newt (*Ichthyosaura alpestris*).

Reptiles

Lateralization has been observed only in a few reptile species. Gekos (*Hoplodactylus duvaucelii*) present an individual-level preference for the right or left hindlimb when releasing off the ground (Seligmann, 2002). Sieg et al. (2010) reported that, during egg laying, leatherback turtle (*Dermochelys coriacea*) females use the right hindlimb to cover the chamber, but this preference is significant only for very few individuals. Instead, terrestrial tortoises (*Testudo hermanni*) display a population-level rightward bias in the righting response, with most individuals retaining the same preferred righting direction up to 10 months later (Stancher et al., 2006). This species also presents a population-level preference for the right forelimb in starting locomotion, while inspecting mirror imaging (Stancher et al., 2018). The limb used to initiate locomotion has been investigated also in sea turtles: Green turtles (*Chelonia mydas*) show an individual-level preference, whereas Olive Ridley turtles (*Lepidochelys olivacea*) do not present any preference to use the right or the left forelimb (Malashichev, 2016). As discussed for amphibians, the species-specific differences in locomotion may explain the differences in the presence, strength, and level of the bias (Malashichev, 2006): sea turtles use all four limbs in scratching, while terrestrial species alternate limbs during locomotion. Thus, the presence of a population-level bias in terrestrial tortoises would be a consequence of the asynchronous locomotion in this species whereas the lack or the presence of bias at the individual level in sea turtles is linked to their synchronous locomotion.

Birds

Lateralized behavior in motor control is common in birds and has mainly been investigated as footedness, i.e., the preferential use of one foot over the other in different tasks. It is important to highlight that in these studies the dominant foot may not necessarily be the one that is used in the task itself, for example, in manipulating food, but the one used to stabilize the posture during the task. Randler (2007) tried to assess whether footedness should be related to the foot used for holding or handling food or to the foot used to stabilize the posture by assessing footedness in wildfowl and waders while roosting on the ground on one foot. While 21 of the species examined did not show any foot preference, avocet (*Recurvirostra avosetta*), northern shoveller (*Anas clypeata*), oystercatcher (*Haematopus ostralegus*), and Eurasian curlew (*Numenius arquata*) showed a preference to roost on the right foot, suggesting that footedness in birds may be linked to postural control (Randler, 2007).

Different bird species show either individual-level or population-level footedness or no footedness depending on the task. The first evidence of preferential foot use was provided in parrots and cockatoos during food manipulation suggesting an overall preference for the left foot in all species (*Cacatua gallerita, Ara macao, Ara severa, Platycercus elegans*) except *P. elegans* which showed a preference for the right foot (Friedman and Davies, 1938; Rogers, 1980; Harris, 1989; Rogers and Workman, 1993). Species such as the great tit (*Parus major*) present instead an individual-level footedness while stabilizing food over a perch and in pulling out some food from a cylinder (Gibb and Hartley, 1957; Yince, 1964). This is also the case of the jungle crows (*Corvus macrorhynchos*) which present a strong individual-level preference for one foot over the other in both beak-scratching and food-holding (Izawa et al., 2005). New Caledonian crows (*Corvus moneduloides*) show a population-level bias while manufacturing tools by making tapered tools from the left edge of leaves even when the spiral direction of the leaves facilitates the access from right side (Hunt, 2000; Hunt et al., 2001). Interestingly, the bias is only exhibited at the individual level in tool use (Rutledge and Hunt, 2004; Weir et al., 2004).

Domestic pigeons (*Columba livia*) do not show any foot preference when removing a piece of tape from their beak (Güntürkün et al., 1988), but show instead an individual-level preference for either the right or left foot to land from a flight (Fisher, 1957; Davies and Green, 1991). Domestic chicks (*G. gallus domesticus*) instead prefer to use the right foot to initiate ground scratching while searching for food, but individuals with only one eye in use prefer to use the foot opposite to it (Rogers and Workman, 1993; Tommasi and Vallortigara, 1999). Interestingly, footedness in this species seems to be influenced by age and type of task, revealing a change in hemispheric dominance from day 5 to day 11 (Dharmaretnam et al., 2002). In fact, consistent with previous studies (Rogers and Workman, 1993; Tommasi and Vallortigara, 1999) chicks preferentially use the right leg to initiate ground scratching on day 5 but revert this preference to the left leg on day 8, to then prefer the right leg again after day 11. For the tape-removing task, instead, the naïve birds did not show any preference, whereas the trained ones preferred to use the right foot on day 8; but on day 11 the trained group reverted the preference to the left foot while the naïve group displayed a right foot bias (Dharmaretnam et al., 2002). Foot preference has been recently investigated also in pukeko (*Porphyrio p. melanotus*), which frequently use their feet to feed, indicating only individual-level footedness in some individuals with roughly the same percentages of birds with a left preference, a right preference, or no preference. Interestingly, the more the foot is used the weaker the preference becomes (Rodgers and Cain, 2019). The type of task also influences leg preferential use in Gyrfalcon (*Falco rusticolus*) as this species presents a population-level preference to stand on the left leg while sleeping, but no preference for placing the head under the left or the right-wing while sleeping and a trend to use the left leg to grasp food (Baciadonna et al., 2022).

These studies suggest that footedness is evident only in species that use their feet for manipulative tasks (Walker, 1980) or that, as the case of the domestic chick, use the feet for feeding in general (Rogers and Workman, 1993).

The potential benefits linked to foot preferences have been investigated in birds. Species of parrots with a stronger preference for one foot while holding food show better performance in both a pebble-grain discrimination task and in a string-pulling task (Magat and Brown, 2009). The strength of the bias seems related to body size and the associated foraging mode in 23 species of Australian parrots, suggesting that parrots with larger bodies benefit from the motor bias in coordinating seed extraction from seedpods, whereas smaller body species may have lost the bias as they shifted to grazing on small seeds (Brown and Magat, 2011). Moreover, a recent study showed that parrot species with stronger foot preferences have larger brains (Kaplan and Rogers, 2021), suggesting a possible link between cognitive performance, brain size, and lateralization. Orange-winged Amazon parrots (*Amazona amazonica*) tested on the Hamilton search task (i.e., a spatial foraging assessment), present overall a left foot preference, and show side preferences contralateral to their preferred limb during location selection. Interestingly, the bias influences learning but has no effect on cognitive flexibility (Cussen and Mench, 2014). Male yellow-bellied tits (*Pardaliparus venustulus*) do not show a population-level preference for one foot over the other to clamp mealworms against a perch, but only at the individual level with right-footed individuals showing better escape performance than other conspecifics, whereas feeding efficiency was not found to be related to footedness (Yu et al., 2020).

The examples above suggest that footedness may be associated with some cognitive abilities or fitness-related measurements. However, only a couple of studies—to our knowledge—have investigated the link between footedness (as a proxy of lateralization) and survival. This is the case of pheasants (*Phasianus colchicus*), in which the strength of footedness seems related to survival as birds with a strong preference to use one foot in a step task once released in the wild are those with low survival (Whiteside et al., 2018). Moreover, footedness in this species is not correlated with cognitive performance in any of the three assays of visual or spatial learning and memory tasks (Whiteside et al., 2020).

Mammals

A recent meta-analysis reported that both rats (*Rattus rattus*) and mice (considering the existing literature on the domestic mouse, *Mus musculus,* and the striped field mouse, *Apodemus agrarius*) present a paw preference but only at the individual level (Manns et al., 2021). Specifically, only the data from the studies employing a food-reaching task to assess paw preference after Collins (1968) were considered and revealed that 81% of mice and 84% of rats show a preference for either the left or the right paw (Manns et al., 2021).

Paw preferences have been widely investigated in domestic dogs (*Canis lupus familiaris*) providing inconsistent evidence. In fact, some studies reported that overall dogs show a population-level preference to use the right paw (Tan, 1987; Tan and Çalşikan, 1987; Tomkins et al., 2010; Laverack et al., 2021; Simon et al., 2022a,b), whereas other studies found a preference only at the individual level (Wells, 2003; Quaranta et al., 2004; Branson and Rogers, 2006; Poyser et al., 2006; McGreevy et al., 2010; Marshall-Pescini et al., 2013; Schneider et al., 2013; Barnard et al., 2018) as supported by a recent meta-analysis indicating that 68% of dogs have an individual-level preference either for the left or the right paw (Ocklenburg et al., 2019). According to some research (Tan, 1987; Tan and Çalşikan, 1987; Laverack et al., 2021; and the meta-analysis on 11 studies by Ocklenburg et al., 2019) dogs show an overall population-level preference for the right paw regardless the sex, but other studies failed to replicate these findings and found instead an effect of sex with a left paw preference in male dogs and a right paw preference in female dogs (Wells, 2003; Quaranta et al., 2004; McGreevy et al., 2010). Importantly, a recent study investigating paw preference in a sample of 17,901 dogs (much bigger than previous studies) found an overall right-paw preference in both sexes, but with more left-paw-biased male dogs than female dogs (Laverack et al., 2021). This sex difference in paw preference has been argued to be related to sex hormones and investigated in neuter and entire dogs revealing an overall population-level preference to use the right paw both while retrieving a piece of food and a toy without a significant effect of sex or neutered status (Duncan et al., 2022). Being dogs a domesticated species that interact a lot with their owners, it has been argued whether owner-handedness could possibly affect paw preference in their dogs and this looks to be the case as left-handed owners seem more likely to own a dog that prefers to use its left paw and vice versa (Charlton and Frasnelli, 2023). Interestingly, in the paw task (i.e., when the owner asked their dog to give a paw), the hand presented to the dog did not significantly predict which paw the dog lifted in response, suggesting that the dog's paw preference does not change in relation to that. The strength of paw preference seems related to other personality traits in dogs as individuals with nonsignificant preference show stronger stress responses to a thunderstorm sound than dogs with a strong preference (Branson and Rogers, 2006). Furthermore, dogs with a nonsignificant bias for one paw produce higher levels of adrenalin, a stress hormone, than dogs with a strong bias at 8 days after vaccination for rabies (Siniscalchi et al., 2010).

Paw preference has also been studied in the domestic cat (*Felis Catus*), providing, as in the case of domestic dogs, conflicting results (Cole, 1955; Tan et al., 1990; Lorincz and Fabre-Thorpe, 1994; Wells and Millsopp, 2009; McDowell et al., 2018; Wells and McDowell, 2019). The first study by Cole (1955) suggested an individual-level paw preference in a food-reaching task with 38.3% of the cats preferentially using the left paw and 20% the right paw. However, another study (Tan et al., 1990) adopting the same methodology as Cole (1955) indicated a higher proportion of cats preferentially using the right paw than those using the left paw. The meta-analysis conducted by Ocklenburg et al. (2019) seems to confirm that paw preference is present in 78% of cats but with differences between males and females. As for dogs, sex seems to influence paw preference also in cats with male cats preferentially using the left paw and female cats the right paw (Tan et al., 1990; Lorincz and Fabre-Thorpe, 1994; McDowell et al., 2018; Ocklenburg et al., 2019; Wells and McDowell, 2019). This sex effect has been further investigated in 44 neutered cats, in a food-reaching task and in three spontaneous behaviors (lying on their side, descending the stairs, and stepping into the litter tray): male cats preferentially used the left paw preference and female cats the right paw in all tasks but lying on their side (McDowell et al., 2018). The effect of neutered status on paw preference has been investigated in food- and toy-reaching tasks indicating a population-level bias to use the right paw in both tasks (Duncan et al., 2022). Moreover, the strength of paw preference (i.e., regardless of the direction) was significantly influenced by an interaction between neuter status and life stage, with senior entire cats having the stronger preference (Duncan et al., 2022). An effect of age on paw preference was previously reported in domestic cats with cats being more ambilateral at 12 weeks of age and showing a paw preference at 1 year (Wells and Millsopp, 2012).

Footedness has been assessed in different breeds of horses (*Equus ferus caballus*). Thoroughbreds (bred to race at the gallop), Standardbreds (bred for pacing) preferentially graze with the left foreleg in front of the other but Quarter Horses (bred for so-called "cutting work" which involves maneuvering individual cattle in and out of herds) do not show any preference, suggesting that training and/or selection may have an influence on motor bias (McGreevy and Thomson, 2006). When started moving off from a standing position, horses that prefer to initially use the right forelimb seem to be more optimistic as they are more likely to expect a neutral stimulus to be positive than individuals preferentially using the left forelimb (Marr et al., 2018). Quarter Horse tested in four motor tasks (walking, stepping on and off a step, truck loading, and unloading) show a population-level preference to use the left forelimb while stepping off a step and during truck loading (Siniscalchi et al., 2014). When observed during reining competitions, in which American Quarter Horses perform the maneuvers such as spins, circles, and stops directed to either the left or right and are judged for symmetric performance, horses do not show any population-level motor bias in the left vs right direction of the maneuvers, suggesting a possible effect of training on forelimb preferences (Whishaw, 2015).

Donkeys (*Equus asinus*) prefer to stand with the right forelimb advanced over the left forelimb when they are in a large enclosure, but this population-level bias disappears when they move to a small enclosure, suggesting that constraints in space availability is an important source of variation in motor biases (Zucca et al., 2011b). In another study, donkeys were assessed for the leading limb used in a motionless posture and for personality using the Donkey Temperament Questionnaire (French, 1993). No population-level bias was observed but limb preference was shown to significantly predict the trait *difficult to handle*: as donkeys that preferentially used the right forelimb were harder to handle (Díaz et al., 2021). Among ungulates, goats (*Capra hircus*) do not show a population-level foreleg preference in the first foreleg used to step off a board after standing with both forelimbs but 36.6% of the individuals tested presented an individual-level preference for a specific foreleg (Baruzzi et al., 2018).

Lions present a population-level preference to place the right forelimb in front of the other while standing with a stronger bias in sick lions compared to healthy ones (Zucca et al., 2011a). In a field study, a weak left motor bias was reported in plains zebras (*Equus burchellii*) but not in impalas (*Aepyceros melampus*) (McGreevy et al., 2007).

Among marsupial species, a population-level forelimb preference in unimanual behaviors has been reported for gray short-tailed opossums (*Monodelphis domestica*) and sugar gliders (*Petaurus breviceps*) with a sex difference: males tend to use the right forelimb and females the left one (Giljov et al., 2013). Red (*Macropus rufus*) and Eastern gray (*Macropus giganteus*) kangaroos also present a population-level preference toward the left forelimb for unimanual feeding and grooming (Giljov et al., 2015). Interestingly, also red-necked wallabies (*Macropus rufogriseus*) prefer to use the left forelimb for auto-grooming in bipedal position (Giljov et al., 2012), but not Goodfellow's tree-kangaroos (*Dandrolagus goodfellowi*) a mainly arboreal species (Giljov et al., 2015). A population-level preference to use the left forelimb for manipulating food has been reported also in the red-necked wallaby (*M. rufogriseus*) and the eastern gray kangaroo (*M. giganteus*) (Giljov et al., 2017). Interestingly, this preference is expressed at the early age, i.e., at about 6–7 months for the red-necked wallaby and at 7–9 months for the gray kangaroo, and no effect of age was observed in relation to the transition to the bipedal posture (Giljov et al., 2017). In fact, when Giljov et al. (2015) compared bipedal and quadrupedal marsupial species, striking differences were observed suggesting that bipedality played an important role in the evolution of limb preference among mammals.

Finally, among aquatic mammals, the only evidence of limb preference has been provided in bottlenose dolphins (*Tursiops aduncus*), which use preferentially left flipper in social rubbing behavior. However, this cannot be considered a purely motor behavior because, as we will see in the next section, the social context plays an important role in the manifestation of lateralization (Sakai et al., 2006).

Invertebrates

Several studies have demonstrated lateralized behavior in invertebrates, including cephalopods, crustaceans, and insects (for a review see Frasnelli, 2013). Common octopuses (*Octopus vulgaris*) show an individual-level preference for the anterior arm used for reaching and exploring objects, as well as for the arm used to reach into a maze and retrieve a food item in 50% of the individuals tested (Byrne et al., 2006). However, they do not show any significant bias when exploring a T-maze in the absence of any specific intra- or extra-maze cues (Frasnelli et al., 2019). Instead, common cuttlefish (*Sepia officinalis*) when tested in a T-maze present a rightward bias in 53% of the individuals and a leftward bias in 20% of them, whereas 27% of them have no preference (Alves et al., 2007). Interestingly, individuals with a significant side bias consistently chose the side of their initial choice (Alves et al., 2007).

Atyid shrimps (*Limnocaridina latipes* and *Neocaridina denticulata*) present a directional bias to escape in a specific direction, a bias that seems to be genetically derived as the behavioral asymmetry of the parents predicts that of the offspring (Takeuchi et al., 2008). Moreover, the left/right direction of the escape response is determined by a morphologic asymmetry in the angle of the abdominal curvature with half of the individuals presenting a leftward curvature and half of them a rightward curvature (Takeuchi et al., 2008). Similarly, wild crayfish (*Procambarus clarkia*) show a bimodal distribution in the direction of escape behaviors with half of the individuals jumping leftward and the other half jumping rightward which correlates with their carapace asymmetry (Tobo et al., 2012). Crab species such as the American lobster crustacean species (*Homarus americanus*) present claw morphologic and behavioral dominance (Govind, 1984, 1992; Govind and Pearce, 1992). Interestingly, it is the behavioral asymmetry in this case that induces and orients the morphologic asymmetry as claw use in juvenile lobsters induces one claw to transform into a specialized crusher claw, whereas induced insufficient stimulation during a critical developmental stage causes no specialization (Govind, 1992; Govind and Pearce, 1992). Also, the male fiddler crab (*Uca vocans vomeris*) presents distinctively asymmetric claws with 98.6% of the individuals being right-clawed (Backwell et al., 2007). In this species being left-clawed seems to be disadvantageous as left-clawed males tend to retreat into their burrow to observe and avoid fights, whereas right-clawed males are more likely to initiate fights, engage in significantly more fights and win more fights than left-clawed individuals (Backwell et al., 2007).

Many spiders tend to present mainly lesions on the left leg than on the right leg (Heuts et al., 2003). Similarly, spitting spiders (*Scytodes globula*) use their left anterior legs more often than the right ones during prey handling (Ades and Ramires, 2002).

Lateral motor biases have been documented in social and solitary insects (e.g., ants, honeybees, fruit flies, cockroaches, and giant water bugs—see Frasnelli et al., 2012a; Niven and Frasnelli, 2018). Ladybirds (*Coccinella septempunctata*) present an individual-level bias toward either the left or the right arm of a Y-maze (Girling et al., 2007), whereas giant water bugs (*Belostoma flumineum*) show a population-level bias to turn leftwards within a T-maze (Kight et al., 2008). A population-level bias has also been reported in American cockroaches (*Periplaneta americana*) in choosing preferentially the right arm of a T-maze, a bias that is likely to be pure motor and not sensory driven as it remains after ablation of the antennae—the antennae of insects are rich of special structures called sensilla with olfactory, taste, CO_2, mechanoreceptors, and hygroreceptors (Cooper et al., 2011). Fruit flies *Drosophila melanogaster* and *Drosophila paulistorum* instead present only individual-level biases in their circling motor behavior, in tapping, and wing extension, but not for the side they preferentially fold their wings or when choosing the left or the right arm of a Y-maze (Perelle et al., 1978). This last result is inconsistent with a more recent study in *D. melanogaster* which found individual-level motor biases in the Y-maze and confirmed the individual-level bias in circling behavior (Buchanan et al., 2015). These motor biases are not correlated with any morphologic traits (e.g., laterality of gut-twisting, leg length asymmetry, or wing-folding preference) nor genetic strains, but seem generated centrally as their strength increases when neurons of the central complex are silenced (Buchanan et al., 2015).

Some ant species present motor biases during their foraging trips with 12 *Lasius* species keeping mainly to the right side of their foraging "streets" and one species to the left (Heuts et al., 2003). Another ant species (*Formica pratensis*) which does not use trail pheromones also shows a population-level bias to keep the left to avoid encountering other workers when walking back to the nest with food but not when they leave the nest to look for food (Hönicke et al., 2015). Red wood ants (*Formica rufa*) belonging to different colonies show population-level biases in different directions when tested for forelimb preference during a gap-crossing task, suggesting that social pressures act to coordinate individuals within the same colony and not necessarily at the species level (Calcraft et al., 2016). Ants (*Temnothorax albipennis*) show a population-level leftward bias when exploring unknown nest sites (Hunt et al., 2014). As ants, honeybees (*A. mellifera*) present a population-level motor bias when exploring unknown open cavities but rightward (instead of leftward as in ants—Hunt et al., 2014), with shorter decision times for strongly right-biased individuals (O'Shea-Wheller, 2019). When individual honeybee workers are provided with a pair of identical apertures through which they have to fly they show an individual-level bias to choose either the right or the left aperture (with some individuals equally preferring the right and the left aperture) (Ong et al., 2017). Moreover, in a recent study (Liga et al., 2024), honeybees trained to associate a visual stimulus (blue or yellow target) to a reward/punishment, tested in a Y-maze to assess the directional choice for one of the two identical targets placed in the left and right arms of the maze showed no population-level directional bias, but only at the individual level in some individuals in line to the result by Ong et al. (2017). However, the data revealed an influence of direction on the flight times as overall bees took less time to choose the stimulus in the left arm (Liga et al., 2024).

Some bumblebee species (*Bombus lapidarius, Bombus lucorum, and Bombus pascuorum*) but not buff-tailed bumblebees (*Bombus terrestris*) prefer to circle in a specific direction (clockwise or anticlockwise) when visiting florets arranged around an upright inflorescence (Kells and Goulson, 2001). Free-flying bumblebees (*B. terrestris*) trained in a T-maze to associate a color with a sugar reward, when presented in unrewarded tests with the trained color at the end of both arms of the maze, overall prefer the right arm, suggesting a visuomotor bias in this direction (Waite and Frasnelli, 2021).

Large white butterflies (*Pieris brassicae)* present intra- and interindividual variation in flight direction with highly mobile and longed-winged individuals flying in similar directions and less mobile and short-winged individuals flying in divergent directions (Larranaga et al., 2013). Similarly, melon flies (*Zeugodacus cucurbitae*) present individual-level flight directional biases in 83% of the individuals to continue "forward" movement. When the effect of wind was analyzed, this was shown to be not the only potential factor influencing the motor bias as the mean flight directions coincided with the wind direction for only 36% of the flies (Miller et al., 2022).

Adult desert locusts (*Schistocerca gregaria*) exhibit an individual-level preference to use the left or the right leg in response to stimulation (Rowell, 1964). This species also presents an individual-level bias (either clockwise or anticlockwise) in their rolling behavior, and this is linked to an asymmetric electrical activity of the muscles (Wilson, 1968). Moreover, many individuals of *Sanicula gregaria* show an individual-level preferential use of the first forelimb used to reach across a gap, with strongly biased locusts outperforming weakly biased individuals (Bell and Niven, 2014, 2016). Analogously, the individuals of antlion larvae (*Myrmeleon bore*) that show an individual-level bias in righting from a supine position perform better than other larvae in a subsequent learning task (Miler et al., 2017).

PERCEPTUAL AND COGNITIVE ASYMMETRIES

In the previous section, we analyzed lateralized *spontaneous* motor behavior. Here, we will consider the motor biases resulting from an asymmetric use of the eyes in visualizing and processing specific stimuli (especially in the case of animals with laterally placed eyes that mainly rely on monocular vision). We will also analyze the asymmetries exhibited in the use of other paired sensory organs (e.g., nostrils, ears, antennae) as well as in cognitive processes (e.g., learning and memory) and discuss the similarity in the direction of the pattern in vertebrate and invertebrate animals.

Fish

A common test used to assess visually induced motor lateralization in fish is the detour test, i.e., a test in which the single individual swims along a runway until it faces a barrier behind which a visual stimulus such as a group of conspecifics or a dummy predator is located. When presented with such a test, Western mosquitofish (*G. holbrooki*) and goldbelly topminnows (*Girardinus falcatus*) show a preferential use of the right eye to monitor a predator and consequently a preference to turn leftwards, but do not in the absence of a predator (Bisazza et al., 1997a,b). Postnatal testosterone seems to influence eye preference when inspecting a predator as the population-level bias for the right eye is present only in fish treated with testosterone (Schaafsma and Groothuis, 2011). Experiments on goldbelly topminnows selected for strength and direction of the eye preference while inspecting a predator in a detour task show that this behavioral asymmetry is inherited in both strength and direction (Bisazza et al., 2000b) and it extends to other tasks (Bisazza et al., 2001a,b, 2005; Dadda et al., 2007, 2009, 2012). The same leftward bias observed for a dummy predator is triggered in male fish of the same species by a group of females, but not by male conspecifics or in the absence of stimuli (Bisazza et al., 1998; Vallortigara and Bisazza, 2002). Instead, female mosquitofish and gold belly topminnow preferentially turn rightward to reach conspecifics of the same sex (Bisazza et al., 1998) and, when sexually deprived, consistently look at males with the right eye. Similarly, sexually deprived female guppies (*Poecilia reticulata*) exhibit a stronger bias to turn leftward when viewing orange-colored than when they view drab males (Kaarthigeyan and Dharmaretnam, 2005). The detour test has recently been challenged to be an appropriate method to assess behavioral asymmetries (Roche et al., 2020). In fact, when the repeatability of lateral biases in a detour test was analyzed in four previously studied fish species (*Ctenolabrus rupestris*, *Neopomacentrus azysron*, *Pomacentrus* amboinensis, and *D. rerio*), both the strength and the direction of lateralization were found to be highly nonrepeatable in all species (Roche et al., 2020).

Turning responses have been studied in the framework of the potential benefits they provide to single individuals when in a school (for example, when escaping from a predator—see Ghirlanda and Vallortigara, 2004). Out of 16 fish species, 10 had a population-level preference to turn either left or right—and all the gregarious species were within those 10 species together with 40% of the nongregarious species tested (Bisazza et al., 2000a,b)—suggesting that the gregarious nature of a species may push lateralization to align in the same direction in most individuals. The strength of lateralization seems instead subject to risk levels as high predation pressure causes important lateralization increases in juvenile coral reef damselfish (Ferrari et al., 2015a,b) and in yellow-and-blueback fusiliers (*C. teres*). Moreover, in this last species, right-biased fish show improved escape performance compared to the left-biased conspecifics (Chivers et al., 2016).

Behavioral asymmetries have been widely assessed also in feeding behavior supporting the idea that, in vertebrates, the right hemisphere (left eye) responds to unexpected stimuli (such as a predator), whereas the left hemisphere (right eye) oversees routine tasks such as feeding (Rogers et al., 2013). When approaching a target to bite, zebrafish do so by using preferentially the right eye (Miklósi and Andrew, 1999), while Australian lungfish bend their bodies rightwards while feeding (Lippolis et al., 2009). Insights into the evolution of the left hemisphere specialization for feeding come from the lancelet *Branchiostoma* as the larvae of this species have the mouth located on the left side of the body indicating that the neural circuitry involved in feeding is likely located on the left side of the brain. This neural connection is maintained also in adults when the mouth moves frontally and becomes symmetric (Jeffries and Lewis, 1978). In some other species such as the *Perissodus* scale-eating cichlids, the behavioral bias to attack a prey on the left or the right side seems related to a morphologic asymmetry in the jaws. In fact, fish that open the mouth on the right side attack their prey on the left side and vice versa (Hori, 1993; Takeuchi et al., 2012). Although this mouth dimorphism seems to be determined by the Mendelian one-locus-two-alleles system, in which dextrality is dominant (Seki et al., 2000), the jaw asymmetry however is likely not the cause of the behavioral bias to attack the prey on one side but a consequence of it as *Perissodus microlepis* juveniles with a symmetric mouth present a strong preference to attack their prey on one side and this seems to induce the development of the morphologic asymmetry in the jaws (Van Dooren et al., 2010; Lee et al., 2012). A population-level bias in prey catching may be an advantage for species that adopt cooperative strategies while feeding but not necessarily in species that catch their prey individually (Ghirlanda and Vallortigara, 2004). In fact, in this case, we expect an individual-level bias to be more advantageous for the predator than a population-level directional bias, since population-level bias would also mean predictability (Vallortigara and Rogers, 2005). This is the case of some predators such as sailfish, which are lateralized at the individual level in attacking schooling sardines on one side (and the stronger they are lateralized, the more successful they are at capturing their prey), but that overall do not show a population-level bias (Kurvers et al., 2017).

Visual lateralization in response to conspecifics has been investigated not only using the detour task but also the mirror test. In this test, a mirror is put along the tank walls to give the swimming fish the impression—through its reflection—of the presence of a conspecific. Females but not males of different species of fish belonging to the Osteoglossiformes, Cypriniformes,

Cyprinodontiformes, Beloniformes orders preferentially look at their own reflection using the left eye (Sovrano, 1999; Sovrano et al., 2001; Sovrano, 2004; Sovrano and Andrew, 2006). The presence of population-level use of the left eye in females but not males is likely due to the fact that females are more social than males (Sovrano, 1999). However, in *Neolamprologus pulcher*, a highly social species of cichlid fish that breed cooperatively, males preferentially use the right eye in the mirror test, but females do not show any preference (Reddon and Balshine, 2010). Moreover, when a predator is presented at the end of the tank, female mosquitofish tend to approach the predator more closely when the mirror is on their left side (and thus giving the impression of swimming together with a conspecific on their left side) than on the right side (Bisazza et al., 1999; De Santi et al., 2001). Although the left eye (right hemisphere) seems the one preferentially used to look at conspecifics, the pattern is a bit more complex as familiarity with the conspecific has been shown to influence the direction of the bias. After ablation of the left side of the brain, juvenile soldierfish (*Myripristis pralinia*) do not prefer conspecifics over nonconspecifics anymore but this preference remains when the right brain side is ablated, providing evidence to the idea that the left hemisphere oversees conspecifics recognition (and more in general familiar stimuli) (Roux et al., 2016).

The mirror test can be used also to investigate laterality biases in relation to aggressive behavior when fish look at their own reflection as a potential competitor. In such a test, male Siamese fighting fish (*Betta splendens*) show an individual-level preference in the eye used to look at the mirror image (Cantalupo et al., 1996; Clotfelter and Kuperberg, 2007) linked to an asymmetry in head morphology (i.e., craniovertebral angle) that induce the fish to incline the head toward a specific direction: fish with a left-inclined head use the left eye and the opposite was true for fish with a right-inclined head (Takeuchi et al., 2010). In three teleost species (*G. holbrooki, Xenotoca eiseni,* and *B. splendens)* a population-level bias to use the right eye has been reported not only when the fish attacked their own mirror image but also toward a live rival (Bisazza and De Santi, 2003). In nest-holding males fighting fish (*B. splendens*), this right-eye bias is influenced by the reproduction stage and paternal care with a shift toward the left eye after spawning (Forsatkar et al., 2015). The same shift can also be induced in *B. splendens* through fluoxetine exposure, whereas cortisol strengthens the right-eye bias (Hedayati Rad et al., 2017). Not only pharmacologic manipulation but also exposure to pesticides influences lateralization in fish. In fact, larval coral reef fish (*Acanthurus triostegus*) exposed to chlorpyrifos, a pesticide often inadvertently added to coral reef waters, show an impaired visual-lateralization (Besson et al., 2017).

The cerebral correlates of behavioral lateralization in fish have been found at the level of the habenular nuclei, i.e., bilateral structures involved in neurotransmitter regulation, motivational states, and stimuli encoding (Bianco and Wilson, 2009). The habenular nuclei are neither morphologic nor functional symmetric as in many fish species the right and left nuclei differ in sizes (Bradshaw and Rogers, 1993) as well as in their neuronal connections, and gene expression (Roberson and Halpern, 2017). In zebrafish larvae, the left dorsal habenula plays a fundamental role in input-dependent light-preference behavior (Zhang et al., 2017), and in reducing fear reactions as it induces recovery after exposure to an aversive event (Duboué et al., 2017).

Amphibians

Lateralized behavior in response to a simulated predator has been investigated in three species of toads (*B. bufo*, *Bufo viridis,* and *B. marinus*). When a simulated snake was presented in the left or right lateral visual hemifield escape responses were elicited more strongly when the stimulus was on the toad's left side compared to its right side, suggesting a specialization of the right hemisphere (left visual hemifield) in response to predators (Lippolis et al., 2002). The left hemisphere is instead specialized in prey-catching behavior as toads respond more with predatory tongue-striking responses when the prey is visualized in their right visual hemifield, and when conspecifics are visualized in the left hemifield (Vallortigara et al., 1998). This complementary specialization of functions in the two hemispheres has been supported by further evidence in *B. marinus* presented with novel and familiar prey stimuli: toads prefer to view and strike a novel prey stimulus with the left eye (right hemisphere) and a familiar prey with the right eye (left hemisphere) (Robins and Rogers, 2006a), in line with the findings in fish (Roux et al., 2016). The behavioral asymmetry to respond to prey stimuli has been supported by electrophysiologic recordings in the forebrain and midbrain of music frogs (*Nidirana daunchina*) while attacking prey with higher evoked responses when the prey was presented in the right visual hemifield (Liu et al., 2020).

The same left-eye preference/rightward turning behavior observed in female fish when looking at conspecifics of the same sex (Bisazza et al., 1998) has been documented in *B. bufo* tadpoles approaching a conspecific to make social aggregation (Dadda et al., 2003) with the strength of such bias decreasing with age (Dadda, 2005). Moreover, the left eye is also the one preferentially used in the mirror test in different species of anuran amphibians (Bisazza et al., 2002). When toads respond aggressively to a conspecific such as a rival male, they do so by preferentially using the left eye and thus attacking the conspecific on their left side (Vallortigara et al., 1998; Robins and Rogers, 2006b). This suggests a specialization of the right hemisphere (left visual hemifield) for attack, a specialization that—as we will see in the next paragraphs—is shared by other vertebrate species and is likely connected to the specialization of the right hemisphere in reacting to negative emotions (Rogers et al., 2013). Side biases have been investigated in interactions with conspecifics, heterospecifics, a prey, and a simulated predator in the Ecuador poison frog (*Ameerega bilinguis)* and the sanguine poison frog (*Allobates zaparo*) (Anderson et al., 2021). The study revealed that only *Al. zaparo* present individual-level lateralization in prey catching and in reaction to the simulated predator but not in the other two tasks, whereas no evidence for lateralization was found in the Ecuador poison frog (Anderson et al., 2021).

The examples above suggest a benefit of behavioral asymmetries in tasks such as feeding, social interactions, and response to predators; however, the link between lateralization and cognition has not been explored much. Blackiston and Levin (2013) investigated the ability to learn colored stimuli in wild-type, left–right-randomized and left–right-reversed *Xenopus* tadpoles, finding that animals with either randomization or reversal of somatic left–right pattern learned more slowly than wild-type siblings.

As for brain asymmetries, morphologic analyses of the habenular nuclei revealed that the left one is more lobate than the right one in *Triturus cristatus*, *Rana esculenta* (Braitenberg and Kemali, 1970), and *R. temporaria* adults and larvae (Morgan et al., 1973). However, recordings of cerebral activity in *R. temporaria* show that the left habenula is characterized by less spontaneous as well as a light-evoked activity than the right habenula (Morgan et al., 1973).

Reptiles

A left-eye bias in attending to predators has been documented in the common wall lizard (*Podarcis muralis*) when tested under monocular conditions of vision (i.e., using eye patches). Individuals with the left eye covered made frequent and fast C-bending responses trying to monitor the predator with the left eye although it was patched (Bonati et al., 2013).

Side biases in aggression have been well documented in reptiles demonstrating a population-level left-eye preference in male tree lizards *(Urosaurus ornatus)* (Hews and Worthington, 2001) *and in* green anoles (*Anolis carolinensis)* (Deckel, 1995, 1997, 1998; Deckel and Jevitts, 1997; Deckel et al., 1998), supporting the idea that the right hemisphere is specialized for aggression. As in males, also females of the striped plateau lizard (*Sceloporus virgatus*) preferentially use the left eye for conspecific aggression (Hews et al., 2004). In anoles, the lateralized aggressive behavior is mediated by stress as males subjected to mild handling stress show a decreased left eye bias compared to nonstressed control subjects (Deckel, 1998). A similar decrease is caused by alcohol withdrawal after 19-days daily consumption (Deckel, 1997). As alcohol exposure induces an increase in serotonin levels in the raphe nucleus, the reduction in the aggressive bias may be explained through the influence of alcohol on the naturally occurring asymmetry in endogenous 5-HT levels observed in the control group (Deckel et al., 1998). This is further supported by a following study showing that quipazine, a serotonin agonist, induces a decreased left-eye aggressive bias compared to controls (Deckel and Fuqua, 1998).

Birds

Visual lateralization has been widely studied in birds not only because they possess laterally placed eyes (as many other vertebrate species) but also because the optic nerves decussate almost completely, revealing asymmetric contralateral projection between specific brain areas and thus making them fundamental to understanding the function of each brain hemisphere (Takatsuji et al., 1983; Ehrlich and Mark, 1984; Weidner et al., 1985; Rogers and Sink, 1988; Remy and Güntürkün, 1991; Güntürkün, 1997).

The left-eye (right hemisphere) functional specialization for predator detection has been documented both in young chicks (Rogers, 2000) and adult chickens (Evans et al., 1993). The left eye is also the preferred one in domestic chicks when they have to discriminate a familiar from an unfamiliar conspecific (Vallortigara and Andrew, 1991, 1994; Vallortigara, 1992), in a detour task to reach an imprinting object (Vallortigara et al., 1999), and in aggressive responses (Vallortigara et al., 2001) as chicks preferentially attack a rival male on their left side.

As for the other vertebrate species analyzed so far, the left hemisphere is instead specialized in feeding behavior as demonstrated by the right-eye bias chicks show in a pebble-grain discrimination task where the animal has to peck on the grains among pebbles stuck to the floor (Mench and Andrew, 1986) but also pigeons (Gunturkun and Kesch, 1987; Güntürkün et al., 2000) and quails (Valenti et al., 2003). In different species of cockatoos, the left foot bias in handling the foot has been associated with a left eye lateralized behavior and shown to be correlated with higher ganglion cell densities in the left compared with the right perifoveal region (Coimbra et al., 2014). Footedness is correlated to eye preference in discriminating food items in several species of Australian parrots but also with brain size: species with higher absolute brain mass have stronger foot biases and the nidopallium (a brain area recruited for higher cognitive tasks) is larger in left-footed parrots (Kaplan and Rogers, 2021).

The domestic chick has been a great model for understanding the neuroanatomic asymmetries reflected in the behavioral asymmetries described above as well as their development. By injecting cycloheximide, the protein synthesis inhibitor, into the chick's left hemisphere (but not the right one) on day 2 after hatching their visual discrimination abilities in a pebble-grain discrimination task is disrupted (Rogers and Anson, 1979). The left hemisphere region responsible for this discrimination function is the hyperpallial (Wulst) region (Deng and Rogers, 1997).

In relation to the development of visual lateralization, a crucial effect of exposure to light at different developmental stages has been initially documented in chicks and later in other bird species. During the last 3 days of incubation, the chick embryo is turned in the egg so that its left eye is occluded by its body and only the right eye can be stimulated by light passing through the shell. This asymmetry of light stimulation leads to asymmetric development of the visual pathways and visually guided behavior. Indeed, chicks incubated in the dark during the final days of incubation lack these asymmetries, and chicks that have been manipulated so that the left eye of the embryo is exposed to light while the right eye is occluded have reversed asymmetry in the use of their eyes (and hemispheres) (Rogers, 1982, 1990, 2008, 2014; Casey and Karpinski, 1999; Chiandetti and Vallortigara, 2009, 2019; Vallortigara et al., 2011; Chiandetti et al., 2013, 2017). Similar effects of light on the development of visual lateralization have been shown in pigeons (Güntürkün, 1993; Skiba et al., 2002; for a review see Chiandetti, 2017), in bobwhite quail (Casey and Lickliter, 1998), and in Japanese quail chicks (*Coturnix coturnix japonica*) (George et al., 2021) as well as zebrafish (Andrew et al., 2009). Furthermore, the development of behavioral and brain asymmetries is also influenced by hormones as the injection of testosterone and corticosterone *in ovo* altered it at the level of the thalamofugal visual pathway in chicks (Schwarz and Rogers, 1992; Rogers and Deng, 2005).

Mammals

Rats have been a useful model to investigate cerebral lateralization and its development revealing that brain asymmetries are present in several brain regions at birth, change during lifespan, and are sexually dimorphic (Glick and Ross, 1981; Denenberg, 1983). Moreover, lateralization in the rat brain shares lots of similarities with the human brain in the dopaminergic asymmetry in basal ganglia function underlying spatial processes (Glick and Ross, 1981; Denenberg, 1983) but also in affective processes as lesions of the right parietal cortex in the rat increase activity in the open field compared with left parietal lesions, suggesting that right parietal cortex has a role in emotionality (Crowne et al., 1987). Mice have also been used to understand the neurobiology of lateralization especially in relation to disorders such as ASD in prenatal zinc deficient (PZD) mice, revealing changes in hemisphere-specific expression of marker genes and in the lateralized expression of dopamine receptor 1 in PZD mice, and thus highlighting the importance of zinc signaling in establishing brain lateralization (Grabrucker et al., 2018). Left-turning biases associated with aggressive behavior have been recently reported in both captive and wild solitary Cape-mole rats (*Georychus capensis*) with a higher number of lateralized individuals and a stronger bias in the wild population, suggesting that the captivity may reduce aggression and the lateralized bias associated with it (Jacobs Paul and Oosthuizen, 2021). A recent review analyzed the existing studies investigating anxiety, fear, addiction, stress, and atypical hemispheric asymmetries in rats and mice (Mundorf and Ocklenburg, 2023), indicating an overall increase in the right-hemispheric brain activity and consequently left-sided behavioral biases in animals experiencing such emotional states. The review also highlighted that sex and developmental stage are important factors affecting the expression of behavioral and brain asymmetries (Mundorf and Ocklenburg, 2023).

A right hemisphere dominance in fearful as well as aggressive behavior has been reported also in dogs which preferentially turn their head leftward in response to thunderstorms (Siniscalchi et al., 2008) and attack sheep livestock more frequently when they circle it in a counterclockwise direction and thus keeping them in the left visual hemifield (Siniscalchi et al., 2019). Dogs have been an important model in the study of lateralization not only in relation to fear and aggression but also in emotional processing and display (i.e., in their asymmetric way of looking at stimuli and to wail their tail in response to them), providing important support for the valence theory according to which the right and the left hemisphere respond respectively to negative and positive emotions (for a review see Siniscalchi et al., 2017; Simon et al., 2022a,b).

Similar lateralization patterns to those reported in domestic dogs have been documented in horses. In fact, also horses present a left-eye bias in aggressive display (Austin and Rogers, 2007, 2012) and in scanning and viewing either a familiar or an unfamiliar person (Farmer et al., 2010), a right- and left-eye bias in assessing respectively a novel and a negative object (De Boyer Des Roches et al., 2008) with this last left-eye preference becoming stronger as the emotionality becomes higher (Larose et al., 2006). Horses also display a preference to use the right nostril while sniffing the feces of an unfamiliar stallion (McGreevy and Rogers, 2005) and stress situations seem to affect the direction and the strength of both sensory and motor biases (Marr et al., 2020). On the contrary, asymmetric handling in early life influences alert levels in horses as foals touched by humans on their right side become warier of humans than those touched on the left side or not touched (De Boyer Des Roches et al., 2011).

The right hemisphere seems involved not only in agonistic interactions with conspecifics but also in face recognition as observed in sheep (Kendrick, 2006; Versace et al., 2007) and more in general in social interactions. Indeed, in 11 species of marine and terrestrial mammals (including whales and kangaroos), juveniles have a strong preference for keeping their mother on their left side/left visual field (Karenina et al., 2017).

As for parrots (Kaplan and Rogers, 2021), the possible link between limb preference (as a proxy for lateralization), brain mass, and neuron number has been analyzed in preregistered cross-species meta-regressions revealing positive associations of brain mass and neuron number with right limb preferences but negative associations with left limb preferences (Ocklenburg et al., 2023).

Invertebrates

Cuttlefish tested in a T-maze show a left-turning bias resulting from an eye use preference. Cerebral correlates of this visual lateralization have been found in adult cuttlefish: the left optic lobe has a larger peduncle lobe and higher monoamine concentrations compared to the right one (Jozet-Alves et al., 2012). Furthermore, this left-turning bias develops at 15 days post-hatch, and it is modulated by embryonic exposure to predator odor (Jozet-Alves and Hébert, 2013), suggesting that the physical–chemical environment affects the development of lateralization in vertebrates as well as in invertebrates. Moreover, cuttlefish exhibit a right eye visual lateralization in their camouflage behavior for brightness matching (Schnell et al., 2018).

Asymmetric use of antennae in social interactions has been documented at the population level in highly social ants (Frasnelli et al., 2012a), honeybees (Rogers et al., 2013), social stingless bees *T. carbonaria* (Rogers and Frasnelli, 2016) but also in solitary mason bees (*O. rufa*) when engaging in aggressive displays (Rogers et al., 2016), suggesting that engagement in specific interindividual interactions rather than "sociality" in general may generate population-level lateralization (Frasnelli and Vallortigara, 2018). Population-level asymmetries have been observed in aggressive and mating displays in so-called solitary insects (e.g., tephrid flies, Benelli et al., 2015; earwigs, Kamimura et al., 2021; for a review see Niven and Frasnelli, 2018). Moreover, social context has been shown to modulate individual-level behavioral asymmetries in five different genotypes of *D. melanogaster* (Versace et al., 2020). This implies that lateralization is not a static feature of the neural organization but it is modulated by the functional context (for a commentary also see Frasnelli, 2021). For example, the nematode *C. elegans* shows a distinct motor handedness preference: at the population level, males show a pronounced right-hand turning bias during mating. Interestingly, this motor preference is also observed in worms with mirror-reversed anatomic handedness, suggesting that it is independent of overall anatomic asymmetry and it may be driven by epigenetic factors rather than by genetic variation (Downes et al., 2012). The nematode has been an extremely important model for uncovering the mechanisms of left and right asymmetries at the single-neuron level. Indeed, in *C. elegans* 104 out of 302 neurons are located on or very close to the midline and have no contralateral analog, whereas the other 198 neuros are present as bilaterally symmetric pairs (Hobert et al., 2002; Hobert, 2014). In particular, the olfactory and the taste neurons of the nematode are great examples of individual- and population-level asymmetries respectively (Ortiz et al., 2006, 2009; Suzuki et al., 2008). Comparative studies with *D. melanogaster* and the zebrafish *D. rerio* suggest some shared commonalities in the mechanisms involved in asymmetric neural development between vertebrates and invertebrates (Betschinger and Knoblich, 2004; Taylor et al., 2010).

Honeybees have been instead a good model to investigate the brain correlates of lateralization (Frasnelli et al., 2014). Their behavioral asymmetry observed in the learning and recall of olfactory memories (Letzkus et al., 2006; Rogers and Vallortigara, 2008; Frasnelli et al., 2011, 2012b) and the corresponding asymmetries in the electroantennographic responses of the neurons in the right and left antenna (Anfora et al., 2010) may be partially explained by a morphologic asymmetry at the peripheral level: the right antenna has about 5% more olfactory sensilla than the left antenna (Letzkus et al., 2006; Frasnelli et al., 2010a,b). Moreover, protein and gene expression in the honeybee brain during learning is also asymmetric (Biswas et al., 2010; McNeill and Robinson, 2015; Guo et al., 2016). Although the functional activity patterns (Galizia et al., 1998) and the volumes (Rigosi et al., 2011) are bilaterally symmetric in the right and left antennal lobes of the same honeybee, odor representations are different suggesting a functional asymmetry at the level of the antennal lobes (Rigosi et al., 2015).

CONCLUSION

As this chapter demonstrates, behavioral asymmetries are a widespread phenomenon across different animal groups, implying specialization of the left and right sides of the brain for motor, sensory, and cognitive control. The findings provide evidence for a long evolutionary history of lateralization in the animal kingdom and support the importance of studying this phenomenon from a comparative perspective to provide insight into the mechanisms, development, and function of lateralized nervous systems. Comparative studies in different vertebrate and invertebrate species indicate that sociality may have been an important factor in determining the evolution of population-level lateralization as the alignment of behavioral biases among individuals arises when social interactions (either synergistic or antagonistic) are present. At the same time, if lateralization has been thought to be a static feature of nervous systems for many years, more recent studies investigating behavioral asymmetries in different contexts and under different environmental conditions suggest that more research should be done to investigate its plasticity and how it can be influenced by environmental stimuli over an individual lifespan.

REFERENCES

Ades C, Ramires EN (2002). Asymmetry of leg use during prey handling in the spider Scytodes globula (Scytodidae). J Insect Behav 15: 563–570. https://doi.org/10.1023/A:1016337418472.

Alves C, Chichery R, Boal JG et al. (2007). Orientation in the cuttlefish Sepia officinalis: response versus place learning. Anim Cogn 10: 29–36. https://doi.org/10.1007/s10071-006-0027-6.

Anderson HM, Fisher DN, McEwen BL et al. (2021). Episodic correlations in behavioural lateralization differ between a poison frog and its mimic. Anim Behav 174: 207–215. https://doi.org/10.1016/j.anbehav.2021.01.011.

Andrew RJ, Osorio D, Budaev S (2009). Light during embryonic development modulates patterns of lateralization strongly and similarly in both zebrafish and chick. Philos Trans R Soc B 364: 983–989. https://doi.org/10.1098/rstb.2008.0241.

Anfora G, Frasnelli E, Maccagnani B et al. (2010). Behavioural and electrophysiological lateralization in a social (Apis mellifera) but not in a non-social (Osmia cornuta) species of bee. Behav Brain Res 206: 236–239. https://doi.org/10.1016/j.bbr.2009.09.023.

Annett M (2002). Handedness and brain asymmetry: the right shift theory, Psychology Press.

Austin NP, Rogers LJ (2007). Asymmetry of flight and escape turning responses in horses. Laterality 12: 464–474. https://doi.org/10.1080/13576500701495307.

Austin NP, Rogers LJ (2012). Limb preferences and lateralization of aggression, reactivity and vigilance in feral horses, Equus caballus. Anim Behav 83: 239–247. https://doi.org/10.1016/j.anbehav.2011.10.033.

Baciadonna L, Zucca P, Samour J (2022). Laterality preferences at rest and predatory behaviour of the gyrfalcon (Falco rusticolus): an alpha predator of the sky. Laterality 27: 86–100. https://doi.org/10.1080/1357650X.2021.1958831.

Backwell PRY, Matsumasa M, Double M et al. (2007). What are the consequences of being left-clawed in a predominantly right-clawed fiddler crab? Proc R Soc B 274: 2723–2729.

Barnard S, Wells D, Hepper P (2018). Laterality as a predictor of coping strategies in dogs entering a rescue shelter. Symmetry 10: 538.

Baruzzi C, Nawroth C, McElligott AG et al. (2018). Motor asymmetry in goats during a stepping task. Laterality 23: 599–609.

Bell ATA, Niven JE (2014). Individual-level, context-dependent handedness in the desert locust. Curr Biol 24: R382–R383. https://doi.org/10.1016/j.cub.2014.03.064.

Bell ATA, Niven JE (2016). Strength of forelimb lateralization predicts motor errors in an insect. Biol Lett 12: 20160547. https://doi.org/10.1098/rsbl.2016.0547.

Benelli G, Donati E, Romano D et al. (2015). Lateralisation of aggressive displays in a tephritid fly. Sci Nat 102: 1. https://doi.org/10.1007/s00114-014-1251-6.

Besson M, Gache C, Bertucci F et al. (2017). Exposure to agricultural pesticide impairs visual lateralization in a larval coral reef fish. Sci Rep 7: 9165. https://doi.org/10.1038/s41598-017-09381-0.

Betschinger J, Knoblich JA (2004). Dare to be different: asymmetric cell division in Drosophila, *C. elegans* and vertebrates. Curr Biol 14: R674–R685. https://doi.org/10.1016/j.cub.2004.08.017.

Bianco IH, Wilson SW (2009). The habenular nuclei: a conserved asymmetric relay station in the vertebrate brain. Philos Trans R Soc B 364: 1005–1020. https://doi.org/10.1098/rstb.2008.0213.

Bisazza A, De Santi A (2003). Lateralization of aggression in fish. Behav Brain Res 141: 131–136. https://doi.org/10.1016/S0166-4328(02)00344-3.

Bisazza A, Vallortigara G (1996). Rotational Bias in mosquitofish (*Gambusia holbrooki*): the role of laterality and sun-compass navigation. Laterality 1: 161–175. https://doi.org/10.1080/713754231.

Bisazza A, Cantalupo C, Robins A et al. (1996). Right-pawedness in toads. Nature 379: 408.

Bisazza A, Pignatti R, Vallortigara G (1997a). Detour tests reveal task- and stimulus-specific behavioural lateralization in mosquitofish (Gambusia holbrooki). Behav Brain Res 89: 237–242. https://doi.org/10.1016/S0166-4328(97)00061-2.

Bisazza A, Pignatti R, Vallortigara G (1997b). Laterality in detour behaviour: interspecific variation in poeciliid fish. Anim Behav 54: 1273–1281. https://doi.org/10.1006/anbe.1997.0522.

Bisazza A, Rogers LJ, Vallortigara G (1998). The origins of cerebral asymmetry: a review of evidence of behavioural and brain lateralization in fishes, reptiles and amphibians. Neurosci Biobehav Rev 22: 411–426. https://doi.org/10.1016/S0149-7634(97)00050-X.

Bisazza A, De Santi A, Vallortigara G (1999). Laterality and cooperation: mosquitofish move closer to a predator when the companion is on their left side. Anim Behav 57: 1145–1149. https://doi.org/10.1006/anbe.1998.1075.

Bisazza A, Cantalupo C, Capocchiano M et al. (2000a). Population lateralisation and social behaviour: a study with 16 species of fish. Laterality 5: 269–284. https://doi.org/10.1080/713754381.

Bisazza A, Facchin L, Vallortigara G (2000b). Heritability of lateralization in fish: concordance of right-left asymmetry between parents and offspring. Neuropsychologia 38: 907–912.

Bisazza A, Lippolis G, Vallortigara G (2001a). Lateralization of ventral fins use during object exploration in the blue gourami (Trichogaster trichopterus). Physiol Behav 72: 575–578. https://doi.org/10.1016/S0031-9384(01)00417-6.

Bisazza A, Sovrano VA, Vallortigara G (2001b). Consistency among different tasks of left–right asymmetries in lines of fish originally selected for opposite direction of lateralization in a detour task. Neuropsychologia 39: 1077–1085. https://doi.org/10.1016/S0028-3932(01)00034-3.

Bisazza A, De Santi A, Bonso S et al. (2002). Frogs and toads in front of a mirror: lateralisation of response to social stimuli in tadpoles of five anuran species. Behav Brain Res 134: 417–424. https://doi.org/10.1016/S0166-4328(02)00055-4.

Bisazza A, Dadda M, Cantalupo C (2005). Further evidence for mirror-reversed laterality in lines of fish selected for leftward or rightward turning when facing a predator model. Behav Brain Res 156: 165–171. https://doi.org/10.1016/j.bbr.2004.05.022.

Biswas S, Reinhard J, Oakeshott J et al. (2010). Sensory regulation of Neuroligins and Neurexin I in the honeybee brain. PLoS One 5: e9133. https://doi.org/10.1371/journal.pone.0009133.

Blackiston DJ, Levin M (2013). Inversion of left–right asymmetry alters performance of Xenopus tadpoles in nonlateralized cognitive tasks. Anim Behav 86: 459–466. https://doi.org/10.1016/j.anbehav.2013.05.043.

Bolis A, Gazzola A, Pellitteri-Rosa D et al. (2020). Exposure during embryonic development to Roundup® Power 2.0 affects lateralization, level of activity and growth, but not defensive behaviour of marsh frog tadpoles. Environ Pollut 263: 114395. https://doi.org/10.1016/j.envpol.2020.114395.

Bonati B, Csermely D, Sovrano VA (2013). Looking at a predator with the left or right eye: asymmetry of response in lizards. Laterality 18: 329–339.

Bradshaw JL, Rogers LJ (1993). The evolution of lateral asymmetries, language, tool use and intellect, Academic Press, San Diego.

Braitenberg V, Kemali M (1970). Exceptions to bilateral asymmetry in the epithalamus of lower vertebrates. J Comp Neurol 138: 137–146.

Branson NJ, Rogers LJ (2006). Relationship between paw preference strength and noise phobia in Canis familiaris. J Comp Psychol 120: 176–183. https://doi.org/10.1037/0735-7036.120.3.176.

Briggs-Gonzalez VS, Gonzalez SC (2016). Lateralized turning biases in two neotropical tadpoles. Ethology 122: 582–587. https://doi.org/10.1111/eth.12503.

Brown C, Magat M (2011). Cerebral lateralization determines hand preferences in Australian parrots. Biol Lett 7: 496–498.

Buchanan SM, Kain JS, De Bivort BL (2015). Neuronal control of locomotor handedness in *Drosophila*. Proc Natl Acad Sci U S A 112: 6700–6705. https://doi.org/10.1073/pnas.1500804112.

Byrne RA, Kuba MJ, Meisel DV et al. (2006). Does Octopus vulgaris have preferred arms? J Comp Psychol 120: 198–204. https://doi.org/10.1037/0735-7036.120.3.198.

Calcraft PRT, Bell AT, Husbands P et al. (2016). The evolution of handedness: why are ant colonies left- and right-handed? In: Biomath communications: proceedings (BIOMATH'16), vol. 3. ISSN 2367-5233.

Cantalupo C, Bisazza A, Vallortigara G (1996). Lateralization of displays during aggressive and courtship behaviour in the Siamese fighting fish (*Betta splendens*). Physiol Behav 60: 249–252. https://doi.org/10.1016/0031-9384(96)00015-7. PMID: 8804671.

Casey MB, Karpinski S (1999). The development of postnatal turning bias is influenced by prenatal visual experience in domestic chicks (Gallus gallus). Psychol Rec 49: 67–74. https://doi.org/10.1007/BF03395307.

Casey MB, Lickliter R (1998). Prenatal visual experience influences the development of turning bias in bobwhite quail chicks (Colinus virginianus). Dev Psychobiol 32: 327–338. https://doi.org/10.1002/(SICI)1098-2302(199805)32:4<327::AID-DEV7>3.0.CO;2-P.

Charlton K, Frasnelli E (2023). Does owner handedness influence paw preference in dogs? Anim Cogn 26: 425–433. https://doi.org/10.1007/s10071-022-01673-x.

Chiandetti C (2017). Manipulation of strength of cerebral lateralization via embryonic light stimulation in birds. In: L Rogers, G Vallortigara (Eds.), Lateralized brain functions: methods in human and non-human species. Humana Press, New York, pp. 611–631.

Chiandetti C, Vallortigara G (2009). Effects of embryonic light stimulation on the ability to discriminate left from right in the domestic chick. Behav Brain Res 198: 240–246. https://doi.org/10.1016/j.bbr.2008.11.018.

Chiandetti C, Vallortigara G (2019). Distinct effect of early and late embryonic light-stimulation on chicks' lateralization. Neuroscience 414: 1–7. https://doi.org/10.1016/j.neuroscience.2019.06.036. PMID: PMID: 31279824; Epub 2019 Jul 4.

Chiandetti C, Galliussi J, Andrew RJ et al. (2013). Early-light embryonic stimulation suggests a second route, via gene activation, to cerebral lateralization in vertebrates. Sci Rep 3: 2701. https://doi.org/10.1038/srep02701.

Chiandetti C, Lemaire B, Versace E et al. (2017). Early- and late-light embryonic stimulation modulates similarly chicks' ability to filter out distractors. Symmetry 9: 84. https://doi.org/10.3390/sym9060084.

Chivers DP, McCormick MI, Allan BJM et al. (2016). At odds with the group: changes in lateralization and escape performance reveal conformity and conflict in fish schools. Proc R Soc B 283: 20161127. https://doi.org/10.1098/rspb.2016.1127.

Clotfelter ED, Kuperberg ES (2007). Cerebral lateralization and its relationship to phylogeny and aggression in Anabantoid fishes. Brain Behav Evol 69: 169–175. https://doi.org/10.1159/000096985.

Coimbra JP, Collin SP, Hart NS (2014). Topographic specializations in the retinal ganglion cell layer correlate with lateralized visual behavior, ecology, and evolution in cockatoos: retinal specializations in cockatoos. J Comp Neurol 522: 3363–3385. https://doi.org/10.1002/cne.23637.

Cole J (1955). Paw preference in cats related to hand preference in animals and men. J Comp Physiol Psychol 48: 137–140.

Collins RL (1968). On the inheritance of handedness. J Hered 59: 9–12.

Cooper R, Nudo N, González JM et al. (2011). Side-dominance of Periplaneta americana persists through antenna amputation. J Insect Behav 24: 175–185. https://doi.org/10.1007/s10905-010-9246-4.

Crowne DP, Richardson CM, Dawson KA (1987). Lateralization of emotionality in right parietal cortex of the rat. Behav Neurosci 101: 134–138.

Cussen VA, Mench JA (2014). Performance on the Hamilton search task, and the influence of lateralization, in captive orange-winged Amazon parrots (Amazona amazonica). Anim Cogn 17: 901–909. https://doi.org/10.1007/s10071-013-0723-y.

Dadda M (2005). Social aggregation and lateralised response to social stimuli in tadpoles (Bufo bufo): influence of developmental stage. Laterality 10: 345–352. https://doi.org/10.1080/13576500442000139.

Dadda M, Sovrano VA, Bisazza A (2003). Temporal pattern of social aggregation in tadpoles and its influence on the measurement of lateralised response to social stimuli. Physiol Behav 78: 337–341. https://doi.org/10.1016/S0031-9384(02)01001-6.

Dadda M, Zandonà E, Bisazza A (2007). Emotional responsiveness in fish from lines artificially selected for a high or low degree of laterality. Physiol Behav 92: 764–772. https://doi.org/10.1016/j.physbeh.2007.06.001.

Dadda M, Zandona E, Agrillo C et al. (2009). The costs of hemispheric specialization in a fish. Proc Roy Soc Ser B, Biol Sci 276: 4399–4407.

Dadda M, Koolhaas WH, Domenici P (2010). Behavioural asymmetry affects escape performance in a teleost fish. Biol Lett 6: 414–417.

Dadda M, Nepomnyashchikh VA, Izvekov EI et al. (2012). Individual-level consistency of different laterality measures in the goldbelly topminnow. Behav Neurosci 126: 845–849. https://doi.org/10.1037/a0030319.

Davies MNO, Green PR (1991). Footedness in pigeons, or simply sleight of foot? Anim Behav 42: 311–312. https://doi.org/10.1016/S0003-3472(05)80562-7.

de Lussanet MH, Osse JW (2015). Decussation as an axial twist: a comment on Kinsbourne (2013). Neuropsychology 29: 713–714. https://doi.org/10.1037/neu0000163.

De Boyer Des Roches A, Richard-Yris M-A, Henry S et al. (2008). Laterality and emotions: visual laterality in the domestic horse (Equus caballus) differs with objects' emotional value. Physiol Behav 94: 487–490. https://doi.org/10.1016/j.physbeh.2008.03.002.

De Boyer Des Roches A, Durier V, Richard-Yris M-A et al. (2011). Differential outcomes of unilateral interferences at birth. Biol Lett 7: 177–180.

De Santi A, Sovrano VA, Bisazza A et al. (2001). Mosquitofish display differential left- and right-eye use during mirror image scrutiny and predator inspection responses. Anim Behav 61: 305–310. https://doi.org/10.1006/anbe.2000.1566.

Deckel AW (1995). Laterality of aggressive responses in Anolis. J Exp Zool 272: 194–200.

Deckel AW (1997). Effects of alcohol consumption on lateralized aggression in Anolis carolinensis. Brain Res 756: 96–105. https://doi.org/10.1016/S0006-8993(97)00125-X.

Deckel AW (1998). Hemispheric control of territorial aggression in *Anolis carolinensis:* effects of mild stress. Brain Behav Evol 51: 33–39. https://doi.org/10.1159/000006527.

Deckel AW, Fuqua L (1998). Effects of serotonergic drugs on lateralized aggression and aggressive displays in Anolis carolinensis. Behav Brain Res 95: 227–232. https://doi.org/10.1016/S0166-4328(98)00048-5.

Deckel AW, Jevitts E (1997). Left vs. right-hemisphere regulation of aggressive behaviors in Anolis carolinensis: effects of eye-patching and fluoxetine administration. J Exp Zool 278: 9–21. https://doi.org/10.1002/(SICI)1097-010X(19970501)278:1<9::AID-JEZ2>3.0.CO;2-T.

Deckel AW, Lillaney R, Ronan PJ et al. (1998). Lateralized effects of ethanol on aggression and serotonergic systems in Anolis carolinensis. Brain Res 807: 38–46. https://doi.org/10.1016/S0006-8993(98)00718-5.

Denenberg VH (1983). Lateralization of function in rats. Am J Physiol Regul Integr Comp Physiol 245: R505–R509.

Deng C, Rogers LJ (1997). Differential contributions of the two visual pathways to functional lateralization in chicks. Behav Brain Res 87: 173–182. https://doi.org/10.1016/S0166-4328(97)02276-6.

Dharmaretnam M, Vijitha V, Priyadharshini K et al. (2002). Ground scratching and preferred leg use in domestic chicks: changes in motor control in the first two weeks post-hatching. Laterality 7: 371–380.

Díaz S, Murray L, Rodway P (2021). Limb preference and personality in donkeys (Equus asinus). Laterality 26: 186–200.

Domenici P, Blake RW (1997). The kinematics and performance of fish fast-start swimming. J Exp Biol 200: 1165–1178. https://doi.org/10.1242/jeb.200.8.1165.

Domenici P, Allan B, McCormick MI et al. (2012). Elevated carbon dioxide affects behavioural lateralization in a coral reef fish. Biol Lett 8: 78–81.

Domenici P, Allan BJM, Watson S-A et al. (2014). Shifting from right to left: the combined effect of elevated CO_2 and temperature on behavioural lateralization in a coral reef fish. PLoS One 9: e87969. https://doi.org/10.1371/journal.pone.0087969.

Downes JC, Birsoy B, Chipman KC et al. (2012). Handedness of a motor program in C. elegans is independent of left-right body asymmetry. PLoS One 7: e52138. https://doi.org/10.1371/journal.pone.0052138.

Duboué ER, Hong E, Eldred KC et al. (2017). Left habenular activity attenuates fear responses in larval zebrafish. Curr Biol 27: 2154–2162. e3. https://doi.org/10.1016/j.cub.2017.06.017.

Duncan A, Simon T, Frasnelli E (2022). Investigating the influence of neuter status on paw preference in dogs and cats. Laterality 27: 359–378. https://doi.org/10.1080/1357650X.2022.2086563.

Eaton RC, Lee RKK, Foreman MB (2001). The Mauthner cell and other identified neurons of the brainstem escape network of fish. Prog Neurobiol 63: 467–485. https://doi.org/10.1016/S0301-0082(00)00047-2.

Ehrlich D, Mark R (1984). An atlas of the primary visual projections in the brain of the chick Gallus gallus. J Comp Neurol 223: 592–610.

Evans CS, Evans L, Marler P (1993). On the meaning of alarm calls: functional reference in an avian vocal system. Anim Behav 46: 23–38.

Farmer K, Krueger K, Byrne RW (2010). Visual laterality in the domestic horse (Equus caballus) interacting with humans. Anim Cogn 13: 229–238. https://doi.org/10.1007/s10071-009-0260-x.

Ferrari MCO, McCormick MI, Allan BJM et al. (2015a). The effects of background risk on behavioural lateralization in a coral reef fish. Funct Ecol 29: 1553–1559. https://doi.org/10.1111/1365-2435.12483.

Ferrari MCO, McCormick MI, Allan BJM et al. (2015b). Living in a risky world: the onset and ontogeny of an integrated antipredator phenotype in a coral reef fish. Sci Rep 5: 15537. https://doi.org/10.1038/srep15537.

Ferrari MCO, McCormick MI, Mitchell MD et al. (2017). Daily variation in behavioural lateralization is linked to predation stress in a coral reef fish. Anim Behav 133: 189–193. https://doi.org/10.1016/j.anbehav.2017.09.020.

Fisher HI (1957). Footedness in domestic pigeons. Wilson Bull 67: 170–177.

Forsatkar MN, Dadda M, Nematollahi MA (2015). Lateralization of aggression during reproduction in male Siamese fighting fish. Ethology 121: 1039–1047. https://doi.org/10.1111/eth.12418.

Frasnelli E (2013). Brain and behavioral lateralization in invertebrates. Front Psychol 4. https://doi.org/10.3389/fpsyg.2013.00939.

Frasnelli E (2021). Looking at lateralization as a dynamic and plastic feature of nervous systems. Laterality 26: 323–326. https://doi.org/10.1080/1357650X.2021.1876083.

Frasnelli E, Vallortigara G (2018). Individual-level and population-level lateralization: two sides of the same coin. Symmetry 10: 739. https://doi.org/10.3390/sym10120739.

Frasnelli E, Anfora G, Trona F et al. (2010a). Morpho-functional asymmetry of the olfactory receptors of the honeybee (Apis mellifera). Behav Brain Res 209: 221–225. https://doi.org/10.1016/j.bbr.2010.01.046.

Frasnelli E, Vallortigara G, Rogers LJ (2010b). Response competition associated with right–left antennal asymmetries of new and old olfactory memory traces in honeybees. Behav Brain Res 209: 36–41. https://doi.org/10.1016/j.bbr.2010.01.014.

Frasnelli E, Vallortigara G, Rogers LJ (2011). Origins of brain asymmetry: lateralization of odour memory recall in primitive Australian stingless bees. Behav Brain Res 224: 121–127. https://doi.org/10.1016/j.bbr.2011.05.026.

Frasnelli E, Iakovlev I, Reznikova Z (2012a). Asymmetry in antennal contacts during trophallaxis in ants. Behav Brain Res 232: 7–12. https://doi.org/10.1016/j.bbr.2012.03.014.

Frasnelli E, Vallortigara G, Rogers LJ (2012b). Left–right asymmetries of behaviour and nervous system in invertebrates. Neurosci Biobehav Rev 36: 1273–1291. https://doi.org/10.1016/j.neubiorev.2012.02.006.

Frasnelli E, Haase A, Rigosi E et al. (2014). The bee as a model to investigate brain and behavioural asymmetries. Insects 5: 120–138. https://doi.org/10.3390/insects5010120.

Frasnelli E, Ponte G, Vallortigara G et al. (2019). Visual lateralization in the cephalopod mollusk Octopus vulgaris. Symmetry 11: 1121. https://doi.org/10.3390/sym11091121.

French JM (1993). Assessment of donkey temperament and the influence of home environment. Appl Anim Behav Sci 36: 249–257. https://doi.org/10.1016/0168-1591(93)90014-G.

Friedman H, Davies M (1938). 'Left handedness' in parrots. Auk 80: 478–480.

Galizia CG, Nägler K, Hölldobler B et al. (1998). Odour coding is bilaterally symmetrical in the antennal lobes of honeybees (*Apis mellifera*): symmetrical odour representation. Eur J Neurosci 10: 2964–2974. https://doi.org/10.1111/j.1460-9568.1998.00303.x.

George I, Lerch N, Jozet-Alves C et al. (2021). Effect of embryonic light exposure on laterality and sociality in quail chicks (Coturnix coturnix japonica). Appl Anim Behav Sci 236: 105270. https://doi.org/10.1016/j.applanim.2021.105270.

Ghirlanda S, Vallortigara G (2004). The evolution of brain lateralization: a game-theoretical analysis of population structure. Proc R Soc Lond B 271: 853–857. https://doi.org/10.1098/rspb.2003.2669.

Ghirlanda S, Frasnelli E, Vallortigara G (2009). Intraspecific competition and coordination in the evolution of lateralization. Philos Trans R Soc B 364: 861–866. https://doi.org/10.1098/rstb.2008.0227.

Gibb J, Hartley PHT (1957). Bird foods and feeding-habits as subjects for amateur research. Brit Birds 50: 278–291 .

Giljov A, Karenina K, Malashichev Y (2012). Limb preferences in a marsupial, Macropus rufogriseus: evidence for postural effect. Anim Behav 83: 525–534. https://doi.org/10.1016/j.anbehav.2011.11.031.

Giljov A, Karenina K, Malashichev Y (2013). Forelimb preferences in quadrupedal marsupials and their implications for laterality evolution in mammals. BMC Evol Biol 13: 61. https://doi.org/10.1186/1471-2148-13-61.

Giljov A, Karenina K, Ingram J et al. (2015). Parallel emergence of true handedness in the evolution of marsupials and placentals. Curr Biol 25: 1878–1884. https://doi.org/10.1016/j.cub.2015.05.043.

Giljov A, Karenina K, Ingram J et al. (2017). Early expression of manual lateralization in bipedal marsupials. J Comp Psychol 131: 225–230. https://doi.org/10.1037/com0000073.

Girling RD, Hassall M, Turner JG (2007). Do turning biases by the 7-spot ladybird, Coccinella septempunctata, increase their foraging efficiency? Behaviour 144: 143–163.

Glick SD, Ross DA (1981). Lateralization of function in the rat brain: basic mechanisms may be operative in humans. Trends Neurosci 4: 196–199.

Goree BD, Wassersug RJ (2001). Are Archeobatrachian anurans ambidextrous? Assessing handedness in Bombina orientalis. J Herpetol 35: 538. https://doi.org/10.2307/1565980.

Govind CK (1984). Development of asymmetry in the neuromuscular system of lobster claws. Biol Bull 167: 94–119.

Govind CK (1992). Claw asymmetry in lobsters: case study in developmental neuroethology. J Neurobiol 23: 1423–1445. https://doi.org/10.1002/neu.480231006.

Govind CK, Pearce J (1992). Mechanoreceptors and minimal reflex activity determining claw laterality in developing lobsters. J Exp Biol 171: 149–162. https://doi.org/10.1242/jeb.171.1.149.

Grabrucker S, Haderspeck JC, Sauer AK et al. (2018). Brain lateralization in mice is associated with zinc signaling and altered in prenatal zinc deficient mice that display features of autism Spectrum disorder. Front Mol Neurosci 10: 450. https://doi.org/10.3389/fnmol.2017.00450.

Green AJ (1997). Asymmetrical turning during spermatophore transfer in the male smooth newt. Anim Behav 54: 343–348. https://doi.org/10.1006/anbe.1996.0461.

Green L, Jutfelt F (2014). Elevated carbon dioxide alters the plasma composition and behaviour of a shark. Biol Lett 10: 20140538. https://doi.org/10.1098/rsbl.2014.0538.

Güntürkün O (1993). The ontogeny of visual lateralization in pigeons. German J Psychol 17: 276–287.

Güntürkün O (1997). Morphological asymmetries of the tectum opticum in the pigeon. Exp Brain Res 116: 561–566. https://doi.org/10.1007/PL00005785.

Gunturkun O, Kesch S (1987). Visual lateralization during feeding in pigeons. Behav Neurosci 101: 433–435.

Güntürkün O, Kesch S, Delius JD (1988). Absence of footedness in domestic pigeons. Anim Behav 36: 602–604. https://doi.org/10.1016/S0003-3472(88)80032-0.

Güntürkün O, Diekamp B, Manns M et al. (2000). Asymmetry pays: visual lateralization improves discrimination success in pigeons. Curr Biol 10: 1079–1081. https://doi.org/10.1016/S0960-9822(00)00671-0.

Guo Y, Wang Z, Li Y et al. (2016). Lateralization of gene expression in the honeybee brain during olfactory learning. Sci Rep 6: 34727.

Harris LJ (1989). Footedness in parrots: three centuries of research, theory, and mere surmise. Can J Psychol 43: 369–396. https://doi.org/10.1037/h0084228.

Hedayati Rad M, Nematollahi MA, Forsatkar MN et al. (2017). Prozac impacts lateralization of aggression in male Siamese fighting fish. Ecotoxicol Environ Saf 140: 84–88. https://doi.org/10.1016/j.ecoenv.2017.02.027.

Heuts BA (1999). Lateralization of trunk muscle volume, and lateralization of swimming turns of fish responding to external stimuli. Behav Processes 47: 113–124. https://doi.org/10.1016/S0376-6357(99)00056-X.

Heuts BA, Cornelissen P, Lambrechts DYM (2003). Different attack modes of Formica species in interspecific one-on-one combats with other ants [Hymenoptera: Formicidae]. Ann Zool 53.

Hews DK, Worthington RA (2001). Fighting from the right side of the brain: left visual field preference during aggression in free-ranging male tree lizards *(Urosaurus ornatus)*. Brain Behav Evol 58: 356–361. https://doi.org/10.1159/000057576.

Hews DK, Castellano M, Hara E (2004). Aggression in females is also lateralized: left-eye bias during aggressive courtship rejection in lizards. Anim Behav 68: 1201–1207. https://doi.org/10.1016/j.anbehav.2003.11.024.

Hobert O (2014). Development of left/right asymmetry in the Caenorhabditis elegans nervous system: from zygote to postmitotic neuron: neuronal asymmetry in *C. elegans*. Genesis 52: 528–543. https://doi.org/10.1002/dvg.22747.

Hobert O, Johnston RJ, Chang S (2002). Left–right asymmetry in the nervous system: the Caenorhabditis elegans model. Nat Rev Neurosci 3: 629–640. https://doi.org/10.1038/nrn897.

Hönicke C, Bliss P, Moritz RFA (2015). Effect of density on traffic and velocity on trunk trails of Formica pratensis. Sci Nat 102: 17. https://doi.org/10.1007/s00114-015-1267-6.

Hori M (1993). Frequency-dependent natural selection in the handedness of scale-eating cichlid fish. Science 260: 216–219. https://doi.org/10.1126/science.260.5105.216.

Horstick EJ, Bayleyen Y, Burgess HA (2020). Molecular and cellular determinants of motor asymmetry in zebrafish. Nat Commun 11: 1170. https://doi.org/10.1038/s41467-020-14965-y.

Hunt GR (2000). Human–like, population–level specialization in the manufacture of pandanus tools by new Caledonian crows Corvus moneduloides. Proc R Soc Lond B Biol Sci 267: 403–413.

Hunt GR, Corballis MC, Gray RD (2001). Laterality in tool manufacture by crows. Nature 414: 707. https://doi.org/10.1038/414707a.

Hunt ER, O'Shea-Wheller T, Albery GF et al. (2014). Ants show a leftward turning bias when exploring unknown nest sites. Biol Lett 10: 20140945. https://doi.org/10.1098/rsbl.2014.0945.

Izawa E-I, Kusayama T, Watanabe S (2005). Foot-use laterality in the Japanese jungle crow (Corvus macrorhynchos). Behav Processes 69: 357–362. https://doi.org/10.1016/j.beproc.2005.02.001.

Izvekov EI, Nepomnyashchikh VA (2010). Comparison of two kinds of functional asymmetry in the roach Rutilus rutilus (Teleostei: Cyprinidae). J Evol Biochem Phys 46: 69–76. https://doi.org/10.1134/S0022093010010081.

Izvekov EI, Kuternitskaya EA, Pankova NA et al. (2014). Lateralisation of rotational swimming but not fast escape response in the juvenile sterlet sturgeon, Acipenser ruthenus (Chondrostei: Acipenseridae). Laterality 19: 302–324. https://doi.org/10.1080/1357650X.2013.804080.

Jacobs Paul J, Oosthuizen MK (2021). Laterality in the cape mole-rat, Georychus capensis. Behav Processes 185: 104346. https://doi.org/10.1016/j.beproc.2021.104346.

Jeffries RPS, Lewis DN (1978). The English Silurian fossil Placocystites forbesianus and the ancestry of vertebrates. Philos Trans 282: 205–323.

Jozet-Alves C, Hébert M (2013). Embryonic exposure to predator odour modulates visual lateralization in cuttlefish. Proc R Soc Lond B Biol Sci 280: 20122575.

Jozet-Alves C, Romagny S, Bellanger C et al. (2012). Cerebral correlates of visual lateralization in Sepia. Behav Brain Res 234: 20–25. https://doi.org/10.1016/j.bbr.2012.05.042.

Kaarthigeyan J, Dharmaretnam M (2005). Relative levels of motivation and asymmetries of viewing and detour task in guppies (Poecilia reticulata). Behav Brain Res 159: 37–41. https://doi.org/10.1016/j.bbr.2004.09.024.

Kamimura Y, Matsumura Y, Yang C-CS et al. (2021). Random or handedness? Use of laterally paired penises in *Nala* earwigs (Insecta: Dermaptera: Labiduridae). Biol J Linn Soc 134: 716–731. https://doi.org/10.1093/biolinnean/blab111.

Kaplan G, Rogers LJ (2021). Brain size associated with foot preferences in Australian parrots. Symmetry 13: 867. https://doi.org/10.3390/sym13050867.

Karenina K, Giljov A, Ingram J et al. (2017). Lateralization of mother–infant interactions in a diverse range of mammal species. Nat Ecol Evol 1: 0030.

Kells AR, Goulson D (2001). Evidence for handedness in bumblebees. J Insect Behav 14: 47–55.

Kendrick KM (2006). Brain asymmetries for face recognition and emotion control in sheep. Cortex 42: 96–98. https://doi.org/10.1016/S0010-9452(08)70328-9.

Kight SL, Steelman L, Coffey G et al. (2008). Evidence of population level in giant water bugs, Belostoma flumineum say (Heteroptera: Belostomatidae): T-maze turning is left biased. Behav Processes 79: 66–69.

Korn H, Faber DS (2005). The Mauthner cell half a century later: a neurobiological model for decision-making? Neuron 47: 13–28. https://doi.org/10.1016/j.neuron.2005.05.019.

Kurvers RHJM, Krause S, Viblanc PE et al. (2017). The evolution of lateralization in group hunting sailfish. Curr Biol 27: 521–526. https://doi.org/10.1016/j.cub.2016.12.044.

Larbi MC, Messa G, Jalal H et al. (2022). An early midbrain sensorimotor pathway is involved in the timely initiation and direction of swimming in the hatchling Xenopus laevis tadpole. Front Neural Circuits 16: 1027831. https://doi.org/10.3389/fncir.2022.1027831.

Larose C, Rogers LJ, Ricard-Yris M-A et al. (2006). Laterality of horses associated with emotionality in novel situations. Laterality 11: 355–367.

Larranaga N, Baguette M, Calvez O et al. (2013). Intra- and inter-individual variations in flight direction in a migratory butterfly co-vary with individual mobility. J Exp Biol. https://doi.org/10.1242/jeb.082883.

Laverack K, Pike TW, Cooper Jonathan J et al. (2021). The effect of sex and age on paw use within a large sample of dogs (Canis familiaris). Appl Anim Behav Sci 238: 105298. https://doi.org/10.1016/j.applanim.2021.105298.

Lee HJ, Kusche H, Meyer A (2012). Handed foraging behavior in scale-eating cichlid fish: its potential role in shaping morphological asymmetry. PLoS One 7: e44670. https://doi.org/10.1371/journal.pone.0044670.

Letzkus P, Ribi WA, Wood JT et al. (2006). Lateralization of olfaction in the honeybee *Apis mellifera*. Curr Biol 16: 1471–1476. https://doi.org/10.1016/j.cub.2006.05.060.

Liga D, Stancher G, Frasnelli E (2024). Visuo-motor lateralization in *Apis mellifera*: flight speed differences in foraging choices. Sci Rep 14: 660. https://doi.org/10.1038/s41598-023-51141-w.

Lippolis G, Bisazza A, Rogers LJ et al. (2002). Lateralisation of predator avoidance responses in three species of toads. Laterality 7: 163–183.

Lippolis G, Joss JMP, Rogers LJ (2009). Australian lungfish *(Neoceratodus forsteri)*: a missing link in the evolution of complementary side biases for predator avoidance and prey capture. Brain Behav Evol 73: 295–303. https://doi.org/10.1159/000230674.

Liu Y, Shen J, Fang K et al. (2020). Behavior and approximate entropy of right-eye lateralization during predation in the music frog. Asian Herpetol Res 11: 115–124.

Lopes AF, Morais P, Pimentel M et al. (2016). Behavioural lateralization and shoaling cohesion of fish larvae altered under ocean acidification. Mar Biol 163: 243. https://doi.org/10.1007/s00227-016-3026-4.

Lorincz E, Fabre-Thorpe M (1994). Effect of practice on paw preference in a reaching task in cats. C R Acad Sci III 317: 1081–1088.

Magat M, Brown C (2009). Laterality enhances cognition in Australian parrots. Proc Roy Soc Ser B, Biol Sci 276: 4155–4162.

Malashichev YB (2002). Asymmetries in amphibians: a review of morphology and behaviour. Laterality 7: 197–217.

Malashichev YB (2006). One-sided limb preference is linked to alternating-limb locomotion in anuran amphibians. J Comp Psychol 120: 401–410. https://doi.org/10.1037/0735-7036.120.4.401.

Malashichev Y (2016). Asymmetry of righting reflexes in sea turtles and its behavioral correlates. Physiol Behav 157: 1–8. https://doi.org/10.1016/j.physbeh.2016.01.001.

Malashichev YB, Nikitina NG (2002). Preferential limb use in relation to epicoracoid overlap in the shoulder girdle of toads. Laterality 7: 1–18.

Manns M, Basbasse YE, Freund N et al. (2021). Paw preferences in mice and rats: meta-analysis. Neurosci Biobehav Rev 127: 593–606. https://doi.org/10.1016/j.neubiorev.2021.05.011.

Marr I, Farmer K, Krüger K (2018). Evidence for right-sided horses being more optimistic than left-sided horses. Animals 8: 219. https://doi.org/10.3390/ani8120219.

Marr I, Preisler V, Farmer K et al. (2020). Non-invasive stress evaluation in domestic horses (*Equus caballus*): impact of housing conditions on sensory laterality and immunoglobulin a. R Soc Open Sci 7: 191994. https://doi.org/10.1098/rsos.191994.

Marshall-Pescini S, Barnard S, Branson NJ et al. (2013). The effect of preferential paw usage on dogs' (Canis familiaris) performance in a manipulative problem-solving task. Behav Processes 100: 40–43. https://doi.org/10.1016/j.beproc.2013.07.017.

McDowell L, Wells D, Hepper P (2018). Lateralization of spontaneous behaviours in the domestic cat, Felis silvestris. Anim Behav 135: 37–43.

McGreevy PD, Rogers LJ (2005). Motor and sensory laterality in thoroughbred horses. Appl Anim Behav Sci 92: 337–352. https://doi.org/10.1016/j.applanim.2004.11.012.

McGreevy PD, Thomson PC (2006). Differences in motor laterality between breeds of performance horse. Appl Anim Behav Sci 99: 183–190. https://doi.org/10.1016/j.applanim.2005.09.010.

McGreevy PD, Landrieu JP, Malou PFJ (2007). A note on motor laterality in plains zebras (Equus burchellii) and impalas (Aepyceros melampus). Laterality 12: 449–457.

McGreevy PD, Brueckner A, Thomson PC et al. (2010). Motor laterality in 4 breeds of dog. J Vet Behav 5: 318–323. https://doi.org/10.1016/j.jveb.2010.05.001.

McNeill MS, Robinson GE (2015). Voxel-based analysis of the immediate early gene, *c-jun*, in the honey bee brain after a sucrose stimulus: *c-jun* responds to sucrose in the bee brain. Insect Mol Biol 24: 377–390. https://doi.org/10.1111/imb.12165.

Mench JA, Andrew RJ (1986). Lateralization of a food search task in the domestic chick. Behav Neural Biol 46: 107–114. https://doi.org/10.1016/S0163-1047(86)90570-4.

Mikhailova GZ, Tiras NR, Pavlik VD et al. (2006). Morphological parameters of mauthner neurons of goldfishes with modified asymmetry of motor behavior. Neurophysiology 38: 15–26. https://doi.org/10.1007/s11062-006-0021-2.

Miklósi A, Andrew RJ (1999). Right eye use associated with decision to bite in zebrafish. Behav Brain Res 105: 199–205.

Miler K, Kuszewska K, Woyciechowski M (2017). Larval antlions with more pronounced behavioural asymmetry show enhanced cognitive skills. Biol Lett 13: 20160786. https://doi.org/10.1098/rsbl.2016.0786.

Miletto Petrazzini ME, Sovrano VA, Vallortigara G et al. (2020). Brain and behavioral asymmetry: a lesson from fish. Front Neuroanat 14: 11. https://doi.org/10.3389/fnana.2020.00011.

Miller ND, Yoder TJ, Manoukis NC et al. (2022). Harmonic radar tracking of individual melon flies, Zeugodacus cucurbitae, in Hawaii: determining movement parameters in cage and field settings. PLoS One 17: e0276987. https://doi.org/10.1371/journal.pone.0276987.

Morgan MJ, O'Donnell JM, Oliver RF (1973). Development of left–right asymmetry in the habenular nuclei of Rana temporaria. J Comp Neurol 149: 203–214.

Moulton JM, Barron SE (1967). Asymmetry in the Mauthner cells of the goldfish brain. Copeia 1967: 836–837.

Mundorf A, Ocklenburg S (2023). Hemispheric asymmetries in mental disorders: evidence from rodent studies. J Neural Transm. https://doi.org/10.1007/s00702-023-02610-z.

Niven JE, Frasnelli E (2018). Insights into the evolution of lateralization from the insects. In: Progress in brain research, Elsevier3–31.

O'Shea-Wheller TA (2019). Honeybees show a context-dependent rightward bias. Biol Lett 15: 20180877. https://doi.org/10.1098/rsbl.2018.0877.

Ocklenburg S, Isparta S, Peterburs J et al. (2019). Paw preferences in cats and dogs: meta-analysis. Laterality 24: 647–677.

Ocklenburg S, El Basbasse Y, Ströckens F et al. (2023). Hemispheric asymmetries and brain size in mammals. Commun Biol 6: 521. https://doi.org/10.1038/s42003-023-04894-z.

Ong M, Bulmer M, Groening J et al. (2017). Obstacle traversal and route choice in flying honeybees: evidence for individual handedness. PLoS One 12: e0184343. https://doi.org/10.1371/journal.pone.0184343.

Ortiz CO, Etchberger JF, Posy SL et al. (2006). Searching for neuronal left/right asymmetry: genomewide analysis of nematode receptor-type guanylyl Cyclases. Genetics 173: 131–149. https://doi.org/10.1534/genetics.106.055749.

Ortiz CO, Faumont S, Takayama J et al. (2009). Lateralized gustatory behavior of C. elegans is controlled by specific receptor-type guanylyl cyclases. Curr Biol 19: 996–1004. https://doi.org/10.1016/j.cub.2009.05.043.

Perelle IB, Saretsky S, Ehrman L (1978). Lateral consistency in Drosophila. Anim Behav 27: 622–623.

Poyser F, Caldwell C, Cobb M (2006). Dog paw preference shows lability and sex differences. Behav Processes 73: 216–221. https://doi.org/10.1016/j.beproc.2006.05.011.

Quaranta A, Siniscalchi M, Frate A et al. (2004). Paw preference in dogs: relations between lateralised behaviour and immunity. Behav Brain Res 153: 521–525. https://doi.org/10.1016/j.bbr.2004.01.009.

Randler C (2007). Foot preferences during resting in wildfowl and waders. Laterality 12: 191–197.

Reddon AR, Balshine S (2010). Lateralization in response to social stimuli in a cooperatively breeding cichlid fish. Behav Processes 85: 68–71. https://doi.org/10.1016/j.beproc.2010.06.008.

Remy M, Güntürkün O (1991). Retinal afferents to the tectum opticum and the nucleus opticus principalis thalami in the pigeon. J Comp Neurol 305: 57–70.

Rigosi E, Frasnelli E, Vinegoni C et al. (2011). Searching for anatomical correlates of olfactory lateralization in the honeybee antennal lobes: a morphological and behavioural study. Behav Brain Res 221: 290–294. https://doi.org/10.1016/j.bbr.2011.03.015.

Rigosi E, Haase A, Rath L et al. (2015). Asymmetric neural coding revealed by *in vivo* calcium imaging in the honey bee brain. Proc R Soc B 282: 20142571. https://doi.org/10.1098/rspb.2014.2571.

Roberson S, Halpern ME (2017). Convergence of signaling pathways underlying habenular formation and axonal outgrowth. Development, dev147751. https://doi.org/10.1242/dev.147751.

Robins A, Rogers LJ (2002). Limb preference and skeletal asymmetry in the cane toad, Bufo marinus (Anura: Bufonidae). Laterality 7: 261–275.

Robins A, Rogers L (2006a). Complementary and lateralized forms of processing in Bufo marinus for novel and familiar prey. Neurobiol Learn Mem 86: 214–227. https://doi.org/10.1016/j.nlm.2006.03.002.

Robins A, Rogers LJ (2006b). Lateralized visual and motor responses in the green tree frog, Litoria caerulea. Anim Behav 72: 843–852. https://doi.org/10.1016/j.anbehav.2006.01.022.

Robins A, Lippolis G, Bisazza A et al. (1998). Lateralized agonistic responses and hindlimb use in toads. Anim Behav 56: 875–881. https://doi.org/10.1006/anbe.1998.0877.

Roche DG, Amcoff M, Morgan R et al. (2020). Behavioural lateralization in a detour test is not repeatable in fishes. Anim Behav 167: 55–64. https://doi.org/10.1016/j.anbehav.2020.06.025.

Rodgers RMR, Cain KE (2019). No evidence for side preference or handedness in a New Zealand rail species with habitual foot-usage, the pukeko. Behav Processes 167: 103910. https://doi.org/10.1016/j.beproc.2019.103910.

Rogers LJ (1980). Lateralisation in the avian brain. Bird Behav 2: 1–12.

Rogers LJ (1982). Light experience and asymmetry of brain function in chickens. Nature 297: 223–225.

Rogers LJ (1990). Light input and the reversal of functional lateralization in the chicken brain. Behav Brain Res 38: 211–221.

Rogers LJ (2000). Evolution of hemispheric specialization: advantages and disadvantages. Brain Lang 73: 236–253. https://doi.org/10.1006/brln.2000.2305.

Rogers LJ (2002). Lateralised brain function in anurans: comparison to lateralisation in other vertebrates. Laterality 7: 219–239.

Rogers LJ (2008). Development and function of lateralization in the avian brain. Brain Res Bull 76: 235–244. https://doi.org/10.1016/j.brainresbull.2008.02.001.

Rogers LJ (2014). Asymmetry of brain and behavior in animals: its development, function, and human relevance: laterality development in animal models. Genesis 52: 555–571. https://doi.org/10.1002/dvg.22741.

Rogers LJ, Anson JM (1979). Lateralisation of function in the chicken fore-brain. Pharmacol Biochem Behav 10: 679–686.

Rogers LJ, Deng C (2005). Corticosterone treatment of the chick embryo affects light-stimulated development of the thalamofugal visual pathway. Behav Brain Res 159: 63–71. https://doi.org/10.1016/j.bbr.2004.10.003.

Rogers LJ, Frasnelli E (2016). Antennal asymmetry in social behavior of the Australian stingless bee, Tetragonula carbonaria. J Insect Behav 29: 491–499. https://doi.org/10.1007/s10905-016-9575-z.

Rogers LJ, Sink HS (1988). Transient asymmetry in the projections of the rostral thalamus to the visual hyperstriatum of the chicken, and reversal of its direction by light exposure. Exp Brain Res 70: 378–384.

Rogers LJ, Vallortigara G (2008). From antenna to antenna: lateral shift of olfactory memory recall by honeybees. PLoS One 3: e 2340. https://doi.org/10.1371/journal.pone.0002340.

Rogers LJ, Workman L (1989). Light exposure during incubation affects competitive behaviour in domestic chicks. Appl Anim Behav Sci 23: 187–198.

Rogers LJ, Workman L (1993). Footedness in birds. Anim Behav 45: 409–411.

Rogers LJ, Vallortigara G, Andrew RJ (2012). Divided brains: the biology and behaviour of brain asymmetries, Cambridge University Press, Cambrige, [England]; New York.

Rogers LJ, Rigosi E, Frasnelli E et al. (2013). A right antenna for social behaviour in honeybees. Sci Rep 3: 2045. https://doi.org/10.1038/srep02045.

Rogers LJ, Frasnelli E, Versace E (2016). Lateralized antennal control of aggression and sex differences in red mason bees, *Osmia bicornis*. Sci Rep 6: 29411. https://doi.org/10.1038/srep29411.

Roux N, Duran E, Lanyon RG et al. (2016). Brain lateralization involved in visual recognition of conspecifics in coral reef fish at recruitment. Anim Behav 117: 3–8. https://doi.org/10.1016/j.anbehav.2016.04.011.

Rowell CHF (1964). Central control of an insect segmental reflex: I. Inhibition by different parts of the central nervous system. J Exp Biol 41: 559–572.

Rutledge R, Hunt GR (2004). Lateralized tool use in wild New Caledonian crows. Anim Behav 67: 327–332. https://doi.org/10.1016/j.anbehav.2003.07.002.

Sakai M, Hishii T, Takeda S et al. (2006). Laterality of flipper rubbing behaviour in wild bottlenose dolphins (Tursiops aduncus): caused by asymmetry of eye use? Behav Brain Res 170: 204–210. https://doi.org/10.1016/j.bbr.2006.02.018.

Schaafsma SM, Groothuis TGG (2011). Sex-specific effects of postnatal testosterone on lateralization in cichlid fish. Anim Behav 81: 283–288. https://doi.org/10.1016/j.anbehav.2010.10.019.

Schneider LA, Delfabbro PH, Burns NR (2013). Temperament and lateralization in the domestic dog (Canis familiaris). J Vet Behav 8: 124–134. https://doi.org/10.1016/j.jveb.2012.06.004.

Schnell AK, Bellanger C, Vallortigara G et al. (2018). Visual asymmetries in cuttlefish during brightness matching for camouflage. Curr Biol 28: R925–R926. https://doi.org/10.1016/j.cub.2018.07.019.

Schwarz IM, Rogers LJ (1992). Testosterone: a role in the development of brain asymmetry in the chick. Neurosci Lett 146: 167–170.

Seki S, Kohda M, Hori M (2000). Asymmetry of mouth morph of a freshwater goby, *Rhinogobius flumineus*. Zoolog Sci 17: 1321–1325. https://doi.org/10.2108/zsj.17.1321.

Seligmann H (2002). Behavioural and morphological asymmetries in hindlimbs of Hoplodactylus duvaucelii (Lacertilia: Gekkonomorpha: Gekkota: Diplodactylinae). Laterality 7: 277–283.

Sieg AE, Zandonà E, Izzo VM et al. (2010). Population level "flipperedness" in the eastern Pacific leatherback turtle. Behav Brain Res 206: 135–138. https://doi.org/10.1016/j.bbr.2009.08.023.

Simon T, Frasnelli E, Guo K et al. (2022a). Is there an association between paw preference and emotionality in pet dogs? Animals 12: 1153. https://doi.org/10.3390/ani12091153.

Simon T, Guo K, Frasnelli E et al. (2022b). Testing of behavioural asymmetries as markers for brain lateralization of emotional states in pet dogs: a critical review. Neurosci Biobehav Rev 143: 104950. https://doi.org/10.1016/j.neubiorev.2022.104950.

Siniscalchi M, Quaranta A, Rogers LJ (2008). Hemispheric specialization in dogs for processing different acoustic stimuli. PLoS One 3: e3349. https://doi.org/10.1371/journal.pone.0003349.

Siniscalchi M, Sasso R, Pepe AM et al. (2010). Catecholamine plasma levels following immune stimulation with rabies vaccine in dogs selected for their paw preferences. Neurosci Lett 476: 142–145. https://doi.org/10.1016/j.neulet.2010.04.016.

Siniscalchi M, Padalino B, Lusito R et al. (2014). Is the left forelimb preference indicative of a stressful situation in horses? Behav Processes 107: 61–67. https://doi.org/10.1016/j.beproc.2014.07.018.

Siniscalchi M, D'Ingeo S, Quaranta A (2017). Lateralized functions in the dog brain. Symmetry 9: 71. https://doi.org/10.3390/sym9050071.

Siniscalchi M, Bertino D, d'Ingeo S et al. (2019). Relationship between motor laterality and aggressive behavior in sheepdogs. Symmetry 11: 233. https://doi.org/10.3390/sym11020233.

Skiba M, Diekamp B, Güntürkün O (2002). Embryonic light stimulation induces different asymmetries in visuoperceptual and visuomotor pathways of pigeons. Behav Brain Res 134: 149–156. https://doi.org/10.1016/S0166-4328(01)00463-6.

Sovrano V (1999). Roots of brain specializations: preferential left-eye use during mirror-image inspection in six species of teleost fish. Behav Brain Res 106: 175–180. https://doi.org/10.1016/S0166-4328(99)00105-9.

Sovrano VA (2004). Visual lateralization in response to familiar and unfamiliar stimuli in fish. Behav Brain Res 152: 385–391. https://doi.org/10.1016/j.bbr.2003.10.022.

Sovrano VA (2007). A note on asymmetric use of the forelimbs during feeding in the European green toad (Bufo viridis). Laterality 12: 458–463.

Sovrano V, Andrew R (2006). Eye use during viewing a reflection: behavioural lateralisation in zebrafish larvae. Behav Brain Res 167: 226–231. https://doi.org/10.1016/j.bbr.2005.09.021.

Sovrano VA, Bisazza A, Vallortigara G (2001). Lateralization of response to social stimuli in fishes: a comparison between different methods and species. Physiol Behav 74: 237–244. https://doi.org/10.1016/S0031-9384(01)00552-2.

Stancher G, Clara E, Regolin L et al. (2006). Lateralized righting behavior in the tortoise (Testudo hermanni). Behav Brain Res 173: 315–319. https://doi.org/10.1016/j.bbr.2006.06.023.

Stancher G, Sovrano VA, Vallortigara G (2018). Motor asymmetries in fishes, amphibians, and reptiles. In: Progress in brain research, Elsevier33–56.

Stennett CR, Strauss RE (2010). Behavioural lateralization in zebrafish and four related species of minnows (Osteichthyes: Cyprinidae). Anim Behav 79: 1339–1342. https://doi.org/10.1016/j.anbehav.2010.03.007.

Suzuki H, Thiele TR, Faumont S et al. (2008). Functional asymmetry in Caenorhabditis elegans taste neurons and its computational role in chemotaxis. Nature 454: 114–117. https://doi.org/10.1038/nature06927.

Takatsuji K, Ito H, Masai H (1983). Ipsilateral retinal projections in Japanese quail, Coturnix coturnix japonica. Brain Res Bull 10: 53–56.

Takeuchi Y, Tobo S, Hori M (2008). Morphological asymmetry of the abdomen and behavioral laterality in Atyid shrimps. Zoolog Sci 25: 355–363. https://doi.org/10.2108/zsj.25.355.

Takeuchi Y, Hori M, Myint O et al. (2010). Lateral bias of agonistic responses to mirror images and morphological asymmetry in the Siamese fighting fish (Betta splendens). Behav Brain Res 208: 106–111. https://doi.org/10.1016/j.bbr.2009.11.021.

Takeuchi Y, Hori M, Oda Y (2012). Lateralized kinematics of predation behavior in a Lake Tanganyika scale-eating cichlid fish. PLoS One 7: e29272. https://doi.org/10.1371/journal.pone.0029272.

Tan Ü (1987). Paw preferences in dogs. Int J Neurosci 32: 825–829.

Tan Ü, Çalşikan S (1987). Allometry and asymmetry in the dog brain: the right hemisphere is heavier regardless of paw preference. Int J Neurosci 35: 189–194.

Tan Ü, Yaprak M, Kutlu N (1990). Paw preference in cats: distribution and sex differences. Int J Neurosci 50: 195–208.

Taylor RW, Hsieh Y-W, Gamse JT et al. (2010). Making a difference together: reciprocal interactions in *C. elegans* and zebrafish asymmetric neural development. Development 137: 681–691. https://doi.org/10.1242/dev.038695.

Tobo S, Takeuchi Y, Hori M (2012). Morphological asymmetry and behavioral laterality in the crayfish, Procambarus clarkii. Ecol Res 27: 53–59. https://doi.org/10.1007/s11284-011-0867-7.

Tomkins LM, Thomson PC, McGreevy PD (2010). First-stepping test as a measure of motor laterality in dogs (Canis familiaris). J Vet Behav 5: 247–255. https://doi.org/10.1016/j.jveb.2010.03.001.

Tommasi L, Vallortigara G (1999). Footedness in binocular and monocular chicks. Laterality 4: 89–95. https://doi.org/10.1080/713754325.

Valenti A, Sovrano VA, Zucca P et al. (2003). Visual lateralisation in quails (Coturnix coturnix). Laterality 8: 67–78.

Vallortigara G (1992). Right hemisphere advantage for social recognition in the chick. Neuropsychologia 30: 761–768. https://doi.org/10.1016/0028-3932(92)90080-6.

Vallortigara G (2006). The evolutionary psychology of left and right: costs and benefits of lateralization. Dev Psychobiol 48: 418–427. https://doi.org/10.1002/dev.20166.

Vallortigara G, Andrew RJ (1991). Lateralization of response by chicks to change in a model partner. Anim Behav 41: 187–194. https://doi.org/10.1016/S0003-3472(05)80470-1.

Vallortigara G, Andrew RJ (1994). Differential involvement of right and left hemisphere in individual recognition in the domestic chick. Behav Processes 33: 41–57. https://doi.org/10.1016/0376-6357(94)90059-0.

Vallortigara G, Bisazza A (2002). How ancient is brain lateralisation? In: LJ Rogers, RJ Andrew (Eds.), Comparative vertebrate lateralisation. Cambridge University Press, Cambridge, pp. 9–69.

Vallortigara G, Rogers LJ (2005). Survival with an asymmetrical brain: advantages and disadvantages of cerebral lateralization. Behav Brain Sci 28: 575–589. https://doi.org/10.1017/S0140525X05000105.

Vallortigara G, Rogers LJ, Bisazza A et al. (1998). Complementary right and left hemifield use for predatory and agonistic behaviour in toads. Neuro Report 9: 3341–3344.

Vallortigara G, Regolin L, Pagni P (1999). Detour behaviour, imprinting and visual lateralization in the domestic chick. Cogn Brain Res 7: 307–320. https://doi.org/10.1016/S0926-6410(98)00033-0.

Vallortigara G, Cozzutti C, Tommasi L et al. (2001). How birds use their eyes: opposite left-right specialization for the lateral and frontal visual hemifield in the domestic chick. Curr Biol 11: 29–33.

Vallortigara G, Chiandetti C, Sovrano VA (2011). Brain asymmetry (animal). WIREs Cogn Sci 2: 146–157. https://doi.org/10.1002/wcs.100.

Van Dooren TJM, Van Goor HA, Van Putten M (2010). Handedness and asymmetry in scale-eating cichlids: antisymmetries of different strength. Evolution. https://doi.org/10.1111/j.1558-5646.2010.00977.x.

Versace E, Morgante M, Pulina G et al. (2007). Behavioural lateralization in sheep (Ovis aries). Behav Brain Res 184: 72–80. https://doi.org/10.1016/j.bbr.2007.06.016.

Versace E, Caffini M, Werkhoven Z et al. (2020). Individual, but not population asymmetries, are modulated by social environment and genotype in Drosophila melanogaster. Sci Rep 10: 4480. https://doi.org/10.1038/s41598-020-61410-7.

Waite L, Frasnelli E (2021). Visuo-motor biases in buff-tailed bumblebees (*Bombus terrestris*). Laterality 26: 55–70. https://doi.org/10.1080/1357650X.2020.1826503.

Walker SF (1980). Lateralization of functions in the vertebrate brain: a review. Br J Psychol 71: 329–367.

Wassersug RJ, Naitoh T, Yamashita M (1999). Turning Bias in tadpoles. J Herpetol 33: 543. https://doi.org/10.2307/1565570.

Weidner C, Reperant J, Miceli D et al. (1985). An anatomical study of ipsilateral retinal projections in the quail using radioautographic, horseradish peroxidase, fluorescence and degeneration techniques. Brain Res 340: 99–108.

Weir AA, Kenward B, Chappell J et al. (2004). Lateralization of tool use in New Caledonian crows (*Corvus moneduloides*). Proc R Soc Lond B Biol Sci 271: S344–S346.

Wells DL (2003). Lateralised behaviour in the domestic dog, Canis familiaris. Behav Processes 61: 27–35. https://doi.org/10.1016/S0376-6357(02)00161-4.

Wells DL, McDowell LJ (2019). Laterality as a tool for assessing breed differences in emotional reactivity in the domestic cat, *Felis silvestris* catus. Animals 9: 647. https://doi.org/10.3390/ani9090647.

Wells DL, Millsopp S (2009). Lateralized behaviour in the domestic cat, Felis silvestris catus. Anim Behav 78: 537–541. https://doi.org/10.1016/j.anbehav.2009.06.010.

Wells DL, Millsopp S (2012). The ontogenesis of lateralized behavior in the domestic cat, Felis silvestris catus. J Comp Psychol 126: 23–30. https://doi.org/10.1037/a0026522.

Whishaw IQ (2015). Absence of population asymmetry in the American quarter horse (Equus ferus caballus) performing skilled left and right manoeuvres in reining competition. Laterality 20: 604–617.

Whiteside MA, Bess MM, Frasnelli E et al. (2018). Low survival of strongly footed pheasants may explain constraints on lateralization. Sci Rep 8: 13791. https://doi.org/10.1038/s41598-018-32066-1.

Whiteside MA, Bess MM, Frasnelli E et al. (2020). No evidence that footedness in pheasants influences cognitive performance in tasks assessing colour discrimination and spatial ability. Learn Behav 48: 84–95. https://doi.org/10.3758/s13420-019-00402-8.

Wilson DM (1968). Inherent asymmetry and reflex modulation of the locust flight motor pattern. J Exp Biol 48: 631–641. https://doi.org/10.1242/jeb.48.3.631.

Yamashita M, Naitoh T, Wassersug RJ (2000). Startle response and turning Bias in Microhyla tadpoles. Zoolog Sci 17: 185–189. https://doi.org/10.2108/zsj.17.185.

Yince MA (1964). Use of the feet in feeding by the great tit Parus major. Ibis 106: 508–529.

Yu G, Guo J, Xie W et al. (2020). Footedness predicts escape performance in a passerine bird. Ecol Evol 10: 4251–4260. https://doi.org/10.1002/ece3.6193.

Zhang B, Yao Y, Zhang H et al. (2017). Left Habenula mediates light-preference behavior in zebrafish via an asymmetrical visual pathway. Neuron 93: 914–928. e4 https://doi.org/10.1016/j.neuron.2017.01.011.

Zottoli SJ (1977). Correlation of the startle reflex and Mauthner cell auditory responses in unrestrained goldfish. J Exp Biol66: 243–254. https://doi.org/10.1242/jeb.66.1.243.

Zucca P, Baciadonna L, Masci S et al. (2011a). Illness as a source of variation of laterality in lions (Panthera leo). Laterality 16: 356–366. https://doi.org/10.1080/13576501003690025.

Zucca P, Cerri F, Carluccio A et al. (2011b). Space availability influence laterality in donkeys (Equus asinus). Behav Processes 88: 63–66. https://doi.org/10.1016/j.beproc.2011.06.012.

Further reading

Baciadonna L, Nawroth C, Briefer EF et al. (2019). Perceptual lateralization of vocal stimuli in goats. Curr Zool 65: 67–74. https://doi.org/10.1093/cz/zoy022.

Bisazza A, Cantalupo C, Robins A et al. (1997). Pawedness and motor asymmetries in toads. Laterality 2: 49–64.

Boulinguez-Ambroise G, Aychet J, Pouydebat E (2022). Limb preference in animals: new insights into the evolution of manual laterality in hominids. Symmetry 14: 96. https://doi.org/10.3390/sym14010096.

Messina A, Boiti A, Vallortigara G (2021). Asymmetric distribution of pallial-expressed genes in zebrafish (*Danio rerio*). Eur J Neurosci 53: 362–375. https://doi.org/10.1111/ejn.14914.

Rogers LJ (2021). Brain lateralization and cognitive capacity. Animals 11: 1996. https://doi.org/10.3390/ani11071996.

Rogers LJ (2023a). Development of hand and paw preferences and their association with other patterns of behaviour and cognition. Symmetry 15: 926. https://doi.org/10.3390/sym15040926.

Rogers LJ (2023b). Unfolding a sequence of sensory influences and interactions in the development of functional brain laterality. Front Behav Neurosci 16: 1103192. https://doi.org/10.3389/fnbeh.2022.1103192.

Handbook of Clinical Neurology, Vol. 208 (3rd series)
Cerebral Asymmetries
C. Papagno and P. Corballis, Editors
https://doi.org/10.1016/B978-0-443-15646-5.00026-9

Chapter 15

Hemispheric asymmetries, paleoneurology, and the evolution of the human genus

EMILIANO BRUNER*

Departamento de Paleobiología, Museo Nacional de Ciencias Naturales, CSIC, Madrid, Spain

Reina Sofía Alzheimer Center, CIEN Foundation, ISCIII, Madrid, Spain

Abstract

Brain asymmetries are a distinctive feature of *Homo sapiens* and are associated with key evolutionary functions including language and handedness. Nonetheless, differences between humans and apes could be just a matter of degree and size and not the expression of unique traits of our species. In this chapter, I introduce paleoneurology and the study of brain morphology in fossil hominids, reviewing the anatomic factors that can influence the main asymmetries of the endocranial cavity (cortical volumes, sulcal patterns, and craniovascular features). The paleoneurological evidence suggests that most extinct human species displayed a pattern of gross endocranial asymmetries similar to modern humans. In addition, the behavioral information on handedness also points to a similar degree of laterality in archaic species of the human genus and in Neandertals. At present, there is therefore no evidence suggesting that the brain asymmetries in *H. sapiens* are part of a derived set of features. Of course, even a simple proportional change due to brain size increase can anyway prompt crucial cognitive changes, mostly if threshold effects are considered. Nonetheless, we still lack much information in basic anatomy to support consistent hypotheses on the biologic factors involved in endocranial asymmetries in fossil hominids. This missing information concerns endocranial morphogenesis and topology, spatial conflicts and constraints, the biomechanical balance between cerebral tissues, and the actual histologic changes associated with encephalization.

BRAIN EVOLUTION AND PREHISTORY

The human brain is patently asymmetric, in both its structure and functions, at macroscopic and microscopic scales, and, in the last century, this has inevitably called the attention of many generations of neurobiologists, psychiatrists, psychologists, and anthropologists (Thiebaut de Schotten and Beckmann, 2022). In humans, brain asymmetries are recognizable since the earliest fetal stages (Corballis, 2013), and they have been associated with complex and derived human features spanning from language to tool-use (Toga and Thompson, 2003; Corballis, 2019). Most theories, in this sense, suggest reciprocal influences between speech and gesture, and many cerebral asymmetries would be the anatomic result of this coevolutionary relationship. Human cognitive complexity can be indeed linked both to hemispheric specialization and to the possibility (or even the necessity) to differentiate the role of each hemisphere through parallel processing, allowing a fine coordination between regions involved, for example, in language, social cognition, or attention (Hervé et al., 2013; Hartwigsen et al., 2021). This generated a pronounced interest in hemispheric specialization not only in neurobiology but also in evolutionary anthropology, and this topic soon became a key issue to investigate the derived cognitive abilities of *H. sapiens*.

Evolutionary biology and comparative neuroanatomy are fields that must inevitably integrate analyses and results from both neontological and paleontological samples (Bruner, 2019). The neontological data refer to living species, in particular other primates and especially apes, which can provide excellent experimental and quantitative evidence. Nonetheless, living zoological groups represent the actual product of evolution and cannot hence provide information on the process of change itself. Paleontological data deal with extinct species, which can supply direct information on the changes that occurred along ancestral–descendant lineages. Nonetheless, fossil specimens can only provide scarce and fragmented evidence in terms of biologic, taxonomical, and

*Correspondence to: Emiliano Bruner, Museo Nacional de Ciencias Naturales, CSIC, Calle J.G. Abascal 2, 28006 Madrid, Spain. E-mail: emiliano.bruner@mncn.csic.es

statistical information. However, such scarce and fragmented evidence is the only one we have on our own phylogenetic history, and surely deserves attention.

Living and extinct taxonomic units (*taxa*; sing.: *taxon*) must be investigated through the lens of comparative biology, considering that their phenotype is an admixture of *plesiomorphic* and *apomorphic* traits. The former are primitive features retained from ancestral groups, while the latter are new (derived) features, associated with a specific evolutionary lineage. In some cases, an evolutionary trait can be strictly limited to one single species (*autapomorphy*). Comparative biology must also consider whether adaptations or traits are *homologous* (namely, derived from the same embryological or developmental element) or *analogous* (i.e., derived from distinct embryological or developmental elements). In the latter case, some features can display a similar phenotype or function in two species but have evolved independently, through parallel or convergent processes, and are not inherited from a common ancestor. All these concepts and principles are habitually applied in zoology when dealing with traditional biologic features like wings, hairs, or metabolites. Nonetheless, they should also be applied to neuroanatomy, behavior, and any cognitive trait or abilities.

This view should be necessarily considered when developing fields that integrate neuroscience and prehistory, focusing on brain anatomy in fossil species (*paleoneurology*), brain functions associated with specific behaviors inferred from the archaeological record (*neuroarchaeology*), and psychologic models applied to prehistoric cultures (*cognitive archaeology*) (Bruner, 2022, 2024). Through a comparative perspective, we should take into account that each taxon is a unique combination of specific characters. If we apply the comparative approach to the human genus (namely, the genus *Homo*—Fig. 15.1), this means, for example, that we should consider whether extinct human species might have had cognitive abilities that we have lost or that we have never evolved (plesiomorphy vs apomorphy). Conversely, other human species might have evolved similar cognitive skills although relying on distinct biologic foundations (homology vs analogy).

Of course, investigating brain and cognition in extinct species is undoubtedly a challenging task, taking into consideration that fossils cannot provide any direct evidence of soft tissues and behavior. Unfortunately, all too often, this limitation is associated with two extreme reactions. On one side, there are those who think that cerebral or cognitive issues cannot be investigated in prehistory. In this case, brain and cognition would simply stand out of the mandate of paleoanthropology. This perspective is not really useful to the development of these kinds of inquiries, because it denies the possibility of any advance in this sense. Such a rejection is not scientific at all, because it relegates research only to those aspects that are certain and well defined, prompting skepticism toward the exploration of the new and unknown. The exploration of these indefinite frontiers should be precisely the role of science itself.

On the other side, there are those who think that the absence of direct evidence in prehistory can be supplied by a noticeable degree of speculation and conclusions based on personal opinions. In this case, the methods are not those of science but those of philosophy, following logical reasoning, the principle of possibility, and some informed background coming from neuroscience. This approach can supply interesting perspectives, but it is particularly sensitive to excesses of narrative. In the absence of prudence, hypotheses and theories can undergo a never-ending multiplication and never-ending unfruitful debates. Without the possibility of making a proper selection of hypotheses through quantitative and experimental paradigms, any debate will be inevitably unresolvable and, as such, out of the realm of science. At last, an excess of speculation and narrative would generate a discredit of the disciplines involved in these fields.

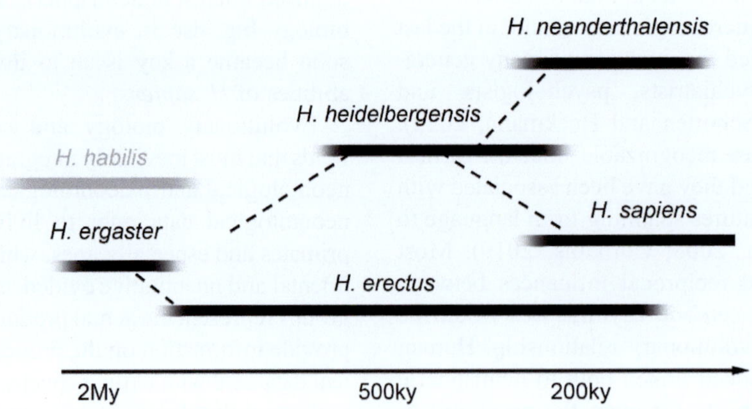

Fig. 15.1. The human genus probably originated in Africa 2 million years ago. This figure displays the key human (*Homo*) species on a general chronological scale (*My*, million years ago; *ky*: thousands of years ago). *Dashed lines* refer to possible phylogenetic relationships. The species *H. habilis* is at present not regarded as a consistent taxon. Encephalization (namely, an increase of brain size relative to body size) probably occurred in *H. ergaster*, *H. heidelbergensis*, *H. sapiens*, and *H. neanderthalensis*. *H. ergaster* and *H. erectus* are affine species generally associated with Africa and Asia, respectively. *H. heidelbergensis*, mostly known for its African and European specimens, is a very heterogeneous taxon which could be split into different groups. *H. sapiens* and *H. neanderthalensis* are the species with the largest brain size and more complex culture. The human family tree is probably much larger, although other species are currently not well described, mostly when dealing with features associated with brain and behavior (see Wood and Richmond, 2000; Bruner and Beaudet, 2023).

As usual, the best option is probably in the middle of these two extreme positions, namely in a cautious interpretation of multiple and independent quantitative data, trying to frame these evolutionary questions into proper hypotheses that, albeit difficult to test, are nevertheless reliable and in agreement with experimental results. Of course, such an aim is difficult, but, precisely, effortlessness is not a traditional feature of science.

With this in mind, hemispheric asymmetry has represented a very standard topic in human paleoneurology since the early stages of this field (LeMay, 1982). The aim, in this sense, was twofold: to investigate the evolution of the hemispheric specialization in hominids, and to consider to what extent such features could be actually investigated in fossil remains.

BRAINS AND ENDOCASTS

The field of paleoneurology (or, more precisely, *paleoneurobiology*) concerns the study of brain morphology in extinct species (Holloway et al., 2004; Falk, 2012; Bruner, 2015, 2017; Bruner and Beaudet, 2023). Although making inferences on cognition and behavior in paleoanthropology is a constant temptation, this field is strictly an anatomic one, and it deals with considering macroscopic features like brain size, the morphology of sulci and gyri, the meningeal vascular channels, or the functional relationships between skull and brain. Therefore, analyses, results, and conclusions in paleoneurology should be, as far as possible, limited to some aspects of brain morphology, and this requires a comprehensive knowledge and competence on the processes and mechanisms involved in brain morphogenesis. The foundation of the field is indeed pretty simple: the endocranial cavity is molded by the growing brain, and its morphology can hence inform on the general geometry and superficial features of the cerebral cortex. Therefore, paleoneurology is but the application to the fossil record of what is called *functional craniology* (Moss and Young, 1960). The rough correspondence between the brain form and the cranial cavity allows making neuroanatomic inferences in those cases in which the brain is no longer available, like in extinct species or museum osteological collections. As primates, however, our senses are not really comfortable with "a cavity," namely with an empty space, and in paleoneurology we therefore fill the cavity and generate a positive cast of that empty space, called *endocast*, resembling the outer and volumetric appearance of the original brain (Fig. 15.2). Such casts were historically made by resins and plastic materials, but, in the last decades, biomedical imaging and virtual reconstructions have represented the golden standard of the field (Gunz et al., 2009; Neubauer, 2014). The correspondence between an endocast and a brain is pretty good, and it allows the recognition of many cortical traits (De Jager et al., 2019, 2022; Dumoncel et al., 2021; Labra et al., 2024). However, an endocranial cast is not a brain and, obviously, it cannot be employed to investigate many crucial aspects of brain biology or to consider the anatomy and morphology of any deep cerebral element. Therefore, working in paleoneurology would require a good amount of caution, trying to make inferences only for those brain features that can be detectable on the endocranial cavity. Indeed, it would be prudent to employ the term "endocast" instead of the term "brain" when dealing with fossil anatomy, to avoid generalizations and misunderstandings.

The information available from the endocasts is not homogeneous, and one needs both biologic competence and experience to provide reliable inferences on the correspondent brain anatomy. For example, there are regions in which, during ontogeny and evolution, the brain molds the skull, but also regions where the skull molds the brain through spatial constraints, biomechanical signaling, and topological balance (Lieberman et al., 2000; Bastir et al., 2006; Bruner et al., 2019; Schuurman and Bruner, 2024). In morphology, the term *topology* refers to the reciprocal position of the anatomic elements, and to the properties of the anatomic system associated with their spatial organization. This spatial organization determines structural and functional constraints, and channels the morphogenetic process, being a key aspect in ontogeny and phylogeny. For what concerns the endocranial cavity, for example, the vault and the cranial base represent two distinct developmental environments. The morphology of the

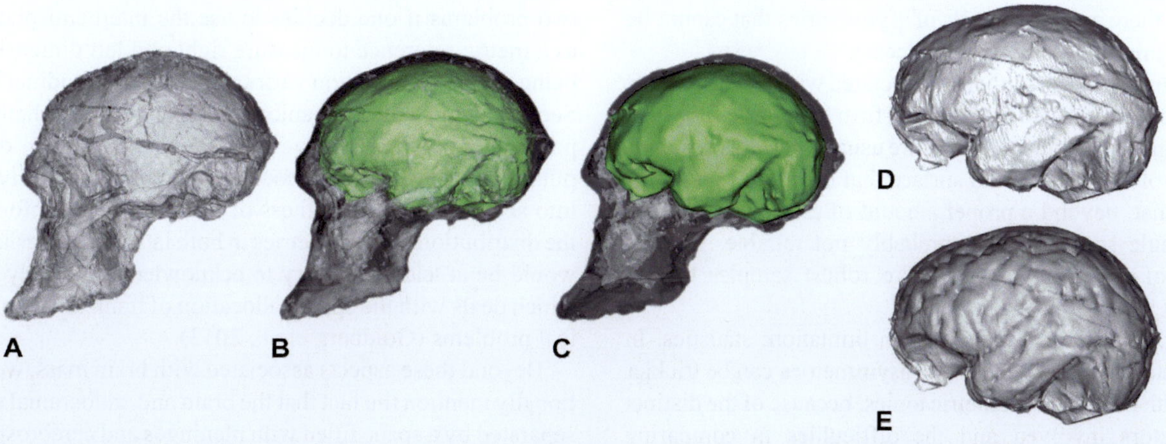

Fig. 15.2. Computed tomography allows the digital reconstruction of fossil specimens (here, a specimen of *Australopithecus africanus*, STS5, from Sterkfontein, South Africa). By using the digital replica of the skull (A), the endocranial cavity is filled through pixel segmentation (B, C) to generate a 3D virtual endocast (D). Sulci and gyri are tentatively identified on the endocast surface, according to personal experience or automatic procedures (E).

lower regions of the brain (like the orbital, temporal, and cerebellar cortices), in contact with the three cranial fossae, is sensitive to multiple morphogenetic factors that have little to do with the brain structure and functions, and their variations can be the result of secondary spatial changes due to cranial architecture. In contrast, the dorsal brain regions (i.e., the frontal and parietal gyri that are in contact with the vault) are freer from cranial constraints, and hence their morphology more likely reflects real variations of the cortical proportions. However, in this case, we have a distinct problem: probably because of the gravity, their surface is more separated from the endocranial table and, consequently, the imprints left on the bone are much feebler. Therefore, we have a paradoxical situation: the sulcal marks on the cranial base are more pronounced but less easy to interpret, while the marks on the vault are more informative but less visible.

With these premises, the question is whether endocasts can be employed to investigate brain asymmetries in fossil species. In this sense, we can identify at least two groups of traits: the overall volumetric asymmetries (i.e., the gross proportions of the right and left regions), and the differences in the morphology of specific sulcal patterns (in terms of size or shape of sulci and gyri).

BRAIN ASYMMETRIES AND ENDOCRANIAL MORPHOLOGY

Brain asymmetries are of various kinds and can refer to gray matter development, cytoarchitecture, connections, neurotransmitters, blood supply, and many other biologic parameters. Clearly, paleoneurology can only investigate the gross anatomic differences between hemispheres, namely the differences in overall dimensions or in the expression of some sulcal scheme. This limitation is an important restriction to the wide concept of brain asymmetries and must be constantly taken into account when investigating this topic in extinct species. Therefore, the first assumption in paleoneurology should be of general caution: the absence of evidence is not the evidence of absence. This means that the presence of macroscopic asymmetries reveals the effect of some underlying neurobiologic factors, but the lack of visible asymmetries does not mean that the two hemispheres did the same job. Namely, there are many kinds of asymmetries that cannot be detected by simply measuring an endocast.

A second limitation is implicitly associated with the difficulties in paleoneurology in detecting clear and firm boundaries between cortical regions. Such identifications are usually based on grooves and imprints of the endocranial surface that are at best tentative. This means that, beyond a proper amount of caution, inferences based on single specimens are probably not reliable. Instead, these kinds of analyses would require robust samples that in paleontology are rarely available.

Moreover, this leads to a third main limitation: statistics. In general, the statistics associated with asymmetries can be trickier than those with other morphometric topics, because of the distinct kinds of factors involved and the difficulties in comparing specular matching elements (Klingenberg et al., 2002; Eckert et al., 2019). Additionally, brain asymmetries are generally associated with minor and subtle metric differences. The anatomic uncertainties associated with endocasts, and the elusive differences, would require very large samples to be properly handled in terms of a quantitative approach. In paleontology, instead, all we have are scattered, incomplete, fragmented, and deformed specimens, which hardly provide sufficient statistical power to test or reject precise hypotheses on this topic. This limitation is not only statistical but also zoological, because the specimens available are not necessarily representative of the phylogenetic radiation they belong to, adding a further bias on the possibility to investigate the intra- and interspecific diversity. With these premises, it is clearly challenging in fossils to discriminate directional from fluctuating asymmetries, in which additional phenotypic effects are influenced by factors like environmental and developmental instability (Leamy and Klingenberg, 2005).

A fourth inevitable limitation deals with the nature of the asymmetries itself. The gross metric differences between hemispheres, in fact, can only reveal the amount of differences, but not their structural causes. Superficially, we are tempted to state that any differences in this sense is due to the number of neurons, but there is no reason to defend this possibility when compared with others. The volumetric changes we can observe in an endocast can be due to the number of neurons but also to differences in connectivity, supporting cells (the glia), cerebrospinal fluid, or even blood supply. Also, a specific cortical difference can be a primary consequence of some local changes, or a secondary consequence of changes in some neighboring regions, displacing the adjacent elements. Finally, many studies on asymmetries require a firm criterion to separate the right and left sides, and in general the interhemispheric fissure is used as a spatial reference. The interhemispheric fissure is the space occupied by the *falx cerebri*, a connective invagination of the meninges that anchors the brain to the cranial cavity, acting as a main biomechanical tensor during growth and development (Moss and Young, 1960). Because of this structural role, it influences the cerebral topology, and at the same time it is influenced by the cerebral topology. Because of the pronounced brain asymmetries in humans, the midline is therefore curved and tilted. Hence, although it is ideally represented as a straight line, often it is not. This generates at least two problems if one decides to use the interhemispheric fissure as a metric reference to measure right and left dimensions. First, being curved, its position varies along the longitudinal brain axis. Second, being a biomechanical tensor, we ignore whether a given part of the curvature is due to elements pushing on one side or pulling from the other. Indeed, the difference is not trivial, taking into account the low stiffness of the brain mass. Information on the distribution of asymmetries in both lateral and mesial elements would be at least necessary to acknowledge properly this issue, which deals with the spatial allocation of brain areas and topological problems (Goldberg et al., 2013).

Beyond these aspects associated with brain mass, we can additionally mention the fact that the brain and endocranial surface are separated by a space filled with meninges and cerebrospinal fluid, and therefore the asymmetries we observe on an endocast are not those of the brain, but of the endocranial cavity, with these elements interposed. However, at present, the limited evidence in this sense suggests that the thickness of this interposed space does vary (most of all in the vertical dimension), but without a patent asymmetrical pattern (Fournier et al., 2011; Dumoncel et al., 2021). Therefore, to date, we can conclude that the interposed space may introduce a systemic bias throughout the endocranial surface but does not influence the local expression of the general pattern of asymmetries.

Most of the structural factors, variables, and parameters mentioned above are, at present, scarcely known in hominoids, in terms of variation and homology, between and within species, and are still scarcely investigated. This should add further caution when speculations on brain asymmetries are provided according to gross endocranial metrics.

BRAIN ASYMMETRIES IN FOSSIL HOMINIDS

Recognizing the importance of brain asymmetries in cognitive evolution, and recognizing that endocasts display asymmetric traits, hemispheric specialization was soon included in the basic set of features investigated in human paleoneurology (LeMay, 1982; Holloway et al., 2004). For half a century, any new report of fossil endocasts was inevitably accompanied by a precise description of the global and local asymmetries, in terms of volumetric or sulcal aspects (e.g., Holloway, 1981a,b; Tobias, 1987).

The main traits that were considered were the gross volumetric differences between the right and left hemispheres and the sulcal patterns associated with the language areas, namely Broca's and Wernicke's regions (Fig. 15.3). For what concerns the global morphology, human brains are frequently characterized by the so-called "torque": the frontal region is larger on the right side and the occipital region is larger on the left side, "twisting" the brain's longitudinal axis (Fig. 15.4). Such gross volumetric asymmetries of the brain are called *petalias*, and the common human pattern is termed right-frontal left-occipital petalia. The more the fossil specimens that were described, the more the evidence that this same pattern can be frequently found in any human (*Homo*) species since the most archaic forms of our genus. Much later, comprehensive quantitative analyses confirmed this general result: all human species present a similar overall pattern of endocranial asymmetry (Balzeau et al., 2012a,b). The fact that all human taxa share a similar pattern of gross asymmetries is surely remarkable, suggesting that hemispheric differences might have had a role during the whole hominid radiation. Nonetheless, it also demotes the appeal of these traits if one is particularly interested in the evolution of *H. sapiens*, a species in which the expression of these features is deceptively similar to the other members of the family.

The paleoneurological interest in endocranial asymmetries further decreased when extending the analysis to great apes. There are still several doubts and disagreements on the possibility of detecting homologous comparisons for hemispheric asymmetries between humans and apes, and if such uncertainty persists for living taxa, we can imagine that concluding results for fossil species would be even more unlikely. Early observations suggested that the human pattern of petalia was less frequent in apes, both in terms of expression and magnitude (Holloway and De La Coste-Lareymondie, 1982). However, the pronounced differences in brain size introduce, in this case, size-related issues that must be acknowledged (Williams et al., 2022). In fact, most anatomic features display complex scaling rules when size does change in ontogeny or evolution (Klingenberg, 1998, 2010). Because of the necessity to maintain structural, functional, and developmental balance, the shape of an anatomic element changes according to changes in its overall size. *Allometry* is commonly defined in biology as the size-related changes of shape, or the study of size and its consequences (Gould, 1966). As expected, size-related scaling is not isometric when dealing with primate brains (Rilling, 2006), so the problem is whether the amount of asymmetry in humans is the one expected for an ape with a similar brain size (Kyriacou and Bruner, 2011). Later quantitative analyses, in fact, did not confirm significant differences among hominoids, suggesting that humans and great apes share the same global volumetric pattern of hemispheric dominance in terms of volume, mostly when the differences in overall brain size are

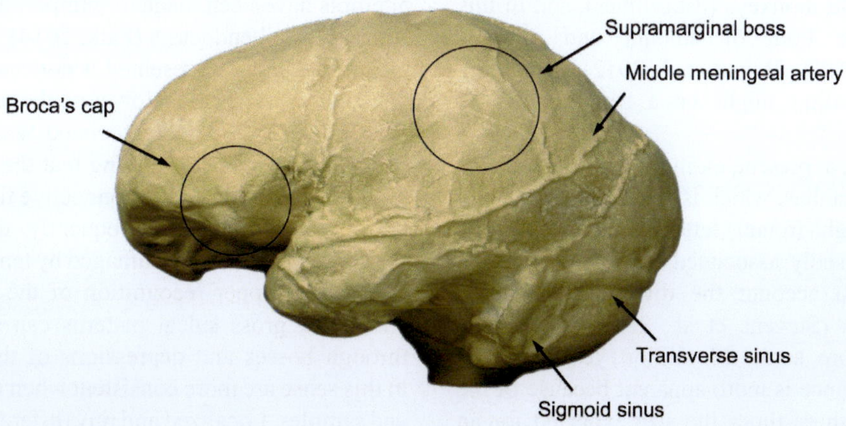

Fig. 15.3. Before the digital era, endocasts were created with plaster, resins, or other plastic materials. Here the endocast of a specimen of *Homo erectus* from Zhoukoudian, China. Traditional features commonly described in paleoneurology included the expression of Broca's cap and the supramarginal region (generally more bulging or shaped on the *left side*) and the course of the transverse and sigmoid sinuses (generally larger on the *right side*). The middle meningeal artery does not display directional asymmetries, although presenting many random or fluctuating asymmetries.

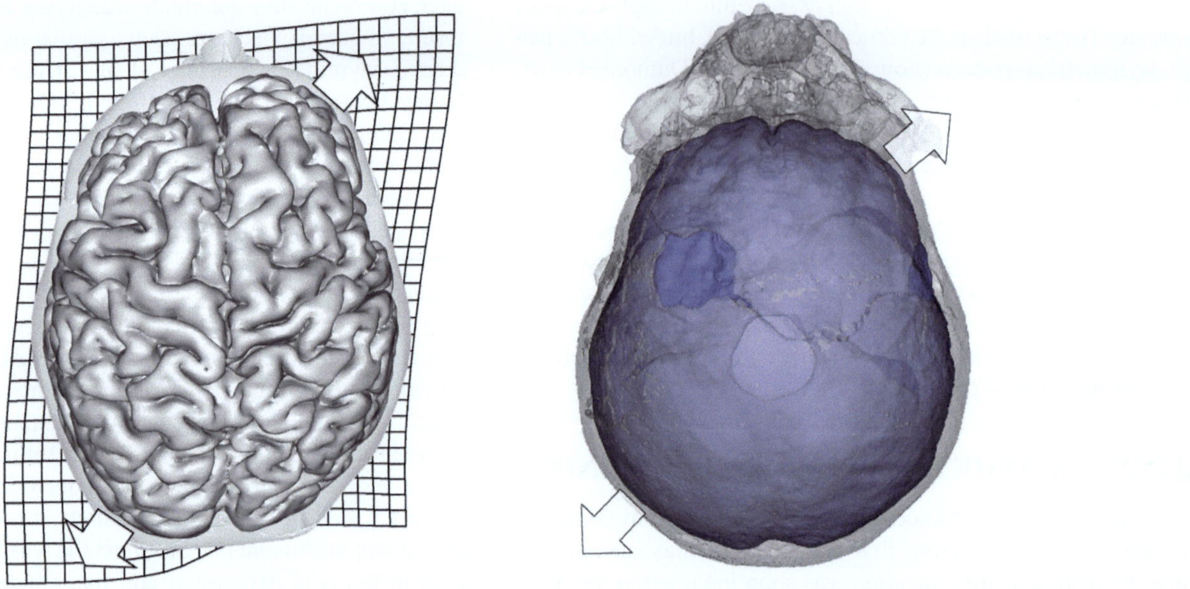

Fig. 15.4. *Left*: most human species display a similar gross asymmetry pattern (*superior view*) in which the frontal region is larger on the *right side* and the occipital region is larger on the *left side*. This pattern generates the so-called brain torque, which may involve a curvature of the midsagittal plane, corresponding to the interhemispheric fissure and *falx cerebri*. *Right*: the digital reconstruction of the Neandertal specimen Saccopastore 1 from Italy (*superior view*), in which the right frontal and left occipital region are wider than the counterpart.

taken into account (Balzeau and Gilissen, 2010; Balzeau et al., 2012a; Gómez-Robles et al., 2013; Neubauer et al., 2020). Indeed, the same surveys also evidenced that humans display, in terms of asymmetries, a more pronounced individual variability, which suggests more fluctuating asymmetries probably due to the increased neural plasticity. Interestingly, an analysis in New World monkeys (platyrrhines) showed that in these primates the *pattern* of asymmetry is influenced by ecology and social factors (namely, it is associated with a specific cognitive niche), but the *amount* of asymmetry is related to brain size (Gonzalez et al., 2022). In this sense, hominoids are probably less diverse than platyrrhines, and their differences seem more of degree than of organization. Indeed, apes and humans display much more asymmetries than Old World monkeys (catarrhines), and in this case differences may be both of amount and patterns (Heilbroner and Holloway, 1989; Bogart et al., 2012), suggesting that hemispheric specialization might be a stem feature of hominoids.

In sum, we can state that, at present, extinct and living humans share a similar pattern of petalias, which is also probably shared with living apes. The right-frontal left-occipital volumetric dominance is hence supposedly associated with their common ancestors, and taking into account the divergence between catarrhines and hominoids (Stevens et al., 2013), the torque may have evolved therefore some 20 million years ago. In humans, volumetric dominance is more apparent because of the large brain size, which is three times the size expected for an ape of our body size. Human specialization would in this case deal more with the individual variation and high neural plasticity than with the pattern of asymmetry itself. The noticeable idiosyncratic variability in humans decreases further the importance of individual assessments when working with specific fossils, stressing the necessity to count with large samples when looking for general trends and average differences. Also, we have to remind once more that this absence of evidence is not evidence of absence. In fact, more than stating that extinct and extant humans share a similar pattern of brain asymmetry, we should better state that there is at present no evidence suggesting the opposite. Nowadays, general comments on the petalias are still part of the standard pack of traits scored in traditional paleoneurological surveys. However, the usefulness of this information is, as far as we know, scarce.

For what concerns possible asymmetries in the sulcal patterns, the current information is probably even more blurry. Many attempts have been made to interpret the morphology of specific sulci on fossil endocasts (Falk, 2014), and digital and automatic techniques have represented a noticeable advance in this sense (Beaudet et al., 2016). However, the recognition of minor sulcal details—such as those associated with asymmetries—is surely challenging when considering that the brain and the endocranial table are separated by the connective tissues of the meninges and the cerebrospinal fluid. Frequently, the endocranial surface in fossil specimens is also damaged by taphonomic processes, which hampers a proper recognition of the original morphology. On endocasts, gross sulcal patterns can sometimes be recognized through bosses and depressions of the surface, but statements in this sense are more consistent when dealing with broad regions and samples. Localized and tiny differences between right and left sides in single specimens are definitely hard to interpret, and even harder to defend on a population or statistical ground. Interestingly, sulcal features of Broca's region (which is usually identified on endocasts as "Broca's cap," namely a bulging boss on the posterior region of the third frontal circumvolution—see Fig. 15.3) are better identified precisely on the hemisphere that is not dominant in terms of volume (that is, the left hemisphere). This raises the question of whether the sharper sulcal pattern is the result of localized cortical development or if, being the region smaller than the contralateral one, the sulci are simply less amassed and more recognizable to a visual inspection. In contrast, parietal and temporal regions are more identifiable as bosses on the left side, where the *planum temporale* and the perisylvian cortex are usually dominant (Toga and Thompson, 2003). However, taking into account that these morphologic and functional correspondences are difficult to resolve even for living humans (e.g., Amunts et al., 1999; Keller et al., 2009a,b), it is unlikely that concluding statements could be provided for fossil specimens.

Some additional cues can be found inside of the skull but outside of the brain, namely within the pyramids of the temporal bones, that protect and seclude the anatomic elements of the middle and inner ear. Auditory features and language are the functions more frequently associated with brain asymmetries, in terms of causes and adaptations (Toga and Thompson, 2003), and there is neuroanatomic evidence of human-derived traits, in this sense (Rilling, 2014). Interestingly, archaic human species like *Homo heidelbergensis*, the possible ancestor of both modern humans and Neandertals, probably displayed a middle ear with the same auditory sensitivity as our species, apparently adapted to speech frequencies (Martínez et al., 2004). Once more, this may suggest that our modern hemispheric specialization was shared with other species of the human genus. At least, this common hearing sensitivity is not in disagreement with this hypothesis.

CRANIOVASCULAR TRAITS

A complementary comment on fossils and endocranial asymmetries concerns the craniovascular features, namely those imprints left on the bone by the vascular network (Píšová et al., 2017) (see Fig. 15.3). The middle meningeal artery, the venous sinuses, the diploic channels, the emissary veins, and the endocranial microvessels are involved in blood flow, endocranial thermal regulation, and brain immunologic response (Bruner and Sherkat, 2008; Bruner et al., 2011; Bruner and Eisová, 2024). For most of these features, however, the pattern is roughly symmetrical on both sides, and asymmetries are apparently random or fluctuating, not associated with directional or consistent differences (Eisová et al., 2019). The only exception is the confluence of the sinuses, in which the right transverse sinus is often larger than the left one, receiving most of the blood from the superior sagittal sinus. Accordingly, the right sigmoid sinus and the right jugular vein are also larger than their left counterpart. This asymmetry can be due to the fact that the occipital cortex is larger on the left side (the occipital petalia), and hence the main blood flow runs on the opposite side because of minor mechanical resistance. Neandertals frequently display a more pronounced asymmetry from the midsagittal plane for this vascular drainage (Rosas et al., 2008), possibly because of the increased width of their endocranial cavity, stressing the rightward deviation of the superior sagittal sinus since the vault. Although different human species displayed differences in the prevalence or distribution of the craniovascular features, the general absence of patent asymmetries is a common rule. At present, despite these traits having crucial biologic functions, they do not seem to show any visible asymmetrical distribution, which is, of course, relevant information too. As a cautionary note, we should remark that also in this case subtle differences would be hard to detect because, beyond the limitations associated with fossil remains, the irregular and fuzzy vascular morphology is very difficult to quantify in terms of metrics.

FOSSIL SPECIES, COGNITION, AND BEHAVIOR

Most information on hemispheric specialization in extinct species comes from the study of fossil endocasts and concerns the gross volumetric dominance of specific cortical regions. Nonetheless, a few but relevant cues can also deal with the behavior of extinct human taxa, as inferred from the archaeological record (Wynn and Coolidge, 2016). In cognitive archaeology, cultural or behavioral traces are contrasted with current models in psychology to test specific skills as well as narrow and broad cognitive abilities (Bruner, 2022, 2024). In the context of cerebral asymmetries, a partial correlation with behavior can be found, for example, by investigating handedness in extinct hominids (Uomini, 2009). Brain asymmetries and handedness do not display a straightforward association (Toga and Thompson, 2003), but indeed hemispheric specialization and hand preference share evolutionary and functional factors (Cashmore et al., 2008). Of course, determining hand preference and handedness in fossil species can be a hard task, because of uncertainties associated with functional inferences and statistical limitations (Cashmore, 2009). Some information can be obtained by skeletal markers, interpreting bone asymmetries in terms of behavioral asymmetries (Lazenby, 2002). Stone tools can also provide specific cues in this sense (Pobiner, 1999). However, the most conclusive result probably came from unsuspected anatomic elements: the teeth. Robust evidence was associated with marks that can be found on the labial surface of incisors and canines, whose inclination depends on the hand used to manipulate the tool (Lozano et al., 2009). In fact, most extinct human species consistently employed their teeth for tool-use and tool-making, using their mouth as a "third hand," possibly because of a lack of cognitive specialization in the body-tool integration capacity (Bruner and Lozano, 2014). The use of the mouth and teeth to grasp leather or other materials while cutting or scratching with stone tools can frequently lead to mark imprints on the front side of the teeth, with a slope that depends upon the hand used for such a task (Fig. 15.5). These traces in modern humans are much more infrequent, even in hunter-gatherers. As far as we know from these kinds of imprints, all human species for which this evidence is available display a ratio of right-handed subjects that is similar to modern humans, and compatible with a population-based right-handedness (Lozano et al., 2017). Accordingly, the conclusion is the same as we have from paleoneurology: the pattern of behavioral asymmetry in fossil hominids is apparently similar to the figure we observe in *H. sapiens*. This can mean that all human species displayed the same scheme of laterality and behavioral specialization, or

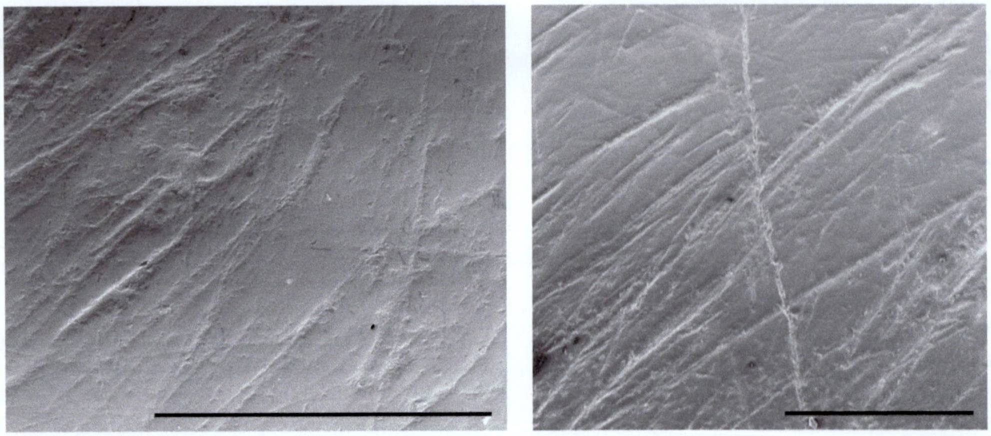

Fig. 15.5. Most extinct human species display many labial scratches on the front teeth due to the habitual (or even obligatory) use of the mouth as a third hand. The direction of the imprints reveals handedness and laterality. Scale bar 1 mm. Images courtesy of Marina Lozano.

else that differences, if any, cannot be properly detected through current paleontological (bones) and archaeological (culture) evidence.

FINAL REMARKS AND SOME OPEN QUESTIONS

At present, we have not identified consistent species-specific differences in the anatomic or behavioral asymmetries among extinct human species, or between extinct and extant humans. Discrete or outstanding gross anatomic differences between humans and apes are not apparent, and most hominoid variations in this sense can be easily interpreted through allometric patterns and scaling rules. It remains to be evaluated whether the size and patterns of the language areas and connections also fit a common allometric trend within hominoids, or else represent some autoapomorphic human traits (Rilling, 2014). In the case of living apes, a major statistical limitation to test this hypothesis may concern the gap between our brain size and the cerebral dimension of the other living species. With such a gap, regression models are difficult to handle, because humans are outliers, whose predicted values must be extrapolated out of the other primate's range. In this situation, subtle changes in the sample composition or metrics employed can profoundly influence the regression parameters. Also the fact that living apes are represented by a small number of species does not help to generate consistent quantitative models. In the case of fossil species, the first limitation is that we can only deal with indirect evidence (endocasts and behavioral traces), and the second concerns the scarce statistical power of a poor sample size. An additional issue deals with the fact that many interspecific differences in anatomy are a matter of averages for traits that, in general, display a pronounced degree of interspecific overlap. Mean differences (mostly when elusive) can support evolutionary hypotheses based on subtle effects of natural selection, but are more difficult to interpret when dealing with individual variation. For example, we give much importance to brain size in human evolution, forgetting that many individuals of *H. sapiens* have brain dimensions that are below the average of *H. heidelbergensis* or *Homo erectus*. In the case of brain asymmetries, differences can be particularly small and with uncertain associations with functional or cognitive aspects, a condition that, at best, requires caution.

On the one hand, we should take into consideration the hypothesis that hemispheric specialization in modern humans (*H. sapiens*) underwent some distinctive changes that we have been not able to disclose yet. Even if only by degree, an allometric effect can be determinant in terms of the amount of change (size matters!) or when trespassing some thresholds associated with emerging functions. The neuropsychologic literature is full of examples in which tiny biologic variations can generate outstanding cognitive or behavioral differences. In this case, we should anyway remember that extinct species are not necessarily primitive versions of modern humans, and they might have had their own specializations that we have lost or never evolved (Bruner, 2024). Therefore, it would be a mistake to think of fossil species as "equally" or "less" evolved than modern humans, because some of them could have been "diversely" evolved. This is particularly true when dealing with species that are not our direct ancestors, like Neandertals or *H. erectus* (see Fig. 15.1). Namely, fossil taxa could have had patterns of asymmetries different from our own, in terms of both anatomy and behaviors.

On the other hand, we should consider the alternative hypothesis, namely that our asymmetries are part of a plesiomorphic evolutionary package shared with all humans, or even with all hominoids. In this case, hemispheric specialization should be interpreted as a primitive (albeit crucial) feature, whose neurobiologic importance has long evolutionary roots.

What kind of features can be further investigated in the paleontological and archaeological record? It is unlikely that at present new evidence can provide noticeable changes to the current conclusions. Of course, in science, this is not the reason to give up searching. Unexpected results are the fuel of scientific progress and they are, tautologically, unexpected. As mentioned above, major cues on language or lateralization in extinct human species came not from brain morphology but from surprising anatomic elements: teeth and middle ear. For what concerns fossils, surely the main and general scope of paleontology is to increase the sample size, to provide more comprehensive samples, in terms of biogeographical, phylogenetic, and statistical representation. Although more difficult to quantify, the asymmetries in some regions (like the dorsal frontal and parietal cortex) have been less investigated and may merit further attention. The relationship between cerebral and cranial asymmetries may also be a topic of interest, for example, when dealing with the vascular system and blood supply.

Nonetheless, all these features can be hardly explored if we miss basic anatomic information on their variability, homology, and functions, in living taxa. At present, we need more research on macroscopic features such as sulcal organization, vessels, and bone layers. Such research is still insufficiently developed for our species, despite its possible implications for health and medicine (Bruner et al., 2014). In the case of petalia, for example, we should investigate the biomechanical, topological, and histological factors associated with endocranial asymmetries, before making inferences on their evolutionary or functional meaning. A commitment to investigate many basic aspects of our own anatomic and functional variations is probably a key requirement to assess what is right (or wrong) and what is left (to do) in terms of research on brain asymmetries and evolution.

ACKNOWLEDGMENTS

I am grateful to François Boller for the invitation to participate in this volume, and to Marina Lozano for her help with Fig. 15.5. This article is supported by the grant PID2021-122355NB-C33 financed by MCIN/AEI/10.13039/501100011033/ FEDER, UE. I declare no conflict of interest.

References

Amunts K, Schleicher A, Bürgel U et al. (1999). Broca's region revisited: cytoarchitecture and intersubject variability. J Comp Neurol 412: 319–341.

Balzeau A, Gilissen E (2010). Endocranial shape asymmetries in *Pan paniscus*, *Pan troglodytes* and *Gorilla gorilla* assessed via skull based landmark analysis. J Hum Evol 59: 54–69.

Balzeau A, Gilissen E, Grimaud-Hervé D (2012a). Shared pattern of endocranial shape asymmetries among great apes, anatomically modern humans, and fossil hominins. PLoS One 7: e29581.

Balzeau A, Holloway RL, Grimaud-Hervé D (2012b). Variations and asymmetries in regional brain surface in the genus *Homo*. J Hum Evol 62: 696–706.

Bastir M, Rosas A, O'Higgins P (2006). Craniofacial levels and the morphological maturation of the human skull. J Anat 209: 637–654.

Beaudet A, Dumoncel J, De Beer F et al. (2016). Morphoarchitectural variation in South African fossil cercopithecoid endocasts. J Hum Evol 101: 65–78.

Bogart SL, Mangin JF, Schapiro SJ et al. (2012). Cortical sulci asymmetries in chimpanzees and macaques: a new look at an old idea. Neuroimage 61: 533–541.

Bruner E (2015). Human paleoneurology, Springer, Cham.

Bruner E (2017). The fossil evidence of human brain evolution. In: J Kaas (Ed.), Evolution of nervous systems, vol. 4. Elsevier, Oxford, pp. 63–92.

Bruner E (2019). Human paleoneurology: shaping cortical evolution in fossil hominids. J Comp Neurol 527: 1753–1765.

Bruner E (2022). Prehistory, neuroscience, and evolutionary anthropology: a personal journey. J Anthropol Sci 100: 173–192.

Bruner E (2024). Cognitive archaeology, and the psychological assessment of extinct minds. J Comp Neurol 532: e25583.

Bruner E, Beaudet A (2023). The brain of *Homo habilis*: three decades of paleoneurology. J Hum Evol 174: 103281.

Bruner E, Eisová S (2024). Vascular microforamina and endocranial surface: normal variation and distribution in adult humans: vascular biology. Anat Rec. https://doi.org/10.1002/ar.2542.

Bruner E, Lozano M (2014). Extended mind and visuo-spatial integration: three hands for the Neandertal lineage. J Anthropol Sci 92: 273–280.

Bruner E, Sherkat S (2008). The middle meningeal artery: from clinics to fossils. Child's Nerv Sys 24: 1289–1298.

Bruner E, Mantini S, Musso F et al. (2011). The evolution of the meningeal vascular system in the human genus: from brain shape to thermoregulation. Am J Hum Biol 23: 35–43.

Bruner E, De la Cuétara JM, Masters M et al. (2014). Functional craniology and brain evolution: from paleontology to biomedicine. Front Neuroanat 8: 19.

Bruner E, Ogihara N, Tanabe HC (2019). Digital endocasts: from skulls to brains, Springer, Tokyo.

Cashmore L (2009). Can hominin "handedness" be accurately assessed? Ann Hum Biol 36: 624–641.

Cashmore L, Uomini N, Chapelain A (2008). The evolution of handedness in humans and great apes: a review and current issues. J Anthropol Sci 86: 7–35.

Corballis MC (2013). Early signs of brain asymmetry. Trends Cogn Sci 17: 554–555.

Corballis MC (2019). Evolution of cerebral asymmetry. Prog Brain Res 250: 153–178.

De Jager EJ, Van Schoor AN, Hoffman JW et al. (2019). Sulcal pattern variation in extant human endocasts. J Anat 235: 803–810.

De Jager EJ, Risser L, Mescam M et al. (2022). Sulci 3D mapping from human cranial endocasts: a powerful tool to study hominin brain evolution. Hum Brain Map 43: 4433–4443.

Dumoncel J, Subsol G, Durrleman S et al. (2021). Are endocasts reliable proxies for brains? A 3D quantitative comparison of the extant human brain and endocast. J Anat. 238: 480–488.

Eckert MA, Vaden Jr KI, Dyslexia Data Consortium (2019). A deformation-based approach for characterizing brain asymmetries at different spatial scales of resolution. J Neurosci Meth 322: 1–9.

Eisová S, Píšová H, Velemínský P et al. (2019). Normal craniovascular variation in two modern European adult populations. J Anat 235: 765–782.

Falk D (2012). Hominin paleoneurology: where are we now? Prog Brain Res 195: 255–272.

Falk D (2014). Interpreting sulci on hominin endocasts: old hypotheses and new findings. Front Hum Neurosci 8: 134.

Fournier M, Combès B, Roberts N et al. (2011). Mapping the distance between the brain and the inner surface of the skull and their global asymmetries. In: Medical imaging 2011: image processing, 79620Y.

Goldberg E, Roediger D, Kucukboyaci NE et al. (2013). Hemispheric asymmetries of cortical volume in the human brain. Cortex 49: 200–210.

Gómez-Robles A, Hopkins WD, Sherwood CC (2013). Increased morphological asymmetry, evolvability and plasticity in human brain evolution. Proc R Soc B 280: 20130575.

Gonzalez PN, Vallejo-Azar M, Aristide L et al. (2022). Endocranial asymmetry in New World monkeys: a comparative phylogenetic analysis of morphometric data. Brain Struct Funct 227: 469–477.

Gould SJ (1966). Allometry and size in ontogeny and phylogeny. Biol Rev 41: 587–638.
Gunz P, Mitteroecker P, Neubauer S et al. (2009). Principles for the virtual reconstruction of hominin crania. J Hum Evol 57: 48–62.
Hartwigsen G, Bengio Y, Bzdok D (2021). How does hemispheric specialization contribute to human-defining cognition? Neuron 109: 2075–2090.
Heilbroner PL, Holloway RL (1989). Anatomical brain asymmetry in monkeys: frontal, temporoparietal, and limbic cortex in *Macaca*. Am J Phys Anthropol 80: 203–211.
Hervé PY, Zago L, Petit L et al. (2013). Revisiting human hemispheric specialization with neuroimaging. Trends Cogn Sci 17: 69–80.
Holloway RL (1981a). The Indonesian *Homo erectus* brain endocasts revisited. Am J Phys Anthropol 55: 503–521.
Holloway RL (1981b). Volumetric and asymmetry determinations on recent hominid endocasts: Spy I and II, Djebel Ihroud I, and the Salé *Homo erectus* specimens, with some notes on neandertal brain size. Am J Phys Anthropol 55: 385–393.
Holloway RL, De La Coste-Lareymondie MC (1982). Brain endocast asymmetry in pongids and hominids: some preliminary findings on the paleontology of cerebral dominance. Am J Phys Anthropol 58: 101–110.
Holloway RL, Broadfiel DC, Yuan M (2004). The human fossil record, volume three: brain endocasts—the paleoneurological evidence, Alan R. Liss, New York.
Keller SS, Crow T, Foundas A et al. (2009a). Broca's area: nomenclature, anatomy, typology and asymmetry. Brain Lang 109: 29–48.
Keller SS, Roberts N, Hopkins W (2009b). A comparative magnetic resonance imaging study of the anatomy, variability, and asymmetry of Broca's area in the human and chimpanzee brain. J Neurosci 29: 14607–14616.
Klingenberg CP (1998). Heterochrony and allometry: the analysis of evolutionary change in ontogeny. Biol Rev 73: 79–123.
Klingenberg CP (2010). Evolution and development of shape: integrating quantitative approaches. Nat Rev Gen 11: 623–635.
Klingenberg CP, Barluenga M, Meyer A (2002). Shape analysis of symmetric structures: quantifying variation among individuals and asymmetry. Evolution 56: 1909–1920.
Kyriacou A, Bruner E (2011). Brain evolution, innovation, and endocranial variations in fossil hominids. Paleoanthropology 2011: 130–143.
Labra N, Mounier A, Leprince Y et al. (2024). What do brain endocasts tell us? A comparative analysis of the accuracy of sulcal identification by experts and perspectives in palaeoanthropology. J Anat 244: 274–296.
Lazenby RA (2002). Skeletal biology, functional asymmetry and the origins of "handedness". J Theor Biol 218: 129–138.
Leamy LJ, Klingenberg C (2005). The genetics and evolution of fluctuating asymmetry. Ann Rev Ecol Evol Syst 36: 1–21.
LeMay M (1982). Morphological aspects of human brain asymmetry: an evolutionary perspective. Trends Neurosci 5: 273–275.
Lieberman DE, Ross CF, Ravosa MJ (2000). The primate cranial base: ontogeny, function, and integration. Am J Phys Anthropol 113: 117–169.
Lozano M, Mosquera M, Bermúdez de Castro JM et al. (2009). Right handedness of *Homo heidelbergensis* from Sima de los Huesos (Atapuerca, Spain) 500,000 years ago. Evol Hum Behav 30: 369–376.
Lozano M, Estalrrich A, Bondioli L et al. (2017). Right-handed fossil humans. Evol Anthropol 26: 313–324.
Martínez I, Rosa M, Arsuaga JL et al. (2004). Auditory capacities in Middle Pleistocene humans from the Sierra de Atapuerca in Spain. Proc Natl Acad Sci USA 101: 9976–9981.
Moss ML, Young RW (1960). A functional approach to craniology. Am J Phys Anthropol 18: 281–292.
Neubauer S (2014). Endocasts: possibilities and limitations for the interpretation of human brain evolution. Brain Behav Evol 84: 117–134.
Neubauer S, Gunz P, Scott NA et al. (2020). Evolution of brain lateralization: a shared hominid pattern of endocranial asymmetry is much more variable in humans than in great apes. Sci Adv 6: eaax9935.
Píšová H, Rangel de Lázaro G, Velemínský P et al. (2017). Craniovascular traits in anthropology and evolution: from bones to vessels. J Anthropol Sci 95: 35–65.
Pobiner BL (1999). The use of stone tools to determine handedness in hominids. Curr Anthropol 40: 90–92.
Rilling JK (2006). Human and nonhuman primate brains: are they allometrically scaled versions of the same design? Evol Anthropol 15: 65–77.
Rilling JK (2014). Comparative primate neurobiology and the evolution of brain language systems. Curr Opin Neurobiol 28: 10–14.
Rosas A, Peña-Melián A, García-Tabernero A et al. (2008). Endocranial occipito-temporal anatomy of SD-1219 from the Neandertal El Sidrón site (Asturias, Spain). Anat Rec 291: 502–512.
Schuurman T, Bruner E (2024). An inclusive anatomical network analysis of human craniocerebral topology. J Anat. https://doi.org/10.1111/joa.14068.
Stevens NJ, Seiffert ER, O'Connor PM et al. (2013). Palaeontological evidence for an Oligocene divergence between Old World monkeys and apes. Nature 497: 611–614.
Thiebaut de Schotten M, Beckmann CF (2022). Asymmetry of brain structure and function: 40 years after Sperry's Nobel Prize. Brain Struct Funct 227: 1–4.
Tobias PV (1987). The brain of *Homo habilis*: a new level of organization in cerebral evolution. J Hum Evol 16: 741–761.
Toga AW, Thompson PM (2003). Mapping brain asymmetry. Nat Rev Neurosci 4: 37–48.
Uomini NT (2009). The prehistory of handedness: archaeological data and comparative ethology. J Hum Evol 57: 411–419.
Williams CM, Peyre H, Toro R et al. (2022). Comparing brain asymmetries independently of brain size. Neuroimage 254: 119118.
Wood B, Richmond BG (2000). Human evolution: taxonomy and paleobiology. J Anat 197: 19–60.
Wynn T, Coolidge FL (2016). Archeological insights into hominin cognitive evolution. Evol Anthropol 25: 200–213.

Handbook of Clinical Neurology, Vol. 208 (3rd series)
Cerebral Asymmetries
C. Papagno and P. Corballis, Editors
https://doi.org/10.1016/B978-0-443-15646-5.00029-4

Chapter 16

Large-scale genetic mapping for human brain asymmetry

ZHIQIANG SHA[1] AND CLYDE FRANCKS[1,2*]

[1]*Language & Genetics Department, Max Planck Institute for Psycholinguistics, Nijmegen, The Netherlands*

[2]*Department of Cognitive Neuroscience & Donders Community for Medical Neuroscience, Radboud University Medical Center, Nijmegen, The Netherlands*

Abstract

Left–right asymmetry is an important aspect of human brain organization for functions including language and hand motor control, which can be altered in some psychiatric traits. The last 5 years have seen rapid advances in the identification of specific genes linked to variation in asymmetry of the human brain and/or handedness. These advances have been driven by a new generation of large-scale genome-wide association studies, carried out in samples ranging from roughly 16,000 to over 1.5 million participants. The implicated genes tend to be most active in the embryonic and fetal brain, consistent with early developmental patterning of brain asymmetry. Several of the genes encode components of microtubules or other microtubule-associated proteins. Microtubules are key elements of the internal cellular skeleton (cytoskeleton). A major challenge remains to understand how these genes affect, or even induce, the brain's left–right axis. Several of the implicated genes have also been associated with psychiatric or neurologic disorders, and polygenic dispositions to autism and schizophrenia have been associated with structural brain asymmetry. Knowledge of developmental mechanisms that lead to hemispheric specialization may ultimately help to define etiologic subtypes of brain disorders.

INTRODUCTION

Population-level asymmetries of anatomy and function arise in the human brain during fetal development, and behavioral asymmetry has been reported as early as 10 weeks of gestational age (Hepper et al., 1998, 2005; Kasprian et al., 2011; Abu-Rustum et al., 2013; Duboc et al., 2015; Taymourtash et al., 2023). Prenatal asymmetries include left-lateralized functional connectivity of the temporal lobe (Taymourtash et al., 2023) and right-lateralized predominance of arm movements (Hepper, 2013). These developmental asymmetries are likely to be precursors of left-hemispheric language dominance and right-handedness. The early appearance of brain and behavioral asymmetries indicates a genetically regulated program of left–right axis formation in the central nervous system (Francks, 2015; Gunturkun and Ocklenburg, 2017; de Kovel et al., 2019; Francks, 2019; Sha et al., 2021b).

Consistent with a genetic origin, twin- and family-based analyses have reported heritability of up to roughly 30% for measures of brain structural or functional asymmetry, particularly those linked to regions or networks important for language (Guadalupe et al., 2017; Kong et al., 2018; Somers et al., 2015; Labache et al., 2023), although heritability of hemispheric language dominance has not always been found (Bishop and Bates, 2019). Left-handedness also has a heritability of roughly 25% in twin-based analysis (Medland et al., 2009) and 1%–6% in population studies that have examined the contribution of common genetic variants specifically (de Kovel and Francks, 2019; Cuellar-Partida et al., 2021).

In addition, studies have measured molecular asymmetries in small numbers of postmortem tissue samples from matched left and right brain regions of the prenatal or adult central nervous system, in terms of messenger RNA, protein, or DNA methylation (Sun et al., 2005; Karlebach and Francks, 2015; de Kovel et al., 2017, 2018; Muntané et al., 2017; Ocklenburg et al., 2017; Zhao et al., 2022b). These postmortem studies have been reviewed previously (Schmitz et al., 2019), and we will not cover them again here. They have provided snapshots of potential molecular asymmetries in the human brain at specific life stages, including the possibility of left–right differences of maturation rates during fetal development, and asymmetries of glutamate receptor expression in the adult cerebral cortex.

In this chapter, we will instead review recent studies that have associated particular genetic variants with interindividual differences in brain asymmetry or handedness. After roughly two decades during which these kinds of studies produced tentative findings in

*Correspondence to: Clyde Francks, Language & Genetics Department, Max Planck Institute for Psycholinguistics, Nijmegen, The Netherlands. Tel: +31-24-3521929, E-mail: clyde.francks@mpi.nl

limited sample sizes (reviewed by Schmitz et al., 2019; de Kovel and Francks, 2019), progress was suddenly accelerated during the past 5 years by large-scale genome-wide association analyses, carried out in general population samples ranging from roughly 16,000 to over 1.5 million participants (de Kovel and Francks, 2019; Wiberg et al., 2019; Carrion-Castillo et al., 2020; Le Guen et al., 2020; Cuellar-Partida et al., 2021; Kong et al., 2021; Sha et al., 2021a,b; Zhao et al., 2022a). We will focus on the findings from this new generation of large-scale genetic association studies.

Altered brain asymmetries have been reported in various psychiatric disorders (DeLisi et al., 1997; Shenton et al., 2001; De Fosse et al., 2004; Herbert et al., 2005; Kawasaki et al., 2008; Schizophrenia Working Group of the Psychiatric Genomics Consortium, 2014; Carper et al., 2016; Okada et al., 2016; Markou et al., 2017; Ravichandran et al., 2017; Sun et al., 2017; Grove et al., 2019; Postema et al., 2019; Floris et al., 2021; Kong et al., 2022), including by recent consortium studies that have helped to overcome some of the statistical uncertainty of earlier literature (Postema et al., 2019; Kong et al., 2020; Postema et al., 2021; Schijven et al., 2023). The consortium studies have found various aspects of brain structural asymmetry to be subtly altered in autism (Postema et al., 2019), schizophrenia (Schijven et al., 2023), obsessive-compulsive disorder (Kong et al., 2020), and attention deficit-hyperactivity disorder (Postema et al., 2021). In addition, meta-analyses have indicated that left-handedness occurs at increased rates in neurodevelopmental and psychiatric conditions, including intellectual disability (Papadatou-Pastou and Tomprou, 2015), autism (Markou et al., 2017), schizophrenia (Hirnstein and Hugdahl, 2014), and reading or language impairment (Abbondanza et al., 2023). These associations suggest that population-typical asymmetries are linked to optimal brain function and raise the possibility of shared genetic contributions to brain asymmetry and disorders. Toward the end of this chapter, we will summarize findings from recent large-scale genetic studies that have supported such a link between brain asymmetry and disorders.

GENOME-WIDE ASSOCIATION STUDIES OF BRAIN STRUCTURAL ASYMMETRY

In 2020, the first two large-scale genome-wide association studies of specific aspects of human brain structural asymmetry were published (Carrion-Castillo et al., 2020; Le Guen et al., 2020). These studies made use of the UK Biobank database (Alfaro-Almagro et al., 2018; Bycroft et al., 2018), which at that time included brain imaging and genetic data from roughly 16,000–18,000 participants, depending on study-specific details of MRI processing and quality control. The UK Biobank is a general population dataset of middle-aged to older adults, and the imaging genetics dataset has since roughly doubled in size (and further data collection and releases are on-going). It became a mainstay for many of the studies described further below, as it is unique in terms of homogeneous data collection from such a large sample size.

The study by Carrion-Castillo et al. (2020) targeted asymmetry of the *planum temporale*, a region of the superior temporal auditory and language cortex whose asymmetry has long been studied in relation to possible associations with hemispheric language dominance, reading- and language-related impairments, and other brain disorders (Sommer et al., 2001; Knaus et al., 2010; Altarelli et al., 2014; Tzourio-Mazoyer et al., 2018). *Planum temporale* volume asymmetry was found to have a significant heritability of roughly 14%, and two genomic loci were associated with its variability, one of which was a protein-altering variant within the gene ITIH5 (Carrion-Castillo et al., 2020). The functions of this gene are not well understood, but it may be involved in stabilization of the extracellular matrix (the network of proteins and other molecules that surround and support cells within tissues), and it has high expression in endothelial cells, which suggests a role in the blood–brain barrier. The mechanisms by which *ITIH5* might affect asymmetry of the superior temporal cortex are not known, especially as this gene is relatively uniformly expressed across the cortex. Interactions of the ITIH5 protein with other, as-yet-unidentified proteins that show more regional and hemisphere-specific expression patterns may underlie its association with *planum temporale* asymmetry. In principle, such processes may occur developmentally—for example, during fetal stages—and/or later in life with respect to age-related changes of asymmetry during adulthood (Roe et al., 2021). We will return to these possibilities further below.

The study by Le Guen et al. (2020) focused on the anatomy of the superior temporal asymmetric pit, a part of the superior temporal sulcus that is located beneath the primary auditory cortex of Heschl's gyrus. The superior temporal asymmetric pit may be a human-specific feature that was involved in the evolution of language. Its depth on the left hemisphere was found to associate with a locus near to the gene *DACT1,* which was also the case on the right hemisphere, but to a lesser extent than the left. This pattern resulted in an association of the *DACT1* locus with asymmetry. *DACT1* is involved in regulation of intracellular signaling pathways during development, which may include very early embryonic morphogenesis as well as later neuronal differentiation.

Unlike *ITIH5*, where a protein-altering genetic variant was associated with structural brain asymmetry, the genetic variant close to *DACT1* does not alter a protein sequence and is more likely to exert a regulatory effect on gene expression at the level of messenger RNA. This means that there is some uncertainty about whether the association of the *DACT1* locus with brain asymmetry is in fact mediated through *DACT1* itself, or through another nearby gene in the genome whose expression might also be affected by the same genetic variant. In fact, most of the genetic associations found in the studies described in this chapter are of this latter type, i.e., where there is some uncertainty about the causal gene pertaining to any given genetic association. The effect at ITIH5 is relatively unusual in this respect, as it directly implicates a specific gene through altering its protein sequence. The uncertainty at most loci can be partly mitigated through the use of gene set (or pathway) analysis, as will be seen below.

In 2021, a genome-wide association study of human structural brain asymmetry was published (Sha et al., 2021b) that had three important features that enabled more extensive discovery than previously. First, the study was based on an updated release of the UK Biobank data that yielded a sample size of over 32,000 individuals—roughly twice that of previous studies—with a consequent boost to statistical power to detect genetic effects. Second, the study considered structural asymmetries of gray matter thickness and surface area for regions spanning the entire cerebral cortex, as well as volumetric asymmetries of subcortical structures (including the thalamus, hippocampus, and striatal structures), rather than focusing on just one region of the brain. Third, the study used multivariate genetic association mapping, where each variant in the genome was tested for its simultaneous association with all heritable regional asymmetry measures. This approach explicitly modeled how each genetic locus can affect asymmetry to a greater or lesser extent across multiple brain regions, which again likely enhanced the statistical power to detect genetic effects.

Sha et al. (2021b) calculated hemispheric asymmetry indexes as (left–right)/((left+right)/2) for each pair of corresponding left-to-right structural measures in the UK Biobank data and found that 42 regional asymmetries showed significant heritability based on common single nucleotide variants in the genome (i.e., with population frequencies above 1%). The heritability ranged from only 2.2% for the asymmetry of entorhinal cortical thickness to 9.4% for the asymmetry of the superior temporal surface area. The overall pattern across regional asymmetry measures was consistent with previous twin-based heritability analyses (Guadalupe et al., 2017; Kong et al., 2018), although twin-based heritability had been reported as somewhat higher, up to roughly 25% for certain brain regional asymmetries. Twin-based heritability is often measured to be higher than that based on common single nucleotide variants alone, which is unsurprising because common single nucleotide variants are just one class of genetic variation (others include rarer mutations and larger genomic structural variants). In addition, twin studies can overestimate heritability when certain assumptions are not fully met (Young, 2019).

In multivariate genome-wide association testing, Sha et al. (2021b) found 27 independent genetic variants that were significantly associated with different aspects of structural brain asymmetry (Fig. 16.1). Strikingly, almost half of these genetic loci implicate genes that either code for components of microtubules called *tubulins*, or else other proteins that directly interact with, or modify, microtubules. Microtubules are prominent parts of the cytoskeleton—the framework of protein filaments internal to cells—that contribute to a wide range of processes including cellular growth, division, migration, shape and axis formation, axon outgrowth and intracellular transport (Janke and Magiera, 2020).

Gene-set enrichment analysis confirmed that genetic variants associated with structural brain asymmetry tend to be located in the genome close to, or within, microtubule-related genes more often than expected by chance. Therefore, even though the causal gene might be uncertain at any single locus (see above), the overall tendency for the loci to be within or near microtubule-related genes means that the causal gene at any given locus is likely to be the nearest microtubule-related gene. Specific microtubule-related genes implicated in structural brain asymmetry by Sha et al. (2021b) code for the tubulins TUBB, TUBB3, TUBB4A, and TUBA1A/TUBA1B/TUBA1C (the genes for these latter three proteins are clustered together on chromosome 12), as well as microtubule-associated proteins MAPRE3, AGBL5, MAP2, DAAM1, and MAPT.

Other loci associated with structural brain asymmetry by Sha et al. (2021b), but which did not directly implicate microtubule components or associated proteins, included the DACT1 and ITIH5 loci that were previously reported (see above), as well as nine others known to affect brain phenotypes or development. The DACT1 locus was now shown to associate with not only asymmetry of the language-related superior temporal sulcus and transverse temporal (Heschl's) gyrus, but also asymmetry of the pericalcarine cortex, which is located medially in the occipital lobe and best known for its functions in the visual system. In fact, almost all of the loci identified by Sha et al. (2021b) were associated with asymmetries of more than one brain region, and the regions were often spatially separated for any given genetic locus (Fig. 16.1). This suggests that the asymmetries of spatially distinct regions might share common ontological mechanisms, and/or that some mechanisms which underlie adult age-related changes of asymmetry are shared across distinct brain regions.

Given the increased sample size of Sha et al. (2021b) compared to previous studies, as well as the multivariate approach of Sha et al. (2021b), the ITIH5 locus was even more significantly associated with brain structural asymmetry than before, with a point-wise P value of 4.75×10^{-38}. Note that such a low P value does not indicate a large effect—rather, it reflects a robust, small-effect association in a sample size of over 32,000 individuals. Other genes implicated by Sha et al. (2021b) in structural brain asymmetry included *ZIC4,* which is involved in visual and auditory pathway development, *NR2F1* involved in neural activity during cortical patterning, *BBS9* which causes retinopathy and intellectual disability when mutated, *FLRT3* which regulates axon guidance and excitatory synapse development, and *COL18A1* involved in neural tube closure during embryonic development.

When analyzing their results jointly with postmortem mRNA expression data from human brain samples across the lifespan, Sha et al. (2021b) discovered that genes associated with interindividual variation in structural brain asymmetry in adults tend to be most strongly expressed during early and mid-prenatal stages. This suggests that at least some of the heritable variation in adult structural brain asymmetry is established early in life, and therefore that it remains partly stable throughout life. Indeed, some of the implicated genes—including the microtubule-related genes—may be involved in developmental mechanisms that create the left–right axis in the central nervous system, possibly within the first month after conception. Others of the implicated genes may then help to implement a downstream program of asymmetric brain development and/or lifespan change, after the initial establishment of the left–right axis. We will return to these possibilities further below.

In addition to the above studies that focused on the genetics of regional measures of structural brain asymmetry (in univariate or multivariate analytic contexts), two recent studies have focused on the genetics of global "torque" of the human brain (Kong et al., 2021; Zhao et al., 2022a). Torque refers to a population-level tendency for the right cerebral hemisphere to be situated slightly

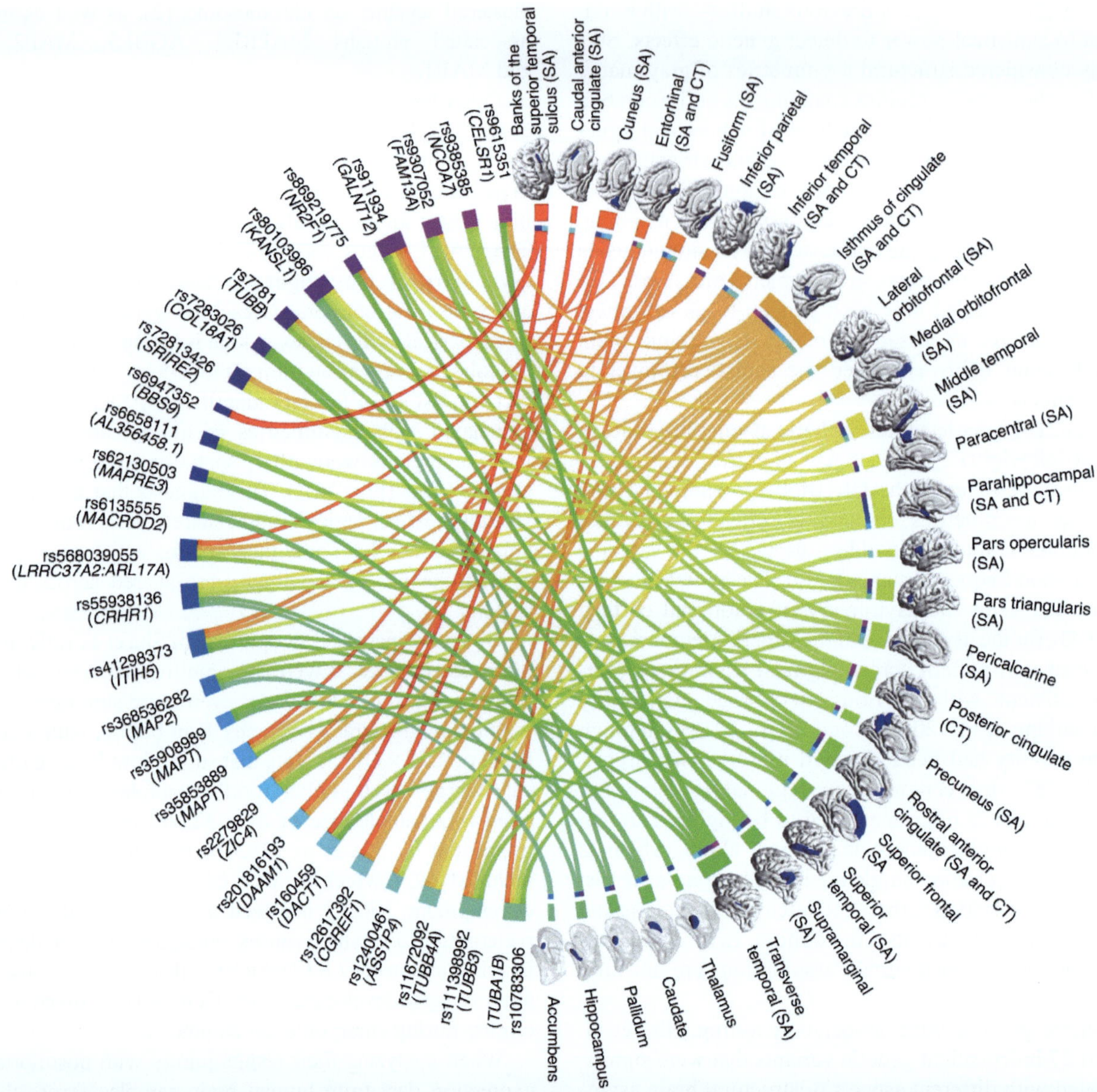

Fig. 16.1. Overview of genetic variants identified by genome-wide scanning for human structural brain asymmetry in 32,256 adult individuals. The genetic variants are indicated on the left of the circle, and they are connected by lines to the specific regional asymmetries with which they showed multivariate associations. Different colors indicate different variants or regional asymmetries. Lines linking variants to regional asymmetries are colored according to the regions. The nearest gene to each variant is shown. SA, surface area; CT, cortical thickness; SUB, subcortical volume. Figure reproduced from Sha Z, Schijven D, Carrion-Castillo A, et al. (2021b). The genetic architecture of structural left–right asymmetry of the human brain. Nat Hum Behav 5: 1226–1239.

anteriorly to the left, sometimes with the anterior right and occipital left hemispheres crossing over the midline. This pattern results in a counterclockwise skew when visualized from above (Toga and Thompson, 2003).

Kong et al. (2021) worked with the UK Biobank data to measure brain skew in the horizontal plane, using an approach based on registration to an artificially symmetrized atlas. They also confirmed a related skew in the vertical plane, whereby the right hemisphere is positioned slightly dorsally relative to the left in the majority of people. Both horizontal and vertical skews showed considerable but largely uncorrelated interindividual variation, and each was independently associated with handedness (the average skew in left-handers was reduced in the horizontal plane, but increased in the vertical plane). These average patterns and associations were also confirmed by Kong et al. (2021) in other smaller datasets. However, although the two skew measures showed SNP-based heritabilities of 4%–13%, they also showed substantial polygenicity in causal mixture model analysis, and no individually significant loci were found in genome-wide association analysis based on nearly 34,000 individuals from the UK Biobank. An even larger sample will probably be needed to reveal individually significant loci for brain torque as measured in this way.

Zhao et al. (2022a) worked with an earlier (smaller) release of the UK Biobank data, in combination with other datasets, for a total of over 24,000 individuals. They applied three-dimensional brain shape analysis to measure multiple aspects of torque, which included frontal and occipital protrusion and midline crossing, bending, dorsoventral shift, tissue distribution asymmetries, and cerebral cortical surface positional asymmetries. All together, this resulted in hundreds of different structural brain measures, some of which showed associations with age, sex, handedness or verbal-numerical reasoning. Zhao et al. (2022a) then carried out univariate genome-wide association scanning for each of these measures separately, which produced numerous individual loci at a lenient threshold for multiple testing correction (i.e., when correcting for multiple testing across the genome but not for the number of brain measures). Only two loci were significant after strict multiple testing correction, which were not in the immediate vicinity of protein-coding genes. Again, an even larger sample will be needed to firmly implicate specific loci in brain torque.

GENOME-WIDE ASSOCIATION STUDIES OF HANDEDNESS

Roughly 90% of the population is right-handed and 10% left-handed, and this bias has been broadly consistent across cultures, ethnicities, and human history, with a limited degree of geographic and temporal variation (Coren and Porac, 1977; Perelle and Ehrman, 1994; Faurie and Raymond, 2004; McManus, 2009; Papadatou-Pastou et al., 2020). Handedness is a prominent behavioral manifestation of brain asymmetry, because right-handedness reflects left-hemisphere dominance for hand preference, and vice versa (Willems et al., 2014). As handedness is relatively easily measured by questionnaire, the sample sizes available for genetic studies have been much larger than MRI-based studies of brain asymmetry.

In 2019, the first two large-scale genome-wide association studies of human handedness were published (de Kovel and Francks, 2019; Wiberg et al., 2019), both based on the UK Biobank data, which permitted sample sizes of over 30,000 left-handers and 300,000 right-handers. As both studies were based on the same dataset, they both obtained similar results, although the findings varied slightly due to differences in quality control parameters, methodological choices, confound variables to include, and so on. These studies identified three or four loci that surpassed the threshold for statistical significance in the context of genome-wide multiple testing. The implicated genes included *TUBB*, which encodes a tubulin subunit, and *MAP2* and *MAPT*, which encode microtubule-associated proteins. The exact number of loci depended on the inclusion or exclusion of around 5000 individuals who reported using both hands equally. [This "ambidextrous" phenotype was found to be unstable in individuals who reported their handedness on multiple visits to a UK Biobank assessment center (de Kovel et al., 2019)].

Then, in 2021, a genome-wide association meta-analysis study was published based on data from the UK Biobank in addition to many other datasets from around the world, for a total of 194,198 left-handed and 1,534,836 right-handed participants (Cuellar-Partida et al., 2021). The greater statistical power of this larger study resulted in 41 genomic loci being associated with left-handedness, including at least eight that implicated microtubule-related genes and others involved in axon development and neurogenesis. A separate analysis of 37,637 ambidextrous vs 1,422,823 right-handed individuals produced 7 genomic loci, with some overlap with those from the left-handed vs right-handed analysis.

Like de Kovel and Francks (2019), the larger study by Cuellar-Partida et al. (2021) found no support for genes that were implicated by very much smaller studies over the preceding decades. This has been a feature of human genetics research in general, i.e., that large-scale genome-wide association studies have often not supported earlier findings from smaller samples. For some common diseases, the paradigm shift in scope and scale of genetic studies began already around 2007 (Burton et al., 2007), whereas for handedness and brain asymmetry the period 2019–2021 was the turning point. Genome-wide association studies in large samples have at least two crucial benefits for finding robust results. First, a large sample size improves statistical power, which is important for detecting effects and also increases the likelihood that statistically significant effects are true (Button et al., 2013). Second, a genome-wide approach is unconstrained by prior hypotheses about trait etiology. This is advantageous when studying complex, multifactorial traits like handedness or brain asymmetry, for which no single mechanism is likely to underlie the heritable variance.

GENE-BRAIN-HANDEDNESS ASSOCIATIONS

The findings summarized above refer to gene-brain or gene-handedness associations. It is also of interest to identify three-way associations—between genetic variants, brain structural/functional measures, and handedness as a behavioral variable. Such gene-brain-behavior associations can be especially informative about molecular and neurobiologic mechanisms underlying lateralized behavior.

In the UK Biobank data, Wiberg et al. (2019) examined temporal correlation data derived from resting-state functional MRI and found that left-handedness was associated with (i) decreased functional connectivity between core regions of the left-hemisphere language network, (ii) increased functional connectivity between the homologous regions of the right hemisphere, (iii) increased functional connectivity between the left and right counterparts. A similar pattern has also been reported in independent data (Labache et al., 2020). Wiberg et al. (2019) then found that a genetic variant associated with left-handedness in their genome-wide scan—rs199512—was also associated with white matter structural connectivity in tracts linking core regions of the language network. In addition, this genetic variant is associated with the expression level of the gene that encodes microtubule-associated protein MAPT, although the variant may also influence the expression of other genes, and is located within the genomic locus of *WNT3*. In fact, this locus on chromosome 17q21 is unusual in having especially long-range linkage disequilibrium (nonindependent co-occurrence of alleles at different polymorphic sites), which relates partly to a common inversion polymorphism that spans almost one million DNA bases and several genes (Alves et al., 2015). The *MAPT* locus is particularly complex in terms of pinpointing the causal gene, or genes, that might mediate the effects of genetic variants on brain and behavioral traits. This locus is also associated with a multitude of different structural and functional brain traits, not limited to asymmetry or language regions/connections (Smith et al., 2021). We will revisit the *MAPT* locus later when discussing genetic links between brain asymmetry and brain disorders.

The largest genome-wide study of structural brain asymmetry (Sha et al., 2021b) and that of handedness (Cuellar-Partida et al., 2021) identified several genetic loci in common, which included the *MAPT* locus and several other microtubule-related genes: *MAP2*, *TUBA1A/TUBA1B/TUBA1C*, *TUBB*, *TUBB3*, and *TUBB4A*. These concordant findings firmly established microtubule biology in the heritability of both brain asymmetry and behavioral asymmetry. They also motivated a subsequent study in the UK Biobank that integrated all three levels of analysis: genomic variation, brain structural asymmetry, and handedness (Sha et al., 2021a), which we will now discuss in some detail.

Despite decades of previous research, the cerebral cortical structural correlates of handedness have remained uncertain, likely due to varying methods and limited sample sizes (as summarized by Sha et al., 2021a). Even a relatively large study of 608 left-handers vs 7243 right-handers did not find clear evidence for altered cerebral cortical anatomy in left-handedness (Kong et al., 2018). The latter study screened the cerebral cortex using atlas-defined regions (34 regions per hemisphere), but the possibility remained that structural correlates of handedness were too focal to be captured by the relatively large parcellations defined in that brain atlas.

Therefore, Sha et al. (2021a) mapped cerebral cortical structural asymmetry with respect to handedness using a higher-resolution, atlas-free approach in 3062 left-handers and 28,802 right-handers from the UK Biobank. They measured asymmetries of cortical surface area and thickness at each of 8681 "vertices" in each hemisphere, where vertex-wise correspondence between the left and right hemispheres was achieved through resampling each individual's cortical surface model to a symmetric template (Greve et al., 2013; Maingault et al., 2016). This approach aligned cortical folding patterns across individuals and hemispheres.

Sha et al. (2021a) found that left-handed people had very subtle but statistically significant group-average alterations of cortical surface area asymmetry, or cortical thickness asymmetry, of quite focal "clusters" located within the fusiform, anterior insula, precentral, postcentral, anterior-middle-cingulate and inferior occipital cortex (Fig. 16.2). Strikingly, for all of these clusters, the alterations were consistent with a relative shift of neural resources to the right hemisphere in left-handers, i.e., to the hemisphere that controls the preferred hand in left-handers. Functional annotation of the regions with altered *surface area* asymmetry in left-handers, based on meta-analyzed functional MRI data, pointed to their involvements in executive functions including working memory, as well as language and reading, mood and pain perception. In contrast, functional annotation of the cortical regions with altered *thickness* asymmetry in left-handers indicated hand sensorimotor functions (driven by a cluster in the primary sensorimotor cortex of the postcentral gyrus) and vision (driven by a cluster located in the occipital lobe).

Sha et al. (2021a) then showed that 18 of the 41 handedness-associated genomic loci reported by Cuellar-Partida et al. (2021)

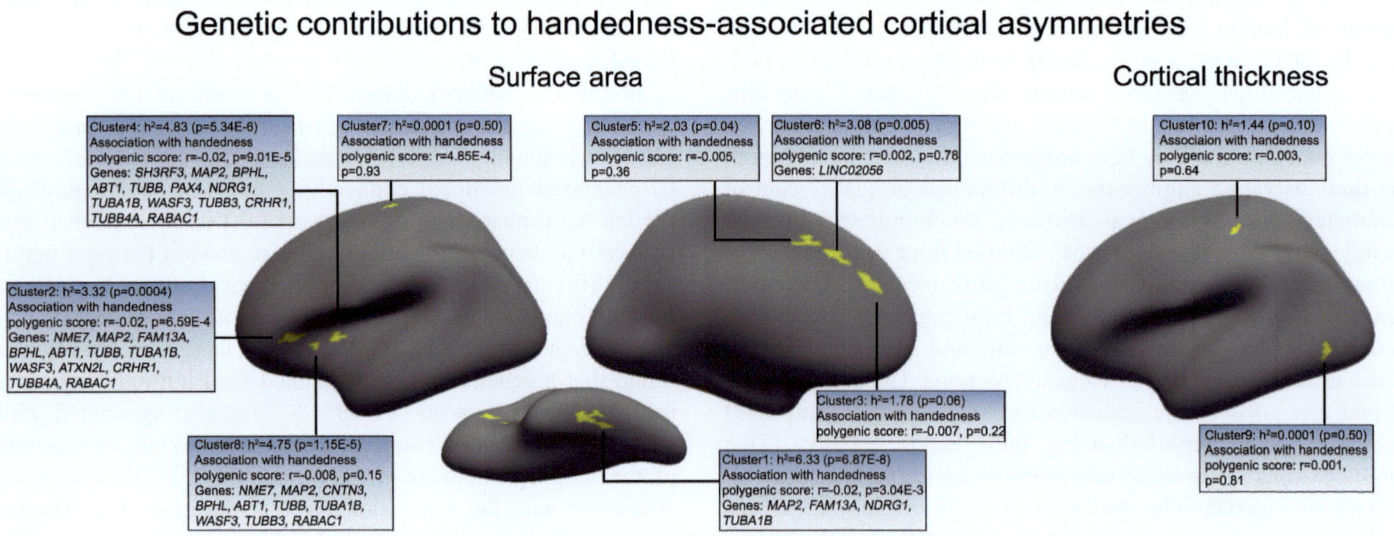

Fig. 16.2. Genetic effects on human handedness also associate with specific cerebral cortical asymmetries that are linked to handedness in 31,864 adult individuals. Ten distinct cortical clusters showed rightward shifts of surface area or thickness asymmetry in left-handers. For surface area asymmetry, five of the clusters showed significant heritability after false-discovery-rate correction (i.e., three clusters in the anterior insular cortex [clusters 2, 4, and 8], one in the fusiform cortex [cluster 1], and one in the anterior-middle-cingulate cortex [cluster 6]). Other handedness-associated asymmetry clusters were not significantly heritable, including the pre- and postcentral clusters of the sensorimotor cortex (clusters 7 and 10) and the inferior occipital cortex (cluster 9). The associations of cortical asymmetries with polygenic disposition to left-handedness are indicated, as well as individual handedness-associated genes implicated in their variabilities. Figure reproduced from Sha Z, Pepe A, Schijven D et al. (2021a). Handedness and its genetic influences are associated with structural asymmetries of the cerebral cortex in 31,864 individuals. Proc Natl Acad Sci U S A 118: e2113095118.

were also associated with at least one of the cortical surface area asymmetry clusters that are linked to left-handedness, thereby identifying a set of gene-brain-behavior associations (Fig. 16.2). The implicated genes again included several that encode components of microtubules, TUBB, TUBA1A/TUBA1B/TUBA1C, TUBB3, and TUBB4A, as well as microtubule-associated proteins MAP2, MAPT, and NME7. Interestingly, recessive mutations in *NME7* cause *situs inversus totalis*, a rare condition in which the visceral organs (heart, liver, etc.) are reversed in their orientation on the left–right axis (Reish et al., 2016). We will discuss visceral laterality further below (see also Chapter 5 in this volume). Other genes implicated by Sha et al. (2021a) to affect both handedness and its cerebral cortical correlates included *CNTN3*, which mediates cell surface interactions during nervous system development (Mohebiany et al., 2014), and *ATXN2L*, implicated in macrocephaly (Alzahrani et al., 2021).

For each of the UK Biobank participants with brain MRI and genetic data, Sha et al. (2021a) also calculated their polygenic disposition to left-handedness, based on the combined effects of variants all over their genomes. Polygenic disposition to left-handedness was associated with the handedness-linked asymmetries of surface area in the fusiform cortex and anterior insular cortex (Fig. 16.2). Specifically, higher polygenic disposition to left-handedness was associated with a relatively rightward shift of *surface area* asymmetry in these cortical regions—the same as for left-handedness itself, whereas regional cortical *thickness* asymmetries that were associated with hand preference (within the postcentral gyrus and inferior occipital cortex) were not significantly heritable and showed no associations with polygenic disposition to left-handedness (Fig. 16.2).

Causal mediation modeling (a statistical approach that seeks to explain the observed relationship between two variables via the inclusion of a third variable) was compatible with fusiform and anterior insula surface area asymmetries partly mediating the effect of polygenic disposition on hand preference. However, mediation models could also be fitted the other way around, i.e., with hand preference partly mediating an effect of polygenic disposition to left-handedness on these particular brain asymmetries. Both models can be considered possible, because the brain's structure may be both causal to behavioral outcomes and shaped plastically by repeated behavior [or correlated due to other shared underlying factors (Bishop, 2013)]. The cortical asymmetries that share genetic influences with hand preference, such as asymmetry of the language-related anterior insula surface area, suggest genetic, developmental, and perhaps evolutionary links between language and handedness. Such links have been envisaged before (Corballis, 2003). In contrast, nonheritable cortical correlates of hand preference, such as asymmetry of primary sensorimotor cortical thickness, may reflect plastic "use-based" brain responses that are relatively distal to primary mechanisms of asymmetry formation in the developing brain.

A ROLE FOR MICROTUBULES IN LEFT–RIGHT AXIS FORMATION?

In seeking the developmental mechanisms of human brain asymmetry, it is worth considering what is known of mechanisms that lead to visceral organ asymmetries in the body cavity (the heart, liver, etc.). An important mechanism involved in vertebrate visceral development involves motile cilia, which are cellular organelles that project from the cellular surface and can beat/rotate to produce an extracellular fluid flow (Norris, 2012). Life on earth is based on L-form amino acids rather than the mirror D-form, and this chirality (an object is chiral if it cannot be superposed on its mirror image) carries through to the macromolecular scale to influence the structure and movement of motile cilia (Fliegauf et al., 2007; Norris, 2012), i.e., they tend to move in one particular orientation. This results in unilateral fluid flow that can trigger asymmetric gene expression in early mammalian embryos (Norris, 2012). After asymmetry of key signaling molecules and transcription factors is established (transcription factors regulate the expression of other genes), lateralized downstream developmental programs can proceed, that are ultimately translated into asymmetries of organ morphology and positioning.

What might this mean for the brain? As reviewed above, microtubule-related and embryonically expressed genes have now been robustly associated with variation in adult human brain asymmetry and handedness by large-scale genetic studies. Microtubules are essential components of motile cilia, supporting their hairlike structure and contributing to their motility. Therefore, embryonic motile cilia might seem to provide a common mechanism for the developments of brain and visceral asymmetry. However, left-hemisphere dominances for language and hand articulation do not usually reverse in people with the rare condition of *situs inversus,* i.e., reversal of the visceral organs on the left–right axis, when caused by genetic mutations that impair motile ciliary function (Tanaka et al., 1999; McManus et al., 2004; Vingerhoets et al., 2018; Postema et al., 2020). Also, mutations in tubulin genes themselves are not known as causes of *situs inversus* of the viscera. These observations suggest at least a partial disconnect between brain and visceral organ asymmetry development and also mean that the fundamental mechanisms of brain left–right axis formation remain unknown in our species, or indeed in any mammal.

As essential components of the cytoskeleton, microtubules can also contribute to *chirality* at the whole-cell scale, i.e., create unidirectional biases in the morphology, position, rotation, or migration of cells (Tee et al., 2015), or the intracellular distributions of organelles (Fan et al., 2019). Studies in invertebrates and frog embryos have indicated that the cytoskeleton can induce asymmetric morphology of various organs through affecting cellular chirality within embryonic precursors of the organs concerned (McNiven and Porter, 1988; Steinhauer and Kalderon, 2006; Okumura et al., 2008; Lobikin et al., 2012; Tee et al., 2015; Davison et al., 2016; Inaki et al., 2016; McDowell et al., 2016). These mechanisms can be organ-intrinsic, i.e., arise independently of other developing organs or systems. Thus, the nonciliary cytoskeleton can be a source of left–right axis creation, in addition to microtubules within cilia. An organ-intrinsic, microtubule-based, but nonmotile-ciliary mechanism of brain left–right axis formation would not only fit the genetic findings pertaining to brain asymmetry and handedness but also explain the degree of apparent disconnect of brain asymmetries from visceral asymmetries (Tanaka et al., 1999; McManus et al., 2004;

Vingerhoets et al., 2018; Postema et al., 2020; Sha et al., 2021b). Such a mechanism may occur very early in development, for example, during formation of the neural tube, which in humans begins in the third week postconception (Sadler, 2005).

It is particularly relevant that Sha et al. (2021a) implicated the gene *NME7* in affecting hand preference and handedness-associated cortical surface area asymmetries. Recessive mutations in this gene cause *situs inversus* of the viscera, but ciliary functions remain intact in the affected people (Reish et al., 2016). The NME7 protein associates with γ-tubulin and the microtubules within cilia (Reish et al., 2016), which means that it can not only potentially influence cilia but also the microtubule cytoskeleton more broadly. Again, this could fit with a mechanism of brain asymmetric development that is based on cellular chirality and partly independent from ciliary-based mechanisms of visceral laterality formation. Alternatively, if motile cilia are involved in creating human brain asymmetry, then perhaps those in the embryonic nervous system have a somewhat distinct composition from those in other organs or developmental stages. Primary (non-motile) cilia may also be involved.

It is also likely that microtubule-related genes influence inter-individual variation in adult brain asymmetry in other ways, apart from a putative role in embryonic left–right axis creation. As mentioned further above, microtubules are ubiquitous components of cells throughout the lifespan and they affect a myriad of cellular properties. Mutations in some tubulin genes cause rare neurologic disorders that are likely to involve altered neuronal migration, differentiation, axon guidance, and maintenance (Tischfield et al., 2011). Mutations in *TUBB2B* can cause asymmetric polymicrogyria (many and small folds) of the cerebral cortex (Jaglin et al., 2009), while mutations in *TUBB3* can cause asymmetric cortical dysplasia and unilateral hypohidrosis (reduced sweating on one side of the body, thought to be linked to disrupted function of the cortex, brain stem, and spinal cord) (Fukumura et al., 2016; Shimojima et al., 2016).

Some of the effects of microtubule-related genes on brain asymmetry may be secondary to more general effects on, e.g., age-related cortical thinning. However, enrichment for microtubule-related genes was not reported in a genome-wide association study of bilaterally averaged cortical surface area and thickness measures in 51,665 individuals (Grasby et al., 2020). This suggests a particular involvement of microtubule genes in hemispheric asymmetry rather than bilateral measures. In addition, Sha et al. (2021b) found no significant overlap between genetic variants associated with structural brain asymmetry and genetic variants associated with intracranial volume [the latter were identified in a study of 32,438 participants (Adams et al., 2016)]. This again indicates that the genetic architecture of brain asymmetry is largely distinct from brain size.

SHARED GENETIC EFFECTS ON BRAIN ASYMMETRY AND PSYCHIATRIC TRAITS

As part of their genetic study of handedness in the UK Biobank, Wiberg et al. (2019) tested for genetic correlations between left-handedness and various neurodegenerative and psychiatric traits, making use of publicly available summary statistics from genome-wide association studies of those traits. Genetic correlation indicates the extent to which the same genetic variants, considered in combination all over the genome, are associated with two different traits. Therefore, significant genetic correlation between a pair of traits can indicate a degree of shared etiology. Wiberg et al. (2019) found that left-handedness showed a significant genetic correlation with schizophrenia, which is consistent with an elevated rate of left-handedness in people with schizophrenia (Hirnstein and Hugdahl, 2014). They also found a significant genetic correlation of left-handedness with Parkinson disease, although this may have been disproportionately influenced by the *MAPT* locus mentioned above, due to its unusual genomic properties. *MAPT* mutations are a known cause of frontotemporal dementia with parkinsonism, and various other neurodegenerative diseases can involve aberrant aggregation within neurons of the protein encoded by this gene—microtubule-associated protein *tau* (Strang et al., 2019). This locus may therefore be a particularly likely candidate for having an age-related and/or neurodegenerative effect on brain asymmetry in older adults, in addition to a possible developmental effect on brain asymmetry and handedness.

For the specific genomic loci that Wiberg et al. (2019) found to be associated with left-handedness, they also tested associations with various clinical phenotypes in the UK Biobank. Significant associations for the *MAPT* locus included neuroticism, mood swings, anxious feelings, and miserableness. The *TUBB* locus also showed associations with traits related to anxiety and depression.

In their larger genetic study of left-handedness, Cuellar-Partida et al. (2021) noted that three of the implicated genomic loci had previously been associated with schizophrenia at a statistically significant level, after adjusting for genome-wide multiple testing. At each of these three loci, the allele that increased the odds of left-handedness also increased the risk of schizophrenia. The nearest genes to these loci were *FURIN* that encodes a membrane-bound protease enzyme, *SLC39* that encodes a membrane-bound zinc-influx transporter, and *ABT1* that encodes a transcription activator.

Sha et al. (2021b) used the results from their multivariate genome-wide study of structural brain asymmetry to examine whether genetic loci associated with brain asymmetry are also associated with neurodevelopmental disorders or other behavioral and psychologic traits. When collapsing their genetic association data across multiple variants within individual genes, they found that 43 genes implicated in brain structural asymmetry had previously been reported to associate with educational attainment (Lee et al., 2018), and 15 with intelligence (Savage et al., 2018). Considering common genetic variants across the whole genome in combination, there was statistically significant evidence for a genetic overlap between brain asymmetries and autism, educational attainment and schizophrenia. Since Sha et al. (2021b) had found that brain asymmetry-related genes tend to be especially highly expressed in the embryonic and fetal brain, the genetic overlap between brain asymmetry and psychiatric or behavioral traits may reflect a shared susceptibility to atypical neurodevelopment in utero. However, shared genetic influences on brain asymmetry and traits such as autism and schizophrenia may also be expressed later, for example, during infancy, childhood, or adolescence.

Although autism and schizophrenia are rare in the UK Biobank general population sample, each individual can be more or less genetically predisposed, based on their genome-wide combination of common genetic variants. In a further study of over 32,000 adult individuals from the UK Biobank (Sha et al., 2021c), polygenic scores for autism and schizophrenia were tested for relations with brain structural asymmetries spanning the cerebral cortex and subcortical structures. Sha et al. (2021c) found that polygenic scores for autism and schizophrenia were weakly positively correlated with each other, and each score was significantly associated with its own distinct set of brain regional asymmetries (Fig. 16.3).

The single regional asymmetry that was noticeably concordant in its associations with polygenic scores for both autism and schizophrenia was asymmetry of the *pars opercularis* surface area. The *pars opercularis* forms part of Broca's classically defined inferior frontal language cortex (Broca, 1861). This region showed reduced leftward asymmetry with higher polygenic risk for both autism and schizophrenia (Fig. 16.3), consistent with leftward asymmetry being the optimal organization to support left-lateralized functions such as language. Therefore, autism and schizophrenia may share an etiologic link through altered structure and function of the inferior frontal language cortex, which would be consistent with impaired social communication in both disorders. Altered hemispheric asymmetry of regions important for language has been proposed to contribute to auditory verbal hallucinations in schizophrenia (DeLisi et al., 1997; Ocklenburg et al., 2015), although with a focus primarily on the temporal rather than frontal cortex (Schijven et al., 2023).

Sha et al. (2021c) found no evidence that the patterns of brain asymmetry associated with polygenic scores for autism and schizophrenia were negatively correlated across brain regions, and therefore the results did not support a concept of autism and schizophrenia as opposing disorders on a single neurobiologic dimension (Crespi and Badcock, 2008; Ciaramidaro et al., 2015). Sha et al. (2021c) also found no evidence that polygenic scores for autism or schizophrenia were associated with more male-like or more female-like average patterns of brain asymmetry, in contrast to what might be expected according to the "extreme male brain" theory of autism (Baron-Cohen, 2002).

The association effect sizes between structural brain asymmetries and polygenic scores for autism and schizophrenia were very small, with multivariate canonical correlations not >0.04 (Sha et al., 2021c). A high degree of statistical significance arose because of the sample size of over 32,000 individuals. It is possible that polygenic dispositions to autism and schizophrenia might show stronger associations with brain measures at levels that are more directly relevant for lateralized functions than macrostructural asymmetries—for example, in asymmetric abundances of certain cell types or gene expression levels.

CONCLUSIONS AND FUTURE PERSPECTIVES

In this chapter, we have seen how recent large-scale studies have enabled the robust identification of genetic loci associated with variation in brain asymmetry and handedness. This has been a major step forward in understanding the genetics of human brain asymmetry. The genes involved tend to be most strongly expressed in the embryonic and fetal brain, compared to brain tissue from later stages of the lifespan. There is a significant functional enrichment for microtubule-related genes. Together, these observations suggest a microtubule-mediated mechanism of left–right axis formation in the developing human brain. Such a mechanism may involve cellular chirality and be distinct from motile cilia-dependent processes of visceral asymmetry development.

However, despite their relatively high expression in the embryonic and fetal brain, many of the implicated genes continue to be expressed at robust levels in the brain throughout the lifespan. Brain structural asymmetry continues to develop throughout the lifespan (Zhou et al., 2013; Roe et al., 2021, 2023), and the UK

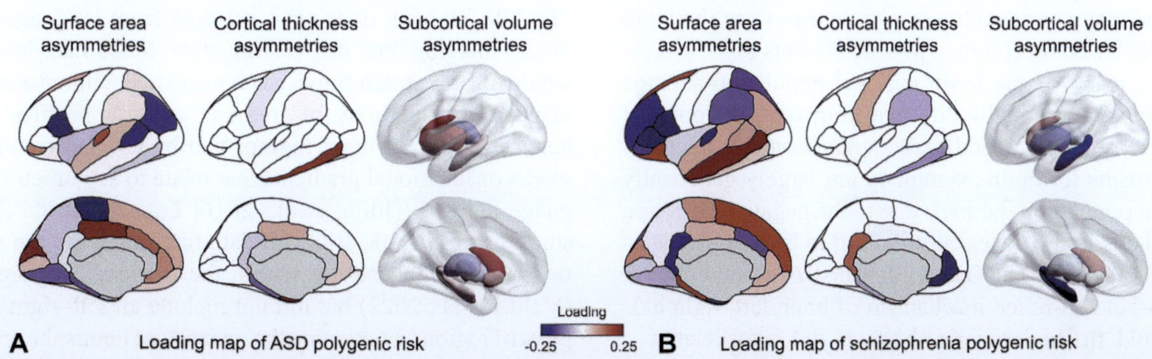

Fig. 16.3. Map of associations of regional brain asymmetries with polygenic disposition to autism or schizophrenia in 32,256 adult individuals from the general population (UK Biobank). Loadings of regional brain asymmetries derived from canonical correlation analyses with (A): autism spectrum disorder (ASD) polygenic disposition and (B): schizophrenia polygenic disposition. A positive loading (red) for a given region indicates a leftward shift of asymmetry associated with increased polygenic disposition. Conversely, a negative loading (blue) indicates a rightward shift of asymmetry associated with increased polygenic disposition. Figure reproduced from Sha Z, Schijven D, Francks C (2021c). Patterns of brain asymmetry associated with polygenic risks for autism and schizophrenia implicate language and executive functions but not brain masculinization. Mol Psychiatry 26: 7652–7660.

Biobank—a mainstay for many of the studies reviewed above—consists of middle-aged to older adults. Furthermore, microtubules are ubiquitous cellular components that are fundamental to many processes throughout cellular life, while many of the genes implicated in interindividual variation in adult brain asymmetry or handedness are not overtly microtubule-related. It is therefore possible that some of the implicated genes exert their influences on brain asymmetry subsequent to neurodevelopment, perhaps by affecting age-related changes in older adults, and neurodegenerative processes.

In order to shed light on these different possibilities, gene-functional approaches will be required, for example, in cellular or animal models. Mice have shown evidence for molecular and neurophysiologic brain asymmetries (Levy et al., 2019; Jordan, 2020) and may be useful as model organisms. One possibility is that microtubule-mediated cellular chirality occurs when the embryonic neural tube is forming and the embryo undergoes axial rotation (Faisst et al., 2002), a process related to left–right asymmetry. Postmortem human embryonic samples are especially rare from this developmental period (2–4 weeks post conception) due to legal and ethical restrictions, which makes studies in model organisms especially important.

Such studies may reveal asymmetry-linked subtypes of genetic susceptibility to psychiatric traits such as autism and schizophrenia, which could contribute to understanding disorder pathogenesis, as well as diagnosis and ultimately personalized treatment. Some forms of these conditions are likely to arise from altered embryonic and fetal neurodevelopment and may be largely irreversible by pharmacologic treatment when diagnosed later in life. Nonetheless, it might still be useful to distinguish such subtypes from drug-tractable forms when making treatment choices. Another crucial question remains on the causal relations that underlie associations between genes, brain asymmetry, and behavior. It may be possible to apply Mendelian randomization (Davey Smith and Ebrahim, 2003), a form of causal modeling suited to cross-sectional genome-wide association data, to understand whether altered brain asymmetries mediate gene-disorder associations.

In the large-scale genetic studies reviewed in this chapter, left-handedness and many regional brain asymmetries were shown to have heritability based on common genetic variants of <10%. At the same time, many of the brain regional population-average asymmetries were substantially different from zero (the point of bilateral symmetry). This overall pattern suggests that developmental mechanisms for brain asymmetry are largely genetically invariant in the population, perhaps due to the negative selection of genetic variants in the genes contributing to brain asymmetry (McManus, 1985; Bishop and Bates, 2019; de Kovel and Francks, 2019). A cytoskeleton-based mechanism of brain left–right axis formation would fit this scenario, because the cytoskeleton is essential for a host of other fundamental functions in cellular biology (Geiger et al., 2001; Janke and Bulinski, 2011). Genetic variation affecting cytoskeletal components may therefore not be well tolerated, for reasons apart from their functions in brain asymmetry.

Twin studies have not found effects of shared environment on brain asymmetries (Guadalupe et al., 2017; Kong et al., 2018), and left-handedness has shown only limited associations with the environmental and early life factors studied so far (de Kovel et al., 2019; McManus, 2021), in which case early developmental randomness may in fact cause most variation in brain and behavioral asymmetries (Mitchell, 2020). Regardless, there is enough heritable variance to have supported successful gene mapping studies, as reviewed in this chapter. Identifying specific genes leads to new insights into trait biology.

Importantly, all of the large-scale studies reviewed in this chapter were based on common genetic variation, i.e., with allele population frequencies of at least half of 1%. This leaves other classes of genetic variation yet to be explored in large-scale studies of human brain asymmetry. For example, rare protein-altering genetic variants may also play a role in interindividual variability. Such variants may have larger effects on brain asymmetry than the common variants studied so far—not necessarily at the population level, but on those rare individuals who carry them. One study in a large consanguineous family with multiple left-handers did not identify such effects using whole exome sequencing (Kavaklioglu et al., 2016). However, an exploratory study of right-hemisphere language dominance—a trait with roughly 1% frequency in the population that occurs mostly in left-handers—found evidence for the involvement of rare variants in genes related to the actin cytoskeleton (Carrion-Castillo et al., 2019). That finding remains tentative as it was based on a sample size below 100. The roles of rare genetic variants in handedness and brain asymmetry may start to be clarified through the recent release of whole exome sequence data in the UK Biobank. [Note added in April 2024: an exome-wide study of rare protein-altering variants in 38,043 left-handed and 313,271 right-handed individuals from the UK Biobank indicated that the heritability of left-handedness due to this class of genetic variant was roughly 1% and implicated another beta-tubulin gene, *TUBB4B* (Schijven et al., 2024)].

The large-scale studies reviewed in this chapter were focused on brain structural asymmetry or handedness. Functional hemispheric language dominance has also shown significant heritability in some studies (Somers et al., 2015; Labache et al., 2020), but large samples for genome-wide association analysis are lacking. The UK Biobank does not have data from language task functional imaging but has resting-state functional imaging, i.e., where intrinsic brain activity was measured without a specific task demand. Correlations of intrinsic activity profiles over time between different brain regions can be used to reconstruct networks or functional gradients that relate to asymmetry of the language network (Joliot et al., 2016; Labache et al., 2023). One study using the UK Biobank data has examined the genetics of functionahl connectivity within the bilateral language network (Mekki et al., 2022) but did not include all left–right homotopic pairs of regions or examine the genetics of hemispheric functional differences. This remains an area for further investigation.

Similarly, the genetics of white matter microstructure and nerve fiber connectivity has been examined in the UK Biobank using diffusion tensor data imaging data (Zhao et al., 2021; Sha et al., 2023), and one of these studies examined connectivity specifically within the core left-hemisphere language network (Sha et al., 2023). However, left–right asymmetry with regard to the contralateral counterpart regions was not examined. This may be a fruitful avenue for future gene mapping studies of brain asymmetry.

In conclusion, large-scale genetic studies carried out during the last 5 years have greatly advanced our understanding of the genetic contributions to human brain asymmetry. The field now has a substantial list of high-confidence findings to be examined in gene-functional and genetic epidemiologic studies, which may help to understand the roles of the implicated genes in human brain asymmetry and related disorders.

ACKNOWLEDGMENTS

Large-scale genetic studies from the authors' research group that were cited in this chapter were supported by the Max Planck Society (Germany) and Netherlands Organization for Scientific Research (Language in Interaction consortium: Gravitation grant number 024.001.006). Several of the studies discussed in this chapter made use of data from the UK Biobank.

References

Abbondanza F, Dale PS, Wang CA et al. (2023). Language and reading impairments are associated with increased prevalence of non-right-handedness. Child Dev 94: 970–984.

Abu-Rustum RS, Ziade MF, Abu-Rustum SE (2013). Reference values for the right and left fetal choroid plexus at 11 to 13 weeks: an early sign of "developmental" laterality? J Ultrasound Med 32: 1623–1629.

Adams HH, Hibar DP, Chouraki V et al. (2016). Novel genetic loci underlying human intracranial volume identified through genome-wide association. Nat Neurosci 19: 1569–1582.

Alfaro-Almagro F, Jenkinson M, Bangerter NK et al. (2018). Image processing and quality control for the first 10,000 brain imaging datasets from UK Biobank. Neuroimage 166: 400–424.

Altarelli I, Leroy F, Monzalvo K et al. (2014). Planum temporale asymmetry in developmental dyslexia: revisiting an old question. Hum Brain Mapp 35: 5717–5735.

Alves JM, Lima AC, Pais IA et al. (2015). Reassessing the evolutionary history of the 17q21 inversion polymorphism. Genome Biol Evol 7: 3239–3248.

Alzahrani F, Albatti TH, Alkuraya FS (2021). A de novo ATXN2L variant in a child with developmental delay and macrocephaly. Am J Med Genet A 185: 949–951.

Baron-Cohen S (2002). The extreme male brain theory of autism. Trends Cogn Sci 6: 248–254.

Bishop DVM (2013). Cerebral asymmetry and language development: cause, correlate, or consequence? Science 340: 1230531.

Bishop DVM, Bates TC (2019). Heritability of language laterality assessed by functional transcranial Doppler ultrasound: a twin study. Wellcome Open Res 4: 161.

Broca P (1861). Remarques sur le siège de la faculté du langage articulé, suivies d'une observation d'aphémie (perte de la parole). Bull Mem Soc Anatomique de Paris 6: 330–357.

Burton PR, Clayton DG, Cardon LR et al. (2007). Genome-wide association study of 14,000 cases of seven common diseases and 3,000 shared controls. Nature 447: 661–678.

Button KS, Ioannidis JPA, Mokrysz C et al. (2013). Power failure: why small sample size undermines the reliability of neuroscience. Nat Rev Neurosci 14: 365–376.

Bycroft C, Freeman C, Petkova D et al. (2018). The UK Biobank resource with deep phenotyping and genomic data. Nature 562: 203–209.

Carper RA, Treiber JM, DeJesus SY et al. (2016). Reduced hemispheric asymmetry of white matter microstructure in autism Spectrum disorder. J Am Acad Child Adolesc Psychiatry 55: 1073–1080.

Carrion-Castillo A, Van der Haegen L, Tzourio-Mazoyer N et al. (2019). Genome sequencing for rightward hemispheric language dominance. Genes Brain Behav 18: e12572.

Carrion-Castillo A, Pepe A, Kong X-Z et al. (2020). Genetic effects on planum temporale asymmetry and their limited relevance to neurodevelopmental disorders, intelligence or educational attainment. Cortex 124: 137–153.

Ciaramidaro A, Bölte S, Schlitt S et al. (2015). Schizophrenia and autism as contrasting minds: neural evidence for the hypo-hyper-intentionality hypothesis. Schizophr Bull 41: 171–179.

Corballis MC (2003). From mouth to hand: gesture, speech, and the evolution of right-handedness. Behav Brain Sci 26: 199–208.

Coren S, Porac C (1977). Fifty centuries of right-handedness: the historical record. Science 198: 631–632.

Crespi B, Badcock C (2008). Psychosis and autism as diametrical disorders of the social brain. Behav Brain Sci 31: 241–261.

Cuellar-Partida G, Tung JY, Eriksson N et al. (2021). Genome-wide association study identifies 48 common genetic variants associated with handedness. Nat Hum Behav 5: 59–70.

Davey Smith G, Ebrahim S (2003). 'Mendelian randomization': can genetic epidemiology contribute to understanding environmental determinants of disease? Int J Epidemiol 32: 1–22.

Davison A, McDowell GS, Holden JM et al. (2016). Formin is associated with left-right asymmetry in the pond snail and the frog. Curr Biol 26: 654–660.

De Fosse L, Hodge SM, Makris N et al. (2004). Language-association cortex asymmetry in autism and specific language impairment. Ann Neurol 56: 757–766.

de Kovel CGF, Francks C (2019). The molecular genetics of hand preference revisited. Sci Rep 9: 5986.

de Kovel CGF, Lisgo S, Karlebach G et al. (2017). Left-right asymmetry of maturation rates in human embryonic neural development. Biol Psychiatry 82: 204–212.

de Kovel CGF, Lisgo SN, Fisher SE et al. (2018). Subtle left-right asymmetry of gene expression profiles in embryonic and foetal human brains. Sci Rep 8: 12606.

de Kovel CGF, Carrión-Castillo A, Francks C (2019). A large-scale population study of early life factors influencing left-handedness. Sci Rep 9: 584.

DeLisi LE, Sakuma M, Kushner M et al. (1997). Anomalous cerebral asymmetry and language processing in schizophrenia. Schizophr Bull 23: 255–271.

Duboc V, Dufourcq P, Blader P et al. (2015). Asymmetry of the brain: development and implications. Annu Rev Genet 49: 647–672.

Faisst AM, Alvarez-Bolado G, Treichel D et al. (2002). Rotatin is a novel gene required for axial rotation and left–right specification in mouse embryos. Mech Dev 113: 15–28.

Fan J, Zhang H, Rahman T et al. (2019). Cell organelle-based analysis of cell chirality. Commun Integr Biol 12: 78–81.

Faurie C, Raymond M (2004). Handedness frequency over more than ten thousand years. Proc Roy Soc London, Ser B: Biol Sci. 271: S43–S45.

Fliegauf M, Benzing T, Omran H (2007). When cilia go bad: cilia defects and ciliopathies. Nat Rev Mol Cell Biol 8: 880–893.

Floris DL, Wolfers T, Zabihi M et al. (2021). Atypical brain asymmetry in autism—a candidate for clinically meaningful stratification. Biol Psychiatry: Cogn Neurosci Neuroimaging 6: 802–812.

Francks C (2015). Exploring human brain lateralization with molecular genetics and genomics. Ann N Y Acad Sci 1359: 1–13.

Francks C (2019). In search of the biological roots of typical and atypical human brain asymmetry: comment on "phenotypes in hemispheric functional segregation? Perspectives and challenges" by guy Vingerhoets. Phys Life Rev 30: 22–24.

Fukumura S, Kato M, Kawamura K et al. (2016). A mutation in the tubulin-encoding TUBB3 gene causes complex cortical malformations and unilateral hypohidrosis. Child Neurol Open 3. https://doi.org/10.1177/2329048x16665758.

Geiger B, Bershadsky A, Pankov R et al. (2001). Transmembrane crosstalk between the extracellular matrix—cytoskeleton crosstalk. Nat Rev Mol Cell Biol 2: 793–805.

Grasby KL, Jahanshad N, Painter JN et al. (2020). The genetic architecture of the human cerebral cortex. Science 367: 6484.

Greve DN, Van der Haegen L, Cai Q et al. (2013). A surface-based analysis of language lateralization and cortical asymmetry. J Cogn Neurosci 25: 1477–1492.

Grove J, Ripke S, Als TD et al. (2019). Identification of common genetic risk variants for autism spectrum disorder. Nat Genet 51: 431–444.

Guadalupe T, Mathias SR, van Erp TGM et al. (2017). Human subcortical brain asymmetries in 15,847 people worldwide reveal effects of age and sex. Brain Imaging Behav 11: 1497–1514.

Gunturkun O, Ocklenburg S (2017). Ontogenesis of lateralization. Neuron 94: 249–263.

Hepper PG (2013). The developmental origins of laterality: fetal handedness. Dev Psychobiol 55: 588–595.

Hepper PG, McCartney GR, Shannon EA (1998). Lateralised behaviour in first trimester human foetuses. Neuropsychologia 36: 531–534.

Hepper PG, Wells DL, Lynch C (2005). Prenatal thumb sucking is related to postnatal handedness. Neuropsychologia 43: 313–315.

Herbert MR, Ziegler DA, Deutsch CK et al. (2005). Brain asymmetries in autism and developmental language disorder: a nested whole-brain analysis. Brain 128: 213–226.

Hirnstein M, Hugdahl K (2014). Excess of non-right-handedness in schizophrenia: meta-analysis of gender effects and potential biases in handedness assessment. Br J Psychiatry 205: 260–267.

Inaki M, Liu J, Matsuno K (2016). Cell chirality: its origin and roles in left-right asymmetric development. Philos Trans R Soc Lond B Biol Sci 371: 20150403.

Jaglin XH, Poirier K, Saillour Y et al. (2009). Mutations in the β-tubulin gene TUBB2B result in asymmetrical polymicrogyria. Nat Genet 41: 746–752.

Janke C, Bulinski JC (2011). Post-translational regulation of the microtubule cytoskeleton: mechanisms and functions. Nat Rev Mol Cell Biol 12: 773–786.

Janke C, Magiera MM (2020). The tubulin code and its role in controlling microtubule properties and functions. Nat Rev Mol Cell Biol 21: 307–326.

Joliot M, Tzourio-Mazoyer N, Mazoyer B (2016). Intra-hemispheric intrinsic connectivity asymmetry and its relationships with handedness and language lateralization. Neuropsychologia 93: 437–447.

Jordan JT (2020). The rodent hippocampus as a bilateral structure: a review of hemispheric lateralization. Hippocampus 30: 278–292.

Karlebach G, Francks C (2015). Lateralization of gene expression in human language cortex. Cortex 67: 30–36.

Kasprian G, Langs G, Brugger PC et al. (2011). The prenatal origin of hemispheric asymmetry: an in utero neuroimaging study. Cereb Cortex 21: 1076–1083.

Kavaklioglu T, Ajmal M, Hameed A et al. (2016). Whole exome sequencing for handedness in a large and highly consanguineous family. Neuropsychologia 93: 342–349.

Kawasaki Y, Suzuki M, Takahashi T et al. (2008). Anomalous cerebral asymmetry in patients with schizophrenia demonstrated by voxel-based morphometry. Biol Psychiatry 63: 793–800.

Knaus TA, Silver AM, Kennedy M et al. (2010). Language laterality in autism spectrum disorder and typical controls: a functional, volumetric, and diffusion tensor MRI study. Brain Lang 112: 113–120.

Kong XZ, Mathias SR, Guadalupe T et al. (2018). Mapping cortical brain asymmetry in 17,141 healthy individuals worldwide via the ENIGMA Consortium. Proc Natl Acad Sci U S A 115: E5154–E5163.

Kong X-Z, Boedhoe PSW, Abe Y et al. (2020). Mapping cortical and subcortical asymmetry in obsessive-compulsive disorder: findings from the ENIGMA Consortium. Biol Psychiatry 87: 1022–1034.

Kong X-Z, Postema M, Schijven D et al. (2021). Large-scale phenomic and genomic analysis of brain asymmetrical skew. Cereb Cortex 31: 4151–4168.

Kong X-Z, Postema MC, Guadalupe T et al. (2022). Mapping brain asymmetry in health and disease through the ENIGMA Consortium. Hum Brain Mapp 43: 167–181.

Labache L, Mazoyer B, Joliot M et al. (2020). Typical and atypical language brain organization based on intrinsic connectivity and multitask functional asymmetries. Elife 9: e58722.

Labache L, Ge T, Yeo BTT et al. (2023). Language network lateralization is reflected throughout the macroscale functional organization of cortex. Nat Commun 14: 3405.

Le Guen Y, Leroy F, Philippe C et al. (2020). Enhancer locus in ch14q23.1 modulates brain asymmetric temporal regions involved in language processing. Cereb Cortex 30: 5322–5332. https://doi.org/10.1093/cercor/bhaa112.

Lee JJ, Wedow R, Okbay A et al. (2018). Gene discovery and polygenic prediction from a genome-wide association study of educational attainment in 1.1 million individuals. Nat Genet 50: 1112–1121.

Levy RB, Marquarding T, Reid AP et al. (2019). Circuit asymmetries underlie functional lateralization in the mouse auditory cortex. Nat Commun 10: 2783.

Lobikin M, Wang G, Xu J et al. (2012). Early, nonciliary role for microtubule proteins in left–right patterning is conserved across kingdoms. Proc Natl Acad Sci U S A 109: 12586–12591.

Maingault S, Tzourio-Mazoyer N, Mazoyer B et al. (2016). Regional correlations between cortical thickness and surface area asymmetries: a surface-based morphometry study of 250 adults. Neuropsychologia 93: 350–364.

Markou P, Ahtam B, Papadatou-Pastou M (2017). Elevated levels of atypical handedness in autism: meta-analyses. Neuropsychol Rev 27: 258–283.

McDowell G, Rajadurai S, M., Levin (2016). From cytoskeletal dynamics to organ asymmetrya nonlinear, regulative pathway underlies left-right patterning. Philos Trans R Soc Lond B Biol Sci 371: 20150409.

McManus IC (1985). Handedness, language dominance and aphasia: a genetic model. Psychol Med Monogr Suppl 8: 1–40.

McManus IC (2009). The history and geography of human handedness. In: IEC Sommer, RS Kahn (Eds.), Language lateralization and psychosis. Cambridge University Press, Cambridge.

McManus C (2021). Is any but a tiny fraction of handedness variance likely to be due to the external environment? Laterality 26: 310–314.

McManus IC, Martin N, Stubbings GF et al. (2004). Handedness and situs inversus in primary ciliary dyskinesia. Proc Biol Sci 271: 2579–2582.

McNiven MA, Porter KR (1988). Organization of microtubules in centrosome-free cytoplasm. J Cell Biol 106: 1593–1605.

Medland SE, Duffy DL, Wright MJ et al. (2009). Genetic influences on handedness: data from 25,732 Australian and Dutch twin families. Neuropsychologia 47: 330–337.

Mekki Y, Guillemot V, Lemaître H et al. (2022). The genetic architecture of language functional connectivity. Neuroimage 249: 118795.

Mitchell KJ (2020). Innate: how the wiring of our brains shapes who we are, Princeton University Press.

Mohebiany AN, Harroch S, Bouyain S (2014). New insights into the roles of the contactin cell adhesion molecules in neural development. Adv Neurobiol 8: 165–194.

Muntané G, Santpere G, Verendeev A et al. (2017). Interhemispheric gene expression differences in the cerebral cortex of humans and macaque monkeys. Brain Struct Funct 222: 3241–3254.

Norris DP (2012). Cilia, calcium and the basis of left-right asymmetry. BMC Biol 10: 102.

Ocklenburg S, Güntürkün O, Hugdahl K et al. (2015). Laterality and mental disorders in the postgenomic age—a closer look at schizophrenia and language lateralization. Neurosci Biobehav Rev 59: 100–110.

Ocklenburg S, Schmitz J, Moinfar Z et al. (2017). Epigenetic regulation of lateralized fetal spinal gene expression underlies hemispheric asymmetries. Elife 6: e22784.

Okada N, Fukunaga M, Yamashita F et al. (2016). Abnormal asymmetries in subcortical brain volume in schizophrenia. Mol Psychiatry 21: 1460–1466.

Okumura T, Utsuno H, Kuroda J et al. (2008). The development and evolution of left-right asymmetry in invertebrates: lessons from Drosophila and snails. Dev Dyn 237: 3497–3515.

Papadatou-Pastou M, Tomprou D-M (2015). Intelligence and handedness: meta-analyses of studies on intellectually disabled, typically developing, and gifted individuals. Neurosci Biobehav Rev 56: 151–165.

Papadatou-Pastou M, Ntolka E, Schmitz J et al. (2020). Human handedness: a meta-analysis. Psychol Bull 146: 481–524.

Perelle IB, Ehrman L (1994). An international study of human handedness: the data. Behav Genet 24: 217–227.

Postema MC, van Rooij D, Anagnostou E et al. (2019). Altered structural brain asymmetry in autism spectrum disorder in a study of 54 datasets. Nat Commun 10: 4958.

Postema MC, Carrion Castillo A, Fisher SE et al. (2020). The genetics of situs inversus without primary ciliary dyskinesia. Sci Rep 10: 3677.

Postema MC, Hoogman M, Ambrosino S et al. (2021). Analysis of structural brain asymmetries in attention-deficit/hyperactivity disorder in 39 datasets. J Child Psychol Psychiatry 62: 1202–1219.

Ravichandran C, Shinn AK, Ongur D et al. (2017). Frequency of non-right-handedness in bipolar disorder and schizophrenia. Psychiatry Res 253: 267–269.

Reish O, Aspit L, Zouella A et al. (2016). A homozygous Nme7 mutation is associated with situs Inversus Totalis. Hum Mutat 37: 727–731.

Roe JM, Vidal-Piñeiro D, Sørensen Ø et al. (2021). Asymmetric thinning of the cerebral cortex across the adult lifespan is accelerated in Alzheimer's disease. Nat Commun 12: 721.

Roe JM, Vidal-Piñeiro D, Amlien IK et al. (2023). Tracing the development and lifespan change of population-level structural asymmetry in the cerebral cortex. Elife 12: e84685.

Sadler TW (2005). Embryology of neural tube development. Am J Med Genet C Semin Med Genet 135C: 2–8.

Savage JE, Jansen PR, Stringer S et al. (2018). Genome-wide association meta-analysis in 269,867 individuals identifies new genetic and functional links to intelligence. Nat Genet 50: 912–919.

Schijven D, Postema MC, Fukunaga M et al. (2023). Large-scale analysis of structural brain asymmetries in schizophrenia via the ENIGMA consortium. Proc Natl Acad Sci U S A 120: e2213880120.

Schijven D, Soheili-Nezhad S, Fisher SE et al. (2024). Exome-wide analysis implicates rare protein-altering variants in human handedness. Nat Commun 15: 2632.

Schizophrenia Working Group of the Psychiatric Genomics Consortium (2014). Biological insights from 108 schizophrenia-associated genetic loci. Nature 511: 421–427.

Schmitz J, Güntürkün O, Ocklenburg S (2019). Building an asymmetrical brain: the molecular perspective. Front Psychol 10: 982.

Sha Z, Pepe A, Schijven D et al. (2021a). Handedness and its genetic influences are associated with structural asymmetries of the cerebral cortex in 31,864 individuals. Proc Natl Acad Sci U S A 118: e2113095118.

Sha Z, Schijven D, Carrion-Castillo A et al. (2021b). The genetic architecture of structural left–right asymmetry of the human brain. Nat Hum Behav 5: 1226–1239.

Sha Z, Schijven D, Francks C (2021c). Patterns of brain asymmetry associated with polygenic risks for autism and schizophrenia implicate language and executive functions but not brain masculinization. Mol Psychiatry 26: 7652–7660.

Sha Z, Schijven D, Fisher SE et al. (2023). Genetic architecture of the white matter connectome of the human brain. Sci Adv 9: eadd2870.

Shenton ME, Dickey CC, Frumin M et al. (2001). A review of MRI findings in schizophrenia. Schizophr Res 49: 1–52.

Shimojima K, Okamoto N, Yamamoto T (2016). A novel TUBB3 mutation in a sporadic patient with asymmetric cortical dysplasia. Am J Med Genet A 170: 1076–1079.

Smith SM, Douaud G, Chen W et al. (2021). An expanded set of genome-wide association studies of brain imaging phenotypes in UK Biobank. Nat Neurosci 24: 737–745.

Somers M, Ophoff RA, Aukes MF et al. (2015). Linkage analysis in a Dutch population isolate shows no major gene for left-handedness or atypical language lateralization. J Neurosci 35: 8730–8736.

Sommer I, Ramsey N, Kahn R et al. (2001). Handedness, language lateralisation and anatomical asymmetry in schizophrenia: meta-analysis. Br J Psychiatry 178: 344–351.

Steinhauer J, Kalderon D (2006). Microtubule polarity and axis formation in the Drosophila oocyte. Dev Dyn 235: 1455–1468.

Strang KH, Golde TE, Giasson BI (2019). MAPT mutations, tauopathy, and mechanisms of neurodegeneration. Lab Invest 99: 912–928.

Sun T, Patoine C, Abu-Khalil A et al. (2005). Early asymmetry of gene transcription in embryonic human left and right cerebral cortex. Science 308: 1794–1798.

Sun Y, Chen Y, Collinson SL et al. (2017). Reduced hemispheric asymmetry of brain anatomical networks is linked to schizophrenia: a connectome study. Cereb Cortex 27: 602–615.

Tanaka S, Kanzaki R, Yoshibayashi M et al. (1999). Dichotic listening in patients with situs inversus: brain asymmetry and situs asymmetry. Neuropsychologia 37: 869–874.

Taymourtash A, Schwartz E, Nenning K-H et al. (2023). Fetal development of functional thalamocortical and cortico-cortical connectivity. Cereb Cortex 33: 5613–5624.

Tee YH, Shemesh T, Thiagarajan V et al. (2015). Cellular chirality arising from the self-organization of the actin cytoskeleton. Nat Cell Biol 17: 445–457.

Tischfield MA, Cederquist GY, Gupta ML et al. (2011). Phenotypic spectrum of the tubulin-related disorders and functional implications of disease-causing mutations. Curr Opin Genet Dev 21: 286–294.

Toga AW, Thompson PM (2003). Mapping brain asymmetry. Nat Rev Neurosci 4: 37–48.

Tzourio-Mazoyer N, Crivello F, Mazoyer B (2018). Is the planum temporale surface area a marker of hemispheric or regional language lateralization? Brain Struct Funct 223: 1217–1228.

Vingerhoets G, Li X, Hou L et al. (2018). Brain structural and functional asymmetry in human situs inversus totalis. Brain Struct Funct 223: 1937–1952.

Wiberg A, Ng M, Al Omran Y et al. (2019). Handedness, language areas and neuropsychiatric diseases: insights from brain imaging and genetics. Brain 142: 2938–2947.

Willems RM, der Haegen LV, Fisher SE et al. (2014). On the other hand: including left-handers in cognitive neuroscience and neurogenetics. Nat Rev Neurosci 15: 193–201.

Young AI (2019). Solving the missing heritability problem. PLoS Genet 15: e1008222.

Zhao B, Li T, Yang Y et al. (2021). Common genetic variation influencing human white matter microstructure. Science 372: eabf3736.

Zhao L, Matloff W, Shi Y et al. (2022a). Mapping complex brain torque components and their genetic architecture and Phenomic associations in 24,112 individuals. Biol Psychiatry 91: 753–768.

Zhao X, Liang W, Wang W et al. (2022b). Changes in and asymmetry of the proteome in the human fetal frontal lobe during early development. Commun Biol 5: 1031.

Zhou D, Lebel C, Evans A et al. (2013). Cortical thickness asymmetry from childhood to older adulthood. Neuroimage 83: 66–74.

Handbook of Clinical Neurology, Vol. 208 (3rd series)
Cerebral Asymmetries
C. Papagno and P. Corballis, Editors
https://doi.org/10.1016/B978-0-443-15646-5.00014-2

Chapter 17

Sex/gender differences in hemispheric asymmetries

SOPHIE HODGETTS AND MARKUS HAUSMANN*

Department of Psychology, Durham University, England, United Kingdom

Abstract

This chapter will critically review evidence across 40 years of research, covering both early and contemporary studies that have investigated sex/gender differences in hemispheric asymmetries, including both structural and functional asymmetries. We argue that small sex/gender differences in hemispheric asymmetry reliably exist, but there is significant overlap between the sexes and considerable within-sex variation. Furthermore, we argue that research to date is limited in its consideration of sex/gender-related factors, such as sex hormones and gender roles. Moreover, we highlight a critical limitation stemming from the lack of universal agreement on the definitions of "sex" and "gender," resulting in the majority of studies interested in sex/gender differences in hemispheric asymmetry involving the separation of participants into dichotomous male/female groups based solely on self-identified sex. Future research involving sex/gender should adopt a biopsychosocial approach whenever possible, to ensure that nonbinary psychologic, biologic, and environmental/social factors related to sex/gender, and their interactions, are routinely accounted for. Finally, we argue that while the human brain is not sexually dimorphic, sex/gender differences in the brain are not trivial and likely have several clinically relevant implications, including for the development of stratified treatment approaches for both neurologic and psychiatric patient populations.

INTRODUCTION

The number of studies investigating the influence of sex/gender on the brain has increased substantially in recent years. Nevertheless, many cognitive neuroscientists continue to debate the existence and importance of such interindividual differences. For example, while some researchers argue for the existence of sex/gender differences in brain structure and function (Cahill, 2017; Jäncke, 2018; Hodgetts and Hausmann, 2020b, 2022), including hemispheric asymmetry (Hirnstein et al., 2019; Hirnstein and Hausmann, 2021), others argue that such differences either do not exist or are small enough to be considered trivial (Eliot et al., 2021). Despite this disagreement, many researchers agree that human brains (and the brains of other species) are not "sexually dimorphic" (i.e., "male brains" and "female brains") but can be better conceptualized as "mosaics" of features, existing on a masculine-feminine continuum (Joel et al., 2015; Hodgetts and Hausmann, 2020b; Eliot et al., 2021; Hirnstein and Hausmann, 2021; Zhang et al., 2021; Rauch and Eliot, 2022). Moreover, findings from several recent studies support taking a biopsychosocial approach to the study of sex/gender differences in the brain, such that both the influence of nonbinary, sex/gender-relevant biologic, psychologic, and environmental factors and the complex and dynamic interaction(s) between these factors are investigated (Cahill, 2017; Jäncke, 2018; Hausmann, 2020; Hodgetts and Hausmann, 2020b; Joel, 2021).

This chapter will review the findings from historical and contemporary research into sex/gender differences in brain lateralization, before presenting a possible new approach for future research in this area. We will argue that sex/gender differences in hemispheric asymmetries exist. Even though some of these differences are relatively small, they are not necessarily trivial, and they are not "sexual dimorphisms" (see also Galea, 2021; Hirnstein and Hausmann, 2021). We will argue that there is a high level of inter- and intraindividual variability resulting in a considerable amount of overlap between the sexes at the population level. We will also argue that future research should adopt a nonbinary, biopsychosocial approach (Hausmann, 2020), to ensure that the full spectrum of sex/gender-relevant environmental factors and interindividual variations are considered more routinely and appropriately.

In general, the term "sex" is used when referring to the biologic classification of an individual as male or female, based on the overt appearance of their reproductive system (e.g., genitalia), and "gender" is used when referring to the psychosocial constructs typically associated with one's sex (e.g., masculinity, femininity). However, the two terms have been frequently and misleadingly conflated in previous research and as such, it is

*Correspondence to: Dr. Markus Hausmann, Department of Psychology, Durham University, South Road, Durham, DH1 3LE, United Kingdom. E-mail: markus.hausmann@durham.ac.uk

not always clear whether a study has investigated sex, gender, or both. To account for this inconsistency, and in line with the previous literature (e.g., Joel et al., 2015; Jäncke, 2018), this chapter will use sex/gender, rather than "sex" or "gender" separately. It should also be noted that we are discussing data from averages of groups, not individuals. This chapter will also focus primarily on large-scale studies and meta-analyses, meaning several high-quality studies will not be reviewed in detail here. Finally, we explicitly avoid the term "sexual dimorphism" when referring to sex/gender differences as this term implies two distinct brains in a binary sense: one male and one female, as all empiric evidence suggests that such dichotomous categorization is incorrect and should be abandoned from the neuroscientific literature.

EVIDENCE REGARDING SEX/GENDER DIFFERENCES IN HEMISPHERIC ASYMMETRIES

Hemispheric asymmetry is a fundamental principle of functional brain organization referring to the asymmetric representation of a specific cognitive process to either the left or right cerebral hemisphere. For example, in the healthy adult human brain, the left hemisphere is typically dominant for many language processes (Kimura, 1967), while the right hemisphere is dominant for some visuospatial processes (Hellige, 1993; Hugdahl and Westerhausen, 2013). Sex/gender differences in both structural and functional hemispheric asymmetries have been investigated at many levels of analysis, and in the following section, we will review the relevant evidence, with a focus on meta-analyses, systematic reviews, and large-scale studies.

Structural Cerebral Asymmetries

Due to the robust sex/gender difference in specific language processes (e.g., verbal fluency) (Hirnstein et al., 2019), most studies of sex/gender differences in structural cerebral asymmetries have focused on cortical areas known to be involved in language processes such as the planum temporale, a region that overlaps with Wernicke's area and is involved in language comprehension (Shapleske et al., 1999). In an early study involving 24 healthy participants, the left planum temporale was shown to be larger in men compared to women (Kulynych et al., 1994). Moreover, only men showed an asymmetry in this region (i.e., larger in the left hemisphere compared to the right). Similar results were yielded by a study of three large structural magnetic resonance imaging (MRI) datasets ($n = 2337$, 935, and 888), which found a stronger leftward asymmetry in the planum temporale for men compared to women (Guadalupe et al., 2015). However, this finding is inconsistent. For example, a meta-analysis of 13 studies ($n = 807$) investigating structural asymmetry of the planum temporale revealed no significant sex/gender differences (Sommer et al., 2008). Finally, a recent structural MRI study used measures of cortical thickness to create "hemispheric morphologic networks" in a sample of 285 participants (Choi et al., 2020). Results showed that the patterns of asymmetry differed between men and women in several cortical regions, with men showing larger rightward asymmetry in the cingulate and superior parietal gyrus, and women showing larger leftward asymmetry in the temporal pole. Taken together, these findings suggest that (1) a sex/gender difference in the structure of the planum temporale might exist, (2) the effect size is small, and (3) a significant difference will only be detected with larger samples.

Functional Cerebral Asymmetries

Evidence From Early Behavioral Studies

Sex/gender influences have been extensively investigated with regard to functional cerebral asymmetries (FCAs). McGlone (1980) provides a comprehensive review of the early evidence concerning sex/gender differences in FCAs from clinical and nonclinical populations, including behavioral, electrophysiologic (e.g., electroencephalography and event-related potentials), developmental, and animal studies. After synthesizing the available findings, McGlone (1980) argued that behavioral studies of clinical populations revealed a greater dependence on the left hemisphere for language in male brain injury patients (i.e., stronger asymmetry in men), evidenced by more severe language impairment in male patients with left-hemispheric lesions compared to female patients with comparable damage (e.g., (McGlone and Kertesz, 1973). Moreover, behavioral data from studies of language FCAs in healthy samples also demonstrated increased asymmetry in men compared to women, for example, dichotic listening studies (Harshman and Remington, 1976; Lake and Bryden, 1976). However, results across the studies included in this review were inconsistent, with some studies suggesting increased asymmetry in women (Dorman and Porter, 1975), while still further studies reported no significant effect of sex/gender (McGlone and Davidson, 1973; McKeever and VanDeventer, 1977; Scott et al., 1979). As such, McGlone (1980) drew several tentative conclusions and provided suggestions for further research. First, it was argued that sex/gender differences in brain organization exist but are small and subject to the influence of other factors including task type, participant's age, and extent of brain injury. Second, while many studies suggested that men were more lateralized than women, this finding was inconsistent, leading McGlone (1980) to suggest that brains are more similar between sexes/genders than different. However, McGlone (1980) also notes that small sex/gender differences are not unimportant and/or trivial. Finally, McGlone notes that most of the studies included here are focused on language lateralization, and studies of other lateralized processes (e.g., spatial skills) were underrepresented.

Several systematic reviews and meta-analyses have been produced in the decades following McGlone's (1980) review and have incorporated evidence from studies of sex/gender effects on lateralized processes other than language (e.g., visuospatial processing). First, a series of reviews produced by Hiscock et al. provided a comprehensive overview of behavioral evidence from auditory (Hiscock et al., 1994), visual (Hiscock et al., 1995), and tactile (Hiscock et al., 1999) laterality paradigms, as well as dual-task interference studies (Hiscock et al., 2001). Most of the studies included in these reviews assessed hemispheric asymmetry using behavioral measures, such as reaction times, accuracy rates, and laterality quotients derived from visual half-field tasks, dichotic listening tasks, and dichaptic paradigms. The authors concluded that there is "a small but reliable population-level sex difference" (Hiscock et al., 2001, p. 137), with men typically demonstrating larger asymmetries compared to women. Subsequently, Voyer (1996) produced the first meta-analysis of studies examining sex/gender differences in functional cerebral asymmetries. This analysis covered two decades of research and considered multiple lateralized processes, including visual, auditory, and tactile behavioral tasks. The results indicated that sex/gender differences in functional cerebral asymmetries were most prominent in the visual and auditory modalities, with men consistently yielding greater degrees of asymmetry than women. However, it should be noted that Voyer (1996) also argued for caution when interpreting these findings, as the meta-analysis results also suggested that studies of sex/gender differences in hemispheric asymmetries are subject to the file-drawer problem.

Evidence From Functional Neuroimaging

Further reviews and meta-analyses were able to include evidence from neuroimaging, alongside behavioral data, to investigate sex/gender differences in FCAs for a range of lateralized processes. For example, Vogel et al. (2003) conducted a meta-analysis of studies using both neuroimaging (e.g., functional magnetic resonance imaging) and behavioral measures (e.g., mental rotation) to investigate sex/gender differences in FCAs for visuospatial processing. The results revealed a significant sex/gender difference, characterized by a stronger right-hemispheric bias (i.e., greater asymmetry) in men compared to women. A further meta-analysis focused on studies of sex/gender differences in language FCAs, assessed using dichotic listening, including both studies with only behavioral data, and studies combining behavioral data with MRI and/or fMRI (Voyer, 2011) reported a similar pattern of sex/gender differences (i.e., greater asymmetry in men compared to women). A separate series of two meta-analyses conducted by Sommer et al. (2008, 2004) further investigated sex/gender differences in language FCAs, with a focus on neuroimaging. The first meta-analysis (Sommer et al., 2004) focused on functional neuroimaging studies (e.g., fMRI, PET, fTCD), while the second focused on structural asymmetries in the planum temporale, and FCAs assessed via dichotic listening (Sommer et al., 2008). No sex/gender differences were found across the neuroimaging studies, but stronger FCAs in men were found across the dichotic listening studies. However, it is important to note that in their meta-analysis of fMRI studies, Sommer et al. (2004) included studies investigating a range of language processes, some of which are not considered to be lateralized such as semantic decision, phonologic processing (Hickok and Poeppel, 2007). As such, it is possible that the effects of sex/gender on FCAs were masked. Nevertheless, Sommer (2010) argued that as the studies showing stronger FCAs in men yielded small effect sizes, they were likely subject to publication biases and thus, in line with the argument previously made by Voyer (1996), that research in this area is subject to the file-drawer problem.

Following these meta-analyses, several large-scale studies using resting state fMRI (rs-fMRI) have been conducted to investigate the effect of sex/gender on FCAs (e.g., Liu et al., 2009; Tomasi and Volkow, 2012; Nielsen et al., 2013; Choi et al., 2020). Liu et al. (2009) found small but significant sex/gender differences for both left- and right-lateralized resting-state networks, with men showing stronger FCAs than women. However, both sexes showed strong FCAs in this study, and there was much overlap between them. In a larger rs-fMRI study, FCAs for short-range connectivity in specific regions of the superior temporal ($n = 913$), Tomasi and Volkow (2012) reported that men had stronger rightward inferior frontal, and inferior occipital cortices, while women had stronger leftward FCAs for long-range connectivity in the inferior frontal cortex. Furthermore, in another large rs-fMRI study ($n = 1011$), Nielsen et al. (2013) identified several resting-state networks characterized by lateralization (e.g., the DMN and language network included several left-lateralized regions), but the patterns of FCAs did not differ between men and women.

In sum, although inconsistencies exist, several meta-analyses and large-scale studies suggest that sex/gender differences in hemispheric asymmetries, both structural and functional, do exist. However, while most studies suggest that men generally demonstrate greater asymmetries compared to women across multiple lateralized processes, effect sizes are small and likely only detectable using large samples. This is in line with several contemporary views of sex/gender differences in the human brain more generally and will be discussed further in the following section.

CONTEMPORARY VIEWS OF SEX/GENDER DIFFERENCES IN HEMISPHERIC ASYMMETRY

Evidence to date suggests that men demonstrate larger asymmetries compared to women, but the effect size is small (Hirnstein et al., 2019). This suggests that small sex/gender differences in FCAs do exist at the population level but may only be detected in studies with larger samples. These findings have prompted a shift amongst researchers away from the view that human brains can be conceptualized as having either a prototypical "male" or "female" pattern of hemispheric asymmetry, that is that hemispheric asymmetries are sexually dimorphic. Instead, many researchers now advocate for a more nuanced view of sex/gender differences in hemispheric asymmetries, noting that there is both considerable overlap between the sexes/genders and a significant amount of within-sex/gender variability (Joel et al., 2015; Sanchis-Segura et al., 2019; Hodgetts and Hausmann, 2020b; Joel, 2021). Indeed, while it has been shown that FCAs are relatively stable over time (e.g., Vingerhoets, 2019), several sex/gender-relevant factors have been shown to contribute to variations and dynamic changes in FCAs (e.g., Hausmann, 2019) such as sex hormones (e.g., Hausmann, 2017; Hodgetts and Hausmann, 2018), suggesting that the sex/gender as a factor should be viewed as "an imperfect, temporary proxy for yet unknown factors, such as hormones or sex-linked genes, that explain variation better than sex" (Maney, 2016; p. 220).

In fact, multiple studies have shown that an individual's distinct sex-related hormonal profile can be an important factor in the generation and maintenance of sex/gender differences in hemispheric asymmetry (Weis and Hausmann, 2010; Hodgetts and Hausmann, 2018, 2020b). However, sex hormone levels are characterized by considerable degrees of fluctuation both across short time intervals (e.g., during the menstrual cycle), and across the lifespan (e.g., reducing after menopause). Indeed, many studies investigating menstrual cycle-related effects of sex hormones on FCAs have revealed reduced asymmetries during the late follicular phase (high levels of estradiol), and/or during the postovulatory luteal phase, and increased asymmetries during menstruation (lowest levels of estradiol and progesterone, for reviews see Hausmann, 2017; Hodgetts and Hausmann, 2018). In contrast, other studies showed larger FCAs during the follicular and/or luteal phase in combination with reduced FCAs during menstruation (e.g., Mead and Hampson, 1996; Sanders and Wenmoth, 1998; Wadnerkar et al., 2008; Cowell et al., 2011). Such inconsistencies, sometimes occurring even in the same study (e.g., Mead and Hampson, 1996; Sanders and Wenmoth, 1998), may suggest that some studies are reporting false positives. Alternatively, these inconsistent results may reflect differences between studies with regard to the specific task and test modality under investigation, suggesting that sex-hormonal effects on FCAs are not uniform across different lateralized processes (Hausmann and Bayer, 2010; Hjelmervik et al., 2012; Hodgetts et al., 2015, 2017).

Although different explanations of how sex hormones might influence hemispheric asymmetries have been proposed, one explanation takes the neuromodulatory properties of sex hormones into account. The hypothesis of progesterone-mediated interhemispheric decoupling (Hausmann and Güntürkün, 2000) stated that high levels of progesterone lead to a reduction in interhemispheric inhibition (i.e., activity in the nondominant hemisphere is no longer suppressed by the dominant hemisphere). Specifically, this hypothesis states that higher levels of progesterone can reduce interhemispheric inhibition by suppressing the excitatory responses of neurons to glutamate (Smith et al., 1987) and by enhancing their inhibitory responses to GABA (Smith, 1991), leading to increased bilateral activation and reduced FCAs (e.g., Cook, 1984; Regard et al., 1994). This mechanism presumes that FCAs arise because the hemisphere dominant in a particular task inhibits the nondominant hemisphere via the corpus callosum (Cook, 1984; Chiarello et al., 1989). Although cortico-cortical transmission is primarily excitatory, callosal projections terminate on pyramidal neurons, which subsequently activate GABAergic interneurons (Toyama and Matsunami, 1976), inhibiting the contralateral hemisphere (Innocenti, 1980). Moreover, it has been shown that callosal projections terminate directly on GABAergic interneurons (Conti and Manzoni, 1994). Both mechanisms would result in widespread inhibition of homotopic regions of the nondominant hemisphere by the dominant hemisphere. This hypothesis was supported empirically by several behavioral studies of FCAs (e.g., Hausmann and Güntürkün, 2000; Hausmann et al., 2002, 2013; Bayer et al., 2008), transcranial magnetic stimulation experiments (Hausmann et al., 2006), and fMRI (Weis et al., 2008, 2011).

Subsequent research revealed that high estradiol levels may also be capable of reducing FCAs via an effect on interhemispheric inhibition (Hausmann, 2005; Hausmann et al., 2006; Weis et al., 2008; Hausmann and Bayer, 2010). For example, in an fMRI study, Weis et al. (2008) investigated FCAs in naturally cycling women as they completed a word-matching task during different menstrual cycle phases. Results revealed that high levels of estradiol during the follicular phase were associated with reduced interhemispheric inhibition, and in turn, reduced language-related FCAs compared to other cycle phases. Hausmann et al. (2013) used electroencephalography to directly measure interhemispheric connectivity by using visual-evoked potentials to estimate interhemispheric transfer time (IHTT). The results showed that IHTT from right to left was increased during the luteal phase as compared to the menstrual phase and that this effect was related to high levels of estradiol, rather than progesterone. Further research has investigated the effect of sex hormones on intrahemispheric activity. In one such study, Weis et al. (2011) used fMRI to scan naturally cycling women as they completed a figure-matching task. The results revealed cycle-phase-related reduced functional connectivity within right-hemispheric networks during the luteal phase, as compared to both the menstrual and the follicular phase. Consequently, the authors suggested that sex hormones modulate not only interhemispheric inhibition between homotopic areas (e.g., as seen in Weis et al., 2008) but can also influence intrahemispheric integration, and interhemispheric connectivity between heterotopic brain regions.

Taken together, the findings discussed thus far suggest that like other indices of human brain structure (e.g., total brain size/volume, callosal size/shape) and function (e.g., functional connectivity, task-related activity), hemispheric asymmetries are not sexually dimorphic. Similar arguments have recently been made by Eliot et al. (2021), following a comprehensive review of the literature on sex/gender differences in the human brain. Specifically, Eliot et al. (2021) stated that sex/gender differences in the brain (and this explicitly included hemispheric asymmetries) are either "trivial," "negligible," "very weak," or "unlikely to be meaningful." Like many other researchers in the field, we agree with the rejection of the "sexual dimorphism" concept in its extreme form. However, the latter points made by Eliot et al. (2021) are similarly extreme and risk overlooking the evidence that some sex/gender differences, while not dimorphisms, do reliably exist and are not necessarily trivial. Indeed, Hirnstein and Hausmann (2021) note that while small effect sizes have little meaning for single events (e.g., comparing one individual with another), they can become meaningful across repeated events. For example, when multiple small sex/gender differences are combined into multivariate analyses using machine learning, classification algorithms can predict an individual's sex/gender with an accuracy between 60% and 80% (Chekroud et al., 2016; Sanchis-Segura et al., 2020; Weis et al., 2020). However, as noted by Sanchis-Segura et al. (2022), caution is needed when interpreting these so-called sex prediction studies as the accuracy of such predictions is dependent on the specific machine learning algorithm used. That is, different algorithms may produce different outcomes from the same dataset because of differences in their internal operations. Thus, such predictions should not be considered generalizable, and the methods used to achieve predictions should not be considered separately from the results (Sanchis-Segura et al., 2022). Moreover, Sanchis-Segura et al. (2022) argued that the accuracy of "sex prediction" should not be taken as evidence of a sexual dimorphism, as the results do not stem from the existence of two sex-typical profiles but rather from a distribution of features that exist as a male-female continuum. Finally, while results across studies are inconsistent and some studies report no sex/gender differences in hemispheric asymmetry, this does not imply that there is no sex/gender effect. Several factors can impact sex/gender differences in the brain such as age, socioeconomic status, education, sexual orientation and identity, gender roles, and sex hormones. For example, it is likely that FCAs are double-coded by stable characteristics (e.g., genetics) and temporary situational aspects, including those related to sex/gender (e.g., sex hormonal environment) (e.g., Hausmann, 2019). As such, our understanding of sex/gender differences in the brain is unlikely to meaningfully progress if research does not take sex/gender-related biologic, psychologic, and social factors (as well as their complex interactions) into account using a biopsychosocial approach.

THE BIOPSYCHOSOCIAL APPROACH

Although the evidence presented above demonstrates that sex hormones can significantly influence hemispheric asymmetries, it is unlikely they are the sole influencing factor related to the generation and maintenance of sex/gender differences in hemispheric asymmetries. Indeed, studies have revealed that environmental factors associated with sex/gender, such as gender roles and stereotypes, can also influence sex/gender differences in brain structure and function (Jäncke, 2018; Hodgetts and Hausmann, 2020b). Moreover, it is likely that this influence is due to a complex interactive relationship between such factors and sex/gender-relevant biologic factors, which can only be fully understood via studies that adopt a biopsychosocial approach.

While the number of studies adopting a biopsychosocial approach toward sex/gender differences is limited in general, those that exist support the notion that gender roles and gender stereotypes can interact with biologic factors in a way that contributes to the generation and maintenance of sex/gender differences in brain structure and function. For example, Belfi et al. (2014) used MRI to investigate the relationship between gender roles and brain structure in 108 children (56 boys and 52 girls). The results showed that masculinity positively correlated with white matter volumes in the frontal lobe, while femininity positively correlated with gray matter volumes in the temporal lobe, even after biologic sex was controlled for. Although the causality of this relationship is not clear, the authors suggested that this effect may represent an environmental influence on sex/gender differences in the brain. Regarding brain function, Wraga et al. (2006) used fMRI to investigate the effects of gender stereotyping on the neural underpinnings of visuospatial performance in 54 women. Significantly poorer performance was found in women exposed to a negative gender stereotype when compared to women given a positive gender stereotype. Additionally, poorer performance in the negative stereotype group was characterized by greater activation in brain regions associated with emotional processing, including the orbital and medial frontal gyri, and the anterior cingulate cortex. These results suggest that gender stereotypes can influence the efficiency of neural activity in sex/gender-sensitive tasks. Hausmann et al. (2009) investigated the effect of gender stereotypes in both men and women using a battery of sex/gender-sensitive cognitive tasks (including mental rotation and verbal fluency) and hormone assays to directly measure testosterone levels in all participants. The expected sex/gender differences favoring men and women were found for mental rotation and verbal fluency, respectively. However, the sex/gender difference in mental rotation was significantly driven by the gender-primed group. Moreover, testosterone levels in the gender stereotype group were 60% higher than those in the control group. These results suggest that the effect of gender priming on cognitive performance can be mediated by the accompanying changes in sex hormone levels, probably depending on whether participants interpreted the testing situation after priming as challenging or threatening. Similar results were also reported by Pletzer et al. (2019) who showed that high levels of testosterone, in conjunction with high levels of self-reported masculinity, yielded highly accurate mental rotation scores regardless of biologic sex. Taken together, these studies show that an interactive relationship exists between sex/gender relevant biologic, psychologic, and social factors that are capable of influencing sex/gender differences in cognition. As such, it follows that the same interactive relationship could influence sex/gender differences in FCAs (e.g., as assessed using lateralized tasks and/or neuroimaging). However, to date, no studies using the biopsychosocial approach to directly test this relationship have been conducted.

Crucially, to adopt a truly biopsychosocial approach to the study of sex/gender differences in the brain (including the study of sex/gender effects on asymmetries), the distinction between "sex" and "gender" should be acknowledged, and the effects of both should be accounted for in future studies. In doing so, it is also important to recognize that neither biologic sex nor gender can accurately be conceptualized as binary (Karkazis, 2019). However, research to date has been limited by a lack of universal agreement on the definitions of "sex" and "gender," resulting in the majority of studies interested in sex/gender differences involving the separation of participants into dichotomous male/female groups based on self-identified sex (Rauch and Eliot, 2022). This results in an approach that imposes a binary onto constructs that are inherently nonbinary, which in turn, cannot accurately investigate the multitude of sex/gender-related factors (e.g., masculinity, femininity) that interact to produce or maintain sex/gender differences. Essentially, studies of the brain to date have not appropriately acknowledged the impact of gender, and "binary sex may not be a precise enough independent variable for partitioning subjects when it is often gendered behavior we seek to understand" (Rauch and Eliot, 2022, p. 1). Studies of sex/gender differences in FCAs are also subject to this limitation, with the exception of studies comparing FCAs between transgender and cisgender participants (though such studies also treat gender as a binary construct, e.g., Cohen-Kettenis et al., 1998; Wisniewski et al., 2005; Beking et al., 2020). However, recent advances in the operationalization and understanding of gender have resulted in a small number of neuroimaging studies that attempted to incorporate a nonbinary, spectrum-based approach to gender in their analyses of sex/gender differences in the brain. In one such study, Zhang et al. (2021), combined connectivity data from four large, international databases ($n=9620$) to create a "brain-gender continuum" of resting state networks. This continuum was then mapped onto measures of internalizing and externalizing behaviors from one of the databases ($n=719$), which are known to present differently according to an individual's masculinity/femininity. This mapping resulted in a U-shaped curve according to internalizing behavior, with "androgynous" (i.e., those in the middle of the continuum) brains being associated with the lowest levels of internalization. This finding, in conjunction with those from similar studies (Phillips et al., 2019; Vosberg et al., 2021), supports the notion of adopting nonbinary sex/gender measures as independent variables, as opposed to separating participants by self-identified binary sex. Moreover, adopting such methods may allow for a truly biopsychosocial approach to the study of sex/gender effects in hemispheric asymmetry.

CLINICAL IMPLICATIONS

Thus far, in this chapter, we have argued that some sex/gender differences in hemispheric asymmetries exist, but there is a high level of inter- and intraindividual variability (due to multiple sex/gender-related factors). Moreover, we have argued that these differences are not trivial. As such, in the following section, we will discuss the potential clinical implications associated with sex/gender differences in FCAs and highlight how future research (adopting a biopsychosocial approach) may contribute to the development of evidence-based, sex/gender-sensitive treatment for a range of diagnoses.

The potential for sex/gender differences in recovery following a unilateral stroke or brain injury has been a particular area of research focus concerning the clinical implications of sex/gender differences in FCAs. Such studies have particularly focused on sex/gender differences in the impact of and recovery from left-hemispheric lesions (due to stroke or injury), with regard to language functioning. For example, early researchers argued that if language FCAs are smaller in women (i.e., more bilaterally organized) compared to men, then they should show increased resilience to unilateral left-hemispheric damage compared to men (Lansdell, 1961, 1962; McGlone, 1980). That is, following left-hemispheric damage, patients with a more bilaterally organized brain should be capable of compensating via the right hemisphere, resulting in less severe language impairment and faster recovery compared to patients with asymmetrically organized brains. However, evidence in support of this notion is inconsistent. Several meta-analyses have suggested that aphasia following left-hemispheric damage is not more likely in male patients (Sommer et al., 2004; Plowman et al., 2012; Watila and Balarabe, 2015), leading some researchers to largely dismiss the notion of a clinically relevant relationship between sex/gender and language FCAs (Wallentin, 2020).

The clinical implications of sex/gender differences in FCAs and the effect of sex/gender-related factors on FCAs have recently been extended to psychiatric conditions, particularly those characterized by psychosis (Hodgetts and Hausmann, 2020a). Indeed, a large body of evidence suggests that the brains of patients with schizophrenia are characterized by atypical FCAs when compared to healthy controls (Oertel-Knochel and Linden, 2011; Crow et al., 2013; Ocklenburg et al., 2015). Of particular relevance are the multiple studies demonstrating reduced activation of the left hemisphere during language processing and/or production using fMRI (Weiss et al., 2004, 2006; Dollfus et al., 2005; Razafimandimby et al., 2007; Zhang et al., 2008; Oertel et al., 2010; Bleich-Cohen et al., 2012; Alary et al., 2013; Leroux et al., 2015) electroencephalography (Angrilli et al., 2009; Jalili et al., 2010) and positron emission tomography (Artiges et al., 2000). Moreover, a separate stream of research has demonstrated the neuroleptic effects of sex hormones, with estrogens (e.g., estradiol) consistently being shown to have therapeutic effects for women with schizophrenia (Kulkarni et al., 2001; Liao et al., 2002; Akhondzadeh et al., 2006; Bergemann et al., 2007; Kulkarni et al., 2008; Kulkarni, 2009; Ghafari et al., 2013; Riecher-Rössler et al., 2018, for a review see Begemann et al., 2012). Therefore, an open question remains regarding the potential for a functional link between FCAs, sex hormones, and psychotic symptoms. Although it is currently not clear whether atypical FCAs are the cause or consequence of psychotic symptoms, one might speculate that sex-hormonal modulations of FCAs will also modulate psychotic symptoms. Indeed, Bayer and Hausmann (2011) suggested that estradiol may produce an antipsychotic effect in schizophrenia by modulating the atypical FCAs demonstrated in such patients. However, given that patients with schizophrenia typically demonstrate reduced FCAs (Løberg et al., 1999, 2004; Hahn et al., 2011; meta-analysis by Ocklenburg et al., 2013) and atypical interhemispheric interaction, such as excessive callosal information (e.g., David, 1993; Narr et al., 2003) or longer interhemispheric transfer (IHTT) (Henshall et al., 2012), compared to healthy controls, it seems unlikely that high levels of estradiol reduce psychotic symptoms by "normalizing" FCAs (i.e., increasing lateralization to a level comparable with healthy controls). Instead, estradiol might support bilateral brain organization in patients with schizophrenia, thereby increasing the difference in FCAs seen between patients and controls. Indeed, several studies now suggest that reduced FCAs (i.e., more bilateral brain organization) are a marker of neurocompensatory functioning in specific populations including healthy elderly people (Cabeza, 2002; Bracco et al., 2011; for a review see Wingfield and Grossman, 2006) and people with neurodevelopmental conditions (e.g., dyslexia, Shaywitz et al., 2007; Illingworth and Bishop, 2009; Bishop, 2013). However, the precise nature of the relationship between FCAs, sex/gender-related factors, and schizophrenia is unclear, as it is yet to be directly investigated.

Crucially, most of the clinical research thus far that has included sex/gender as an independent variable with respect to FCAs and/or clinical outcomes is subject to the same limitations discussed in the previous section. That is, they either adhere to a binary conceptualization of sex/gender (i.e., grouping participants according to self-identified sex/gender) or do not consider other sex/gender relevant biologic factors (e.g., sex hormones) or they do not consider gender/sex-relevant psychologic and environmental influences at all. For example, studies investigating the potential relationship between sex/gender differences in FCAs and language outcomes following unilateral stroke have not considered the probable role of sex hormones in such a relationship. This is surprising given that sex hormones, including estradiol and progesterone, have been shown to influence both language-related FCAs (Wadnerkar et al., 2008; Cowell et al., 2011; Hodgetts et al., 2015) and recovery from brain injury (e.g., stroke, Stein, 2001; Spychala et al., 2017; Sohrabji et al., 2019). Moreover, studies that have investigated the effect of sex hormones on outcomes following brain injury or stroke have not accounted for sex/gender-related factors beyond sex assigned at birth, such as gender identity, stereotypes, or behaviors associated with masculinity/femininity that are relevant to clinical outcomes (e.g., internalizing vs externalizing). This is important because, as discussed in Section "The Biopsychosocial Approach," such factors have been shown to interact with the biologic components of sex/gender (e.g., sex hormones Hausmann et al., 2009). Thus, future research investigating the potential relationship between sex/gender, FCAs, clinical outcomes (particularly language recovery following unilateral stroke), and psychiatric treatment (particularly estrogen adjunctive treatment in schizophrenia) should also aim to adopt a truly biopsychosocial approach by adopting nonbinary sex/gender measures as independent variables.

CONCLUSIONS

In sum, research to date supports the notion that sex/gender differences in the brain and behavior exist at multiple levels, including structural and functional cerebral asymmetries. However, it is not accurate to conceptualize the brain as "sexually dimorphic" and an oversimplified binary (e.g., "male brain," "female brain") is misleading and should be avoided in future research. Although many of these sex/gender differences are characterized by small effect sizes, some of these differences are reliably found and should not be considered trivial. Indeed, it is possible that the small effect sizes are driven by studies to date have not appropriately acknowledged the impact of gender, instead relying on self-identified sex and imprecise male-female binary groups. Recent advances in the operationalization and understanding of gender will allow future studies to incorporate a nonbinary, spectrum-based approach to gender in their analyses of sex/gender differences in hemispheric asymmetries. Such advances are in line with the biopsychosocial approach, which calls for the acknowledgment of include biologic factors such as levels of sex hormones and sex-linked genes and environmental factors such as gender roles and stereotypes. Finally, acknowledgment of the full range of sex/gender-related factors will allow future research to directly investigate the potential relationship between sex/gender, hemispheric asymmetries, and treatment outcomes, thus facilitating the development of evidence-based, sex/gender-sensitive treatment for a range of neurologic and psychiatric diagnoses.

References

Akhondzadeh S, Rezaei F, Larijani B et al. (2006). Correlation between testosterone, gonadotropins and prolactin and severity of negative symptoms in male patients with chronic schizophrenia. Schizophr Res 84: 405–410. https://doi.org/10.1016/j.schres.2006.02.008.

Alary M, Delcroix N, Leroux E et al. (2013). Functional hemispheric lateralization for language in patients with schizophrenia. Schizophr Res 149: 42–47. https://doi.org/10.1016/j.schres.2013.06.003.

Angrilli A, Spironelli C, Elbert T et al. (2009). Schizophrenia as failure of left hemispheric dominance for the phonological component of language. PLoS One 4. https://doi.org/10.1371/journal.pone.0004507.

Artiges E, Martinot J-L, Verdys M et al. (2000). Altered hemispheric functional dominance during word generation in negative schizophrenia. Schizophr Bull 26: 709–721. https://doi.org/10.1093/oxfordjournals.schbul.a033488.

Bayer U, Hausmann M (2011). Sex hormone therapy and functional brain plasticity in postmenopausal women. Neuroscience 191: 118–128. https://doi.org/10.1016/j.neuroscience.2011.03.034.

Bayer U, Kessler N, Güntürkün O et al. (2008). Interhemispheric interaction during the menstrual cycle. Neuropsychologia 46: 2415–2422. https://doi.org/10.1016/j.neuropsychologia.2008.02.028.

Begemann MJH, Dekker CF, van Lunenburg M et al. (2012). Estrogen augmentation in schizophrenia: a quantitative review of current evidence. Schizophr Res 141: 179–184. https://doi.org/10.1016/j.schres.2012.08.016.

Beking T, Burke SM, Geuze RH et al. (2020). Testosterone effects on functional amygdala lateralization: a study in adolescent transgender boys and cisgender boys and girls. Psychoneuroendocrinology 111: 104461. https://doi.org/10.1016/j.psyneuen.2019.104461.

Belfi AM, Conrad AL, Dawson J et al. (2014). Masculinity/femininity predicts brain volumes in normal healthy children. Dev Neuropsychol 39: 25–36. https://doi.org/10.1080/87565641.2013.839681.

Bergemann N, Abu-Tair F, Strowitzki T (2007). Estrogen in the treatment of late-onset schizophrenia. J Clin Psychopharmacol 27: 717–720. https://doi.org/10.1097/jcp.0b013e31815a6af9.

Bishop DVM (2013). Cerebral asymmetry and language development: cause, correlate or consequence? Why are some children poor at language learning? Science 340. https://doi.org/10.1126/science.1230531.

Bleich-Cohen M, Sharon H, Weizman R et al. (2012). Diminished language lateralization in schizophrenia corresponds to impaired interhemispheric functional connectivity. Schizophr Res 134: 131–136.

Bracco L, Bessi V, Alari F et al. (2011). Cerebral hemodynamic lateralization during memory tasks as assessed by functional transcranial Doppler (fTCD) sonography: effects of gender and healthy aging. Cortex 47: 750–758. https://doi.org/10.1016/j.cortex.2010.03.007.

Cabeza R (2002). Hemispheric asymmetry reduction in older adults: the HAROLD model. Psychol Aging 17: 85–100. https://doi.org/10.1037/0882-7974.17.1.85.

Cahill L (2017). Sex influences exist at all levels of human brain function, principles of gender-specific medicine. 121–128. https://doi.org/10.1016/B978-0-12-803506-1.00034-6.

Chekroud AM, Ward EJ, Rosenberg MD et al. (2016). Patterns in the human brain mosaic discriminate males from females. Proc Natl Acad Sci U S A 113: E1968. https://doi.org/10.1073/pnas.1523888113.

Chiarello C, McMahon MA, Schaefer K (1989). Visual cerebral lateralization over phases of the menstrual cycle: a preliminary investigation. Brain Cogn 11: 18–36. https://doi.org/10.1016/0278-2626(89)90002-X.

Choi YH, Yun JY, Kim BH et al. (2020). Gender-related and hemispheric effects in cortical thickness-based hemispheric brain morphological network. Biomed Res Int 2020. https://doi.org/10.1155/2020/3560259.

Cohen-Kettenis P, van Goozen SHM, Doorn CD et al. (1998). Cognitive ability and cerebral lateralisation in transsexuals. Psychoneuroendocrinology 23: 631–641. https://doi.org/10.1016/S0306-4530(98)00033-X.

Conti F, Manzoni T (1994). The neurotransmitters and postsynaptic actions of callosally projecting neurons. Behav Brain Res 64: 37–53.

Cook ND (1984). Homotopic callosal inhibition. Brain Lang 23: 116–125. https://doi.org/10.1016/0093-934X(84)90010-5.

Cowell PE, Ledger WL, Wadnerkar MB et al. (2011). Hormones and dichotic listening: evidence from the study of menstrual cycle effects. Brain Cogn 76: 256–262. https://doi.org/10.1016/j.bandc.2011.03.010.

Crow TJ, Chance SA, Priddle TH et al. (2013). Laterality interacts with sex across the schizophrenia/bipolarity continuum: an interpretation of meta-analyses of structural MRI. Psychiatry Res 210: 1232–1244. https://doi.org/10.1016/j.psychres.2013.07.043.

David AS (1993). Callosal transfer in schizophrenia: too much or too little? J Abnorm Psychol 102: 573.

Dollfus S, Razafimandimby A, Delamillieure P et al. (2005). Atypical hemispheric specialization for language in right-handed schizophrenia patients. Biol Psychiatry 57: 10202–11028. https://doi.org/10.1016/j.biopsych.2005.01.009.

Dorman MF, Porter RJ (1975). Hemispheric lateralization for speech perception in stutterers. Cortex 11: 181–185. https://doi.org/10.1016/S0010-9452(75)80042-6.

Eliot L, Ahmed A, Khan H et al. (2021). Dump the "dimorphism": comprehensive synthesis of human brain studies reveals few male-female differences beyond size. Neurosci Biobehav Rev 125: 667–697. https://doi.org/10.1016/j.neubiorev.2021.02.026.

Galea LAM (2021). Chasing red herrings and wild geese: sex differences versus sex dimorphism. Front Neuroendocrinol 63: 100940. https://doi.org/10.1016/j.yfrne.2021.100940.

Ghafari E, Fararouie M, Shirazi HG et al. (2013). Combination of estrogen and antipsychotics in the treatment of women with chronic schizophrenia. Clin Schizophr Relat Psychoses 6: 172–176. https://doi.org/10.3371/CSRP.GHFA.01062013.

Guadalupe T, Zwiers MP, Wittfeld K et al. (2015). Asymmetry within and around the human planum temporale is sexually dimorphic and influenced by genes involved in steroid hormone receptor activity. Cortex 62: 41–55. https://doi.org/10.1016/j.cortex.2014.07.015.

Hahn C, Neuhaus AH, Pogun S et al. (2011). Smoking reduces language lateralization: a dichotic listening study with control participants and schizophrenia patients. Brain Cogn 76: 300–309. https://doi.org/10.1016/j.bandc.2011.03.015.

Harshman R, Remington R (1976). Sex, language and the brain. Part I: a review of the literature on adult sex differences in lateralization. UCLA Working Papers in Phonetics 31: 86–103.

Hausmann M (2005). Hemispheric asymmetry in spatial attention across the menstrual cycle. Neuropsychologia 43: 1559–1567. https://doi.org/10.1016/j.neuropsychologia.2005.01.017.

Hausmann M (2017). Why sex hormones matter for neuroscience: a very short review on sex, sex hormones, and functional brain asymmetries. J Neurosci Res 95: 40–49. https://doi.org/10.1002/jnr.23857.

Hausmann M (2019). Variations of hemispheric functional segregation in the laterality spectrum: comment on "Phenotypes in hemispheric functional segregation? Perspectives and challenges" by Guy Vingerhoets. Phys Life Rev. https://doi.org/10.1016/j.plrev.2019.08.006.

Hausmann M (2020). Sex/gender differences in brain activity—its time for a biopsychosocial approach to cognitive neuroscience. Cogn Neurosci1–2. https://doi.org/10.1080/17588928.2020.1853087.

Hausmann M, Bayer U (2010). Sex hormonal effects on hemispheric asymmetry and interhemispheric interaction. The two halves of the brain: information processing in the cerebral hemispheres, MIT Press, Cambridge, MA, pp. 253–285. https://doi.org/10.7551/mitpress/9780262014137.003.0176.

Hausmann M, Güntürkün O (2000). Steroid fluctuations modify functional cerebral asymmetries: the hypothesis of progesterone-mediated interhemispheric decoupling. Neuropsychologia 38: 1362–1374. https://doi.org/10.1016/S0028-3932(00)00045-2.

Hausmann M, Becker C, Gather U et al. (2002). Functional cerebral asymmetries during the menstrual cycle: a cross-sectional and longitudinal analysis. Neuropsychologia 40: 808–816. https://doi.org/10.1016/S0028-3932(01)00179-8.

Hausmann M, Tegenthoff M, Sänger J et al. (2006). Transcallosal inhibition across the menstrual cycle: a TMS study. Clin Neurophysiol 117: 26–32. https://doi.org/10.1016/j.clinph.2005.08.022.

Hausmann M, Schoofs D, Rosenthal HES et al. (2009). Interactive effects of sex hormones and gender stereotypes on cognitive sex differences—a psychobiosocial approach. Psychoneuroendocrinology 34: 389–401. https://doi.org/10.1016/j.psyneuen.2008.09.019.

Hausmann M, Hamm JP, Waldie KE et al. (2013). Sex hormonal modulation of interhemispheric transfer time. Neuropsychologia 51: 1734–1741. https://doi.org/10.1016/j.neuropsychologia.2013.05.017.

Hellige JB (1993). Hemispheric asymmetry: whats right and whats left, Harvard University Press, Cambridge, MA (Perspectives in cognitive neuroscience.) pp. xiii, 396.

Henshall KR, Sergejew AA, McKay CM et al. (2012). Interhemispheric transfer time in patients with auditory hallucinations: an auditory event-related potential study. Int J Psychophysiol 84: 130–139.

Hickok G, Poeppel D (2007). The cortical organization of speech understanding. Nature 8: 393–402.

Hirnstein M, Hausmann M (2021). Sex/gender differences in the brain are not trivial—A commentary on Eliot et al. (2021). Neurosci Biobehav Rev 130: 408–409. https://doi.org/10.1016/j.neubiorev.2021.09.012.

Hirnstein M, Hugdahl K, Hausmann M (2019). Cognitive sex differences and hemispheric asymmetry: a critical review of 40 years of research. Laterality 24: 204–252. https://doi.org/10.1080/1357650X.2018.1497044.

Hiscock M, Inch R, Jacek C et al. (1994). Is there a sex difference in human laterality? I. An exhaustive survey of auditory laterality studies from six neuropsychology journals. J Clin Exp Neuropsychol 16: 423–435. https://doi.org/10.1080/01688639408402653.

Hiscock M, Israelian M, Inch R et al. (1995). Is there a sex difference in human laterality? II. An exhaustive survey of visual laterality studies from six neuropsychology journals. J Clin Exp Neuropsychol 17: 590–610. https://doi.org/10.1080/01688639508405148.

Hiscock M, Inch R, Hawryluk J et al. (1999). Is there a sex difference in human laterality? III. An exhaustive survey of tactile laterality studies from six neuropsychology journals. J Clin Exp Neuropsychol 17–28. https://doi.org/10.1076/jcen.21.1.17.944. United Kingdom: Taylor and Francis.

Hiscock M, Perachio N, Inch R (2001). Is there a sex difference in human laterality? IV. An exhaustive survey of dual-task interference studies from six neuropsychology journals. J Clin Exp Neuropsychol 23: 137–148. https://doi.org/10.1076/jcen.23.2.137.1206.

Hjelmervik H, Westerhausen R, Osnes B et al. (2012). Language lateralization and cognitive control across the menstrual cycle assessed with a dichotic-listening paradigm. Psychoneuroendocrinology 37: 1866–1875. https://doi.org/10.1016/j.psyneuen.2012.03.021.

Hodgetts S, Hausmann M (2018). The neuromodulatory effects of sex hormones on functional cerebral asymmetries and cognitive control: an update. Z fur Neuropsychol 29. https://doi.org/10.1024/1016-264X/a000224.

Hodgetts S, Hausmann M (2020a). Antipsychotic effects of sex hormones and atypical hemispheric asymmetries. Cortex 127: 313–332. https://doi.org/10.1016/j.cortex.2020.02.016.

Hodgetts S, Hausmann M (2020b). Sex/gender differences in the human brain, in reference module in neuroscience and biobehavioral psychology, Elsevier. https://doi.org/10.1016/b978-0-12-809324-5.24103-5.

Hodgetts S, Hausmann M (2022). Sex/gender differences in brain lateralisation and connectivity. https://doi.org/10.1007/7854_2022_303.

Hodgetts S, Weis S, Hausmann M (2015). Sex hormones affect language lateralisation but not cognitive control in normally cycling women. Horm Behav 74: 194–200. https://doi.org/10.1016/j.yhbeh.2015.06.019.

Hodgetts S, Weis S, Hausmann M (2017). Estradiol-related variations in top-down and bottom-up processes of cerebral lateralization. Neuropsychology 31: 319–327. https://doi.org/10.1037/neu0000338.

Hugdahl K, Westerhausen R (2013). In: K Hugdahl, R Westerhausen (Eds.), The two halves of the brain. MIT Press, Cambridge, MA. https://doi.org/10.7551/mitpress/9780262014137.001.0001.

Illingworth S, Bishop DVM (2009). Atypical cerebral lateralisation in adults with compensated developmental dyslexia demonstrated using functional transcranial Doppler ultrasound. Brain Lang 111: 61–65. https://doi.org/10.1016/j.bandl.2009.05.002.

Innocenti GM (1980). The primary visual pathway through the corpus callosum: morphological and functional aspects in the cat. Arch Ital Biol 118: 124–188.

Jalili M, Meuli R, Do KQ et al. (2010). Attenuated asymmetry of functional connectivity in schizophrenia: a high-resolution EEG study. Psychophysiol 47: 706–716. https://doi.org/10.1111/j.1469-8986.2009.00971.x.

Jäncke L (2018). Sex/gender differences in cognition, neurophysiology, and neuroanatomy [version 1; referees: 3 approved]. F1000Research 7: 1–10. https://doi.org/10.12688/f1000research.13917.1.

Joel D (2021). Beyond the binary: rethinking sex and the brain. Neurosci Biobehav Rev 122: 165–175. https://doi.org/10.1016/j.neubiorev.2020.11.018.

Joel D, Berman Z, Tavor I et al. (2015). Sex beyond the genitalia: the human brain mosaic. Proc Natl Acad Sci U S A 112: 15468–15473. https://doi.org/10.1073/pnas.1509654112.

Karkazis K (2019). The misuses of "biological sex". Lancet 394: 1898–1899. https://doi.org/10.1016/S0140-6736(19)32764-3.

Kimura D (1967). Functional asymmetry of the brain in dichotic listening. Cortex 3: 163–178. https://doi.org/10.1016/S0010-9452(67)80010-8.

Kulkarni J (2009). Oestrogen: a new treatment approach for schizophrenia? Med J Aust 190: 16–17.

Kulkarni J, Riedel A, de Castella AR et al. (2001). Estrogen: a potential treatment for schizophrenia. Schizophr Res 48: 137–144.

Kulkarni J, de Castella A, Fitzgerald PB et al. (2008). Estrogen in severe mental illness: a potential new treatment approach. Arch Gen Psychiatry 65: 955–960. https://doi.org/10.1001/archpsyc.65.8.955.

Kulynych JJ, Vladar K, Jones DW et al. (1994). Gender differences in the normal lateralization of the supratemporal cortex: MRI surface-rendering morphometry of Heschls gyrus and the planum temporale. Cereb Cortex 4: 107–118. https://doi.org/10.1093/cercor/4.2.107.

Lake DA, Bryden MP (1976). Handedness and sex differences in hemispheric asymmetry. Brain Lang 3: 266–282. https://doi.org/10.1016/0093-934X(76)90022-5.

Lansdell H (1961). The effect of neurosurgery on a test of proverbs. Am Psychol 16: 448.

Lansdell H (1962). A sex difference in effect of temporal lobe neurosurgery on design preference. Nature 194: 852–854. https://doi.org/10.1038/194852a0.

Leroux E, Delcroix N, Dollfus S (2015). Left-hemisphere lateralization for language and interhemispheric fiber tracking in patients with schizophrenia. Schizophr Res 165. https://doi.org/10.1016/j.schres.2015.03.028.

Liao DL, Chen H, Lee SM et al. (2002). Estrogen supplementation for female schizophrenics treated with atypical antipsychotics. Gen Hosp Psychiatry 24: 357–359.

Liu H, Stufflebeam SM, Sepulcre J et al. (2009). Evidence from intrinsic activity that asymmetry of the human brain is controlled by multiple factors. Proc Natl Acad Sci U S A 106: 20499–20503. https://doi.org/10.1073/pnas.0908073106.

Løberg EM, Hugdahl K, Green MF (1999). Hemispheric asymmetry in schizophrenia: a "dual deficits" model. Biol Psychiatry 45: 76–81. https://doi.org/10.1016/S0006-3223(98)00219-4.

Løberg EM, Jørgensen HA, Hugdahl K (2004). Dichotic listening in schizophrenic patients: effects of previous vs. ongoing auditory hallucinations. Psychiatry Res 128: 167–174. https://doi.org/10.1016/j.psychres.2004.01.011.

Maney DL (2016). Perils and pitfalls of reporting sex differences. Philos Trans R Soc Lond B Biol Sci 371. https://doi.org/10.1098/rstb.2015.0119.

McGlone J (1980). Sex differences in human brain asymmetry: a critical survey. Behav Brain Sci 3: 215–227. https://doi.org/10.1017/S0140525X00004398.

McGlone J, Davidson W (1973). The relation between cerebral speech laterality and spatial ability with special reference to sex and hand preference. Neuropsychologia 11: 105–113. https://doi.org/10.1016/0028-3932(73)90070-5.

McGlone J, Kertesz A (1973). Sex differences in cerebral processing of visuospatial tasks. Cortex 9: 313–320. https://doi.org/10.1016/s0010-9452(73)80009-7.

McKeever WF, VanDeventer AD (1977). Visual and auditory language processing asymmetries: influences of handedness, familial sinistrality, and sex. Cortex 13: 225–241. https://doi.org/10.1016/S0010-9452(77)80033-6.

Mead LA, Hampson E (1996). Asymmetric effects of ovarian hormones on hemispheric activity: evidence from dichotic and tachistoscopic tests. Neuropsychology 10: 578–587. https://doi.org/10.1037/0894-4105.10.4.578.

Narr KL, Green MF, Capetillo-Cunliffe L et al. (2003). Lateralized lexical decision in schizophrenia: hemispheric specialization and interhemispheric lexicality priming. J Abnorm Psychol 112: 623.

Nielsen JA, Zielinski BA, Ferguson MA et al. (2013). An evaluation of the left-brain vs. right-brain hypothesis with resting state functional connectivity magnetic resonance imaging. PLoS One 8. https://doi.org/10.1371/journal.pone.0071275.

Ocklenburg S, Westerhausen R, Hirnstein M et al. (2013). Auditory hallucinations and reduced language lateralization in schizophrenia: a meta-analysis of dichotic listening studies. J Int Neuropsychol Soc 19: 410–418. https://doi.org/10.1017/S1355617712001476.

Ocklenburg S, Gunturkun O, Hugdahl K et al. (2015). Laterality and mental disorders in the postgenomic age: a closer look at schizophrenia and language lateralization. Neurosci Biobehav Rev 59: 100–110. https://doi.org/10.1016/j.neubiorev.2015.08.019.

Oertel V, Knochel C, Rotarska-Jagiela A et al. (2010). Reduced laterality as a trait marker of schizophrenia—evidence from structural and functional neuroimaging. J Neurosci 30: 2289–2299. https://doi.org/10.1523/jneurosci.4575-09.2010.

Oertel-Knochel V, Linden DEJ (2011). Cerebral asymmetry in schizophrenia. Neuroscientist 17: 456–467. https://doi.org/10.1177/1073858410386493.

Phillips OR, Onopa AK, Hsu V et al. (2019). Beyond a binary classification of sex: an examination of brain sex differentiation, psychopathology, and genotype. J Am Acad Child Adolesc Psychiatry 58: 787–798. https://doi.org/10.1016/j.jaac.2018.09.425.

Pletzer B, Jäger S, Hawelka S (2019). Sex hormones and number processing: progesterone and testosterone relate to hemispheric asymmetries during number comparison. Horm Behav 115: 104553. https://doi.org/10.1016/j.yhbeh.2019.07.001.

Plowman E, Hentz B, Ellis C (2012). Post-stroke aphasia prognosis: a review of patient-related and stroke-related factors. J Eval Clin Pract 18: 689–694. https://doi.org/10.1111/j.1365-2753.2011.1650.x.

Rauch JM, Eliot L (2022). Breaking the binary: gender versus sex analysis in human brain imaging. Neuroimage 264: 119732. https://doi.org/10.1016/j.neuroimage.2022.119732.

Razafimandimby A, Maiza O, Herve PY et al. (2007). Stability of functional language lateralization over time in schizophrenia patients. Schizophr Res 94: 197–206. https://doi.org/10.1016/j.schres.2007.04.011.

Regard M, Cook ND, Wieser HG et al. (1994). The dynamics of cerebral dominance during unilateral limbic seizures. Brain 117: 91–104. https://doi.org/10.1093/brain/117.1.91.

Riecher-Rössler A, Butler S et al. (2018). Sex and gender differences in schizophrenic psychoses: a critical review. Archives of Women's Mental. Health 21: 627–648.

Sanchis-Segura C, Ibañez-Gual MV, Adrián-Ventura J et al. (2019). Sex differences in gray matter volume: How many and how large are they really? Biol Sex Differ 10: 1–19. https://doi.org/10.1186/s13293-019-0245-7.

Sanchis-Segura C, Ibañez-Gual MV, Aguirre N et al. (2020). Effects of different intracranial volume correction methods on univariate sex differences in grey matter volume and multivariate sex prediction. Sci Rep 10: 12953.

Sanchis-Segura C, Aguirre N, Cruz-Gómez AJ et al. (2022). Beyond "sex prediction": estimating and interpreting multivariate sex differences and similarities in the brain. Neuroimage 257: 119343. https://doi.org/10.1016/j.neuroimage.2022.119343.

Sanders G, Wenmoth D (1998). Verbal and music dichotic listening tasks reveal variations in functional cerebral asymmetry across the menstrual cycle that are phase and task dependent. Neuropsychologia 36: 869–874. https://doi.org/10.1016/S0028-3932(98)00022-0.

Scott S, Hynd GW, Hunt L et al. (1979). Cerebral speech lateralization in the Native American Navajo. Neuropsychologia 17: 89–92. https://doi.org/10.1016/0028-3932(79)90026-5.

Shapleske J, Rossell SL, Woodruff PWR et al. (1999). The planum temporale: a systematic, quantitative review of its structural, functional and clinical significance. Brain Res Rev 29: 26–49. https://doi.org/10.1016/S0165-0173(98)00047-2.

Shaywitz BA, Skudlarski P, Holahan JM et al. (2007). Age-related changes in reading systems of dyslexic children. Ann Neurol 61: 363–370. https://doi.org/10.1002/ana.21093.

Smith SS (1991). Progesterone administration attenuates excitatory amino acid responses of cerebellar Purkinje cells. Neuroscience 42: 309–320.

Smith SS, Waterhouse BD, Woodward DJ (1987). Sex steroid effects on extrahypothalamic CNS. II. Progesterone, alone and in combination with estrogen, modulates cerebellar responses to amino acid neurotransmitters. Brain Res 422: 52–62. https://doi.org/10.1016/0006-8993(87)90539-7.

Sohrabji F, Okoreeh A, Panta A (2019). Sex hormones and stroke: beyond estrogens. Horm Behav 111: 87–95. https://doi.org/10.1016/j.yhbeh.2018.10.010.

Sommer IEC (2010). Sex differences in handedness, brain asymmetry, and language lateralization. In: The two halves of the brain: information processing in the cerebral hemispheres, MIT Press, Cambridge, MA, pp. 287–312. https://doi.org/10.7551/mitpress/9780262014137.003.0211.

Sommer IEC, Aleman A, Bouma A et al. (2004). Do women really have more bilateral language representation than men? A meta-analysis of functional imaging studies. Brain 127: 1845–1852. https://doi.org/10.1093/brain/awh207.

Sommer IE, Aleman A, Somers M et al. (2008). Sex differences in handedness, asymmetry of the Planum Temporale and functional language lateralization. Brain Res 1206: 76–88. https://doi.org/10.1016/j.brainres.2008.01.003.

Spychala MS, Honarpisheh P, McCullough LD (2017). Sex differences in neuroinflammation and neuroprotection in ischemic stroke. J Neurosci Res 95: 462–471. https://doi.org/10.1002/jnr.23962.

Stein DG (2001). Brain damage, sex hormones and recovery: a new role for progesterone and estrogen? Trends Neurosci 24: 386–391. https://doi.org/10.1016/S0166-2236(00)01821-X.

Tomasi D, Volkow ND (2012). Laterality patterns of brain functional connectivity: gender effects. Cereb Cortex 22: 1455–1462. https://doi.org/10.1093/cercor/bhr230.

Toyama K, Matsunami K (1976). Convergence of specific visual and commissural impulses upon inhibitory interneurones in cats visual cortex. Neuroscience 1: 107–112. https://doi.org/10.1016/0306-4522(76)90004-X.

Vingerhoets G (2019). Phenotypes in hemispheric functional segregation? Perspectives and challenges. Phys Life Rev. https://doi.org/10.1016/j.plrev.2019.06.002.

Vogel JJ, Bowers CA, Vogel DS (2003). Cerebral lateralization of spatial abilities: a meta-analysis. Brain Cogn 52: 197–204. https://doi.org/10.1016/S0278-2626(03)00056-3.

Vosberg DE, Syme C, Parker N et al. (2021). Sex continuum in the brain and body during adolescence and psychological traits. Nat Hum Behav 5: 265–272. https://doi.org/10.1038/s41562-020-00968-8.

Voyer D (1996). On the magnitude of laterality effects and sex differences in functional lateralities. Laterality 1: 51–84. https://doi.org/10.1080/713754209.

Voyer D (2011). Sex differences in dichotic listening. Brain Cogn 76: 245–255. https://doi.org/10.1016/j.bandc.2011.02.001.

Wadnerkar MB, Whiteside SP, Cowell PE (2008). Dichotic listening asymmetry: sex differences and menstrual cycle effects. Laterality 13: 297–309. https://doi.org/10.1080/13576500701821106.

Wallentin M (2020). Gender differences in language are small but matter for disorders. Handb Clin Neurol 175: 81–102. https://doi.org/10.1016/B978-0-444-64123-6.00007-2.

Watila MM, Balarabe B (2015). Factors predicting post-stroke aphasia recovery. J Neurol Sci 352: 12–18. https://doi.org/10.1016/j.jns.2015.03.020.

Weis S, Hausmann M (2010). Sex hormones: modulators of interhemispheric inhibition in the human brain. Neuroscientist 16: 132–138. https://doi.org/10.1177/1073858409341481.

Weis S, Hausmann M, Stoffers B et al. (2008). Estradiol modulates functional brain organization during the menstrual cycle: an analysis of interhemispheric inhibition. J Neurosci 28: 13401–13410. https://doi.org/10.1523/jneurosci.4392-08.2008.

Weis S, Hausmann M, Stoffers B et al. (2011). Dynamic changes in functional cerebral connectivity of spatial cognition during the menstrual cycle. Hum Brain Mapp 32: 1544–1556. https://doi.org/10.1002/hbm.21126.

Weis S, Patil KR, Hoffstaedter F et al. (2020). Sex classification by resting state brain connectivity. Cereb Cortex 30: 824–835. https://doi.org/10.1093/cercor/bhz129.

Weiss EM, Hofer A, Golaszewski S et al. (2004). Brain activation patterns during a verbal fluency test—a functional MRI study in healthy volunteers and patients with schizophrenia. Schizophr Res 70: 287–291. https://doi.org/10.1016/j.schres.2004.01.010.

Weiss EM, Hofer A, Golaszewski S et al. (2006). Language lateralization in unmedicated patients during an acute episode of schizophrenia: a functional MRI study. Psychiatry Res Neuroimaging 146: 185–190. https://doi.org/10.1016/j.pscychresns.2005.11.003.

Wingfield A, Grossman M (2006). Language and the aging brain: patterns of neural compensation revealed by functional brain imaging. J Neurophysiol 96: 2830–2839. https://doi.org/10.1152/jn.00628.2006.

Wisniewski AB, Prendeville MT, Dobs AS (2005). Handedness, functional cerebral hemispheric lateralization, and cognition in male-to-female transsexuals receiving cross-sex hormone treatment. Arch Sex Behav 34: 167–172. https://doi.org/10.1007/s10508-005-1794-x.

Wraga M, Helt M, Jacobs E et al. (2006). Neural basis of stereotype-induced shifts in women's mental rotation performance. 12–19. https://doi.org/10.1093/scan/nsl041.

Zhang Z, Shi J, Yuan Y et al. (2008). Relationship of auditory verbal hallucinations with cerebral asymmetry in patients with schizophrenia: an event-related fMRI study. J Psychiatr Res 42: 477–486. https://doi.org/10.1016/j.jpsychires.2007.04.003.

Zhang Y, Luo Q, Huang CC et al. (2021). The human brain is best described as being on a female/male continuum: evidence from a neuroimaging connectivity study. Cereb Cortex 31: 3021–3033. https://doi.org/10.1093/cercor/bhaa408.

Section VI

Language

Handbook of Clinical Neurology, Vol. 208 (3rd series)
Cerebral Asymmetries
C. Papagno and P. Corballis, Editors
https://doi.org/10.1016/B978-0-443-15646-5.00020-8

Chapter 18

Functional and structural brain asymmetries in language processing

PATRICK C. TRETTENBREIN[1,2,3] AND ANGELA D. FRIEDERICI[1*]

[1]*Department of Neuropsychology, Max Planck Institute for Human Cognitive and Brain Sciences, Leipzig, Germany*

[2]*International Max Planck Research School on Neuroscience of Communication: Structure, Function, and Plasticity (IMPRS NeuroCom), Leipzig, Germany*

[3]*Experimental Sign Language Laboratory (SignLab), Department of German Philology, University of Göttingen, Göttingen, Germany*

Abstract

The lateralization of language to the left hemisphere of the human brain constitutes one of the classic examples of asymmetry in biology. At the same time, it is also commonly understood that damage to the left hemisphere does not lead to a complete loss of all linguistic abilities. These seemingly contradictory findings indicate that neither our cognitive capacity for language nor its neural substrates are monolithic. This chapter reviews the functional and structural lateralization of the neural substrates of different aspects of language as revealed in the past decades by neuroimaging research. Most aspects of language processing indeed tend to be functionally lateralized to the left hemisphere in the adult human brain. Nevertheless, both hemispheres exhibit a certain equipotentiality with regard to some aspects of language processing, especially with regard to processing meaning and sound. In contrast, the so-called "core language network" in the left hemisphere constitutes a functional and structural asymmetry: This network (i) is crucial for a core aspect of language processing, namely syntax, which refers to the generation of hierarchically structured representations of utterances linking meaning and sound, (ii) matures in accordance with a genetically determined biologic matrix, and (iii) its emergence may have constituted a prerequisite for the evolution of the human language capacity.

INTRODUCTION

Humans are bilateria, a clade of animals whose body plans as embryos are essentially symmetric (i.e., the left and the right side of the body are largely mirror images of each other). This symmetric organization of our bodies is also reflected in a generally very high degree of symmetry in the brain, especially in sensorimotor regions. However, structural asymmetries, such as the imbalance of the left lung having fewer lobes to accommodate the heart, or functional dissymmetries, such as the preferential use of one hand over the other, emerge at different points in time during development. This emergence of asymmetries during development seems to be genetically determined (Corballis, 2020; Wan et al., 2022), though some research suggests that only the presence (or absence) of an asymmetry is genetically encoded and not its directionality (Arning et al., 2013). Importantly, asymmetries of body and brain emerge independently, as documented by patients with situs inversus totalis (i.e., a mirror-inversed organization of organs in the chest and abdomen) who nevertheless usually are right-handed (Matsumoto et al., 1997) and exhibit left-hemispheric dominance for language (Kennedy et al., 1999; Vingerhoets et al., 2018).

The insight that asymmetries in biology and human physiology are not limited to the abovementioned structural asymmetry of the left lung or the functional asymmetry of hand preference, but also extend to the human brain and cognitive functions such as language, dates back to the 19th century. The French physician Marc Dax was the first to link lesions to the left hemisphere of the adult human brain to both an impaired command of language and a disturbance of speech (Dax, 1863, 1865). A few decades later, in a usually much more widely known paper, fellow Frenchman and physician Paul Broca first claimed that lesions specifically to the third convolution of the left frontal cortex cause what he called aphemia, a permanent apraxia of speech (Broca, 1861). Yet, reexaminations of the conserved brains of some of Broca's patients using neuroimaging methods have since shown that their lesions actually extended far into neighboring regions of what is now called Broca's area, as well as into the white matter (Dronkers et al., 2007).

*Correspondence to: Angela D. Friederici, Department of Neuropsychology, Max Planck Institute for Human Cognitive and Brain Sciences, Stephanstraße 1a, 04103, Leipzig, Germany. Tel: +49-3419940112, E-mail: friederici@cbs.mpg.de

In this chapter, we take Dax' and Broca's observations about the lateralization of language and speech as well as early work in neuropsychology, as a starting point for a review of the functional and structural lateralization of the neural substrates of language as revealed by the past decades of research using a variety of neuroimaging methods. Our focus will be on studies using (functional) magnetic resonance imaging ([f]MRI) but we include findings using other methods wherever relevant. We start by briefly considering the language–brain relationship as well as brain asymmetries in general. Next, we discuss the functional and structural properties of the language network in the left hemisphere as well as other regions and networks in both hemispheres that are frequently recruited in tandem with the language network—always in the context of lateralization. This description of language-related structures in the adult human brain is complemented by discussions of our current understanding of the development of the so-called "core language network" in the left hemisphere, which is crucial for hierarchical syntactic processing. Lastly, note that this review is primarily based on studies of spoken and written language processing (see Trettenbrein et al., 2025, this volume, for a review focusing on sign language).

What is "language"?

Modern linguistics distinguishes between individual *languages* as commonly understood, such as English, Swahili, or Japanese Sign Language; and *language* as a neurally implemented cognitive capacity that enables humans to produce and comprehend an, in principle, infinite amount of different hierarchically structured expressions (Chomsky, 1965, 1986, 1995; Everaert et al., 2015; Friederici et al., 2017). On this view, the language system uses individual lexical items (i.e., roughly "words"; but see Krauska and Lau, 2023 for a discussion of this notion) to generate structured representations of utterances, which are mapped to the sensorimotor system during production and perception and to the conceptual–intentional system for thought. *Language* in this technical sense then constitutes a structured mapping from meaning (i.e., semantics) to sound (phonetics/phonology) and vice versa; and the core functionality of the human language system is to generate representations that enable this mapping (Chomsky, 2011). Notice, however, that the mapping to the sensorimotor system is not fixed but develops during language acquisition. This explains why humans can readily acquire *languages* that differ with regard to their phonology, lexicon, grammar, and even in their modality, as in the case of sign languages (Klima et al., 1979; Emmorey, 2015; Trettenbrein et al., 2021).

A peculiarity of the human language system, to date not observed in other animals, appears to be that the representations it generates are structured hierarchically (Berwick et al., 2013; Everaert et al., 2015; Friederici et al., 2017). For example, the English sentence in Fig. 18.1 can be produced and perceived in different modalities (e.g., speech or writing), yet at the representational level, different signals rely on the same abstract hierarchically structed representation. Notice, however, that unlike speech or sign, the different writing systems for spoken languages constitute a relatively recent cultural invention (pictographic and ideographic systems emerged about 6000, alphabetic systems about 3000 years ago) and, unlike natural language acquisition, require explicit instruction (Friederici, 2017). This makes it unlikely that the human brain has adapted to reading and writing in the same way as it has for language. Instead, already existing neural circuitry appears to have been exapted for processing written language (Dehaene et al., 2005; Dehaene and Cohen, 2007). In this chapter, we equally consider studies carried out using spoken and written stimuli, as they have been shown to differentially recruit (primary) sensory cortex but not association cortex (Jobard et al., 2007; Vigneau et al., 2011; Walenski et al., 2019; Uddén et al., 2022).

Evidence for the hypothesis that representations of utterances are indeed structured hierarchically and do not simply constitute a flat concatenation of lexical items into a string can be observed when considering that certain elements of an utterance, such as "an apple," behave as functional units (i.e., constituents), whereas others does not. That is, in a more complex version of our example sentence such as (2) the unit "an apple" can be focused (i.e., moved to the more prominent position at the beginning of the sentence to express emphasis) in a so-called cleft construction, such as (3), whereas "apple in" cannot, as shown in (4):

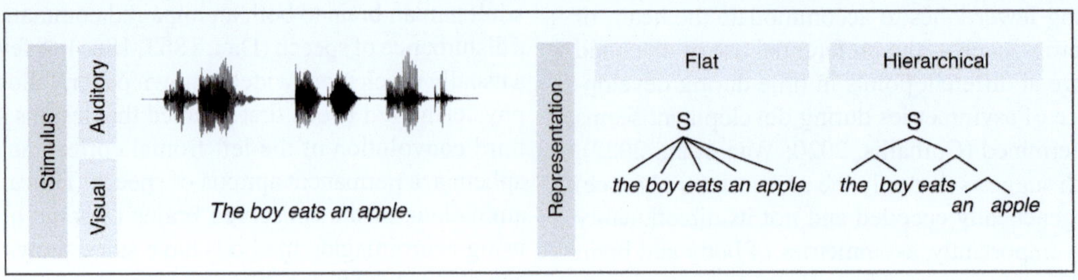

Fig. 18.1. Common linguistic signals and possible syntactic representations. The left panel shows two examples of linguistic signals frequently used as stimuli in experimental studies on language processing, a spectrogram of the recorded speech signal as well as the written version of the sentence *The boy eats an apple*. The right panel illustrates two possible ways of how the syntactic structure of either of these stimuli may be represented in linguistic analysis: Either as a flat structure in which all elements in the tree diagram are connected directly to the root, or as a hierarchical structure where elements form constituents which can be embedded in each other. Evidence from both linguistics and cognitive neuroscience strongly suggests that human brains automatically recover the hierarchical structure from linguistic signals, despite the fact that this structure is not contained in the strictly sequential signal in speech or writing. Illustration is loosely based on materials from Zaccarella and Trettenbrein (2021) and is subject to a Creative Commons (CC-BY 4.0) license.

(1) *The boy eats an apple.*
(2) *The boy eats an apple in the park.*
(3) *It is an apple that the boy eats __ in the park.*
(4) **It is apple in that the boy eats an __ the park.*

This indicates that "an apple" behaves as a functional unit in the examples above, whereas "apple in" does not. See Zaccarella and Trettenbrein (2021) for a more detailed discussion of hierarchy in language and its relevance to research in cognitive neuroscience.

The emerging picture in the cognitive sciences has been that the cognitive machinery that enables the generation of hierarchical syntactic structures linking meaning to a phonologic form is rooted in a universal computational mechanism that is biologically determined, implemented neurally, and specific to our species (Lenneberg, 1969; Hauser et al., 2002; Berwick et al., 2013; Friederici et al., 2017). Accordingly, the cognitive specifics of the language system are understood to be determined by a "biologic matrix" (Lenneberg, 1967) that enables and, at the same time, constrains the development of the language system from infancy into adulthood (Lenneberg, 1964, 1967, 1969; Crain et al., 2016; Yang et al., 2017). How such formal descriptions of the language system and its development relate to neurobiology remains, to an extent, an open question as there is no one-to-one mapping between the foundational elements of linguistic theory (e.g., phonemes, lexical items, etc.) and those of neuroscience (e.g., cell assemblies, brain regions, etc.; Poeppel and Embick, 2013; Embick and Poeppel, 2015). Consequently, it is important to keep in mind that the vast majority of studies in the cognitive neuroscience of language are correlational in nature and that their interpretation depends greatly on the operationalization of linguistic constructs in a particular experiment (van der Burght et al., 2023).

Asymmetries in the brain

Upon first look, the human brain may appear to be a rather symmetrically organized organ (Kandel and Hudspeth, 2013), yet asymmetries may manifest themselves in two major ways (Hervé et al., 2013): (i) An asymmetry is structural in nature when a certain part of the brain [e.g., a lobe, region, Brodmann area (BA), etc.] is more pronounced in one hemisphere than in the other with regard to its volume, size, surface, neuron density, predominant neurotransmitter, or a similar measure. (ii) An asymmetry is functional in nature when the activation observed with a method that indexes neural activity in a certain part of the brain [e.g., the blood oxygen level-dependent (BOLD) in functional magnetic resonance imaging (fMRI)] is more pronounced in one hemisphere than in the other. Intuitively, one may assume that structural and functional asymmetries are linked in a straightforward sense so that, for example, a bigger volume of one region implies also more activation in this region. However, there is no a priori reason for this assumption, and it is likely that structure–function correlations in the context of asymmetry are substantially influenced by the chosen task, measurement, and method. Indeed, structure–function correlations in speech- and language-related asymmetries seem ubiquitous (Toga and Thompson, 2003; Hervé et al., 2013; Bradshaw et al., 2017), but have not always been observed (Keller et al., 2011).

As already alluded to above, the neural substrates of speech and language have a long history of inspiring theories of hemispheric specialization and asymmetry which, to an extent, can be confirmed by contemporary large-scale meta-analyses of neuroimaging studies (Fig. 18.2) based on data from the BrainMap database (Fox and Lancaster, 2002; Laird et al., 2011). Both, speech production and language comprehension, are significantly left-lateralized and recruit regions in superior temporal cortex bilaterally, but differ in their involvement of frontal regions: Speech production recruits bilateral pre-motor and especially motor cortices, whereas spoken and written language comprehension always shows strong and lateralized recruitment of the entire left inferior frontal gyrus (including Broca's area), extending dorsally also into premotor regions. Notice, however, that these meta-analyses do not necessarily consider that there are actually two dichotomies involved here: (i) the difference between speech production and comprehension and (ii) the difference between speech and language (Friederici et al., 2017). The former refers to the production and perception of auditory signals, which form only one way in which language can be externalized (another possibility being the visuo-gestural modality of sign language; Trettenbrein et al., 2021). The latter refers to the processing of the abstract information reconstructed from and, in part, even imposed onto different linguistic signals with regard to, for example, their phonologic, semantic, and syntactic content.

Beyond the expected difference between production and comprehension, different subcomponents of the language system also recruit available neural resources differently during language processing, depending on the task (Bradshaw et al., 2017) and, even more importantly, the nature of the stimulus (i.e., the type of linguistic information being processed). This is evidenced by additional large-scale meta-analyses of studies that focused on the neural resources underlying phonologic, semantic, and syntactic processing in language also based on data from BrainMap (Fig. 18.2, central panels). Across hundreds of fMRI studies with several thousand participants, the general stark and significant left-lateralization of language processing is also confirmed for processing phonology, semantics, and syntax, essentially corroborating older meta-analytical work on lateralization of subcomponents of language (Vigneau et al., 2011). Among the three analyzed subcomponents, phonologic processing is the least left-lateralized. In contrast, syntactic processing (i.e., the neural substrates of the cognitive machinery that enables the generation of hierarchical syntactic representations linking sound to meaning) is almost completely left-lateralized and recruits large portions of the inferior frontal gyrus (including Broca's area) and the posterior temporal cortex. Notice, however, that this differentiation into phonology, semantics, and syntax constitutes a widely used classification highlighting the processing of these different types of linguistic information yet, at the same time, also disregards others (e.g., prosody, morphology, pragmatics, etc.; Everaert et al., 2015; Friederici, 2002, 2017; Hagoort, 2017).

From the meta-analyses of different linguistic domains summarized in Fig. 18.2 we can conclude that there generally is a clear

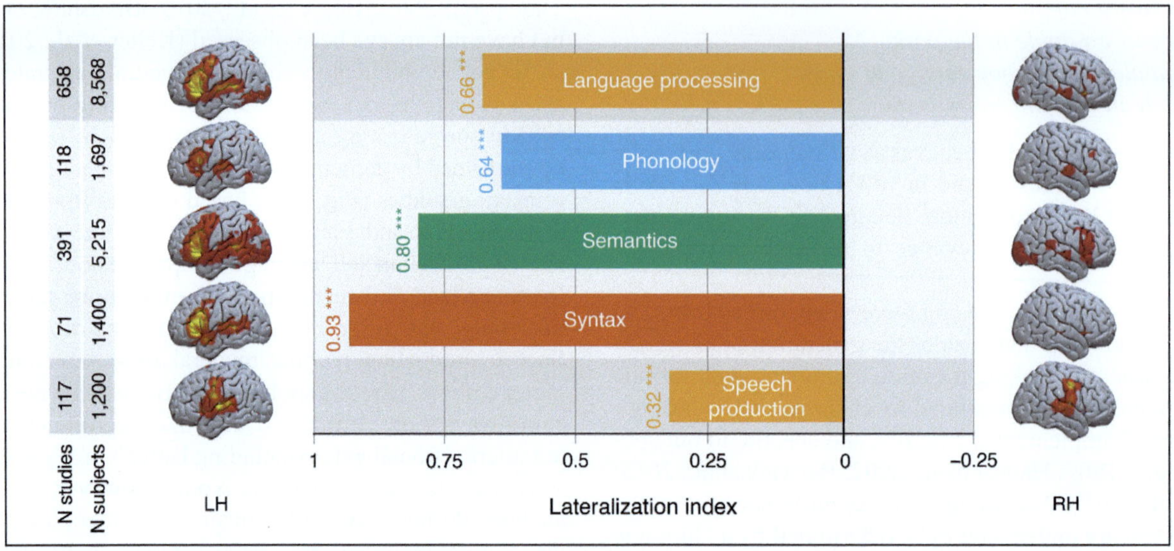

Fig. 18.2. Lateralization in meta-analyses of language processing and speech production. Five different meta-analyses of different aspects of language processing and speech production based on data from the BrainMap database (Fox and Lancaster, 2002; Laird et al., 2011) analyzed using Activation Likelihood Estimation (Eickhoff et al., 2009, 2012, 2017; Turkeltaub et al., 2012) and a weighted lateralization index (Matsuo et al., 2012). Each row contains the following data for one of the different meta-analyses: The number (N) of studies included in the analysis, the total number of study participants in the input data set, an illustration of the convergence across studies for the left hemisphere (LH), a plot of the lateralization index across hemispheres (ranging from 1 indicating complete left-lateralization to −1 indicating complete right-lateralization; significant differences from the mean was performed using permutation tests described in Trettenbrein et al. (2021) and are indicated using common significance levels: * for $P<0.05$, ** for $P<0.01$, and *** for $P<0.001$), and an illustration of the convergence across studies in the right hemisphere (RH). A short description of the data displayed in the five rows from top to bottom: The top row (color-coded in orange) displays information for a general meta-analysis of "language processing," that is all studies in the BrainMap database tagged as "Cognition.Language." The second to fourth row contain information for three different subcomponents of the language systems as identified in the taxonomy of the BrainMap database: Meta-analyses of phonologic ("Cognition.Language.Phonology"; color-coded in blue), semantic ("Cognition.Language.Semantics"; color-coded in green), and syntactic processing ("Cognition.Language.Syntax"; color-coded in brown). The last row (also color-coded in orange) contains another general meta-analysis of speech production including all studies tagged as "Action.Execution.Speech" in the BrainMap database. (The complete details as to how these data were retrieved and how the different analyses were performed are available as online supplementary material to this chapter via Figshare: https://10.6084/m9.figshare.21836079.)

functional leftward lateralization for language processing in the adult brain, with some involvement of the right hemisphere for semantics and phonology (i.e., meaning and sound) and only marginal involvement for syntax (i.e., grammatical structure). Such meta-analytic findings, generalizing about large sets of studies with different stimuli, can provide a useful general picture but should nevertheless be interpreted with some caution against the background of results from individual experiments: For example, in the domain of phonology, it has been shown that segmental aspects reflected in phonetic features are processed in the left hemisphere, whereas suprasegmental aspects reflected in prosodic features are processed in the right hemisphere (Meyer et al., 2002, 2004; Plante et al., 2002). That is, the hemispheres exhibit differential specializations for different types of phonologic information. Significantly, the functional activation associated with the processing of prosodic information shifts from the right to the left hemisphere as soon as this information is relevant for syntactic phrasing (van der Burght et al., 2019).

COMPONENTS OF THE FUNCTIONAL LANGUAGE NETWORK

Even though linguistic signals can be produced and perceived in different modalities (e.g., auditory or visual), the linguistic information (i.e., phonologic, semantic, or syntactic) recovered from or imposed on different signals is nevertheless processed predominantly in left-hemispheric networks in the perisylvian cortex in the adult brain (Fig. 18.3). Meta-analyses of functional studies on language processing using a variety of comprehension paradigms consistently implicate the left inferior frontal gyrus, especially the pars opercularis, superior and middle temporal gyrus, as well as posterior superior temporal sulcus as key regions subserving language processing (Vigneau et al., 2006, 2011; Price, 2010; Walenski et al., 2019; Stefaniak et al., 2021; see also Fig. 18.2, top panel). These regions identified in functional studies largely correspond to the two major canonical language regions first identified in the literature on aphasia: First, the so-called Broca's area, consisting of the cytoarchitectonic regions (i.e., Brodmann areas; Brodmann, 1909) BA 44 and BA 45 in the left inferior frontal gyrus, located ventrally and medially to the frontal operculum (Friederici, 2011; Tremblay and Dick, 2016). Second, the so-called Wernicke's area is in the left superior temporal gyrus, corresponding mostly to cytoarchitectonic BA 22 (though some researchers' definitions have also included BA 42 and BA 41, which are part of the primary auditory cortex or extended the region into BA 39 and BA 40 in the inferior parietal lobule; Bogen and Bogen, 1976; Tremblay and Dick, 2016; Wernicke, 1881).

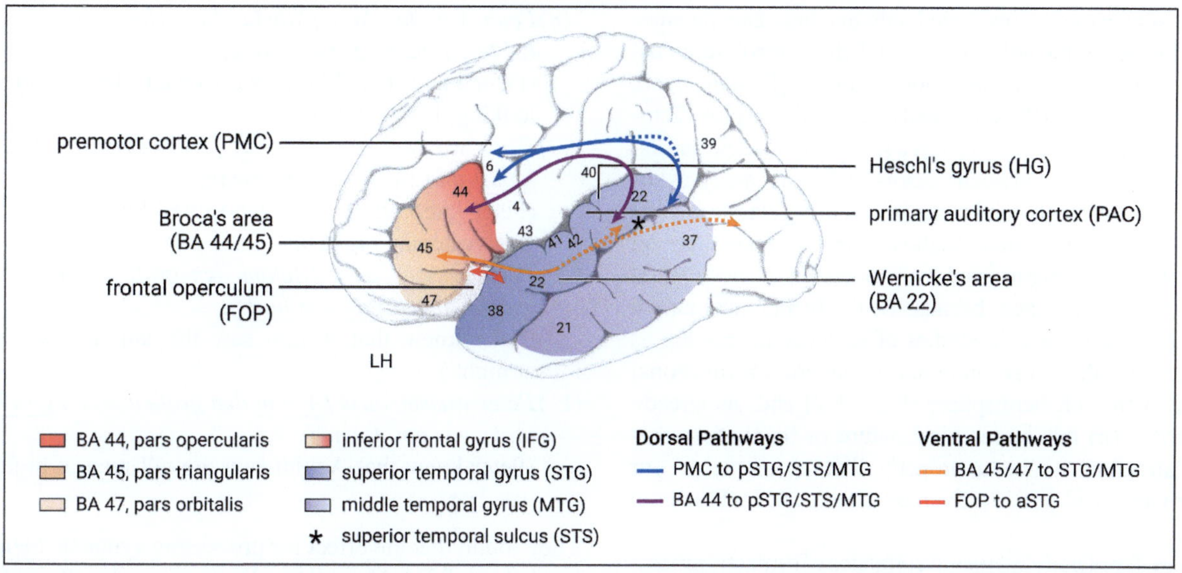

Fig. 18.3. Schematic of the language network. Schematic depiction of the left hemisphere (LH) showing anatomic landmarks and cytoarchitectonic details of language-relevant regions of the cortex. Major gyri that are part of the language network are color-coded: Inferior frontal gyrus (IFG) and its subregions Brodmann areas (BA) 44, BA 45, and BA 47 are depicted in shades of red; superior temporal gyrus (STG) and middle temporal gyrus (MTG) are shown in shades of blue. The location of the language-relevant portion of the superior temporal sulcus (STS) in between STG and MTG is indicated using an asterisk (*). Numbers on the schematic indicate the respective cytoarchitectonic label of a region (i.e., BA) as defined on the basis of histologic studies by Brodmann (1909). Together, BA 44 (pars opercularis) and BA 45 (pars triangularis) form Broca's area. The pars orbitalis (BA 47) is located anteriorly to Broca's area (BA 44 and BA 45) while the frontal operculum (FOP) is located ventrally and medially to it. Premotor cortex (PMC) is located in BA 6. Wernicke's area is located in BA 22, though some definitions also include BA 42 which is part of primary auditory cortex (PAC) together with Heschl's gyrus (HG). The main white-matter fiber tracts connecting these language-relevant regions of cortex dorsally and ventrally are indicated using color-coded arrows: Blue for the pathway connecting PMC to posterior STG (pSTG), STS, and MTG; purple for the pathway connecting BA 44 to pSTG, STS, and MTG; orange for the pathway connecting BA 45 and BA 47 to STG and MTG; and red for the pathway connecting FOP to aSTG. Adapted from Friederici AD (2011). The brain basis of language processing: from structure to function. Physiol Rev 91: 1357–1392. https://doi.org/10.1152/physrev.00006.2011.

In addition, the advent of functional neuroimaging has revealed that other left (perisylvian) regions besides the canonical Broca's area and Wernicke's area are also involved in different aspects of language processing. For example, the frontal operculum has frequently been implicated in studies of syntactic processing (Friederici et al., 2000). An extensive body of work by Pylkkänen and colleagues has demonstrated the involvement of the left anterior temporal lobe (i.e., BA 38) in different aspects of conceptual (i.e., semantic) combination (Bemis and Pylkkanen, 2011; Pylkkänen, 2019). In a synthesis of several fMRI studies, Kuhnke et al. (2023) have recently shown that the left angular gyrus may constitute a multimodal convergence zone for semantic processing. A meta-analysis of semantic processing by Binder et al. (2009) has revealed primarily left-lateralized effects located in posterior inferior parietal lobe, middle temporal gyrus, fusiform and parahippocampal gyri, dorsomedial prefrontal cortex, inferior frontal gyrus, ventromedial prefrontal cortex, and posterior cingulate gyrus. Lastly, large-scale study of sentence processing in the auditory and visual domain with more than 200 participants indicates that left inferior frontal cortex (i.e., BA 44 and 45), bilateral anterior temporal lobe, posterior middle temporal gyrus, and left inferior parietal lobule as major hubs for language processing (Uddén et al:, 2022).

The observation that language processing recruits a wide set of regions primarily in the left perisylvian cortex is not surprising when considering that this notion is unlikely to capture a single monolithic cognitive operation. Instead, "language processing" involves several subcomponents, which have been shown to differentially recruit different parts of language-relevant networks. Accordingly, there is widespread agreement among researchers in the cognitive neuroscience of language that the primarily left-hemispheric frontotemporal networks in the human brain depicted in Fig. 18.3 are involved in, and, to an extent, specialized for processing linguistic information, whereas the exact details of functional associations within these regions and networks continue to be subject of extensive debate (Hickok, 2009; Fedorenko et al., 2011; Friederici, 2011; Friederici and Gierhan, 2013; Fedorenko and Thompson-Schill, 2014; Friederici et al., 2017; Hagoort, 2017, 2019; Matchin and Hickok, 2020). Such disagreements are not surprising because even a notion such as "syntax" is likely still too vague to be mapped to one particular cortical region or network, as syntactic processing at the very least entails the generation of a hierarchical representation and the mapping of this representation to the sensorimotor system during production and perception as well to the conceptual-intentional system for thought.

In our view, experiments using careful and well-controlled manipulations of linguistic constructs have the potential to reveal a more fine-grained picture of the functional correspondences of different parts of the brain, while we focus here on the cortex.

Traditionally, such experiments varied only one linguistic parameter between the experimental stimulus and the control stimulus, following the logic of "pure insertion" (but see Friston et al., 1996). Contemporary fMRI experiments frequently employ factorial experimental designs, which follow a similar basic logic but can also detect interactions between factors, whereas analyses of data from such experiments can be carried out using classic univariate or more recent multivariate methods. We will focus here on experiments that have investigated the processing of hierarchical structures in human language, because (i) the hierarchical nature of language is most evident in studies of sentence processing as a subsystem with the most pronounced pattern of functional lateralization to the left hemisphere (Fig. 18.2) and, as already discussed above, (ii) this hierarchical nature of linguistic representations also seems to be a property that is unique to our species (Berwick et al., 2013; Everaert et al., 2015; Friederici et al., 2017).

One strategy for identifying neural correlates of syntactic processing has been to systematically vary, step by step, the complexity of syntactic structures (but leaving the number of words identical) as done by Friederici et al. (2005). All example sentences from this study in (5)–(7) translate into English as *Today, the grandfather gifted the boy a lollipop.* Yet, the mandatory case marking of functional elements (i.e., determiners) in German allows elements of the sentence to be "scrambled," meaning that phrases can be moved from their canonical position, making their processing more complex and, presumably, more effortful with regard to cognitive resources.

(5) *Heute hat [der Opa]*$_1$ *[dem Jungen]*$_2$ *[den Lutscher]*$_3$ *geschenkt.*

(6) *Heute hat [dem Jungen]*$_2$ *[der Opa]*$_1$ __$_2$ *[den Lutscher]*$_3$ *geschenkt.*

(7) *Heute hat [dem Jungen]*$_2$ *[den Lutscher]*$_3$ *[der Opa]*$_1$ __$_2$ __$_3$ *geschenkt.*

The authors found that the necessary reordering of non-canonical (i.e., scrambled) sentences activated BA 44. Interestingly, a look at the percent signal change in this region revealed that the activation systematically increased as a function of the number of scrambled elements, with sentences like (5) with the lowest and sentences like (7) with the highest activity. Other studies using similar designs have also observed activation of BA 44 (Meyer et al., 2012) or the inferior frontal gyrus (Röder et al., 2002; Grewe et al., 2005).

A possible confound in these studies is the increased working memory demand, which is why some researchers who have sought to investigate syntactic complexity turned to a related phenomenon, that is, sentences with hierarchically embedded structures. Makuuchi et al. (2009) systematically varied the amount of hierarchical embedding in German sentences as listed in (8)–(11). Example sentences (8) and (9) are high and (10) and (11) are low in syntactic complexity, whereas (8) and (10) are high and (9) and (11) are low in working memory requirements due to the differences in syntactic distance (illustrated by underlining the dependent elements).

(8) *Peter wusste, dass [Maria, [die Hans, [der gut aussah], liebte], Johann geküsst hatte].*
(Peter knew that Maria, who loved Hans, who was good looking, kissed Johann.)

(9) *Peter wusste, dass [Maria, [die weinte], Johann geküsst hatte] und zwar gestern Abend.*
(Peter knew that Maria, who cried kissed Johann and that was yesterday night.)

(10) *Peter wusste, dass [Achim den großen Mann gestern am späten Abend gesehen hatte].*
(Peter knew that Achim saw the tall man yesterday late at night.)

(11) *Peter wusste, dass [Achim den großen Mann gesehen hatte] und zwar am Abend.*
(Peter knew that Achim saw the tall man at night and that was late.

They found that the effect for processing syntactic hierarchy was located in left BA 44, and was distinct from the effect of sentence-related working memory processes, which was localized to the inferior frontal sulcus (Makuuchi et al., 2009). These findings have since been conceptually replicated with an adaptation of the basic experimental design to Japanese observing an effect for syntactic hierarchy in left BA 44 and left posterior middle temporal gyrus (Iwabuchi et al., 2019).

Because processing sentences usually entails the simultaneous constructing and processing of meaning (i.e., semantics) in addition to the hierarchical relations between lexical items, researchers have used experimental paradigms relying on the so-called artificial grammars (e.g., Bahlmann et al., 2008; Chen et al., 2021a,b). Usually, such artificial grammars use meaningless syllables (e.g., /de/, /bo/, etc.), which can be combined according to rules of varying grammatical complexity and participants are taught these associations prior to scanning. For example, a study by Friederici et al. (2006) taught participants two different types of artificial grammar: First, a so-called finite-state grammar where rules specify only the local linear relationship to other elements (i.e., category A is always followed by category B, resulting in ABABAB structures). Second, a so-called phrase-structure grammar where rules can entail long-distance dependencies (i.e., n elements from category A are always followed by n elements from category B, resulting in A^nB^n structures). The authors found that processing of finite-state grammars activated the left frontal operculum and right superior temporal sulcus, whereas processing of phrase-structure grammars activated left BA 44 and left middle temporal gyrus, in addition to the left frontal operculum and right superior temporal sulcus (Friederici et al., 2006). These results clearly implicate the core language network in the left hemisphere for processing grammars that rely on recursive rules (i.e., somewhat similar to natural language) and suggest that while both the frontal operculum and the posterior portion of Broca's area (i.e., BA 44) are involved in "syntactic" processing in the broader sense, processing hierarchical representations recruits also the latter in combination with posterior superior temporal gyrus and sulcus (Friederici et al., 2009; Goucha et al., 2017).

Lastly, yet another strategy adopted by many researchers to isolate the hierarchical relations of words processed in a sentential context has been to contrast grammatically correct sentences with mere word lists. In a meta-analysis reviewing studies that have employed such paradigms, Zaccarella et al. (2017) found that the meta-analytic convergence mass across studies was strongly influenced by the linguistic properties of the word lists that were used as part of the respective contrast analyses: When the word list conditions also included function words that enabled the build-up of local phrase structure (e.g., *money, the, [the, client], washed]*), the convergence mass was located in the anterior portion of Broca's area (i.e., left BA 45), the left posterior temporal gyrus and sulcus, as well as in left anterior temporal lobe. However, when the word list conditions used in the experiments contained only content words and thus did not enable the build-up of local phrase structure, the convergence mass was located in the posterior portion of Broca's area (i.e., left BA 44) as well as in the left posterior temporal gyrus und sulcus. Accordingly, the authors suggest that these two regions in the left hemisphere constitute the core system for the generation of hierarchical representations in language.

From the data reviewed here we can conclude that there is a strong left lateralization of the functional language network recruited for processing syntactic information in the typically developed mature brain.

COMPONENTS OF THE STRUCTURAL LANGUAGE NETWORK

The functional connectivity between the different perisylvian regions forming the core and extended language network in the left hemisphere described in the previous section is enabled structurally by myelinated fiber pathways in the white matter. A lot of work has focused on describing the direct connections between two regions, yet it should be mentioned that indirect tracts with a gray matter "relay station" in between are also feasible (Gierhan, 2013). For example, the posterior portion of Broca's area (i.e., BA 44) projects directly to cortical regions in the posterior temporal lobe via the arcuate fasciculus. At the same time, there also is an indirect connection from the premotor cortex and the dorsal portion of BA 44 to the inferior parietal cortex via the superior longitudinal fasciculus, which is then, in turn, connected to the posterior temporal cortex. Notice that structural imaging cannot detect the directions in which information flows in white-matter tracts, so that labeling a connection as "frontotemporal" does not imply any directionality. In general, the information flow in language-relevant networks is currently not well understood and will not be considered further here due to our focus on lateralization (but see Friederici, 2011, 2017 for discussion and a first model of information flow in language-relevant networks based on data from electroencephalography, magnetoencephalography, and fMRI connectivity analses).

For different aspects of language processing, at least four major pathways interconnecting inferior frontal and temporal regions dorsally and ventrally have been identified (Fig. 18.3). The already mentioned arcuate fasciculus and superior longitudinal fasciculus dorsally connect inferior and superior frontal regions to posterior temporal as well as inferior parietal regions. Due to methodological limitations in tracking the horizontal parts of both pathways, some researchers have sometimes grouped both pathways together (Friederici and Gierhan, 2013). However, both pathways show different functional associations, with the arcuate fasciculus having been shown to support the processing of complex syntax by connecting the posterior portion of Broca's area (i.e., BA 44) to the posterior temporal cortex, and the superior longitudinal fascicle supporting speech repetition by connecting premotor cortex to posterior temporal cortex via the inferior parietal cortex and can be subdivided into several substructures (Friederici and Gierhan, 2013; Gierhan, 2013; Janelle et al., 2022). In addition, at least two pathways connect the inferior frontal cortex to the temporal cortex ventrally: One pathway assumed to support basic syntactic processes connects the anterior inferior frontal cortex and the frontal operculum to the anterior temporal cortex via the uncinate fascicle. Another pathway hypothesized to support semantic processing connects the inferior frontal cortex including the anterior portion of Broca's area (i.e., BA 45) with the posterior temporal, occipital, and parietal cortex via what is either called the extreme capsule fiber system or the (inferior) fronto-occipital fasciculus (Friederici and Gierhan, 2013; Gierhan, 2013; Meola et al., 2015).

Among these four pathways, most studies have reported the arcuate fasciculus to show a clear left-lateralization in the majority of the population (Paus et al., 1999; Pujol et al., 2002; Glasser and Rilling, 2008), whereas the amount of lateralization observed seems to depend on the exact measurement and analysis method used. Some studies suggested that the arcuate fasciculus is strongly left-lateralized only in half of the population (e.g., Catani et al., 2007), yet recent studies with large samples of several hundred participants have found the arcuate fasciculus to be strongly left-lateralized in two-thirds of participants (Gallardo et al., 2020; Warrington et al., 2020). For example, in a sample of 600 subjects, Gallardo et al. (2020) found that 70% of participants had at least 1.5 times more streamlines for the arcuate fasciculus in the left hemisphere than in the right hemisphere. In contrast, the uncinate fasciculus and certain parts of the superior longitudinal fasciculus (i.e., the so-called SLF II and SLF III; Friederici and Gierhan, 2013; Janelle et al., 2022) have been shown to be strongly right-lateralized in most of the population (Gallardo et al., 2020; Warrington et al., 2020). The data and analyses of both groups are less clear for the (inferior) fronto-occipital fasciculus, which, however, also tends to be slightly right-lateralized.

Significantly, these patterns of lateralization of language-relevant fiber pathways, especially the left-lateralization of the arcuate fasciculus, seem to be independent of the modality of language use and have also been observed in deaf native signers who acquired a sign language early in life and use it as their primary means of communication (Cheng et al., 2019; Finkl et al., 2019). Despite the obvious modality differences between sign language and auditory language production and perception, which was evidenced in structural differences in production-related fiber tracts, Finkl et al. (2019) did not observe any differences between both groups with regard to the structural integrity of the arcuate fasciculus. This indicates that this pathway is relevant for processing linguistic information regardless of whether language is spoken or signed. Similarly, in another group comparison of different samples, Cheng et al. (2019) also found no differences in the structural integrity of the arcuate fasciculus between deaf signers and hearing controls, as well as a clear left-lateralization of this core language pathway. Interestingly, the authors also report data from three deaf individuals who grew up with severe language deprivation and only acquired a sign language late in life. These individuals showed altered white-matter microstructure in the arcuate fasciculus and presented with severe problems in comprehension of morphosyntax and complex sentences (Cheng and Mayberry, 2019; see Trettenbrein et al., 2025, this volume, for an in-depth discussion).

Structural differences between the left and the right hemisphere in language-related networks are not limited to the white matter but can also be observed using a number of other measures, such as cortical surface area derived using MRI data or the cytoarchitectonic structure of a region determined in ex vivo brains. Broca's area, as one of the canonical language regions and parts of the core language network, has repeatedly been described as an asymmetric region. However, studies using a variety of structural imaging approaches have not consistently reported a left-over-right asymmetry for Broca's area or its subregions (for extensive reviews see Keller et al., 2009; Sprung-Much et al., 2022). In contrast to these mixed findings from neuroimaging, studies investigating the microstructure of Broca's area post mortem paint the clearest picture and indicate a leftward asymmetry, especially for the cytoarchitecturally defined posterior portion of Broca's area (i.e., BA 44), over its right-hemispheric homolog (Amunts et al., 1999, 2003; Sprung-Much et al., 2022). These discrepancies between different kinds of data may either be related to methodological differences or, as some researchers have hypothesized, be related to the documented high individual variability in these regions (Sprung-Much and Petrides, 2020; Sprung-Much et al., 2022).

With regard to the posterior temporal cortex, which generally shows higher white-matter volume in the left hemisphere (Anderson et al., 1999), a number of structural asymmetries have been reported. A classic finding is that the upper portion of the superior temporal gyrus (i.e., the planum temporale) located within the Sylvian fissure is larger in the left hemisphere (Geschwind and Levitsky, 1968; Binder et al., 1996; Shapleske et al., 1999). Similarly, the superior temporal sulcus is deeper in the right hemisphere than in the left hemisphere close to the area of Heschl's gyrus (Leroy et al., 2015; Bodin et al., 2018), and generally exhibits a genetically constrained more chaotic morphology on the left side than on the right side (Le Guen et al., 2018). While the superior temporal sulcus has been shown to be involved in auditory language comprehension, establishing a structure–function correlation between these structural asymmetries and cognitive functions has not been met with success in experiments (Specht and Wigglesworth, 2018). Lastly, with regard to microstructure, Galuske et al. (2000) used postmortem neuronal tract tracing in posterior BA 22 to show that this region contained about 30% more distinct columnar subsystems in the left hemisphere than in the right hemisphere, suggesting increased processing capacity.

The data reviewed here suggests a structural lateralization of the arcuate fasciculus as one of the main fiber tracts connecting the temporal cortex to Broca's area in the inferior frontal cortex in the typically developed adult brain.

DEVELOPMENT OF THE LANGUAGE NETWORK

Spoken language acquisition starts before birth, because the auditory system is already functioning in the fetus in utero. However, the acoustic information perceivable in the uterus is filtered by the surrounding water and tissue. Therefore, the acoustic information that arrives at the auditory system of the fetus is filtered by about 400 Hz, leaving prosodic information of speech (i.e., the patterns of stress and intonation of a particular language) intact, but not its phonetic details (i.e., the actual speech sounds as such). The presumption that the prosody of a given language is processed even before birth is supported by newborns' behavior (for reviews see Gervain, 2018; Ghio et al., 2021), as it has been shown that newborns' first cries are already language-specific: For example, French babies and German babies cry with a different melody, each mirroring the speech melody of their respective prospective native language (i.e., the majority language spoken by the mother and used in her environment; Mampe et al., 2009). Moreover, newborns also react specifically and preferably when listening to their native language in comparison to a foreign language that is not familiar to them, even when these are filtered by 400 Hz to mimic the conditions previously experienced in utero (Mehler et al., 1988).

These are interesting observations given that it takes several months before children utter the first words in their native language, indicating that comprehension and production abilities and their neural basis do not necessarily develop in tandem despite their reliance on the development of linguistic competence as such. Over the past decades, systematic research in the field of language acquisition has advanced our knowledge not only concerning the language abilities of young infants and children but also concerning the neural networks supporting this process both with respect to its functional specifics and its structural basis. Here, we review the development of the core language network previously identified in adults and already discussed above, and particularly consider its hemispheric lateralization in the developing brain. The available literature on the development of the lateralization of language processing in the developing brain comes from studies that, at times, used very different methodological approaches and therefore allow us to draw a rather clear, converging picture: At birth, there is bilateral involvement for speech processing, then lateralization of language toward the left hemisphere increases during development leading to a dominance of the left hemisphere for language processing in the vast majority of the adult population.

Functional lateralization during development

Taking the language network in the newborn brain as a starting point, a seminal study by Perani et al. (2011) provided functional and structural data in the context of language lateralization in the developing brain. Functional brain data were acquired in a setting in which 2-day-old Italian newborns heard different acoustic sequences: Either normal Italian speech or speech with the intonational contour was taken out, leading to a flattened speech version, or the phonetic details were filtered out, leading to a hummed speech version. The functional brain activation in these newborns was mainly located in the posterior superior temporal gyrus in both hemispheres. A region-of-interest analysis of the activation in the auditory cortices revealed larger activation in the right compared to the left hemisphere for normal speech and no stimulus-specific activation for the altered speech conditions. This indicates that only normal speech triggers significant brain activity in newborns and that the right hemisphere shows a stronger involvement, possibly due to a perceptual preference for prosodic information, which in the adult brain is primarily processed in the right hemisphere.

Thus, the functional brain response to speech in newborns appears to be bilateral with a slight lateralization to the right, whereas a seeming left-hemispheric dominance has been reported as early as 3 months of age. Investigating the functional connectivity of language-relevant brain regions in these newborns studied by Perani et al. (2011) compared to adults revealed a clear difference: While the group of typically developed adults showed an intrahemispheric functional connectivity between Broca's area in the left frontal cortex and the left posterior superior temporal gyrus, healthy newborns displayed an interhemispheric functional connectivity between Broca's area and its right-hemispheric homolog as well as between left superior temporal gyrus and its right-hemispheric homolog (Lohmann et al., 2010; Perani et al., 2011; see Fig. 18.4). At the age of 3 months, infants already seem to show a somewhat different activity pattern related to speech: Two fMRI studies reported speech-related activation in the superior temporal cortex bilaterally, but with a dominance in the left hemisphere (Dehaene-Lambertz, 2002; Dehaene-Lambertz et al., 2006). However, when directly contrasting natural forward speech to speech played backward, the effect in the temporal cortex disappeared and activation was found in the left angular gyrus and the precuneus.

Functional neuroimaging studies on language processing and its lateralization in early childhood are rare due to the experimental constraint that the participant has to avoid movements in the scanner, which especially for young children has proven to be difficult. Alternative approaches to reduce scanning time often involve testing language function behaviorally outside the scanner and performing only structural scans, which allow for the analysis of the gray matter in language-relevant brain regions and the white matter connecting these regions and then correlating these structural data with behavioral data. One such study investigated the cortical thickness of the regions in the neural language network relevant in 5-year-old and 6-year-old children and found that an increase in language performance was associated with changes of cortical thickness asymmetry in the left inferior frontal gyrus, in particular in Broca's area (Qi et al., 2019). The cortical thinning of this region between the ages of 5 and 6 years observed in this study was larger in the left hemisphere than in its right-hemispheric homolog and furthermore associated with the observed language performance at the age of 7 years.

Another approach that aims to minimize scanning time and the requirement of children to comply with performing an experimental task in the scanner has been to use resting state MRI and behavioral language tests. With this approach, it was found that 5-year-old children displayed an interhemispheric correlation

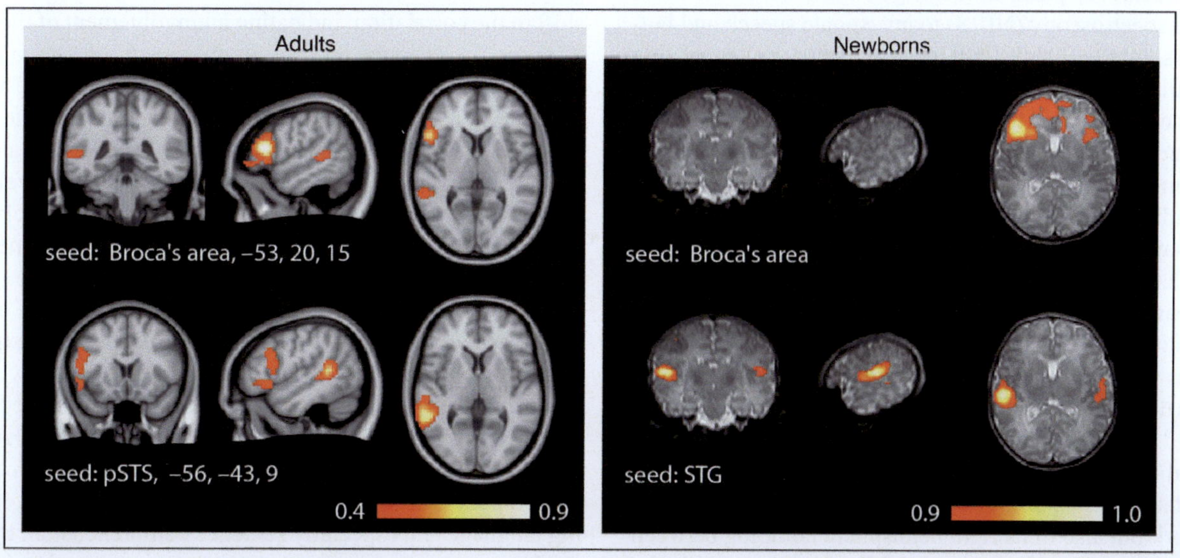

Fig. 18.4. Functional brain connectivity in adults and newborns. Functional connectivity results comparing the BOLD response to speech of adults and newborns. The left panel shows correlation values of low-pass–filtered residuals of language experiments in adults and the right panel shows corresponding data for newborns, in both cases with seeds in Broca's area (upper part of panels) and in the posterior superior temporal sulcus (pSTS) and superior temporal gyrus (STG) shown in the lower part of both panels. For adults in the left panel, coordinates are given in Talairach space (Talairach and Tournoux, 1988). As no such coordinates are available for newborns' brains, the neuroanatomic location of the seed is given instead in the right panel. In the upper and lower rows in both panels, slices are shown in the following order (from left to right): Coronal view, sagittal view, and axial view. Adapted from Perani D, Saccuman MC, Scifo P, et al. (2011). Neural language networks at birth. Proc Natl Acad Sci U S A 108: 16056–16061. https://doi.org/10.1073/pnas.1102991108.

of the left inferior frontal gyrus with its right homolog region. When relating resting state functional connectivity data to sentence processing, the local connectivity within the left inferior frontal gyrus was found to be associated with processing syntactically simple sentences, while the processing of syntactically complex sentences was associated with the long-range connectivity between the inferior frontal gyrus and the posterior superior temporal sulcus in the left hemisphere (Xiao et al., 2016). These data suggest that a selective left frontotemporal connectivity in the neural network is already in place at the age of 5 years when measured in a so-called "task-free" setting of resting state MRI.

However, it takes some more time before the neural language network in the left hemisphere reacts specifically to different aspects of the language input. An fMRI study that directly investigated syntactic processes during language comprehension demonstrated strong functional connectivity between left posterior superior temporal gyrus and left BA 44 in 6-year-olds, which was not observed in 3-year-olds, suggesting a clear development step (Vissiennon et al., 2017). These data, in combination with the infant data, suggest that the brain response to speech early in life differs from that in the adult brain, whereas the lateralization of speech and language processing seems to emerge during brain maturation. Hence, while the functional connectivity patterns observed during speech processing in newborns and adults shift from an initially bilateral to a mature left-hemispheric functional response, it should be considered that this developmental change does not merely reflect a change in the response to speech but instead evidences the emergence and functional segregation of a network that processes linguistic information (e.g., semantics, syntax, etc.) driven by the spoken input.

A number of developmental language studies used functional near-infrared spectroscopy (NIRS) to investigate speech and language processing in early life. This measurement does not provide the same spatial precision as fMRI, but allows to differentiate the involvement of the left and the right hemisphere. On the basis of a review of functional NIRS data, it was concluded that the lateralization for language emerges out of a bilateral left–right involvement for auditory processes to a left-hemispheric dominance during language learning (Minagawa-Kawai et al., 2011). Significantly, similar lateralization for processing linguistic input is also present in deaf children who acquire a sign language (Payne et al., 2019), again indicating that the observed left-hemispheric dominance relates to processing linguistic information and not just speech. This developmental pattern is confirmed by a recent study that analyzed fMRI language activation in children from 4 to 13 years and adults (Olulade et al., 2020): The study revealed language-related activation not only in the left-hemispheric language network but also in its right hemisphere homolog between the age of 4 and 6 years. During later childhood, a decrease in the involvement of the right hemisphere and an increase in the involvement of the left hemisphere were observed. This again suggests that language becomes more and more left-lateralized throughout life.

There are also several developmental language studies that only looked at functional activation in the left hemisphere and can therefore not speak directly to the asymmetry of language in the developing brain. Some of these studies, however, provide interesting information concerning the involvement of Broca's area and the posterior temporal cortex during language development. A review of studies on the development of the cortical language network by Skeide and Friederici (2016) summarized the literature, including region-of-interest studies, and proposed that there are two development stages: a first stage from birth to 3 years and a second stage continuing into adolescence with a possible closure roughly at the age of puberty (Lenneberg, 1967; Friederici, 2017). During the first stage, language processing is mainly supported by the temporal cortex, whereas the second stage involves the inferior frontal cortex including Broca's area and the dorsal fiber tract connection to the posterior temporal cortex (i.e., the core language network; also see Skeide and Friederici, 2016). It seems that phonologic, prosodic-, and word-based processes can be dealt with by both the left and right temporal cortices. Broca's area in the left inferior frontal gyrus, in particular, comes into play later when the demands on processing syntax increase.

This raises the question as to how children process sentences and syntactic information at the age of 3 years before this developmental shift for language processing in the core language network occurs. One fMRI study investigated sentence processing in 3- to 10-year-old children and varied semantic plausibility and syntactic complexity of the sentences to answer this question (Skeide et al., 2014): Although activation in Broca's area was observed in 3- to 4-year-old children, the main effects of activation for semantic and syntax were found only in the posterior temporal cortex, and similarly so in 6- to 7-year-olds. That is, main effects for syntax in Broca's area were observed in 9- to 10-year-old children as well as adults. More recently, a combined gray matter and behavioral correlation study reported a regional shift in the correlation, indicating an involvement of the posterior STG in 3-year-olds and an involvement of Broca's area in 4-year-old children (Klein et al., 2022). This suggests that, in addition to a developmental change in laterality of the language network from initially more bilateral involvement to a clear leftward preference, there appears to be an additional developmental shift within the left hemisphere, from the primary involvement of posterior temporal cortex to a preferential involvement of the inferior frontal cortex.

Lastly, a recent meta-analysis of functional fMRI studies on language comprehension, including 27 independent experiments involving children between the ages 4 and 12 years, with a mean age of 8.9 years, allows to further specify this assumed developmental shift in preschool children (Enge et al., 2020). As already discussed above, typically developed adults display activation in the left pars opercularis (BA 44) and the left temporal cortex during language and especially sentence comprehension. In contrast, the brain responses during language comprehension in children shift from bilateral temporal activation and the left pars triangularis (BA 45) toward the adult pattern involving the left posterior temporal cortex and the left pars opercularis (BA 44) during development (Fig. 18.5). Given that the maturation of the arcuate fasciculus as one of the main fiber pathways connecting these regions of the core language network has been linked to the performance during tasks involving processing complex sentences

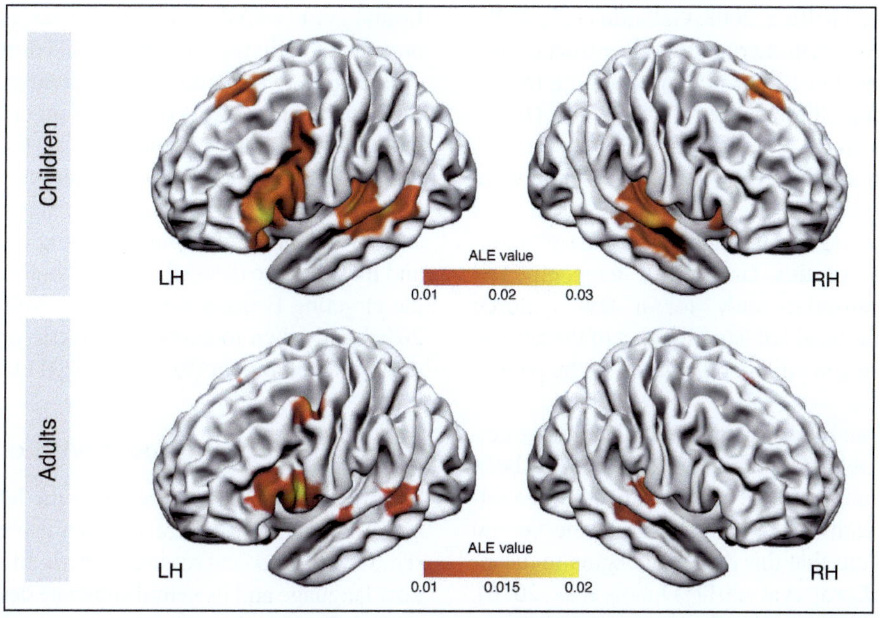

Fig. 18.5. Meta-analyses of language comprehension in children and adults. Results from meta-analyses of language comprehension in children and adults by Enge et al. (2020). The top panel shows an activation likelihood estimation (ALE) map of significant clusters associated with language comprehension in children, superimposed onto a standard cortical surface. Activations reported in 27 experiments that showed above-chance overlap ($P < 0.05$, cluster-wise family-wise error [FWE] corrected) are shown. The color bar represents the ALE value of any given voxel, that is, its degree of nonrandom convergence in activation between experiments. The bottom panel shows an ALE map of significant clusters associated with language comprehension in adults. These data were reproduced using the sample of studies reported in a previous meta-analysis by Rodd et al. (2015). Maps depict clusters with above-chance overlap ($P < 0.05$, cluster-wise FWE-corrected) and their associated ALE value (color bar), that is, the degree of nonrandom convergence in activation between experiments at any given voxel. Both panels show the left hemisphere (LH) as well as the right hemisphere (RH). Adapted from Enge A, Friederici AD, Skeide MA (2020). A meta-analysis of fMRI studies of language comprehension in children. Neuroimage 215: 116858. https://doi.org/10.1016/j.neuroimage.2020.116858.

(Skeide et al., 2014; Skeide and Friederici, 2016; Klein et al., 2022), it stands to reason that this developmental shift may be related to the increasing functional specialization of the brain for processing syntactic information.

Structural lateralization during development

The description of the lateralization of the structural language network in newborns and children has been of scientific interest for more than 50 years. Early research focused on the gray matter of the temporal cortex and found a larger left than right lateralization already in the fetus (Chi et al., 1977) and in the newborn (Witelson and Pallie, 1973). This is an important finding given that the asymmetry of the planum temporale in the posterior temporal gyrus has long been associated with the human lateralization of language (Binder et al., 1996); also see the discussion above. However, later studies showed that the leftward structural asymmetry of the planum temporale observed in many studies does not seem to relate systematically to language lateralization. One study examined the relation between the asymmetry of the planum temporale and language laterality in 287 healthy adults and concluded that the asymmetry of the planum temporale appears to be associated with local functional lateralization only in auditory areas, but is not a marker of interindividual variability of language lateralization (Tzourio-Mazoyer and Mazoyer, 2017). An earlier study by Dorsaint-Pierre et al. (2006) similarly found that it was not the planum temporale but rather the posterior part of the inferior frontal gyrus (i.e., Broca's area) for which a direct relationship to language lateralization was observed.

A structural analysis of Broca's area in the inferior frontal gyrus was performed using a cytoarchitectonic approach (Amunts et al., 2003). This approach quantifies the number of different types of neurons in the six layers of the cortex (Brodmann, 1909). It was discovered that the anterior part of Broca's area (BA 45) and the posterior part of Broca's area (BA 44) followed different development trajectories to achieve the asymmetry usually observed in the adult brain: BA 45, a region known to primarily support semantic processes, achieved a leftward asymmetry in its cytoarchitectonic profile by the age of 5 years, whereas BA 44, a region known to primarily support syntactic processes, achieved a leftward asymmetry in its cytoarchitectonic profile much later around the age of 10–11 years. This result indicates that different regions of the language network area show distinct lateralization patterns to the left hemisphere at different stages during development.

Another set of studies looked at the white-matter fiber tracts connecting the different regions of the language network and their lateralization. As discussed above, there are dorsally and ventrally located fiber tracts that connect the inferior frontal gyrus with the temporal cortex, among which the arcuate fasciculus has been shown to be stronger (e.g., in terms of the number of streamlines generated in diffusion MRI analyses, etc.) in the left hemisphere than in the right hemisphere in the adult brain (Paus et al., 1999;

Pujol et al., 2002; Glasser and Rilling, 2008; Gallardo et al., 2020; Warrington et al., 2020). The asymmetry of this fiber tract during development was investigated in 183 participants ranging from 5 to 30 years of age, including children from 5 to 13 years (Lebel and Beaulieu, 2009). This study reported that fractional anisotropy and the number of streamlines of the arcuate fascicle were higher in the left hemisphere than in the right hemisphere in most participants, independent of age. A subanalysis of the children revealed left lateralization of this fascicle for most children (N = 53) and right lateralization only for a few children (N = 15), suggesting that the usual left lateralization of the arcuate fascicle is already established in childhood, similar to the pattern observed in adults.

In a review of the functional and structural neural language network in the infant brain (Dehaene-Lambertz, 2017), it has been argued that during the maturation of the perisylvian regions, a relative delay of the dorsal pathway in comparison to the ventral pathway can be observed, but that this disparity begins to disappear after 3 months of age (Leroy et al., 2011; Dubois et al., 2016). The authors argue that this maturational step is related to the increase in vocalizations at this age. However, in the analysis of the dorsal pathway we have to consider that it may not constitute an indivisible entity: The arcuate fascicle is often taken to be a dorsally located fiber bundle that directly connects the inferior frontal gyrus with the temporal cortex. However, it has been demonstrated that in the adult brain, this dorsal pathway actually subdivides into two bundles: one targeting Broca's area and another targeting the premotor cortex (Perani et al., 2011). In the newborn brain, only the fiber bundle targeting the premotor cortex is myelinated, whereas the fiber bundle targeting Broca's area is not (Fig. 18.6). The fiber bundle targeting the premotor cortex supports auditory-to-motor mapping allowing early vocalization in infants and may well be detectable in 3-month-olds. In contrast, the bundle targeting Broca's area is only myelinated later (Skeide et al., 2016) and taken to support syntactic processing coming in later during development (Friederici, 2011).

Language development with only one hemisphere

Relatively rare cases where either the left or right hemisphere has been damaged or where one hemisphere had to be surgically removed for medical reasons early in life make it possible to study how language and its neural substrate develop in the absence of the typical functional and structural lateralization. Clinical findings from brain-injured individuals suggest that the left hemisphere and the right hemisphere, in principle, appear to be equipotential and therefore equally involved in speech and language processing early in life with a gradual increase of left hemisphere involvement

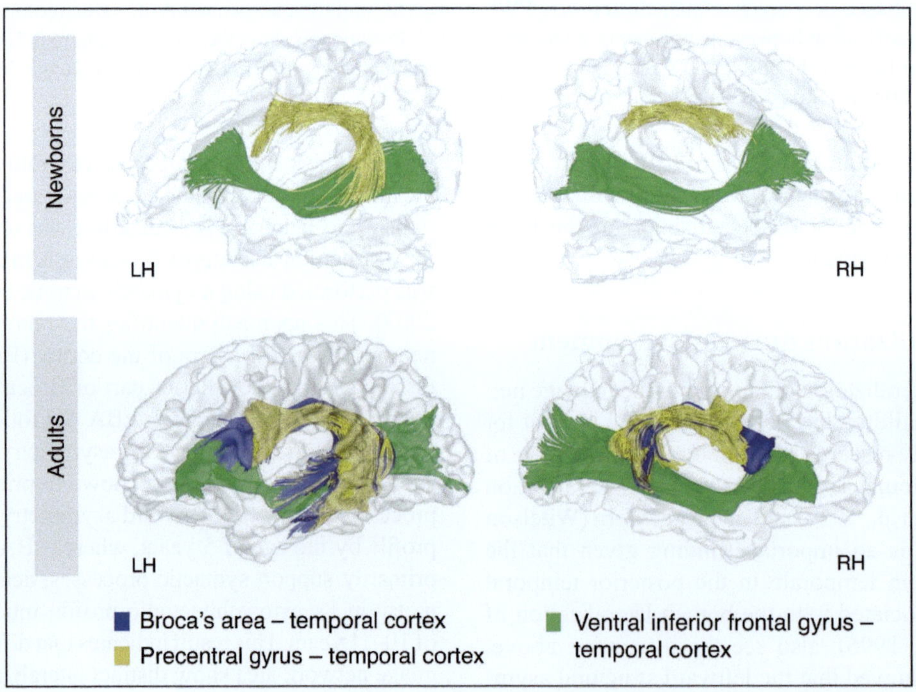

Fig. 18.6. Ontogeny of language-relevant white-matter tracts in the human brain. The top panel shows fiber tracking results of diffusion tensor imaging data seeding in Broca's area and the precentral gyrus/premotor cortex in the left hemisphere (LH) and right hemisphere (RH) of the brains of newborns (maximum age = 3 days). The bottom panel shows fiber tracking results with the same seed regions in a group of adults. In both newborns and adults, the pathway connecting the ventral inferior frontal gurus to the temporal cortex can be detected (color-coded in green). Only in adults, two dorsal pathways can be detected: The arcuate and superior longitudinal fasciculus connecting the temporal cortex to Broca's area (color-coded in blue) and the premotor cortex (color-coded in yellow). This indicates that the dorsal connection to Broca's area matures only later during development, whereas the ventral connection via the extreme capsule is present in both groups. A certain degree of structural asymmetry in the dorsal white-matter pathways is already visible early in life and becomes more pronounced in adulthood. Adapted from Perani D, Saccuman MC, Scifo P, et al. (2011). Neural language networks at birth. Proc Natl Acad Sci U S A 108: 16056–16061. https://doi.org/10.1073/pnas.1102991108.

during childhood (Basser, 1962; Bates et al., 2001). In a clinical study, 102 individuals with early unilateral brain lesions were examined (Basser, 1962): In children whose lesions occurred prior to speech onset (between 18 and 24 months), left hemisphere lesions and right hemisphere lesions equally resulted in delayed language development. Children who received a hemispherectomy before the age of 13 years only developed transitory aphasia in a few cases, whereas hemispherectomy in adults resulted in permanent aphasia when the left hemisphere was taken out. For those patients in whom the right hemisphere was surgically removed no aphasia was observed. This pattern has been confirmed by more recent studies, which show that adolescents and young adults who suffered a perinatal stroke to the left hemisphere nevertheless developed normal abilities for processing complex sentences in the right-hemispheric homolog of the language network (Newport et al., 2017, 2022).

The data reviewed here suggest a certain equipotentiality of the left and the right hemisphere for developing speech and language processing systems during the first months of life, as well as the increasing reliance of the language system on the left hemisphere during development, which usually results in the dominance of left hemisphere for language in the vast majority of typically developed adults.

DISCUSSION

Our review of functional and structural asymmetries in language-relevant regions of the cortex and the white matter has confirmed that the human language system in the typically developed adult brain exhibits a leftward asymmetry, whereas it remains an open question whether and how functional and structural asymmetries in the same regions and networks ultimately relate to each other (Keller et al., 2011). The core language network subserving the generation of hierarchically structured representations during syntactic processing is made up of the posterior portion of Broca's area (i.e., left BA 44) and the posterior superior temporal gyrus and sulcus. This network also showed the most pronounced functional leftward asymmetry in studies focusing on the processing of syntactic hierarchy even independent of meaning (e.g., in so-called artificial grammar studies). A left-hemispheric preference for processing linguistic information is already present early in life, however, the functional dissociation of semantic and syntactic processing on the cortical level emerges only gradually as part of brain maturation during childhood. The main fiber pathway connecting the regions of the core language network via the arcuate fascicle targeting BA 44 in Broca's area also shows a clear leftward asymmetry in the vast majority of the adult population and the maturation of this fiber pathway is correlated with the emergence of complex syntactic abilities.

Establishing consistent links between structure and function may have proven difficult to date because different definitions of "language" that confound language processing with speech production have sometimes been employed in the literature. For example, the widely cited number of more than 96% of right hander showing left-lateralized language functions is derived from studies using only production tasks (Rasmussen and Milner, 1977; Knecht, 2000). In contrast, meta-analyses of language processing, usually also showing strong left-lateralization, tend to rely on data stemming from studies using comprehension paradigms (Price, 2010; Vigneau et al., 2011; Walenski et al., 2019; see also Fig. 18.2). Hence, while both speech production and language comprehension necessarily access the same linguistic competence (Chomsky, 1965; van der Burght et al., 2023) and thus recruit the language network (Friederici, 2017; Friederici et al., 2017), this does not imply that all subcomponents of language always need to be lateralized to the same hemisphere.

Recent work investigating a large sample of 287 participants (Labache et al., 2020) has revealed a task-specific intra-individual differentiation of lateralization patterns in the core and extended language network: In this study, most participants showed a left-lateralized BOLD response in the core language network and its right-hemispheric homolog for speech production (92.34%), auditory (86.76%), and written (84.67%) sentence comprehension, whereas a small subgroup exhibited what the authors called a "crossed" pattern for different tasks (i.e., some of the tasks showed left lateralization and others right lateralization). This "crossed" pattern was observed in 5% of the participants for whom speech production was lateralized to the left hemisphere and auditory sentence comprehension to the right hemisphere in the same subject. The "crossed" pattern was most pronounced when the analysis was limited to the core language network and its right-hemispheric homolog. Participants exhibiting such a "crossed" activation pattern moreover showed a larger corpus callosum volume, indicating stronger inter-hemispheric connectivity and allowing for fast information transfer between the two hemispheres. The observed difference in lateralization for speech production in the left hemisphere and comprehension in the right hemisphere is interesting, since the right hemisphere is known to support prosodic processes, which appear to be most relevant for auditory language comprehension (Meyer et al., 2002; Sammler et al., 2010; van der Burght et al., 2019). The reported increased volume of the corpus callosum is crucial for the interaction of prosodic information and syntactic information during sentence comprehension (Friederici et al., 2007). Future work should examine whether these "crossed" individuals may have relied more on prosodic information during auditory sentence comprehension and therefore involved the right hemisphere.

From the point of view of evolutionary biology, both symmetry and asymmetry can constitute an advantage in fitness (Corballis, 2020). On the one hand, the natural world we live in does not have a systematic left–right bias, which is why sensorimotor systems can be symmetrically organized without any obvious negative consequences. By design, such a symmetric organization also allows for a certain degree of redundancy. On the other hand, an asymmetric organization, in the case of the brain, drastically increases the available neural resources (Rogers, 2021), as different regions or networks can be dedicated to distinct cognitive operations by means of differential functional specialization of the two hemispheres. At the same time, the typical strong left-lateralization of speech, language, and especially of the processing of complex syntax also illustrates that a high degree of asymmetry increases the probability of a design including a so-called "single point of failure" (i.e., a point of a system that, if it stops working, will stop the entire system or at least a major part of it from functioning). This is evidenced by the well-known aphasic syndromes that follow only left-hemispheric lesions but not right-hemispheric lesions to the perisylvian cortex and connecting white-matter tracts (for detailed discussions, see Raymer and Gonzalez Rothi, 2015).

Accordingly, while even semantic and phonologic processing tend to show left-lateralized patterns in functional neuroimaging studies (Vigneau et al., 2011; see also Fig. 18.2), the literature on development, as well as on aphasia and other neuropsychologic syndromes, makes it clear that the right hemisphere can also process these types of linguistic information, though maybe somewhat less efficiently or in a more restricted matter (Vigneau et al., 2011; Stefaniak et al., 2021; Wilson et al., 2023). For example, so-called "split-brain" patients, in which the left and right hemispheres have been surgically disconnected for medical reasons (Sperry, 1961), confirm such supposed limitations with regard to processing semantic and phonologic information, because their left hemisphere exhibits an advantage over the right hemisphere (Zaidel, 1978). In the only reported case of a split-brain patient with alleged right-hemispheric dominance for language, both hemispheres actually succeeded in oral naming, reading, and auditory comprehension tasks using single words, yet failed at a task testing passive voice syntax (Lutsep et al., 1995). Both hemispheres performed above chance in an active voice syntax task (i.e., the canonical word order of English). In the context of our discussion, such a pattern is difficult to interpret with regard to the impairment of syntactic abilities, but it provides evidence for both hemispheres' capabilities in semantics and phonology for comprehension and production.

The left-lateralization of language in typically developed persons is not merely an epiphenomenon of language acquisition, but reflects the maturation of a network dedicated to generating hierarchically structured representations in tandem with other language-relevant regions supporting the processing of semantics and phonology as well as speech production. However, relatively rare cases of so-called "crossed" lateralization patterns for sentence production and auditory sentence comprehension (Labache et al., 2020) confirm the theoretical stipulation that speech and language functions are principally independent also in terms of their neurobiologic implementation. At the same time, genetically determined biologic matrices of the core and extended language networks (Hervé et al., 2013; Kong et al., 2020; Sha et al., 2021; Wan et al., 2022) exhibit sufficient flexibility early in life for speech production and syntactic processing as usually highly left-lateralized processes to shift to the right hemisphere in cases of perinatal stroke (Newport et al., 2017, 2022). Significantly, this plasticity is highly constrained and has only been observed after injuries during the critical period for language acquisition early in life (Lenneberg, 1967; Friederici, 2017) and is always limited to the right-hemispheric regions homologous to the core and extended language network (Newport et al., 2022). Whether this reflects the plasticity and flexibility of the developing brain or a general equipotentiality of both hemispheres early in life for developing fully fledged language functions including syntactic processing and speech production remains an open question.

CONCLUSION

Our review of the neuroimaging literature in this chapter has confirmed the centuries-old assumption that the left hemisphere of the human brain is usually specialized for language processing in typically developed adults and, at the same time, has allowed us to draw the more nuanced conclusion that observed functional leftward asymmetries are most pronounced for components of the language system subserving syntactic processing. These functional asymmetries emerge during development and are, to an extent, paralleled in structural asymmetries, although with some individual variation. Both hemispheres display a weak equipotentiality with regard to processing meaning (i.e., semantics) and sound (i.e., phonetics/phonology), but even these processes tend to show a left-hemispheric dominance during most experimental tasks.

The processing of hierarchical syntactic structure is primarily supported by a left-hemispheric frontotemporal network connecting the posterior inferior frontal gyrus to the posterior temporal cortex via the arcuate fasciculus. This network constitutes both functional and structural asymmetries. It subserves the generation of hierarchically structured representations linking meaning and sound, develops in accordance with a genetically determined biologic matrix, and its emergence may have constituted a prerequisite for the evolution of the capacity for language in our species. In the future, studies that use linguistically informed experimental manipulations and consider individual variation will likely provide a more fine-grained picture of the lateralization of language and its different subsystems.

ACKNOWLEDGMENTS

This work was funded by the Max Planck Society. While completing revisions of this chapter Patrick C. Trettenbrein was funded by the Deutsche Forschungsgemeinschaft (DFG, German Research Foundation)—501984557 which is part of the DFG SPP 2392 "Visual Communication" (ViCom).

CONFLICTS OF INTEREST

We have no known conflicts of interest to disclose.

References

Amunts K, Schleicher A, Bürgel U et al. (1999). Broca's region revisited: cytoarchitecture and intersubject variability. J Comp Neurol 412: 319–341. https://doi.org/10.1002/(SICI)1096-9861(19990920)412:2<319::AID-CNE10>3.0.CO;2-7.

Amunts K, Schleicher A, Ditterich A et al. (2003). Broca's region: cytoarchitectonic asymmetry and developmental changes. J Comp Neurol 465: 72–89. https://doi.org/10.1002/cne.10829.

Anderson B, Southern BD, Powers RE (1999). Anatomic asymmetries of the posterior superior temporal lobes: a postmortem study. Neuropsychiatry Neuropsychol Behav Neurol 12: 247–254.

Arning L, Ocklenburg S, Schulz S et al. (2013). PCSK6 VNTR polymorphism is associated with degree of handedness but not direction of handedness. PLoS One 8: e67251. https://doi.org/10.1371/journal.pone.0067251.

Bahlmann J, Schubotz RI, Friederici AD (2008). Hierarchical artificial grammar processing engages Broca's area. Neuroimage 42: 525–534. https://doi.org/10.1016/j.neuroimage.2008.04.249.

Basser LS (1962). Hemiplegia of early onset and the faculty of speech with special reference to the effects of hemispherectomy. Brain 85: 427–460. https://doi.org/10.1093/brain/85.3.427.

Bates E, Reilly J, Wulfeck B et al. (2001). Differential effects of unilateral lesions on language production in children and adults. Brain Lang 79: 223–265. https://doi.org/10.1006/brln.2001.2482.

Bemis DK, Pylkkanen L (2011). Simple composition: a magnetoencephalography investigation into the comprehension of minimal linguistic phrases. J Neurosci 31: 2801–2814. https://doi.org/10.1523/JNEUROSCI.5003-10.2011.

Berwick RC, Friederici AD, Chomsky N et al. (2013). Evolution, brain, and the nature of language. Trends Cogn Sci 17: 89–98. https://doi.org/10.1016/j.tics.2012.12.002.

Binder JR, Frost JA, Hammeke TA et al. (1996). Function of the left planum temporale in auditory and linguistic processing. Brain 119: 1239–1247. https://doi.org/10.1093/brain/119.4.1239.

Binder JR, Desai RH, Graves WW et al. (2009). Where is the semantic system? A critical review and meta-analysis of 120 functional neuroimaging studies. Cereb Cortex 19: 2767–2796. https://doi.org/10.1093/cercor/bhp055.

Bodin C, Takerkart S, Belin P et al. (2018). Anatomo-functional correspondence in the superior temporal sulcus. Brain Struct Funct 223: 221–232. https://doi.org/10.1007/s00429-017-1483-2.

Bogen JE, Bogen GM (1976). Wernicke's region—where is it? Ann N Y Acad Sci 280: 834–843. https://doi.org/10.1111/j.1749-6632.1976.tb25546.x.

Bradshaw AR, Thompson PA, Wilson AC et al. (2017). Measuring language lateralisation with different language tasks: a systematic review. PeerJ 5: e3929. https://doi.org/10.7717/peerj.3929.

Broca P (1861). Remarques sur le siége de la faculté du langage articulé, suivies d'une obsevation d'aphémie (perte de la parole). Bull de La Soc Anatomique 6: 330–357.

Brodmann K (1909). Beiträge zur histologischen Lokalisation der Grosshirnrinde: VI. Die Cortexgliederung des Menschen. J Für Psychol Neurol 10: 231–246.

Catani M, Allin MPG, Husain M et al. (2007). Symmetries in human brain language pathways correlate with verbal recall. Proc Natl Acad Sci U S A 104: 17163–17168. https://doi.org/10.1073/pnas.0702116104.

Chen L, Goucha T, Männel C et al. (2021a). Hierarchical syntactic processing is beyond mere associating: functional magnetic resonance imaging evidence from a novel artificial grammar. Hum Brain Mapp 42: 3253–3268. https://doi.org/10.1002/hbm.25432.

Chen L, Wu J, Hartwigsen G et al. (2021b). The role of a critical left fronto-temporal network with its right-hemispheric homologue in syntactic learning based on word category information. J Neurolinguistics 58: 100977. https://doi.org/10.1016/j.jneuroling.2020.100977.

Cheng Q, Mayberry RI (2019). Acquiring a first language in adolescence: the case of basic word order in American sign language. J Child Lang 46: 214–240. https://doi.org/10.1017/S0305000918000417.

Cheng Q, Roth A, Halgren E et al. (2019). Effects of early language deprivation on brain connectivity: language pathways in deaf native and late first-language learners of American sign language. Front Hum Neurosci 13: 320. https://doi.org/10.3389/fnhum.2019.00320.

Chi JG, Dooling EC, Gilles FH (1977). Left-right asymmetries of the temporal speech areas of the human fetus. Arch Neurol 34: 346–348. https://doi.org/10.1001/archneur.1977.00500180040008.

Chomsky N (1965). Aspects of the theory of syntax, The MIT Press.

Chomsky N (1986). In: RN Anshen (Ed.), Knowledge of language: its nature, origins, and use. Praeger.

Chomsky N (1995). The minimalist program, MIT Press.

Chomsky N (2011). Language and other cognitive systems: what is special about language? Lang Learn Dev 7: 263–278. https://doi.org/10.1080/15475441.2011.584041.

Corballis MC (2020). Bilaterally symmetrical: to be or not to be? Symmetry 12: 326. https://doi.org/10.3390/sym12030326.

Crain S, Koring L, Thornton R (2016). Language acquisition from a biolinguistic perspective. Neurosci Biobehav Rev. https://doi.org/10.1016/j.neubiorev.2016.09.004.

Dax G (1863). M. dax soumet au jugement de l'Académie Un Mémoire Intitulé: 'Observations tendant à prouver la coïncidence constante des dérangements de la parole avec une lésion de l'hémisphère gauche du cerveau. Compt Rend HebdomSéanl'Acad Scien 56: 536.

Dax M (1865). Lésions de la moitié gauche de l'encéphale coïncident avec l'oubli des signes de la pensée (lu à Montpellier En 1836). Gaz Heb Méd Chir 2: 259–262.

Dehaene S, Cohen L (2007). Cultural recycling of cortical maps. Neuron 56: 384–398. https://doi.org/10.1016/j.neuron.2007.10.004.

Dehaene S, Cohen L, Sigman M et al. (2005). The neural code for written words: a proposal. Trends Cogn Sci 9: 335–341. https://doi.org/10.1016/j.tics.2005.05.004.

Dehaene-Lambertz G (2002). Functional neuroimaging of speech perception in infants. Science 298: 2013–2015. https://doi.org/10.1126/science.1077066.

Dehaene-Lambertz G (2017). The human infant brain: a neural architecture able to learn language. Psychon Bull Rev 24: 48–55. https://doi.org/10.3758/s13423-016-1156-9.

Dehaene-Lambertz G, Hertz-Pannier L, Dubois J et al. (2006). Functional organization of perisylvian activation during presentation of sentences in preverbal infants. Proc Natl Acad Sci U S A 103: 14240–14245. https://doi.org/10.1073/pnas.0606302103.

Dorsaint-Pierre R, Penhune VB, Watkins KE et al. (2006). Asymmetries of the planum temporale and Heschl's gyrus: relationship to language lateralization. Brain 129: 1164–1176. https://doi.org/10.1093/brain/awl055.

Dronkers NF, Plaisant O, Iba-Zizen MT et al. (2007). Paul Broca's historic cases: high resolution MR imaging of the brains of Leborgne and Lelong. Brain 130: 1432–1441. https://doi.org/10.1093/brain/awm042.

Dubois J, Poupon C, Thirion B et al. (2016). Exploring the early organization and maturation of linguistic pathways in the human infant brain. Cereb Cortex 26: 2283–2298. https://doi.org/10.1093/cercor/bhv082.

Eickhoff SB, Laird AR, Grefkes C et al. (2009). Coordinate-based activation likelihood estimation meta-analysis of neuroimaging data: a random-effects approach based on empirical estimates of spatial uncertainty. Hum Brain Mapp 30: 2907–2926. https://doi.org/10.1002/hbm.20718.

Eickhoff SB, Bzdok D, Laird AR et al. (2012). Activation likelihood estimation meta-analysis revisited. Neuroimage 59: 2349–2361. https://doi.org/10.1016/j.neuroimage.2011.09.017.

Eickhoff SB, Laird AR, Fox PM et al. (2017). Implementation errors in the GingerALE software: description and recommendations: errors in the GingerALE software. Hum Brain Mapp 38: 7–11. https://doi.org/10.1002/hbm.23342.

Embick D, Poeppel D (2015). Towards a computational (ist) neurobiology of language: correlational, integrated and explanatory neurolinguistics. Lang Cogn Neurosci 30: 357–366. https://doi.org/10.1080/23273798.2014.980750.

Emmorey K (2015). The neurobiology of sign language. In: AW Toga, P Bandettini, P Thompson, K Friston (Eds.), Brain mapping: an encyclopedic reference. vol. 3. Academic Press, pp. 475–479.

Enge A, Friederici AD, Skeide MA (2020). A meta-analysis of fMRI studies of language comprehension in children. Neuroimage 215: 116858. https://doi.org/10.1016/j.neuroimage.2020.116858.

Everaert MBH, Huybregts MAC, Chomsky N et al. (2015). Structures, not strings: linguistics as part of the cognitive sciences. Trends Cogn Sci 19: 729–743. https://doi.org/10.1016/j.tics.2015.09.008.

Fedorenko E, Thompson-Schill SL (2014). Reworking the language network. Trends Cogn Sci 18: 120–126. https://doi.org/10.1016/j.tics.2013.12.006.

Fedorenko E, Behr MK, Kanwisher N (2011). Functional specificity for high-level linguistic processing in the human brain. Proc Natl Acad Sci U S A 108: 16428–16433. https://doi.org/10.1073/pnas.1112937108.

Finkl T, Hahne A, Friederici AD et al. (2019). Language without speech: Segregating distinct circuits in the human brain. Cereb Cortex 30: 812–823. https://doi.org/10.1093/cercor/bhz128.

Fox PT, Lancaster JL (2002). Mapping context and content: the BrainMap model. Nat Rev Neurosci 3: 319–321. https://doi.org/10.1038/nrn789.

Friederici AD (2002). Towards a neural basis of auditory sentence processing. Trends Cogn Sci 6: 78–84. https://doi.org/10.1016/S1364-6613(00)01839-8.

Friederici AD (2011). The brain basis of language processing: from structure to function. Physiol Rev 91: 1357–1392. https://doi.org/10.1152/physrev.00006.2011.

Friederici AD (2017). Language in our brain: the origins of a uniquely human capacity, MIT Press.

Friederici AD, Gierhan SME (2013). The language network. Curr Opin Neurobiol 23: 250–254. https://doi.org/10.1016/j.conb.2012.10.002.

Friederici AD, Meyer M, von Cramon DY (2000). Auditory language comprehension: an event-related fMRI study on the processing of syntactic and lexical information. Brain Lang 74: 289–300. https://doi.org/10.1006/brln.2000.2313.

Friederici AD, Fiebach CJ, Schlesewsky M et al. (2005). Processing linguistic complexity and grammaticality in the left frontal cortex. Cereb Cortex 16: 1709–1717. https://doi.org/10.1093/cercor/bhj106.

Friederici AD, Bahlmann J, Heim S et al. (2006). The brain differentiates human and non-human grammars: functional localization and structural connectivity. Proc Natl Acad Sci U S A 103: 2458–2463. https://doi.org/10.1073/pnas.0509389103.

Friederici AD, Von Cramon DY, Kotz SA (2007). Role of the Corpus callosum in speech comprehension: interfacing syntax and prosody. Neuron 53: 135–145. https://doi.org/10.1016/j.neuron.2006.11.020.

Friederici AD, Makuuchi M, Bahlmann J (2009). The role of the posterior superior temporal cortex in sentence comprehension. Neuroreport 20: 563–568. https://doi.org/10.1097/WNR.0b013e3283297dee.

Friederici AD, Chomsky N, Berwick RC et al. (2017). Language, mind and brain. Nat Hum Behav. https://doi.org/10.1038/s41562-017-0184-4.

Friston KJ, Price CJ, Fletcher P et al. (1996). The trouble with cognitive subtraction. Neuroimage 4: 97–104. https://doi.org/10.1006/nimg.1996.0033.

Gallardo G, Wassermann D, Anwander A (2020). Bridging the gap: from neuroanatomical knowledge to tractography of brain pathways. Neuroscience. https://doi.org/10.1101/2020.08.01.232116.

Galuske RAW, Schlote W, Bratzke H et al. (2000). Interhemispheric asymmetries of the modular structure in human temporal cortex. Science 289: 1946–1949. https://doi.org/10.1126/science.289.5486.1946.

Gervain J (2018). The role of prenatal experience in language development. Curr Opin Behav Sci 21: 62–67. https://doi.org/10.1016/j.cobeha.2018.02.004.

Geschwind N, Levitsky W (1968). Human brain: left-right asymmetries in temporal speech region. Science 161: 186–187. https://doi.org/10.1126/science.161.3837.186.

Ghio M, Cara C, Tettamanti M (2021). The prenatal brain readiness for speech processing: a review on foetal development of auditory and primordial language networks. Neurosci Biobehav Rev 128: 709–719. https://doi.org/10.1016/j.neubiorev.2021.07.009.

Gierhan SME (2013). Connections for auditory language in the human brain. Brain Lang 127: 205–221. https://doi.org/10.1016/j.bandl.2012.11.002.

Glasser MF, Rilling JK (2008). DTI tractography of the human brain's language pathways. Cereb Cortex 18: 2471–2482. https://doi.org/10.1093/cercor/bhn011.

Goucha T, Zaccarella E, Friederici AD (2017). A revival of the Homo loquens as a builder of labeled structures: neurocognitive considerations. Neurosci Biobehav Rev. https://doi.org/10.1016/j.neubiorev.2017.01.036.

Grewe T, Bornkessel I, Zysset S et al. (2005). The emergence of the unmarked: a new perspective on the language-specific function of Broca's area. Hum Brain Mapp 26: 178–190. https://doi.org/10.1002/hbm.20154.

Hagoort P (2017). The core and beyond in the language-ready brain. Neurosci Biobehav Rev. https://doi.org/10.1016/j.neubiorev.2017.01.048.

Hagoort P (2019). The neurobiology of language beyond single-word processing. Science 366: 55–58. https://doi.org/10.1126/science.aax0289.

Hauser MD, Chomsky N, Fitch WT (2002). The faculty of language: what is it, who has it, and how did it evolve? Science 298: 1569–1579. https://doi.org/10.1126/science.298.5598.1569.

Hervé P-Y, Zago L, Petit L et al. (2013). Revisiting human hemispheric specialization with neuroimaging. Trends Cogn Sci 17: 69–80. https://doi.org/10.1016/j.tics.2012.12.004.

Hickok G (2009). The functional neuroanatomy of language. Phys Life Rev 6: 121–143. https://doi.org/10.1016/j.plrev.2009.06.001.

Iwabuchi T, Nakajima Y, Makuuchi M (2019). Neural architecture of human language: hierarchical structure building is independent from working memory. Neuropsychologia 132: 107137. https://doi.org/10.1016/j.neuropsychologia.2019.107137.

Janelle F, Iorio-Morin C, D'amour S et al. (2022). Superior longitudinal fasciculus: a review of the anatomical descriptions with functional correlates. Front Neurol 13: 794618. https://doi.org/10.3389/fneur.2022.794618.

Jobard G, Vigneau M, Mazoyer B et al. (2007). Impact of modality and linguistic complexity during reading and listening tasks. Neuroimage 34: 784–800. https://doi.org/10.1016/j.neuroimage.2006.06.067.

Kandel ER, Hudspeth AJ (2013). The brain and behavior. In: ER Kandel, JH Schwartz, TM Jessell, SA Siegelbaum, AJ Hudspeth (Eds.), Principles of neural science, fifth edn. McGraw-Hill, pp. 5–21.

Keller SS, Crow T, Foundas A et al. (2009). Broca's area: nomenclature, anatomy, typology and asymmetry. Brain Lang 109: 29–48. https://doi.org/10.1016/j.bandl.2008.11.005.

Keller SS, Roberts N, García-Fiñana M et al. (2011). Can the language-dominant hemisphere be predicted by brain anatomy?

J Cogn Neurosci 23: 2013–2029. https://doi.org/10.1162/jocn.2010.21563.

Kennedy DN, O'Craven KM, Ticho BS et al. (1999). Structural and functional brain asymmetries in human situs inversus totalis. Neurology 53: 1260. https://doi.org/10.1212/WNL.53.6.1260.

Klein CC, Berger P, Goucha T et al. (2022). Children's syntax is supported by the maturation of BA44 at 4 years, but of the posterior STS at 3 years of age. Cereb Cortex 33: 5426–5435. https://doi.org/10.1093/cercor/bhac430.

Klima ES, Bellugi U, Battison R et al. (1979). The signs of language, Harvard University Press.

Knecht S (2000). Handedness and hemispheric language dominance in healthy humans. Brain 123: 2512–2518. https://doi.org/10.1093/brain/123.12.2512.

Kong X-Z, Tzourio-Mazoyer N, Joliot M et al. (2020). Gene expression correlates of the cortical network underlying sentence processing. Neurobiol Lang 1: 77–103. https://doi.org/10.1162/nol_a_00004.

Krauska A, Lau E (2023). Moving away from lexicalism in psycho- and neuro-linguistics. Front Lang Sci 2: 1125127. https://doi.org/10.3389/flang.2023.1125127.

Kuhnke P, Chapman CA, Cheung VKM et al. (2023). The role of the angular gyrus in semantic cognition: a synthesis of five functional neuroimaging studies. Brain Struct Funct 228: 273–291. https://doi.org/10.1007/s00429-022-02493-y.

Labache L, Mazoyer B, Joliot M et al. (2020). Typical and atypical language brain organization based on intrinsic connectivity and multitask functional asymmetries. Elife 9: e58722. https://doi.org/10.7554/eLife.58722.

Laird AR, Eickhoff SB, Fox PM et al. (2011). The BrainMap strategy for standardization, sharing, and meta-analysis of neuroimaging data. BMC Res Notes 4. https://doi.org/10.1186/1756-0500-4-349.

Le Guen Y, Leroy F, Auzias G et al. (2018). The chaotic morphology of the left superior temporal sulcus is genetically constrained. Neuroimage 174: 297–307. https://doi.org/10.1016/j.neuroimage.2018.03.046.

Lebel C, Beaulieu C (2009). Lateralization of the arcuate fasciculus from childhood to adulthood and its relation to cognitive abilities in children. Hum Brain Mapp 30: 3563–3573. https://doi.org/10.1002/hbm.20779.

Lenneberg EH (1964). The capacity for language acquisition. In: JA Fodor, JJ Katz (Eds.), The structure of language: readings in the philosophy of language. Prentice-Hall, pp. 579–603.

Lenneberg EH (1967). Biological foundations of language, Wiley.

Lenneberg EH (1969). On explaining language. Science 164: 635–643. https://doi.org/10.1126/science.164.3880.635.

Leroy F, Glasel H, Dubois J et al. (2011). Early maturation of the linguistic dorsal pathway in human infants. J Neurosci 31: 1500–1506. https://doi.org/10.1523/JNEUROSCI.4141-10.2011.

Leroy F, Cai Q, Bogart SL et al. (2015). New human-specific brain landmark: the depth asymmetry of superior temporal sulcus. Proc Natl Acad Sci U S A 112: 1208–1213. https://doi.org/10.1073/pnas.1412389112.

Lohmann G, Hoehl S, Brauer J et al. (2010). Setting the frame: the human brain activates a basic low-frequency network for language processing. Cereb Cortex 20: 1286–1292. https://doi.org/10.1093/cercor/bhp190.

Lutsep HL, Wessinger CM, Gazzaniga MS (1995). Cerebral and callosal organisation in a right hemisphere dominant "split brain" patient. J Neurol Neurosurg Psychiatry 59: 50–54. https://doi.org/10.1136/jnnp.59.1.50.

Makuuchi M, Bahlmann J, Anwander A et al. (2009). Segregating the core computational faculty of human language from working memory. Proc Natl Acad Sci U S A 106: 8362–8367. https://doi.org/10.1073/pnas.0810928106.

Mampe B, Friederici AD, Christophe A et al. (2009). Newborns' cry melody is shaped by their native language. Curr Biol 19: 1994–1997. https://doi.org/10.1016/j.cub.2009.09.064.

Matchin W, Hickok G (2020). The cortical organization of syntax. Cereb Cortex 30: 1481–1498. https://doi.org/10.1093/cercor/bhz180.

Matsumoto T, Kuriya N, Akagi T et al. (1997). Handedness and laterality of the viscera. Neurology 49: 1751. https://doi.org/10.1212/WNL.49.6.1751.

Matsuo K, Chen S-HA, Tseng W-YI (2012). AveLI: a robust lateralization index in functional magnetic resonance imaging using unbiased threshold-free computation. J Neurosci Methods 205: 119–129. https://doi.org/10.1016/j.jneumeth.2011.12.020.

Mehler J, Jusczyk P, Lambertz G et al. (1988). A precursor of language acquisition in young infants. Cognition 29: 143–178. https://doi.org/10.1016/0010-0277(88)90035-2.

Meola A, Comert A, Yeh F et al. (2015). The controversial existence of the human superior fronto-occipital fasciculus: connectome-based tractographic study with microdissection validation. Hum Brain Mapp 36: 4964–4971. https://doi.org/10.1002/hbm.22990.

Meyer M, Alter K, Friederici AD et al. (2002). FMRI reveals brain regions mediating slow prosodic modulations in spoken sentences. Hum Brain Mapp 17: 73–88. https://doi.org/10.1002/hbm.10042.

Meyer M, Steinhauer K, Alter K et al. (2004). Brain activity varies with modulation of dynamic pitch variance in sentence melody. Brain Lang 89: 277–289. https://doi.org/10.1016/S0093-934X(03)00350-X.

Meyer L, Obleser J, Anwander A et al. (2012). Linking ordering in Broca's area to storage in left temporo-parietal regions: the case of sentence processing. Neuroimage 62: 1987–1998. https://doi.org/10.1016/j.neuroimage.2012.05.052.

Minagawa-Kawai Y, Cristià A, Dupoux E (2011). Cerebral lateralization and early speech acquisition: a developmental scenario. Dev Cogn Neurosci 1: 217–232. https://doi.org/10.1016/j.dcn.2011.03.005.

Newport EL, Landau B, Seydell-Greenwald A et al. (2017). Revisiting Lenneberg's hypotheses about early developmental plasticity: language organization after left-hemisphere perinatal stroke. Biol Theory 11: 407–422. https://doi.org/10.5964/bioling.9105.

Newport EL, Seydell-Greenwald A, Landau B et al. (2022). Language and developmental plasticity after perinatal stroke. Proc Natl Acad Sci U S A 119: e2207293119. https://doi.org/10.1073/pnas.2207293119.

Olulade OA, Seydell-Greenwald A, Chambers CE et al. (2020). The neural basis of language development: changes in lateralization over age. Proc Natl Acad Sci U S A 117: 23477–23483. https://doi.org/10.1073/pnas.1905590117.

Paus T, Zijdenbos A, Worsley K et al. (1999). Structural maturation of neural pathways in children and adolescents: in vivo study. Science 283: 1908–1911. https://doi.org/10.1126/science.283.5409.1908.

Payne H, Gutierrez-Sigut E, Woll B et al. (2019). Cerebral lateralisation during signed and spoken language production in children born deaf. Dev Cogn Neurosci 36: 100619. https://doi.org/10.1016/j.dcn.2019.100619.

Perani D, Saccuman MC, Scifo P et al. (2011). Neural language networks at birth. Proc Natl Acad Sci U S A 108: 16056–16061. https://doi.org/10.1073/pnas.1102991108.

Plante E, Creusere M, Sabin C (2002). Dissociating sentential prosody from sentence processing: activation interacts with task demands. Neuroimage 17: 401–410. https://doi.org/10.1006/nimg.2002.1182.

Poeppel D, Embick D (2013). Defining the relation between linguistics and neuroscience. In: A Cutler (Ed.), Twenty-first century psycholinguistics: four cornerstones. Psychology Press, pp. 103–118.

Price CJ (2010). The anatomy of language: a review of 100 fMRI studies published in 2009. Ann N Y Acad Sci 1191: 62–88. https://doi.org/10.1111/j.1749-6632.2010.05444.x.

Pujol J, López-Sala A, Deus J et al. (2002). The lateral asymmetry of the human brain studied by volumetric magnetic resonance imaging. Neuroimage 17: 670–679. https://doi.org/10.1006/nimg.2002.1203.

Pylkkänen L (2019). The neural basis of combinatory syntax and semantics. Science 366: 62–66. https://doi.org/10.1126/science.aax0050.

Qi T, Schaadt G, Friederici AD (2019). Cortical thickness lateralization and its relation to language abilities in children. Dev Cogn Neurosci 39: 100704. https://doi.org/10.1016/j.dcn.2019.100704.

Rasmussen T, Milner B (1977). The role of early left-brain injury in determining lateralization of cerebral speech functions. Ann N Y Acad Sci 299: 355–369. https://doi.org/10.1111/j.1749-6632.1977.tb41921.x.

Raymer AM, Gonzalez Rothi LJ (Eds.), (2015). The Oxford handbook of aphasia and language disorders. vol. 1. Oxford University Press. https://doi.org/10.1093/oxfordhb/9780199772391.001.0001.

Rodd JM, Vitello S, Woollams AM et al. (2015). Localising semantic and syntactic processing in spoken and written language comprehension: an activation likelihood estimation meta-analysis. Brain Lang 141: 89–102. https://doi.org/10.1016/j.bandl.2014.11.012.

Röder B, Stock O, Neville H et al. (2002). Brain activation modulated by the comprehension of normal and pseudo-word sentences of different processing demands: a functional magnetic resonance imaging study. Neuroimage 15: 1003–1014. https://doi.org/10.1006/nimg.2001.1026.

Rogers LJ (2021). Brain lateralization and cognitive capacity. Animals 11: 1996. https://doi.org/10.3390/ani11071996.

Sammler D, Kotz SA, Eckstein K et al. (2010). Prosody meets syntax: the role of the corpus callosum. Brain 133: 2643–2655. https://doi.org/10.1093/brain/awq231.

Sha Z, Schijven D, Carrion-Castillo A et al. (2021). The genetic architecture of structural left–right asymmetry of the human brain. Nat Hum Behav. https://doi.org/10.1038/s41562-021-01069-w.

Shapleske J, Rossell SL, Woodruff PWR et al. (1999). The planum temporale: a systematic, quantitative review of its structural, functional and clinical significance. Brain Res Rev 29: 26–49. https://doi.org/10.1016/S0165-0173(98)00047-2.

Skeide MA, Friederici AD (2016). The ontogeny of the cortical language network. Nat Rev Neurosci 17: 323–332. https://doi.org/10.1038/nrn.2016.23.

Skeide MA, Brauer J, Friederici AD (2014). Syntax gradually segregates from semantics in the developing brain. Neuroimage 100: 106–111. https://doi.org/10.1016/j.neuroimage.2014.05.080.

Skeide MA, Brauer J, Friederici AD (2016). Brain functional and structural predictors of language performance. Cereb Cortex 26: 2127–2139. https://doi.org/10.1093/cercor/bhv042.

Specht K, Wigglesworth P (2018). The functional and structural asymmetries of the superior temporal sulcus. Scand J Psychol 59: 74–82. https://doi.org/10.1111/sjop.12410.

Sperry RW (1961). Cerebral organization and behavior: the split brain behaves in many respects like two separate brains, providing new research possibilities. Science 133: 1749–1757. https://doi.org/10.1126/science.133.3466.1749.

Sprung-Much T, Petrides M (2020). Morphology and spatial probability maps of the horizontal ascending ramus of the lateral fissure. Cereb Cortex 30: 1586–1602. https://doi.org/10.1093/cercor/bhz189.

Sprung-Much T, Eichert N, Nolan E et al. (2022). Broca's area and the search for anatomical asymmetry: commentary and perspectives. Brain Struct Funct 227: 441–449. https://doi.org/10.1007/s00429-021-02357-x.

Stefaniak JD, Alyahya RSW, Lambon Ralph MA (2021). Language networks in aphasia and health: a 1000 participant activation likelihood estimation meta-analysis. Neuroimage 233: 117960. https://doi.org/10.1016/j.neuroimage.2021.117960.

Talairach J, Tournoux P (1988). Co-planar stereotaxic atlas of the human brain: 3-dimensional proportional system—an approach to cerebral imaging, Thieme Medical Publisher.

Toga AW, Thompson PM (2003). Mapping brain asymmetry. Nat Rev Neurosci 4: 37–48. https://doi.org/10.1038/nrn1009.

Tremblay P, Dick AS (2016). Broca and Wernicke are dead, or moving past the classic model of language neurobiology. Brain Lang 162: 60–71. https://doi.org/10.1016/j.bandl.2016.08.004.

Trettenbrein PC, Papitto G, Friederici AD et al. (2021). Functional neuroanatomy of language without speech: an ALE meta-analysis of sign language. Hum Brain Mapp 42: 699–712. https://doi.org/10.1002/hbm.25254.

Trettenbrein PC, Zaccarella E, Friederici AD (2025). Functional and structural brain asymmetries in sign language processing. In: C Papagno, P Corballis (Eds.), Handbook of clinical neurology, pp. 321–350. Elsevier.

Turkeltaub, Eickhoff SB, Laird AR et al. (2012). Minimizing within-experiment and within-group effects in activation likelihood estimation meta-analyses. Hum Brain Mapp 33: 1–13. https://doi.org/10.1002/hbm.21186.

Tzourio-Mazoyer N, Mazoyer B (2017). Variations of planum temporale asymmetries with Heschl's gyri duplications and association with cognitive abilities: MRI investigation of 428 healthy volunteers. Brain Struct Funct 222: 2711–2726. https://doi.org/10.1007/s00429-017-1367-5.

Uddén J, Hultén A, Schoffelen J-M et al. (2022). Supramodal sentence processing in the human brain: FMRI evidence for the influence of syntactic complexity in more than 200 participants. Neurobiol Lang 3: 575–598. https://doi.org/10.1162/nol_a_00076.

van der Burght CL, Goucha T, Friederici AD et al. (2019). Intonation guides sentence processing in the left inferior frontal gyrus. Cortex 117: 122–134. https://doi.org/10.1016/j.cortex.2019.02.011.

van der Burght CL, Friederici AD, Maran M et al. (2023). Cleaning up the brickyard: how theory and methodology shape experiments in cognitive neuroscience of language. J Cogn Neurosci1–22. https://doi.org/10.1162/jocn_a_02058.

Vigneau M, Beaucousin V, Hervé PY et al. (2006). Meta-analyzing left hemisphere language areas: phonology, semantics, and sentence processing. Neuroimage 30: 1414–1432. https://doi.org/10.1016/j.neuroimage.2005.11.002.

Vigneau M, Beaucousin V, Hervé P-Y et al. (2011). What is right-hemisphere contribution to phonological, lexico-semantic, and sentence processing? Neuroimage 54: 577–593. https://doi.org/10.1016/j.neuroimage.2010.07.036.

Vingerhoets G, Li X, Hou L et al. (2018). Brain structural and functional asymmetry in human situs inversus totalis. Brain Struct Funct. https://doi.org/10.1007/s00429-017-1598-5.

Vissiennon K, Friederici AD, Brauer J et al. (2017). Functional organization of the language network in three- and six-year-old children. Neuropsychologia 98: 24–33. https://doi.org/10.1016/j.neuropsychologia.2016.08.014.

Walenski M, Europa E, Caplan D et al. (2019). Neural networks for sentence comprehension and production: an ALE-based meta-analysis of neuroimaging studies. Hum Brain Mapp. https://doi.org/10.1002/hbm.24523.

Wan B, Bayrak Ş, Xu T et al. (2022). Heritability and cross-species comparisons of human cortical functional organization asymmetry. Elife 11: e77215. https://doi.org/10.7554/eLife.77215.

Warrington S, Bryant KL, Khrapitchev AA et al. (2020). XTRACT-standardised protocols for automated tractography in the human and macaque brain. Neuroimage 217: 116923. https://doi.org/10.1016/j.neuroimage.2020.116923.

Wernicke C (1881). Lehrbuch der Gehirnkrankheiten für Aerzte und Studirende, vols. 1–2–3. Verlag von Theodor Fischer.

Wilson SM, Entrup JL, Schneck SM et al. (2023). Recovery from aphasia in the first year after stroke. Brain 146: 1021–1039. https://doi.org/10.1093/brain/awac129.

Witelson SF, Pallie W (1973). Left hemisphere specialization for language in the newborn: neuroanatomical evidence of asymmetry. Brain 96: 641–646. https://doi.org/10.1093/brain/96.3.641.

Xiao Y, Friederici AD, Margulies DS et al. (2016). Development of a selective left-hemispheric fronto-temporal network for processing syntactic complexity in language comprehension. Neuropsychologia 83: 274–282. https://doi.org/10.1016/j.neuropsychologia.2015.09.003.

Yang C, Crain S, Berwick RC et al. (2017). The growth of language: universal grammar, experience, and principles of computation. Neurosci Biobehav Rev. https://doi.org/10.1016/j.neubiorev.2016.12.023.

Zaccarella E, Trettenbrein PC (2021). Neuroscience and syntax. In: N Allott, T Lohndal, G Rey (Eds.), A companion to Chomsky. Wiley-Blackwell, pp. 325–347. https://doi.org/10.1002/9781119598732.ch20.

Zaccarella E, Schell M, Friederici AD (2017). Reviewing the functional basis of the syntactic merge mechanism for language: a coordinate-based activation likelihood estimation meta-analysis. Neurosci Biobehav Rev 80: 646–656. https://doi.org/10.1016/j.neubiorev.2017.06.011.

Zaidel E (1978). The elusive right hemisphere of the brain. Eng Sci 42: 10–32.

Handbook of Clinical Neurology, Vol. 208 (3rd series)
Cerebral Asymmetries
C. Papagno and P. Corballis, Editors
https://doi.org/10.1016/B978-0-443-15646-5.00013-0

Chapter 19

Brain asymmetries in figurative language comprehension

COSTANZA PAPAGNO*

CIMeC, Center for Mind/Brain Sciences, University of Trento, Rovereto, Italy

Abstract

This chapter reviews the literature concerning the neural basis of three types of figurative expressions, namely, idioms, metaphors, and irony. Besides these three forms of language, which are the most investigated, many other types exist, differing in their linguistic structure and, consequently, in the corresponding comprehension processes.

After defining the most common figurative expressions and showing how they differ in terms of linguistic properties, the chapter presents early studies that focused on the role of the right hemisphere (RH) in figurative language comprehension in general, without a clear distinction among different forms. Later literature shows how evidence has been accumulated, suggesting that both hemispheres are involved in figurative language processing. Therefore, a sharp distinction between the left hemisphere (LH) and the RH, the first involved in literal language and the second in figurative language, is not tenable. Idioms, metaphors, and irony will be considered separately, demonstrating that different expressions rely on the LH and RH to a different degree.

INTRODUCTION

Traditionally, language has been distinguished into literal and figurative. Figurative language is a way of expressing oneself using a word outside a strict or realistic meaning. People use about 6 nonliteral expressions per minute of discourse, specifically 1.8 new metaphors and 4.08 conventionalized metaphors (Pollio et al., 1977; Cacciari and Papagno, 2012), giving us an idea of how pervasive figurative language is. Besides metaphors, there are many other different forms of figurative language (i.e., idioms, clichés, proverbs, sarcasm, hyperbole, oxymorons, metonymies, etc.) (see Table 19.1).

The distinction between literal and nonliteral language has led to a dichotomy between LH and RH, with the consequence that the RH has been viewed as equally engaged in the processing of all sorts of nonstrictly denotative linguistic materials, including prosodic cues (typically carrying emotional as well as linguistic information), metaphors, idioms, proverbs, and different types of speech acts (Burgess and Chiarello, 1996). Accordingly, linguistic examinations have been developed to investigate the so-called RH language abilities, which implies testing idiom or metaphor comprehension, or in better terminology, formulaic language (see, e.g., Kempler and Van Lancker, 1996; Rinaldi et al., 2008).

On the one hand, however, the distinction between figurative and nonfigurative language is not totally clear; on the other hand, various figurative forms are likely to call for different comprehension processes, and the involvement of the RH in all these processes is neither the same nor exclusive.

In the final years of the 20th century, many researchers started to seriously doubt that such a distinction could be drawn, proposing a more realistic view: namely, to consider language on a continuum from not conventional at all to fully conventional. At the fully conventional pole, one could place idioms (see below), while literal (or propositional) language would be placed at the nonconventional end (see, for example, Cacciari and Papagno, 2012).

DEFINING FIGURES OF SPEECH

The definition of figurative language, in fact, covers several different figures of speech based on various cognitive processes, as reported in Table 19.1. For example, **metaphors** rely on categorization processes (Glucksberg, 2001), in that they make a comparison between two entities that share a common feature. They contain a "topic" or "tenor" for comment, which is the subject of the metaphor and a "vehicle" for making the comment (Richards, 1936). The vehicle is the semantic domain from which certain attributes are selected and assigned to the tenor. Metaphors can be classified in various ways: they can diverge in the grammatical form of the vehicle (Chen et al., 2008; Cacciari et al., 2011; Sakamoto and Utsumi, 2014), and we can distinguish nominal metaphors (noun-based: "This man is an elephant."), predicate metaphors (verb-based: "She ran for president."), prepositional

*Correspondence to: Costanza Papagno, CIMeC, Center for Mind/Brain Sciences, University of Trento, CeRiN, via Matteo del Ben 5/b, 30068 Rovereto, Italy. Tel: +39-0464808165, E-mail: costanza.papagno@unitn.it

Table 19.1

Different figures of speech

Figures	Definition	Example
Metaphor	A word or phrase applied to an object or action to which it is not literally applicable.	This woman is a rabbit
Simile	The two elements of the comparison need to be connected with a like	This dress fits like a glove
Metonymy	The name of an object is replaced with a word closely related	I am studying Dante
Idiom	A group of words having a meaning not deducible from those of the individual words	Kick the bucket
Proverb	A short, well-known saying, stating a general truth	All that glitters is not gold
Hyperbole	A deliberate exaggeration	I have thousands of things to do
Irony	A description which is the opposite of the reality	What a wonderful day
Sarcasm	Use of irony to mock, insult someone	This cork smells the wine
Oxymoron	A phrase with two contradictory words	A peaceful war
Cliché	A fixed, over-repeated expression	The perfect storm
Understatement	Intentional downplaying of a situation	Nothing serious, I had only a heart attack
Euphemism	To say something in an understated way	He passed away
Personification	Assigning human attributes to things	The road runs along the river

metaphors (preposition-based: "Chandelier earrings are out."), and adjective metaphors (adjective-based: "He has a warm heart.").

Another possible distinction relies on the level of conventionality, that is, how much their usage is well-established and recurrent in a culture; therefore, metaphors can be either highly conventional (e.g., "Marriage is a jail.") or novel (e.g., "She is a loudspeaker."). A slightly different criterion of categorization is the degree of salience (see also section "The neural correlates of idioms"): salient metaphor meaning is very familiar to the individual and already stored in his/her mental lexicon, whereas unfamiliar metaphor meaning is unknown or less prominent to him/her (Giora, 2002; Bowdle and Gentner, 2005).

Third, metaphors differ in their level of aptness, that is, the degree to which the vehicle captures salient attributes of the tenor. Apt metaphors sound like well-constructed expressions because of the clear analogy between the two semantic domains (e.g., "Memory is a warehouse."); conversely, nonapt metaphors may appear inappropriate or less intuitive (e.g., "His spectacles were sandpaper.") (Giora, 2002; Thibodeau and Durgin, 2011).

Metaphors, even those most conventionalized, do not have a unique standardized meaning, and in this aspect, they differ from another type of figurative language, idioms (see below). Indeed, when we say, as in the previous example of a nominal metaphor, "This man is an elephant," we might be referring to his dimensions, to his clumsiness, or to some other feature; on the other hand, idioms have a unique meaning (even though many of them derive from metaphors) that cannot change depending on the context. We can create new metaphors, just as we can use different forms of irony, but we cannot create new idioms simply by combining words online. To become an idiom, a linguistic expression must be conventionalized for an entire linguistic community.

A figure of language close to metaphor is a **simile**, in that it is also a comparison, but the two elements of the comparison need to be connected through a "like" (This shirt fits like a glove.), reducing the amount of "imagination" required for metaphor comprehension. The distinction between simile and metaphor is among the oldest and most widely recognized in rhetorical theory. It is also one of the most tenuous (Israel et al., 2004).

When one makes an obvious reference to a concept by referring to it as one of its attributes, a **metonymy** is produced, i.e., "In that class, the students read Cicerone." This is also a sort of comparison-based figure of speech; however, instead of making a direct connection, people are simply letting one word or phrase stand in for another (in this example, the author's name for the work itself). Metaphors are based on analogies, while in metonymies, the word is substituted because there is already a clear association between the new word and the original. Saying "Give me a hand" to ask for help is another example. Metonymies are found to be easier to process in a sentence context than metaphors (Bambini et al., 2013).

Irony is more interpretive and complex than metaphor, in that it involves a relationship between the thoughts of the speaker and the thoughts of someone else (e.g., by alluding to the failure of shared expectations), therefore involving theory-of-mind (ToM) mechanisms. In irony, what is said is, in general, the opposite of what is intended (e.g., "What a nice day." when it is raining heavily) (Grice, 1975). Understanding irony relates to ambiguity resolution, and irony is ambiguous per se (Rapp et al., 2013). Indeed, irony is comprehended at a later age than metaphor (Szücs, 2013). **Sarcasm** is close to irony, but it is considered more caustic and is usually directed against an individual (Gibbs, 1986). Sarcasm is used to indirectly convey the speaker's criticism of a victim and is deployed in a hurtful way.

Differently from metaphors, the meaning of an **idiom** must be retrieved from semantic memory, as idioms are unique to a language, culture, or even region. In these types of phrases, the literal meaning differs from the figurative one, so only a native speaker or a person otherwise familiar with the phrase can understand what is actually being expressed.

According to Nunberg et al. (1994), the class of idiom is not unitary but rather varies in syntactic and semantic dimensions. Some idioms have literal meanings that denote concrete things and their relations (spilling the beans), while others do not (second thoughts). Idioms can involve metaphors (take the bull by the horns), metonymies (lend a hand), hyperbole (not worth the paper it's printed on), or other kinds of figuration not necessarily clearly perceived by the listener. Yet some idioms involve no other figuration at all (e.g., "by dint of").

An idiom can be semantically opaque when the meaning can be derived from neither the image evoked (kick the bucket) nor the constituent words. A transparent idiom, on the other hand, shows a clear relationship between metaphorical and literal sense: "to skate on thin ice" clearly suggests a dangerous situation. Another way to distinguish idioms is compositionality: the degree to which the phrasal meaning, once known, can be analyzed in terms of the contribution of the idiom's parts. In "*spill the beans*," for example, there is a clear correspondence between "spill" and "beans and the relevant parts of its figurative meaning of divulge information" (Papagno et al., 2004*)*. Idioms also vary in the extent to which they can be syntactically transformed while retaining their idiomatic meaning (Gibbs and Gonzales, 1985). Finally, some idioms do not have any well-formed literal counterpart, while others do but have an ambiguous correspondence to idiomatic meaning (e.g., "break the ice").

Finally, a figure of speech which, like idiom, is typical of a culture, is a **proverb**, which is often used as a general comment on shared communicative situations and consists of generally true statements, both literally and figuratively. They are full sentences, signaled by specific phonetic and/or rhetorical patterns. Proverb interpretation requires two general cognitive stages: a denotative stage, in which the specific meaning of the individual words of the proverb is retrieved from some semantic knowledge, and a connotative stage, in which a coherent, abstract principle or concept is formed on the bases of the meanings of the individual words (Kaiser et al., 2013).

It is evident from this short review that these figures of language involve common but also different processes, and therefore, that the networks recruited for their interpretations overlap only partially. Of course, they all involve the comprehension of individual words and the processing of semantic and syntactic features, and, accordingly, they must thus require at least part of the same brain networks involved in literal language. Indeed, the psycholinguistic models on figurative-language processing are based on language-unimpaired participants, and they imply lexical integrity as a prerequisite.

THE RIGHT-HEMISPHERE HYPOTHESIS

Following what has been described in the previous section, damage to the LH that results in aphasia should also affect the ability to comprehend figurative language. However, a widely accepted view in the neuropsychologic literature assumes that damage to the LH may have no effect on figurative language, while damage to the nondominant RH would be threatening to the processing of figurative expressions (Van Lancker Sidtis and Postman, 2006).

This theory that considers RH as crucial to any form of figurative language originated with a misinterpretation of the results of the first classic lesion study on metaphor comprehension (Winner and Gardner, 1977). Winner and Gardner found that people with aphasia (PWA) correctly associated a metaphoric sentence with a picture, while this was not the case for right-brain-damaged patients, who chose the picture representing the literal meaning. However, when the same right-brain-damaged patients were required to produce an oral explanation of the figurative expression, they responded correctly, while those with PWA, who despite their language impairment, were able to produce an oral answer and provided the literal meaning. Nonetheless, Winner and Gardner's results have been considered an example of how LH and RH are responsible for different aspects of language: literal and figurative, respectively. What really emerges from this study is that right-brain-damaged people were unable to select the appropriate picture. One possibility could be that these patients, as often happens after RH damage, had trouble with pictorial material (e.g., Papagno et al., 2006) but no information is available to support or disconfirm this explanation.

A similar study—but on idioms—was conducted by Stachowiak et al. (1977), who investigated the comprehension of spoken texts in four subgroups of PWA, in normal controls, and in right-brain-damaged patients. Participants were presented with short texts and were required to choose the appropriate picture for the story from a multiple-choice set of five: one picture showing the main event, another picture depicting the literal sense of a metaphorical comment, and the others misrepresenting semantic functions (subject-noun phrase, verb, verb complement phrase) expressed in the text. Half of the stories were commented with transparent idioms and half with opaque idioms. The performance of the aphasic groups was not poorer than that of the other two groups, being the performance is better for opaque idioms for all groups. Nevertheless, Wernicke's aphasics showed difficulties in dealing with metaphorical meanings and tended to point to the picture representing the literal interpretation.

Some years later, Van Lancker and Kempler (1987), and Kempler et al. (1999) reported a double dissociation between a series of linguistic expressions including idioms, proverbs, and formulaic expressions (I'll get back to you), and novel sentence comprehension: right-brain-damaged participants performed correctly on novel (literal) sentences and poorly on familiar (figurative) sentences, while left-brain-damaged aphasic patients showed the opposite pattern.

Similarly, Brownell et al. (1990) found that right brain-damaged patients experienced difficulties in retrieving the metaphoric meaning of a word. However, RHD patients' performance suggested a general deficit in appreciation of alternative meanings that extended beyond nonliteral interpretations of words and phrases to include homonymous nouns whose secondary meaning was not metaphoric. Therefore, the result cannot be interpreted as a specific deficit in comprehending figurative language.

Importantly, Tompkins et al. (1992) pointed out that, in the abovementioned studies, the tasks of providing a definition or choosing a picture—therefore offline paradigms—require several cognitive operations, making it difficult to determine at which level the deficit was located.

To overcome this problem, Tompkins et al. tested idiom comprehension using an online word-monitoring task, in which participants listened to a sentence, and when they heard a specific target word, they had to press a button as quickly as possible. Target words were concrete nouns that were the final constituent of ambiguous familiar idioms, and they appeared in three different contexts: idiomatic, inside an ambiguous idiom but prompting a literal interpretation, and in a control condition in which the word appeared in a sentence that was not an idiom. Right- and left-brain-damaged patients performed similarly to controls, all being faster with idiomatic and literal contexts, while in an offline task (definition of an idiomatic string), patients were equally impaired with respect to controls, suggesting impairment in a later stage of processing. However, left-brain-damaged patients showed a high-performance variability in the online task, which could be accounted for by the fact that some were aphasic, and their reduced language abilities diminished performance.

This was the first study that challenged the RH hypothesis. Later, a series of papers appeared in which figurative language comprehension resulted impaired in left brain-damaged patients (Nenonen et al., 2002; Hillert, 2004; Papagno et al., 2004; see Papagno and Romero Lauro, 2010, for a review), However, when the lesion involved the frontal lobe, right brain-damaged patients' performance was pathologic as well (Papagno et al., 2006).

Starting from these last studies, the different forms of figurative language were investigated separately, with a major interest in three categories, namely, idioms, metaphors, and irony. In particular, several neuroimaging studies have shed light on the anatomic structures involved in processing figurative language, pointing to an asymmetrically distributed network involving both hemispheres, with predominance for LH.

THE NEURAL CORRELATES OF IDIOMS

Information concerning the neural circuits involved in idiom comprehension comes from neuropsychologic, neurophysiological, and neuroimaging studies. Several neuropsychologic studies on PWA and right-brain-damaged patients have been conducted, in most cases distinguishing opaque (Papagno et al., 2004, 2006) and ambiguous idioms (Cacciari et al., 2006), leading to the conclusion that PWA is impaired in the comprehension of both types when they are presented out of context. They can, however, benefit from contextual cues, which indeed improve performance. Right-brain-damaged patients are impaired as well, but crucially, only when the lesion involves the prefrontal lobe. Therefore, it is possible that LH activity is required in the interpretation of single elements within an expression, while RH (and more specifically, the prefrontal cortex) is responsible for the activation of contextual cues. In particular, neuropsychologic studies have highlighted two specific cortico-subcortical regions that, when damaged, yield impaired idiom comprehension: the left temporal and bilateral prefrontal regions (Cacciari et al., 2006; Papagno et al., 2006). These results are confirmed by TMS studies (Oliveri et al., 2004; Fogliata et al., 2007; Rizzo et al., 2007; Häuser et al., 2016), in which inhibition of the left temporal cortex decreased literal and idiomatic comprehension to the same level, while modulation of the prefrontal cortex affects idiomatic comprehension alone. More specifically, when the time course of opaque-idiom processing was investigated, it was found that a selective impairment in accuracy for idioms was found when rTMS was applied to the left prefrontal and temporal cortex 80 ms after picture presentation, confirming the role of these regions in this task. Moreover, rTMS to the prefrontal cortex, but not to the temporal cortex, continued to affect performance on idiomatic sentences at the later time of 120 ms, suggesting that the prefrontal region is involved in both the retrieval of figurative meaning from semantic memory and the monitoring of the response by inhibiting alternative interpretations when a picture-matching task is used. This was, however, an offline task, with all its limitations. Häuser et al. (2016) also stimulated the left prefrontal cortex, specifically the ventrolateral, as in Fogliata et al. (2007), and found that low-familiarity idioms were particularly affected by TMS disruption, and they explained this result suggesting a higher semantic conflict between a salient literal and a less-known figurative meaning. Rizzo et al. (2007) obtained the same results, namely a disruption of unambiguous idiom comprehension, by stimulating the bilateral dorsolateral prefrontal cortex. Similarly, Kurada et al. (2021) applied rTMS to the same regions to explore the role of the two hemispheres depending on the idiom type. They demonstrated that the left dorsolateral prefrontal cortex was more critical to comprehension of opaque rather than transparent idioms.

The prefrontal cortex can have different roles in idiom comprehension: specifically, after the linguistic analysis of a string is performed, the choice of the correct response from among alternatives requires a selection process coupled with monitoring of the response. Metzler (2001) obtained a similar result with ambiguous words: in a semantic priming task on ambiguous words, patients with left frontal lesions did not show semantic facilitation for contextually appropriate meaning compared to controls. Alternative interpretations could be suggested, for example, a tendency to give concrete responses resulting from premature selection of an interpretation, as found with proverbs by Burgess and Chiarello (1996) and also observed in schizophrenic patients (Schettino et al., 2010). Another possibility could be a lack of suppression of the literal meaning, which suppression would allow the activation of the figurative one (Papagno et al., 2003).

Neuroimaging studies have confirmed the involvement of both hemispheres in idiom comprehension—LH to a greater extent—challenging the notion of selective RH involvement in processing idioms. The network that is specifically activated by the nonliteral condition (opaque-idiom processing) involves the left temporal cortex, the left superior medial frontal gyrus (Brodmann area 9), and the left inferior frontal gyrus. Activations in RH have been found in the right superior and middle temporal gyri, temporal pole, and the right inferior frontal gyrus (Lauro et al., 2008). A similar activation has been described in a different study (Zempleni et al., 2007) in which figurative comprehension of both ambiguous and unambiguous idioms was found to be supported by bilateral inferior frontal gyri and left middle temporal gyrus. The right middle temporal gyrus was also involved but apparently only to process ambiguous idioms (Zempleni et al., 2007). Mashal et al. (2008), in a further experiment with ambiguous idioms, found that processing the idiomatic interpretation of idioms and the literal interpretations of literal sentences involved LH regions, whereas processing the literal interpretation of idioms was associated with increased activity in the right brain regions, suggesting that these RH areas are involved in semantic ambiguity resolution and in processing nonsalient meanings of idiomatic expressions. More recently, an fMRI study in a different language, Spanish, was run to assess comprehension of ambiguous and nonambiguous idioms (Bendersky et al., 2021), and again, a bilateral activation in the frontotemporal regions was observed, showing consistency across languages. Findings from the processing of Chinese idioms provide additional evidence that both LH and RH are involved in idiom processing, although in different ways, and that different types of idioms produce different patterns of activation. Significantly, idioms engaged more activation in the right parietal cortex, and the activation strength decreased as a function of the semantic transparency of phrases (Yang et al., 2016). Gender differences are also observed: women showed greater overall activation than men when comprehending literal and idiomatic sentences; whereas men had significantly greater functional connectivity between LIFG left inferior frontal gyrus and LMTG left middle temporal gyrus than women across tasks (Kana et al., 2012).

To sum up, all data converge to indicate a bilateral involvement of the hemispheres in idiom processing, with an asymmetry due to higher and more constant activation of LH, in sharp contradiction to the seminal studies from 1977 to the 1990s. RH seems to be "asked for help," either when an idiom maintains two meanings, the literal and the figurative, or when the figurative meaning has low familiarity. Alternatively, right inferior frontal gyrus activation has been attributed to enhanced attention to the salience of familiar idioms (Bohrn et al., 2012).

THE NEURAL CORRELATES OF METAPHOR

As already reported, the RH hypothesis originates from a misinterpretation of the results of the study on metaphor comprehension by Winner and Gardner (1977) and was initially supported by studies on healthy participants using different paradigms, such as a divided-visual-field paradigm (see Anaki et al., 1998). It has already been stressed that subsequent studies did not confirm this special involvement of RH neither in patient studies—right-brain-damaged patients performed just as well as healthy participants with conventional metaphors (Giora et al., 2000; Zaidel et al., 2002)—nor in activation studies (e.g., Rapp et al., 2004; Lee and Dapretto, 2006; Stringaris et al., 2007). However, a fundamental aspect of metaphors is that they can differ in their level of conventionalization. According to this distinction, several hypotheses have been made to account for the different roles of the two hemispheres. The graded-salience hypothesis (GSH) (Giora, 1997) suggests that the main variable affecting process lateralization is salience, referring to the degrees of familiarity, conventionality, and predictability of a metaphor. Therefore, LH is required for the processing of salient meanings, while RH, for the processing of nonsalient meanings: a novel metaphor would (also) be processed by RH. Accordingly, the conflicting evidence from previous fMRI research studying the role of RH in metaphor processing could be related to the confounding effects of salience. The findings of another fMRI study (Mashal et al., 2005) support this explanation, as does a meta-analysis on figurative language, where novel and conventional metaphors were contrasted to test the GSH, and RH involvement was found only for novel metaphors (Bohrn et al., 2012). Similarly, a TMS study (Pobric et al., 2008) investigated a group of healthy participants who were presented with four types of word pairs—literal, conventional metaphoric, and novel metaphoric expressions, and unrelated word pairs—and were then asked to perform a semantic judgment task. While novel metaphor processing was disrupted by rTMS over the right posterior superior temporal sulcus, conventional metaphors, as well as literal word pairs, were impaired when rTMS was applied over the left inferior frontal gyrus. The findings suggest that RH may be critically involved in processing nonsalient meanings, supporting the GSH described above but also a relatively similar hypothesis: the fine-coarse semantic coding theory (CSCT, Beeman et al., 1994; Beeman, 1998; Jung-Beeman, 2005), which also assumes that it is not figurativeness itself that determines hemispheric lateralization. According to this theory, RH and LH are specialized in two different processes of semantic-spreading activation and integration: coarse semantic coding and fine semantic coding, respectively. When a word is heard, LH rapidly triggers a focused semantic field composed of closely related meanings; meanwhile, RH activates a wide semantic field comprising both close and distant meanings. According to this model, conventional metaphor interpretation depends mostly on the LH fine coding since it allows the quick retrieval of conventional metaphoric association between target and vehicle. On the other hand, the RH coarse coding contributes crucially to novel metaphor comprehension by integrating the unusually associated, distant terms of the unfamiliar metaphor.

A novel metaphor evolves, however, over time, into a familiar or conventional one. The career metaphor model (Bowdle and Gentner, 2005) suggests that nonconventional metaphors call for an online comparison, i.e., an evaluation of the highest number of attributes of both the vehicle and the tenor to discover similarities among concepts. Conventional metaphors require relating a certain target attribute to a stored abstract schema of meaning directly elicited by the vehicle (categorization), therefore, allowing faster access to the meaning because it no longer requires the online comparison of semantic domains, but rather, the retrieval of a previously learned meaning stored in semantic memory. However, a novel metaphor can become familiar when we are repeatedly exposed to it: the gradual process by which a word acquires additional and easily accessible figurative meaning represents the "career of metaphor." Cardillo et al. (2012) performed an fMRI study to test the process of conventionalization of novel metaphors and found that it modulates the activity within the bilateral inferior prefrontal cortex, left posterior middle temporal gyrus, and right posterolateral occipital cortex, producing a decreased neural load within the semantic networks, rather than a hemispheric or regional shift across brain areas.

A recent study on brain-damaged patients (Cardillo et al., 2018) demonstrated that both LH and RH patients performed worse on metaphoric than literal sentences, and the degree of this difficulty varied for the different types of metaphor (the nominal and predicate), but the two patient groups did not differ. When single cases were examined to look for dissociations between literal and metaphoric sentences, it was found that there were either general comprehension deficits of metaphors and literal sentences to equal extents, and selective metaphor impairments that were specific to certain types of metaphor (impairment of nominal but not predicate metaphors). Patients with selective metaphor deficits had an injury to LH, once again providing an argument against the hypothesis of a specific or necessary contribution of RH in understanding metaphoric language. More specifically, the overlay map of the lesions revealed an involvement of the left prefrontal region. Cardillo et al. (2018) concluded that the RH hypothesis of metaphor does not hold up when considering the neuroimaging results (see below) or lesion studies, and they interpret the role of RH as supportive but not necessary, probably engaging when more demanding stimuli require additional resources. In line with the cognitive resource hypothesis (Tompkins, 2012), Monetta et al. (2006) showed that when healthy participants had to process alternative metaphorical and nonmetaphorical meanings of words in an isolated condition or in a dual-task paradigm, their performances were more affected by the dual-task condition with the metaphorical meaning than with the nonmetaphorical task, suggesting that greater cognitive resources are required for a metaphorical linguistic task than for a nonmetaphorical task.

Generally, nominal metaphors are tested with some reported exceptions (see, for example, Cardillo et al., 2018). A study that specifically considered novel predicate metaphors using auditory presentation in a brief context revealed bilateral activation of parietal areas, including the left angular and right inferior parietal gyri and right precuneus (Obert et al., 2014). The involvement of RH could be attributed to the novelty of the metaphor (GSH), to attentional resources (Bohrn et al., 2012) or to deeper semantic processes required by the lack of familiarity. No general RH advantage for metaphor processing could be found in a recent meta-analysis of neuroimaging studies concerning figurative language. On the contrary, significant clusters of activation for metaphor conditions were lateralized mostly to LH.

Metaphor processing is also vulnerable to neurodegenerative diseases, essentially probable Alzheimer disease (Papagno, 2001; Papagno et al., 2003; Fujimoto et al., 2019; Klooster et al., 2020). A recent study of patients with LH neurodegeneration affected by logopenic aphasia and demographically matched healthy participants (Klooster et al., 2020) showed that LH participants performed disproportionately poorly on moderately familiar metaphor comprehension. A voxel-based morphometry analysis showed a relationship between metaphor accuracy and atrophy in the left middle and superior temporal gyri, as well as the left inferior frontal gyrus, in line with several neuroimaging studies. Finally, Hartung et al. (2020), in an fMRI study, demonstrated that the neural correlates of processing metaphoric sentences and their literal counterparts are indistinguishable when embedded in a narrative: both conditions activate bilateral areas along the anterior temporal poles, middle temporal gyri, superior temporal sulci, and the angular gyri.

THE NEURAL CORRELATES OF IRONY

As already reported, irony is a form of nonliteral language in which the speaker makes a statement meaning something different, in general, the opposite of what has been said (Gibbs, 1986). The cognitive processes that correlate with irony detection seem to be illuminated by the tasks that evaluate ToM, inhibitory control, and verbal abilities. Indeed, the comprehension of irony should involve brain areas specialized in social cognition, since the listener must be aware that the intended meaning is different from what is said literally. Deficits of ToM, particularly second-order ToM, can co-occur with difficulties in understanding irony in many brain-damaged patients, and several studies using neuroimaging methods have shown the activation of a neural ToM network—left and right temporal-parietal junction, medial prefrontal cortex, and precuneus—when performing a task of irony understanding (Eviatar and Just, 2006). Indeed, the most severe disturbances in irony comprehension have been found in patients who are insensitive to the context, that is, they do not perceive an incongruity between the statement and the context and are affected by a joint impairment of ToM and executive functions. In other patients, who are sensitive to the context, an isolated deficit in executive function can be sufficient to produce a deficit in irony if the executive function demand of the irony task is increased. However, markers helping the comprehension of irony, such as hyperbolic statements, may eliminate the effect of task demand (Cordonier et al., 2020).

A lack of sensitivity to contextual incongruity has already been demonstrated in right brain-damaged patients by Champagne-Lavau et al. (2018), which leads to an impairment in irony comprehension and a consequent literal interpretation.

Apparently, irony has never been tested in patients with damage to LH, but at least one study (Giora et al., 2000) assessed comprehension of sarcasm, which is a form of irony used in a critical way, as explained above. Both LH and RH brain-damaged patients were impaired, but the RH patients performed worse than patients with LH damage when aphasia was controlled for. Considering the lesion site, anterior regions on the left (corresponding to the middle and inferior frontal gyri, including Broca's area) correlated negatively with sarcasm comprehension, while no correlation with any specific area was found in the case of RH. Therefore, even though RH contribution seemed greater, it is not localized to any specific region, whereas LH contribution is associated with specific left frontal regions. Thus, it is possible that LH contributes to sarcasm comprehension because of its linguistic format and not due to the particular type of expression, while RH is involved because of that specific category of utterance, clearly showing that the processing of sarcasm in the two hemispheres is qualitatively different.

A further neuropsychologic study of sarcasm (Shamay-Tsoory et al., 2005) demonstrated that patients with prefrontal lesions were impaired, but the most severe deficit in sarcasm comprehension was found in those with right ventromedial lesions. ToM and emotion recognition were related to the ability to understand sarcasm.

Additional information about the differential contribution of the two hemispheres in irony comprehension comes from neuroimaging studies. In general, fMRI studies on irony contrasted it with one of the other types of figurative expressions, such as metaphor. Eviatar and Just (2006) required participants to read brief, three-sentence stories that concluded with either a literal, metaphoric, or ironic sentence, acquiring fMRI images during the reading of the critical sentence. A selective neural response to metaphor and irony was found. In particular, metaphoric expressions produced significantly higher activation in the left inferior frontal gyrus and in the bilateral inferior temporal cortex than literal and ironic utterances. On the other hand, ironic expressions resulted in significantly higher activation than literal statements in the right superior and middle temporal gyri, with metaphoric statements generating intermediate levels in these regions. This result shows that RH is differentially involved in irony comprehension, in line with the study on sarcasm reported above. In addition, the activation patterns in the right temporal lobe were in different regions for irony and metaphor, supporting the hypothesis that different types of figurative language recruit different areas.

In contrast with these findings, a recent meta-analysis based on the mean of irony studies (Reyes-Aguilar et al., 2018) revealed increased activation in the dorsomedial prefrontal cortex and right medial prefrontal cortex (as was found for sarcasm in brain-damaged patients), inferior frontal gyrus and superior temporal gyrus for irony comprehension, with no difference in irony relative to other pragmatic forms.

Bosco et al. (2017) contrasted irony with deceit and, at odds with the previously cited studies, found that the recognition of ironic communicative intention activated a left frontotemporal network, involving the left temporoparietal junction, the left inferior frontal gyrus, the left middle frontal gyrus, the left middle temporal gyrus, and the left dorsolateral prefrontal cortex.

Comprehension of deceitful communicative intention activated the same areas with the exception of the middle temporal gyrus, thus suggesting its specific role in irony recognition.

Bohrn et al. (2012) combined data from 354 participants across 22 fMRI studies and one positron emission tomography study to identify the differences in neural correlates of figurative and literal language processing and to investigate the role of RH in figurative-language processing. As in previously reported studies, they demonstrated that figurative language activates a bilateral distributed network, involving the left and right inferior frontal gyrus, large parts of the left temporal lobe, the bilateral medial frontal gyri, and an area around the left amygdala. Subgroup comparisons between experiments on metaphor, idiom, and irony/sarcasm revealed shared activations in left frontotemporal regions for idiom and metaphor processing. Irony/sarcasm processing was correlated with activations in anterior midline structures such as the medial frontal gyrus, anterior cingulate cortex, the right superior frontal gyrus, and the right precentral gyrus, these regions being distinct from those involved in idiom and metaphor comprehension. Irony processing depends on regions distinct from metaphor/idiom processing and barely overlapped, other than a small part of the medial prefrontal cortex in BA 9/10. Not a single region could be identified in which all three types of figurative-language processing overlapped.

All neuroimaging data on irony/sarcasm processing point toward an important role of the mentalizing network—especially the medial prefrontal cortex—in irony processing (see also Akimoto et al., 2014), consistently showing activation of semantic networks and theory-of-mind networks during irony comprehension (see, e.g., Akimoto et al., 2014).

CONCLUSIONS

This short review of three types of figurative language supports the statement at the beginning of this chapter, that different figures of language should not be treated as one, because they rely on different processes and, consequently, on different neural correlates (see Fig. 19.1). It also explicates the different contributions of the two hemispheres in processing the various forms of figuration. Even two, almost overlapping, figures, such as irony and sarcasm, present specific activations (Filik et al., 2019). In a recent meta-analysis, idioms showed convergence in the left inferior frontal gyrus (BA 45) (Reyes-Aguilar et al., 2018), but previous studies demonstrated that idiom activations differ depending on their characteristics (ambiguous, opaque, decomposable, etc.); in the case of metaphor, the level of conventionalization determines the involvement of RH, but also their status as nominal or predicate can change the processing areas; irony involves the ToM network and seems to be the figure of speech with the highest involvement of RH. Otherwise, in general, figurative language shows a bias for LH, possibly because *it is* language, and as such, it requires the same perisylvian areas as literal language, along with the medial prefrontal cortex, due to higher-level processes of inference.

Some authors have also suggested differences among languages since frontotemporal regions were recruited by Germanic-language speakers, while only left frontal areas were recruited by Romance-language speakers, and the right medial prefrontal cortex by Japanese speakers (Reyes-Aguilar et al., 2018). However, this variation seems unlikely, since idioms, for example, were studied in speakers of Hebrew, Italian, Spanish, and Chinese—all with the same results (see above, section "The neural correlates of idioms").

Inconsistent results are due to various methodological issues, such as a lack of differentiation of lesion sites within RH, the absence of a clear definition of the neuropsychologic profiles of patients, types of stimuli, use of pictorial material (Oliveri et al., 2004; Papagno and Caporali, 2007), and different paradigms in activation studies.

The bottom line is that figurative language is a fascinating topic that effectively illustrates how the two hemispheres must communicate with each other internally to allow successful social communication externally.

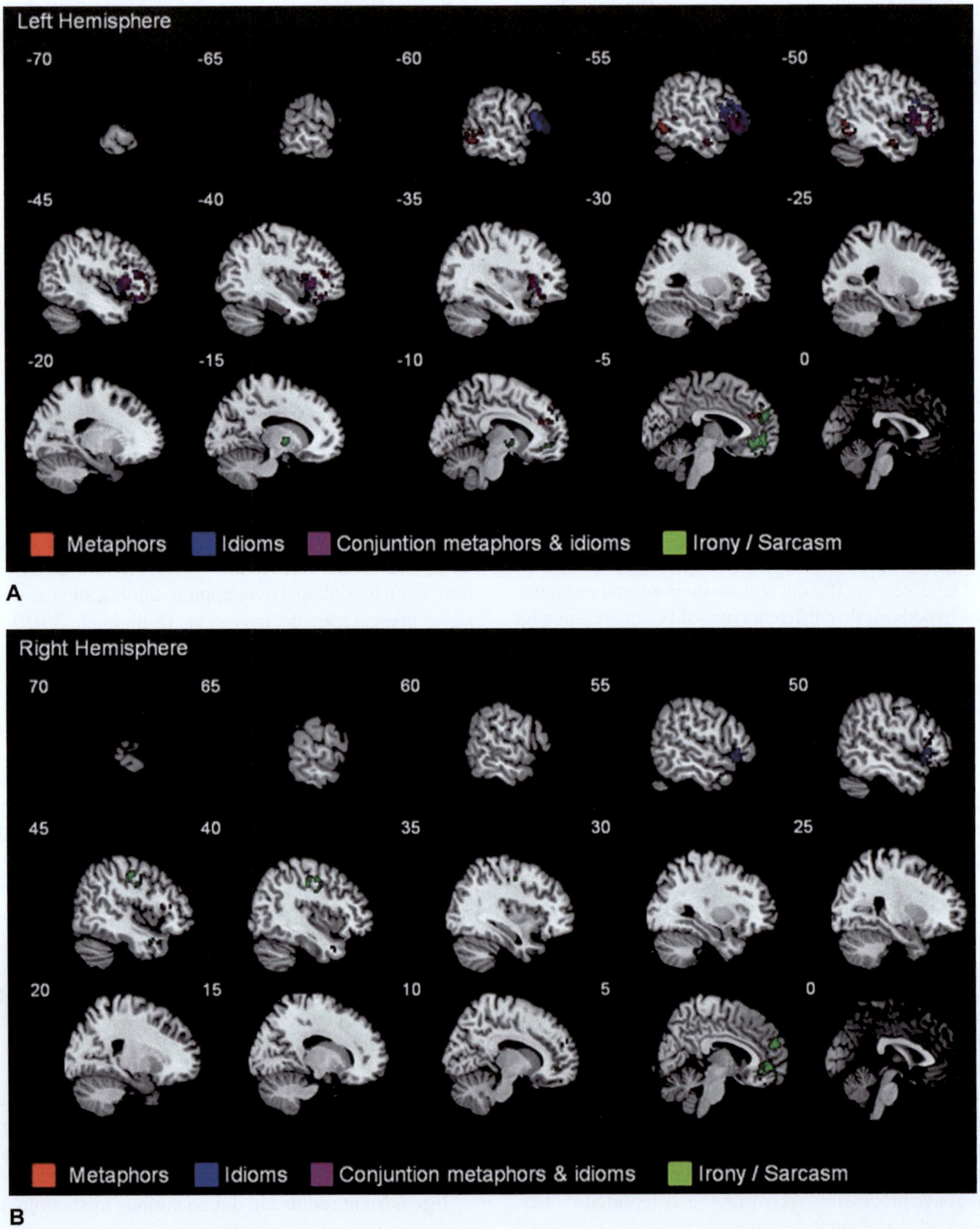

Fig. 19.1. Representation of the different areas activated for metaphor, idiom, or irony processing, compared to literal language: (A) left hemisphere and (B) right hemisphere. The left inferior frontal region is activated by both, idiom and metaphor. Irony processing activates the medial frontal cortex, the right superior frontal, and the middle and superior temporal gyri. Modified from fig. 2, Bohrn IC, Altmann U, Jacobs AM (2012). Looking at the brains behind figurative language—a quantitative meta-analysis of neuroimaging studies on metaphor, idiom, and irony processing. Neuropsychologia 50: 2669–2683.

References

Anaki D, Faust M, Kravetz S (1998). Cerebral hemispheric asymmetries in processing lexical metaphors. Neuropsychologia 36: 353–362. https://doi.org/10.1016/s0028-3932(97)00110-3.

Akimoto Y, Sugiura M, Yomogida Y et al. (2014). Irony comprehension: social conceptual knowledge and emotional response. Hum Brain Mapp 35: 1167–1178.

Bambini V, Ghio M, Moro A et al. (2013). Differentiating among pragmatic uses of words through timed sensicality judgments. Front Psychol 4: 938. https://doi.org/10.3389/fpsyg.2013.00938.

Beeman M (1998). Coarse semantic coding and discourse comprehension. In: M Beeman, C Chiarello (Eds.), Right hemisphere language comprehension. Perspective from cognitive neuroscience. Erlbaum, Mahwah, NJ, pp. 255–284.

Beeman M, Friedman RB, Grafman J et al. (1994). Summation priming and coarse semantic coding in the right hemisphere. J Cogn Neurosci 6: 26–45. https://doi.org/10.1162/jocn.1994.6.1.26.

Bendersky M, Lomlomdjian C, Abusambra V et al. (2021). Functional anatomy of idiomatic expressions. Brain Topogr 34: 489–503. https://doi.org/10.1007/s10548-021-00843-3.

Bohrn IC, Altmann U, Jacobs AM (2012). Looking at the brains behind figurative language—A quantitative meta-analysis of neuroimaging studies on metaphor, idiom, and irony processing. Neuropsychologia 50: 2669–2683.

Bosco FM, Parola A, Valentini MC et al. (2017). Neural correlates underlying the comprehension of deceitful and ironic communicative intentions. Cortex 94: 73–86.

Bowdle BF, Gentner D (2005). The career of metaphor. Psychol Rev 112: 193–216. https://doi.org/10.1037/0033-295X.112.1.193.

Brownell HH, Simpson TL, Bihrle AM et al. (1990). Appreciation of metaphoric alternative word meanings by left and right brain-damaged patients. Neuropsychologia 28: 375–383.

Burgess C, Chiarello C (1996). Neurocognitive mechanisms underlying metaphor comprehension and other figurative language. Metaphor Symb Activ 11 (1): 67–84. https://doi.org/10.1207/s153278 motion68ms1101_4.

Cacciari C, Papagno C (2012). Neuropsychological and neurophysiological correlates of idiom understanding: how many hemispheres are involved? In: M Faust (Ed.), The handbook of the neuropsychology of language. Wiley-Blackwell 1: 368–385.

Cacciari C, Bolognini N, Senna I et al. (2011). Literal, fictive, and metaphorical motion sentences preserve the motion component of the verb: a TMS study. Brain Lang 119: 149–157. https://doi.org/10.1016/j.bandl.2011.05.004.

Cacciari C, Reati F, Colombo MR et al. (2006). The comprehension of ambiguous idioms in aphasic patients. Neuropsychologia 44: 1305–1314.

Cardillo ER, McQuire M, Chatterjee A (2018). Selective metaphor impairments after left, not right, hemisphere injury. Front Psychol 9: 2308. https://doi.org/10.3389/fpsyg.2018.02308.

Cardillo ER, Watson CE, Schmidt GL et al. (2012). From novel to familiar: tuning the brain for metaphors. Neuroimage 59: 3212–3221. https://doi.org/10.1016/j.neuroimage.2011.11.079.

Champagne-Lavau M, Cordonier N, Bellmann A et al. (2018). Context processing during irony comprehension in right-frontal brain-damaged individuals. Clin Linguist Phon 32: 721–738. https://doi.org/10.1080/02699206.2018.1430851.

Chen E, Widick P, Chatterjee A (2008). Functional-anatomical organization of predicate metaphor processing. Brain Lang 107: 194–202. https://doi.org/10.1016/j.bandl.2008.06.007.

Cordonier N, Fossard M, Champagne-Lavau M (2020). Differential impairments in irony comprehension in brain-damaged individuals: insight from contextual processing, theory of mind, and executive functions. Neuropsychology 34: 750. https://doi.org/10.1037/neu0000682.

Eviatar Z, Just MA (2006). Brain correlates of discourse processing: an fMRI investigation of irony and conventional metaphor comprehension. Neuropsychologia 44 (12): 2348–2359.

Filik R, Ţurcan A, Ralph-Nearman C et al. (2019). What is the difference between irony and sarcasm? An fMRI study. Cortex 115: 112–122. https://doi.org/10.1016/j.cortex.2019.01.025.

Fogliata A, Rizzo S, Reati F et al. (2007). The time course of idiom processing. Neuropsychologia 45 (14): 3215–3222. https://doi.org/10.1016/j.neuropsychologia.2007.06.009.

Gibbs R (1986). On the psycholinguistic of sarcasm. J Exp Psychol Gen 115: 3–15.

Fujimoto N, Nakamura H, Tsuda T et al. (2019). Impaired comprehension of metaphorical expressions in very mild Alzheimer's disease. Neuropsychiatr Dis Treat 15: 713–720. https://doi.org/10.2147/NDT.S193645.

Gibbs RW, Gonzales GP (1985). Syntactic frozenness in processing and remembering idioms. Cognition 20: 243–259.

Giora R (1997). Understanding figurative and literal language: the graded salience hypothesis. Cogn Linguist 7: 183–206.

Giora R (2002). Literal vs. figurative language: different or equal? J Pragmat 34: 487–506.

Giora R, Zaidel E, Soroker N et al. (2000). Differential effect of right- and left-hemisphere damage on understanding sarcasm and metaphor. Metaphor Symb 15: 63–83.

Glucksberg S (2001). Understanding figurative language, Oxford University Press, Oxford.

Grice HP (1975). Logic and conversation. In: P Cole, J Morgan (Eds.), Syntax and semantics, 3: speech acts. Academic Press, New York, NY.

Hartung F, Kenett YN, Cardillo ER et al. (2020). Context matters: novel metaphors in supportive and non-supportive contexts. Neuroimage 212: 116645. https://doi.org/10.1016/j.neuroimage.2020.116645.

Häuser KI, Titone DA, Baum SR (2016). The role of the ventrolateral prefrontal cortex in idiom comprehension: an rTMS study. Neuropsychologia 91: 360–370. https://doi.org/10.1016/j.neuropsychologia.2016.09.003.

Hillert DG (2004). Spared access to idiomatic and literal meanings: a single-case approach. Brain Lang 89: 207–215.

Jung-Beeman M (2005). Bilateral brain processes for comprehending natural language. Trends Cogn Sci 9: 512–518. https://doi.org/10.1016/j.tics.2005.09.009.

Israel M, Harding JR, Tobin V (2004). On simile. In: Language, culture, and mind, 123–135.

Kaiser NC, Lee GJ, Lu PH et al. (2013). What dementia reveals about proverb interpretation and its neuroanatomical correlates. Neuropsychologia 51: 1726–1733.

Kana RK, Murdaugh DL, Wolfe KR et al. (2012). Brain responses mediating idiom comprehension: gender and hemispheric differences. Brain Res 1467: 18–26. https://doi.org/10.1016/j.brainres.2012.05.031.

Kempler D, Van Lancker D, Marchman V et al. (1999). Idiom comprehension in children and adults with unilateral brain damage. Dev Neuropsychol 15: 327–349.

Kempler D, Van Lancker D (1996). The formulaic and novel language comprehension test (FANL-C), Copyright, Los Angeles, CA.

Klooster N, McQuire M, Grossman M et al. (2020). The neural basis of metaphor comprehension: evidence from left hemisphere degeneration. Neurobiol Lang 1: 474–491. https://doi.org/10.1162/nol_a_00022.

Kurada HZ, Arıca-Akkök E, Özaydın-Aksun Z et al. (2021). The impact of transparency on hemispheric lateralization of idiom comprehension: an rTMS study. Neuropsychologia 163: 108062. https://doi.org/10.1016/j.neuropsychologia.2021.108062.

Lauro LJ, Tettamanti M, Cappa SF et al. (2008). Idiom comprehension: a prefrontal task? Cereb Cortex 18: 162–170. https://doi.org/10.1093/cercor/bhm042.

Lee SS, Dapretto M (2006). Metaphorical vs. literal word meanings: fMRI evidence against a selective role of the right hemisphere. Neuroimage 29: 536–544. https://doi.org/10.1016/j.neuroimage.2005.08.003.

Mashal N, Faust M, Hendler T (2005). The role of the right hemisphere in processing nonsalient metaphorical meanings: application of principal components analysis to fMRI data. Neuropsychologia 43 (14): 2084–2100. https://doi.org/10.1016/j.neuropsychologia.2005.03.019.

Mashal N, Faust M, Hendler T et al. (2008). Hemispheric differences in processing the literal interpretation of idioms: converging evidence from behavioral and fMRI studies. Cortex 44: 848–860. https://doi.org/10.1016/j.cortex.2007.04.004.

Metzler C (2001). Effects of left frontal lesions on the selection of context-appropriate meanings. Neuropsychology 15: 315.

Monetta L, Ouellet-Plamondon C, Joanette Y (2006). Simulating the pattern of right-hemisphere-damaged patients for the processing of the alternative metaphorical meanings of words: evidence in favor of a cognitive resources hypothesis. Brain Lang 96: 171–177. https://doi.org/10.1016/j.bandl.2004.10.014.

Nenonen M, Niemi J, Laine M (2002). Representation and processing of idioms: evidence from aphasia. J Neurolinguistics 15: 43–58.

Nunberg G, Sag IA, Wasow T (1994). Idioms. Language 70: 491–538.

Obert A, Gierski F, Calmus A et al. (2014). Differential bilateral involvement of the parietal gyrus during predicative metaphor processing: an auditory fMRI study. Brain Lang 137: 112–119. https://doi.org/10.1016/j.bandl.2014.08.002.

Oliveri M, Romero L, Papagno C (2004). Left but not right temporal lobe involvement in opaque idiom comprehension: a repetitive transcranial stimulation study. J Cogn Neurosci 16: 848–855.

Papagno C (2001). Comprehension of metaphors and idioms in patients with Alzheimer's disease: a longitudinal study. Brain 124: 1450–1460. https://doi.org/10.1093/brain/124.7.1450.

Papagno C, Caporali A (2007). Testing idiom comprehension in aphasic patients: the effects of task and idiom type. Brain Lang 100: 208–220. https://doi.org/10.1016/j.bandl.2006.01.002.

Papagno C, Curti R, Rizzo S et al. (2006). Is the right hemisphere involved in idiom comprehension? A neuropsychological study. Neuropsychology 20: 598–606. https://doi.org/10.1037/0894-4105.20.5.598.

Papagno F, Lucchelli S, Muggia SR (2003). Idiom comprehension in Alzheimer's disease: the role of the central executive. Brain 126: 2419–2430.

Papagno C, Romero Lauro L (2010). The neural basis of idiom processing: neuropsychological, neurophysiological and neuroimaging evidence. Italian J Linguist 22 (1): 21–24.

Papagno C, Tabossi P, Colombo M et al. (2004). Idiom comprehension in aphasic patients. Brain Lang 89: 226–234. https://doi.org/10.1016/S0093-934X(03)00398-5.

Pobric G, Mashal N, Faust M et al. (2008). The role of the right cerebral hemisphere in processing novel metaphoric expressions: a transcranial magnetic stimulation study. J Cogn Neurosci 20: 170–181. https://doi.org/10.1162/jocn.2008.20005.

Pollio H, Barlow J, Fine H et al. (1977). Metaphor and the poetics of growth: figurative language in psychology, psychotherapy and education, Erlbaum, Hillsdale, NJ.

Rapp AM, Leube DT, Erb M et al. (2004). Neural correlates of metaphor processing. Brain Res 20: 395–402. https://doi.org/10.1016/j.cogbrainres.2004.03.017.

Rapp AM, Langohr K, Mutschler DE et al. (2013). Isn't it ironic? Neural correlates of irony comprehension in schizophrenia. PLoS One 8: e74224. https://doi.org/10.1371/journal.pone.0074224.

Reyes-Aguilar A, Valles-Capetillo E, Giordano MA (2018). Quantitative meta-analysis of neuroimaging studies of pragmatic language comprehension: in search of a universal neural substrate. Neuroscience 395: 60–88. https://doi.org/10.1016/j.neuroscience.2018.10.043.

Richards IA (1936). The philosophy of rhetoric, Oxford University Press, New York.

Rinaldi MC, Marangolo P, Lauriola M (2008). BLED Batteria del Linguaggio dell'Emisfero destro. Giunti Psychomterics.

Rizzo S, Sandrini M, Papagno C (2007). The dorsolateral prefrontal cortex in idiom interpretation: an rTMS study. Brain Res Bull 71: 523–528. https://doi.org/10.1016/j.brainresbull.2006.11.006.

Sakamoto M, Utsumi A (2014). Adjective metaphors evoke negative meanings. PLoS One 9: e89008. https://doi.org/10.1371/journal.pone.0089008.

Schettino A, Lauro L, Crippa F et al. (2010). The comprehension of idiomatic expressions in schizophrenic patients. Neuropsychologia 48: 1032–1040. https://doi.org/10.1016/j.neuropsychologia.2009.11.030.

Shamay-Tsoory SG, Tomer R, Aharon-Peretz J (2005). The neuroanatomical basis of understanding sarcasm and its relationship to social cognition. Neuropsychology 19: 288–300. https://doi.org/10.1037/0894-4105.19.3.288.

Stachowiak F, Huber W, Poeck K et al. (1977). Text comprehension in aphasia. Brain Lang 4: 177–195.

Stringaris AK, Medford NC, Giampietro V et al. (2007). Deriving meaning: distinct neural mechanisms for metaphoric, literal, and non-meaningful sentences. Brain Lang 100: 150–162. https://doi.org/10.1016/j.bandl.2005.08.001.

Szücs M (2013). The role of theory of mind, age, and reception of grammar in metaphor and irony comprehension of preschool children. Proceedings of the third central European conference in linguistics for postgraduate students. Pázmány Péter Catholic University, Budapest.

Thibodeau PH, Durgin FK (2011). Metaphor aptness and conventionality: a processing fluency account. Metaphor Symb 26: 206–226. https://doi.org/10.1080/10926488.2011.583196.

Tompkins C (2012). Rehabilitation for cognitive-communication disorders in right hemisphere brain damage. Arch Phys Med Rehabil 1: S61–S69. https://doi.org/10.1016/j.apmr.2011.10.015.

Tompkins CA, Boada R, McGarry K (1992). The access and processing of familiar idioms by brain-damaged and normally aging adults. J Speech Hear Res 35: 626–637.

Van Lancker D, Kempler D (1987). Comprehension of familiar phrases by left but not by right hemisphere damaged patients. Brain Lang 32: 265–277.

Van Lancker Sidtis D, Postman WA (2006). Formulaic expressions in spontaneous speech of left and right hemisphere damaged subjects. Aphasiology 20: 411–426.

Yang J, Li P, Fang X et al. (2016). Hemispheric involvement in the processing of Chinese idioms: an fMRI study. Neuropsychologia 87: 12–24. https://doi.org/10.1016/j.neuropsychologia.2016.04.029.

Winner E, Gardner H (1977). The comprehension of metaphor in brain-damaged patients. Brain 100: 717–729.

Zaidel E, Kasher A, Soroker N et al. (2002). Effects of right and left hemisphere damage on performance of the "Right Hemisphere Communication Battery". Brain Lang 80: 510–535.

Zempleni MZ, Haverkort M, Renken R et al. (2007). Evidence for bilateral involvement in idiom comprehension: an fMRI study. Neuroimage 34: 1280–1291.

Handbook of Clinical Neurology, Vol. 208 (3rd series)
Cerebral Asymmetries
C. Papagno and P. Corballis, Editors
https://doi.org/10.1016/B978-0-443-15646-5.00012-9

Chapter 20

The lateralization of reading

JASON J.S. BARTON[1*], ANDREA ALBONICO[1,2], AND RANDI STARRFELT[3]

[1]*Departments of Medicine (Neurology), Ophthalmology and Visual Sciences, University of British Columbia, Vancouver, BC, Canada*
[2]*Department of Psychology, University of the Fraser Valley, Abbotsford, BC, Canada*
[3]*Department of Psychology, Center for Visual Cognition, Copenhagen University, Copenhagen, Denmark*

Abstract

Reports in the 1890s described reading disorders from left hemisphere damage. Subsequent work converging from a variety of research approaches have confirmed a strong dependence of reading on the left ventral occipitotemporal cortex, though there is also evidence for some reading capacity of the right hemisphere. The development of this leftward bias parallels reading acquisition in children and adults and is blunted in developmental dyslexia. Several structural and functional hypotheses have been advanced to explain why reading lateralizes to the left. In the second half of this review we explore the extension of these findings to other forms of reading. Most reading studies used the alphabetic scripts of Europe but there are many writing systems. Comparisons with logographic scripts such as Chinese and kanji have revealed subtle differences. Also, while we often think of reading as the extraction of verbal language from written text, it can be broadened to other types of information extraction from symbols. Reading can occur with visual stimuli that are not written text, as with sign language in the deaf and lip-reading, and with non-visual stimuli that are textual, as with Braille. Musical notation and number reading are other text-based visual forms of reading that do not involve words. Overall, most studies show that the left ventral occipitotemporal cortex is involved in processing these diverse types of reading, with variable contributions from the right hemisphere.

Reading is a recent development, about 5000 years old, and because it is relatively new in evolutionary terms, it is highly unlikely that the human brain could have evolved specialized regions for the artificial stimuli that are written texts (Henderson, 2008, 2019). Nevertheless, over 130 years ago Déjerine described selective reading deficits from left posterior strokes (Dejerine, 1891, 1892). Since then, work with tachistoscopic techniques, functional neuroimaging, and split-brain and alexic patients has confirmed this lateralization and identified left-sided reading networks. In the first part of this review, we describe this evidence along with studies of how lateralization develops as subjects learn to read, reports of its reduction in developmental dyslexia, hypotheses about why reading lateralizes to the left, and what reading capacity remains in the right hemisphere.

In the second part, we review the data for different types of reading. How do the findings with alphabetic scripts like English apply to the many different writing systems in the world? Most remain unexplored, but there is a substantial literature on the logographic scripts of Chinese and *kanji*. Then there is the issue of what exactly constitutes reading. Whether the script is Chinese or English, most think of "reading" as what you are doing right now, extracting language in terms of both meaning and sound from a series of complicated line segments on a page, symbols that often have an arbitrary relation to what is being communicated. However, there are other forms of symbolic communication that we might also call reading, including those with visual stimuli that are not written text, such as sign language and lip-reading, or nonvisual stimuli that are textual, such as Braille. There are also texts that do not correspond to spoken language, such as musical notation. Finally, there is the question of number reading, of particular interest since some of the earliest writing systems may have been developed to do Sumerian accounting (Chrisomalis, 2009). For all of these, there have been studies examining their lateralization and similarities to the reading of written text.

*Correspondence to: Jason J S Barton, MD PhD FRCPC, Neuro-ophthalmology, Section K, VGH Eye Care Centre, 2550 Willow Street, Vancouver, BC, Canada. Tel: +1-604-875-4339, Fax: +1-604-875-4302, E-mail: jasonbarton@shaw.ca

HISTORY OF LATERALIZATION IN READING

The earliest observations of lateralized cognitive functions involved language. Broca (1861) and Dax (1865) noted the effect of left frontal lesions on spoken language. Next, Exner (Exner, 1881; Roux et al., 2010) reported impaired writing with damage to the posterior part of the left middle frontal gyrus. Then, in 1891, a left-sided lesion affecting reading was described.

That year, Déjerine reported a patient with impaired reading and writing (*alexia with agraphia*) but relatively preserved spoken language, whose autopsy showed a lesion of the left angular gyrus (Dejerine, 1891). In 1892, he reported Monsieur C, who had lost the ability to read words or even single letters after a stroke (Dejerine, 1892). While this patient had a right hemianopia, his speech production, comprehension, and object recognition were normal, and most intriguingly he could write, even if he could not read what he wrote. Four years later, a second stroke eliminated Monsieur C's writing ability, and he died shortly after. As with the 1891 patient, the autopsy showed a recent stroke of the left angular gyrus. However, the older lesions had damaged the splenium and left medial occipital cortex, corresponding to the time when he had *pure word blindness* (now called *pure alexia* or *alexia without agraphia*) (Fig. 20.1).

From these two cases, Déjerine argued that the left angular gyrus was the "visual center for words." He explained the *pure alexia* of Monsieur C after his first stroke as a disconnection of this center from its visual inputs. The fact that Monsieur C could still write at first was attributed to the integrity of the left angular gyrus. However, the right hemianopia meant that reading depended on visual information from the left hemifield, which would be processed by the right visual cortex. Loss of callosal fibers from the right to the left hemisphere eliminated the transfer of this visual information to the left angular gyrus for reading. While current concepts of the role of the angular gyrus differ from Déjerine's conjecture (e.g., Junker et al., 2023), his concepts of *pure alexia* and *alexia with agraphia* both posit the existence of a visual word processor in the left hemisphere.

There were occasional writings on alexia through the first half of the 20th century (e.g., Holmes, 1950), but the idea of specialized cerebral centers for language lost traction for a while. It was not until the 1960s that the interest in the cerebral substrate for alexia increased again, due in part to Geschwind's (1965) paper on disconnection syndromes. Geschwind, like Déjerine, interpreted *pure alexia* as a disconnection of visual input to language processors in the left hemisphere. At the same time, more cognitively oriented papers about acquired reading deficits were published (e.g., Warrington and Zangwill, 1957; Kinsbourne and Warrington, 1962), arguing that the posterior left hemisphere was critical for visual word recognition.

In 1980, Shallice and Warrington suggested a division between the *central* and *peripheral* alexias. The central alexias are reading deficits associated with the more general language problems of aphasia. While many early cognitive studies were not interested in localization, it logically followed from their link with aphasia that central alexias were caused by damage to the language-dominant hemisphere, typically the left. Peripheral alexias are reading disorders related to impaired attention or perceptual processing, affecting the input to language systems. These too are most often linked to unilateral lesions, typically but not exclusively on the left.

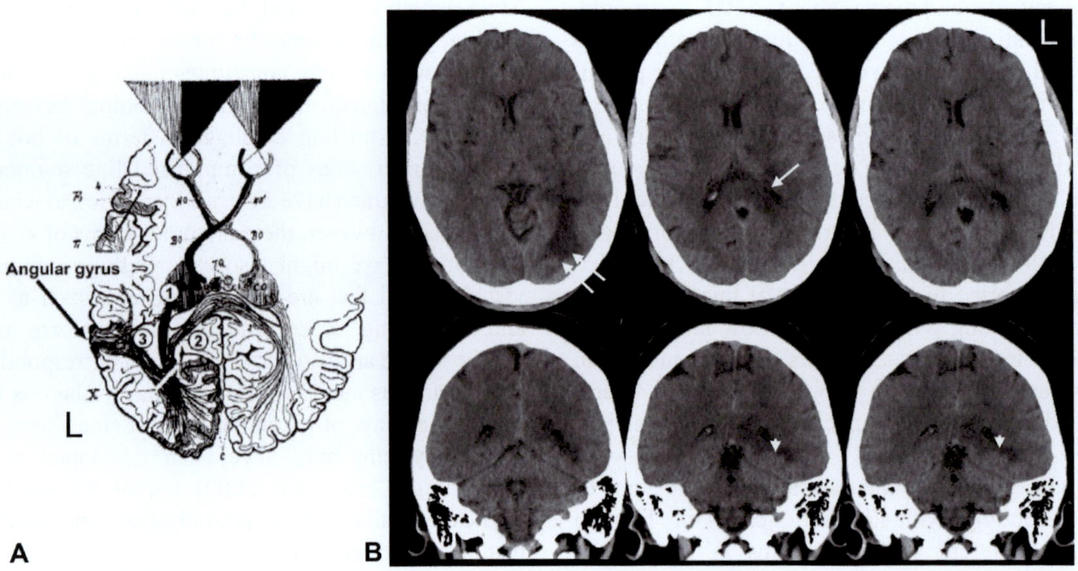

Fig. 20.1. *Pure alexia* as a disconnection. (A) The original axial diagram from Dejerine (1892), where the shaded cortical ribbon at the bottom indicates damage to the left occipital cortex, and the white bar marked 'X' shows the disruption of fibers from this cortex and the splenium, disconnecting the reading center Déjerine located in the angular gyrus. (B) A modern example. Axial CT images of GB, a 63-year-old woman, 1 day after a stroke causing *pure alexia* and right hemianopia. Double arrows show the damage to the left occipital optic radiations causing the right hemianopia; single arrow indicates the infarction of the splenium, while arrowheads show the approximate location of the VWFA in the left fusiform gyrus, which appears to be spared. L = left side of images: note that this is reversed between A and B.

THE ALEXIAS

Given the prominent role of alexia in the history of reading lateralization, we begin with the neuropsychologic evidence. Alexia is an impairment in reading acquired from a cerebral injury in a person who was literate before its onset (Leff and Starrfelt, 2014).[a]

There are many kinds of alexia, which are divided into two categories. Peripheral alexias are due to defects in perception or attention. These defects can be general, as with hemianopia and neglect, in which case any lateralizing effect is more variable and depends on other factors. When they are more reading-selective, as with hemialexia or *pure alexia*, they show lateralizations that reflect a left visual word processor. Because peripheral alexias are not linguistic in origin, writing, spoken language, and comprehension are generally preserved. In contrast, the central alexias are due to disruption of linguistic processes and are typically encountered with verbal language impairments in semantics or phonology.

Peripheral alexias: Perception

Hemianopic alexia refers to the impact on reading when a homonymous hemianopia encroaches on the 5° surrounding the center of vision (Trauzettel-Klosinski and Brendler, 1998; Leff et al., 2006; Rubino et al., 2016). Reading is slow rather than inaccurate. This impact differs between right and left hemianopia, not because of the lateralization of word processing, but because of the direction in which we read. Patients whose language is read from left to right have an asymmetric reading span, from 4 letters to the left to 15 letters to the right of their point of fixation (Rayner et al., 2010). When reading sentences, they preview the upcoming text to the right to plan the optimal placement of the next fixation (Schotter et al., 2012). Hence, a right hemianopia slows down reading. It creates a slightly longer "word-length effect," the increase in reading time as a function of the number of letters in a word (Sheldon et al., 2012), and results in many more fixations on a line of text (Trauzettel-Klosinski and Brendler, 1998). In contrast, left hemianopic alexia only causes hesitation and searching for the beginning of the next line.

Left hemi-alexia refers to the rare loss of reading for words in the left hemifield only (Gazzaniga and Freedman, 1973; Suzuki et al., 1998; Molko et al., 2002). Damage to the splenium disconnects the right occipital cortex from the visual word processor in the left hemisphere. Because all left hemispheric structures are intact, they can read words in their right hemifield. Since the right occipital cortex is unaffected, other visual processing in the left hemifield is preserved. This includes the ability to copy letters and words in that hemifield by writing with their left hand.

Right hemi-alexia has only one well-documented case (Barton et al., 2022). This was attributed to an intra-hemispheric disconnection between the left occipital cortex and the left reading network, which retained its callosal input from the right occipital cortex (Fig. 20.2).

Pure alexia is also referred to as alexia without agraphia. When extreme, the patient is unable to read (*global alexia*), but more commonly patients become slow and inefficient readers, with an elevated word-length effect (see Barton et al., 2014 for a review). They may spell words out loud to read them (*letter-by-letter alexia*). It has been inferred from these observations that *pure alexia* reflects a breakdown of fast, automatic parallel letter processing (Behrmann et al., 1998; Starrfelt et al., 2009), so that patients rely on serial letter identification to read words (Rayner and Johnson, 2005; Cumming et al., 2006; Shallice, 2014; Bormann et al., 2015). *Pure alexia* may be due to damage to or disconnection of the visual word form area (VWFA) in the left fusiform gyrus (Leff et al., 2001; Cohen et al., 2003; Gaillard et al., 2006; Dehaene and Cohen, 2011) (Fig. 20.3). The VWFA is the first clearly lateralized structure in the reading process. It is thought to process abstract letter identities for word recognition, regardless of letter size, case, font, or location in the visual field (Cohen et al., 2000, 2002; Dehaene and Cohen, 2011).

Peripheral alexias: Attention

Attentional alexia is rare. It affects the reading of text or sentences, rather than single words. Its defining mistake is the *migration error*, in which letters "migrate" from one word to another—e.g., WIN-FED read as FIN-FED (Shallice and Warrington, 1977). Patients with attentional alexia may have trouble identifying the individual letters of words, while able to read the word and also identify single letters (Shallice and Warrington, 1977; Mayall and Humphreys, 2002). They struggle particularly with words written in unconventional formats, such as vertically or rotated (Vinckier et al., 2006) or in alternating case (Hall et al., 2001). Attentional alexia follows bilateral lesions causing Bálint's syndrome (Baylis et al., 1994) or simultanagnosia (Coslett and Saffran, 1991; Hall et al., 2001; Vinckier et al., 2006), but can occur with left parietal lesions (Shallice and Warrington, 1977; Price and Humphreys, 1993). A related disorder is *letter position alexia*, in which patients make migration errors within words rather than between them. It has been described mainly with abjad scripts such as Hebrew and Arabic, which do not represent vowels (Friedmann and Gvion, 2001). It occurs with left occipito-parietal lesions.

Neglect alexia is related to visuospatial neglect, and like that disorder, it is more common after right-sided lesions. In some cases, the neglect is limited to reading (Caramazza and Hillis, 1990; Moore and Demeyere, 2017). *Allocentric* ("object-centered") errors are made on the left side of words, including

[a]While disturbances of central or paracentral vision (e.g., macular degeneration, cataracts) or eye movements (e.g., nystagmus, saccadic dysmetria) can also affect reading (Rodriguez and Barton, 2015), these are not called alexias.

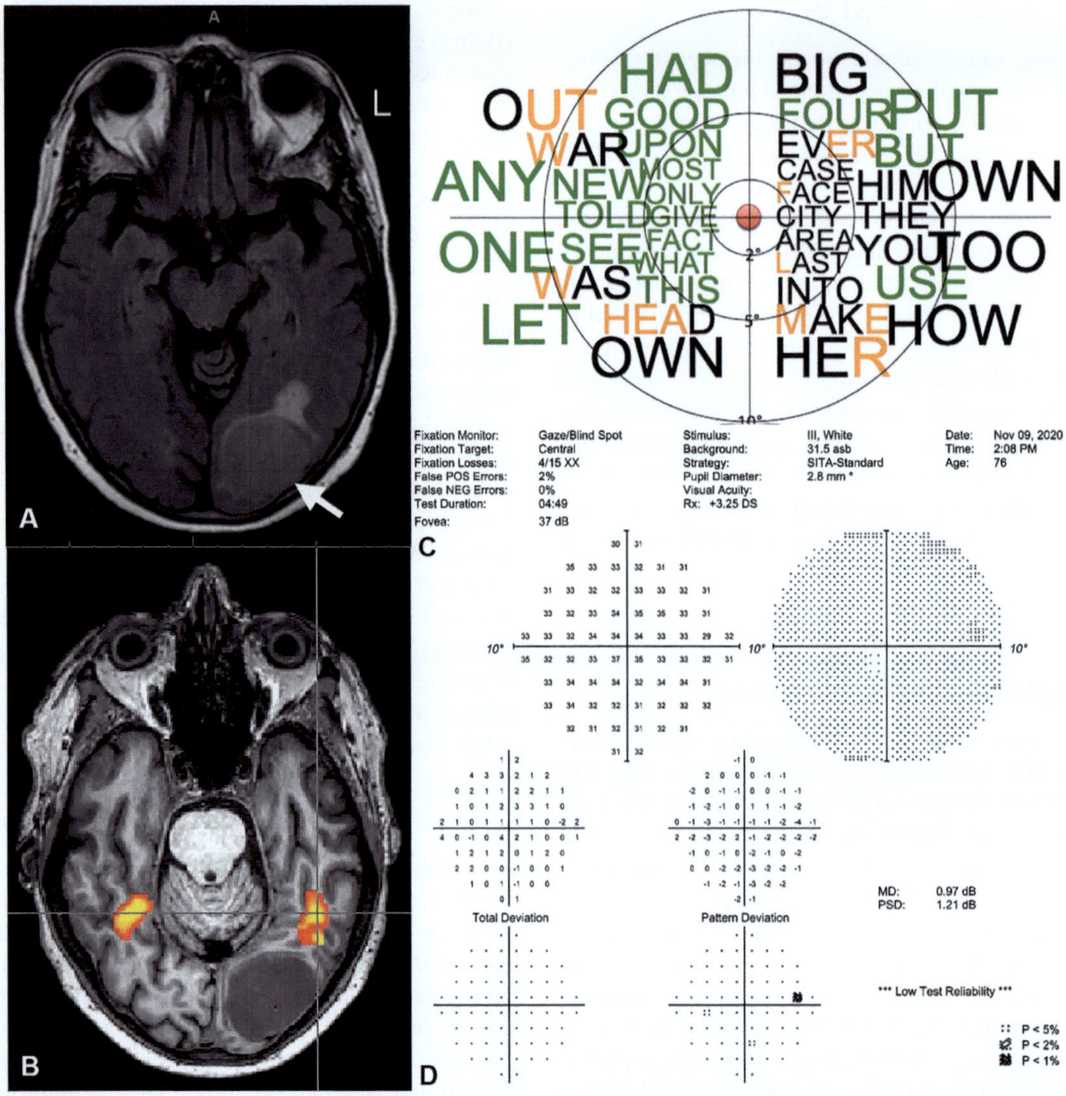

Fig. 20.2. *Right hemi-alexia* in PH, a 76-year-old woman. (A) Axial image from her structural MRI scan showed a left occipital meningioma (arrow) surrounded by edema. (B) Functional MRI for word perception showed an intact left visual word form area at the intersection of the crosshairs. (C) Perimetry with words as stimuli showed that she could read most words in the left hemifield but not in the right (green = correctly read, orange = some letters read, black = not read). Outer circle indicates 10° eccentricity. (D) Automated perimetry of the central 10 degrees of vision was normal. Adapted from Barton JJ, Rubino C, Albonico A et al. (2022). Right hemi-alexia. Cortex 157: 288–303. https://doi.org/10.1016/j.cortex.2022.09.015.

omissions (e.g., OAT read as AT); additions (OAT read as BOAT); and substitutions (OAT read as CAT) (Vallar et al., 2010). With *egocentric* (viewer-centered) errors, patients omit whole words on the left side of a column of text (Beschin et al., 2014) (Fig. 20.4). Patients can show one or both types of errors. Neglect alexia occurs with damage to the right inferior parietal and posterior temporal lobes (Ptak et al., 2012), sometimes to the right anterior temporal lobe (Boukrina et al., 2020). There are rare cases of neglect alexia after left-sided lesions, with allocentric errors for the right side of words (Hillis and Caramazza, 1995; Vallar and Calzolari, 2018; Moore and Demeyere, 2019; Moore et al., 2023).

Central alexias

The central alexias include surface alexia, phonologic alexia, and deep alexia. These have contributed to the *dual route model of reading* (Coltheart et al., 2001). This postulates the existence of a lexical route for reading, in which visual text is read by accessing an internal lexicon, and a phonologic route, in which the graphemes of written words are mapped to their corresponding phonology first. All central alexias are caused by left hemispheric injury, often strokes in the territory of the left middle cerebral artery, sometimes degenerative disorders.

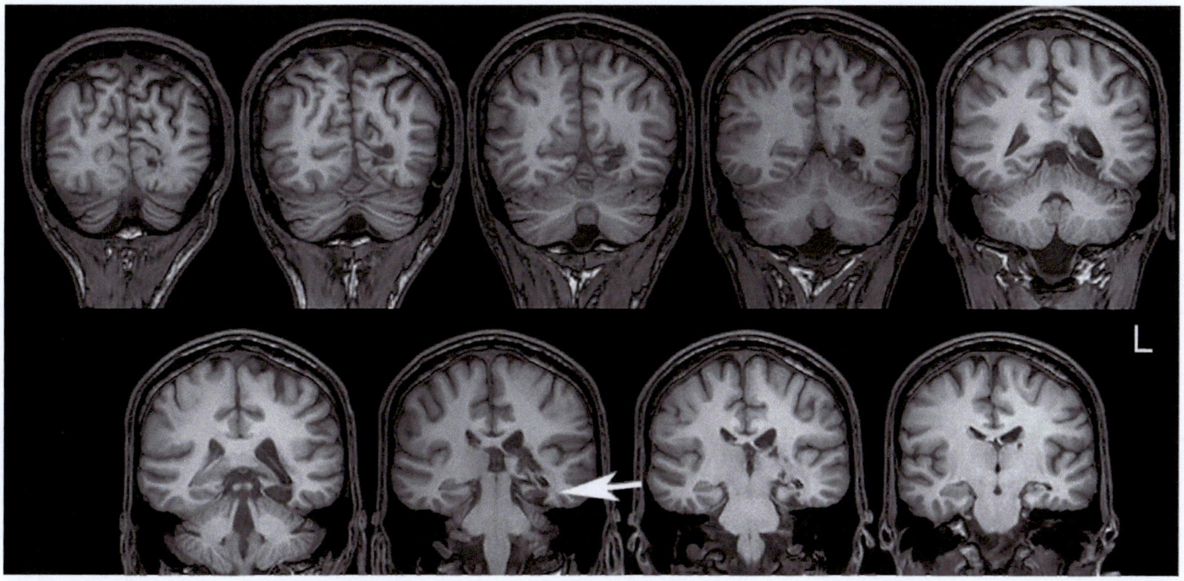

Fig. 20.3. *Pure alexia* from damage to the VWFA. CJ, a 41-year-old man, suffered a stroke 15 months prior, causing pure alexia with right hemianopia. Coronal structural T1-weighted MRI images show a small infarct of the left inferior occipital cortex, extending to the approximate location of the VWFA in the fusiform gyrus (arrow). L = left side of scans. Adapted from Albonico A, Barton JJS (2017). Face perception in pure alexia: complementary contributions of the left fusiform gyrus to facial identity and facial speech processing. Cortex 96: 59–72.

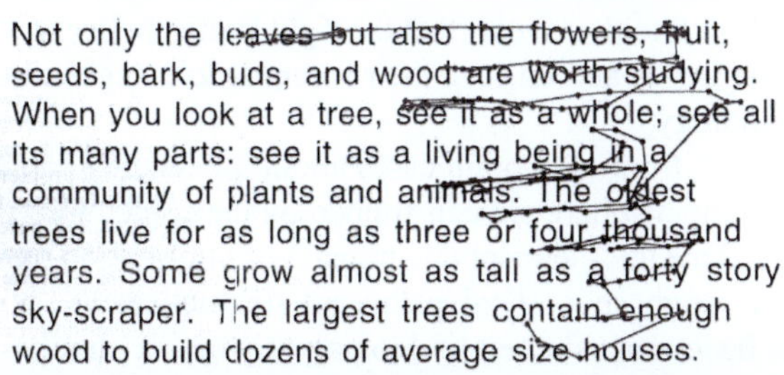

Fig. 20.4. *Neglect dyslexia*, egocentric errors. Eye movements are depicted as dots for fixations and lines for the scan path between successive fixations during the reading of the paragraph. The patient fails to read words on the left side of the page, a problem that gets progressively worse with each following line. From Rodriguez AR, Barton JJ (2015). The 20/20 patient who can't read. Can J Ophthalmol 50: 257–264. https://doi.org/10.1016/j.jcjo.2015.05.013.

Surface alexia arises from damage to the lexical route. Reading of irregularly spelled real words is impaired, but reading of regular words and pseudo-words[b] is relatively preserved (Marshall and Newcombe, 1973). Patients make *regularization errors*—e.g., pronouncing PINT to rhyme with MINT. Surface alexia is most common with semantic dementia from degeneration of the anterior temporal lobes (Mummery et al., 2000; Nestor et al., 2006; Woollams, 2014; Woollams et al., 2007). Lesion mapping links surface alexia to damage of the lateral and ventral left temporal lobe (Ripamonti et al., 2014; Binder et al., 2016; Aguilar et al., 2018; Tomasino et al., 2020).

As its name implies, *phonologic alexia* arises from damage to the phonologic route. Patients can pronounce real words they know but not unfamiliar words or pseudo-words (Beauvois and Derouesne, 1979). They make *lexicalization errors*, reading pseudo-words as a visually similar real word—e.g., NOOM read as ROOM. Reading of real words with low imageability can be affected: for instance, function words (e.g., BUT, THUS, FOR) are read inaccurately or not at all. It is associated with damage to a variety of left-sided structures, including left inferior parietal, inferior frontal, and superior temporal cortices (Rapcsak et al., 2009; Ripamonti et al., 2014; Beeson et al., 2022). These regions are involved in phonologic processing and patients often have a more general phonologic impairment (Crisp and Lambon Ralph, 2006; Rapcsak et al., 2009).

Deep alexia is seen in patients with severe aphasia from large lesions of the territory of the left middle cerebral artery. These patients have a nearly complete loss of reading ability. When they do read, their classic mistake is a *semantic reading error,* a substitution of a related real word—e.g., CHICKEN read as EGG (Marshall and Newcombe, 1973). They do not make phonologic errors, substitutions that sound similar to the desired word, but sometimes make visual errors—e.g., BIT for BET—derivational errors and, most typically, bizarre unclassifiable errors. Similar to patients with phonologic alexia, they struggle with function words and pseudo-words, leading to suggestions that phonologic and deep alexia might form a continuum of severity (Friedman, 1996; Crisp and Lambon Ralph, 2006).

TACHISTOSCOPIC STUDIES

Typical subjects

Mishkin and Forgays (1952) first found that English words are more accurately read when flashed in the right than the left

[b]Pseudo-words are pronounceable letter strings that aren't real words.

hemifield. Subsequently many studies have replicated a right-field advantage for word reading, with several explanations advanced (Pirozzolo, 1977).

The "directional scanning effect" proposed that reading from left to right results in better training and sensitivity of neurons in the right hemifield. This was supported by a small left-field advantage for Yiddish, which is read from right to left (Mishkin and Forgays, 1952). However, other studies of Jewish reading found either that its hemifield advantage depended on which language was learned first (Orbach, 1952), or that both English and Jewish had the same right-field advantage (Barton et al., 1965).

"Attentional bias" held that a linguistic task primes the language-dominant left hemisphere, creating a right-field advantage (Kinsbourne, 1970). Others suggested that, while both hemispheres have orthographic lexicons, those of the left are better (Bonandrini et al., 2023). The most influential proposal is the "direct access theory" (Kimura, 1961; Barton et al., 1965). While right-field stimuli are processed by the left occipital cortex and thus have a short route to language processors in the left hemisphere, left-field information is received by the right occipital cortex and requires a more arduous interhemispheric transfer.

Some factors that modulate these field asymmetries point to superior reading processes on the left. Orthographic regularity, which means a consistent correspondence between text and sound, enhances the right-field advantage (Beaton et al., 2007), suggesting that phonologic coding is better in the left hemisphere. The right field shows less impact of the number of letters in the word, implying better whole-word processing by the left too (Bub and Lewine, 1988). More efficient connections between orthography and lexical/semantic meaning in the left hemisphere may be the reason that infrequently encountered words are read more easily in the right field (Bonandrini et al., 2023).

On the other hand, tasks that stress visual rather than linguistic processing of visual words can create a left-field advantage. This was found for simply matching words (Gibson et al., 1972), a task that can be done even if one doesn't know the language. Similarly, matching of two letters shows a left-field advantage if the letters have the same case, but a right-field advantage when their case differs (Geffen et al., 1972). Text in a more irregular script shows a left-field advantage (Bryden and Allard, 1976), attributed to a need for more visual preprocessing. A similar reduction in the right-field advantage occurs with handwritten words (Hellige and Adamson, 2007).

Split-brain studies

The most dramatic tachistoscopic evidence for hemispheric differences comes from those rare subjects who have lost the callosal fibers that connect the two cerebral hemispheres. In such cases, stimuli in one visual hemifield can only be processed by the contralateral hemisphere. Callosal loss can be complete, as with some surgical callosotomies for epilepsy, or partial, as with natural lesions of the splenium, which is the key part of the corpus callosum for interhemispheric transfer of written text (Funnell et al., 2000b).

In general, split-brain subjects have no trouble processing written text in the right hemifield. Nevertheless, there is some reading capability of the right hemisphere for words in the left field, as long as reading is evaluated by manual rather than verbal responses. Early work showed that two of three patients could read numbers, letters, and short words in the left field (Gazzaniga and Sperry, 1967).

The right hemisphere is not the same at reading as the left, though. It is slow. Split-brain patients show a much larger right-field advantage in latency than typical subjects (Cohen et al., 2000). While split-brain subjects can derive the semantic meaning of words in the left field (Zaidel and Peters, 1981; Gazzaniga et al., 1984)—being able to point to drawings that correspond to the written word, for example—they don't know the sound of the word, suggesting that the right hemisphere processes words like ideograms (Zaidel and Peters, 1981). One subject with left hemi-alexia from a splenial lesion could slowly name words in the left field but made occasional semantic errors, suggesting similarities to deep alexia (Cohen and Dehaene, 2000). Some split-brain subjects can make accurate lexical decisions about pseudo-words or consonant strings in their left field (Sergent, 1987; Cohen et al., 2000), which has been attributed to differences in visual familiarity between real words and non-words. All told, it seems that the right hemisphere can recognize whole words and retrieve their meaning to a degree, but not their phonology.

Inferences from split-brain cases about the reading capability of the right hemisphere are constrained by the contentious possibility that some interhemispheric transfer of information remains, perhaps through subcortical pathways (Funnell et al., 2000a). This has been most studied by looking at whether simultaneous stimuli in both hemifields are integrated, with results both for (Sergent, 1987) and against (Gazzaniga et al., 1984; Kingstone and Gazzaniga, 1995; Miller and Kingstone, 2005). Less contentious is the possibility that callosal resection is sometimes less complete than assumed, allowing some residual cortical interhemispheric transfer. Remnant fibers have been revealed by MRI scans (Seymour et al., 1994), raising doubts about many of the early cases. Newer studies have tried to exclude transfer by showing that stimuli in the left field generate a lexical effect in the right but not the left N170 potential, or do not activate the left visual word form area on fMRI (Cohen et al., 2000).

LATERALIZATION IN NEUROIMAGING

Like all complex tasks, reading involves an extensive network (Fiez and Petersen, 1998; Price, 2012). Beyond the bilateral striate and peri-striate cortex, this includes mainly left-lateralized visual and linguistic regions in the occipitotemporal and perisylvian cortex, as well as ocular motor areas involved in generating saccades when reading sentences and paragraphs.

One key area with a relatively selective response to visual words is located in the fusiform gyrus, the VWFA (Fig. 20.5), which shows a strong leftward lateralization (McCandliss et al., 2003; Cohen and Dehaene, 2004; Dehaene and Cohen, 2011). A review of over 25 studies located the peak activation of the VWFA at approximately [$x=-43$, $y=-54$, $z=-12$]

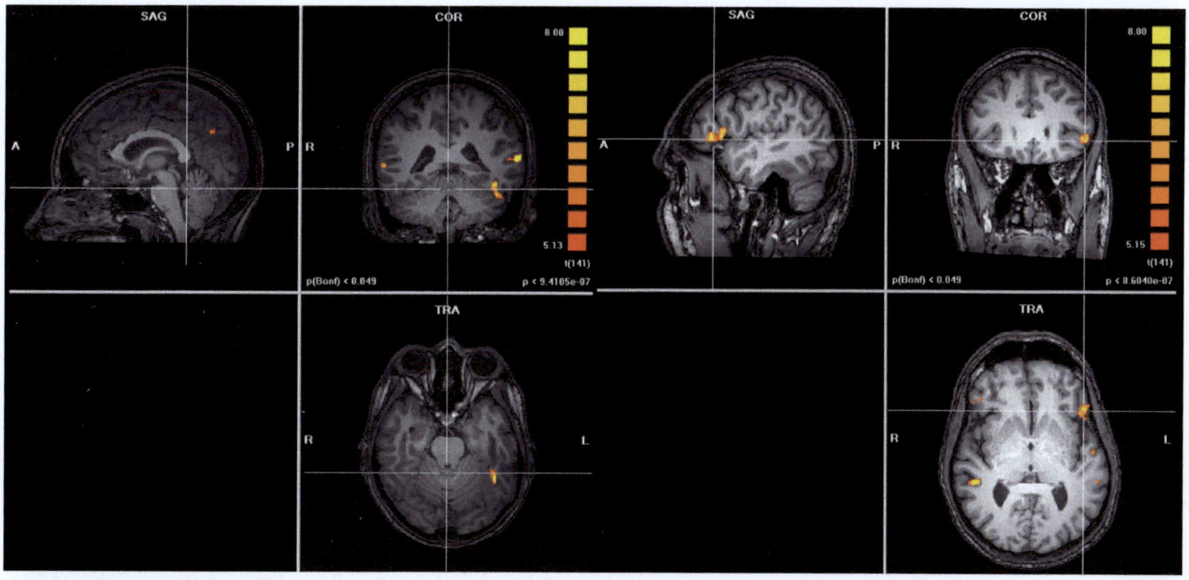

Fig. 20.5. Functional MRI of the reading network. The contrast between reading English and reading Korean in the first author, who does not read Korean, with activated areas shown in orange-yellow. Left half shows the left VWFA and the middle temporal gyrus. Right half: Crosshairs are aligned on the inferior frontal gyrus. Note: Left side of the brain is on the right side of coronal and axial scans. From Barton JJS, Fox CJ, Sekunova A, Iaria G (2010). Encoding in the visual word form area: an fMRI-adaptation study of words versus handwriting. J Cogn Neurosci 22: 1649–1661.

(Cohen et al., 2002). This word-sensitive area has also been located with intracranial recordings (Nobre et al., 1994) and magnetoencephalography (Tarkiainen et al., 2002). Event-related potentials show that word-selective activity emerges here in the N170, a negative peak occurring 150–200 ms after stimulus presentation (Bentin et al., 1999). Although the VWFA is strongly activated by words and letter strings, this selectivity may be relative rather than absolute, as it has some responses to colors and visual objects (Price and Devlin, 2003; Starrfelt, 2007) as well as to symbolic visual material, such as musical notation, logos, and even text from a language that the subject cannot read (Muayqil et al., 2015).

The properties of the left VWFA have been extensively studied. It responds to visual but not spoken words (Dehaene et al., 2002) and shows invariance for visual features irrelevant for reading, such as font, case, and size (Polk and Farah, 2002; Dehaene et al., 2004), whether text is printed or handwritten (Qiao et al., 2010), and the hemifield in which the word is located (Cohen et al., 2000). On the other hand, it distinguishes between words and their mirror images (Pegado et al., 2011). It is sensitive to linguistic factors. It is activated more by real words than pronounceable pseudo-words or letter strings (Price et al., 1996). Pseudo-words that conform to the orthographic rules of the language elicit a more robust response than ones that do not (Cohen et al., 2002). While some claim that it is not sensitive to word frequency (Fiebach et al., 2002), others note an effect of the frequency of letter strings (Kronbichler et al., 2004) and the two-letter combinations called bigrams (Binder et al., 2006).

Word-selective areas of the left ventral occipitotemporal cortex are not homogenous. Imaging studies suggest a high degree of functional organization (Vinckier et al., 2007; Caffarra et al., 2021), and the degree of lateralization may vary across different components (Seghier and Price, 2011). This organization is a posterior-to-anterior progression of increasing selectivity for higher-order aspects of words, from letters to bigrams, quadrigrams, and real words (Vinckier et al., 2007), a functional gradient that develops as children mature (Olulade et al., 2013).

THE ROLE OF THE RIGHT HEMISPHERE

Given the converging evidence from tachistoscopic, split-brain, alexic, and neuroimaging work for lateralization of reading to left hemisphere structures, what if anything does the right hemisphere contribute? After all, neuroimaging studies consistently show modest activation of right hemisphere homologs of the left reading network (Nestor et al., 2013).

On the one hand, there may be an early perceptual contribution of the right hemisphere to reading. As stated, tachistoscopic work shows that tasks that stress the visual aspect of word processing have a left-field advantage (Geffen et al., 1972; Gibson et al., 1972; Bryden and Allard, 1976). Functional imaging studies report that handwritten words activate right visual areas more than printed ones, though priming suggests that it is the left VWFA that displays word activity that is invariant for textual style (Qiao et al., 2010; Longcamp et al., 2011).

On the other hand, the right hemisphere may perform some semantic processing of text. Split-brain studies suggest that the right hemisphere can extract some semantic meaning but not the phonology of visual words (Zaidel and Peters, 1981; Gazzaniga et al., 1984). In support, functional imaging studies have shown that the semantic processing of visual words activates frontal and temporal sites in both hemispheres (Taylor and Regard, 2003). A DTI tractography study claimed a role for the right frontal lobe in reading comprehension, based on a correlation of that skill with the fractional anisotropy of the right arcuate fasciculus (Horowitz-Kraus et al., 2014).

It is controversial whether the data from alexia support residual semantic processing by the right hemisphere. In patients with severe aphasia from large left hemispheric strokes, deep alexia with its characteristic semantic errors and inability to read pseudo-words suggests a pattern of reading based on approximate semantics but not phonology, somewhat reminiscent of the left-field reading of split-brain subjects, and therefore perhaps a type of right hemisphere reading (Coltheart, 1980; Saffran et al., 1980). Against this, an imaging study argued that activity in the surviving left hemisphere regions explains this residual reading (Price et al., 1998). The covert abilities of patients with *pure alexia* to make lexical or semantic decisions might be based on semantic representations in the right hemisphere (Coslett and Saffran, 1998; Goodglass et al., 2000). This too has been challenged by claims that the same damaged reading system performs both explicit and implicit reading by these patients (Lambon Ralph et al., 2004).

Right-hemispheric reading may contribute little to the spontaneous improvement of alexia, which correlates more with residual perfusion in the left reading network than with that in the homologous regions on the right (Boukrina et al., 2019). While the rehabilitation of alexia remains difficult, studies of the effects of training show increased activity in the left hemisphere rather than recruitment of the right (Woodhead et al., 2013, 2014). On the other hand, improvement in the reading of children at risk for dyslexia correlates with increased fractional anisotropy in the right superior longitudinal fasciculus, which differs from the left-sided changes seen in typical children (Wang et al., 2017).

There is support for some right-hemispheric readings from large group studies of stroke patients. Word, object, and face perception were tested in 64 patients with right, left, or bilateral strokes of the posterior cerebral artery (Rice et al., 2021). Most commonly, patients had deficits in all three, indicating damage to general visual processes. Six patients had selective impairments in word perception only, but this occurred with either left- or right-sided strokes. A large survey of 731 stroke patients using the brief reading tests of the Birmingham Cognitive Screen found reading impairment in 68% of those with left-hemisphere strokes, as expected, but also in 54% of those with right-sided ones (Balani and Bickerton, 2023). Reading patterns differed, though. Left-stroke patients had more difficulty with irregular words and words with low frequency and low imageability, while right-stroke patients struggled more with regular words and pseudo-words.

Beyond reading, the right hemisphere may be engaged in recognizing nonlinguistic aspects of text. Studies have shown that prosopagnosic patients with right occipitotemporal damage cannot recognize handwriting or font (Rentschler et al., 1994; Hills et al., 2015).

LATERALIZATION IN THE DEVELOPMENT OF READING

Reading is a highly developed visual skill, but it is not innate. No one is born able to read. Learning to read provides an opportunity to study the neural changes that accompany the acquisition of reading proficiency, which occurs during childhood in most, but sometimes in adults.

Typical development

Many studies show that learning to read is accompanied by lateralized changes in the left hemisphere. One of the earliest observations was the emergence of the right-field advantage in tachistoscopic studies of reading as children become more proficient readers (Forgays, 1953).

In children, left hemisphere structures both predict and change with reading proficiency. The gray matter volume of the left superior temporal cortex predicts reading gains over the first 2 years of school, and these gains correlate with reductions in the volume of other left frontal and parietal regions (Linkersdörfer et al., 2015). Reading gains correlate with white matter changes too, such as increases in connectivity between the left VWFA and the inferior parietal lobe (Moulton et al., 2019) and in the fractional anisotropy of the left arcuate fasciculus (Wang et al., 2017). Similar increases in fractional anisotropy of the left arcuate and inferior longitudinal fasciculi were reported in superior readers (Yeatman et al., 2012).

In terms of function, learning to read creates more lateralized activity to visual words. Such activation increases in the left but not the right occipitotemporal cortex from ages 7 to 13 and correlates with measures of reading speed (Ben-Shachar et al., 2011). This increase can be detected in the first year of school, not just in the left VWFA but also in other components of the reading network (Dehaene-Lambertz et al., 2018). The lateralizing effects of learning to read are surprisingly rapid. A longitudinal fMRI/EEG study of kindergarten children showed that 8 weeks of grapheme-to-phoneme training enhanced the response to letter strings in the left fusiform gyrus (Brem et al., 2010).

Learning to read can also be studied in cross-sections of adults with varying degrees of literacy. An fMRI study compared illiterate adults, late-literate adults, and adults who learned to read in childhood, and found that activation of the left VWFA by written text increased across these three groups (Dehaene et al., 2010). Compared to illiterate subjects, late-literate adults have more white matter in the splenium of the corpus callosum and more gray matter in the left supramarginal and left superior temporal gyri (Carreiras et al., 2009).

As with children, learning to read by adults is accompanied rapidly by left-lateralized activity for text. When subjects learn to read a new script, the left VWFA develops selective responses to that script after just a few training sessions (Hashimoto and Sakai, 2004; Song et al., 2010; Yoncheva et al., 2010). This effect requires specific attention to the correspondences between print and speech sounds, supporting the role of the VWFA in translating graphemes to phonemes (Dehaene and Cohen, 2011).

Why is the VWFA on the left? Developmental hypotheses

The fusiform gyri are involved in expert visual processes that require fine levels of discrimination (Weiner and Zilles, 2016), the two best examples being visual word and face recognition. Why does the VWFA develop on the left and not the right? There are both structural and functional hypotheses, which are not mutually exclusive.

Structural arguments focus on an anatomic advantage for the left over the right fusiform gyrus in relation to the verbal language networks. The *many-to-many hypothesis* proposes that learning to read leads to competition between word and face representations in the fusiform cortex (Behrmann and Plaut, 2013). Space constraints lead to a drive to minimize axonal length, favoring connections between regions near each other (Behrmann and Plaut, 2020). This "principle of proximity" results in visual word processing co-lateralizing with the left language network (Gerrits et al., 2019), while face processing shifts to the right fusiform gyrus, though some argue that there is no conclusive evidence for the latter dynamic (Rossion and Lochy, 2022).

Another structural argument is made by the *bias connectivity hypothesis* (Hannagan et al., 2015). This proposes that the left VWFA region is primed for word processing because it has a higher density of white matter connections to the left perisylvian language network, even before the onset of reading. Indeed, studies in children before and after they learn to read show that the location of the VWFA they develop at age 8 is predicted by the connectivity patterns of their left ventral occipitotemporal cortex at age 5 (Saygin et al., 2016).

A functional argument is that there may be some intrinsic visual processing skill that makes the left fusiform gyrus better suited to analyzing written text. The *neuronal recycling hypothesis* (Dehaene and Cohen, 2007) notes that human writing systems have certain common features, such as the use of three strokes per character on average and of non-accidental intersections of contours, and conjectures that there may be a "functional bias" for such features in the left fusiform gyrus. Visual word processing capitalizes on this sensitivity and "recycles" it for reading. In support, one study found that alphanumeric characters already generate more activity than faces in the left fusiform cortex of children at age 5, before reading begins (Cantlon et al., 2011). However, another study found that the region destined to become the left VWFA at age 8 does not prefer visual words over faces or false fonts at age 5 (Saygin et al., 2016).

A related functional proposal is the *shape hypothesis*, which argues that the shape sensitivity of the inferotemporal cortex explains why it is activated by either visual text or tactile Braille reading (Hannagan et al., 2015). This proposal did not develop an explanation for lateralization, however. Finally, there are older functional arguments that the left hemisphere is better suited to word processing—and the right to face processing—because the left emphasizes part-based rather than holistic processing, high rather than low spatial frequencies, or categorical rather than coordinate relations (Farah, 1990; Plaut and Behrmann, 2011).

Developmental dyslexia

The cause of developmental dyslexia remains unclear. There are pathogenetic hypotheses centered on cerebellar function, sensory processing, and the magnocellular stream (Peterson and Pennington, 2012), but the most prominent current proposal is that it is related to a phonologic deficit (Ramus et al., 2003). One can ask whether any of these proposed mechanisms involve or cause altered lateralization of reading in the brain. Indeed, older anatomic studies proposed that learning disabilities such as dyslexia were associated with fundamental disruptions in brain asymmetry, particularly an enlarged planum temporale in the right hemisphere (Galaburda et al., 1978).

Reduced left-lateralization is suggested by tachistoscopic studies showing that dyslexic subjects have an anomalous left-field advantage (Gerlach et al., 2022) or loss of the right-field advantage for words (Collins et al., 2017; Gabay et al., 2017). Functional imaging studies of reading have reported reduced activation of the left extrastriate cortex (Maisog et al., 2008; Langer et al., 2015) and lower amplitude MEG signals in this region (Salmelin et al., 1996). A cross-sectional study comparing preteen and late teen children suggested that with age, word-related activity in the anterior fusiform cortex lateralizes more to the left in typical subjects, whereas dyslexic children develop less lateralized activity in more posteromedial regions (Shaywitz et al., 2007). Word responses in the left VWFA may also be anomalous in dyslexia, with less evidence of the posterior-to-anterior gradient of responses to word components and fewer effects of word familiarity (Van der Mark et al., 2009).

There are many more such studies. A meta-analysis concluded that words activate the left ventral occipitotemporal region less in both children and adults with dyslexia, and the left superior temporal region less in the adults (Richlan et al., 2011). Other reviews concluded that the entire left reading network of occipitotemporal, inferior frontal, and inferior parietal regions is less active during reading in dyslexic subjects, though reductions in the ventral occipitotemporal cortex may precede the rest (Maisog et al., 2008; Richlan, 2012).

Longitudinal studies of dyslexic children are harder to come by (Chyl et al., 2021). These can both reveal predictors of later reading difficulties as well as characterize anomalous developmental trajectories. Various structural precursors of dyslexia have been reported, including thinner gray matter in the left perisylvian network (Clark et al., 2014) and white matter alterations in the left arcuate fasciculus (Kraft et al., 2016; Borchers et al., 2019; Kuhl et al., 2020). Among functional imaging findings, the most promising markers may be reduced activation of the left ventral occipitotemporal cortex by letters (Centanni et al., 2019) and poor convergence of print- and speech-related activation in the left superior temporal gyrus (Preston et al., 2016; Chyl et al., 2018).

Learning to read is also accompanied by aberrant functional changes in the brains of dyslexic children, often with abnormal lateralization (Fig. 20.6). While typical children develop left temporoparietal activation over 3 months of reading, dyslexic children show bilateral temporoparietal and frontal activity in response to words (Yamada et al., 2011). Failure of reading-related activity to lateralize to the left over the first 2 years of schooling was seen in an MEG study (Simos et al., 2005). Similarly, fMRI studies show reduced word activation of the left

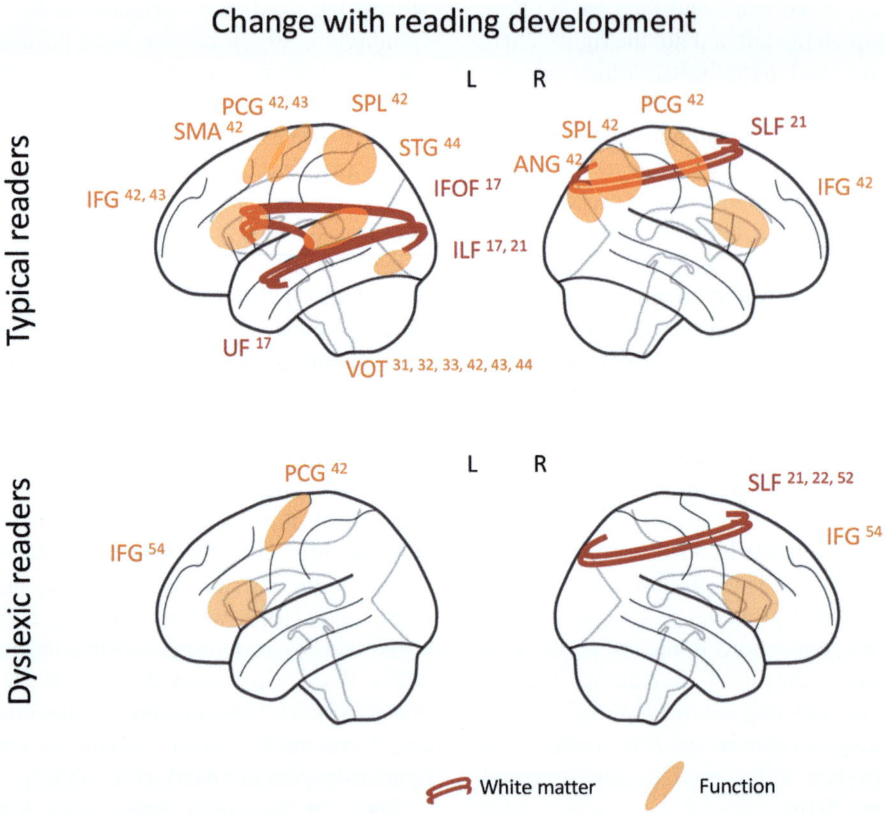

Fig. 20.6. Summary diagram of cerebral changes reported in longitudinal studies of children learning to read. Structural changes in white matter (red lines) and alterations in functional activity (orange zones) are depicted for typical and dyslexic children, for the left and right hemispheres. ANG, angular gyrus; IFG, Inferior Frontal Gyrus; IFOF, inferior fronto-occipital fasciculus; ILF, inferior longitudinal fasciculus; PCG, pre- and postcentral gyri; SLF, superior longitudinal fasciculus; SMA, supplemental motor area; SPL, superior parietal lobule; STG, superior temporal gyrus; UF, uncinate fasciculus; VOT, ventral occipitotemporal cortex. Numbers refer to the references in the original. Adapted from Chyl K, Fraga-Gonzalez G, Brem S, et al. (2021). Brain dynamics of (a)typical reading development-a review of longitudinal studies. NPJ Sci Learn 6: 4. https://doi.org/10.1038/s41539-020-00081-5.

ventral occipitotemporal cortex and left inferior frontal gyrus over those 2 years (Chyl et al., 2019), with reduced coupling between the left ventral occipitotemporal cortex, left superior temporal gyrus, and left inferior frontal gyrus (Wang et al., 2020).

All told, there is substantial evidence of structural and functional abnormalities in the left reading network of dyslexic subjects, particularly in the left ventral occipitotemporal cortex. Some of these may be correlative effects of poor reading rather than its cause, though at least a few anomalies are evident even before children start to read.

Can training reverse these effects? A number of reviews have addressed this question (Barquero et al., 2014; Braid and Richlan, 2022; Perdue et al., 2022). As for all intervention studies, such overviews are limited by variations in the number and type of subjects studied, the type and duration of training, and imaging methods. Perhaps as a result, meta-analyses find no consistent effects across training studies (Braid and Richlan, 2022; Perdue et al., 2022).

LOGOGRAPHIC VS ALPHABETIC SCRIPT

Many writing systems exist. These differ in the aspects of the language they represent. Segmental scripts have written elements (graphemes) that correspond to the sounds of auditory language (phonemes), either just its consonants (abjad scripts) or both consonants and vowels, as in alphabetic scripts like English. Syllabaries are similar but their graphemes represent syllables. The characters of logographic scripts like Chinese represent words or their components (morphemes). Because they often represent some phonetic elements as well, they are also called morphosyllabic scripts. Finally, ideographic systems are pictorial codes that represent concepts or ideas with no reference to sound, the best modern example being emojis.

The lateralization of most nonalphabetic scripts is largely unexplored. Comparisons have been made mainly between alphabetic and logographic scripts, primarily Chinese or Japanese. The Chinese logographic system was imported and then abandoned in Vietnam and North Korea, but continues as *kanji* in Japan and South Korea. These last two languages have additional writing systems. Japanese has a syllabary known as *kana*, comprising both *hiragana* and *katakana*—the latter used for words imported from other languages—while Korean has an alphabet called *hangul*.

Within-subject contrasts between logographic and alphabetic reading systems have often used bilingual subjects, though this requires control for which language was learned first. Studies

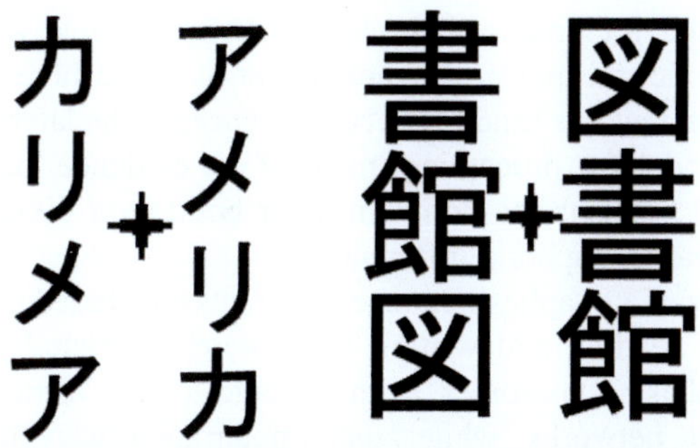

Fig. 20.7. Examples of words and non-words in *kana* (left) and *kanji* script (right). The two differ in complexity. Adapted from Ino T, Nakai R, Azuma T et al. (2009). Recognition and reading aloud of kana and kanji word: an fMRI study. Brain Res Bull 78: 232–239.

comparing *kanji* with *kana* or *hangul* are preferable, but there are still potential confounds. These include differences in the timing and intensity of learning, with Japanese children learning *kana* first but then spending more time learning *kanji* (Hasuike et al., 1986), as well as the amount of learning required, given that there are 71 letters for *kana* and around 40,000 for *kanji*. There are also linguistic differences, with *kanji* used for nouns and verb stems and *hiragana* more for tenses, particles, and conjunctions, as well as elementary visual differences (Fig. 20.7), with *kanji* characters being more complex than *kana* symbols (Horie et al., 2012).

There are speculations that the processes involved in logographic and alphabetic reading are lateralized differently. One is that *kana* reading depends on the left hemisphere and *kanji* reading on the right (Bradshaw et al., 1981). Reasons advanced include the fact that logographic script is less tied to phonology, derives historically from pictograms, and possibly involves more configurational processing.

Tachistoscopic studies of logographic reading

English words are better recognized when flashed in the right than the left visual field (Bryden and Rainey, 1963). Do logographic scripts show the opposite? Early studies of Chinese did not find this. Bilingual Chinese subjects had a similar right-field advantage for both English and Chinese (Kershner and Jeng, 1972) and for Arabic and Chinese numbers (Besner et al., 1982), while two experiments with bilingual English or Chinese subjects found no hemifield advantage for either English or Chinese words (Hardyck et al., 1978). Two later studies claimed an interesting twist: a left-field advantage for single Chinese characters, but a right-field advantage for words made of two characters (Tzeng et al., 1979; Cheng and Yang, 1989). This was attributed to the holistic processing of single characters by the right hemisphere and sequential analytic processing of multicharacter words by the left. However, this "Chinese character-word dissociation hypothesis" has not received support from functional imaging of Chinese (Tan et al., 2000) or event-related potentials for *kanji* (Maurer et al., 2008).

Japanese reports show a consistent right-field superiority for the syllabic script of *kana* (Hatta, 1976; Endo et al., 1978; Shimizu and Endo, 1981a,b; Shimizu et al., 1981). For *kanji* the results are mixed. Some reported the same right-field superiority as for *kana* (Ino et al., 2009), but others claimed a left-field superiority (Hatta, 1977). Tests of *kana* and *kanji* in the same subjects found a right-field advantage for *kana* but no hemifield difference for *kanji* (Sasanuma et al., 1977; Endo et al., 1981; Shimizu et al., 1981).

Hangul can serve as a control "nonverbal" stimulus in Japanese subjects who cannot read Korean. One study showed a right-field advantage for *kana* but a left-field advantage for *hangul*, the latter disappearing after subjects learned the meanings and pronunciations of the *hangul* words (Shimizu and Endo, 1981a,b). A second study contrasted *kana*, *kanji*, and *hangul* in Japanese subjects vs Korean subjects whose first language was Japanese but who later learned Korean (Endo et al., 1981). Both groups had a right-field superiority for *kana* and a small left-field advantage for *kanji*. The group that could not read Korean showed a large left-field advantage for *hangul*, but the subjects who could read it showed the same right-field advantage for *hangul* as for *kana*.

In summary, tachistoscopic studies consistently show a strong right-field/left hemisphere superiority for alphabetic scripts or syllabaries such as English, *hangul*, and *kana* once subjects have learned to read them. For logographic scripts, the results vary. A few find the same right-field superiority as for alphabetic script but most report either no hemifield difference or a weak left-field advantage. Some of the inconsistencies in logograms may be related to issues like exposure duration and word complexity (Hasuike et al., 1986), or the "concreteness" of the word (Shimizu and Endo, 1981a,b).

Event-related potentials of logographic reading

Visual word perception generates three electrophysiologic components: a bilateral occipital P100 linked to basic visual feature analysis, a left-lateralized occipitotemporal N170 linked to visual word processing in the VWFA, and a central N400 that reflects semantic processing. Some studies of bilingual Asians show strong left-lateralization of the N170 for English or *hangul*, but more bilateral activation for Chinese logograms (Liu and Perfetti, 2003; Kim et al., 2004). Others report similar left-lateralization of the N170 for both *kana* and *kanji* (Maurer et al., 2008; Horie et al., 2012). The N170 with Asian scripts shows either no or slightly right-biased lateralization in English subjects who cannot read them (Wong et al., 2005; Maurer et al., 2008). One study of Japanese subjects with limited English experience found bilateral N170 patterns for *kana* and *kanji* but a more right-lateralized potential for English (Ishiwatari et al., 2002). Overall, the impression is similar to that from the tachistoscopic work.

Functional neuroimaging of logographic reading

An fMRI study of Chinese found activation of the left fusiform and the right lingual gyrus (Tan et al., 2000). One study asked if Chinese characters were processed more like pictograms or linguistic units and found that their activation resembled that of English text more than that of pictures, with a greater left temporal bias (Chee et al., 2000).

Two fMRI studies contrasted Chinese with *pinyin*, which uses English letters as phonetic symbols to teach Chinese. One found little evidence overall for hemispheric lateralization, but greater activation in the left fusiform gyrus for Chinese than for *pinyin* (Chen et al., 2002). The other found more left hemispheric activity for both, but curiously extrastriate activation was lateralized more to the right for *pinyin* and more to the left for Chinese (Fu et al., 2002).

For Japanese, an early MEG study claimed no lateralization for either *kana* or *kanji* (Koyama et al., 1998). Subsequent PET and fMRI studies have shown that both *kana* and *kanji* activate the left fusiform gyrus (Sakurai et al., 2000; Nakamura et al., 2005a,b) but differ in the additional areas they activate. *Kanji* activates the right fusiform gyrus, while *kana* recruits either bilateral medial and lateral occipital gyri (Sakurai et al., 2000) or the left posterior middle temporal cortex (Nakamura et al., 2005b).

Does it matter which language is learned first? One study of the fusiform region reported that native Chinese readers have bilateral, slightly left-biased activation for both English and Chinese, while English readers learning Chinese show strong left-lateralization for English, but also recruit the right fusiform area for Chinese (Nelson et al., 2009). However, a Japanese study of bilingual English and Japanese subjects found that, regardless of which was learned first, *kanji* showed weaker left-lateralization in the lateral occipital complex than either *kana* or English (Koyama et al., 2014).

There may be a relationship between lateralization and the proficiency of second-language learning. Among Chinese subjects taught *hangul*, those with stronger left-lateralization of *hangul* activity in the fusiform gyrus were better at learning it (Xue et al., 2006). In Japanese subjects, English proficiency correlated with more left-lateralized activation in the temporoparietal junction for either *kana* or English (Koyama et al., 2014). In contrast, the *kanji* proficiency of English subjects correlated with more right-lateralized activation of the lateral occipital cortex when viewing either *kanji* or Tibetan pictograms they could not read.

All told, neuroimaging studies suggest that the processing of Chinese or *kanji* logograms still emphasizes left fusiform activity, although these recruit more occipital or fusiform structures in the right hemisphere than alphabetic or syllabic scripts do, a conclusion supported by two meta-analyses (Bolger et al., 2005; Tan et al., 2005).

Neuropsychologic studies of logographic reading

Are there dissociable deficits in logographic and alphabetic processing? Both aphasic and alexic subjects have been studied, with early cases reporting relative dissociations. One Japanese patient had greater difficulty reading and writing *kana* than *kanji* (Asayama, 1914), while a Chinese man with a left parieto-occipital tumor had the converse, more difficulty with Chinese than English (Lyman et al., 1938). Since then, evidence has accumulated for double dissociations between *kana* and *kanji* in both alexic and aphasic Japanese patients. While there has been speculation that this might reflect differences in the lateralization of *kana* and *kanji* processing, most of the evidence indicates a strong left hemispheric dominance for both, with dissociations due to intra-hemispheric differences.

A study of 378 aphasic Japanese patients concluded that problems with reading and writing *kana* were associated with the classic aphasic syndromes from damage to the Wernicke and Broca area networks, while *kanji* impairments were linked to transcortical or semantic aphasias (Sasanuma, 1975). However, a study of 50 patients noted that most reading impairments affected both *kana* and *kanji* and refuted any link between the reading impairment and the type of aphasia (Sugishita et al., 1992). A different basis for dissociation was forwarded in a review of alexia with or without agraphia (Iwata, 1984). This claimed that *kanji* reading was disrupted by lesions in a ventral semantic pathway using the inferior and middle temporal gyri, while *kana* reading was affected by damage to a dorsal phonologic route involving the angular gyrus.

Subsequent cases of alexia with agraphia have confirmed an association of *kanji* deficits with left ventral temporal damage (Jibiki and Yamaguchi, 1993; Sakurai, 2004). Of seven patients with pure alexia from left occipitotemporal lesions, four were impaired for both *kana* and *kanji*, while three had more difficulty with *kanji*, though in two the difference disappeared when the scripts were equated for pictoral complexity (Sugishita et al., 1992). Sakurai et al. (2006) reported that a left fusiform gyrus lesion caused severe difficulty reading *kanji*, slowness with *kana*, and some difficulty writing *kanji* (Fig. 20.8), reminiscent of the surface dysgraphia in cases of pure alexia for English (Rapcsak et al., 1990).

Selective *kana* deficits were not found in any of the seven pure alexic patients of Sugishita et al. (1992), but in one patient with alexia and agraphia from left occipitoparietal damage. Others have reported selective *kana* impairments with left posterior temporal (Shiota et al., 1986) or posterior occipital strokes (Sakurai, 2004; Sakurai et al., 2006) (Fig. 20.8). As summarized in a recent review (Sakurai, 2019), the current model attributes dissociations not to a difference between phonologic and semantic processing, but to a contrast between whole-word reading for *kanji* in the visual word form area, and letter identification for *kana* in the inferior occipital cortex just posterior to the visual word form area.

Tellingly, alexia for *kanji*, *kana*, or both is rare after right hemispheric damage. A review of all Japanese reports from 1901 to 1985 found right hemispheric lesions in 5 of 69 cases of alexia (Paradis et al., 1985). In these five, only one, a left-handed man, showed the result others had predicted, namely more difficulty with *kanji*. In fact, three of the five struggled more with *kana*.

Thus, even though tachistoscopic, neuroimaging, and some electrophysiologic studies suggest more right occipitotemporal involvement with logograms than alphabetic or syllabic scripts, this does not seem to be a critical contribution to reading. In support, tachistoscopic studies of three patients with section of

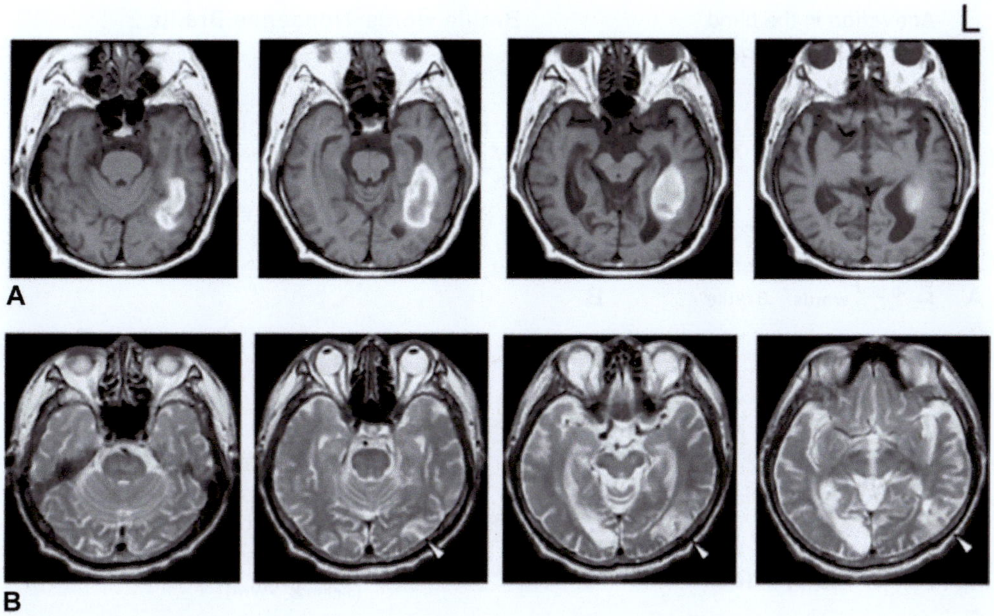

Fig. 20.8. (A) Axial T1-weighted (top row) and coronal T2-weighted (bottom row) MRI images of a patient with alexia affecting *kanji* more than *kana* after a hemorrhage involving the left fusiform gyrus. (B) Axial (top row) and coronal (bottom row) T2-weighted images of a patient with a first stroke of the right lingual and cuneate gyri, who developed alexia for *kana* after a second stroke of the left inferior and middle occipital gyri (arrow). Adapted from Sakurai Y, Yagishita A, Goto Y et al. (2006). Fusiform type alexia: pure alexia for words in contrast to posterior occipital type pure alexia for letters. J Neurol Sci 247: 81–92.

the splenium during the treatment of pineal tumors showed intact reading aloud and comprehension of both *kana* and *kanji* in the right visual field, but difficulty in the left visual field for both. This essentially amounts to left hemi-alexia for both *kana* and *kanji*—though *kanji* reading in the left field improved with time (Sugushita et al., 1978).

TACTILE (NONVISUAL) TEXT: BRAILLE READING BY THE BLIND

What sort of network is involved in reading when sensory input is tactile rather than visual? This would be an interesting question even in sighted individuals, but it is the rare sighted person who learns Braille. Rather, studies of tactile reading are done almost exclusively in the blind, which confounds the effects of tactile reading inputs with those of visual deprivation.

Early functional neuroimaging investigations focused on showing that the striate and peri-striate cortex could be re-purposed for nonvisual functions such as Braille reading in the congenitally blind (Sadato et al., 1996), sometimes even in those who lost vision later (Büchel et al., 1998). This occipital neuroplasticity is not limited to Braille, as spoken language can also activate this cortex in the blind (Röder et al., 2002; Tian et al., 2022). This is consistent with claims that network reorganization in early blindness leads to a merger of language and visual modules (Hasson et al., 2016). Two observations support the critical role of the striate cortex in tactile reading by the blind. First, repetitive transcranial magnetic stimulation over the occipital lobe disrupts the reading of Braille or embossed letters in blind but not sighted subjects (Cohen et al., 1997). Second, bilateral striate infarcts abolished the ability of a congenitally blind woman to read Braille but did not affect other tactile perceptions (Hamilton et al., 2000).

Are these occipital neuroplastic effects lateralized? There are no data yet about the laterality of occipital activity during Braille reading. However, one study found that auditory activation of the occipital cortex co-lateralized with that of the perisylvian language networks in the blind (Lane et al., 2017).

Several reports have studied the fate of the left reading network. What happens to the VWFA, given that it is a *visual* area? Is it co-opted by tactile reading like the striate cortex? Or, since reading Braille is tactile, does reading selectivity develop in the parietal cortex? The answer may be yes to both. The left VWFA is active during Braille reading in subjects with congenital or adult-onset blindness (Büchel et al., 1998; Reich et al., 2011) (Fig. 20.9), but a recent study also found posterior parietal areas that were more activated by Braille reading than other tactile shapes (Tian et al., 2022).

There are some differences in the VWFA of the blind and the sighted, though. First, the anatomic inputs to the VWFA differ. Despite the neuroplastic changes in the striate cortex, low-vision subjects show stronger functional connectivity of the VWFA with the parietal than with the occipital cortex (Zhou et al., 2020). This is particularly true in subjects who can read Braille, in whom the strength of the parietal connection correlates with the age at which they had learned Braille. Increased functional connectivity between the somatosensory cortex and the left VWFA can develop quickly, though, as it is also seen in sighted subjects after only 9 months of learning Braille (Siuda-Krzywicka et al., 2016).

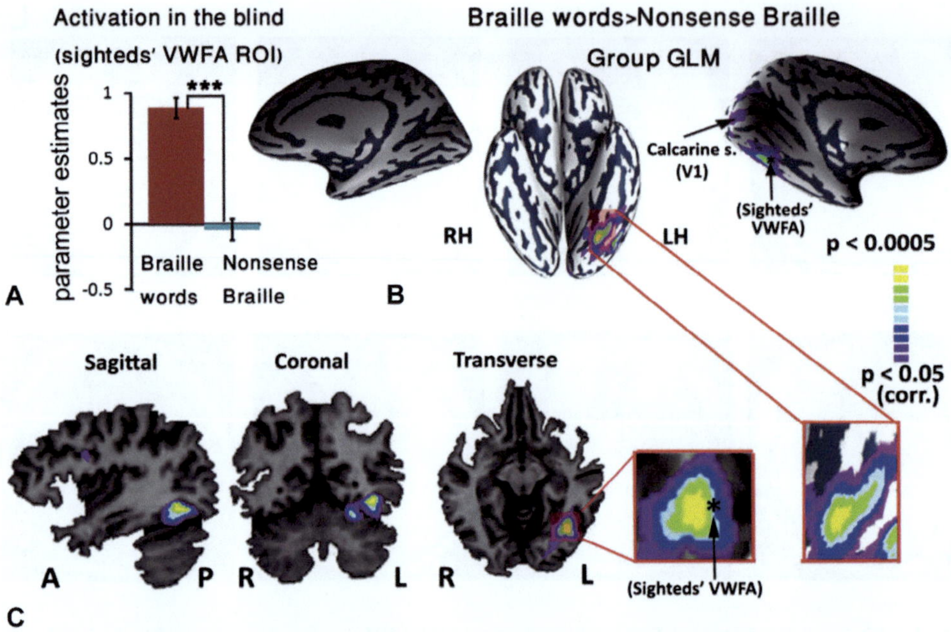

Fig. 20.9. Braille-reading activation in the congenitally blind. (A) Activation for Braille vs nonsense Braille in a region corresponding to the location of the VWFA in sighted subjects. Statistical parametric map across the whole brain for the contrast of Braille vs nonsense Braille, on an inflated brain (B) and three brain sections (C). Activation peaks in the VWFA. From Reich L, Szwed M, Cohen L et al. (2011). A ventral visual stream Reading center independent of visual experience. Curr Biol 21: 363–368.

Second, while the ventral occipitotemporal cortex of sighted subjects has a posterior-to-anterior hierarchy (Vinckier et al., 2007), one study failed to find a similar anatomic progression from shapes to consonants to words for Braille in the blind (Tian et al., 2022).

Third, just like the responses in the striate cortex, activation in the left VWFA may be less modality-specific in the blind. In these subjects, the VWFA and other occipitotemporal areas can be activated by spoken words (Tian et al., 2022) or auditory sentence comprehension (Kim et al., 2017).

Fourth, while visual reading is strongly co-localized with other language functions to the left hemisphere in sighted subjects, the lateralization of Braille reading in the blind may not be as consistent—but see Reich et al. (2011). The peri-sylvian network for spoken language is less strongly lateralized to the left in blind subjects, with more individuals having right or bilateral processing (Röder et al., 2000, 2002; Lane et al., 2017; Tian et al., 2022). Why this occurs is not yet known. One study found that Braille-selective responses in the VWFA co-lateralize strongly with auditory language in blind subjects, with the result being that Braille-reading in the blind is less consistently left-lateralized than the visual reading of sighted subjects (Tian et al., 2022).

Another point regarding lateralization concerns the sensory input. Visual reading is seldom confined to one visual hemifield, and thus both the right and left striate cortex contribute information when sighted individuals read. However, Braille reading is often done with one hand, and therefore the initial sensory input is strongly lateralized to the somatosensory cortex contralateral to that hand. While one study found a consistent left-lateralization of the VWFA regardless of the hand preferred for reading (Reich et al., 2011), another found a modest effect of the preferred reading hand on lateralization of Braille reading in the VWFA (Tian et al., 2022).

In summary, there is strong evidence that the tactile reading of blind individuals recruits not only the occipital cortex but also the VWFA and other components of the network used for reading in sighted individuals, as well as tactile association areas in the parietal cortex. However, reading-selective responses are less consistently localized to the left hemisphere in the blind, in parallel with the more variable lateralization of spoken language in these subjects, and there may be a mild effect of the preferred reading hand on lateralization.

Finally, as noted already, tactile reading is rarely studied in sighted subjects. However, one study trained sighted subjects daily in Braille for 9 months (Siuda-Krzywicka et al., 2016). Training increased activation from Braille in the left VWFA and other frontal and parietal regions, but not in the somatosensory cortex. It also increased the resting-state functional connectivity between the left primary somatosensory cortex and the left VWFA. This confirms that the adaptation of the VWFA to use tactile inputs can occur without visual deprivation.

VISUAL NON-TEXT READING

Sign language in the deaf

Some conjecture that there are reasons why sign language might be less biased to the left hemisphere. Watching a person making signs may recruit visuospatial and social cognitive processes, for which the right hemisphere is dominant (Trettenbrein et al., 2021).

As for the striate cortex in the blind reading Braille, the primary auditory cortex of both hemispheres is recruited during sign language in the deaf (Petitto et al., 2000). Also, like the striate cortex in the blind, this neuroplasticity is non-specific, since activation occurs also with signing in deaf subjects who do not know sign language (Cardin et al., 2013) and with nonlinguistic visual or tactile stimuli in deaf signers (Finney et al., 2001; Bola et al., 2017). More specific to linguistic issues is the degree to which signing involves the left perisylvian network for spoken language (Campbell et al., 2008). Here, we discuss effects in areas considered "visual."

How about lateralization in the VWFA? In deaf signers, perception of sign language generates more left-sided activity in the fusiform gyri than perception of nonlinguistic gestures (Newman et al., 2015). However, this lateralization of mid-fusiform activity is less with sign language in the deaf than it is with written language in hearing nonsigners (Waters et al., 2007).

Unlike the case for Braille, where few sighted individuals acquire proficiency in tactile reading, sign language can be studied in both deaf and hearing native signers, the latter often being the children of deaf parents. This comparison gives a clearer picture of the neuroplastic responses to deafness. One such study found more activation of the left superior temporal gyrus and secondary auditory cortex in the deaf compared to hearing native signers (MacSweeney et al., 2002). Comparisons between deaf signers, deaf nonsigners, and hearing nonsigners can separate the effects of acquiring sign language from the effects of deafness. One such study concluded that deafness drove plastic changes more in the right than the left superior temporal cortex, while acquiring sign language led to plastic changes in the left superior temporal sulcus (Cardin et al., 2013).

To summarize, the evidence suggests that, as expected, sign language is mediated in the deaf both by visual areas and the left perisylvian language network. Just as with written text, reading sign language involves the left VWFA, though the fusiform lateralization may not be as strong as for written text.

Lip-reading

Lip-reading uses visual cues about facial movement to infer speech sounds. Since auditory language is lateralized to the left but the perception of dynamic facial expressions is processed more by the right (Sato et al., 2019), the lateralization of lip-reading is uncertain.

An early tachistoscopic study found a left-field advantage for matching facial speech to sounds, though it expressed reservations that this might reflect general perceptual factors (Campbell, 1986). A later study requiring matching of facial speech across viewpoint changes found a right-field advantage (Campbell et al., 1996).

Several fMRI studies exist. One showed bilateral activation of the auditory cortex (MacSweeney et al., 2000), while another claimed to isolate a temporal visual speech area in the junction of the posterior superior temporal sulcus and middle temporal gyrus bilaterally (Bernstein et al., 2011). These studies did not take lip-reading proficiency into account, though. Others showed that lip-reading skill correlates with activity in the left posterior superior temporal region (Hall et al., 2005), and that this may be more significant for deaf than for hearing subjects (Capek et al., 2008). A study that included only proficient lip readers showed that speaking faces activate a left-sided network of auditory and visual association areas as well as language regions (Ruytjens et al., 2006). A study of the McGurk effect, in which incongruent auditory and visual displays of syllables lead to the illusion of a blended sound (McGurk and MacDonald, 1976), found that those who report the illusion show more activity in the left superior temporal sulcus, and that this activity correlates with the frequency of reporting the illusion (Nath and Beauchamp, 2012). Finally, an electrophysiologic study using the visual mismatch negativity inferred that the left posterior temporal cortex is involved in discriminating facial speech patterns (Files et al., 2013).

Studies of small numbers of patients have supported this left-lateralization of lip-reading. Lip-reading was abnormal in one alexic patient with a left hemispheric lesion, but normal in a patient with a left middle cerebral artery stroke who had aphasia that resolved into a "word meaning deafness," as well as in two prosopagnosic patients with right-sided lesions (Campbell et al., 1986, 1990). A later study showed that facial speech identification and audiovisual integration in the McGurk effect was reduced in four alexic subjects with left fusiform damage, but unaffected in a prosopagnosic subject with right fusiform damage (Albonico and Barton, 2017) (Fig. 20.10).

NON-PHONOLOGIC READING OF SYMBOLS: MUSICAL NOTATION

Reading music also involves the processing of a temporal sequence of symbols in written text. Unlike written language, musical notation is not linked to phonology, and unlike aphasia, the various acquired deficits in music perception are lateralized to the right hemisphere (Stewart et al., 2006). Nevertheless, studies consistently show that the reading of musical notation is lateralized to the left, in close proximity but not identical with the network for reading language.

Tachistoscopic studies have shown a right-field advantage for reading single notes (Oscar-Berman et al., 1976) or chords (Segalowitz et al., 1979; Salis, 1980). An early study with functional imaging showed that reading music activates the left parieto-occipital lobe (Sergent et al., 1992). Later work with fMRI showed activation of the left occipitotemporal cortex near the VWFA (Mongelli et al., 2017; Bouhali et al., 2020). Music literacy is associated with increased left-lateralization of activity associated with reading music (Bouhali et al., 2017) and an increase in the size of this left-sided activation and its network (Mongelli et al., 2017).

Neuropsychologic cases are rare, as only a portion of the populace is literate in music and there are no standardized tests or easy methods to adjust for premorbid proficiency.

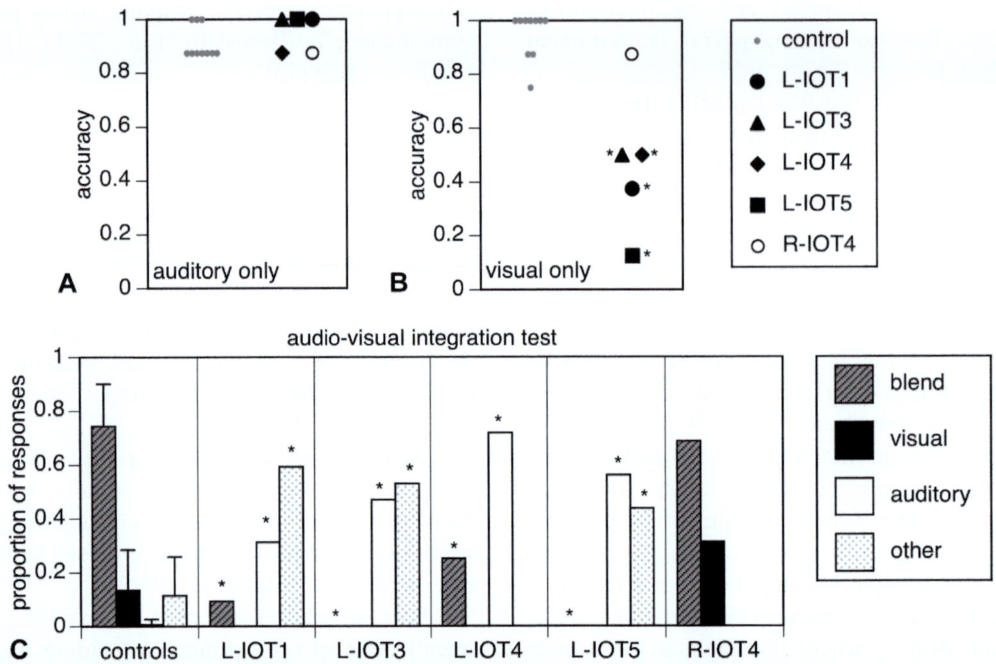

Fig. 20.10. Lip-reading by five patients with unilateral fusiform lesions. Four have alexia with left-sided lesions (black symbols) while one has prosopagnosia with a right-sided lesion (white disc). (A) All five can identify speech sounds in auditory-only clips. (B) All four with left fusiform lesions are impaired in identifying facial speech patterns in visual-only clips. (C) McGurk effect, with audiovisual clips of incongruent stimuli, in which the speech sound differs from the facial speech pattern. Subjects can report perceiving a blend of the two (blend), the facial speech pattern only (visual), the speech sound only (auditory), or none of the above (other). The four alexic patients rarely reported the blend and never reported the facial speech pattern only, indicating a failure to perceive facial speech patterns, consistent with their performance shown in (B), with visual-only clips. Asterisks mark abnormal scores. Error bars indicate one standard error for controls. From Albonico A, Barton, JJS (2017). Face perception in pure alexia: complementary contributions of the left fusiform gyrus to facial identity and facial speech processing. Cortex 96: 59–72.

Nevertheless, a review of 16 cases with loss of music literacy noted that all had left posterior damage, and most had alexia as well (Hébert and Cuddy, 2006). In musicians being evaluated for resection of focal brain lesions, music reading could be disrupted by direct current stimulation of various left-hemisphere sites in five patients, but not in any right-hemisphere sites of two others (Roux et al., 2007).

READING NUMBERS

Digits and numbers have had an uncertain relationship with letters and words from the beginning. Monsieur C could read numbers better than letters or words, though not perfectly, as he read numbers slowly, "digit-by-digit" (Bub et al., 1993). This led some to think that number reading was lateralized to the right.

Most patient work since then has shown that, while there may be some visual number processing by the right hemisphere, the left side is better. An early split-brain study (Teng and Sperry, 1973) found better identification of digits than letters in the left hemifield, but both stimuli still showed a large right-field advantage. Later split-brain studies showed that the right hemisphere can make magnitude judgments about numbers but struggles with adding or naming numbers, whereas all tasks are done well by the left hemisphere (Seymour et al., 1994; Cohen and Dehaene, 1996). This pattern was reproduced in two patients with pure alexia from left occipital strokes: they had intact magnitude judgments but impaired addition and poor naming of numbers with multiple digits (Cohen and Dehaene, 1995). A review of 90 cases of pure alexia (Starrfelt and Behrmann, 2011) concluded that none had firm evidence of spared number reading and speculated that any trend to easier reading of numbers than letters was due to easier discriminability of the former.

Some studies in healthy subjects have argued for a right dominance in number reading. An fMRI (Park et al., 2012) and an event-related potential study (Park et al., 2014) reported more number-related activity in the right ventral visual cortex and more letter-related activity in the left. A direct electrocorticographic study showed bilateral inferior temporal activity that was greater on the right during the identification of Arabic numerals than of letters (Shum et al., 2013). However, all of these studies relied on subtractions between number and letter activity, so the apparent right hemisphere superiority for numbers could simply reflect a marked right inferiority for letters combined with more bilateral representation of numbers, in keeping with proposals that each hemisphere has a "number form area" (Hannagan et al., 2015).

CONCLUSION

Reading is an activity that draws on both perceptual and linguistic operations (Fig. 20.11). While there is bilateral activity in the brain during reading, there is strong evidence that left

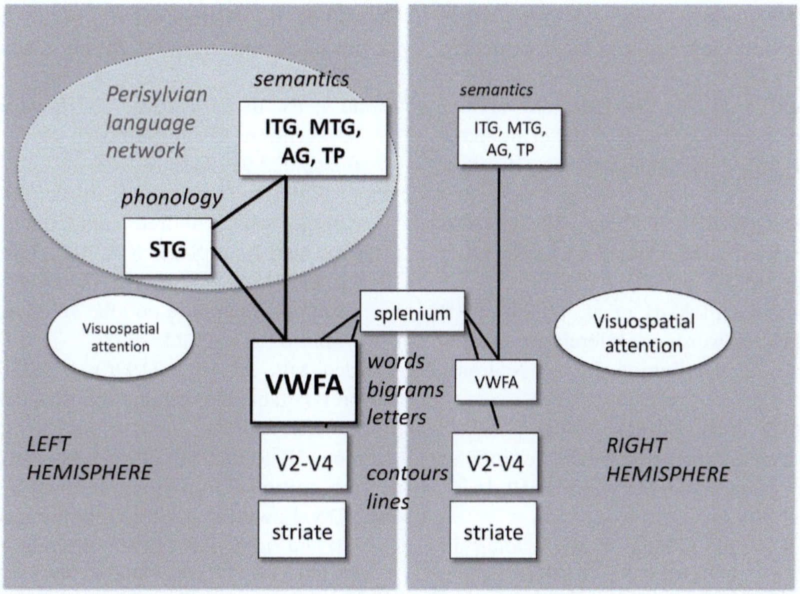

Fig. 20.11. Summary structural schematic of reading. Cortical perceptual processing begins at the bottom, in the left and right striate cortex, with early visual areas V2–4 extracting line segments and increasingly complex contours. The VWFA in the ventral occipitotemporal cortex performs reading selective operations in extracting letters, then bigrams, and words, in a hierarchical stream processing posterior to anterior. This is lateralized more to the left. Contributions from right hemispheric perceptual processes are received via the splenium. Perceptual processing interacts with linguistic processing, with phonologic and semantic operations in the left perisylvian language network (light gray ring), though the right hemisphere is capable of some semantic processing. Attentional processes in both right and left hemispheres are also involved. Adapted from Boukrina O, Chen P, Budinoska T et al. (2020). Exploratory examination of lexical and neuroanatomic correlates of neglect dyslexia. Neuropsychology 34: 404–419. https://doi.org/10.1037/neu0000619.

occipitotemporal structures dominate and are critical for the perceptual aspects of this function. Leftward lateralization of visual word-related activity is a marker for the acquisition of reading skill and is reduced in developmental dyslexia, though it is not clear how much of this reduction is cause or effect. The left fusiform gyrus may have special functional attributes that predispose it to visual word processing, as well as superior connectivity to the left language network. On its own, the right hemisphere is capable of some semantic processing but shows no access to the phonology of words. Work with other forms of reading generally shows a similar emphasis on left visual areas, often the ventral occipitotemporal cortex, though the degree of lateralization may be less, more variable, or affected by other lateralizing factors.

References

Aguilar OM, Kerry SJ, Crinion JT et al. (2018). Dorsal and ventral visual stream contributions to preserved reading ability in patients with central alexia. Cortex 106: 200–212. https://doi.org/10.1016/j.cortex.2018.06.003.

Albonico A, Barton JJS (2017). Face perception in *pure alexia*: complementary contributions of the left fusiform gyrus to facial identity and facial speech processing. Cortex 96: 59–72.

Asayama T (1914). Über die Aphasie bein Japanern. Deutch Arch f klin Med 113: 523.

Balani AB, Bickerton WL (2023). Acquired reading impairment following brain injury. Appl Neuropsychol Adult 1–19. https://doi.org/10.1080/23279095.2023.2165923.

Barquero LA, Davis N, Cutting LE (2014). Neuroimaging of reading intervention: a systematic review and activation likelihood estimate meta-analysis. PLoS One 9: e83668. https://doi.org/10.1371/journal.pone.0083668.

Barton MI, Goodglass H, Shai A (1965). Differential recognition of tachistoscopically presented English and Hebrew words in right and left visual fields. Percept Mot Skills 21: 431–437. https://doi.org/10.2466/pms.1965.21.2.431.

Barton JJS, Hanif HM, Eklinder Björnström L et al. (2014). The word-length effect in reading: a review. Cogn Neuropsychol 31: 378–412. https://doi.org/10.1080/02643294.2014.895314.

Barton JJ, Rubino C, Albonico A et al. (2022). Right hemi-alexia. Cortex 157: 288–303. https://doi.org/10.1016/j.cortex.2022.09.015.

Baylis GC, Driver J, Baylis LL et al. (1994). Reading of letters and words in a patient with Balint's syndrome. Neuropsychologia 32: 1273–1286. https://doi.org/10.1016/0028-3932(94)90109-0.

Beaton AA, Suller S, Workman L (2007). Visual laterality effects in readers of a deep and a shallow orthography. Laterality 12: 199–215. https://doi.org/10.1080/13576500601112283.

Beauvois MF, Derouesne J (1979). Phonological alexia: three dissociations. J Neurol Neurosurg Psychiatry 42: 1115–1124. https://doi.org/10.1136/jnnp.42.12.1115.

Beeson PM, Rising K, Sachs A et al. (2022). Common predictors of spoken and written language performance in aphasia, alexia, and agraphia. Front Hum Neurosci 16: 1025468. https://doi.org/10.3389/fnhum.2022.1025468.

Behrmann M, Plaut DC (2013). Distributed circuits, not circumscribed centers, mediate visual recognition. Trends Cogn Sci 17: 210–219.

Behrmann M, Plaut DC (2020). Hemispheric organization for visual object recognition: a theoretical account and empirical evidence. Perception 49: 373–404.

Behrmann M, Plaut DC, Nelson J (1998). A literature review and new data supporting an interactive account of letter-by-letter reading. Cogn Neuropsychol 15: 7–51. https://doi.org/10.1080/026432998381212.

Ben-Shachar M, Dougherty RF, Deutsch GK et al. (2011). The development of cortical sensitivity to visual word forms. J Cogn Neurosci 23: 2387–2399. https://doi.org/10.1162/jocn.2011.21615.

Bentin S, Mouchetant-Rostaing Y, Giard M-H et al. (1999). ERP manifestations of processing printed words at different psycholinguistic levels: time course and scalp distribution. J Cogn Neurosci 11: 235–260.

Bernstein LE, Jiang J, Pantazis D et al. (2011). Visual phonetic processing localized using speech and nonspeech face gestures in video and point-light displays. Hum Brain Mapp 32: 1660–1676. https://doi.org/10.1002/hbm.21139.

Beschin N, Cisari C, Cubelli R et al. (2014). Prose reading in neglect. Brain Cogn 84: 69–75. https://doi.org/10.1016/j.bandc.2013.11.002.

Besner D, Daniels S, Slade C (1982). Ideogram reading and right hemisphere language. Brit J Psychol 73: 21–28.

Binder JR, Medler DA, Westbury CF et al. (2006). Tuning of the human left fusiform gyrus to sublexical orthographic structure. Neuroimage 33: 739–748.

Binder JR, Pillay SB, Humphries CJ et al. (2016). Surface errors without semantic impairment in acquired dyslexia: a voxel-based lesion–symptom mapping study. Brain 139: 1517–1526. https://doi.org/10.1093/brain/aww029 %J Brain.

Bola Ł, Zimmermann M, Mostowski P et al. (2017). Task-specific reorganization of the auditory cortex in deaf humans. Proc Natl Acad Sci U S A 114: E600–E609.

Bolger DJ, Perfetti CA, Schneider W (2005). Cross-cultural effect on the brain revisited: universal structures plus writing system variation. Hum Brain Mapp 25: 92–104. https://doi.org/10.1002/hbm.20124.

Bonandrini R, Paulesu E, Traficante D et al. (2023). Lateralized reading in the healthy brain: a behavioral and computational study on the nature of the visual field effect. Neuropsychologia 180: 108468. https://doi.org/10.1016/j.neuropsychologia.2023.108468.

Borchers LR, Bruckert L, Dodson CK et al. (2019). Microstructural properties of white matter pathways in relation to subsequent reading abilities in children: a longitudinal analysis. Brain Struct Funct 224: 891–905. https://doi.org/10.1007/s00429-018-1813-z.

Bormann T, Wolfer S, Hachmann W et al. (2015). Fast word reading in *pure alexia*: "fast, yet serial". Neurocase 21: 251–267. https://doi.org/10.1080/13554794.2014.890732.

Bouhali F, Mongelli V, Cohen L (2017). Musical literacy shifts asymmetries in the ventral visual cortex. Neuroimage 156: 445–455. https://doi.org/10.1016/j.neuroimage.2017.04.027.

Bouhali F, Mongelli V, Thiebaut de Schotten M et al. (2020). Reading music and words: the anatomical connectivity of musicians' visual cortex. Neuroimage 212: 116666. https://doi.org/10.1016/j.neuroimage.2020.116666.

Boukrina O, Barrett AM, Graves WW (2019). Cerebral perfusion of the left reading network predicts recovery of reading in subacute to chronic stroke. Hum Brain Mapp 40: 5301–5314.

Boukrina O, Chen P, Budinoska T et al. (2020). Exploratory examination of lexical and neuroanatomic correlates of neglect dyslexia. Neuropsychology 34: 404–419. https://doi.org/10.1037/neu0000619.

Bradshaw JL, Nettleton NC, Taylor MJ (1981). Right hemisphere language and cognitive deficit in sinistrals? Neuropsychologia 19: 113–132.

Braid J, Richlan F (2022). The functional neuroanatomy of reading intervention. Front Neurosci 16: 921931. https://doi.org/10.3389/fnins.2022.921931.

Brem S, Bach S, Kucian K et al. (2010). Brain sensitivity to print emerges when children learn letter–speech sound correspondences. Proc Natl Acad Sci U S A 107: 7939–7944.

Broca PP (1861). Perte de la parole; ramollisement chronique et destruction partielle du lobe antérieur gauche du cerveau. Bull Soc Anthrop Paris 2: 235–238.

Bryden MP, Allard F (1976). Visual hemifield differences depend on typeface. Brain Lang 3: 191–200. https://doi.org/10.1016/0093-934x(76)90016-x.

Bryden MP, Rainey CA (1963). A left-right difference in tachistoscopic recognition. J Exp Psychol 66: 568–571.

Bub DN, Lewine J (1988). Different modes of word recognition in the left and right visual fields. Brain Lang 33: 161–188. https://doi.org/10.1016/0093-934x(88)90060-0.

Bub DN, Arguin M, Lecours AR (1993). Jules Dejerine and his interpretation of *pure alexia*. Brain Lang 45: 531–559. https://doi.org/10.1006/brln.1993.1059.

Büchel C, Price C, Friston K (1998). A multimodal language region in the ventral visual pathway. Nature 394: 274–277. https://doi.org/10.1038/28389.

Caffarra S, Karipidis II, Yablonski M et al. (2021). Anatomy and physiology of word-selective visual cortex: from visual features to lexical processing. Brain Struct Funct 226: 3051–3065.

Campbell R (1986). The lateralization of lip-read sounds: a first look. Brain Cogn 5: 1–21. https://doi.org/10.1016/0278-2626(86)90059-x.

Campbell R, Landis T, Regard M (1986). Face recognition and lipreading. A neurological dissociation. Brain 109: 509–521. https://doi.org/10.1093/brain/109.3.509.

Campbell R, Garwood J, Franklin S et al. (1990). Neuropsychological studies of auditory-visual fusion illusions. Four case studies and their implications. Neuropsychologia 28: 787–802. https://doi.org/10.1016/0028-3932(90)90003-7.

Campbell R, De Gelder B, De Haan E (1996). The lateralization of lip-reading: a second look. Neuropsychologia 34: 1235–1240. https://doi.org/10.1016/0028-3932(96)00046-2.

Campbell R, MacSweeney M, Waters D (2008). Sign language and the brain: a review. J Deaf Stud Deaf Educ 13: 3–20.

Cantlon JF, Pinel P, Dehaene S et al. (2011). Cortical representations of symbols, objects, and faces are pruned back during early childhood. Cereb Cortex 21: 191–199.

Capek CM, Macsweeney M, Woll B et al. (2008). Cortical circuits for silent speechreading in deaf and hearing people. Neuropsychologia 46: 1233–1241. https://doi.org/10.1016/j.neuropsychologia.2007.11.026.

Caramazza A, Hillis AE (1990). Where do semantic errors come from? Cortex 26: 95–122. https://doi.org/10.1016/s0010-9452(13)80077-9.

Cardin V, Orfanidou E, Rönnberg J et al. (2013). Dissociating cognitive and sensory neural plasticity in human superior temporal cortex. Nat Commun 4: 1473.

Carreiras M, Seghier ML, Baquero S et al. (2009). An anatomical signature for literacy. Nature 461: 983–986. https://doi.org/10.1038/nature08461.

Centanni TM, Norton ES, Ozernov-Palchik O et al. (2019). Disrupted left fusiform response to print in beginning kindergartners is associated with subsequent reading. Neuroimage Clin 22: 101715. https://doi.org/10.1016/j.nicl.2019.101715.

Chee MWL, Weekes B, Lee KM et al. (2000). Overlap and dissociation of semantic processing of Chinese characters, English words, and pictures: evidence from fMRI. Neuroimage 12: 392–403.

Chen Y, Fu S, Iversen SD et al. (2002). Testing for dual brain processing routes in Reading: a direct contrast of Chinese character and pinyin Reading using fMRI. J Cogn Neurosci 14: 1088–1098.

Cheng C-M, Yang M-J (1989). Lateralization in the visual perception of Chinese characters and words. Brain Lang 36: 669–689.

Chrisomalis S (2009). The origins and co-evolution of literacy and numeracy. In: DR Olson, N Torrance (Eds.), The Cambridge handbook of literacy. Cambridge University Press, Cambridge, U.K.

Chyl K, Kossowski B, Debska A et al. (2018). Prereader to beginning reader: changes induced by reading acquisition in print and speech brain networks. J Child Psychol Psychiatry 59: 76–87. https://doi.org/10.1111/jcpp.12774.

Chyl K, Kossowski B, Debska A et al. (2019). Reading acquisition in children: developmental processes and dyslexia-specific effects. J Am Acad Child Adolesc Psychiatry 58: 948–960. https://doi.org/10.1016/j.jaac.2018.11.007.

Chyl K, Fraga-Gonzalez G, Brem S et al. (2021). Brain dynamics of (a) typical reading development-a review of longitudinal studies. NPJ Sci Learn 6: 4. https://doi.org/10.1038/s41539-020-00081-5.

Clark KA, Helland T, Specht K et al. (2014). Neuroanatomical precursors of dyslexia identified from pre-reading through to age 11. Brain 137: 3136–3141. https://doi.org/10.1093/brain/awu229.

Cohen L, Dehaene S (1995). Number processing in *pure alexia*: the effect of hemispheric asymmetries and task demands. Neurocase 1: 121–137. https://doi.org/10.1080/13554799508402356.

Cohen L, Dehaene S (1996). Cerebral networks for number processing: evidence from a case of posterior Callosal lesion. Neurocase 2: 155–174. https://doi.org/10.1080/13554799608402394.

Cohen L, Dehaene S (2000). Calculating without reading: unsuspected residual abilities in *pure alexia*. Cogn Neuropsychol 17: 563–583. https://doi.org/10.1080/02643290050110656.

Cohen L, Dehaene S (2004). Specialization within the ventral stream: the case for the visual word form area. Neuroimage 22: 466–476.

Cohen LG, Celnik P, Pascual-Leone A et al. (1997). Functional relevance of cross-modal plasticity in blind humans. Nature 389: 180–183.

Cohen L, Dehaene S, Naccache L et al. (2000). The visual word form area: spatial and temporal characterization of an initial stage of reading in normal subjects and posterior split-brain patients. Brain 123: 291–307.

Cohen L, Lehericy S, Chochon F et al. (2002). Language-specific tuning of visual cortex? Functional properties of the visual word form area. Brain 125: 1054–1069.

Cohen L, Martinaud O, Lemer C et al. (2003). Visual word recognition in the left and right hemispheres: anatomical and functional correlates of peripheral alexias. Cereb Cortex 13: 1313–1333. https://doi.org/10.1093/cercor/bhg079.

Collins E, Dundas E, Gabay Y et al. (2017). Hemispheric organization in disorders of development. Vis Cogn 25: 416–429. https://doi.org/10.1080/13506285.2017.1370430.

Coltheart M (1980). The semantic error: types and theories. In Deep dyslexia, pp. 146–159.

Coltheart M, Rastle K, Perry C et al. (2001). DRC: a dual route cascaded model of visual word recognition and reading aloud. Psychol Rev 108: 204–256.

Coslett HB, Saffran EM (1998). Reading and the right hemisphere: evidence from acquired dyslexia. In: Right hemisphere language comprehension: perspectives from cognitive neuroscience, pp. 105–132.

Coslett HB, Saffran E (1991). Simultanagnosia. To see but not two see. Brain 114: 1523–1545. https://doi.org/10.1093/brain/114.4.1523.

Crisp J, Lambon Ralph MA (2006). Unlocking the nature of the phonological–deep dyslexia continuum: the keys to reading aloud are in phonology and semantics. J Cogn Neurosci 18: 348–362. https://doi.org/10.1162/jocn.2006.18.3.348.

Cumming TB, Patterson K, Verfaellie M et al. (2006). One bird with two stones: abnormal word length effects in *pure alexia* and semantic dementia. Cogn Neuropsychol 23: 1130–1161. https://doi.org/10.1080/02643290600674143.

Dax M (1865). Lésions de la moitié gauche de l'encéphale coïncidant avec l'oubli des signes de la pensée. Gazette Hebdomadaire de Médecine et de Chirurgie 2: 259–262.

Dehaene S, Cohen L (2007). Cultural recycling of cortical maps. Neuron 56: 384–398.

Dehaene S, Cohen L (2011). The unique role of the visual word form area in reading. Trends Cogn Sci 15: 254–262.

Dehaene S, Le Clec'H G, Poline J-B et al. (2002). The visual word form area: a prelexical representation of visual words in the fusiform gyrus. Neuroreport 13: 321–325.

Dehaene S, Jobert A, Naccache L et al. (2004). Letter binding and invariant recognition of masked words: behavioral and neuroimaging evidence. Psychol Sci 15: 307–313.

Dehaene S, Pegado F, Braga LW et al. (2010). How learning to read changes the cortical networks for vision and language. Science 330: 1359–1364.

Dehaene-Lambertz G, Monzalvo K, Dehaene S (2018). The emergence of the visual word form: longitudinal evolution of category-specific ventral visual areas during reading acquisition. PLoS Biol 16: e2004103.

Dejerine J (1891). Sur un cas de cécité verbale avec agraphie suivi d'autopsie. Mém Soc Biol 3: 197–201.

Dejerine J (1892). Contribution à l'étude anatomo-pathologique et clinique des différentes variétés cécité verbale. C R Hebd Séances Mémoires de la Société de Biologie 4: 61–90.

Endo M, Shimizu A, Hori T (1978). Functional asymmetry of visual fields for Japanese words in kana (syllable-based) writing and random shape-recognition in Japanese subjects. Neuropsychologia 16: 291–297. https://doi.org/10.1016/0028-3932(78)90022-2.

Endo M, Shimizu A, Nakamura I (1981). Laterality differences in recognition of Japanese and Hangul words by mono- linguals and bilinguals. Cortex 17: 391–400.

Exner S (1881). Untersuchungen uber die Localisation der Functionen in der Grosshirnrinde des Menschen, Braumüller, Wien.

Farah M (1990). Visual agnosia: disorders of visual recognition and what they tell us about normal vision, MIT Press, Cambridge.

Fiebach CJ, Friederici AD, Müller K et al. (2002). fMRI evidence for dual routes to the mental lexicon in visual word recognition. J Cogn Neurosci 14: 11–23.

Fiez JA, Petersen SE (1998). Neuroimaging studies of word reading. Proc Natl Acad Sci U S A 95: 914–921.

Files BT, Auer Jr ET, Bernstein LE (2013). The visual mismatch negativity elicited with visual speech stimuli. Front Hum Neurosci 7: 371. https://doi.org/10.3389/fnhum.2013.00371.

Finney EM, Fine I, Dobkins KR (2001). Visual stimuli activate auditory cortex in the deaf. Nat Neurosci 4: 1171–1173.

Forgays DG (1953). The development of differential word recognition. J Exp Psychol 45: 165–168. https://doi.org/10.1037/h0053888.

Friedman RB (1996). Recovery from deep alexia to phonological alexia: points on a continuum. Brain Lang 52: 114–128. https://doi.org/10.1006/brln.1996.0006.

Friedmann N, Gvion A (2001). Letter position dyslexia. Cogn Neuropsychol 18: 673–696. https://doi.org/10.1080/02643290143000051.

Fu S, Chen Y, Smith S et al. (2002). Effects of word form on brain processing of written Chinese. Neuroimage 17: 1538–1548.

Funnell MG, Corballis PM, Gazzaniga MS (2000a). Cortical and subcortical interhemispheric interactions following partial and complete callosotomy. Arch Neurol 57: 185–189. https://doi.org/10.1001/archneur.57.2.185.

Funnell MG, Corballis PM, Gazzaniga MS (2000b). Insights into the functional specificity of the human corpus callosum. Brain 123: 920–926. https://doi.org/10.1093/brain/123.5.920.

Gabay Y, Dundas E, Plaut D et al. (2017). Atypical perceptual processing of faces in developmental dyslexia. Brain Lang 173: 41–51. https://doi.org/10.1016/j.bandl.2017.06.004.

Gaillard R, Naccache L, Pinel P et al. (2006). Direct intracranial, FMRI, and lesion evidence for the causal role of left inferotemporal cortex in reading. Neuron *50*: 191–204. https://doi.org/10.1016/j.neuron.2006.03.031.

Galaburda AM, LeMay M, Kemper TL et al. (1978). Right-left asymmetries in the brain: structural differences between the hemispheres may underlie cerebral dominance. Science 199: 852–856.

Gazzaniga MS, Freedman H (1973). Observations on visual processes after posterior callosal section. Neurology 23: 1126–1130. https://doi.org/10.1212/wnl.23.10.1126.

Gazzaniga MS, Sperry RW (1967). Language after section of the cerebral commissures. Brain 90: 131–148. https://doi.org/10.1093/brain/90.1.131.

Gazzaniga MS, Smylie CS, Baynes K et al. (1984). Profiles of right hemisphere language and speech following brain bisection. Brain Lang 22: 206–220. https://doi.org/10.1016/0093-934x(84)90090-7.

Geffen G, Bradshaw JL, Nettleton NC (1972). Hemispheric asymmetry: verbal and spatial encoding of visual stimuli. J Exp Psychol 95: 25–31. https://doi.org/10.1037/h0033265.

Gerlach C, Kuhn CD, Poulsen M et al. (2022). Lateralization of word and face processing in developmental dyslexia and developmental prosopagnosia. Neuropsychologia 170: 108208. https://doi.org/10.1016/j.neuropsychologia.2022.108208.

Gerrits R, Van der Haegen L, Brysbaert M et al. (2019). Laterality for recognizing written words and faces in the fusiform gyrus covaries with language dominance. Cortex 117: 196–204. https://doi.org/10.1016/j.cortex.2019.03.010.

Geschwind N (1965). Disconnexion syndromes in animals and man. Brain 88: 17–294.

Gibson AR, Dimond SJ, Gazzaniga MS (1972). Left field superiority for word matching. Neuropsychologia 10: 463–466. https://doi.org/10.1016/0028-3932(72)90009-7.

Goodglass H, Lindfield KC, Alexander MP (2000). Semantic capacities of the right hemisphere as seen in two cases of pure word blindness. J Psycholinguist Res 29: 399–422. https://doi.org/10.1023/a:1005155228509.

Hall DA, Humphreys GW, Cooper AC (2001). Neuropsychological evidence for case-specific reading: multi-letter units in visual word recognition. Q J Exp Psychol A 54: 439–467. https://doi.org/10.1080/713755978.

Hall D, Fussell C, Summerfield AQ (2005). Reading fluent speech fromtalking faces: typical brain networks and individual differences. J Cogn Neurosci 17: 939–953. https://doi.org/10.1162/0898929054021175.

Hamilton R, Keenan JP, Catala M et al. (2000). Alexia for braille following bilateral occipital stroke in an early blind woman. Neuroreport 11: 237–240. https://doi.org/10.1097/00001756-200002070-00003.

Hannagan T, Amedi A, Cohen L et al. (2015). Origins of the specialization for letters and numbers in ventral occipitotemporal cortex. Trends Cogn Sci 19: 374–382.

Hardyck C, Tzeng OJL, Wang WS-Y (1978). Cerebral lateralization of function and bilingual decision processes: is thinking lateralized? Brain Lang 5: 56–71.

Hashimoto R, Sakai KL (2004). Learning letters in adulthood: direct visualization of cortical plasticity for forming a new link between orthography and phonology. Neuron 42: 311–322.

Hasson U, Andric M, Atilgan H et al. (2016). Congenital blindness is associated with large-scale reorganization of anatomical networks. Neuroimage 128: 362–372.

Hasuike R, Tzeng O, Hung D (1986). Script effects and cerebral lateralization: the case of chinese characters: language processing in bilinguals. In: J Vaid (Ed.), Psycholoinguistic and neuropsychological perspectives. Lawrence Erlbaum Associates, Hillsdale, New Jersey, pp. 275–288.

Hatta T (1976). Asynchrony of lateral onset as a factor in difference in visual field. Percept Mot Skills 42: 163–166.

Hatta T (1977). Recognition of Japanese *kanji* in the left and right visual fields. Neuropsychologia 15: 685–688.

Hébert S, Cuddy LL (2006). Music-reading deficiencies and the brain. Adv Cogn Psychol 2: 199–206.

Hellige JB, Adamson MM (2007). Hemispheric differences in processing handwritten cursive. Brain Lang 102: 215–227. https://doi.org/10.1016/j.bandl.2006.11.003.

Henderson VW (2008). Alexia and agraphia. *Contrasting perspectives of J.-M. Charcot and J. Hughlings Jackson.* Neurology 70: 391–400. https://doi.org/10.1212/01.wnl.0000298680.47382.61.

Henderson VW (2019). Alexia and Agraphia from 1861 to 1965. Front Neurol Neurosci 44: 39–52. https://doi.org/10.1159/000494951.

Hillis AE, Caramazza A (1995). A framework for interpreting distinct patterns of hemispatial neglect. Neurocase 1: 189–207. https://doi.org/10.1080/13554799508402364.

Hills CS, Pancaroglu R, Duchaine B et al. (2015). Word and text processing in acquired prosopagnosia. Ann Neurol 78: 258–271. https://doi.org/10.1002/ana.24437.

Holmes G (1950). Pure word blindness. Folia Psychiatr Neurol Neurochir Neerl 53: 279–288.

Horie S, Yamasaki T, Okamoto T et al. (2012). Differential roles of spatial frequency on reading processes for ideograms and phonograms: a high-density ERP study. Neurosci Res 72: 68–78.

Horowitz-Kraus T, Wang Y, Plante E et al. (2014). Involvement of the right hemisphere in reading comprehension: a DTI study. Brain Res 1582: 34–44.

Ino T, Nakai R, Azuma T et al. (2009). Recognition and reading aloud of kana and kanji word: an fMRI study. Brain Res Bull 78: 232–239.

Ishiwatari K, Koga Y, Nagata K (2002). Different reading processes in different writing systems-ERP analysis using LORETA. Int Congr Ser 1232: 63–69.

Iwata M (1984). Kanji versus kana, neuropsychological correlates of the Japanese writing system. Trends Neurosci 7: 290–293.

Jibiki I, Yamaguchi N (1993). The Goi (word-meaning) syndrome with impaired kanji processing: alexia with agraphia. Brain Lang 45: 61–69.

Junker FB, Schlaffke L, Lange J et al. (2023). The angular gyrus serves as an interface between the non-lexical reading network and the semantic system: evidence from dynamic causal modeling. Brain Struct Funct. https://doi.org/10.1007/s00429-023-02624-z.

Kershner JR, Jeng AG-R (1972). Dual functional hemispheric asymmetry in visual perception: effects of ocular dominance and postexposural processes. Neuropsychologia 10: 437–445.

Kim KH, Yoon HW, Park HW (2004). Spatiotemporal brain activation pattern during word/picture perception by native Koreans. Neuroreport 15: 1099–1103.

Kim JS, Kanjlia S, Merabet LB et al. (2017). Development of the visual word form area requires visual experience: evidence from blind braille readers. J Neurosci 37: 11495–11504.

Kimura D (1961). Cerebral dominance and the perception of verbal stimuli. Can J Psychol 15: 166–171.

Kingstone A, Gazzaniga M (1995). Subcortical transfer of higher order information: more illusory than real? Neuropsychology 9: 321–328.

Kinsbourne M (1970). The cerebral basis of lateral asymmetries in attention. Acta Psychol (Amst) 33: 193–201. https://doi.org/10.1016/0001-6918(70)90132-0.

Kinsbourne M, Warrington EK (1962). A disorder of simultaneous form perception. Brain 85: 461–486.

Koyama S, Kakigi R, Hoshiyama M et al. (1998). Reading of Japanese Kanji (morphograms) and Kana (syllabograms): a magnetoencephalographic study. Neuropsychologia 36: 83–98.

Koyama MS, Stein JF, Stoodley CJ et al. (2014). A cross-linguistic evaluation of script-specific effects on fMRI lateralization in late second language readers. Front Hum Neurosci *8*: 249. https://doi.org/10.3389/fnhum.2014.00249.

Kraft I, Schreiber J, Cafiero R et al. (2016). Predicting early signs of dyslexia at a preliterate age by combining behavioral assessment with structural MRI. Neuroimage 143: 378–386. https://doi.org/10.1016/j.neuroimage.2016.09.004.

Kronbichler M, Hutzler F, Wimmer H et al. (2004). The visual word form area and the frequency with which words are encountered: evidence from a parametric fMRI study. Neuroimage 21: 946–953.

Kuhl U, Neef NE, Kraft I et al. (2020). The emergence of dyslexia in the developing brain. Neuroimage 211: 116633. https://doi.org/10.1016/j.neuroimage.2020.116633.

Lambon Ralph MA, Hesketh A, Sage K (2004). Implicit recognition in pure alexia: the Saffran effect-a tale of two systems or two procedures? Cogn Neuropsychol 21: 401–421. https://doi.org/10.1080/02643290342000384.

Lane C, Kanjlia S, Richardson H et al. (2017). Reduced left lateralization of language in congenitally blind individuals. J Cogn Neurosci 29: 65–78.

Langer N, Benjamin C, Minas J et al. (2015). The neural correlates of reading fluency deficits in children. Cereb Cortex 25: 1441–1453.

Leff A, Starrfelt R (2014). Alexia: diagnosis, treatment and theory, Springer Science+Business Media, London.

Leff AP, Crewes H, Plant GT et al. (2001). The functional anatomy of single-word reading in patients with hemianopic and *pure alexia*. Brain 124: 510–521. https://doi.org/10.1093/brain/124.3.510.

Leff AP, Spitsyna G, Plant GT et al. (2006). Structural anatomy of pure and hemianopic alexia. J Neurol Neurosurg Psychiatry 77: 1004–1007. https://doi.org/10.1136/jnnp.2005.086983.

Linkersdörfer J, Jurcoane A, Lindberg S et al. (2015). The association between gray matter volume and reading proficiency: a longitudinal study of beginning readers. J Cogn Neurosci 27: 308–318.

Liu Y, Perfetti CA (2003). The time course of brain activity in reading English and Chinese: an ERP study of Chinese bilinguals. Hum Brain Mapp 18: 167–175.

Longcamp M, Hlushchuk Y, Hari R (2011). What differs in visual recognition of handwritten vs. printed letters? An fMRI study. Hum Brain Mapp 32: 1250–1259.

Lyman R, Kwan S, Chao W (1938). Left occipito-parietal brain tumor with observations on alexia and agraphia in Chinese and English. Chin Med J (Engl) 54: 491–516.

MacSweeney M, Amaro E, Calvert GA et al. (2000). Silent speechreading in the absence of scanner noise: an event-related fMRI study. Neuroreport 11: 1729–1733. https://doi.org/10.1097/00001756-200006050-00026.

MacSweeney M, Woll B, Campbell R et al. (2002). Neural systems underlying British sign language and audio-visual English processing in native users. Brain 125: 1583–1593.

Maisog JM, Einbinder ER, Flowers DL et al. (2008). A meta-analysis of functional neuroimaging studies of dyslexia. Ann N Y Acad Sci 1145: 237–259.

Marshall JC, Newcombe F (1973). Patterns of paralexia: a psycholinguistic approach. J Psycholinguist Res 2: 175–199.

Maurer U, Zevin JD, McCandliss BD (2008). Left-lateralized N170 effects of visual expertise in reading: evidence from Japanese syllabic and logographic scripts. J Cogn Neurosci 20: 1878–1891.

Mayall K, Humphreys GW (2002). Presentation and task effects on migration errors in attentional dyslexia. Neuropsychologia 40: 1506–1515. https://doi.org/10.1016/S0028-3932(01)00175-0.

McCandliss BD, Cohen L, Dehaene S (2003). The visual word form area: expertise for reading in the fusiform gyrus. Trends Cogn Sci 7: 293–299.

McGurk H, MacDonald J (1976). Hearing lips and seeing voices. Nature 264: 746–748. https://doi.org/10.1038/264746a0.

Miller MB, Kingstone A (2005). Taking the high road on subcortical transfer. Brain Cogn 57: 162–164. https://doi.org/10.1016/j.bandc.2004.08.037.

Mishkin M, Forgays DG (1952). Word recognition as a function of retinal locus. J Exp Psychol 43: 43–48. https://doi.org/10.1037/h0061361.

Molko N, Cohen L, Mangin JF et al. (2002). Visualizing the neural bases of a disconnection syndrome with diffusion tensor imaging. J Cogn Neurosci 14: 629–636. https://doi.org/10.1162/08989290260045864.

Mongelli V, Dehaene S, Vinckier F et al. (2017). Music and words in the visual cortex: the impact of musical expertise. Cortex 86: 260–274. https://doi.org/10.1016/j.cortex.2016.05.016.

Moore M, Demeyere N (2017). Neglect dyslexia in relation to unilateral visuospatial neglect: a review. AIMS Neuroscience 4: 148–168. https://doi.org/10.3934/Neuroscience.2017.4.148.

Moore MJ, Demeyere N (2019). Neglect dyslexia as a word-centred impairment: a single case study. Cortex 119: 543–554. https://doi.org/10.1016/j.cortex.2018.10.024.

Moore MJ, Milosevich E, Mattingley JB et al. (2023). The neuroanatomy of visuospatial neglect: a systematic review and analysis of lesion-mapping methodology. Neuropsychologia 180: 108470. https://doi.org/10.1016/j.neuropsychologia.2023.108470.

Moulton E, Bouhali F, Monzalvo K et al. (2019). Connectivity between the visual word form area and the parietal lobe improves after the first year of reading instruction: a longitudinal MRI study in children. Brain Struct Funct 224: 1519–1536.

Muayqil T, Davies-Thompson J, Barton JJ (2015). Representation of visual symbols in the visual word processing network. Neuropsychologia 69: 232–241. https://doi.org/10.1016/j.neuropsychologia.
2015.01.045.

Mummery CJ, Patterson K, Price CJ et al. (2000). A voxel-based morphometry study of semantic dementia: relationship between temporal lobe atrophy and semantic memory. Ann Neurol 47: 36–45. https://doi.org/10.1002/1531-8249(200001)47:1<36::AID-ANA8>3.0.CO;2-L.

Nakamura K, Dehaene S, Jobert A et al. (2005a). Subliminal convergence of Kanji and Kana words: further evidence for functional parcellation of the posterior temporal cortex in visual word perception. J Cogn Neurosci 17: 954–968.

Nakamura K, Oga T, Okada T et al. (2005b). Hemispheric asymmetry emerges at distinct parts of the occipitotemporal cortex for objects, logograms and phonograms: a functional MRI study. Neuroimage 28: 521–528.

Nath AR, Beauchamp MS (2012). A neural basis for interindividual differences in the McGurk effect, a multisensory speech illusion. Neuroimage 59: 781–787. https://doi.org/10.1016/j.neuroimage.2011.07.024.

Nelson JR, Liu Y, Fiez J et al. (2009). Assimilation and accommodation patterns in ventral occipitotemporal cortex in learning a second writing system. Hum Brain Mapp 30: 810–820.

Nestor PJ, Fryer TD, Hodges JR (2006). Declarative memory impairments in Alzheimer's disease and semantic dementia. Neuroimage 30: 1010–1020. https://doi.org/10.1016/j.neuroimage.2005.10.008.

Nestor A, Behrmann M, Plaut DC (2013). The neural basis of visual word form processing: a multivariate investigation. Cereb Cortex 23: 1673–1684. https://doi.org/10.1093/cercor/bhs158.

Newman AJ, Supalla T, Fernandez N et al. (2015). Neural systems supporting linguistic structure, linguistic experience, and symbolic communication in sign language and gesture. Proc Natl Acad Sci U S A 112: 11684–11689.

Nobre AC, Allison T, McCarthy G (1994). Word recognition in the human inferior temporal lobe. Nature 372: 260–263.

Olulade OA, Flowers DL, Napoliello EM et al. (2013). Developmental differences for word processing in the ventral stream. Brain Lang 125: 134–145.

Orbach J (1952). Retinal locus as a factor in the recognition of visually perceived words. Am J Psychol 65: 555–562.

Oscar-Berman M, Blumstein S, De Luca D (1976). Iconic recognition of musical symbols in the lateral visual fields. Cortex 12: 241–248. https://doi.org/10.1016/s0010-9452(76)80005-6.

Paradis M, Hagiwara H, Hildebrandt N (1985). Neurolinguistic aspects of the Japanese writing system, Academic Press, New York and London.

Park J, Hebrank A, Polk TA et al. (2012). Neural dissociation of number from letter recognition and its relationship to parietal numerical processing. J Cogn Neurosci 24: 39–50. https://doi.org/10.1162/jocn_a_00085.

Park J, Chiang C, Brannon EM et al. (2014). Experience-dependent hemispheric specialization of letters and numbers is revealed in early visual processing. J Cogn Neurosci 26: 2239–2249. https://doi.org/10.1162/jocn_a_00621.

Pegado F, Nakamura K, Cohen L et al. (2011). Breaking the symmetry: mirror discrimination for single letters but not for pictures in the visual word form area. Neuroimage 55: 742–749.

Perdue MV, Mahaffy K, Vlahcevic K et al. (2022). Reading intervention and neuroplasticity: a systematic review and meta-analysis of brain changes associated with reading intervention. Neurosci Biobehav Rev 132: 465–494. https://doi.org/10.1016/j.neubiorev.2021.11.011.

Peterson RL, Pennington BF (2012). Developmental dyslexia. The Lancet 379: 1997–2007.

Petitto LA, Zatorre RJ, Gauna K et al. (2000). Speech-like cerebral activity in profoundly deaf people processing signed languages: implications for the neural basis of human language. Proc Natl Acad Sci U S A 97: 13961–13966.

Pirozzolo FJ (1977). Lateral asymmetries in visual perception: a review of tachistoscopic visual half-field studies. Percept Mot Skills 45: 695–701. https://doi.org/10.2466/pms.1977.45.3.695.

Plaut DC, Behrmann M (2011). Complementary neural representations for faces and words: a computational exploration. Cogn Neuropsychol 28: 251–275. https://doi.org/10.1080/02643294.2011.609812.

Polk TA, Farah MJ (2002). Functional MRI evidence for an abstract, not perceptual, word-form area. J Exp Psychol Gen 131: 65.

Preston JL, Molfese PJ, Frost SJ et al. (2016). Print-speech convergence predicts future reading outcomes in early readers. Psychol Sci 27: 75–84. https://doi.org/10.1177/0956797615611921.

Price CJ (2012). A review and synthesis of the first 20 years of PET and fMRI studies of heard speech, spoken language and reading. Neuroimage 62: 816–847.

Price CJ, Devlin JT (2003). The myth of the visual word form area. Neuroimage 19: 473–481.

Price CJ, Humphreys GW (1993). Attentional dyslexia: the effect of co-occurring deficits. Cogn Neuropsychol 10: 569–592. https://doi.org/10.1080/02643299308253474.

Price CJ, Wise R, Frackowiak R (1996). Demonstrating the implicit processing of visually presented words and pseudowords. Cereb Cortex 6: 62–70.

Price CJ, Howard D, Patterson K et al. (1998). A functional neuroimaging description of two deep dyslexic patients. J Cogn Neurosci 10: 303–315. https://doi.org/10.1162/089892998562753.

Ptak R, Di Pietro M, Schnider A (2012). The neural correlates of object-centered processing in reading: a lesion study of neglect dyslexia. Neuropsychologia 50: 1142–1150. https://doi.org/10.1016/j.neuropsychologia.2011.09.036.

Qiao E, Vinckier F, Szwed M et al. (2010). Unconsciously deciphering handwriting: subliminal invariance for handwritten words in the visual word form area. Neuroimage 49: 1786–1799.

Ramus F, Rosen S, Dakin SC et al. (2003). Theories of developmental dyslexia: insights from a multiple case study of dyslexic adults. Brain 126: 841–865. https://doi.org/10.1093/brain/awg076.

Rapcsak SZ, Rubens AB, Laguna JF (1990). From letters to words: procedures for word recognition in letter-by-letter reading. Brain Lang 38: 504–514. https://doi.org/10.1016/0093-934X(90)90134-3.

Rapcsak SZ, Beeson PM, Henry ML et al. (2009). Phonological dyslexia and dysgraphia: cognitive mechanisms and neural substrates. Cortex 45: 575–591. https://doi.org/10.1016/j.cortex.2008.04.006.

Rayner K, Johnson RL (2005). Letter-by-letter acquired dyslexia is due to the serial encoding of letters. Psychol Sci 16: 530–534. https://doi.org/10.1111/j.0956-7976.2005.01570.x.

Rayner K, Slattery TJ, Belanger NN (2010). Eye movements, the perceptual span, and reading speed. Psychon Bull Rev 17: 834–839. https://doi.org/10.3758/PBR.17.6.834.

Reich L, Szwed M, Cohen L et al. (2011). A ventral visual stream Reading center independent of visual experience. Curr Biol 21: 363–368.

Rentschler I, Treutwein B, Landis T (1994). Dissociation of local and global processing in visual agnosia. Vision Res 34: 963–971.
Rice GE, Kerry SJ, Robotham RJ et al. (2021). Category-selective deficits are the exception and not the rule: evidence from a case-series of 64 patients with ventral occipito-temporal cortex damage. Cortex 138: 266–281. https://doi.org/10.1016/j.cortex.2021.01.021.
Richlan F (2012). Developmental dyslexia: dysfunction of a left hemisphere reading network. Front Hum Neurosci 6: 120.
Richlan F, Kronbichler M, Wimmer H (2011). Meta-analyzing brain dysfunctions in dyslexic children and adults. Neuroimage 56: 1735–1742. https://doi.org/10.1016/j.neuroimage.2011.02.040.
Ripamonti E, Aggujaro S, Molteni F et al. (2014). The anatomical foundations of acquired reading disorders: a neuropsychological verification of the dual-route model of reading. Brain Lang 134: 44–67. https://doi.org/10.1016/j.bandl.2014.04.001.
Röder B, Rösler F, Neville HJ (2000). Event-related potentials during auditory language processing in congenitally blind and sighted people. Neuropsychologia 38: 1482–1502.
Röder B, Stock O, Bien S et al. (2002). Speech processing activates visual cortex in congenitally blind humans. Eur J Neurosci 16: 930–936.
Rodriguez AR, Barton JJ (2015). The 20/20 patient who can't read. Can J Ophthalmol 50: 257–264. https://doi.org/10.1016/j.jcjo.2015.05.013.
Rossion B, Lochy A (2022). Is human face recognition lateralized to the right hemisphere due to neural competition with left-lateralized visual word recognition? A critical review. Brain Struct Funct 227: 599–629.
Roux FE, Lubrano V, Lotterie JA et al. (2007). When "abegg" is read and ("A, B, E, G, G") is not: a cortical stimulation study of musical score reading. J Neurosurg 106: 1017–1027. https://doi.org/10.3171/jns.2007.106.6.1017.
Roux FE, Draper L, Köpke B et al. (2010). Who actually read Exner? Returning to the source of the frontal "writing centre" hypothesis. Cortex 46: 1204–1210. https://doi.org/10.1016/j.cortex.2010.03.001.
Rubino C, Yeung SC, Barton JJS (2016). The impact of central sparing on the word-length effect in hemianopia. Cogn Neuropsychol 33: 353–361. https://doi.org/10.1080/02643294.2016.1232707.
Ruytjens L, Albers F, van Dijk P et al. (2006). Neural responses to silent lipreading in normal hearing male and female subjects. Eur J Neurosci 24: 1835–1844. https://doi.org/10.1111/j.1460-9568.2006.05072.x.
Sadato N, Pascual-Leone A, Grafman J et al. (1996). Activation of the primary visual cortex by Braille reading in blind subjects. Nature 380: 526–528. https://doi.org/10.1038/380526a0.
Saffran E, Bogyo L, Schwartz M et al. (1980). Does deep dyslexia reflect right-hemisphere reading. In: Deep dyslexia, pp. 381–406.
Sakurai Y (2004). Varieties of alexia from fusiform, posterior inferior temporal and posterior occipital gyrus lesions. Behav Neurol 15: 35–50.
Sakurai Y (2019). Kanji (morphogram) and kana (phonogram) problem in Japanese alexia and agraphia. Front Neurol Neurosci 44: 53–63.
Sakurai Y, Momose T, Iwata M et al. (2000). Different cortical activity in reading of Kanji words, Kana words and Kana nonwords. Cogn Brain Res 9: 111–115.
Sakurai Y, Yagishita A, Goto Y et al. (2006). Fusiform type alexia: *pure alexia* for words in contrast to posterior occipital type *pure alexia* for letters. J Neurol Sci 247: 81–92.
Salis DL (1980). Laterality effects with visual perception of musical chords and dot patterns. Percept Psychophys 28: 284–292. https://doi.org/10.3758/bf03204387.
Salmelin R, Kiesilä P, Uutela K et al. (1996). Impaired visual word processing in dyslexia revealed with magnetoencephalography. Ann Neurol 40: 157–162.
Sasanuma S (1975). Kana and Kanji processing in Japanese aphasics. Brain Lang 2: 369–383.
Sasanuma S, Itoh M, Mori K et al. (1977). Tachistoscopic recognition of *Kana* and *Kanji* words. Neuropsychologia 15: 547–553.
Sato W, Kochiyama T, Uono S et al. (2019). Widespread and lateralized social brain activity for processing dynamic facial expressions. Hum Brain Mapp 40: 3753–3768. https://doi.org/10.1002/hbm.24629.
Saygin ZM, Osher DE, Norton ES et al. (2016). Connectivity precedes function in the development of the visual word form area. Nat Neurosci 19: 1250–1255. https://doi.org/10.1038/nn.4354.
Schotter ER, Angele B, Rayner K (2012). Parafoveal processing in reading. Atten Percept Psychophys 74: 5–35. https://doi.org/10.3758/s13414-011-0219-2.
Segalowitz SJ, Bebout LJ, Lederman SJ (1979). Lateralization for reading musical chords: disentangling symbolic, analytic, and phonological aspects of reading. Brain Lang 8: 315–323. https://doi.org/10.1016/0093-934x(79)90059-2.
Seghier ML, Price CJ (2011). Explaining left lateralization for words in the ventral occipitotemporal cortex. J Neurosci 31: 14745–14753. https://doi.org/10.1523/JNEUROSCI.2238-11.2011.
Sergent J (1987). A new look at the human split brain. Brain 110: 1375–1392. https://doi.org/10.1093/brain/110.5.1375.
Sergent J, Zuck E, Terriah S et al. (1992). Distributed neural network underlying musical sight-reading and keyboard performance. Science 257: 106–109.
Seymour SE, Reuter-Lorenz PA, Gazzaniga MS (1994). The disconnection syndrome. Basic findings reaffirmed. Brain 117: 105–115. https://doi.org/10.1093/brain/117.1.105.
Shallice T (2014). On compensatory strategies and computational models: the case of *pure alexia*. Cogn Neuropsychol 31: 529–543. https://doi.org/10.1080/02643294.2014.914023.
Shallice T, Warrington EK (1977). The possible role of selective attention in acquired dyslexia. Neuropsychologia 15: 31–41. https://doi.org/10.1016/0028-3932(77)90112-9.
Shaywitz BA, Skudlarski P, Holahan JM et al. (2007). Age-related changes in reading systems of dyslexic children. Ann Neurol 61: 363–370. https://doi.org/10.1002/ana.21093.
Sheldon CA, Abegg M, Sekunova A et al. (2012). The word-length effect in acquired alexia, and real and virtual hemianopia. Neuropsychologia 50: 841–851. https://doi.org/10.1016/j.neuropsychologia.2012.01.020.
Shimizu A, Endo M (1981a). Laterality differences in recognition of Kana and Kanji words. Acta Neurol (Napoli) 3: 705–720.
Shimizu A, Endo M (1981b). Tachistoscopic recognition of Kana and Hangul words, handedness and shift of laterality difference. Neuropsychologia 19: 665–673.
Shimizu A, Endo M, Egami M (1981). Does sex influence laterality differences in recognition of visually-presented Kana or Kanji words? Folia Psychiatr Neurol Jpn 35: 417–424.
Shiota J, Kawamura M, Isono O et al. (1986). Alexia with agraphia produced by localized infarction in the inferior posterior region of the left temporal lobe. No To Shinkei 38: 1051–1055.
Shum J, Hermes D, Foster BL et al. (2013). A brain area for visual numerals. J Neurosci 33: 6709–6715. https://doi.org/10.1523/JNEUROSCI.4558-12.2013.
Simos PG, Fletcher JM, Sarkari S et al. (2005). Early development of neurophysiological processes involved in normal reading and reading

disability: a magnetic source imaging study. Neuropsychology 19: 787–798. https://doi.org/10.1037/0894-4105.19.6.787.

Siuda-Krzywicka K, Łukasz B, Paplinska M et al. (2016). Massive cortical reorganization in sighted braille readers. Elife 5: e10762.

Song Y, Hu S, Li X et al. (2010). The role of top-down task context in learning to perceive objects. J Neurosci 30: 9869–9876.

Starrfelt R (2007). Selective alexia and agraphia sparing numbers-a case study. Brain Lang 102: 52–63. https://doi.org/10.1016/j.bandl.2006.09.005.

Starrfelt R, Behrmann M (2011). Number reading in *pure alexia*—a review. Neuropsychologia 49: 2283–2298. https://doi.org/10.1016/j.neuropsychologia.2011.04.028.

Starrfelt R, Habekost T, Leff AP (2009). Too little, too late: reduced visual span and speed characterize *pure alexia*. CerebCortex 19: 2880–2890. https://doi.org/10.1093/cercor/bhp059.

Stewart L, von Kriegstein K, Warren JD et al. (2006). Music and the brain: disorders of musical listening. Brain 129: 2533–2553. https://doi.org/10.1093/brain/awl171.

Sugishita M, Otomo K, Kabe S et al. (1992). A critical appraisal of neuropsychological correlates of Japanese ideogram (KANJI) and phonogram (KANA) reading. Brain 42: 2207–2210.

Sugushita M, Iwata M, Toyokura Y et al. (1978). Reading of ideograms and phonograms in Japanese patients after partial commissurotomy. Neuropsychologia 16: 417–426.

Suzuki K, Yamadori A, Endo K et al. (1998). Dissociation of letter and picture naming resulting from callosal disconnection. Neurology 51: 1390–1394. https://doi.org/10.1212/wnl.51.5.1390.

Tan LH, Spinks JA, Gao J-H et al. (2000). Brain activation in the processing of Chinese characters and words: a functional MRI study. Hum Brain Mapp 10: 16–27.

Tan LH, Laird AR, Li K et al. (2005). Neuroanatomical correlates of phonological processing of Chinese characters and alphabetic words: a meta-analysis. Hum Brain Mapp 25: 83–91.

Tarkiainen A, Cornelissen PL, Salmelin R (2002). Dynamics of visual feature analysis and object-level processing in face versus letter-string perception. Brain 125: 1125–1136.

Taylor KI, Regard M (2003). Language in the right cerebral hemisphere: contributions from reading studies. News Physiol Sci 18: 257–261. https://doi.org/10.1152/nips.01454.2003.

Teng EL, Sperry RW (1973). Interhemispheric interaction during simultaneous bilateral presentation of letters or digits in commissurotomized patients. Neuropsychologia 11: 131–140. https://doi.org/10.1016/0028-3932(73)90001-8.

Tian M, Saccone EJ, Kim JS et al. (2022). Sensory modality and spoken language shape reading network in blind readers of braille. Cereb Cortex 1–15. https://doi.org/10.1093/cercor/bhac216.

Tomasino B, Ius T, Skrap M et al. (2020). Phonological and surface dyslexia in individuals with brain tumors: performance pre-, intra-, immediately post-surgery and at follow-up. Hum Brain Mapp 41: 5015–5031.

Trauzettel-Klosinski S, Brendler K (1998). Eye movements in reading with hemianopic field defects: the significance of clinical parameters. Graefes Arch Clin Exp Ophthalmol 236: 91–102. https://doi.org/10.1007/s004170050048.

Trettenbrein PC, Papitto G, Friederici AD et al. (2021). Functional neuroanatomy of language without speech: an ALE meta-analysis of sign language. Hum Brain Mapp 42: 699–712.

Tzeng O, Hung D, Cotton B et al. (1979). Visual lateralization effect in reading Chinese characters. Nature 282: 499–501.

Vallar G, Calzolari E (2018). Unilateral spatial neglect after posterior parietal damage. Handb Clin Neurol 151: 287–312. https://doi.org/10.1016/b978-0-444-63622-5.00014-0.

Vallar G, Burani C, Arduino LS (2010). Neglect dyslexia: a review of the neuropsychological literature. Exp Brain Res 206: 219–235. https://doi.org/10.1007/s00221-010-2386-0.

Van der Mark S, Bucher K, Maurer U et al. (2009). Children with dyslexia lack multiple specializations along the visual word-form (VWF) system. Neuroimage 47: 1940–1949.

Vinckier F, Naccache L, Papeix C et al. (2006). "What" and "where" in word Reading: ventral coding of written words revealed by parietal atrophy. J Cogn Neurosci 18: 1998–2012. https://doi.org/10.1162/jocn.2006.18.12.1998.

Vinckier F, Dehaene S, Jobert A et al. (2007). Hierarchical coding of letter strings in the ventral stream: dissecting the inner organization of the visual word-form system. Neuron 55: 143–156.

Wang Y, Mauer MV, Raney T et al. (2017). Development of tract-specific white matter pathways during early reading development in at-risk children and typical controls. Cereb Cortex 27: 2469–2485.

Wang F, Karipidis II, Pleisch G et al. (2020). Development of print-speech integration in the brain of beginning readers with varying Reading skills. Front Hum Neurosci 14: 289. https://doi.org/10.3389/fnhum.2020.00289.

Warrington E, Zangwill OL (1957). A study of dyslexia. J Neurol Neurosurg Psychiatry 20: 208–215. https://doi.org/10.1136/jnnp.20.3.208.

Waters D, Campbell R, Capek CM et al. (2007). Fingerspelling, signed language, text and picture processing in deaf native signers: the role of the mid-fusiform gyrus. Neuroimage 35: 1287–1302.

Weiner KS, Zilles K (2016). The anatomical and functional specialization of the fusiform gyrus. Neuropsychologia 83: 48–62. https://doi.org/10.1016/j.neuropsychologia.2015.06.033.

Wong AC, Gauthier I, Woroch B et al. (2005). An early electrophysiological response associated with expertise in letter perception. Cogn Affect Behav Neurosci 5: 306–318.

Woodhead ZV, Penny W, Barnes GR et al. (2013). Reading therapy strengthens top-down connectivity in patients with pure alexia. Brain 136: 2579–2591. https://doi.org/10.1093/brain/awt186.

Woodhead ZV, Barnes GR, Penny W et al. (2014). Reading front to back: MEG evidence for early feedback effects during word recognition. Cereb Cortex 24: 817–825. https://doi.org/10.1093/cercor/bhs365.

Woollams AM (2014). Connectionist neuropsychology: uncovering ultimate causes of acquired dyslexia. Philos Trans R Soc Lond B Biol Sci 369: 20120398. https://doi.org/10.1098/rstb.2012.0398.

Woollams AM, Ralph MA, Plaut DC et al. (2007). SD-squared: on the association between semantic dementia and surface dyslexia. Psychol Rev 114: 316–339. https://doi.org/10.1037/0033-295X.114.2.316.

Xue G, Chen C, Jin Z et al. (2006). Cerebral asymmetry in the fusiform areas predicted the efficiency of learning a new writing system. J Cogn Neurosci 18: 923–931.

Yamada Y, Stevens C, Dow M et al. (2011). Emergence of the neural network for reading in five-year-old beginning readers of different levels of pre-literacy abilities: an fMRI study. Neuroimage 57: 704–713. https://doi.org/10.1016/j.neuroimage.2010.10.057.

Yeatman JD, Dougherty RF, Ben-Shachar M et al. (2012). Development of white matter and reading skills. Proc Natl Acad Sci U S A 109: E3045–E3053. https://doi.org/10.1073/pnas.1206792109.

Yoncheva YN, Blau VC, Maurer U et al. (2010). Attentional focus during learning impacts N170 ERP responses to an artificial script. Dev Neuropsychol 35: 423–445.

Zaidel E, Peters AM (1981). Phonological encoding and ideographic reading by the disconnected right hemisphere: two case studies. Brain Lang 14: 205–234. https://doi.org/10.1016/0093-934x(81)90077-8.

Zhou W, Pang W, Zhang L et al. (2020). Altered connectivity of the visual word form area in the low-vision population: a resting-state fMRI study. Neuropsychologia 137: 107302.

Handbook of Clinical Neurology, Vol. 208 (3rd series)
Cerebral Asymmetries
C. Papagno and P. Corballis, Editors
https://doi.org/10.1016/B978-0-443-15646-5.00021-X

Chapter 21

Functional and structural brain asymmetries in sign language processing

PATRICK C. TRETTENBREIN[1,2,3*], EMILIANO ZACCARELLA[1], AND ANGELA D. FRIEDERICI[1]

[1]*Department of Neuropsychology, Max Planck Institute for Human Cognitive and Brain Sciences, Leipzig, Germany*

[2]*International Max Planck Research School on Neuroscience of Communication: Structure, Function, and Plasticity (IMPRS NeuroCom), Leipzig, Germany*

[3]*Experimental Sign Language Laboratory (SignLab), Department of German Philology, University of Göttingen, Göttingen, Germany*

Abstract

The capacity for language constitutes a cornerstone of human cognition and distinguishes our species from other animals. Research in the cognitive sciences has demonstrated that this capacity is not bound to speech but can also be externalized in the form of sign language. Sign languages are the naturally occurring languages of the deaf and rely on movements and configurations of hands, arms, face, and torso in space. This chapter reviews the functional and structural organization of the neural substrates of sign language, as identified by neuroimaging research over the past decades. Most aspects of sign language processing in adult deaf signers markedly mirror the well-known, functional left-lateralization of spoken and written language. However, both hemispheres exhibit a certain equipotentiality for processing linguistic information and the right hemisphere seems to specifically support processing of some constructions unique to the signed modality. Crucially, the so-called "core language network" in the left hemisphere constitutes a functional and structural asymmetry in typically developed deaf and hearing populations alike: This network is (i) pivotal for processing complex syntax independent of the modality of language use, (ii) matures in accordance with a genetically determined biologic matrix, and (iii) may have constituted an evolutionary prerequisite for the emergence of the human capacity for language.

INTRODUCTION

The capacity for language serves as a cornerstone of human cognition and is one of the key properties that distinguishes our species from other animals. The first known attempts to establish a possible neural basis for this capacity date back to the 19th century and were based on the observations of Marc Dax, who determined that damage to the left hemisphere of the brain would lead to either a disturbance of speech or impaired language abilities (Dax, 1863, 1865). This was later followed by the often much more widely known work of Broca (1861) suggesting that especially lesions to the left inferior frontal gyrus of the cortex affected expressive language. This cortical region is nowadays known as Broca's area, even though a reexamination of the conserved brains of some of Broca's patients using modern neuroimaging has shown that their lesions extended into neighboring regions and the white matter tracts originating from them (Dronkers et al., 2007). Ample lesion studies in combination with the advent of neuroimaging have since confirmed Dax' and Broca's very first notions that certain aspects of speech and language processing are asymmetrically organized in the adult human brain (see Trettenbrein and Friederici, 2025, for a review based primarily on studies of spoken and written language).

Since the 1960s, research in the cognitive sciences has accumulated evidence which strongly suggests that the human-specific language capacity (Hauser et al., 2002; Bolhuis et al., 2014) is not bound to speech but may be perceived and externalized in different modalities (Stokoe, 1960; Klima et al., 1979; Chomsky, 1986a): auditory-oral for spoken language, visuospatial for sign language, and tactuo-spatial for tactile sign language. Sign languages are the natural languages of the deaf and, just like spoken languages, exhibit complex linguistic organization on different levels such as phonology or syntax (Sandler and Lillo-Martin, 2008; Mathur and Rathmann, 2014; Cecchetto, 2017). (Note: In English, uppercase "Deaf" is sometimes used to refer to people, who identify as members of a linguistic and cultural

*Correspondence to: Patrick C. Trettenbrein, Department of Neuropsychology, Max Planck Institute for Human Cognitive and Brain Sciences, Stephanstraße 1a, 04103 Leipzig, Germany. Tel: +49-341-99402625, E-mail: trettenbrein@cbs.mpg.de

minority, lowercase "deaf" is then seen to refer to people's audiologic status. Following Caselli et al. (2017), we use lowercase "deaf" throughout here because we discuss studies with diverse groups of participants from laboratories around the world, who may differ in whether and how they identify with the respective community.) Intriguingly, emerging research on tactile sign languages of the deafblind has revealed a similarly complex linguistic organization (Checchetto et al., 2018; Edwards and Brentari, 2020). In contrast to speech, sign, and tacticle sign, the different writing systems for spoken languages constitute cultural inventions that have to be acquired by explicit instruction. Written language only emerged relatively recently (about 6000 years for pictographic and ideographic systems and about 3000 years for alphabetic systems), which makes it unlikely that the human brain has evolutionarily adapted to reading and writing as it has for language processing as such (Dehaene et al., 2005; Dehaene and Cohen, 2007; Friederici, 2017).

First neuropsychologic evidence for a left-lateralization of sign language, similar to spoken language, came from case studies of deaf signers, who exhibited a variety of aphasic symptoms in their signing following lesions to the left but not the right hemisphere (Klima et al., 1979; Poizner et al., 1987; Bellugi et al., 1989; Hickok et al., 1996a, 1998; Corina, 1998). For example, Poizner et al. (1987) documented case studies of two deaf signers with agrammatic and paragrammatic behavioral profiles following left-hemispheric lesions to the perisylvian cortex. Similarly, mirroring the picture of Broca's classic cases for speech production, Hickok et al. (1996b) report a deaf signer with a lesion to the left inferior frontal gyrus extending into neighboring regions and white matter, who exhibited acute expressive aphasia. These data are complemented by a case study of a hearing bimodal bilingual that underwent presurgical Wada testing (i.e., left intracarotid injection of Amytal to render left-hemispheric language areas inoperative), which led to a temporary aphasia in both American Sign Language (ASL) and spoken English (Damasio et al., 1986). Collectively, these studies provided the first, suggestive evidence that the left hemisphere of the human brain may not be specialized for processing the fast temporal structure of speech (Zatorre et al., 2002; Schönwiesner et al., 2005), but is specialized for processing language independent of modality.

In this chapter, we take these early neuropsychologic observations about the lateralization of language across different modalities as a starting point for a review of the functional and structural lateralization of the neural substrates of sign language as revealed by the past decades of research using a range of neuroimaging methods. Throughout the discussion, we will assume that readers are already familiar with the fundamentals of the functional and structural neuroanatomy of spoken and written language processing (for an overview see, e.g., Friederici, 2017; Hagoort, 2017). While we aim to incorporate findings using a variety of neuroimaging methods, our focus here will be on studies using (functional) magnetic resonance imaging ([f]MRI). We start by briefly considering the nature of language and its relation to the brain in general, including a primer on the linguistic analysis of sign languages, as well as sign language acquisition. Next, we review case studies of sign language aphasia before laying out the functional and structural neural basis for processing sign language in the brains of adult deaf signers (i.e., individuals with congenital or early onset hearing loss, who primarily use a sign language to communicate). We end with a summarizing discussion as well as an attempt to sketch a model of the functional neuroanatomy of sign language processing.

What is "language"?

Different from everyday use, modern linguistics distinguishes between *language* as a neurally implemented cognitive capacity and individual *languages* such as, for example, English, Japanese Sign Language (JSL), and Nahuatl. *Language* as a cognitive capacity enables humans to assemble individual lexical items (i.e., a technical notion from linguistics roughly corresponding to "words" that also includes signs) into structured representations of utterances, which are mapped to the sensorimotor system for production and comprehension and the conceptual-intentional system for thought (Chomsky, 1965, 1986b, 1995; Everaert et al., 2015; Friederici et al., 2017). In a technical sense, the human language system therefore enables a structured mapping from meaning to a physical signal (i.e., signs or sound) and vice versa (Chomsky, 2011). This architecture of *language* enables humans to flexibly produce and comprehend an in principle infinite number of different utterances, which they have never encountered before. At the same time, this cognitive architecture also accounts for why *language* can readily be externalized in different modalities and why individual *languages*, regardless of whether they are spoken or signed, differ in their phonology, lexicon, and grammar: The specifics of the mapping to the sensorimotor system appear not to be fixed in human biology but instead develop during language acquisition (Lenneberg, 1964; Crain et al., 2016; Yang et al., 2017).

The representations generated by the human language system are evolutionarily distinctive because they are not merely sequences of lexical items but instead specify the relations between individual lexical items hierarchically, a property that seems to be shared across modalities (spoken, signed, etc.) but has not yet been observed in the communication systems of other animals (Berwick et al., 2013; Everaert et al., 2015; Friederici et al., 2017). The German Sign Language (DGS) sentence schematically depicted in Fig. 21.1 (top panel) can be used to illustrate two different ways in which the grammatical relationship between the individual signs making up a sentence can be analyzed linguistically: We might assume that the signs are merely a sequence of lexical items, which can be represented as a flat structure in which all elements in the diagram are connected directly to the root (Fig. 21.1, bottom-left panel). Alternatively, we may assume that lexical items are grouped into constituents, which are embedded in each other hierarchically (Fig. 21.1, bottom-right panel). Significantly, despite the fact that spoken languages and sign languages are perceived and produced in radically different modalities, linguistic analysis has demonstrated that they exhibit deep structural similarities on several levels of description, including the hierarchical syntactic

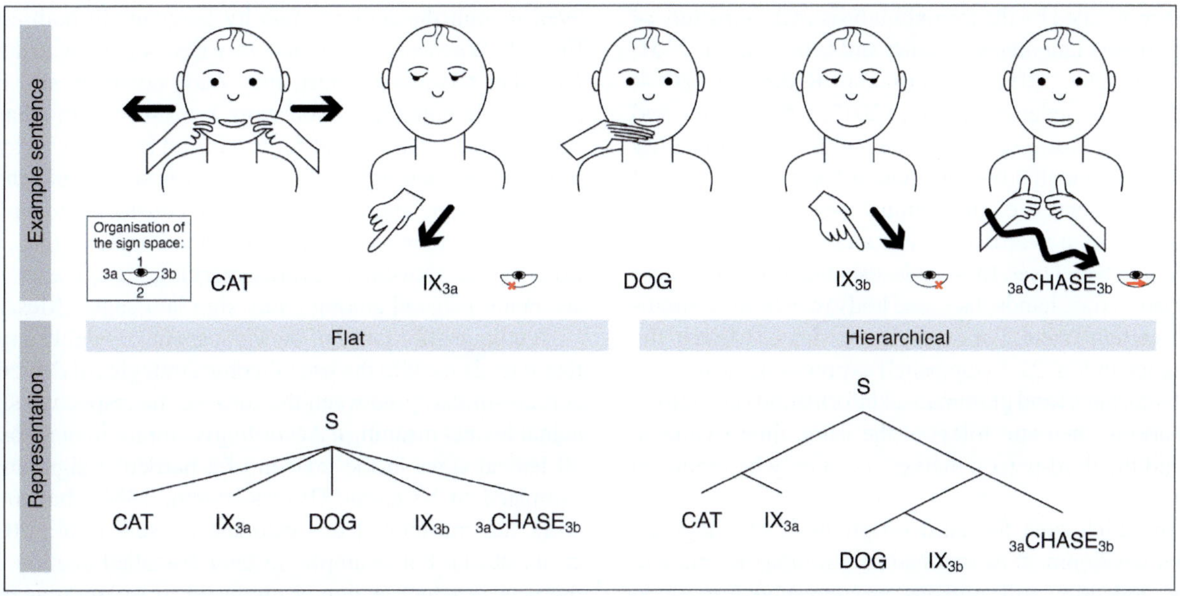

Fig. 21.1. Example sentence in German sign language (DGS) and possible syntactic representations. The top panel shows a schematic depiction of an example sentence in German Sign Language (DGS), which roughly translates into English as "The cat chased the dog." The schematics indicate handshape, location, and movement of manual signs, known nonmanual components, as well as mouthings and are accompanied by English glosses. Glossing amounts to writing one language (i.e., DGS) using another (i.e., English) and is therefore necessarily an imprecise yet widely used tool in sign language linguistics (see Pfau et al., 2012 for an in-depth explanation of notational conventions). Index signs are glossed using the abbreviation IX and locations in sign space are given as subscripts for the respective signs, indicating start (and endpoint) of a syntax-relevant movement in sign space. Notice that the sentence in this example primarily serves illustrative purposes; in natural conversations, indexing can also occur by producing a sign in the respective location in sign space and is also frequently at least partially or even completely omitted if relying only on word order patterns [the default word order for DGS is subject-object-verb (SOV)] may be sufficient for expressing a proposition (Mathur and Rathmann, 2012; Pfau et al., 2018). The bottom panel illustrates two possible ways in which the syntactic structure of this example sentence may be represented in linguistic analysis: The bottom-left panel depicts an analysis of the sentence of a flat structure in which all elements in the tree diagram are directly connected to the root. In contrast, the bottom-right panel shows an analysis of the sentence highlighting its hierarchical structure, where lexical items combine to form constituents, which can be embedded in each other. While research on spoken and written language has, by now, established that human brains automatically attribute hierarchical structure to a linguistic signal despite the fact that this structure is not contained in the signal itself (see, e.g., Friederici et al., 2017; Zaccarella and Trettenbrein, 2021), only a handful of studies have investigated syntactic processing in sign language so far (e.g., MacSweeney et al., 2006; Stroh et al., 2019; Matchin et al., 2021), which is why the relevance of hierarchical structure during sign language processing requires further psycho- and neurolinguistic research. However, formal linguistic analyses strongly suggests that sign languages, just like spoken languages, rely on such hierarchical representations (Sandler and Lillo-Martin, 2001, 2008; Tang and Lau, 2012; Mathur and Rathmann, 2014; Cecchetto, 2017).

structure of grammatical representations (Sandler and Lillo-Martin, 2001, 2008; Tang and Lau, 2012; Mathur and Rathmann, 2014; Cecchetto, 2017).

The, at least, more than 200 different sign languages of the deaf around the world (Hammarström et al., 2022) as well as the various tactile sign languages of the deafblind, are testament to the contemporary understanding among cognitive scientists that the capacity for language is not bound to speech, but that the neurocognitive machinery for language flexibly interfaces with the sensorimotor system depending on individual developmental circumstances. Accordingly, the emerging picture in the cognitive sciences has been that the language system essentially enables the generation of hierarchically structured syntactic representations linking meaning (i.e., semantics) to a physical signal (i.e., sign or sound; respectively, representations thereof) and vice versa, whereas this capacity is rooted in a universal computational mechanism that is biologically determined, implemented neurally, and specific to humans (Lenneberg, 1969; Hauser et al., 2002; Berwick et al., 2013; Friederici et al., 2017). That is, the "biologic matrix" (Lenneberg, 1967), which determines and, at the same time, also constrains the cognitive specifics and development of the language system from infancy into adulthood (Lenneberg, 1964, 1967, 1969; Crain et al., 2016; Yang et al., 2017) appears not to specify the modality in which linguistic information can be perceived and externalized (Lenneberg, 1970; Klima et al., 1979; Poizner et al., 1987; Meier, 2016).

Once this perspective on the language system is adopted, it is not surprising that sign languages and spoken languages share deep similarities on the level of formal description but, at the same time, also exhibit features, which specifically reflect the different constraints (or lack thereof) imposed by the respective modality in which language is externalized and perceived (i.e., visuospatial for sign language and auditory-oral for spoken language). For example, the hierarchical grouping of lexical items into constituents is motivated by the formal analysis of the structure of both sign and spoken languages (Sandler and Lillo-Martin, 2001, 2008; Tang and Lau, 2012; Mathur and Rathmann, 2014; Cecchetto, 2017; Friederici et al., 2017; Zaccarella and Trettenbrein, 2021). However, while the auditory-oral modality imposes a strictly sequential organization of lexical items as well as morphologic elements (i.e., the minimal meaning-bearing elements that make up a word, e.g., *eat* and *-s* in *eats*), these constrains seem to be relaxed for the visuospatial modality where several articulators (i.e., hands, face, and body) can be used simultaneously to a certain extent. For example, the sign CHASE in the example sentence in Fig. 21.1 (top panel) expresses lexical information (i.e., handshape) and grammatical information (i.e., movement path marking thematic roles) at the same time (while in English *eat* and the third-person marker *-s* can only be produced sequentially).

The way in which such formal descriptions of the language system and its development as outlined here ultimately relate to neurobiology generally still remains an unresolved issue in cognitive neuroscience, whereas researchers working on sign language face some additional challenges. Like other areas of cognitive neuroscience, the neuroscience of sign language faces the issue that there is no established linking theory for mapping the basic elements of linguistic theory (e.g., lexical items, syntactic operations, etc.) to the basic elements of neuroscience (e.g., cell assemblies, brain regions, etc.; Poeppel and Embick, 2013; Embick and Poeppel, 2015). Accordingly, the vast majority of studies in the cognitive neuroscience of language in any modality are correlational in nature and their interpretation depends greatly on the way in which formal constructs from linguistics have been operationalized in an experiment (van der Burght et al., 2023). In addition, because sign languages are minority languages primarily used by the respective deaf communities, they are frequently understudied in comparison to the primarily Western spoken languages commonly used in experiments (Henrich et al., 2010). As a consequence, relatively little previous research and formal analyses of many phenomena may be available (Baker et al., 2016; Quer and Steinbach, 2019) and resources that are routinely used in research on spoken languages (e.g., corpus data) are currently not available for the vast majority of sign languages.

A PRIMER ON SIGN LANGUAGE LINGUISTICS

As already alluded to above, the study of sign language over the past decades and the therefrom emerging understanding that sign languages are fully fledged natural languages with complex organization on all levels of linguistic analysis have revolutionized how cognitive scientists attempt to understand the human capacity for language. Pioneering work carried out from the 1960s of the past century onward (e.g., Stokoe, 1960; Klima et al., 1979) demonstrated that sign languages such as, for example, ASL are not in any way "translations" of the spoken majority language in a region (i.e., English in case of the United States) into a gestural system but instead have their own and independent grammar as well as lexicon. This is best demonstrated by the fact that ASL and British Sign Language (BSL) are not mutually intelligible, even though the spoken majority language in both countries is English. When an ASL signer and a BSL signer, who do not know the other's language meet, they may, however, still be able to make basic communication work by resorting to what is sometimes called International Sign. This ad hoc sign system used in international settings heavily relies on signs from the signers' respective sign languages that can be considered to have a strong iconic motivation as well as grammatical features (e.g., modifications of path movement overtly marking the agreement of verbs; see below) shared among many sign languages (Mesch, 2010).

A unique affordance of the visuospatial modality appears to be that it lends itself to the use of iconic strategies of depiction (i.e., a certain similarity between the form of the respective sign and its actual lexical meaning). Accordingly, approximately one-third of all lexical signs in the lexicon of a particular sign language are estimated to be iconic (Boyes-Braem, 1986), but more recent work has questioned this estimate (Caselli et al., 2017; Sehyr et al., 2021). For example, in their so-called citation form (i.e., the way in which a sign is produced in isolation as opposed to a sentence or discourse context), the DGS signs in Fig. 21.2 (top-right panel) are clearly iconic: The sign TREE depicts the ground as well as stem and crown of a tree, whereas the sign for SLEEP depicts the head of the signer being laid down on a cushion. Because the latter sign overlaps with a similar-looking gesture for "sleeping," its meaning is rather transparent also to nonsigners (Trettenbrein et al., 2021b). Despite its clear iconic motivation this is not true for the sign TREE, because comprehending the sign requires knowledge of the structured mapping from the strategy of iconic depiction to the lexical meaning (Emmorey, 2014; McGarry et al., 2023). Importantly, iconic strategies differ between sign languages: Whereas the DGS sign iconically depicts a tree with stem and crown using the arm and hand, Chinese Sign Language depicts only the outline of the stem (Bellugi and Klima, 1976).

While sign language relies on a set of articulators that is radically different from those used for spoken language, signs and spoken words nevertheless exhibit familiar internal organizational principles. The visuospatial modality allows for the simultaneous use of different articulators (i.e., hands, face, and body), which can combine in different ways to articulate a sign and convey linguistic information: Usually, a distinction between manual and nonmanual components of signs is made, which can be accompanied by so-called mouthings or mouth gestures (Fig. 21.2, left panel). Just like lexical items and morphemes in spoken languages can be broken down into individual phonemes (i.e., speech sounds), the manual components of individual signs are also not holistic units but consist of sub-lexical parameters (i.e., phonemes; Stokoe, 1960; Brentari, 2012; Baker et al., 2016; Fenlon et al., 2017): These parameters include hand form, hand position, contact area, and movement (from the starting point to end point). Sign languages differ with regard to the phonemes they use. The relevance of sub-lexical parameters for sign formation is evidenced by the existence of minimal pairs (i.e., words, which differ only in one phonologic parameter such as *grammar* and *glamour* in English) in sign languages. For example, the DGS signs TEACHER and LECTURER differ in one

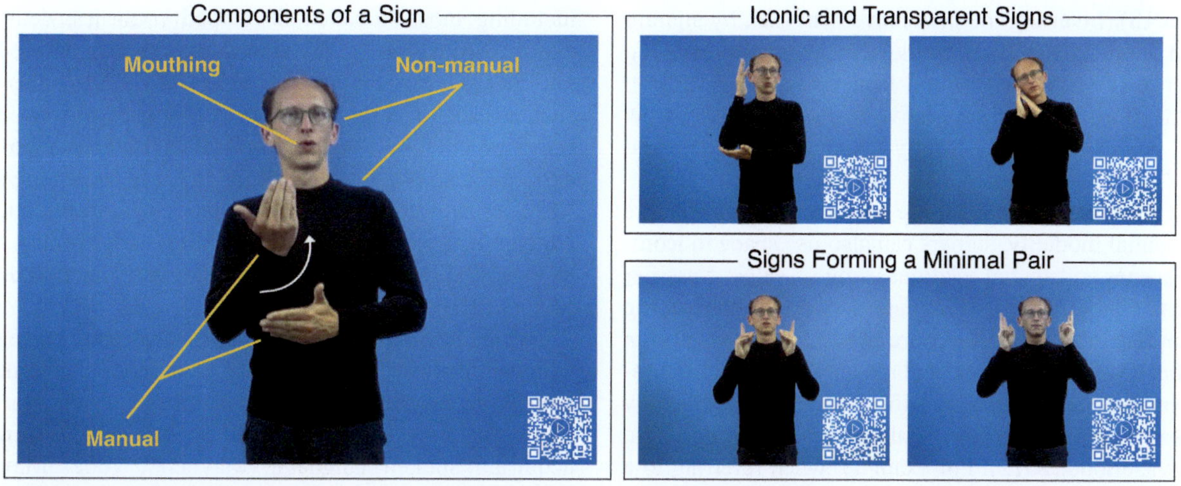

Fig. 21.2. Linguistic analysis of individual signs. The left panel shows the different manual, and nonmanual components of a sign. In general, non-manual components include the direction of gaze, movement of eyebrows, movement of head and torso, facial expression, mouthings [i.e., forming (part of) a spoken word with the lips], and mouth gestures. Manual components include the hand form, hand position, contact area, and movement (starting point to end point). The example shown here is the DGS sign NEW (video: https://doi.org/10.6084/m9.figshare.21711794), which is signed with flat hands with the nondominant (left) hand steady in front of the body, whereas the dominant (right) performs an upward movement behind the nondominant hand. These manual components are accompanied by a mouthing roughly corresponding to the spoken German nɔɪ̯ as well as momentarily raised eyebrows. The right top panel shows the DGS signs TREE (left; video https://doi.org/10.6084/m9.figshare.21712619) and SLEEP (right; video: https://doi.org/10.6084/m9.figshare.21713123), which are both iconic signs yet have different levels of transparency. While the iconic motivation of the sign TREE (i.e., the dominant right hand depicting the stem whereas the fingers represent the treetop) is obvious once the meaning of the sign is revealed; the meaning of the sign is, nevertheless, accurately identified by only 10% of non-signing participants in a guessing task (Trettenbrein et al., 2021b), rendering the sign nontransparent to nonsigners despite its iconic motivation. In contrast, the iconic motivation for the lexical sign SLEEP overlaps with a gesture also frequently used by nonsigners to indicate "sleep," "sleeping," or "being tired," thereby rendering the sign highly transparent also to nonsigners. The right-bottom panel shows the DGS signs LECTURER (left; video https://doi.org/10.6084/m9.figshare.21711575) and TEACHER (right; https://doi.org/10.6084/m9.figshare.21708107), which form a so-called minimal pair, as their manual components differ only in hand form but share the hand position, and movement from the starting to the end point. In addition, the lexical distinction may be emphasized when both signs are accompanied by different mouthings. Representative frames of signs in this illustration and the accompanying videos available online via a QR code or DOI have been adapted from a publicly available dataset by Trettenbrein et al. (2021b) under a Creative Commons (CC-BY 4.0) license to which this illustration is also subject.

manual parameter, namely hand form (Fig. 21.2, bottom-right panel; notice though that, in many cases, these DGS signs will also be accompanied by different mouthings).

Besides phonology, sign languages also exhibit a rich morphologic structure, which closely interacts with syntax (i.e., the grammatical structure of the entire sentence). In the visuospatial modality, morphologic processes can be realized in two ways: First, one or more phonemes of a sign may be altered (e.g., the DGS sign TREE depicted in Fig. 21.2 may be produced with a slight left-to-right movement to mark plurality and change the meaning to TREES or FOREST). Such alterations of the movement path, in combination with repetitions of (part) of a sign, yield complex and language-specific systems of derivational morphology (Bellugi and Klima, 1976; Bellugi et al., 1989). Second, an affix may be added to a sign to derive a related, yet different meaning (e.g., in the DGS sign SCIENCE + PERSON a lexicalized person marker is suffixed to the sign SCIENCE to derive the meaning "scientist"). Yet, because producing signs sequentially generally takes longer than producing speech sounds (Klima et al., 1979), affixation appears to be relatively rare in sign languages. That is, the signing rate is generally lower than the speaking rate when only sequentially produced manual signs are considered. However, the simultaneous morphology and articulation of signs ultimately results in an overall similar information density for sign and speech (Klima et al., 1979).

Moving to the level of sentences, many sign languages including DGS can express grammatical relations using a combination of two different syntactic strategies: The relation of lexical elements in a sentence may either be marked by relying on word order [subject-object-verb (SOV) being the default word order for DGS, different from the subject-verb-object (SVO) order of German] or via the grammatical use of sign space. For example, the proposition "the cat is chasing the dog" can be expressed by signing CAT DOG CHASE. In this case, the order in which signs are produced indicates their grammatical role in the sentence (i.e., CAT is the subject and DOG is the object). Alternatively, the same proposition can be expressed using the spatial grammar that is unique to the visuospatial modality (Fig. 21.1, top panel): In the example sentence, subject and object are first assigned positions in sign space (the sign CAT location 3a and the sign DOG location 3b) and the sign CHASE is then produced in morphologically modified form (glossed as ${}_{3a}$CHASE${}_{3b}$) where the start and end of the altered movement path indicate the position of the subject and object in sign space (Mathur and Rathmann, 2012;

Pfau et al., 2018). Inverting the path movement (i.e., by signing the inflected form $_{3b}$CHASE$_{3a}$) would also invert the meaning of the sentence (i.e., the dog would then chase the cat). Sign languages therefore provide suggestive evidence for the view that lexical processes are inherently syntactic (Krauska and Lau, 2023).

Lastly, besides the grammatical use of space, which is unique to the visuospatial modality, signers can also use space to iconically depict scenes and spatial relations using so-called classifier constructions. Classifiers are morphemes with a nonspecific meaning expressed using the hands, are attested in almost all sign languages that have been studied, and use a particular configuration of the hand to represent a specific category of entities by depicting a salient property shared by all entities belonging to that category (Zwitserlood, 2012). For example, when referring to a car slowly driving along a road, in a classifier construction a deaf signer using DGS would use a flat handshape (i.e., ) with the palm pointing downward. This is a classifier for vehicles with four wheels that can then be used to iconically depict the slow movement in sign space (i.e., the hand would be moving forward slowly). However, if the signer was referring to a motorcycle they would use the same handshape but with the palm pointing sideways, a classifier that is used in DGS for all vehicles, which have only two wheels. Classifiers are lexicalized and differ between sign languages (e.g., ASL uses the  handshape for vehicles regardless of the number of wheels; Zwitserlood, 2012; Baker et al., 2016).

Interim summary

Sign languages are natural languages with complex organization on all levels of linguistic analysis (e.g., phonology, morphology, and syntax) which are externalized in the visuospatial modality. This indicates that the mapping of the hierarchically structured representations of the language system to the sensorimotor system is not biologically specified but emerges during development. The principally iconic motivation of many signs, grammatical use of space, as well as flexible use of classifiers in sign space are affordances specific to the visuospatial modality.

SIGN LANGUAGE ACQUISITION AND DEVELOPMENT

The language experience of deaf children is highly variable and this developmental variability is also reflected both behaviorally and neurally. Language acquisition in the deaf population depends on several factors: (i) the degree and age of onset of hearing loss (i.e., either congenital, during early childhood, or later), (ii) the use of aided technology (if desired and suitable), and (iii) the language primarily used by a child's caregivers. It has been estimated that 90%–95% of infants with profound hearing loss are born to hearing parents, who do not know any sign language (Mitchell and Karchmer, 2004; Meier, 2016). Consequently, the number of deaf children, who naturally acquire the sign language used by their parents, is remarkably small with the vast majority of the deaf population exhibiting heavily varying language acquisition profiles. Indeed, in large parts of the world, including most so-called developed countries, deaf children's exposure to a sign language is still frequently delayed—a situation that is completely preventable (Mayberry, 2002). Crucially, such a lack of early access to a linguistic system has lasting developmental consequences: Deaf individuals who were only exposed to an established sign language in late childhood or adolescence show considerable limitations in grammatical knowledge and performance but not nonlinguistic abilities (Emmorey et al., 1995a; Mayberry, 2002; Mayberry and Squires, 2006; Meier, 2016).

Typically developing children of deaf parents, regardless of whether they are deaf or hearing, undergo the same developmental milestones during sign language acquisition that have previously been identified based on studies of hearing children acquiring spoken language (Mayberry, 2002; Mayberry and Squires, 2006; Meier, 2016). This similarity is evidenced by the near-identical growth curves observed during early development of the productive vocabulary of a group of deaf children acquiring BSL and a larger cohort of hearing children acquiring British English (Fig. 21.3, left panel). The development of deaf children's syntactic abilities also proceeds in similar stages and at a similar pace like that of their hearing peers (Fig. 21.3, right panel), despite the fact that common measures like the mean length of utterance are prone to underestimate the complexity of signed utterances (Terrace et al., 1979). While recent research suggests that lexical development during the early phase of sign language acquisition of a deaf child born to hearing parents can be aided by the parent's rudimentary learning and usage of the local sign language (Berger et al., 2023), early exposure to a sign language in its full form is, nevertheless, crucial for the typical development of the language system (Mayberry, 2002; Meier, 2016).

The deep similarities between the development of the language system in deaf children acquiring a sign language and hearing children acquiring a spoken language indicate that language constitutes a modality-independent capacity for manipulating abstract units and rules, but this does not imply that there are no effects of modality. A widely known example is the observation that deaf children, on average, seem to produce their first signs several months before their hearing peers produce their first words (Newport and Meier, 1985; Meier, 2016). This is predicted by the fact that motoric systems relevant for producing signs mature before those for controlling speech, in combination with the greater perspicuity and recognizability of signs to caretakers (Newport and Meier, 1985). However, even in cases where one might expect an impact of modality, for example, due to an overlap between sign and gesture, the independent maturation of the language system can be observed: Even though the ASL signs for the pronouns ME and YOU superficially resemble paralinguistic gestures (i.e., pointing with the index finger at oneself or a second person), which are commonly used prelinguistically by deaf and hearing children already at 9 months of age, deaf children undergo a later developmental phase during language acquisition in which they systematically confuse these pronominal signs at the same age when hearing children acquiring a spoken language make similar pronoun-reversal errors (Petitto, 1987).

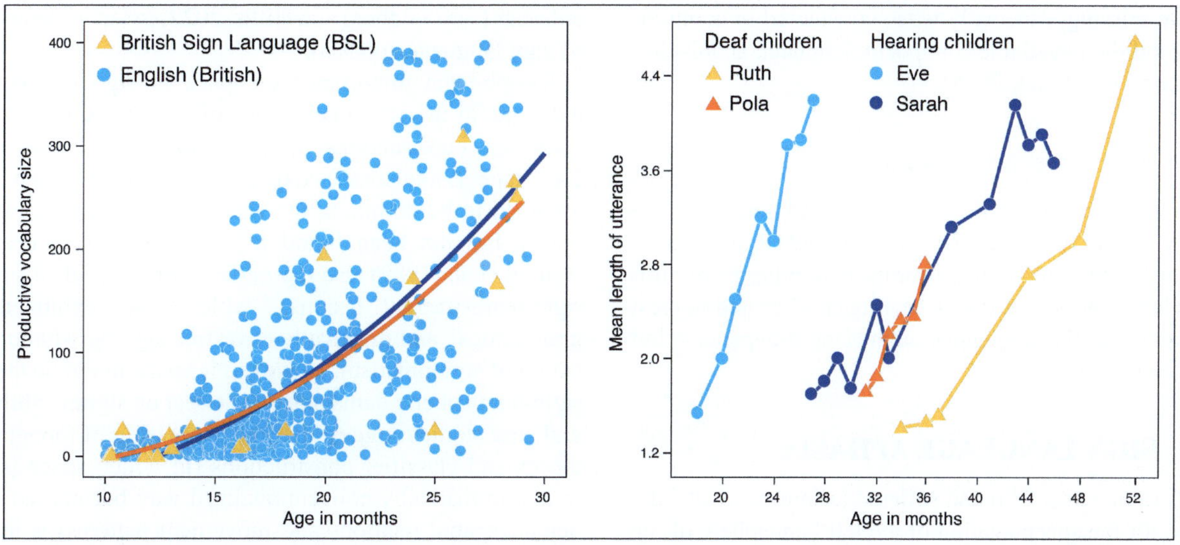

Fig. 21.3. Lexical and syntactic development in deaf children acquiring a sign language. Two plots showing growth curves for the lexical (left panel) and syntactic (right panel) development of deaf children acquiring a sign language in comparison to hearing peers acquiring a spoken language, both measured by analyzing produced utterances. The graph in the left panel shows the development of the productive vocabulary size for a group of typically developing deaf children acquiring British Sign Language (BSL) (color-coded in orange) in comparison to a group of typically developing hearing children acquiring British English (color-coded in blue). The figure clearly indicates that the size of the productive vocabulary grows at a similar speed in both groups. The illustration is based upon data freely available from the Wordbank database as of June 20, 2023 (Frank et al., 2017); specifically the graph shows quadratic regression models fit to the BSL data (Wordbank identifier "WG") by Woolfe et al. (2010) as well as the data for British English (Wordbank identifier "Oxford CDI") by Floccia (2017), which includes data from Hamilton et al. (2000). The right panel shows the development of mean length of utterance (MLU) of two deaf children acquiring American Sign Language (ASL) in comparison to two hearing children acquiring American English. These data are reproduced here from Terrace et al. (1979) in an adapted form and should be treated with some caution because MLU was measured here based on definitions developed for research on spoken languages, as the authors of the original paper also remark in a footnote. While spoken utterances are often broken down not just into words but also sequentially produced morphemes (e.g., *drive there* vs *drive-ing*), many sign languages (e.g., ASL or DGS) will rely on simultaneity to express the same propositions using a single morphologically modified sign (see section "A Primer on Sign Language Linguistics").

Additional evidence for the notion that humans are born with a modality-independent propensity for symbolic communication (Lenneberg, 1964) comes from studies uncovering the generative and rule-based nature of so-called homesign systems developed by deaf children, whose hearing parents have not exposed them to a sign language (Goldin-Meadow, 2003, 2017; Goldin-Meadow and Yang, 2017; Flaherty et al., 2021). Deaf children, who do not have access to a conventionalized sign language, have been found to develop such homesign systems derived ad hoc from iconic gestures. That is, these children develop individual homesign systems, which exhibit many of the properties of natural languages (e.g., morphologic patterns, basic syntactic structure of utterances, etc.), despite the fact that these properties are not contained in the gestural input they receive from caretakers (Goldin-Meadow, 2003, 2017; Flaherty et al., 2021). Analyzing the statistical profile of observed grammatical patterns in homesign confirms that, while homesign systems fall short of the conventionalized lexical and syntactic complexity of natural languages, they are nevertheless based on productive grammars (Goldin-Meadow and Yang, 2017). This strongly suggests that children bring a structural bias to the process of language acquisition, which determines them to impose discrete units that are combined in a rule-based fashion onto any kind of potentially linguistic input.

Whereas the cognitive prerequisites and developmental milestones of sign language acquisition in deaf children have, as of now, been studied for decades and with veritable success, the neural basis of sign language during infancy and childhood and its development remain severely understudied. A relatively recent study by Payne et al. (2019) is an interesting exception to this general lack of neural data: The authors used functional transcranial Doppler sonography (fTCD), a method that measures cerebral perfusion changes in response to neural activation, in a sample of 19 typically developing deaf children, who were on average 8 years old, and found that language production was significantly left-lateralized, independent of the modality of language use (i.e., whether the children were producing BSL, spoken English, or a combination of the two). The authors conclude that the lateralization of language in the tested sample reflects a left-hemispheric specialization for processing linguistic representations. While this interpretation is in line with the data of typically developed deaf adults, as we will discuss below, studies of how this usually left-lateralized pattern for (sign) language processing emerges in deaf signers during (sign) language acquisition are missing, especially when considered in comparison to the neural data available for spoken language acquisition (Trettenbrein and Friederici, 2025).

Interim summary

Most deaf children are born to hearing parents and their language experience is highly variable. The developmental milestones during sign language acquisition mirror those familiar from studies of spoken language acquisition in hearing children and suggest that the maturation of the language system is input-dependent but modality-independent.

SIGN LANGUAGE APHASIA

While reports of patients, who had suffered lesions, provided the first evidence for the specialization of the left hemisphere of the human brain for spoken language processing already more than 150 years ago (Broca, 1861; Dax, 1863, 1865), the first detailed reports of deaf signers with aphasia following left-hemispheric brain damage surfaced only in the 1980s of the past century (e.g., Poizner et al., 1987; Bellugi et al., 1989). Until then, it was still commonly assumed that the left hemisphere and especially the posterior portion of the left inferior frontal gyrus (i.e., Broca's area) are specialized for processing speech. A case report by Damasio et al. (1986) about a hearing, bimodal, bilingual patient provided the first evidence for a left-hemispheric specialization for language and not just speech: The patient was a native speaker of English but had acquired ASL for professional reasons, when she was 18 years of age, and used ASL daily for her work. An examination of the patient's linguistic performance, while undergoing a so-called Wada test (i.e., the injection of barbiturate into the left carotis), revealed that the patient exhibited a marked aphasia in both English and ASL when left-hemispheric, language-relevant areas of the cortex were rendered inoperative by the procedure.

This first suggestive evidence for a general specialization of the left hemisphere for language independent of modality, based on data from a hearing late signer, was soon complemented by detailed and linguistically sophisticated case studies of deaf signers with left-hemispheric damage, who exhibited different impairments in their signing depending on the location of the lesion within the perisylvian regions of the left hemisphere (Poizner et al., 1987; Bellugi et al., 1989; Hickok et al., 1996b). For example, Bellugi et al. (1989) reported data from a deaf patient (GD), who had suffered a left frontal lesion comprising most of the frontal lobe, including Broca's area, in the inferior frontal gyrus. After the stroke, the patient's signing exhibited clear signs of agrammatism as it was generally slow and effortful, consisted mostly of one-sign utterances, and lacked the grammaticalized morphologic and syntactic markings of ASL. Generally speaking, the aphasic syndromes observed in deaf signers with left-hemispheric damage mirror those known from research with hearing users of a spoken language insofar as lesions affecting frontal perisylvian regions and white matter cause non-fluent aphasia, whereas lesions to posterior perisylvian regions lead to fluent aphasias (see Goldberg and Hillis, 2022, for an in-depth treatment of sign language aphasia).

Despite the importance of space in sign language as, for example, in the grammatical use of sign space (Fig. 21.1) or in classifier constructions (see section "A Primer on Sign Language Linguistics"), damage to the RH, which is widely assumed to be dominant for visuospatial processing (Corballis, 2003), has not been found to cause aphasia in deaf signers. Bellugi et al. (1989) also report data from three deaf signers with right-hemispheric lesions, which all exhibited fluent, grammatical, and practically error-free signing without any indication of agrammatism. However, it seems not to be the case that right-hemispheric damage has no effect on signing abilities: Several case studies suggest that the processing of topographic sentences and classifier constructions (in which space is not used in a grammatically conventionalized way but instead iconically depicts spatial relations and movement patterns) is impaired in deaf signers following right-hemispheric lesions (Hickok et al., 1996c; Atkinson et al., 2005). These observations are in line with data from so-called split-brain patients indicating that the left hemisphere retains relatively sophisticated visuospatial abilities (Corballis, 2003), which can support the processing of spatial aspects of sign language in the context of generating syntactic representations (i.e., grammatical use of sign space) but not the generation of spatial representations iconically depicting spatial relations (Bellugi et al., 1989; Emmorey et al., 1995b; Hickok et al., 1996c; Atkinson et al., 2005).

In contrast to sign language aphasias following damage to left but not right perisylvian regions, two independent case studies have documented a unique pattern of sign language aphasia following lesions to the left occipital cortex. Hickok et al. (1995) documented a case of a deaf signer with a lesion spanning the left medial temporal and occipital lobe, the left occipital pole, and the white matter giving rise to the splenium of the corpus callosum. Behaviorally, the patient presented with right hemianopia (i.e., loss of half of the visual field), alexia (i.e., inability to recognize written words), and a severe inability to comprehend signs in spite of seemingly intact sign production with the exception of occasional paraphasic errors (i.e., incorrect substitution of a sign or a part of it). The authors interpreted this as a case of so-called "sign blindness" where the spared visual areas of the RH were disconnected from the spared left perisylvian language regions by the lesion. Saito et al. (2007) describe a similar case of a deaf signer with a lesion to the left occipital lobe, medial temporal lobe and hippocampus, as well as the corpus callosum. Interestingly, this patient presented also with severely impaired sign production abilities in addition to impaired comprehension. In sum, these cases (i) show that the RH is unable to fully support sign language comprehension and (ii) suggest that posterior regions of the left hemisphere may undergo some plastic changes, which make them more language-relevant as the result of deafness and/or sign language acquisition.

Given that both hands are usually used as the primary articulators during sign language production, the observed lateralization patterns for sign language aphasia seem to confirm the dissociation between the primarily left-hemispheric processing of linguistic representations generated by the language system as opposed to the respective modality-specific (i.e., speech or sign) processing of the bilateral production system during externalization. That is, even though sign language imposes articulatory demands that are radically different from those of spoken language, processing sign language seems to primarily involve the left hemisphere for processing linguistic representations independent of whether the left or right hand is being used as the primary articulator during sign production (Corina et al., 2003). In this context, it is also interesting to note that Gutiérrez-Sigut et al. (2015) compared spoken and sign language production in a group of hearing, native, bimodal bilingual users of English and BSL and found that the left-hemispheric lateralization was actually more pronounced for producing signs in comparison to producing speech. Because the lateralization was not primarily driven by motor (i.e., production) factors, the authors argue that the visuospatial modality imposes increased processing demands with regard to somatosensory feedback and the simultaneous processing of phonologic parameters (i.e., handshape, location, and path movement; Gutiérrez-Sigut et al., 2015).

Additional evidence for a dissociation of the language system from the sensorimotor system as well as other cognitive systems used during the externalization of language and communication stems from the observation that the production of signs and the production of gestures can be differentially impaired in aphasic signers despite both being produced in the same modality. A suggestive case study by Corina et al. (1992) describes a deaf signer with a marked aphasia disrupting comprehension and production of ASL following left-hemispheric damage, while their ability to communicate using nonlinguistic gestures was mostly preserved. The patient would substitute lexical signs with pantomimic gestures iconically related to the object such as, for example, miming the bouncing of a ball instead of using the lexical ASL sign, BALL. Interestingly, these substitutions were produced regardless of the iconicity of the target sign, hinting at the lexicalized status even of highly iconic signs (also see Marshall et al., 2004, for a similar case report). Likewise, Bellugi et al. (1989) found that all three aphasic deaf signers with left-hemispheric damage showed strong dissociations between their linguistic, nonlinguistic gestural, and general motoric capacities. As of now, there is also neuroimaging evidence indicating that the processing of gestures as opposed to lexicalized signs differentially engages perisylvian regions (Newman et al., 2015), thereby supporting the idea that sign and gesture rely on distinct representations.

Lastly, independent of the interesting data from the variety of case studies discussed here it must also be mentioned that researchers and clinicians working with deaf patients with aphasic symptoms frequently face the problem that suitable diagnostic tools may simply not be available for the particular sign language under study, a subsisting issue that also has serious consequences for the quality of care that deaf patients may receive. In such cases, researchers have to create a suitable test either by adapting an existing standardized aphasia test battery for a spoken language to the sign language under study (e.g., the adaptation of the Boston Diagnostic Aphasia Examination to ASL available from the Salk Institute; https://lcn.salk.edu/ASLNew/SAT.htm) or by creating a suitable test battery de novo (e.g., the BSL Aphasia Assessment Battery developed by the Deafness Cognition and Language Research Centre; https://dcalportal.org/tests/aab). For nonlinguistic assessment, researchers or clinicians, who are not proficient in the respective sign language used by the participant, may opt to work closely with interpreters and rely on available tests standardized using data from the hearing, nonsigning population. However, tests that depend on knowledge of the spoken majority language used in the respective environment are likely not appropriate (Miranda et al., 2022).

Interim summary

In adult signers, the left hemisphere is usually dominant for sign language processing. The patterns of documented aphasic syndromes in deaf signers with damage to different left perisylvian regions align with those known from studies of hearing aphasics. In contrast, lesions to the left occipital cortex that also affect the splenium of the corpus callosum may cause primarily receptive aphasia in deaf signers by disconnecting left perisylvian regions from the visual system. Moreover, deaf signers' difficulties with processing classifier constructions following right-hemispheric lesions constitute a modality-specific syndrome.

FUNCTIONAL NEUROIMAGING STUDIES OF SIGN LANGUAGE

While case studies of deaf signers with aphasia provided important, first insights into the neural basis of sign language, the advent of neuroimaging in the 1990s of the past century brought about the first, functional neuroimaging studies of sign language processing in deaf signers using noninvasive methods such as electroencephalography (EEG), positron emission tomography (PET), and functional magnetic resonance imaging (fMRI). Whereas the methodological foundations and respective limitations of these research methods apply to sign language research in the same way as they apply to research on spoken and written language, the specifics of the visuospatial modality nevertheless require adaptations such as, for example, the reliance on videos for presenting stimulus materials and a variety of associated potential issues for presentation as well as analysis (Capek and Neville, 2015). We will begin by briefly reviewing EEG studies of sign language processing in deaf signers in comparison to insights from similar studies in hearing users of spoken languages, but will then focus on studies of the neural basis of sign language processing using PET and fMRI as these are potentially more informative in the context of lateralization. Studies carried out with different functional methods [e.g., magnetoencephalography (MEG)] or different populations (e.g., hearing signers) will be considered wherever relevant.

Electrophysiologic studies

The major event-related potentials (ERPs) observed during sign language comprehension parallel those reported for auditory language comprehension (Fig. 21.4, top panel). Early ERPs peaking

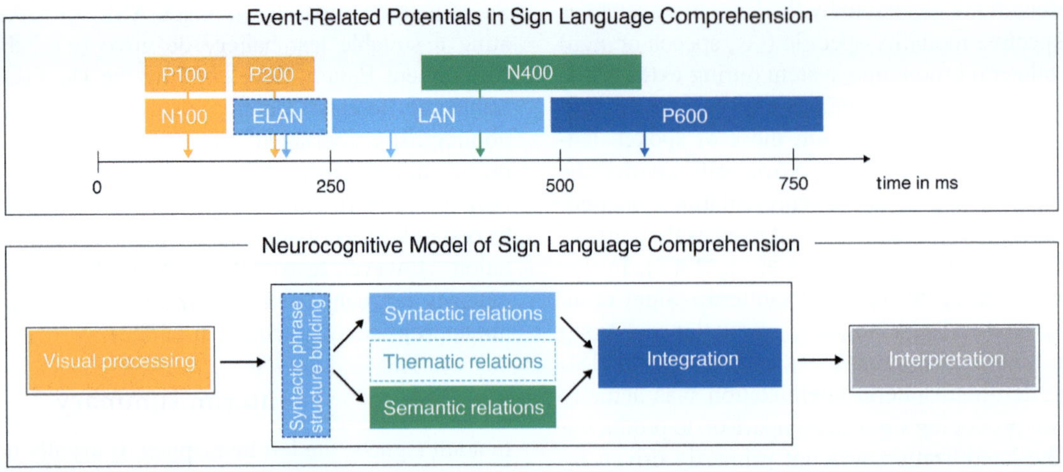

Fig. 21.4. Event-related potentials and a neurocognitive model of sign language comprehension. The top panel depicts the time course of ERPs that have been observed in studies of sign language comprehension. The colored boxes indicate different ERPs, whereas the width of boxes roughly indicates associated time-windows reported for different components in the literature. Arrows for each box indicate the canonical peak of a component in milliseconds (ms). Notice that part of the reported variability with regard to the timing of peaks reported for different components may reflect researchers' degrees of freedom (Luck and Gaspelin, 2017) with regard to the time-locking of their analyses: ERPs can be time-locked to the onset of stimulus videos, the actual onset of the first sign (Emmorey et al., 2022), or the so-called recognition or uniqueness point. However, determining the recognition or uniqueness point is complicated by the fact that signers use transitional movements between signs as cues during comprehension (Hosemann et al., 2013). Abbreviations used in the top panel: positivity (P), negativity (N), early left anterior negativity (ELAN), and left anterior negativity (LAN). The bottom panel describes the major consecutive processing steps of a tentative neurocognitive model of sign language comprehension. Different processing stages are indicated by different boxes. Arrows depict the temporal relationship between processing stages. The color coding is consistent between the two panels and links the different ERPs to their assumed processing stage: Visual processing is depicted in orange, syntax-related processing in light blue, semantic processing in green, and integration and repair processes in dark blue. The diagram in the bottom panel is loosely based on the neurocognitive model for auditory language comprehension depicted in Friederici (2017).

at around 100 ms (i.e., P100 and N100) and 200 ms (i.e., N200) have been attributed to visual processing (Emmorey et al., 2022). Yet, some studies report differential topographic distributions of ERPs related to open and closed class signs (e.g., nouns vs pronouns) already as early as 100 ms (Neville et al., 1997). During the processing of sign language sentences, an early left anterior negativity (ELAN) peaking at 140–200 ms assumed to reflect syntactic phrase structure building (Capek et al., 2009; Friederici, 2011) and (left) anterior negativities in the time-window of 250–600 ms related to the manipulation of syntactic information (Capek et al., 2009; Hänel-Faulhaber et al., 2014) have been reported. Following semantic manipulations on the sentence- and single-sign level, negativities peaking at approximately 400 ms (i.e., N400) have been observed (Gutiérrez et al., 2012b; Hosemann et al., 2013; Hänel-Faulhaber et al., 2014), whereas their amplitude and latency can be modulated by priming (Gutiérrez et al., 2012a; Emmorey et al., 2022) and psycholinguistic variables such as lexical frequency (Emmorey et al., 2020). Lastly, late positivities assumed to reflect integration and repair processes occur after 500 ms (i.e., the so-called post-N400 positivities or P600s; Neville et al., 1997; Capek et al., 2009; Hänel-Faulhaber et al., 2014).

Interpreting the topographic distribution of ERPs observed during sign language processing has proven difficult because (i) EEG generally allows only coarse and approximate estimation of lateralization due to the so-called "inverse problem" (Grech et al., 2008), and (ii) some language-related components and their corresponding canonical topographies have actually not consistently been observed. In their study of violations of so-called agreement verbs (see Fig. 21.1) in ASL, Capek et al. (2009) found an ELAN when the agreement between subject and object was reversed (a manipulation possible due to the SVO word order of ASL) and a more right-lateralized anterior negativity when the agreement relation was unspecified (i.e., the verb agreed with a locus in space that had not previously been assigned). Hänel-Faulhaber et al. (2014) employed another violation of verb agreement where signs did not agree with either the subject or the object but moved from a neutral point in sign space toward the signer. Interestingly, they did not observe an ELAN but a relatively late LAN (400–600 ms). Lastly, using unspecified agreement violations in DGS, Hosemann et al. (2018) observed neither an ELAN, nor a LAN, nor a P600, but two independent effects (one mostly right posterior positivity at 220–570 ms, followed by a left anterior effect at 300–600 ms).

This diversity of ERP findings related to sign language comprehension indicates that a first model of the time course of differ rent processing steps sketched in Fig. 21.4 (bottom panel) will require refinement in the future, including adjustments reflecting possible modality-specific aspects (Gutiérrez et al., 2012b). A possible explanation for the observed variation is that all sign language studies relating to ELAN and LAN to date have used manipulations of agreement verbs: This likely constitutes a confound because the spatial nature of sign language agreement requires the recruitment of additional cognitive resources. Moreover, almost all paradigms mentioned above using agreement manipulations relied on morphologic alteration of the verb sign's movement parameter (e.g., moving in sign space from point 3a to 2 instead of from 3a to 3b; see Fig. 21.1). Recall, however, that the ELAN in auditory language comprehension is elicited by word category violations (Friederici, 2011, 2017). For example, the words *he butterfly* cannot be integrated into a phrase but *her butterfly* can (Maran et al., 2022a,b). Yet, verb agreement violations in sign language do not necessarily always also constitute word category violations, which block local phrase structure building, because alternative and less likely yet nevertheless grammatical interpretations of an utterance may be possible (for extended discussion, also see Hosemann et al., 2018).

Magnetic resonance and radionuclide imaging studies

The first fMRI and PET studies of sign language were carried out in the late 1990s of the past century and yielded conflicting results which, on the one hand, confirmed the stronger involvement of the left hemisphere and left perisylvian regions during sign language processing similar to spoken language but, on the other hand, also implicated right-hemispheric perisylvian regions as well as regions not typically reported in studies of spoken language processing. One of the first direct comparisons between hearing non-signers, deaf signers, and hearing signers processing videos in ASL revealed that both deaf and hearing signers showed activation in left perisylvian regions as well as their right-hemispheric homologs (Bavelier et al., 1998; Neville et al., 1998). This pattern was absent in all groups when processing written English. A study by Söderfeldt et al. (1997) was among the first to report the largest response to sign language in the bilateral visual association cortex (BA 37 and BA 19) in a sample of so-called children of deaf adults (CODAs), who were native signers, regions that are not usually observed in studies of spoken language processing. Lastly, studies by McGuire et al. (1997) and Petitto et al. (2000) suggested the primary involvement of left perisylvian regions in sign language processing, overlapping with frontal and temporal sites previously identified in studies of spoken and written language processing.

A recent meta-analysis of PET and fMRI studies of sign language comprehension in deaf signers confirmed the larger left-hemispheric involvement especially in left frontal regions, but also found extensive right-hemispheric convergence across all 23 neuroimaging studies in their dataset (Trettenbrein et al., 2021a; Fig. 21.5). Specifically, the authors observed seven clusters of convergence mass across studies in (1) Broca's area in the left inferior frontal gyrus (IFG) with peaks in both anatomic subregions, the anterior Brodmann area (BA) 45 and the more posterior BA 44; (2) left precentral and middle frontal gyrus (with peaks in BA 6 and BA 8); (3) left anterior insula; (4) left middle occipital gyrus (BA 19) bordering middle temporal gyrus; (5) right pars triangularis (i.e., BA 45); (6) right superior temporal gyrus (STG; BA 22); and (7) posterior portions of right middle- and inferior temporal gyrus spanning into the occipital gyrus (BA 37). Notice that the studies included in the dataset varied with regard to the experimental paradigms (e.g., semantic judgments, etc.), baseline conditions (e.g., fixation cross, videos of reversed signing, etc.), and stimulus complexity (e.g., single signs, sentences, etc.). Accordingly, while this analysis identified brain regions that are engaged by deaf signers during sign language comprehension, the functional characteristics of these regions remain underspecified.

To better characterize the observed pattern of convergence across studies of sign language processing, Trettenbrein et al. (2021a) performed two additional analyses which revealed that (i) the convergence mass was generally left-lateralized and especially so within Broca's area and (ii) part of the convergence mass

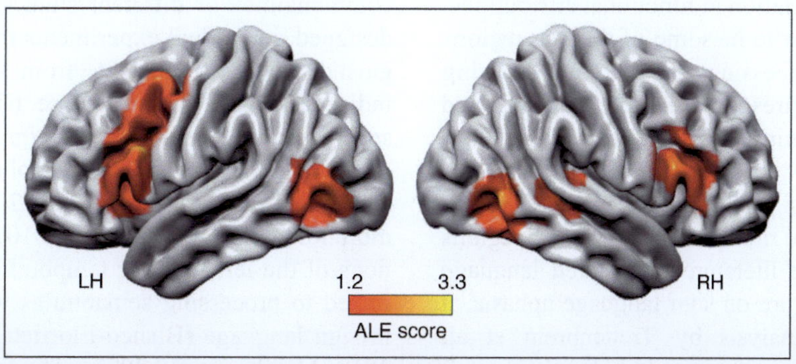

Fig. 21.5. Meta-analysis of sign language comprehension. Meta-analytical convergence across different fMRI and PET studies of significant clusters for sign language comprehension over baseline contrasts from 23 different neuroimaging studies with deaf signers and (total number of participants = 316). Color bar at the bottom indicates the Activation Likelihood Estimation (ALE; Eickhoff et al., 2009, 2012, 2017; Turkeltaub et al., 2012) score of any given voxel, which represents the degree of nonrandom convergence in activation between contrasts in the dataset. In the left hemisphere (LH), convergence was observed in the inferior frontal gyrus, middle occipital gyrus, precentral gyrus, middle frontal gyrus, and insula (ordered by descending size of clusters). In the RH, convergence was observed in middle and inferior temporal gyrus extending into occipital cortex, inferior frontal gyrus, and superior temporal gyrus. Based on data from Trettenbrein PC, Papitto G, Friederici AD, et al. (2021a). Functional neuroanatomy of language without speech: an ALE meta-analysis of sign language. Hum Brain Mapp 42: 699–712. https://doi.org/10.1002/hbm.25254.

in Broca's area but not its right-hemispheric homolog was specific to sign language processing in deaf signers and overlapped with spoken and written language processing in hearing non-signers. In a first step, the authors computed weighted lateralization indices using hemispheric and anatomic masks of Broca's area, which revealed a slight global left-lateralization [lateralization index (LI): 0.24, P 0.001] as well as a strong left-lateralization in Broca's area and it's right-hemispheric homolog (LI: 0.68, P 0.001). The left lateralization was especially pronounced in the posterior portion of Broca's area (BA 44, LI: 0.78, P 0.001). In a second step, the authors contrasted their meta-analysis of sign language comprehension studies with an independent meta-analysis of studies in which hearing participants observed sign-like actions. This revealed that only a cluster in Broca's area (peak in BA 44), the right superior temporal gyrus (peaks in BA 22) and left middle frontal and precentral gyrus (peaks in BA 6, BA 44, and BA 8) were uniquely involved in sign language processing but not in processing sign-like actions.

Based on the three clusters in IFG, right STG, and left middle frontal and precentral gyrus identified as sign-language specific by the contrast analysis with an independent meta-analysis of the processing of sign-like actions, Trettenbrein et al. (2021a) used the BrainMap database to characterize the functional associations of these clusters using a purely data-driven approach. This analysis revealed that voxels within the sign-language specific cluster in Broca's area were associated with studies that investigated cognition and, more specifically, with studies of language processing in hearing nonsigners across the several thousand studies listed in the database (Fig. 21.6). Broca's area can, therefore, be considered a modality-independent hub for language processing that is involved in processing linguistic information both in deaf signers as well as in hearing users of a spoken language. The sign-language-specific right STG cluster as well as left middle frontal and precentral cluster were also mostly associated with different aspects of language and speech processing in hearing nonsigners in the BrainMap database, though to a lesser extent and with larger variation in functional attributions. These regions, therefore, appear to be some of the key regions that deaf signers recruit for processing visual stimuli showing manual, facial, and bodily gestures that carry conventionalized linguistic structure and meaning in their respective sign language.

Because an overlap between sign and spoken language processing in the left inferior frontal and right posterior temporal regions is expected on the basis of the literature for spoken language processing as well as the literature on sign language aphasia, it is surprising that the meta-analysis by Trettenbrein et al. (2021a) did not observe convergence across studies in the left posterior temporal cortex, which houses Wernicke's area as another major hub for language processing. The involvement of posterior STG, posterior superior temporal sulcus (STS), and posterior middle temporal gyrus (MTG) in sign language processing is well documented by individual studies (Bavelier et al., 1998; Neville et al., 1998; San José-Robertson et al., 2004; MacSweeney et al., 2006; Moreno et al., 2018; Matchin et al., 2021). A possible explanation for this lack of convergence in the meta-analysis is that the different articulators (Fig. 21.2, left panel) were not used uniformly across studies in the data set: Stimuli including mouthings or mouth gestures have been found to activate the middle portion of the superior temporal cortex, whereas stimulus videos using only manual actions activate more posterior and inferior parts of lateral temporal lobe (Capek et al., 2008). The left and right posterior temporal cortex may, therefore, not subserve the same functional role during sign language processing, similar to the usually left-hemispheric linguistic advantage in spoken language processing (Vigneau et al., 2011; Wilson et al., 2023).

Similarly, given the important role of space in most sign languages, it is also surprising that Trettenbrein et al. (2021a) did not observe any convergence in either the left or right parietal cortex, after all a region, which is known to generally support spatial processing and has been found relevant for encoding phonologic information in sign languages (Cardin et al., 2013, 2016). The most likely explanation for this lack of convergence across studies in the meta-analysis is that involvement of parietal regions in sign language processing is, in many cases, only observed when the experimental paradigm explicitly taps into a property of sign language that requires the kind of information processing for which parietal regions seem to be specialized (Söderfeldt et al., 1997). For example, in a study of ASL locative classifier constructions, Emmorey et al. (2021) found bilateral involvement of the superior parietal lobule. Additional studies have found involvement of the left parietal cortex in processing topographic sentences (MacSweeney et al., 2002a), as well as bilateral involvement of the supramarginal gyrus and right superior parietal lobule for the processing of classifier constructions that use space iconically (Emmorey et al., 2002; Jednoróg et al., 2015). Thus, the neuroimaging literature as well as lesion studies (Bellugi et al., 1989; Emmorey et al., 1995b; Hickok et al., 1996c; Atkinson et al., 2005) demonstrate the relevance of parietal regions for sign language processing.

While a meta-analytic approach has the potential to reveal the commonalities of different studies, it is evident that carefully designed, individual experiments that pay close attention to linguistic details and insights from sign language linguistics are indispensable for gaining a closer understanding of the functional specifics of regions involved in sign language processing. In such studies, Broca's area has been implicated in processing mouthing and mouth gestures (Capek et al., 2008), as well as sign language morphology (Newman et al., 2010). Anterior and posterior portions of the left superior temporal cortex and sulcus have been linked to processing semantic as well as syntactic information in sign language (Blanco-Elorrieta et al., 2018; Moreno et al., 2018; Matchin et al., 2021). The left posterior middle temporal gyrus, which houses the motion-sensitive area MT/V5 and borders on the extrastriate regions of the occipital cortex has been found to be involved in processing movement in signed stimuli (Levanen, 2001; McCullough et al., 2012) and in lexical processing (Capek et al., 2008). Processing mouth gestures involves the fusiform face area located bilaterally at the bottom of the inferior temporal cortex in the fusiform gyrus (Capek et al., 2008). The left supramarginal gyrus has been linked to processing sign language

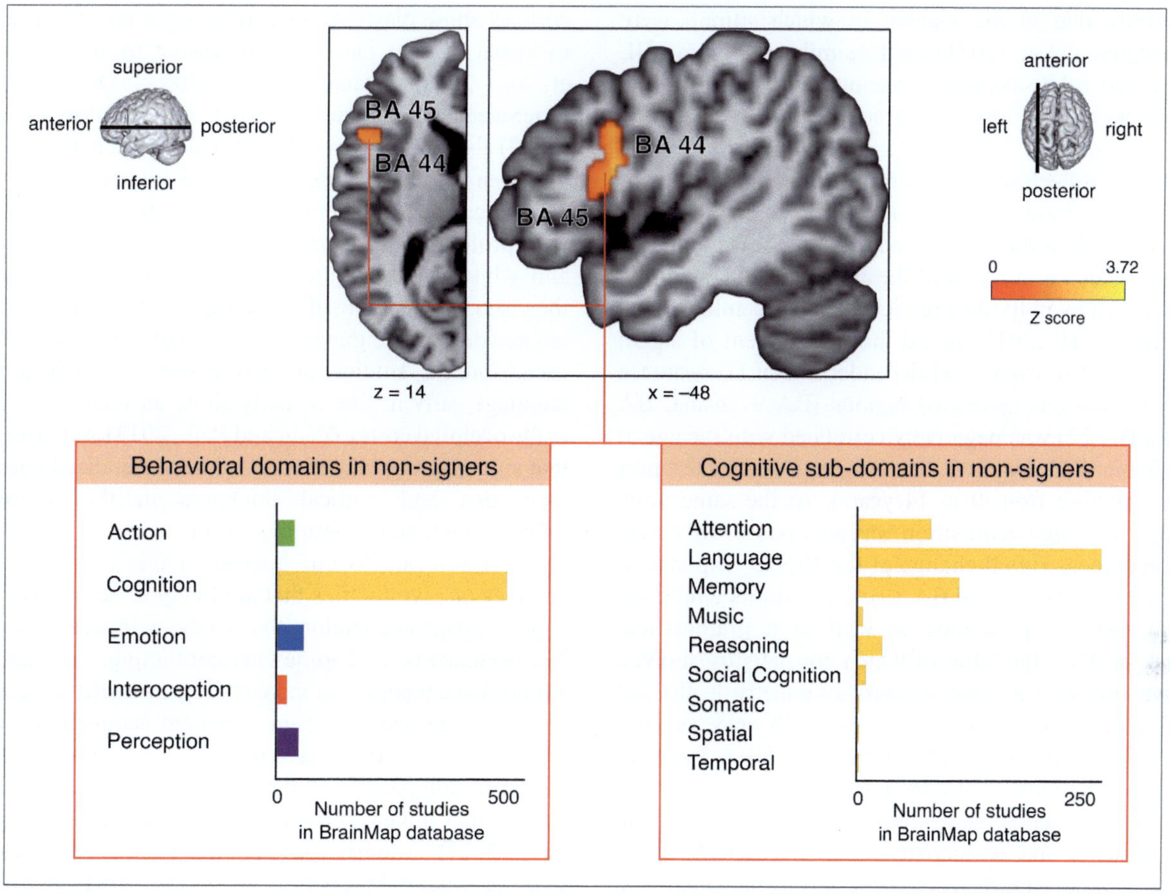

Fig. 21.6. Sign language-specific voxels in Broca's area and their function in non-signers. Functional attributions according to the BrainMap database (Fox and Lancaster, 2002) for sign language-specific voxels in Broca's area resulting from a contrast analysis comparing a meta-analysis of sign language comprehension in deaf signers and an independent meta-analysis of the observation of sign-like actions by hearing non-signers. The top panel shows a transverse and sagittal plane of the left hemisphere with the significant cluster in Broca's area [spanning Brodmann areas (BA) 44 and 45] surviving the contrast analysis of sign language comprehension and sign-like actions. The left box shows the number of studies with nonsigners in the BrainMap database that report peaks in voxels of the cluster specific to sign language comprehension in the left inferior frontal gyrus (Broca's area) organized by behavioral domain. Most studies in the database clearly have associated voxels in this cluster with cognitive processing. The right box shows the number of studies with nonsigners in the BrainMap database that reported peaks in voxels of the sign language-specific cluster in Broca's area organized by behavioral subdomains within the domain of cognition, according to the BrainMap taxonomy. Most studies in the database clearly associated voxels in the sign-language specific cluster with language processing in hearing non-signers, as well as to a lesser extent with attention and well as memory-related processes according to the BrainMap taxonomy. This meta-analytic overlap suggest that Broca's area constitutes a modality-independent hub in the (core) language network that processes linguistic information regardless of whether it was originally perceived as sign, speech, or writing. Adapted from Trettenbrein PC, Papitto G, Friederici AD, et al. (2021a). Functional neuroanatomy of language without speech: an ALE meta-analysis of sign language. Hum Brain Mapp 42: 699–712. doi:https://doi.org/10.1002/hbm.25254.

phonology as well as classifier constructions (Emmorey et al., 2002, 2021; MacSweeney et al., 2002a; Jednoróg et al., 2015; Cardin et al., 2016).

Given that the generation of hierarchically structured representations of phrases and sentences is a key function of the language network, it is somewhat unsatisfying to observe that syntactic processing remains one of the relatively understudied areas of neurocognitive research on sign language (Emmorey, 2021). An early study by MacSweeney et al. (2006) used a contrast of sentences and mere lists of unconnected signs in BSL and found that the presence of syntactic structure reliably recruited posterior portions of the left middle and superior temporal gyrus and left inferior frontal cortex. A more recent study used a semantic and syntactic violation paradigm in DGS and found that semantic processing recruited the posterior portion of Broca's area (i.e., BA 45), whereas syntactic processing of DGS sentences using spatial syntax increased activation in the right supramarginal gyrus (Stroh et al., 2019). Two independent studies in French Sign Language (LSF) and ASL attempted to implement a parametric design similar to Pallier et al. (2011), where the neural response is thought to increase systematically as a function of constituent size: Moreno et al. (2018) found main effects for this manipulation in subcortical structures such as the basal ganglia and only a smaller effect was found in the left perisylvian regions—possibly due to the manner in which stimuli were constructed. Matchin et al. (2021) used a similar design in ASL but only observed a left-lateralized parametric response in the anterior and posterior superior temporal sulcus in the syntactically most complex condition.

Lastly, directly reflecting insights from behavioral studies of sign language acquisition, neuroimaging data also suggests that the observed activations during sign language processing in deaf signers are modulated as a factor of the age at which a sign language was acquired by individual participants. In a seminal fMRI study, Mayberry et al. (2011) found that the percent of signal change of the blood oxygen level dependent (BOLD) response in left perisylvian language-related regions (BA 9, insula, BA 44, BA 6, and BA 22) was negatively correlated with the age at which participants had acquired sign language (with the onset of acquisition ranging from 0 to 14 years). At the same time, the age of sign language acquisition was also positively correlated with the percent signal change of the BOLD response in the occipital cortex (BA 18 and BA 19). The authors employed two different tasks, a phonemic as well as a grammatical judgment task in ASL, the latter of which consistently showed bilateral activations with a clear leftward asymmetry in the left IFG—specifically, the posterior portion of Broca's area (i.e., BA 44)—and the frontal operculum. Mayberry et al. (2011) hypothesized that these findings could be reflective of a modality-independent developmental shift during language acquisition, where more posterior and sensory/signal-based processing shifts to a more anterior adult pattern of automatic and abstract linguistic processing (Trettenbrein and Friederici, 2025).

Interim summary

The deep formal similarities between sign and spoken language are mirrored on the neural level. Major ERP components observed during sign language processing parallel those reported for spoken language. Studies with PET and fMRI have shown that sign language processing primarily recruits left perisylvian regions, which have previously been identified as relevant for spoken language processing. In particular, Broca's area has been identified as a modality-independent hub for language processing. Bilateral visual association areas as well as parietal regions subserve modality-specific aspects of sign language processing. Several brain regions including Broca's area show modulations in their response to sign language, according to the age of onset of sign language acquisition.

STRUCTURAL NEUROIMAGING OF SIGN LANGUAGE

A primary interest in the context of structural imaging has been to identify whether and where structural changes in the brain's gray and white matter can be observed as the result of deafness or sign language acquisition, whereas it is not clear whether these factors can actually be disentangled. In a systematic review and meta-analysis, Grégoire et al. (2022) identified a volumetric decrease in gray matter around the auditory cortex in deaf signers. This is not surprising, even though it is understood that auditory cortices show plasticity related changes and seem to participate in visual or sign language processing in deaf signers (Finney et al., 2001; Sadato et al., 2004; Ding et al., 2015; MacSweeney and Cardin, 2015). Beyond this, Grégoire et al. (2022) also identified a volumetric decrease in the gray and white matter in the visual cortex and a gray matter increase in the right cerebellum as consistent findings in structural imaging studies with profoundly deaf participants. The general decrease in gray and white matter volume in visual areas is unexpected given the visuospatial nature of sign language, but it may reflect the varied and delayed language experience of the majority of deaf children. Indeed, studies of deaf signers, who acquired a sign language early in life, actually show an increase in gray matter in the occipital cortex (Pénicaud et al., 2013) and there is evidence that sign language experience leads to lateralized changes in surface area and cortical thickness in the occipital cortex (McCullough and Emmorey, 2021).

More relevant to our interest in this chapter are structural changes or asymmetries that are likely to be the consequence of sign language acquisition, though the evidence for such changes has been scarce and somewhat conflicting. For example, some studies have reported an increase in gray matter volume in the left insula as the result of sign language acquisition (Allen et al., 2008), whereas others have reported a bilateral decrease in gray matter in comparison to hearing controls and a decrease in the right insula linked to the acquisition of sign language (Olulade et al., 2014). Another approach has been to specifically investigate the structural properties of regions, which have already been identified as relevant for spoken language processing: In a recent paper, Cheng et al. (2023) attempted to link age of acquisition to differences in structural data for the group of participants for which they previously reported an effect of age of sign language acquisition on the BOLD response (Mayberry et al., 2011). Significantly, delayed language exposure in early childhood was associated with negative changes in gray matter volume and cortical thickness in bilaterial perisylvian regions. No such differences were observed in deaf signers with early exposure to ASL (Cheng et al., 2023).

With regard to the white matter pathways connecting the inferior frontal and posterior temporal perisylvian language regions, no differences between deaf signers and hearing users of a spoken language have been observed in the pathways that have been studied to date. Cheng et al. (2019) studied the arcuate fasciculus connecting the inferior frontal to posterior temporal regions, inferior fronto-occipital fasciculus, inferior longitudinal fasciculus connecting the anterior temporal to extrastriate cortex and occipital lobe, and the uncinate fasciculus connecting the anterior inferior frontal cortex and the frontal operculum to the anterior temporal cortex in a group of native deaf signers using ASL and a control group of hearing nonsigners using English and observed no group differences. Significantly, the arcuate and inferior longitudinal fascicles constituted structural asymmetries and were left-lateralized in both groups, thereby reproducing a pattern already reported in hearing speakers (Gallardo et al., 2020; Warrington et al., 2020). The inferior fronto-occipital fasciculus as well as the uncinate fasciculus have been reported to be right-lateralized in the literature (Gallardo et al., 2020; Warrington et al., 2020), but this pattern was not observed by Cheng et al. (2019) for either group. A possible reason could be that this pattern has been reported in studies with very large samples with several hundred participants so that the analysis of a necessarily smaller sample of deaf signers may not have sufficient power to detect these effects.

Due to its role as a part of the core language network, Finkl et al. (2019) specifically investigated the arcuate fasciculus using diffusion tensor imaging (DTI) and probabilistic tractography and observed no difference in the connectivity profiles of this pathway between a group of deaf signers, who had acquired DGS early in life, and matched controls. However, some differences between groups were observed depending on the respective seed region: When seeding in the posterior portion of Broca's area (i.e., BA 44), the authors found differences in the pathways connecting this region to the pre-supplementary motor area as well as to the thalamus (Fig. 21.7, left panel). When seeding in the posterior superior temporal gyrus (i.e., BA 22), the authors observed a difference in the pathway connecting this region to the ipsilateral parietal cortex and the precuneus (Fig. 21.7, right panel). Lastly, the transcallosal connection between the auditory cortices was lower in the deaf group. The lack of a difference between groups indicates that a typically developed arcuate fascicle does not require auditory language input. Furthermore, the authors interpret the remaining observed group differences as relating to different requirements of the mapping to the sensorimotor system imposed by the visuospatial modality.

While the arcuate fasciculus exhibits no difference between typically developed early deaf signers and hearing users of a spoken language, the development of this pathway is nevertheless shaped by language experience early in life. Cheng et al. (2019) reported data from three deaf individuals, who experienced only minimal language during childhood and acquired a conventionalized sign language only later in life after the onset of puberty and thus outside the so-called critical period for language acquisition (Lenneberg, 1967, 1969). Significantly, when compared to a control group of deaf early signers, all three deaf late signers showed alterations of white matter microstructure in all four fiber pathways investigated in this study; yet, the difference was most pronounced for the arcuate fasciculus. The authors argued that the decreased values in fractional anisotropy in the left doral pathway that they observed when comparing each of the three participants, who were deaf late signers to a group of deaf controls, who were early signers, are due to the reduced laterality of the pathway (Cheng et al., 2019). These data suggest that adequate early language exposure is required for the emergence of the usually left-lateralized structural pattern of the dorsal pathway that is part of the core language network in typically developed adults.

In this context, it is suggestive to note that the behavioral linguistic profiles for all three deaf late signers in the study by Cheng et al. (2019) reported in independent publications indicate limited syntactic abilities in ASL (Ramírez et al., 2013; Ferjan Ramirez et al., 2014; Mayberry et al., 2018). Moreover, all three participants also exhibited atypical neural response patterns to ASL signs that primarily recruited the bilateral dorsolateral superior parietal and occipital cortex and minimally involved perisylvian language regions (Ferjan Ramirez et al., 2014; Mayberry et al., 2018). A follow-up study conducted 15 months after data collection for two of the three deaf late signers showed an altered response to highly familiar ASL signs, which was more concentrated in left perisylvian language regions (Ferjan Ramirez et al., 2016). Notice that these case studies provide only indirect

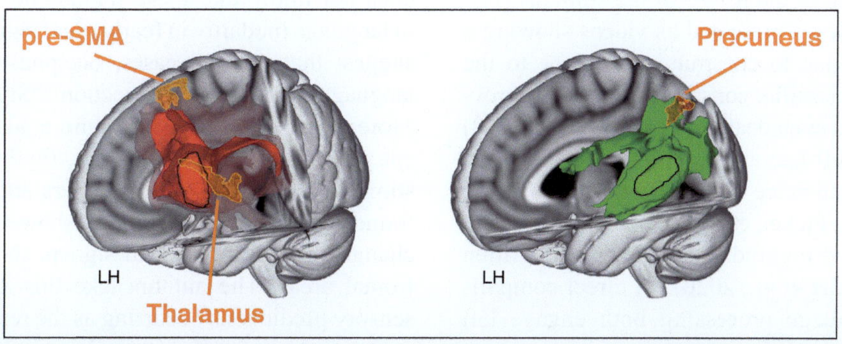

Fig. 21.7. Fiber-tracking of the core language network in the deaf brain. Visualization of fiber-tracking results with two hubs of the so-called "core" language network in the left hemisphere (LH) as seed regions in a sample of deaf signers. Both illustrations show results based on a group comparison of deaf signers compared to a sample of matched hearing controls without any knowledge of sign language. Parts shown in orange indicate regions where connectivity was weaker in the group of deaf signers compared to the hearing controls. Left panel: Using the posterior portion of Broca's area, left Brodmann area 44 (BA 44; shown in red and indicated using a black outline), as a seed region demonstrates the strong frontal connectivity of this core language region, as well as its connection to the posterior temporal cortex via the arcuate fasciculus. The group of deaf signers shows weaker connectivity to pre-supplementary motor area (pre-SMA) as well as the thalamus (both indicated in orange). Right panel: Using the left posterior temporal gyrus (pSTG; shown in green and indicated using a black outline) as a seed region confirms the connection of this core language region to other parts of temporal as well as parietal cortex, including also the canonical connection to frontal language-relevant regions via the arcuate fasciculus. The connection of pSTG to the precuneus is weaker in the LH for the deaf group when compared to hearing controls (indicated in orange). Adapted from Finkl T, Hahne A, Friederici AD, et al. (2019). Language without speech: Segregating distinct circuits in the human brain, Cereb Cortex 30: 812–823. doi:https://doi.org/10.1093/cercor/bhz128, under a Creative Commons (CC-BY 4.0) license.

evidence for the relevance of a typically developed dorsal pathway for processing complex syntactic structures in sign language. However, as studies of spoken language processing have shown that the structural integrity of the arcuate fasciculus during maturation correlates with the ability for processing syntactically complex sentences (Friederici and Gierhan, 2013; Skeide and Friederici, 2016), it seems reasonable to take these case studies as indicative of the important role that the arcuate fasciculus and its development may play in the deaf brain during sign language acquisition.

Interim summary

The core language network connecting the left posterior inferior frontal and posterior temporal regions shows no differences between deaf signers, who acquired a sign language early in life, and hearing speakers, pointing at its potentially modality-independent nature. Observed structural brain differences between deaf signers and hearing controls all pertain to the modality of language use as they seem to either be the consequence of deafness or sign language acquisition. Several studies have observed an effect of the age of sign language acquisition on the structural properties of the brains of deaf signers, thereby underlining the importance of an early exposure to a sign language for the typical development of the language system.

DISCUSSION

Our review of the functional and structural asymmetries in cortical regions and the white matter relevant for sign language processing has confirmed that the human language system in the typically developed adult deaf brain is functionally and, with regard to the core language network also structurally left-lateralized, though the RH appears to be more engaged during sign language processing due to the nature of the stimulus (i.e., sign language stimuli are always presented as videos showing a person signing) as well as due to constructions unique to the visuospatial modality (e.g., classifier constructions). While prevalent in the deaf, hearing, and even deafblind (Obretenova, 2010) population, this functional left-lateralization is not absolute and cases of reversed cerebral dominance for sign language have been reported in the literature (e.g., Pickell et al., 2005), similar to what has previously been observed for studies of spoken and written language processing (Labache et al., 2020). In direct comparison, sign and spoken language processing both engage left perisylvian regions, yet rely on different mappings to the sensorimotor system determined by the modality of language use. In addition, the spatial nature of sign language is also reflected in the engagement of bilateral parietal regions, though it seems that the left and right parietal cortex differentially support either the construction of grammatical or spatial representations (Hickok et al., 1996c).

The neural substrates of sign, spoken, and written language processing in the left perisylvian cortex show significant overlap in the inferior frontal and posterior temporal regions (MacSweeney et al., 2002b; Sakai et al., 2005; Inubushi and Sakai, 2013; Trettenbrein et al., 2021a; Uddén et al., 2022), though the exact activation patterns in the (posterior) temporal regions appear to be modulated by modality-specific properties (Evans et al., 2019), including the use of different articulators (Capek et al., 2008) but not iconicity (Emmorey et al., 2004; Klann et al., 2005). These data are compatible with the notion that the language network primarily processes linguistic information in a modality-independent fashion and interacts with modality-specific regions and networks depending on the requirements of the modality of language use as well as task demands. For example, the greater self-monitoring demands of the visuospatial modality lead to a more pronounced left-lateralization during sign than speech production (Gutiérrez-Sigut et al., 2015, 2016). Moreover, comprehension and production tasks may induce different activation patterns (Emmorey, 2021); but, while many studies have treated Broca's area as a primarily production-related brain region (Braun, 2001; Horwitz et al., 2003; Emmorey, 2006), the identification of Broca's area as a modality-independent hub for language comprehension speaks against a clear dichotomy between production and comprehension (Trettenbrein et al., 2021a).

The engagement of the bilateral occipital cortex in sign language is, of course, expected because sign language is perceived visually (Fig. 21.5) and due to the neuroanatomic properties of the visual system, however, it seems that sign language acquisition does not simply lead to a differential recruit of the visual system but instead triggers lasting structural (Pénicaud et al., 2013) and possibly also functional adaptions within the visual system (Brookshire et al., 2017; Stroh et al., 2022). While the bilateral clusters located on the border of the posterior middle temporal and extrastriate cortex observed in the meta-analysis of sign language comprehension in deaf signers by Trettenbrein et al. (2021a) did not survive the comparison with an independent meta-analysis of sign-like actions in hearing non-signers, this does not imply that these regions do not show any adaptation to language modality in (early) deaf signers. In fact, lesion studies suggest that, in rare cases, occipital lesions can lead to sign language aphasias (see section "Sign Language Aphasia"). Moreover, a study by Brookshire et al. (2017) looked at cortical entrainment to oscillations (i.e., rhythmic modulation) of ASL stimuli in a group of deaf signers and hearing nonsigners and found that, while both groups showed phase-locking to visual changes in ASL, only deaf signers showed strong coherence in frontal areas. The authors take this as evidence for top-down sensory predictions occurring as the result of knowledge of ASL.

A number of different explanations for the seemingly larger right-hemispheric involvement observed in neuroimaging studies of sign language processing in deaf signers have been put forward in the literature, with most researchers arguing for either a modality-specific or stimulus-specific difference (Peperkamp and Mehler, 1999). That is, if the greater right-hemispheric involvement was due to specific requirements or properties of the signed modality, then some regions in the RH should show a stronger response to modality-specific constructions. This is actually the case for the seemingly right-hemispheric specialization for processing classifier constructions, which can also be selectively impaired by lesions (Hickok et al., 1996c; Atkinson,

2005). Similarly, if the greater right-hemispheric involvement was due to specific properties of stimuli used in sign language experiments (i.e., videos showing a person signing) as well as their linguistic properties, then we should expect that processing this additional (social) information about the signer performing manual actions, which is part of the sign language stimuli but not necessarily of auditory or written stimuli will recruit more right-hemispheric networks in signers and nonsigners alike (Campbell et al., 2011). The meta-analysis by Trettenbrein et al. (2021a) seems to confirm this in line with studies attempting to disentangle sentence processing from processing prosody and discourse-level information (Atkinson et al., 2004; Inubushi and Sakai, 2013; Newman et al., 2015). Hence, larger involvement of the RH during signing than during auditory or written language processing appears to be driven by both modality and stimulus-specific properties.

The role that left perisylvian language regions, including Broca's area, take on in language processing appears to be dependent on exposure to adequate linguistic input already early in life, independent of the modality of language use. That is, delayed exposure to a sign language has a severe and lasting impact on the development of the core and extended language network in the LH of deaf signers, which is evidenced in structural properties of the brain, functional neural correlates, as well as on the behavioral level (see sections "Structural Neuroimaging of Sign Language" and "Sign Language Acquisition and Development"). The different studies of the effects that delayed exposure to a sign language has on neural and linguistic development reviewed here provide strong support for the notion that the growth of language in the individual is determined and also constrained by a biologic matrix, whereas this matrix does not specify the modality in which language can or should be perceived or externalized. Similarly, the limited syntactic abilities of individuals who were exposed to a conventionalized sign language only later in life after the onset of puberty supports the notion of a critical period for language acquisition (Lenneberg, 1967, 1969), independent of the modality in which language is acquired. In sum, these observations underpin the importance of providing early access to a sign language to deaf children to foster the typical development of the language system.

Lastly, it should be mentioned that the studies reviewed here suggest that the usual left-lateralization of language processing in the typically developed adult brain does not merely reflect a specialization for processing the fast temporal structure of speech (Zatorre et al., 2002; Schönwiesner et al., 2005); instead, we take this pattern to indicate that the LH is usually specialized for the processing of linguistic information and, specifically, the generation of the hierarchically structured representations underlying sentences independent of the modality of language use. Ursula Bellugi and colleagues had already reached a similar conclusion based on their extensive studies of deaf aphasics (Poizner et al., 1987; Emmorey et al., 1995b; Hickok et al., 1996a, 1998), though it has been questioned whether sign language processing indeed does not impose similar temporal processing demands like speech because it is not clear that a greater reliance on space in the signed signal does not impose demands on temporal processing (Corina, 1999). A middle ground between both views may be supplied by the literature on so-called split-brain patients who were hearing and non-signers: Based on his work with these patients, especially in the domain of visual processing, Zaidel (1978) proposed that, while the RH can recognize units (e.g., spoken or printed words) as whole patterns, the LH always decomposes words and by extension signs as well as sentences into smaller units by feature analysis.

TOWARD A FUNCTIONAL NEUROANATOMY OF SIGN LANGUAGE

Against the background of the data discussed in this chapter, we can attempt a sketch of the neural basis of sign language processing including relevant cortical regions and white-matter pathways (Fig. 21.8). It seems clear now that the so-called "classical" model of language processing consisting only of Broca's and Wernicke's area familiar from text books was severely underspecified not just in the context of spoken language but even more so when we consider sign language processing (Poeppel et al., 2012; Emmorey, 2015, 2021). While sign language and spoken language processing overlap in their recruitment of the bilateral posterior temporal cortex and a strong left-lateralization in the IFG, sign language processing additionally recruits higher parts of the bilateral visual system as well as bilateral parietal cortices for the processing of modality-specific phenomena such as classifier constructions in which space is used iconically. We take this to indicate that the core language system exhibits similar functional and structural asymmetries, independent of the modality of language use, yet interfaces with different networks depending on modality-specific processing demands. This is also reflected in overall similar electrophysiologic response patterns and processing stages (Fig. 21.4). The white-matter pathways connecting regions implicated in sign language processing are best understood for the perisylvian cortex and remain to be explored in future studies in the context of regions recruited specifically during sign language processing.

In direct comparison to models of spoken and written language processing (e.g., Friederici, 2011, 2017; Hagoort, 2017), our current understanding of exact functional attributions of regions or possible functional dissociations of different linguistic subsystems (e.g., semantics vs syntax) for sign language processing remains somewhat limited and requires further research. Broca's area has been identified as a modality-independent hub for language processing and has also been implicated in studies of grammatical processing in ASL, in processing mouthing and mouth gestures, as well as processing sign morphology. Both the anterior and posterior portions of the left superior temporal cortex and sulcus have been linked to semantic and syntactic processing. The left posterior middle temporal gyrus (i.e., area V5/MT) bordering on occiptial cortical regions such as the extrastriate body area has been found to be involved in lexical processing, sentence comprehension, and motion processing. Regions on the underside of the temporal cortex such as the fusiform face area are involved in processing mouth gestures. The bilateral parietal cortex has been linked to processing phonologic and

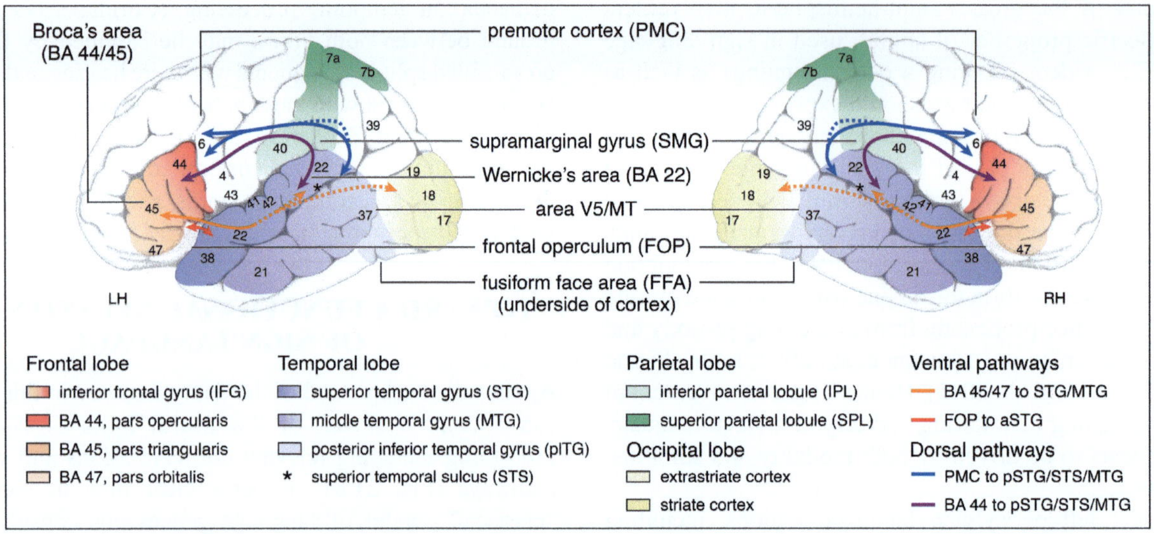

Fig. 21.8. Schematic of the neural basis of sign language. Schematic depiction of the LH and RH displaying anatomic landmarks and cytoarchitectonic details of cortical regions relevant for processing sign language. Major gyri that are involved in sign language processing are color-coded: The IFG and its subregions, Brodmann areas (BA) 44, BA 45, and BA 47, are depicted in shades of red; superior temporal gyrus (STG), middle temporal gyrus (MTG), and posterior inferior temporal gyrus (pITG) are shown in shades of blue. The location of the superior temporal sulcus (STS) in between STG and MTG is indicated using an asterisk (*). Inferior parietal lobule (IPL) and superior parietal lobule (SPL) are depicted in shades of green. Extrastriate and striate portions of the visual cortex are indicated in shades of yellow. The different numbers on the schematic indicate the respective cytoarchitectonic label of a region (i.e., BA), as defined by Brodmann (1909), on the basis of histologic studies. BA 44 (pars opercularis) and BA 45 (pars triangularis) in the LH jointly form Broca's area. BA 47 (pars orbitalis) in the LH is located anteriorly to Broca's area, whereas the frontal operculum (FOP) is located ventrally and medially to it. The premotor cortex (PMC) is located in BA 6. Wernicke's area is located in BA 22 in the LH, though definitions based on studies of spoken language sometimes also include BA 42, which is part of primary auditory cortex and adjacent to Heschel's gyrus (BA 41). The supramarginal gyrus is located dorsally in the IPL in BA 40. The extrastriate body area (EBA) is located in BA 19. The fusiform face area (FFA) is not shown in the diagram as it is located at the bottom of the inferior temporal cortex in the fusiform gyrus (BA 37). The main white matter fiber tracts connecting regions of the cortex relevant for processing sign language are located dorsally and ventrally and indicated using color-coded arrows: Blue for the pathway connecting PMC to the posterior STG (pSTG), STS, and MTG; purple for the pathway connecting BA 44 to pSTG, STS, and MTG; orange for the pathway connecting BA 45 and BA 47 to STG and MTG; and red for the pathway connecting FOP to aSTG. Based on a similar schematic based on studies of spoken and written language processing by Friederici AD (2011). The brain basis of language processing: from structure to function. Physiol Rev 91: 1357–1392. https://doi.org/10.1152/physrev.00006.2011.

spatial information in sign language: The left supramarginal gyrus has been linked to lexical phonologic processing, whereas its right-hemispheric counterpart has been linked to form-based phonologic processing. The right superior parietal lobule may be functionally more relevant for processing classifier constructions.

CONCLUSION

Our review of the neuroimaging literature on sign language in this chapter has confirmed the general insight that the LH of the typically developed adult human brain is specialized for processing language independent of the modality of language use. This functionally asymmetric organization emerges during development and is, to an extent, paralleled in structural asymmetries. Both functional and structural asymmetries are subject to individual variation and directly impacted by delays in sign language acquisition. Generally speaking, both hemispheres display a certain equipotentiality and are involved in processing sign language, especially with regard to processing visual and spatial information in the occipital and the parietal cortex as well as linguistic information in the posterior temporal cortex. The RH seems to specifically support processing of some constructions unique to the signed modality (i.e., classifier constructions that use space iconically and not grammatically). A key difference between our current state of knowledge about sign and spoken language processing concerns our lack of understanding of syntactic processing in the visuospatial modality.

The studies reviewed in this chapter support the speculative notion that the processing of sentences in both spoken and sign language in typically developed adults is supported by a primarily left-hemispheric fronto-temporal network connecting the IFG to the posterior temporal cortex. This network appears to subserve the generation of hierarchically structured representations linking meaning to the representations of a physical signal (i.e., sign or sound) independent of the modality of language use and develops in accordance with a genetically determined biologic matrix, suggesting that its emergence may have constituted a prerequisite for the evolution of the human language capacity. However, the precise functional roles of the major frontal and temporal hubs in this network during sign language processing remain underspecified, at least in comparison to work on spoken and written language processing. Future work using linguistically informed experimental manipulations and methodologic advances will likely provide a more fine-grained picture of the functional specialization and lateralization of sign language and its subsystems.

ACKNOWLEDGMENTS

The authors would like to thank (in alphabetical order by last name) Thomas A. Finkbeiner, Cheslie C. Klein, Nina-Kristin Meister, Joëlle A. M. Schroën, and Markus Steinbach for consultation, discussion, and/or feedback on a first draft of this manuscript. We are grateful to Heike Schmidt-Duderstedt and Andrea Gast-Sandmann for their artistic assistance with creating illustrations, to Jens-Michael Cramer for recording the original set of videos from which the stills and videos in Fig. 21.2 are derived, as well as to Alfred Anwander and Theresa Finkl for providing materials for Fig. 21.7. The handshape fonts used in this chapter were created by the Centre for Sign Linguistics and Deaf Studies at the Chinese University of Hong Kong. This work was funded by the Max Planck Society. While completing revisions of this chapter, Patrick C. Trettenbrein was funded by the Deutsche Forschungsgemeinschaft (DFG, German Research Foundation)—501984557, which is part of the DFG SPP 2392 "Visual Communication" (ViCom).

CONFLICTS OF INTEREST

We have no known conflicts of interest to disclose.

References

Allen JS, Emmorey K, Bruss J et al. (2008). Morphology of the insula in relation to hearing status and sign language experience. J Neurosci 28: 11900–11905. https://doi.org/10.1523/JNEUROSCI.3141-08.2008.

Atkinson JR (2005). The perceptual characteristics of voice-hallucinations in deaf people: insights into the nature of subvocal thought and sensory feedback loops. Schizophr Bull 32: 701–708. https://doi.org/10.1093/schbul/sbj063.

Atkinson JR, Campbell R, Marshall J et al. (2004). Understanding "not": neuropsychological dissociations between hand and head markers of negation in BSL. Neuropsychologia 42: 214–229. https://doi.org/10.1016/S0028-3932(03)00186-6.

Atkinson JR, Marshall J, Woll B et al. (2005). Testing comprehension abilities in users of British sign language following CVA. Brain Lang 94: 233–248. https://doi.org/10.1016/j.bandl.2004.12.008.

Baker A, van den Bogaerde B, Pfau R et al. (2016). The linguistics of sign language, John Benjamins.

Bavelier D, Corina D, Jezzard P et al. (1998). Hemispheric specialization for English and ASL: left invariance-right variability. Neuroreport 9: 1537–1542.

Bellugi U, Klima ES (1976). Two faces of sign: iconic and abstract. Ann N Y Acad Sci 280: 514–538. https://doi.org/10.1111/j.1749-6632.1976.tb25514.x.

Bellugi U, Poizner H, Klima ES (1989). Language, modality and the brain. Trends Neurosci 12: 380–388. https://doi.org/10.1016/0166-2236(89)90076-3.

Berger L, Pyers J, Lieberman A et al. (2023). Parent American sign language skills correlate with child–but not toddler–ASL vocabulary size. Lang Acquis 31: 85–99. https://doi.org/10.1080/10489223.2023.2178312.

Berwick RC, Friederici AD, Chomsky N et al. (2013). Evolution, brain, and the nature of language. Trends Cogn Sci 17: 89–98. https://doi.org/10.1016/j.tics.2012.12.002.

Blanco-Elorrieta E, Kastner I, Emmorey K et al. (2018). Shared neural correlates for building phrases in signed and spoken language. Sci Rep 8. https://doi.org/10.1038/s41598-018-23915-0.

Bolhuis JJ, Tattersall I, Chomsky N et al. (2014). How could language have evolved? PLoS Biol 12: e1001934. https://doi.org/10.1371/journal.pbio.1001934.

Boyes-Braem P (1986). Two aspects of psycholinguistic research: iconicity and temporal structure. Proceedings of the second European congress on sign language research; signs of life. University of Amsterdam.

Braun AR (2001). The neural organization of discourse: an $H_2{}^{15}O$-PET study of narrative production in English and American sign language. Brain 124: 2028–2044. https://doi.org/10.1093/brain/124.10.2028.

Brentari D (2012). Phonology. In: R Pfau, M Steinbach, B Woll (Eds.), Sign language: an international handbook. de Gruyter, pp. 21–54.

Broca P (1861). Remarques sur le siége de la faculté du langage articulé, suivies d'une obsevation d'aphémie (perte de la parole). Bull Soc Anatomique 6: 330–357.

Brodmann K (1909). Beiträge zur histologischen Lokalisation der Grosshirnrinde: VI. Die Cortexgliederung des Menschen. J Psychol Neurol 10: 231–246.

Brookshire G, Lu J, Nusbaum HC et al. (2017). Visual cortex entrains to sign language. Proc Natl Acad Sci U S A 114: 6352–6357. https://doi.org/10.1073/pnas.1620350114.

Campbell R, Capek CM, Gazarian K et al. (2011). The signer and the sign: cortical correlates of person identity and language processing from point-light displays. Neuropsychologia 49: 3018–3026. https://doi.org/10.1016/j.neuropsychologia.2011.06.029.

Capek CM, Neville HJ (2015). Studying sign language processing using functional neuroimaging techniques: fMRI, ERP, MEG and TMS. In: E Orfanidou, B Woll, G Morgan (Eds.), Research methods in sign language studies: a practical guide. Wiley-Blackwell, pp. 321–335.

Capek CM, Waters D, Woll B et al. (2008). Hand and mouth: cortical correlates of lexical processing in British sign language and speechreading English. J Cogn Neurosci 20: 1220–1234. https://doi.org/10.1162/jocn.2008.20084.

Capek CM, Grossi G, Newman AJ et al. (2009). Brain systems mediating semantic and syntactic processing in deaf native signers: biological invariance and modality specificity. Proc Natl Acad Sci U S A 106: 8784–8789. https://doi.org/10.1073/pnas.0809609106.

Cardin V, Orfanidou E, Rönnberg J et al. (2013). Dissociating cognitive and sensory neural plasticity in human superior temporal cortex. Nat Commun 4. https://doi.org/10.1038/ncomms2463.

Cardin V, Orfanidou E, Kästner L et al. (2016). Monitoring different phonological parameters of sign language engages the same cortical language network but distinctive perceptual ones. J Cogn Neurosci 28: 20–40. https://doi.org/10.1162/jocn_a_00872.

Caselli NK, Sehyr ZS, Cohen-Goldberg AM et al. (2017). ASL-LEX: a lexical database of American sign language. Behav Res Methods 49: 784–801. https://doi.org/10.3758/s13428-016-0742-0.

Cecchetto C (2017). The syntax of sign language and Universal Grammar. In: I Roberts (Ed.), The Oxford handbook of universal grammar. Oxford University Press.

Checchetto A, Geraci C, Cecchetto C et al. (2018). The language instinct in extreme circumstances: the transition to tactile Italian sign language (LISt) by deafblind signers. Glossa 3: 65. https://doi.org/10.5334/gjgl.357.

Cheng Q, Roth A, Halgren E et al. (2019). Effects of early language deprivation on brain connectivity: language pathways in deaf native and late first-language learners of American sign language. Front Hum Neurosci 13: 320. https://doi.org/10.3389/fnhum.2019.00320.

Cheng Q, Roth A, Halgren E et al. (2023). Restricted language access during childhood affects adult brain structure in selective language regions. Proc Natl Acad Sci U S A 120: e2215423120. https://doi.org/10.1073/pnas.2215423120.

Chomsky N (1965). Aspects of the theory of syntax, The MIT Press.

Chomsky C (1986a). Analytic study of the Tadoma method: language abilities of three deaf-blind subjects. J Speech Lang Hear Res 29: 332. https://doi.org/10.1044/jshr.2903.347.

Chomsky N (1986b). In: RN Anshen (Ed.), Knowledge of language: its nature, origins, and use. Praeger.

Chomsky N (1995). The minimalist program, MIT Press.

Chomsky N (2011). Language and other cognitive systems: what is special about language? Lang Learn Dev 7: 263–278. https://doi.org/10.1080/15475441.2011.584041.

Corballis PM (2003). Visuospatial processing and the right-hemisphere interpreter. Brain Cogn 53: 171–176. https://doi.org/10.1016/S0278-2626(03)00103-9.

Corina DP (1998). The processing of sign language: evidence from aphasia. In: B Stemmer, HA Whitaker (Eds.), Handbook of neurolinguistics. Academic Press 313–329.

Corina DP (1999). On the nature of left hemisphere specialization for signed language. Brain Lang 69: 230–240. https://doi.org/10.1006/brln.1999.2062.

Corina DP, Poizner H, Bellugi U et al. (1992). Dissociation between linguistic and nonlinguistic gestural systems: a case for compositionality. Brain Lang 43: 414–447. https://doi.org/10.1016/0093-934X(92)90110-Z.

Corina DP, San Jose-Robertson L, Guillemin A et al. (2003). Language lateralization in a bimanual language. J Cogn Neurosci 15: 718–730. https://doi.org/10.1162/089892903322307438.

Crain S, Koring L, Thornton R (2016). Language acquisition from a biolinguistic perspective. Neurosci Biobehav Rev. https://doi.org/10.1016/j.neubiorev.2016.09.004.

Damasio A, Bellugi U, Damasio H et al. (1986). Sign language aphasia during left-hemisphere amytal injection. Nature 322: 363–365. https://doi.org/10.1038/322363a0.

Dax G (1863). M. Dax soumet au jugement de l'Académie un Mémoire intitulé: 'Observations tendant à prouver la coïncidence constante des dérangements de la parole avec une lésion de l'hémisphère gauche du cerveau. Compt Rend HebdomSéanl'Acad Scien 56: 536.

Dax M (1865). Lésions de la moitié gauche de l'encéphale coïncident avec l'oubli des signes de la pensée (lu à montpellier En 1836). Bull Hebd Médecine Chir 2: 259–262.

Dehaene S, Cohen L (2007). Cultural recycling of cortical maps. Neuron 56: 384–398. https://doi.org/10.1016/j.neuron.2007.10.004.

Dehaene S, Cohen L, Sigman M et al. (2005). The neural code for written words: a proposal. Trends Cogn Sci 9: 335–341. https://doi.org/10.1016/j.tics.2005.05.004.

Ding H, Qin W, Liang M et al. (2015). Cross-modal activation of auditory regions during visuo-spatial working memory in early deafness. Brain 138: 2750–2765. https://doi.org/10.1093/brain/awv165.

Dronkers NF, Plaisant O, Iba-Zizen MT et al. (2007). Paul Broca's historic cases: high resolution MR imaging of the brains of Leborgne and Lelong. Brain 130: 1432–1441. https://doi.org/10.1093/brain/awm042.

Edwards T, Brentari D (2020). Feeling phonology: the conventionalization of phonology in protactile communities in the United States. Language. https://doi.org/10.1353/lan.0.0248.

Eickhoff SB, Laird AR, Grefkes C et al. (2009). Coordinate-based activation likelihood estimation meta-analysis of neuroimaging data: a random-effects approach based on empirical estimates of spatial uncertainty. Hum Brain Mapp 30: 2907–2926. https://doi.org/10.1002/hbm.20718.

Eickhoff SB, Bzdok D, Laird AR et al. (2012). Activation likelihood estimation meta-analysis revisited. Neuroimage 59: 2349–2361. https://doi.org/10.1016/j.neuroimage.2011.09.017.

Eickhoff SB, Laird AR, Fox PM et al. (2017). Implementation errors in the GingerALE software: description and recommendations: errors in the GingerALE software. Hum Brain Mapp 38: 7–11. https://doi.org/10.1002/hbm.23342.

Embick D, Poeppel D (2015). Towards a computational (ist) neurobiology of language: correlational, integrated and explanatory neurolinguistics. Lang Cogn Neurosci 30: 357–366. https://doi.org/10.1080/23273798.2014.980750.

Emmorey K (2006). The role of Broca's area in sign language. In: Y Grodzinsky, K Amunts (Eds.), Broca's region. Oxford University Press, pp. 169–184.

Emmorey K (2014). Iconicity as structure mapping. Philos Trans R Soc Lond B Biol Sci 369. https://doi.org/10.1098/rstb.2013.0301.

Emmorey K (2015). The neurobiology of sign language. In: AW Toga, P Bandettini, P Thompson, K Friston (Eds.), Brain mapping: an encyclopedic reference. vol. 3. Academic Press 475–479.

Emmorey K (2021). New perspectives on the neurobiology of sign languages. Front Commun 6: 748430. https://doi.org/10.3389/fcomm.2021.748430.

Emmorey K, Bellugi U, Friederici A et al. (1995a). Effects of age of acquisition on grammatical sensitivity: evidence from on-line and off-line tasks. Appl Psycholinguist 16: 1–23. https://doi.org/10.1017/S0142716400006391.

Emmorey K, Corina DP, Bellugi U (1995b). Differntial processing of topographic and referential functions of space. In: Language, gesture, and space, Lawrence Erlbaum Associates 43–62.

Emmorey K, Damasio H, McCullough S et al. (2002). Neural systems underlying spatial language in American sign language. Neuroimage 17: 812–824. https://doi.org/10.1006/nimg.2002.1187.

Emmorey K, Grabowski T, McCullough S et al. (2004). Motor-iconicity of sign language does not alter the neural systems underlying tool and action naming. Brain Lang 89: 27–37. https://doi.org/10.1016/S0093-934X(03)00309-2.

Emmorey K, Winsler K, Midgley KJ et al. (2020). Neurophysiological correlates of frequency, concreteness, and iconicity in American sign language. Neurobiol Lang 1: 249–267. https://doi.org/10.1162/nol_a_00012.

Emmorey K, Brozdowski C, McCullough S (2021). The neural correlates for spatial language: perspective-dependent and -independent relationships in American sign language and spoken English. Brain Lang 223: 105044. https://doi.org/10.1016/j.bandl.2021.105044.

Emmorey K, Midgley KJ, Holcomb PJ (2022). Tracking the time course of sign recognition using ERP repetition priming. Psychophysiology 59. https://doi.org/10.1111/psyp.13975.

Evans S, Price CJ, Diedrichsen J et al. (2019). Sign and speech share partially overlapping conceptual representations. Curr Biol 29: 3739–3747. https://doi.org/10.1016/j.cub.2019.08.075.

Everaert MBH, Huybregts MAC, Chomsky N et al. (2015). Structures, not strings: linguistics as part of the cognitive sciences. Trends Cogn Sci 19: 729–743. https://doi.org/10.1016/j.tics.2015.09.008.

Fenlon J, Cormier K, Brentari D (2017). The phonology of sign languages. In: SJ Hannahs, A Bosch (Eds.), The Routledge handbook of phonological theory. Routledge.

Ferjan Ramirez N, Leonard MK, Torres C et al. (2014). Neural language processing in adolescent first-language learners. Cereb Cortex 24: 2772–2783. https://doi.org/10.1093/cercor/bht137.

Ferjan Ramirez N, Leonard MK, Davenport TS et al. (2016). Neural language processing in adolescent first-language learners: longitudinal case studies in American sign language. Cereb Cortex 26: 1015–1026. https://doi.org/10.1093/cercor/bhu273.

Finkl T, Hahne A, Friederici AD et al. (2019). Language without speech: segregating distinct circuits in the human brain. Cereb Cortex 30: 812–823. https://doi.org/10.1093/cercor/bhz128.

Finney EM, Fine I, Dobkins KR (2001). Visual stimuli activate auditory cortex in the deaf. Nat Neurosci 4: 1171–1173. https://doi.org/10.1038/nn763.

Flaherty M, Hunsicker D, Goldin-Meadow S (2021). Structural biases that children bring to language learning: a cross-cultural look at gestural input to homesign. Cognition 211: 104608. https://doi.org/10.1016/j.cognition.2021.104608.

Floccia C (2017). *Data collected with the Oxford CDI over a course of 5 years in Plymouth Babylab, UK*. [dataset]. https://wordbank.stanford.edu.

Fox PT, Lancaster JL (2002). Mapping context and content: the BrainMap model. Nat Rev Neurosci3: 319–321. https://doi.org/10.1038/nrn789.

Frank MC, Braginsky M, Yurovsky D et al. (2017). Wordbank: an open repository for developmental vocabulary data. J Child Lang 44: 677–694. https://doi.org/10.1017/S0305000916000209.

Friederici AD (2011). The brain basis of language processing: from structure to function. Physiol Rev 91: 1357–1392. https://doi.org/10.1152/physrev.00006.2011.

Friederici AD (2017). Language in our brain: the origins of a uniquely human capacity, MIT Press.

Friederici AD, Gierhan SME (2013). The language network. Curr Opin Neurobiol 23: 250–254. https://doi.org/10.1016/j.conb.2012.10.002.

Friederici AD, Chomsky N, Berwick RC et al. (2017). Language, mind and brain. Nat Hum Behav. https://doi.org/10.1038/s41562-017-0184-4.

Gallardo G, Wassermann D, Anwander A (2020). Bridging the gap: from neuroanatomical knowledge to tractography of brain pathways. Neuroscience. https://doi.org/10.1101/2020.08.01.232116.

Goldberg EB, Hillis AE (2022). Sign language aphasia. In: Handbook of clinical neurology, vol. 185. Elsevier 297–315. https://doi.org/10.1016/B978-0-12-823384-9.00019-0.

Goldin-Meadow S (2003). The resilience of language: what gesture creation in deaf children can tell us about how all children learn language, Psychology Press.

Goldin-Meadow S (2017). What the hands can tell us about language emergence. Psychon Bull Rev 24: 213–218. https://doi.org/10.3758/s13423-016-1074-x.

Goldin-Meadow S, Yang C (2017). Statistical evidence that a child can create a combinatorial linguistic system without external linguistic input: implications for language evolution. Neurosci Biobehav Rev 81: 150–157. https://doi.org/10.1016/j.neubiorev.2016.12.016.

Grech R, Cassar T, Muscat J et al. (2008). Review on solving the inverse problem in EEG source analysis. J Neuroeng Rehabil 5: 25. https://doi.org/10.1186/1743-0003-5-25.

Grégoire A, Deggouj N, Dricot L et al. (2022). Brain morphological modifications in congenital and acquired auditory deprivation: a systematic review and coordinate-based meta-analysis. Front Neurosci 16: 850245. https://doi.org/10.3389/fnins.2022.850245.

Gutiérrez E, Müller O, Baus C et al. (2012a). Electrophysiological evidence for phonological priming in Spanish sign language lexical access. Neuropsychologia 50: 1335–1346. https://doi.org/10.1016/j.neuropsychologia.2012.02.018.

Gutiérrez E, Williams D, Grosvald M et al. (2012b). Lexical access in American sign language: an ERP investigation of effects of semantics and phonology. Brain Res 1468: 63–83. https://doi.org/10.1016/j.brainres.2012.04.029.

Gutiérrez-Sigut E, Daws R, Payne H et al. (2015). Language lateralization of hearing native signers: a functional transcranial Doppler sonography (fTCD) study of speech and sign production. Brain Lang 151: 23–34. https://doi.org/10.1016/j.bandl.2015.10.006.

Gutiérrez-Sigut E, Payne H, MacSweeney M (2016). Examining the contribution of motor movement and language dominance to increased left lateralization during sign generation in native signers. Brain Lang 159: 109–117. https://doi.org/10.1016/j.bandl.2016.06.004.

Hagoort P (2017). The core and beyond in the language-ready brain. Neurosci Biobehav Rev. https://doi.org/10.1016/j.neubiorev.2017.01.048.

Hamilton A, Plunkett K, Schafer G (2000). Infant vocabulary development assessed with a British communicative development inventory. J Child Lang 27: 689–705. https://doi.org/10.1017/S0305000900004414.

Hammarström H, Forkel R, Haspelmath M et al. (2022). *Glottolog database* (v4.7) [dataset], Zenodo. https://doi.org/10.5281/ZENODO.7398962.

Hänel-Faulhaber B, Skotara N, Kügow M et al. (2014). ERP correlates of German sign language processing in deaf native signers. BMC Neurosci 15: 62. https://doi.org/10.1186/1471-2202-15-62.

Hauser MD, Chomsky N, Fitch WT (2002). The faculty of language: what is it, who has it, and how did it evolve? Science 298: 1569–1579. https://doi.org/10.1126/science.298.5598.1569.

Henrich J, Heine SJ, Norenzayan A (2010). The weirdest people in the world? Behav Brain Sci 33: 61–83. https://doi.org/10.1017/S0140525X0999152X.

Hickok G, Klima E, Kritchevsky M et al. (1995). A case of 'sign blindness' following left occipital damage in a deaf signer. Neuropsychologia 33: 1597–1606. https://doi.org/10.1016/0028-3932(95)00111-5.

Hickok G, Bellugi U, Klima ES (1996a). The neurobiology of sign language and its implications for the neural basis of language. Nature 381: 699–702. https://doi.org/10.1038/381699a0.

Hickok G, Kritchevsky M, Bellugi U et al. (1996b). The role of the left frontal operculum in sign language aphasia. Neurocase 2: 373–380. https://doi.org/10.1080/13554799608402412.

Hickok G, Say K, Bellugi U et al. (1996c). The basis of hemispheric asymmetries for language and spatial cognition: clues from focal brain damage in two deaf native signers. Aphasiology 10: 577–591. https://doi.org/10.1080/02687039608248438.

Hickok G, Bellugi U, Klima ES (1998). The neural organization of language: evidence from sign language aphasia. Trends Cogn Sci 2: 129–136. https://doi.org/10.1016/S1364-6613(98)01154-1.

Horwitz B, Amunts K, Bhattacharyya R et al. (2003). Activation of Broca's area during the production of spoken and signed language: A combined cytoarchitectonic mapping and PET analysis. Neuropsychologia 41: 1868–1876. https://doi.org/10.1016/S0028-3932(03)00125-8.

Hosemann J, Herrmann A, Steinbach M et al. (2013). Lexical prediction via forward models: N400 evidence from German sign language. Neuropsychologia 51: 2224–2237. https://doi.org/10.1016/j.neuropsychologia.2013.07.013.

Hosemann J, Herrmann A, Sennhenn-Reulen H et al. (2018). Agreement or no agreement. ERP correlates of verb agreement violation in

German sign language. Lang Cogn Neurosci 1–21. https://doi.org/10.1080/23273798.2018.1465986.
Inubushi T, Sakai KL (2013). Functional and anatomical correlates of word-, sentence-, and discourse-level integration in sign language. Front Hum Neurosci 7. https://doi.org/10.3389/fnhum.2013.00681.
Jednoróg K, Bola Ł, Mostowski P et al. (2015). Three-dimensional grammar in the brain: dissociating the neural correlates of natural sign language and manually coded spoken language. Neuropsychologia 71: 191–200. https://doi.org/10.1016/j.neuropsychologia.2015.03.031.
Klann J, Kastrau F, Huber W (2005). Lexical decision with no iconicity effect in German sign language: an fMRI-study. Brain Lang 95: 167–169. https://doi.org/10.1016/j.bandl.2005.07.091.
Klima ES, Bellugi U, Battison R et al. (1979). The signs of language, Harvard University Press.
Krauska A, Lau E (2023). Moving away from lexicalism in psycho- and neuro-linguistics. Front Lang Sci 2: 1125127. https://doi.org/10.3389/flang.2023.1125127.
Labache L, Mazoyer B, Joliot M et al. (2020). Typical and atypical language brain organization based on intrinsic connectivity and multitask functional asymmetries. eLife 9: e58722. https://doi.org/10.7554/eLife.58722.
Lenneberg EH (1964). The capacity for language acquisition. In: JA Fodor, JJ Katz (Eds.), The structure of language: readings in the philosophy of language. Prentice-Hall 579–603.
Lenneberg EH (1967). Biological foundations of language, Wiley.
Lenneberg EH (1969). On explaining language. Science 164: 635–643. https://doi.org/10.1126/science.164.3880.635.
Lenneberg EH (1970). What is meant by a biological approach to language? Am Ann Deaf 115: 67–72.
Levanen S (2001). Cortical representation of sign language: comparison of deaf signers and hearing non-signers. Cereb Cortex 11: 506–512. https://doi.org/10.1093/cercor/11.6.506.
Luck SJ, Gaspelin N (2017). How to get statistically significant effects in any ERP experiment (and why you shouldn't): how to get significant effects. Psychophysiology 54: 146–157. https://doi.org/10.1111/psyp.12639.
MacSweeney M, Cardin V (2015). What is the function of auditory cortex without auditory input? Brain 138: 2468–2470. https://doi.org/10.1093/brain/awv197.
MacSweeney M, Woll B, Campbell R et al. (2002a). Neural correlates of British sign language comprehension: spatial processing demands of topographic language. J Cogn Neurosci 14: 1064–1075. https://doi.org/10.1162/089892902320474517.
MacSweeney M, Woll B, Campbell R et al. (2002b). Neural systems underlying British sign language and audio-visual English processing in native users. Brain 125: 1583–1593. https://doi.org/10.1093/brain/awf153.
MacSweeney M, Campbell R, Woll B et al. (2006). Lexical and sentential processing in British sign language. Hum Brain Mapp 27: 63–76. https://doi.org/10.1002/hbm.20167.
Maran M, Friederici AD, Zaccarella E (2022a). Syntax through the looking glass: a review on two-word linguistic processing across behavioral, neuroimaging and neurostimulation studies. Neurosci Biobehav Rev 142: 104881. https://doi.org/10.1016/j.neubiorev.2022.104881.
Maran M, Numssen O, Hartwigsen G et al. (2022b). Online neurostimulation of Broca's area does not interfere with syntactic predictions: a combined TMS-EEG approach to basic linguistic combination. Front Psychol 13: 968836. https://doi.org/10.3389/fpsyg.2022.968836.
Marshall J, Atkinson JR, Smulovitch E et al. (2004). Aphasia in a user of British sign language: dissociation between sign and gesture. Cogn Neuropsychol 21: 537–554. https://doi.org/10.1080/02643290342000249.
Matchin W, İlkbaşaran D, Hatrak M et al. (2021). The cortical organization of syntactic processing is supramodal: evidence from American sign language. J Cogn Neurosci 34: 224–235. https://doi.org/10.1162/jocn_a_01790.
Mathur G, Rathmann C (2012). Verb agreement. In: R Pfau, M Steinbach, B Woll (Eds.), Sign language: an international handbook. de Gruyter 136–157.
Mathur G, Rathmann C (2014). The structure of sign languages. In: MA Goldrick, VS Ferreira, M Miozzo (Eds.), The Oxford handbook of language production. Oxford University Press 379–392.
Mayberry RI (2002). Cognitive development in deaf children: the interface of language and perception in neuropsychology. In: SJ Segalowitz, I Rapin (Eds.), Handbook of neuropsychology, vol. 8. Elsevier71–107.
Mayberry RI, Squires B (2006). Sign language: acquisition. In: Encyclopedia of language and linguistics, Elsevier, 291–296. https://doi.org/10.1016/B0-08-044854-2/00854-3.
Mayberry RI, Chen J-K, Witcher P et al. (2011). Age of acquisition effects on the functional organization of language in the adult brain. Brain Lang 119: 16–29. https://doi.org/10.1016/j.bandl.2011.05.007.
Mayberry RI, Davenport T, Roth A et al. (2018). Neurolinguistic processing when the brain matures without language. Cortex 99: 390–403. https://doi.org/10.1016/j.cortex.2017.12.011.
McCullough S, Emmorey K (2021). Effects of deafness and sign language experience on the human brain: voxel-based and surface-based morphometry. Lang Cogn Neurosci 36: 422–439. https://doi.org/10.1080/23273798.2020.1854793.
McCullough S, Saygin AP, Korpics F et al. (2012). Motion-sensitive cortex and motion semantics in American sign language. Neuroimage 63: 111–118. https://doi.org/10.1016/j.neuroimage.2012.06.029.
McGarry ME, Midgley KJ, Holcomb PJ et al. (2023). How (and why) does iconicity effect lexical access: an electrophysiological study of American sign language. Neuropsychologia 183: 108516. https://doi.org/10.1016/j.neuropsychologia.2023.108516.
McGuire PK, Robertson D, Thacker A et al. (1997). Neural correlates of thinking in sign language. Neuroreport 8: 695–698.
Meier RP (2016). Sign language acquisition, Oxford University Press. https://doi.org/10.1093/oxfordhb/9780199935345.013.19.
Mesch (2010). Perspectives on the concept and definition of international sign, World Federation of the Deaf. http://wfdeaf.org/wp-content/uploads/2016/11/Perspectives-on-the-Concept-and-Definition-of-IS_Mesch-FINAL.pdf.
Miranda M, Arias F, Arain A et al. (2022). Neuropsychological evaluation in American sign language: a case study of a deaf patient with epilepsy. Epilepsy Behav Rep 19: 100558. https://doi.org/10.1016/j.ebr.2022.100558.
Mitchell RE, Karchmer MA (2004). Chasing the mythical ten percent: parental hearing status of deaf and hard of hearing students in the United States. Sign Lang Stud 4: 138–163. https://doi.org/10.1353/sls.2004.0005.
Moreno A, Limousin F, Dehaene S et al. (2018). Brain correlates of constituent structure in sign language comprehension. Neuroimage 167: 151–161. https://doi.org/10.1016/j.neuroimage.2017.11.040.
Neville HJ, Coffey SA, Lawson DS et al. (1997). Neural systems mediating American sign language: effects of sensory experience and age of acquisition. Brain Lang 57: 285–308. https://doi.org/10.1006/brln.1997.1739.

Neville HJ, Bavelier D, Corina D et al. (1998). Cerebral organization for language in deaf and hearing subjects: biological constraints and effects of experience. Proc Natl Acad Sci U S A 95: 922–929. https://doi.org/10.1073/pnas.95.3.922.

Newman AJ, Supalla T, Hauser P et al. (2010). Dissociating neural subsystems for grammar by contrasting word order and inflection. Proc Natl Acad Sci U S A 107: 7539–7544. https://doi.org/10.1073/pnas.1003174107.

Newman AJ, Supalla T, Fernandez N et al. (2015). Neural systems supporting linguistic structure, linguistic experience, and symbolic communication in sign language and gesture. Proc Natl Acad Sci U S A 112: 11684–11689. https://doi.org/10.1073/pnas.1510527112.

Newport EL, Meier RP (1985). The acquisition of American sign language. In: DI Slobin (Ed.), The crosslinguistic study of language acquisition. Psychology Press881–938.

Obretenova S (2010). Neuroplasticity associated with tactile language communication in a deaf-blind subject. Front Hum Neurosci 3. https://doi.org/10.3389/neuro.09.060.2009.

Olulade OA, Koo DS, LaSasso CJ et al. (2014). Neuroanatomical profiles of deafness in the context of native language experience. J Neurosci 34: 5613–5620. https://doi.org/10.1523/JNEUROSCI.3700-13.2014.

Pallier C, Devauchelle A-D, Dehaene S (2011). Cortical representation of the constituent structure of sentences. Proc Natl Acad Sci U S A 108: 2522–2527. https://doi.org/10.1073/pnas.1018711108.

Payne H, Gutierrez-Sigut E, Woll B et al. (2019). Cerebral lateralisation during signed and spoken language production in children born deaf. Dev Cogn Neurosci 36: 100619. https://doi.org/10.1016/j.dcn.2019.100619.

Pénicaud S, Klein D, Zatorre RJ et al. (2013). Structural brain changes linked to delayed first language acquisition in congenitally deaf individuals. NeuroImage 66: 42–49. https://doi.org/10.1016/j.neuroimage.2012.09.076.

Peperkamp S, Mehler J (1999). Signed and spoken language: a unique underlying system? Lang Speech 42: 333–346. https://doi.org/10.1177/00238309990420020901.

Petitto LA (1987). On the autonomy of language and gesture: evidence from the acquisition of personal pronouns in American sign language. Cognition 27: 1–52. https://doi.org/10.1016/0010-0277(87)90034-5.

Petitto LA, Zatorre RJ, Gauna K et al. (2000). Speech-like cerebral activity in profoundly deaf people processing signed languages: implications for the neural basis of human language. Proc Natl Acad Sci 97: 13961–13966. https://doi.org/10.1073/pnas.97.25.13961.

Pfau R, Steinbach M, Woll B (Eds.), (2012). Sign language: an international handbook. de Gruyter.

Pfau R, Salzmann M, Steinbach M (2018). The syntax of sign language agreement: common ingredients, but unusual recipe. Glossa 3. https://doi.org/10.5334/gjgl.511.

Pickell H, Klima E, Love T et al. (2005). Sign language aphasia following right hemisphere damage in a left-hander: a case of reversed cerebral dominance in a deaf signer? Neurocase 11: 194–203. https://doi.org/10.1080/13554790590944717.

Poeppel D, Embick D (2013). Defining the relation between linguistics and neuroscience. In: A Cutler (Ed.), Twenty-first century psycholinguistics: four cornerstones. Psychology Press 103–118.

Poeppel D, Emmorey K, Hickok G et al. (2012). Towards a new neurobiology of language. J Neurosci 32: 14125–14131. https://doi.org/10.1523/JNEUROSCI.3244-12.2012.

Poizner H, Klima ES, Bellugi U (1987). What the hands reveal about the brain, MIT Press.

Quer J, Steinbach M (2019). Handling sign language data: the impact of modality. Front Psychol 10: 483. https://doi.org/10.3389/fpsyg.2019.00483.

Ramírez NF, Lieberman AM, Mayberry RI (2013). The initial stages of first-language acquisition begun in adolescence: when late looks early. J Child Lang 40: 391–414. https://doi.org/10.1017/S0305000911000535.

Sadato N, Yamada H, Okada T et al. (2004). Age-dependent plasticity in the superior temporal sulcus in deaf humans: a functional MRI study. BMC Neurosci 5: 56. https://doi.org/10.1186/1471-2202-5-56.

Saito K, Otsuki M, Ueno S (2007). Sign language aphasia due to left occipital lesion in a deaf signer. Neurology 69: 1466–1468. https://doi.org/10.1212/01.wnl.0000277455.08067.9c.

Sakai KL, Tatsuno Y, Suzuki K et al. (2005). Sign and speech: Amodal commonality in left hemisphere dominance for comprehension of sentences. Brain 128: 1407–1417. https://doi.org/10.1093/brain/awh465.

San José-Robertson L, Corina DP, Ackerman D et al. (2004). Neural systems for sign language production: mechanisms supporting lexical selection, phonological encoding, and articulation: neural systems for sign language production. Hum Brain Mapp 23: 156–167. https://doi.org/10.1002/hbm.20054.

Sandler W, Lillo-Martin DC (2001). Natural sign languages. In: M Aronoff, J Rees-Miller (Eds.), The handbook of linguistics. Blackwell 533–562.

Sandler W, Lillo-Martin DC (2008). Sign language and linguistic universals, Cambridge University Press.

Schönwiesner M, Rübsamen R, Von Cramon DY (2005). Hemispheric asymmetry for spectral and temporal processing in the human antero-lateral auditory belt cortex: spectro-temporal processing in human auditory cortex. Eur J Neurosci 22: 1521–1528. https://doi.org/10.1111/j.1460-9568.2005.04315.x.

Sehyr ZS, Caselli N, Cohen-Goldberg AM et al. (2021). The ASL-LEX 2.0 project: a database of lexical and phonological properties for 2,723 signs in American sign language. Jf Deaf Stud Deaf Edu 26: 263–277. https://doi.org/10.1093/deafed/enaa038.

Skeide MA, Friederici AD (2016). The ontogeny of the cortical language network. Nat Rev Neurosci 17: 323–332. https://doi.org/10.1038/nrn.2016.23.

Söderfeldt B, Ingvar M, Rönnberg J et al. (1997). Signed and spoken language perception studied by positron emission tomography. Neurology 49: 82–87. https://doi.org/10.1212/WNL.49.1.82.

Stokoe WC (1960). Sign language structure: an outline of the visual communication systems of the American deaf. In: Studies in linguistics, University of Buffalo.

Stroh A-L, Rösler F, Dormal G et al. (2019). Neural correlates of semantic and syntactic processing in German sign language. Neuroimage 200: 231–241. https://doi.org/10.1016/j.neuroimage.2019.06.025.

Stroh A, Grin K, Rösler F et al. (2022). Developmental experiences alter the temporal processing characteristics of the visual cortex: evidence from deaf and hearing native signers. Eur J Neurosciejn.15629. https://doi.org/10.1111/ejn.15629.

Tang G, Lau P (2012). Coordination and subordination. In: R Pfau, M Steinbach, B Woll (Eds.), Sign language: an international handbook. de Gruyter 340–365.

Terrace H, Petitto L, Sanders R et al. (1979). Can an ape create a sentence? Science 206: 891–902. https://doi.org/10.1126/science.504995.

Trettenbrein PC, Friederici AD (2025). Functional and structural brain asymmetries in language processing. In: C Papagno, P Corballis (Eds.), Handbook of clinical neurology, pp. 269–287.

Trettenbrein PC, Papitto G, Friederici AD et al. (2021a). Functional neuroanatomy of language without speech: an ALE meta-analysis of sign language. Hum Brain Mapp 42: 699–712. https://doi.org/10.1002/hbm.25254.

Trettenbrein PC, Pendzich N-K, Cramer J-M et al. (2021b). Psycholinguistic norms for more than 300 lexical signs in German sign language (DGS). Behav Res Methods 53: 1817–1832. https://doi.org/10.3758/s13428-020-01524-y.

Turkeltaub PE, Eickhoff SB, Laird AR et al. (2012). Minimizing within-experiment and within-group effects in activation likelihood estimation meta-analyses. Hum Brain Mapp 33: 1–13. https://doi.org/10.1002/hbm.21186.

Uddén J, Hultén A, Schoffelen J-M et al. (2022). Supramodal sentence processing in the human brain: fMRI evidence for the influence of syntactic complexity in more than 200 participants. Neurobiol Lang 3: 575–598. https://doi.org/10.1162/nol_a_00076.

van der Burght CL, Friederici AD, Maran M et al. (2023). Cleaning up the brickyard: how theory and methodology shape experiments in cognitive neuroscience of language. PsyArXiv. https://doi.org/10.31234/osf.io/6zpjq.

Vigneau M, Beaucousin V, Hervé P-Y et al. (2011). What is right-hemisphere contribution to phonological, lexico-semantic, and sentence processing? Neuroimage 54: 577–593. https://doi.org/10.1016/j.neuroimage.2010.07.036.

Warrington S, Bryant KL, Khrapitchev AA et al. (2020). XTRACT-standardised protocols for automated tractography in the human and macaque brain. Neuroimage 217: 116923. https://doi.org/10.1016/j.neuroimage.2020.116923.

Wilson SM, Entrup JL, Schneck SM et al. (2023). Recovery from aphasia in the first year after stroke. Brain 146: 1021–1039. https://doi.org/10.1093/brain/awac129.

Woolfe T, Herman R, Roy P et al. (2010). Early vocabulary development in deaf native signers: a British sign language adaptation of the communicative development inventories. J Child Psychol Psychiatry 51: 322–331. https://doi.org/10.1111/j.1469-7610.2009.02151.x.

Yang C, Crain S, Berwick RC et al. (2017). The growth of language: universal grammar, experience, and principles of computation. Neurosci Biobehav Rev. https://doi.org/10.1016/j.neubiorev.2016.12.023.

Zaccarella E, Trettenbrein PC (2021). Neuroscience and syntax. In: N Allott, T Lohndal, G Rey (Eds.), A companion to Chomsky. Wiley-Blackwell, 325–347. https://doi.org/10.1002/9781119598732.ch20.

Zaidel E (1978). The elusive right hemisphere of the brain. Eng Sci 42: 10–32. https://resolver.caltech.edu/CaltechES:42.1.hemisphere.

Zatorre RJ, Belin P, Penhune VB (2002). Structure and function of auditory cortex: music and speech. Trends Cogn Sci6: 37–46. https://doi.org/10.1016/S1364-6613(00)01816-7.

Zwitserlood I (2012). Classifiers. In: R Pfau, M Steinbach, B Woll (Eds.), Sign language: an international handbook. de Gruyter, 158–185.

Handbook of Clinical Neurology, Vol. 208 (3rd series)
Cerebral Asymmetries
C. Papagno and P. Corballis, Editors
https://doi.org/10.1016/B978-0-443-15646-5.00008-7

Chapter 22

Unveiling the hemispheric specialization of language: Organization and neuroplasticity

SONJA BANJAC AND MONICA BACIU*

Université Grenoble Alpes, Université Savoie Mont Blanc, CNRS, LPNC, Grenoble, France

Abstract

The advancements in understanding hemispheric specialization of language (HSL) have been following two primary avenues: the development of neuroimaging techniques and the study of its reorganizations in patients with various neuropathologic conditions. Hence, the objectives of this chapter are twofold. First, to provide an overview of the key neuroimaging techniques employed to investigate HSL, along with the notable findings derived from them in the healthy population. Second, it focuses on the reorganization of HSL in physiologic (healthy aging) and pathologic (poststroke aphasia and temporal lobe epilepsy) conditions. The chapter emphasizes the importance of employing multimodal methodologies to comprehend the complex relationship between underlying HSL mechanisms affected by disease and resulting language impairments. Combining the neuroimaging techniques can help us understand how different characteristics of language networks combine into general mechanisms that support their plasticity. Nevertheless, it highlights the need for standardized HSL metrics, as the absence of such metrics poses challenges in synthesizing findings across studies. Additionally, while HSL findings are being accumulated, albeit multimodal, there is a lack of integration within a robust theoretical framework. In conclusion, there is a need for novel models acknowledging multimodal aspects of HSL while positioning it within the context of other cognitive functions.

GENERAL CONSIDERATIONS

Functional hemispheric specialization underscores the significant differences between the two brain hemispheres regarding cognitive processing (Hellige, 2002; Hervé et al., 2013; Tzourio-Mazoyer et al., 2017; Vingerhoets, 2019). It highlights that the hemispheres are not functionally equivalent, as one hemisphere tends to support specific cognitive functions preferentially. This phenomenon reveals a fascinating aspect of our cognitive architecture, emphasizing the specialized and asymmetric nature of our hemispheres (Hellige, 2002; Badzakova-Trajkov et al., 2016; Tzourio-Mazoyer et al., 2017).

The notion of hemispheric specialization emerges from the clinical neuropsychologic perspective, wherein a specific cognitive function impairment occurs when a brain lesion is localized within one hemisphere (Corballis, 2021). However, it is essential to note that this definition is relative, as cognitive functions are primarily supported by one hemisphere but not exclusively. Consequently, "hemispheric complementarity" acknowledges that both hemispheres contribute to complex functions such as language (Lindell, 2006), which was also corroborated by studies (Vigneau et al., 2011). The right and left hemisphere participation in a specialized cognitive function follows a continuum of interhemispheric dynamics (Jung-Beeman, 2005). Nonetheless, one hemisphere is predominantly favored over the other. A typical example is the language function. The estimates of typical left hemispheric predominance show variable percentages ranging from 83% to 90% in right-handers and 64% to 80% in left-handers (Knecht, 2000; Carey and Johnstone, 2014; Mazoyer et al., 2014; Thornton et al., 2016). There are various explanations for left predominance (e.g., Bishop, 2013; Corballis, 2019; Olulade et al., 2020; Bisiacchi and Cainelli, 2022), but to present it in detail is not the objective of this chapter.

The study of language function has traditionally focused on the specialized regions within the left hemisphere involved in various neurolinguistic operations or intrahemispheric specialization. With the development of functional neuroimaging methods, the understanding of language networks has moved from several "eloquent" regions, such as the left inferior frontal gyrus (IFG, commonly referred to as Broca's area), primarily responsible

*Correspondence to: Monica Baciu, Université Grenoble Alpes, Université Savoie Mont Blanc, CNRS, LPNC, Grenoble, France. Tel: +33 (0)4 76 74 81 45, E-mail: monica.baciu@univ-grenoble-alpes.fr

for language production and the left posterior temporal cortex (known as Wernicke's area) responsible for language comprehension, toward a more distributed representation of language, emphasizing the interactions between various cognitive processes (Price, 2012; Middlebrooks et al., 2017; Hertrich et al., 2020; Monroy-Sosa et al., 2021).

Overall, the hemispheric specialization of language (HSL), as well as its regional representation in the predominant left hemisphere, have been the subject of extensive investigation (Mazoyer et al., 2014; Baciu and Perrone-Bertolotti, 2015; Tzourio-Mazoyer et al., 2017; Vingerhoets, 2019; Gurunandan et al., 2020; Malik-Moraleda et al., 2022). However, although leftward HSL is well established (Hausmann, 2019; Hausmann et al., 2019), it exhibits remarkable neural plasticity in various situations and conditions, including learning, aging, and brain damage (Gurunandan et al., 2020). Moreover, the interindividual variability of the HSL is multifactorial, including factors such as age, gender, and genetic basis (Tzourio-Mazoyer et al., 2004, 2017; Labache et al., 2023). We will only briefly touch upon handedness, the factor most frequently related to HSL, and only in healthy participants, given that pathologic conditions will be the focus of the second part of this chapter. Although strong, the relationship between handedness and HSL is nonlinear and nonsymmetrical (Kroliczak et al., 2021). For instance, a recent well-powered study found that despite a weak correlation between HSL assessed via dichotic listening and handedness, there were significantly more left-handed participants with atypical language HSL on the categorical level (Packheiser et al., 2020). While right-handed individuals have proportionally increased left HSL (Carey and Johnstone, 2014), most left-handers also exhibit left-lateralization for language. However, more left-handers than right-handers show bilateral or right lateralization (Knecht, 2000; Mazoyer et al., 2014; Woodhead et al., 2021). Another study has shown that left-handed participants have weaker HSL and that their HSL covariance is weaker between different language tasks than right-handed participants (Woodhead et al., 2021).

Neuroimaging techniques have played a crucial role in advancing our understanding of HSL. Each neuroimaging approach provides a unique perspective, contributing to a more comprehensive understanding of this complex phenomenon (D'Esposito et al., 2016). Just as initial understandings of HSL were gained through patient studies (Rutten, 2017; Corballis, 2021) exploring the reorganization of HSL in different healthy and pathologic conditions using diverse modern neuroimaging techniques can further progress in unraveling the intricacies of HSL. Hence, the aims of this chapter are twofold. First, it will provide an overview of the principal neuroimaging techniques and methodologies used for studying HSL and some general HSL findings obtained with each of them in healthy individuals. Second, it will delve into the reorganization of HSL in physiologic (healthy aging) and pathologic (poststroke aphasia and epilepsy) conditions. Finally, we will conclude this chapter by summarizing the main conclusions drawn from this overview and highlighting potential directions for future research in the field of HSL.

NEUROIMAGING EVIDENCE AND HEMISPHERIC PREDOMINANCE FOR LANGUAGE

HSL through the functional brain mapping

Measuring regional brain activity by blood oxygen level-dependent (BOLD) contrast during cognitive task performance, fMRI allows interferences about the association between localized brain activity and specific processes (D'Esposito et al., 2016). This observational, correlative technique is frequently utilized as the noninvasive substitution for the Wada test, even more so given the high correspondence between these two methods (Dym et al., 2011; Campbell et al., 2022). It is often used to identify an individual's specific functional network (e.g., language), primarily used in work with patients with epilepsy and aphasia, as we will later see.

Decades of using fMRI for studying HSL showed that it depends on methodological choices (Black et al., 2017; Bradshaw et al., 2017a,b). The first crucial aspect is how HSL is measured and expressed. Generally, the hemispheric dominance is expressed as the laterality index (LI), which expresses the relative activation difference between the left hemisphere (LH) and right hemisphere (RH) in task-related activity (Seghier, 2008, 2019; Agarwal et al., 2019). This classic LI ratio captures the general idea of hemispheric domination and is still the most widely used method (Bradshaw et al., 2017a). However, the classic LI encounters challenges such as disregarding interindividual signal strength variability, poor data quality, and, particularly, dependence on thresholds (Suarez et al., 2009; Seghier, 2019). There are numerous proposed solutions like threshold-independent methods such as curve LI (Abbott et al., 2010), average LI (Matsuo et al., 2012), weighted histogram (Suarez et al., 2009), and the area under the curve (Brumer et al., 2020). Although there is no widely accepted method, on their systematic review, Bradshaw et al. (2017a) recommend the use of bootstrapping (Wilke and Lidzba, 2007) as the preferred method for assessing HSL, which continues to be validated and supported by the literature (Johnstone et al., 2020).

The second aspect is that language encompasses multiple levels of processing (e.g., phonology, semantics, syntax), which despite some common regions, engage specific cerebral networks (Middlebrooks et al., 2017; Hertrich et al., 2020) and can exhibit different HSL (Bradshaw et al., 2017b; see Fig. 22.1). Employing multiple tasks when assessing an individual's HSL is recommended given that one task cannot simultaneously activate all language processes and dimensions (Black et al., 2017; Arantes and Cendes, 2020). Indeed, based on the Neurosynth database (Yarkoni et al., 2011), Lee et al. (2021) verified that tasks associated with semantics, naming, and syntax are strongly expressive and show higher engagement of the IFG, the tasks associated with listening and speech perception are weighted toward posterior and middle temporal gyri and Heschl's gyrus, while the tasks such as lexical, phonologic, sentence, and word generation are balanced across language network. The American Society of Functional Neuroradiology

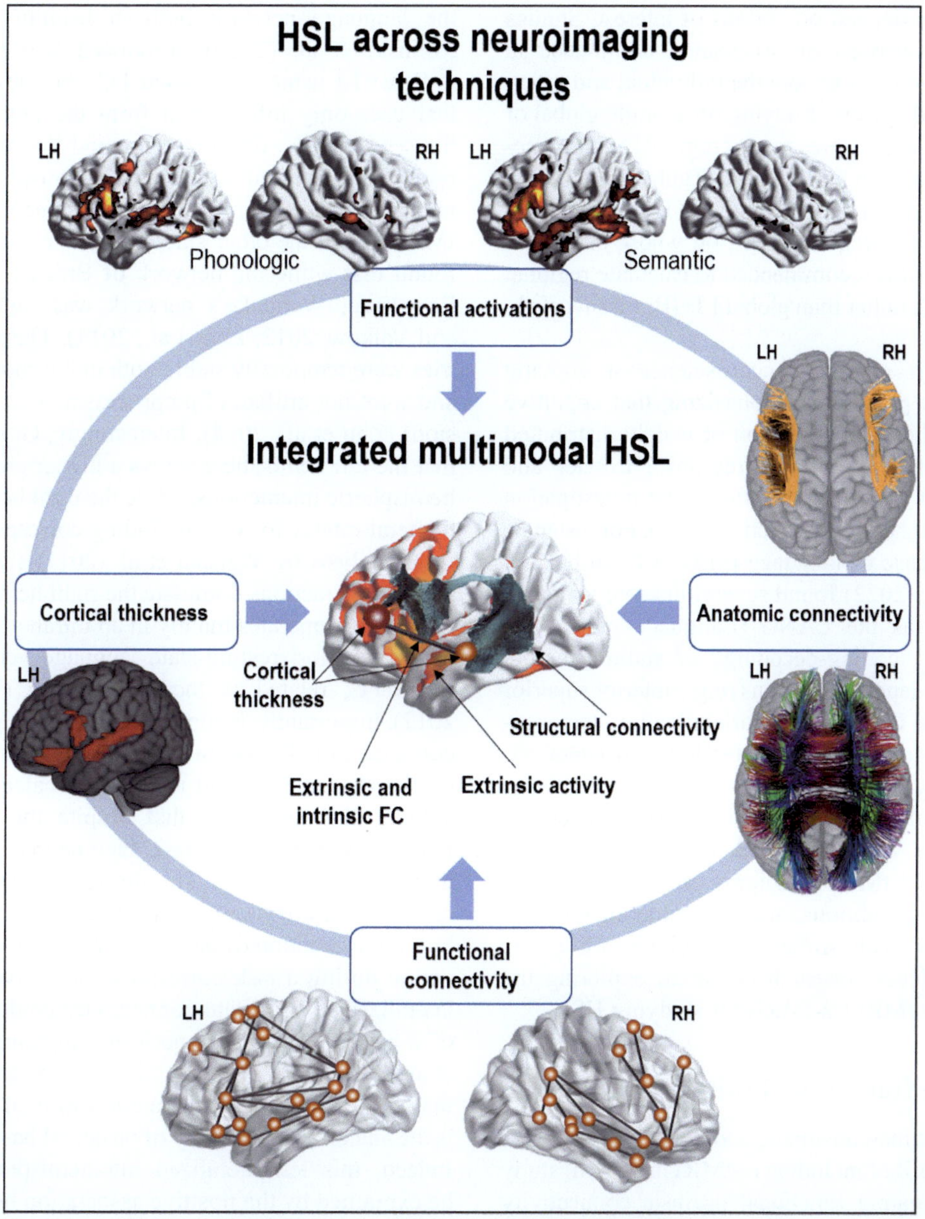

Fig. 22.1. Schematic illustration of HSL across neuroimaging techniques. Each neuroimaging approach offers a unique perspective, contributing to a more comprehensive understanding of the intricate nature of language lateralization. Functional brain mapping showed mostly left-lateralized language networks, albeit with variations in lateralization observed across different language tasks, particularly within the temporal regions. Here we present meta-analytic activation maps derived from the Neurosynth database (http://neurosynth.org; Yarkoni et al., 2011) for the terms "phonologic" and "semantic," which yielded 377 and 1031 studies, respectively. Studies on HSL based on FC did not identify specific lateralization patterns, yet they show concordance with task-based HSL and allow identifying relatively segregated left-lateralized networks showing left lateralization. Therefore, here we presented a schematic illustration of FC. Studies suggest that the relationship between cortical thickness and HSL is complex and nonlinear, possibly multifaced. Furthermore, it is suggested that VBM asymmetry measures are more accurate than unilateral cortical measures. These leftward asymmetries were found for IFG and are particularly replicated for the superior temporal gyrus. Differences in white matter were proposed to be underlying functional HSL. Although its relationship with HSL is not straightforward, ARC was found to have leftward lateralization in both macro and microstructure levels. Illustrations are based on the Population-Averaged Tractography Atlas of Yeh (2022). Finally, future studies should aim to find new composite or fusion metrics that would allow grasping the relationship between the different aspects of HSL revealed by various neuroimaging techniques. *LH*, left hemisphere; *RH*, right hemisphere.

highlighted that silent word generation and rhyming are more helpful in identifying lateralization than sentence completion or nonverb association (Black et al., 2017). Similarly, Bradshaw et al. (2017b) reported that the verbal fluency task exhibits robust lateralization, but its variants (semantic, verb, and phonemic) induce different lateralization patterns in language areas. Moreover, LI values can vary depending on whether the task is active or passive. Additionally, another study showed that overt speech could also influence the LI and prevent identifying HSL reliably (Berro et al., 2021).

Finally, LI values also depend on regions of interest. Studies have reported cases of crossed or dissociated dominance in different cortical language areas at both the individual and group levels, indicating the inadequacy of relying on a single global or regional LI and emphasizing the presence of regional heterogeneity in hemispheric lateralization for language (Tailby et al., 2017). Therefore, the third important aspect of studying HSL using task-based fMRI is the choice of regions based on whose activation the LI will be calculated. It is recommended to calculate regional LIs from predefined ROIs rather than global LIs (Bradshaw et al., 2017a).

Nevertheless, modern cognitive neuroscience is moving toward the connectomic paradigm, emphasizing that cognitive functions are based on the interconnection of widely distributed cortical networks rather than isolated regions (Bressler and Menon, 2010; Fornito et al., 2016), leading to the investigation of HSL by focusing on functional connectivity (FC). For instance, using an extensive database of language fMRI tasks in healthy participants, Roger et al. (2022) found several discrete language subnetworks regrouped as the LANG (Language) atlas. The network specialized for "coding-decoding" of auditory-verbal signals composed of key language regions (e.g., anterior superior temporal gyrus, posterior IFG, inferior parietal cortex, temporo-parietal junction, etc.) showed the strongest leftward lateralization, while the sensory-motor network was bilateral. Other regions in their LANG atlas showed more balanced connectivity between the hemispheres, and hence weaker leftward lateralization. However, task-based fMRI paradigms often demand time and additional technical conditions, and can sometimes be too demanding for patients with different cognitive difficulties (Cirillo et al., 2022). Hence, researchers started exploring the potential of resting-state fMRI (rs-fMRI) for studying HSL.

HSL through the functional brain connectivity

In addition to being less time-consuming and not cognitively burdening, the primary benefit of including rs-fMRI in an HSL study is that it allows investigating functional intrinsic connectivity (Labache et al., 2020), which some researchers suggest underlies left language lateralization in the majority of healthy participants (Wang et al., 2019). For instance, Wang et al. (2019) found that the differences in typical and atypical HSL in left-handers are reflected in the FC, particularly the strength of connections between the left IFG and other language regions. Atypical lateralization is associated with the global topology of brain networks, suggesting that stronger atypical lateralization is associated with better brain network organization.

Moreover, studies using rs-fMRI demonstrated that it can identify reproducible and relatively segregated left-lateralized networks related to articulatory, phonologic, and semantic processes (Battistella et al., 2020). Similar to task-based fMRI, rs-fMRI still does not have an accepted method for lateralization assessment. For instance, Mbwana et al. (2021) proposed FC hemispheric contrast based on the number and strength of connections within a hemisphere compared to those across hemispheres. Using this index, they identified the expected left-lateralization of the language network both in healthy adults and children. Rolinski et al. (2020) introduced two methods to calculate rs-fMRI LI using seed-based FC, the intrahemispheric method that uses only information from the hemisphere ipsilateral to the seed and the difference method that uses information from both ipsilateral and contralateral hemisphere to the seed. Both methods showed significant, but fair concordance with task-based fMRI LI. On the other side, using a seed-based FC, two studies found that while the network of Broca's area was mainly left-lateralized, Wernicke's network was right-lateralized (Tomasi and Volkow, 2012; Zhu et al., 2014). These functional asymmetries were temporally stable both in the short term and long term, and were not artifact of preprocessing (i.e., global signal regression) (Zhu et al., 2014). Interestingly, Gotts et al. (2013) found that the left hemisphere shows a higher preference toward intrahemispheric interactions, while the right hemisphere shows more bilateral interactions. This finding corresponds to a well-known meta-analysis by Vigneau et al. (2011) showing that interhemispheric interactions dominate the right hemisphere, while the left hemisphere operates mainly in an intrahemispheric manner. Furthermore, most resting-state language networks rely on short-range FC, related to local processing (Tomasi and Volkow, 2012). Importantly, Rolinski et al. (2020) highlighted that the concordance of HSL assessed with task-based and rs-fMRI depends on the chosen ROI and rs-fMRI LI calculation method. These authors also suggested that despite the relationship between HSL at rest and during a task, their degree of functional asymmetry may not be identical. This discrepancy between task-based and rs-fMRI lateralization might reveal different aspects of the cerebral representation of language. While more left-lateralized activation during a task corresponds to active processing, the less lateralized resting-state connectivity could reflect the readiness of a wider network to perform language processing (Doucet et al., 2017; Mbwana et al., 2021). It has long been hypothesized that the basis of functional lateralization of the language function is the anatomic asymmetry of its neural basis (Keller et al., 2011). Indeed, this left-lateralized intrahemispheric organization can be explained by the positive association between leftward lateralization of speech processing and brain volume since brain volume can increase distance, influencing transfer time and favoring intrahemispheric organization (i.e., strong lateralization) for high-speed processes (Gotts et al., 2013; Hervé et al., 2013). The following two sections provide more details on HSL association with gray and white matter asymmetries.

HSL through the anatomic cortical thickness assessed with VBM-MRI (voxel-based morphometry)

The relationship between structural asymmetry and functional lateralization of language remains to be determined, with inconsistent findings across different brain regions. While some studies suggested that the cortical thickness or brain volume might correspond to the language functional asymmetry (Josse et al., 2006; Plessen et al., 2014), other studies (Eckert et al., 2006; Keller et al., 2011; Greve et al., 2013; Gerrits et al., 2022) and a meta-analysis did not replicate this finding (Kong et al., 2018). Nevertheless, in the mentioned meta-analysis, Kong et al. (2018) found that the IFG and transverse temporal gyrus showed the largest leftward asymmetries in terms of surface area, while the other language regions, such as the triangular part of the IFG and inferior parietal gyrus showed a strong rightward asymmetry. Indeed, the leftward anatomic asymmetry of planum temporale has been replicated many times, although its association with functional lateralization remains unclear (Greve et al., 2013; Gerrits et al., 2022). Interestingly, another study (Bain et al., 2019) validated the anatomic asymmetries of the superior temporal sulcus across different independent datasets showing a leftward asymmetry of surface area but a rightward asymmetry of cortical thickness and sulcal depth. The rightward asymmetry was attributed to interruptions in the left STS sulcus, possibly caused by white matter. The most robust relationship with functional lateralization was found for the insula (Keller et al., 2011; Greve et al., 2013; Gerrits et al., 2022). A previous study also pointed out that subcortical structures should be given attention when investigating HSL (Guadalupe et al., 2022). This study combining VBM and dichotic listening showed that the asymmetry index of dichotic listening had the strongest association with gray matter asymmetry of the amygdala and lobule VI. However, the authors warn that although differences in anatomic asymmetry of language regions can be found between individuals with typical and atypical language HSL on the group level, this cannot be used as a predictor of hemispheric specialization on the individual level (Gerrits et al., 2022). These results indicate that the structural basis of the functional HSL is more complex and nonlinear, possibly multifaced (Dos Santos et al., 2006; Eckert et al., 2006; Kong et al., 2022). It was also suggested that the asymmetry measures are more accurate than unilateral cortical measures for differentiating between patients and healthy participants (Kong et al., 2018). Furthermore, it was also proposed that the lack of evidence for the relationship between anatomic asymmetry and HSL comes from methodological differences across studies and that in future studies, multiple metrics need to be integrated (Sprung-Much et al., 2022). Moreover, Sprung-Much et al. (2022) also open the possibility that structural asymmetries that underly functional HSL cannot be found on the gross cortical anatomy level but rather on the level of microanatomy.

HSL through the anatomic connectivity

Diffusion tensor imaging (DTI), a neuroimaging technique allowing the identification and microstructural integrity assessment of the white matter pathways (Catani et al., 2007), showed that the white matter of perisylvian language networks differs between LH and RH (Catani et al., 2007; Catani and Budisavljević, 2014). For instance, a study combining DTI and graph theory (Caeyenberghs and Leemans, 2014) found that LH is more performant than RH and that especially regions related to language in LH displayed increased nodal efficiency. These differences in intrahemispheric white matter connectivity were proposed to impact functional lateralization during language processing (Ocklenburg et al., 2016). Principal DTI measures used in HSL studies are mean diffusivity (MD), which represents the magnitude of water diffusion, and fractional anisotropy (FA), which indicates directionality. An increased MD is associated with the loss of myelin sheaths, cell membranes, or axons, while FA measures microstructural organization and integrity, reflecting the alignment of neuronal fibers (Perlaki et al., 2013; Catani and Budisavljević, 2014; Ocklenburg et al., 2016). For instance, Powell et al. (2006) found leftward lateralization of FA and tract volume. Although the volume did not correlate with HSL, the FA asymmetry correlated positively with lateralization of frontal lobes during verb generation and temporal lobes during reading comprehension, indicating that a more significant leftward FA asymmetry corresponded to greater functional activation asymmetry.

Moreover, some studies were also investigating the relationship between HSL and individual fibers. One prominent perspective suggests that the structural characteristics of the corpus callosum (CC) play a crucial role in HSL (Ocklenburg et al., 2016). A larger CC was found to be associated with a more pronounced leftward HSL in the IFG and posterior temporal regions, resulting from stronger left-sided functional activation in the IFG and posterior region and reduced activation in the right posterior temporal region with the increase of the CC (Josse et al., 2008). Atypical HSL was associated with enhanced interhemispheric connectivity manifested as higher anisotropic diffusion through the CC (Häberling et al., 2011). However, Ocklenburg et al. (2016) emphasize that it is necessary to consider the intrahemispheric white matter asymmetries to comprehend HSL. The arcuate fasciculus (ARC), connecting the ventrolateral frontal cortex to the posterior superior temporal lobe and closely associated with language processing, is often investigated in relation to HSL (see Fig. 22.1). ARC was found to have leftward lateralization in both macro and microstructure (Piervincenzi et al., 2016) using a voxel-based whole-brain approach (Ocklenburg et al., 2013), tract-specific DTI (Bain et al., 2019), and fixed-based analysis (Verhelst et al., 2021). Given this structural asymmetry of the ARC and the prevalence of left HSL, it was proposed that it could represent an anatomic substrate for HSL, although the data suggest that this relationship is not straightforward (Catani et al., 2007; Ocklenburg et al., 2014). While some studies found that this leftward ARC structural lateralization is associated with HSL (Powell et al., 2006; Ocklenburg et al., 2014, 2016), other studies did not find this relationship (Verhelst et al., 2021; Gerrits et al., 2022). For instance, Vernooij et al. (2007) reported that leftward asymmetries in fiber density of ARC were positively associated with left HSL during verb generation in right-handers, while this association was not found in left-handers. However, Verhelst et al. (2021) found no difference between participants with left and right HSL regarding the structural asymmetry of ARC, suggesting that there is a predominant leftward ARC asymmetry regardless of HSL or handedness.

However, the lateralization pattern of other tracts and its association with HSL has been less investigated (Verhelst et al., 2021). For instance, Perlaki et al. (2013) found in a sample of left-handed women that right HSL was associated with reduced microstructural integrity of superior longitudinal fasciculus (SLF) and white matter of the left parietal lobe, suggesting that decreased microstructural integrity of the language-related tracts in the LH can be linked to the rightward HSL. Moreover, the uncinate fasciculus (UF) connecting the anterior temporal lobe with the orbitofrontal cortex (Dick et al., 2014; Monroy-Sosa et al., 2021) was also found to exhibit left–right asymmetries and to be related to temporal HSL during the dichotic listening task (Ocklenburg et al., 2014).

Paradoxically, despite this predominant leftward structural lateralization, results suggest that bilateral structural representation and not extreme lateralization might be more beneficial for language performance (Catani et al., 2007; Piervincenzi et al., 2016).

NEUROPLASTICITY OF THE HSL

Neuroplasticity is a continuous process that allows the remodeling of brain networks over the short, medium, and long term to optimize their functioning (Duffau, 2006). This notion is closely related to the broad concept of homeostasis applied to the brain (Turrigiano and Nelson, 2000) since neural plasticity optimizes and stabilizes (Turrigiano and Nelson, 2004) cognitive functions. While neural changes occurring during development can be viewed as a natural form of plasticity that underlies learning and memory, brain lesions give rise to pathologically induced plasticity (Duffau, 2006). Interestingly, similar neural mechanisms underpin both natural and pathologically induced plasticity. Still, neuroplasticity or reorganization can be adaptive (efficient) or maladaptive (poor recovery due to ineffective "reprogramming") (Aghaz et al., 2018). Language reorganization, including at its hemispheric level, is dynamic and nonlinear (Saur, 2006; Kiran and Thompson, 2019). The HSL reorganization in healthy aging, poststroke aphasia, and temporal lobe epilepsy (TLE) will be now described. Our selection of language network plasticity models was motivated by several reasons. First, patients with TLE offer a unique opportunity to investigate HSL reorganization patterns associated with a chronic abnormal activity that can manifest from an early age. Second, poststroke patients with aphasia exhibit language impairments resulting from brain lesions, which can be attributed to both stroke-related factors and aging effects. Therefore, we included healthy aging as the third model to explore the reorganization of the left hemisphere language network that occurs throughout a typical lifespan. Importantly, language difficulties vary across these three models, enabling us to examine the associations between different forms of HSL plasticity and specific language impairments.

Reorganization of the HSL in healthy aging

Brain and cognitive functions change as individuals age, although the nature of these changes varies among older adults. Aging is a complex process characterized by different dynamics, including lifespan aging (LA) that encompasses the effects of chronological age and reserve aging (RA), which involves additional compensatory strategies employed by older adults with greater cognitive reserve capacity, as described by Barulli and Stern (2013) and Stern (2009). When considering language abilities, it is observed that language function experiences less decline compared to other cognitive functions. However, focusing on two primary dimensions, namely language production and language comprehension, studies by Baciu et al. (2016, 2021) and Wlotko et al. (2010) demonstrate that language production shows an earlier and more pronounced decline, while language comprehension remains relatively stable.

We will present the most common models and findings on aging-related HSL reorganization patterns (see Fig. 22.2). The PASA model (posterior–anterior shift in aging; Davis et al., 2008; Dennis and Cabeza, 2008) suggests a reduction in the activity of posterior occipitotemporal regions with increased frontal activity in older adults compared to younger adults. The HAROLD model (hemispheric asymmetry reduction in older adults; Cabeza, 2002) is based on the observation that older participants show reduced hemispheric asymmetry during the execution of various cognitive tasks compared to younger adults. This profile is mainly observed in higher-performing older adults (Manenti et al., 2013). The DECHA model (default-executive coupling hypothesis of aging; Spreng et al., 2018; Spreng and Turner, 2019) postulates that older adults fail to modulate the FC between regions of the executive control network (frontoparietal, FPN) typically recruited when task demands increase. In this case, older adults would rely more on their semantic knowledge stored in semantic memory, reflected by additional recruitment of the default mode network (DMN), which reflects semantic processes. This would decrease DMN deactivation during task performance and alter the connectivity within the FPN. Additionally, an increase in connectivity between the FPN and DMN is observed to facilitate task performance. Furthermore, He et al. (2020) demonstrated that age-related changes in connectomic properties of brain networks, specifically segregation and integration, occur particularly in intrinsic rs-fMRI brain networks. Specifically, there is a reduction in functional segregation (specificity or specialization of brain networks) with age and an increase in internetwork integration of intrinsic brain networks. A dedifferentiation process (reduced network specificity) occurs with age, necessitating improved network integration and connectivity (King et al., 2018; Chen et al., 2019). Furthermore, some aspects of language, such as phonology, decline with the aging of white matter, while other aspects, like semantics, remain resistant to these structural changes (Chen et al., 2021), although studies found that the aging of semantic fluency is influenced by gray matter volume in frontotemporal regions and underlying white matter (Zhu et al., 2022). Indeed, when older and young adults were compared regarding the white matter integrity, the effects of aging were observed within CC and UF (Fan et al., 2019). Similarly, the integrity of left but not right ARC was found to decline with age, suggesting asymmetry in the aging of this fiber tract (Ikuta et al., 2020). Additionally, consistent age effects were observed in cortical thickness in language-related regions such as IFG, superior and middle temporal gyri, fusiform gyrus, and temporoparietal junction (Fjell et al., 2009).

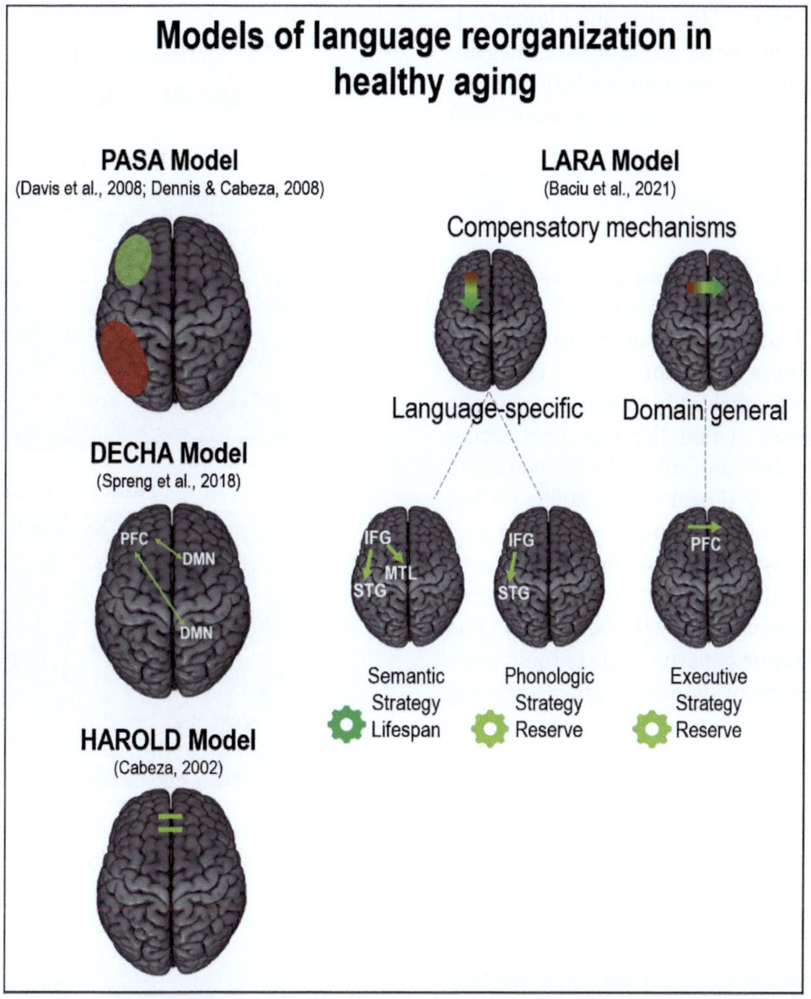

Fig. 22.2. Models of language reorganization in healthy aging. The PASA model proposes that there is an anterior–posterior language reorganization in older adults since a decline in regional occipitotemporal activity is coupled with increased frontal activity (Davis et al., 2008; Dennis and Cabeza, 2008). According to the DECHA model, language reorganization in the elderly is reflected as increased connectivity between lateral prefrontal regions and the default network, which on the cognitive level, manifests as a reduction of control resources and greater influence of the semantic knowledge (Spreng et al., 2018). The HAROLD model suggests that there is a reduction of left HSL in prefrontal regions in older adults during cognitive performance (Cabeza, 2002). The LARA model proposes that there are two interrelated mechanisms in older adults: language-specific and domain-general (Baciu et al., 2021). Language-specific mechanism refers to a functional shift from frontal to posterior regions and encompasses two strategies: (i) a semantic strategy that is associated with universal aging (or lifespan) and manifested as over-recruitment of temporoparietal and mesial temporal regions reflected in increased access to semantic memory. (ii) A phonologic strategy that mitigates phonologic deficits in aging by increasing connectivity between left IFG and superior temporal gyrus. This strategy is related to cognitive reserve capacity. Domain general mechanisms, dependent on the cognitive reserve, include recruitment of the contralateral hemisphere, especially in the prefrontal regions. *DMN*, default mode network; *IFG*, inferior frontal gyrus; *MTL*, mesial temporal lobe; *PFC*, prefrontal cortex; *STG*, superior temporal gyrus.

Compensatory mechanisms are implemented so that older adults can perform accurately like younger adults. They also explain the abovedescribed reorganization of HSL. These mechanisms can be language-specific, either lexico-semantic or phonologic (Dorot and Mathey, 2013; Valente and Laganaro, 2015; Boudiaf et al., 2018; Hoyau et al., 2018; Baciu et al., 2021) or can be nonlinguistic such as a supplementary involvement of the cognitive control (Salthouse, 1996; Murphy et al., 2000; LaGrone and Spieler, 2006; Baciu et al., 2021). In a previous work (Baciu et al., 2021), global findings on language production with aging, were formalized as a multifunctional compensatory model named LARA for lexical access and retrieval in aging, presenting complementary mechanisms that occur in healthy aging to palliate the decline of language production. LARA postulates a domain-general (executive functioning) and a language-specific compensatory domain according to the two dimensions of aging, LA and RA. While language-specific strategies rely on increasing cooperation between language subprocesses, domain-general mechanisms reflect executive strategies that depend on the cognitive reserve. Interestingly, a previous study (Brichko et al., 2022) showed that the relationship between cognitive reserve and white matter integrity changes during aging. Namely, higher levels of education and intellectual attainment in late middle adulthood were associated with better overall white matter microstructure, indicating the neuroprotective effects. However, the higher cognitive reserve was associated with poorer white matter integrity in older adults, possibly indicating a greater ability to compensate for age-related white matter declines among individuals with higher cognitive reserve.

Reorganization of the HSL in poststroke aphasia

The recovery of language function following a stroke exhibits considerable variability, with diverse cognitive and linguistic outcomes (van Oers et al., 2010). It is important to note that poststroke recovery encompasses different types, including spontaneous recovery and neurorehabilitation-based recovery. Studies usually focus on patients within the first year (acute and subacute stages) following the onset of stroke as the period of maximum spontaneous recovery (Lazar et al., 2008; Elkana et al., 2013). However, functional changes continue to occur during the chronic stage, primarily related to speech recovery (Meinzer et al., 2004; Cramer, 2008), although some elements of spontaneous recovery mechanisms may still be present. The evaluation of cerebral reorganization patterns in the chronic stage is typically conducted in the context of speech therapy, attributing reorganizational changes solely to intensive intervention. Such studies compared brain activation before and after the direct intervention, which generally included 2 weeks of speech therapy at most (Abo et al., 2004; Breier et al., 2006; Richter et al., 2008; Menke et al., 2009). In general, studies using functional imaging yielded highly heterogeneous recovery patterns across individual patients, making it difficult to draw general conclusions about the brain areas involved in the success of intensive interventions compared to regions recruited during spontaneous language rehabilitation (Peck et al., 2004; Crosson et al., 2005; Fridriksson et al., 2007; Menke et al., 2009).

In general, the prevailing view among researchers emphasizes the dominant (left) hemisphere role in language recovery (Léger et al., 2002; Breier et al., 2004; Fernandez et al., 2004, see Fig. 22.3). However, other authors suggest a compensatory role for the RH (Musso et al., 1999; Blasi et al., 2002). Some authors propose that the RH cannot fully support linguistic functions as effectively as the LH (Rosen et al., 2000; Naeser et al., 2004). Instead, the increased activation in the LH associated with language recovery may indicate a "disinhibition" of the RH resulting from impaired interhemispheric coordination due to stroke, along with a reduction in the normal inhibitory influence of the LH on the RH, which is typically observed in HC and responsible for hemispheric specialization (Thiel et al., 2006). However, conflicting results have been reported by Menke et al. (2009) and Elkana et al. (2013) suggesting that RH activation may be associated with better recovery. Consistent with these findings, Richter et al. (2008) showed that the initial activation in the right IFG and other RH regions positively correlates with language improvement. They propose that RH activation before aphasia therapy serves as a strong predictor of therapeutic success. Heiss and Thiel (2006) suggest that functional recovery following focal brain damage relies on the adaptive plasticity of unaffected networks. Nevertheless, studies have not found a consistent direct

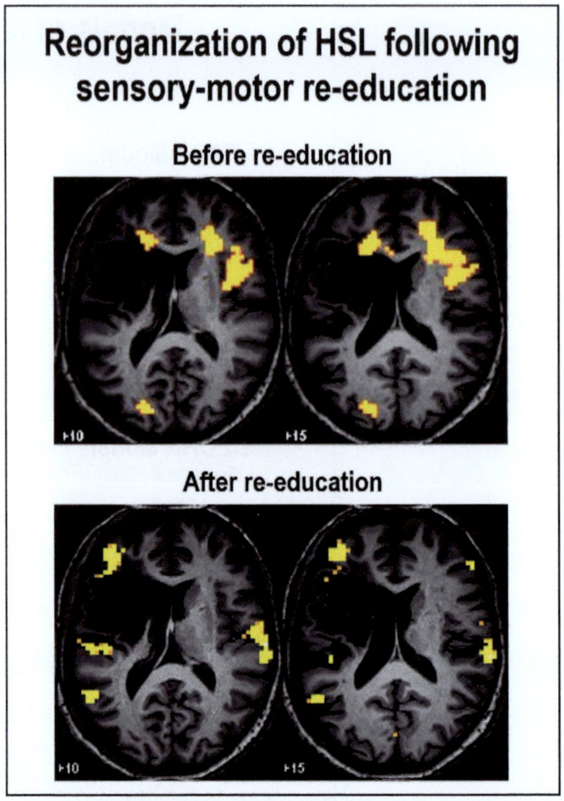

Fig. 22.3. Reorganization of HSL following sensory-motor reeducation. Here we illustrate unpublished data of our team showing how HSL during lexical production task changes in a patient with poststroke aphasia after having a 3-month reeducation based on the sensory-motor interaction.

association between RH activation and functional recovery (Price and Crinion, 2005; Heiss and Thiel, 2006; Crosson et al., 2007). Engagement of RH and LH changes across the recovery period (Hartwigsen and Saur, 2019). The enhanced RH activation is typically observed within the first 2 weeks after stroke but may return to control levels, characterized by LH predominance, after 1 year. In contrast, LH activity gradually increases over the course of months to years (Saur, 2006). This suggests that the early and temporary increase in RH activity does not reflect a functionally relevant and gradual reorganization process. Some studies have suggested that the early activation of RH areas reflects alterations in transcallosal inhibition and may even be maladaptive (Rosen et al., 2000). Additionally, a meta-analysis found only modest evidence for the recruitment of RH homologs in patients with aphasia (Wilson and Schneck, 2020). The functional significance of activation in homologous RH language areas in stroke patients requires further clarification (Hartwigsen and Saur, 2019).

Rs-fMRI is especially promising in studies on people with aphasia as participants do not need to perform a language task, making it accessible even for patients with the most severe language impairments. Indeed, reviewing rs-fMRI studies in poststroke aphasia, Klingbeil et al. (2019) concluded that they show connectivity disruption within language networks related to performance which tends to resolve with recovery.

Nevertheless, these studies still have small and heterogeneous samples and diverse analytic approaches, impeding the identification of consistent rs-FC reorganization patterns in poststroke aphasia. DTI studies also suggest the asymmetry effects both for performance and recovery. For instance, the integrity of the left language pathways, such as ARC, inferior fronto-occipital fasciculus (IFOF), and inferior longitudinal fasciculus (ILF), were found to be the best correlates of language performance, while no such association was found for the RH (Keser et al., 2020). Additionally, the spontaneous recovery after stroke was predicted by the left–right asymmetry of the ARC integrity (Hosomi et al., 2009) and by the possibility of reconstructing the left ARC regardless of the left ARC integrity (Kim and Jang, 2013). However, the better outcome of the therapy was associated with better pretreatment WM integrity of left temporal tracts (Bonilha et al., 2016).

Reorganization of the HSL in TLE

Epilepsy, particularly TLE, a highly prevalent (Fiest et al., 2017) neurologic disorder characterized by recurrent seizures and cognitive difficulties (Schoenberg et al., 2011; Barr, 2015), has significant implications for understanding the plasticity of HSL given that it affects the temporal lobe and that it is often associated with language networks reorganization (Thom and Bertram, 2012; Barr, 2015; Jaimes-Bautista et al., 2015). TLE patients, often drug-resistant and undergoing surgery, systematically undergo a presurgical evaluation to assess language, providing rich databases for understanding the structure–function base of language (Drane and Pedersen, 2019). FMRI data reveal a higher prevalence of atypical language representation in epilepsy patients (20–50%) compared to HC (2.5%) (Goldmann and Golby, 2005; Oh and Koh, 2009; Berl et al., 2014; Tailby et al., 2017), with a tendency toward bilaterality in TLE patients (Thornton et al., 2016; Tailby et al., 2017). Three types of language network reorganization are identified: interhemispheric, crossed interhemispheric, and intrahemispheric (Baciu and Perrone-Bertolotti, 2015; Balter et al., 2019). These reorganizations manifest as a "shift" of the typical left language networks to their right homologs (Baciu and Perrone-Bertolotti, 2015; Chang et al., 2017; Tzourio-Mazoyer et al., 2017; Balter et al., 2019; Foesleitner et al., 2021), a partial "shift" of the typical language regions to the RH (Baciu and Perrone-Bertolotti, 2015; Gonzálvez et al., 2016; Balter et al., 2019), or recruitment of nonspecialized regions in the dominant hemisphere (Cousin et al., 2008; Baciu and Perrone-Bertolotti, 2015; Gonzálvez et al., 2016; Drane and Pedersen, 2019), respectively. However, using a data-driven hierarchical clustering approach to LI obtained with semantic fMRI task, Berl et al. (2014) found six patterns of atypical HSL in epilepsy patients: symmetrically bilateral, two unilaterally crossed, and three right dominant patterns. These findings highlighted that atypical HSLs are more variable and that they should be evaluated based on the regional level. It is important to note that while TLE patients can have atypical neural substrates, often there is no substantial evidence of reorganization (Drane and Pedersen, 2019), and some studies found comparable rates of atypical HSL in TLE patients and controls for naming and semantic tasks (Gonzálvez et al., 2016; Lopes et al., 2019). Doucet et al. (2015) observed no difference in LI between TLE patients and HC, with less than 4% of patients showing right HSL. Authors associated this with the age of onset, as right HSL was found to be strongly related to an earlier age of seizure onset. Indeed, the probability and degree of cerebral reorganization in TLE are variable and related to various factors, such as the location of the epileptogenic zone in the language-dominant hemisphere, hippocampal atrophy (HA), and epilepsy duration (Stewart et al., 2014; Baciu and Perrone-Bertolotti, 2015; Thornton et al., 2016; Chang et al., 2017; Marcelle et al., 2022). LTLE patients are more likely to exhibit atypical HSL (e.g., bilateral) than RTLE patients, who usually show left HSL (Rosazza et al., 2013; Neudorf et al., 2020). Decreased left frontal lateralization during verbal fluency tasks is linked to better performance and the absence of periictal aphasia in LTLE patients, suggesting that engagement of contralateral regions can help avoid language obstruction (Rosazza et al., 2013). HA is associated with a higher degree of atypical HSL in both frontal and temporal language regions (Thornton et al., 2016), but it can also be associated with intrahemispheric reorganization, such as wider dispersion of posterior language sites (Benjamin et al., 2020). Nevertheless, Lopes et al. (2019) found that although HA has a negative impact on the language network, it does not affect HSL.

Previous studies in TLE have also used fMRI data to investigate language network reorganization based on FC. Indeed, some report that task-based fMRI FC is more sensitive to language network reorganization than activation analyses in TLE patients (Foesleitner et al., 2021). An altered language connectome has been observed in TLE patients, with decreased frontotemporal and fronto-hippocampal FC (Foesleitner et al., 2021). Importantly, He et al. (2018) demonstrated that combining task-based with rs-fMRI FC reveals dynamic properties of language network, providing a more accurate prediction of verbal fluency than traditional activation-based measures. Resting-state fMRI studies have also provided valuable insights into TLE-related HSL reorganization. For instance, regardless of similar activation patterns of TLE and HC, LTLE patients had reduced FC between language regions compared to HC (Waites et al., 2006; Pravatà et al., 2011). Moreover, stronger leftward HSL in LTLE was associated with the increased FC between the epileptogenic mesial temporal lobe and left IFG. Conversely, in RLTE, stronger leftward HSL was associated with reduced FC between the epileptogenic hippocampus and left IFG (Doucet et al., 2015). Resting-state FC also revealed the intrahemispheric reorganization pattern as epilepsy patients with atypical HSL had a more dispersed resting-state language network compared to those with typical HSL (DeSalvo et al., 2014), and both LTLE and RTLE showed higher leftward lateralization of resting-state language network than HC (Doucet et al., 2017). There is also evidence for crossed interhemispheric reorganization, as LTLE patients showed more rightward lateralization of rs-FC than RTLE patients in temporo-parietal regions only (Whitten et al., 2021).

Studies in TLE patients suggest that anatomic asymmetries can be related to HSL. Volumetric MRI studies support the role of left lateral frontotemporal and medial temporal lobe structures in language performances and their impairment in TLE, as well as that atypical HSL can be related to structural shifts to RH homologous language regions in LTLE patients supporting more efficient language reorganization (Balter et al., 2019). Keller et al. (2018) reported greater rightward asymmetry in the mesial temporal lobe and superior temporal gyrus in right-handed patients with left HSL compared to HC. Direct comparison of LTLE patients with left and right HSL revealed differences in structural asymmetries of language regions such as pars triangularis and orbitofrontal cortex, among others (Keller et al., 2018). Although not focusing on HSL, a mega-analytic study found that language-related hubs, such as precentral, superior temporal cortices, ipsilateral hippocampus, and thalamus, are among atrophy-prone regions in TLE (Larivière et al., 2020). However, studies do not consistently report an association between HSL or language performance in LTLE patients and cortical thickness measures but rather with changes in the white matter tracts (Kucukboyaci et al., 2014; Chang et al., 2017; Kaestner et al., 2019).

Studies suggest that atypical HSL and altered extrinsic frontotemporal connectivity in TLE may be associated with microstructural alterations in perisylvian tracts (Chang et al., 2017; Reyes et al., 2019; Trimmel et al., 2021; Caciagli et al., 2023). Indeed, compared to HC and RTLE, patients with LTLE demonstrate reduced LH and increased RH structural connections, reflecting altered HSL (Powell et al., 2007; Thornton et al., 2016; Balter et al., 2019). Specifically, LTLE patients show decreased asymmetry in functional and microstructural measures in the ARC (Chang et al., 2017; Balter et al., 2019). Language difficulties in TLE patients relate to impaired left ARC and ILF, but LTLE patients without language difficulties still display tract damage, suggesting that it is necessary but insufficient language impairment conditions (Kaestner et al., 2019). Furthermore, LTLE patients show greater tract integrity in the right ILF, IFOF, and UF (Neudorf et al., 2020). This contralateral reorganization suggests an adaptive compensatory mechanism supporting more bilateral language processing (Thornton et al., 2016; Tailby et al., 2017; Neudorf et al., 2020).

CONCLUSIONS AND PERSPECTIVES

Understanding HSL through patient studies is not a novel concept (Rutten, 2017; Corballis, 2021) but advancements in neuroimaging techniques have provided a more nuanced understanding. These techniques shed light on how various conditions impact specific aspects of language networks, highlighting their plasticity and facilitating comprehension of structure–function relationships. Indeed, findings in patients have inspired various models of language reorganization in aphasia (Chang and Lambon Ralph, 2020; Stefaniak et al., 2020), aging (Reuter-Lorenz and Park, 2014; Spreng and Turner, 2019; Baciu et al., 2021), and TLE (Baciu and Perrone-Bertolotti, 2015). Moreover, studies in TLE patients provided important information about the role of the hippocampus in language and HSL (Tracy and Boswell, 2008; Hamberger and Cole, 2011; Baciu and Perrone-Bertolotti, 2015) although some highlight that the findings are inconclusive (see Drane and Pedersen, 2019).

Nevertheless, we have noticed a growing trend in the field toward conducting multimodal and multicenter studies (Kong et al., 2018; Larivière et al., 2020; Sisodiya et al., 2020), aiming to validate and expand upon the findings of previous studies with small sample sizes by incorporating diverse methodologies. Several authors (Piervincenzi et al., 2016; Chang et al., 2017; Doucet et al., 2017; Hartwigsen and Saur, 2019) suggested that multimodal methodology is needed to understand the association between the underlying mechanisms that are affected by the disease and language impairment. By integrating these approaches, we can better understand how various aspects of the language network (WM, GM, FC) combine to form overarching mechanisms that support language plasticity. For instance, contradictory results regarding the compensatory role of RH (Musso et al., 1999; Blasi et al., 2002) or LH dominance (Léger et al., 2002; Breier et al., 2004; Fernandez et al., 2004) in poststroke aphasia can be related to DTI studies finding the left perisylvian regions to be the best correlates of language performance (Keser et al., 2020). However, we firmly believe that the next crucial step is to dedicate efforts toward developing models and theories that can establish connections between these multimodal findings, offering a solid theoretical framework and enabling predictions in the field (see Fig. 22.1). Although assessing HSL from a connectomic perspective is on the rise with the augmentation of the number of FC (task-based or rs-fMRI) and DTI studies, the association between HSL and other functions remains to be explored.

Finally, as we showed, even with highly used methods such as fMRI, there is no standardized approach for calculating different HSL metrics (although there are recommendations, Bradshaw et al., 2017a), particularly pronounced in rs-fMRI and DTI studies. Consequently, consolidating results across different studies and identifying HSL phenotypes in these conditions pose significant challenges.

In conclusion, unifying methods and analyses are crucial to advance our understanding of HSL while focusing on those that would allow the fusion of multimodal datasets. This initiative should be followed by efforts to integrate obtained findings through models and theories that would acknowledge multimodal aspects of language organization while positioning it within the broader context of other cognitive functions.

FUNDING SOURCES

This work was funded by the French program "AAP GENERIQUE 2017" run by the "Agence Nationale pour la Recherche" grant "REORG" [grant number ANR-17-CE28-0015-01]; and by NeuroCoG IDEX UGA within the "Investissements d'avenir" program [grant number ANR-15-IDEX-02]. IRMaGe MRI/Neurophysiology facility was partly funded by the French program "Investissement d'Avenir" run by the "Agence Nationale pour la Recherche"; grant "Infrastructure d'avenir en Biologie Santé" [grant number ANR-11-INBS-0006].

References

Abbott DF, Waites AB, Lillywhite LM et al. (2010). fMRI assessment of language lateralization: an objective approach. Neuroimage 50: 1446–1455.

Abo M, Senoo A, Watanabe S et al. (2004). Language-related brain function during word repetition in post-stroke aphasics. Neuroreport 15: 1891–1894.

Agarwal S, Sair HI, Gujar S et al. (2019). Language mapping with fMRI: current standards and reproducibility. Top Magn Reson Imaging 28: 225–233.

Aghaz A, Hemmati E, Ghasisin L (2018). Types of neuroplasticity and factors affecting language recovery in patients with aphasia: a systematic review. Arch Neurosci 5: e62265.

Arantes ME, Cendes F (2020). In search of a new paradigm for functional magnetic resonance experimentation with language. Front Neurol 11: 588.

Baciu M, Perrone-Bertolotti M (2015). What do patients with epilepsy tell us about language dynamics? A review of fMRI studies. Rev Neurosci 26323–341.

Baciu M, Boudiaf N, Cousin E et al. (2016). Functional MRI evidence for the decline of word retrieval and generation during normal aging. Age 38: 3.

Baciu M, Banjac S, Roger E et al. (2021). Strategies and cognitive reserve to preserve lexical production in aging. GeroScience 43: 1725–1765.

Badzakova-Trajkov G, Corballis MC, Häberling IS (2016). Complementarity or independence of hemispheric specializations? A brief review. Neuropsychologia 93: 386–393.

Bain JS, Filo S, Mezer AA (2019). The robust and independent nature of structural STS asymmetries. Brain Struct Funct 224: 3171–3182.

Balter S, Lin G, Leyden KM et al. (2019). Neuroimaging correlates of language network impairment and reorganization in temporal lobe epilepsy. Brain Lang 193: 31–44.

Barr WB (2015). Neuropsychological assessment of patients with epilepsy. In: WB Barr, C Morrison (Eds.), Handbook on the neuropsychology of epilepsy. Springer.

Barulli D, Stern Y (2013). Efficiency, capacity, compensation, maintenance, plasticity: emerging concepts in cognitive reserve. Trends Cogn Sci 17: 502–509.

Battistella G, Borghesani V, Henry M et al. (2020). Task-free functional language networks: reproducibility and clinical application. J Neurosci 40: 1311–1320.

Benjamin CFA, Gkiatis K, Matsopoulos GK et al. (2020). Presurgical language fMRI in epilepsy: an introduction. In: GPD Argyropoulos (Ed.), Translational neuroscience of speech and language disorders. Springer International Publishing, Cham.

Berl MM, Zimmaro LA, Khan OI et al. (2014). Characterization of atypical language activation patterns in focal epilepsy: language activation patterns. Ann Neurol 75: 33–42.

Berro DH, Lemée J-M, Leiber L-M et al. (2021). Overt speech critically changes lateralization index and did not allow determination of hemispheric dominance for language: an fMRI study. BMC Neurosci 22: 74.

Bishop DVM (2013). Cerebral asymmetry and language development: cause, correlate, or consequence? Science 340: 1230531.

Bisiacchi P, Cainelli E (2022). Structural and functional brain asymmetries in the early phases of life: a scoping review. Brain Struct Funct 227: 479–496.

Black DF, Vachha B, Mian A et al. (2017). American society of functional neuroradiology–recommended fMRI paradigm algorithms for presurgical language assessment. Am J Neuroradiol 38: E65–E73.

Blasi V, Young AC, Tansy AP et al. (2002). Word retrieval learning modulates right frontal cortex in patients with left frontal damage. Neuron 36: 159–170.

Bonilha L, Gleichgerrcht E, Nesland T et al. (2016). Success of anomia treatment in aphasia is associated with preserved architecture of global and left temporal lobe structural networks. Neurorehabil Neural Repair 30: 266–279.

Boudiaf N, Laboissière R, Cousin É et al. (2018). Behavioral evidence for a differential modulation of semantic processing and lexical production by aging: a full linear mixed-effects modeling approach. Neuropsychol Dev Cogn B Aging Neuropsychol Cogn 25: 1–22.

Bradshaw AR, Bishop DVM, Woodhead ZVJ (2017a). Methodological considerations in assessment of language lateralisation with fMRI: a systematic review. Peer J 5: e3557.

Bradshaw AR, Thompson PA, Wilson AC et al. (2017b). Measuring language lateralisation with different language tasks: a systematic review. Peer J 5: e3929.

Breier JI, Castillo EM, Boake C et al. (2004). Spatiotemporal patterns of language-specific brain activity in patients with chronic aphasia after stroke using magnetoencephalography. Neuroimage 23: 1308–1316.

Breier JI, Maher LM, Novak B et al. (2006). Functional imaging before and after constraint-induced language therapy for aphasia using magnetoencephalography. Neurocase 12: 322–331.

Bressler SL, Menon V (2010). Large-scale brain networks in cognition: emerging methods and principles. Trends Cogn Sci 14: 277–290.

Brichko R, Soldan A, Zhu Y et al. (2022). Age-dependent association between cognitive reserve proxy and longitudinal white matter microstructure in older adults. Front Psychol 13: 859826.

Brumer I, De Vita E, Ashmore J et al. (2020). Implementation of clinically relevant and robust fMRI-based language lateralization: choosing the laterality index calculation method. PLoS One 15: e0230129.

Cabeza R (2002). Hemispheric asymmetry reduction in older adults: the HAROLD model. Psychol Aging 17: 85–100.

Caciagli L, Paquola P, He X et al. (2023). Disorganization of language and working memory systems in frontal versus temporal lobe epilepsy. Brain 146: 935–953

Caeyenberghs K, Leemans A (2014). Hemispheric lateralization of topological organization in structural brain networks: hemispheric lateralization in structural brain networks. Hum Brain Mapp 35: 4944–4957.

Campbell JM, Kundu B, Lee JN et al. (2022). Evaluating the concordance of functional MRI-based language lateralization and Wada testing in epilepsy patients: a single-center analysis. Interv Neuroradiol 29: 599–604.

Carey DP, Johnstone LT (2014). Quantifying cerebral asymmetries for language in dextrals and adextrals with random-effects meta analysis. Front Psychol 5: 1128.

Catani M, Budisavljević S (2014). Contribution of diffusion tractography to the anatomy of language. Diffusion MRI, Elsevier.

Catani M, Allin MPG, Husain M et al. (2007). Symmetries in human brain language pathways correlate with verbal recall. Proc Natl Acad Sci U S A 104: 17163–17168.

Chang Y-N, Lambon Ralph MA (2020). A unified neurocomputational bilateral model of spoken language production in healthy participants and recovery in poststroke aphasia. Proc Natl Acad Sci U S A 117: 32779–32790.

Chang Y-HA, Kemmotsu N, Leyden KM et al. (2017). Multimodal imaging of language reorganization in patients with left temporal lobe epilepsy. Brain Lang 170: 82–92.

Chen Q, Xia Y, Zhuang K et al. (2019). Decreased inter-hemispheric interactions but increased intra-hemispheric integration during typical aging. Aging 11: 10100–10115.

Chen P-Y, Chen C-L, Tseng H-M et al. (2021). Differential associations of white matter brain age with language-related mechanisms in word-finding ability across the adult lifespan. Front Aging Neurosci 13: 701565.

Cirillo S, Battistella G, Castellano A et al. (2022). Comparison between inferior frontal gyrus intrinsic connectivity network and verb-generation task fMRI network for presurgical language mapping in healthy controls and in glioma patients. Brain Imaging Behav 16: 2569–2585.

Corballis MC (2019). Evolution of cerebral asymmetry. Prog Brain Res 250: 153–178.

Corballis MC (2021). Humanity and the left hemisphere: the story of half a brain. Laterality 26: 19–33.

Cousin E, Baciu M, Pichat C et al. (2008). Functional MRI evidence for language plasticity in adult epileptic patients: preliminary results. Neuropsychiatr Dis Treat 4: 235–246.

Cramer SC (2008). Repairing the human brain after stroke: I. Mechanisms of spontaneous recovery. Ann Neurol 63: 272–287.

Crosson B, Moore AB, Gopinath K et al. (2005). Role of the right and left hemispheres in recovery of function during treatment of intention in aphasia. J Cogn Neurosci 17: 392–406.

Crosson B, McGregor K, Gopinath KS et al. (2007). Functional MRI of language in aphasia: a review of the literature and the methodological challenges. Neuropsychol Rev 17: 157–177.

D'Esposito M, Kayser A, Chen A (2016). Functional MRI: applications in cognitive neuroscience. In: M Filippi (Ed.), fMRI techniques and protocols, second edn. Humana Press, New York, NY.

Davis SW, Dennis NA, Daselaar SM et al. (2008). Qué PASA? The posterior–anterior shift in aging. Cereb Cortex 18: 1201–1209.

Dennis NA, Cabeza R (2008). Neuroimaging of healthy cognitive aging. In: C FIM, TA Salthouse (Eds.), The handbook of aging and cognition. Psychology Press, New York, NY, pp. 1–54.

DeSalvo MN, Douw L, Takaya S et al. (2014). Task-dependent reorganization of functional connectivity networks during visual semantic decision making. Brain Behav 4: 877–885.

Dick AS, Bernal B, Tremblay P (2014). The language connectome: new pathways, new concepts. Neuroscientist 20: 453–467.

Dorot D, Mathey S (2013). Accès aux représentations sémantiques et phonologiques chez des adultes jeunes et âgés : une étude des mots sur le bout de la langue. Psychol FR 58: 1–16.

Dos Santos SS, Woerner W, Walter C et al. (2006). Handedness, dichotic-listening ear advantage, and gender effects on planum temporale asymmetry—a volumetric investigation using structural magnetic resonance imaging. Neuropsychologia 44: 622–636.

Doucet GE, Pustina D, Skidmore C et al. (2015). Resting-state functional connectivity predicts the strength of hemispheric lateralization for language processing in temporal lobe epilepsy and normals: RsFC predicts language specialization. Hum Brain Mapp 36: 288–303.

Doucet GE, He X, Sperling MR et al. (2017). From "rest" to language task: task activation selects and prunes from broader resting-state network: language network spatial configuration change. Hum Brain Mapp 38: 2540–2552.

Drane DL, Pedersen NP (2019). Knowledge of language function and underlying neural networks gained from focal seizures and epilepsy surgery. Brain Lang 189: 20–33.

Duffau H (2006). Brain plasticity: from pathophysiological mechanisms to therapeutic applications. J Clin Neurosci 13: 885–897.

Dym RJ, Burns J, Freeman K et al. (2011). Is functional MR imaging assessment of hemispheric language dominance as good as the wada test?: a meta-analysis. Radiology 261: 446–455.

Eckert MA, Leonard CM, Possing ET et al. (2006). Uncoupled leftward asymmetries for planum morphology and functional language processing. Brain Lang 98: 102–111.

Elkana O, Frost R, Kramer U et al. (2013). Cerebral language reorganization in the chronic stage of recovery: a longitudinal fMRI study. Cortex 49: 71–81.

Fan Y-T, Fang Y-W, Chen Y-P et al. (2019). Aging, cognition, and the brain: effects of age-related variation in white matter integrity on neuropsychological function. Aging Ment Health 23: 831–839.

Fernandez B, Cardebat D, Demonet J-F et al. (2004). Functional MRI follow-up study of language processes in healthy subjects and during recovery in a case of aphasia. Stroke 35: 2171–2176.

Fiest KM, Sauro KM, Wiebe S et al. (2017). Prevalence and incidence of epilepsy. Neurology 88: 296–303.

Fjell AM, Westlye LT, Amlien I et al. (2009). High consistency of regional cortical thinning in aging across multiple samples. Cereb Cortex 19: 2001–2012.

Foesleitner O, Sigl B, Schmidbauer V et al. (2021). Language network reorganization before and after temporal lobe epilepsy surgery. J Neurosurg 134: 1694–1702.

Fornito A, Zalesky A, Bullmore ET (2016). Fundamentals of brain network analysis, Elsevier/Academic Press, Amsterdam; Boston.

Fridriksson J, Moser D, Bonilha L et al. (2007). Neural correlates of phonological and semantic-based anomia treatment in aphasia. Neuropsychologia 45: 1812–1822.

Gerrits R, Verhelst H, Dhollander T et al. (2022). Structural perisylvian asymmetry in naturally occurring atypical language dominance. Brain Struct Funct 227: 573–586.

Goldmann RE, Golby AJ (2005). Atypical language representation in epilepsy: implications for injury-induced reorganization of brain function. Epilepsy Behav 6: 473–487.

Gonzálvez GG, Trimmel K, Haag A et al. (2016). Activations in temporal areas using visual and auditory naming stimuli: a language fMRI study in temporal lobe epilepsy. Epilepsy Res 128: 102–112.

Gotts SJ, Jo HJ, Wallace GL et al. (2013). Two distinct forms of functional lateralization in the human brain. Proc Natl Acad Sci U S A 110: E3435–E3444.

Greve DN, Van der Haegen L, Cai Q et al. (2013). A surface-based analysis of language lateralization and cortical asymmetry. J Cogn Neurosci 25: 1477–1492.

Guadalupe T, Kong X-Z, Akkermans SEA et al. (2022). Relations between hemispheric asymmetries of grey matter and auditory processing of spoken syllables in 281 healthy adults. Brain Struct Funct 227: 561–572.

Gurunandan K, Arnaez-Telleria J, Carreiras M et al. (2020). Converging evidence for differential specialization and plasticity of language systems. J Neurosci 40: 9715–9724.

Häberling IS, Badzakova-Trajkov G, Corballis MC (2011). Callosal tracts and patterns of hemispheric dominance: a combined fMRI and DTI study. Neuroimage 54: 779–786.

Hamberger MJ, Cole J (2011). Language organization and reorganization in epilepsy. Neuropsychol Rev 21: 240–251.

Hartwigsen G, Saur D (2019). Neuroimaging of stroke recovery from aphasia—Insights into plasticity of the human language network. Neuroimage 190: 14–31.

Hausmann M (2019). Variations of hemispheric functional segregation in the laterality spectrum. Phys Life Rev 30: 27–29.

Hausmann M, Brysbaert M, van der Haegen L et al. (2019). Language lateralisation measured across linguistic and national boundaries. Cortex 111: 134–147.

He X, Bassett DS, Chaitanya G et al. (2018). Disrupted dynamic network reconfiguration of the language system in temporal lobe epilepsy. Brain 141: 1375–1389.

He L, Wang X, Zhuang K et al. (2020). Decreased dynamic segregation but increased dynamic integration of the resting-state functional networks during normal aging. Neuroscience 437: 54–63.

Heiss WD, Thiel A (2006). A proposed regional hierarchy in recovery of post-stroke aphasia. Brain Lang 98: 118–123.

Hellige JB (2002). Hemispheric specialization: contemporary issues. Rev Neuropsychol 12: 7–49.

Hertrich I, Dietrich S, Ackermann H (2020). The margins of the language network in the brain. Front Commun 5: 519955.

Hervé P-Y, Zago L, Petit L et al. (2013). Revisiting human hemispheric specialization with neuroimaging. Trends Cogn Sci 17: 69–80.

Hosomi A, Nagakane Y, Yamada K et al. (2009). Assessment of arcuate fasciculus with diffusion-tensor tractography may predict the prognosis of aphasia in patients with left middle cerebral artery infarcts. Neuroradiology 51: 549–555.

Hoyau E, Roux-Sibilon A, Boudiaf N et al. (2018). Aging modulates fronto-temporal cortical interactions during lexical production. A dynamic causal modeling study. Brain Lang 184: 11–19.

Ikuta T, Gollnick HM, Rutledge AN (2020). Age associated decline in the arcuate fasciculus and IQ. Brain Imaging Behav 14: 362–367.

Jaimes-Bautista AG, Rodríguez-Camacho M, Martínez-Juárez IE et al. (2015). Semantic processing impairment in patients with temporal lobe epilepsy. Epilepsy Res Treat 2015: 1–8.

Johnstone LT, Karlsson EM, Carey DP (2020). The validity and reliability of quantifying hemispheric specialisation using fMRI: evidence from left and right handers on three different cerebral asymmetries. Neuropsychologia 138: 107331.

Josse G, Hervé P-Y, Crivello F et al. (2006). Hemispheric specialization for language: brain volume matters. Brain Res 1068: 184–193.

Josse G, Seghier ML, Kherif F et al. (2008). Explaining function with anatomy: language lateralization and corpus callosum size. J Neurosci 28: 14132–14139.

Jung-Beeman M (2005). Bilateral brain processes for comprehending natural language. Trends Cogn Sci 9: 512–518.

Kaestner E, Reyes A, Macari AC et al. (2019). Identifying the neural basis of a language-impaired phenotype of temporal lobe epilepsy. Epilepsia 60: 1627–1638.

Keller SS, Roberts N, García-Fiñana M et al. (2011). Can the language-dominant hemisphere be predicted by brain anatomy? J Cogn Neurosci 23: 2013–2029.

Keller SS, Roberts N, Baker G et al. (2018). A voxel-based asymmetry study of the relationship between hemispheric asymmetry and language dominance in Wada tested patients. Hum Brain Mapp 39: 3032–3045.

Keser Z, Meier EL, Stockbridge MD et al. (2020). The role of microstructural integrity of major language pathways in narrative speech in the first year after stroke. J Stroke Cerebrovasc Dis 29: 105078.

Kim SH, Jang SH (2013). Prediction of aphasia outcome using diffusion tensor tractography for arcuate fasciculus in stroke. Am J Neuroradiol 34: 785–790.

King BR, van Ruitenbeek P, Leunissen I et al. (2018). Age-related declines in motor performance are associated with decreased segregation of large-scale resting state brain networks. Cereb Cortex 28: 4390–4402.

Kiran S, Thompson CK (2019). Neuroplasticity of language networks in aphasia: advances, updates, and future challenges. Front Neurol 10: 295.

Klingbeil J, Wawrzyniak M, Stockert A et al. (2019). Resting-state functional connectivity: an emerging method for the study of language networks in post-stroke aphasia. Brain Cogn 131: 22–33.

Knecht S (2000). Handedness and hemispheric language dominance in healthy humans. Brain 123: 2512–2518.

Kong X-Z, Mathias SR, Guadalupe T et al. (2018). Mapping cortical brain asymmetry in 17,141 healthy individuals worldwide via the ENIGMA Consortium. Proc Natl Acad Sci U S A 115: E5154–E5163.

Kong X-Z, Postema MC, Guadalupe T et al. (2022). Mapping brain asymmetry in health and disease through the ENIGMA consortium. Hum Brain Mapp 43: 167–181.

Kroliczak G, Buchwald M, Kleka P et al. (2021). Manual praxis and language-production networks, and their links to handedness. Cortex 140: 110–127.

Kucukboyaci NE, Kemmotsu N, Leyden KM et al. (2014). Integration of multimodal MRI data via PCA to explain language performance. Neuroimage Clin 5: 197–207.

Labache L, Mazoyer B, Joliot M et al. (2020). Typical and atypical language brain organization based on intrinsic connectivity and multitask functional asymmetries. Elife 9: e58722.

Labache L, Ge T, Yeo BTT et al. (2023). Language network lateralization is reflected throughout the macroscale functional organization of cortex. Nat Commun 14: 3405.

LaGrone S, Spieler DH (2006). Lexical competition and phonological encoding in young and older speakers. Psychol Aging 21: 804–809.

Larivière S, Rodríguez-Cruces R, Royer J et al. (2020). Network-based atrophy modeling in the common epilepsies: a worldwide ENIGMA study. Sci Adv 6: eabc6457.

Lazar RM, Speizer AE, Festa JR et al. (2008). Variability in language recovery after first-time stroke. J Neurol Neurosurg Psychiatry 79: 530–534.

Lee JJ, Luckett P, Fakhri MM et al. (2021). Resting state functional MR imaging of language function. Neuroimaging Clin N Am 31: 69–79.

Léger A, Démonet JF, Ruff S et al. (2002). Neural substrates of spoken language rehabilitation in an aphasic patient: an fMRI study. Neuroimage 17: 174–183.

Lindell AK (2006). In your right mind: right hemisphere contributions to language processing and production. Neuropsychol Rev 16: 131–148.

Lopes TM, Campos BM, Zanão TA et al. (2019). Hippocampal atrophy disrupts the language network but not hemispheric language lateralization. Epilepsia 60: 744–755.

Malik-Moraleda S, Ayyash D, Gallée J et al. (2022). An investigation across 45 languages and 12 language families reveals a universal language network. Nat Neurosci 25: 1014–1019.

Manenti R, Brambilla M, Petesi M et al. (2013). Compensatory networks to counteract the effects of ageing on language. Behav Brain Res 249: 22–27.

Marcelle M, You X, Fanto EJ et al. (2022). Impact of development and recent-onset epilepsy on language dominance. Epilepsia 63: 2637–2649.

Matsuo K, Chen S-HA, Tseng W-YI (2012). AveLI: a robust lateralization index in functional magnetic resonance imaging using unbiased threshold-free computation. J Neurosci Methods 205: 119–129.

Mazoyer B, Zago L, Jobard G et al. (2014). Gaussian mixture modeling of hemispheric lateralization for language in a large sample of healthy individuals balanced for handedness. PLoS One 9: e101165.

Mbwana JS, You X, Ailion A et al. (2021). Functional connectivity hemispheric contrast (FC-HC): a new metric for language mapping. Neuroimage Clin 30: 102598.

Meinzer M, Elbert T, Wienbruch C et al. (2004). Intensive language training enhances brain plasticity in chronic aphasia. BMC Biol 2: 1–9.

Menke R, Meinzer M, Kugel H et al. (2009). Imaging short- and long-term training success in chronic aphasia. BMC Neurosci 10: 118.

Middlebrooks EH, Yagmurlu K, Szaflarski JP et al. (2017). A contemporary framework of language processing in the human brain in the context of preoperative and intraoperative language mapping. Neuroradiology 59: 69–87.

Monroy-Sosa A, Chakravarthi SS, Cortes-Contreras AP et al. (2021). The evolution of cerebral language localization: historical analysis and current trends. World Neurosurg 145: 89–97.

Murphy DR, Craik FIM, Li KZH et al. (2000). Comparing the effects of aging and background noise on short-term memory performance. Psychol Aging 15: 323–334.

Musso M, Weiller C, Kiebel S et al. (1999). Training-induced brain plasticity in aphasia. Brain 122: 1781–1790.

Naeser MA, Martin PI, Baker EH et al. (2004). Overt propositional speech in chronic nonfluent aphasia studied with the dynamic susceptibility contrast fMRI method. Neuroimage 22: 29–41.

Neudorf J, Kress S, Gould L et al. (2020). Language lateralization differences between left and right temporal lobe epilepsy as measured by overt word reading fMRI activation and DTI structural connectivity. Epilepsy Behav 112: 107467.

Ocklenburg S, Hugdahl K, Westerhausen R (2013). Structural white matter asymmetries in relation to functional asymmetries during speech perception and production. Neuroimage 83: 1088–1097.

Ocklenburg S, Schlaffke L, Hugdahl K et al. (2014). From structure to function in the lateralized brain: how structural properties of the arcuate and uncinate fasciculus are associated with dichotic listening performance. Neurosci Lett 580: 32–36.

Ocklenburg S, Friedrich P, Güntürkün O et al. (2016). Intrahemispheric white matter asymmetries: the missing link between brain structure and functional lateralization? Rev Neurosci 27: 465–480.

Oh Y-M, Koh E-J (2009). Language lateralization in patients with temporal lobe epilepsy: a comparison between volumetric analysis and the Wada test. J Korean Neurosurg Soc 45: 329.

Olulade OA, Seydell-Greenwald A, Chambers CE et al. (2020). The neural basis of language development: changes in lateralization over age. Proc Natl Acad Sci U S A 117: 23477–23483.

Packheiser J, Schmitz J, Arning L et al. (2020). A large-scale estimate on the relationship between language and motor lateralization. Sci Rep 10: 13027.

Peck KK, Moore AB, Crosson BA et al. (2004). Functional magnetic resonance imaging before and after aphasia therapy: shifts in hemodynamic time to peak during an overt language task. Stroke 35: 554–559.

Perlaki G, Horvath R, Orsi G et al. (2013). White-matter microstructure and language lateralization in left-handers: a whole-brain MRI analysis. Brain Cogn 82: 319–328.

Piervincenzi C, Petrilli A, Marini A et al. (2016). Multimodal assessment of hemispheric lateralization for language and its relevance for behavior. Neuroimage 142: 351–370.

Plessen KJ, Hugdahl K, Bansal R et al. (2014). Sex, age, and cognitive correlates of asymmetries in thickness of the cortical mantle across the life span. J Neurosci 34: 6294–6302.

Powell HWR, Parker GJM, Alexander DC et al. (2006). Hemispheric asymmetries in language-related pathways: a combined functional MRI and tractography study. Neuroimage 32: 388–399.

Powell HWR, Parker GJM, Alexander DC et al. (2007). Abnormalities of language networks in temporal lobe epilepsy. Neuroimage 36: 209–221.

Pravatà E, Sestieri C, Mantini D et al. (2011). Functional connectivity MR imaging of the language network in patients with drug-resistant epilepsy. Am J Neuroradiol 32: 532–540.

Price CJ (2012). A review and synthesis of the first 20years of PET and fMRI studies of heard speech, spoken language and reading. Neuroimage 62: 816–847.

Price CJ, Crinion J (2005). The latest on functional imaging studies of aphasic stroke. Curr Opin Neurol 18: 429–434.

Reuter-Lorenz PA, Park DC (2014). How does it STAC up? Revisiting the scaffolding theory of aging and cognition. Neuropsychol Rev 24: 355–370.

Reyes A, Kaestner E, Bahrami N et al. (2019). Cognitive phenotypes in temporal lobe epilepsy are associated with distinct patterns of white matter network abnormalities. Neurology 92: e1957–e1968.

Richter M, Miltner WHR, Straube T (2008). Association between therapy outcome and right-hemispheric activation in chronic aphasia. Brain 131: 1391–1401.

Roger E, Rodrigues De Almeida L, Loevenbruck H et al. (2022). Unraveling the functional attributes of the language connectome: crucial subnetworks, flexibility and variability. Neuroimage 263: 119672.

Rolinski R, You X, Gonzalez-Castillo J et al. (2020). Language lateralization from task-based and resting state functional MRI in patients with epilepsy. Hum Brain Mapp 41: 3133–3146.

Rosazza C, Ghielmetti F, Minati L et al. (2013). Preoperative language lateralization in temporal lobe epilepsy (TLE) predicts peri-ictal, pre- and post-operative language performance: an fMRI study. Neuroimage Clin 3: 73–83.

Rosen HJ, Petersen SE, Linenweber MR et al. (2000). Neural correlates of recovery from aphasia after damage to left inferior frontal cortex. Neurology 55: 1883–1894.

Rutten G-J (2017). The Broca-Wernicke doctrine: a historical and clinical perspective on localization of language functions, Springer.

Salthouse TA (1996). The processing-speed theory of adult age differences in cognition. Psychol Rev 103: 403–428.

Saur D (2006). Dynamics of language reorganization after stroke. Brain 129: 1371–1384.

Schoenberg MR, Werz MA, Drane DL (2011). Epilepsy and seizures. The little black book of neuropsychology: a syndrome-based approach, Springer.

Seghier ML (2008). Laterality index in functional MRI: methodological issues. Magn Reson Imaging 26: 594–601.

Seghier ML (2019). Categorical laterality indices in fMRI: a parallel with classic similarity indices. Brain Struct Funct 224: 1377–1383.

Sisodiya SM, Whelan CD, Hatton SN et al. (2020). The ENIGMA-epilepsy working group: mapping disease from large data sets. Hum Brain Mapp 43: 113–128.

Spreng RN, Turner GR (2019). The shifting architecture of cognition and brain function in older adulthood. Perspect Psychol Sci 14: 523–542.

Spreng RN, Lockrow AW, DuPre E et al. (2018). Semanticized autobiographical memory and the default—executive coupling hypothesis of aging. Neuropsychologia 110: 37–43.

Sprung-Much T, Eichert N, Nolan E et al. (2022). Broca's area and the search for anatomical asymmetry: commentary and perspectives. Brain Struct Funct 227: 441–449.

Stefaniak JD, Halai AD, Lambon Ralph MA (2020). The neural and neurocomputational bases of recovery from post-stroke aphasia. Nat Rev Neurol 16: 43–55.

Stern Y (2009). Cognitive reserve. Neuropsychologia 47: 2015–2028.

Stewart CC, Swanson SJ, Sabsevitz DS et al. (2014). Predictors of language lateralization in temporal lobe epilepsy. Neuropsychologia 60: 93–102.

Suarez RO, Whalen S, Nelson AP et al. (2009). Threshold-independent functional MRI determination of language dominance: a validation study against clinical gold standards. Epilepsy Behav 16: 288–297.

Tailby C, Abbott DF, Jackson GD (2017). The diminishing dominance of the dominant hemisphere: language fMRI in focal epilepsy. Neuroimage Clin 14: 141–150.

Thiel A, Habedank B, Herholz K et al. (2006). From the left to the right: how the brain compensates progressive loss of language function. Brain Lang 98: 57–65.

Thom M, Bertram EH (2012). Temporal lobe epilepsy. Handb Clin Neurol 107: 225–240, Elsevier.

Thornton RC, van Graan LA, Powell RHW (2016). fMRI in epilepsy. In: M Filippi (Ed.), fMRI techniques and protocols, second edn. Humana Press, New York, NY.

Tomasi D, Volkow ND (2012). Resting functional connectivity of language networks: characterization and reproducibility. Mol Psychiatry 17: 841–854.

Tracy JI, Boswell SB (2008). Mesial temporal lobe epilepsy: a model for understanding the relationship between language and memory. In: B Stemmer, HA Whitaker (Eds.), Handbook of the neuroscience of language. Elsevier, Oxford, pp. 319–328.

Trimmel K, Vos SB, Caciagli L et al. (2021). Decoupling of functional and structural language networks in temporal lobe epilepsy. Epilepsia 62: 2941–2954.

Turrigiano GG, Nelson SB (2000). Hebb and homeostasis in neuronal plasticity. Curr Opin Neurobiol 10: 358–364.

Turrigiano GG, Nelson SB (2004). Homeostatic plasticity in the developing nervous system. Nat Rev Neurosci 5: 97–107.

Tzourio-Mazoyer N, Josse G, Crivello F et al. (2004). Interindividual variability in the hemispheric organization for speech. Neuroimage 21: 422–435.

Tzourio-Mazoyer N, Perrone-Bertolotti M, Jobard G et al. (2017). Multi-factorial modulation of hemispheric specialization and plasticity for language in healthy and pathological conditions: a review. Cortex 86: 314–339.

Valente A, Laganaro M (2015). Ageing effects on word production processes: an ERP topographic analysis. Lang Cogn Neurosci 30: 1259–1272.

van Oers CAMM, Vink M, van Zandvoort MJE et al. (2010). Contribution of the left and right inferior frontal gyrus in recovery from aphasia. A functional MRI study in stroke patients with preserved hemodynamic responsiveness. Neuroimage 49: 885–893.

Verhelst H, Dhollander T, Gerrits R et al. (2021). Fibre-specific laterality of white matter in left and right language dominant people. Neuroimage 230: 117812.

Vernooij MW, Smits M, Wielopolski PA et al. (2007). Fiber density asymmetry of the arcuate fasciculus in relation to functional hemispheric language lateralization in both right- and left-handed healthy subjects: a combined fMRI and DTI study. Neuroimage 35: 1064–1076.

Vigneau M, Beaucousin V, Hervé P-Y et al. (2011). What is right-hemisphere contribution to phonological, lexico-semantic, and sentence processing? Neuroimage 54: 577–593.

Vingerhoets G (2019). Phenotypes in hemispheric functional segregation? Perspectives and challenges. Phys Life Rev 30: 1–18.

Waites AB, Briellmann RS, Saling MM et al. (2006). Functional connectivity networks are disrupted in left temporal lobe epilepsy. Ann Neurol 59: 335–343.

Wang S, Van der Haegen L, Tao L et al. (2019). Brain functional organization associated with language lateralization. Cereb Cortex 29: 4312–4320.

Whitten A, Jacobs ML, Englot DJ et al. (2021). Resting-state hippocampal networks related to language processing reveal unique patterns in temporal lobe epilepsy. Epilepsy Behav 117: 107834.

Wilke M, Lidzba K (2007). LI-tool: a new toolbox to assess lateralization in functional MR-data. J Neurosci Methods 163: 128–136.

Wilson SM, Schneck SM (2020). Neuroplasticity in post-stroke aphasia: a systematic review and meta-analysis of functional imaging studies of reorganization of language processing. Neurobiol Lang 2: 22–82.

Wlotko EW, Lee C-L, Federmeier KD (2010). Language of the aging brain: event-related potential studies of comprehension in older adults. Lang Linguist Compass 4: 623–638.

Woodhead ZVJ, Thompson PA, Karlsson EM et al. (2021). An updated investigation of the multidimensional structure of language lateralization in left- and right-handed adults: a test–retest functional transcranial Doppler sonography study with six language tasks. R Soc Open Sci 8: rsos.200696,-200696.

Yarkoni T, Poldrack RA, Nichols TE et al. (2011). Large-scale automated synthesis of human functional neuroimaging data. Nat Methods 8: 665–670.

Yeh F-C (2022). Population-based tract-to-region connectome of the human brain and its hierarchical topology. Nat Commun 13: 4933.

Zhu L, Fan Y, Zou Q et al. (2014). Temporal reliability and lateralization of the resting-state language network. PLoS One 9: e85880.

Zhu Z, Deng J, Li M et al. (2022). Processing speed mediates the relationship between brain structure and semantic fluency in aging. Neurosci Lett 788: 136838.

Section VII

Movement/motor asymmetries/cerebellum

Handbook of Clinical Neurology, Vol. 208 (3rd series)
Cerebral Asymmetries
C. Papagno and P. Corballis, Editors
https://doi.org/10.1016/B978-0-443-15646-5.00005-1

Chapter 23

Cerebellar asymmetries

CAROLINE NETTEKOVEN[1,2*] AND JÖRN DIEDRICHSEN[1,2,3]

[1]*Western Institute for Neuroscience, Western University, London, ON, Canada*

[2]*Department of Computer Science, Western University, London, ON, Canada*

[3]*Department of Statistical and Actuarial Sciences, Western University, London, ON, Canada*

Abstract

The cerebellum is a subcortical structure tucked underneath the cerebrum that contains the majority of neurons in the brain, despite its small size. While it has received less attention in the study of brain asymmetries than the cerebrum, structural asymmetries in the cerebellum have been found in cerebellar volume that mirror cerebral asymmetries. Larger cerebellar structures have been reported on the right compared to the left, either for the whole cerebellar hemisphere or the anterior part of the cerebellum, with the latter accompanied by a left increase in the posterior cerebellum. Cerebellar asymmetries are considered evolutionary recent and have been observed prenatally and in early development. Both asymmetries in anterior–posterior divisions and specific lobules have been linked to handedness and cognitive abilities, in particular language. Functional lateralization in the cerebellum varies across motor and cognitive functions, with language activation predominantly localized in the right hemisphere, contralateral to cerebral activation. Meanwhile, working memory and executive functions are not lateralized to one hemisphere. New neuroimaging methods and resources, including a symmetric functional atlas of the cerebellum that enables precision mapping, open novel avenues for exploring cerebellar asymmetries and answering questions about the developmental timeline, relationships to behavior, and clinical relevance.

INTRODUCTION

The cerebellum, or "little brain," is a small structure situated underneath the much larger cerebrum. Despite only making up 10% of the total brain mass, the cerebellar cortex holds 80% of all neurons (Azevedo et al., 2009) and when fully unfolded has a surface area of 78% of the neocortex (Sereno et al., 2020). The cerebellum supports a multitude of functions, including motor, executive, social, linguistic, and emotional processes. Studies of functional lateralization, particularly of higher-order cognitive functions such as language, often prioritize the cerebrum over the cerebellum. Similarly, studies of structural asymmetries have largely focused on the cerebrum, leaving many questions about cerebellar structural and circuitry asymmetries unanswered. This is partially driven by difficulties in imaging the cerebellum and a lack of tools necessary for careful investigations of cerebellar functional asymmetries. Still, studies to date provide a picture of cerebellar organization that is broadly symmetric, but includes fine hemispheric differences, some of which mirror cerebral asymmetries. Here, we review asymmetries in cerebellar structure, function, and neurochemistry. We also consider technical challenges in studying cerebellar asymmetries and outline recent advances addressing these limitations, enabling future in-depth studies of cerebellar asymmetries.

STRUCTURAL ASYMMETRIES

Broad cerebellar structure

The cerebellum consists of two hemispheres that are connected through the vermis, a worm-like midline structure. The cerebellum sits within the posterior fossa and connects to the brainstem via three pairs of peduncles: the superior, middle, and inferior peduncle. The inferior cerebellar peduncle and middle cerebellar peduncle contain mainly afferent projections to the cerebellum from the spinal cord and the cerebrum, respectively. The inferior peduncle also contains efferent projections from the cerebellum to the spinal cord. The superior peduncle primarily transports efferent fibers from the cerebellum, projecting to the cerebrum. Cerebellar afferents and efferents cross over to the contralateral cerebral hemisphere in the brainstem. In general, the cerebellum therefore connects to the contralateral cerebral hemisphere and is concerned with inputs from the ipsilateral side of the body.

*Correspondence to: Caroline Nettekoven, Department of Computer Science, Western Institute for Neuroscience, Western University, London, ON, Canada. Tel: +1-519-6612111x86057, Fax: +1-519-661-3613, E-mail: cr.nettekoven@gmail.com

The tightly folded cerebellar cortex is commonly divided along broad anatomic divisions, the fissures, into 10 lobules and denoted I–X according to the Larsell nomenclature (Larsell and Jansen, 1973; Schmahmann et al., 2000). Lobules I–V form the anterior cerebellum, lobules VI–IX the posterior cerebellum, and lobule X forms the flocculonodular lobe of the cerebellum. Because lobules VII and VIII are the two largest lobules in the human cerebellum (Diedrichsen and Zotow, 2015), many studies split them into two sections (VIIA and VIIB, as well as VIIIA and VIIIB), and lobule VIIA is split again into Crus I and Crus II.

Neuroimaging studies have identified gross asymmetries in cerebellar structure in children (Holland et al., 2014) and across the adult lifespan (Herve et al., 2006; Fan et al., 2010; Bernard and Seidler, 2013; Kang et al., 2015). While some found an overall larger right hemisphere of the cerebellum (Herve et al., 2006; Fan et al., 2010; Kang et al., 2015), others reported this volume increase on the right for only some cerebellar lobules (Kavaklioglu et al., 2017). However, some studies fail to show any structural asymmetries in the cerebellum (Gocmen-Mas et al., 2009; Ertekin et al., 2013). The earliest report of cerebellar structural asymmetry divided the cerebellum into an anterior and a posterior section according to coronal MRI slices acquired in 15 right-handed and 8 left-handed participants (Snyder et al., 1995). The authors found that in the anterior cerebellum, the right hemisphere was significantly larger than the left, whereas in the posterior cerebellum the left was bigger than the right. A similar pattern had previously been described for the neocortex (LeMay, 1976), with right anterior structures (e.g., frontal pole) extending more anteriorly and left posterior structures (e.g., occipital pole) protruding more posteriorly. This is despite the cerebellum primarily projecting to the contralateral cerebral hemisphere, which could suggest an opposite pattern of asymmetry.

Handedness

When first reporting cerebellar structural asymmetry (Snyder et al., 1995), also found a link to handedness: while all 23 participants showed an overall similar asymmetry pattern, it was more pronounced in right-handed people. A later study of 19 monozygotic female adult twins confirmed some of the structural asymmetries in the cerebellum, with the anterior cerebellum (lobules I–V) being larger on the right, but near symmetry in the posterior cerebellum (lobules VI–VIIIB) (Rosch et al., 2018). The anterior cerebellar asymmetry was stronger in right-handed than left-handed twins, potentially reflecting an experience-dependent maintenance of higher growth rates in the right hemispheric cerebellum. However, differences in the definition of regions of interest between the first report, where sections were divided by coronal slice, and the more precise definition in the later study based on a lobular atlas (Diedrichsen, 2006), make a direct comparison difficult. Furthermore, the purported relationship between cerebellar asymmetry and handedness was not replicated in a large-scale study of 2226 participants (Kavaklioglu et al., 2017), but a recent investigation of over 37,000 participants associated handedness with whole-brain asymmetry patterns that included the cerebellum. Directly comparing left- and right-handed participants revealed that right-handers showed larger lobules VIIIA and VIIIB in the right cerebellar hemisphere and larger Crus I and Crus II in the left cerebellar hemisphere (Saltoun et al., 2023).

Evolutionary and developmental studies

Cerebellar structural asymmetry seems to have emerged recently in evolution, as it is only observed in modern humans—*Homo sapiens*—but not in *Homo erectus* or *Homo neanderthalensis* (Zhang and Wu, 2021). This is in contrast to cerebral asymmetries, which emerged earlier, appearing already in *Homo erectus* and *Homo neanderthalensis* (Holloway and De La Costelareymondie, 1982; Li et al., 2018).

During gestation, the cerebellum grows more quickly than any other brain region, increasing 34-fold from weeks 18–39. Early in gestation, the left cerebellar hemisphere is larger than the right, but the right shows an accelerated growth curve, nearly overtaking the left by week 39 (Andescavage et al., 2017). During the first 3 months after birth, the cerebellum again shows the fastest growth, more than doubling in size (Holland et al., 2014). During this time, the right cerebellar hemisphere is larger than the left. Cerebellar asymmetry does not depend on sex in the first 3 months of life (Holland et al., 2014), though there is some evidence for an interaction between sex and cerebellar structural asymmetries in children (Isıklar et al., 2023) and young adults (Fan et al., 2010).

Relationship with movement, language, and cognition

Several studies have related cerebellar structural asymmetries to language, motor, and cognitive functions. Neonatal whole-cerebellar asymmetry has been found to predict later language skills (Vassar et al., 2020), with more cerebellar asymmetry being linked to lower language scores as assessed with the Bayley Scales of Infant-Toddler Development-III (Bayley, 2006). However, cerebellar asymmetries in this study were based on rating scale assessments by radiologists, which differs substantially from the quantitative techniques used in previous studies (Holland et al., 2014; Andescavage et al., 2017). In 10-year-old children, whole cerebellar volume showed no association with language content, but correlated positively with right gray matter volume in Crus I/Crus II (Stipdonk et al., 2021). In a cohort of 48 autistic and non-autistic boys aged 6–13 with and without language impairment (autistic language impairment or specific language impairment), only those with impaired language function showed leftward lobule VIIIA asymmetry, irrespective of autism diagnosis. Non-impaired boys showed larger right lobule VIIIA volume, and language performance correlated with lobule VIIIA asymmetry across groups, with increased volume on the right relating to higher language scores (Hodge et al., 2010). Decreased cerebellar volume in the right hemisphere of 32 children with cerebellar malformations (aged 1–6) has been linked to impaired expressive language, as well as cognitive and motor impairments (Bolduc et al., 2012).

In adults, working memory task performance has been found to positively correlate with gray matter volume in left Crus I (Ding et al., 2012), and improved timing in musical performance has been associated with smaller right lobule VI volume (Baer et al., 2015). Meanwhile, increased volume of the cerebellar cortex was associated with mild cognitive impairment in a study of 400 randomly selected older adults (aged 64–70) (Cherbuin et al., 2010). Asymmetric gray matter reductions and torque have also been linked to neuropsychiatric disorders such as autism spectrum disorder, developmental dyslexia (Stoodley, 2014), and schizophrenia (Szeszko et al., 2003).

Cerebellar damage from injury, stroke, or degeneration has previously been shown to cause, in addition to the classic motor impairments, a range of non-motor symptoms. These include language deficits, anagrammatism, executive function deficits, and social and affective impairments that were termed the "Cerebellar Cognitive Affective Syndrome" (Schmahmann and Sherman, 1998). To our knowledge, there are so far no reports of Cerebellar Cognitive Affective Syndrome manifesting as lateralized to a particular hemisphere, although functions that belong to the symptom set of the syndrome certainly appear lateralized (see section Task-based asymmetries). Notably, cognitive symptoms after cerebellar damage in adults appear milder than in young children (Glickstein, 1994; Fabbro et al., 2004; Ronconi et al., 2017) and often result in a decrease in function rather than a full loss of it (Olson et al., 2023). For example, cerebellar damage to right Crus I/II appears to not result in aphasia or absence of speech, but in dysarthria (Ackermann, 2008; Stoodley et al., 2016) or agrammatism (Silveri et al., 1994).

Deep cerebellar nuclei

The cerebellar white matter envelops the deep cerebellar nuclei, which lie close to the midline in each hemisphere and receive the output generated from the cerebellar cortex. The deep cerebellar nuclei then project to the brainstem nuclei and the cerebral cortex via the thalamus. Only the flocculonodular cerebellar cortex, lobule X, deviates from this by projecting directly to the vestibular nuclei in the brainstem. As such, the deep cerebellar nuclei and the vestibular nuclei are the sole transmitters of the output from the cerebellum. The deep cerebellar nuclei include in each hemisphere along the medial-to-lateral axis, the dentate nucleus, the interpositus (consisting of globose and emboliform nuclei), and the fastigial nucleus.

Few neuroimaging studies have examined the structure of deep cerebellar nuclei due to their small size and their near invisibility in standard T1-weighted anatomic images (Diedrichsen et al., 2011). Only susceptibility-weighted images, which are not part of standard MRI protocols, provide suitable contrast for localizing the deep cerebellar nuclei, as these images are sensitive to their high iron content. A first volumetric comparison of left and right dentate nuclei reported increased volume on the right in 9 of 10 examined participants (Deoni and Catani, 2007), though this was not replicated in a later study of 23 participants (Diedrichsen et al., 2011). In postmortem sections of human brains, visual comparison of the dentate ribbon revealed left–right differences in the folding pattern of a single case, but no systematic differences.

Cytoarchitecture

While the cerebral cortex is dividable according to its cytoarchitecture, the cerebellar cortical makeup is generally remarkably uniform. The single output cell of the cerebellar cortex is the Purkinje cell. Each Purkinje cell receives inputs via two streams. First, they receive multiple inputs from multiple parallel fibers. Parallel fibers arise from the densely packed granule cells. Granule cells receive information via mossy fibers that originate in the spinal cord, medulla oblongata, and most massively from the pontine nuclei. The majority of axons exiting the neocortex for the brainstem either project directly to the pons or send a collateral there, which projects onto the granule cells (Tomasch, 1969). Granule cells then innervate Purkinje cells via parallel fibers. Second, each Purkinje cell receives input from a single climbing fiber, originating in the inferior olivary nucleus, though a single climbing fiber branches to innervate one to seven Purkinje cells.

Detailed descriptions of the cerebellar circuit have led to the development of the Marr-Albus-Ito model of cerebellar function (Eccles, 1967; Marr, 1969; Albus, 1971; Ito and Kano, 1982). This model is based on the convergence of the two input types at the Purkinje cell. First, the many inputs from parallel fibers are proposed to represent a detailed context. Second, the single climbing fiber input is proposed to represent an error, or "teaching" signal. When a Purkinje cell receives climbing fiber activation (error signal) shortly after activation of a set of parallel fibers (context), then the synaptic weights of this particular set of parallel fibers change through long-term depression (Ito and Kano, 1982). When this set of parallel fibers is activated again at this Purkinje cell, the Purkinje cell firing is reduced compared to before. The change in Purkinje cell firing leads to a change in behavior, a change in output to the deep cerebellar nuclei, and suppression of the climbing fiber input (Medina and Lisberger, 2008).

The Marr-Albus-Ito model explains a wealth of experimental data, particularly in the motor domain, and has been highly influential. However, recent studies cast doubt on a universal learning principle within the cerebellar cortex by painting a more complex picture of cerebellar circuitry (Beckinghausen and Sillitoe, 2019; Fujita et al., 2020; Busch and Hansel, 2023). In humans, Busch et al. (Busch and Hansel, 2023) showed in sagittal slices of three postmortem human brains that only 4% of human Purkinje cells have the stereotypic single dendritic branch, while 96% show a multibranched dendritic branch that bifurcates into multiple primary dendrites close to the neuron body. In mice, 15% of multibranched Purkinje cells also receive multiple climbing fiber inputs to the different dendritic branches, and some branches seemed to have locally distinct functional responses to whisker stimulation. These results show a clear divergence from the purported regularity of the cerebellar circuitry. The authors also described an anterior-to-posterior gradient in dendritic branching patterns, where the majority of Purkinje cells in lobules I–VI had normative, single-branched dendrites and the multibranched dendrites occurred in the majority of lobule VII–X. However, left–right asymmetries in Purkinje cell branching could not be detected, as the investigation focused on sagittal slices of the mid-section of only one hemisphere, precluding a between-hemispheric comparison.

Summary

Studies of structural asymmetries in the cerebellum have largely focused on the cerebellar cortex or overall cerebellar volume rather than the cerebellar nuclei, due to its relative accessibility in neuroimaging data. Generally, studies find rightward asymmetry in the cerebellum, with increased cerebellar volume on the right anterior cerebellum and increased left volume in the posterior cerebellum. Cerebellar asymmetries seem to be evolutionary recent and appear before birth. Some links with handedness, cognitive functions, and clinical diagnosis have been reported, although little consensus has been found. This may in part be driven by differences in analysis methods and techniques used to assess cerebellar asymmetry. Some studies have divided the cerebellum along the anterior-to-posterior axis for asymmetry analysis (Snyder et al., 1995), others group lobules into cerebellar zones (Bolduc et al., 2012; Rosch et al., 2018) or use ratings from radiologists to assess cerebellar asymmetry (Stipdonk et al., 2021). Establishing consensus on cerebellar asymmetry across the heterogeneous literature is therefore difficult.

As tools for cerebellar segmentation and atlases for cerebellar parcellation have been developed, most studies now parcellate the cerebellum along lobular boundaries (Fan et al., 2010; Bernard et al., 2015; Kavaklioglu et al., 2017; Isıklar et al., 2023; Saltoun et al., 2023) based on a probabilistic group atlas of cerebellar lobules (Diedrichsen, 2006). This makes comparing results across neuroimaging studies easier. However, most researchers are interested in linking observed asymmetries to cerebellar function, relying on the assumption that different cerebellar-dependent functions can be localized to different lobules. Mapping cerebellar activity across motor and cognitive domains revealed this is not the case (King et al., 2019). Indeed, cerebellar activity usually spans several lobules, and one lobule often contains multiple functional regions. Future studies of structural asymmetry in the cerebellar cortex should therefore use symmetric regions of interest that capture functional regions when attempting to link structural asymmetries to function. For a discussion of recent advances in this realm, see section Technical considerations for studying cerebellar lateralization.

FUNCTIONAL LATERALIZATION

The cerebellum has traditionally been associated with motor function, though neuroimaging studies show cerebellar activity during a broad range of behavioral tasks, including those probing language (Petersen et al., 1989), social cognition (Van Overwalle et al., 2015), attention (Allen et al., 1997), and working memory (Marvel and Desmond, 2010). While the cerebellum has been researched extensively in motor control, the study of the cerebellar role in higher-order cognitive functions such as language is in its relative infancy.

Task-based asymmetries

A function is considered lateralized if it engages one hemisphere of the brain more than the other, of which language function is a prime example. Right-lateralization of language activation in the cerebellum, mirroring left lateralization in the cerebrum, is well established (Amunts et al., 1996; Stoodley and Schmahmann, 2009; Fedorenko et al., 2010; Haberling and Corballis, 2016; LeBel and D'Mello, 2023). Though most language studies do not focus on the cerebellum or do not cover the cerebellum during data acquisition or analysis, there are early reports of functional lateralization in the right cerebellum during language processing (Petersen et al., 1989; Raichle et al., 1994).

Using positron emission tomography, Petersen et al. (Petersen et al., 1989) were the first to show clear lateralized cerebellar activation during language processing, even when motor demands were accounted for. Subsequent studies confirmed right cerebellar activation during language processes such as semantic retrieval and prediction (Fiez et al., 1996; Moberget and Ivry, 2016; Lesage et al., 2017).

A meta-analysis of task-based fMRI studies showed that language activity primarily occurs in the right cerebellum in lobule VI, Crus I/Crus II as well as vermal lobule VIIA (Stoodley and Schmahmann, 2009). On the other hand, left-lateralized activation was found primarily for spatial tasks, located in lobule VI. However, both language and spatial tasks showed some weaker, but consistent bilateral activation, occupying a small cluster in opposite lobule VI.

To examine the relationship between language lateralization in the cerebellum and neocortex, Wang et al. (Wang et al., 2013) analyzed fMRI data acquired while participants performed a semantic decision task. In the cerebellum, participants showed strong language lateralization in the right Crus I/Crus II and in the cerebrum, in the left prefrontal cortex. Language activity lateralization in the cerebellum was correlated with language lateralization in the cerebrum: right cerebellar task lateralization correlated with left cerebral task-based language lateralization. In line with lateralization of language and spatial function, patients with left-sided cerebellar lesions showed larger impairments in attention and visuospatial tasks, but not language, fluency, or motor function (Starowicz-Filip et al., 2021). Right-sided cerebellar lesions, on the other hand, led to impaired language and verbal fluency, but also general cognitive deficits, including memory, attention, and visuospatial functions (Silveri et al., 1994; Marien et al., 1996; Leggio et al., 2000; Marien et al., 2000). A detailed analysis of cerebellar lateralization was recently provided by a voxel-wise comparison of functional profiles across 417 tasks probed in seven fMRI datasets (Nettekoven et al., 2024) (1A). Social-linguistic-spatial regions located in Crus I and Crus II (2A) showed low correlations of functional profiles, indicating strongly lateralized functional responses. On the left, these regions activated while viewing social and emotional stimuli, whereas the right regions showed far less activity in those tasks. On the right, the cerebellum responded primarily to tasks involving linguistic information, such as a word reading and theory of mind story reading task or a verb generation task. In contrast, cerebellar working memory regions in lobules VI and VII

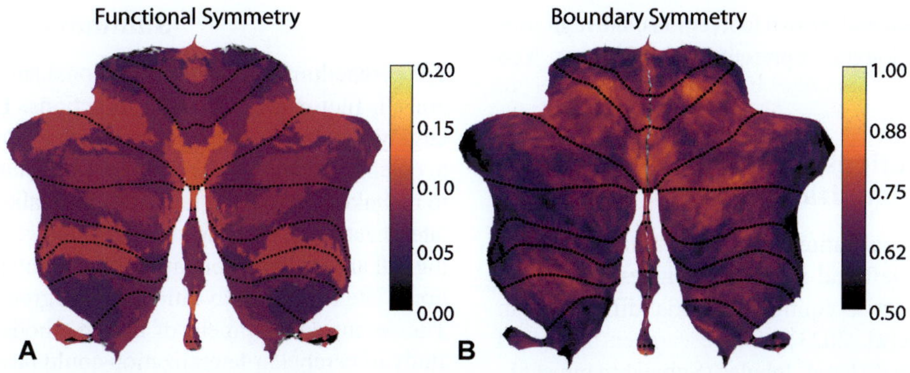

Fig. 23.1. Functional lateralization and boundary symmetry of cerebellar functional regions. Functional symmetry calculated as the correlations between the functional responses of anatomically corresponding voxel in the left and right hemispheres, averaged across participants and within each functional region (A) reveals bilateral response profiles of multiple demand regions and lateralized responses of social-linguistic-spatial regions. Boundary symmetry calculated as the correlations of the probabilistic voxel assignments between the symmetric and asymmetric versions of the atlas (B) shows symmetric boundaries in the motor regions, and asymmetric boundaries in social-linguistic-spatial regions. Reproduced from Nettekoven C, Zhi D, Shahshahani L, et al. (2024). A hierarchical atlas of the human cerebellum for functional precision mapping. Nat Commun 15: 8376. https://doi.org/10.1038/s41467-024-52371-w.

showed largely bilateral responses (Fig. 23.1A), as indicated by highly correlated functional profiles across tasks. This was despite the variety of executive function and working memory tasks in the task set, including those probing verbal working memory. Hence, while there might be some lateralization of working memory and executive function, the majority of tasks activate cerebellar working memory regions bilaterally.

Neuroimaging studies of motor and sensory tasks show sensorimotor homunculi on each side of lobules III–VI and lobule VIII (Stoodley and Schmahmann, 2009). Cerebellar activity is largely confined to the hemisphere ipsilateral to the moving effector (Saadon-Grosman et al., 2022), but bilateral for tongue and eyes (Buckner et al., 2011; Nettekoven et al., 2024). A voxel-wise comparison of the functional profiles confirmed this pattern of lateralization in cerebellar motor regions. The left and right cerebellar hand regions (M3) show low correlations of functional profiles (Fig. 23.1A). Finally, activity in the oculomotor vermis, responsible for eye movements and saccades, showed strongly bilateral responses. Therefore, while some functions are clearly lateralized in the cerebellum, such as language and social cognition, others are consistently bilateral like working memory and executive function, or exhibit laterality that appears effector-dependent such as motor functions.

In addition to having different functional profiles, boundaries between functional regions themselves can be asymmetric. The recent development of a symmetric functional atlas of the cerebellum (see section Technical considerations for studying cerebellar lateralization) allowed for the first systematic comparison of boundary symmetry in the cerebellum (Nettekoven et al., 2024). Examining the correspondence between the symmetric and asymmetric version of the atlas (Fig. 23.1A and B) across the different regions revealed that regions involved in movement and working memory functions have highly symmetric boundaries, i.e., their boundaries fall into the same place on the left and right. Among the motor regions, the oculomotor vermis M1 and the hand region M3 (Fig. 23.1A) fall particularly similar in left and right cerebellar hemisphere. Meanwhile, the social-linguistic-spatial regions have more dissimilar region boundaries between left–right region pairs, indicated by the low overall boundary symmetry in social-linguistic-spatial regions (Nettekoven et al., 2024).

Resting-state connectivity asymmetries

fMRI data acquired at rest has been used to quantify cortico-cerebellar functional connectivity and examine cerebellar lateralization in the context of cerebral lateralization (O'Reilly et al., 2010; Buckner et al., 2011; Wang et al., 2013; Wang et al., 2014). In the cerebellum, resting-state fMRI data from 1000 participants demonstrated an approximately homotopic map of cerebral connectivity networks (Buckner et al., 2011). Here, each cerebellar voxel was assigned to the cerebral resting-state network its resting-state time course correlated with the most (Buckner et al., 2011). Consistent with viral tracing studies in monkeys demonstrating that cerebral cortical areas project to the contralateral cerebellum (Kelly and Strick, 2003), cortico-cerebellar connectivity is strongest between contralateral cerebellar and cerebral regions, particularly in the motor regions (Krienen and Buckner, 2009; O'Reilly et al., 2010; Buckner et al., 2011). Resting-state and task-based fMRI data demonstrated a somatomotor map of the body including foot, hand, and tongue representations in the anterior cerebellum (Buckner et al., 2011). The task-based maps revealed that lateralization of somatomotor representations in the cerebellum varies with body part. While tongue representation is bilateral, consistent with the bilateral nature of the performed tongue movement, right foot and hand representations are lateralized, with right hand activation showing the strongest lateralization. Whether this lateralization depends on handedness to our knowledge has so far not been investigated.

Resting-state connectivity between ipsilateral and contralateral corticocerebellar regions followed a similar pattern: Hand and foot regions showed much stronger coupling contralaterally than ipsilaterally, whereas tongue regions showed no difference between ipsi- and contralateral connectivity. This echoes observations for the cerebral cortex showing the lowest inter-hemispheric connectivity between the two hand regions and the two foot regions, and high coupling for left and right tongue representations (Thomas Yeo et al., 2011).

Technical considerations for studying cerebellar lateralization

Studies of functional lateralization often necessitate defining regions of interest in the left and right hemispheres, which are matched in location and size to control for spatial differences in signal-to-noise ratio (Yan et al., 2023). Most researchers have used anatomic subdivisions into different lobules (Schmahmann et al., 2000; Diedrichsen et al., 2009) to define regions of interest, taking advantage of the anatomic symmetry of lobules. However, lobular boundaries do not reflect functional boundaries in the cerebellum (King et al., 2019), rendering them unsuitable for functional lateralization studies. Recently, a symmetric functional atlas has been developed, using a machine learning model that learns the functional organization of the cerebellum using over 100 participants across seven fMRI studies (Nettekoven et al., 2024) (Fig. 23.2A and B). The model was constrained to learn corresponding symmetric regions, while the functional responses of the matching regions could vary between the left and right hemispheres (Zhi et al., 2025). The symmetric group atlas captured functional organization 5% less accurately than the asymmetric group atlas, which represents a small trade-off between the validity of the region boundaries and practical utility for lateralization studies.

The cerebellum poses many challenges to functional neuroimaging, but the study of cerebellar functional lateralization holds promise for answering fundamental questions about brain asymmetries. Though it suffers from lower signal-to-noise ratio, the sources of noise in the cerebellum differ from those in the cerebrum, driven by different anatomic and technical constraints. By contrasting and comparing cerebral and cerebellar asymmetries, some consensus on brain asymmetries can be established, independently of the different noise sources (Wang et al., 2013).

Summary

The cerebellum shows clear functional lateralization for some, but not all, motor and cognitive functions. Language activation is largely located in the right cerebellar hemisphere, contralateral to cerebral activation. Cerebellar functional lateralization appears to be linked to cerebral functional lateralization, handedness, and lateralization of language function. However, difficulties in aligning left and right cerebellar regions of interest for studies of functional lateralization have hindered progress in this line of research. The recent development of new methods and resources for the study of cerebellar lateralization could pave the way to more precise studies of functional asymmetries in the cerebellum.

NEUROCHEMICAL ASYMMETRIES

Few studies of neurochemical asymmetries in the cerebellum exist, primarily due to the technical challenges in measuring neurochemicals from the human cerebellum in vivo. Neurochemicals in the brain can be measured noninvasively using magnetic resonance spectroscopy (MRS). However, the use of MRS in the cerebellum is impeded by the close proximity of the cerebellum to two large sources of noise: the brainstem on the anterior side and the neck on the posterior side. The fat tissue in the neck emits a signal over 100 times stronger than brain metabolites (Rothman et al., 1993) and when MRS measurements are collected close to this fatty tissue, imperfect slice selection pulses can lead to lipid contaminations of the measured spectrum (Kreis, 2004). Placing an MRS voxel in the small, curved cerebellar hemispheres at a sufficient distance from fat tissue while maximizing the gray matter content of the voxel is therefore difficult. Similarly, areas adjacent to the brainstem can suffer from high levels of physiologic noise, driven by the cardiac and respiratory cycles (Brooks et al., 2013), which induce field inhomogeneities and reduce the quality of the MRS spectra. However, with the advent of high-field imaging and improved shimming techniques for reducing field

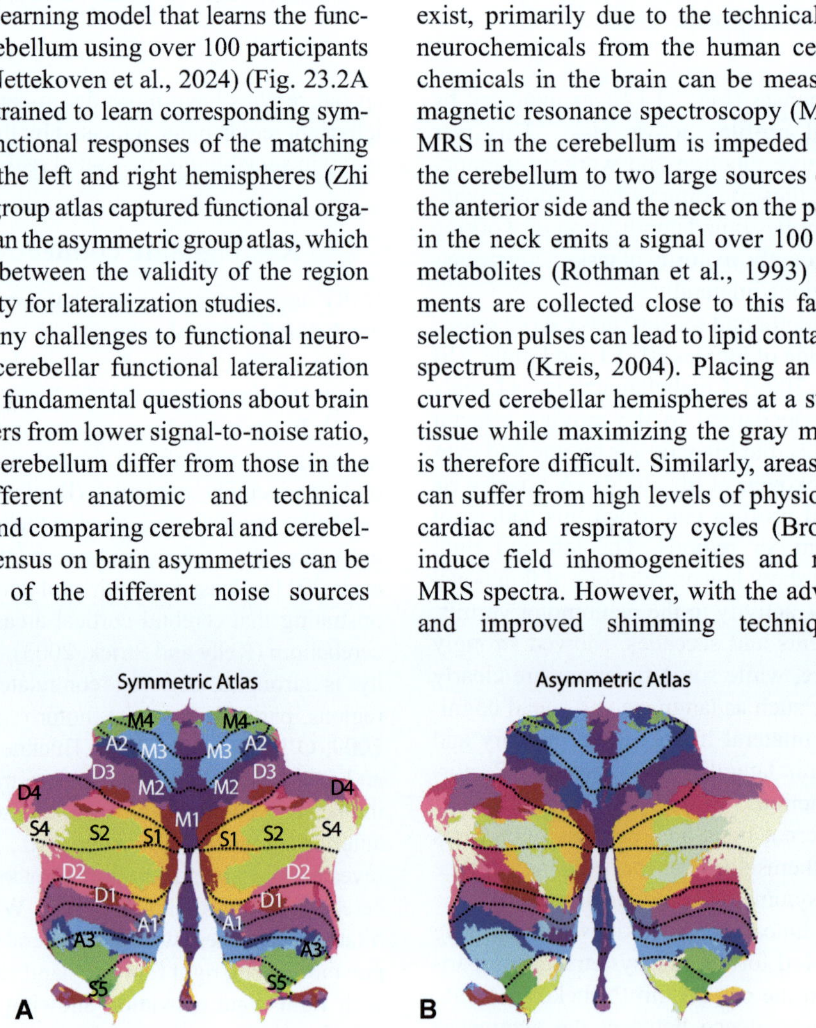

Fig. 23.2. Symmetric functional atlas of the cerebellum allows investigations of lateralization. (A) Functional atlas of the cerebellum with 32 symmetric regions. (B) Functional atlas of the cerebellum with 32 corresponding asymmetric regions. The atlas (A and B) across the different regions revealed that regions involved in movement and working memory functions have highly symmetric boundaries, i.e., their boundaries fall into the same place on the left and right. Among the motor regions, the oculomotor vermis M1 and the hand region M3 (A) fall particularly similar in the left and right cerebellar hemispheres. Meanwhile, the social-linguistic-spatial regions have more dissimilar region boundaries between left–right region pairs, indicated by the low overall boundary symmetry in social-linguistic-spatial regions (Nettekoven et al., 2024). Reproduced from Nettekoven C, Zhi D, Shahshahani L, et al. (2024). A hierarchical atlas of the human cerebellum for functional precision mapping. Nat Commun 15: 8376. https://doi.org/10.1038/s41467-024-52371-w.

inhomogeneities, there have been some investigations of cerebellar neurochemistry and their hemispheric differences.

In a first investigation study of cerebellar neurochemistry, Rae et al. (Rae et al., 1998) examined hemispheric differences in cerebellar neurochemical ratios in 14 dyslexic men and compared them to 15 male controls. In the right cerebellum of dyslexic men, the authors found decreases in the ratio of choline (Cho) to *N*-acetylaspartate (NAA), a commonly used reference metabolite (Bachtiar and Stagg, 2013) due to the high concentrations of NAA in the brain. Dyslexic men also showed significantly lower levels of creatine (Cr)/NAA in the right cerebellum compared to controls and a significant hemispheric difference in Cr/NAA concentration, whereas controls showed no hemispheric difference. The dyslexic group showed contralateral alterations in the neocortex. In the left temporoparietal lobe, dyslexic men showed significantly reduced Cho/NAA levels compared to controls and compared to their right Cho/NA levels, where controls showed equal levels.

Although motor impairments in dyslexia have been linked to a role of the cerebellum in this developmental disorder (Nicolson et al., 1995), in line with the hypothesis that cerebellar dysfunction could manifest in incoordination of eye movements during reading, no association between cerebellar neurochemistry and handedness or performance on a peg-board test was found. Nevertheless, the study provided compelling evidence for asymmetrically altered levels of cerebellar neurochemicals in dyslexia.

Tracking cerebellar neurochemical concentration over time in healthy participants has been of recent interest as a means to understanding cerebellar plasticity in healthy human function. In particular, the major inhibitory neurotransmitter γ-aminobutyric acid (GABA) has been suggested to play a fundamental role in cerebellar plasticity, since it signals information from the sole output neuron of the cerebellar cortex, the Purkinje cell, and has been implicated in parallel fiber synaptic plasticity (Orts-Del'Immagine and Pugh, 2018). Jalali et al. (Jalali et al., 2018) were the first to probe the responsiveness of the GABAergic system in the cerebellum and attempt to link it to individual differences on a task probing motor adaptation, where the cerebellum plays a key role (Diedrichsen, 2005; Graydon et al., 2005; Miall and Jenkinson, 2005). Based on previous work linking anodal transcranial direct current stimulation (tDCS) to reductions in GABA in the primary motor cortex and motor learning (Stagg et al., 2009; Kim et al., 2014), the authors tested for changes in GABA concentration in the right cerebellar hemisphere in response to right cerebellar tDCS and attempted to link these changes to adaptation performance. There was no significant change in GABA, but exploratory analyses revealed a correlation between late adaptation and stimulation-driven GABA decrease. The authors noted as one of the reasons for the lack of GABA change in response to adaptation the long MRS acquisition time of 25 min, which might have rendered transient changes in GABAergic levels undetectable.

In a subsequent investigation of GABA changes occurring in the cerebellum during adaptation, MRS measurements were acquired in 9-min blocks from both the left and right cerebellar hemispheres (see Fig. 23.3A) while participants performed a right-hand adaptation task in the scanner (Nettekoven et al., 2022). Isolating adaptation-driven GABA changes revealed diverging GABA concentration at the right and left cerebellar nuclei (see Fig. 23.3B) with left cerebellar GABA increasing and right cerebellar GABA decreasing. The extent of the early GABA change at the right cerebellar nuclei showed a relationship with adaptation performance (Fig. 23.3C). Those participants who showed greater GABA decrease also adapted better. This relationship was specific to right cerebellar GABA change and adaptation, as there was no relationship with left cerebellar GABA change. Though these results present the first evidence for lateralized neurochemical signatures of cerebellar-dependent motor behavior of the right hand, further MRSI studies are necessary to provide a full picture of cerebellar neurochemistry and its role in motor and non-motor functions.

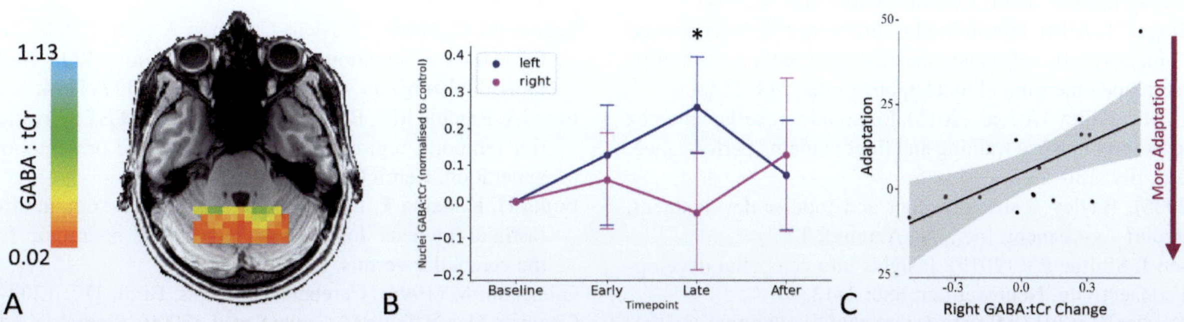

Fig. 23.3. Asymmetric neurochemical changes at cerebellar nuclei during adaptation. Magnetic resonance spectroscopic imaging (MRSI) can be used to quantify the major inhibitory neurotransmitter GABA in several cerebellar voxels in vivo, as shown for a representative participant (A). Isolating neurochemical changes in response to righthand adaptation reveals diverging GABA at the left and right cerebellar nuclei (B) and a relationship between early GABA change at the right cerebellar nuclei and adaptation performance (C). Reproduced from Nettekoven C, Mitchell L, Clarke WT, et al. (2022). Cerebellar GABA change during visuomotor adaptation relates to adaptation performance and cerebellar network connectivity: a magnetic resonance spectroscopic imaging study. J Neurosci 42: 7721–7732.

CONCLUSIONS

Asymmetries have been documented in the structure, function, and neurochemistry of the cerebellum. Functional asymmetries are largely confined to the cerebellar hemisphere contralateral to the cerebrum, while broad structural asymmetries show the same pattern in the cerebellum and cerebrum. The extent of cerebellar asymmetries has also been linked to handedness and performance on lateralized functions, such as language. However, technical challenges in measuring and analyzing cerebellar asymmetries have so far hindered thorough studies of cerebellar lateralization. Hence, several questions regarding the function and clinical consequences of cerebellar asymmetries remain open. For example, does the development of functional asymmetries in the cerebellum precede, parallel, or follow the development of corresponding cerebral functional asymmetries? To what extent does cerebellar asymmetry vary with biologic factors such as age, sex, handedness, and heredity? And what is the relationship between cerebellar asymmetry and disorders that involve the cerebellum such as schizophrenia, and how can we use this knowledge to improve diagnosis and treatment? The development of specialized imaging techniques and new tools for the study of the cerebellum now opens the opportunity to answer these questions.

REFERENCES

Ackermann H (2008). Cerebellar contributions to speech production and speech perception: psycholinguistic and neurobiological perspectives. Trends Neurosci 31: 265–272.

Albus JS (1971). A theory of cerebellar function. Math Biosci 10: 25–61.

Allen G, Buxton RB, Wong EC et al. (1997). Attentional activation of the cerebellum independent of motor involvement. Science 275: 1940–1943.

Amunts K, Schlaug G, Schleicher A et al. (1996). Asymmetry in the human motor cortex and handedness. Neuroimage 4: 216–222.

Andescavage NN, Du Plessis A, McCarter R et al. (2017). Complex trajectories of brain development in the healthy human fetus. Cereb Cortex 27: 5274–5283.

Azevedo FA, Carvalho LR, Grinberg LT et al. (2009). Equal numbers of neuronal and nonneuronal cells make the human brain an isometrically scaled-up primate brain. J Comp Neurol 513: 532–541.

Bachtiar V, Stagg CJ (2013). Interindividual differences in behavior and plasticity. In: Magnetic resonance spectroscopy: tools for neuroscience research and emerging clinical applications, 243–253.

Baer LH, Park MT, Bailey JA et al. (2015). Regional cerebellar volumes are related to early musical training and finger tapping performance. Neuroimage 109: 130–139.

Bayley N (2006). Bayley scales of infant and toddler development, 3rd edHarcourt Assessment, Inc., San Antonio, TX.

Beckinghausen J, Sillitoe RV (2019). Insights into cerebellar development and connectivity. Neurosci Lett 688: 2–13.

Bernard JA, Seidler RD (2013). Relationships between regional cerebellar volume and sensorimotor and cognitive function in young and older adults. Cerebellum 12: 721–737.

Bernard JA, Leopold DR, Calhoun VD et al. (2015). Regional cerebellar volume and cognitive function from adolescence to late middle age. Hum Brain Mapp 36: 1102–1120.

Bolduc ME, Du Plessis AJ, Sullivan N et al. (2012). Regional cerebellar volumes predict functional outcome in children with cerebellar malformations. Cerebellum 11: 531–542.

Brooks JC, Faull OK, Pattinson KT et al. (2013). Physiological noise in brainstem fMRI. Front Hum Neurosci 7: 1–13.

Buckner RL, Krienen FM, Castellanos A et al. (2011). The organization of the human cerebellum estimated by intrinsic functional connectivity. J Neurophysiol 106: 2322–2345.

Busch SE, Hansel C (2023). Climbing fiber multi-innervation of mouse Purkinje dendrites with arborization common to human. Science 381: 420–427.

Cherbuin N, Reglade-Meslin C, Kumar R et al. (2010). Mild cognitive disorders are associated with different patterns of brain asymmetry than normal aging: the PATH through life study. Front Psych 1: 1–9.

Deoni SC, Catani M (2007). Visualization of the deep cerebellar nuclei using quantitative T1 and ρ magnetic resonance imaging at 3 Tesla. Neuroimage 37: 1260–1266.

Diedrichsen J (2005). Neural correlates of reach errors. J Neurosci 25: 9919–9931.

Diedrichsen J (2006). A spatially unbiased atlas template of the human cerebellum. Neuroimage 33: 127–138.

Diedrichsen J, Zotow E (2015). Surface-based display of volume-averaged cerebellar imaging data. PLoS One 10: 1–18.

Diedrichsen J, Balsters JH, Flavell J et al. (2009). A probabilistic MR atlas of the human cerebellum. Neuroimage 46: 39–46.

Diedrichsen J, Maderwald S, Kuper M et al. (2011). Imaging the deep cerebellar nuclei: a probabilistic atlas and normalization procedure. Neuroimage 54: 1786–1794.

Ding H, Qin W, Jiang T et al. (2012). Volumetric variation in subregions of the cerebellum correlates with working memory performance. Neurosci Lett 508: 47–51.

Eccles JC (1967). The cerebellum as a neuronal machine, Springer-Verlag, Berlin.

Ertekin T, Acer N, Icer S et al. (2013). Comparison of two methods for the estimation of subcortical volume and asymmetry using magnetic resonance imaging: a methodological study. Surg Radiol Anat 35: 301–309.

Fabbro F, Tavano A, Corti S et al. (2004). Long-term neuropsychological deficits after cerebellar infarctions in two young adult twins. Neuropsychologia 42: 536–545.

Fan L, Tang Y, Sun B et al. (2010). Sexual dimorphism and asymmetry in human cerebellum: an MRI-based morphometric study. Brain Res 1353: 60–73.

Fedorenko E, Hsieh PJ, Nieto-Castanon A et al. (2010). New method for fMRI investigations of language: defining ROIs functionally in individual subjects. J Neurophysiol 104: 1177–1194.

Fiez JA, Raichle ME, Balota DA et al. (1996). PET activation of posterior temporal regions during auditory word presentation and verb generation. Cereb Cortex 6: 1–10.

Fujita H, Kodama T, Du Lac S (2020). Modular output circuits of the fastigial nucleus for diverse motor and nonmotor functions of the cerebellar vermis. Elife 9: 1–36.

Glickstein M (1994). Cerebellar agenesis. Brain 117: 1209–1212.

Gocmen-Mas N, Pelin C, Canan S et al. (2009). Stereological evaluation of volumetric asymmetry in healthy human cerebellum. Surg Radiol Anat 31: 177–181.

Graydon FX, Friston KJ, Thomas CG et al. (2005). Learning-related fMRI activation associated with a rotational visuo-motor transformation. Cogn Brain Res 22: 373–383.

Haberling IS, Corballis MC (2016). Cerebellar asymmetry, cortical asymmetry and handedness: two independent networks. Laterality 21: 397–414.

Herve PY, Crivello F, Perchey G et al. (2006). Handedness and cerebral anatomical asymmetries in young adult males. Neuroimage 29: 1066–1079.

Hodge SM, Makris N, Kennedy DN et al. (2010). Cerebellum, language, and cognition in autism and specific language impairment. J Autism Dev Disord 40: 300–316.

Holland D, Chang L, Ernst TM et al. (2014). Structural growth trajectories and rates of change in the first 3 months of infant brain development. JAMA Neurol 71: 1266–1274.

Holloway RL, De La Costelareymondie MC (1982). Brain endocast asymmetry in pongids and hominids: some preliminary findings on the paleontology of cerebral dominance. Am J Phys Anthropol 58: 101–110.

Isıklar S, Demir I, Ozdemir ST et al. (2023). Examination of the development and asymmetry of the cerebellum and its lobules in individuals aged 1–18 years: a retrospective MRI study. Brain Topogr 901–925.

Ito M, Kano M (1982). Long-lasting depression of parallel fiber-Purkinje cell transmission induced by conjunctive stimulation of parallel fibers and climbing fibers in the cerebellar cortex. Neurosci Lett 33: 253–258.

Jalali R, Chowdhury A, Wilson M et al. (2018). Neural changes associated with cerebellar tDCS studied using MR spectroscopy. Exp Brain Res 236: 997–1006.

Kang X, Herron TJ, Ettlinger M et al. (2015). Hemispheric asymmetries in cortical and subcortical anatomy. Laterality 20: 658–684.

Kavaklioglu T, Guadalupe T, Zwiers M et al. (2017). Structural asymmetries of the human cerebellum in relation to cerebral cortical asymmetries and handedness. Brain Struct Funct 222: 1611–1623.

Kelly RM, Strick PL (2003). Cerebellar loops with motor cortex and prefrontal cortex of a nonhuman primate. J Neurosci 23: 8432–8444.

Kim S, Stephenson MC, Morris PG et al. (2014). TDCS induced alterations in GABA concentration within primary motor cortex predict motor learning and motor memory: a 7T magnetic resonance spectroscopy study. Neuroimage 99: 237–243.

King M, Hernandez-Castillo CR, Poldrack RA et al. (2019). Functional boundaries in the human cerebellum revealed by a multi-domain task battery. Nat Neurosci 22: 1371–1378.

Kreis R (2004). Issues of spectral quality in clinical 1H-magnetic resonance spectroscopy and a gallery of artifacts. NMR Biomed 17: 361–381.

Krienen FM, Buckner RL (2009). Segregated fronto-cerebellar circuits revealed by intrinsic functional connectivity. Cereb Cortex 19: 2485–2497.

Larsell O, Jansen J (1973). The comparative anatomy and histology of the cerebellum—review. Comp Gen Pharmacol 3: 467–468.

LeBel A, D'Mello AM (2023). A seat at the (language) table: incorporating the cerebellum into frameworks for language processing. Curr Opin Behav Sci 53: 101310.

Leggio MG, Silveri MC, Petrosini L et al. (2000). Phonological grouping is specifically affected in cerebellar patients: a verbal fluency study. J Neurol Neurosurg Psychiatry 69: 102–106.

LeMay M (1976). Morphological cerebral asymmetries of modern man, fossil man, and nonhuman primate. Ann N Y Acad Sci 280: 349–366.

Lesage E, Hansen PC, Miall RC (2017). Right lateral cerebellum represents linguistic predictability. J Neurosci 37: 6231–6241.

Li X, Crow TJ, Hopkins WD et al. (2018). Human torque is not present in chimpanzee brain. Neuroimage 165: 285–293.

Marien P, Saerens J, Nanhoe R et al. (1996). Cerebellar induced aphasia: case report of cerebellar induced prefrontal aphasic language phenomena supported by SPECT findings. J Neurol Sci 144: 34–43.

Marien P, Engelborghs S, Pickut BA et al. (2000). Aphasia following cerebellar damage: fact or fallacy? J Neurolinguistics 13: 145–171.

Marr JE (1969). A theory of cerebellar cortex. J Physiol 202: 437–470.

Marvel CL, Desmond JE (2010). Functional topography of the cerebellum in verbal working memory. Neuropsychol Res 20: 271–279.

Medina JF, Lisberger SG (2008). Links from complex spikes to local plasticity and motor learning in the cerebellum of awake-behaving monkeys. Nat Neurosci 11: 1185–1192.

Miall RC, Jenkinson EW (2005). Functional imaging of changes in cerebellar activity related to learning during a novel eye-hand tracking task. Exp Brain Res 166: 170–183.

Moberget T, Ivry RB (2016). Cerebellar contributions to motor control and language comprehension: searching for common computational principles. Ann N Y Acad Sci 1369: 154–171.

Nettekoven C, Mitchell L, Clarke WT et al. (2022). Cerebellar GABA change during visuomotor adaptation relates to adaptation performance and cerebellar network connectivity: a magnetic resonance spectroscopic imaging study. J Neurosci 42: 7721–7732.

Nettekoven C, Zhi D, Shahshahani L et al. (2024). A hierarchical atlas of the human cerebellum for functional precision mapping. Nat Commun 15: 8376. https://doi.org/10.1038/s41467-024-52371-w.

Nicolson RI, Fawcett AJ, Dean P (1995). Time estimation deficits in developmental dyslexia: evidence of cerebellar involvement. Proc Roy Soc London, Ser B Biol Sci 259: 43–47.

O'Reilly JX, Beckmann CF, Tomassini V et al. (2010). Distinct and overlapping functional zones in the cerebellum defined by resting state functional connectivity. Cereb Cortex 20: 953–965.

Olson IR, Hoffman LJ, Jobson KR et al. (2023). Little brain, little minds: the big role of the cerebellum in social development. Dev Cogn Neurosci 60: 101238.

Orts-Del'Immagine A, Pugh JR (2018). Activity-dependent plasticity of presynaptic GABAB receptors at parallel fiber synapses. Synapse.

Petersen SE, Fox PT, Posner MI et al. (1989). Positron emission tomographic studies of the processing of single words. J Cogn Neurosci 1: 153–170.

Rae C, Lee MA, Dixon RM et al. (1998). Metabolic abnormalities in developmental dyslexia detected by 1H magnetic resonance spectroscopy. Lancet 351: 1849–1852.

Raichle ME, Fiez JA, Videen TO et al. (1994). Practice-related changes in human brain functional anatomy during nonmotor learning. Cereb Cortex 4: 8–26.

Ronconi L, Casartelli L, Carna S et al. (2017). When one is enough: impaired multisensory integration in cerebellar agenesis. Cereb Cortex 27: 2041–2051.

Rosch RE, Cowell PE, Gurd JM (2018). Cerebellar asymmetry and cortical connectivity in monozygotic twins with discordant handedness. Cerebellum 17: 191–203.

Rothman DL, Petroff OA, Behar KL et al. (1993). Localized 1H NMR measurements of γ-aminobutyric acid in human brain in vivo. Proc Natl Acad Sci U S A 90: 5662–5666.

Saadon-Grosman N, Angeli PA, DiNicola LM et al. (2022). A third Somatomotor representation in the human cerebellum. J Neurophysiol 128: 1051–1073.

Saltoun K, Adolphs R, Paul LK et al. (2023). Dissociable brain structural asymmetry patterns reveal unique phenome-wide profiles. Nat Hum Behav 7: 251–268.

Schmahmann JD, Sherman JC (1998). The cerebellar cognitive affective syndrome. Brain 121: 561–579.

Schmahmann J, Doyon J, Toga A et al. (2000). MRI atlas of the human cerebellum, Elsevier.

Sereno MI, Diedrichsen J, Tachrount M et al. (2020). The human cerebellum has almost 80 of the surface area of the neocortex. Proc Natl Acad Sci U S A 117: 19538–19543.

Silveri MC, Leggio MG, Molinari M (1994). The cerebellum contributes to linguistic production: a case of agrammatic speech following a right cerebellar lesion. Neurology 44: 2047–2050.

Snyder PJ, Bilder RM, Wu H et al. (1995). Cerebellar volume asymmetries are related to handedness: a quantitative MRI study. Neuropsychologia 33: 407–419.

Stagg CJ, Best JG, Stephenson MC et al. (2009). Polarity-sensitive modulation of cortical neurotransmitters by transcranial stimulation. J Neurosci 29: 5202–5206.

Starowicz-Filip A, Prochwicz K, Klosowska J et al. (2021). Cerebellar functional lateralization from the perspective of clinical neuropsychology. Front Psychol 12: 1–12.

Stipdonk LW, Boumeester M, Pieterman KJ et al. (2021). Cerebellar volumes and language functions in school-aged children born very preterm. Pediatr Res 90: 853–860.

Stoodley CJ (2014). Distinct regions of the cerebellum show gray matter decreases in autism, ADHD, and developmental dyslexia. Front Syst Neurosci 8: 1–17.

Stoodley CJ, Schmahmann JD (2009). Functional topography in the human cerebellum: a meta-analysis of neuroimaging studies. Neuroimage 44: 489–501.

Stoodley CJ, MacMore JP, Makris N et al. (2016). Location of lesion determines motor vs. cognitive consequences in patients with cerebellar stroke. NeuroImage Clin 12: 765–775.

Szeszko PR, Gunning-Dixon F, Ashtari M et al. (2003). Reversed cerebellar asymmetry in men with first-episode schizophrenia. Biol Psychiatry 53: 450–459.

Thomas Yeo BT, Krienen FM, Sepulcre J et al. (2011). The organization of the human cerebral cortex estimated by intrinsic functional connectivity. J Neurophysiol 106: 1125–1165.

Tomasch J (1969). The numerical capacity of the human corticoponto-cerebellar system. Brain Res 13: 476–484.

Van Overwalle F, D'aes T, Marien P (2015). Social cognition and the cerebellum: a meta-analytic connectivity analysis. Hum Brain Mapp 36: 5137–5154.

Vassar R, Schadl K, Cahill-Rowley K et al. (2020). Neonatal brain microstructure and machine-learning-based prediction of early language development in children born very preterm. Pediatr Neurol 108: 86–92.

Wang D, Buckner RL, Liu H (2013). Cerebellar asymmetry and its relation to cerebral asymmetry estimated by intrinsic functional connectivity. J Neurophysiol 109: 46–57.

Wang D, Buckner RL, Liu H et al. (2014). Functional specialization in the human brain estimated by intrinsic hemispheric interaction. J Neurosci 34: 12341–12352.

Yan X, Kong R, Xue A et al. (2023). Homotopic local-global parcellation of the human cerebral cortex from resting-state functional connectivity. Neuroimage 273: 120010. https://doi.org/10.1016/j.neuroimage.2023.120010 Epub 2023 Mar 12, PMID: 36918136; PMCID: PMC10212507.

Zhang Y, Wu X (2021). Asymmetries of cerebellar lobe in the genus homo. Symmetry 13.

Zhi D, Shahshahani L, Nettekoven C et al. (2025). A hierarchical Bayesian brain parcellation framework for fusion of functional imaging datasets. Imaging Neurosci 3. https://doi.org/10.1162/imag_a_00408.

Handbook of Clinical Neurology, Vol. 208 (3rd series)
Cerebral Asymmetries
C. Papagno and P. Corballis, Editors
https://doi.org/10.1016/B978-0-443-15646-5.00016-6

Chapter 24

Handedness

SEBASTIAN OCKLENBURG[1,2*] AND ONUR GÜNTÜRKÜN[2,3]

[1]*ICAN Institute for Cognitive and Affective Neuroscience, MSH Medical School Hamburg, Hamburg, Germany*

[2]*Institute of Cognitive Neuroscience, Ruhr University Bochum, Bochum, Germany*

[3]*Research Center One Health Ruhr, Research Alliance Ruhr, Ruhr University Bochum, Bochum, Germany*

Abstract

This chapter offers an overview of the literature on human handedness and its assessment in clinical neurologic practice and research. There are two major forms of handedness: hand preference, which describes a subjective preference to use one hand over the other for skilled motor activities like writing, and hand skill, which describes objectively measured mother skill. This chapter gives an overview of widely used questionnaires and tests to assess hand preference and hand skill, as well as suggestions on how to determine handedness categories such as left-handed, right-handed, and mixed-handed based on the results of these questionnaires and tests. Handedness is just one form of hemispheric asymmetry in the human motor system, and the chapter also provides an overview of its association with other motor asymmetries such as footedness. Moreover, the associations of handedness with functional brain activation as well as with structural markers on the cortical, subcortical, cerebellar, and spinal levels are discussed. Furthermore, the potential relevance of handedness retraining for clinical neurologic research and the association of handedness and cognitive abilities are discussed. The chapter concludes with an outlook on the critical importance of including handedness in clinical neurologic research and practice.

INTRODUCTION: WHAT IS HANDEDNESS?

Compared to the hands of nonhuman primate species, human hands are more efficient in performing firm one-handed precision grips and fine precision movements. This enables humans to make and use tools much more efficiently than monkeys and apes (Marzke and Marzke, 2000). During tool use, an important functional feature of the human hand becomes evident: most people strongly prefer to use one specific hand over the other for fine precision movements such as using a pen to sign their name under a document. This strong motor preference is called handedness (Corballis, 1997, 2003; McManus, 2004, 2019, 2022; Häberling et al., 2016). Handedness is a form of functional hemispheric asymmetries, i.e., a functional difference between the left and the right hemisphere of the brain (Badzakova-Trajkov et al., 2016; Corballis and Häberling, 2017; Ocklenburg and Güntürkün, 2018; Corballis, 2019; Gerrits, 2022; Zatorre, 2022). Specifically, handedness is the functional hemispheric asymmetry of the brain's motor networks for hand use (Amunts et al., 1996, 1997, 2000; Corballis, 1998; Sha et al., 2021). This neural origin of handedness also becomes evident when looking at the left and right hands of people. Typically, the two hands are largely symmetric in terms of bones, muscles, and other anatomic features. It is not possible to tell whether a person is left-handed or right-handed just by looking at their hands (see Fig. 24.1).

Typically, two different ways of assessing handedness can be distinguished (Palmer and Corballis, 1996; Callahan et al., 2016; Mundorf et al., 2023). First, there is hand preference, which indicates the subjective preference of the person to use one hand over the other for a given task, independent of the skill level of the hands (e.g., someone can prefer to use the left hand for writing even though they are equally fast and accurate with both hands). Second, there is hand skill (sometimes also called hand dexterity), which refers to objectively measured skill independent of subjective preference (e.g., somebody can write faster and more accurately with the right hand, even though they prefer to use the left hand for writing). In the following sections, commonly used questionnaires and tests to assess both hand skills and hand preference will be discussed, as well as how to interpret their results.

*Correspondence to: Sebastian Ocklenburg, ICAN Institute for Cognitive and Affective Neuroscience, MSH Medical School Hamburg, Hamburg, Germany. Tel: +49-40-361-22649269. E-mail: sebastian.ocklenburg@medicalschool-hamburg.de.

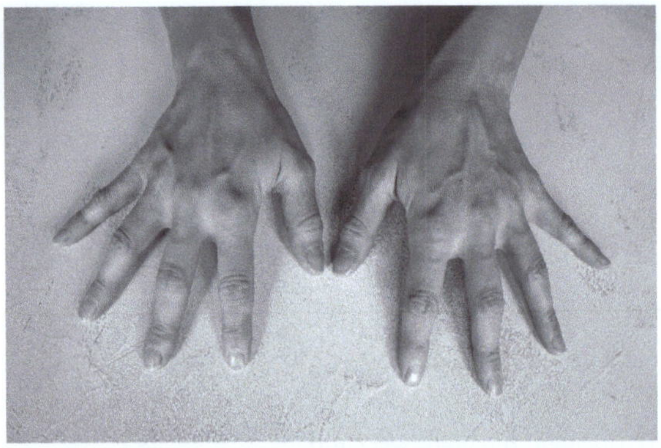

Fig. 24.1. A left and a right human hand. Just by looking at the picture, it is not possible, whether the person is left handed or right handed.

The assessment of handedness 1: Hand preference

While it is not uncommon to see older studies that used a single question to assess handedness (e.g., "Do you prefer the left or the right hand for writing?"), it is now considered best practice to use multiple-item questionnaires to assess hand preference (Vingerhoets et al., 2023). The first widely used multiple-item questionnaire to assess hand preference is the Edinburgh Handedness Inventory (EHI) (Oldfield, 1971). Fig. 24.2 shows the 10 different activities participants are asked to indicate their preferred hand in the EHI [writing, drawing, throwing, using scissors, using a toothbrush, using a knife (without a fork), using a spoon, using a broom (upper hand), striking a match (the hand that holds the match), and opening a box (the hand that holds the lid)].

In the EHI, participants must indicate the strength of their individual preferences for each of the 10 items. There are five possible answer options. If a participant prefers one hand so strongly that they would never try to use the other hand unless forced to, they must write two pluses in the box for the respective hand. If a participant prefers one hand, they write a single plus in the box for the respective hand. If a participant is completely indifferent about which hand to use for a specific activity, they must write a plus in both the box for the left hand and the box for the right hand. Based on the sum of pluses the participant made for the right hand (R) and the left hand (L), a lateralization quotient (LQ) is determined using the formula:

$$LQ = [(R - L)/(R + L)] * 100$$

The LQ reflects both the strength and the direction of hand preference in a single value. It ranges from −100 (consistently left-handed) to 100 (consistently right-handed). A value of 0 indicates complete mixed-handedness.

The original EHI paper by Oldfield has been cited more than 11,000 times by now, making the EHI the most widely used questionnaire to assess hand preference (Oldfield, 1971). While originally published in English, it has been translated into many different languages, including Portuguese (Espírito-Santo et al., 2017), Spanish (Albayay et al., 2019), Chinese (Yang et al., 2018), and Italian (Viggiano et al., 2001). Despite its widespread use, the EHI is not without problems. A comprehensive review of

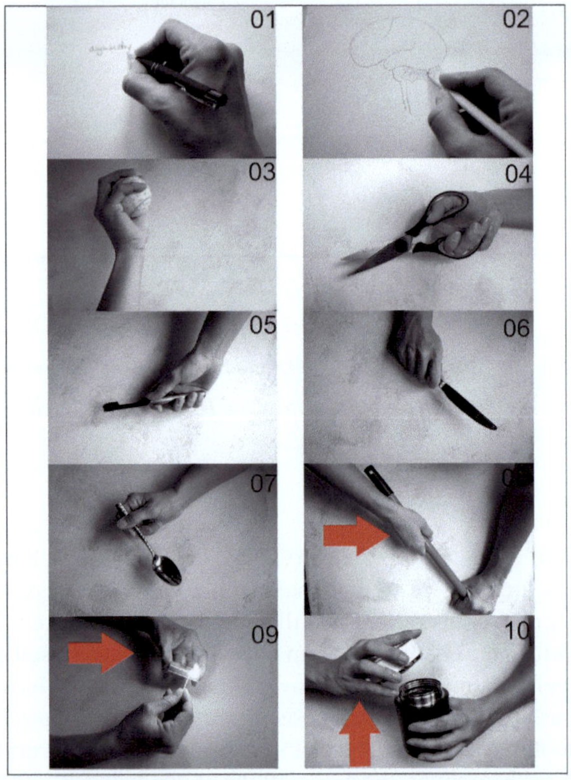

Fig. 24.2. The 10 activities used to determine hand preference in the original version of the Edinburgh Handedness Inventory (EHI): writing (01), drawing (02), throwing (03), using scissors (04), using a toothbrush (05), using a knife without a fork (06), using a spoon (07), using a broom (upper hand) (08), striking a match (match) (09), and opening a box (lid) (10). Red arrows indicate the relevant hand when two hands are in the picture.

899 scientific articles that used the EHI published between 1998 and 2012 investigated to what extent researchers modified the original EHI questionnaire in modern research (Edlin et al., 2015). The analysis revealed substantial diversity between studies regarding test items, response format, and schemes for classifying individuals as right-handed or left-handed. Moreover, there was evidence that researchers sometimes had not been aware that they had been using a non-original variant of the EHI, and the uniformity of variants was low. Together, these factors may have negative consequences for the meta-analytic integration of EHI data and replicability of handedness results. A more recent survey of the use of the EHI in functional magnetic resonance imaging studies since 2013 also found substantial evidence for nontransparent usage and reporting of EHI data and many changes to response format or cut-off scores that were not justified by any references (Yeung and Wong, 2022). Moreover, the EHI has been criticized for being outdated and having a response format that is frequently misunderstood (Fazio et al., 2013).

Over the decades, many alternatives to the EHI have been developed, but none have ever reached comparable citation rates. These include both updated versions of the EHI like the EHI Short Form (Veale, 2014) and many independent questionnaires. Table 24.1 gives an overview of some of the most widely used

Table 24.1

Commonly used questionnaires and behavioral tests to determine hand preference.

Name	Publication	Number of citations
Annett Hand Preference Questionnaire (AHPQ)	Annett (1970a)	11
Edinburgh Handedness Inventory (EHI)—Original version	Oldfield (1971)	11,264
EHI Short Form	Veale (2014)	199
Beukelaar and Kroonenberg questionnaire	Beukelaar and Kroonenberg (1983)	63
Waterloo Handedness Questionnaires (WHQ)	Steenhuis and Bryden (1989)	43
Waterloo Handedness Questionnaires Revised (WHQ-R)	Elias et al. (1998)	170
The lateral preference inventory	Coren (1993)	351
EHI-WHQ (combination of EHI and WHQ)	Stone et al. (2013)	36
WatHand Box Test (WBT)[a]/WatHand Cabinet Test (WHCT)[a]	Bryden et al. (2000) and Bryden et al. (2007)	9/10
Inventory of global lateral preference (IPLAG)	Marim et al. (2011)	5
The Flinders Handedness Survey (FLANDERS)	Nicholls et al. (2013)	72
The Fazio Laterality Inventory (FLI)	Fazio et al. (2013)	9
Home Handedness Questionnaire (HHQ)[a]	Nelson et al. (2019)	5

[a]Observational behavioral preference test.
Citations were determined using https://pubmed.ncbi.nlm.nih.gov/ in January 2024. Citation rates may be underestimated in some cases, in which more than one paper can be cited for a specific test or there are additional books or book chapters that can be cited like in the case of the Annett Hand Preference Questionnaire (Annett, 1970a).

hand preference questionnaires and tests. Given the large amount of hand preference questionnaires developed over the years, it does not give a full overview of every handedness test ever developed. For a complete overview, the interested reader is referred to a recent systematic review on inventories of human lateral preference (Bazo et al., 2023).

While hand preference is typically assessed using questionnaires, it is also possible to assess it using behavioral tests (e.g., by asking participants to perform certain activities in real life or pantomime and write down which hand they used). In contrast to behavioral tests of hand skill (see below), behavioral tests of hand preferences typically do not include measurement of reaction times and accuracy. Instead, only the preferred hand for a set of activities is determined.

Behavioral tests of hand preferences are typically used in children since self-report and parental reports are not sufficient indicators of hand preference in this group (Vingerhoets et al., 2023). For example, in one study, children between 7 and 11 years were asked to perform the following seven activities (Ocklenburg et al., 2010):

1. Write their name on a piece of paper.
2. Throw a ball.
3. Cut out a nametag using scissors.
4. Use a toothbrush.
5. Cut a piece of play dough with a knife.
6. Use a spoon to eat puffed rice.
7. Play a C-major scale on a xylophone.

Thus, the items partly overlapped with those of the EHI but some unsuitable items like striking a match were substituted and the overall number of items was reduced to keep the stress level low. Comparable to the EHI, an LQ was determined based on the observed hand preferences.

A validated preference test specifically designed for younger children (e.g., preschoolers) is the Home Handedness Questionnaire (HHQ) (Nelson et al., 2019; Gonzalez and Nelson, 2021). In this task, parents must place an object on a table or countertop within the child's reach. Then, the parent needs to ask the child to perform 15 unilateral actions (such as picking up a bottle cap or picking up a sheet of paper off the table) and 15 role-differentiated bimanual manipulations, which are tasks that must be performed with both hands, but each hand has a specific role (such as holding a bag and retrieving a toy from inside). Each of the 30 actions is performed twice. Based on whether the child performed the actions with the left or the right hand, two separate laterality scores are calculated for the two types of actions. Individual preferences are then determined based on binomial z-scores ($z < -1.96$: left-handedness, $z > 1.96$: right-handedness, and z between -1.96 and 1.96: no preference).

Two further observational tests to behaviorally assess hand preferences in children are the WatHand Box Test (WBT) (Bryden et al., 2000) and the WatHand Cabinet Test (WHCT) (Bryden et al., 2007). In the WBT, participants have to perform several different unimanual tasks, including opening a lock with a key, using a hammer, and tossing a ball (Bryden et al., 2000). The WHCT is a cabinet with several features that allow for the assessment of hand preferences, such as a Velcro bull's-eye target and ball to assess hand preference for throwing a ball or a padlock hung from a hook that allows to assess hand preference for opening a lock with a key (Bryden et al., 2007; Scharoun and Bryden, 2015).

The assessment of handedness 2: Hand skill

In contrast to hand preference, hand skill (sometimes also called hand performance) refers to objectively measured motor skill or performance with the hands. While both hand preference and hand skill typically point in the same direction (most people who prefer to write with their left hand are also faster and more accurate when writing with their left hand than with their right hand), this is not necessarily the case in 100% of individuals. It has been estimated in a simulation that the correlation between hand preference and hand skill is significant but not 1.0, with an estimated value of 0.70 for a five-preference scale, and an estimated value of 0.72 for a nine-item scale (Bishop, 1989). In contrast to hand preference questionnaires, hand skill tests typically involve measuring reaction time and acquiring accuracy data that are used to determine the individual skill level.

Several different families of tests have been developed to test hand skills (Doyen and Carlier, 2002). One widely used form of hand skill tests is pegboard tests such as the Anett pegboard test (Annett, 1970b) or the classic Purdue pegboard test (Tiffin and Asher, 1948). In this type of test, participants must move several wooden or plastic pegs (often 10) that are located in holes on a wooden or plastic board as fast and accurately as possible to a second set of empty holes on the same board. The time it takes to move all pegs with each hand is measured several times (often two or three times). Afterwards, the reaction times for both hands are typically integrated into a single value that indicates the direction of the dominant hand, e.g., by using the following formula (Scharoun and Bryden, 2015):

$$\text{Pegboard test laterality quotient} = [(\text{left hand time} - \text{right hand time}) / (\text{left hand time} + \text{right hand time})] * 100$$

Note that, unlike the LQ for EHI, the average time for the left hand (not the right hand) is used as the minuend in the formula, and the average time for the right hand as the subtrahend. This is done since for reaction times, larger values indicate worse performance. This way, negative values indicate left-handedness, and positive values right-handedness just like for the EHI LQ.

One alternative to the pegboard task that does not necessarily require physical objects to be manipulated is the finger-tapping task, which can be conducted online with digital devices (Parker et al., 2021). For example, in a recent preregistered study focused on assessing the reliability of an online laterality test battery, the authors asked participants to press a sequence of keys with the forefinger of their left hand, and a different sequence of keys with the forefingers of their right hand as many times as possible within 30 s (Parker et al., 2021). By comparing left-hand and right-hand performance over four blocks, the more skilled hand can be identified.

A further commonly used test for hand skill is the Tapley and Bryden dot-filling task (Tapley and Bryden, 1985; Bryden et al., 1994; McManus et al., 2016). In this task, participants receive a sheet of paper on which several circles are printed. Participants must place dots in these circles with a pen as fast and accurately as possible. By comparing the performance of the two hands, the dominant hand can be determined.

The skill tests described so far focus on simple and repetitive motor actions. These actions typically are easier to perform than the tasks usually assessed in hand preference questionnaires (e.g., writing). Therefore, it can make sense to also include a hand skill test for more complicated motor actions like writing in a handedness test battery. To this end, the handwriting speed test can be used (Provins and Magliaro, 1993). In this test, participants must write the alphabet as one word three times as quickly as possible with each hand. By comparing the mean number of letters written per minute per hand, the dominant hand can be identified. Fig. 24.3 shows a participant performing the handwriting speed test.

A test that combines preference and skills aspects of handedness is Bishop's Card Reaching Task (Bishop et al., 1996; Doyen and Carlier, 2002; Musalek et al., 2016). In this task, participants must pick up cards that can be located at different positions. There are no time constraints and the hand used to pick up the card is noted on each trial.

Best practice for handedness assessment in clinical neurologic practice and research

To ensure high between-study comparability and foster large-scale meta-analyses on handedness, it is important to develop best practices or gold standards for handedness assessment in clinical neurologic practice and research. Unfortunately, there is no consensus in the field, but a recent Delphi expert survey report on recommendations for laterality assessment (Vingerhoets et al., 2023) included several important recommendations.

Most importantly, it should be avoided to assess handedness with a single question (e.g., "Are you left-handed or right-handed?), and multi-item questionnaires or multi-trial tests should be used. Only by including multiple items, both direction and strength of handedness can be assessed. Moreover, both a hand preference (e.g., the EHI) and a hand skill (e.g., the pegboard test) test should be used to get a full picture of handedness. Also, cultural biases in hand use should be considered. For example, if a culture has biases against the use of the left hand because it is perceived as less skillful and less powerful (Zverev, 2006), lower rates of left-handedness may be expected. Moreover, based on the literature on problems with the EHI (Fazio et al., 2013;

Fig. 24.3. The handwriting speed test to determine hand skill (Provins and Magliaro, 1993). The participant has to write the whole alphabet as a single word as fast and as accurately as possible.

Edlin et al., 2015; Yeung and Wong, 2022), it becomes clear that transparent reporting of test items, scoring procedures, and cut-off scores is very important. If any modifications to the original form of a used handedness test are made, these should be supported by references and made transparent.

Handedness categories for hand preference

Based on the results of hand preference tests, individuals can be classified into different handedness categories, and different categorization schemes have been developed over the years (McManus, 2019). A recent systematic review of inventories used to assess lateral preference (Bazo et al., 2023) identified three major ways to classify hand preference:

1. dichotomous classification (left-handed and right-handed)
2. three categories (left-handed, right-handed, and mixed-handed)
3. five categories (consistently left-handed, inconsistently left-handed, mixed-handed, inconsistently right-handed, and consistently right-handed)

The most basic one is to distinguish between left-handers and right-handers. A recent meta-analysis on human handedness found that the best prevalence estimate for left-handedness using this dichotomous classification is 10.6%, with the remaining 89.4% of people being right-handed (Papadatou-Pastou et al., 2020). The second most common handedness classification is the three-category scheme (left-handed, right-handed, and mixed-handed). While the dichotomous handedness classification is probably the most widely used one, it may not be optimal in clinical research, and the three-category scheme may be a better choice for clinical neurology research. Specifically, some studies and meta-analyses have found effects for mixed-handers, but not left handers. For example, a recent meta-analysis on handedness in post-traumatic stress disorder (PTSD) reported a significantly higher frequency of mixed-handedness but not left-handedness in PTSD (Borawski et al., 2023). Thus, there may be pathology-relevant neurobiologic differences between left-handed, right-handed, and mixed-handed individuals. However, there is a major methodological issue when it comes to the definition of mixed-handedness that has not been resolved so far. There is an enormous heterogeneity regarding the LQ cut-offs used to define whether an individual is mixed-handed, and no consensus on a best practice regarding specific LQ cut-offs has been reached so far. Table 24.2 gives an overview of some of the different LQ cut-off values that have been used to determine mixed-handedness.

Presently, there is no consensus on which of the various possible mixed-handedness LQ cut-offs reflects the biologic mechanism underlying a categorization into three hand preference types in an optimal way. There is also the possibility that hand preference truly is a continuum and that any form of categorization into three types is artificial. Still, attempts have been made to determine the optimal LQ cut-off for mixed-handedness. For example, one study used latent-class analysis on handedness data from 354 individuals and identified a three-category solution with

Table 24.2

Different LQ cut-offs that have been used to determine mixed handedness.

Study	Mixed-handedness cut-off
Arning et al. (2015) and Mundorf et al. (2023)	LQ: −40 to 40
Orlando et al. (2023)	LQ: −50 to 50
Veale (2014) and Thomas and Fitch (2023)	LQ: −60 to 60
Deep-Soboslay et al. (2010)	LQ: −70 to 70
Christman et al. (2007) and Cheng et al. (2020)	LQ: −80 to 80
Cannon et al. (1995)	LQ: −99 to 99

Note that the minimum of the LQ is −100 and the maximum is 100.

the optimal cut-off point for mixed-handedness between LQ = ±50 and LQ = ±70 (Dragovic, 2004).

The five-category handedness scheme (consistently left-handed, inconsistently left-handed, mixed-handed, inconsistently right-handed, and consistently right-handed) is used less commonly than the other two schemes but is often applied when there is a research question that relates to handedness consistency. In general, individuals who always prefer one hand for all activities are considered consistently handed (e.g., an LQ of −100 or 100 in the EHI). People who prefer the nondominant hand for at least one activity are considered inconsistently handed. As there is evidence that people with consistent and inconsistent handedness differ in some aspects of memory, decision-making, and others (Prichard et al., 2013, 2023), it may be worthwhile to assess this variable in clinical neurology studies focused on handedness.

Handedness categories for hand skill

Comparable to hand preference, different handedness categories can be determined for hand skill (Todor and Doane, 1978; Vingerhoets et al., 2023). The simplest categorization is to differentiate left-handers (faster and more accurate with the left hand) from right-handers (faster and more accurate with the right hand). In addition, ambidexter has been suggested as a third category for people who are fast and accurate with both hands (Vingerhoets et al., 2023). Ambidexterity is not the same as mixed-handedness. While mixed-handedness refers to an ambiguity in hand preferences across different tasks (e.g., someone prefers to write and draw with their left hand, but to throw and open lids with their right hand), ambidexterity refers to equal hand skill in one specific task (Vingerhoets et al., 2023). Therefore, someone can be ambidextrous in one task but not ambidextrous in a different task.

While the concept of ambidexterity seems to be generally accepted in the laterality research community and frequently is mentioned in peer-reviewed research papers, there is also a fourth hand skill category that is less commonly used. The term "ambisinistrality" has been used to describe individuals who are slow and inaccurate with both hands (Todor and Doane, 1978). Table 24.3 gives an overview of the four hand skill categories.

Table 24.3

Overview of the different categories for hand skills based on speed and accuracy in hand skill tests (RH: right hand; LH: left hand).

Category	Speed	Accuracy
Right handed	RH fast/LH slow	RH high/LH low
Left handed	LH fast/RH slow	LH high/RH low
Ambidextrous	RH fast/LH fast	RH high/LH high
Ambisinistral	RH slow/LH slow	RH low/LH low

How can the individual hand skill category be determined? Different methods have been suggested such as determining congruency scores between left- and right-hand performance (Lombana et al., 2022), but there is currently no agreed-upon best practice. In general, ambidexterity should indicate little or no performance difference between the two hands (Vingerhoets et al., 2023). The only study that included all four categories (Todor and Doane, 1978) first determined a handedness index (HI) following the formula:

$$\mathrm{HI} = (\mathrm{right-hand\ performance} - \mathrm{left-hand\ performance}) / (\mathrm{right-hand\ performance} + \mathrm{left-hand\ performance})$$

Participants with a HI of 0.05 or higher were classified as right-handed and participants with a HI of −0.05 or lower were classified as left-handed. Participants with an HI between −0.05 and 0.05 were classified as ambilateral. Ambilateral participants were categorized as ambidextrous if their performance with both hands was not more than one standard deviation below the mean of the performance that the right-handed group showed with their dominant right hand. Ambilateral participants who did not fulfill this criterion were classified as ambisinistral.

The association between handedness and other forms of motor asymmetries

Handedness is not the only form of behavioral asymmetry in the motor system (Elias, 2022). Other commonly investigated forms of motor asymmetries in humans include footedness (Elias et al., 1998; Packheiser et al., 2020; Bazo et al., 2023), head-turning asymmetries during kissing and other activities (Güntürkün, 2003; Ocklenburg and Güntürkün, 2009), left–right asymmetries during cradling (Packheiser et al., 2019b; Malatesta et al., 2022; Vauclair, 2022; van der Meer, 2023), and asymmetries during embracing another person (Ocklenburg et al., 2018; Packheiser et al., 2019a). Moreover, there is a bias to turn the left cheek forward when posing for a portrait (McManus and Humphrey, 1973; Lindell, 2017, 2019; Hew and Lindell, 2022), there are whole-body rotational and turning biases (Bradshaw and Bradshaw, 1988), and chewing side preferences (Martinez-Gomis et al., 2009)—just to name a few examples. Thus, it is evident that behavioral asymmetries are ubiquitous on many different levels in the motor system of the human brain.

Importantly, many (but not all) of these motor asymmetries are associated with handedness, suggesting that motor asymmetries are not independent of each other. The most widely investigated form of motor asymmetry apart from handedness is footedness (e.g., a preference to perform activities such as picking up a marble with one specific foot). Foot preference is commonly assessed with the Waterloo Footedness Questionnaire (Elias and Bryden, 1998; Elias et al., 1998). This questionnaire works similarly to the EHI in that participants have to indicate their preferences for several activities, and in the end, a lateralization quotient is calculated. Fig. 24.4 shows the performance of one example item from the WFQ.

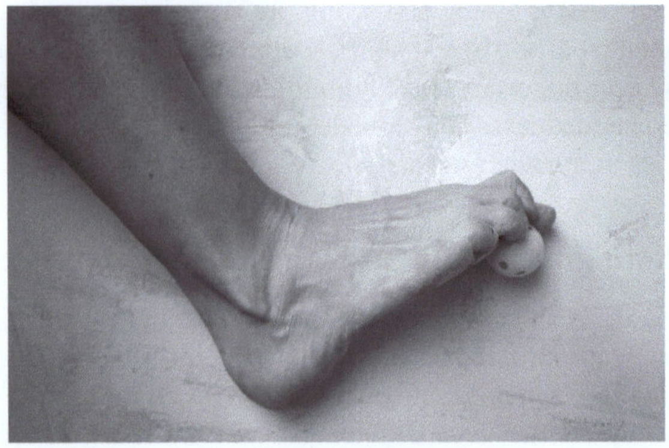

Fig. 24.4. A volunteer performing an example item from the Waterloo footedness Questionnaire ("If you wanted to pick up a marble with your toes, which foot would you use?"), which is used to assess footedness (Elias et al., 1998).

Footedness is associated with handedness. A recent meta-analysis of 164 studies on footedness found that about 60.1% of left-handers were left-footed but only 3.2% of right-handers were left-footed (Packheiser et al., 2020). Similarly, handedness and cradling asymmetry are associated. A recent meta-analysis found a robust leftward cradling bias that was less pronounced in left-handers (Packheiser et al., 2019b). For head-turning laterality during kissing, it was found that right-kissers also had a more rightward hand preference (Ocklenburg and Güntürkün, 2009). Thus, handedness has high, but not perfect, predictive value for other asymmetries in the human motor system.

Handedness and asymmetries in cortical structure

As explained in the beginning of this chapter, handedness is not caused by any anatomic differences between the left and right hands of left-handers and right-handers. Instead, it is a form of hemispheric asymmetry and originates in the central nervous system. This has led researchers to investigate differences in cortical structure between left- and right-handers, with a focus on the motor cortex (e.g., primary motor cortex, premotor cortex, and supplementary motor area) located in the precentral gyrus anterior to the central sulcus (Hatta, 2007). Several studies have suggested differences between left- and right-handers in cortical areas relevant for motor processing but the result patterns across studies are rather heterogenous and there is no single brain area that shows clear-cut differences between handedness groups across several studies.

One early study found that the left central sulcus is deeper than the right in right-handers, but the right central sulcus is deeper than the left in left-handers (Amunts et al., 1996). A further study found that handedness was associated with asymmetries in cortical structure only in male but not in female participants (Amunts et al., 2000). In this study, the left central sulcus was significantly deeper than the right, but only in consistent right-handers. Moreover, one study found that the motor hand area of the precentral gyrus showed a significant leftward asymmetry in right-handers, but there were no left–right differences in this area in left-handers (Foundas et al., 1998). However, other studies did not find significant differences in cortical anatomy between left-handers and right-handers (Ocklenburg et al., 2016) or only nonsignificant differences. For example, one study found a difference between left-handers and right-handers in the left precentral sulcus that reached nominal significance but did not survive correction for multiple comparisons (Guadalupe et al., 2014). The largest study on handedness and asymmetries in cortical structure included 28,802 right-handers and 3062 left-handers from the UK Biobank dataset (Sha et al., 2021). The authors found significant differences between left- and right-handers in cortical thickness asymmetries in clusters in the postcentral gyrus and the inferior occipital cortex, with left-handers showing less asymmetry than right-handers. Functional annotation of the observed clusters revealed a relevance for somatosensory, tactile, and motor tasks, but also to vision. The authors suggested that this relationship may be explained by the need for hand-eye coordination during manual tasks.

In addition to gray matter asymmetries, a few studies have also investigated the link between handedness and cortical white matter, but there are no robust findings that replicated across many studies. A recent review article concluded that the existing evidence is incomplete and partly inconsistent, but suggests that left- and right-handers show differences in white matter pathways relevant to visuomotor abilities (e.g., superior longitudinal fasciculus), but not in projection tracts involved in the motor system itself (e.g., corticospinal tract) (Budisavljevic et al., 2021). For example, a DTI study found that there were no differences between left-handers and right-handers in size, mean diffusion, and anisotropy of the corticospinal tract (Westerhausen et al., 2007).

Handedness and the cerebellum

The cerebellum plays a crucial role in motor control and motor learning, but it is also likely involved in various cognitive functions (Koziol et al., 2014; Peterburs and Desmond, 2016; Peterburs et al., 2018, 2021; Jacobi et al., 2021). Due to its important role in the motor system, several studies have investigated the potential role of the cerebellum in handedness (McManus, 2022). For example, a neurologic single-case study reported that hemicerebellitis during childhood can lead to a switch in handedness (Mascalchi et al., 2017). Just like the cortex, the cerebellum shows structural and functional hemispheric asymmetries (Hu et al., 2008; Kavaklioglu et al., 2017; Schutter, 2019), and some studies have found associations between asymmetries in the cerebellum and handedness. For example, an early MRI study found that right-handers showed stronger cerebellar asymmetries than left-handers (Snyder et al., 1995). Moreover, an MRI study in twins discordant for handedness reported larger rightward asymmetry in the anterior cerebellum in right-handed twins compared to left-handed twins (Rosch et al., 2018). However, the largest MRI study on volumetric gray matter asymmetries in the cerebellum ($n = 2226$) did not report any significant associations between handedness and cerebellar asymmetries (Kavaklioglu et al., 2017). This led the authors to conclude that there is no association between hand motor control and volumetric gray matter asymmetries in the cerebellum.

Interestingly, functional asymmetries in the cerebellum have also been linked to handedness. For example, one resting-state fMRI study found that right-handers showed stronger asymmetries in both right-lateralized cerebellar and left-lateralized cerebellar regions (Wang et al., 2013). Moreover, a recent study on whole-brain connectivity found that functional connectivity of the cerebellum may be more relevant for handedness than asymmetries in cerebellar structure (Tejavibulya et al., 2022). In this study, the authors found that left-handers and right-handers showed differences in functional connectivity with cortical motor, somatosensory, and language areas, but especially strong differences in functional connectivity within the cerebellum or between the cerebellum and the cortex. Overall, right-handers showed more functional cerebellar connectivity to frontal regions, while left-handers showed more functional cerebellar connectivity to more posterior regions, including the motor and somatosensory cortex. This suggests that future clinical neurology studies on handedness should take cerebral and cerebellar functional connectivity into account.

Handedness and the spinal cord

In addition to the cerebral cortex and the cerebellum, another structure in the central nervous system that has been suggested to be relevant for handedness is the spinal cord (Isa, 2019). Specifically, the cervical segments C5–C8 of the spinal cord are relevant for motor control of arms and hands (Isa et al., 2007). The idea that the spinal cord may be relevant for handedness largely is founded on a developmental perspective. Ultrasound studies have shown lateral preferences for arm movements as soon as independent arm movements are possible, at about 10 weeks after gestation (Hepper, 2013). These early preferences for arm movement show strong correlations with handedness later in life. For example, thumb-sucking preferences as a fetus have been shown to predict handedness at ages 10–12 years (Hepper et al., 2005). As the motor cortex is not yet functionally connected to the spinal cord before week 15 after gestation (ten Donkelaar et al., 2004), these findings suggest that it is unlikely that brain asymmetries are involved in the formation of these early precursors to handedness. Instead, the spinal cord may play a pivotal role in the development of handedness.

Indeed, developmental studies suggest asymmetries in spinal cord development that may be relevant for handedness. For example, an RNA sequencing study in which gene expression levels in the left and right spinal cords of postmortem human embryos 4–8 weeks after gestation were measured found that the left side of the spinal cord matured faster than the right side (de Kovel et al., 2017). A further study showed development-dependent gene expression asymmetries between the left and right segments of the spinal cord, which are relevant for hand and arm movements. These asymmetries were epigenetically regulated by both lateralized methylation of CpG islands (specific gene regions that regulate gene expression) and miRNA expression asymmetries in the TGF-b signaling pathway (an important signaling pathway that is relevant for various processes in the developing embryo, such as cell growth and migration) (Ocklenburg et al., 2017).

Handedness and asymmetries in brain activity

Handedness is a form of hemispheric asymmetries for the execution of skilled motor abilities like writing. Therefore, researchers have investigated differences in brain activity between left- and right-handers using tasks such as finger tapping while in the MRI scanner. For example, two large-scale fMRI studies (Tzourio-Mazoyer et al., 2015, 2021) have used data from the BIL&GIN (Brain Imaging of Lateralization by the Groupe d'Imagerie Fonctionnelle) database (Mazoyer et al., 2016). The more recent study (Tzourio-Mazoyer et al., 2021) analyzed finger-tapping data from 143 left-handers and 144 right-handers to identify brain areas involved in manual preference. They found that handedness could be best explained by brain activity differences between movements of the dominant compared to the nondominant hands in 11 regions of interest involved in hand movements. These 11 regions showed activation profiles that depended on whether the dominant or the nondominant hand moved and included areas in the central sulcus, the supramarginal gyrus, the postcentral gyrus, the basal ganglia, the cerebellum, the thalamus, and others. The author identified two major mechanisms relevant for handedness. First, there was stronger contralateral activity of motor areas when the right hand was moved, but this effect was modulated by handedness, with some areas showing stronger asymmetries in left-handers while others showed stronger asymmetries in right-handers. Second, there was stronger deactivation in ipsilateral areas when the dominant hand was used than when the nondominant hand was used. For this effect, the pattern in left-handers mirrored that in right-handers. This study suggests that functional asymmetries in brain activation between left- and right-handers are complex and involve the activation of many parts of the motor system.

Interestingly, differences in brain activation between left-handers and right-handers have also been observed outside the motor domain. For example, a fMRI study that compared face, word, and spatial processing in left- and right-handers showed that right-handers showed a significantly stronger leftward brain activation bias for processing words than left-handers (Badzakova-Trajkov et al., 2010).

Moreover, right-handers showed a significantly stronger rightward brain activation bias for processing faces than left-handers. No handedness group differences were found for the spatial task. Thus, handedness affects the typically leftward asymmetry for processing of words and the typically rightward asymmetry for processing of faces, with left-handers showing more atypical asymmetries than right-handers. A more recent fMRI study in left- and right-handers also reported that left-handers showed weaker lateralization for processing language, faces, and bodies (Johnstone et al., 2021). Therefore, handedness, language lateralization, and lateralization for face and body processing are not wholly independent but are to some extent related, suggesting a partly shared ontogenesis. The finding that left-handers showed weaker leftward language lateralization has also been supported by a recent functional transcranial Doppler sonography study (Woodhead et al., 2021) and a recent behavioral study with dichotic listening (Karlsson et al., 2023).

Handedness and cognitive abilities

Since differences in both brain structure and brain function have been reported between left-handers and right-handers, it is conceivable that there are differences in cognitive abilities between handedness groups. Indeed, a vast number of studies have been conducted that investigated the link between handedness and cognitive abilities, and several meta-analyses have been carried out that integrated these findings (Papadatou-Pastou, 2018).

A meta-analysis on the cognitive benefits of right-handedness (Somers et al., 2015) found no difference between handedness groups for verbal abilities but reported a small but significant advantage for right-handers in spatial abilities. A meta-analysis on handedness and intelligence that included data from intellectually disabled, typically developing, and especially intelligent people found that out of five meta-analyses, only the comparison between intellectually disabled and typically developing individuals yielded a robust effect (Papadatou-Pastou and Tomprou, 2015). Specifically, it was found that individuals with intellectual disability had a higher prevalence of non-right-handedness than typically developing individuals. A more recent meta-analysis from the same group on handedness and IQ scores revealed that IQ differences between left- and right-handers are negligible (Ntolka and Papadatou-Pastou, 2018). While a significant effect in favor of right-handers was observed in the meta-analysis, the authors highlighted that the magnitude of the effect was marginal ($d = -0.07$) and hardly had any real-life relevance. Regarding mathematical abilities, a meta-analysis on handedness and mathematical learning difficulties (Papadatou-Pastou et al., 2021) revealed no differences between handedness groups. One cognitive domain for which some stereotypes suggest that left-handers are better than right-handers is artistic ability. While no meta-analysis has been published on this topic yet, the largest study so far suggests that left-handers and right-handers are equally artistically creative (van der Feen et al., 2020).

Thus, taken together there is no evidence for any pronounced differences in cognitive abilities between left-handers. If differences exist on the meta-analytic level, they are typically so small that they have very low relevance in real life.

Handedness retraining

One factor that can complicate the interpretation of handedness data is the fact that sometimes people have been retrained to use their previously nondominant hand as their dominant hand. Handedness retraining can have various reasons:

1. Forced handedness retraining: In several countries, left-handers used to be retrained to write with their right hand instead of their left, sometimes with force (Porac and Friesen, 2000; Porac and Searleman, 2002; Porac and Martin, 2007; Kushner, 2011, 2012; Stetter et al., 2023). The practice has been abolished in many countries, making forced handedness retraining less common than it was in the 1970s and in preceding decades and centuries. However, there are reports that forced handedness retraining is still ongoing in some countries. For example, a 2004 study found that 17% of primary school teachers asked in the study strongly urged left-handed students to switch to the right hand for writing (Holder and Kateeba, 2004). Moreover, left-handedness rates in China are much lower than worldwide (about 1% compared to 10.6%), and attempts to switch left-handers to write with the right hand are still common (Kushner, 2013).
2. Handedness retraining due to amputation: If the dominant hand of an individual needs to be amputated after an accident, the person has no choice but to retrain their previously nondominant hand (Xiaoli et al., 2015).
3. Voluntary handedness retraining to enhance athletic performance: Professional athletes sometimes deliberately practice the use of the nondominant hand to become more ambidextrous to gain an advantage in sports like tennis or combat sports (Marcori et al., 2019).

A central aspect of handedness retraining for research in clinical neurology is that retraining a left-hander to write with their right hand does make performance similar (but often not 100% alike) to that of a person who is born right-handed, but the neuronal processes underlying hand use are different between born right-handers and retrained left-handers (Marcori et al., 2019). For example, an fMRI study on hand motor control in left-handers, right-handers, and people who were born left-handed but had been retrained to mainly use their right hand found that switched individuals showed a functional motor architecture that contained elements of both left-handedness and right-handedness (Grabowska et al., 2012). Therefore, when assessing handedness in a clinical neurology study, including an item in the general questionnaire on whether a person has been retrained to use their previously nondominant hand as their dominant hand can be helpful for data interpretation.

Conclusion: The critical importance of handedness assessment in clinical neurology and basic neuroscience

In his 2023 book "How Handedness Shapes Lived Experience, Intersectionality, and Inequality—Hand and World" philosopher Peter Westmoreland coined the term "right-hand bias" (Westmoreland, 2023). According to Westmoreland (2023), a right-hand bias exists because the vast majority of people are right-handed and privilege the right hand to interact with the world, which is, for example, evident in human-made tools like scissors and pens.

A bias toward right-handedness is also evident in empiric research in both clinical neurology and basic neuroscience. A recent study aimed at quantifying handedness-based recruitment biases in neuroimaging research reported clear results (Bailey et al., 2020). In this study, the authors assessed the handedness of participants from more than 1000 published neuroimaging papers. From the more than 30,000 volunteers included in these studies, about 3%–4% were non-right-handed (either left- or mixed-handed). This percentage is considerably lower than what would be expected based on a recent meta-analysis on handedness reporting the best estimate for left-handedness to be 10.6% (Papadatou-Pastou et al., 2020). Therefore, compared to their actual percentage in the population, left-handers and mixed-handers are severely underrepresented in empiric neuroimaging research. The reasons for this are complex and mostly include the wish to reduce variance in the data in studies with smaller sample sizes that may reduce the probability of obtaining significant results. Moreover, targeted recruitment of left-handers and mixed-handers may be challenging due to their rareness, prolonging the time researchers need to recruit all participants and complete data collection. Despite these issues, it has been argued that it is important to include left-handed participants in neuroscience and neurogenetics as they represent an important part of the population, and left-handedness falls within the normal range of brain diversity (Willems et al., 2014). It is therefore crucial to include left-handers and mixed-handers in neurology studies as well. For basic science studies in neurology, including left-handers and mixed-handers ensures capturing the full range of diversity in brain structure and function. For clinical studies in neurology, including left-handers and mixed-handers is important to assess whether different symptoms can be expected after specific brain lesions or other pathologic processes. Also, they may react differently to treatments. Therefore, the assessment of handedness should be established as a standard procedure in clinical neurology.

References

Albayay J, Villarroel-Gruner P, Bascour-Sandoval C et al. (2019). Psychometric properties of the Spanish version of the Edinburgh handedness inventory in a sample of Chilean undergraduates. Brain Cogn 137: 103618.

Amunts K, Schlaug G, Schleicher A et al. (1996). Asymmetry in the human motor cortex and handedness. Neuroimage 4: 216–222.

Amunts K, Schlaug G, Jäncke L et al. (1997). Motor cortex and hand motor skills: structural compliance in the human brain. Hum Brain Mapp 5: 206–215.

Amunts K, Jäncke L, Mohlberg H et al. (2000). Interhemispheric asymmetry of the human motor cortex related to handedness and gender. Neuropsychologia 38: 304–312.

Annett M (1970a). A classification of hand preference by association analysis. Br J Psychol 61: 303–321.

Annett M (1970b). The growth of manual preference and speed. Br J Psychol 61: 545–558.

Arning L, Ocklenburg S, Schulz S et al. (2015). Handedness and the X chromosome: the role of androgen receptor CAG-repeat length. Sci Rep 5: 8325.

Badzakova-Trajkov G, Häberling IS, Roberts RP et al. (2010). Cerebral asymmetries: complementary and independent processes. PLoS One 5: e9682.

Badzakova-Trajkov G, Corballis MC, Häberling IS (2016). Complementarity or independence of hemispheric specializations? A brief review. Neuropsychologia 93: 386–393.

Bailey LM, McMillan LE, Newman AJ (2020). A sinister subject: quantifying handedness-based recruitment biases in current neuroimaging research. Eur J Neurosci 51: 1642–1656.

Bazo NS, Marcori AJ, Guimarães AN et al. (2023). Inventories of human lateral preference: a systematic review. Percept Mot Skills 130: 2505–2529.

Beukelaar LJ, Kroonenberg PM (1983). Towards a conceptualization of hand preference. Br J Psychol 74: 33–45.

Bishop DV (1989). Does hand proficiency determine hand preference? Br J Psychol 80: 191–199.

Bishop DV, Ross VA, Daniels MS et al. (1996). The measurement of hand preference: a validation study comparing three groups of right-handers. Br J Psychol 87: 269–285.

Borawski J, Papadatou-Pastou M, Packheiser J et al. (2023). Handedness in post-traumatic stress disorder: a meta-analysis. Neurosci Biobehav Rev 145: 105009.

Bradshaw JL, Bradshaw JA (1988). Rotational and turning tendencies in humans: an analog of lateral biases in rats? Int J Neurosci 39: 229–232.

Bryden MP, Singh M, Steenhuis RE et al. (1994). A behavioral measure of hand preference as opposed to hand skill. Neuropsychologia 32: 991–999.

Bryden PJ, Pryde KM, Roy EA (2000). A developmental analysis of the relationship between hand preference and performance: II. A performance-based method of measuring hand preference in children. Brain Cogn 43: 60–64.

Bryden PJ, Roy EA, Spence J (2007). An observational method of assessing handedness in children and adults. Dev Neuropsychol 32: 825–846.

Budisavljevic S, Castiello U, Begliomini C (2021). Handedness and white matter networks. Neuroscientist 27: 88–103.

Callahan D, de Virgilio C, Tillou A et al. (2016). Medical student hand preference, perceived dexterity, and impact of handedness on training. J Surg Res 204: 114–117.

Cannon M, Byrne M, Cassidy B et al. (1995). Prevalence and correlates of mixed-handedness in schizophrenia. Psychiatry Res 59: 119–125.

Cheng Y, Hegarty M, Chrastil ER (2020). Telling right from right: the influence of handedness in the mental rotation of hands. Cogn Res: Princ Implic 5: 25.

Christman SD, Jasper JD, Sontam V et al. (2007). Individual differences in risk perception versus risk taking: handedness and interhemispheric interaction. Brain Cogn 63: 51–58.

Corballis MC (1997). The genetics and evolution of handedness. Psychol Rev 104: 714–727.

Corballis MC (1998). Cerebral asymmetry: motoring on. Trends Cogn Sci 2: 152–157.

Corballis MC (2003). From mouth to hand: gesture, speech, and the evolution of right-handedness. Behav Brain Sci 26: 199–208. discussion 208-60.

Corballis MC (2019). Evolution of cerebral asymmetry. Prog Brain Res 250: 153–178.

Corballis MC, Häberling IS (2017). The many sides of hemispheric asymmetry: a selective review and outlook. J Int Neuropsychol Soc 23: 710–718.

Coren S (1993). The lateral preference inventory for measurement of handedness, footedness, eyedness, and earedness: norms for young adults. Bull Psychon Soc 31: 1–3.

de Kovel CGF, Lisgo S, Karlebach G et al. (2017). Left-right asymmetry of maturation rates in human embryonic neural development. Biol Psychiatry 82: 204–212.

Deep-Soboslay A, Hyde TM, Callicott JP et al. (2010). Handedness, heritability, neurocognition and brain asymmetry in schizophrenia. Brain 133: 3113–3122.

Doyen A-L, Carlier M (2002). Measuring handedness: a validation study of Bishop's reaching card test. Laterality 7: 115–130.

Dragovic M (2004). Categorization and validation of handedness using latent class analysis. Acta Neuropsychiatrica 16: 212–218.

Edlin JM, Leppanen ML, Fain RJ et al. (2015). On the use (and misuse?) of the Edinburgh handedness inventory. Brain Cogn 94: 44–51.

Elias LJ (2022). Side effects: how left-brain right-brain differences shape everyday behaviour, Dundurn, Toronto.

Elias LJ, Bryden MP (1998). Footedness is a better predictor of language lateralisation than handedness. Laterality 3: 41–51.

Elias LJ, Bryden MP, Bulman-Fleming MB (1998). Footedness is a better predictor than is handedness of emotional lateralization. Neuropsychologia 36: 37–43.

Espírito-Santo H, Pires CF, Garcia IQ et al. (2017). Preliminary validation of the Portuguese Edinburgh handedness inventory in an adult sample. Appl Neuropsychol Adult 24: 275–287.

Fazio R, Dunham KJ, Griswold S et al. (2013). An improved measure of handedness: the Fazio laterality inventory. Appl Neuropsychol Adult 20: 197–202.

Foundas AL, Hong K, Leonard CM et al. (1998). Hand preference and magnetic resonance imaging asymmetries of the central sulcus. Neuropsychiatry Neuropsychol Behav Neurol 11: 65–71.

Gerrits R (2022). Variability in hemispheric functional segregation phenotypes: a review and general mechanistic model. Neuropsychol Rev. https://doi.org/10.1007/s11065-022-09575-y.

Gonzalez SL, Nelson EL (2021). Factor analysis of the home handedness questionnaire: unimanual and role differentiated bimanual manipulation as separate dimensions of handedness. Appl Neuropsychol Adult 28: 173–184.

Grabowska A, Gut M, Binder M et al. (2012). Switching handedness: fMRI study of hand motor control in right-handers, left-handers and converted left-handers. Acta Neurobiol Exp 72: 439–451.

Guadalupe T, Willems RM, Zwiers MP et al. (2014). Differences in cerebral cortical anatomy of left- and right-handers. Front Psychol 5: 261.

Güntürkün O (2003). Human behaviour: adult persistence of head-turning asymmetry. Nature 421: 711.

Häberling IS, Corballis PM, Corballis MC (2016). Language, gesture, and handedness: evidence for independent lateralized networks. Cortex 82: 72–85.

Hatta T (2007). Handedness and the brain: a review of brain-imaging techniques. Magn Reson Med Sci 6: 99–112.

Hepper PG (2013). The developmental origins of laterality: fetal handedness. Dev Psychobiol 55: 588–595.

Hepper PG, Wells DL, Lynch C (2005). Prenatal thumb sucking is related to postnatal handedness. Neuropsychologia 43: 313–315.

Hew VY, Lindell AK (2022). Putting your best face forward: posing biases in psychologists' online portraits. Laterality 27: 406–414.

Holder MK, Kateeba D (2004). Hand preference survey of 5136 school children in Western Uganda. Laterality 9: 201–207.

Hu D, Shen H, Zhou Z (2008). Functional asymmetry in the cerebellum: a brief review. Cerebellum 7: 304–313.

Isa T (2019). Dexterous hand movements and their recovery after central nervous system injury. Annu Rev Neurosci 42: 315–335.

Isa T, Ohki Y, Alstermark B et al. (2007). Direct and indirect cortico-motoneuronal pathways and control of hand/arm movements. Physiology (Bethesda) 22: 145–152.

Jacobi H, Faber J, Timmann D et al. (2021). Update cerebellum and cognition. J Neurol 268: 3921–3925.

Johnstone LT, Karlsson EM, Carey DP (2021). Left-handers are less lateralized than right-handers for both left and right hemispheric functions. Cerebral Cortex 31: 3780–3787.

Karlsson EM, Hugdahl K, Hirnstein M et al. (2023). Analysis of distributions reveals real differences on dichotic listening scores between left- and right-handers. Cereb Cortex Commun 4: tgad009.

Kavaklioglu T, Guadalupe T, Zwiers M et al. (2017). Structural asymmetries of the human cerebellum in relation to cerebral cortical asymmetries and handedness. Brain Struct Funct 222: 1611–1623.

Koziol LF, Budding D, Andreasen N et al. (2014). Consensus paper: the cerebellum's role in movement and cognition. Cerebellum 13: 151–177.

Kushner HI (2011). Retraining the King's left hand. Lancet 377: 1998–1999.

Kushner HI (2012). Retraining left-handers and the aetiology of stuttering: the rise and fall of an intriguing theory. Laterality 17: 673–693.

Kushner HI (2013). Why are there (almost) no left-handers in China? Endeavour 37: 71–81.

Lindell AK (2017). Celebrity chefs put their left cheek forward: cover image orientation in celebrity cookbooks. Laterality 22: 515–520.

Lindell AK (2019). Left cheek poses garner more likes: the effect of pose orientation on Instagram engagement. Laterality 24: 600–613.

Lombana NF, Naeger PA, Padilla PL et al. (2022). Ambidexterity in left-handed and right-handed individuals and implications for surgical training. Proc (Bayl Univ Med Cent) 35: 176–179.

Malatesta G, Marzoli D, Tommasi L (2022). Environmental and genetic determinants of sensorimotor asymmetries in mother-infant interaction. Front Behav Neurosci 16: 1080141.

Marcori AJ, Monteiro PHM, Okazaki VHA (2019). Changing handedness: what can we learn from preference shift studies? Neurosci Biobehav Rev 107: 313–319.

Marim EA, Lafasse R, Okazaki VHA (2011). Global lateral preference inventory. Braz J Mot Behav 6: 14–23.

Martinez-Gomis J, Lujan-Climent M, Palau S et al. (2009). Relationship between chewing side preference and handedness and lateral asymmetry of peripheral factors. Arch Oral Biol 54: 101–107.

Marzke MW, Marzke RF (2000). Evolution of the human hand: approaches to acquiring, analysing and interpreting the anatomical evidence. J Anat 197: 121–140.

Mascalchi M, Lenge M, Bianchi A et al. (2017). Hemicerebellitis can drive handedness shift. Cerebellum Ataxias 4: 14.

Mazoyer B, Mellet E, Perchey G et al. (2016). BIL&GIN: a neuroimaging, cognitive, behavioral, and genetic database for the study of human brain lateralization. Neuroimage 124: 1225–1231.

McManus C (2004). Right hand, left hand: the origins of asymmetry in brains, bodies, atoms and cultures, Phoenix, London.

McManus C (2019). Half a century of handedness research: myths, truths; fictions, facts; backwards, but mostly forwards. Brain Neurosci Adv 3. https://doi.org/10.1177/2398212818820513.

McManus C (2022). Cerebral polymorphisms for lateralisation: modelling the genetic and phenotypic architectures of multiple functional modules. Symmetry 14: 814.

McManus C, Humphrey NK (1973). Turning the left cheek. Nature 243: 271–272.

McManus C, van Horn JD, Bryden PJ (2016). The Tapley and Bryden test of performance differences between the hands: the original data, newer data, and the relation to pegboard and other tasks. Laterality 21: 371–396.

Mundorf A, Getzmann S, Gajewski PD et al. (2023). Stress exposure, hand preference, and hand skill: a deep phenotyping approach. Laterality 28: 209–237.

Musalek M, Scharoun SM, Bryden PJ (2016). Using Bishop's card reaching task to assess hand preference in 8- to 10-year-old Czech children. PLoS One 11: e0166337.

Nelson EL, Gonzalez SL, El-Asmar JM et al. (2019). The home handedness questionnaire: pilot data from preschoolers. Laterality 24: 482–503.

Nicholls MER, Thomas NA, Loetscher T et al. (2013). The Flinders handedness survey (FLANDERS): a brief measure of skilled hand preference. Cortex 49: 2914–2926.

Ntolka E, Papadatou-Pastou M (2018). Right-handers have negligibly higher IQ scores than left-handers: systematic review and meta-analyses. Neurosci Biobehav Rev 84: 376–393.

Ocklenburg S, Güntürkün O (2009). Head-turning asymmetries during kissing and their association with lateral preference. Laterality 14: 79–85.

Ocklenburg S, Güntürkün O (2018). The lateralized brain: the neuroscience and evolution of hemispheric asymmetries, Academic Press, London.

Ocklenburg S, Bürger C, Westermann C et al. (2010). Visual experience affects handedness. Behav Brain Res 207: 447–451.

Ocklenburg S, Friedrich P, Güntürkün O et al. (2016). Voxel-wise grey matter asymmetry analysis in left- and right-handers. Neurosci Lett 633: 210–214.

Ocklenburg S, Schmitz J, Moinfar Z et al. (2017). Epigenetic regulation of lateralized fetal spinal gene expression underlies hemispheric asymmetries. Elife 6: e22784.

Ocklenburg S, Packheiser J, Schmitz J et al. (2018). Hugs and kisses—the role of motor preferences and emotional lateralization for hemispheric asymmetries in human social touch. Neurosci Biobehav Rev 95: 353–360.

Oldfield RC (1971). The assessment and analysis of handedness: the Edinburgh inventory. Neuropsychologia 9: 97–113.

Orlando I, Ricci C, Griffanti L et al. (2023). Neural correlates of successful emotion recognition in healthy elderly: a multimodal imaging study. Soc Cogn Affect Neurosci 18: nsad058.

Packheiser J, Rook N, Dursun Z et al. (2019a). Embracing your emotions: affective state impacts lateralisation of human embraces. Psychol Res 83: 26–36.

Packheiser J, Schmitz J, Berretz G et al. (2019b). Handedness and sex effects on lateral biases in human cradling: three meta-analyses. Neurosci Biobehav Rev 104: 30–42.

Packheiser J, Schmitz J, Berretz G et al. (2020). Four meta-analyses across 164 studies on atypical footedness prevalence and its relation to handedness. Sci Rep 10: 14501.

Palmer RE, Corballis MC (1996). Predicting reading ability from handedness measures. Br J Psychol 87: 609–620.

Papadatou-Pastou M (2018). Handedness and cognitive ability: using meta-analysis to make sense of the data. Prog Brain Res 238: 179–206.

Papadatou-Pastou M, Tomprou D-M (2015). Intelligence and handedness: meta-analyses of studies on intellectually disabled, typically developing, and gifted individuals. Neurosci Biobehav Rev 56: 151–165.

Papadatou-Pastou M, Ntolka E, Schmitz J et al. (2020). Human handedness: a meta-analysis. Psychol Bull 146: 481–524.

Papadatou-Pastou M, Panagiotidou D-A, Abbondanza F et al. (2021). Hand preference and mathematical learning difficulties: new data from Greece, the United Kingdom, and Germany and two meta-analyses of the literature. Laterality 26: 485–538.

Parker AJ, Woodhead ZVJ, Thompson PA et al. (2021). Assessing the reliability of an online behavioural laterality battery: a pre-registered study. Laterality 26: 359–397.

Peterburs J, Desmond JE (2016). The role of the human cerebellum in performance monitoring. Curr Opin Neurobiol 40: 38–44.

Peterburs J, Hofmann D, Becker MPI et al. (2018). The role of the cerebellum for feedback processing and behavioral switching in a reversal-learning task. Brain Cogn 125: 142–148.

Peterburs J, Liang Y, Cheng DT et al. (2021). Sensory acquisition functions of the cerebellum in verbal working memory. Brain Struct Funct 226: 833–844.

Porac C, Friesen IC (2000). Hand preference side and its relation to hand preference switch history among old and oldest-old adults. Dev Neuropsychol 17: 225–239.

Porac C, Martin WLB (2007). A cross-cultural comparison of pressures to switch left-hand writing: Brazil versus Canada. Laterality 12: 273–291.

Porac C, Searleman A (2002). The effects of hand preference side and hand preference switch history on measures of psychological and physical well-being and cognitive performance in a sample of older adult right-and left-handers. Neuropsychologia 40: 2074–2083.

Prichard E, Propper RE, Christman SD (2013). Degree of handedness, but not direction, is a systematic predictor of cognitive performance. Front Psychol 4: 9.

Prichard EC, Clarkson EM, Christman SD (2023). Differences between consistent and inconsistent handedness remain consistently interesting: ten years of research on the consistency of handedness with the Edinburgh handedness inventory. Percept Mot Skills. https://doi.org/10.1177/00315125231217624.

Provins KA, Magliaro J (1993). The measurement of handedness by preference and performance tests. Brain Cogn 22: 171–181.

Rosch RE, Cowell PE, Gurd JM (2018). Cerebellar asymmetry and cortical connectivity in monozygotic twins with discordant handedness. Cerebellum 17: 191–203.

Scharoun SM, Bryden PJ (2015). Is strength of handedness reliable over repeated testing? An examination of typical development and autism spectrum disorder. Front Psychol 6: 17.

Schutter DJLG (2019). Hemispheric asymmetries in the human cerebellum. Cortex 115: 352–356.

Sha Z, Pepe A, Schijven D et al. (2021). Handedness and its genetic influences are associated with structural asymmetries of the cerebral cortex in 31,864 individuals. Proc Natl Acad Sci U S A 118: e2113095118.

Somers M, Shields LS, Boks MP et al. (2015). Cognitive benefits of right-handedness: a meta-analysis. Neurosci Biobehav Rev 51: 48–63.

Snyder PJ, Bilder RM, Wu H et al. (1995). Cerebellar volume asymmetries are related to handedness: a quantitative MRI study. Neuropsychologia 33: 407–419.

Steenhuis RE, Bryden MP (1989). Different dimensions of hand preference that relate to skilled and unskilled activities. Cortex 25: 289–304.

Stetter L, Sattler JB, Marquardt C et al. (2023). Handwriting kinematics during learning to write with the dominant left hand in converted left-handers. Sci Rep 13: 2171.

Stone KD, Bryant DC, Gonzalez CLR (2013). Hand use for grasping in a bimanual task: evidence for different roles? Exp Brain Res 224: 455–467.

Tapley SM, Bryden MP (1985). A group test for the assessment of performance between the hands. Neuropsychologia 23: 215–221.

Tejavibulya L, Peterson H, Greene A et al. (2022). Large-scale differences in functional organization of left- and right-handed individuals using whole-brain, data-driven analysis of connectivity. Neuroimage 252: 119040.

ten Donkelaar HJ, Lammens M, Wesseling P et al. (2004). Development and malformations of the human pyramidal tract. J Neurol 251: 1429–1442.

Thomas CL, Fitch SB (2023). Handedness and test anxiety: an examination of mixed-handed and consistent-handed students. Experimental Results 4: e15.

Tiffin J, Asher EJ (1948). The Purdue pegboard; norms and studies of reliability and validity. J Appl Psychol 32: 234–247.

Todor JI, Doane T (1978). Handedness and hemispheric asymmetry in the control of movements. J Mot Behav 10: 295–300.

Tzourio-Mazoyer N, Petit L, Zago L et al. (2015). Between-hand difference in ipsilateral deactivation is associated with hand lateralization: fMRI mapping of 284 volunteers balanced for handedness. Front Hum Neurosci 9: 5.

Tzourio-Mazoyer N, Labache L, Zago L et al. (2021). Neural support of manual preference revealed by BOLD variations during right and left finger-tapping in a sample of 287 healthy adults balanced for handedness. Laterality 26: 398–420.

van der Feen FE, Zickert N, Groothuis TGG et al. (2020). Does hand skill asymmetry relate to creativity, developmental and health issues and aggression as markers of fitness? Laterality 25: 53–86.

van der Meer ALH (2023). Handedness as a major determinant of lateral bias in human functional cradling. Infancy. https://doi.org/10.1111/infa.12572. Epub ahead of print 15 December 2023.

Vauclair J (2022). Maternal cradling bias: a marker of the nature of the mother-infant relationship. Infant Behav Dev 66: 101680.

Veale JF (2014). Edinburgh handedness inventory-short form: a revised version based on confirmatory factor analysis. Laterality 19: 164–177.

Viggiano M, Borelli P, Vannucci M et al. (2001). Hand preference in Italian students. Laterality 6: 283–286.

Vingerhoets G, Verhelst H, Gerrits R et al. (2023). Laterality indices consensus initiative (LICI): a Delphi expert survey report on recommendations to record, assess, and report asymmetry in human behavioural and brain research. Laterality 28: 122–191.

Wang D, Buckner RL, Liu H (2013). Cerebellar asymmetry and its relation to cerebral asymmetry estimated by intrinsic functional connectivity. J Neurophysiol 109: 46–57.

Westerhausen R, Huster RJ, Kreuder F et al. (2007). Corticospinal tract asymmetries at the level of the internal capsule: is there an association with handedness? Neuroimage 37: 379–386.

Westmoreland P (2023). How handedness shapes lived experience, intersectionality, and inequality: hand and world, Palgrave Macmillan, Cham, Switzerland.

Willems RM, van der Haegen L, Fisher SE et al. (2014). On the other hand: including left-handers in cognitive neuroscience and neurogenetics. Nat Rev Neurosci 15: 193–201.

Woodhead ZVJ, Thompson PA, Karlsson EM et al. (2021). An updated investigation of the multidimensional structure of language lateralization in left- and right-handed adults: a test-retest functional

transcranial Doppler sonography study with six language tasks. R Soc Open Sci 8: 200696.
Xiaoli G, Yuanyuan L, Bekrater-Bodmann R et al. (2015). Handedness change after dominant side amputation: evaluation from a hand laterality judgment task. Annu Int Conf IEEE Eng Med Biol Soc 2015: 8002–8005.
Yang N, Waddington G, Adams R et al. (2018). Translation, cultural adaption, and test-retest reliability of Chinese versions of the Edinburgh handedness inventory and Waterloo footedness questionnaire. Laterality 23: 255–273.
Yeung AWK, Wong NSM (2022). The non-transparent usage and reporting of the Edinburgh handedness inventory in functional magnetic resonance imaging literature: a survey of studies published since 2013. Laterality 27: 221–231.
Zatorre RJ (2022). Hemispheric asymmetries for music and speech: Spectrotemporal modulations and top-down influences. Front Neurosci 16: 1075511.
Zverev YP (2006). Cultural and environmental pressure against left-hand preference in urban and semi-urban Malawi. Brain Cogn 60: 295–303.

Handbook of Clinical Neurology, Vol. 208 (3rd series)
Cerebral Asymmetries
C. Papagno and P. Corballis, Editors
https://doi.org/10.1016/B978-0-443-15646-5.00024-5

Chapter 25

Hemispheric asymmetries in the control of upper limb movements

LUIGI TROJANO*

Department of Psychology, University of Campania 'Luigi Vanvitelli', Caserta, Italy

Abstract

This chapter deals with the unique human abilities of using tools, imitating others' gestures, drawing, and building complex items. Herein, after a brief overview of clinical manifestations and assessment of disorders of tool use and imitation (upper limb apraxia) and of the impairments in drawing and assembling multipart objects (constructional apraxia), brain asymmetries are discussed mainly starting from the neuropsychologic studies on patients with focal brain lesions, although both upper limb apraxia and constructional apraxia are often observed during the course of neurodegenerative diseases. Although no room is allowed here for a full discussion of brain–behavior relationships, relevant functional neuroimaging findings in healthy individuals are considered. The data presented in this chapter clearly demonstrate that tool use, gesture imitation, and "formative" activities (i.e., drawing and assembling) require the interplay of several brain areas and neural networks distributed over both hemispheres. Nonetheless, gesture processing exhibits a quite strong lateralization to the left hemisphere in typical right-handers. As regards "formative" activities, the neural networks in both hemispheres seem to provide complementary contributions, although the left inferior parietal lobule might play a specific role. The convergence of gesture processing and "formative activities" in the parietal lobes might be related to their considerable expansion in our species and could suggest that some shared basic, still unknown, computational processes allowed the development of these skills in modern humans.

INTRODUCTION

The ability to perform skilled movements of the upper limbs is a defining feature of human beings, since the time of upright-standing ancestors (Bradshaw and Rogers, 1996). The "explosion" of human uniqueness occurred during the Upper Paleolithic period (from about 50,000 years ago until about 10,000 years ago), which saw modern humans producing and using a variety of sophisticated tools, creating paintings and sculptures, and likely nonmaterial art (see Zaidel, 2018).

This chapter will deal with these "unique" human abilities: tool use (and imitation of others' gestures), drawing, and building complex items. Several kinds of brain lesions can lead to selective acquired impairments in these skills. In this chapter, I will briefly overview clinical manifestations and assessment of disorders of tool use and imitation (upper limb apraxia) and of impairments in drawing and assembling multipart objects (constructional apraxia). This will provide the bases for tackling hemispheric asymmetries from a neuropsychologic viewpoint, although no room is allowed here for a full discussion of brain-behavior relationships in these two domains. Herein, cognitive and behavioral findings in patients with focal brain lesions will be the main source of evidence for constraining hypothesis about hemispheric specialization, although upper limb apraxia and constructional apraxia are often observed during the course of neurodegenerative diseases (for reviews, see Lesourd et al., 2013; Trojano and Gainotti, 2016; Yliranta and Jehkonen, 2020).

UPPER LIMB APRAXIA

Clinical manifestations

Pioneering observations in late 19th century already established that some brain lesions can impair skilled upper limb movements in the absence of elementary motor disturbances. For instance, Steinthal (1881) described an aphasic patient who showed obvious difficulties in manipulating and using common everyday tools (e.g., pen, knife, his own violin), yet had no difficulties in performing simple motor acts. Commenting on this patient, Steinthal (1881) used the term "apraxia" to mean a disturbance in which arm movements were not impeded; but, in his words, the relationships between the movements and the objects were disrupted. Such observations paved the way for the extensive case study of a patient with unilateral disordered control of upper limb

*Correspondence to: Luigi Trojano, MD, Full Professor of Cognitive and Rehabilitative Neuropsychology, Department of Psychology, University of Campania 'Luigi Vanvitelli', Caserta, Italy. E-mail: luigi.trojano@unicampania.it

movements by Liepmann (1900), who eventually performed the first systematic group study on praxic abilities in patients with brain lesions (Liepmann, 1905).

Later on, Liepmann (1920, 1925) elaborated a definition of apraxia as a specific disorder of gesture production, which is still considered clinically sound. Indeed, modern literature converges in considering limb apraxia (LA) as a disturbance of intentional limb actions independent of sensory-motor impairments and of cognitive deficits preventing comprehension of task instructions or stimulus processing (Osiurak and Rossetti, 2017). One peculiarity of LA, already noted by Liepmann, is a huge discrepancy between patients' performance during clinical assessment and spontaneous behavior in ecologic conditions, as contextual cues can greatly improve patients' gesture production (Trojano et al., 2007). Such a discrepancy is one instance of the so-called "automatic-voluntary dissociation" first described by Jackson (1866).

LA can present in quite heterogeneous forms, and already Liepmann distinguished three variants differing for clinical presentation and postulated mechanisms (see Goldenberg, 2003). According to Liepmann (1920, 1925), a first variant was "ideational apraxia" consisting in a failure to create an appropriate image of the intended action ("movement formula"), resulting in faulty use of single objects and omission or misplacement of parts of complex actions; the more complex the actions, the more likely the errors. A second variant was initially termed "motor apraxia" and later "ideo-kinetic apraxia" (but in subsequent literature the term "ideomotor apraxia" from the French tradition prevailed; Goldenberg, 2003). Ideomotor apraxia was conceptualized as the specific inability to convert movement formulas into patterns of motor innervation, leading to impaired imitation of gestures; as in imitation the "movement formula" is provided by the examiner's demonstration, imitation errors would demonstrate the inability to produce appropriate movements, even when the movement formula is available. The third variant, "limb-kinetic apraxia," was thought not to be confined to movements directed by a conscious plan but to affect all movements, because of defective kinematic memories of an extremity. Indeed, limb-kinetic apraxia would affect only the limb contralateral to the lesioned hemisphere, whereas the two main variants, i.e., ideational and ideomotor apraxias, would manifest bilaterally, usually after a left-sided lesion, as it will be discussed later. In Liepmann's view (1920, 1925), the three variants of apraxia would be related to different lesion sites.

Although the general concept of LA and its partition into three main variants are still considered as a valid reference frame, the great expansion of studies in modern literature led to different elaborations of the original definitions. Divergent interpretative frameworks led to a substantial inconsistency in the use of the terms "ideational" and "ideomotor" apraxia, whereas a general agreement exists in defining "limb-kinetic" apraxia as characterized by inaccurate or clumsy distal limb movements in the contralesional arm across all gesture production tasks and under ecologic conditions as well (Osiurak and Rossetti, 2017; Baumard and Le Gall, 2021). Nonetheless, limb-kinetic apraxia could be difficult to disentangle from elementary motor disorders, and some authors even consider it not as a cognitive disorder per se, but rather as a consequence of a breakdown in motor action plans, clearly distinct from the two high-order (cognitive) main disturbances of learned skilled movements (e.g., Foundas and Duncan, 2019).

Current definitions of ideational and ideomotor apraxia differ greatly, in some cases departing from Liepmann's original classification based on impairments in tool use vs imitation tasks. One influential viewpoint mainly refers the dichotomy to the quality of errors, affecting content (e.g., gestures not adequate for the target object, perseveration) vs spatiotemporal execution (e.g., errors in timing, sequencing, amplitude of motor acts, or in hand configuration; Gonzalez Rothi et al., 1988). Within this framework, types of errors (content vs spatiotemporal) would provide cues for ascribing the variants of LA to a disturbance of a praxis conceptual system or of a praxis production system, respectively (e.g., Foundas and Duncan, 2019). However, patients with LA produce a great variety of (often inconsistent) errors in gesture production tasks, sometimes repeatedly trying to achieve the intended goal, and this may make it challenging to obtain definite clues for error classification. Nonetheless, many authors adopt this taxonomy and many papers aimed at comprehending the cognitive and neural correlates of gesture production errors within this theoretical perspective (see later).

The picture is further complicated by the multiplicity of theoretical accounts. Several models of gesture production and comprehension have been proposed (see Buxbaum and Randerath, 2018, for a review), some of which integrated praxis within the general framework of cognitive models put forward for language (e.g., Rothi et al., 1991; Cubelli et al., 2000; Stamenova et al., 2012). According to the basic tenets of these models, praxic abilities are subtended by several functionally independent, hierarchically connected modules. Some modules interact to ensure correct comprehension and production of meaningful (i.e., learned) gestures, both those involved in tool use (transitive gestures, i.e., gestures implying purposeful interactions with objects according to their function) and those shared for communicative purposes (intransitive symbolic gestures, e.g., waving goodbye). A "lexical-semantic route" would thus link gesture processing with acquired conceptual knowledge about objects and their function; an impairment in this route would typically produce not only content errors, but also spatiotemporal distortions. A distinct "direct route" would allow imitation of meaningless (novel) gestures, by mapping seen configurations of hands and arms for informing gesture production. This kind of models has been repeatedly refined to incorporate interaction of gesture processing with working memory or body representation abilities (see Buxbaum and Randerath, 2018), and identified further subtypes of LA, such as "conceptual apraxia" (an impairment in both recognizing and producing well-known gestures) and "conduction apraxia" (a selective defect in imitating novel gestures; Ochipa et al., 1994). These LA subtypes, however, might partly overlap to classic LA types, and need further studies to reach consistent and generally accepted diagnostic criteria (Baumard and Le Gall, 2021).

On these bases, to avoid any possible misunderstanding, for clinical and research purposes it appears strongly advisable to resort to the well-established definition of LA, and to specify the kind of gestures and tasks, which individual patients are impaired on (Buxbaum and Randerath, 2018). This approach could help comprehend LA in terms of the processes implied in each task by different types of gestures, and also identify the neural correlates of upper limb control (Baumard and Le Gall, 2021).

Clinical assessment

In his first case series on patients with brain lesions, Liepmann (1905) examined the hand ipsilateral to the lesion and assessed: use of common objects, "simulation" of object use (i.e., pantomime of object use on verbal command), production of expressive (symbolic) movements on verbal commands, and gesture imitation. All these tasks have been retained in subsequent studies and complemented with further tasks and stimuli, to address the clinical manifestations of LA, and to obtain cues for their interpretation.

Pantomime of object use on a verbal command is perhaps the most common tasks assessing gesture production, is considered one of the "gold standards" for the diagnosis of apraxia (Baumard and Le Gall, 2021), and is included in several test batteries (e.g., Rothi et al., 1997; Power et al., 2010; Vanbellingen et al., 2010a,b). However, pantomiming on verbal command is a complex task, as it implies verbal comprehension abilities, access to semantic knowledge, working memory, manipulation knowledge, and motor simulation (Goldenberg, 2017). Tool-use pantomime appears to involve both communicative and motor-cognitive aspects, which could be affected by dissociable lesions (Finkel et al., 2018). However, the frequent association of LA and aphasia (LA without aphasia occurs quite rarely after LH lesions; e.g., Goldenberg and Randerath, 2015; Schmidt et al., 2022) would advise against using verbal commands (specific verbal instructions may affect performance even in healthy participants; Bartolo et al., 2020). Moreover, it should be considered that pantomiming tool use is quite far from everyday activities, so that its ecologic validity could be questioned. An alternative task is use of actual single or multiple objects (De Renzi and Lucchelli, 1988), which circumvents difficulties in verbal processing and facilitates responses by providing examinees with concrete objects as cues for eliciting correct gestures. In this respect, actual tool use could be less sensitive in identifying impaired performance with respect to pantomiming, but is highly specific and can trigger striking errors (Fig. 25.1). Patients often express strong feelings of bafflement having a well-known and easily recognized object in their hands, correctly recalling its function, manipulating it with reiterated different attempts, and yet failing in performing the correct action. By contrast, pantomiming could be considered more apt to capture subtle LA manifestations, if the contribution of non-apraxic impairments (e.g., in language or working memory; Bartolo et al., 2003) is adequately discounted.

The second most used task for evaluating LA is imitation, which can be assessed both with transitive (tool use) and intransitive (i.e., symbolic) meaningful gestures (Fig. 25.1), and with meaningless gestures. Imitation of meaningful gestures can be performed referring to previous stored knowledge, whereas imitation of meaningless gestures can more directly assess the ability to reproducing a given configuration of limb postures and movements, even though implying several cognitive operations (e.g., body image or body schema, visual–spatial skills; for a review, see Baumard and Le Gall, 2021); when requiring distal finger movements, imitation of meaningless gestures could be demanding even for non-apraxic patients with focal brain lesions (Tessari et al., 2021).

The multifaceted attempts at comprehending the cognitive and neural bases of LA led to proposals of further gesture production tasks (e.g., use of novel tools; Buchmann and Randerath, 2017), and of tasks not requiring gesture production but assessing gesture comprehension and discrimination, particularly in cognitive-oriented research (e.g., Bartolo et al., 2008; Power et al., 2010). For clinical purposes, it seems sufficient, however, to restrict LA assessment to classic gesture production tasks, with the recommendation of assessing pantomimes of transitive and intransitive gestures (preferring pictures of objects instead of verbal commands), actual object use and imitation of meaningful and meaningless gestures to gather sufficient information for describing patients' impairment in upper limb control. Although this procedure might seem painstaking, such a clinical evaluation can be performed with a limited number of stimuli even at the bedside, and is warranted since many dissociations between tasks and types of gestures have been reported in patients with LA, as will be discussed below.

The most feasible task to assess motor deftness of either hand at the bedside is perhaps the coin rotation task (examinees are asked to rotate coins using their thumb, index, and middle fingers, as rapidly as possible; Mendoza et al., 2009). This task could be useful to highlight the clumsy movements characterizing limb-kinetic apraxia (e.g., Falchook et al., 2016), reported also in patients with subcortical degenerative diseases (e.g., Zadikoff and Lang, 2005; Quencer et al., 2007).

Hemispheric asymmetries in upper limb praxic abilities

Liepmann (1900) did not address the issue of hemispheric lateralization in his first description of an apraxic patient with impaired control of the right arm (see later). However, in his first group study Liepmann (1905) obtained clear-cut evidence about left hemispheric dominance in praxic abilities, as approximately 50% of patients with lesions in left hemisphere (LH) failed gesture production tasks with their ipsilesional hand, whereas all patients with right hemisphere (RH) lesions performed the same tasks flawlessly. This and other clinical observations led Liepmann (1925) to maintain that LH was dominant both in the first stages of movement planning and in translation from the "movement formula" to execution of movements with either limb, whereas RH had only a marginal role in motor control in typical right-handed individuals. The strong asymmetry in the prevalence of LA between lesions of either hemisphere has been confirmed in modern studies (see Donkervoort et al., 2000).

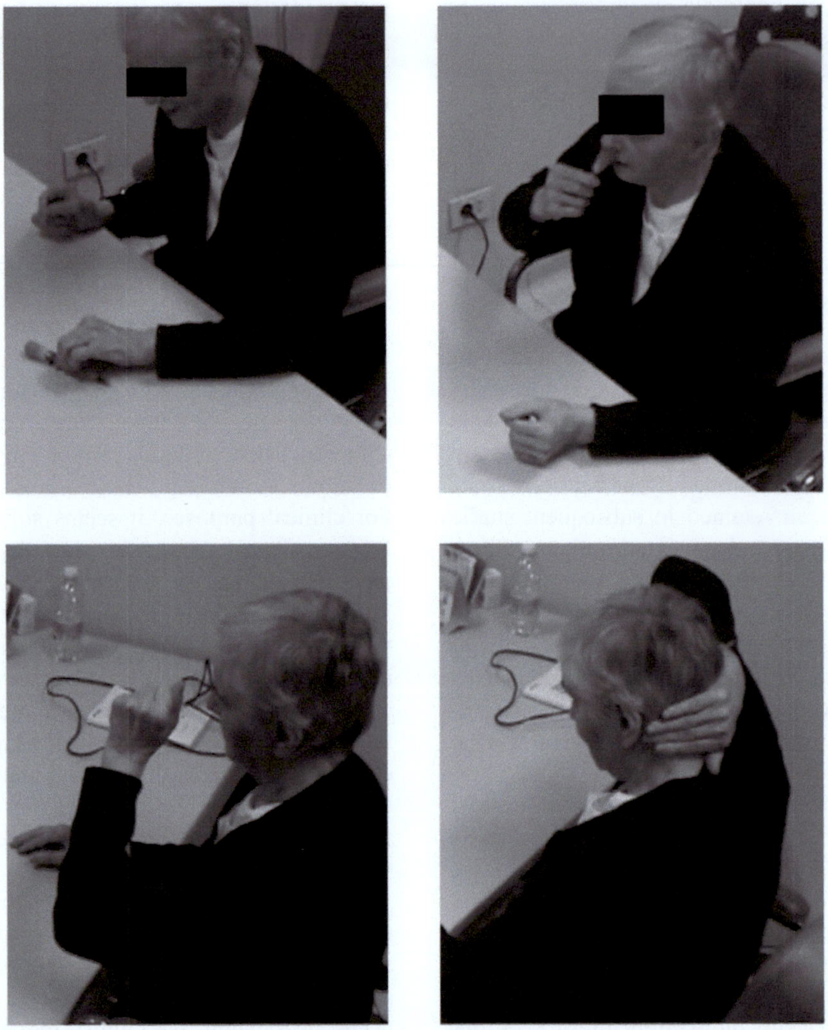

Fig. 25.1. Freeze frames from a video of a patient with a left parietotemporal lesion showing bilateral upper limb apraxia on single object use and gesture imitation tasks. The patient, as most apraxic patients, made many inconsistent types of errors on such tasks with either hand (even with the same hand), and some are depicted here. In the upper panels, the patient's attempts at using a paintbrush (she used it apparently as an eraser with her left hand and brushed her upper lip with her right hand); in the lower panels, the patient's imitation of a military salute (she imitated the salute with her left clenched fist, whereas in her attempt with her right hand she positioned the hand behind the back of her head instead of the forehead).

According to some authors, LH controls even execution of bilateral distal independent and precise movements better than RH, and only rapid movements not requiring precise control (e.g., finger tapping) would not be lateralized (Hanna-Pladdy et al., 2002).

Lateralization of praxic abilities in LH was supported by the observation of left unilateral apraxia after lesion of the anterior two-thirds of the corpus callosum. Liepman himself (Liepmann and Maas, 1907) observed this picture in a patient with an infarction in the territory of the left anterior cerebral artery, who could not write, use objects and perform skilled movements to command with his left hand correctly (the patient had a right paresis preventing assessment of his right limb). Although this patient also had lesions in the superior and medial portions of the left frontal lobe, possibly complicating symptom interpretation (see Heilman and Watson, 2008), further clinical observations by Liepmann's contemporaries helped establish that this clinical picture, defined "callosal apraxia," was related to the lack of transfer of motor representations from the LH to the RH motor system. Callosal apraxia has been reported in modern studies frequently associated with one or more further signs of callosal disconnection (such as left-sided tactile anomia and left-sided agraphia; e.g., Leiguarda et al., 1989). In these cases, an impairment in pantomiming on verbal command could be ascribed to some language-praxis disconnection, but clear evidence of impairments in pantomiming to visual presentation of tools and in meaningless gesture imitation has also been reported (e.g., Watson and Heilman, 1983), consistent with LH dominance for praxic abilities. The persistence of left unilateral LA for a long time after the callosal lesion (Falchook et al., 2016), and the improvement of callosal apraxia (at least in imitation tasks) associated with recovery of callosal connectivity (Morihara et al., 2021), are further data supporting LH lateralization for high-level control of praxic abilities and suggest that RH would be unable to promote recovery.

In patients with surgical callosotomy data are mixed. Left unilateral apraxia can be absent or restricted to pantomiming to verbal command (see Lausberg et al., 2003) and this might suggest that complete callosal section would allow some autonomy between the two hemispheres more than spontaneous disconnection (Goldenberg et al., 2001); however, it is should be considered that patients with severe epilepsy often show substantial cerebral reorganization and have been tested for apraxia several years after surgery. To avoid possible confounding of verbal instructions on praxis evaluation, Lausberg et al. (2003) assessed three patients with complete callosotomy on pantomiming to visual objects and on actual tool use with their either hand. The patients showed no difficulties in dealing with actual objects, but failed in pantomiming, making both concept and execution errors with their left hand. The dissociation between pantomiming to visual objects and using actual objects had already been reported both in patients with spontaneous callosotomy (e.g., Boldrini et al., 1992) and in patients with LH lesions (e.g., in De Renzi et al., 1982, where also two patients with reverse dissociation were identified). On these bases, Lausberg et al. (2003) argued left-hand object use might proceed via mechanisms not dependent on left lateralized functions, at least in surgical split-brain patients; but, it is worth recalling that actual use of objects is a highly specific task for LA, less sensitive than pantomiming in detecting mild disturbances.

The possibility that RH can support at least some praxic competences has been entertained from different perspectives. Building on automatic-voluntary dissociation, some authors put forward the hypothesis that RH could be involved in the context-dependent execution of familiar, well-established action routines, whereas LH would be critical for the control of context-independent movements and for learning novel movement sequences (see Gainotti, 2021). However, the large majority of authors focused on the possible involvement of RH in the gesture production tasks used during neuropsychologic assessment.

Patients, who developed apraxia after RH stroke, have been reported since early studies; but, at that time it was suggested that a lesion in the white matter of the right frontal lobe might damage the anterior callosal fibers, thus reconciling these clinical finding with the basic assumption of LH dominance for praxis (Wilson, 1908). In modern times, Roy et al. (2000) reported that defective performance on pantomime or on imitation of transitive and intransitive gestures can be equally common after damage to either hemisphere, whereas deficits in *both* pantomiming and imitating the same gestures are more common in patients with LH lesions. Focusing on intransitive gestures only, a further study by the same group confirmed that low scores on imitation or pantomime tasks were equally likely after lesions to either hemisphere (Heath et al., 2001). These findings led to the proposal that LH controls transitive gestures while both hemispheres may be involved in the control of *intransitive* gestures (e.g., Buxbaum et al., 2007). This proposal was supported by two later studies by Roy's group (Stamenova et al., 2010, 2012) showing that deficits in pantomime of intransitive, but not transitive gestures, were associated with RH lesions, beyond confirming the well-known notions that pantomime elicits poorer performance than imitation, and that LH lesions determine more impairments than RH lesions, irrespective of task or gesture type. On these bases, the authors concluded that LH is dominant for pantomime and production of tool-related gestures, whereas in some cases RH may contribute to production of *intransitive* gestures. However, other studies were not consistent with this hypothesis. For instance, Hanna-Pladdy et al. (2001) required patients with lesions to LH or RH to execute pantomimes of transitive and intransitive gestures, and observed that patients with LH lesions made a large number of spatiotemporal and conceptual errors for both transitive and intransitive gestures whereas patients with RH lesion made spatial and temporal errors when performing *transitive* gestures. Similarly, Flores-Medina et al. (2014) reported that patients with LH lesions are impaired particularly on intransitive gestures, whereas LH and RH lesions would give rise to equivalent performance in *transitive* gestures; moreover, the patients with LH lesions showed significantly more content errors, and the patients with RH lesions showed more execution errors. This overall inconsistent pattern of results has been tackled by a recent review of structural data in patients with brain lesions and of functional neuroimaging studies on healthy participants (Lesourd et al., 2018). The results showed that imitation of meaningful (as well as meaningless) intransitive gestures relies on a wide neural network in which the left hemisphere, and mainly the inferior parietal cortex, plays a key role. Nonetheless, the authors observed that several areas of the right hemisphere can contribute to intransitive gesture imitation, as shown by several neurofunctional studies focusing on the imitation of finger gestures. Once more, then, contrasting results have been obtained about the possible asymmetries in processing transitive and intransitive gestures, whereas a more consistent pattern seems to suggest that the so-called content errors would be related to the specific praxic competences of LH.

The possible contribution of RH (although lower than that provided by LH) in imitating symbolic intransitive and meaningless gestures had already been proposed in classic studies on patients with unilateral brain lesions (Barbieri and De Renzi, 1988). More recently, several studies by Goldenberg (e.g., Goldenberg, 1996; Goldenberg and Strauss, 2002) suggested that impairments in imitating meaningless gestures might depend on the body parts involved. According to Goldenberg, imitation of hand postures would be rooted in LH, whereas imitation of finger postures could be affected to similar degrees in patients with lesions either in LH or RH, thus suggesting that RH could contribute to spatial coding of the relationships of fingers between each other. However, subsequent studies failed to replicate these findings (e.g., Della Sala et al., 2006), and eventually Goldenberg et al. (2009) observed that the number of errors in imitating meaningless gestures in patients with RH damage was correlated with the severity of left unilateral neglect, and could be ascribed not only to the spatial exploration bias, but also to a general narrowing of the focus of attention and to reduced visual processing abilities. These data have been not only recently confirmed by a study on a large sample of patients with RH lesions (Dressing et al., 2020), and support the lateralization of praxic abilities to LH, but also imply that apparently simple tasks such as gesture imitation can be affected by other non-praxic cognitive impairments. This consideration could be consistent with findings from a recent detailed analysis of errors made by patients with unilateral brain lesions in imitating transitive and intransitive gestures (Scandola et al., 2021). The authors found that apraxic patients with LH lesions made content and configuration errors correlated to lesions of a wide frontoparietal network; content errors were also associated with fronto-insular lesions and movement errors with lesions in precentral and postcentral gyri; spatial errors were instead associated with ventral frontal lesions. In the same study, non-apraxic patients with RH lesions only made spatial errors, which therefore cannot be considered apraxic in nature.

Praxic abilities thus appear to be lateralized to LH as well as language, whereas RH would play a role only in controlling nonspecific aspects of movement planning in the vast majority of right-handers. However, some individuals may show a different cerebral organization, as demonstrated by reports of selective "crossed" apraxia, i.e., of LA without aphasia after RH stroke in right-handed patients (e.g., Marchetti and Della Sala, 1997). These observations suggest that control of upper limb movements could dissociate from both language and handedness. This issue has been addressed in a study systematically assessing language and upper limb praxis in a series of 50 left-handers with unilateral brain lesions (Goldenberg, 2013). The study identified three left-handed patients with LH lesions showing aphasia and pervasive apraxia, i.e., a dissociation of language and praxis from handedness, and three patients with RH lesions showing pervasive apraxia without aphasia, i.e., a dissociation of apraxia from aphasia. Moreover, left-handed patients showed less severe praxis and language impairments than those shown by a control group of right-handed individuals with brain lesions, thus partially supporting the idea that praxis is more bilaterally organized in left handers than in right handers. Regarding single praxis tasks, Goldenberg (2013) observed a frequent association of defective pantomime and actual tool use with aphasia, whereas defective imitation of hand postures was more frequently associated with the occurrence of neglect, consistent with the idea that the role of RH in praxis is eminently related to spatial processing. A recent functional neuroimaging study of healthy individuals (Kroliczak et al., 2021) showed that atypical (not lateralized to LH) organization of praxis was present in some right-handed participants but was more frequent in left-handers, whereas atypical organization of language was far less frequent, though with the same discrepancy between right- and left-handers. Notwithstanding several dissociations between praxis and language, computation of asymmetries in brain activation showed a strict association between pantomime of object use and language, regardless of handedness and typicality. Kroliczak et al. (2021) suggested that the developmental order of handedness, praxis, and speech acquisition could contribute to determine hemispheric asymmetries of functions; but, the observed relationships between praxis and language could be contingent on a common cerebral specialization, possibly related to the ability to construct symbolic representations, and/or to represent action sequences. However, it is important to underline that these conclusions could apply specifically to pantomiming; that is considered a task more complex than actual object use since classic studies (e.g., Goodglass and Kaplan, 1963) and is currently held a semantic-based task, relying on specific LH neural correlates in right-handers (Osiurak et al., 2021).

In synthesis, currently available evidence suggests a strong lateralization of praxic abilities to LH in most right-handers, maybe weaker than lateralization of language. The most recent research led to identification of different neurocognitive correlates of single praxic tasks, of processing of different kinds of gestures, and of production of different errors. This could be considered true both for studies on patients with brain lesions and for neuroimaging functional studies in healthy individuals, in which, however, small variations in paradigms and tasks can yield to different findings (for some reviews on the extensive neurofunctional literature on praxic abilities, see: Lesourd et al., 2018, 2021; O'Neal et al., 2021). Some cues have been gathered about the possible contribution of RH at least for some tasks and gestures (and errors); but, at the moment it could be possible to maintain that such a contribution mainly pertains to nonspecific aspects of gesture processing.

A last facet of gesture processing that could be lateralized is related to selection (and inhibition) of gestures. Informative data on this issue come from patients with brain lesions, who develop apparently purposeful, goal-directed, and well-executed movements that, however, are beyond their control and in contrast with their own will. This rare condition, formerly denoted as "alien hand" syndrome, is better conceived of as "anarchic hand," because usually patients do not report a sense of alienation or loss of ownership toward their affected limb, and instead tend to ascribe an autonomous volition to it (Della Sala et al., 1991). The syndrome is, however, heterogeneous in its clinical presentations, some of which are related to posterior (parietal) lesions and can be indeed complemented by feeling of disownership toward the affected limb (Hassan and Josephs, 2016). Moreover, unwilled, and yet purposeful, gestures are often observed in neurodegenerative diseases, and notably in cortico-basal degeneration (see Di Pietro et al., 2021); but, "proper" anarchic hand related to focal anterior lesions is more relevant for the present issue. Interestingly, the first patient with unilateral LA described by Liepmann (1900) showed unintended actions of his right hand, thus raising the possibility that he was affected by a right-sided, contralesional, anarchic hand related to complete damage of the anterior two-thirds of the corpus callosum and to left frontal and parietal subcortical lesions (Goldenberg, 2003). Feinberg et al. (1992) proposed the existence of two variants of this condition related to anterior lesions. The first would be determined by an anterior callosal damage, causing an intermanual conflict during behaviors requiring LH control; the second would occur exclusively in the dominant limb after left medial frontal lesions (often involving the corpus callosum), and would consist in groping and compulsive manipulation of tools, due to an increased exploratory behavior mediated by LH, released from contralateral inhibitory control. A different classificatory framework (Della Sala et al., 1991) posits that fleeting, involuntary movements can appear short after callosal disconnection, whereas the stabilized chronic form of the anarchic hand can occur on either hand after contralateral medial frontal lesions combined with lesions of anterior callosal fibers (Trojano et al., 1993). Indeed, a case series of patients with an anarchic hand confirmed that the disturbance can affect either limb after combined lesions of contralateral cortical and subcortical medial frontal regions and of the genu or anterior rostrum of the corpus callosum (Brainin et al., 2008). Further studies pointed to the anterior part of the supplementary motor area (pre-SMA) as the medial frontal region of either hemisphere whose lesion, combined with a lesion of the anterior corpus callosum, can give rise to the specific signs of contralesional anarchic hand (Brugger et al., 2015). A very rare, bilateral lesion of SMA, with sparing of the corpus callosum, has been reported to determine persistent bilateral utilization behavior, considered as a bilateral anarchic hand (Boccardi et al., 2002). Goldberg et al. (1981) posited that the anarchic hand can be ascribed to faulty initiation, execution, and inhibition of preexisting intact motor subroutines, but several theoretical accounts have eventually been put forward (see Wolpe et al., 2020). Functional neuroimaging studies in healthy participants provided contrasting findings, but would suggest that SMA is involved in several aspects of action monitoring (see Charalampaki et al., 2022). Independently from these considerations, the occurrence of a persistent anarchic hand at either limb after a stroke in the anterior cerebral artery territory of either side seems to imply that, although the planning of upper limb movements is strongly lateralized to LH, the inhibition and selection of gestures is based on medial prefrontal regions bilaterally.

CONSTRUCTIONAL APRAXIA

Clinical manifestations

Early descriptions of patients with LA signaled that the impairment in controlling upper limb movements could also affect the ability of building predefined patterns by sticks (Rieger, 1909). A few years later, it became clear that activities such as drawing, assembling, or cutting a shape with scissors specifically require that upper limbs act under strict control of vision, thus leading to the construct of "optic apraxia" (Poppelreuter, 1917). Because of the broad range of impairments that could be included under this heading, Kleist (1934) proposed the term "constructional apraxia" (CA) to designate a specific disturbance of "formative" activities (such as assembling and drawing) in which the spatial form of the product is missed. According to Kleist's original definition (1934), CA would result from a disconnection between visual–spatial functions and the kinestetic engrams implied in manual activity, thus excluding from CA visuo-perceptual impairments and the different variants of LA identified by Liepmann (1920, 1925). Indeed, one of Kleist's pupil, Strauss (1924) systematically assessed constructional abilities in 14 patients with mainly vascular focal brain lesions, and observed selective impairments in "visual-related formative activities" (i.e., CA) without associated LA or relevant visual disorders in five patients. Incidentally, it should be noted that Strauss formally credited Kleist for having coined the definition of CA, although Strauss' paper preceded publication of Kleist's (1934) famous book (Trojano, 2020).

In their papers, Kleist and Strauss maintained that the features of errors in "constructional" tasks might help distinguish CA from other disorders: errors in CA are spatial in nature, whereas errors in LA would consist of difficult or misguided movements producing an awkward but recognizable final output. CA would be characterized by spatial displacements or incorrect proportions of objects or their parts, rotations (usually by 90° or 180°), and omissions, to the point that apraxics' drawings might resemble children's drawings. Strauss (1924) endorsed Kleist's proposal that spatial errors distinguish selective CA from apraxic disorders, and further suggested that a complete disintegration of patients' productions would instead denote co-occurrence of complex visual disorders, such as Lissauer's apperceptive agnosia. Later, in his authoritative book on parietal lobes, Critchley (1953) maintained that errors typical of CA in drawing and assembling sticks would include wrong orientation (e.g., tilting horizontal lines, making vertical lines oblique), dimensions and placement of single elements, tendency to produce the copy close to or superimposed on the model, perseverations, omissions, mirror reversals of parts or whole figures, and loss of perspective. However, the analysis of constructional errors is not a straightforward procedure, even in spite of attempts at quantitative analyses (e.g., Carson et al., 2019). The amount and the type of errors may greatly vary within the same individual patient and errors are hardly ascribed to specific mechanisms unequivocally. For instance, the reproduction of a line in the wrong orientation could be due to faulty visuospatial perceptual processes, or to impaired motor execution, or to defects in planning drawing (Trojano and Conson, 2008). Another instance of this difficulty is the tendency to produce responses close to the given model or to overlap it. Mayer-Gross (1935) specifically investigated this phenomenon in six patients with CA and proposed the definition of "closing-in." According to Mayer-Gross, the phenomenon became more manifest as the difficulty of spatial tasks increased, and was related to a "primary biologic protective mechanism" in patients with spatial disorders. Several later authors investigated different aspects of "closing-in," studying its qualitative features in children and in patients with unilateral brain lesions (e.g., Gainotti, 1972) or assessing its sensitivity and specificity as a neuropsychologic marker of dementia (e.g., Gainotti et al., 1998). On the basis of these data, several interpretations of closing-in were advanced (for a review, see Ambron and Della Sala, 2017), but accumulating evidence supported the idea that closing-in could be the consequence of an impaired attentional control of spatial actions in patients with vascular or degenerative brain diseases (e.g., De Lucia et al., 2014, 2016).

Notwithstanding Kleist's and Strauss' specific clinical observations, in subsequent years the original definition was substantially disregarded, since some authors considered CA as an executive disorder restricted to complex manipulation, in the absence of visuo-perceptive disturbances, whereas others reckoned CA to result from a basic visuo-perceptive impairment (for a thorough critical review, see De Renzi, 1982). This "dispute" also led to contrasting proposals about neuroanatomic correlates of CA (see later), that ultimately imply that constructional abilities rely on widely distributed neural networks. Analogously, attempts at defining the cognitive processes involved in constructional abilities within comprehensive models (Roncato et al., 1987; Van Sommers, 1989; Grossi, 1991; Guérin et al., 1999) share the idea that visuospatial processes, dedicated planning abilities, and general control processes contribute to drawing and assembling tasks; but, no model gained general acceptance, also because of the difficulties in error analysis and of paucity of data on patients making consistent and clearly interpretable errors (for a review, see Gainotti and Trojano, 2018). It is worth mentioning that Grossi (1991), exploiting the distinction between lexical and sub-lexical routes proposed for language and praxic abilities, suggested the existence of two copying procedures: a "lexical" route predominantly involving activation of familiar constructional schemata (for example, in drawing a square or a face), and a "line-by-line" procedure, based on a spatial analysis without resorting to constructional representations (e.g., when reproducing novel stimuli).

Gradually, the term CA began to be used (until nowadays) as an operational definition to refer to all disturbances observed during drawing and assembling and building complex models without any implication about the putative causal mechanisms (for a review, see Gainotti and Trojano, 2018). In this perspective, CA is associated with a large variety of cognitive disturbances, spanning from visual agnosia, to optic ataxia and other visuospatial impairments, to unilateral neglect, LA and dysexecutive syndrome, among others, in patients with focal or diffuse (vascular or degenerative) brain diseases (see Trojano and Conson, 2008). In modern literature, limited instances of "pure" CA have been reported (e.g., Trojano and Grossi, 1998).

Clinical assessment

Kleist's assumption that all "formative" activities producing a "spatially formed output" share a common specific brain mechanism, rested mainly on paper-and-pencil tasks. In his systematic assessment of CA, though, Strauss (1924) used a wider range of tasks: assembling blocks or sticks to reproduce familiar forms (e.g., a cross) or novel abstract figures; spontaneous drawing; and copying geometric figures and object shapes. However, drawing and assembling cannot be considered equivalent: for instance, Strauss himself (1924) observed impairments in assembling but not in drawing in two of his patients with pure CA. For this reason, some authors recommend complementing drawing tasks with two- and three-dimensional assembling tasks (e.g., Benton, 1989), which nevertheless are used rarely. In recent years, interest in two- and three-dimensional assembling tasks has been revived by some clinical studies (Capruso and Hamsher, 2011; De Lucia et al., 2020; Martins-Rodrigues et al., 2021); but, it is necessary to keep in mind that assembling items of different forms or colors (e.g., as in Kohs' cubes) demands general problem-solving abilities, above all when the examinee is presented with more items than those necessary to reproduce the model (Critchley, 1953). Drawing tasks are those most widely used to assess constructional abilities for their easy administration even at bedside, but copying and free drawing cannot be considered equivalent. Free drawing, requiring patients to draw a named object (e.g., a clock; Fig. 25.2), is apparently the most immediate tool for assessing constructional skills, providing information about the ability to draw complete shapes and to organize the figure as a whole, with its component elements in their correct spatial relationships. However, this task does not easily lend itself to standardization (for instance, see the huge variability in administration and scoring procedures for the clock drawing test; Hazan et al., 2018) and relies on non-constructional cognitive abilities too, particularly lexical-semantic knowledge and mental imagery (see Trojano and Conson, 2008). Instead, copying tasks directly evaluate the patient's ability to reproduce simple, such as circles and squares, or complex figures, such as the Rey-Osterreith Complex Figure (ROCF, Rey, 1941; Osterrieth, 1944). However, copying performance is strongly related to age, educational level, and cultural background (see Rosselli and Ardila, 2003), beyond visuospatial and executive abilities. It is, therefore, crucial for the diagnosis of CA to adopt standardized tasks, with solid normative data, and employ well-known and novel stimuli of graded complexity (De Renzi, 1982).

Hemispheric asymmetries in constructional abilities

In proposing CA as an independent disorder, Kleist (1934) suggested its association with lesions in the left posterior parietal lobe, which would serve as a center for linking visual and kinesthetic functions. In his book, Kleist (1934) was quite prudent about this hypothesis; but, in his attempts at finding a precise localization for cognitive and mental disorders (see Trojano, 2020), in a related paper (Kleist, 1926) he explicitly stated that CA was due to a damage to the left Brodmann area 39 (angular gyrus), differently from Liepmann's LA, to be ascribed to lesions in the left Brodmann area 40 (supramarginal gyrus). Strauss' analysis of patients with pure CA (1924) supported the left parietal localization of CA; but, in subsequent years, the dispute between interpretations of CA as a disorder related to motor planning and execution, or as a consequence of impaired visuospatial perception made clear that disturbances in drawing and assembling tasks could be generated by lesions in either hemisphere (for reviews see De Renzi, 1982; Gainotti and Trojano, 2018). For instance, in their first major systematic survey of the relationships between visuo-constructive disabilities and hemispheric locus of lesion, Piercy et al. (1960) observed that CA was significantly more frequent and more severe in patients with right-sided (and in particular with right parietal) lesions and that errors made by patients with LH or RH parietal lesions differed qualitatively: drawings of patients with LH lesions mainly showed great simplification of the outline, and frequent hesitations and lack of details, but good overall spatial organization, whereas those of patients with RH lesions were characterized by a failure to reproduce the spatial relationships among the various parts of the model.

These observations were consistent with the general idea that RH (and, in particular, the right parietal lobe) is dominant for

Fig. 25.2. Instances of graphic productions on the clock drawing test by patients with left (left panels) or right brain lesions (right panels; a drawing by a patient with unilateral neglect is depicted in the top right panel). In this version of the test, a pre-drawn circle is provided and patients are required to complete the dial and then to draw the clock hands pointing at 2:45.

visual–spatial functions, but a substantial number of subsequent studies of CA failed to find a greater prevalence or severity of CA in patients with RH. For instance, Arena and Gainotti (1978) did not find differences between patients with unilateral RH or LH brain damage on a graphomotor test and on a multiple choice visuoperceptual task. Moreover, as stated above, error analysis is not straightforward for constructional tasks, and the idea of clearly separable kinds of errors after LH or RH lesions (apart from neglect-related errors) has not been supported consistently (see Carson et al., 2019). Therefore, a different approach to hemispheric contribution to "formative" activities tried to identify correlations between constructional performance and other cognitive abilities in patients with lesions in either hemisphere. For instance, Kirk and Kertesz (1989) noted that performance on a free drawing task correlated strongly with scores on a visual-perceptual task in patients with RH lesions, but with tests of verbal comprehension, and with the severity of the motor defect in patients with LH lesions, and concluded that CA originates from a visual–spatial defect after RH damage, while it could be linked to semantic or to elementary motor disorders after LH lesions. However, this proposal should be considered with caution, as free drawing has inherent semantic correlates. Two later studies on copying tasks reported partially conflicting correlations of constructional disturbances: Carlesimo et al. (1993) confirmed that a disorder at the elementary motor level may play a role in patients with LH lesions, whereas Trojano et al. (2004) observed that drawing accuracy was significantly correlated with spatial perceptual and representational tests (as assessed by a battery tapping visuospatial processing; Trojano et al., 2015, 2018) in patients with RH but not LH lesions, without detecting cognitive abilities significantly correlated with drawing in patients with LH lesions. More recently, Biesbroek et al. (2014) examined data from a large cohort of patients with first-ever ischemic stroke, who completed the ROCF copy test and a well-known test for visuospatial perception (Judgment of Line Orientation, JLOT; Benton et al., 1975). The authors observed that the two tests share correlation with several areas in RH (dorsal frontal, superior temporal, and inferior parietal regions), whereas lesions in the right superior parietal lobule, angular gyrus, and middle occipital gyri were associated with poor performance on the ROCF, but not on the JLOT. These findings would provide evidence for a specific role of RH inferior and superior parietal lobules in constructional tasks, independently from visuospatial perception; but, the exclusive association with RH is likely explained by the exclusion of patients with right paresis and/or tool apraxia from the study. The relationships between drawing performance (assessed by copying a ROCF-like complex figure from which several aspects of drawing can be quantified; Humphreys et al., 2012) and brain lesions has also been assessed in a large, consecutive sample of patients with unilateral stroke (Chechlacz et al., 2014). The results showed that areas in the two hemispheres were significantly correlated with different aspects of drawing: overall accuracy and omissions of left details of the stimulus correlated with lesions in RH; spatial errors correlated with a wide range of lesions in either hemisphere, and with lesions in the left precuneus; and errors in reproducing the stimulus' details correlated with RH lesions, whereas errors in reproducing the stimulus' overall configuration correlated with LH lesions. This study demonstrated that lesions in either hemisphere might determine different drawing impairments, and, by implication, that drawing depends on several cognitive processes based on wide, interconnected neural networks. It is worth noting that the left angular gyrus identified by Kleist as the main anatomic correlate of CA would not be included within these neural networks. In a further analysis of data obtained on a subsample of the same cohort of patients, Chen et al. (2016) observed that overall constructional performance was correlated with LA and left neglect: the association between CA and LA was related to lesions of the left post-rolandic gyrus, and of thalamus and right middle frontal gyrus, whereas the association between CA and left neglect was related to right inferior parietal lesions. By means of a principal factor analysis, the author identified three components associated with accuracy in copying a complex figure, which were related to different neural networks present in discrete regions of both hemispheres and putatively involved in high-order motor control, complex visuomotor transformation, and planning and sequential organization of action. While this and the previously cited studies strongly supported the idea that constructional abilities are based on wide, interconnected networks, it must be stressed that they were based on a correlational approach applied to consecutive series of patients in which CA was likely associated to other cognitive disturbances, without focusing on patients with selective CA, as those described by Kleist and Strauss. Indeed, several functional neuroimaging studies on healthy individuals demonstrated that drawing is associated with activation of wide brain networks in either hemisphere, but their findings were quite inconsistent because of differences in control tasks and stimuli employed in experimental paradigms. A recent meta-analysis of these studies supported the role of the left inferior parietal lobule in drawing in healthy individuals, when subtracting the areas involved in elementary motor activities from those recruited by copying tasks (Raimo et al., 2021).

In synthesis, many neuropsychologic and neurofunctional data demonstrate that CA, in its wider acceptation, is related to lesion to either hemisphere with analogous prevalence and severity (if neglect-related errors are not considered). In this perspective, CA has not a strong localizing value, but can be considered as a sensitive marker for cognitive impairments, and is frequently searched for in assessing general cognitive deterioration (Trojano and Gainotti, 2016; Youn et al., 2021). At the moment, the hemispheric asymmetries in the contribution of LH and RH to "formative" activities cannot be inferred by qualitative or quantitative analyses of patients' productions (Carson et al., 2019); but, several experimental studies tried to define possible cognitive and neural correlates of CA in patients with brain lesions. RH's role in constructional abilities appears to be mainly related to lateralized attentional disorders (left neglect), in which stimuli's spatial position and hierarchical structure strongly influence patients' graphic productions (e.g., Behrmann and Plaut, 2001). Independently from neglect-related disorders, some studies put forward further hypotheses about RH's role in constructional abilities. For instance, Russell et al. (2010) suggested that CA in patients with RH frontal and parietal lesions might result from impairments in remembering spatial information across eye fixations, i.e., in remapping stimuli's locations across saccades. This hypothesis would be consistent with the idea that a specialized RH network is responsible for visuospatial remapping that is a prerequisite for subsequent visuospatial processing, which would be symmetrically organized (Pisella et al., 2011). A different viewpoint was assumed by Laeng (2006), according to whom RH might be responsible for computing precise metric ("coordinate") spatial relations among figure components, whereas LH could be involved in appreciation of abstract ("categorical") spatial relations, like those usually captured by prepositions or locatives (e.g., "inside," "between"), in keeping with hypotheses about lateralization of spatial computations (Kosslyn, 1987; see also Trojano et al., 2006). A further proposal, based on the postulated hemispheric asymmetries in attending to local vs global features of hierarchical visual stimuli (Delis et al., 1986), posits that poor complex figure drawing could result from deficits in hierarchical representation of complex stimuli (Chechlacz et al., 2015). In particular, impairments in reproducing local features in figure copying tasks would be related with LH lesions of parietal, occipital, and insular cortex and of their interconnection via the inferior frontooccipital fasciculus, whereas impairments in reproducing global stimuli configuration would be associated with lesions to an RH network interconnected by the third branch of the superior longitudinal fasciculus and the long segment of the perisylvian network. Further studies are needed to substantiate these hypotheses about hemispheric asymmetries in "formative activities" and to ascertain their heuristic values in interpreting neuropsychologic observations on patients with brain lesions.

CONCLUDING REMARKS

This chapter dealt with the unique human abilities of using tools, imitating others' gestures, and drawing and building complex items. The data presented here delineated that all these activities require the interplay of several brain areas and neural networks distributed over both hemispheres. Nonetheless, a quite strong asymmetry is present in the neural processing of gestures, as, in typical right-handers, the areas responsible for gesture-specific processing largely (but not completely) overlap language-related areas and exhibit a similar lateralization to the LH. The same does not apply to "formative activities," in which neural networks in both hemispheres seem to provide crucial contributions, although the left inferior parietal lobule might play a specific role.

The convergence of gesture processing and "formative activities" (and language) in the parietal lobes might be related to their considerable expansion in our species (Bruner et al., 2023), and could suggest that some shared basic, still unknown, computational processes allowed the development of these skills in modern humans.

References

Ambron E, Della Sala S (2017). A critical review of closing-in. Neuropsychology 31: 105–117.

Arena R, Gainotti G (1978). Constructional apraxia and visuoperceptive disabilities in relation to laterality of cerebral lesions. Cortex 14: 463–473.

Barbieri C, De Renzi E (1988). The executive and ideational components of apraxia. Cortex 24: 535–543.

Bartolo A, Cubelli R, Della Sala S et al. (2003). Pantomimes are special gestures which rely on working memory. Brain Cogn 53: 483–494.

Bartolo A, Cubelli R, Della Sala S (2008). Cognitive approach to the assessment of limb apraxia. Clin Neuropsychol 22: 27–45.

Bartolo A, Della Sala S, Cubelli R (2020). Effect of test instructions: the example of the pantomime production task. Brain Cogn 139: 105516.

Baumard J, Le Gall D (2021). The challenge of apraxia: toward an operational definition? Cortex 141: 66–80.

Behrmann M, Plaut DC (2001). The interaction of spatial reference frames and hierarchical object representations: evidence from figure copying in hemispatial neglect. Cogn Affect Behav Neurosci 1: 307–329.

Benton AL (1989). Constructional apraxia. In: F Boller, J Grafman (Eds.), Handbook of neuropsychology. vol. 2. Elsevier, Amsterdam, pp. 387–394.

Benton AL, Hannay HJ, Varney NR (1975). Visual perception of line direction in patients with unilateral brain disease. Neurology 25: 907–910.

Biesbroek JM, van Zandvoort MJ, Kuijf HJ et al. (2014). The anatomy of visuospatial construction revealed by lesion-symptom mapping. Neuropsychologia 62: 68–76.

Boccardi E, Della Sala S, Motto C et al. (2002). Utilisation behaviour consequent to bilateral SMA softening. Cortex 38: 289–308.

Boldrini P, Zanella R, Cantagallo A et al. (1992). Partial hemispheric disconnection syndrome of traumatic origin. Cortex 28: 135–143.

Bradshaw JL, Rogers LJ (1996). Tool use and the evolutionary development of manual asymmetry. In: D Elliott, EA Roy (Eds.), Motor asymmetries in motor performance. CRC Press, Boca Raton, FL, pp. 33–54.

Brainin M, Seiser A, Matz K (2008). The mirror world of motor inhibition: the alien hand syndrome in chronic stroke. J Neurol Neurosurg Psychiatry 79: 246–252.

Brugger F, Galovic M, Weder BJ et al. (2015). Supplementary motor complex and disturbed motor control—a retrospective clinical and lesion analysis of patients after anterior cerebral artery stroke. Front Neurol 6: 209.

Bruner E, Battaglia-Mayer A, Caminiti R (2023). The parietal lobe evolution and the emergence of material culture in the human genus. Brain Struct Funct 228: 145–167.

Buchmann I, Randerath J (2017). Selection and application of familiar and novel tools in patients with left and right hemispheric stroke: psychometrics and normative data. Cortex 94: 49–62.

Buxbaum LJ, Randerath J (2018). Limb apraxia and the left parietal lobe. Handb Clin Neurol 151: 349–363.

Buxbaum LJ, Kyle K, Grossman M et al. (2007). Left inferior parietal representations for skilled hand-object interactions: evidence from stroke and corticobasal degeneration. Cortex 43: 411–423.

Capruso DX, Hamsher K (2011). Constructional ability in two- versus three-dimensions: relationship to spatial vision and locus of cerebrovascular lesion. Cortex 47: 696–705.

Carlesimo GA, Fadda L, Caltagirone C (1993). Basic mechanisms of constructional apraxia in unilateral brain-damaged patients: role of visuo-perceptual and executive disorders. J Clin Exp Neuropsychol 15: 342–358.

Carson L, Filipowicz A, Anderson B et al. (2019). Representational drawing following brain injury. Neuropsychologia 133: 107154.

Charalampaki A, Ninija Karabanov A, Ritterband-Rosenbaum A et al. (2022). Sense of agency as synecdoche: multiple neurobiological mechanisms may underlie the phenomenon summarized as sense of agency. Conscious Cogn 101: 103307.

Chechlacz M, Novick A, Rotshtein P et al. (2014). The neural substrates of drawing: a voxel-based morphometry analysis of constructional, hierarchical, and spatial representation deficits. J Cogn Neurosci 26: 2701–2715.

Chechlacz M, Mantini D, Gillebert CR et al. (2015). Asymmetrical white matter networks for attending to global versus local features. Cortex 72: 54–64.

Chen H, Pan X, Lau JKL et al. (2016). Lesion-symptom mapping of a complex figure copy task: a large-scale PCA study of the BCoS trial. Neuroimage Clin 11: 622–634.

Critchley M (1953). The parietal lobes, Hafner Press, New York.

Cubelli R, Marchetti C, Boscolo G et al. (2000). Cognition in action: testing a model of limb apraxia. Brain Cogn 44: 144–165.

De Lucia N, Grossi D, Trojano L (2014). The genesis of closing-in in Alzheimer disease and vascular dementia: a comparative clinical and experimental study. Neuropsychology 28: 312–318.

De Lucia N, Grossi D, Trojano L (2016). A clinical study on closing-in in focal brain-damaged individuals. J Neurol Sci 363: 170–175.

De Lucia N, Grossi D, Milan G et al. (2020). The closing-in phenomenon in constructional tasks in dementia and mild cognitive impairment. Neuropsychology 34: 168–175.

De Renzi E (1982). Disorders of space exploration and cognition, John Wiley & Sons, Chichester, p. 1982.

De Renzi E, Lucchelli F (1988). Ideational apraxia. Brain 111: 1173–1185.

De Renzi E, Faglioni P, Sorgato P (1982). Modality-specific and supramodal mechanisms of apraxia. Brain 105: 301–312.

Delis DC, Robertson LC, Efron R (1986). Hemispheric specialization of memory for visual hierarchical stimuli. Neuropsychologia 24: 205–214.

Della Sala S, Marchetti C, Spinnler H (1991). Right-sided anarchic (alien) hand: a longitudinal study. Neuropsychologia 29: 1113–1127.

Della Sala S, Faglioni P, Motto C et al. (2006). Hemisphere asymmetry for imitation of hand and finger movements, Goldenberg's hypothesis reworked. Neuropsychologia 44: 1496–1500.

Di Pietro M, Russo M, Dono F et al. (2021). A critical review of alien limb-related phenomena and implications for functional magnetic resonance imaging studies. Front Neurol 12: 661130.

Donkervoort M, Dekker J, van den Ende E et al. (2000). Prevalence of apraxia among patients with a first left hemisphere stroke in rehabilitation centres and nursing homes. Clin Rehabil 14: 130–136.

Dressing A, Martin M, Beume LA et al. (2020). The correlation between apraxia and neglect in the right hemisphere: a voxel-based lesion-symptom mapping study in 138 acute stroke patients. Cortex 132: 166–179.

Falchook AD, Watson RT, Heilman KM (2016). Callosal apraxia: a 34-year follow-up study. Neurocase 22: 306–311.

Feinberg TE, Schindler RJ, Flanagan NG et al. (1992). Two alien hand syndromes. Neurology 42: 19–24.

Finkel L, Hogrefe K, Frey SH et al. (2018). It takes two to pantomime: communication meets motor cognition. Neuroimage Clin 19: 1008–1017.

Flores-Medina Y, Chávez-Oliveros M, Medina LD et al. (2014). Brain lateralization of complex movement: neuropsychological evidence from unilateral stroke. Brain Cogn 84: 164–169.

Foundas AL, Duncan ES (2019). Limb apraxia: a disorder of learned skilled movement. Curr Neurol Neurosci Rep 19: 82.

Gainotti G (1972). A quantitative study of the "closing-in" symptom in normal children and in brain-damaged patients. Neuropsychologia 10: 429–436.

Gainotti G (2021). Is there a causal link between the left lateralization of language and other brain asymmetries? A review of data gathered in patients with focal brain lesions. Brain Sci 11: 1644.

Gainotti G, Trojano L (2018). Constructional apraxia. Handb Clin Neurol 151: 331–348.

Gainotti G, Marra C, Villa GP et al. (1998). Sensitivity and specificity of some neuropsychological markers of Alzheimer's dementia. Alzheimer Dis Assoc Disord 12: 152–162.

Goldberg G, Mayer NH, Toglia JU (1981). Medial frontal cortex infarction and the alien hand sign. Arch Neurol 38: 683–686.

Goldenberg G (1996). Defective imitation of gestures in patients with damage in the left or right hemisphere. J Neurol Neurosurg Psychiatry 61: 176–180.

Goldenberg G (2003). Apraxia and beyond: life and work of Hugo Liepmann. Cortex 39: 509–524.

Goldenberg G (2013). Apraxia in left-handers. Brain 136: 2592–2601.

Goldenberg G (2017). Facets of pantomime. J Int Neuropsychol Soc 23: 121–127.

Goldenberg G, Randerath J (2015). Shared neural substrates of apraxia and aphasia. Neuropsychologia 75: 40–49.

Goldenberg G, Strauss S (2002). Hemisphere asymmetries for imitation of novel gestures. Neurology 59: 893–897.

Goldenberg G, Laimgruber K, Hermsdorfer J (2001). Imitation of gestures by disconnected hemispheres. Neuropsychologia 39: 1432–1443.

Goldenberg G, Münsinger U, Karnath H-O (2009). Severity of neglect predicts accuracy of imitation in patients with right hemisphere lesions. Neuropsychologia 47: 2948–2952.

Gonzalez Rothi LJ, Mack L, Verfaellie M et al. (1988). Ideomotor apraxia: error pattern analysis. Aphasiology 2: 381–387.

Goodglass G, Kaplan E (1963). Disturbance of gesture and pantomime in aphasia. Brain 86: 703–720.

Grossi D (1991). La riabilitazione dei disturbi della cognizione spaziale, Masson, Milano.

Guérin F, Ska B, Belleville S (1999). Cognitive processing of drawing abilities. Brain Cogn 40: 464–478.

Hanna-Pladdy B, Daniels SK, Fieselman MA et al. (2001). Praxis lateralization: errors in right and left hemisphere stroke. Cortex 37: 219–230.

Hanna-Pladdy B, Mendoza JE, Apostolos GT et al. (2002). Lateralised motor control: hemispheric damage and the loss of deftness. J Neurol Neurosurg Psychiatry 73: 574–577.

Hassan A, Josephs KA (2016). Alien hand syndrome. Curr Neurol Neurosci Rep 16: 73.

Hazan E, Frankenburg F, Brenkel M et al. (2018). The test of time: a history of clock drawing. Int J Geriatr Psychiatry 33: e22–e30.

Heath M, Roy EA, Westwood D et al. (2001). Patterns of apraxia associated with the production of intransitive limb gestures following left and right hemisphere stroke. Brain Cogn 46: 165–169.

Heilman KM, Watson RT (2008). The disconnection apraxias. Cortex 44: 975–982.

Humphreys GW, Bickerton WL, Samson D et al. (2012). The Birmingham cognitive screen (BCoS), Psychology Press, London.

Jackson H (1866). Notes on the physiology and pathology of language. Brain 38: 48–58.

Kirk A, Kertesz A (1989). Hemispheric contributions to drawing. Neuropsychologia 27: 881–886.

Kleist K (1926). Gehirnpathologische und gehirnlokalisatorische Ergebnisse, vornehmlich auf Grund von Kriegshirnverletzungen. In: VV.AA., Festschrift für Woldemar Bechterew zum 40. Jahre seiner Lehrtätigkeit. Leningrad, pp. 201–214.

Kleist K (1934). Gehirnpathologie vornehmlich auf Grund der Kriegerfahrungen, J.A. Barth, Leipzig.

Kosslyn SM (1987). Seeing and imagining in the cerebral hemispheres: a computational approach. Psychol Rev 94: 148–175.

Kroliczak G, Buchwald M, Kleka P et al. (2021). Manual praxis and language-production networks, and their links to handedness. Cortex 140: 110–127.

Laeng B (2006). Constructional apraxia after left or right unilateral stroke. Neuropsychologia 44: 1595–1606.

Lausberg H, Cruz RF, Kita S et al. (2003). Pantomime to visual presentation of objects: left hand dyspraxia in patients with complete callosotomy. Brain 126: 343–360.

Leiguarda R, Starkstein S, Berthier M (1989). Anterior callosal haemorrhage. A partial interhemispheric disconnection syndrome. Brain 112: 1019–1037.

Lesourd M, Le Gall D, Baumard J et al. (2013). Apraxia and Alzheimer's disease: review and perspectives. Neuropsychol Rev 23: 234–256.

Lesourd M, Osiurak F, Baumard J et al. (2018). Cerebral correlates of imitation of intransitive gestures: an integrative review of neuroimaging data and brain lesion studies. Neurosci Biobehav Rev 95: 44–60.

Lesourd M, Servant M, Baumard J et al. (2021). Semantic and action tool knowledge in the brain: identifying common and distinct networks. Neuropsychologia 159: 107918.

Liepmann H (1900). Das Krankheitsbild der Apraxie (motorische Asymbolie) auf Grund eines Falles von einseitiger Apraxie. Monatschr Psychiat Neurol 8: 15–44, 102–132, 182–197.

Liepmann H (1905). Die linke Hemisphaere und das Handeln. Muench Med Wochenschr 52: 2322–2326, 2375–2378.

Liepmann H (1920). Apraxia. Ergb Gesamte Med 1: 516–543.

Liepmann H (1925). Apraktische Störungen. In: H Curschmann, F Kramer (Eds.), Lehrbuch der Nervenkrankheiten. Springer, Berlin, pp. 408–416.

Liepmann H, Maas O (1907). Fall von linksseitiger Agraphie und Apraxie bei rechsseitiger Lahmung. Z Psychol Neurol 10: 214–227.

Marchetti C, Della Sala S (1997). On crossed apraxia. Description of a right-handed apraxic patient with right supplementary motor area damage. Cortex 33: 341–354.

Martins-Rodrigues R, da Fonsêca ÉKG, Lucena-Alves SS et al. (2021). Clinical utility of two- and three-dimensional visuoconstructional tasks in mild cognitive impairment and early Alzheimer's disease. Arch Clin Neuropsychol 36: 177–185.

Mayer-Gross W (1935). Some observations on apraxia. Proc R Soc Med 28: 1203–1212.

Mendoza JE, Apostolos GT, Humphreys JD et al. (2009). Coin rotation task (CRT): a new test of motor dexterity. Arch Clin Neuropsychol 24: 287–292.

Morihara K, Kakinuma K, Kobayashi E et al. (2021). Improvement in callosal disconnection syndrome with recovery of callosal connectivity. Neurocase 27: 323–331.

Ochipa C, Rothi LJ, Heilman KM (1994). Conduction apraxia. J Neurol Neurosurg Psychiatry 57: 1241–1244.

O'Neal CM, Ahsan SA, Dadario NB et al. (2021). A connectivity model of the anatomic substrates underlying ideomotor apraxia: a meta-analysis of functional neuroimaging studies. Clin Neurol Neurosurg 207: 106765.

Osiurak F, Rossetti Y (2017). Definition: limb apraxia. Cortex 93: 228.

Osiurak F, Reynaud E, Baumard J et al. (2021). Pantomime of tool use: looking beyond apraxia. Brain Commun 3: fcab263.

Osterrieth PA (1944). Le test de copie d'une figure complexe. Arch Psychol 30: 286–356.

Piercy M, Hécaen H, Ajuriaguerra J (1960). Constructional apraxia associated with unilateral cerebral lesions-left and right sided cases compared. Brain 83: 225–242.

Pisella L, Alahyane A, Blangero A et al. (2011). Right-hemispheric dominance for visual remapping in humans. Philos Trans R Soc Lond B Biol Sci 366: 572–585.

Poppelreuter W (1917). Die Psychischen Schaedigungen durch Kopfschuss im Kriege 1914–1916, Band 1. Die Störungen der niederen und höheren Sehleitstungen durch Verletzungen des Okzipitalhirns. Voss, Leipzig.

Power E, Code C, Croot K et al. (2010). Florida Apraxia Battery–Extended and Revised Sydney (FABERS): design, description, and a healthy control sample. J Clin Exp Neuropsychol 32: 1–18.

Quencer K, Okun MS, Crucian G et al. (2007). Limb-kinetic apraxia in Parkinson disease. Neurology 68: 150–151.

Raimo S, Santangelo G, Trojano L (2021). The neural bases of drawing. A meta-analysis and a systematic literature review of neurofunctional studies in healthy individuals. Neuropsychol Rev 31: 689–702.

Rey A (1941). L'examen psychologique dans les cas d'encephalopathie traumatique. Arch Psychol 28: 215–285.

Rieger C (1909). Über Apparate in dem Hirn, Fisher, Jena.

Roncato S, Sartori G, Rumiati R (1987). Constructional apraxia: an information processing analysis. Cogn Neuropsychol 4: 113–129.

Rosselli M, Ardila A (2003). The impact of culture and education on non-verbal neuropsychological measurements: a critical review. Brain Cogn 52: 326–333.

Rothi LJG, Ochipa C, Heilman KM (1991). A cognitive neuropsychological model of limb praxis. Cogn Neuropsychol 8: 443–458.

Rothi LJG, Raymer AM, Heilman KM (1997). Limb praxis assessment. In: LJG Rothi, KM Heilman (Eds.), Apraxia: the neuropsychology of action. Brain damage, behaviour and cognition series. Psychology Press, Hove, UK.

Roy EA, Heath M, Westwood D et al. (2000). Task demands and limb apraxia in stroke. Brain Cogn 44: 253–279.

Russell C, Deidda C, Malhotra P et al. (2010). A deficit of spatial remapping in constructional apraxia after right-hemisphere stroke. Brain 133: 1239–1251.

Scandola M, Gobbetto V, Bertagnoli S et al. (2021). Gesture errors in left and right hemisphere damaged patients: a behavioural and anatomical study. Neuropsychologia 162: 108027.

Schmidt C, Achilles EIS, Fink GR et al. (2022). Distinct cognitive components and their neural substrates underlying praxis and language deficits following left hemisphere stroke. Cortex 146: 200–215.

Stamenova V, Roy EA, Black SE (2010). Associations and dissociations of transitive and intransitive gestures in left and right hemisphere stroke patients. Brain Cogn 72: 483–490.

Stamenova V, Black SE, Roy EA (2012). An update on the conceptual-production systems model of apraxia: evidence from stroke. Brain Cogn 80: 53–63.

Steinthal H (1881). Abriss der Sprachwissenschaft, Ferd. Dümmlers Verlagsbuchhandlung Harrwitz und Gossmann, Berlin.

Strauss H (1924). Über konstructive apraxie. Mschr Psychiatr 56: 65–124.

Tessari A, Mengotti P, Faccioli L et al. (2021). Effect of body-part specificity and meaning in gesture imitation in left hemisphere stroke patients. Neuropsychologia 151: 107720.

Trojano L (2020). Constructional apraxia from the roots up: Kleist, Strauss, and their contemporaries. Neurol Sci 41: 981–988.

Trojano L, Conson M (2008). Visuospatial and visuoconstructive deficits. Handb Clin Neurol 88: 373–391.

Trojano L, Gainotti G (2016). Drawing disorders in Alzheimer's disease and other forms of dementia. J Alzheimers Dis 53: 31–52.

Trojano L, Grossi D (1998). 'Pure' constructional apraxia—a cognitive analysis of a single case. Behav Neurol 11: 43–49. https://doi.org/10.1155/1998/614728.

Trojano L, Crisci C, Lanzillo B et al. (1993). How many alien hand syndromes? Follow-up of a case. Neurology 43: 2710–2712.

Trojano L, Fragassi NA, Chiacchio L et al. (2004). Relationships between constructional and visuospatial abilities in normal subjects and in focal brain-damaged patients. J Clin Exp Neuropsychol 26: 1103–1112.

Trojano L, Conson M, Maffei R et al. (2006). Categorical and coordinate spatial processing in the imagery domain investigated by rTMS. Neuropsychologia 44: 1569–1574.

Trojano L, Labruna L, Grossi D (2007). An experimental investigation of the automatic/voluntary dissociation in limb apraxia. Brain Cogn 65: 169–176.

Trojano L, Siciliano M, Pedone R et al. (2015). Italian normative data for the battery for visuospatial abilities (TERADIC). Neurol Sci 36: 1353–1361.

Trojano L, Siciliano M, Cristinzio C et al. (2018). Exploring visuospatial abilities and their contribution to constructional abilities and nonverbal intelligence. Appl Neuropsychol Adult 25: 166–173.

Van Sommers P (1989). A system for drawing and drawing-related neuropsychology. Cogn Neuropsychol 6: 117–164.

Vanbellingen T, Kersten B, Van de Winckel A et al. (2010a). A new bedside test of gestures in stroke: the apraxia screen of TULIA (AST). J Neurol Neurosurg Psychiatry 82: 389–392.

Vanbellingen T, Kersten B, Van Hemelrijk B et al. (2010b). Comprehensive assessment of gesture production: a new test of upper limb apraxia (TULIA). Eur J Neurol 17: 59–66.

Watson RT, Heilman KM (1983). Callosal apraxia. Brain 106: 391–403.

Wilson SAK (1908). A contribution to the study of apraxia with a review of the literature. Brain 31: 164–216.

Wolpe N, Hezemans FH, Rowe JB (2020). Alien limb syndrome: a Bayesian account of unwanted actions. Cortex 127: 29–41.

Yliranta A, Jehkonen M (2020). Limb and face apraxias in frontotemporal dementia: a systematic scoping review. Cortex 129: 529–547.

Youn YC, Pyun JM, Ryu N et al. (2021). Use of the clock drawing test and the Rey-Osterrieth complex figure test-copy with convolutional neural networks to predict cognitive impairment. Alzh Res Ther 13: 85.

Zadikoff C, Lang AE (2005). Apraxia in movement disorders. Brain 128: 1480–1497.

Zaidel DW (2018). Culture and art: importance of art practice, not aesthetics, to early human culture. Prog Brain Res 237: 25–40.

Section VIII

Other cognitive and perceptual function

Handbook of Clinical Neurology, Vol. 208 (3rd series)
Cerebral Asymmetries
C. Papagno and P. Corballis, Editors
https://doi.org/10.1016/B978-0-443-15646-5.00002-6

Chapter 26

The arts and hemispheric specialization

D.W. ZAIDEL*

Department of Psychology, University of California, Los Angeles (UCLA), Los Angeles, CA, United States

Abstract

Art was initially thought of as a single function linked mainly to spatial perception and right hemisphere functional specialization. Art was also considered to be diametrically opposed to language, further solidifying the right hemisphere specialization model. This view remained dominant for many decades. However, increase in published observations of artworks by professional artists following acquired unilateral hemispheric damage showed that quality artistic works can be produced when there is damage in either the left or the right hemisphere. With the advent of neuroimaging such as functional magnetic resonance imaging (fMRI), light was shed on the neural underpinning of the esthetics of artworks revealing activation in multiple regions across both hemispheres. The bulk of recent data suggests complementary hemispheric contributions to art production and esthetic evaluation of numerous art expressions. Similarly, creativity and imagination, upon which art expression depends, both appear to recruit interhemispheric processes. Culturally, the early evolutionary origin of art is associated mainly with *Homo sapiens* (HS) but, despite evidence for cerebral asymmetry based on fossil skulls, their emergence did not coincide with visual art making. A significant lag of hundreds of thousands of years intervened before humans produced art consistently and abundantly. By now, the arts are practiced ubiquitously throughout the globe, in all human societies.

INTRODUCTION

The term art covers a wide range of artistic expressions spanning multiple forms in the visual arts, music, dance, and literature. It is considered a uniquely human form of sociocultural communication between artist and viewer, ubiquitously present in all societies, and used as the symbolic representation of cultural identity.

The neural underpinning of art making and esthetics of artworks had been rather elusive before the 21st century. Up until then, the nature of art was largely discussed in philosophical and psychologic terms (with a few exceptions, Alajouanine, 1948; Gardner, 1984). Early assumptions viewed art as a single function, which, like language, was thought to be localized in a specific brain region. Initial understanding of the neural underpinning of art production and similarly, esthetic reactions to art was approached from the hemispheric specialization and functional localization perspectives. In keeping with the localization tradition of deducing function from neurologic patients with unilateral focal brain lesions, neurologists observed post-lesion artistic behavior of established professional artists. This formed the basis for long-lasting scientific thinking about the topic.

THE EARLY DOMINANT MODEL OF ART: HEMISPHERIC SPECIALIZATION AND FUNCTIONAL LOCALIZATION

Language specialization in the left hemisphere is the earliest example of documented functional asymmetry (Finger, 2001). Aphasia, a disorder of various aspects of language, follows damage to specific localized regions in the left hemisphere in most right-handed people. In 1948, the neurologist, Théophile Alajouanine, published the first influential report about art and brain where he described three well-known professional artists with aphasia (Alajouanine, 1948). Each represented a separate domain of art. The painter suffered a left-hemisphere stroke resulting in Wernicke-type aphasia but continued to paint relatively uninterrupted poststroke, expressing the same themes and same *genre* style; no alterations in his artwork were reported. The fact that the patient's painting activity was not curtailed by a left hemisphere lesion suggested at the time that the intact right hemisphere mediated the artistic expression. The second patient was a famous poet and novelist. He, too, had a left-sided stroke, which, based on the symptoms, at first presented as global aphasia and later

*Correspondence to: D. W. Zaidel, Department of Psychology, University of California, Los Angeles (UCLA), Los Angeles, CA, United States. E-mail: dahliaz@ucla.edu.

evolved into a Broca-type aphasia (Boller, 2005). Unlike the painter, he was unable to express his art poststroke, implying the left hemisphere plays an exclusive critical role in the literary arts, a natural conclusion at that time. The third patient was the famous musical composer (Maurice Ravel). Rather than a stroke, he is believed to have developed a type of neurodegenerative dementia with progressive aphasia as the most notable symptom; he did continue to compose a few pieces after symptoms onset, the composition Bolero being the best known from that phase, but ceased composing as additional motoric symptoms developed along with a wide range of disease-affected brain regions. (The absence of tissue biopsy remains a source of continued debate regarding diagnosis; Amaducci et al., 2002; Cavallera et al., 2012.)

At the same time, it was observed that in nonartists, the ability to draw common objects was preserved following unilateral left or right lesions. MacDonald Critchley described severe aphasia following left hemisphere stroke in a fashion designer artist (Critchley, 1953). His artistic skills were preserved poststroke but there was a qualitative difference in his illustrations; specifically, there was emphasis on the global contours rather than on the inside details (see Zaidel, 2005, for similar cases). Subsequently, the work of Warrington and James (1966) with nonartist groups of neurologic patients showed that the spontaneous drawings of familiar daily objects differed only qualitatively depending on the side of the lesion: Patients with left hemisphere lesions drew the overall gestalt, largely leaving out the details while those with right hemisphere damage drew the details, leaving out the general gestalt (see review, Kirk and Kertesz, 1993). Inferring function from lesion, this was interpreted to show that in the healthy brain, the left hemisphere specializes in focusing, remembering, and attending to details, and the right hemisphere specializes in overall gestalt and spatial perception. Further large studies of similar left–right lesioned neurologic cases failed to show qualitative differences in drawings (Gainotti et al., 1985). However, right hemisphere specialization in memory for gestalt configurations and left hemisphere attention to details does apply to perceptual and cognitive functions independently of art (McFie and Zangwill, 1960; Banich, 2009). Thus, performance in standard neuropsychologic tests does not automatically generalize to artistic skills.

NEUROPSYCHOLOGIC LOGIC OF HEMISPHERIC SPECIALIZATION IN ART

Both its creation by artists and its perception by viewers was initially thought to be specialized in the right hemisphere. The bases for this formulation were gleaned from nonartist neurologic patients with unilateral lesions and from commissurotomy patients (see Zaidel, 2013), and the assumption that nonlanguage cognition underpinned artistic endeavors in artist and viewer. This was the prevailing view for many decades, bolstered by the few, largely isolated cases of established artists described in the preceding paragraphs.

(a) Spatial perception was regarded as the principal skill in making and perceiving visual art. This notion was based on the common neuropsychologic clinical assessment consisting of failure to accurately copy on paper a drawn Necker cube following injury to the right hemisphere, particularly when localized in the parietal lobe (McFie et al., 1950), as well as inaccuracies in copying the cube with the right hand (left hemisphere control) but not with the left hand (right hemisphere control) of complete commissurotomy patients (Sperry, 1968; Bogen and Vogel, 1975). Selectively poor memory for the Rey-Osterrieth Complex Figure following right anterior temporal lobectomy (Taylor, 1969) supported the conclusion that spatial cognition is specialized in the right hemisphere (De Renzi, 1982). An example of this is the ability to transform perception of external reality unto a two-dimensional (2-D) surface (canvas, paper, wall), as illustrated in the converging lines in vanishing point depictions of depth emphasized in the Renaissance. This understanding generalized to inferring that the artistic juxtaposition of the parts in a composition, regardless of whether they were familiar objects (realism style), or dots, lines, curves, colors, and so on (abstract style), taps the same function of spatial perception (Zaidel, 2015).

Similar assumptions were made in the mental process of transforming the perception of the real world unto three-dimensional (3-D) constructions, as in sculptures. In the mind of the spectator of visual art, the relationship among the details of the artistic composition was assumed to be neurally supported by the same right hemisphere specialization. No attempt was made initially to differentiate between art making and spectator viewing.

Spatial cognition consists of a range of facets that cannot simply be assessed through cube copying or multiple 2-D tests of the cognition (Colby, 2009). The view of spatial perception described above and its role in art making was too narrow. Spatial knowledge, remembering geographic layouts, within personal (home) and extra-personal (outside the home) space, orienting and navigating applies to nonart activities. On the other hand, transforming real external depth in the mind and depicting it unto 2-D space (canvas, wall), organizing the relationship among details on a surface so that the emerging whole pattern ends up appealing to viewers seems to be applicable to artistic compositions. Similarly, transforming imagined ideas, emotions, and memory of experiences into understandable representations has both art and nonart applications.

(b) Art was viewed basically as a nonverbal function that did not recruit language cognition and thus was naturally inferred to be specialized by what was considered to be the nonlanguage hemisphere, namely the right. This logic was extended to viewers' cognitive and esthetic assessment of art as well (reviewed in Zaidel, 2015). Eventually, with more available published data, both of these initial assumptions proved incomplete in the face of subsequent observations of post-lesion professional artists as well as by the eventual use of neuroimaging studies in healthy participants.

THE ADVANCED PERSPECTIVE: CURRENT UNDERSTANDING OF HEMISPHERIC SPECIALIZATION AND ART

While right hemisphere specialization in spatial perception and cognition still holds, and can provide important clues to lesion localization in both artists and nonartists, multiple published findings and theoretical assumptions have progressed to question its major role in art making and spectator viewing. Professional artists with neurologic lesions, ranging in etiology from unilateral stroke, neurosurgery, and various types of neurodegenerative dementias, helped shed additional significant light on the neural support of art making since Alajouanine's (1948) article (see Jung, 1974; Rose, 2004; Bogousslavsky and Boller, 2005; Zaidel, 2005; Bäzner and Hennerici, 2007; Finger et al., 2013). This is described in the following paragraphs.

VISUAL ART

In reported cases of professional artists with right hemisphere lesions, there is continued artistic activity, despite the occasional presence of perceptual and motoric deficits, which, importantly, are observed in nonartists with similar lesions as well (Zaidel, 2005; Bäzner and Hennerici, 2007). The fact that the post-lesion symptoms do not distinguish between artists and nonartists indicates that they cannot simply be used as clues to the neural underpinning of art making. Hemi-neglect of the left half of external and personal space commonly following right parietal lobe stroke is an example of a perceptual symptom. In many cases, the symptom dissipates approximately 2–3 months post-damage. During that hemineglect phase, artistic abilities do not diminish; they can be observed in the right space of the painting surface. Two such instances are the German artists Anton Räderscheit (Gardner, 1974) and Lovis Corinth (Bäzner and Hennerici, 2007). Another German artist, Otto Dix, suffered a unilateral right hemisphere stroke and nevertheless went on to produce artworks that displayed pre-lesion artistic skills (McKiernan, 2014); they did not lose their artistry following the stroke (Jung, 1974). Similarly, professional artists with left hemisphere lesions, followed by aphasia, continued being engaged unhampered with their artistic works post-lesion (Zaimov et al., 1969).

With right-handed artists, after left-sided lesions there is the added issue of right-hand apraxia or paralysis, and yet these artists were able to execute their craft with the left hand. One would expect some graphic compensation resulting from sheer motoric control issue of holding the paint brush with the nondominant hand but this does not appear to interfere with artistic skill and cognition.

Visual professional artists diagnosed with dementia in old age similarly continue to produce artworks well into their illness, regardless of the disease's etiology. The initial stages of Alzheimer disease (AD) do not hamper painting and drawing nor are major alterations observed. Only with progression of the disease process, when there is increase in affected brain areas and loss of axonal connectivity, obvious changes in stylistic techniques rather than in artistic talent or skill per se become apparent (Maurer and Prvulovic, 2004; Fornazzari, 2005). Both spatial misproportions and narrow color choices can be seen, reflecting extensive damage to sensory, perceptual, and cognitive brain areas in both hemispheres. In the late phases of the disease, artists resort to simplification of patterns, increased use of rounded geometric patterns, making more ovals than squares, and applying a small range of colors (Lee et al., 2015).

Nonartists diagnosed with AD display visuospatial impairments in drawings (Kirk and Kertesz, 1991, 1993) even as the capacity to draw common objects and scenes remains intact (Miller and Johansson, 2016). With artists with AD, visuospatial distortions reflect the presence of right hemisphere disease damage, particularly in the posterior attention-controlling regions, while at the same time sparing object knowledge and memory, with the ability to coordinate the motoric execution demanded by the drawings. As the disease process progresses, the motoric ability to coordinate and execute diminishes substantially (Harrison, 2013).

This outcome is not unique to AD pathology; similar patterns can be observed in artworks of artists with frontotemporal dementia (Miller et al., 1996; Safar and Press, 2011), corticobasal degeneration (Kleiner-Fisman et al., 2003), and Lewy Body disease (Sahlas, 2003; Drago et al., 2006). Severe brain deterioration in multiple regions and worsening memory in the final stages of these dementias lead to form simplification and distortions of scenic space, the human body, and the face.

VIEWERS OF ARTWORKS: NEUROESTHETICS

Contrary to the early assumptions that the functionality of a single brain region underpins art esthetics, recent findings indicate that art esthetic reactions recruit multiple regions across both hemispheres and activate several interconnected pathways (Nadal, 2013). The typical method for measuring the reactions in the brain is with functional magnetic resonance imaging (fMRI). In many experiments, participants choose either beautiful, not beautiful, or neutral upon viewing pictures of works of art while their brain activity is mapped. Meta-analyses revealed consistent activity across widely distributed regions involving both hemispheres, thereby refuting the single-region-within-a-hemisphere assumption (Nadal, 2013; Vartanian and Skov, 2014; Vartanian, 2015; Boccia et al., 2016).

Moreover, detailed analysis in one meta-analysis has shown that maximally activated regions are not art-specific; they are active in nonart cognitive and perceptual tasks as well (Boccia et al., 2016). The subject matter of these artworks determined regional activation: Painted portraits maximally engaged activity in the right fusiform, the right inferior occipital gyrus, and the right amygdala. In nonart studies, the right fusiform and inferior occipital gyri are maximally active in face perception and recognition (Kanwisher, 2010), and the right amygdala is linked to emotional response to faces. Painted landscapes showed maximum activation in the right parahippocampal gyrus, right retrosplenial cortex, right middle temporal gyrus, and left lingual gyrus. Based on four patients with right parahippocampal lesions, impairments were found in the perception of nonart outdoor scenes (Habib and Sirigu, 1987; and see Aguirre et al., 1996).

Reactions to abstract art involved maximal activation in the posterior cingulate gyrus, bilaterally. In nonart studies, the posterior cingulate is active in tasks involving internally directed contemplation, retrieval of episodic and semantic memories, and conceptual thinking (see Maddock et al., 2001; den Ouden et al., 2005; Busler et al., 2019; Fernandino et al., 2022). The fact that this region was active during the esthetic evaluation of artworks suggests that conceptual assessment of art intersects with art esthetics. Taken together, findings from fMRI studies show that esthetics reactions to art draw on intellectual, perceptual, and emotional functions.

DRAWING

Structural MRI investigations of brain gray matter correlating drawings and size in the brains of students with art training to those without the training revealed a relationship between specific brain regions and drawing skills: gray matter area in the right medial frontal gyrus and the left anterior cerebellum correlated with drawing skills in nonart students (Chamberlain et al., 2014). In students with artistic training, the same correlation was found and there was the additional enhanced gray matter region in the right precuneus, a region localized in the parietal lobe (Schlegel et al., 2015). Functional neuroimaging (fMRI) of subjects drawing geometric figures with each hand revealed bilateral hemispheric activation while there was largely left hemisphere activation for writing (Potgieser et al., 2015).

Nonartist patients with aphasia communicate with reasonable clarity through simple line drawings (rarely of scenes), but the quality and organization of those drawings do not add up to artworks (Lyon, 1995; Sacchett, 2002; Sacchett and Lindsay, 2007; Hung and Ostergren, 2019). Similarly, complete commissurotomy patients studied at Caltech (Sperry, 1974; Bogen and Vogel, 1975) drew adequate human figures with the right hand (left hemisphere control) in free vision (not under lateralized conditions). They were nonartists drawing the figures with a pencil without observing a model or copying. The absence of interhemispheric communication in these cases, as well as absence of feedback from the right hemisphere, is consistent with the ability of the left hemisphere to mediate drawings of familiar daily objects. In the healthy brain, under conditions of intact interhemispheric communication, one would expect drawing to recruit skills controlled by both hemispheres.

Hyperschematia

Right hemisphere specialization in spatial cognition and orientation to the left half of space plays a role in attention to the left vs the right half of the drawing surface, influencing size and volume of the depicted objects. In drawings and paintings by nonartist neurologic patients with right hemisphere lesions, objects drawn or painted in the left half appear enlarged compared to the right half, contributing to left–right asymmetry in the composition. The exaggeration is attributed to excessive attention to that left side. This is known as hyperschematia, first described by French neurologist Pierre Bonnier (Vallar and Papagno, 2003). The asymmetry can be expressed in both 2-D and 3-D (Rode et al., 2006). Visual neglect (hemiinattention) phenomena are unrelated to this condition. Rode and colleagues noted that even healthy professional artists sometimes exhibit exaggerated emphasis of objects through enlargement in the left half, albeit as an unconscious artistic choice; they provide illustration of two works of the Cubist painter Braque as examples (Rode et al., 2018). Gaining insights and understanding for the neural underpinning of artistic choices will benefit from further observations of this tendency.

In explaining hyperschematia, a strong argument is that developmentally learned cultural habits, such as reading direction, play a role in left-half drawing enlargements in artists whose native language is read left-to-right, unlike habits originating with right-to-left reading direction as in Hebrew, Arabic, and so on (Girelli et al., 2017). Another argument concerns aging, specifically that increased age affects the balance of left–right attentional processes, with attention biased leftward (Learmonth and Papadatou-Pastou, 2022). Generalizing to artists with advanced age, left–right asymmetric attentional bias might emerge, directing more attention to the left half than the right of space. This, too, remains to be investigated further in future studies.

LEFT HEMISPHERE: COLOR VISION AND MEANING OF COLORS

Colors in artworks require deliberate choices and knowledge of objects. Deficits in color perception known as color blindness (color vision deficiency) can originate with impaired color-sensitive receptors in the retina, specifically the cones, but does not lessen the artistry of color-blind artists (Marmor, 2016). On the other hand, deficit in color perception arising from damage to posterior brain regions is known as achromatopsia. The condition does not impair vision, it impairs perception of colors. Patients with this condition report seeing the world in gray-scale. They know which colors normally go with familiar objects, but they are no longer able to access wavelength signals normally processed in the anterior occipital and posterior regions of the temporal lobes. One such artist described by the neurologist Oliver Sacks saw his world and art in shades of gray following a car accident, and at first included the altered perception of the world in his art but later, with time, started to add some colors (Lawley, 1996; Sacks, 1996). The lesion localization in his brain went undetermined.

The left lingual gyrus is associated with functions related to the meaning of colors, in knowing the significance of blue skies as opposed to gray skies, green as opposed to purple grass, ripe compared to unripe bananas, and so on (Meadows, 1974; Hsu et al., 2011; Oishi et al., 2021). Color agnosia is defined in terms of loss of knowledge of previously known color meaning, specifically the linking of colors with familiar objects (Kinsbourne and Warrington, 1964). An artist with this acquired condition described loss of the ability to use colors in his art, reporting that their meaning was lost to him (Holmes, 1945). He was not color blind nor did he have achromatopsia, rather he was unable to process the meaning and concepts of colors due to the damage. In the healthy brain, knowledge of colors includes their meaning, achieved by mapping the perception arising from wavelengths unto semantic color knowledge (Woodward et al., 1999). In creating art, artists incorporate this knowledge in their works to make the composition appealing to viewers, to attract them to consider the totality of the work.

LITERARY ARTS

One would naturally expect left hemisphere and aphasia to prevent further expressions by literary artists. Writing has always been regarded as an aspect of expressive language. Impairments, such as agraphia and dysgraphia, have been associated with left hemisphere damage. However, a striking dissociation between post-lesion language impairments and literary output has been observed following left-hemisphere lesions and aphasia (Iniesta, 2012, 2013). The latter deepened the mystery of how art is created and controlled by the brain.

A few such established writers that became aphasic following left-sided stroke have been described (Zaidel, 2005). The poet, Tomas Tranströmer, suffered a stroke in left frontal regions resulting in Broca-type aphasia and right-hand paralysis (Iniesta, 2013). Typically, patients with non-fluent aphasia display hesitant, slow speech with some anomia, and, in addition, also have grammatic deficits in expressing themselves and comprehending speech. His poststroke poetry did not reflect alterations in style, they continued to express conceptual thought, and were received with praise. Despite the location of the lesion, Tranströmer continued to write poems poststroke and was awarded the Nobel prize in Literature. Another established poet and writer, William Carlos Williams, suffered a series of left hemisphere strokes, which resulted in speech impairments (including inability to speak), anomia, and right arm paralysis. Yet, he continued his literary productivity for 11 years poststroke, his poems were well-received and no alterations were noted in style, structure, originality, and skill. A Pulitzer Prize was awarded to one of the poststroke poems. And another writer, Agnes de Mille, following a stroke in the left thalamus resulting in short-lived anomia and acalculia, whose artistic life was devoted to the dancing arts and choreography, was able to write several books poststroke (Iniesta, 2013). In total, contrary to what one would intuitively predict, aphasia due to left hemisphere damage does not necessarily eliminate the capacity to engage in literary (creative) writing, presumably because conceptual thinking and word knowledge regions remain accessible.

On the other hand, three other cases of well-known professionally established writers with left-sided stroke followed by aphasia ceased to write altogether (Vitturi and Gagliardi, 2021), as did Alajounine's patient (described in the opening paragraphs of this chapter). One is led to the conclusion that the extent of damage can explain the difference between the two groups. The damage in the second group might have been large and widely spread, involving regions subserving conceptual thinking, semantic ideation, focused attention, memory, and pathways accessing the mental lexicon. Although this scenario is highly likely judging from the symptoms, without brain imaging or postmortem examination the lesion extent cannot be ascertained.

An example of the effects of multiregional brain damage is the case of the novelist, Iris Murdoch, whose failed attempt to produce a readable, coherent novel at the end of her creative and productive life, is attributed to Alzheimer Disease (Garrard et al., 2005). In the disease, diffused areas in the brain, bilaterally, are affected by neuronal loss accompanied by impaired axonal connectivity. Critically, memory deterioration due to hippocampal damage would have interfered with organization of the story elements and coherent composition of the story.

MUSICAL ARTS

Post-lesion musical compositions have been reported for established professional composers with hemispheric lesions due to stroke or to other conditions. Unlike with visual artists, the number of such cases is small. With approximately10 published cases, what is known is that 7 composers with left hemisphere lesions continued to write well-accepted musical pieces, of the three with right hemisphere lesions, 2 did go on to compose, and 1 did not (Zaidel, 2005). The full extent of the lesion in each one of the composers is not known. Mazzucchi et al. (2017) studied the post-stroke musical compositions by 4 of the left hemisphere stroke composers and observed some changes in style, structure, and harmony. These post-lesion works demonstrate that the ability to create compositions resembling pre-stroke musical level relies on both left and right hemispheric functions. (One case with dementia of unknown etiology, Maurice Ravel, was described in the introductory paragraph in this chapter.)

Music and language share acoustic properties with asymmetric functions matching hemispheric specialization (Zatorre, 2022). Musical abilities were thought to be specialized in the right hemisphere following the early published studies of anterior temporal lobectomy patients from the Montreal Neurological Institute (Milner, 1962). Doreen Kimura confirmed this conclusion when her results with the dichotic listening method showed that healthy subjects recognized melodies better in the left than in the right ear (Kimura, 1964). Kimura did caution, however, that not all musical abilities are represented in melody recognition. Music was nevertheless assumed to represent a nonverbal domain, that is, not depending on linguistic abilities, and despite paucity of observations of post-lesion musical composers, the Milner and Kimura findings formed the basis for conceiving of the right hemisphere as specializing in music.

Since then, multiple studies revealed that music is not a unitary function, but rather consists of multiple components with asymmetric hemispheric representations (Janata, 2015). The primary auditory cortex is in the superior temporal gyrus as is Heschel's gyrus, in both hemispheres. The left primary auditory cortex is larger and has a greater gray and white matter than the right; there is asymmetry in cytoarchitectonic organization as well (Dorsaint-Pierre et al., 2006).

The temporal lobes are largely viewed as the principal areas for music processing. Some musical components were found to be lateralized to the left hemisphere and others to the right hemisphere: Perception of timing, intervals, and rhythm are specialized by the left while pitch, timbre, and melodic contour recognition are specialized in the right hemisphere (Warrier and Zatorre, 2004). However, studies have also shown that global musical contour processing can include information about timing and intervals and neurologic patients with right hemisphere lesions can be impaired in both components (Liégeois-Chauvel et al., 1998).

Neuroanatomic connectivity, specifically with regard to fine differences along spectral frequencies, explains specialization for pitch sensitivity in the right temporal lobe (Zatorre, 2022). Functional axonal connectivity in the left hemisphere is linked to the reliance of language on temporal components rather than on pitch (Giroud et al., 2020).

CREATIVITY AND ART

The general definition of creativity in all fields of human endeavor regards it as something new that did not exist previously, imparting positive value to society (Zaidel, 2014). Creative ideas in art are innovations, extensions of previously established knowledge, and shaped by socially based factors (cultural trends, political, religious, and so on). The well-known Renaissance artists were influenced by the realism of ancient Greek artists. Some of the artworks of Van Gogh, Degas, Monet, Picasso, and many more artists from the late 1800s and early 1900s were influenced by Japanese and African art (Van Tilborgh et al., 2018). Their artistic creativity is the innovative incorporation and blending of that influence into their own new style.

In 1988, neurosurgeon J. E. Bogen suggested a significant role for the corpus callosum in interhemispheric integration in the process of creativity (Bogen and Bogen, 1988). The brain's neural circuitry giving rise to creative ideas is currently not fully understood (Benedek and Fink, 2019; Ovando-Tellez et al., 2022). Knowledge, memory, attention, and imagination together with skill and personality play a major role in creativity. No one hemisphere has thus far been reliably identified with "creativity specialization." fMRI studies suggest a major involvement of specific areas in the left hemisphere as well areas in the right hemisphere in creativity. A meta-analysis of 45 studies of idea generation found multiple active regions in both the left and right hemispheres, although more pronounced activation was seen in left hemisphere regions (Kleibeuker et al., 2013).

Memory processes contribute significantly to creativity. Memory consists both of consolidation of sensory input and of retrieval from long-term memory. It plays an important role in linking associations in stored semantic knowledge and in highly rehearsed skills involving expertise; it is a critical link in the functional circuitry underpinning creative art expressions. Artistic decisions draw on knowledge of the world. Remembering ideas and past established and highly rehearsed styles make art creation possible. Neuroanatomically, parietal regions of the precuneus and posterior cingulate gyrus are associated with conceptual thinking, in both hemispheres, and have rich axonal connections to the hippocampus. One would expect that, under conditions where the hippocampus and its projections to the rest of the brain are intact, retrieval of knowledge stored in long-term memory would be functional (i.e., not damaged) and would facilitate creative writing, music composing, painting, drawing, and so on.

HEMISPHERIC ASYMMETRIES: IMAGINATION AND IMAGERY IN THE MIND OF THE ARTIST

Conceiving a plan in the artist's mind for a piece of artwork recruits brain regions giving rise to imagination and imagery, two essential components of artistic compositions. Imagination refers to the intellectual initiation in the mind of the conceptual idea for the composition while visual mental imagery refers to the mind's eye, to visualizing mentally the individual components not externally visually present, that is, retrieving the image from memory to enable bringing the composition to fruition. Originally, researchers assumed from clinical neuropsychologic tests that spatial perception was the same as generating visual mental images, and that the right hemisphere specialized in the generation. However, based on neuropsychologic and electrophysiologic observations Farah proposed (Farah et al., 1988; Farah, 1989) that the formation of the image itself in the mind's eye (retrieving it from memory) is supported by posterior regions in the left inferior occipital gyrus. Corballis and Sergent (1988) studied this issue in a complete commissurotomy case at Caltech, Pasadena, and confirmed the separation between forming mental visual images with the left hemisphere and spatial localization with the right hemisphere. Recent neuroimaging studies using fMRI argue for the critical involvement of the left hemisphere in formation of mental imagery, specifically "frontoparietal networks initiate, modulate, and maintain activity in a core temporal [lobe] network centered on the fusiform imagery node, a high-level visual region in the left fusiform gyrus." (Spagna et al., 2021), abstract and see Spagna (2022) for a review.

The neural underpinning of imagination, as defined above, has been difficult to tease apart mostly because imagination involves multiple functions. Imagination reflects, at the very least, memory retrieval from long-term semantic and episodic memory, hypothesis testing, reconstruction of existing knowledge, problem solving, working memory, and processes involved with creativity (Abraham, 2016). Furthermore, imagination does not necessarily co-occur with visual mental imagery, as evidenced by mathematicians and some successful scientists, who describe being able to solve problems and achieve creative solutions while not being able to mentally visualize the solution (Zaidel, 2020). Functional specialization in the left and right hemispheres could be involved in the generation of intellectual conceptual ideas, integrated via interhemispheric transfer, with no valid reason (as far as we know) to assume a difference in this regard between artists and nonartists.

ART'S EVOLUTIONARY BACKGROUND: LANGUAGE LATERALITY, HANDEDNESS, AND ARTMAKING

To fully understand the practice of art in human societies, we need to consider its evolutionary purpose. When did humans start making art and why? What was its purpose? The early origin of art making is tied to cultural group identification (Lewis-Williams, 2002; Zaidel, 2017). Consistent, abundant, and varied art is mainly associated with HS. Judging from the fossil record available now, the earliest HS emerged about ~315,000 years ago in Africa (Hublin et al., 2017; Richter et al., 2017; Schlebusch et al., 2017).

Bone skull endocasts reveal structural asymmetries that imply functional language and cognitive lateralization in early HS (Holloway et al., 2004, 117; Balzeau et al., 2012; Neubauer et al., 2020). There is a strong correlation between cerebral asymmetry and handedness in humans, whereby left hemisphere specialization for language is linked to right-handedness (Corballis, 2009, 2019). Significantly, unlike humans' close primate relatives, the great majority of humans throughout the world are right-handed. The predominant use of the right hand is assumed to have gradually evolved in the common hominin ancestor of HS and the Neanderthals based on evidence from stone tool knapping and dental markings (Bermúdez de Castro et al., 1988; Uomini, 2009; Frayer et al., 2016; and see, Fitch and Braccini, 2013), alongside the neural support of language functions.

Human communication systems recruit underlying semantic cognition not exclusively specialized for deriving meaning from language. Cultural rules, social norms, and symbols tap into meaning systems as well (Zaidel, 2019). Those systems recruit symbolic and abstract cognition, that is, they rely on attaining meaning agreed upon by the social group (Deacon, 1998, 2011). The biologic roots of language have a long evolutionary history, expressed through vocalization and body signals (Corballis, 2017), and can be traced to brain regional markers in nonhuman primates going back many millions of years (Aboitiz and Garcia, 1997; Aboitiz, 2017). Visual art, on the other hand, is practiced only in human cultures and appears to have emerged much after vocal language (Zaidel, 2019).

Conceptual thinking capability is considered to have coevolved with the cognition-supporting language (Marques et al., 2008; Schoenemann, 2012), but not restricted exclusively to language operations. Evidence for semantic concept formation has been found in both the left and right hemispheres (Zaidel, 2000; Zhang et al., 2020). Critically, concepts symbolically representing group belongingness identity are linked to art's evolutionary adaptive emergence in human culture, their function utilized and aimed at promoting group cohesiveness, which, in turn, optimized survival of early socially oriented existence (Zaidel, 2017, 2018). Tracing this early beginning, archeologists discovered in multiple HS camp sites in Africa, the currently presumed region in the world where HS evolved, increasingly advanced manufacture of stone tools (McBrearty and Brooks, 2000). The tools became refined, fashioned not only from different stone material, at times also from other sources (McBrearty and Brooks, 2000; Stout, 2011) attesting to intelligence and successful survival strategies (Brooks et al., 2018). And, yet, no physical evidence for visual art appeared in Africa around the time of HS emergence and not for over 150,000 years afterwards, even as stone techniques and their refinements showed advances. One would have expected the finger dexterity and bimanual hand control exercised in tool making to lead to engravings, sculpting, and stone etchings but this clearly did not happen for an exceptionally long stretch of time.

However, given evolutionary adaptive reliance on socially oriented group life, it is possible that HS nevertheless engaged in art activities inherently lacking physical archeological traces such as body paintings, communal dancing, and group chorus singing (Zaidel, 2017, 2018, 2019). These art formats would have symbolically promoted the concept of group unity and cohesiveness so critical for survival and could well have predated the physical artifacts of art.

The earliest art-related visual artifacts found in Africa were perforated seashell beads with evidence suggesting they were strung together into a type of a necklace and possibly served as visual image of identity; they were discovered in Blombos cave, South Africa, and dated to around 77,000 years ago (Henshilwood, 2007; Henshilwood et al., 2011). Evidence for what is taken as symbolic cognition was also found in the cave, dated to around 100,000 years ago, in the form of differently pigmented heated ochre clay resulting in assorted colors (Henshilwood et al., 2011). The consistency of the colors resulting from heating could have aided in smearing deliberate paints on the body, a likely practice given unearthed artifacts suggesting sophisticated cognition prior to the seashell beads (McBrearty and Brooks, 2000; McBrearty and Stringer, 2007; Mellars et al., 2007).

While a trickle of visual art commenced in Africa, human art-making advanced to regular practice around ~45,000–35,000 years ago but only after groups of HS migrated away from Africa; it became consistent and abundant in Western Europe (Bar-Yosef, 2002; McBrearty, 2013) and recently, found in cave wall paintings in Indonesia, dated to 45,000 years ago (Aubert et al., 2019). Today, visual art, music, dance, literature, storytelling, and theater are the broad categories practiced in both modern and traditional societies. Art has become an integral part of human cultural identity.

SUMMARY AND CONCLUSIONS

The early working hypothesis for a neural basis for art making and art esthetics was led off by the hemispheric specialization and functional laterality approach. Two main notions dominated the thinking about the neural basis, right hemisphere specialization for spatial perception and observations of neurologic cases of established artists. Clinical neuropsychologic tests of spatial perception (poor copying of Necker cube and other tests) led to the conclusion that functional specialization in art is lateralized to the right hemisphere. Supported by complete commissurotomy findings, art was viewed as a nonlanguage function, since the main expressive and receptive language centers were found lateralized to the left hemisphere. Music, too, was viewed as a nonlanguage function and, based on little evidence, was "relegated" to the right hemisphere. However, studies of spatial perception have shown that it is not a unitary function, that each hemisphere contributes selectively to the cognition. It, thus, became obvious that spatial perception cannot singly play a central role in art making.

Mentally visualizing external space is useful for realistic depictions and does not exclusively activate right hemisphere regions.

Eventually, increase in published observations of established professional artists with neurologic lesions modified and expanded the early assumptions. Knowledge of artistic skills appears to be stored bilaterally and can be tapped following unilateral damage in either hemisphere. The great majority of reported established artists continue to produce their skillful art regardless of lesion laterality. In general, no striking alterations in style or genre that can be attributed to brain neural damage in the artists have been noted. In dementia cases of artists, of various etiologies, late disease stages are associated with characteristic form and content alterations.

Functional asymmetry and right-handedness are inferred from early HS skull fossil findings and stone tool knapping direction. Yet, the early origins of HS do not coincide with archeological evidence for visual art (the only physical form with traces). Rather, a lag of many hundreds of thousands of years intervened before signs of the art became evident. Brain capability for creating it may nevertheless have been present and cultural norms and survival strategies could have initially favored nonphysical art forms (body painting, communal dance, and singing), useful for expressions of unity and harmony concepts. Conceptual thinking is one of the hallmarks of human cognition. The ubiquitous worldwide presence of art making and art esthetics shows that art evolved into a system of sociocultural communication recruiting functions subserved by multiple brain regions in both cerebral hemispheres.

References

Aboitiz F (2017). A brain for speech: a view from evolutionary neuroanatomy, Pelgrave Macmillan.

Aboitiz F, Garcia VR (1997). The evolutionary origin of the language areas in the human brain. A neuroanatomical perspective. Brain Res Rev 25: 381–396.

Abraham A (2016). The imaginative mind. Hum Brain Mapp 37: 4197–4211.

Aguirre GK, Detre JA, Alsop DC et al. (1996). The parahippocampus subserves topographical learning in man. Cereb Cortex 6: 823–829.

Alajouanine T (1948). Aphasia and artistic realization. Brain 71: 229–241.

Amaducci L, Grassi E, Boller F (2002). Maurice ravel and right-hemisphere musical creativity: influence of disease on his last musical works? Eur J Neurol 9: 75–82.

Aubert M, Lebe R, Oktaviana AA et al. (2019). Earliest hunting scene in prehistoric art. Nature 576: 442–445.

Balzeau A, Gilissen E, Grimaud-Hervé D (2012). Shared pattern of endocranial shape asymmetries among great apes, anatomically modern humans, and fossil hominins. PLoS One 7: e29581.

Banich MT (2009). Hemispheric specialization and cognition. In: LR Squire (Ed.), Encyclopedia of neuroscience. Academic Press, Oxford.

Bar-Yosef O (2002). The upper paleolithic revolution. Ann Rev Anthropol 31: 363–393.

Bäzner H, Hennerici MG (2007). Painting after right-hemisphere stroke-case studies of professional artists. Front Neurol Neurosci 22: 1–13.

Benedek M, Fink A (2019). Toward a neurocognitive framework of creative cognition: the role of memory, attention, and cognitive control. Curr Opin Behav Sci 27: 116–122.

Bermúdez de Castro J, Bromage TG, Jalvo YF (1988). Buccal striations on fossil human anterior teeth: evidence of handedness in the middle and early upper Pleistocene. J Hum Evol 17: 403–412.

Boccia M, Barbetti S, Piccardi L et al. (2016). Where does brain neural activation in aesthetic responses to visual art occur? Meta-analytic evidence from neuroimaging studies. Neurosci Biobehav Rev 60: 65–71.

Bogen JE, Bogen GM (1988). Creativity and the corpus callosum. Psychiatr Clin North Am 11: 293–301.

Bogen JE, Vogel PJ (1975). Neurologic status in the long term following complete cerebral commissurotomy. In: F Michel, B Schott (Eds.), Les syndromes de disconnexion calleuse chez l'home. Lyon, Hospital Neurologie.

Bogousslavsky J, Boller F (Eds.), (2005). Neurological disorders in famous artists. Karger, Basel.

Boller F (2005). Valery Larbaud. In: J Bogousslavsky, F Boller (Eds.), Neurological disorders in famous artists. Karger, Basel.

Brooks AS, Yellen JE, Potts R et al. (2018). Long-distance stone transport and pigment use in the earliest middle stone age. Science 360: 90–94.

Busler JN, Yanes JA, Bird RT et al. (2019). Differential functional patterns of the human posterior cingulate cortex during activation and deactivation: a meta-analytic connectivity model. Exp Brain Res 237: 2367–2385.

Cavallera GM, Giudici S, Tommasi L (2012). Shadows and darkness in the brain of a genius: aspects of the neuropsychological literature about the final illness of Maurice ravel (1875-1937). Med Sci Monit 18: Mh1-8.

Chamberlain R, McManus IC, Brunswick N et al. (2014). Drawing on the right side of the brain: a voxel-based morphometry analysis of observational drawing. Neuroimage 96: 167–173.

Colby CL (2009). Spatial cognition. In: LR Squire (Ed.), Encyclopedia of neuroscience. Academic Press, Oxford.

Corballis MC (2009). The evolution and genetics of cerebral asymmetry. Philos Trans R Soc Lond B Biol Sci 364: 867–879.

Corballis MC (2017). Language evolution: a changing perspective. Trends Cogn Sci 21: 229–236.

Corballis MC (2019). Evolution of cerebral asymmetry. Prog Brain Res 250: 153–178.

Corballis MC, Sergent J (1988). Imagery in a commissurotomized patient. Neuropsychologia 26: 13–26.

Critchley M (1953). The parietal lobes, Hafner, New York.

De Renzi E (1982). Disorders of space exploration and cognition, Wiley, New York.

Deacon TW (1998). The symbolic species: the co-evolution of language and the brain, W. W. Norton and Company, New York.

Deacon TW (2011). The symbol concept. In: M Tallerman, K Gibson (Eds.), Oxford handbook of language evolution. Oxford University Press, Oxford.

den Ouden HE, Frith U, Frith C et al. (2005). Thinking about intentions. Neuroimage 28: 787–796.

Dorsaint-Pierre R, Penhune VB, Watkins KE et al. (2006). Asymmetries of the planum temporale and Heschl's gyrus: relationship to language lateralization. Brain 129: 1164–1176.

Drago V, Crucian GP, Foster PS et al. (2006). Lewy body dementia and creativity: case report. Neuropsychologia 44: 3011–3015.

Farah MJ (1989). The neural basis of mental imagery. Trends Neurosci 12: 395–399.

Farah MJ, Hammond KM, Levine DN et al. (1988). Visual and spatial mental imagery: dissociable systems of representation. Cogn Psychol 20: 439–462.

Fernandino L, Tong JQ, Conant LL et al. (2022). Decoding the information structure underlying the neural representation of concepts. Proc Natl Acad Sci U S A 119.

Finger S (2001). Origins of neuroscience: a history of explorations into brain function, Oxford University Press, Oxford.

Finger S, Zaidel DW, Boller F et al. (Eds.), (2013). The fine arts, neurology, and neuroscience: history and modern perspectives: new discoveries and changing landscapes. Elsevier, Oxford.

Fitch WT, Braccini SN (2013). Primate laterality and the biology and evolution of human handedness: a review and synthesis. Ann N Y Acad Sci 1288: 70–85.

Fornazzari LR (2005). Preserved painting creativity in an artist with Alzheimer's disease. Eur J Neurol 12: 419–424.

Frayer DW, Clarke RJ, Fiore I et al. (2016). OH-65: the earliest evidence for right-handedness in the fossil record. J Hum Evol 100: 65–72.

Gainotti G, D'Erme P, Diodato S (1985). Are drawing errors different in right-sided and left-sided constructional apraxics? Ital J Neurol Sci 6: 495–501.

Gardner H (1974). The shattered mind, Vintage, New York.

Gardner H (1984). Art, mind, and brain, Basic Books, New York.

Garrard P, Maloney LM, Hodges JR et al. (2005). The effects of very early Alzheimer's disease on the characteristics of writing by a renowned author. Brain 128: 250–260.

Girelli L, Marinelli C, Grossi G et al. (2017). Cultural and biological factors modulate spatial biases over development. Laterality 22: 1–15.

Giroud J, Trébuchon A, Schön D et al. (2020). Asymmetric sampling in human auditory cortex reveals spectral processing hierarchy. PLoS Biol 18: e3000207.

Habib M, Sirigu A (1987). Pure topographical disorientation: a definition and anatomical basis. Cortex 23: 73–85.

Harrison EM (2013). Understanding suffering: Utermohlen's self-portraits and Alzheimer's disease. Nurse Educ 38: 20–25.

Henshilwood CS (2007). Fully symbolic sapiens behaviour: innovation in the Middle Stone Age at Blombos Cave, South Africa. In: C Stringer, P Mellars (Eds.), Rethinking the human revolution: new behavioural and biological perspectives on the origins and dispersal of modern humans. Cambridge University Press, Cambridge.

Henshilwood CS, d'Errico F, van Niekerk KL et al. (2011). A 100,000-year-old ochre-processing workshop at Blombos cave, South Africa. Science 334: 219–222.

Holloway RL, Broadfield DC, Yuan MS (2004). The human fossil record: brain endocasts, paleoneurological evidence, Wiley-Liss, New York.

Holmes G (1945). The organization of the visual cortex in man. Proc R Soc Lond B Biol Sci 132: 348–361.

Hsu NS, Kraemer DJ, Oliver RT et al. (2011). Color, context, and cognitive style: variations in color knowledge retrieval as a function of task and subject variables. J Cogn Neurosci 23: 2544–2557.

Hublin J-J, Ben-Ncer A, Bailey SE et al. (2017). New fossils from Jebel Irhoud, Morocco and the pan-African origin of Homo sapiens. Nature 546: 289–292.

Hung P-F, Ostergren J (2019). A comparison of drawing and writing on facilitating word retrieval in individuals with aphasia. Aphasiology 33: 1462–1481.

Iniesta I (2012). The Iatroversalia (doctor poems) of William Carlos Williams. Clin Med (Lond) 12: 92–93.

Iniesta I (2013). Tomas Tranströmer's stroke of genius. Clin Med (Lond) 13: 109–110.

Janata P (2015). Neural basis of music perception. Handb Clin Neurol 129: 187–205.

Jung R (1974). Neuropsychologie und neurophysiologie des kontur- und formschens in zeichnung und malerie. In: HH Wieck (Ed.), Psychopathologie musischer Gestaltungen. FK Schattauer, Stuttgart.

Kanwisher N (2010). Functional specificity in the human brain: a window into the functional architecture of the mind. Proc Natl Acad Sci U S A 107: 11163–11170.

Kimura D (1964). Left-right differences in the perception of melodies. Q J Exp Psychol 16: 355–358.

Kinsbourne M, Warrington EK (1964). Observations on colour agnosia. J Neurol Neurosurg Psychiatry 27: 296–299.

Kirk A, Kertesz A (1991). On drawing impairment in Alzheimer's disease. Arch Neurol 48: 73–77.

Kirk A, Kertesz A (1993). Subcortical contributions to drawing. Brain Cogn 21: 57–70.

Kleibeuker S, Koolschijn PC, Jolles D et al. (2013). The neural coding of creative idea generation across adolescence and early adulthood. Front Hum Neurosci 7: 905–917.

Kleiner-Fisman G, Black SE, Lang AE (2003). Neurodegenerative disease and the evolution of art: the effects of presumed corticobasal degeneration in a professional artist. Mov Disord 18: 294–302.

Lawley J (1996). Vision is not black and white: the colourless case of Mr I, Rapport.

Learmonth G, Papadatou-Pastou M (2022). A meta-analysis of line bisection and landmark task performance in older adults. Neuropsychol Rev 32: 438–457.

Lee Y-T, Tsai Y-C, Chen L-K (2015). Progressive changes in artistic performance of a Chinese master with Alzheimer's disease. Cortex 69: 282–283.

Lewis-Williams D (2002). The mind in the cave: consciousness and the origins of art, Thames and Hudson, London.

Liégeois-Chauvel C, Peretz I, Babaï M et al. (1998). Contribution of different cortical areas in the temporal lobes to music processing. Brain 121: 1853–1867.

Lyon JG (1995). Communicative drawing: an augmentative mode of interaction. Aphasiology 9: 84–94.

Maddock RJ, Garrett AS, Buonocore MH (2001). Remembering familiar people: the posterior cingulate cortex and autobiographical memory retrieval. Neuroscience 104: 667–676.

Marmor MF (2016). Vision, eye disease, and art. Eye (Lond) 30: 287–303.

Marques JF, Canessa N, Siri S et al. (2008). Conceptual knowledge in the brain: fMRI evidence for a featural organization. Brain Res 1194: 90–99.

Maurer K, Prvulovic D (2004). Paintings of an artist with Alzheimer's disease: visuoconstructural deficits during dementia. J Neural Transm 111: 235–245.

Mazzucchi A, Fanticini F, Bellocchio MG et al. (2017). The influence of brain lesion on musical masterpieces of famous composers. J Neurol Neurosci 8: 236–248.

McBrearty S (2013). Advances in the study of the origin of humanness. J Anthropol Res 69: 7–31.

McBrearty S, Brooks AS (2000). The revolution that wasn't: a new interpretation of the origin of modern human behavior. J Hum Evol 39: 453–563.

McBrearty S, Stringer C (2007). The coast in colour. Nature 449: 793.

McFie J, Zangwill OL (1960). Visual-constructive disabilities associated with lesions of the left cerebral hemisphere. Brain 83: 243–260.

McFie J, Piercy MF, Zangwill OL (1950). Visual spatial agnosia association with lesions of the right cerebral hemisphere. Brain 73: 167–190.

McKiernan M (2014). Otto dix, the match seller 1920. Occup Med 64: 148–149.

Meadows JC (1974). Disturbed perception of colors associated with localised cerebral lesions. Brain 97: 615–632.

Mellars P, Boyle K, Bar-Yosef O et al. (2007). Rethinking the human revolution: new behavioural and biological perspectives on the origin and dispersal of modern humans, McDonald Institute for Archaeological Research, Cambridge.

Miller E, Johansson B (2016). Capability to paint and Alzheimer's disease: relationship to disease stages and instructions. SAGE Open 6. https://doi.org/10.1177/2158244016631799.

Miller BL, Ponton M, Benson FD et al. (1996). Enhanced artistic creativity with temporal lobe degeneration. Lancet 348: 1744–1745.

Milner B (1962). Laterality effects in audition. In: VB Mountcastle (Ed.), Interhemispheric relations and cerebral dominance. Johns Hopkins University Press, Baltimore.

Nadal M (2013). The experience of art: insights from neuroimaging. Prog Brain Res 204: 135–158.

Neubauer S, Gunz P, Scott NA et al. (2020). Evolution of brain lateralization: a shared hominid pattern of endocranial asymmetry is much more variable in humans than in great apes. Sci Adv 6: eaax9935.

Oishi Y, Nagasawa H, Hirayama K et al. (2021). Neural bases of color-specific semantic loss: two cases of object-color knowledge impairment. Cortex 141: 211–223.

Ovando-Tellez M, Benedek M, Kenett YN et al. (2022). An investigation of the cognitive and neural correlates of semantic memory search related to creative ability. Commun Biol 5: 604.

Potgieser ARE, van der Hoorn A, de Jong BM (2015). Cerebral activations related to writing and drawing with each hand. PLoS One 10: e0126723.

Richter D, Grün R, Joannes-Boyau R et al. (2017). The age of the hominin fossils from Jebel Irhoud, Morocco, and the origins of the middle stone age. Nature 546: 293–296.

Rode G, Michel C, Rossetti Y et al. (2006). Left size distortion (hyperschematia) after right brain damage. Neurology 67: 1801–1808.

Rode G, Vallar G, Chabanat E et al. (2018). What do spatial distortions in Patients' drawing after right brain damage teach us about space representation in art? Front Psychol 9: 1058.

Rose FC (Ed.), (2004). Neurology of the arts: painting, music, literature. Imperial College Press, London.

Sacchett C (2002). Drawing in aphasia: moving towards the interactive. Int J Hum Comput Stud 57: 263–277.

Sacchett C, Lindsay J (2007). Revealing competence and rethinking identity in severe aphasia using drawing and a communication book, Psychology Press.

Sacks O (1996). An anthropologist on Mars, Vintage, NY.

Safar L, Press DZ (2011). Art and the brain: effects of dementia on art production in art therapy. Art Ther 28: 96–103.

Sahlas DJ (2003). Dementia with Lewy bodies and the neurobehavioral decline of Mervyn Peake. Arch Neurol 60: 889–892.

Schlebusch CM, Malmström H, Günther T et al. (2017). Southern African ancient genomes estimate modern human divergence to 350,000 to 260,000 years ago. Science 358: 652–655.

Schlegel A, Alexander P, Fogelson SV et al. (2015). The artist emerges: visual art learning alters neural structure and function. Neuroimage 105: 440–451.

Schoenemann PT (2012). Evolution of brain and language. Prog Brain Res 195: 443–459.

Spagna A (2022). Chapter 8: Visual mental imagery: Inside the mind's eyes. In: G Miceli, P Bartolomeo, V Navarro (Eds.), Handbook of clinical neurology. Elsevier.

Spagna A, Hajhajate D, Liu J et al. (2021). Visual mental imagery engages the left fusiform gyrus, but not the early visual cortex: a meta-analysis of neuroimaging evidence. Neurosci Biobehav Rev 122: 201–217.

Sperry RW (1968). Hemisphere deconnection and unity in conscious awareness. Am Psychol 23: 723–733.

Sperry RW (1974). Lateral specialization in the surgically separated hemispheres. In: FO Schmitt, FG Worden (Eds.), Neuroscience third study program. MIT Press, Boston.

Stout D (2011). Stone toolmaking and the evolution of human culture and cognition. Philos Trans R Soc Lond B Biol Sci 366: 1050–1059.

Taylor LB (1969). Localisation of cerebral lesions by psychological testing. Clin Neurosurg 16: 269–287.

Uomini NT (2009). The prehistory of handedness: archaeological data and comparative ethology. J Hum Evol 57: 411–419.

Vallar G, Papagno C (2003). In: C Code, C Wallesch, Y Joanette et al. (Eds.), Classic cases in neuropsychology. Psychology Press, Hove.

Van Tilborgh L, Bakker N, Homburg C et al. (2018). Van Gogh and Japan, Mercatorfonds, Amsterdam.

Vartanian O (2015). Neuroimaging studies of making aesthetic products. In: JP Huston, M Nadal, F Mora et al. (Eds.), Art, aesthetics, and the brain. Oxford University Press, Oxford.

Vartanian O, Skov M (2014). Neural correlates of viewing paintings: evidence from a quantitative meta-analysis of functional magnetic resonance imaging data. Brain Cogn 87: 52–56.

Vitturi BK, Gagliardi RJ (2021). Post-stroke aphasia in famous writers: when neurology left geniuses speechless. Arq Neuropsiquiatr 79: 251–253.

Warrier CM, Zatorre RJ (2004). Right temporal cortex is critical for utilization of melodic contextual cues in a pitch constancy task. Brain 127: 1616–1625.

Warrington EK, James M (1966). Drawing disability in relation to laterality of cerebral lesion. Brain 89: 53–82.

Woodward T, Dixon M, Mullen K et al. (1999). Analysis of errors in color agnosia: a single-case study. Neurocase 5: 95–108.

Zaidel DW (2020). Imagination, symbolic cognition, and human evolution: The early arts facilitated group survival. In: J Carroll, M Clasen, E Jonsson (Eds.), Evolutionary perspectives on imaginative culture. Switzerland, Springer Nature.

Zaidel DW (2000). Different concepts and meaning systems in the left and right hemispheres. Psychol Learn Motiv 40: 1–21.

Zaidel DW (2005). Neuropsychology of art: neurological, cognitive, and evolutionary perspectives, Psychology Press, UK.

Zaidel DW (2013). Split-brain, the right hemisphere, and art: fact and fiction. Prog Brain Res 204: 3–17.

Zaidel DW (2014). Creativity, brain, and art: biological and neurological considerations. Front Hum Neurosci 8: 1–9.

Zaidel DW (2015). Hemispheric specialization, art, and aesthetics. In: JP Huston, M Nadal, F Mora et al. (Eds.), Art, aesthetics, and the brain. Oxford University Press, Oxford.

Zaidel DW (2017). Art in early human evolution: socially driven art forms versus material art evolutionary studies in imaginative. Culture 1: 149–157.

Zaidel DW (2018). Culture and art: importance of art practice, not aesthetics, to early human culture. Prog Brain Res 237: 25–40.

Zaidel DW (2019). Co-evolution of language and symbolic meaning: co-opting meaning underlying the initial arts in early human culture. WIREs. Cognit Sci 11: e1520.

Zaimov K, Kitov D, Kolev N (1969). Aphasie chez un peintre. Encéphale 58: 377–417.

Zatorre RJ (2022). Hemispheric asymmetries for music and speech: spectrotemporal modulations and top-down influences. Front Neurosci 16: 1075511.

Zhang Y, Han K, Worth R et al. (2020). Connecting concepts in the brain by mapping cortical representations of semantic relations. Nat Commun 11: 1877.

Handbook of Clinical Neurology, Vol. 208 (3rd series)
Cerebral Asymmetries
C. Papagno and P. Corballis, Editors
https://doi.org/10.1016/B978-0-443-15646-5.00001-4

Chapter 27

Emotion: An evolutionary model of lateralization in the human brain

GUIDO GAINOTTI*

Institute of Neurology, Università Cattolica del Sacro Cuore, Fondazione Policlinico A. Gemelli, IRCCS, Rome, Italy

Abstract

Since several reviews have recently discussed the lateralization of emotions, this chapter will take into account the possible evolutionary meaning of this lateralization. The organization of the chapter will be based on the following steps. I will first propose that emotions must be considered as a complex adaptive system, complementary to the more phylogenetically advanced cognitive system. Second, I will remind historical aspects and consolidated results on the lateralization of emotions. Then I will discuss the phylogenetic aspects of the problem, trying to evaluate if emotional asymmetries concern only humans and some nonhuman primates or are part of a continuum between humans and many phylogenetically distant animal species. After having reviewed various aspects of emotional lateralization across different animal species and (more specifically) in nonhuman primates, I will propose a general model of hemispheric asymmetries in the human brain, based on theoretical models and empiric data. Theoretical models stem from the influence that the presence or the absence of language can have on concomitant hemispheric functions, whereas supporting neuropsychologic data have been gathered in patients with unilateral brain damage.

INTRODUCTION

The lateralization of emotions is a well-established aspect of neuropsychology and of neurosciences which has triggered in recent years several general and specific reviews (e.g., Gainotti, 1989, 2001, 2012, 2020, 2021a; Davidson, 1992, 1998; Gainotti et al., 1993; Borod et al., 1998; Borod, 2000; Demaree et al., 2005; Ocklenburg and Güntürkün, 2018; Stankovic, 2021; Ocklenburg et al., 2022; Palomero-Gallagher and Amunts, 2022).

For this reason, instead of considering this chapter as an update or discussion of previous reviews on the subject, I thought it appropriate to rely on evolutionary considerations akin to those that have inspired the contribution of Mike Corballis to the study of hemispheric asymmetries (e.g., Corballis, 1989, 1991, 2003, 2009, 2012, 2017; Corballis et al., 2002). The line of thought that I intend to follow in the organization of this chapter will therefore is follows.

I will start with some theoretical considerations concerning the nature of emotions, considered as a multicomponent, hierarchical adaptive system, complementary to the philogenetically more advanced cognitive system.

I will then shortly remind the historical development of studies on emotional lateralization, starting from the clinical observations that have triggered these studies and from the subsequent experimental investigations dealing with the communicative and vegetative components of emotions. The results of these experimental investigations will be discussed within the framework of theoretical models advanced to explain the hemispheric asymmetries for emotions.

In the core section of this chapter, I will try to evaluate if human emotional lateralization is part of a continuum between humans and many phylogenetically distant animal species, addressing this problem in two steps. The first will consist of a critical survey (Gainotti, 2021a) of investigations assuming that emotional asymmetries may concern not only humans and some nonhuman primates but also many other vertebrates (see Bisazza et al., 1998 for review) and even invertebrates (see Frasnelli et al., 2012 for review). The second step will consist of discussing the results of a systematic review of various aspects of emotional lateralization in nonhuman primates (Gainotti, 2022), conducted to clarify if emotional lateralization only concerns humans (and nonhuman primates), or extends along all the phylogenetic evolution from invertebrates to humans. In the last part of my contribution, I will address the more general problem of the nature of hemispheric lateralization, trying to evaluate if asymmetries observed in animals and humans are structurally or only apparently similar and proposing a general model of hemispheric asymmetries in the human brain. This model will be based on the theoretical implications of the different lateralization of language in the left hemisphere (LH)

*Correspondence to: Guido Gainotti, Institute of Neurology, Università Cattolica del Sacro Cuore, Policlinico Gemelli, Largo Agostino Gemelli 8, 00168 Rome, Italy. Tel: +39 06 3015 6435, Fax: + 39 06 3015 1909, E-mail: guido.gainotti@unicatt.it and of emotions in the right hemisphere (RH) (Gainotti, 2023) and will be supported by neuropsychologic data gathered in unilateral brain-damaged patients.

THE NATURE OF EMOTIONS

Even if the nature of emotions remains controversial (e.g., James, 1884; Cannon, 1927; Schachter and Singer, 1962; Plutchik, 1980; Ekman, 1984, 1992; Oatley, 1994; Cabanac, 2002) several authors (e.g., Ekman, 1984; Gainotti, 1989, 2000) have considered emotions as a complex adaptive system, more primitive, but complementary to the more phylogenetically advanced cognitive system.

Similarities and differences between the emotional and the cognitive system

Due to their complex multicomponent structure and their general adaptive functions, emotions form a true adaptive system, whose functional architecture is similar to that of the cognitive system, despite the different goals pursued by these two functional organizations. From the structural point of view, both the emotional and the cognitive system require the integration of components that must: (a) orient attention toward the most relevant stimuli, (b) analyze the sensory data to appraise their significance, (c) select an appropriate response, and (d) memorize the most relevant data. On the other hand, the goals of the two functional systems are different. According to Oatley and Johnson-Laird (1987, 2014), emotions can, indeed, be considered as an emergency system, capable of interrupting with an urgent procedure the ongoing action, to rapidly and automatically select a new operative scheme, whereas cognition forms a more advanced adaptive system, capable of analyzing and solving complex situations, but requiring much more time to respond appropriately. The main characteristics of the different components of the emotional and cognitive system have been summarized in Table 27.1.

Hierarchical models of emotions

Besides the similar architecture and the different goals of the two systems, several authors (e.g., Papez, 1937; Nauta, 1971; Izard, 1971, 1990; Zajonc, 1980, 1984; Lazarus, 1982, 1984; Drevets and Raichle, 1998; Pessoa, 2008) have stressed the interactions existing between them and, in particular, the hierarchical structure of human emotions. Two reasons suggest a hierarchical organization of emotions. The first is that it can help to solve conflicts that can arise between the responses selected by two different independent adaptive systems. The second is that only basic emotions (joy, sadness, fear, anger, surprise, and disgust), which appear during the earliest stages of child development, have the characteristics of an emergency system, whereas complex emotions, such as vanity, proudness, remorse, or nostalgia, require much more elaborated cognitive evaluations and flexible response options. Now, since complex emotions derive from the basic ones, thanks to the increasing interactions existing between the emotional and the cognitive systems, this difference between elementary and complex emotions requires the construction of hierarchically organized psychologic and neuroanatomic models of emotions. A developmental model consistent with this logic has been formulated by Leventhal (1979, 1987), who has proposed that human emotions may derive from the activity of a hierarchical multicomponent system, based on three functional levels (the sensorimotor, the schematic, and

Table 27.1

Main characteristics of the different components of the emotional and of the cognitive system.

Common functions	Manner in which this function is accomplished by the Emotional system	Cognitive system
1) Attentional orienting	Automatic orienting elicited by external stimuli	Intentional orienting partly determined by internal representations
2) Analysis of sensory data	a) Quick computation of poorly processed sensory data leading to a direct motor response	Exhaustive but slow computation of highly processed information
3) Computation of stimulus significance	Concerns subjective, self-referential meaning	Concerns objective, social meaning
4) Response activated by the processing of sensory data	a) Is quickly selected from a small number of innate operative patterns b) Is produce with an emergency procedure c) Consists of a limited number of automatic reactions d) Includes communicative (facial or vocal) expressive components and a strong involvement of the autonomic system	a) Is selected after a thorough evaluation of the stimulus features b) Often requires the inhibition of the most spontaneous motor response c) Consists of a very large number of controlled strategic plans d) Does not usually include expressive and autonomic components
Learning mechanisms	Are based on conditioned, automatic and unconscious forms of learning	Are based on the conscious and controlled acquisition of new information in declarative memory

the conceptual level) According to this model, each person disposes since birth of a set of **inborn expressive-motor programs**, which constitute the basic interactive schemata of the human species. During individual development, these innate emotional programs are linked, thanks to mechanisms of conditioned learning, to **emotional schemata,** which differ from the expressive-motor programs of the sensorimotor level, because they are different for each individual, whereas the inborn expressive-motor programs are innate and universal. Since the automatic activation of these emotional schemata is accompanied by the evocation of the corresponding subjective experiences, these emotional schemata correspond to real emotions. Finally, the third, **conceptual level** of emotions is based on mechanisms of conscious declarative memory and does not include concrete representation of true emotions, but abstract notions of what emotions are and of how to deal appropriately with them: at this level are, therefore, represented the **social "display rules,"** which establish "who can show what emotion to whom and when they can show it" (Ekman, 1984).

HISTORICAL DEVELOPMENTS OF THE CONSTRUCT OF EMOTIONAL LATERALIZATION

The first clinical observations

According to Harrington (1985) and Harris (1999), the hypothesis assuming that emotions may be lateralized was first suggested by Luys (1879), who reported differences between the emotional behavior of right and left brain-damaged patients and proposed the existence of an "emotion" center in the RH, complementary to the "intellectual" centers in the LH. Partly similar observations were made, in the following years by Mills (1912), Babinski (1914), and Schilder (1935) in unilateral brain-damaged patients. Mills (1912) noticed that damage to the right side of the brain can cause a decrease in emotional expression, Babinski (1914) observed that patients with RH lesions are sometimes indifferent ("anosodiaphoria") to their left-sided hemiplegia, and Schilder (1935) suggested that the RH might play a special role in autonomic functions. All these observations went, however, unnoticed, and only in the second half of the last century the hypothesis of an emotional lateralization was reintroduced by new empiric data. Clinical observations consistent with this assumption were, in fact, made by Terzian and Cecotto (1958), Alemà and Donini (1960), Perria et al. (1961), Terzian (1964), and Rossi and Rosadini (1967) observing the emotional behavior of patients submitted to pharmacologic inactivation of the RH and LH and by Gainotti (1969, 1972) analyzing the emotional reactions shown by right and left brain-damaged patients. The interpretation that these authors gave of their observations was, however, quite different. The emotional reactions to the pharmacologic inactivation of the RH and LH were, in fact, considered by Perria et al. (1961) and Rossi and Rosadini (1967) as due to a different specialization of the LH and RH for positive and, respectively, for negative emotions. On the contrary, Gainotti (1972) considered the "catastrophic reactions" of left brain-damaged patients as a dramatic but psychologically appropriate form of emotional response, contrasting with the abnormal and emotionally inappropriate "indifference reactions" of right brain-damaged patients. This Gainotti's interpretation was due both to a qualitative analysis of the observed reactions and to the other paradoxical emotional patterns, such as "anosognosia" (Babinski, 1914), "misoplegia" and "personification of the paralyzed limbs" (Critchley, 1955, 1957) often observed in patients with right-sided lesions. Gainotti (1972), therefore, assumed that the abnormal emotional reaction of right brain-damaged patients may stem from a major involvement of the RH in emotional functions. Thus, the first clinical observations that raised the problem of hemispheric asymmetries in emotion representation also prompted the two models of emotional lateralization that are still considered the most important ones in this area of inquiry, namely the different specializations of the LH and RH for positive and negative emotions (also named **the "valence" hypothesis**) and the RH dominance for emotions (also named **"the right hemisphere dominance" hypothesis**).

Experimental studies that have tried to clarify the meaning of the previous clinical observations

In the following years, many experimental studies were carried out to clarify the meaning of the different emotional reactions observed after RH and LH damage. Most of these investigations contrasted the "valence" with the "RH dominance" hypothesis, focusing attention on two typical aspects of the emotional response, namely the expressive-communicative and the vegetative components. The former has the social function of alerting of a possible danger to other members of the group, whereas the latter increases the efficacy of the behavioral response. Furthermore, both contribute to the generation of the emotional experience, which is a marker of felt, "schematic" emotions (e.g., Scherer, 1984; Leventhal, 1987).

INVESTIGATIONS CONCERNING THE COMMUNICATIVE ASPECTS OF EMOTIONS

The nonverbal **communicative aspects of emotions** were studied at both the expressive and receptive levels, in healthy subjects and in unilateral brain-damaged patients. Investigations that studied in healthy subjects the lateralized comprehension of emotional stimuli in the auditory modality usually used the dichotic listening technique, whereas those that investigated the monohemispheric recognition of facial emotional expressions used a lateralized presentation of emotional faces to the right and left visual fields. Studies using dichotic stimuli have generally supported the hypothesis of RH dominance for emotion perception (see Erhan et al., 1998 for review) and similar results were obtained by studies using the tachistoscopic lateralized presentation of emotional faces (see Tamietto et al., 2006 and Gainotti, 2023 for reviews), even if, in this case, some exceptions were found. For instance, Reuter-Lorenz and Davidson (1981) and Reuter-Lorenz et al. (1983) documented an RH advantage for negative emotions and an LH advantage for positive emotions and Asthana and Mandal (2001) found RH superiority for the recognition of sad facial expressions, but not of happy expressions. Results supporting the RH dominance hypothesis were also obtained by studying facial asymmetries with video records of spontaneous emotions or using chimeric faces (i.e., composite photographs of posed positive or negative emotions, created using two left or two right half faces). Several authors (reviewed by Asthana and Mandal, 1998; Borod et al., 2001; Nicholls et al., 2002; Gainotti, 2023) found, indeed, that the left half face appears more emotionally expressive than the right half-face and that both positive and negative left–left composites of chimeric faces are judged as more emotionally expressive than the corresponding right–right composites.

Results similar to those obtained in healthy subjects were gathered by investigations that studied the communicative aspects of emotions in unilateral brain-damaged patients, comparing the ability of right and left brain-damaged patients to recognize or express emotions with voice (emotional prosody) and facial expression. Some authors showed, in fact, that right brain-damaged patients can easily understand the meaning of a sentence, but are unable to recognize the emotion expressed by the voice of a speaker, and similar results were obtained studying the capacity of right and left-brain-damaged patients to express emotions through the tone of voice (see Schmitt et al., 1997 and Ross and Monnot, 2008 for reviews). Other authors showed that patients with RH lesions are seriously impaired in recognizing facial emotional expressions and in expressing emotions through facial expressions (see Blonder et al., 1993; Mandal et al., 1993; Borod et al., 1998; Gainotti, 2019b for reviews). Therefore, most data obtained in healthy subjects and in unilateral brain-damaged patients strongly promote the hypothesis of a general RH dominance for emotional comprehension and expression, even if results of some investigations conducted in normal subjects could support the "valence" more than the "RH dominance" hypothesis. An interpretation of these last results divergent from the "valence hypothesis" was, however, advanced by Davidson (1983, 1992, 1998) on the basis of electroencephalographic (EEG) asymmetries observed at the level of the frontal lobes in depressed individuals and in normal subjects, during the expression of positive and negative emotions. Davidson claimed that these asymmetries are not related to the valence but to the motivational system engaged by the emotional stimulus and proposed that the left prefrontal cortex (PFC) may be involved in a system-inducing approach to appetitive stimuli and the right PFC in a system promoting withdrawal from aversive stimuli. This "**approach/withdrawal hypothesis**" became very popular and can be considered as a third general model of emotional laterality followed in the literature.

Investigations concerning the autonomic components of emotions

The first investigation explicitly designed to study this issue in unilateral brain-damaged patients was undertaken by Heilman et al. (1978), who studied the galvanic skin response to painful stimuli applied to the hand ipsilateral to the lesioned hemisphere and observed a flattened vegetative response in patients with RH lesions. After this seminal research, other investigations studied: (a) the psychophysiologic correlates of emotional activation in unilateral brain-damaged patients, (b) hemispheric asymmetries in cardiac autonomic control, and (c) the psychophysiologic correlates of the selective emotional stimulation of the RH or LH in normal subjects. All studies conducted along the first line of research showed a significant reduction in galvanic skin response and other indices of sympathetic activation in patients with right brain lesions, whereas the vegetative response of left brain-damaged patients was much more variable (see Hagemann et al., 2003 and Gainotti, 2020 for review). Studies investigating hemispheric asymmetries in autonomic heart control showed that physiologic heart rate variability (HRV) is reduced more by right than by left brain lesions, and that in patients with hemispheric brain infarction, changes in circadian blood pressure patterns and cardiovascular parameters indicate a right lateralization of sympathetic activation. Other authors confirmed the leading role played by the RH in sympathetic activities but suggested a left lateralization of parasympathetic functions (see Guo et al., 2016 and Sturm et al., 2018 for review). A strong similarity can, therefore, be found between results obtained studying hemispheric asymmetries for autonomic functions and those obtained from studying the lateralization of emotional communication in the human brain. In fact, just as an RH superiority for negative emotions is almost universally acknowledged and an LH prevalence for positive emotions is supported only by some studies, an RH dominance for sympathetic functions is much more supported than an LH prevalence for parasympathetic activities.

Inconsistent results obtained by neuroimaging studies concerning the laterality of emotions

Results obtained by neuroimaging studies concerning the laterality of emotions have been much less consistent than those concerning the communicative aspects and the vegetative components of emotions. As a matter of fact, results obtained by two recent meta-analyses, concerning respectively the discrete neural correlates of basic emotions (Vytal and Hamann, 2010) and the large-scale neural networks underlying emotion regulation (Morawetz et al., 2020), are inconsistent and contradictory. Vytal and Hamann (2010) showed that a right lateralization could be associated only with the processing of fear, whereas for all the other (positive or negative) basic emotions, the largest activation clusters were found in both the left and right hemispheres. Analogously, Morawetz et al. (2020) showed that perception, experience, and expression of emotion are each subserved by a distinct large-scale network, composed of left-lateralized, right-lateralized, or bilaterally activated areas. In our opinion, the contrast between the consistent results obtained by studies concerning the communicative aspects and the inconsistent outcomes of neuroimaging studies regarding the laterality of emotions could be due to the automatic nature of the emotional processing. This hypothesis stems from the fact that the analysis of asymmetries in emotional communication directly probes the laterality question, whereas the study of metabolic changes is an indirect marker of it and is also linked to the complexity and the amount of attentional resources requested by the task. Now, since the latter are usually small during automatic tasks, results might be more sensitive to the influence of uncontrolled variables during the analysis of automatically processed emotional stimuli.

IS LATERALIZATION OF EMOTIONS A PHYLOGENETICALLY RECENT OR OLD PHENOMENON?

A critical survey of the literature assuming that emotional asymmetries may have evolved along all vertebrate species

Although hemispheric asymmetries have been considered a uniquely human feature since the discovery of the left lateralization of language, this idea has been challenged in more recent times by an increasing number of studies describing a lateralization of different behavioral functions in several nonhuman species. These studies were summarized in a book by Bradshaw and Rogers (1993), who, in contrast to Corballis' (1991) views assuming that lateralization is a uniquely human property that has evolved quite recently, argued that hemispheric asymmetries have a long evolutionary history and that the present state of affairs in *Homo sapiens* is the final product of this coevolution. Two different meanings of the term "laterality" underpinned the views of Corballis (1991) and Bradshaw and Rogers (1993), because the former viewed lateralization as the latest stage of the process of functional localization, whereas the latter simply reported empiric evidence showing that lateralization of certain functions can contribute to improve performance on related tasks. Vallortigara et al. (1999) expressed this position with this statement: "We believe that the essence of brain lateralization lies in the fact that one side of the brain performs certain types of computational operations and the other side performs other, different computational operations. When faced with complex computational problems, brains might perform better if they segregate them into subproblems that differ in a typical ways." This computational acceptance of the term "laterality" can help to understand why laterality effects have been described not only in all nonhuman vertebrates (see Rogers and Andrew, 2002 and MacNeilage et al., 2009 for review), but also in invertebrates (see Frasnelli et al., 2012 for review), suggesting that, even in the absence of a neocortex, animal brains can perform better on complex problems if lateralization helps them to segregate the task into different subproblems. This acceptation of the term "laterality" raises, in my opinion the problem of evaluating if brain asymmetries documented in animals and humans are quite comparable (and the present state of affairs in *H. sapiens* is the final product of this coevolution) or if a qualitative difference exists between these two kinds of lateralization. Two requirements can, in my opinion, help answer to this question. The first requirement concerns the population level (rather than the individual level) of the observed asymmetries, because only population-level lateralizations may reflect evolutionarily stable strategies that can develop across species (Corballis, 2003). In partial contrast with this position, Frasnelli and Vallortigara (2018) and Vallortigara and Rogers (2020) used the mathematical theory of game to clarify the relationships between lateral bias at the individual level and at the population level. They have argued that strategies of lateralization at the individual or population level, depend on ecologic contexts and that individual-level and population-level lateralization should be considered as two aspects of the same continuum. The second requirement is the existence of a common basic pattern of asymmetries, evolving from lower vertebrates to humans because this common pattern is necessary to assume that the present form of human lateralization may be the final product of coevolution across different species. Irrespective of these different acceptations of the term "laterality," and the scattered or structured nature of human brain asymmetries (to which I will return in section Are the Main Characteristics of the Right Hemisphere Partly Determined by Evolutionary Factors Linked to its Involvement in Emotional Functions? of this review), an aspect of animal lateralization that has attracted particular attention has concerned the hemispheric asymmetries for emotions. According to Désiré et al. (2002) and Dawkins (2006), this interest was due to the fact that most investigations dealing with this subject were based on observational data or physiologic measurements that are particularly appropriate for the study of animal behavior. In any case, these investigations have revealed important homologies between human and animal asymmetries for various aspects of emotional behavior, even if many problems concerning theoretical and methodological issues remain open in this area of research.

Theoretical problems in studies dealing with emotional lateralization in animals

A first theoretical problem is that some data gathered in studies dealing with emotional lateralization in animals contrast with the computational meaning of the term "laterality" proposed by Bradshaw and Rogers (1993) and developed by more recent reviews (e.g., Vallortigara et al., 1999; Vallortigara and Rogers, 2020). This acceptation maintains, in fact, that the direction of the lateralization is less important than the fact that the two sides of the brain perform different computational operations, since a similar direction of the lateralization can only be explained on the basis of general principles. However, some studies on emotional lateralization in animals have concerned just the direction of the lateralization, because these studies have shown that in several vertebrate species, most individuals react faster when a predator approaches from their left than from the right side. Following this line of thought, MacNeilage et al. (2009) suggested that hemispheric specializations might have evolved from a basic lateralization pattern, common to all vertebrates, in which the RH is specialized in detecting and responding to unexpected stimuli in the environment, whereas the LH is involved in the control of well-established patterns of behavior under ordinary and familiar settings. These observations are paradoxically consistent with the two criteria required by the coevolution model of animal and human asymmetries, but inconsistent with the computational meaning of the term "laterality."

A second theoretical problem concerns the multicomponent nature of emotions because if their lateralization is a phylogenetically old phenomenon, it could be logical to expect that the basic visceral autonomic components of emotions may be more widespread than the more recent expressive-communicative components, linked to the development of the social aspects of emotions. To our knowledge, however, no studies dealing with emotional lateralization in animals have taken into account the problem of the lateralization of the autonomic components of emotions. The last theoretical problem concerns the models of emotional lateralization used in different studies, because in many investigations results have been discussed with reference to the "RH hypothesis," in other studies with regard to the "emotional-valence hypothesis" and, in still other research, in relation to the "approach-withdrawal hypothesis."

Methodological problems met in these studies

Equally or even more relevant are the methodological questions, concerning both the techniques used to assess a functional lateralization and the procedures adopted to document the existence of a laterality effect.

An example of a questionable method used to assess brain lateralization, is the preferential use of the left or right eye for looking at relevant stimuli, which, according to Vallortigara et al. (1999) has provided important insights into the lateralization of brain functions. However, even if Bishop et al. (1953) have suggested that the crossed fibers of the optic nerve may transmit sensory information faster than uncrossed fibers, no preferential relation has been proven between an eye and the contralateral hemisphere, since each visual cortex receives visual information from both eyes. The use of eye preference as an indicator of contralateral visual dominance seems therefore quite questionable.

As for the procedures adopted to document the existence of a laterality effect, many studies have not specified if the observed lateralization has been found at the individual- or at the population level, in spite of the fact that asymmetries at the individual level cannot be considered as forms of brain lateralization reflecting evolutionarily stable strategies.

A systematic review of investigations that have studied asymmetries for emotions in nonhuman primates

Since a systematic review of all studies that have investigated aspects of emotional lateralization across all the vertebrate species was not feasible, I thought, that a first step toward a clarification of this complex issue could consist of a systematic review of investigations that have studied different aspects of hemispheric asymmetries for emotions in nonhuman primates. These animals are, indeed, more similar to humans than all the other animal species not only from the cognitive but also from the physical point of view, allowing us infer that methodologies used to study their asymmetries in emotional behavior should be more similar than those used in investigations conducted in other animal species to those used for human studies. Furthermore, within the nonhuman primates, phylogenetic differences exist between different species of prosimians, monkeys, and apes (e.g., Chatterjee et al., 2009; Herlyn, 2016; Yousaf et al., 2021). It could, therefore, be predicted that patterns of emotional lateralization within nonhuman primates should be clear and consistent if emotional lateralization has a long evolutionary history, whereas they could be less clear and consistent if emotional lateralization is a much more recent process.

Purposes of the systematic review

The main purposes of the review were therefore following:

(1) To check, with an explorative study, if methodologies used to investigate various aspects of emotional lateralization in nonhuman primates could help to evaluate the reliability of data concerning the emotional lateralization of animals along all the phylogenetic evolution. It could, indeed, be hypothesized that methodologies used in studies conducted in cognitively well-developed animals such as primates could be more complex and reliable than those used in research conducted in animals, characterized by brain and body structures very different from the human ones.

(2) To take separately into account results obtained in studies conducted in great apes and in monkeys concerning different components and stages of emotions to evaluate if some of these components point to a continuum between humans and many other animal species, whereas other components suggest a similarity only between humans and nonhuman primates or between humans and great apes. The hypothesis of a long-standing evolution could, indeed, lead to predicting that right lateralization of sympathetic functions (involved in the more primitive emotions of fear and rage) could be observed across many animal species, whereas asymmetries in emotional communication (prevalent in more evolutionary recent social emotions) could be documented only in primates and, above all, in humans and great apes.

(3) To evaluate if data obtained in the more methodologically sound areas of research mainly endorsed the "RH model," the "valence hypothesis" or the "approach–withdrawal hypothesis."

Methodological aspects of the review

Papers considered as appropriate to be included in the review were subdivided into three broad categories, corresponding to different components of the emotional system: (a) lateralization of emotional expression; (b) lateralization in the processing of emotional stimuli; (c) lateralization of markers of emotional activation. These broad categories were subsequently broken into different subheadings, to evaluate results obtained in more specific lines of research and the reliability of methodologies used to investigate various aspects of emotional lateralization. The subheadings concerning the *lateralization of emotional expression* took separately into account studies regarding the asymmetric expression of emotions at the face and at the whole-body level.

Those concerning the *asymmetries observed in the perceptual processing of emotions* took separately into account visual asymmetries documented in split-brain monkeys, gaze asymmetries observed while primates looked to emotional faces (including the left-side cradling preference) or to other emotional stimuli, visual investigations based on the eye preference methodology and asymmetries observed in the auditory modality. More disappointing, was the review of investigations concerning the *laterality of markers of emotional activation*, because only the asymmetry of the tympanic membrane temperature (TMT) has been used in the literature as an indicator of emotional activation in primates.

Main results of the systematic review

Overall, the results of the review were difficult to evaluate, due to the fragmentary and heterogeneous nature of topics taken into account, the small number of papers included in each category, and the presence of methodological inconsistencies. A rather general problem concerned the individual or population level of the reported asymmetries, because in most studies these asymmetries had been investigated at the population level, but in some, they had observed at the individual level, and in still other studies this information could not be obtained due to the small number of animals studied.

The most consistent and methodologically sound line of research considered in the review concerned the asymmetry in emotional expression at the facial level, which pointed to a RH lateralization of emotional expressions not influenced by the valence of the emotional stimuli. A right lateralization of the perceptual processing of visual emotional cues was also observed in split-brain apes and monkeys, in which visual stimuli are independently processed by each hemisphere and in studies that have investigated gaze asymmetries while looking at emotional stimuli. This category of investigations included the left-side cradling preference, that, according to Salk (1973), Manning et al. (1994), and Boulinguez-Ambroise et al. (2020), could be due to a "left gaze bias" (i.e., prevalent use of the RH to look to emotionally laden puppies). It must also be noted that the greatest evidence of lateralization at the population level of facial emotional expressions and cradling preference was found in great apes, rather than in monkeys. Much less consistent and methodologically criticizable were: asymmetries in the production of bodily emotional expressions; results obtained assessing hemispheric asymmetries in the processing of auditory emotional stimuli; and results obtained using the eye preference as a marker of lateralized treatment of emotional stimuli.

Equally disappointing were results obtained using asymmetries in TMT as a marker of lateralized emotional activation. These results were, in fact, highly inconsistent and had been criticized from the methodological point of view by Propper and Brunyé (2013), who had noticed that two different interpretations had been given of data obtained with this methodology. Asymmetric TMT had been, indeed, attributed in some studies (e.g., Boyce et al., 2002; Gunnar and Donzella, 2004) to an ipsilaterally increased activity and in other studies (e.g., Boyce et al., 1996; Helton, 2010) to a contralaterally decreased activity.

Overall, the greater consistency of results obtained studying the communicative aspects of emotions and the prevalence of these asymmetries in great apes support models assuming a similarity only between humans and nonhuman primates rather than suggesting a continuum in emotion lateralization across many animal species.

As for the problem of the models of emotional lateralization, data of the review, mainly supported the "RH model," because in the most methodologically sound lines of research, the right lateralization of emotional expression was not influenced by the valence of the emotional situation. Results supporting the "valence hypothesis" were instead found in the more questionable lines of research, such as those concerning eye preference (and even in these last investigations some results were at odds with the same model). Poorly supported was also the "approach–withdrawal hypothesis," which predicts an LH activation during aggression/approach and an RH activation during fear/withdrawal, because most results showed an RH activation during both aggressive and fear-inducing conditions.

ARE THE MAIN CHARACTERISTICS OF THE RIGHT HEMISPHERE PARTLY DETERMINED BY EVOLUTIONARY FACTORS LINKED TO ITS INVOLVEMENT IN EMOTIONAL FUNCTIONS?

Since the aim of this chapter was to draw on the strong evolutionary beliefs that have oriented the contribution of Corballis to the study of hemispheric asymmetries (e.g., Corballis, 1991, 2017; Corballis et al., 2002), I will propose in the last section of this review a model assuming that the main characteristics of the RH may have been partly determined by evolutionary factors. This model will also try to answer to the question, raised in section Is Lateralization of Emotions a Phylogenetically Recent or Old Phenomenon? of the present review, concerning the "scattered and arbitrary" or "organized and purposeful" structure of the human brain asymmetries. The rationale of this model is that, since emotions constitute a primitive and preverbal adaptive system, some functional features of this system may also characterize the nonverbal cognitive functioning of the RH. The model is based on cognitive considerations and is supported by neuropsychologic data gathered in unilateral brain-damaged patients.

The cognitive considerations on which the model is based stem from three principles: (a) the "crowding model" (Kosslyn, 1987), which assumes that to avoid interferences, functions (such as emotions and language) that could select different responses options in front of external situations, must be crowded in different hemispheres; (b) the assumption that the presence or the absence of language must have important implications for the general functioning of the left and right hemispheres; and (c) the competition for cortical space, which implies that the left lateralization of language may have led to a greater representation of perceptual modalities in the RH. The neuropsychologic data supporting the model have been obtained through a general review of disorders shown by unilateral brain-damaged patients (Gainotti, 2021b, 2023) and of the brain mechanisms underlying these disorders.

To match cognitive predictions with neuropsychologic data, a preliminary distinction was made between three kinds of brain activities, that could: (a) take advantage of the capacity of language to shape cognitive activities, such as its contribution to solving nonverbal cognitive tasks, and to increase the levels of consciousness and intentionality; (b) characterize a cognitive system not equipped with these positive effects of language, and therefore prone to a prevalence of "automatic over intentional," or of "unconscious over conscious" ways of functioning; and (c) be less represented in the hemisphere where language had emerged, at the cost of preexisting simplest activities, due to competition for cortical space.

The decision to focus our attention (within the activities that could take advantage of the shaping influence of language) on the solution of nonverbal cognitive tasks and on the levels of consciousness and intentionality was due to the fact these activities have been studied from both the theoretical and the neuropsychologic point of view. The contribution of language to the solution of nonverbal cognitive task, indeed, has been proven by studies conducted in children and in adults, which had shown that solving nonverbal tasks is associated with the use of inner speech (e.g., Sokolov, 1972; Winsler and Naglieri, 2003; Lupyan, 2012). Analogously, discussing the influence that language could have on the levels of consciousness and intentionality, Piaget (1967) stated, that "thought becomes conscious to the degree to which the child is able to communicate it," Carruthers (2000) said that a mental state is conscious only when it is represented by a reflexive conceptual thought and Brandom (2014) acknowledged that discursive intentionality is only exhibited by subjects provided with language and concepts.

The correspondence between predictions based on the theoretical impact of language on other brain functions and neuropsychologic data obtained in unilateral brain-damaged patients was then assessed and a good correspondence was found between predictions and neuropsychologic data. Several investigations (reviewed by Gainotti, 2014, 2021b) have, indeed, confirmed the influence of language on the solution of nonverbal tasks, showing that some nonverbal cognitive disorders of aphasic patients could be due to the inability to solve the task through an intermediate propositional use of covert language. For instance, in one of these studies, Baldo et al. (2015) found that left brain-damaged patients with language disorders are not impaired in cognitive tasks that can be performed with simple perceptual activity, but are defective when an intermediate verbal activity is required by the task. Equally confirmed was the influence that language can have on the levels of intentionality and consciousness both in left and right brain-damaged patients. As for the intentional/automatic distinction, several authors have reported patients, who, in the context of global aphasia, could utter automatic but not intentional speech, whereas other authors have documented a prevalent role of the intact RH in automatic speech production (see Gainotti, 2021b for review). Furthermore, a disruption of automatic aspects of behavior has been observed in many syndromes typical of RH lesions, such as unilateral spatial neglect and acquired prosopagnosia. Many authors (e.g., Gainotti et al., 1991; Gainotti, 1996; Bartolomeo and Chockron, 2002; Corbetta and Shulman, 2002, 2011) have, indeed, shown that the conscious and controlled ("endogenous") components of contralateral orienting of attention are relatively spared in the unilateral neglect syndrome, whereas the automatic ("exogenous") components are much more impaired. A similar disruption of the automatic components of a task subsumed by the RH has been repeatedly documented in prosopagnosia. In this defect of familiar people identification typical of the RH lesions, a selective defect of the (automatically raised) "face familiarity feelings" has been found by Gainotti and Marra (2011) in patients with right temporal damage and by Borghesani et al. (2019), studying various components of prosopagnosia in a large sample of patients with neurodegenerative diseases.

A last hemispheric asymmetry that is usually neglected, but could apply to the third category of brain activities listed in our schematic distinction, concerns the greater representation of perceptual functions in the RH. This asymmetry was first documented De Renzi and Spinnler (1966, 1967), De Renzi et al. (1969), and Faglioni et al. (1969) in unilateral brain-damaged patients and has been afterward confirmed by Rizzolatti et al. (1971) in healthy subjects and by Corballis et al. (2002) in split-brain patients. Gazzaniga (2000) has suggested that this asymmetry may be due to the fact that: "While language emerged in the left hemisphere at the cost of pre-existing perceptual systems, the critical features of the bilaterally present perceptual system were spared in the opposite half-brain." This mechanism, therefore, assumes that the cortical areas involved in language processing in the LH had remained (at least in part) dedicated to the perceptual processing in the homologous areas of the RH. It could be objected that these mild interhemispheric differences in perceptual processing are probably not provided of biologic significance, but this objection overlooks a central tenet of the embodied cognition framework (Barsalou, 1999, 2008; Simmons and Barsalou, 2003) which states that semantic knowledge and other representational activities are grounded in sensorimotor activities. On the basis of these theoretical premises, it is possible to assume that the greater involvement of the RH in sensorimotor functions may have led to the development in this hemisphere of "higher-order convergence zones" (Damasio, 1989, 1990) supporting nonverbal forms of person recognition and conceptual knowledge. Results consistent with this hypothesis have been obtained by Snowden et al. (2004) and Gainotti (2007, 2012), administering verbal and nonverbal tests of conceptual thinking and of famous people identification to semantic dementia patients with prevalent degeneration of the left or right anterior temporal lobes (ATL). Subjects with a predominant left ATL atrophy were, indeed, more impaired on verbal tests of conceptual thinking and person identification, whereas those with a predominant right ATL atrophy were more impaired on nonverbal tests of conceptual thinking and person identification.

CONCLUDING REMARKS

The main reason why I have proposed this model, concerning the hemispheric asymmetries of the human brain, is not that I reject computational models advanced to explain laterality effects observed in animals, but that these models could be inadequate to explain the distribution of functions between the human left and the right hemispheres. In agreement with the positions of Corballis, who assumed that a strong link should exists between the LH dominance for language and for hand gestures, our model assumes that brain asymmetries are not randomly distributed, but follow a logic intrinsic to the brain organization. Both models are, therefore, strongly based on evolutionary factors and assume that human brain asymmetries may have emerged either by interactions between the individuals and the external milieu (association between vocalizations and gestures) or by factors intrinsic to the brain organization. Among these factors, we can include the "crowding principle" (Kosslyn, 1987), which assumes that to avoid interference, functions critical for survival must be crowded in different hemispheres and the competition for cortical space within each hemisphere, which can explain the greater involvement of the nonverbal hemisphere in perceptual functions, or in abilities based on these sensory-motor activities. These principles are quite consistent with empiric data showing that the degree of lateralization is greater for language than for emotions, perceptual activities, and other cognitive functions for which a right lateralization is usually acknowledged. Emotions and other right-sided functions are, indeed bilaterally represented at the hemispheric level, whereas language is almost only represented in this hemisphere. Furthermore, the more concrete (semantic-lexical) levels of language are partly represented in the right hemisphere (e.g., Gainotti et al., 1981; Chiarello, 1998), whereas the abstract grammatical levels are only represented in the left hemisphere.

On the basis of the abovementioned considerations, I think that the few similarities between the animals' and human hemispheric asymmetries could be only phenomenological, because these asymmetries are scattered and unsystematic in animals, whereas they are systematic and congruent in humans. Problematic is, therefore, the hypothesis of an evolutional continuum between humans and animals with respect to the specific problem of emotional asymmetries, because in humans the different distribution of emotional functions in the RH and of the highest cognitive functions in the LH seems linked to the emergence of language (and of the related cognitive functions) in the LH. However, since in animals the only adaptive system is constituted by emotions and there are no reasons that suggest a lateralization of this adaptive system, there are also no reasons to suggest a continuum between humans and animals with respect to the problem of emotional laterality.

References

Alemà G, Donini G (1960). Sulle modificazioni cliniche ed elettroencefalografiche da introduzione intracarotidea di iso-amil-etil-barbiturato di sodio nell'uomo. Boll Soc Ital Biol Sper 36: 900–904.

Asthana HS, Mandal MK (1998). Hemifacial asymmetry in emotion expression. Behav Modif 22: 177–183. https://doi.org/10.1177/01454455980222005.

Asthana HS, Mandal MK (2001). Visual-field bias in the judgment of facial expression of emotion. J Gen Psychol 128: 21–29. https://doi.org/10.1080/00221300109598895.

Babinski J (1914). Contribution à l'étude des troubles mentaux dans l'hémiplégie organique cérébrale (Anosognosie). Rev Neurol (Paris) 27: 845–848.

Baldo JV, Paulraj SR, Curran BC et al. (2015). Impaired reasoning and problem-solving in individuals with language impairment due to aphasia or language delay. Front Psychol 6: 1523. https://doi.org/10.3389/fpsyg.2015.01523.

Barsalou LW (1999). Perceptual symbol systems. Behav Brain Sci 22: 577–660. https://doi.org/10.1017/s0140525x99002149.

Barsalou LW (2008). Grounded cognition. Annu Rev Psychol 59: 617–645. https://doi.org/10.1146/annurev.psych.59.103006.093639.

Bartolomeo P, Chockron S (2002). Orienting of attention in left unilateral neglect. Neurosci Biobehav Rev 26: 217–234. https://doi.org/10.1016/s0149-7634(01)00065-3.

Bisazza A, Rogers LJ, Vallortigara G (1998). The origins of cerebral asymmetry: a review of evidence of behavioural and brain lateralization in fishes, reptiles and amphibians. Neurosci Biobehav Rev 22: 411–426. https://doi.org/10.1016/s0149-7634(97)00050-x.

Bishop PO, Jeremy D, Lance JW (1953). The optic nerve: properties of a central tract. J Physiol 121: 415–432. https://doi.org/10.1113/jphysiol.1953.sp004955.

Blonder LX, Burns A, Bowers D et al. (1993). Right hemisphere facial expressivity during natural conversation. Brain Cogn 21: 44–56. https://doi.org/10.1006/brcg.1993.1003.

Borghesani V, Narvid J, Battistella G et al. (2019). "Looks familiar, but I do not know who she is": the role of the anterior right temporal lobe in famous face recognition. Cortex 115: 72–85. https://doi.org/10.1016/j.cortex.2019.01.006. Epub 2019 Jan 23.

Borod JC (Ed.), (2000). The neuropsychology of emotions. Oxford University Press, New York.

Borod JC, Cicero BA, Obler LK et al. (1998). Right hemisphere emotional perception: evidence across multiple channels. Neuropsychology 12: 446–458. https://doi.org/10.1037//0894-4105.12.3.446.

Borod JC, Zgaljardic D, Tabert MH et al. (2001). Asymmetries of emotional perception and expression in normal adults. In: G Gainotti (Ed.), Emotional behavior and its disorders, F Boller, J Grafman (Eds.), Handbook of neuropsychology. second edn. vol. 5. Elsevier, Amsterdam, pp. 181–205.

Boulinguez-Ambroise G, Pouydebat E, Disarbois È et al. (2020). Human-like maternal left-cradling bias in monkeys is altered by social pressure. Sci Rep 10: 11036. https://doi.org/10.1038/s41598-020-68020-3.

Boyce WT, Higley JD, Jemerin JJ et al. (1996). Tympanic temperature asymmetryand stress behavior in rhesus macaques and children. Arch Pediatr Adolesc Med 150: 518–523. https://doi.org/10.1001/archpedi.1996.02170300072014.

Boyce WT, Essex MJ, Alkon A et al. (2002). Temperament, tympanum, and temperature: four provisional studies of the biobehavioral correlates of tympanic membrane temperature asymmetries. Child Dev 73: 718–733. https://doi.org/10.1111/1467-8624.00434.

Bradshaw JL, Rogers LJ (1993). The evolution of lateral asymmetries, language, tool use, and intellect, Academic Press, New York.

Brandom B (2014). Intentionality and language. In: NJ Enfield, P Kockelman, J Sidnell (Eds.), The Cambridge handbook of linguistic anthropology. Cambridge University Press, Cambridge, UK.

Cabanac M (2002). What is emotion? Behav Processes 60: 69–83. https://doi.org/10.1016/S0376-6357(02)00078-5.

Cannon WB (1927). The James–Lange theory of emotions: a critical examination and an alternation. Am J Psychol 39: 106–124.

Carruthers P (2000). Phenomenal consciousness: a naturalistic theory, Cambridge University Press, Cambridge UK.

Chatterjee HJ, Ho SY, Barnes I et al. (2009). Estimating the phylogeny and divergence times of primates using a supermatrix approach. BMC Evol Biol 9: 259. https://doi.org/10.1186/1471-2148-9-259.

Chiarello C (1998). Right hemisphere contributions to lexical semantics, Springer-Verlag.

Corballis MC (1989). Laterality and human evolution. Psychol Rev 96: 492–505. https://doi.org/10.1037/0033-295X.96.3.492.

Corballis MC (1991). The lopsided ape, Oxford University Press, New York.

Corballis MC (2003). From hands to mouth: the origins of language, Princeton University Press, Princeton, New Jersey.

Corballis MC (2009). The evolution and genetics of cerebral asymmetry. Philos Trans R Soc Lond B Biol Sci 364: 867–879. https://doi.org/10.1098/rstb.2008.0232.

Corballis MC (2012). Lateralization of the human brain. Prog Brain Res 195: 103–121. https://doi.org/10.1016/B978-0-444-53860-4.00006-4.

Corballis MC (2017). The evolution of lateralized brain circuits. Front Psychol 8. https://doi.org/10.3389/fpsyg.2017.01021.

Corballis MC, Funnell MG, Gazzaniga MS (2002). Hemispheric asymmetries for simple visual judgments in the split brain. Neuropsychologia 40: 401–410. https://doi.org/10.1016/s0028-3932(01)00100-2.

Corbetta MN, Shulman GL (2002). Control of goal-directed and stimulus-driven attention in the brain. Nat Rev Neurosci 3: 201–215. https://doi.org/10.1038/nrn755.

Corbetta MN, Shulman GL (2011). Spatial neglect and attention networks. Annu Rev Neurosci 34: 569–599. https://doi.org/10.1146/annurev-neuro-061010-113731.

Critchley M (1955). Personification of paralysed limbs in hemiplegics. Br Med J 30: 284–286.

Critchley M (1957). Observations on anosodiaphoria. Encéphale 46: 540–546.

Damasio AR (1989). Time-locked multiregional retroactivation: a systems level proposal for the neural substrates of recall and recognition. Cognition 33: 25–62. https://doi.org/10.1016/0010-0277(89)90005-X.

Damasio AR (1990). Category-related recognition defects as a clue to the neural substrates of knowledge. Trends Neurosci 13: 95–98. https://doi.org/10.1016/0166-2236(90)90184-C.

Davidson RJ (1983). Hemispheric specialization for cognition and affect. In: A Gale, J Edwards (Eds.), Physiological correlates of human Behavior. Academic Press, London, pp. 203–216.

Davidson RJ (1992). Anterior cerebral asymmetry and the nature of emotion. Brain Cogn 20: 125–151. https://doi.org/10.1016/0278-2626(92)90065-t.

Davidson RJ (1998). Affective style and affective disorders: perspectives from affective neuroscience. Cogn Emot 12: 307–330. https://doi.org/10.1080/026999398379628.

Dawkins MS (2006). Through animal eyes: what behaviour tells us. Appl Anim Beha Sci 100: 4–10.

De Renzi E, Spinnler H (1966). Visual recognition in patients with unilateral cerebral disease. J Nerv Ment Dis 142: 515–525. https://doi.org/10.1097/00005053-196606000-00002.

De Renzi E, Spinnler H (1967). Impaired performance on color tasks in patients with hemispheric damage. Cortex 3: 194–217. https://doi.org/10.1016/S0010-9452(67)80012-1.

De Renzi E, Scotti G, Spinnler H (1969). Perceptual and associative disorders of visual recognition. Their relation to the side of cerebral lesion. Neurology 19: 634–642. https://doi.org/10.1212/WNL.19.7.634.

Demaree HA, Everhart DE, Youngstrom EA et al. (2005). Brain lateralization of emotional processing: historical roots and a future incorporating "dominance". Behav Cogn Neurosci Rev 4: 3–20.

Désiré L, Boissy A, Veissier I (2002). Emotions in farm animals. Behav Processes 60: 165–180. https://doi.org/10.1016/s0376-6357(02)00081-5.

Drevets WC, Raichle ME (1998). Reciprocal suppression of regional cerebral blood flow during emotional versus higher cognitive processes: implications for interactions between emotion and cognition. Cognit Emot 12: 353–385. https://doi.org/10.1080/026999398379646.

Ekman P (1984). Expression and the nature of emotion. In: K Scherer, P Ekman (Eds.), Approachs to emotion. Erlbaum, Hillsdale, NJ, pp. 319–344.

Ekman P (1992). An argument for basic emotions. Cognit Emot 6: 169–200. https://doi.org/10.1080/02699939208411068.

Erhan H, Borod JC, Tenke CE et al. (1998). Identification of emotion in a dichotic listening task: event-related brain potential and behavioral findings. Brain Cogn 37: 286–307. https://doi.org/10.1006/brcg.1998.0984.

Faglioni P, Spinnler H, Vignolo LA (1969). Contrasting behavior of right and left hemisphere-damaged patients on a discriminative and a semantic task of auditory recognition. Cortex 5: 3166–3389. https://doi.org/10.1016/s0010-9452(69)80014-6.

Frasnelli E, Vallortigara G (2018). Individual-level and population-level lateralization: two sides of the same coin. Symmetry 10: 739. https://doi.org/10.3390/sym10120739.

Frasnelli E, Vallortigara G, Rogers LJ (2012). Left-right asymmetries of behavioural and nervous system in invertebrates. Neurosci Biobehav Rev 36: 1273–1291. https://doi.org/10.1016/j.neubiorev.2012.02.006.

Gainotti G (1969). Réaction catastrophiques et manifestations d'indifferénce au cours des atteintes cérébrales. Neuropsychologia 7: 195–204. https://doi.org/10.1016/0028-3932(69)90017-7.

Gainotti G (1972). Emotional behavior and hemispheric side of the lesion. Cortex 8: 41–55. https://doi.org/10.1016/s0010-9452(72)80026-1.

Gainotti G (1989). The meaning of emotional disturbances resulting from unilateral brain injury. In: G Gainotti, C Caltagirone (Eds.), Emotions and the dual brain. Springer, Heidelberg, pp. 147–167.

Gainotti G (1996). Lateralization of brain mechanisms underlying automatic and controlled forms of spatial orienting of attention. Neurosci Biobehav Rev 20: 617–622. https://doi.org/10.1016/0149-7634(95)00074-7.

Gainotti G (2000). Neuropsychological theories of emotions. In: JC Borod (Ed.), The neuropsychology of emotions. Oxford University Press, New York, pp. 147–167.

Gainotti G (2001). Components and levels of emotion disrupted in patients with unilateral brain damage. In: G Gainotti (Ed.), Emotional behavior and its disorders, Handbook of neuropsychology. second edn. vol. 5. Elsevier, Amsterdam, pp. 161–180.

Gainotti G (2007). Different patterns of famous people recognition disorders in patients with right and left anterior temporal lesions: a systematic review. Neuropsychologia 45: 1591–1607. https://doi.org/10.1016/j.neuropsychologia.2006.12.013.

Gainotti G (2012). The format of conceptual representations disrupted in semantic dementia: a position paper. Cortex 48: 521–529. https://doi.org/10.1016/j.cortex.2011.06.019.

Gainotti G (2014). Old and recent approaches to the problem of nonverbal conceptual disorders in aphasic patients. Cortex 2014: 78–89. https://doi.org/10.1016/j.cortex.2014.01.009.

Gainotti G (2019b). A historical review of investigations on laterality of emotions in the human brain. J Hist Neurosci 28: 23–41. https://doi.org/10.1080/0964704X.2018.1524683.

Gainotti G (2020). Emotions and the right side of the brain, Springer Nature, Switzerland AG.

Gainotti G (2021a). The difficult integration between human and animal studies on emotional lateralization: a perspective article. Brain Sci 11: 975. https://doi.org/10.3390/brainsci11080975.

Gainotti G (2021b). Is there a causal link between the left lateralization of language and other brain asymmetries? A review of data gathered in patients with focal brain lesions. Brain Sci 11: 1644. https://doi.org/10.3390/brainsci11121644.

Gainotti G (2022). Hemispheric asymmetries for emotion in non-human primates: a systematic review. Neurosci Biobehav Rev 141C: 104830. https://doi.org/10.1016/j.neubiorev.2022.104830.

Gainotti G (2023). Some historical notes orienting towards brain mechanisms that could underlie hemispheric asymmetries. Cortex 163: 26–41. https://doi.org/10.1016/j.cortex.2023.03.001.

Gainotti G, Marra C (2011). Differential contribution of right and left temporo-occipital and anterior temporal lesions to face recognition disorders. Front Hum Neurosci 5: 55. https://doi.org/10.3389/fnhum.2011.00055. eCollection 2011.

Gainotti G, Caltagirone C, Miceli G et al. (1981). Selective semantic-lexical impairment of language comprehension in right-brain-damaged patients. Brain Lang 13: 201–211. https://doi.org/10.1016/0093-934X(81)90090-0.

Gainotti G, D'Erme P, Bartolomeo P (1991). Early orientation of attention toward the half space ipsilateral to the lesion in patients with unilateral brain damage. J Neurol Neurosurg Psychiatry 54: 1082–1089. https://doi.org/10.1136/jnnp.54.12.1082.

Gainotti G, Caltagirone C, Zoccolotti P (1993). Left/right and cortical/subcortical dichotomies in the neuropsychological study of human emotions. Cognit Emot 7: 71–93. https://doi.org/10.1080/02699939308409178.

Gazzaniga MS (2000). Cerebral specialization and interhemispheric communication: does the corpus callosum enable the human condition? Brain 123: 1293–1326. https://doi.org/10.1093/brain/123.7.1293.

Gunnar MR, Donzella B (2004). Tympanic membrane temperature and emotional dispositions in preschool-aged children: a methodological study. Child Dev 75: 497–504. https://doi.org/10.1111/j.1467-8624.2004.00690.x.

Guo CC, Sturm VE, Zhou Jet al. (2016). Dominant hemisphere lateralization of cortical parasympathetic control as revealed by frontotemporal dementia. Proc Natl Acad Sci U S A 113: E2430–E2439. https://doi.org/10.1073/pnas.1509184113.

Hagemann D, Waldstein SR, Thayer JF (2003). Central and autonomic nervous system integration in emotion. Brain Cogn 52: 79–87. https://doi.org/10.1016/S0278-2626(03)00011-3.

Harrington A (1985). Nineteenth-century ideas on hemispheric differences and duality of mind: target article and commentaries. Behav Brain Sci 8: 617–660.

Harris LJ (1999). Early theory and research on hemispheric specialization. Schizophr Bull 25: 11–40. https://doi.org/10.1093/oxfordjournals.schbul.a033359.

Heilman KM, Schwartz HD, Watson RT (1978). Hypoarousal in patients with the neglect syndrome and emotional indifference. Neurology 28: 229–232. https://doi.org/10.1212/wnl.28.3.229.

Helton WS (2010). The relationship between lateral differences in tympanic membrane temperature and behavioral impulsivity. Brain Cogn 74: 75–78. https://doi.org/10.1016/j.bandc.2010.06.008. Epub 2010 Jul 24.

Herlyn H (2016). The phylogenetic system of primates—character evolution in the light of a consolidated tree. Org Divers Evol 16: 689–713. https://doi.org/10.1007/s13127-016-0279-1.

Izard CE (1971). The face of emotion, Appleton-Century-Crofts, New York.

Izard CE (1990). Facial expression and the regulation of emotions. Pers Soc Psychol 58: 487–498. https://doi.org/10.1037//0022-3514.58.3.487.

James W (1884). What is an emotion? Mind 9: 188–205.

Kosslyn SM (1987). Seeing and imagining in the cerebral hemispheres: a computational approach. Psychol Rev 94: 148–175.

Lazarus RS (1982). Thoughts on relations between emotion and cognition. Am Psychol 37: 1019–1024.

Lazarus RS (1984). On the primacy of cognition. Am Psychol 39: 124–129.

Leventhal H (1979). A perceptual-motor processing model of emotion. In: P Pliner, K Blankestein, IM Spiegel (Eds.), Perception of emotion in self and others. vol. 5. Plenum, New York, pp. 1–46.

Leventhal H (1987). A perceptual motor theory of emotion. In: L Berkowitz (Ed.), Advances in experimental social psychology. vol. 17. Academic Press, New York, pp. 117–182.

Lupyan G (2012). Linguistically modulated perception and cognition: the label-feedback hypothesis. Front Psychol 2012: 54. https://doi.org/10.3389/fpsyg.2012.00054.

Luys JB (1879). Etudes sur le dédoublement des opérations cérébrales et sur le rôle isolé de chaque hémisphère dans les phénemènes de la pathologie mentale. Bull Acad Natl Méd 8: 516–534. 547–65.

MacNeilage PF, Rogers LJ, Vallortigara G (2009). Origins of the left and right brain. Sci Am 301: 60–67. https://doi.org/10.1038/scientificamerican0709-60.

Mandal MK, Asthana HS, Tandon SC (1993). Judgment of facial expression of emotion in unilateral brain-damaged patients. Arch Clin Neuropsychol 8: 171–183.

Manning JT, Heaton R, Chamberlain AT (1994). Left-side cradling: similarities and differences between apes and humans. J Hum Evol 26: 77–83. https://doi.org/10.1006/jhev.1994.1005.

Mills CK (1912). The cerebral mechanisms of emotional expression. Trans Stud Coll Physicians Phila 34: 381–390.

Morawetz C, Riedel MC, Salo T et al. (2020). Multiple large-scale neural networks underlying emotion regulation. Neurosci Biobehav Rev 116: 382–395. https://doi.org/10.1016/j.neubiorev.2020.07.001.

Nauta WJH (1971). The problem of the frontal lobe: a reinterpretation. J Psychiatr Res 8: 167–187.

Nicholls ME, Wolfgang BJ, Clode D et al. (2002). The effect of left and right poses on the expression of facial emotion. Neuropsychologia 40: 1662–1665. https://doi.org/10.1016/s0028-3932(02)00024-6.

Oatley K (1994). Emotion. In: MW Eysenck (Ed.), The Blackwell dictionary of cognitive psychology. Blackwell, Oxford, pp. 129–134.

Oatley K, Johnson-Laird P (1987). Toward a cognitive theory of emotions. Cognit Emot 1: 29–50. https://doi.org/10.1080/02699938708408362.

Oatley K, Johnson-Laird P (2014). Cognitive approaches to emotions. Trends Cogn Sci 18: 134–140. https://doi.org/10.1016/j.tics.2013.12.004.

Ocklenburg S, Güntürkün O (2018). Hemispheric asymmetries in emotion processing. In: S Ocklenburg, O Güntürkün (Eds.), The lateralized brain: the neuroscience and evolution of hemispheric asymmetries. Academic Press, New York, pp. 213–238.

Ocklenburg S, Peterburs J, Mundorf A (2022). Hemispheric asymmetries in the amygdala: a comparative primer. Prog Neurobiol 2022: 102283. https://doi.org/10.1016/j.pneurobio.2022.102283.

Palomero-Gallagher N, Amunts K (2022). A short review on emotion processing: a lateralized network of neuronal networks. Brain Struct Funct 227: 673–684. https://doi.org/10.1007/s00429-021-02331-7.

Papez JW (1937). A proposed mechanism of emotion. Arch Neurol Psychiatry 79: 217–224.

Perria L, Rosadini G, Rossi GF (1961). Determination of side of cerebral dominance with amobarbital. Arch Neurol 4: 173–181.

Pessoa L (2008). On the relation between emotion and cognition. Nat Rev Neurosci 9: 148–158. https://doi.org/10.1038/nrn2317.

Piaget J (1967). In: D Elkind (Ed.), Six psychological studies. Vintage, NewYork.

Plutchik R (1980). Emotion: a psychobioevolutionary synthesis, Harper and Row, New York.

Propper RE, Brunyé TT (2013). Lateralized difference in tympanic membrane temperature: emotion and hemispheric activity. Front Psychol 4: 104. https://doi.org/10.3389/fpsyg.2013.00104.

Reuter-Lorenz P, Davidson RJ (1981). Differential contribution of the two cerebral hemispheres to the perception of happy and sad faces. Neuropsychologia 19: 609–613. https://doi.org/10.1016/0028-3932(81)90030-0.

Reuter-Lorenz P, Givis RP, Moscovitch M (1983). Hemispheric specialization and the perception of emotion: evidence from right-handers and from inverted and noninverted left handers. Neuropsychologia 21: 687–692. https://doi.org/10.1016/0028-3932(83)90068-4.

Rizzolatti G, Umilta C, Berlucchi G (1971). Opposite superiorities of the right and left cerebral hemispheres in discriminative reaction time to physiognomical and alphabetical material. Brain 94: 431–442. https://doi.org/10.1093/brain/94.3.43.

Rogers LJ, Andrew JR (2002). Comparative vertebrate lateralization, Cambridge University Press, Cambridge, U.K.

Ross ED, Monnot M (2008). Neurology of affective prosody and its functional-anatomic organization in right hemisphere. Brain Lang 104: 51–74. https://doi.org/10.1016/j.bandl.2007.04.007.

Rossi GF, Rosadini G (1967). Experimental analysis of cerebral dominance in man. In: CJ Millikan, FL Darly (Eds.), Brain mechanisms underlying speech and language. Grune and Stratton, New York.

Salk L (1973). The role of the heartbeat in the relations between mother and infant. Sci Am 228: 24–29.

Schachter S, Singer JE (1962). Cognitive, social, and physiological determinants of emotional state. Psychol Rev 69: 379–399.

Scherer KR (1984). On the nature and function of emotion: a component process approach. In: KR Scherer, P Ekman (Eds.), Approaches to emotion. Erlbaum, Hillsdale, NJ, pp. 293–317.

Schilder P (1935). The image and appearance of the human body. K Paul, London: UK.

Schmitt JJ, Hartje W, Willmes K (1997). Hemispheric asymmetry in the recognition of emotional attitude conveyed by facial expression, prosody and propositional speech. Cortex 33: 65–81. https://doi.org/10.1016/s0010-9452(97)80005-6.

Simmons WK, Barsalou LW (2003). The similarity-in-topography principle: reconciling theories of conceptual deficit. Cogn Neuropsychol 20: 451–486. https://doi.org/10.1080/02643290342000032.

Snowden JS, Thompson JC, Neary D (2004). Knowledge of famous faces and names in semantic dementia. Brain 127: 860–872. https://doi.org/10.1093/brain/awh099.

Sokolov A (1972). Inner speech and thought, Plenum Press, New York, NY, USA.

Stankovic M (2021). A conceptual critique of brain lateralization models in emotional face perception: toward a hemispheric functional-equivalence (HFE) model. Int J Psychophysiol 160: 57–70. https://doi.org/10.1016/j.ijpsycho.2020.11.001.

Sturm VE, Sible IJ, Datta S et al. (2018). Resting parasympathetic dysfunction predicts prosocial helping deficits in behavioral variant frontotemporal dementia. Cortex 109: 141–155. https://doi.org/10.1016/j.cortex.2018.09.006.

Tamietto M, Latini Corazzini L, de Gelder B et al. (2006). Functional asymmetry and interhemispheric cooperation in the perception of emotions from facial expressions. Exp Brain Res 171: 389–404. https://doi.org/10.1007/s00221-005-0279-4.

Terzian H (1964). Behavioural and EEG effects of intracarotid sodium amytal injection. Acta Neurochir 12: 230–239.

Terzian H, Cecotto S (1958). Su un nuovo metodo per la determinazione e lo studio della dominanza emisferica. G Psichiatr Neuropatol 87: 889–924.

Vallortigara G, Rogers LJ (2020). A function for the bicameral mind. Cortex 124: 274–285. https://doi.org/10.1016/j.cortex.2019.11.018.

Vallortigara G, Rogers LJ, Bisazza A (1999). Possible evolutionary origins of cognitive brain lateralization. Brain Res Brain Res Rev 30: 164–175. https://doi.org/10.1016/s0165-0173(99)00012-0.

Vytal K, Hamann S (2010). Neuroimaging support for discrete neural correlates of basic emotions: a voxel-based meta-analysis. J Cogn Neurosci 22: 2864–2885. https://doi.org/10.1162/jocn.2009.21366.

Winsler A, Naglieri J (2003). Overt and covert verbal problem-solving strategies: developmental trends in use, awareness, and relations with task performance in children aged 5 to 17. Child Dev 74: 659–678. https://doi.org/10.1111/1467-8624.00561.

Yousaf A, Liu J, Ye S et al. (2021). Current progress in evolutionary comparative genomics of great apes. Front Genet 12: 657468. https://doi.org/10.3389/fgene.2021.657468.

Zajonc RB (1980). Feeling and thinking: preferences need no inferences. Am Psychol 2: 151–176.

Zajonc RB (1984). On the primacy of affect. In: KR Scherer, P Ekman (Eds.), Approaches to emotion. Erlbaum, Hillsdale, NJ, pp. 239–246.

Further reading

Gainotti G (2019a). Emotions and the right hemisphere: can new data clarify old models? Neuroscientist 25: 258–270. https://doi.org/10.1177/1073858418785342.

Gainotti G (2021c). Unconscious processing of emotions and the right hemisphere. In: KM HeilmanSE Nadeau (Eds.), Disorders of emotion in neurologic disease. Handbook of clinical neurology. vol. 183. Elsevier, Amsterdam, Chap. 2, pp. 27–46.

LeDoux J (1996). The emotional brain. The mysterious underpinnings of emotional life, Simon and Schuster, New York.

Handbook of Clinical Neurology, Vol. 208 (3rd series)
Cerebral Asymmetries
C. Papagno and P. Corballis, Editors
https://doi.org/10.1016/B978-0-443-15646-5.00010-5

Chapter 28

Hemispheric asymmetries in face recognition in health and dysfunction

MARLENE BEHRMANN*

Department of Ophthalmology, University of Pittsburgh School of Medicine, Pittsburgh, PA, United States

Abstract

A defining characteristic of the human brain is that, notwithstanding the clear anatomic similarities, the two cerebral hemispheres have several different functional superiorities. The focus of this chapter is on the hemispheric asymmetry associated with the function of face identity processing, a finely tuned and expert behavior for almost all humans that is acquired incidentally from birth and continues to be refined through early adulthood.

The first section lays out the well-accepted doctrine that face perception is a product of the right hemisphere, a finding based on longstanding behavioral data from healthy adult human observers. Data are then presented from neuropsychologic studies conducted with individuals with prosopagnosia, which is either acquired after a lesion to the right hemisphere or is developmental in nature with no obvious lesion. The second section reviews data on the neural correlates of face perception, gathered using a host of imaging methodologies all the way from electroencephalography (EEG) through functional magnetic resonance imaging (fMRI) and positron emission tomography (PET) studies to transcranial magnetic stimulation and intracranial depth recording. The penultimate section reviews empirical findings that track the emergence of the hemispheric asymmetry for faces, and offers a theoretical proposal that lays out possible origins of the adult asymmetry profile. Lastly, the hemispheric asymmetry associated with the perception of emotional face expression is considered. While considerable progress has been made in understanding the functional organization of the human cerebral cortex and its biases and asymmetries, much remains to be determined and the many inconsistencies remain to be reconciled in future research.

INTRODUCTION

One defining characteristic of the human brain is that, despite the anatomic and structural equivalence, the two cerebral hemispheres have different functional superiorities. This hemispheric lateralization of function is thought to be phylogenetically recent given its absence in close relatives of humans such as old-world monkeys, as documented in studies of population-level behavioral, neuroanatomic, and functional asymmetries (Corballis, 1982; Ettlinger, 1988). Consistent with this, no significant differences between face patches of the left hemisphere (LH) and right hemisphere (RH) in macaque monkeys were uncovered in a study using transient pharmacologic inactivation coupled with functional magnetic resonance imaging (fMRI) to explore the circuitry and directional connectivity underlying face processing (Liu et al., 2022). Other studies, however, have uncovered hemispheric differences in nonhuman primates perhaps primarily for the perception of emotional stimuli (Parr and Hopkins, 2000) [for a review, see Hopkins and Cantalupo, 2008 and Hopkins et al., 2015] (Fig. 28.1).

The discussion of hemispheric superiorities in humans and the origin and potential malleability of these lateralized differences remains ongoing and continues to motivate hundreds, if not thousands, of investigations of human brain functional organization. As will be evident from this brief review, notwithstanding the massive effort to elucidate the functional arrangement of the cerebral cortices, the exact nature and extent of their lateralization remain unsettled. Here, we focus specifically on the lateralized profile or hemispheric superiority of one particular function, namely face processing, which has long been attributed to the RH in human adults. While some researchers have argued that face perception is solely a function of the RH, others have argued for a RH advantage or a bias that coexists with a more limited left hemisphere (LH) contribution, rather than a purely binary assignment of face processing to the RH [for a review, see Petunia et al., 2024]. Findings that challenge the RH dominance view

*Correspondence to: Marlene Behrmann, PhD, Department of Ophthalmology, University of Pittsburgh School of Medicine, Pittsburgh, PA, 15213, United States. E-mail: mbehrmann@pitt.edu

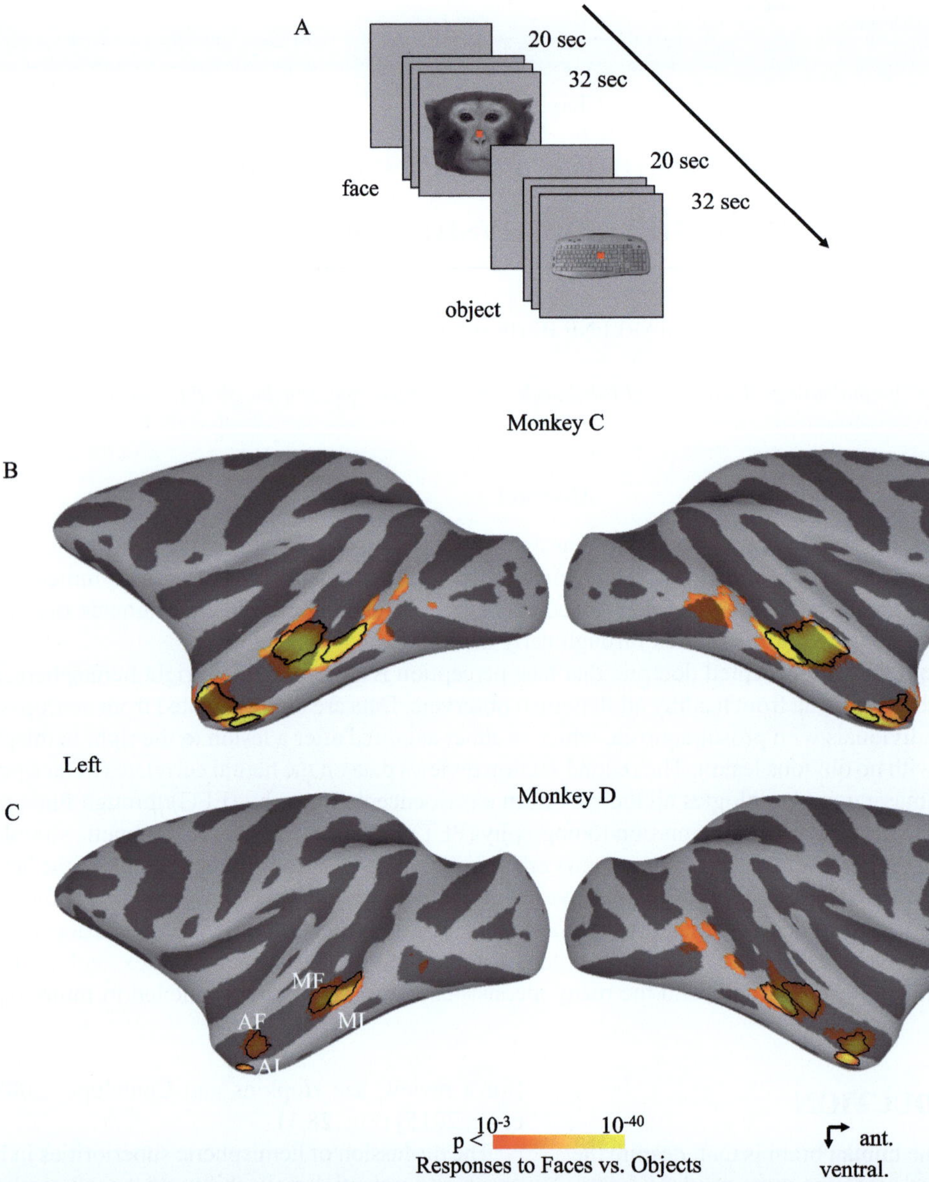

Fig. 28.1. fMRI experiment and face-selective patches in two monkeys, C and D. (A) An illustrative example of a sequence of monkey faces and objects displayed to monkeys while fMRI data were collected. (B), (C) Face-selective (neutral monkey faces vs objects, $P < 0.005$ uncorrected) activation maps shown on lateral views of the inflated cortex in monkeys C and D. The selected voxels in each ROI are encircled by black lines. MF middle fundal face patch, ML middle lateral face patch, AF anterior fundal face patch, and AL anterior lateral face patch. From Liu N, Behrmann M, Turchi JN, et al. (2022). Bidirectional and parallel relationships in macaque face circuit revealed by fMRI and causal pharmacological inactivation. Nat Commun 13: 6787; an open access article distributed under the terms of the Creative Commons CC BY license).

of behavior-brain correspondence have demonstrated that laterality may fluctuate and, possibly, even be reversed (Wu et al., 2022). Wu et al. (2022) used dynamic functional connectivity measures to quantify laterality fluctuations, defined as the standard deviation of the laterality time series, and laterality reversal, defined as the number of zero crossings in the laterality time series. They used dynamic changes of lateralization to map resting state data from the Human Connectome Project (http://www.humanconnectome.org/data) and noted that variations in laterality were primarily in lower order brain regions, including in visual and other sensory areas, but not in higher order brain regions. This pattern of fluctuation is conceptually consistent with the fact that lower order regions have shorter intrinsic time scales and, bilaterally, are required to respond to rapidly changing input, which is then integrated to yield a coherent, more stable percept. This variability obviously and directly challenges the central assumption that laterality is a stable trait (Liu et al., 2009; Gotts et al., 2013) that optimizes neural function (Rogers and Vallortigara, 2017).

The question to be addressed here then is whether, how, and to what extent, face perception emerges from computations mediated solely by the RH and if so, what mechanisms and pressures drive this lateralized function in humans. First, we describe current evidence supporting the now-longstanding claim of the RH superiority for face perception, largely drawing on findings from behavioral studies of typical individuals, and we address some possible properties of the RH that predisposes the RH to be well suited for computations underlying face perception. We then provide findings favoring a RH advantage for face processing based on results from neural studies, starting with neuropsychologic evidence from individuals with prosopagnosia, either of the acquired variety or of the type that appears lifelong and apparent from birth. We follow this with a presentation of data aggregated from neuroimaging methodologies, including electroencephalography (EEG) and fMRI. We also include results from newer data acquisition methods like transcranial magnetic stimulation and intracranial depth electrodes, and review both spatial as well as latency differences for face perception compared to the perception of other stimulus classes in the two cerebral hemispheres. We then review the literature on the development of face perception over human ontogeny [for a review of face perception over phylogeny, see Behrmann and Avidan, 2022], present an account of the origin of this functional organization, and briefly consider the mechanisms associated with the perception of emotional expression on faces.

EVIDENCE FOR RH SUPERIORITY FOR FACE PERCEPTION IN TYPICAL OBSERVERS

The longstanding claim that the RH has greater involvement than the LH in face perception comes from studies largely conducted in the 1970s and 1980s using the half-field or tachistoscopic procedure. This method entails having an observer seated in front of a screen while maintaining fixation on the center of the screen; ideally, the observer's eye movements are monitored to ensure maintenance of fixation. A stimulus is presented briefly (too brief to permit a saccade so ideally, for 150 ms or less) to either the left or right visual field and, depending on the task, the participant names the displayed item or makes a decision about it (for example, is the image a face or an object? Is the face smiling or frowning?). Because there is, at least initially, preferential processing of a visual stimulus in the hemisphere contralateral to the field, one can determine whether performance is better for left (RH) or right (LH) visual field stimuli. In this way, one can determine which hemisphere has the advantage for face processing, and the findings consistently point to the RH as having the advantage or superiority for face perception.

Another common paradigm included the use of chimeric stimuli or composite symmetric faces made by, for example, combining the left or right half of a face with its corresponding mirror image (or halves of two different faces). In this latter paradigm, participants fixate the center of the image such that each half of the composite falls in one visual field. Participants indicate which composite (right or left visual field) more closely resembles the original or target face usually displayed prior to the presentation of the chimera. Regardless of the upright or inverted orientation of the image, a preference for the composite half in the left visual field was demonstrated in typical observers, consistent with a RH advantage in face identification (Kolb et al., 1983). These studies also showed that the RH was more engaged for the recognition of previously unfamiliar faces (Levy et al., 1972) and for the recognition of familiar faces, although this latter result is more controversial (Levine and Koch-Weser, 1982). One possible concern with some of these studies, however, is that the RH has other functions that might have supported this advantage: for example, attentional function is generally attributed to the RH (Shulman et al., 2010), and, so, the left field advantage might have been attentional rather than face-selective in nature (a comparison or control from a different stimulus class showing no left visual field advantage was not always tested). These studies also tended to place emphasis and interpret the left visual field/RH advantage in binary terms but there was also, almost always, some selection of the composite face that appeared in the right visual field.

This demonstrable RH face advantage has been well complemented by many recent studies. For example, in a study with a large and heterogeneous sample of 122 participants all of whom completed multiple tasks with unilateral or bilateral stimulus presentation, group-level analyses confirmed the predicted RH lateralization by revealing correlations between performance on faces, global forms, low spatial frequency processing, and spatial attention tasks, and the predicted LH lateralization through correlations between visual word and local feature processing (but not high spatial frequency processing) (Brederoo et al., 2020). These authors attributed the lateralized asymmetries to the causal complementarity principle according to which different processes recruiting similar brain regions come to be lateralized to homolog areas in opposite hemispheres [for related ideas, see the following studies that demonstrate neural competition over development with fine-tuning the LH lateralization for visual word processing during literacy acquisition and its correlation with the RH lateralization for face processing (Dundas et al., 2013)].

For a long time, the RH superiority for face perception was assumed to be a product of the general superiority in visuospatial tasks including configural or holistic processing [but see Gerlach and Poirel, 2020 for critical analysis of divided field measures] but alternative hypotheses have also been proposed. Among these alternative hypotheses is the notion that, whereas the LH processes visual input at a more local or featural level, the RH processes information at a more holistic or global level. For example, compound stimuli (Navon, 1977) made of the letters "H" and "S" whose identities are the same (consistent) or different (inconsistent) at the local and global level (see Fig. 28.2A) have been shown to the right visual field/LH or left visual field/RH of typical observers. In different blocks of trial with appropriate instructions to identify the letter at either the local or global level, participants indicated the letter identity (for example, by pushing either a "H" or "S" button, Fig. 28.2A). Identifying the local letters was faster for stimuli presented in the right visual field/LH than in the left visual field/RH, and the converse was true for the global letter identification trials (Van Kleeck, 1989; Ivry and Robertson, 1998).

According to Sergent (1982), this local–global hemispheric asymmetry may be a manifestation of a more fundamental spatial frequency sensitivity, with the LH biased toward high spatial frequencies and, hence, responding better to the local elements, and the RH evincing a bias to processing low spatial frequency

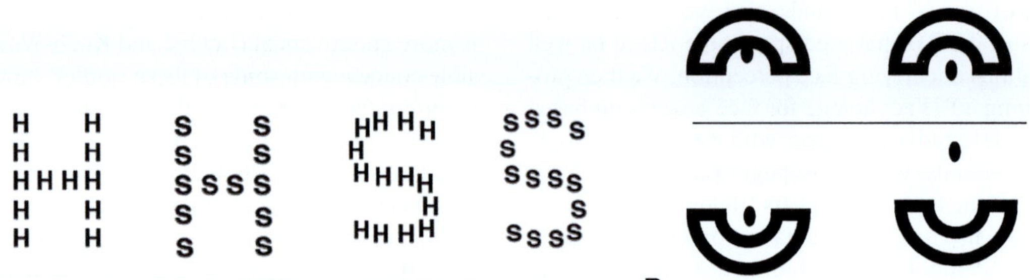

A B

Fig. 28.2. Schematic depiction of stimuli to illustrate experimental conditions. (A) Example of "Navon" or compound letters made of "H" and "S" stimuli, with "consistent" stimuli, i.e., same letter identity at the local and global level (first and fourth examples) and "inconsistent" stimuli, i.e., different letter identities at the global and local level (the second and third examples). (B) Two kinds of judgments were made on stimuli such as these: (left column) is the dot on or off the contour of a line drawing of a blob (the on/off task; categorical) or (right column) is the dot within 2 mm of the contour (the distance task; coordinate).

information and, hence, the global identity (Sergent, 1982; Ivry and Robertson, 1998; Robertson and Ivry, 2000). Proposals such as these dovetail well with the Brederoo et al. (2020) report mentioned earlier that documented the laterality-specific correlations between spatial frequency, global vs local processing, and word vs face perception.

The specific link to face perception comes from the claim that the LH high spatial frequency superiority permits the processing of local face features while the low spatial frequency RH advantage exists for and enables processing a face as a global whole. Indeed, in a study using positron emission tomography (PET) scanning (Rossion et al., 2000), the right middle fusiform gyrus was more activated when participants matched whole faces than parts of faces and the left homologous region was more activated for the part than for the whole condition. This held only for faces but not for a control condition of houses and led to the conclusion of lateralized localization of mechanisms supporting face processing. In a related fMRI study, the right fusiform gyrus signal profile was correlated with categorical face vs non-face judgments (Is this a "face"?) (Meng et al., 2012). The left fusiform gyrus activation, however, was correlated with image-level face-semblance, implicating processing of face resemblance through a more featural process, which, then, provides this earlier emerging signal to deeper, categorical analyses by the RH.

Over time, other claims regarding fundamental but general differences between the hemispheres have also been offered to account for the RH face advantage, although these other proposals are not necessarily mutually exclusive with the hypotheses about spatial frequency and feature or part/whole distinctions. One example hypothesis is that there are two kinds of spatial relations, one tuned more for categorical relations and one for specific coordinate representations. When human participants had to decide whether the dot was on or off (categorical judgment) the object shape (see Fig. 28.2B), performance was better for stimuli presented to the right visual field/LH than left visual field/RH. When the decision required was the coordinate or metric distance (e.g., the dot within 2 mm of the shape boundary), responses were faster for stimuli presented to the left visual field/RH than the right visual field/LH (Kosslyn et al., 1989).

Another related account explains the hemispheric differences as a result of the extent to which there is coarse vs fine encoding of the input. The claim is that neurons in the RH and LH have differently sized receptive fields, with larger fields in the RH than LH (although this account was formulated primarily for parietal cortex differences in the two hemispheres) (Monaghan and Shillcock, 2004). Unsurprisingly, if this were the case, more coarse larger-scale processing would be a property of the RH and smaller, finer-grain processing would be a property of the LH.

A last, different perspective comes from the claim that hemispheric asymmetries operate at an encoding stage beyond the sensory level. The proposal is that the asymmetries arise as a result of differences in within-hemisphere anatomic connections with more distal connection structure between columns in language areas of the LH and more local connection structure between columns in the RH homotopic region (Hsiao et al., 2013). To explore the implications of this proposal, Hsiao et al. (2013), in a series of neurocomputational simulations, demonstrated the emergence of both processing differences and differential frequency filtering, thereby replicating a host of empirical findings reviewed above. These simulations led to the suggestion that it is not frequency tuning per se that drives the hemispheric asymmetries but, rather, the anatomic differences in connection structure are key [see also Gomez et al., 2019 for changes in receptive field size over development in the two hemispheres].

NEURAL EVIDENCE FOR THE ROLE OF THE RH IN FACE PERCEPTION

Neuropsychologic investigations

Studies of typical observers who have delineated behavioral asymmetries, while intriguing and highly suggestive, do not provide a causal link between face perception and the RH [see the transient "lesion model" in Liu et al., 2022 at the beginning of this chapter]. This link in humans has long been inferred on the basis of neuropsychologic studies of individuals with a deficit in recognizing faces termed "prosopagnosia" [from the Greek prosopon (face) and gnosis (knowledge)]. Prosopagnosia in premorbidly normal face recognizers can occur following brain injury from stroke, trauma, or tumor. A second type of prosopagnosia, termed "congenital" (CP) or "developmental" prosopagnosia, assessed over the past few decades, can manifest in individuals with a clearly disproportionate difficulty in recognizing faces but in the absence of neurologic damage and after ample exposure to face stimuli. We first describe similarities and differences in the two forms of prosopagnosia and then consider their underlying etiologies.

In both instances of prosopagnosia, with the former being more rare than the latter, the deficit cannot be accounted for by a loss in sensory vision or abnormal low-level visual perception nor by a deficit such as anomia or a semantic impairment. Individuals with prosopagnosia are impaired at identifying individual people by face, even close family members, and are often, but perhaps not always, also impaired at discriminating between individuals (Behrmann and Avidan, 2005; Behrmann et al., 2005a). In prosopagnosia, recognition is usually accomplished using other visual cues such as gait and body posture [but see Biotti et al., 2017] or salient clues about hairstyle or clothing and nonvisual cues, predominantly voice. Several studies have noted the specificity of the face perception impairment and have demonstrated that other facial information, such as facial expression and emotions, may be spared (Lahiri, 2020; Bell et al., 2023; Bennetts et al., 2024) (see section on Facial expression perception in the following).

Whether the recognition of other objects is also affected remains highly controversial. Whereas some have argued that more complex high-level regions of the visual cortex participate in processing multiple object types and is not domain-specific (Behrmann and Plaut, 2015), others have proposed that regions associated with face recognition are domain-specific, more modular in nature, and dedicated to processing only one object type, namely the face (McKone and Kanwisher, 2005). In one study that examines the generality of the visual recognition deficits in prosopagnosia, individuals with developmental prosopagnosia showed deficits in the recognition of both human and animal faces, and, close inspection at the single case level revealed that of the 60% of cases in whom face recognition was impaired, there was a concurrent deficit in the recognition of animal faces (Epihova et al., 2023). A similar conclusion was reached in a review paper that examined published papers on CP cases from 1976 to 2016. Of the 238 CP cases whose data permitted a satisfactory evaluation of the face vs non-face comparison, 80.3% of the individuals were impaired at face and object recognition, and only 19.7% were impaired at face perception alone (Geskin and Behrmann, 2018). Other studies have also pointed out that this congenital form of prosopagnosia is often accompanied by other developmental comorbidities such as an impairment in memory or ADHD or dyslexia (Svart and Starrfelt, 2022) although some dissociations are also present (Kuhn et al., 2021) and some findings are more mixed (Bate et al., 2019; Gray et al., 2019).

Acquired prosopagnosia

Acquired prosopagnosia is one consequence of a lesion to the RH occipitotemporal lobe or middle fusiform gyrus cortex. A lesion on the left side may also result in prosopagnosia (Meadows, 1974; Grüsser and Landis, 1991), albeit less often and less severely (Barton, 2008; Behrmann and Plaut, 2014). For an historical overview of prosopagnosia and its relationship to the lesion site, see Gainotti and Marra (2011) and Gainotti (2023) and for a recent review of the unilateral right and left and bilateral lesions in those with prosopagnosia, see Barton (2022). One recent interesting source of neuropsychologic data comes from the Back-of-the-Brain project, a detailed neuroimaging and neuropsychologic examination of 64 patients with unilateral Posterior Cerebral Artery stroke (Rice et al., 2021; Robotham et al., 2023). Of the patients, 32 had a LH lesion, 23 had a RH lesion and 9 had bilateral lesions. The results are of particular interest in that very few patients (6 out of 64) showed a selective deficit in only one domain, and this was in the domain of word processing. About one-third of the patients showed a significant impairment across domains, irrespective of side of unilateral lesion or bilateral lesions and, surprisingly, a further third of the patients showed no deficits whatsoever. While total lesion volume had the strongest relationship with behavioral performance, face recognition impairment was not related to lesion laterality per se but, rather, in a more equipotential way, was related to lesion volume across both hemispheres. These findings offer clear evidence that both hemispheres are involved in higher order processing of faces (and of words too) and deficits can be observed after damage to these extrastriate regions in either hemisphere even in adulthood; see also Behrmann and Plaut (2014). [For a review of hemispheric variability in EEG studies, see Petunia et al., 2024].

In addition to the debate on lateralization, there is ongoing discussion regarding the lesion site, whether more posterior or more anterior just within the RH (Cohen et al., 2019; Barton, 2022). Notably, some individuals who have more anterior temporal lesions (Evans et al., 1995; Barton, 2003) present with a concurrent deficit in voice recognition (Gainotti, 2013, 2023). As illustrated in Fig. 28.1, just as there are multiple face-selective or face "patches" in monkeys, multiple face-selective regions are evident in humans too in both hemispheres albeit to a greater degree in the right hemisphere (Gobbini and Haxby, 2007). Although there may not be direct homology between humans and nonhuman primates (Rossion and Taubert, 2019), the host of face-selective regions in both species indicates that face perception likely requires the joint activity of multiple brain regions distributed throughout the "core" occipital and temporal lobes (and "extended" regions as well), typically with a RH advantage, and some patches/regions contributing slight different computations [e.g., whether patch has viewpoint invariance or not (Dubois et al., 2015)]. Some theories consider this face network a processing hierarchy beginning with the inferior occipital gyrus and fusiform gyrus and projecting to more anterior temporal and even frontal lobe regions while others have suggested a multiple-route network with nonhierarchical components and no strict posterior to anterior directionality. For example, a well-studied prosopagnosic patient, PS, who has a right inferior occipital gyrus lesion, still showed typical face-selectivity more anteriorly in the right fusiform gyrus, implicating direct connectivity from the early visual cortex and the fusiform area and circumventing the proposed intermediate occipital face area (Rossion, 2022).

Congenital or developmental prosopagnosia

Having reviewed the neural characteristics of acquired prosopagnosia, we now consider the neural basis of the congenital (CP) or developmental form of the deficit. CP is not accounted for by any obvious neurologic concomitant and no disease is readily identifiable on conventional MRI neuroimaging. However, multiple, differing hypotheses regarding the etiology have been postulated. One early study of the morphometry and volume of the occipitotemporal cortex in a small group of individuals with CP and matched controls revealed no significant group differences in the depth or deviation from the midline of the occipitotemporal lobe or collateral sulci. In contrast, a larger anterior and posterior middle temporal gyrus and a significantly smaller anterior fusiform (aF) gyrus in the RH was revealed in the CP individuals, and this anatomic change was correlated with the extent of the face recognition deficit (Behrmann et al., 2007). Of course, the question, then, is what gives rise to this particular anatomic alteration. Subsequent research using diffusion tensor imaging offered a possible explanation. In a group of individuals with CP, a relationship was observed between the structural integrity of the tracts that project through the ventral temporal occipital cortex, especially in the RH, and the severity of the face recognition deficit (Thomas et al., 2009; Zemmoura et al., 2021) [and see Lohse et al., 2016]. Specifically, the data revealed that the white-matter fibers of the inferior longitudinal fasciculus and inferior fronto-occipital fasciculus showed a reduction in microstructural integrity (fractional anisotropy measure). No group differences, either macro- or microstructural, were evident when other tracts such as the forceps major and forceps minor were quantified [for similar results, see Zhao et al., 2022].

Note that some of these structural MRI results have not been replicated; whether different criteria are used for participant inclusion, whether there is substantial variability in the population itself, or whether the variability is a product of different methodological approaches is not yet clear. For example, in one study of CP and age-matched control participants, the group differences emerged not in long-range microstructural differences as noted above but, rather, in the differences in local structural changes to face-specific regions of interest, and these changes were correlated with face perception success (Song et al., 2015) [see also Zhang et al., 2015].

Findings from fMRI have also shown differences between CP individuals and controls. Consistent with the structural MRI above, the interruption of signal propagation in the RH from more posterior to more anterior regions is evident in the reduction of functional connectivity during the resting state between more posterior sites and more anterior temporal lobe regions (Avidan et al., 2014). Also, whereas the RH anterior temporal lobe typically serves as the major network hub for face processing in controls, the CPs evinced hyper-connectivity in the posterior regions of the visual cortex, mostly associated with the lateral occipital and the inferior temporal cortices and not with the anterior temporal lobe (Rosenthal et al., 2017) [also see a theoretical framework for explaining eye movement differences, population receptive fields (pRFs), and connectivity (Avidan and Behrmann, 2021)]. In a study using multi-voxel pattern analysis showing the wide extent of the cortical perturbations, relative to controls, reduced discriminability between faces and objects was identified in the CP individuals, in both the core (occipital and temporal face areas) as well as in the extended region (anterior temporal lobe) (Rivolta et al., 2014).

Although still in its relative infancy, there have been multiple attempts to develop rehabilitation strategies, especially for those with developmental prosopagnosia. Hypothetically, such studies might also contribute to a further understanding of hemispheric differences. The findings to date, however, mostly indicate that compensation for the deficit is not easily achievable [for a review, see DeGutis et al., 2014]. Left to their own devices, patients typically rely on cues such as voice, haircut, or clothing of other individuals but these cues are not foolproof and do not always suffice. Attempts at intervention have included training holistic processing with non-face stimuli (Behrmann et al., 2005b), enhancing sensitivity to spatial distances between features for faces, for example, increasing sensitivity to the between-eye or eye-nose distances or even non-face stimuli, or relying on developing mnemonic cues for faces (Bate and Bennetts, 2014; Davies-Thompson et al., 2017); but, none has yielded extraordinary results. Techniques used to train novice observers with emphasis on featural arrangement (Towler et al., 2021), on eye movement strategies (Bobak et al., 2016), or even with external intervention with oxytocin (Bate et al., 2014) and simultaneous stimulation [for an example of approach, see Willis et al., 2019] might also be a possible avenue to explore in future, and the hemispheric impact of these strategies will be of interest, as well.

MORE GENERAL REVIEW OF NEUROIMAGING STUDIES

Decades of studies have used EEG with measurement of responses to faces evoked by the presence of visual input (evoked response potentials, ERP) (Petunia et al., 2024). Seminal findings by many investigators have used a target detection task for examining the scalp EEG waveform potentials. One such study involved noting how many butterflies were embedded in a stream of unfamiliar human faces or among isolated face components, or even among inverted faces or other nonface stimuli. The major finding was the appearance of a negative potential at around 170 ms, which was the largest over the posterior temporal scalp and greater over the RH than LH (Bentin et al., 1996). This EEG approach has now been employed extensively in the mapping of face processing and has also demonstrated that a later negative potential around 250 ms can differentiate learned target faces from objects and from other faces, and that one's own face and one's own objects can elicit this potential too, even when not identified a priori as targets (Pierce et al., 2011). Much traction has been leveraged recently from coupling EEG measurement with a novel and implicit visual discrimination paradigm (FPVS: fast periodic visual stimulation) with high signal-to-noise ratio. In this approach, images (e.g., of faces) are displayed at a periodic frequency rate between, for example, 6 Hz. For faces, the largest response is observed over the right occipitotemporal cortex with a 6-Hz stimulation rate (corresponding to ~170 ms cycle duration) (Rossion, 2014) [see also Yan et al., 2019].

Consistent with the neuropsychologic data, the ascription of face perception to the RH found strong support following the advent of human neuroimaging. From the seminal papers of Sergent and colleagues using PET (Sergent et al., 1992; Sergent and Signoret, 1992) and of Kanwisher and colleagues using functional magnetic resonance (Kanwisher et al., 1997), the greater activation of the RH in response to faces in the "fusiform face area" or "FFA" was robustly confirmed and documented. Further support for the RH face perception advantage was also provided by correlational evidence that the strength of the left visual field bias during face perception, as measured using fMRI data, covaried with lateralization indices of the FFA (Yovel et al., 2008). Hundreds of fMRI studies have now replicated these findings but, interestingly, with close scrutiny of the data, many studies uncover activation in response to faces in homologous regions in the LH as well (e.g., Hasson et al., 2002; Wang et al., 2019). There are now several reviews offering a broad perspective of studies using fMRI to characterize the neural basis of face processing, and the interested reader may consult them for further details (e.g., Rossion et al., 2018; Rossion and Lochy, 2022).

NEW APPROACHES FOR EXAMINING RIGHT-HEMISPHERE FACE SUPERIORITY

A growing approach that has generated increasing interest is one in which face selectivity can be directly assayed from intracerebral electrophysiologic signals measured in individuals, who are being clinically evaluated for surgery for the treatment of drug-resistant epilepsy. In the early 1990s, Allison and colleagues (Allison et al., 1994a,b; McCarthy et al., 1999) reported findings of responses to face stimuli from strips of electrodes placed over ventral temporal cortex for the localization of seizure activity. Across many studies (Puce et al., 1999), they replicated the finding that a discrete region of the occipitotemporal cortex was activated in response to faces. More recently, but along similar lines, recordings from intracranial depth electrodes, as depicted in Fig. 28.3A, have indicated widespread neural activity in response to faces (using FPVS in some cases), extending broadly even as far as the anterior regions of the ventral occipitotemporal cortex (Rossion et al., 2023). There is, however, clustering of the contacts with greater amplitude of face selectivity to the RH in the vicinity of the middle section of the lateral fusiform gyrus, consistent with much of the existing neuroimaging data.

Intracranial data has also permitted examination of the medial to lateral loci of activation for faces relative to other stimuli such as words in adjoining cortical regions (see Fig. 28.3B) and, for the first time, has enabled researchers to uncover the relative temporal dynamics and location of regions that respond to centrally presented faces (and words), as is common in daily life to ensure high-resolution foveal input. Face-selective responses were initially evident bilaterally in posterior fusiform regions bilaterally, and then were propagated to more anterior regions. Studies such as these begin to offer glimpses into the complex and dynamic bihemispheric, albeit RH weighted, distributed circuits involved in face processing.

Last, transcranial magnetic stimulation (TMS), a method that creates a refractory period or temporal loss of function has been used as has the more recent variant, transcranial direct current stimulation (tDCS). Studies have, for example, uncovered

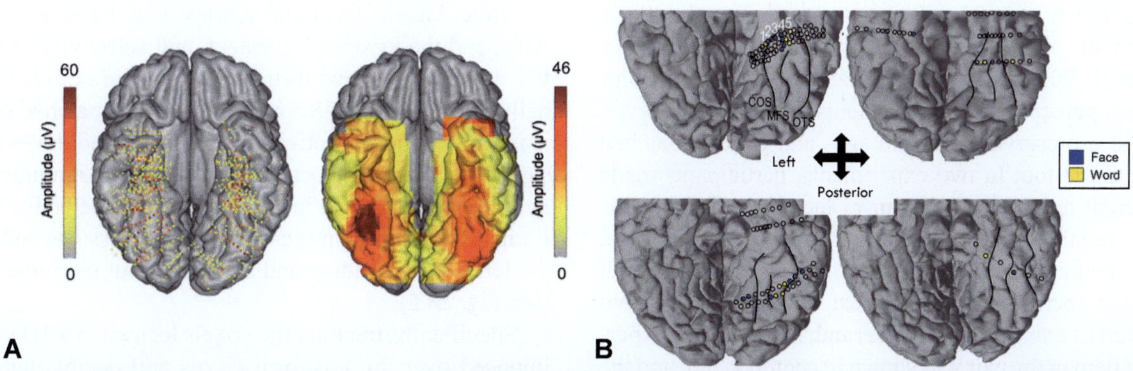

Fig. 28.3. (A) Quantification of face selectivity from intracranial depth electrodes at the group level in a common space. Significant contacts across 84 subjects were color-coded according to their relative face-selective amplitude and displayed in the Talairach space (left panel). Amplitude across contacts are smoothed and projected on the cortical surface by plotting the mean face-selective amplitude within 15 mm × 15 mm voxels (right panel). (B) Multiple adjoining word- and face-selective patches in four individual participants. See alternating bands of word- and face-selectivity along the left fusiform gyrus. Major VTC sulci (collateral sulcus [COS], midfusiform sulcus [MFS], and occipitotemporal sulcus [OTS]) have been outlined for clarity. Shaded electrodes are those selective to words (yellow) and faces (blue). Open circles represent ventral temporal electrodes that did not reach the selectivity criterion for either of these categories. Panel (A) adapted from Rossion B, Jacques C, Jonas J (2023). Intracerebral electrophysiological recordings to understand the neural basis of human face recognition. Brain Sci 13, with the author's permission. Panel (B) adapted with the author's permission from Boring MJ, Silson EH, Ward MJ, et al. (2021). Multiple adjoining word- and face-selective regions in ventral temporal cortex exhibit distinct dynamics. J Neurosci 41: 6314–6327.

evidence that same-different judgments or part-whole matching of faces, but not houses, can be disrupted by TMS to the right occipital face area (OFA), which is close to the lateral surface and more accessible from the scalp than the FFA (Pitcher et al., 2007; Bona and Silvanto, 2014; Bona et al., 2018). Likewise, when a sham version of tDCS was applied to the locus of electrode P08 in the RH (where face inversion effects are typically found using EEG), in participants viewing upright vs inverted faces of the same or different race from the participant, a large face inversion was observed especially for own-race faces, as would be expected. However, when anodal-tDCS was applied at this same location, this large own race inversion was reduced as performance for upright faces was adversely affected, offering insight into face perception in the RH (Civile et al., 2021; Civile and McLaren, 2022).

THE EMERGENCE OF LATERALIZED SUPERIORITY FOR FACES AND HEMISPHERIC PLASTICITY

Thus far, the focus of this review has been almost exclusively on the mature face recognition system, its neural bases and ways in which hemispheric asymmetry has been demonstrated. Of course, one might ask how this mature system and its remarkably replicable topography arises across individuals. Although face processing is clearly superior over the RH in adults, this does not seem to be the case for younger individuals. Provocatively, in individuals aged 5–17 years, a posterior injury (pre-, peri-, or post-natal) involving the temporal regions of either hemisphere resulted in a face processing deficit (Mancini et al., 1994; de Schonen et al., 2005).

Relatedly, a recent study in a relatively large group of 39 individuals (median age 16.7 years at testing, median age at surgery 5.5–6 years) all of whom had hemispherectomy during childhood for the treatment of pharmacologically intractable epilepsy (Granovetter et al., 2022) also showed bilateral hemispheric competence for face perception. Of the patients, 15 had a preserved LH and 24 had a preserved RH, and age- and gender-matched controls were tested too. In two experiments, participants made the same/different judgments to a pair of faces or pair of words presented sequentially (see Fig. 28.4 on top for the experimental procedure). In the first experiment, all participants viewed stimuli in central vision (permitting the contrast between having two hemispheres versus a single hemisphere) and, in the second experiment, the first item in the pair was shown in central vision and the second item was shown in the periphery (the intact visual field for the hemianopic patients), permitting the contrast between just a single hemisphere in the patients versus controls. The results from the two studies did not differ and, surprisingly, independent of central or peripheral presentation, both the left and right resection patients averaged approximately 85% accuracy (see Fig. 28.4 bottom). It is especially interesting that the preservation of just the LH or just the RH yielded equivalent results and that this held for both the face and word conditions. While the patients performed statistically more poorly than controls (with accuracies around 95%), the patients' performance was still impressive given the extent of the cortical resection (hemispherectomy or hemispherotomy). These findings suggest that, perhaps earlier in development, either hemisphere may be sufficient to assume representation of both stimulus classes, which then become increasing lateralized over the course of typical development. With age, then, the lateralization of functions (although not totally dichotomous) becomes more apparent. But, the finding that the mean accuracy of the patients does not reach normal performance still demands explanation. One possibility is that the relative accuracy decrement in the patients may result from increased competition or "neural crowding" in a limited neural space (Lidzba et al., 2006; Krageloh-Mann et al., 2017; Danguecan and Smith, 2019). Thus, cortical territory cannot fully accommodate both face- and word-processing functions in a single hemisphere, in contrast with controls, who have somewhat lateralized functions distributed over the two hemispheres.

Despite the consensus on mechanisms such as "neural crowding," little direct evidence for the competition between functions during the process of reorganization exists. Because longitudinal neuroimaging and behavioral data must be tracked to uncover the microgenesis of reorganization, there has not been much attention to this question. One study, which has addressed this dynamic process, acquired fMRI data from a single child, UD, who had his right occipitotemporal region resected at age 6 years 9 months to manage drug-resistant epilepsy resulting from a low-grade, slow-growing, dysembryoplastic neuroepithelial brain tumor (Liu et al., 2018). UD showed normal performance on face and word experiments (like those shown in Fig. 28.4 top). Data from functional MRI acquired over four time points postsurgically (see stimuli employed in Fig. 28.5A), roughly 6 months apart, confirmed the presence of an immutable left visual field hemianopia. In extrastriate cortex, however, even at the first postsurgical scan, object- and scene-selective responses were observed in the typical category selective Lateral Occipital Cortex and Parahippocampal Place Area, and the topography, extent, and selectivity of these regions fell within the normal distribution and did not change longitudinally as seen in Fig. 28.5 compared to the matched control participants. In contrast, activation for faces and for words showed significant change in location as well as in magnitude of activation over a session, with the location of the word-selective "Visual Word Form Area" shifting more laterally over sessions and the extent of the left FFA expanding and abutting it but more medially located (see Fig. 28.5).

Specifically, tracking the voxels located in a 3D volume superimposed over the Fusiform Gyrus and occipitotemporal sulcus revealed the shift in category selectivity from a response that, initially, was only mildly selective for one of the two categories, and then gradually becoming more selective for either faces or words. These longitudinal findings reveal the dynamics by which plasticity may come about in the human visual system and implicate processes of cooperation (between voxels with similar selectivities) and competition (between voxels with differing selectivities) in the process of neural crowding and reorganization (Fig. 28.6).

Inspired by findings of initial hemispheric equivalence for faces and then, over development subsequent fine-tuning in the RH,

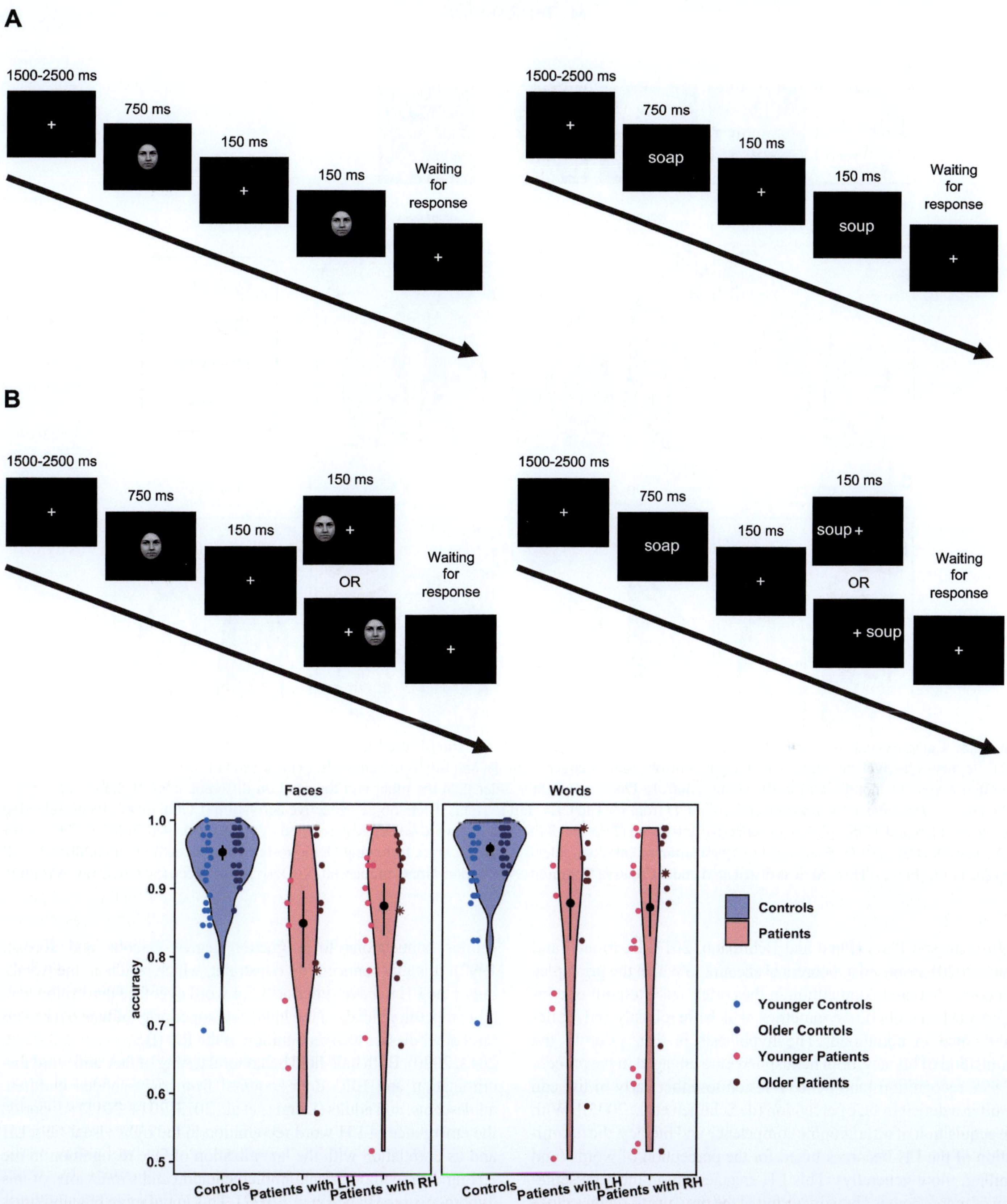

Fig. 28.4. Top: (A) Sequential matching paradigm for faces and words with both pairs of the single trial shown in central vision. (B) Sequential matching paradigm for pairs of faces and words with the "target" shown centrally first followed by the second stimulus shown in either the left or right visual field (to accommodate the hemianopia in the resected patients). (This article is distributed under Creative Commons Attribution-NonCommercial-NoDerivatives License 4.0 (CC BY-NC-ND). Adapted and reprinted from Granovetter MC, Robert S, Ettensohn L, et al. (2022). With childhood hemispherectomy, one hemisphere can support-but is suboptimal for-word and face recognition. Proc Natl Acad Sci U S A 119: e2212936119.

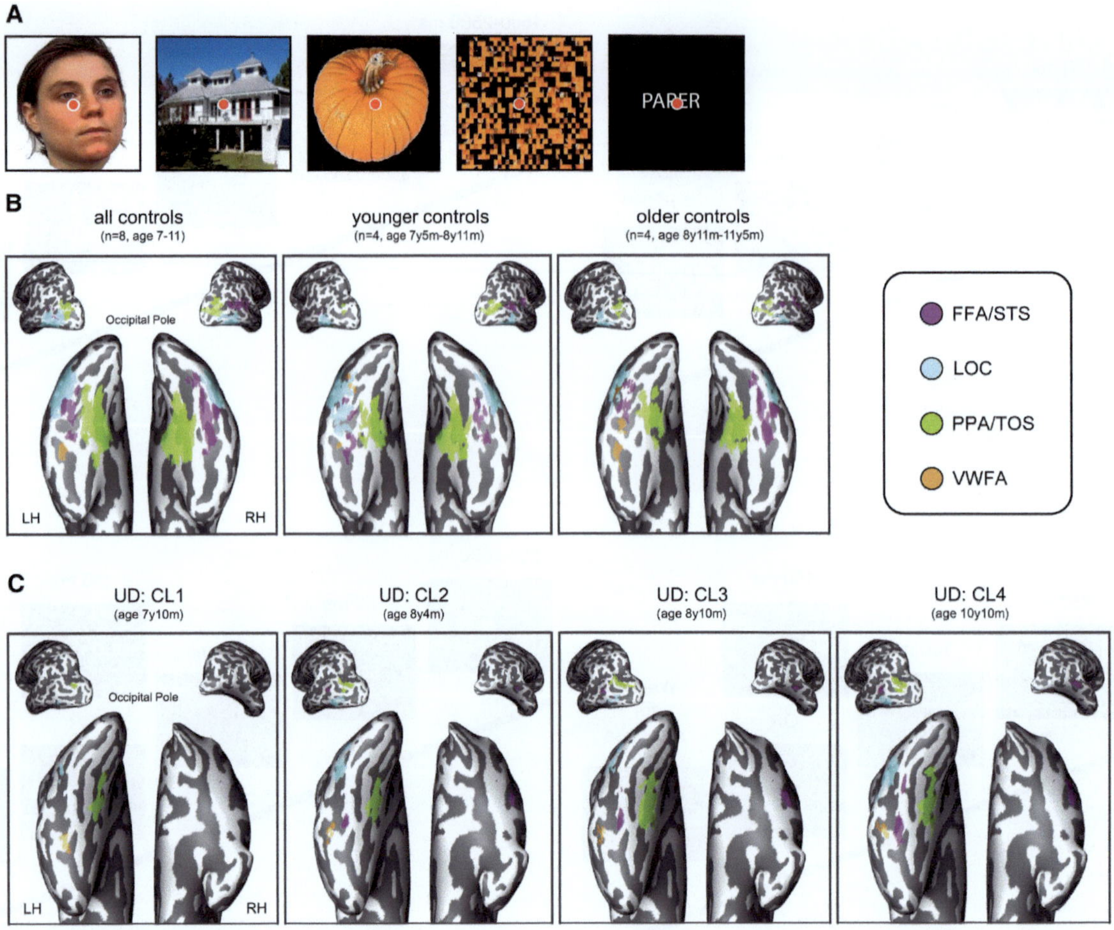

Fig. 28.5. Category-selective activations in controls and UD. (A) Examples of stimuli used in the fMRI experiments of category localizer (CL). (B) Category-selective activations for all eight controls, four younger controls, and four older controls. (This 8 year 11 month child is different from the other 8 year 11 month child in the young controls. One was slightly older than the other and they fell on different sides of the median split.) (C) Category-selective activations visualized in UD from CL1 to CL4 (left to right). Blue: object-selective activation (LOC); green: scene-selective activation (PPA and TOS); pink: face-selective activation (FFA and STS); orange: word-selective activation (VWFA). Adapted from Liu TT, Nestor A, Vida MD, et al. (2018). Successful reorganization of category-selective visual cortex following Occipito-temporal lobectomy in childhood. Cell Rep 24: 1113, (Fig. 2). This article is distributed under Creative Commons Attribution-NonCommercial-NoDerivatives License 4.0 (CC BY-NC-ND).

Behrmann and Plaut (Plaut and Behrmann, 2011; Behrmann and Plaut, 2020) proposed a theoretical account in which the principles of cooperation and competition in the cortex serve to craft or configure the hemispheric asymmetries, which are robustly and replicably noted in adulthood. The hypothesis is that, prior to the acquisition of literacy, both hemispheres are engaged in the process of face recognition [and hence a lesion to either early in life can result in a deficit in face perception (de Schonen et al., 2005)]. With the acquisition of orthographic competence and literacy, the optimization of the LH becomes tuned for the perception of words (and reading, more generally). This LH engagement in the representation of words comes about by virtue of the pressure to retain a close coupling between extrastriate visual areas and language regions in the dominant LH (in the modal right-hander). The pressure to keep the connection length short was inspired by the implementation of neural networks in which error-driven learning was augmented with a spatial loss function penalizing large weights on longer vs shorter connections to a greater degree (Jacobs and Jordan, 1992). It is this connectivity constraint, which results in the tweaking of the LH to have superiority for word over face perception and, by competition, leads to the increased superiority of face over word (and other classes too) recognition in the RH (Behrmann and Plaut, 2013, 2020). Both half-field behavioral testing of face and word discrimination and EEG data acquired from right-handed children, adolescents, and adults (Dundas et al., 2013, 2014, 2015) supported the emergence of LH word recognition in the right visual field/LH and its correlation with the lateralization of face recognition in the RH [and for computational implementation and simulations of this hypothesis, see Blauch et al., 2022]. This initial more bilateral organization for faces, which evolves into more lateralized function over age appears to be true across other functions too—see findings showing bilateral language functions in early childhood with a slight LH bias, which is enhanced with age and a "weak shadow" of language function in the RH (Olulade et al., 2020; Martin et al., 2022).

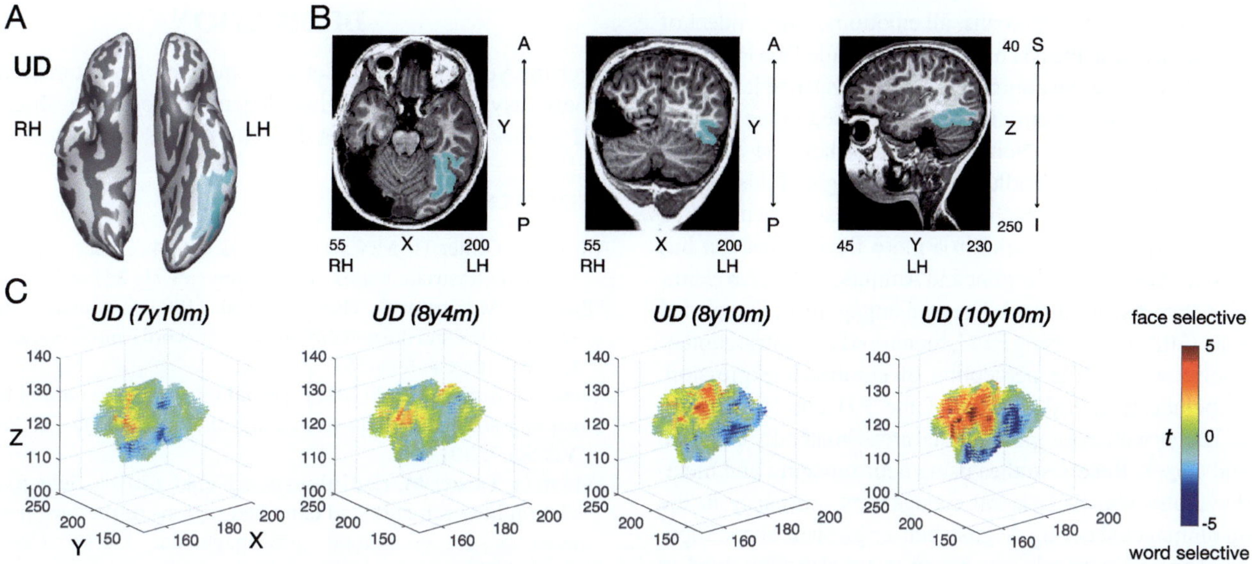

Fig. 28.6. (A) Fusiform Gyrus (FG) and OccipitoTemporal Suclus (OTS) hand-drawn in surface space. (B) FG and OTS in the corresponding volume space (coordinates in native space). (C) Change over time in each voxel's response to faces over words within the FG/OTS region, where the XYZ coordinates and the t score [$t_{(face-word)}$] are plotted for each voxel. Color: darker in red denotes higher sensitivity to faces and darker in blue denotes higher sensitivity to words. Adapted from Liu TT, Nestor A, Vida MD, et al. (2018). Successful reorganization of category-selective visual cortex following Occipito-temporal lobectomy in childhood. Cell Rep 24: 1113, (Fig. 4). This article is distributed under Creative Commons Attribution-NonCommercial-NoDerivatives License 4.0 (CC BY-NC-ND).

There are ongoing debates about the exact nature of the competition and continued discussion about the exact nature of the lateralized competition or neural recycling (Dehaene and Cohen, 2007; Dehaene et al., 2015; Dehaene-Lambertz et al., 2018) for faces vs words or for words vs other categories like limbs (Nordt et al., 2021). Also, one study reported that the word-selective cortex did not recycle the previously face-selective cortex (Dehaene-Lambertz et al., 2018). Rather, word-selective voxels appeared to show weak selectvity before the onset of reading, some showing mild preferences for tools, but not for faces.

Further challenges to the view that the RH face advantage is a consequence of the acquisition of word recognition in the LH comes from findings showing a left visual field gaze (but not necessarily perceptual processing) bias in 6-month-old children (Donati et al., 2020) and even in some nonhuman species (Guo et al., 2009). These findings contest the claim that literacy per se is the trigger for hemispheric differentiation for faces and for words (Lochy et al., 2020; Rossion and Lochy, 2022). Evidently, much remains to be learned about the inherent biases of the two hemispheres and the nonlinearities in hemispheric asymmetry, which seem to be present as well—for example, whereas 4–6 month-old infants evinced a RH lateralization for faces in an EEG fast period visual stimulation study (de Heering and Rossion, 2015), 5-year-old children exhibited a bilateral response under the exact same conditions (Lochy et al., 2017). Obviously, understanding the process by which the mature pattern of hemispheric lateralization emerges requires further investigation and consideration [see also Davies-Thompson et al., 2016].

FACIAL EXPRESSION PERCEPTION

This chapter, thus far, has focused on mechanisms that support face identification and its hemispheric underpinnings. Although there has been less research into the hemispheric differences in the perception of facial expression, there is, nevertheless, interesting work that warrants consideration. There is a general consensus that emotional facial expression is better perceived when the face is viewed in the observer's left visual field than the right, supporting the claim that facial expression is computed superiorly in the right hemisphere (Bourne, 2010; Blom et al., 2020). Consistently, the left of a chimeric face is considered to be more expressive than the right hemiface of the chimeric (Sackeim et al., 1978; Sackeim and Gur, 1980). Last, neuropsychologic data have shown that RH brain damage results in difficulties in the perception and processing of emotional expressions of faces (Kucharska-Pietura et al., 2003; Kucharska-Pietura and David, 2003).

As noted in the introduction, studies of emotional face perception in nonhuman primates have been fairly extensive. In one of the key papers examining the neural underpinning of this behavior in nonhuman primates, monkeys viewed images of neutral faces or faces showing aggression, fear, or submission expressions. Both the amygdala and the inferior temporal were modulated by facial expressions, especially in response to fear and the regions of activation were essentially independent of maps of facial identity perception (Hadj-Bouziane et al., 2008). No obvious hemispheric differences were apparent as is typical for studies like this [for a reviews, see Hopkins and Cantalupo, 2008, Leopold and Rhodes, 2010 and Hopkins et al., 2015].

The central claim that, in humans, all emotions, independent of valence, are processed in the RH has come to be called the RH dominance account. This can be contrasted with the valence lateralization view in which positive emotions are processed in the LH and negative emotions in the RH. Neither hypothesis has iron-clad evidence (Fridlund et al., 1985; Fridlund, 1988). In light of this latter, a more nuanced approach, in which emotion expression is mediated by a host of interrelated networks in a more flexible fashion has been proposed (Palomero-Gallagher and Amunts, 2022). An example of such networks and their differential engagement is reported by Prete et al. (2022) who, using EEG, monitored the distribution of signals associated with the perception of emotional expression. Networks include temporal regions of the RH and, to a lesser degree, the LH, as well as temporoparietal areas bilaterally and, last, right frontal cortex. Recent studies using neuroimaging and meta-analysis have also shown bilateral engagement, primarily in the processing of images showing disgust; but, a right-lateralized superior temporal-frontal network was found to be more involved in social disgust and social cognition (Gan et al., 2022).

Clearly, the jury is still out (and the jurors do not seem ready to appear any time soon) and much research is needed. Perhaps the most intriguing mission to interested researchers at present is to consider the recent findings of networks and distributed functions with multiple regions involved in emotion processing. Attention to the amygdala and its role in emotional face expression, especially fear, is also important (Wang et al., 2023). The move toward using big data to uncover the full complement of neural tissues that underlie complex social and emotional processing is noteworthy, and several relevant atlases and big data analyses are helpful in this regard (Fusar-Poli et al., 2009a,b).

CONCLUSION

As is apparent from even this brief review, face identity and emotion perception and their underlying neural correlates have been the foci of probably thousands of studies by now. And still, notwithstanding the plethora of studies, which have yielded progress, there are still no definitive conclusions. While there is growing recognition that there are likely complex networks in each hemisphere and multiple regions whose joint activity determine the behavioral outcome, there is much that we still do not know. The increasingly refined methodological approaches as well as the novel cutting-edge analytic techniques will surely advance our understanding, and perhaps a less hemisphere-specific and more interactive and dynamic neural mechanism will, in time, be revealed and confirmed.

ACKNOWLEDGMENTS

The writing of this chapter was funded by a grant from the National Science Foundation (BCS2123069) and from the National Institutes of Health (NEI R01EY027018). Marlene Behrmann also acknowledges support from P30 CORE award EY08098 from the National Eye Institute, NIH, and unrestricted supporting funds from The Research to Prevent Blindness Inc., NY, and the Eye & Ear Foundation of Pittsburgh.

DEDICATION

Over the years, Michael Corballis inspired my interest in hemispheric asymmetries with his elegant and creative observations and studies. This chapter is dedicated to his memory.

References

Allison T, Ginter H, McCarthy G et al. (1994a). Face recognition in human extrastriate cortex. J Neurophysiol 71: 821–825.

Allison T, McCarthy G, Nobre A et al. (1994b). Human extrastriate visual cortex and the perception of faces, words, numbers, and colors. Cereb Cortex 4: 544–554.

Avidan G, Behrmann M (2021). Spatial integration in normal face processing and its breakdown in congenital prosopagnosia. Annu Rev Vis Sci 7: 301–321.

Avidan G, Tanzer M, Hadj-Bouziane F et al. (2014). Selective dissociation between core and extended regions of the face processing network in congenital prosopagnosia. Cereb Cortex 24: 1565–1578.

Barton JJS (2003). Disorders of face perception and recognition. Neurol Clin 21: 521–548.

Barton JJ (2008). Prosopagnosia associated with a left occipitotemporal lesion. Neuropsychologia 46: 2214–2224.

Barton JJS (2022). Face processing in the temporal lobe. Handb Clin Neurol 187: 191–210.

Bate S, Bennetts RJ (2014). The rehabilitation of face recognition impaiments: a critical review and future directions. Front Hum Neurosci 8: 1–17.

Bate S, Cook SJ, Duchaine BC et al. (2014). Intranasal inhalation of oxytocin improves face processing in developmental prosopagnosia. Cortex 50: 55–63.

Bate S, Bennetts RJ, Tree JJ et al. (2019). The domain-specificity of face matching impairments in 40 cases of developmental prosopagnosia. Cognition 192: 104031.

Behrmann M, Avidan G (2005). Congenital prosopagnosia: face-blind from birth. Trends Cogn Sci 9: 180–187.

Behrmann M, Avidan G (2022). Face perception: computational insights from phylogeny. Trends Cogn Sci 26: 350–363.

Behrmann M, Plaut DC (2013). Distributed circuits, not circumscribed centers, mediate visual recognition. Trends Cogn Sci 17: 210–219.

Behrmann M, Plaut DC (2014). Bilateral hemispheric processing of words and faces: evidence from word impairments in prosopagnosia and face impairments in pure alexia. Cereb Cortex 24: 1102–1118.

Behrmann M, Plaut DC (2015). A vision of graded hemispheric specialization. Ann N Y Acad Sci 1359: 30–46.

Behrmann M, Plaut DC (2020). Hemispheric organization for visual object recognition: a theoretical account and empirical evidence. Perception 49: 373–404.

Behrmann M, Avidan G, Marotta JJ et al. (2005a). Detailed exploration of face-related processing in congenital prosopagnosia: 1. Behavioral findings. J Cogn Neurosci 17: 1130–1149.

Behrmann M, Marotta J, Gauthier I et al. (2005b). Behavioral change and its neural correlates in visual agnosia after expertise training. J Cogn Neurosci 17: 554–568.

Behrmann M, Avidan G, Gao F et al. (2007). Structural imaging reveals anatomical alterations in inferotemporal cortex in congenital prosopagnosia. Cereb Cortex 17: 2354–2363.

Bell L, Duchaine B, Susilo T (2023). Dissociations between face identity and face expression processing in developmental prosopagnosia. Cognition 238: 105469.

Bennetts RJ, Gregory NJ, Bate S (2024). Both identity and non-identity face perception tasks predict developmental prosopagnosia and face recognition ability. Sci Rep 14: 6626.

Bentin S, Allison T, Puce A et al. (1996). Electrophysiological studies of face perception in humans. J Cogn Neurosci 8: 551–565.

Biotti F, Gray KLH, Cook R (2017). Impaired body perception in developmental prosopagnosia. Cortex 93: 41–49.

Blauch NM, Behrmann M, Plaut DC (2022). A connectivity-constrained computational account of topographic organization in primate high-level visual cortex. Proc Natl Acad Sci U S A 119: e2112566119.

Blom S, Aarts H, Semin GR (2020). Lateralization of facial emotion processing and facial mimicry. Laterality 25: 259–274.

Bobak AK, Parris BA, Gregory NJ et al. (2016). Eye-movement strategies in developmental prosopagnosia and "super" face recognition. Q J Exp Psychol (Hove) 70: 201–217. https://doi.org/10.1080/17470218.2016.1161059.

Bona S, Silvanto J (2014). Accuracy and confidence of visual short-term memory do not go hand-in-hand: behavioral and neural dissociations. PLoS One 9: e90808.

Bona S, Silvanto J, Cattaneo Z (2018). TMS over right OFA affects individuation of faces but not of exemplars of objects. Neuropsychologia117:364–370. https://doi.org/10.1016/j.neuropsychologia.2018.06.024.

Bourne VJ (2010). How are emotions lateralised in the brain? Contrasting existing hypotheses using the chimeric faces test. Cognit Emot 24: 903–911.

Brederoo SG, Van der Haegen L, Brysbaert M et al. (2020). Towards a unified understanding of lateralized vision: a large-scale study investigating principles governing patterns of lateralization using a heterogeneous sample. Cortex 133: 201–214.

Civile C, McLaren IPL (2022). Transcranial direct current stimulation (tDCS) eliminates the other-race effect (ORE) indexed by the face inversion effect for own versus other-race faces. Sci Rep 12: 12958.

Civile C, Quaglia S, Waguri E et al. (2021). Using transcranial direct current stimulation (tDCS) to investigate why faces are and are not special. Sci Rep 11: 4380.

Cohen AL, Soussand L, Corrow SL et al. (2019). Looking beyond the face area: lesion network mapping of prosopagnosia. Brain 142: 3975–3990.

Corballis MC (1982). The lopsided brain: evolution of the generative mind, Oxford University Press.

Danguecan AN, Smith ML (2019). Re-examining the crowding hypothesis in pediatric epilepsy. Epilepsy Behav 94: 281–287.

Davies-Thompson J, Johnston S, Tashakkor Y et al. (2016). The relationship between visual word and face processing lateralization in the fusiform gyri: a cross-sectional study. Brain Res 1644: 88–97.

Davies-Thompson J, Fletcher K, Hills C et al. (2017). Perceptual learning of faces: a rehabilitative study of acquired prosopagnosia. J Cogn Neurosci 29: 573–591.

de Heering A, Rossion B (2015). Rapid categorization of natural face images in the infant right hemisphere. Elife 4: e06564.

de Schonen S, Mancini J, Camps R et al. (2005). Early brain lesions and face-processing development. Dev Psychobiol 46: 184–208.

DeGutis JM, Chiu C, Grosso ME et al. (2014). Face processing improvements in prosopagnosia: successes and failures over the last 50 years. Front Hum Neurosci 8: 561.

Dehaene S, Cohen L (2007). Cultural recycling of cortical maps. Neuron 56: 384–398.

Dehaene S, Cohen L, Morais J et al. (2015). Illiterate to literate: behavioral and cerebral changes induced by reading acquisition. Nat Rev Neurosci 16: 234–244.

Dehaene-Lambertz G, Monzalvo K, Dehaene S (2018). The emergence of the visual word form: longitudinal evolution of category-specific ventral visual areas during reading acquisition. PLoS Biol 16: e2004103.

Donati G, Davis R, Forrester GS (2020). Gaze behaviour to lateral face stimuli in infants who do and do not gain an ASD diagnosis. Sci Rep 10: 13185.

Dubois J, de Berker AO, Tsao DY (2015). Single-unit recordings in the macaque face patch system reveal limitations of fMRI MVPA. J Neurosci 35: 2791–2802.

Dundas EM, Plaut DC, Behrmann M (2013). The joint development of hemispheric lateralization for words and faces. J Exp Psychol 142: 348–358.

Dundas EM, Plaut DC, Behrmann M (2014). An ERP investigation of the co-development of hemispheric lateralization of face and word recognition. Neuropsychologia 61C: 315–323.

Dundas EM, Plaut DC, Behrmann M (2015). Variable left-hemisphere language and orthographic lateralization reduces right-hemisphere face lateralization. J Cogn Neurosci 27: 913–925.

Epihova G, Cook R, Andrews TJ (2023). Recognition of animal faces is impaired in developmental prosopagnosia. Cognition 237: 105477.

Ettlinger GF (1988). Hand preference, ability, and hemispheric specialization: in how far are these factors related in the monkey? Cortex 24: 389–398.

Evans JJ, Heggs AJ, Antoun N et al. (1995). Progressive prosopagnosia associated with selective right temporal atrophy. Brain 118: 1–13.

Fridlund AJ (1988). What can asymmetry and laterality in EMG tell us about the face and brain? Int J Neurosci 39: 53–69.

Fridlund AJ, Gilbert AN, Izard CE et al. (1985). Emotions and facial expression. Science 230: 607–608, 610, 687.

Fusar-Poli P, Placentino A, Carletti F et al. (2009a). Laterality effect on emotional faces processing: ALE meta-analysis of evidence. Neurosci Lett 452: 262–267.

Fusar-Poli P, Placentino A, Carletti F et al. (2009b). Functional atlas of emotional faces processing: a voxel-based meta-analysis of 105 functional magnetic resonance imaging studies. J Psychiatry Neurosci 34: 418–432.

Gainotti G (2013). Is the right anterior temporal variant of prosopagnosia a form of 'associative prosopagnosia' or a form of 'multimodal person recognition disorder'? Neuropsychol Rev 23: 99–110.

Gainotti G (2023). Some historical notes orienting towards brain mechanisms that could underlie hemispheric asymmetries. Cortex 163: 26–41.

Gainotti G, Marra C (2011). Differential contribution of right and left temporo-occipital and anterior temporal lesions to face recognition disorders. Front Hum Neurosci 5: 55.

Gan X, Zhou X, Li J et al. (2022). Common and distinct neurofunctional representations of core and social disgust in the brain: coordinate-based and network meta-analyses. Neurosci Biobehav Rev 135: 104553.

Gerlach C, Poirel N (2020). Who's got the global advantage? Visual field differences in processing of global and local shape. Cognition 195: 104131.

Geskin J, Behrmann M (2018). Congenital prosopagnosia without object agnosia? A literature review. Cogn Neuropsychol 35: 4–54.

Gobbini MI, Haxby JV (2007). Neural systems for recognition of familiar faces. Neuropsychologia 45: 32–41.

Gomez J, Drain A, Jeska B et al. (2019). Development of population receptive fields in the lateral visual stream improves spatial coding amid stable structural-functional coupling. Neuroimage 188: 59–69.

Gotts SJ, Jo HJ, Wallace GL et al. (2013). Two distinct forms of functional lateralization in the human brain. Proc Natl Acad Sci U S A 110: E3435–E3444.

Granovetter MC, Robert S, Ettensohn L et al. (2022). With childhood hemispherectomy, one hemisphere can support-but is suboptimal for-word and face recognition. Proc Natl Acad Sci U S A 119: e2212936119.

Gray KLH, Biotti F, Cook R (2019). Evaluating object recognition ability in developmental prosopagnosia using the Cambridge Car Memory Test. Cogn Neuropsychol 36: 89–96.

Grüsser OJ, Landis T (1991). Visual agnosias and other disturbances of visual perception and cognition, MacMillian, London.

Guo K, Meints K, Hall C et al. (2009). Left gaze bias in humans, rhesus monkeys and domestic dogs. Anim Cogn 12: 409–418.

Hadj-Bouziane F, Bell AH, Knusten TA et al. (2008). Perception of emotional expressions is independent of face selectivity in monkey inferior temporal cortex. Proc Natl Acad Sci U S A 105: 5591–5596.

Hasson U, Levy I, Behrmann M et al. (2002). Eccentricity Bias as an organizing principle for human high-order object areas. Neuron 34: 479–490.

Hopkins WD, Cantalupo C (2008). Theoretical speculations on the evolutionary origins of hemispheric specialization. Curr Dir Psychol Sci 17: 233–237.

Hopkins WD, Misiura M, Pope SM et al. (2015). Behavioral and brain asymmetries in primates: a preliminary evaluation of two evolutionary hypotheses. Ann N Y Acad Sci 1359: 65–83.

Hsiao JH, Cipollini B, Cottrell GW (2013). Hemispheric asymmetry in perception: a differential encoding account. J Cogn Neurosci 25: 998–1007. https://doi.org/10.1162/jocn_a_00377.

Ivry R, Robertson LC (1998). The two sides of perception, MIT Press, Cambridge, MA.

Jacobs RA, Jordan MI (1992). Computational consequences of a bias toward short connections. J Cogn Neurosci 4: 323–336.

Kanwisher N, McDermott J, Chun MM (1997). The fusiform face area: a module in human extrastriate cortex specialized for face perception. J Neurosci 17: 4302–4311.

Kolb B, Sutherland RJ, Whishaw IQ (1983). A comparison of the contributions of the frontal and parietal association cortex to spatial localization in rats. Behav Neurosci 97: 13–27.

Kosslyn SM, Koenig O, Barrett A et al. (1989). Evidence for two types of spatial representations: hemispheric specialization for categorical and coordinate relations. J Exp Psychol Hum Percept Perform 15: 723–735.

Krageloh-Mann I, Lidzba K, Pavlova MA et al. (2017). Plasticity during early brain development is determined by ontogenetic potential. Neuropediatrics 48: 66–71.

Kucharska-Pietura K, David AS (2003). The perception of emotional chimeric faces in patients with depression, mania and unilateral brain damage. Psychol Med 33: 739–745.

Kucharska-Pietura K, Phillips ML, Gernand W et al. (2003). Perception of emotions from faces and voices following unilateral brain damage. Neuropsychologia 41: 1082–1090.

Kuhn CD, Gerlach C, Andersen KB et al. (2021). Face recognition in developmental dyslexia: evidence for dissociation between faces and words. Cogn Neuropsychol 38: 107–115. https://doi.org/10.1080/02643294.2020.1847060.

Lahiri D (2020). Prosopagnosia. Cortex 132: 479.

Leopold DA, Rhodes G (2010). A comparative view of face perception. J Comp Psychol 124: 233–251.

Levine SC, Koch-Weser MP (1982). Right hemisphere superiority in the recognition of famous faces. Brain Cogn 1: 10–22.

Levy J, Trevarthen C, Sperry RW (1972). Reception of bilateral chimeric figures following hemispheric deconnexion. Brain 95: 61–78.

Lidzba K, Staudt M, Wilke M et al. (2006). Visuospatial deficits in patients with early left-hemispheric lesions and functional reorganization of language: consequence of lesion or reorganization? Neuropsychologia 44: 1088–1094.

Liu H, Stufflebeam SM, Sepulcre J et al. (2009). Evidence from intrinsic activity that asymmetry of the human brain is controlled by multiple factors. Proc Natl Acad Sci U S A 106: 20499–20503.

Liu TT, Nestor A, Vida MD et al. (2018). Successful reorganization of category-selective visual cortex following occipito-temporal lobectomy in childhood. Cell Rep 24: 1113.

Liu N, Behrmann M, Turchi JN et al. (2022). Bidirectional and parallel relationships in macaque face circuit revealed by fMRI and causal pharmacological inactivation. Nat Commun 13: 6787.

Lochy A, de Heering A, Rossion B (2017). The non-linear development of the right hemispheric specialization for human face perception. Neuropsychologia.

Lochy A, Schiltz C, Rossion B (2020). The right hemispheric dominance for face perception in preschool children depends on the visual discrimination level. Dev Sci 23: e12914.

Lohse M, Garrido L, Driver J et al. (2016). Effective connectivity from early visual cortex to posterior occipitotemporal face areas supports face selectivity and predicts developmental prosopagnosia. J Neurosci 36: 3821–3828.

Mancini J, de Schonen S, Deruelle C et al. (1994). Face recognition in children with early right or left brain damage. Dev Med Child Neurol 36: 156–166.

Martin KC, Seydell-Greenwald A, Berl MM et al. (2022). A weak shadow of early life language processing persists in the right hemisphere of the mature brain. Neurobiol Lang (Camb) 3: 364–385.

McCarthy G, Puce A, Belger A et al. (1999). Electrophysiological studies of human face perception. II: response properties of face-specific potentials generated in occipitotemporal cortex. Cereb Cortex 9: 431–444.

McKone E, Kanwisher N (2005). Does the human brain process objects of expertise like faces? A review of the evidence. In: S Dehaene, JR Duhamel, M Hauser et al. (Eds.), From monkey brain to human brain. MIT Press, Cambridge, MA.

Meadows JC (1974). The anatomical basis of prosopagnosia. J Neurol Neurosurg Psychiatry 37: 489–501.

Meng M, Cherian T, Singal G et al. (2012). Lateralization of face processing in the human brain. Proc Biol Sci 279: 2052–2061.

Monaghan P, Shillcock R (2004). Hemispheric asymmetries in cognitive modeling: connectionist modeling of unilateral visual neglect. Psychol Rev 111: 283–308.

Navon D (1977). Forest before trees: the precedence of global features in visual perception. Cogn Psychol 9: 353–383.

Nordt M, Gomez J, Natu VS et al. (2021). Cortical recycling in high-level visual cortex during childhood development. Nat Hum Behav 5: 1686–1697. https://doi.org/10.1038/s41562-021-01141-5.

Olulade OA, Seydell-Greenwald A, Chambers CE et al. (2020). The neural basis of language development: changes in lateralization over age. Proc Natl Acad Sci U S A 117: 23477–23483. https://doi.org/10.1073/pnas.1905590117.

Palomero-Gallagher N, Amunts K (2022). A short review on emotion processing: a lateralized network of neuronal networks. Brain Struct Funct 227: 673–684.

Parr LA, Hopkins WD (2000). Brain temperature asymmetries and emotional perception in chimpanzees, pan troglodytes. Physiol Behav 71: 363–371.

Petunia R, Deneke L, Ocklenburg S (2024). Asymmetries in event-related-potentials part 1: a systematic review of face processing studies. Int J Psychophysiol 202: 112386.

Pierce LJ, Scott LS, Boddington S et al. (2011). The n250 brain potential to personally familiar and newly learned faces and objects. Front Hum Neurosci 5: 111.

Pitcher D, Walsh V, Yovel G et al. (2007). TMS evidence for the involvement of the right occipital face area in early face processing. Curr Biol 17: 1568–1573.

Plaut DC, Behrmann M (2011). Complementary neural representations for faces and words: a computational exploration. Cogn Neuropsychol 28: 251–275.

Prete G, Croce P, Zappasodi F et al. (2022). Exploring brain activity for positive and negative emotions by means of EEG microstates. Sci Rep 12: 3404.

Puce A, Allison T, McCarthy G (1999). Electrophysiological studies of human face perception. III: effects of top-down processing on face-specific potentials. Cereb Cortex 9: 445–458.

Rice GE, Kerry SJ, Robotham RJ et al. (2021). Category-selective deficits are the exception and not the rule: evidence from a case-series of 64 patients with ventral occipito-temporal cortex damage. Cortex 138: 266–281.

Rivolta D, Woolgar A, Palermo R et al. (2014). Multi-voxel pattern analysis (MVPA) reveals abnormal fMRI activity in both the "core" and "extended" face network in congenital prosopagnosia. Front Hum Neurosci 8: 925.

Robertson LC, Ivry R (2000). Hemispheric asymmetries: attention to visual and auditory primitives. Curr Dir Psychol Sci 9: 59–64.

Robotham RJ, Rice GE, Leff AP et al. (2023). Systematic evaluation of high-level visual deficits and lesions in posterior cerebral artery stroke. Brain Commun 5: fcad050.

Rogers LJ, Vallortigara G (2017). Lateralized brain functions, Springer.

Rosenthal G, Tanzer M, Simony E et al. (2017). Altered topology of neural circuits in congenital prosopagnosia. Elife 6. https://doi.org/10.7554/eLife.25069.

Rossion B (2014). Understanding individual face discrimination by means of fast periodic visual stimulation. Exp Brain Res 232: 1599–1621.

Rossion B (2022). Twenty years of investigation with the case of prosopagnosia PS to understand human face identity recognition. Part II: neural basis. Neuropsychologia 173: 108279.

Rossion B, Lochy A (2022). Is human face recognition lateralized to the right hemisphere due to neural competition with left-lateralized visual word recognition? A critical review. Brain Struct Funct 227: 599–629.

Rossion B, Taubert J (2019). What can we learn about human individual face recognition from experimental studies in monkeys? Vision Res 157: 142–158.

Rossion B, Dricot L, Devolder A et al. (2000). Hemispheric asymmetries for whole-based and part-based face processing in the human fusiform gyrus. J Cogn Neurosci 12: 793–802.

Rossion B, Jacques C, Jonas J (2018). Mapping face categorization in the human ventral occipitotemporal cortex with direct neural intracranial recordings. Ann N Y Acad Sci. https://doi.org/10.1111/nyas.13596.

Rossion B, Jacques C, Jonas J (2023). Intracerebral electrophysiological recordings to understand the neural basis of human face recognition. Brain Sci 13. https://doi.org/10.3390/brainsci13020354.

Sackeim HA, Gur RC (1980). Asymmetry in facial expression. Science 209: 834–836.

Sackeim HA, Gur RC, Saucy MC (1978). Emotions are expressed more intensely on the left side of the face. Science 202: 434–436.

Sergent J (1982). The cerebral balance of power: confrontation or cooperation. J Exp Psychol Hum Percept Perform 8: 253–273.

Sergent J, Signoret J-L (1992). Functional and anatomical decomposition of face processing; evidence from prosopagnosia and PET study of normal subjects. Phil Trans R Soc Lond B 335: 55–62.

Sergent J, Ohta S, MacDonald B (1992). Functional neuroanatomy of face and object processing. Brain 115: 15–36.

Shulman GL, Pope DL, Astafiev SV et al. (2010). Right hemisphere dominance during spatial selective attention and target detection occurs outside the dorsal frontoparietal network. J Neurosci 30: 3640–3651.

Song S, Garrido L, Nagy Z et al. (2015). Local but not long-range microstructural differences of the ventral temporal cortex in developmental prosopagnosia. Neuropsychologia 78: 195–206.

Svart N, Starrfelt R (2022). Is it just face blindness? Exploring developmental comorbidity in individuals with self-reported developmental prosopagnosia. Brain Sci 12: 230.

Thomas C, Avidan G, Humphreys K et al. (2009). Reduced structural connectivity in ventral visual cortex in congenital prosopagnosia. Nat Neurosci 12: 29–31.

Towler A, Keshwa M, Ton B et al. (2021). Diagnostic feature training improves face matching accuracy. J Exp Psychol Learn Mem Cogn 47: 1288–1298.

Van Kleeck MH (1989). Hemispheric differences in global versus local processing of hierarchical visual stimuli by normal subjects: new data and a meta-analysis of previous studies. Neuropsychologia 27: 1165–1178.

Wang L, Baumgartner F, Kaule FR et al. (2019). Individual face- and house-related eye movement patterns distinctively activate FFA and PPA. Nat Commun 10: 5532.

Wang Y, Luo L, Chen G et al. (2023). Rapid processing of invisible fearful faces in the human amygdala. J Neurosci 43: 1405–1413. https://doi.org/10.1523/JNEUROSCI.1294-22.2022.

Willis ML, Costantino AI, Nitsche MA et al. (2019). Anodal tDCS and high-frequency tRNS targeting the occipitotemporal cortex do not always enhance face perception. Front Neurosci 13: 78.

Wu X, Kong X, Vatansever D et al. (2022). Dynamic changes in brain lateralization correlate with human cognitive performance. PLoS Biol 20: e3001560.

Yan X, Liu-Shuang J, Rossion B (2019). Effect of face-related task on rapid individual face discrimination. Neuropsychologia 129: 236–245.

Yovel G, Tambini A, Brandman T (2008). The asymmetry of the fusiform face area is a stable individual characteristic that underlies the left-visual-field superiority for faces. Neuropsychologia 46: 3061–3068.

Zemmoura I, Burkhardt E, Herbet G (2021). The inferior longitudinal fasciculus: anatomy, function and surgical considerations. J Neurosurg Sci 65: 590–604. https://doi.org/10.23736/S0390-5616.21.05391-1.

Zhang J, Liu J, Xu Y (2015). Neural decoding reveals impaired face configural processing in the right fusiform face area of individuals with developmental prosopagnosia. J Neurosci 35: 1539–1548.

Zhao Y, Tian F, Song Y et al. (2022). Multiple-stage impairments of unfamiliar face learning in developmental prosopagnosia: evidence from fMRI repetition suppression and multi-voxel pattern stability. Neuropsychologia 176: 108370.

Handbook of Clinical Neurology, Vol. 208 (3rd series)
Cerebral Asymmetries
C. Papagno and P. Corballis, Editors
https://doi.org/10.1016/B978-0-443-15646-5.00023-3

Chapter 29

Hemispheric asymmetries in episodic memory

GIAN DANIELE ZANNINO[1] AND GIOVANNI AUGUSTO CARLESIMO[1,2]*

[1]*Laboratory of Neuropsychology of Memory, IRCSS Santa Lucia Foundation, Rome, Italy*

[2]*Department of Systems Medicine, Tor Vergata University, Rome, Italy*

Abstract

The term "episodic memory" refers to our ability to remember past personal experiences. This ability is severely disrupted following bilateral damage to a dedicated neural substrate located symmetrically in the mesial temporal lobes. Milder deficits are also observed following unilateral damage to the same structures. In this chapter, we contrast memory deficits after left and right mesiotemporal damage and review some of the hypotheses proposed to account for the observed differences. As proposed by other authors (e.g., Binder et al., 2010), the observed lesion side effects in memory profiles after unilateral brain damage do not directly reflect specialization across the two symmetric memory substrates. Rather, they depend on the different kinds of information each of the two mesiotemporal structures receives from the ipsilateral hemisphere where other non-memory-specific cognitive systems are asymmetrically housed. In particular, this chapter outlines the role of language, working memory, and visuospatial processes in accounting for side effects in memory profiles. This is because all these systems greatly contribute to the functioning of episodic memory and also show a variable extent of lateralization in the human brain.

INTRODUCTION

In a broad sense, memory refers to the effects of previous experiences on the way an organism reacts to stimuli it encounters in the future. The physical counterparts of memories are changes in the neural circuitry that sustain the interaction between an organism and the environment. In this chapter, we will focus on a particular type of memory in a particular kind of organism, that is, episodic memory in humans. Our topic will be even narrower as we will address the more specific issue of hemispheric specialization in the episodic memory of humans.

Before we get into the heart of the topic, we must establish some general, sufficiently well agreed upon assumptions about episodic memory to serve as a starting point for further discussion. In particular, we will provide a differential characterization of episodic memory in contrast to the other forms of human memory (particularly semantic memory).

Episodic memory is the form of memory that is more akin to the ability the word "memory" refers to in everyday language, that is, the ability to remember a past experience we have taken part in. Following a well agreed upon psychologic taxonomy of human memory, episodic memory is characterized as being a "declarative," "long-term" repository. Both these features also apply to semantic memory, which is approximatively what we envision as our knowledge about word meanings and encyclopedic facts. When we say that episodic memory is "declarative," we mean that its content is consciously accessible and can be expressed explicitly. This occurs typically in a verbal report (such as when we tell someone about our last vacation), but can also occur otherwise (such as when we draw the floor plan of the apartment for sale we have just visited). This feature does not apply to other forms of "nondeclarative" memory, such as procedural memory. Its contents cannot be "declared" by the owner, but need to be demonstrated by a corresponding skill or behavior. Thus, for example, if I learn to drive a car I cannot demonstrate this by describing the actions required to drive a car (also someone who is not able to drive cars might be able to produce such a description). In fact, the acquisition of this procedural form of memory could only be ascertained by evaluating my actual driving ability.

As for their duration, both declarative and nondeclarative memories have the potential to last a lifetime, and even if this is not the case (as occurs for priming and classic conditioning), their decay is much slower than what is observed for so-called "short-term" memories that fade within seconds. As we will see later, among the different forms of short-term memory systems, working memory is of particular importance for this discussion

*Correspondence to: Giovanni Augusto Carlesimo, IRCCS Santa Lucia Foundation, Via Ardeatina 306, 00179, Rome, Italy. E-mail: memolab@hsantalucia.it.

(Baddeley and Hitch, 1974; Baddeley, 2000). Working memory is a limited capacity system that allows for the short-term storage of verbal acoustic (in the phonologic loop), visuospatial (in the visuospatial sketchpad), and supramodal information (in the episodic buffer) for further elaboration (by the central executive) in the service of complex cognitive tasks.

In addition to its content and duration, episodic memory can be differentiated from other forms of memory with regard to its functional architecture. Usually, we subdivide memory processes into four different phases: encoding, retrieval, consolidation, and storage. For the encoding phase, for an episode to be subsequently retrieved, the incoming experience has to be processed online at the moment it occurs, that is, the memory trace must be *encoded*. The encoding phase also has specific characteristics according to the type of memory being considered. In particular, although episodic memories are as a rule acquired during a single learning episode (i.e., real live events do not repeat themselves over time, so I can only participate in them once), semantic memories, such as word meanings and encyclopedic contents, are typically learned through multiple exposures to the relevant content. For example, think of a child who acquires word meanings through repeated encounters with a lexical label associated with the corresponding objects, or the way we can acquire encyclopedic knowledge through repeated readings of a textbook.

The differences in the encoding phase of episodic and semantic memories are mirrored in the retrieval phase. Tulving (2002) proposed the term "autonoetic" to describe the particular kind of conscious state of mind associated with traveling back in time and reliving a personal experience. The *autonoetic* consciousness that characterizes remembering episodic memories is contrasted with the *noetic* experience of knowing semantic content. In fact, semantic memories cannot be remembered because we are unable to engage in backward mental time travel to the particular time point where the learning experience occurred (I *know* what "apple" means but I cannot *remember* the specific episode when I learned it).

Finally, the consolidation and storage of memories also differ according to the different memory systems. There is widespread agreement that the mental representations generated during the online processing of events (i.e., pictures, thoughts, emotions, etc.) need to be bound together to produce a memory trace, which in turn needs to be consolidated and stored for further retrieval. Consolidation is usually envisaged as a process of strengthening the links among single pieces of multimodal information comprising the memory trace. Interestingly, one of the mechanisms through which an episodic memory trace is consolidated is its reactivation over time. According to the very influential multiple trace theory (Nadel and Moscovitch, 1997), every time we remember a past experience, a new memory trace is laid down that corresponds to the content of our recollection. The new trace might be very similar to the previous ones (including the one laid down when the event first occurred) but not identical; the presence of multiple, grossly overlapping traces is thought to facilitate further memory retrieval while also protecting them from decay (consolidation). Importantly, the repeated reactivation of an episodic memory trace loosens its links to the spatiotemporal context in which it was originally acquired, resulting in a semanticization process (Moscovitch et al., 2005). Through semanticization, the episodic memory trace becomes a kind of common denominator across the representations generated during several retrieval episodes, in the same way that semantic memories comprise information "averaged" across multiple learning episodes and thus loosen their links with the spatiotemporal context of each single learning episode.

In addition to being consolidated to prevent decline, memory traces must also be stored somewhere in the nervous system so that they can subsequently be retrieved. Prima facie it is not easy to provide a generally agreed upon picture regarding this point. The so-called declarative model (Squire, 1986) assumes that medial temporal lobe structures only have a temporary role in consolidating, storing, and retrieving episodic *and* semantic memories; once the consolidation process is over, the storage and retrieval of memories rely exclusively on neocortical sites. On the other hand, according to the multiple trace theory (Nadel and Moscovitch, 1997), only semantic and semanticizated memories eventually become hippocampal-independent, whereas autonoetic retrieval (which only applies to truly episodic memories) will always be dependent on the memory-dedicated medial temporal lobe structures. In fact, these two points of view seem much less distant when one considers that the boundaries between episodic and semantic memories are far less clear cut than they might seem. We have already seen that episodic memories might become semanticized through repeated recovery. On the other hand, semantic memories (such as encyclopedic facts) might also be acquired through a limited number of learning episodes, or even a single one (I still remember when and where I was when I learned from my father that Enzo Ferrari had a son, Dino, who died young and to whom he dedicated a Ferrari model). Thus, the strength of the link between a memory trace and a specific spatiotemporal context has considerable variability in the domains of both episodic and semantic memory traces as a function of the number of learning/rehearsal episodes that characterize that trace. At one extreme of this continuum, we find events that occurred in a particular spatiotemporal context that we never rehearsed; at the other extreme, we will probably find the meanings of words we acquired in infancy. Somewhere in between, we will find rehearsed personal episodes and public events to which we were only marginally exposed. The difference regarding the role of the hippocampus in storing and retrieving different kinds of memories according to the time that has elapsed is probably due to the different weights of the two models in the above-described continuum. The declarative model probably pays little attention to the relatively small number of old but little rehearsed personal events (which likely cannot be retrieved without hippocampus-dependent mental time travel, even though they occurred a long time ago). On the other hand, the multiple trace theory probably underestimates the importance of facts, such as public events, that were acquired through only one or a few learning episodes (which, even though they are semantic in nature, probably rely on hippocampal activity to be reactivated).

THE ROLE OF LANGUAGE, VISUOSPATIAL PERCEPTION, AND WORKING MEMORY IN EPISODIC MEMORY

In the following sections of this chapter, we will revise the observed lesion side effects in episodic memory deficits following unilateral brain damage and some of the possible underlying mechanisms that have been advanced in the literature to explain them. We will organize our review of the literature by consistently taking into account the interplay between memory-specific functions/substrates and three out of many possible nonmemory-specific functions/substrates, that is, language, visuospatial perception, and working memory.[a] The choice of focusing on these nonmemory functions is justified by their chief role in the phenomenology of episodic memory, which in turn is reflected by the fact that most of the explanatory mechanisms proposed for the observed lesion side effects can be traced back to the particular contribution of each of these functions to episodic memory.

In the rest of this section, we will sketch the differential contributions of language, visuospatial processes, and working memory to the different phases of episodic memory processing, particularly to the encoding and retrieval phases. We will start by focusing on the encoding phase. Since Paivio's seminal work (Paivio, 1986), we consider the visual and the verbal formats as the two main coding formats of episodic memory traces.[b] Along with the visual and verbal information present in the incoming stimulation, information generated by the subject while taking part in an event also becomes integrated in the corresponding memory trace in a verbal (e.g., verbalized thoughts) or visual (e.g., experienced mental imagery) format. In service of the online conscious processing of incoming visuospatial and verbal stimuli and for their further generative elaboration, *in addition* to specialized language and visuospatial systems, a short-term memory system is also needed to make information available to our awareness for as long as necessary to carry out these operations. It is generally agreed upon that this role is accomplished by the working memory system (Baddeley and Hitch, 1974). During the encoding phase, not only information corresponding to environmental stimulation enters our working memory to be consciously perceived and further elaborated. Often, an event we take part in needs to be processed in light of past autobiographic experiences or general world knowledge, which also has to be available to our consciousness in working memory. Thus, for example, the memory trace of my visit to Omaha Beach this summer is strictly conditioned by what I already knew about the Normandy landings. The original model of working memory (Baddeley and Hitch, 1974) comprised two "slave systems" devoted to the short-term storage of phonologic (the phonologic loop) and visuospatial (the visuospatial sketchpad) information in the service of a central executive that was in charge of performing complex cognitive tasks on it (including the encoding of new memory traces). In a subsequent refinement of this model, a third source of temporarily held information, the episodic buffer, was added (Baddeley, 2000). The role of this "back-up store" was to make all kinds of modality-specific information available to the central executive "in some integrated form" (Baddeley, 2000, p. 419). In our opinion, the best candidate for this "integrated" supramodal code is the semantic one (note that semantic information must be made available in working memory as general world knowledge. Also, as we will see, autobiographic knowledge is in part stored in this format). There is widespread agreement about the role of the human semantic system in recoding modality-specific information in a supramodal format. See, for example, the hub-and-spoke model of semantic memory (Patterson et al., 2007), which assumes the existence of a semantic hub where our modality-specific knowledge about object features (e.g., the taste, color, and shape of apples) is coded in a supramodal format.[c] By contrast, there is much less agreement about the relationship between semantics and language. Is the supramodal semantic code identical with the meaning of words (as we maintain, see Zannino et al., 2014, 2015) or are word meanings only one of several kinds of modality-specific information that needs to be translated into an integrated semantic format (as postulated by the hub-and-spoke model, see Patterson et al., 2007)? This point is relevant to the present discussion because, due to the strong lateralization of language, the perspective we assume in this regard will heavily constrain our interpretation of laterality effects in memory. We will come back to this issue at the end of this section.

The role of conceptual knowledge (i.e., semantics) and working memory in the retrieval phase of episodic memories is not less important than the role they play in the encoding phase. As proposed by Conway (Conway, 2005; see also Moscovitch and Melo, 1997), the autonoetic experience of vividly reliving a past experience can be brought about automatically ("direct retrieval" in Conway's terminology) through the encounter of an external or even internal stimulus that matches part of the information in the memory trace. More typically, however, especially for experiences that are not very recent, episodic memories are accessed through a generative retrieval process that ends with the production of an elaborated cue, which in turn is able to bring the episodic memory into awareness. According to Conway (2005), during the generative retrieval process, subjects search for a suitable elaborated cue in their *conceptual* autobiographic knowledge where only the *meaning* of the experience is retained without visual imagery, which instead predominates in episodic memories (for the visuospatial content of remembered episodes, also see Brewer, 1986 and Rubin et al., 2003). In other words, a strategic search does not directly address the visuospatial contents of episodic memories because information is coded in a semantic format within the autobiographic conceptual store (Conway, 2005, p. 612; see also

[a]We are aware that defining working memory as a "nonmemory-specific function" might sound odd, but for the sake of brevity we have adopted this wording instead of the more correct and longer "nonepisodic-memory-specific function."

[b]For the sake of simplicity, and also because language and visuospatial processing are two of the nonmemory substrates we will focus on, our discussion here is limited to the comparison between language and vision. It must be said, however, that most of the prerogatives of language compared to vision are also true when comparing language to other modality-specific information formats, such as the haptic or the emotional format.

[c]Coding information in a supramodal format has many advantages that we cannot discuss here without losing sight of the focus of our chapter (see Zannino et al., 2015, for an in-depth discussion of this point). Perhaps the most obvious of these is the possibility of comparing different exemplars by taking into account simultaneously all the possible dimensions (i.e., visual, verbal, haptic, acoustic, emotional, etc.) with respect to which may be similar or dissimilar.

Kopelman et al., 1989; Binder et al., 2009, p. 2781). Exactly like the encoding process, this strategic search is a typical example of a "complex cognitive task" (Baddeley, 2000) that can be performed by the central executive in collaboration with the episodic buffer. Not surprisingly, Gabrieli and coworkers (Gabrieli et al., 1998) assign the same task to the "domain-specific semantic working memory" system. Once a well-suited, semantically coded, elaborated cue has been found, the experience of remembering can take place with the contribution of other working memory systems, particularly the visuospatial sketchpad, which makes possible the conscious awareness of vivid visuospatial details during the autonoetic experience of remembering.

Thus, semantic and visuospatial contents and the nonmemory systems dealing with them are critical for the functioning of episodic memory in both the encoding and retrieval phases. Finally, we want to come back to the issue of the relationship between language and semantics in the context of the encoding and retrieving of episodic memories. Most of the arguments provided by advocates of a supramodal nonverbal format for semantic information were proposed in the domain of single concept representations. In this context, it is easy to argue, for example, that my knowledge of the concept of a lemon comprises what lemons look like, how they taste and smell, and what they are called. Note, however, that the semantic contents of episodic memories (apart from the case of some simplified laboratory experiments) resemble a narrative much more than a list of single concepts/words. A narrative codes the logical relations between words through syntax to construct higher-order semantic contents. Thus, given that our natural languages are well suited for coding modality-specific *contents and their logical relationships* in an "integrated form" (thanks to the *lexicon* and *syntax*) and considering that language is also best suited for communicating (i.e., "declaring") these contents to others, why should we rely on a different supramodal code for dealing with the semantic contents of our episodic memories? In light of this consideration, we can summarize the point as follows: language (not only the lexicon but also syntax) is critical for providing a supramodal code to deal with the *semantic content of episodic memories*. The latter comprises: (1) verbal meanings that were already present in the episode I took part in or verbal contents I generated at encoding (e.g., what a friend *told* me yesterday and what I *thought* about it); (2) originally modality-specific, nonverbal (e.g., visuospatial) contents, which were "translated" to access my episodic buffer; and (3) sematicized episodic memories in my long-term store.

On the other hand, the semantic content of episodic memories *does not* comprise: (1) visuospatial (or emotional and other modality-specific) contents that might be present in my working memory (not in the episodic buffer, however) during online encoding/remembering of episodes; or (2) the same kind of contents comprising the nonsemantic part of episodic memories in my long-term store.

WHERE DO HEMISPHERIC ASYMMETRIES IN EPISODIC MEMORY COME FROM?

The role of the medial temporal lobe structures in sustaining episodic memory, particularly of the hippocampus, has been universally agreed upon since Scoville and Milner's report of the case of H.M. (Scoville and Milner, 1957), who become densely amnesic after the bilateral removal of these structures for relief from intractable epilepsy. The hippocampus, however, is necessary but not sufficient to ensure the functioning of episodic memory.

As we mentioned in the previous two sections of this chapter, the phenomenology of episodic memory can be viewed as the end product of the interplay among several memory-specific and nonmemory-specific mental processes. The hippocampus, which is the most important substrate underpinning memory-specific processes, is a symmetric structure housed bilaterally in the medial temporal lobes and connected throughout the brain by means of mainly ipsilateral tracts (Maller et al., 2019). In addition to being primarily ipsilateral, hippocampal connections to neocortical sites are asymmetric, as demonstrated by Huang et al. (2021) using diffusion tractography imaging in a large-scale study that enrolled >170 participants. This means that the ipsilateral hippocampal connection with the homotopic areas in the left and right hemispheres might differ in strength. Of particular interest for the following discussion (see below) are two findings from Huang et al. (2021), see comments about these findings on pp. 4661, 4664: that of a left > right hippocampal connectivity with the ipsilateral anterior temporal lobe cortex that is critical for semantic processing and that of a right > left hippocampal connectivity with the ipsilateral parietal areas involved in spatial processing.

Regarding the nonmemory-specific mental activities involved in the memory phenomenon, they also rely on specialized neural substrates that exhibit a variable degree of lateralization across the cerebral hemispheres. This latter point, as well as the fact that most of the hippocampal connections are ipsilateral, is likely the key to understanding the quality and extent of observed lesion side effects in episodic memory functioning following unilateral brain damage. If we extend the statement from Binder and coworkers beyond the particular case of material-specific deficits (see below), we can say that it seems likely that the underlying cause of side effects in episodic memory "is not a property of the medial temporal lobe itself, but is rather a function of the type of information received by the medial temporal lobe from the ipsilateral hemisphere" (Binder et al., 2010, p. 623).

In the remainder of this section, we will provide a sketch of the neural implementation of language, visuospatial perception, and working memory and their connections to the mesiotemporal structures that are in the service of episodic memory.

The evolution of human language is at the base of hemispheric specialization because not only left lateralized substrates evolved in the service of language, but older cognitive functions that were once bilaterally represented became right-skewed to leave room for language in the left hemisphere (Gazzaniga, 2000). Although language is by far the most lateralized function, the degree of interhemispheric specialization is not the same across all levels of language organization. As shown by many decades of neuropsychologic research devoted to the description of people who became aphasic following left hemispheric damage or developed subtler impairments following commissurotomy (for relief of intractable epilepsy), *expressive* speech (both at the level of single words and sentence production) is almost an exclusive endeavor of the left hemisphere. By contrast, the right hemisphere, as demonstrated in commissurotomized subjects, is *sometimes* capable of a limited degree of language *comprehension* at the level of single words and basic short sentences (Sperry et al., 1969; Volz and Gazzaniga, 2017). The reduced language comprehension abilities of the right hemisphere are not only due to its limited ability to process words. Also abstracting from difficulties in processing word form, comprehension is limited in the right hemisphere due to a reduced ability to process meanings per se, as suggested by a limited ability to interpret single pictures and logical combinations of pictures exhibited by left brain-damaged individuals (Chertkow et al., 1997) as well as by "split-brain" subjects when pictorial material is presented only to their right hemisphere (Sperry et al., 1969). Finally, the neuroanatomy of semantic dementia, a degenerative disease that leads to the selective impairment of semantic knowledge, is particularly relevant to this discussion. It is widely agreed upon that this pathology is associated with a left-skewed atrophy of the anterolateral temporal lobes (Mummery et al., 2000). This finding clearly argues in favor of a left hemisphere dominance also for semantics (not only for word form and syntax) and points to the anterior temporal cortex as a critical substrate for this kind of knowledge (see also Chapter 20 of this volume).

Visuospatial abilities are a good example of those functions that are likely to become right skewed as a consequence of the development of language in the left hemisphere (Corballis et al., 2000). Regarding the human neuropsychology of vision, there is a clear division between retinotopic deficits that affect acuity, color, and shape discrimination (e.g., being able to distinguish a square from a rectangle) in the contralesional visual field and apperceptive agnosia, which equally affects vision in the right and left visual fields. The first are sensory deficits that originate from damage to a neural substrate symmetrically located in the occipital lobes. The latter is a disorder of object perception due to lesions of a lateralized substrate housed in the right parietal lobe (McCarthy and Warrington, 1990). Impaired object recognition in apperceptive agnosia is a consequence of the inability to access long-term stored structural descriptions (Riddoch and Humphreys, 1987a) from incoming visual stimulation. As a result, the patient might be able to perform slavish copies of familiar objects (e.g., a car) but still be unable to identify them (Riddoch and Humphreys, 1987b).[d] Another lateralized disorder of visual perception is prosopagnosia. In this disorder, when the recognition deficit is limited to familiar faces (sparing objects) and is not due to more elementary sensory processes, the clinical presentation is associated with bilateral or right unilateral temporo-occipital damage (McCarthy and Warrington, 1990). In keeping with neuropsychologic evidence, functional magnetic resonance imaging (FMRI) investigations have confirmed the existence of a right-skewed "fusiform face area" that is more active when subjects view faces than when they view common objects (Kanwisher et al., 1997; Rossion et al., 2003). To conclude this brief review of the lateralization of visuospatial abilities, we mention the superiority of the right hemisphere in exploring space through vision, as demonstrated by the strong association between right parietal damage and two neuropsychologic syndromes, unilateral spatial neglect (see Chapter 20 of this volume) and topographic disorientation. In the first case, damage to the inferior parietal lobe causes a reduced awareness of stimuli on the contralesional side of space (Parton et al., 2004). In topographic disorientation, right lesions of the posterior parietal lobe (Aguirre and D'Esposito, 1999) result in the "inability to define the spatial location of salient elements in the environment (i.e., environmental landmarks) with respect to the individual's spatial position in the environment" (Iaria and Burles, 2021).

Regarding the last nonmemory-specific function involved in episodic memory, there is general agreement that the lateral prefrontal cortex is a crucial substrate for working memory. It is also generally agreed upon that just as the content of episodic memory is *retrieved* by the hippocampus but *stored* elsewhere in the neocortex (Moscovitch et al., 2005), the representations made available to our consciousness in working memory are also stored in the posterior neocortical sites (Prabhakaran et al., 2000; Ruchkin et al., 2003; Ricker et al., 2010). Thus, according to some authors (Wager and Smith, 2003), the labor division within the lateral prefrontal cortex is more constrained by the type of executive demand than by the type of material held in memory. Nevertheless, neuroimaging studies using delayed recognition tasks have provided empiric support for a major role of a right-skewed lateral prefrontal substrate in maintaining spatial (McCarthy et al., 1994) as well as perceptual (Gazzaley et al., 2004) information during the delay periods. Furthermore, neuropsychologic findings argue strongly in favor of a major involvement of the right frontal lobe in visuospatial working memory (Bonni et al., 2014). Overall, these results are not surprising given the above-mentioned superiority of the right hemisphere in the long-term storage of the same visuospatial information made available in the short-term in working memory.

In the same vein, we expect a left-skewed representation of prefrontal substrates devoted to the maintenance of both word-form and semantic information in working memory.[e] There is a widespread consensus (based on neuropsychologic, neuroimaging, and brain stimulation studies, see Romero Lauro et al., 2020, for review) that the left prefrontal cortex constitutes the critical substrate of phonologic working memory. Neuroimaging experiments also provide a great deal of evidence arguing for a substrate in the left inferior prefrontal cortex dedicated to the manipulation of semantic information (Gabrieli et al., 1998). Compelling evidence of the relationship between semantic working memory and language (see section The Role of Language, Visuospatial Perception, and Working Memory in episodic

[d]The identification deficit can become apparent not only in patients' inability to name pictures of familiar objects, but also in their inability to match pictures of the same object seen from different viewpoints (Warrington and Taylor, 1978).

[e]See above (section The Role of Language, Visuospatial Perception, and Working Memory in Episodic Memory) for our assumption about a strong relationship between language and semantics in general and, in particular, between language and the semantic contents of episodic memories.

Memory) was provided by Desmond and colleagues (Desmond et al., 1995). In an FMRI experiment, these authors found that in their subject sample, lateralization of the prefrontal area active during semantic (as compared to perceptual) judgments to visually presented words was in keeping with the language lateralization established definitively through Wada evaluation.

Fig. 29.1 provides a schematic representation of the implementation of memory-specific and nonmemory-specific neural systems we sketched in this section. The relevant connections between the different substrates in the service of encoding and strategic retrieval of episodic memories (see section The Role of language, Visuospatial Perception, and Working Memory in episodic Memory) are also depicted.

In the remainder of this chapter, we will provide a selective revision of the literature aimed at exemplifying the above-outlined interaction between vision and language/semantic as well as memory processes. In particular, we will apply our perspective on the empiric evidence derived from two areas of research, that is, on persons suffering from temporal lobe epilepsy, which led to the formulation of the material-specific hypothesis (see section Medial Temporal Lobe Epilepsy and the Material-Specific Hypothesis), and that based on neuroimaging studies of healthy subjects performing cognitive tasks, which led to the formulation of the hemispheric encoding retrieval asymmetry (HERA) model (see section The Neuroimaging of Memory and the HERA Model).

MEDIAL TEMPORAL LOBE EPILEPSY AND THE MATERIAL-SPECIFIC HYPOTHESIS

The study of memory deficits in patients who underwent an unilateral anterior temporal lobectomy for relief from intractable epilepsy is closely intertwined with the so-called material-specific hypothesis first proposed by Brenda Milner more than 50 years ago (Milner, 1966, 1970). At variance with the profound memory loss observed after a bilateral resection of the mesial temporal lobes observed in patient H.M. (Scoville and Milner, 1957, see above), when surgical resection was limited to the side of the seizure onset, postoperative neuropsychologic testing revealed only a mild memory deficit. Based on her previous research on interhemispheric specialization, Milner hypothesized that unilateral damage to the left mesial temporal lobe would result in greater impairment on verbal memory tasks, whereas right-sided damage would result in a greater deficit with visual material (Xia, 2006, see also Saling, 2009). Since then, the so-called material-specific hypothesis has been greatly investigated and partially revised. In particular, it became clear that visuoverbal dissociation is not as symmetric as originally proposed. In fact, the status of the left mesial temporal lobe not only has an impact on verbal memories but also on visual memories, whereas the reverse does not seem to be the case. As we will see in the remainder of this section, this asymmetry is in keeping with the disproportionate contribution of language, as compared to vision, to episodic memory functioning, as outlined in sections The Role of

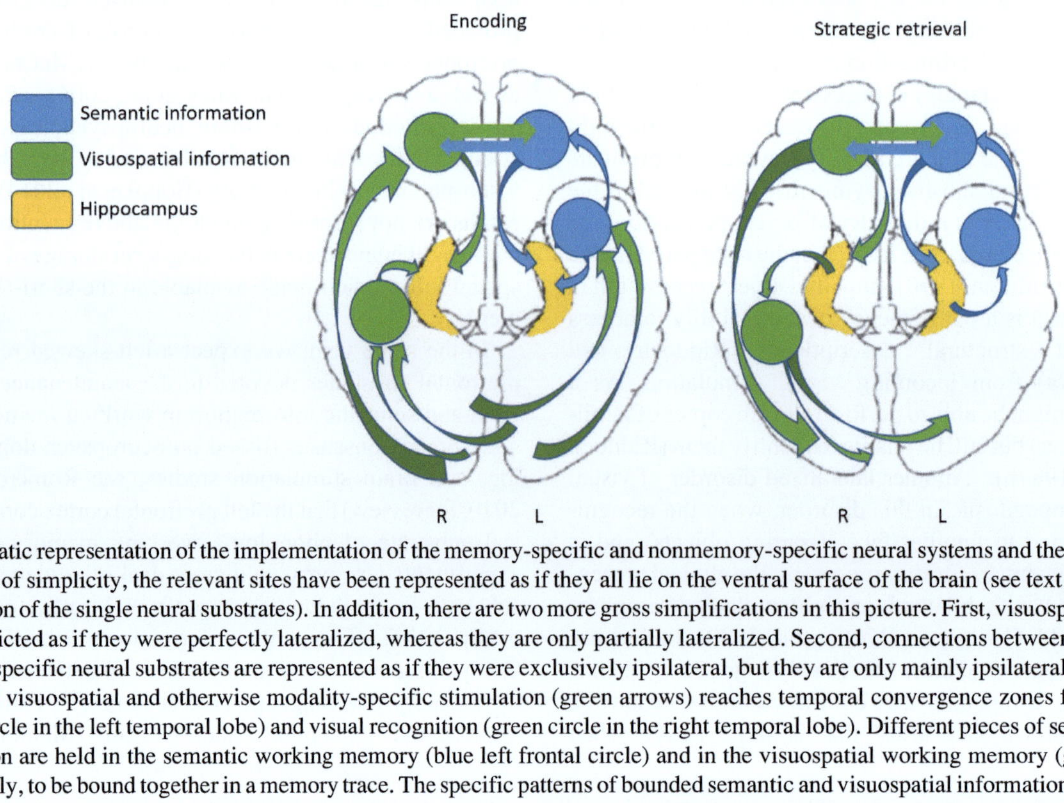

Fig. 29.1. Schematic representation of the implementation of the memory-specific and nonmemory-specific neural systems and their functional interplay. For the sake of simplicity, the relevant sites have been represented as if they all lie on the ventral surface of the brain (see text for some detail on the real localization of the single neural substrates). In addition, there are two more gross simplifications in this picture. First, visuospatial and semantic substrates are depicted as if they were perfectly lateralized, whereas they are only partially lateralized. Second, connections between the hippocampus and nonmemory-specific neural substrates are represented as if they were exclusively ipsilateral, but they are only mainly ipsilateral. At encoding (left panel), peripheral visuospatial and otherwise modality-specific stimulation (green arrows) reaches temporal convergence zones for semantic interpretation (blue circle in the left temporal lobe) and visual recognition (green circle in the right temporal lobe). Different pieces of semantic and visuospatial information are held in the semantic working memory (blue left frontal circle) and in the visuospatial working memory (green right frontal circle), respectively, to be bound together in a memory trace. The specific patterns of bounded semantic and visuospatial information are sent to the left and right hippocampus, respectively, for further reactivation during the retrieval phase. During strategic retrieval (right panel), an elaborated cue, which is comprised of a piece of semantic information bound to a particular memory trace, is retrieved through the combined action of the left hippocampus and the ipsilateral semantic substrates. This cue interacts with the right lateralized visuospatial substrates (green circles) and eventually leads to the reactivation of the original visuospatial experience in the posterior neocortex during remembering.

Language, Visuospatial Perception, and Working Memory in Episodic Memory and Where do Hemispheric Asymmetries in episodic Memory Come From? of this chapter.

Fig. 29.2 shows an idealized picture of the group performance of a sample of people who suffered from mesial temporal lobe epilepsy on a typical neuropsychologic memory test battery. Accuracy levels are displayed according to lesion side (bilateral, left, right) and stimulus material (verbal, visual). Although Fig. 29.2 does not represent any particular subjects' sample performance in any particular neuropsychologic battery, it summarizes several commonly observed findings in the literature.[f] Subsequently, we will revise and comment upon these findings in light of the theoretical framework we described in the previous sections.

The first evidence depicted in Fig. 29.2 regards the point that bilateral damage to the mesial temporal structures results in incomparably more severe memory deficits than unilateral injuries, irrespective of the stimulus material (Saling, 2009; Massot-Tarrús et al., 2019). This feature is in keeping with two characteristics of the functional architecture and neural underpinnings of episodic memory and related nonmemory functions. As we discussed in section Where Do Hemispheric Asymmetries in Episodic Memory Come From? language (including semantics, phonology, and grammar), working memory (for semantic and visuospatial material), and visuospatial processing substrates are represented bilaterally, although with a variable degree of left/right skewness. Moreover, mesial temporal structures show *mainly* ipsilateral connections but are also connected to the contralateral hemisphere. Thus, this indicates a considerable redundancy in memory processing within the mesial temporal structures of each hemisphere, which prevents very severe memory loss in the case of unilateral damage.

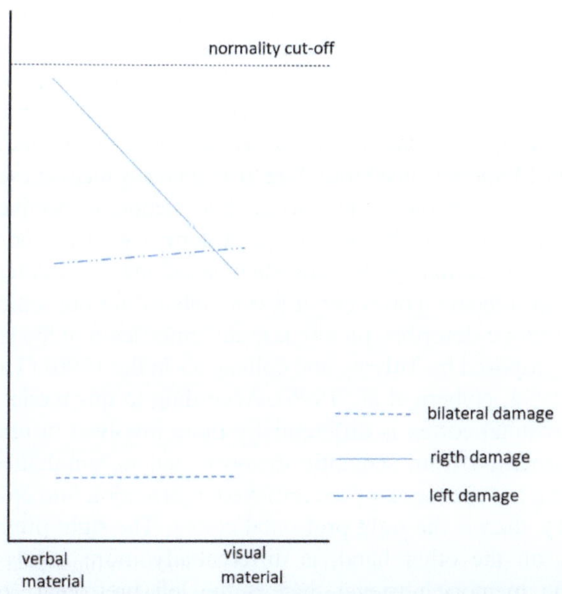

Fig. 29.2. Idealized picture of the lesion side by material type interaction in performance on a typical episodic memory battery administered to people suffering from mesial temporal lobe epilepsy. See text for details.

The second highly agreed upon observation represented in Fig. 29.2 is that of a stronger diagnostic power of verbal memory tasks than visual memory tasks for discriminating between left and right mesial temporal damage (Alessio et al., 2006; Binder et al., 2008). In particular, although as a rule individuals with left temporal lobe foci/resections[g] score worse than those with right temporal lobe pathology on verbal memory tasks, visual memory tasks are often inappropriate for discriminating between the two groups (Lee et al., 2002; Saling, 2009; Barnett et al., 2015; Bentvelzen et al., 2022). The lesion side by material type interaction depicted in Fig. 29.2 is also in keeping with another commonly observed finding, that of a more severe memory impairment associated with left compared to right mesial temporal damage (Binder et al., 2008, 2010; Conradi et al., 2021).

These characteristics of a lesion side by material type interaction are correctly predicted by the interplay between differently lateralized functions in the service of episodic memory that we described in the previous sections. In particular, assuming that semantics rely strictly on language, as we did, and given the straightforward left-skewed lateralization of this latter function, the impact of left-sided mesial lesions on visuospatial (as well as verbal) memory tasks is in keeping with the cardinal role we ascribed to semantics in memory functioning. This role is twofold. First, semantics represent the "integrated" code in which both verbal and extraverbal modality-specific information converges for further elaboration at the time of encoding (in the semantic buffer). Consequently, although the autonoetic experience of remembering relies on the ability to relive modality-specific (mostly visual) experiences, a significant amount of information regarding the *episode* to be remembered (also a simple learning episode in a laboratory set-up) is semantically encoded. Second, in the retrieval phase, the strategic search for episodic memory traces, also when visual stimuli were studied, starts by targeting a semantically elaborated cue that eventually leads to an autonoetic experience of remembering a visual representation. Thus, the retrieval of visual stimuli also relies on language-based processes that are housed in the left hemisphere. Recently, we addressed this issue in an experimental investigation (Zannino et al., 2020) in which we enrolled people who had to undergo a unilateral anterior temporal lobectomy for epilepsy treatment. We used a memory task that had been expressly devised to require participants to retrieve visually encoded memory traces without the possibility of relying on an external cue. In the study phase, the subjects had to copy line drawings of the silhouettes of real objects that were partially occluded so as to make them unrecognizable (otherwise the task could have been solved by simply retrieving the

[f]See, for example, (St-Laurent et al., 2014, Table 1) and (Loring et al., 1995, Table 1) for experimental data that strictly match the group by lesion side interaction depicted in Fig. 29.2.

[g]It should be noted that in temporal lobe epilepsy, surgery exacerbates the symptomatology associated presurgically with left vs right epileptogenic foci (Barnett et al., 2015).

corresponding verbal "translation" from memory). In the test phase, participants had to choose one of three photographs that depicted an object that could have been hidden in the studied drawing. None of the photographs perceptually matched the drawing that had to be retrieved; however, once the drawing was retrieved from memory, it would be easy to match it with the corresponding photograph. In keeping with the hypothesis of the reliance on a language-based process in the strategic search for purely visual memoranda, we found no performance difference associated with lesion side.

As we have seen, semantic-coded information is crucial for strategically searching memory traces irrespective of the kind of material in the memorandum. However, the primacy of the semantic over the visual code also applies when the focus is limited to the encoding phase. Imagine the case of a simple picture recognition task. When the picture of a real item is shown at study, the subjects cannot refrain from interpreting it and, accordingly, the memory trace will be dually coded as a visual and a verbal/semantic representation. However, the reverse is not necessarily true, that is, verbal stimuli are not also invariably coded in a visual format. According to Paivio (1971), this asymmetry gives rise to the so-called picture superiority effect, that is, "the highly consistent empirical finding that stimuli presented for study in picture form are more likely to be recalled in a subsequent free recall test (…) and to be discriminated from unstudied stimuli in a subsequent recognition memory test (…) than stimuli presented for study in word form" (Mintzer and Snodgrass, 1999, p. 113). Coming back to the focus of our discussion, the above-described asymmetry in dually coding pictures and words clearly predicts that the limited availability of visual representations due to right temporal epilepsy should impair word retrieval to a lesser extent than the limited availability of verbal/semantic representations due to left temporal epilepsy should impair picture retrieval.

The group (left vs right) by material (verbal vs visual) interaction depicted in Fig. 29.2 can be made more symmetric through the choice of visual tasks expressly devised to minimize the contribution of the left hemisphere and/or maximize reliance on the right one. Face recognition is a well-suited task for both these purposes. As to the first point, according to the perspective we adopted in this chapter, the testing format of recognition is better suited than that of recall for minimizing the contribution of the left hemisphere. This is true for several reasons. First, retrieval in this testing format is automatic and nonstrategic, so it does not require the left-sided elaboration of a semantic cue. Second, faces are hard to verbalize, that is, when unknown faces are presented at study, it is difficult to generate different semantic representations for each of them. Finally, the balance that favors a right hemisphere contribution to solving this task is further warranted by the fact that processing faces shows a relatively high degree of right skewedness in visual processing areas compared to other visual stimuli such as common objects (see section Where Do Hemispheric Asymmetries in Episodic Memory Come From?). In keeping with the picture outlined above, Bentvelzen and colleagues (Bentvelzen et al., 2021), applied a metaanalytic approach to a sample of 158 experimental studies that involved patients with temporal lobe epilepsy. They found that the only factors reliably associated with the effect size of the disadvantage for right temporal lobe patients compared to left temporal lobe patients in visual memory tasks were testing format (recognition > recall) and stimulus type (faces > designs).

Among memory tasks with visual material, those that stress the retrieval of spatial information are also well-suited for discriminating between left and right mesial temporal damage, as demonstrated by Kessels and colleagues in a metaanalysis that included 27 studies (Kessels et al., 2001). Also, in this case, the relevant kind of information is hard to verbalize and the substrate for processing it is strongly right skewed (see section Where Do Hemispheric Asymmetries in Episodic Memory Come From?).

THE NEUROIMAGING OF MEMORY AND THE HERA MODEL

According to the theoretical framework we outlined in the first three sections of this chapter, an episodic memory trace always comprises two kinds of information, semantic and modality-specific information. Semantic information concerns our interpretation of the episode. It comprises the *meaning* of verbal and nonverbal stimuli we were exposed to or we generated at the time we took part in the episode. Modality-specific information consists of an uninterpreted record of the episode. It contains modality-specific (mainly visuospatial) information in a non "integrated" format. Semantic information is more comprehensive than modality-specific information because more modality-specific information is "copied" in the semantic format at encoding than vice versa (see section Medial Temporal Lobe Epilepsy and the Material-Specific Hypothesis). In this framework, the roles of the left and right dorsolateral prefrontal cortices (see section Where Do Hemispheric Asymmetries in Episodic Memory Come From?) are to temporarily maintain semantic and visuospatial modality-specific information, respectively, in conscious awareness in the service of current goals. Among all the possible "current goals," encoding, searching for, and remembering episodic memories are of special interest for our topic.

The above-described picture is quite reminiscent of the HERA model proposed by Tulving and colleagues in the 1990s (Tulving et al., 1994; Nyberg et al., 1996). According to this model, "the left prefrontal cortex is differentially more involved in retrieval of information from semantic memory, and in simultaneously encoding novel aspects of the retrieved information into episodic memory, than is the right prefrontal cortex. The right prefrontal cortex, on the other hand, is differentially more involved in episodic memory retrieval than is the left prefrontal cortex" (Nyberg et al., 1996, p.138). These authors came to their conclusions about the role of the left prefrontal cortex based on the observation that deep encoding of incoming stimuli (as compared to shallow encoding of the same stimuli) leads to two correlated phenomena at the behavioral and neuroimaging levels of description. Regarding the behavioral level, it was already known that "the durability of the memory trace is a function of depth of processing" (Craik and Tulving, 1975, p. 270), that is, a stimulus is better remembered at encoding the more we move "from early sensory processing to later semantic associative operations" (Craik and Tulving, 1975, p. 270). The new observation, which led to the formulation of the HERA model, was that in several neuroimaging studies, the subtraction of shallow encoding conditions from deep encoding conditions[h] almost invariably resulted in the selective activation of the left prefrontal cortex.

Regarding the role of the right prefrontal cortex, the critical observation for the formulation of the HERA model was that as a rule, subtracting recall or recognition tasks from a reference task involving the same kind of material gives rise to a selective right, or at least right-skewed, activation of the prefrontal cortex. Thus, in summary, the literature reviewed by Tulving and colleagues suggests a strong left lateralization of encoding and a slightly weaker right lateralization of retrieval processes.

Also, the theoretical framework we adopted in this chapter is compatible with the observations at the base of the HERA model, but their interpretation is somewhat different. According to the picture sketched in the first three sections of this chapter, the observed asymmetries at the level of the prefrontal cortex do not rely on different processes (encoding vs retrieval) but on differences in the kinds of processed information (integrated, semantic vs unimodal, mainly visuospatial). As for the left-skewed neuroimaging activation observed by Tulving and colleagues, we maintain that these reflect the priority of semantic (over modality-specific) information in the encoding process. Remember that left-sided activation is associated with deep encoding (as compared to shallow encoding) and that the difference between deep and shallow encoding regards precisely the level of reliance on semantic vs "early sensory" information, respectively.

As for the right-skewed activation associated with retrieval tasks in the neuroimaging studies reviewed by Tulving and colleagues, the issue is a bit more complex. In our framework, retrieval is associated with a strategic search eventually followed by the autonoetic experience of remembering. Alternatively, the strategic search might be replaced by the sudden popping up of memories in reaction to a (mainly external) cue. A strategic search relies strongly on semantic information because, as we saw in sections The Role of Language, Visuospatial Perception, and Working Memory in Episodic Memory and Where Do Hemispheric Asymmetries in Episodic Memory Come From?, this is the kind of information that is suitable for generating an elaborated cue that, in turn, can bring about the retrieval of the entire episodic trace. The episodic trace, in addition to the information stored in an integrated semantic format, also contains a wealth of modality-specific information ensuring to us the possibility of performing backward mental time travel to relive past experiences. In this view, the observation of right-skewed prefrontal activation is the correlate of holding retrieved visuospatial information in working memory during the autonoetic experience of remembering. Importantly, the asymmetric activation that favors the right prefrontal cortex during remembering is far less straightforward than the one that favors the left prefrontal cortex at encoding (as can be seen when comparing tables 1 and 2 in Nyberg et al. (1996), which show prefrontal positron emission tomography (PET) activations associated with episodic memory at encoding and retrieval, respectively). This imbalance is expected in our framework. In fact, we expect that the contribution of semantic working memory will be variable when remembering, according to both, the quote of semantic information in the memory trace itself *and* the need to perform a strategic search to retrieve it. At one extreme, we can imagine a difficult word recall task (strategic search of verbal stimuli) and at the other extreme, an easy face recognition task (automatic retrieval of visual stimuli).

In conclusion, we would like to draw attention to the fact that these arguments parallel those we used in section Medial Temporal Lobe Epilepsy and the Material-Specific Hypothesis to account for the weak association between right temporal damage and impaired visual memory. In both cases. The critical role of the right hemisphere in episodic memory is more elusive than that of the left hemisphere because of the almost ubiquitous additional involvement of semantics. By contrast, some aspects of memory processing, such as strategic search and retrieval of semanticized memories, can be carried out with little or no contribution of modality-specific information, resulting in a more easily recognizable selective involvement of the left hemisphere in the right circumstances.

CONCLUSIONS

Since the material-specific hypothesis was first proposed more than 50 years ago (Milner, 1970, see section Medial Temporal Lobe Epilepsy and the Material-Specific Hypothesis) to account for the relatively subtle memory impairment that occurs after unilateral medial temporal damage, the central tenet of this hypothesis–that the specialization of the left and right temporal lobes in episodic memory processes regards the verbal vs nonverbal nature of the material to be retained, has progressively given way to alternative interpretations based on task or process specificity (Saling, 2009). One of these is the above-revised HERA model (see section The Neuroimaging of Memory and the HERA Model), which contrasts a left-skewed encoding process with a right-skewed retrieval process. In this chapter, we have attempted to review the laterality effects in episodic memory in light of a new divide, that is, between left-skewed processes targeting *supramodal* (semantic) information and the right-skewed processes dealing with *modality-specific* (mainly visuospatial) information. Perhaps the perspective we applied to the empiric

[h] An example of deep encoding as compared to shallow encoding is the verb generation task (producing verbs in response to nouns; e.g., hear "car," say "drive") as compared to a simple noun repetition task (hear "car," say "car"). Cue words ("car" in the example above) encoded through verb generation are recalled better later on than those encoded through repetition. Importantly, deep/semantic encoding can be applied also to nonverbal stimuli provided that we are able to translate the memorandum into the semantic/verbal code (e.g., when we recognize a picture shown at study as "the picture of a dog") and/or to associate semantically coded information with the memorandum (e.g., looking at a face we might think, "this guy looks like an old schoolmate of mine.")

evidence has one major merit. It highlights an intriguing paradox regarding episodic memory. The autonoetic experience of vividly remembering past experiences is primarily based on the ability to retrieve modality-specific uninterpreted information. Nevertheless, in keeping with their declarative nature, the contents of episodic memories must be recoded in a supramodal (i.e., semantic) code to be verbally communicated (and investigated). Moreover, only interpreted experiences (i.e., originally semantic or subsequently semanticized experiences) resist oblivion (as the effect of deep vs shallow encoding demonstrates). In our view, this paradox should be emphasized as it has a more general significance in that it represents one of the many manifestations of the intimate link between language and consciousness in humans.

References

Aguirre GK, D'Esposito M (1999). Topographical disorientation: a synthesis and taxonomy. Brain 122: 1613–1628. https://doi.org/10.1093/brain/122.9.1613. PMID: 10468502.

Alessio A, Bonilha L, Rorden C et al. (2006). Memory and language impairments and their relationships to hippocampal and perirhinal cortex damage in patients with medial temporal lobe epilepsy. Epilepsy Behav 8: 593–600. https://doi.org/10.1016/j.yebeh.2006.01.007. Epub 2006 Mar 6. PMID: 16517214.

Baddeley A (2000). The episodic buffer: a new component of working memory? Trends Cogn Sci 4: 417–423. https://doi.org/10.1016/s1364-6613(00)01538-2. PMID: 11058819.

Baddeley AD, Hitch G (1974). Working memory. In: GA Bower (Ed.), The psychology of learning and motivation. Academic Press, pp. 48–79.

Barnett AJ, Park MT, Pipitone J et al. (2015). Functional and structural correlates of memory in patients with mesial temporal lobe epilepsy. Front Neurol 13: 103. https://doi.org/10.3389/fneur.2015.00103. PMID: 26029159; PMCID: PMC4429573.

Bentvelzen AC, Kessels RPC, Badcock NA et al. (2022). The impact of right temporal lobe epilepsy on nonverbal memory: meta-regression of stimulus- and task-related moderators. Neuropsychol Rev 32: 537–557. https://doi.org/10.1007/s11065-021-09514-3. Epub 2021 Sep 24. Erratum in: Neuropsychol Rev. 2022 Jun 13. PMID: 34559363.

Binder JR, Sabsevitz DS, Swanson SJ et al. (2008). Use of preoperative functional MRI to predict verbal memory decline after temporal lobe epilepsy surgery. Epilepsia 49: 1377–1394. https://doi.org/10.1111/j.1528-1167.2008.01625.x. PMID: 18435753Epub 2008 Apr 24; PMCID: PMC2943848.

Binder JR, Desai RH, Graves WW et al. (2009). Where is the semantic system? A critical review and meta-analysis of 120 functional neuroimaging studies. Cereb Cortex 19: 2767–2796. https://doi.org/10.1093/cercor/bhp055. Epub 2009 Mar 27; PMCID: PMC2774390. PMID: 19329570.

Binder JR, Swanson SJ, Sabsevitz DS et al. (2010). A comparison of two fMRI methods for predicting verbal memory decline after left temporal lobectomy: language lateralization versus hippocampal activation asymmetry. Epilepsia 51: 618–626. https://doi.org/10.1111/j.1528-1167.2009.02340.x. Epub 2009 Oct 8; PMCID: PMC2945221. PMID: 19817807.

Bonnì S, Perri R, Fadda L et al. (2014). Selective deficit of spatial short-term memory: role of storage and rehearsal mechanisms. Cortex 59: 22–32. https://doi.org/10.1016/j.cortex.2014.06.004. Epub 2014 Jul 11. PMID: 25127485.

Brewer WF (1986). What is recollective memory? In: DC Rubin (Ed.), Remembering our past. Studies in autobiographical memory. Cambridge University Press, Cambridge, England, pp. 19–66.

Chertkow H, Bub D, Deaudon C et al. (1997). On the status of object concepts in aphasia. Brain Lang 58: 203–232. https://doi.org/10.1006/brln.1997.1771. PMID: 9182748.

Conradi N, Rosenberg F, Knake S et al. (2021). Wada test results contribute to the prediction of change in verbal learning and verbal memory function after temporal lobe epilepsy surgery. Sci Rep 11: 10979. https://doi.org/10.1038/s41598-021-90376-3.

Conway MA (2005). Memory and the self. J Mem Lang 53: 594–628. https://doi.org/10.1016/j.jml.2005.08.005.

Corballis PM, Funnell MG, Gazzaniga MS (2000). An evolutionary perspective on hemispheric asymmetries. Brain Cogn 43: 112–117. PMID: 10857675.

Craik FIM, Tulving E (1975). Depth of processing and the retention of words in episodic memory. J Exp Psychol Gen 104: 268–294.

Desmond JE, Sum JM, Wagner AD et al. (1995). Functional MRI measurement of language lateralization in Wada-tested patients. Brain 118: 1411–1419. https://doi.org/10.1093/brain/118.6.1411. PMID: 8595473.

Gabrieli JD, Poldrack RA, Desmond JE (1998). The role of left prefrontal cortex in language and memory. Proc Natl Acad Sci U S A 95: 906–913. https://doi.org/10.1073/pnas.95.3.906. PMID: 9448258; PMCID: PMC33815.

Gazzaley A, Rissman J, D'Esposito M (2004). Functional connectivity during working memory maintenance. Cogn Affect Behav Neurosci 4: 580–599. https://doi.org/10.3758/cabn.4.4.580. PMID: 15849899.

Gazzaniga MS (2000). Cerebral specialization and interhemispheric communication: does the corpus callosum enable the human condition? Brain 123: 1293–1326. https://doi.org/10.1093/brain/123.7.1293. PMID: 10869045.

Huang CC, Rolls ET, Hsu CH et al. (2021). Extensive cortical connectivity of the human hippocampal memory system: beyond the "what" and "where" dual stream model. Cereb Cortex 31: 4652–4669. https://doi.org/10.1093/cercor/bhab113. PMID: 34013342; PMCID: PMC8866812.

Iaria G, Burles F (2021). Definition: topographical disorientation. Cortex 137: 330. https://doi.org/10.1016/j.cortex.2021.01.007. Epub 2021 Feb 1. PMID: 33593606.

Kanwisher N, McDermott J, Chun MM (1997). The fusiform face area: a module in human extrastriate cortex specialized for face perception. J Neurosci 17: 4302–4311. https://doi.org/10.1523/JNEUROSCI.17-11-04302.1997. PMID: 9151747; PMCID: PMC6573547.

Kessels RP, de Haan EH, Kappelle LJ et al. (2001). Varieties of human spatial memory: a meta-analysis on the effects of hippocampal lesions. Brain Res Brain Res Rev 35: 295–303. https://doi.org/10.1016/s0165-0173(01)00058-3. PMID: 11423159.

Kopelman MD, Wilson BA, Baddeley AD (1989). The autobiographical memory interview: a new assessment of autobiographical and personal semantic memory in amnesic patients. J Clin Exp Neuropsychol 11: 724–744. https://doi.org/10.1080/01688638908400928. PMID: 2808661.

Lee TM, Yip JT, Jones-Gotman M (2002). Memory deficits after resection from left or right anterior temporal lobe in humans: a meta-analytic review. Epilepsia 43: 283–291. https://doi.org/10.1046/j.1528-1157.2002.09901.x. PMID: 11906514.

Loring DW, Meador KJ, Lee GP et al. (1995). Wada memory asymmetries predict verbal memory decline after anterior temporal lobectomy. Neurology 45: 1329–1333. https://doi.org/10.1212/wnl.45.7.1329. PMID: 7617192.

Maller JJ, Welton T, Middione M et al. (2019). Revealing the hippocampal connectome through super-resolution 1150-direction diffusion MRI. Sci Rep 9: 2418. https://doi.org/10.1038/s41598-018-37905-9. PMID: 30787303; PMCID: PMC6382767.

Massot-Tarrús A, White K, Mirsattari SM (2019). Comparing the Wada test and functional MRI for the presurgical evaluation of memory in temporal lobe epilepsy. Curr Neurol Neurosci Rep 19: 31. https://doi.org/10.1007/s11910-019-0945-8. PMID: 31044310.

McCarthy RA, Warrington EK (1990). Cognitive neuropsychology: a clinical introduction, Academic Press.

McCarthy G, Blamire AM, Puce A et al. (1994). Functional magnetic resonance imaging of human prefrontal cortex activation during a spatial working memory task. Proc Natl Acad Sci U S A 91: 8690–8694. https://doi.org/10.1073/pnas.91.18.8690. PMID: 8078943; PMCID: PMC44672.

Milner B (1966). Amnesia following operation on the temporal lobes. In: CWM Whitty, OL Zangwill (Eds.), Amnesia. Butterworths, London, pp. 109–133.

Milner B (1970). Memory and the medial temporal regions of the brain. In: KH Pribram, DE Broadbent (Eds.), Biology of memory. Academic Press, New York, pp. 29–50.

Mintzer MZ, Snodgrass JG (1999). The picture superiority effect: support for the distinctiveness model. Am J Psychol 112: 113–146. PMID: 10696280.

Moscovitch M, Melo B (1997). Strategic retrieval and the frontal lobes: evidence from confabulation and amnesia. Neuropsychologia 35: 1017–1034. https://doi.org/10.1016/S0028-3932(97)00028-6.

Moscovitch M, Rosenbaum RS, Gilboa A et al. (2005). Functional neuroanatomy of remote episodic, semantic and spatial memory: a unified account based on multiple trace theory. J Anat 207: 35–66. https://doi.org/10.1111/j.1469-7580.2005.00421.x. PMID: 16011544; PMCID: PMC1571502.

Mummery CJ, Patterson K, Price CJ et al. (2000). A voxel-based morphometry study of semantic dementia: relationship between temporal lobe atrophy and semantic memory. Ann Neurol 47: 36–45. PMID: 10632099.

Nadel L, Moscovitch M (1997). Memory consolidation, retrograde amnesia and the hippocampal complex. Curr Opin Neurobiol 7: 217–227. https://doi.org/10.1016/s0959-4388(97)80010-4. PMID: 9142752.

Nyberg L, Cabeza R, Tulving E (1996). PET studies of encoding and retrieval: the HERA model. Psychon Bull Rev 3: 135–148. https://doi.org/10.3758/BF03212412. PMID: 24213861.

Paivio A (1971). Imagery and verbal processes, Holt, Rinehart, and Winston, New YorkReprinted (1979), Erlbaum, Hillsdale, NJ.

Paivio A (1986). Mental representations: a dual-coding approach, Oxford University Press, New York.

Parton A, Malhotra P, Husain M (2004). Hemispatial neglect. J Neurol Neurosurg Psychiatry 75: 13–21. PMID: 14707298; PMCID: PMC1757480.

Patterson K, Nestor PJ, Rogers TT (2007). Where do you know what you know? The representation of semantic knowledge in the human brain. Nat Rev Neurosci 8: 976–987. https://doi.org/10.1038/nrn2277. PMID: 18026167.

Prabhakaran V, Narayanan K, Zhao Z et al. (2000). Integration of diverse information in working memory within the frontal lobe. Nat Neurosci 3: 85–90. https://doi.org/10.1038/71156. PMID: 10607400.

Ricker TJ, AuBuchon AM, Cowan N (2010). Working memory. Wiley Interdiscip Rev Cogn Sci 1: 573–585. https://doi.org/10.1002/wcs.50. Epub 2010 Feb 1. PMID: 26271504.

Riddoch MJ, Humphreys GW (1987a). Visual object processing in optic aphasia: a case of semantic access agnosia. Cogn Neuropsychol 4: 131–185.

Riddoch MJ, Humphreys GW (1987b). A case of integrative visual agnosia. Brain 110: 1431–1462. https://doi.org/10.1093/brain/110.6.1431. PMID: 3427396.

Romero Lauro LJ, Vergallito A, Anzani S et al. (2020). Primary motor cortex and phonological recoding: a TMS-EMG study. Neuropsychologia 2: 107368. https://doi.org/10.1016/j.neuropsychologia.2020.107368. Epub 2020 Feb 1. PMID: 32014451.

Rossion B, Joyce CA, Cottrell GW et al. (2003). Early lateralization and orientation tuning for face, word, and object processing in the visual cortex. Neuroimage 20: 1609–1624.

Rubin DC, Schrauf RW, Greenberg DL (2003). Belief and recollection of autobiographical memories. Mem Cognit 31: 887–901. https://doi.org/10.3758/bf03196443. PMID: 14651297.

Ruchkin DS, Grafman J, Cameron K et al. (2003). Working memory retention systems: a state of activated long-term memory. Behav Brain Sci 26: 709–728. discussion 728–77. https://doi.org/10.1017/s0140525x03000165. PMID: 15377128.

Saling MM (2009). Verbal memory in mesial temporal lobe epilepsy: beyond material specificity. Brain 132: 570–582. https://doi.org/10.1093/brain/awp012. Epub 2009 Feb 27. PMID: 19251757.

Scoville WB, Milner B (1957). Loss of recent memory after bilateral hippocampal lesions. J Neurol Neurosurg Psychiatry 20: 11–21. https://doi.org/10.1136/jnnp.20.1.11. PMID: 13406589; PMCID: PMC497229.

Sperry RW, Gazzaniga MS, e J. E. Bogen. (1969). Interhemispheric relationships: the neocortical commissures; syndromes of hemisphere disconnection. In: Disorders of speech, perception and symbolic behavior, North-Holland Publishing Co, Amsterdam, pp. 273–290.

Squire LR (1986). Mechanisms of memory. Science 232: 1612–1619. https://doi.org/10.1126/science.3086978. PMID: 3086978.

St-Laurent M, McCormick C, Cohn M et al. (2014). Using multivariate data reduction to predict postsurgery memory decline in patients with mesial temporal lobe epilepsy. Epilepsy Behav 31: 220–227. https://doi.org/10.1016/j.yebeh.2013.09.043. Epub 2013 Nov 6. PMID: 24210456.

Tulving E (2002). Episodic memory: from mind to brain. Annu Rev Psychol 53: 1–25. https://doi.org/10.1146/annurev.psych.53.100901.135114. PMID: 11752477.

Tulving E, Kapur S, Craik FI et al. (1994). Hemispheric encoding/retrieval asymmetry in episodic memory: positron emission tomography findings. Proc Natl Acad Sci U S A 91: 2016–2020. https://doi.org/10.1073/pnas.91.6.2016. PMID: 8134342; PMCID: PMC43300.

Volz LJ, Gazzaniga MS (2017). Interaction in isolation: 50 years of insights from split-brain research. Brain 140: 2051–2060. https://doi.org/10.1093/brain/awx139. PMID: 29177496.

Wager TD, Smith EE (2003). Neuroimaging studies of working memory: a meta-analysis. Cogn Affect Behav Neurosci 3: 255–274. https://doi.org/10.3758/cabn.3.4.255. PMID: 15040547.

Warrington EK, Taylor AM (1978). Two categorical stages of object recognition. Perception 7: 695–705. https://doi.org/10.1068/p070695. PMID: 740510.

Xia C (2006). Understanding the human brain: a lifetime of dedicated pursuit. Interview with Dr. Brenda Milner. McGill J Med 9: 165–172. PMID: 18523614; PMCID: PMC2323518.

Zannino GD, Perri R, Monaco M et al. (2014). The special status of verbal knowledge in semantic memory: evidence from performance of semantically impaired subjects on verbalizable and non-verbalizable versions of the object decision task. Brain Lang 128: 9–17. https://doi.org/10.1016/j.bandl.2013.11.003. Epub 2013 Dec 1. PMID: 24300660.

Zannino GD, Caltagirone C, Carlesimo GA (2015). The contribution of neurodegenerative diseases to the modelling of semantic memory: a new proposal and a review of the literature. Neuropsychologia 75: 274–290. https://doi.org/10.1016/j.neuropsychologia.2015.06.023. Epub 2015 Jun 20. PMID: 26102188.

Zannino GD, Murolo R, Grammaldo L et al. (2020). Visuo-verbal distinction revisited: new insights from a study on temporal lobe epilepsy patients in the debate over the lateralization of material-specific and process-specific aspects of memory. J Clin Exp Neuropsychol 42: 1085–1098. https://doi.org/10.1080/13803395.2020.1844868. Epub 2020 Nov 16. PMID: 33198572.

Handbook of Clinical Neurology, Vol. 208 (3rd series)
Cerebral Asymmetries
C. Papagno and P. Corballis, Editors
https://doi.org/10.1016/B978-0-443-15646-5.00017-8

Chapter 30

Brain laterality of numbers and calculation: Complex networks and their development

CARLO SEMENZA[1,2*], SILVIA BENAVIDES-VARELA[2,3], AND ELENA SALILLAS[4]

[1]*Department of Neuroscience, University of Padova, Padova, Italy*

[2]*Padova Neuroscience Center, University of Padova, Padova, Italy*

[3]*Department of Developmental Psychology and Socialisation, University of Padova, Padova, Italy*

[4]*Department of Psychology and Sociology, University of Zaragoza, Zaragoza, Spain*

Abstract

This chapter reviews notions about the lateralization of numbers and calculation in the brain, including its developmental pattern. Such notions have changed dramatically in recent decades. What was once considered a function almost exclusively located in the left hemisphere has been found to be sustained by complex brain networks encompassing both hemispheres. Depending on the specific task, however, each hemisphere has its own role. Much of this progress was determined by the convergency of investigations conducted with different methods. Contrary to traditional wisdom, the right hemisphere is not involved in arithmetic just as far as generic spatial aspects are concerned. Very specific arithmetic functions like remembering the spatial templates for complex operations, or processing of zero in complex numbers, are indeed sustained in specific right-sided areas. The system used in the typical adult appears to be the result of a complex pattern of development. The numerical brain clearly evolved from less mature to more advanced brain networks because of growth and education. Children seem to be equipped with the ability to represent the number nonverbally from a very early age. The bilateral processing of number-related tasks is however a late acquisition.

INTRODUCTION

Number processing and calculation, like language, have been, in the typical right-handed, for a long time believed to be functions of the left hemisphere. Since the first descriptions of acalculia as an independent disorder (Lewandowski and Stadelmann, 1908; Henschen, 1919), this conclusion was supported by studies on focal lesions as well as on the split-brain condition. Math functions in the brain seemed, in fact, to invariably accompany linguistic functions (Goldstein, 1948). Indeed, in the exceptional cases where linguistic functions in the right-handed are in the right hemisphere, math functions are found to be located on the same side. Thus, in the case of crossed aphasia, whereby a lesion in the right hemisphere produces aphasia in the right-handed, signs of acalculia are invariably detected (Semenza et al., 2006). In the study of split-brain, Roger Sperry and colleagues (Sperry et al., 1969) concluded that the contribution of the right hemisphere to calculation was negligible. However, Sperry (1974) later observed how the right hemisphere could, in fact, perform additions up to sums <20, but that was the only exception to a complete left-hemisphere dominance.

Despite this conclusion, signs of acalculia in right-handed individuals were reported after lesions of the right hemisphere (see the historical review by Kahn and Whitaker, 1991). Most studies seem to have assumed that acalculia from a right-hemisphere lesion was secondary to a generalized reduction in cognitive resources or, more specifically, to spatial disorders (see, for reviews, Hartje, 1987; Miceli and Capasso, 1999). Whenever, in fact, a spatial arrangement of a given written operation is required, the right hemisphere was thought to intervene. One typical exception to the left-hemisphere dominance was thus traditionally made for the case of complex, multidigit, written calculation, affected in "spatial acalculia." Spatial acalculia was observed starting from the studies of Marie et al. (1922) and Henschen and Schaller (1925) and related to the right-hemisphere dominance for spatial functions. While in these cases the calculation ability seems to be preserved, errors arise from the inability to respect (Hécaen, 1972) the position of digits and their reciprocal relation. Procedures for complex arithmetic operations indeed include generic spatial components, as well as specific requirements depending on the type of operation.

*Correspondence to: Carlo Semenza, Department of Neuroscience, University of Padova, Padova, Italy. Tel: +39-340-3950526, E-mail: carlo.semenza@unipd.it.

For instance, in the case of complex written multiplication, "move the second partial product one slot to the left." Such procedures would be, in fact, very sensitive to problems with space processing, e.g., typically, unilateral spatial neglect.

The observation of such problems eventually led to the classic tripartite classification of acalculia (Berger, 1926; Hécaen et al., 1961; Hécaen and Angelergues, 1961): (1) "primary acalculia," or "anarithmetia," the disturbance of core math functions, following lesions of the left parietal lobe, (2) secondary acalculia, determined by language problems, mainly reading and writing disorders, following lesions in a language supporting areas of the left hemisphere, and (3) secondary, "spatial" or "right hemisphere" acalculia, a consequence of spatial disorders, for example, unilateral spatial neglect, following lesions of the posterior areas of the right hemisphere.

The classic view was maintained until about the end of the 1970s, when dramatic changes in both theory and methodologies took place. Since then, two factors have contributed to changing our knowledge on brain lateralization of math functions.

The first of these factors was the advent of "cognitive" neuropsychology. Cognitive neuropsychology promoted the splitting math functions and relative tasks into independent, theoretically defensible, functions, and subtasks. Considering each subtask independently from other subtasks enabled researchers to describe in detail very specific disturbances following brain lesions (Nöel, 2001; Semenza, 2008; Cappelletti, 2015; Salillas and Semenza, 2015; Semenza and Salillas, 2021; Priftis, 2023). For example, patients were described who could not retrieve the result of multiplication tables (Warrington, 1982; McCloskey et al., 1985; Hittmair-Delazer et al., 1994; Girelli et al., 1996; Dehaene and Cohen, 1997) while performing perfectly other calculation tasks; other patients were unable to use arithmetic signs except for minus (Lampl et al., 1994; Pesenti et al., 1994) or could surprisingly perform division while unable to perform other operations (Chiarelli et al., 2011). In one particular case, a patient with a left parietal lesion could not deal in any way with numbers >4, despite no evidence of cognitive impairment of other functions, including linguistic ones (Cipolotti et al., 1991). This and other cases definitely show how language and number representation and processing can be totally independent.

The second factor was the dramatic increment in the availability of more and more sophisticated neuroimaging and neurophysiologic techniques. These techniques allowed not only a more precise localization of the lesions that result in selective math disturbances, but also the identification of the brain correlates of specific math tasks in healthy subjects. One of the earliest and most important results of these investigations has been the so-called Triple-Code model (Dehaene and Cohen, 1995), described later in this chapter, the first cognitive model where precise math functions are attributed to specific anatomic areas including both the left and the right hemisphere.

While the observation of clinical cases keeps its heuristic value by revealing unexpected patterns of selective dissociations and paving the way to further investigation toward unforeseen directions, the new technical tools allow more and more sophisticated investigations on localization according to controlled experimental designs. The result is that our understanding of the brain lateralization of numbers and calculation processing made progress hardly paralleled in other domains, leading in recent years to a knowledge of a very fine-grained distribution of functions between the left and the right hemisphere. Our notions of developmental trajectories in the acquisition of math abilities also benefitted from the new theoretical and instrumental tools. This chapter will summarize the present state of knowledge about the laterality of math functions, describing clinical studies, studies on healthy adults, and studies on developmental populations.

CLINICAL STUDIES: MATH ERRORS FOLLOWING LEFT OR RIGHT BRAIN LESIONS

The first systematic studies of acalculia in focal lesions typically reported a non-negligible quantity of calculation errors also when the damage was in the posterior areas of the right hemisphere. The nature of these errors and how they differ from the errors resulting from lesions to the left hemisphere may provide crucial information about the laterality of math functions.

In Hécaen and Angelergues (1961), the proportion of patients committing calculation errors after damage in right-hemisphere areas was 8% for the type "alexia/agraphia," 15% for "anarithmetia," and 75% for the "spatial" type. Hécaen (1972) later observed different but still conspicuous proportions: 2.1% for the alexia/agraphia, 20.2% for anarithmetia and 31% for the spatial type. Hécaen et al. (1961) provided, instead, the incidence of spatial acalculia in patients with right retrorolandic lesions (23.5%), left retrorolandic lesions (2%), and bilateral posterior lesions (17%). They considered spatial acalculia a disturbance whereby patients were unable to respect the order and the position of digits in relation to each other, or when one side of the calculation space was neglected. Interestingly, errors following right-hemisphere damage were anecdotally reported also in mental calculation (Hécaen, 1972), where the possibility that errors originate from a spatial disorder is minimal if any at all (no available theory is there to explain such a possibility other than a lapse of attention due to diminished generic resources). Such errors were otherwise never described, and their nature was never thoroughly discussed. Pure anarithmetia, unaccompanied by other math problems, is relatively rare and follows lesions of the left hemisphere. It may feature interesting dissociations, like the one shown by Benson and Weir (1972) in a person with damage to the posterior parietal lobe, who, while performing normally complex and difficult addition and subtraction as well as multiplication tables, failed with complex multiplication (where however, the addition parts were correct) and division.

In one of the largest investigations, Grafman et al. (1982) studied 76 patients, divided into four groups: left anterior, left posterior, right anterior, and right posterior. The tasks consisted of written additions, subtractions, multiplications, and divisions, including multidigit ones. Posterior patients committed the largest quantity of errors, the left posterior being by far the worst group. Importantly, this study could also show that this result still held even when the influence of more generic factors like impairments of intelligence, visuospatial disorders, and aphasia was considered. Acalculia, the authors concluded, can thus be viewed as an independent symptom. Errors were classified into three categories: (1) "spatial configuration errors," including misalignment (when a number was not in its appropriate column) and rotation (when a written number was rotated within a 180° radius) and (2) "form errors," including size (enlargement or shrinkage in relation to the size of the printed number) and distortion (the production of an unrecognizable number); "attention errors," including omission (of one or more numbers from their proper alignment) and perseveration (inappropriate production of numbers that were present in the immediately preceding operation or space). This classification of errors was aimed at capturing difficulties that did not originate from deficits in arithmetic but from visuo-constructive disorders, thus with "shape-like features of numbers." This type of error was expected to be found mainly in right-hemisphere lesions. The left posterior group, however, committed the largest number of errors for each of these three categories. No interaction of error categories with the group was found, thus no laterality effect was demonstrated in the quality of errors. The nature of the errors was not discussed further. Spatial disorders, it was reported, were present also in patients with Wernicke's aphasia. This classification of errors was never used again, and it is thus hard to fully compare the results of this study with the studies conducted by other authors in the following years.

Very different tasks were used by Warrington and James (1967) for comparing left and right brain-damaged patients. They contrasted tachistoscopic estimation of number of dots (3–7) with untimed counting of dots (up to 15). While this last task did not show lateralization effects, the estimation of the number of dots was worse in the right-hemisphere patients, particularly those with a parietal lesion.

Rosselli and Ardila (1989) compared calculation deficits in the two hemispheres, distinguishing pre- and postrolandic lesions. Moreover, within the left-hemisphere group, they distinguished several groups according to the type of concomitant aphasia. A complex battery was administered including number reading and writing, number transcoding from Arabic code to alphabetical code and vice versa, number size judgments, reading of arithmetic signs, mental and written calculation, successive additions, and subtractions, aligning of numbers for complex additions, and orally presented numerical problems. The left posterior group committed the highest number of errors; no difference was found in the number of errors between the two right-hemisphere groups. Different types of errors seemed to be committed in different types of aphasia: the small size of the aphasia subgroup did not, however, allow a meaningful statistical comparison.

Ardila and Rosselli (1994) analyzed their right-hemisphere group in a later study. Errors appeared particularly in the written calculation, and the same difficulties of spatial nature observed in reading and writing, in general, were observed in reading and writing numbers. The strategies and sequences required to add, subtract, multiply, and divide were preserved, and arithmetic signs were correctly interpreted. Disturbances of arithmetic were however found, and they were strongly associated with two underlying deficits: "reasoning" errors, appearing in one-third of the whole sample, and what the authors called, following Grafman (1988), "loss of calculation automatisms." Calculation automatisms would include what is generally called "fact retrieval," including tables and other simple operations learned by rote. Interestingly, the authors explicitly stated that the loss of calculation automatisms was related to spatial acalculia but did not explain how. The defect was associated with the general loss of automatisms reported by Luria (1966) as a result of right brain damage.

Basso et al. (2000) study on acalculia following lateralized lesions included 26 right brain-damaged participants excluding patients affected by unilateral neglect. Only three patients in this group were found to be affected by acalculia, and no detailed description of their errors is provided. In their left-hemisphere group with aphasia, 15 out of 34 patients did not show signs of acalculia.

Delazer et al. (1999) analyzed the pattern of math errors in a large sample of left-hemisphere patients affected by aphasia. They found that about a quarter of people with aphasia did not show signs of acalculia. As expected, the most severe impairments were found in global aphasia. Broca's and Wernicke's aphasics scored similarly at the quantitative level. Only mild difficulties were reported in amnesic aphasia. Multiplication was especially difficult for Broca's aphasics. Calculation procedures were mainly impaired in Wernicke's aphasia. As previously observed by other authors (e.g., Seron and Deloche, 1983), syntactic errors were more frequent in Broca's aphasia, whereas lexical errors were mainly observed in Wernicke's aphasia.

To summarize thus far, until recent years the analysis of errors was just limited to their classification into those pertaining to one of the three main varieties of acalculia. Important investigations assessing calculation disorders in large cohorts of patients invariably mentioned acalculia errors in people with damage to the right hemisphere, but this finding was just contrasted with the much more conspicuous quantity of errors following damage to the left hemisphere. Some efforts have been made to relate specific math difficulties to specific language problems; most research, however, was devoted to relating math disorders to aphasia categories rather than to specific language tasks. The clinical findings suggestive of the individual role of each hemisphere had to wait until the turn of the millennium.

Dehaene and Cohen (1991, 1997) opened the way to deeper investigations reporting important single cases of acalculia. Patient NAU (Dehaene and Cohen, 1991) could not perform even the simplest exact calculation but was good with approximation tasks. The authors, confirming the earlier findings of Warrington and James (1967) suggested that approximation might be carried out in the right hemisphere. This hypothesis was confirmed in a subsequent study on two acalculia cases (Dehaene and Cohen, 1997). Patient BOO, right-handed, affected by a subcortical lesion of the left hemisphere, had a problem with simple multiplication that was far worse than problems with addition and subtraction. In patient MAR, left-handed, a lesion in the right lower parietal areas affected subtraction but spared multiplication and addition. Dehaene and Cohen (1997) suggested that simple single digit calculation can be solved in two ways: via a "direct" route or via an "indirect," semantic route. In the direct route, mainly used for addition and multiplication (as in tables), operands would elicit the rote memory for the operation. This route would be sustained by a left-sided cortico-subcortical circuit including the basal ganglia and the thalamus. In contrast, in the indirect route, preferentially performed in subtraction, operands represent quantities on which semantically meaningful manipulations can be performed. This route would also monitor the plausibility of a result retrieved by the direct route by referring to approximate magnitude knowledge. The route would be placed bilaterally in the inferior parietal cortex and the left perisylvian network.

Another single case of acalculia, revealing the specific role the right hemisphere has in calculation, was reported by Granà et al. (2006). Their patient PN had a lesion limited to the posterior areas of the right hemisphere (the polar planum, the Heschl gyrus, the temporal planum, the posterior insula, the supramarginal gyrus, and the angular gyrus). This patient's errors affected procedures in multidigit written multiplication. PN, however, was able to describe the procedures in words and showed knowledge of the various steps that are necessary to completely perform the operation. His errors differed from errors determined by a generic inability to deal with spatial material, or from other right-hemisphere deficits like neglect. An example of these errors was adding the carry to the leftmost side of the multiplicand or of the partial product or to the partial product in the row below. He thus knew when and how the carry must be added but he seemed to have forgotten where. According to the authors, these errors could be explained by difficulties in relying on a visuospatial layout representation specific to multiplication: this spatial layout is, in fact, learned at school and guides written multiplication. Granà et al. (2006) concluded that PN, while knowing what, when, and how to carry out the various steps, did not know where to write the digits.

Modern work on right-hemisphere acalculia started testing how specifically neglect influences calculation. Benavides-Varela et al. (2014) compared right-hemisphere patients with and without left-sided neglect. Patients without neglect still showed an impairment in math. Moreover, a difference between patients with neglect and healthy controls was found in tasks such as one-digit mental subtraction and multiplication that do not require written visuospatial abilities. Thus, the authors concluded, unilateral right-hemisphere lesions may produce specific representational deficits that affect simple mental calculation, and not just the spatial arrangement of multidigit written numbers as previously thought. Along the same lines, Dormal et al. (2014) made a comparison between right-hemisphere patients with and without neglect on mental addition and subtraction of two-digit numbers (multiplication was not assessed). Neglect patients performed worse in subtraction, while they performed almost perfectly on matched additions. The authors suggested that this result demonstrated a causal relationship between attending to the left side of space and solving subtractions.

Benavides-Varela et al. (2017) tested 30 right-hemisphere patients administering NADL, a test for numerical abilities (Semenza et al., 2014) and a comprehensive battery of neuropsychologic tests. These patients, as a group, were impaired in number comprehension, transcoding, and written operations (particularly in subtraction and multiplication). Typical spatial errors (misalignment, omissions of the leftmost digits, number inversions, etc.) were linked with extensive lesions in fronto-temporoparietal areas, frequently associated with neglect. Pure arithmetic errors (e.g., failure with number facts, carrying, borrowing, etc.) were associated with lesions in the lower parietal lobe, mostly in the right angular gyrus and in the immediate surrounding areas. Spatial errors were predicted by composite measures of visuospatial attentional and representational abilities. Specific errors of arithmetic nature instead were not predicted by visuospatial abilities. In contrast with traditional views, this group of patients made a much higher proportion of arithmetic errors with respect to spatial errors across various calculation tasks.

Langdon and Warrington (1997) administered a calculation battery and an arithmetic reasoning test to both left- and right-hemisphere patients. Whereas left-hemisphere patients were more impaired than the right-hemisphere lesion group in calculation both groups were equally impaired in arithmetic reasoning. The right hemisphere would be therefore involved in arithmetic reasoning as much as the left hemisphere.

Another function that was found to be in the right hemisphere was processing complex multidigit numbers containing zero in various positions. Benavides-Varela et al. (2016) showed that right-hemisphere patients committed many errors in reading aloud and writing on dictation of such numbers. Interestingly, such errors had no relation with neglect or to generic cognitive impairment. They seemed to derive from damage to mechanisms that are independent of those used in reading and writing numbers without zero. The authors proposed that the basic mechanism would entail parsing and the ability to set up empty-slot structures required for processing *zeros* in complex numbers. Voxel-based lesion-symptom analysis could locate this ability in the right insula and its immediate surroundings. Importantly, such finding was confirmed in a subsequent study on a huge group of stroke patients (Haupt et al., 2017).

A study by Proios et al. (2021) importantly showed for the first time how people with aphasia, damaged in the left hemisphere, might have problems with basic number comprehension in tasks that do not necessarily require verbal abilities. These tasks included number comparison (comparing the number of squares in two panels presented simultaneously), number line marking (marking a given number on a line defined by its endpoints, e.g., "where is 7 in a 1–10 interval?), and digit comprehension (pointing to the appropriate digit corresponding to the number of squares on a panel). These tasks have been previously shown to be disturbed in right-hemisphere lesions, thus they cannot be specifically carried out by the left hemisphere.

On similar lines, Gonzalez et al. (2021) showed, in a very large group (200 patients), how nonverbal abilities could be predictors of calculation tests, whereas agraphia predicted the performance in written calculation. They concluded that calculation abilities can be regarded both as written language-dependent and verbal language-independent.

A step forward in understanding the respective role of the two hemispheres in the retrieval of arithmetic facts and other simple operations came with the use of direct cortical electrostimulation (DCE) in patients undergoing tumor neurosurgery. Yu et al. (2011) found that inhibition of the right parietal areas impaired performance on simple subtraction. These authors related this result to quantity processing rather than to verbal numerical processing. In further studies, Della Puppa and coworkers (Della Puppa et al., 2013, 2015) could show a right-hemisphere involvement also in multiplication (in the angular gyrus, the supramarginal gyrus, the interparietal sulcus, and the superior parietal lobule) and addition (in the supramarginal gyrus).

Analysis of errors committed in DCE investigations added crucial information. Thus Semenza et al. (2017) not only found positive sites in both parietal areas but also found that substitution errors (the majority of omissions were only about 5%) were of a different nature after inhibition of the left or the right hemisphere, unveiling the role of the hemisphere contralateral to the inhibition. Multiplication errors, when inhibition was on the right, mostly consisted of replacing the correct solution with another table solution (e.g., $7 \times 3 = 28$ or $7 \times 3 = 14$). This finding, the authors argued, was because the left hemisphere, where the retrieval of tables is supported, took over the operation. In contrast, most errors following inhibition of the left hemisphere were of the approximation type (e.g., $7 \times 3 = 20$). This would reflect the activity of the right hemisphere, where an approximation mechanism is believed to be acting. Importantly no individual site was positive for both addition and multiplication. This finding ruled out the possibility that errors came from a generalized shortage of processing resources, which would have not resulted in operation-specific deficits. Approximation errors after left-hemisphere inhibition were also found for addition; inhibition led to an underestimation of the correct result in both hemispheres. A larger deviation from correct was however found after right-hemisphere inhibition, reflecting the lack of an approximation mechanism.

NEUROIMAGING STUDIES IN HEALTHY ADULTS

In a well-defined neurofunctional proposal, the Triple Code Model (TCM, Dehaene, 1992; Dehaene and Cohen, 1995) proposed three main numerical representations and a neural localization for them, including possible patterns of lateralization. First, the *analogic magnitude representation* (analogic code) refers to a preverbal representation that carries the numerical meaning, the number sense. This analogic code activates in virtually all numerical processes, and it was proposed to reside bilaterally in the parietal lobe, more specifically, in the bilateral horizontal intraparietal sulcus (hIPS). This precise localization was derived from a later meta-analysis of neuroimaging data (Dehaene et al., 2003). This basic representation would allow for the approximate contrast between sets of objects differing in numerosity. It seems to be common to nonhuman species and preverbal infants (Roitman and Brannon, 2003; see studies in developmental population below) and precludes symbolic processing (Piazza, 2010). This approximate estimation of numerosity has been further accounted for by the approximate number system (ANS; Piazza, 2010). A crucial aspect of this precursor system is that it represents numerosity in a compressed manner, according to Weber's law.

The neural basis of the analogic representation has been explored mostly using adaptation paradigms (e.g., Piazza et al., 2004; Cantlon et al., 2006; Cohen Kadosh et al., 2007; Kadosh et al., 2011) and comparison tasks (e.g., Pinel et al., 1999, 2001) where the ratio, or the distance, between numerosities, needs to be computed. Brain areas that are ratio dependent in the adult brain have been shown to be located bilaterally, in mid-parietal regions (Piazza et al., 2004, 2007). However, a higher reliance on the right parietal areas for analog quantities has also been proposed (Faye et al., 2019), despite the fact that both intraparietal sulci (IPS) play a role in nonsymbolic processing. Concurrently with Piazza studies (Piazza et al., 2004, 2007), Cohen Kadosh et al. (2007) emphasized that the left parietal areas might process magnitude independently of format and that the right parietal areas would process magnitude in a format-dependent manner.

Hence, the approximate system precedes symbolic processing (Nieder and Dehaene, 2009; Odic and Starr, 2018) and the pattern of lateralization might change when symbols are acquired and attached to numerical meanings. For example, Vogel et al. (2015) show hemispheric differences in the IPS using symbolic vs nonsymbolic stimuli. In addition, activations in the left IPS using symbolic material were correlated with age. In a recent meta-analysis (Escobar-Magariño et al., 2022) addressing specific studies using the adaptation paradigm, lateralization indexes suggested right IPS dominance when the stimuli were nonsymbolic, and left IPS dominance when the stimuli were symbolic. Both IPS appeared involved in both formats; however, imbalance to the left IPS or right IPS appeared for symbolic vs nonsymbolic stimuli, respectively.

Any Arabic numerical input, on the other hand, necessarily starts with the decoding of the numerical symbols (i.e., Arabic numbers). A second representation for the TCM, namely the *visual Arabic form* (or Arabic code), carries that decoding for visual numbers. Both hemispheres contribute to this symbolic recognition, interconnected through the corpus callosum and the respective parietal areas of each hemisphere, as well as to the verbal circuit (see below). The TCM locates it in the occipito-temporal cortices. Neuroimaging studies have tried to unveil the specific location of the so-called number form area (NFA) and its differentiation from the homolog area for word recognition, the visual word form area (VWFA) (Hannagan et al., 2015).

A recent meta-analysis aiming to reconcile literature and test for functional specificity on the NFA (Yeo et al., 2017) evidenced a prevalence of the right inferior temporal cortex as specific for the NFA. On the other hand, an Arabic number, unlike letters or letter strings, carries numerical semantic information. Hence, one might have to consider a wider network for a fusion between numerical magnitude processing, ordinality, syntax, lexicon, and symbol-referent associations. This network involves the parietal regions bilaterally and right lateralized superior and inferior frontal regions. Consequently, top-down processes must be considered. Hannagan and collaborators (Hannagan et al., 2015) acknowledge how electrocorticography studies (Shum et al., 2013) show bilaterality, but stress an imbalance favoring the right inferior temporal gyrus, more than the left one. Furthermore, they suggest it is likely that the left NFA could be less differentiated from the visual word form area (VWFA), and hence, less specific to number symbols. The right predominance of NFA could be explained by the default functional connectivity between ITG and IPS in the right hemisphere before learning symbols (Nemmi et al., 2018). Functional connectivity with left parietal areas is shown later in development. Structural data is still missing in this respect.

A third code proposed by the TCM is the *verbal word frame* (or auditory-verbal code), which organizes the numbers as word sequences. This code would support counting and the retrieval of arithmetic facts. Arithmetic facts refer to the stored one-digit operations whereby operands and solutions are stored associatively in long-term memory. This code is verbal because the storage and retrieval of facts is achieved phonologically through rote verbal memory. Due to this verbal nature, fact storage and retrieval would be lateralized in the left (dominant) hemisphere, specifically, in perisylvian areas and the left angular gyrus (AG). Initial neuroimaging research on calculation suggested a key role of the left hemisphere (Burbaud et al., 1995; Dehaene et al., 1999). This network was nicely exemplified in the training study of Delazer et al. (2003): they showed a shift from IPS to left AG, when learning of new arithmetic facts is achieved. Hence, they showed a shift from quantity-based processes in IPS, to automatic retrieval of arithmetic facts in left AG. Additional studies converged with these results: overall, left-hemispheric perisylvian areas and the left AG, as well as the left hippocampus seem to be behind trained problems more than untrained problems (e.g., Delazer et al., 2005; Ischebeck et al., 2006; Grabner et al., 2009; Bloechle et al., 2016).

However, as anticipated in the description of clinical data, the view that only the left hemisphere carries simple calculations is now questioned. Calculation is taken as a much more complex process, requiring a bilateral network. Even within simple calculation, a consideration of several other functions, apart from retrieval, during arithmetic is needed. This view is shared by recent proposals (Arsalidou and Taylor, 2011; Menon, 2014). Importantly, Arsalidou and Taylor (2011) carried out a meta-analysis where lateralization indexes were explicitly calculated for different numerical operations. Again, bilateral parietal areas appear to be relevant for multiplication, but also for addition and subtraction. However, addition showed left-hemisphere dominance, subtraction appeared balanced bilaterally, and surprisingly, a right-hemisphere dominance was found for multiplication. In a modification to the TCM, Arsalidou and Taylor propose a role of the right AG for spatial fact retrieval, among other changes.

Then what is the specific role of each hemisphere during calculation? Using magnetoencephalography (MEG), we were able to specify how the two hemispheres might partially contribute to simple calculation (Salillas et al., 2021). While, as the TCM stated, retrieval relied effectively on a left frontoparietal circuit, a similar right frontoparietal circuit appeared to manage approximation processes during calculation (Semenza et al., 2017; Arcara et al., 2021). In turn, the verification of multiplication solutions unrelated to the operands ($3 \times 2 = 7$) needed a right-hemisphere network, and the verification of multiplication solutions related to the operands ($3 \times 2 = 8$) required a left-hemisphere network (see the EEG study by Kiefer and Dehaene, 1997, for earlier similar findings). The essential difference between these two incorrect problems is the interference of associations within the memory networks for arithmetic that occurs predominantly in the second case and that is irrelevant in the first case. Functional connectivity measures suggested that a loop between the left dorsolateral and the left AG might play a crucial role in resolving this interference in front of incorrect, yet related solutions. These two networks at least, might orchestrate during calculation. Together with bilateral inferior frontal areas probably used for final selection processes. Convergently, Arcara and collaborators (Arcara et al., 2021) showed how during a multiplication production task, easy items (fastest responses) needed of the left AG, again emphasizing the use of retrieval in the first stage. The role of the right dorsolateral prefrontal cortex was then emphasized, as well as a later contribution of right parietal areas, likely helping in the identification of the solution.

Finally, the link between numbers and space needs to be addressed. Certain numerical representations might rely on similar neural roots than spatial cognition. The neural proximity between these two domains would lead to a shared parietal space subserving a culture-dependent cognitive link (Hubbard et al., 2005). There are extensive behavioral and neuropsychologic data on this interaction, but it has been less directly studied using neuroimaging. An early discovered effect, the Spatial Numerical Association of Response Codes (SNARC; Dehaene et al., 1993) prompted the study of the links between numbers and space. It implies that, within a given interval, small numbers are responded faster with the left hand, and larger numbers are responded faster with the right hand. It was therefore suggested that numerical representations would be organized in the form of a "mental number line" with small numbers on the left and big numbers on the right. Using functional near-infrared spectroscopy (fNIRS), Cutini et al. (2014) applied a comparison task and contrasted right and left responses to small and large numbers. Hence, they could contrast the hemodynamic response to SNARC compatible responses vs SNARC-incompatible responses. Bilateral HIPS and left AG appeared related to the SNARC effect and HIPS overlapped with the response to distance between numbers. The distance effect is a marker of the access to magnitude. Thus, these authors suggested a close link between space and the core number process during numerical comparison.

Another recent study addressed the mental number line with a different approach (Liu et al., 2019). A previous work (Zorzi et al., 2002) had shown how neglected patients tend to displace the center of number intervals to the right. That is, when asked to tell the middle number between two numbers (1–5), neglected patients shift the center of the interval to the right (responding 4 instead of 3), in the same way as they do with a line bisection task. Using fMRI, Liu and collaborators (Liu et al., 2019) applied the number bisection task and contrasted it with line bisection in healthy participants. The conjunction test for both tasks showed common activations mainly located bilaterally in parietal–frontal areas. They showed a bias toward the left hemisphere when contrasting number bisection minus physical bisection. This entailed a left-hemisphere network: lingual gyrus, inferior frontal gyrus, precentral, inferior parietal lobule, the supplementary motor area, insula, caudate, inferior temporal gyrus, and thalamus. The authors highlighted a pivotal role of the left SMA which is relevantly coupled with the bilateral frontoparietal networks, especially during number bisection. Hence, this work unveiled bilaterality for number bisection. However, it stressed that number bisection, more than line bisection, requires a left-lateralized network.

Fig. 30.1 depicts a graphical abstract of the neuroimaging studies in adults, focusing on lateralization of the parietal lobes. Overall, the bilateral IPS appears to be used for quantity discrimination with a higher reliance on the left IPS for symbolic stimuli and a more important role of the right IPS for nonsymbolic stimuli. During simple calculation, the view of a unique left parietal involvement needs to be further specified. While it appears that the left parietal lobe is important during fact retrieval, there is a requirement of the right parietal areas for approximation, which would likely complement retrieval. During spatio-numerical processes, we found, again, bilaterality in IPS. Number bisection tasks might differ from other spatial stimuli: number bisection is biased toward the left inferior parietal lobule. More work is needed in this last domain, especially because the emphasis of the reported studies is on the IPL, whereas a role of the posterior superior parietal lobe (PSPL) has been proposed, commonly to other spatial attention functions (Dehaene et al., 2003).

Evidently, the parietal lobes do not work in isolation (Arsalidou and Taylor, 2011). For nonsymbolic quantity contrasts, the IFG bilaterally supports the process. There is predominance for the right hemisphere for these stimuli. The frontal lobe is also important for spatio-numerical processes, bilaterally. As per calculation, the left dorsolateral prefrontal cortex might interact with AG during retrieval and resolution of interference during fact retrieval. The IFG bilaterally, would supplement the parietal areas during selection and decision processes.

As a final remark, it is important to note that a direct contrast between activations in the two hemispheres is highly desirable. Showing that both hemispheres are involved or not in a function is critical. However, their relative importance should also be directly contrasted (i.e., lateralization indexes). We have described two works using this approach, but not all studies that addressed lateralization applied it. This explicit contrast is overall missing for the numerical domain, even when brain lateralization was a research question in a variety of studies.

TRANSCRANIAL MAGNETIC STIMULATION (TMS) AND TRANSCRANIAL DIRECT-CURRENT STIMULATION (tDCS)

With the TMS technique, changing magnetic fields induce changes in the brain tissue noninvasively, altering neuronal activity. With adequate experimental designs, and unlike correlational neuroimaging methods such as fMRI, TMS provides causal information between the stimulated sites and the measured function. Many times, these studies are preceded by neuroimaging techniques, or patient data, which suggest the brain sites to be stimulated, sometimes at the individual level, as well as control sites. Electrophysiologic data are also generally important to determine the timing of stimulation. TMS tells us whether a given brain site is *essential* for a given function: that is, if we transiently alter that site during a given task, and we obtain an observable change in the related behavior (in error rate or reaction times), then that area is considered crucial for the function.

TMS studies focusing on *quantity processing* have almost always focused on a comparison between hemispheres. Not all results have converged with fMRI studies (Salillas and Semenza, 2015). Taken together, the study of quantity using TMS has shown a role for both parietal lobes. More specifically, left IPL (Sandrini et al., 2004), left posterior parietal cortex (Andres et al., 2005), and bilateral IPS (Cappelletti et al., 2007; Dormal et al., 2008; Dormal et al., 2012; Sasanguie et al., 2013) including symbolic and nonsymbolic stimuli. Göbel et al. (2001) measured comparison times between two-digit numbers and a reference while stimulating the left and right AG. A stronger effect for closer distances larger than the references appeared after left AG stimulation. In Andres et al. (2005), the posterior parietal cortices (PPC) appeared to respond differently to close and far digits: while bilateral PPC stimulation

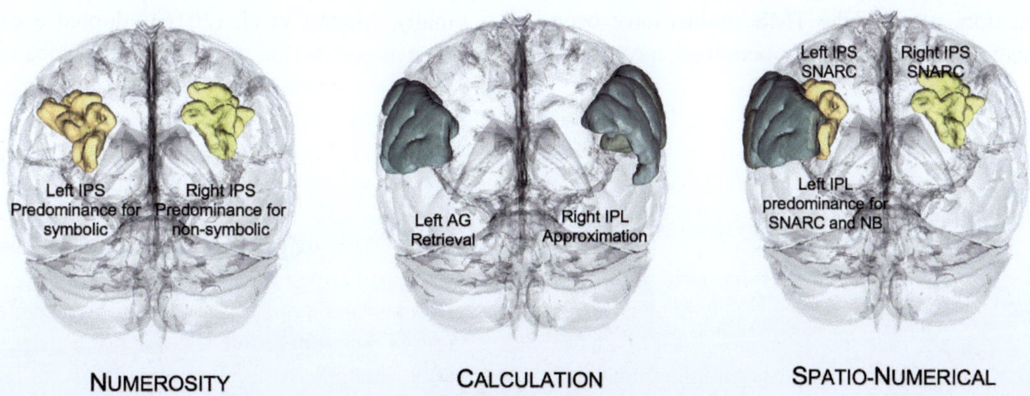

Fig. 30.1. Graphical abstract of the neuroimaging studies in adults restricted to the parietal lobes. Left: for quantity judgments IPS lateralization might be dependent on the use of symbolic (left hemisphere) or nonsymbolic stimuli (right hemisphere). Center: the inferior parietal lobes, bilaterally, coordinate for retrieval (left AG) and for approximation (right IPL) during exact arithmetic. Right: SNARC effects rely on bilateral IPS and left IPL. Number interval bisection tends to recruit left IPL more than line bisection. IPS, intraparietal sulcus; AG, angular gyrus; IPL, inferior parietal lobe; SNARC, Spatial Numerical Association of Response Codes; NB, number bisection.

disrupted the comparison for close numbers, only left PPC stimulation disrupted the comparison between far numbers.

Most studies have used symbolic material (Arabic studies), yet Cappelletti et al. (2007) contrasted symbolic with nonsymbolic stimuli. They showed that the left IPS was crucial for both formats. However, they found facilitation after right IPS stimulation. Sasanguie et al. (2013) on the other hand, showed the crucial role of the left IPS in the mapping between symbols and qualities. This symbol-to-quantity mapping was also explored by Cohen Kadosh et al. (2012), yet they highlighted a role of the right IPS in the facilitation effect during a numerical Stroop task for physical judgments. Regarding the NFA, Grotheer et al. (2016) show causal involvement of the right temporo-occipital cortex in the processing of masked Arabic numbers. They also show an association with Roman letters.

As per the number-space relation, interhemispheric differences have been related specifically to number magnitude, which tends to agree contralaterally with the MNL: right hemisphere with small numbers and left hemisphere with big numbers. For example, Cattaneo et al. (2009) showed a variation in visual cortex excitability caused by the presentation of numbers. Small numbers increased the proportion of trials in which TMS-induced phosphenes after right occipital stimulation. Conversely, large numbers increased the proportion of trials in which TMS-induced phosphenes after left occipital stimulation. Rusconi et al. (2007) explored the SNARC effect in the PPL, showing a bilateral effect. In another study, they remarked on the importance of right IFG and the frontal eye fields in the SNARC effect (Rusconi et al., 2011). Finally, Salillas et al. (2009) (see also Renzi et al., 2011), extended the study of space-number link to motion perception. They showed that the ventral IPS, in the occipital part of IPS and within the network of motion perception (Vanduffel et al., 2002; Orban et al., 2006) was causally linked to motion perception, and to number comparison. Here again, some more important right-hemisphere involvement was shown for small numbers, while large numbers seem to be more bilateral. In turn, this motion-quantity association suggested that attention to motion might be a part of attention along the MNL (McCrink et al., 2007).

Regarding calculation, most of the TMS studies have been addressing lateralization. The explored operations have been addition (Göbel et al., 2006; Salillas et al., 2012; Maurer et al., 2016; Montefinese et al., 2017), subtraction (Maurer et al., 2016; Montefinese et al., 2017; Fresnoza et al., 2020), and multiplication (Andres et al., 2011; Salillas et al., 2012; Maurer et al., 2016; Fresnoza et al., 2020). One study has also included division (Maurer et al., 2016). TCM proposes bilaterality from subtraction but left lateralization for simple addition and multiplication, which would rely on the verbal word frame. Fig. 30.2 unifies the positive sites for each operation based on the mentioned studies. A similar integration of studies was provided (see Salillas and Semenza, 2015) for quantity processing and number-space relations. Since then, more calculation TMS studies have been published and a similar joining picture can be provided for calculation.

For double-digit addition, Göbel et al. (2006) showed a role of the anterior and posterior left IPS and no effect on the right hemisphere. Montefinese et al. (2017) however show a stronger role of the right hIPS for addition and less asymmetry for the ventral IPS (vIPS). But overall, they show that bilateral hIPS and vIPS subserve addition and subtraction to different degrees. Moreover, within the inferior parietal lobe, they showed that bilateral AG and SMG support addition and subtraction. However, the effects on the right SMG were stronger. Finally, while ANG was more important in addition, SMG was slightly more involved in subtraction. Importantly, Fresnoza et al. (2020) differentiate retrieval from calculation strategies. They showed a left AG contribution for retrieval and calculation in multiplication, while it was only involved in retrieval during subtraction. No detrimental effect was found for left hIPS. The study by Fresnoza highlights the importance of differentiating the actual solving process, from the concrete operation under study. They did not stimulate the homolog right parietal areas. Finally, Andres et al. (2011) also contrasted subtraction and multiplication, showing a bilateral role of the hIPS, for both operations. Salillas et al. (2012) explored addition and multiplication and found addition to be supported by the horizontal part of the IPS (hIPS) bilaterally, and multiplication relying on left hIPS and the bilateral part of ventral IPS (vIP).

Finally, Maurer et al. (2016) adopted a different approach, which overcomes the usual extreme locality assumption in

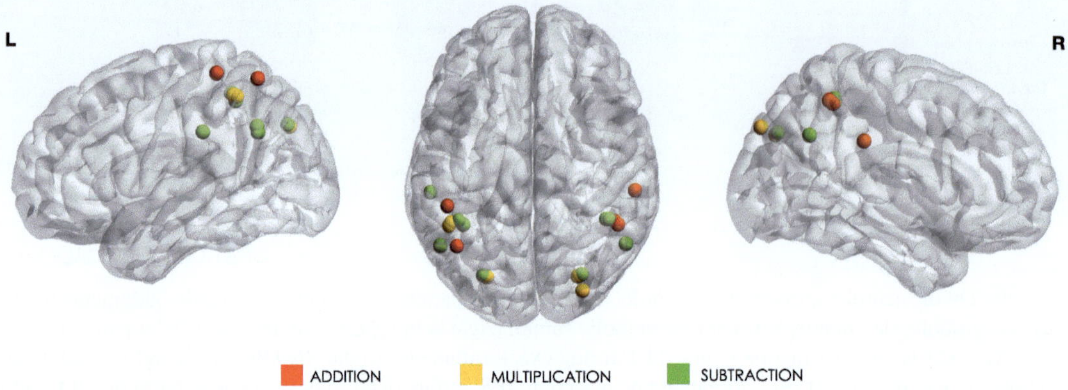

Fig. 30.2. Positive sites from TMS studies on calculation. The different spheres depict the positive sites reported in the studies for addition, multiplication, or subtraction. Table 30.1 details the coordinates in the different studies.

TMS experiments. These authors stimulated more than 50 brain sites; hence, they exposed a distribution of probability of error within division, subtraction, addition, or multiplication. For division, the highest error rates upon stimulation were in the left inferior parietal lobe, left middle frontal gyrus, bilateral posterior superior temporal cortex, and right middle precentral gyrus. For subtraction, the highest error rates appeared after right AG stimulation, as well as the right SG and the middle frontal gyrus. In the addition task, the higher error rates appeared after right middle frontal gyrus stimulation, right AG, and superior temporal gyrus, as well as after anterior superior temporal gyrus. Finally, in multiplication, the highest error rates appeared after left AG stimulation and after middle frontal gyrus stimulation.

Fig. 30.2 shows the localization of positive sites in calculation studies for addition, subtraction, and multiplication, restricted to parietal areas (see Salillas and Semenza, 2015, for other processes). As can be seen, jointly considering the different studies, crucial sites have been shown in both hemispheres for calculation. There are slight proportional differences in the involvement of the right hemisphere between operations (42% in addition tasks, 37% for multiplication, and 40% for subtraction), and not all studies have contrasted both hemispheres (Fresnoza et al., 2020, see Table 30.1).

The most used of a group of similar techniques, transcranial direct-current stimulation (tDCS) is used to either excite (anodal tDCS) or inhibit (cathodal tDCS) the brain cortex. It provided further, albeit limited and largely confirmatory, data on the lateralization of math. Whether tDCS effects are math-specific, however, remains dubious in many cases and no clear patterns of lateralization are consistently found (Mosbacher et al., 2020). Several tDCS studies were aimed at improving learning math. Not all these studies evidenced beneficial effects: depending on the task, detrimental effects were also reported (Mosbacher et al., 2020). All this considered, only pivotal examples of these studies are reported here, and the reader is referred to Schroeder et al. (2017) and Hussain et al. (2024) for more comprehensive reviews.

Table 30.1

Details of the TMS studies on calculation.

Operation			X	Y	Z	Study
Addition	LH	pIPS	−40.75	−59.91	60.74	Göbel et al. (2006)
8 LH vs 6 RH	LH	aIPS	−46.11	−37.63	63.68	
	LH	HIPS	−44	−46.76	52.09	Salillas et al. (2012)
	LH	HIPS	−45.15	−48.5	51.32	
	RH	HIPS	46.87	−46	51.5	
	RH	HIPS	48.89	−47.99	53.18	
	LH	HIPS	−39.77	−45.6	53.09	Montefinese et al. (2017)
	LH	VIPS	−25.97	−76.05	36.23	
	RH	HIPS	42.64	−44.91	52.56	
	RH	VIPS	28.97	−75.59	35.87	
	LH	ANG	−50.34	−59.45	35.23	
	RH	ANG	53.21	−58.58	34.57	
	LH	SMG	−55.73	−30.14	32.15	
	RH	SMG	58.38	−29.19	31.41	
Multiplication	LH	HIPS	−37	−48	49	Andres et al. (2011)
5 LH vs 3 RH	RH	HIPS	44	−45	52	
Studies contrasting both hemispheres: 4 LH vs 3 RH	RH	VIPS	26.86	−76.73	34.85	Salillas et al. (2012)
	LH	VIPS	−23.86	−77.15	35.18	
	LH	HIPS	−44	−46.76	52.09	
	LH	HIPS	−45.15	−48.5	51.32	
	RH	VIPS	29.54	−84.69	38.82	
	LH	AG	−50	−60	32	Fresnoza et al. (2020)
Subtraction	LH	HIPS	−37	−48	49	Andres et al. (2011)
6 LH/4 RH	RH	HIPS	44	−45	54	
Studies contrasting both hemispheres:	LH	HIPS	−39.77	−45.6	53.09	Montefinese et al. (2017)
5 LH vs 4 RH	LH	VIPS	−25.97	−76.05	36.23	
	RH	HIPS	42.64	−44.91	52.56	
	RH	VIPS	28.97	−75.59	35.87	
	LH	ANG	−50.34	−59.45	35.23	
	RH	ANG	53.21	−58.58	34.57	
	LH	SMG	−55.73	−30.14	32.15	
	LH	AG	−50	−60	32	Fresnoza et al. (2020)

All coordinates are in MNI space. In studies in which coordinates were reported in Talairach space, they were transformed to MNI.

Cohen Kadosh et al. (2010) demonstrated that anodal tDCS to the right parietal lobe and cathodal stimulation to the left parietal lobe improved mapping between symbols and magnitude, while cathodal stimulation to the right parietal lobe and anodal stimulation to the left parietal lobe impaired performance. Clemens et al. (2013) applied anodal tDCS over the right parietal cortex but did not find effects on multiplication verification; an fMRI analysis, however, indicated a stronger activation in the angular gyrus during tDCS. Hauser et al. (2013) found that anodal tDCS over left parietal areas reduced calculation times in large subtractions; in contrast, stimulation over right parietal areas, as well as bilateral stimulation, did not show any effect. Left anodal/right cathodal parietal tDCS also seems to improve calculation time in complex multiplication (Kasahara et al., 2013). Anodal stimulation of the left dorsolateral prefrontal cortex improved performance in people with high math anxiety on a math verification task (Sarkar et al., 2014) and in a serial subtraction task (Pope et al., 2015). Klein et al. (2013) showed that bilateral bicephalic tDCS affected mental addition but not a control task (color word Stroop), thus demonstrating a specific influence of tDCS. The numerical distractor distance effect was greater under cathodal than under anodal stimulation. In the same condition, the effect of target identity was not affected. According to Klein et al. (2013), this finding supports the specific claim of the TCM that number magnitude information is represented bilaterally in the intraparietal cortices. Mosbacher et al. (2020) compared the effects of anodal tDCS over the left dorsolateral prefrontal cortex and left posterior parietal cortex, as well as in a sham condition, on arithmetic performance in small and large problems of addition and subtraction, concomitantly recording EEG activity. No general improvement in arithmetic performance was found. In the left hemisphere, a change was found from theta band ERD to ERS in all three conditions; an analogous change in the right hemisphere was only found in the sham condition. Overall, the authors concluded, tDCS did not lead to a general improvement in arithmetic performance: however frontal stimulation seemed to accelerate training gains, while parietal stimulation halted them.

STUDIES IN DEVELOPMENTAL POPULATIONS

How does the "numerical brain" develop? In this section, we briefly review the studies that focus on understanding the brain areas that support numerical representations in the first stages of development or how these networks gradually change throughout brain maturation and enculturation. We will focus on consistent hemispheric asymmetries and the putative roles of each hemisphere in supporting math throughout development. The reader interested in a more comprehensive view of the brain areas involved is referred to these detailed reviews: Arsalidou et al. (2018), Peters and De Smedt (2018), and Vogel and De Smedt (2021).

A wealth of behavioral studies indicate that children are equipped with the ability to represent numbers nonverbally from an early age (e.g., Izard et al., 2009; Benavides-Varela and Reoyo-Serrano, 2021). Two separated, although possibly related abilities can be broadly distinguished when analyzing numerical capacities across development. First, the ability to encode and process nonsymbolic numerical representations (i.e., distinguishing a change in the number of objects/dots in a familiar set). This ability is present since birth and is considered not dependent on language (Gelman and Butterworth, 2005; Dehaene, 2011). Second, the capacity to comprehend and use symbolic representation of numbers, such as Arabic numbers and number words. These abilities are gradually acquired during school age and strongly rely on education and other related capacities such as verbal counting, learning to read, etc.

Right-lateralized parietal activation supports early numerical processing

Studies investigating the neural correlates of nonsymbolic numerical abilities consistently find a developmental continuity with respect to the neural locus, such that the same areas are reported in adults and young children. However, studies also found clear hemispheric asymmetries in the parietal regions that are evident only in the first period of life. Using fMRI Cantlon et al. (2006) tested 4-year-old children and adults with the same paradigm: all participants were exposed to a stream of dot arrays which occasionally changed in numerosity. The results indicated that, while both groups showed stronger BOLD responses to the novel than to familiar numerosities in the IPS, 4-year-old children compared to adults exhibited less activity in the left IPS. Furthermore, a study conducted by Holloway and Ansari (2010) with older children (7- to 9-year-olds) and adults who performed both symbolic and nonsymbolic numerical comparison tasks, showed that only a right-lateralized IPS activity was common to both groups (Holloway and Ansari, 2010). These studies suggest that, while the left IPS does not have a strong involvement in supporting basic numerical abilities early in life, the right IPS readily encodes numerical magnitude independently of the amount of experience, mathematical knowledge, or age. Congruently, studies with younger infants who have not even started to speak, have also reported right-lateralized brain responses to numerosity changes using different techniques. For example, a study with 3-month-old infants compared event-related potentials to changes in numerosity with those to changes in shape (Izard et al., 2008). Then, using a source localization algorithm on scalp ERP data, the authors found that the response to numerical information originated primarily from right frontoparietal regions. The scalp topography was also fully consistent with an EEG study by Berger (2011), which showed right parietal activity related to the evaluation of numerosity in simple arithmetic equations in 6- to 9-month-old infants. Similarly, another EEG study found parametric variations in neural oscillations as a function of numerical ratio in 7-month-old infants. Specifically, Alpha-band (6–8 Hz) oscillations over the midline and right posterior scalp sites were modulated by the ratio between familiar and novel numerosities (Libertus et al., 2009). Two additional studies using fNIRS, also found that only the right parietal cortex of 6-month-olds was sensitive to numerosity, as indicated by an increase in oxygenated hemoglobin when infants were presented with nonsymbolic numerical changes in contrast to shape changes (Hyde et al., 2010; Edwards et al., 2016).

Altogether these studies indicate a right lateralized response of the IPS supporting magnitude representations early in life. Noticeably, the involvement of the right IPS is observed before sophisticated estimation and symbolic numerical abilities emerge in development. This contrasts with the more balanced, bilateral, IPS activation pattern typically observed when healthy adults engaged in similar magnitude comparison tasks (see for example Dehaene et al., 2003) and speaks about the biologic role of the right IPS in supporting numerical processing. As summarized by Bisiacchi and Cainelli (2022), in very young infants the predominance of right-lateralized structures and functions has been observed in several studies outside the number domain, including those focusing on memory functions (Benavides-Varela et al., 2012, 2017), nonspeech auditory stimulation (Telkemeyer et al., 2009), rhythmic visual stimuli (Crowell et al., 1973), and taste (Fox and Davidson, 1986). The dominance of the right hemisphere for many early-life functions is in line with theories stating that the right hemisphere develops earlier and that its development is less subject to external influences because it sustains functions necessary to survival (Geschwind and Galaburda, 1985; Geschwind et al., 2002).

Bilateral parietal activation emerges in childhood

A modification of the hemispheric asymmetry in numerical processing tasks is observed throughout development. Indeed, as children grow up, they usually learn other numerical codes, properties, and functions through education. The effects of instruction were described for the first time in a developmental fMRI study by Rivera et al. (2005). In their study, the authors revealed an age-related increment in the recruitment of the left inferior parietal cortex (specifically the left SMG) during a task with symbolic digits. No such increment was observed in the right parietal lobe. Similarly, in a study by Ansari and Dhital (2006), children (9–11 years old) and adults were presented with a nonsymbolic numerical comparison task. They were presented with two sets of squares side-by-side and were instructed to select the numerically larger group of squares as quickly and as accurately as possible. Their results showed that, in comparison to the group of children, the adult participants exhibited greater effects of numerical distance on the left IPS. Thus, the involvement of left IPS seems to develop as a function of age and experience with representations that entail enculturated symbols, such as Arabic numerals and number words (e.g., Ansari, 2008). The activation of the right IPS, instead, is predominant in nonsymbolic magnitude processing tasks since the earliest stages of development, when the influence of education and instruction is less marked. Emerson and Cantlon (2015) directly tested this hypothesis using fMRI in a longitudinal design with 4- to 9-year-old children. They found that neural responses in the left IPS change systematically as a function of children's numerical discrimination acuity whereas right IPS responses are constant over a 1- to 2-year-period in young children.

The sustained activation of the right IPS and the incremental takeover of the left inferior parietal regions eventually give rise to the more bilateral pattern observed in healthy adults. This bilateral activation supports numerical processing independently of the code use (Piazza et al., 2007) but with an apparent difference in precision. Approximate judgments, for example, correlate with stronger activation in the right than in the left IPS, while exact judgments correlate with more activation in the left vs right IPS (Piazza et al., 2004, 2007). Similarly, a right hemispheric activation in terms of magnitude checking seems to support multiplication tables retrieval in healthy adults, whereas, as mentioned above, fast accurate response selection elicits a stronger activation of the left parietal areas (Arcara et al., 2021). Although evidence of a transition from coarse to accurate representations has been also reported in behavioral studies in children (Berteletti et al., 2010) and theorized based on other number-related asymmetric responses (Felisatti et al., 2020), to our knowledge there are no studies directly linking those transitions with functional changes in the developing brain. It is, however, plausible that learning Arabic symbols contributes to refining the precision of the numerical representation (Verguts and Fias, 2004) and at the same time modifying the hemispheric asymmetry observed in numerical processing tasks early in development. As stated above, while the right IPS keeps a comparable involvement from the first months of life and throughout development, the enculturation process seems to increase the activation in the left parietal areas, progressively supporting more advanced competencies. This explains why children with a stronger neural ratio effect in the left IPS during symbolic number processing also exhibit higher arithmetic competence (Bugden et al., 2012).

Cross-hemispheric integration of parietal areas

Coordination between hemispheres might also be an important feature that gradually increases and supports more efficient numerical processing throughout development. Cantlon et al. (2011) conducted a study comparing adults and young children (6- to 7-year-olds) using fiber-based tractography to examine the relation between the integrity of the fibers of the corpus callosum connecting the left and right parietal lobes and performance on magnitude comparison tasks. This connectivity is reviewed at the structural level by Moeller et al. (2015), covering different ages. Their results showed significantly less integrity of these fibers in children compared to adults and a correlation between integrity and performance on the tasks. This suggests that this interhemispheric connection could play an important role in the maturation of numerical representations. A further fMRI study with even younger children (4- to 6-year-olds) strengthened this view, by showing that the degree of the effective connectivity between the right parietal and the left parietal cortex can predict individual scores in a standardized test of mathematical achievement (Park et al., 2014).

Notably, correlations between white matter tracts that connect the prefrontal cortex and the posterior parietal cortex or the occipitotemporal cortex have been also consistently found across studies. Generally speaking, better white matter integrity of these distant tracts coincides with better arithmetic performance (for a review see Matejko and Ansari, 2015).

Hemispheric asymmetries in other regions supporting basic number processing in children

The neural networks underlying symbolic number processing undergo other developmental changes that imply some levels of lateralization which, however, have not been fully accounted for. For example, a study conducted by (Ansari and Dhital, 2006) with children (9- to 11-year-olds) and adults who performed comparison tasks with Arabic numerals, showed a right-lateralized frontal activity only in the group of children and the typical bilateral IPS activation in adults. While discussing these results, the authors focused on an interesting developmental phenomenon, namely the frontoparietal transfer in the processing of symbolic magnitude processing, possibly due to a shift from more controlled and effortful to more automatic processing of numerical magnitude. This age-dependent frontal-to-parietal shift was replicated by Cantlon et al. (2009) (see Fig. 30.3). Some authors have also hypothesized a stronger involvement of working memory in childhood (Menon, 2016). The lateralization of the effect in the frontal lobe, however, has been often disregarded. In that respect, one could argue that monitoring—the process of continuously evaluating the internal or external contingencies to optimize behavior—which is right-lateralized in the frontal lobe (e.g., Vallesi, 2012), could be enhanced in children but gradually decreases with age or practices. Further studies would be necessary to verify this hypothesis and confirm which specific executive functions are engaged in children when performing even basic numerical (comparison) tasks (Fig. 30.4).

Left- and right-lateralized activation contributes to the typical development of arithmetic skills

Another important aspect for understanding numerical development regards the learning of arithmetic, namely the ability to add, subtract, multiply, and divide symbolic whole numbers. The flourishing body of brain imaging studies regarding calculation

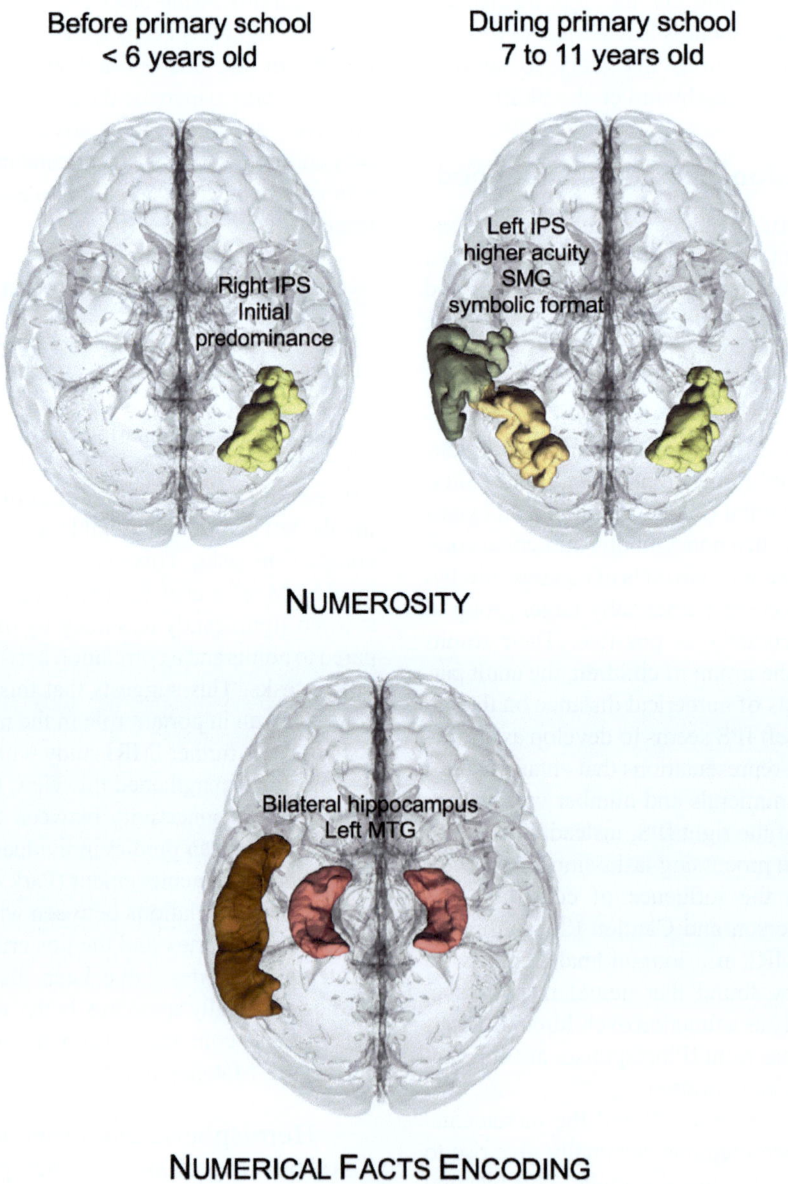

Fig. 30.3. Graphical abstract of developmental neuroimaging studies. Up: evolution with age in the lateralization of the parietal areas involved in numerosity. Down: highlight of the structures recruited in children during numerical facts encoding.

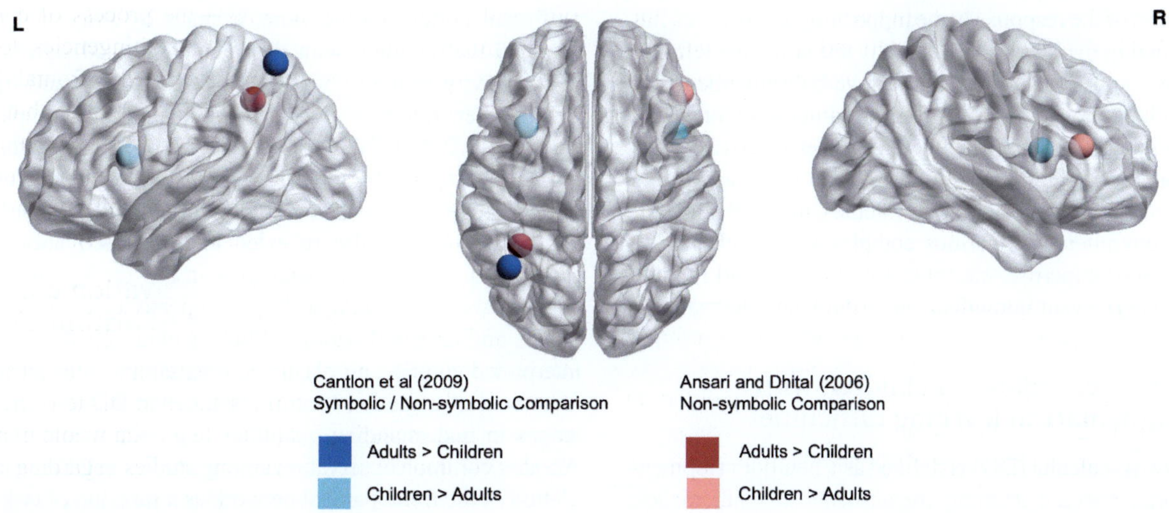

Fig. 30.4. Frontal to parietal shift during number comparison. The studies of Ansari and Dhital (2006) with nonsymbolic stimuli and Cantlon et al. (2009) using symbolic and nonsymbolic stimuli are integrated. Note that only contrasts (adults > children), or (children > adults) area included (i.e., in Ansari and Dhital (2006) adults also showed activations in right IPS). Talairach coordinates have been transformed to MNI space.

in children, has revealed a widespread set of interconnected brain areas which include the IPS as a key region (see previous sections), but also additional areas involved in domain-specific and domain-general processes (i.e., working memory, language, visuospatial abilities, long-term memory, etc.), such as the dorsolateral and ventrolateral prefrontal cortex, the posterior superior parietal lobule (PSPL), the fusiform gyrus, the angular gyrus, the supramarginal gyrus, and the hippocampus.

The right PSPL

Activation in the right posterior superior parietal lobule (PSPL) is often reported in studies investigating the development of arithmetic abilities in children. For example, Rosenberg-Lee et al. (2011) assessed children in second and third grades with fMRI while performing a set of two-digit addition problems. Their study revealed that, compared to second graders, third graders showed greater activity in the right SPL. Similarly, Cho et al. (2012) found that the only region in which brain activity increased with age in second and third-graders was the right PSPL. Furthermore, Prado et al. (2014) assessed second and seventh-grade children while performing a set of one-digit problems. Their study revealed grade-related increases of activity for subtraction, in the right PSPL. Altogether, these studies indicate that activity associated with addition and subtraction increases over development in this region. However, in contrast to the IPS, the right-lateralized PSPL is not considered to be specific to quantity-based processing, but to mechanisms of visuospatial attention, which might be progressively engaged or "recycled" in these tasks during arithmetic education (Hubbard et al., 2005; Dehaene and Cohen, 2007; Knops et al., 2009; Prado et al., 2014). The left PSPL, instead, does not appear to play a significant role in these tasks.

The hippocampus supports encoding numerical facts in the development

Acquiring expertise in other operations, like learning arithmetic facts in multiplications and additions, might rely on different strategies and developmental correlates (Suárez-Pellicioni et al., 2022). Interestingly, although studies in adults often observe activation in language-related areas, particularly the left angular gyrus during the retrieval of multiplication facts (but see also Benavides-Varela et al., 2014, 2016; Semenza and Benavides-Varela, 2017), this is not consistently seen in developmental studies (e.g., Rosenberg-Lee et al., 2011; Prado et al., 2014; Soltanlou et al., 2018; Declercq et al., 2022) (for an exception, see Polspoel et al., 2017). Mastering of single-digit facts in school-age children has been correlated with activity in the left-middle temporal gyrus (Prado et al., 2014) and more consistently with the hippocampus (Rivera et al., 2005; De Smedt et al., 2011; Qin et al., 2014). For example, Qin et al. examined developmental changes in a longitudinal study with 7- to 9-years old children over a 1.2-year interval. The findings indicated reduced brain activity in the prefrontal cortex and increased activity in the right hippocampus along with a shift from the use of counting to memory-based strategies. Evidence for the gradual replacement of procedural strategies in favor of direct retrieval of arithmetic facts was found in the study of Cho et al. (2012). These authors found that children who frequently used a retrieval strategy exhibited enhanced activity and connectivity from the right hippocampus and other areas in the frontal cortex when compared to children who used procedural strategies (e.g., decomposition, counting) more consistently (see Moeller et al. (2015), for a review on structural and functional connectivity). In yet another study with 10- to 12-year-olds, De Smedt et al. reported that the left hippocampus showed increased brain activity during fact retrieval problems (De Smedt et al., 2011). Given the hippocampus's role in memory encoding and learning, this outcome is unsurprising. However, as stated above, the involvement of the hippocampus as opposed to the AG, does not align with the most common findings reported in adults. Thus, it has been suggested that the role of the hippocampus might be to encode new arithmetic facts before their representations can be transferred to language-related regions of the left temporoparietal cortex (De Smedt et al., 2011; Peters and De Smedt, 2018).

The lateralization of the response in the hippocampus, however, has not been clarified in the literature. Generally speaking, the left hippocampus plays a prominent role in storing verbal/semantic memories, while right hippocampus might be more important for spatial memory processing (Burgess et al., 2002; Ezzati et al., 2016). Since both the left and the right hippocampi have been found associated with learning new numerical facts across studies, it is likely that different dimensions interact in various complex ways with participants' age, level of expertise, etc. to enable efficient and flexible processing of the relevant numerical and arithmetic information.

Brain correlates in children with mathematical learning difficulties

Developmental dyscalculia (DD) is defined as a neurodevelopmental disorder with specific learning impairments in mathematics. Children with DD have difficulties in processing numerical information, and learning arithmetic facts, and have poor calculation and math reasoning abilities. Its prevalence is estimated at around 3%–6% of children. It is a neural deficit with a broad social impact, adversely affecting academic and career opportunities (e.g., Vigna et al., 2022). While the etiology of dyscalculia is still debated (Butterworth et al., 2011; Butterworth, 2018; Benavides-Varela, 2020), many efforts have been made in the last decades to support affected children. Efficient interventions often consider individual's needs (e.g., Re et al., 2020) and use both traditional and digital-based tools (e.g., Benavides-Varela et al., 2020, 2023) to help children overcome their deficits.

Like most developmental learning disorders, DD has been linked to abnormalities in how the brain works and how it is structured. However, only a handful of studies have examined the brain activity in children with DD, and their findings remain mixed. For example, differences between typically developing (TD) children and children with DD have been found in the right IPS (Molko et al., 2003; Rotzer et al., 2008) the left IPS (Isaacs et al., 2001) or bilaterally (Rykhlevskaia et al., 2009). Structural differences have also been reported in brain areas that are linked to general learning skills, including the hippocampal circuitry, the SPL, and the prefrontal cortex (Rotzer et al., 2008; Rykhlevskaia et al., 2009). Moreover, children with DD showed reduced white matter integrity of the superior longitudinal fasciculus as compared to controls (Rykhlevskaia et al., 2009). One study also showed that the volume of many of these white matter bundles undergoes abnormal development during the preteenage and teenage years. Specifically, these tracts appeared to increase with age in the TD but not in the DD children (Ranpura et al., 2013). Altogether, these studies indicate that different anatomic alterations of the arithmetic brain network can be linked to severe difficulties in mathematics in children, with no clear differences in the pattern of alterations between the left and the right-hemisphere areas.

Dysfunctional activations in the areas devoted to both numerical and arithmetic processing (e.g., IPS) and to domain-general brain regions (e.g., frontal cortex) have been reported in various studies. For example, Kucian et al. (2006) presented DD and TD children with an approximate arithmetic task. They found that children with DD, compared to TD, had weaker activation in the left IPS as well as the right inferior frontal gyrus and right medial frontal gyrus in the prefrontal cortex. Furthermore, a study by Price and colleagues revealed that children with dyscalculia showed significantly less brain activity in the right IPS compared to age- and IQ-matched TD children during a nonsymbolic magnitude comparison task (Price et al., 2007). Right (Soltész et al., 2007) and bilateral IPS differences compared to controls were also observed in further symbolic comparison tasks (Mussolin et al., 2010). On the other hand, research has also revealed increased activation in the left IPS, frontal cortex, and visual areas in DD during number processing, as well as functional hyperconnectivity in parietal, frontal, visual, and temporal regions (Michels et al., 2018). This has been interpreted considering plastic compensatory mechanisms and the increment of cognitive control resources in DD to overcome challenges in understanding quantities (e.g., Kaufmann et al., 2009). Another common observation among studies in DD is a lack of regulation of the frontoparietal network as a function of task difficulty. Differently from TD children, children with DD recruit this network to the same extent for both easy and hard problems (De Smedt et al., 2011; Ashkenazi et al., 2012).

Identifying which factors determine the lateralization of the functional differences between TD and DD in the single studies would be speculative, given that available research is still rather limited. At the very least, the diversity and level of difficulty of tasks, as well as the children's age should be considered among the factors that possibly result in varying findings. Moreover, small groups of DD children are typically reported within each study, and there is an enormous variation in the inclusion criteria. These factors weaken the power of the results and may confound the conclusions of single studies. A recent meta-analysis comprehensively summarized the findings of 24 studies with 728 children to better understand the neurobiologic basis of math learning difficulties (Tablante et al., 2023). The results revealed that the most consistent dysfunction across studies was found in the right parietal lobe along the intraparietal sulcus (IPS) although atypically upregulated functions in brain regions supporting attention, working memory, and visual processing, were also identified. The authors interpreted these findings considering both the core quantity deficit hypothesis and the multifaceted cognitive impairment hypothesis. Notably, the right-hemispheric deficit in the IPS is suggestive of a core deficit in quantity processing.

Last, detailed qualitative and individual analyses are missing in this field. Close examination of the errors committed constitutes an opportunity for the full appreciation of the condition, the mix of strategies that are used to calculate the answer to a particular problem, their brain correlates, and hemispheric asymmetries in typical and atypical conditions (Benavides-Varela et al., 2017). Error analyses can also reveal insights for designing effective and personalized interventions and, along with longitudinal examinations, constitute a promising avenue for future investigations in the field.

CONCLUSIONS

Notions about the lateralization of math in the brain have changed dramatically in recent decades. What was considered a function almost exclusively sustained by the left hemisphere has been found to be sustained by complex brain networks encompassing both hemispheres.

Much of the progress was determined by the convergency of investigations conducted with different methods. A typical example of this interaction is studies on arithmetic tables. The first finding with TMS showing a role of the right hemisphere (Salillas et al., 2012) thus led to additional clinical investigations (Benavides-Varela et al., 2014, 2016, 2017) including explorations via DCE (Semenza et al., 2017), which in turn prompted research with MEG (Arcara et al., 2021; Salillas et al., 2021). Consequently, we now know that arithmetic tables are not just processed in the left angular gyrus, as previously believed: they are rather sustained by a much more complex, bilaterally located, network, characterized by a complex temporal dynamic.

Depending on the specific task, each hemisphere has been found to have its own role. A simple calculation, again, provides an example. The analysis of errors after DCE, as reported above, revealed that the left-hemisphere retrieves results from a stored repertory, while the right hemisphere contributes by indicating the approximate result. The right hemisphere is not involved in arithmetic just as far as generic spatial aspects are concerned. Very specific arithmetic functions like remembering the spatial templates for complex operations, or processing of zero in complex numbers, are indeed sustained in specific right-sided areas.

Importantly, the system used in the typical adult appears to be the result of a complex pattern of development. The numerical brain clearly evolved from less mature to more advanced brain networks because of growth and education. Children, in fact, seem to be equipped with the ability to represent numbers nonverbally from a very early age. The bilateral processing of a number of related tasks is however a late acquisition. Magnitude representation starts in the right IPS. A dysfunction of this area is related to the core quantity basis deficit at the basis of developmental acalculia. The left IPS intervenes at later stages in relation to age and experience, with exposure to numerical codes like the Arabic code and to number words.

References

Andres M, Seron X, Olivier E (2005). Hemispheric lateralization of number comparison. Cogn Brain Res 25: 283–290. https://doi.org/10.1016/j.cogbrainres.2005.06.002.

Andres M, Pelgrims B, Michaux N et al. (2011). Role of distinct parietal areas in arithmetic: an fMRI-guided TMS study. Neuroimage 54: 3048–3056. https://doi.org/10.1016/j.neuroimage.2010.11.009.

Ansari D (2008). Effects of development and enculturation on number representation in the brain. Nat Rev Neurosci 9: 278–291. https://doi.org/10.1038/NRN2334.

Ansari D, Dhital B (2006). Age-related changes in the activation of the intraparietal sulcus during nonsymbolic magnitude processing: an event-related functional magnetic resonance imaging study. J Cogn Neurosci 18: 1820–1828. https://doi.org/10.1162/JOCN.2006.18.11.1820.

Arcara G, Pezzetta R, Benavides-Varela S et al. (2021). Magnetoencephalography reveals differences in brain activations for fast and slow responses to simple multiplications. Sci Rep 11: 1–13. https://doi.org/10.1038/s41598-021-97927-8.

Ardila A, Rosselli M (1994). Spatial acalculia. Int J Neurosci 78: 177–184. https://doi.org/10.3109/00207459408986056.

Arsalidou M, Taylor MJ (2011). Is 2+2=4? Meta-analyses of brain areas needed for numbers and calculations. Neuroimage 54: 2382–2393. https://doi.org/10.1016/j.neuroimage.2010.10.009.

Arsalidou M, Pawliw-Levac M, Sadeghi M et al. (2018). Brain areas associated with numbers and calculations in children: meta-analyses of fMRI studies. Dev Cogn Neurosci 30: 239–250. https://doi.org/10.1016/J.DCN.2017.08.002.

Ashkenazi S, Rosenberg-Lee M, Tenison C et al. (2012). Weak task-related modulation and stimulus representations during arithmetic problem solving in children with developmental dyscalculia. Dev Cogn Neurosci 2. https://doi.org/10.1016/J.DCN.2011.09.006.

Basso A, Burgio F, Caporali A (2000). Acalculia, aphasia and spatial disorders in left and right brain-damaged patients. Cortex 36: 265–280. https://doi.org/10.1016/S0010-9452(08)70528-8.

Benavides-Varela S (2020). What causes the disorder—theories and perspectives. In: Understanding dyscalculia, pp. 5–22. https://doi.org/10.4324/9780429423581-2.

Benavides-Varela S, Reoyo-Serrano N (2021). Small-range numerical representations of linguistic sounds in 9- to 10-month-old infants. Cognition 213. https://doi.org/10.1016/J.COGNITION.2021.104637.

Benavides-Varela S, Hochmann JR, MacAgno F et al. (2012). Newborn's brain activity signals the origin of word memories. Proc Natl Acad Sci U S A 109: 17908–17913. https://doi.org/10.1073/PNAS.1205413109.

Benavides-Varela S, Pitteri M, Priftis K et al. (2014). Right-hemisphere (spatial?) acalculia and the influence of neglect. Front Hum Neurosci 8. https://doi.org/10.3389/FNHUM.2014.00644.

Benavides-Varela S, Passarini L, Butterworth B et al. (2016). Zero in the brain: a voxel-based lesion-symptom mapping study in right hemisphere damaged patients. Cortex 77: 38–53. https://doi.org/10.1016/J.CORTEX.2016.01.011.

Benavides-Varela S, Piva D, Burgio F et al. (2017). Re-assessing acalculia: distinguishing spatial and purely arithmetical deficits in right-hemisphere damaged patients. Cortex 88: 151–164. https://doi.org/10.1016/J.CORTEX.2016.12.014.

Benavides-Varela S, Zandonella Callegher C, Fagiolini B et al. (2020). Effectiveness of digital-based interventions for children with mathematical learning difficulties: a meta-analysis. Comput Educ 157: 103953. https://doi.org/10.1016/J.COMPEDU.2020.103953.

Benavides-Varela S, Laurillard D, Piperno G et al. (2023). Digital games for learning basic arithmetic at home. Prog Brain Res 276: 35–61. https://doi.org/10.1016/BS.PBR.2023.02.001.

Benson DF, Weir WF (1972). Acalculia: acquired anarithmetia. Cortex 8: 465–472. https://doi.org/10.1016/S0010-9452(72)80008-X.

Berger H (1926). Uber Rechenstorunger bei Herderkraunkunger des Grosshirns. Arch Psychiatr Nervenkr 78: 236–263.

Berger A (2011). Electrophysiological evidence for numerosity processing in infancy. Dev Neuropsychol 36: 668–681. https://doi.org/10.1080/87565641.2010.549878.

Berteletti I, Lucangeli D, Piazza M et al. (2010). Numerical estimation in preschoolers. Dev Psychol 46: 545–551. https://doi.org/10.1037/A0017887.

Bisiacchi P, Cainelli E (2022). Structural and functional brain asymmetries in the early phases of life: a scoping review. Brain Struct Funct 227: 479–496. https://doi.org/10.1007/S00429-021-02256-1.

Bloechle J, Huber S, Bahnmueller J et al. (2016). Fact learning in complex arithmetic-the role of the angular gyrus revisited. Hum Brain Mapp 37: 3061–3079. https://doi.org/10.1002/hbm.23226.

Bugden S, Price GR, McLean DA et al. (2012). The role of the left intraparietal sulcus in the relationship between symbolic number processing and children's arithmetic competence. Dev Cogn Neurosci 2: 448–457. https://doi.org/10.1016/J.DCN.2012.04.001.

Burbaud P, Degreze P, Lafon P et al. (1995). Lateralization of prefrontal activation during internal mental calculation: a functional magnetic resonance imaging study. J Neurophysiol 74: 2194–2200. https://doi.org/10.1152/jn.1995.74.5.2194.

Burgess N, Maguire EA, O'Keefe J (2002). The human hippocampus and spatial and episodic memory. Neuron 35: 625–641. https://doi.org/10.1016/S0896-6273(02)00830-9.

Butterworth B (2018). Dyscalculia: from science to education, Routledge, pp. 1–188. https://doi.org/10.4324/9781315538112.

Butterworth B, Varma S, Laurillard D (2011). Dyscalculia: from brain to education. Science 332: 1049–1053. https://doi.org/10.1126/SCIENCE.1201536.

Cantlon JF, Brannon EM, Carter EJ et al. (2006). Functional imaging of numerical processing in adults and 4-y-old children. PLoS Biol 4: 844–854. https://doi.org/10.1371/JOURNAL.PBIO.0040125.

Cantlon JF, Libertus ME, Pinel P et al. (2009). The neural development of an abstract concept of number. J Cogn Neurosci 21: 2217. https://doi.org/10.1162/JOCN.2008.21159.

Cantlon JF, Davis SW, Libertus ME et al. (2011). Inter-parietal white matter development predicts numerical performance in young children. Learn Individ Differ 21: 672–680. https://doi.org/10.1016/J.LINDIF.2011.09.003.

Cappelletti M (2015). The neuropsychology of acquired number and calculation disorders. In: R Cohen Kadosh, A Dowker (Eds.), The Oxford handbook of numerical cognition. Oxford University Press, pp. 808–836.

Cappelletti M, Barth H, Fregni F et al. (2007). rTMS over the intraparietal sulcus disrupts numerosity processing. Exp Brain Res 179: 631–642. https://doi.org/10.1007/s00221-006-0820-0.

Cattaneo Z, Silvanto J, Battelli L et al. (2009). The mental number line modulates visual cortical excitability. Neurosci Lett 462: 253–256. https://doi.org/10.1016/j.neulet.2009.07.027.

Chiarelli V, Menichelli A, Zadini A et al. (2011). Good division, but bad addition, subtraction and multiplication. A "leftmost-first" bug? Cortex 47: 250–258. https://doi.org/10.1016/j.cortex.2010.08.004.

Cho S, Metcalfe AWS, Young CB et al. (2012). Hippocampal-prefrontal engagement and dynamic causal interactions in the maturation of children's fact retrieval. J Cogn Neurosci 24: 1849–1866. https://doi.org/10.1162/JOCN_A_00246.

Cipolotti L, Butterworth B, Denes G (1991). A specific deficit for numbers in a case of dense acalculia. Brain 114: 2619–2637. https://doi.org/10.1093/brain/114.6.2619.

Clemens B, Jung S, Mingoia G et al. (2013). Influence of anodal transcranial direct current stimulation (tDCS) over the right angular gyrus on brain activity during rest. PLoS One 9: e95984. https://doi.org/10.1371/journal.pone.0095984.

Cohen Kadosh R, Cohen Kadosh K, Kaas A et al. (2007). Notation-dependent and -independent representations of numbers in the parietal lobes. Neuron 53: 307–314. https://doi.org/10.1016/j.neuron.2006.12.025.

Cohen Kadosh K, Henson RN, Cohen Kadosh R et al. (2010). Task-dependent activation of face-sensitive cortex: an fMRI adaptation study. J Cogn Neurosci 22: 903–917. https://doi.org/10.1162/jocn.2009.21224.

Cohen Kadosh R, Bien N, Sack AT (2012). Automatic and intentional number processing both rely on intact right parietal cortex: a combined fMRI and neuronavigated TMS study. Front Hum Neurosci 6. https://doi.org/10.3389/fnhum.2012.00002.

Crowell DH, Jones RH, Kapuniai LE et al. (1973). Unilateral cortical activity in newborn humans: an early index of cerebral dominance? Science 180: 205–207. https://doi.org/10.1126/SCIENCE.180.4082.205.

Cutini S, Scarpa F, Scatturin P et al. (2014). Number-space interactions in the human parietal cortex: enlightening the snarc effect with functional near-infrared spectroscopy. Cereb Cortex 24: 444–451. https://doi.org/10.1093/cercor/bhs321.

De Smedt B, Holloway ID, Ansari D (2011). Effects of problem size and arithmetic operation on brain activation during calculation in children with varying levels of arithmetical fluency. Neuroimage 57: 771–781. https://doi.org/10.1016/J.NEUROIMAGE.2010.12.037.

Declercq M, Bellon E, Sahan MI et al. (2022). Arithmetic learning in children: an fMRI training study. Neuropsychologia 169. https://doi.org/10.1016/J.NEUROPSYCHOLOGIA.2022.108183.

Dehaene S (1992). Varieties of numerical abilities. Cognition 44: 1–42. https://doi.org/10.1016/0010-0277(92)90049-N.

Dehaene S (2011). The number sense: how the mind creates mathematics. Revised and Updated Edition.

Dehaene S, Cohen L (1991). Two mental calculation systems: a case study of severe acalculia with preserved approximation. Neuropsychologia 29: 1045–1074. https://doi.org/10.1016/0028-3932(91)90076-K.

Dehaene S, Cohen L (1995). Towards an anatomical and functional model of number processing. Math Cogn 1: 83–120.

Dehaene S, Cohen L (1997). Cerebral pathways for calculation: double dissociation between rote verbal and quantitative knowledge of arithmetic. Cortex 33: 219–250. https://doi.org/10.1016/S0010-9452(08)70002-9.

Dehaene S, Cohen L (2007). Cultural recycling of cortical maps. Neuron 56: 384–398. https://doi.org/10.1016/J.NEURON.2007.10.004.

Dehaene S, Bossini S, Giraux P (1993). The mental representation of parity and number magnitude. J Exp Psychol Gen 122: 371–396. https://doi.org/10.1037/0096-3445.122.3.371.

Dehaene S, Spelke E, Pinel P et al. (1999). Sources of mathematical thinking: behavioral and brain-imaging evidence. Science 284: 970–974. https://doi.org/10.1126/science.284.5416.970.

Dehaene S, Piazza M, Pinel P et al. (2003). Three parietal circuits for number processing. Cogn Neuropsychol. https://doi.org/10.1080/02643290244000239.

Delazer M, Girelli L, Semenza C et al. (1999). Numerical skills and aphasia. J Int Neuropsychol Soc 5: 213–221. https://doi.org/10.1017/S1355617799533043.

Delazer M, Domahs F, Bartha L et al. (2003). Learning complex arithmetic—an fMRI study. Cogn Brain Res 18: 76–88. https://doi.org/10.1016/j.cogbrainres.2003.09.005.

Delazer M, Ischebeck A, Domahs F et al. (2005). Learning by strategies and learning by drill—evidence from an fMRI study. Neuroimage 25: 838–849. https://doi.org/10.1016/j.neuroimage.2004.12.009.

Della Puppa A, De Pellegrin S, D'Avella E et al. (2013). Right parietal cortex and calculation processing: intraoperative functional mapping of multiplication and addition in patients affected by a brain tumor clinical article. J Neurosurg 119: 1107–1111. https://doi.org/10.3171/2013.6.JNS122445.

Della Puppa A, De Pellegrin S, Salillas E et al. (2015). Functional mapping of left parietal areas involved in simple addition and multiplication. A single-case study of qualitative analysis of errors. J Neuropsychol 9: 330–335. https://doi.org/10.1111/jnp.12043.

Dormal V, Andres M, Pesenti M (2008). Dissociation of numerosity and duration processing in the left intraparietal sulcus: a transcranial magnetic stimulation study. Cortex 44: 462–469. https://doi.org/10.1016/j.cortex.2007.08.011.

Dormal V, Andres M, Pesenti M (2012). Contribution of the right intraparietal sulcus to numerosity and length processing: an

fMRI-guided TMS study. Cortex 48: 623–629. https://doi.org/10.1016/j.cortex.2011.05.019.
Dormal V, Schuller AM, Nihoul J et al. (2014). Causal role of spatial attention in arithmetic problem solving: evidence from left unilateral neglect. Neuropsychologia 60: 1–9. https://doi.org/10.1016/j.neuropsychologia.2014.05.007.
Edwards LA, Wagner JB, Simon CE et al. (2016). Functional brain organization for number processing in pre-verbal infants. Dev Sci 19: 757–769. https://doi.org/10.1111/DESC.12333.
Emerson RW, Cantlon JF (2015). Continuity and change in children's longitudinal neural responses to numbers. Dev Sci 18: 314–326. https://doi.org/10.1111/DESC.12215.
Escobar-Magariño D, Turel O, He Q (2022). Bilateral intraparietal activation for number tasks in studies using adaptation paradigm: a meta-analysis. Neuroscience. https://doi.org/10.1016/j.neuroscience.2022.02.024.
Ezzati A, Katz MJ, Zammit AR et al. (2016). Differential association of left and right hippocampal volumes with verbal episodic and spatial memory in older adults. Neuropsychologia 93: 380–385. https://doi.org/10.1016/J.NEUROPSYCHOLOGIA.2016.08.016.
Faye A, Jacquin-Courtois S, Reynaud E et al. (2019). Numerical cognition: a meta-analysis of neuroimaging, transcranial magnetic stimulation and brain-damaged patients studies. NeuroImage Clin 24. https://doi.org/10.1016/j.nicl.2019.102053.
Felisatti A, Laubrock J, Shaki S et al. (2020). A biological foundation for spatial-numerical associations: the brain's asymmetric frequency tuning. Ann N Y Acad Sci 1477: 44–53. https://doi.org/10.1111/NYAS.14418.
Fox NA, Davidson RJ (1986). Taste-elicited changes in facial signs of emotion and the asymmetry of brain electrical activity in human newborns. Neuropsychologia 24: 417–422. https://doi.org/10.1016/0028-3932(86)90028-X.
Fresnoza S, Christova M, Purgstaller S et al. (2020). Dissociating arithmetic operations in the parietal cortex using 1 Hz repetitive transcranial magnetic stimulation: the importance of strategy use. Front Hum Neurosci 14: 498065. https://doi.org/10.3389/fnhum.2020.00271.
Gelman R, Butterworth B (2005). Number and language: how are they related? Trends Cogn Sci 9: 6–10. https://doi.org/10.1016/J.TICS.2004.11.004.
Geschwind N, Galaburda AM (1985). Cerebral lateralization. Biological mechanisms, associations, and pathology: III. A hypothesis and a program for research. Arch Neurol 42: 634–654. https://doi.org/10.1001/ARCHNEUR.1985.04060070024012.
Geschwind DH, Miller BL, DeCarli C et al. (2002). Heritability of lobar brain volumes in twins supports genetic models of cerebral laterality and handedness. Proc Natl Acad Sci U S A 99: 3176–3181. https://doi.org/10.1073/PNAS.052494999.
Girelli L, Delazer M, Semenza C et al. (1996). The representation of arithmetical facts: evidence from two rehabilitation studies. Cortex 32: 49–66. https://doi.org/10.1016/S0010-9452(96)80016-5.
Göbel S, Walsh V, Rushworth MFS (2001). The mental number line and the human angular gyrus. Neuroimage 14: 1278–1289. https://doi.org/10.1006/NIMG.2001.0927.
Göbel SM, Rushworth MF, Walsh V (2006). Inferior parietal rTMS affects performance in an addition task. Cortex 42: 774–781. https://doi.org/10.1016/S0010-9452(08)70416-7.
Goldstein K (1948). Language and language disturbances, Grune and Stratton, New York, NY.
Gonzalez R, Rojas M, Rosselli M et al. (2021). Acalculia in aphasia. Arch Clin Neuropsychol 36: 455–464. https://doi.org/10.1093/arclin/acaa072.
Grabner RH, Ansari D, Koschutnig K et al. (2009). To retrieve or to calculate? Left angular gyrus mediates the retrieval of arithmetic facts during problem solving. Neuropsychologia 47: 604–608. https://doi.org/10.1016/j.neuropsychologia.2008.10.013.
Grafman J (1988). Acalculia. In: F Boller, J Grafman (Eds.), Handbook of neuropsychology. vol. 1. Elsevier, Amsterdam, pp. 414–430.
Grafman J, Passafiume D, Faglioni P et al. (1982). Calculation disturbances in adults with focal hemispheric damage. Cortex 18: 37–49. https://doi.org/10.1016/S0010-9452(82)80017-8.
Granà A, Hofer R, Semenza C (2006). Acalculia from a right hemisphere lesion. Dealing with "where" in multiplication procedures. Neuropsychologia 44: 2972–2986. https://doi.org/10.1016/j.neuropsychologia.2006.06.027.
Grotheer M, Ambrus GG, Kovács G (2016). Causal evidence of the involvement of the number form area in the visual detection of numbers and letters. Neuroimage 132: 314–319. https://doi.org/10.1016/j.neuroimage.2016.02.069.
Hannagan T, Amedi A, Cohen L et al. (2015). Origins of the specialization for letters and numbers in ventral occipitotemporal cortex. Trends Cogn Sci. https://doi.org/10.1016/j.tics.2015.05.006.
Hartje W (1987). The effect of spatial disorders on arithmetical skills. In: G Deloche, X Seron (Eds.), Mathematical disabilities a cognitive neuropsychological perspective. Lawrence Erlbaum Associates, London, UK, pp. 121–135.
Haupt M, Gillebert CR, Demeyere N (2017). The zero effect: voxel-based lesion symptom mapping of number transcoding errors following stroke. Sci Rep 7: 1–10. https://doi.org/10.1038/s41598-017-08728-x.
Hauser TU, Rotzer S, Grabner RH et al. (2013). Enhancing performance in numerical magnitude processing and mental arithmetic using transcranial Direct Current Stimulation (tDCS). Front Hum Neurosci 7: 244. https://doi.org/10.3389/fnhum.2013.00244.
Hécaen H (1972). Introduction à la neuropsychologie, Larousse, Paris.
Hécaen H, Angelergues R (1961). Etude anatomo-clinique de 280 cas de lesions retrorolandiques unilaterales des hemispheres cerebraux [Anatomo-clinical study of 280 cases of unilateral retrorolandic lesions of the cerebral hemispheres]. Encéphale 50: 533–562.
Hécaen H, Angelergues R, Houiller S (1961). Les varietes cliniques des acalculies au cours des lesions retroronlandiques (Clinical varieties of acalculias following retrorolandic lesions). Rev Neurol 105: 85–103.
Henschen SE (1919). Über sprach-, musik- und rechenmechanismen und ihre lokalisationen im großhirn. Zeitschrift für die gesamte Neurol und Psychiatr 52: 273–298. https://doi.org/10.1007/BF02872428.
Henschen SE, Schaller WF (1925). Clinical and anatomical contributions on brain pathology. Arch Neurol Psychiatry 13: 226–249. https://doi.org/10.1001/archneurpsyc.1925.02200080073006.
Hittmair-Delazer M, Semenza C, Denes G (1994). Concepts and facts in calculation. Brain 117: 715–728. https://doi.org/10.1093/brain/117.4.715.
Holloway ID, Ansari D (2010). Developmental specialization in the right intraparietal sulcus for the abstract representation of numerical magnitude. J Cogn Neurosci 22: 2627–2637. https://doi.org/10.1162/JOCN.2009.21399.
Hubbard EM, Piazza M, Pinel P et al. (2005). Interactions between number and space in parietal cortex. Nat Rev Neurosci 6: 435–448. https://doi.org/10.1038/NRN1684.
Hussain M, Davis NJ, Benn Y (2024). A single tDCS session can enhance numerical competence. Neuropsychologia 193: 108760. https://doi.org/10.1016/j.neuropsychologia.2023.108760.
Hyde DC, Boas DA, Blair C et al. (2010). Near-infrared spectroscopy shows right parietal specialization for number in pre-verbal

infants. Neuroimage 53: 647–652. https://doi.org/10.1016/J.NEUROIMAGE.2010.06.030.

Isaacs EB, Edmonds CJ, Lucas A et al. (2001). Calculation difficulties in children of very low birthweight: a neural correlate. Brain 124: 1701–1707. https://doi.org/10.1093/BRAIN/124.9.1701.

Ischebeck A, Zamarian L, Siedentopf C et al. (2006). How specifically do we learn? Imaging the learning of multiplication and subtraction. Neuroimage 30: 1365–1375. https://doi.org/10.1016/j.neuroimage.2005.11.016.

Izard V, Dehaene-Lambertz G, Dehaene S (2008). Distinct cerebral pathways for object identity and number in human infants. PLoS Biol 6: 0275–0285. https://doi.org/10.1371/JOURNAL.PBIO.0060011.

Izard VR, Sann C, Spelke ES et al. (2009). Newborn infants perceive abstract numbers. Proc Natl Acad Sci U S A 106: 10382–10385. https://doi.org/10.1073/PNAS.0812142106.

Kadosh RC, Bahrami B, Walsh V et al. (2011). Specialization in the human brain: the case of numbers. Front Hum Neurosci. https://doi.org/10.3389/fnhum.2011.00062.

Kahn HJ, Whitaker HA (1991). Acalculia: an historical review of localization. Brain Cogn. https://doi.org/10.1016/0278-2626(91)90071-F.

Kasahara K, Tanaka S, Hanakawa T et al. (2013). Lateralization of activity in the parietal cortex predicts the effectiveness of bilateral transcranial direct current stimulation on performance of a mental calculation task. Neurosci Lett 545: 86–90. https://doi.org/10.1016/j.neulet.2013.04.022.

Kaufmann L, Vogel SE, Starke M et al. (2009). Developmental dyscalculia: compensatory mechanisms in left intraparietal regions in response to nonsymbolic magnitudes. Behav Brain Funct 5. https://doi.org/10.1186/1744-9081-5-35.

Kiefer M, Dehaene S (1997). The time course of parietal activation in single-digit multiplication: evidence from event-related potentials. Math Cogn 3: 1–30. https://doi.org/10.1080/135467997387461.

Klein E, Mann A, Huber S et al. (2013). Bilateral bi-cephalic tDCS with two active electrodes of the same polarity modulates bilateral cognitive processes differentially. PLoS One 8: 1–11. https://doi.org/10.1371/journal.pone.0071607.

Knops A, Thirion B, Hubbard EM et al. (2009). Recruitment of an area involved in eye movements during mental arithmetic. Science 324: 1583–1585. https://doi.org/10.1126/SCIENCE.1171599.

Kucian K, Loenneker T, Dietrich T et al. (2006). Impaired neural networks for approximate calculation in dyscalculic children: a functional MRI study. Behav Brain Funct 2. https://doi.org/10.1186/1744-9081-2-31.

Lampl Y, Eshel Y, Gilad R et al. (1994). Selective acalculia with sparing of the subtraction process in a patient with left parietotemporal hemorrhage. Neurology 44: 1759–1761. https://doi.org/10.1212/wnl.44.9.1759.

Langdon DW, Warrington EK (1997). The abstraction of numerical relations: a role for the right hemisphere in arithmetic? J Int Neuropsychol Soc 3: 260–268. https://doi.org/10.1017/s1355617797002609.

Lewandowski M, Stadelmann E (1908). Über einen bemerkenswerten Fall von Hirnblutung and Über Re-chenstörungen bei Herderkrankung des Gehirns. Zeit Neurol Psychiatry Brain Res 2: 249–265.

Libertus ME, Pruitt LB, Woldorff MG et al. (2009). Induced alpha-band oscillations reflect ratio-dependent number discrimination in the infant brain. J Cogn Neurosci 21: 2398–2406. https://doi.org/10.1162/JOCN.2008.21162.

Liu D, Zhou D, Li M et al. (2019). The neural mechanism of number line bisection: a fMRI study. Neuropsychologia 129: 37–46. https://doi.org/10.1016/j.neuropsychologia.2019.03.007.

Luria AR (1966). Higher cortical functions in man, Basic Books, New York, NY.

Marie P, Bouttier H, Bailey P (1922). La planotopokinesie. Etude sur les erreurs d'exécution de certains mouvements dans leur rapports avec la representation spatiale. Rev Neuro (Paris) 38: 505–512.

Matejko AA, Ansari D (2015). Drawing connections between white matter and numerical and mathematical cognition: a literature review. Neurosci Biobehav Rev 48: 35–52. https://doi.org/10.1016/J.NEUBIOREV.2014.11.006.

Maurer S, Tanigawa N, Sollmann N et al. (2016). Non-invasive mapping of calculation function by repetitive navigated transcranial magnetic stimulation. Brain Struct Funct 221: 3927–3947. https://doi.org/10.1007/s00429-015-1136-2.

McCloskey M, Caramazza A, Basili A (1985). Cognitive mechanisms in number processing and calculation: evidence from dyscalculia. Brain Cogn 4: 171–196. https://doi.org/10.1016/0278-2626(85)90069-7.

McCrink K, Dehaene S, Dehaene-Lambertz G (2007). Moving along the number line: operational momentum in nonsymbolic arithmetic. Percept Psychophys 69: 1324–1333. https://doi.org/10.3758/BF03192949.

Menon V (2014). Arithmetic in the child and adult brain. In: RC Kadosh, A Dowker (Eds.), The Oxford handbook of numerical cognition. Oxford Library of Psycholog p. 1. https://doi.org/10.1093/OXFORDHB/9780199642342.013.041.

Menon V (2016). Working memory in children's math learning and its disruption in dyscalculia. Curr Opin Behav Sci 10: 125–132.

Miceli G, Capasso R (1999). Calculation and number processing. In: G Denes, L Pizzamiglio (Eds.), Handbook of clinical and experimental neuropsychology. Zanichelli, Bologna, pp. 583–612.

Michels L, O'Gorman R, Kucian K (2018). Functional hyperconnectivity vanishes in children with developmental dyscalculia after numerical intervention. Dev Cogn Neurosci 30: 291–303. https://doi.org/10.1016/J.DCN.2017.03.005.

Moeller K, Willmes K, Klein E (2015). A review on functional and structural brain connectivity in numerical cognition. Front Hum Neurosci 9: 227. https://doi.org/10.3389/fnhum.2015.00227.

Molko N, Cachia A, Rivière D et al. (2003). Functional and structural alterations of the intraparietal sulcus in a developmental dyscalculia of genetic origin. Neuron 40: 847–858. https://doi.org/10.1016/S0896-6273(03)00670-6.

Montefinese M, Turco C, Piccione F et al. (2017). Causal role of the posterior parietal cortex for two-digit mental subtraction and addition: a repetitive TMS study. Neuroimage 155: 72–81. https://doi.org/10.1016/j.neuroimage.2017.04.058.

Mosbacher JA, Brunner C, Nitsche MA et al. (2020). Effects of anodal tDCS on arithmetic performance and electrophysiological activity. Front Hum Neurosci 11: 14–17. https://doi.org/10.3389/fnhum.2020.00017.

Mussolin C, De Volder A, Grandin C et al. (2010). Neural correlates of symbolic number comparison in developmental dyscalculia. J Cogn Neurosci 22: 860–874. https://doi.org/10.1162/JOCN.2009.21237.

Nemmi F, Schel MA, Klingberg T (2018). Connectivity of the human number form area reveals development of a cortical network for mathematics. Front Hum Neurosci 12. https://doi.org/10.3389/fnhum.2018.00465.

Nieder A, Dehaene S (2009). Representation of number in the brain. Annu Rev Neurosci. https://doi.org/10.1146/annurev.neuro.051508.135550.

Nöel M (2001). Numerical cognition. In: B Rapp (Ed.), The handbook of cognitive neuropsychology. Psychology Press, Philadelphia, pp. 495–518.

Odic D, Starr A (2018). An introduction to the approximate number system. Child Dev Perspect 12: 223–229. https://doi.org/10.1111/cdep.12288.

Orban GA, Claeys K, Nelissen K et al. (2006). Mapping the parietal cortex of human and non-human primates. Neuropsychologia 44: 2647–2667. https://doi.org/10.1016/j.neuropsychologia.2005.11.001.

Park J, Li R, Brannon EM (2014). Neural connectivity patterns underlying symbolic number processing indicate mathematical achievement in children. Dev Sci 17: 187–202. https://doi.org/10.1111/DESC.12114.

Pesenti M, Seron X, Van Der Linden M (1994). Selective impairment as evidence for mental organisation of arithmetical facts: BB, a case of preserved subtraction? Cortex 30: 661–671. https://doi.org/10.1016/S0010-9452(13)80242-0.

Peters L, De Smedt B (2018). Arithmetic in the developing brain: a review of brain imaging studies. Dev Cogn Neurosci 30: 265–279. https://doi.org/10.1016/J.DCN.2017.05.002.

Piazza M (2010). Neurocognitive start-up tools for symbolic number representations. Trends Cogn Sci. https://doi.org/10.1016/j.tics.2010.09.008.

Piazza M, Izard V, Pinel P et al. (2004). Tuning curves for approximate numerosity in the human intraparietal sulcus. Neuron 44: 547–555. https://doi.org/10.1016/j.neuron.2004.10.014.

Piazza M, Pinel P, Le Bihan D et al. (2007). A magnitude code common to numerosities and number symbols in human intraparietal cortex. Neuron 53: 293–305. https://doi.org/10.1016/J.NEURON.2006.11.022.

Pinel P, Le Clec'H G, Van De Moortele PF et al. (1999). Event-related FMRI analysis of the cerebral circuit for number comparison. Neuroreport 10: 1473–1479. https://doi.org/10.1097/00001756-199905140-00015.

Pinel P, Dehaene S, Rivière D et al. (2001). Modulation of parietal activation by semantic distance in a number comparison task. Neuroimage 14: 1013–1026. https://doi.org/10.1006/nimg.2001.0913.

Polspoel B, Peters L, Vandermosten M et al. (2017). Strategy over operation: neural activation in subtraction and multiplication during fact retrieval and procedural strategy use in children. Hum Brain Mapp 38: 4657–4670. https://doi.org/10.1002/HBM.23691.

Pope PA, Brenton JW, Miall RC (2015). Task-specific facilitation of cognition by anodal transcranial direct current stimulation of the prefrontal cortex. Cereb Cortex 25: 4551–4558. https://doi.org/10.1093/cercor/bhv094.

Prado J, Mutreja R, Booth JR (2014). Developmental dissociation in the neural responses to simple multiplication and subtraction problems. Dev Sci 17: 537–552. https://doi.org/10.1111/DESC.12140.

Price GR, Holloway I, Räsänen P et al. (2007). Impaired parietal magnitude processing in developmental dyscalculia. Curr Biol 17. https://doi.org/10.1016/J.CUB.2007.10.013.

Priftis K (2023). The neuropsychology of mathematical cognition. In: APA handbook of neuropsychology, Neurobehavioral disorders and conditions: accepted science and open questions, volume 1. American Psychological Association, pp. 59–78. https://doi.org/10.1037/0000307-004.

Proios H, Tsakpounidou K, Karapanayiotides T et al. (2021). Aphasia and math: deficits with basic number comprehension and in numerical activities of daily living. J Int Neuropsychol Soc 27: 939–951. https://doi.org/10.1017/S1355617720001368.

Qin S, Cho S, Chen T et al. (2014). Hippocampal-neocortical functional reorganization underlies children's cognitive development. Nat Neurosci 17: 1263–1269. https://doi.org/10.1038/NN.3788.

Ranpura A, Isaacs E, Edmonds C et al. (2013). Developmental trajectories of grey and white matter in dyscalculia. Trends Neurosci Educ 2: 56–64. https://doi.org/10.1016/j.tine.2013.06.007.

Re AM, Benavides-Varela S, Pedron M et al. (2020). Response to a specific and digitally supported training at home for students with mathematical difficulties. Front Psychol 11. https://doi.org/10.3389/FPSYG.2020.02039.

Renzi C, Vecchi T, Silvanto J et al. (2011). Overlapping representations of numerical magnitude and motion direction in the posterior parietal cortex: a TMS-adaptation study. Neurosci Lett 490: 145–149. https://doi.org/10.1016/j.neulet.2010.12.045.

Rivera SM, Reiss AL, Eckert MA et al. (2005). Developmental changes in mental arithmetic: evidence for increased functional specialization in the left inferior parietal cortex. Cereb Cortex 15: 1779–1790. https://doi.org/10.1093/CERCOR/BHI055.

Roitman J, Brannon E (2003). Nonverbal representations of time and number in animals and human infants. In: WH Meck (Ed.), Functional and neural mechanisms of interval timing. CRC Press/Routledge/Taylor & Francis Group, pp. 143–182. https://doi.org/10.1201/9780203009574.ch6.

Rosenberg-Lee M, Barth M, Menon V (2011). What difference does a year of schooling make? Maturation of brain response and connectivity between 2nd and 3rd grades during arithmetic problem solving. Neuroimage 57: 796–808. https://doi.org/10.1016/J.NEUROIMAGE.2011.05.013.

Rosselli M, Ardila A (1989). Calculation deficits in patients with right and left hemisphere damage. Neuropsychologia 27: 607–617. https://doi.org/10.1016/0028-3932(89)90107-3.

Rotzer S, Kucian K, Martin E et al. (2008). Optimized voxel-based morphometry in children with developmental dyscalculia. Neuroimage 39: 417–422. https://doi.org/10.1016/J.NEUROIMAGE.2007.08.045.

Rusconi E, Turatto M, Umiltà C (2007). Two orienting mechanisms in posterior parietal lobule: an rTMS study of the Simon and SNARC effects. Cogn Neuropsychol 24: 373–392. https://doi.org/10.1080/02643290701309425.

Rusconi E, Bueti D, Walsh V et al. (2011). Contribution of frontal cortex to the spatial representation of number. Cortex 47: 2–13. https://doi.org/10.1016/j.cortex.2009.08.005.

Rykhlevskaia E, Uddin LQ, Kondos L et al. (2009). Neuroanatomical correlates of developmental dyscalculia: combined evidence from morphometry and tractography. Front Hum Neurosci 3. https://doi.org/10.3389/NEURO.09.051.2009.

Salillas E, Semenza C (2015). Mapping the brain for math: reversible inactivation by direct cortical electrostimulation and transcranial magnetic stimulation. In: R Cohen Kadosh, A Dowker (Eds.), The Oxford handbook of numerical cognition. Oxford University Press, Oxford, pp. 583–611.

Salillas E, Basso D, Baldi M et al. (2009). Motion on numbers: transcranial magnetic stimulation on the ventral intraparietal sulcus alters both numerical and motion processes. J Cogn Neurosci 21: 2129–2138. https://doi.org/10.1162/jocn.2008.21157.

Salillas E, Semenza C, Basso D et al. (2012). Single pulse TMS induced disruption to right and left parietal cortex on addition and multiplication. Neuroimage 59: 3159–3165. https://doi.org/10.1016/j.neuroimage.2011.10.093.

Salillas E, Piccione F, Di Tomasso S et al. (2021). Neurofunctional components of simple calculation: a magnetoencephalography study. Cereb Cortex 31: 1149–1162. https://doi.org/10.1093/CERCOR/BHAA283.

Sandrini M, Rossini PM, Miniussi C (2004). The differential involvement of inferior parietal lobule in number comparison: a rTMS study. Neuropsychologia 42: 1902–1909. https://doi.org/10.1016/j.neuropsychologia.2004.05.005.

Sarkar A, Dowker A, Cohen Kadosh R (2014). Cognitive enhancement or cognitive cost: trait-specific outcomes of brain stimulation in the case of mathematics anxiety. J Neurosci 34: 16605–16610.

Sasanguie D, Göbel SM, Reynvoet B (2013). Left parietal TMS disturbs priming between symbolic and non-symbolic number representations. Neuropsychologia 51: 1528–1533. https://doi.org/10.1016/j.neuropsychologia.2013.05.001.

Schroeder PA, Dresler T, Bahnmueller J et al. (2017). Cognitive enhancement of numerical and arithmetic capabilities: a mini-review of available transcranial electric stimulation studies. J Cogn Enhanc 1: 39–47. https://doi.org/10.1007/s41465-016-0006-z.

Semenza C (2008). Number processing. In: B Stemmer, H Whitaker (Eds.), Handbook of the neuroscience of language. Elsevier, Oxford, UK, pp. 219–227.

Semenza C, Benavides-Varela S (2017). Reassessing lateralization in calculation. Philos Trans R Soc Lond B Biol Sci 373. https://doi.org/10.1098/RSTB.2017.0044.

Semenza C, Salillas E (2021). Numbers, calculation and acalculia. In: Encyclopedia of behavioural neuroscience, second edn. Elsevier, Oxford, UK, pp. 510–519. https://doi.org/10.1016/B978-0-12-819641-0.00036-0.

Semenza C, Delazer M, Bertella L et al. (2006). Is math lateralised on the same side as language? Right hemisphere aphasia and mathematical abilities. Neurosci Lett 406: 285–288. https://doi.org/10.1016/j.neulet.2006.07.063.

Semenza C, Meneghello F, Arcara G et al. (2014). A new clinical tool for assessing numerical abilities in neurological diseases: numerical activities of daily living. Front Aging Neurosci 6. https://doi.org/10.3389/fnagi.2014.00112.

Semenza C, Salillas E, De Pallegrin S et al. (2017). Balancing the 2 hemispheres in simple calculation: evidence from direct cortical electrostimulation. Cereb Cortex 27: 4806–4814. https://doi.org/10.1093/cercor/bhw277.

Seron X, Deloche G (1983). From 4 to four: A SUPPLEMENT to “from THREE to 3.”. Brain 106: 735–744. https://doi.org/10.1093/brain/106.3.735.

Shum J, Hermes D, Foster BL et al. (2013). A brain area for visual numerals. J Neurosci 33: 6709–6715. https://doi.org/10.1523/JNEUROSCI.4558-12.2013.

Soltanlou M, Artemenko C, Ehlis AC et al. (2018). Reduction but no shift in brain activation after arithmetic learning in children: a simultaneous fNIRS-EEG study. Sci Rep 8. https://doi.org/10.1038/S41598-018-20007-X.

Soltész F, Szucs D, Dékány J et al. (2007). A combined event-related potential and neuropsychological investigation of developmental dyscalculia. Neurosci Lett 417: 181–186. https://doi.org/10.1016/J.NEULET.2007.02.067.

Sperry R (1974). Lateral specialization in the surgically separated hemispheres. In: FO Schmitt, FG Worden (Eds.), Neuroscience third study program, Vol. 3. MIT Press, Cambridge, pp. 5–19.

Sperry R, Gazzaniga M, Bogen J (1969). Interhemispheric relationships: the neocortical commissures; syndromes of hemisphere disconnection. In: Handbook of clinical neurology, 273–290.

Suárez-Pellicioni M, Prado J, Booth JR (2022). Neurocognitive mechanisms underlying multiplication and subtraction performance in adults and skill development in children: a scoping review. Curr Opin Behav Sci 48: 101228. https://doi.org/10.1016/J.COBEHA.2022.101228.

Tablante J, Krossa L, Azimi T et al. (2023). Dysfunctions associated with the intraparietal sulcus and a distributed network in individuals with math learning difficulties: an ALE meta-analysis. Hum Brain Mapp 44. https://doi.org/10.1002/HBM.26240.

Telkemeyer S, Rossi S, Koch SP et al. (2009). Sensitivity of newborn auditory cortex to the temporal structure of sounds. J Neurosci 29: 14726–14733. https://doi.org/10.1523/JNEUROSCI.1246-09.2009.

Vallesi A (2012). Organisation of executive functions: hemispheric asymmetries. J Cogn Psychol 24: 367–386. https://doi.org/10.1080/20445911.2012.678992.

Vanduffel W, Fize D, Peuskens H et al. (2002). Extracting 3D from motion: differences in human and monkey intraparietal cortex. Science 298: 413–415. https://doi.org/10.1126/science.1073574.

Verguts T, Fias W (2004). Representation of number in animals and humans: a neural model. J Cogn Neurosci 16: 1493–1504. https://doi.org/10.1162/0898929042568497.

Vigna G, Ghidoni E, Burgio F et al. (2022). Dyscalculia in early adulthood: implications for numerical activities of daily living. Brain Sci 12. https://doi.org/10.3390/BRAINSCI12030373.

Vogel SE, De Smedt B (2021). Developmental brain dynamics of numerical and arithmetic abilities. NPJ Sci Learn 6. https://doi.org/10.1038/S41539-021-00099-3.

Vogel SE, Goffin C, Ansari D (2015). Developmental specialization of the left parietal cortex for the semantic representation of Arabic numerals: an fMR-adaptation study. Dev Cogn Neurosci 12: 61–73. https://doi.org/10.1016/j.dcn.2014.12.001.

Warrington EK (1982). The fractionation of arithmetical skills: a single case study. Q J Exp Psychol Sect A 34: 31–51. https://doi.org/10.1080/14640748208400856.

Warrington EK, James M (1967). Tachistoscopic number estimation in patients with unilateral cerebral lesions. J Neurol Neurosurg Psychiatry 30: 468–474. https://doi.org/10.1136/jnnp.30.5.468.

Yeo DJ, Wilkey ED, Price GR (2017). The search for the number form area: a functional neuroimaging meta-analysis. Neurosci Biobehav Rev 78: 145–160. https://doi.org/10.1016/J.NEUBIOREV.2017.04.027.

Yu X, Chen C, Pu S et al. (2011). Dissociation of subtraction and multiplication in the right parietal cortex: evidence from intraoperative cortical electrostimulation. Neuropsychologia 49: 2889–2895. https://doi.org/10.1016/j.neuropsychologia.2011.06.015.

Zorzi M, Priftis K, Umiltà C (2002). Brain damage: neglect disrupts the mental number line. Nature 417: 138–139. https://doi.org/10.1038/417138a.

Handbook of Clinical Neurology, Vol. 208 (3rd series)
Cerebral Asymmetries
C. Papagno and P. Corballis, Editors
https://doi.org/10.1016/B978-0-443-15646-5.00019-1

Chapter 31

Seeing and visualizing across the hemispheres

SELENE SCHINTU[1,2] AND PAOLO BARTOLOMEO[3*]

[1]*CIMeC—Center for Mind/Brain Sciences, University of Trento, Trento, Italy*

[2]*Department of Psychology, George Washington University, Washington, DC, United States*

[3]*Sorbonne Université, Institut du Cerveau/Paris Brain Institute—ICM, Inserm, CNRS, APHP, Hôpital de la Pitié Salpêtrière, Paris, France*

Abstract

Despite our subjective experience of a largely symmetric visual world, the human brain exhibits varying patterns and degrees of hemispheric asymmetry in distinct processes of visual cognition. This chapter reviews behavioral and neuroimaging evidence from neurotypical individuals and neurological patients, concerning functional asymmetries between the right hemisphere (RH) and the left hemisphere (LH) in visual object processing and mental imagery. Hierarchical perception shows RH preference for global processing and LH preference for local processing. At later stages of visual object processing, RH-based circuits exhibit a relative advantage in terms of perceptual integration, with a subsequent shift toward LH-based circuits for processing at higher conceptual and semantic levels. In voluntary visual mental imagery, circuits in the LH ventral temporal cortex play a pivotal role in transitioning from object meaning to simulated visualization. These hemispheric asymmetries in visual object processing might, in part, be influenced by the overall need to minimize wiring, coupled with the presence of distinct specialized networks within each hemisphere, such as the RH attention networks and the LH language networks. From a broader viewpoint, the evidence examined in this chapter indicates that visual object processing involves the interactions of large-scale cortical circuits within and between the hemispheres.

INTRODUCTION[1]

The visual system has evolved to navigate and interact with the external world. To this end, a very large surface of the cerebral cortex is dedicated to processing visual information in the mammalian brain (Van Essen, 2004). A fundamental principle of the organization of the visual cortex is its segregation into a ventral cortical visual pathway for object perception and recognition (Kravitz et al., 2013), and a dorsal cortical visual pathway for object localization and interaction (Kravitz et al., 2011). The ventral pathway reaches domain-preferring regions in the ventral temporal cortex (VTC), important for processing perceptual domains such as objects, colors, faces, and letters (Mahon and Caramazza, 2011; Grill-Spector and Weiner, 2014; Bartolomeo, 2022). Nevertheless, recent research suggests that the dorsal pathway, similarly to the ventral one, also contributes to the processing of some object properties (Ayzenberg and Behrmann, 2022a).

Hierarchical feedforward connections in these pathways are consistently accompanied by reciprocal feedback connections that convey information about the behavioral context (Gilbert and Li, 2013). These feedback connections also play a substantial role in visual mental imagery, where specific portions of the visual system simulate perception and predict future scenarios, even in the absence of external visual stimuli (Spagna, 2022). These internal processes emphasize the central role that vision plays in shaping our mental life. In the human brain, distinct processes of visual cognition and visual mental imagery exhibit varying patterns and degrees of hemispheric asymmetry.

VISUAL OBJECT RECOGNITION

Visual perception refers to the brain's capacity to interpret and understand information received through the light that enters the eyes. Object recognition can be conceived as a hierarchical process (Ungerleider and Haxby, 1994), in which posterior regions of the ventral cortical visual pathway process object features and more anterior regions integrate those basic features into more abstract representations, which are necessary for an object to

[1]Abbreviations used in the chapter are listed at the end of the chapter before References section.

*Correspondence to: Paolo Bartolomeo, Institut du Cerveau—Paris Brain Institute—ICM, Hôpital de la Pitié-Salpêtrière, 47 Bd de l'Hôpital, F-75013, Paris, France. Tel: +33157274140. E-mail: paolo.bartolomeo@icm-institute.org

acquire meaning (Ungerleider and Mishkin, 1982; Grill-Spector et al., 2001; Grill-Spector, 2003; Kravitz et al., 2013). Functional MRI (fMRI) unveiled the existence of a bilateral ventral region called the lateral occipital complex (LOC), which seems to encode object shape independently of specific low-level features (contour, luminance, motion, texture, stereoscopic depth cues) that define that shape is invariant to object size and location but does not appear to be engaged in the final, "semantic" stages of the recognition process (Malach et al., 1995; Kourtzi and Kanwisher, 2001). More recently, activity in the dorsal visual pathway [intraparietal sulcus (IPS)] has been proposed to be crucial for processing the global shape of objects (Ayzenberg and Behrmann, 2022a). The human brain hemispheres appear to be biased toward different roles in these processes.

Global and local processing of visual information

Object perception follows a hierarchical structure whereby lower-level objects are nested within higher-level ones. For instance, a forest consists of trees, and trees are composed of branches and leaves. A substantial body of research has revealed a hemispheric bias in hierarchical perception. Processing of global levels (the forest) and of local levels (the trees) tends to rely preferentially on the right hemisphere (RH) and the left hemisphere (LH), respectively. Navon (1977) assessed hierarchical perception by using a task in which stimuli consisted of large letters made of smaller ones (Fig. 31.1).

These hierarchical stimuli are called *congruent* when the identity of the global (large feature) and local (small feature) level matches and *incongruent* when the two levels do not match. By asking healthy participants to identify, in separate blocks, stimuli at the local or global level it was found that when participants were attending the local level had a response time cost for incongruent stimuli compared to congruent stimuli. This suggests the influence of a global level interference, due to participants' difficulty in inhibiting the processing of the global form. On the other hand, when participants attended the stimulus global level the interference of the local level was absent, indicating a precedence of global processing in the intact brain. Martin (1979) documented this overall global precedence effect, as a tendency to process global features before local features. She also described an RH advantage for global attention, indicating that the RH was more adept at processing global information, and an LH advantage for local attention, suggesting that the LH was more sensitive to local details. Sergent (1982) observed similar hemispheric differences, with the LH showing a preference for local details and the RH for global information. However, two subsequent investigations failed to replicate these results (Alivisatos and Wilding, 1982; Boles, 1984). In a further study, Van Kleeck (1989) did not find statistically significant evidence supporting the proposed pattern of hemispheric specialization. However, when he employed meta-analytic techniques to combine these results with findings from previous studies, there was evidence supporting the existence of the hypothesized LH "local" and RH "global" specialization.

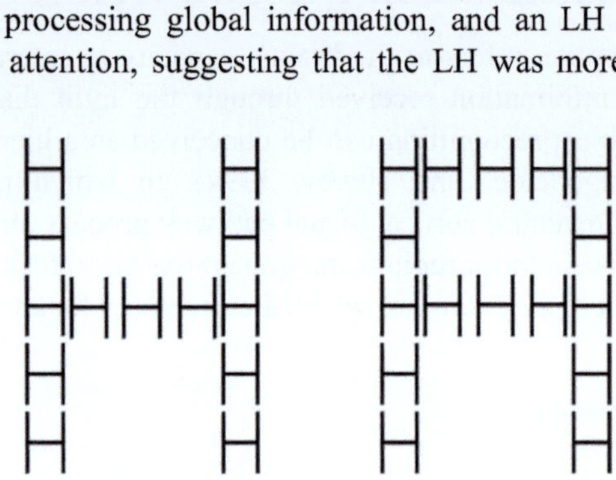

Fig. 31.1. Example of hierarchical stimuli.

Early neuropsychologic studies confirmed the presence of hemispheric asymmetries in hierarchical perception. Before and after undergoing commissurotomy for intractable epilepsy, a patient was asked to draw a series of hierarchical stimuli either with his right hand or with his left hand after having seen them in his central visual field for about 5 seconds. Before surgery, the patient was able to draw global and local level forms with either hand, although he had some difficulty in drawing global forms when using the right hand (controlled by the LH). After surgery, the patient's drawing abilities showed a notable difference depending on the hand used. When using his left hand (controlled by the RH), he displayed higher accuracy in drawing global forms compared to local-level forms. On the contrary, when using his right hand (LH), the opposite pattern emerged, and he showed better accuracy in drawing local over global forms (Delis, 1988). In another study, Delis et al. (1986) asked patients with unilateral strokes to draw hierarchical stimuli from memory and found that patients with RH damage had more difficulty in drawing larger relative to smaller forms, whereas those who had a lesion in the LH found more difficult drawing smaller relative to the larger forms.

A subsequent series of lesion studies demonstrated that the global/local processing advantage relies on the right/left superior temporoparietal junction, respectively. Lesions in this location in the RH led to difficulty in identifying features at the global level, while homolog LH lesions impaired the identification of local features (Delis et al., 1986; Robertson et al., 1988; Lamb and Robertson, 1989; Lamb et al., 1990).

Neuroimaging studies confirmed the advantage for the RH in processing information at the global level and for the LH in processing information at the local level (Martinez et al., 1997). Moreover, they better defined the role of the temporoparietal cortex in hierarchical visual perception (Fink et al., 1996). Using positron emission tomography (PET), Fink and collaborators (1996) tracked the cerebral blood flow of healthy participants while they were processing hierarchical Navon letters in two separate experiments. In the first experiment, participants were instructed to attend and name either the global or the local level of stimuli in separate blocks (directed attention task). Attention to the global level increased blood flow in the RH lingual gyrus, while attention to the local level increased blood flow in the LH inferior occipital cortex. In the second experiment, participants had a target letter, which could appear at the global or local level in varying numbers of consecutive trials, and were asked to identify at which of the two levels the target letter appeared (divided attention task). The number of "attend-to-global" or "attend-to-local" consecutive trials correlated with the increase in

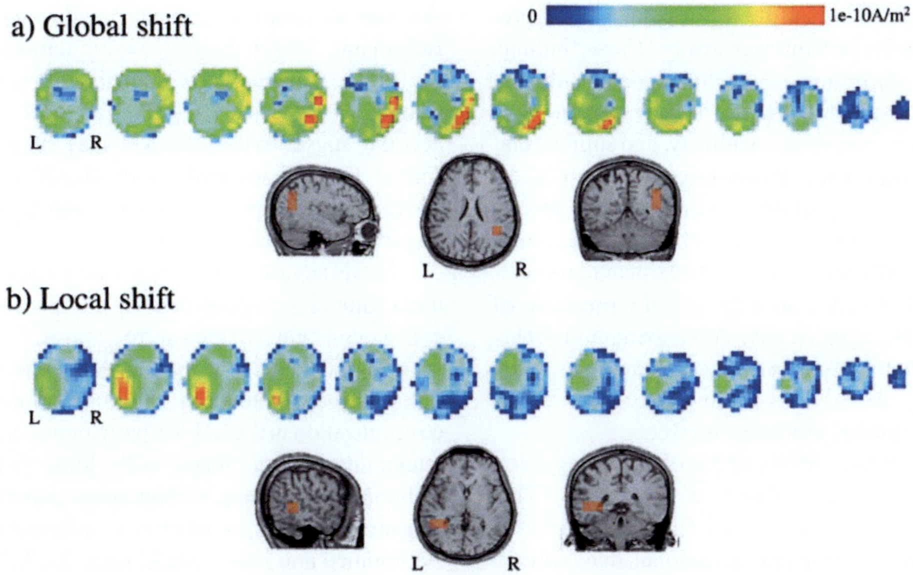

Fig. 31.2. Maps of the current source density of neural activity, analyzed using low-resolution electromagnetic tomography (LORETA; Pascual-Marqui et al., 1994), for attention shifts to (A) global and (B) local levels. For each shift condition, the areas showing a maximal current source are superimposed on a representative anatomic MRI. Figure reproduced from Yamaguchi S, Yamagata S, Kobayashi S (2000). Cerebral asymmetry of the top-down allocation of attention to global and local features. J Neurosci 20: RC72, with authorization. Copyright 2000 Society for Neuroscience.

blood flow in the bilateral temporoparietal cortex but not in extrastriate areas. These findings further support the hemispheric specialization for global vs local processing and indicate that the neural transformations occurring in visual areas are modulated by the attentional control exerted by the temporoparietal regions. Yamaguchi et al. (2000) employed event-related brain potentials (ERPs) to further assess the contributions of the temporoparietal cortex. Participants were instructed to shift their attention between the global and local levels of hierarchically structured stimuli. The findings revealed that the RH temporoparietal cortex exhibited activity when processing the global level, whereas the LH posterior temporal cortex was active during the processing of the local level (Fig. 31.2).

Another PET study conducted on healthy individuals (Fink et al., 1997) tested whether stimulus category influences the hemispheric lateralization of hierarchical perception. Fink and collaborators (1997) tracked participants' cerebral blood flow while they were engaged in a hierarchical perception task. Instead of the Navon letters used by Fink et al. (1996), here the stimuli were large objects made of smaller ones. Blood flow increased in brain areas anatomically similar to the ones identified by the letter-based hierarchical task (Fink et al., 1996), but with an opposite pattern of hemispheric asymmetry. Global processing was associated with increased activity in the LH extrastriate cortex, and local processing with increased activity in the RH extrastriate cortex. It is also worth noting that the respective homolog areas in the contralateral hemispheres also showed an increase in blood flow, which, however, did not reach statistical significance. These findings led the authors to hypothesize that the hemisphere specialized for the specific stimulus category processes the local aspect, which is considered the most demanding aspect of hierarchical stimuli. Thus, when incongruent hierarchical stimuli made of letters are presented, the language-dominant LH processes the local level. On the other hand, when the hierarchical stimuli are objects, the RH processes the local level. When participants are not specifically instructed to direct attention to the global or to the local level, the processing system defaults to the global level.

Integration of the constituent parts of visual objects

To perceive an object, the brain must process the spatial relationships among its constituent parts. Understanding these part relations is essential for creating invariant global shape representations (Biederman, 1987). Recently, Ayzenberg and Behrmann (2022b) investigated how the brain processes the global shape of objects and, thus, computes the relationship between object parts. They proposed a crucial role for parietal regions in the dorsal cortical visual pathway, which deals with several aspects of spatial processing (Mishkin et al., 1983; Kravitz et al., 2011). In the first fMRI experiment, healthy individuals underwent four different localizer scans to quantify voxel activation in response to object-centered part relations, allocentric relations (i.e., spatial relations between objects rather than between object parts), three-dimensional (3D) shape, and tools. They found regions within the IPS that were selectively involved in computing object-centered part relations and distinct from other dorsal regions dealing instead with allocentric relations, 3D shape, and tools. Crucially, such regions exhibited task-dependent functional and effective connectivity with the ventral cortex. In a second experiment, the authors discovered that the multivariate response of the posterior portion of the IPS could be used to decode object categories at levels comparable to ventral object regions. Finally, mediation and multivariate effective connectivity analyses suggested that the IPS mediates the representation of part relations in the ventral pathway. These findings emphasize the specific contributions of the dorsal visual pathway to object recognition, indicating that the dorsal cortex serves as an important source of input for the ventral pathway, and suggest that the dorsal cortex may support the ability to categorize objects based on their global shape. Importantly, not only do the results of Ayzenberg and Behrmann's (2022b) study confirm the strong link between the dorsal and ventral visual cortex in accomplishing object recognition, but they also highlight the presence of hemispheric asymmetries. Specifically, Ayzenberg and Behrmann found that processing of object-part relations was often stronger in the RH (Fig. 31.3). This finding may well reflect the above-described RH global precedence effect.

It seems, therefore, that early levels of the object recognition hierarchy already present hemispheric asymmetries. This hemispheric bias is not absolute and can be influenced by several factors, such as stimulus category and attentional deployment. However, the available evidence suggests a tendency for RH-based circuits to play a prominent role in integrating or grouping individual elements to form a coherent percept. This percept is then further processed until it is recognized (although top-down effects also contribute; see Gilbert and Li, 2013).

The reasons for the RH advantage in global processing remain debated (Kimchi, 1992). One proposed explanation is related to the lateralization of object-based attention (Shomstein and Behrmann, 2006). Indeed, manipulating the focus of attention can either enhance or disrupt the RH global precedence effect (Kimchi and Merhav, 1991; Van Vleet et al., 2011). Another account suggests that the RH may be more responsive to low spatial frequencies (Iidaka et al., 2004; Peyrin et al., 2004; Proverbio and Zani, 2021) and consequently support the ability to perceive global forms.

The RH advantage in global perception may have an evolutionary basis. Global perception enables individuals to quickly grasp a general understanding of their surroundings, even though it may come at the cost of detailed information. A rapid reaction to potentially dangerous visual stimuli in the environment might be facilitated by the co-localization in the RH of a parietotemporal circuit, responsible for integrating global shape with local features (Ayzenberg and Behrmann, 2022b) and of frontoparietal networks crucial for processing urgent or unexpected external information (Corbetta et al., 2008; Bartolomeo and Seidel Malkinson, 2019, 2023).

Object perception and recognition

In 1890, Heinrich Lissauer provided a detailed description of a patient with "mind-blindness" (Lissauer, 1890, 1988; Bartolomeo, 2021). After a stroke in the LH involving the cuneus, lingual, and calcarine gyri, as well as the lower two-thirds of the callosal

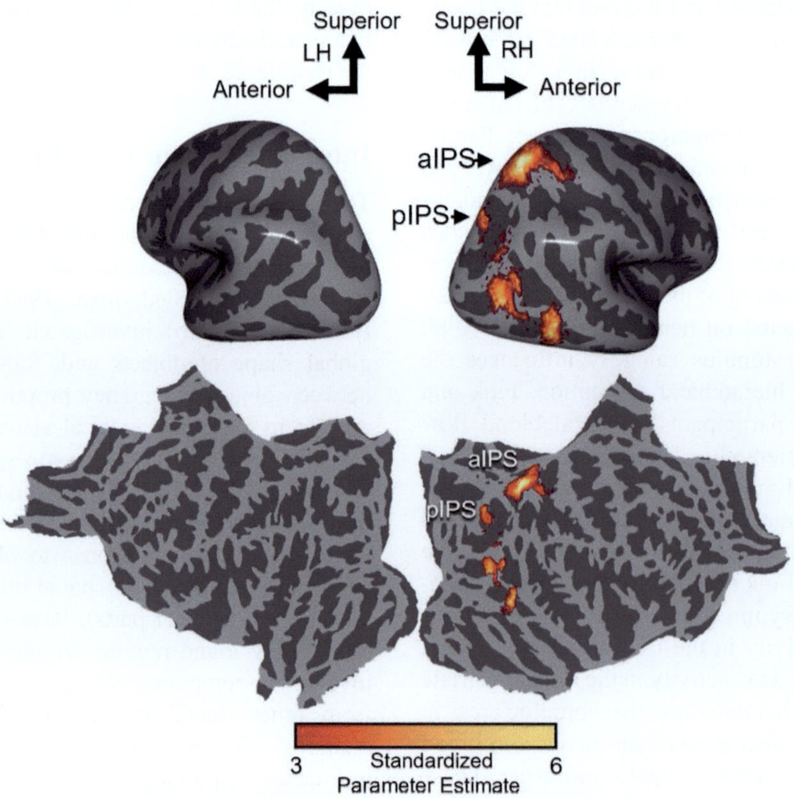

Fig. 31.3. Increased BOLD response for the contrast (part relations > features) for a group average map, inflated (top panel) and flattened (bottom panel). Note the clear asymmetry in favor of the anterior and posterior IPS clusters in the RH. Figure reproduced from Ayzenberg V, Behrmann M (2022b). The dorsal visual pathway represents object-centered spatial relations for object recognition. J Neurosci 42: 4693–4710. https://doi.org/10.1523/JNEUROSCI.2257-21.2022, with authorization. Copyright 2022 the authors.

splenium (Hahn, 1895 cited by Kurachi et al., 1979), the patient had become "quite incapable of visually recognizing the most common objects, although he could recognize everything by touch or hearing." Based on thorough behavioral testing, Lissauer reached the conclusion that his patient was able to "perceive many things without fully comprehending them." In Lissauer's theoretical framework, the patient exhibited normal conscious awareness of sensory impressions, referred to as "apperception," but was unable to connect the content of apperception with stored knowledge. Thus, Lissauer introduced the well-known dichotomy between *apperceptive* and *associative agnosia*, although he maintained that this distinction was not absolute and did not expect to find pure examples of either form of agnosia.

According to De Renzi (1982), posterior occipital and temporal regions in the RH show a preference for discriminating objects' basic features, while more anterior temporal regions in the LH seem to be biased toward assigning meaning to percepts. Support for hemispheric asymmetries in object perception and recognition comes from both neuropsychologic and neuroimaging studies. Right brain-damaged patients generally fail on tasks requiring perceptual categorization, while patients with lesions to the LH mainly fail on tasks requiring semantic categorization and identification of meanings (De Renzi et al., 1969; Warrington and Taylor, 1978). Such asymmetry is also shown in a voxel-based correlational study (Acres et al., 2009), where the behavioral performance of patients with temporal lobe damage was examined in relation to voxel-based measures of neuronal integrity (signal intensities). Each patient's T1-weighted structural MRI image was correlated, on a voxel-by-voxel basis across the entire brain, with the behavioral scores obtained from four different tasks. There were two verbal tasks: (a) picture naming, which involved mapping conceptual knowledge to word forms, and (b) semantic categorization of words, which required mapping word forms to conceptual knowledge. Additionally, two visual object tasks, without a verbal component, were included. These tasks involved retrieving conceptual information from visual objects: (c) visual object categorization and (d) normal and chimera object decisions. The results indicated that there was a correlation between the performance on verbal tasks and the integrity level of temporal regions in the LH. On the other hand, performance on object perception tasks was correlated with the integrity level of corresponding regions in the RH (see Fig. 31.4).

Schintu et al. (2014) also used a data-driven approach to investigate the neural underpinnings of object processing in neurological patients. They assessed performance on the Visual Object Space Perception battery (VOSP; Warrington and James, 1991) in 247 Vietnam veterans, 192 with penetrating traumatic brain injuries and 55 controls. Lesion location was assessed using voxel-based lesion-symptom mapping (Bates et al., 2003). The results indicated that the RH lateral occipital complex (Grill-Spector et al., 1999) plays a critical role in early perceptual

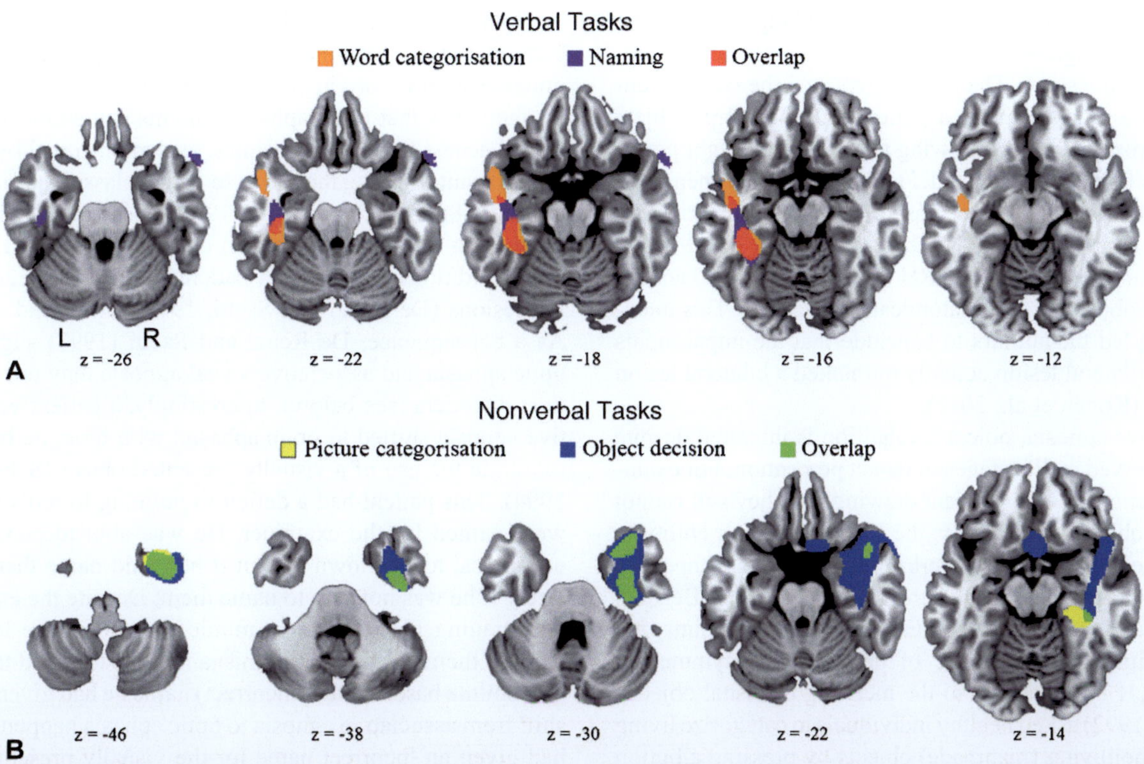

Fig. 31.4. Representative brain slices where voxel signal intensities correlated with behavioral performance on (A) verbal tasks of word categorization (orange) and naming (purple; overlap shown in red) and (B) nonverbal tasks of picture categorization (yellow) and object decision (blue; overlap shown in green). The reported coordinates are in the MNI space. Figure reproduced from Acres K, Taylor KI, Moss HE, et al. (2009). Complementary hemispheric asymmetries in object naming and recognition: a voxel-based correlational study. Neuropsychologia 47: 1836–1843. https://doi.org/10.1016/j.neuropsychologia.2009.02.024, with permission of Elsevier.

discrimination, while more anterior temporal and frontal regions in the LH are essential for more complex forms of object discrimination and recognition. These findings support the idea that both hemispheres contribute to object recognition, with distinct regions along the ventral stream associated with different cognitive processes.

Apperceptive agnosia, also called visual form agnosia or object agnosia, is defined as a deficit in object recognition due to perceptual impairment despite preserved elementary visual functions (Benson and Greenberg, 1969; Farah, 1990). It typically follows bilateral or RH lesions. While some studies described visual form agnosia following a lesion to the RH hemisphere (Humphreys and Riddoch, 1984; Konen et al., 2011), the majority of cases report bilateral lesions to the ventral visual pathway (Goodale et al., 1991; McIntosh et al., 2004; Karnath et al., 2009). Structural MRI showed that patient DF, following carbon monoxide poisoning, had severe damage to the LOC (James et al., 2003), consistent with neuroimaging findings from healthy participants showing bilateral LOC responsible for encoding object shape (Malach et al., 1995; Grill-Spector and Malach, 2004). Patient JS's ischemic brain damage in the medial occipitotemporal cortex was affecting mainly the fusiform and lingual gyri bilaterally, as well as the right parahippocampal cortex (Karnath et al., 2009). Finally, Patient SVD had bilateral damage (due to anoxia) to Brodmann areas (BA) 17, 18, 19, as well as more dorsal areas BA 30, 31, and 7 bleeding into the posterior parietal cortex (Serino et al., 2014).

A study conducted by Konen et al. (2011) helped clarify the reasons why visual agnosia was reported to occur after either unilateral or bilateral lesions. The authors reported the comprehensive single case study of SM, a patient suffering from object agnosia and prosopagnosia following a lesion to the right posterior lateral fusiform gyrus (Fig. 31.5). Using fMRI, Konen et al. showed that the patient had decreased object-related activation (compared to controls) at sites both proximal and distal to the lesion. Importantly, the unilateral RH lesion also affected activation related to objects in the anatomically intact LH. This unexpected finding led the authors to conclude that the impairments following a unilateral lesion actually mimicked a bilateral lesion in this patient (Konen et al., 2011).

In associative agnosia, object recognition is impaired despite relatively preserved earlier stages of object perception. For example, these patients can copy a linear drawing, but they still cannot recognize the object it depicts. On the other hand, their ability to draw from memory based on verbal instructions is impaired. Lesions are typically localized in the LH (Rubens and Benson, 1971; Bauer and Rubens, 1985; Farah, 1990). Neuroimaging findings confirmed the presence of hemispheric asymmetries favoring the LH in the access to the meaning of visual objects. Sergent et al. (1992) asked healthy individuals to categorize living (natural) and nonliving (manmade) objects by pressing a button (i.e., no verbal response required). Meanwhile, they tracked blood flow using PET scans. The results revealed that object categorization was associated with activity in the LH occipitotemporal cortex, with no involvement of RH regions. We can interpret this as evidence of object recognition because, even though the participants were not required to name the stimuli, categorization implies that the stimuli were actually recognized. Price et al. (1996) also conducted a PET study on healthy individuals to investigate the neural processes involved in object recognition and naming. The researchers used stimuli consisting of colored outline drawings of objects and colored nonsense shapes. Participants were asked to either passively view the objects or explicitly name them, as well as to view the colored shapes or name the color of the shape. To differentiate between structural and semantic processing specific to object conditions, the researchers compared both object conditions (naming and viewing) with both shape conditions. This contrast revealed that object recognition is associated with activity in the LH middle occipital and bilateral anterior temporal cortices. The importance of the LH involvement in object recognition was further supported by another study (Stewart et al., 2001), where TMS-induced disruption of the left temporal cortex led to a slowing in participants' picture naming ability without affecting word reading or color naming, when compared to the effect of stimulating the corresponding region on the opposite hemisphere or the vertex.

Several neuroimaging studies have also found that the left posterior temporal cortex is active during the conceptual processing of both pictures and words (Vandenberghe et al., 1996; Bookheimer, 2002; Thompson-Schill, 2003; Xu et al., 2009), likely due to the localization of language circuits in the LH. Lesion location in patients with associative agnosia is typically in the LH occipitotemporal cortex (Bartolomeo, 2022).

Optic aphasia is a condition related to associative agnosia. In both conditions, the naming of visually presented objects is impaired after lesions in the LH occipitotemporal cortex. The core distinction is that optic aphasic patients can show signs of preserved access to object semantics, as demonstrated by nonverbal means (pantomiming the use or semantic classification of objects; Freund, 1889; Beauvois et al., 1982; Poeck, 1984). However, a continuum of visual recognition deficits with mixed features of both conditions may exist in patients who suffer from posterior LH lesions (De Renzi and Saetti, 1997; Kwon and Lee, 2006). As a consequence, De Renzi and Saetti (1997) suggested that optic aphasia and associative visual agnosia may not be two distinct disorders (see below). Interestingly, a patient with associative agnosia shifted to optic aphasia: with time, he became able to mimic the use of a visually presented object (Schnider et al., 1994). This patient had a deficit in pointing to tools when those were named by the examiner. He was able to pantomime the use of real tools shown to him if he could name them and only rarely if he was not able to name them. Despite the experimenter encouraging the patients to mimic the use of the tools before naming them, he frequently misnamed the tool and then made a pantomime based on the (incorrect) name he had given. Since this shift from associative agnosia to optic aphasia happened when he had given an incorrect name for the visually presented objects, Schnider et al. (1994) concluded that the RH must be forced to perform the task for optic aphasia to emerge. According to the authors, optic aphasia occurs when the splenial fibers are damaged in addition to the LH lesion, thus again forcing the RH to take care of the task, i.e., processing stimuli beyond the

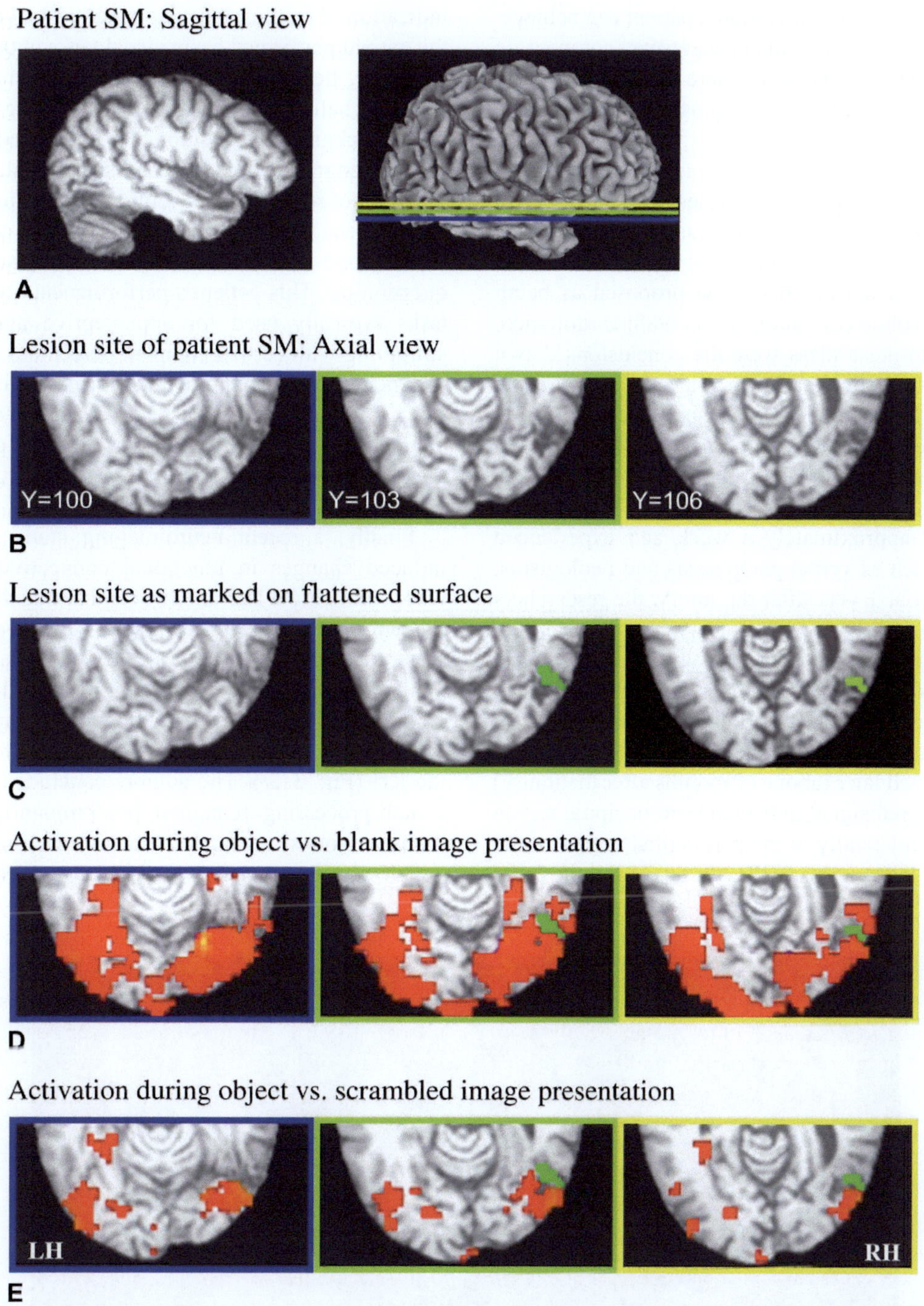

Fig. 31.5. Lesion site in patient SM. (A) Sagittal view of SM's lesioned RH in anatomic space (left panel) and inflated (right panel). The colored horizontal lines correspond to the location of the axial slices shown from B to E, which were cut along the temporal poles for an enlarged representation of occipitotemporal cortex. Note: the bottom, blue, slice is inferior to the lesion, whereas the middle and top ones (green and yellow) cover the lesion site. (B) Axial view of the lesion site. (C) Axial view of the lesion site (green) marked on the flattened surface. (D) Activation during object vs blank image presentations. (E) Activation during object vs scrambled image presentations. Figure reproduced from Konen CS, Behrmann M, Nishimura M, et al. (2011). The functional neuroanatomy of object agnosia: a case study. Neuron 71: 49–60. https://doi.org/10.1016/j.neuron.2011.05.030, with the authorization of Cell Press.

perceptual level. However, De Renzi and Saetti (1997) were skeptical of this anatomic account. They discussed three reported cases of associative agnosia with extensive splenial damage (Feinberg et al., 1994) and a patient who transitioned from associative agnosia to optic aphasia over time (Rubens and Benson, 1971). Based on this evidence, they concluded that this continuum suggested a partial recovery of access to semantic information. Importantly, De Renzi and Saetti also proposed that the same lesion to the LH can lead to either associative agnosia or optic aphasia, depending on varying levels of RH semantic competence in different individuals. According to their hypothesis, the individual's premorbid RH semantic potential is the crucial factor determining the level of recognition a patient can achieve. This also influences the extent to which patients can compensate for the (unavailable) LH resources and where their behavior will fall on the continuum from associative agnosia to optic aphasia (De Renzi and Saetti, 1997).

Importantly, however, most cases of associative agnosia exhibit some degree of perceptual impairment (Lissauer, 1890; Farah, 1990). As noted earlier, Heinrich Lissauer himself did not expect to observe "pure" cases of apperceptive or associative agnosias and considered the distinction he proposed as being relative rather than absolute (Lissauer, 1890, 1988; Bartolomeo, 2021). Consistent with these ideas were the conclusions drawn by De Renzi and Lucchelli (1993) on what they called the "fuzzy boundaries of apperceptive agnosia." De Renzi and Lucchelli conducted extensive testing on a young woman who had a hemorrhagic lesion in the RH frontal and temporal lobes due to a car accident. Following the brain injury, the patient was unconscious for approximately a week and experienced language disorders, such as verbal paraphasias and neologisms, for a few months. About 1 year after the injury, the researchers evaluated the patient and discovered that while the aphasia in spontaneous speech had resolved, a profound impairment in object recognition had emerged. A computed tomography scan conducted around 9 months after the injury confirmed the presence of a hypodensity in the right temporo-occipital cortex. The MRI scan conducted later (about 14 months after the injury) revealed an area of altered signal in the temporo-occipital region of the RH, as well as bilaterally in the parasagittal region of the frontal and parietal lobes. This MRI finding was the only indication of structural damage to the LH. On behavioral testing, the patient performed basic visual tasks within the normal range. However, her deficits became evident when faced with more visually challenging tests, like disentangling a Gestalt from a confused background. Additionally, she failed tests that involved object recognition (even in nonvisual modalities) and visual tasks that did not require the attribution of meaning. Despite these difficulties, she showed preserved knowledge of the functional, contextual, and semantic characteristics of stimuli during verbal questioning. This patient's performance, not failing basic visual tasks typically used for apperceptive agnosia diagnosis but struggling with certain complex perceptual tests, led the authors to conclude that her pattern of performance fell *between* apperceptive and associative agnosia. This discovery challenges the binary distinction between the two types of visual agnosia and supports the idea of a continuum of different degrees of impairment, which is in line with Lissauer's original position.

Finally, a recent neuroimaging study investigated lesion-induced changes in functional connectivity in AL, a patient who, following a lateral LH occipital damage (Fig. 31.6), failed to identify visual objects despite the preserved perception of their structures (Ptak and Lazeyras, 2019). The authors compared the patient's functional connectivity, measured at rest, with that of controls and observed a decrease in connectivity between the lesioned LH lateral occipital cortex and the intact homolog in the RH (Fig. 31.7). The authors concluded that, in this patient, visual processing remained predominantly segregated within the posterior cortex, and could not reach more anterior circuits which encode associative object knowledge. These findings

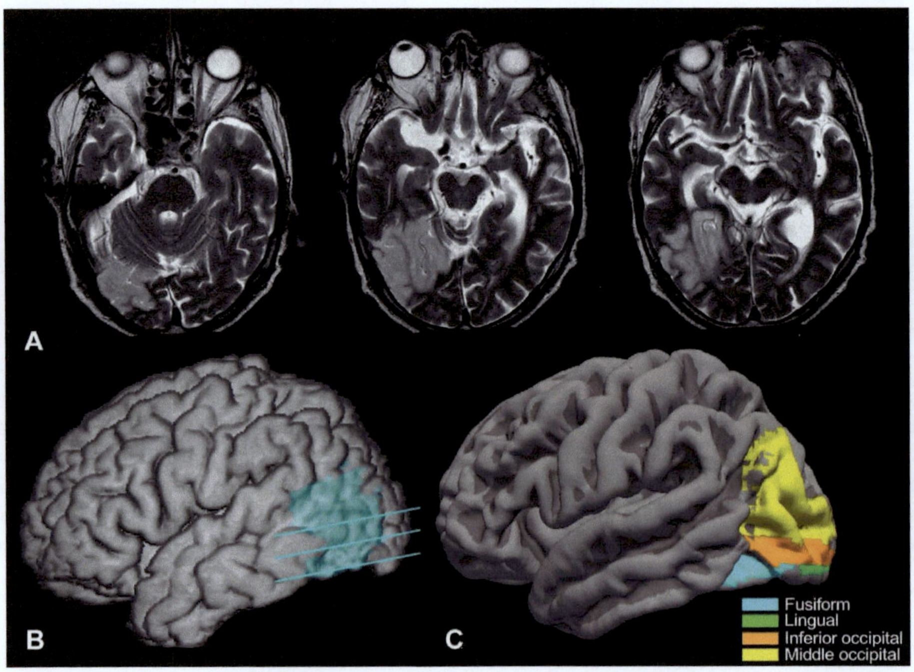

Fig. 31.6. Brain lesion of patient AL. (A) Axial MRI images, showing the ischemic brain lesion located in the left occipitotemporal region. (B) A normalized 3D reconstruction of AL's brain. The lateral occipital lesion is highlighted blue and the parallel blue lines indicate the approximate locations of the three slices shown in panel (A). (C) A lateral view of the template brain displaying the four occipitotemporal regions of interest for the connectivity analyses. Figure reproduced from Ptak R, Lazeyras F (2019). Functional connectivity and the failure to retrieve meaning from shape in visual object agnosia. Brain Cogn 131: 94–101. https://doi.org/10.1016/j.bandc.2018.12.007, with the authorization of Elsevier.

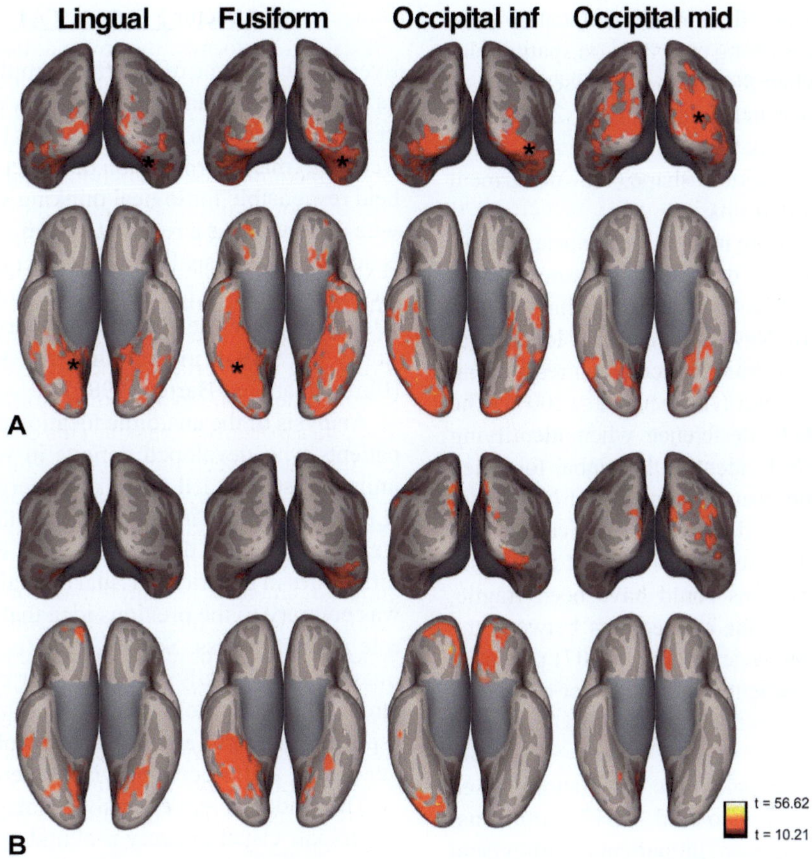

Fig. 31.7. Functional connectivity of four seed regions in the RH with every voxel in the brain in (A) healthy controls and (B) patient AL. Asterisks indicate the approximate location of the seed ROI. Figure reproduced from Ptak R, Lazeyras F (2019). Functional connectivity and the failure to retrieve meaning from shape in visual object agnosia. Brain Cogn 131: 94–101. https://doi.org/10.1016/j.bandc.2018.12.007, with the authorization of Elsevier.

confirm that focal unilateral damage can impact visual object processing in both hemispheres and support the hypothesis that this ability relies on a distributed coding of information across the occipitotemporal cortex bilaterally (Behrmann and Plaut, 2013; Bartolomeo et al., 2015).

However, some patients exhibit a pattern of object agnosia that cannot be easily explained by the apperceptive/associative dichotomy proposed by Lissauer. Riddoch and Humphreys (1987) reported the case of HJA, who, following bilateral vascular lesions of the inferior temporo-occipital gyrus, was able to draw both from copy and from memory quite well but could not identify his own drawings afterward. HJA's performance pattern appeared to reflect a deficit in the integration of local elements into a coherent and global description of the object. This led to the coining of the term "integrative agnosia" as a subtype of visual agnosia, whereby patients struggle with the perceptual integration of local parts into higher-order shapes (Humphreys, 1999).

The majority of other patients with a similar object recognition deficit also had lesions affecting both hemispheres. For example, Madame D experienced integrative agnosia following bilateral lesions at the boundaries between the occipital and temporal lobes (Bartolomeo et al., 1998a). The two strokes occurred sequentially. After the first stroke in the LH, Madame D experienced a slowed letter-by-letter reading but had no issues with recognizing objects and performed the VOSP flawlessly (Bachoud-Lévi and Bartolomeo, 2003). After a year and a half, her reading speed was progressively improving when a second stroke, situated in an almost mirror location in the RH, unfortunately occurred. Following the RH stroke, Madame D became severely impaired in reading not only words but also single letters (Bartolomeo et al., 1998b, 2003) and experienced difficulties recognizing objects (Bartolomeo et al., 1998a), colors (Bartolomeo et al., 1997), and faces. In this patient, impaired recognition was likely a consequence of visual deafferentation of domain-preferring regions in the VTC (Bartolomeo, 2021). Interestingly, however, a unilateral lesion in the LH alone was not enough to cause object recognition problems. Presumably, after the LH stroke visual input was sufficient to access LH semantic and language circuits, thereby supporting object recognition and naming. The RH stroke disrupted this process and impaired communication within high-level visual regions and between the visual and the semantic systems.

Patient SM also had signs of integrative agnosia as a result of damage to the RH inferior temporal lobe, although a previous scan had also revealed lesions in the corpus callosum and left basal ganglia (Behrmann et al., 2006). Although SM did not exhibit significant abnormalities in identifying global and local hierarchical letters (Navon, 1977), he faced difficulties deriving a global form from local components when the number of local components was limited (i.e., requiring strong emphasis on spatial relations between the parts) and when there was insufficient time to bind the local parts together (Behrmann and Kimchi, 2003). The authors concluded that since the patient was sensitive to basic grouping but unable to derive a global shape, the impairment might be in configuring or shape formation.

Another patient, SE, suffered from integrative agnosia due to a bilateral infarct that affected the right dorsal and left ventral visual pathways (Aviezer et al., 2007). Interestingly, SE had trouble identifying the global aspect of Navon hierarchical letters, but he showed normal global-to-local interference when responding to local letters in hierarchical stimuli (Aviezer et al., 2007). The observation that SE had global interference when identifying local letters, despite his inability to identify the global form, led the authors to conclude that the visual system could build integrated representations that enabled covert identification, but lacked the processing articulation required for overt identification. SE's perceptual representations could have been fragile, leading to insufficient stability in the connections between the perceptual and memory systems. Aviezer et al. (2007) proposed that integrative agnosia in this patient might be better characterized as a lack of conscious access to an established integrated perceptual representation.

Although it is difficult to reach anatomic conclusions based on single case reports (see Bartolomeo et al., 2017, for discussion), the pattern of lesion location in patients with visual agnosia clearly suggests that access to LH semantic circuits is crucial for object recognition in most patients, consistent with Lissauer's original definition of associative agnosia. Regarding earlier perceptual stages of processing, the cases of Madame D, who developed visual agnosia only after RH damage added to her previous LH stroke (Bartolomeo et al., 1998a), and patient SM, who experienced visual agnosia after unilateral RH damage with functional consequences on the homolog LH regions (Konen et al., 2011), suggest a crucial role for RH-based circuits. However, rather than the specific hemisphere where the lesion is located, the critical factor seems to be the pattern of anatomic or functional disconnections it creates between the language and semantic systems in the LH and the bilateral perceptual circuits (Freund, 1889; Bartolomeo et al., 2015; Siuda-Krzywicka et al., 2019; Bartolomeo, 2021, 2022; Bartolomeo and Miceli, 2023).

VISUAL MENTAL IMAGERY

Even in the absence of external stimuli, our "mind's eye" enables us to engage in visual mental imagery and simulate the experience of visual perception (Kosslyn et al., 1995; Spagna, 2022). In the *New Age* (mis)interpretation of Roger Sperry's work, the LH was held responsible for logical thinking while the RH was attributed with creativity. As a result, it has often been believed that the RH is dominant for visual mental imagery, which is often regarded as the most creative form of visualization. Already in the early 1980s, however, the evidence supporting the notion of RH superiority in visual mental imagery was found to be lacking (Ehrlichman and Barrett, 1983).

Analysis of the anatomic location of lesions in brain-damaged patients who developed deficits in visual mental imagery after unilateral strokes led to the discovery of two important patterns (Liu et al., 2022). First, it was found that patients with LH lesions were more likely than those with RH damage to experience difficulties in generating visual mental images (Farah, 1984). This was contrary to the previous idea that the RH played a dominant role in this process. Second, deficits in voluntary visual mental imagery are typically observed after extensive damage to the left temporal lobe (Bartolomeo, 2002). This suggests that the LH, and especially circuits in its temporal lobe, plays a critical role in generating and processing visual mental images.

The evidence that extensive damage to the LH temporal lobe can impair visual imagery is abundant (Bartolomeo, 2002, 2008; Moro et al., 2008; Liu et al., 2022). Contrary to the dominant model of visual mental imagery (Pearson, 2019), however, more posterior regions in the occipital lobe do not seem to participate in this ability. For example, Madame D, the abovementioned patient with visual agnosia after bilateral lesions at the borders between the occipital and temporal lobes (see Fig. 31.8), was able to effortlessly build vivid visual mental images of the very same items she could not recognize (Bartolomeo et al., 1998a).

The findings from neurological patients were further specified by a meta-analysis of 27 fMRI studies in healthy participants (Spagna et al., 2021), which revealed imagery-associated activity in a specific region of the LH VTC. Spagna et al. proposed to designate this region as the Fusiform Imagery Node (FIN) (Fig. 31.9A).

The FIN is located within the cytoarchitectonic area FG4 (Lorenz et al., 2015) of the LH mid-fusiform gyrus.

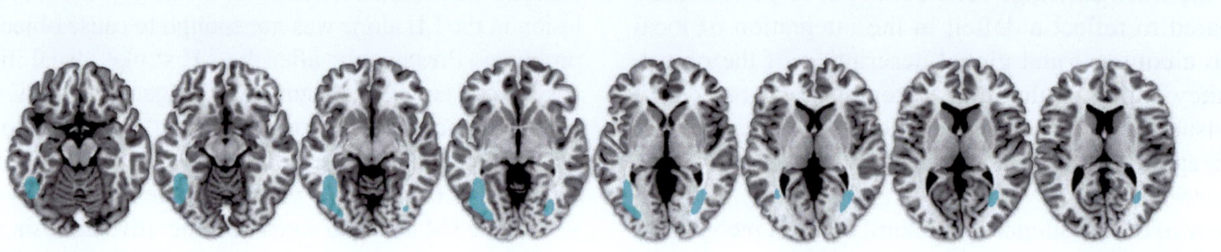

Fig. 31.8. Reconstruction of the lesions of Madame D, who suffered from two consecutive strokes, which occurred first in the LH and then in the RH. Object recognition deficits only emerged after the second (RH) stroke. Image modified from Bartolomeo P, Bachoud-Lévi AC, Thiebaut de Schotten M (2014). The anatomy of cerebral achromatopsia: a reappraisal and comparison of two case reports. Cortex 56: 138–144. https://doi.org/10.1016/j.cortex.2013.01.013, with permission of Elsevier.

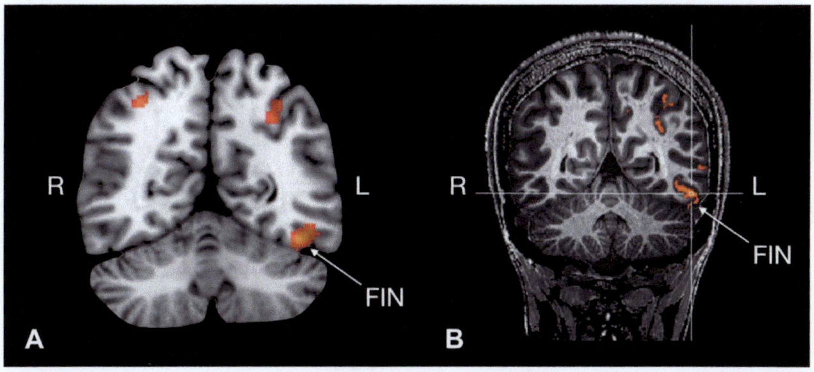

Fig. 31.9. Coronal brain sections showing (A) the results of Spagna et al.'s meta-analysis: brain regions with consistently increased activation associated with the contrast (visual mental imagery > control task), including the FIN in the LH. (B) 7T fMRI results in a typical imager showing the FIN for the contrast (visual mental imagery > control task with abstract words). Voxel size = 1.2 mm, no smoothing, TR = 2 s, $P < 0.001$ uncorrected. Panel (A) reprinted from Spagna A, Hajhajate D, Liu J, et al. (2021). Visual mental imagery engages the left fusiform gyrus, but not the early visual cortex: a meta-analysis of neuroimaging evidence. Neurosci Biobehav Rev 122: 201–217. https://doi.org/10.1016/j.neubiorev.2020.12.029., with Elsevier permission. Panel (B) data from Liu J, Zhan M, Hajhajate D, et al. (2023b). Visual mental imagery in typical imagers and in aphantasia: a millimiter-scale 7-T fMRI study. BioRxiv. https://doi.org/10.1101/2023.06.14.544909, figure courtesy of Jianghao Liu.

However, there are exceptions to the LH preferential engagement in visual mental imagery. First, there is a type of spatial neglect known as imaginal or representational neglect that is preferentially associated with RH damage (Bisiach and Luzzatti, 1978; Bartolomeo et al., 1994). When asked to describe known places from memory, these patients "forget" to mention the left-sided details of their visual mental images. Some patients with LH (Beis et al., 2004; Toba et al., 2022) or bilateral lesions (Weintraub et al., 1996; Bartolomeo et al., 1998; Andrade et al., 2010) may show signs of right-sided *perceptual* neglect; however, the occurrence of right-sided neglect in visual mental imagery appears to be exceedingly rare: only 2 out of the 100 patients with imaginal neglect reported in a recent review (Salvato et al., 2014) displayed signs of right-sided imaginal neglect. A second exception is mental imagery for faces, which might be impaired by bilateral, predominantly RH damage (Etcoff et al., 1991); however, the available evidence is scarce. A case report by Cohen et al. (1992) added to this body of evidence, although the connection to voluntary imagery remains uncertain. The authors described a patient who had a distinct and vivid hallucination of a face, accompanied by a severe headache. MRI revealed a hemorrhage in the anterior two-thirds of the patient's right temporal lobe, whose occurrence was likely to be related to the face hallucination. These exceptions suggest that while LH circuits play an important role in voluntary visual mental imagery for items such as colors, objects, or words (Farah, 1984; Bartolomeo, 2002; Liu et al., 2022), the RH may also contribute, particularly in specific contexts.

On the other hand, a healthy RH may provide the neural substrate for preserved perception in patients with lesion-induced deficits of visual mental imagery. For example, a largely preserved right temporal lobe in the patients described by Moro et al. (2008), who had left temporal damage and impaired visual mental imagery, might have contributed to their normal performance on perceptual tests.

Together with abundant neuroimaging evidence (Spagna et al., 2021; Liu et al., 2022, 2023b), the findings from neurological patients reveal hemispheric asymmetry patterns that contradict the predominant model of visual mental imagery, which emphasizes the involvement of bilateral early visual cortex (Kosslyn, 1994; Pearson, 2019). Instead, evidence suggests that visual mental imagery relies on extensive brain networks including high-level visual circuits, with a notable lateralization pattern in endogenous visual mental imagery, particularly involving the left inferior temporal lobe. The left fusiform gyrus appears to play a crucial role in imagery prompted by words and word-like stimuli such as letters, songs, and national anthems, among others (Liu et al., 2022). In particular, the FIN might act as a bridge between semantic networks and high-level vision (Spagna et al., 2021; Hajhajate et al., 2022; Spagna, 2022). On the other hand, studies using neuroimaging and neurostimulation suggest that the RH inferior temporal lobe is essential for visuospatial aspects of mental imagery, such as mental rotation or navigation (Liu et al., 2022), as well as for face imagery.

To provide a flavor of the patterns of hemispheric lateralization in different domains of visual mental imagery, Fig. 31.10 schematically represents in space the asymmetries reported in different imagery domains in the studies reviewed by Liu et al. (2022). A "voting" procedure established the degree of consistency among studies.

A recent neuroimaging study using ultrahigh field, 7T fMRI (Liu et al., 2023b) further specified the patterns of hemispheric asymmetry in visual mental imagery. Liu et al. asked participants to answer auditorily presented questions pertaining to the visual appearance of items in five domains: object shape, object color, written words, faces, and spatial relationships. A control task minimizing visual mental imagery included questions concerning the meaning of abstract words. The contrast (imagery tasks minus abstract words) demonstrated increased BOLD response in (i) LH frontoparietal areas; (ii) the previously mentioned

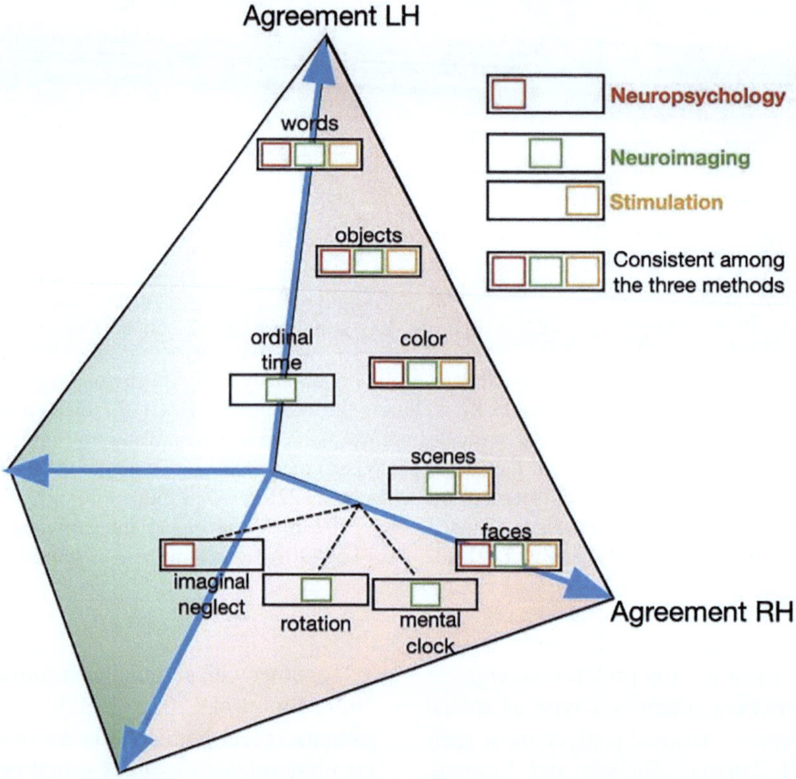

Fig. 31.10. Dimensional spatial embedding of the agreement of laterality patterns of visual mental imagery toward the left hemisphere (LH) or the right hemisphere (RH), across three research approaches: lesion neuropsychology (13 studies, red), neuroimaging (33 studies, green), and direct stimulation (26 studies, orange). The distance along the axes indicates the degree of agreement and was obtained by summing votes from each approach as (left, right) coordinates. A research approach showing clear signs of hemispheric asymmetry yielded one vote. A half vote (0.5) corresponded to evidence of bilateral hemispheric implication; unclear/unknown evidence was given a 0 vote. The closer an item is to the axis, the more hemispheric asymmetry was found (e.g., words for LH). Items showing equal distance from LH and RH axes feature low agreement for lateralization (e.g., color). Figure reproduced from Liu J, Spagna A, Bartolomeo P (2022). Hemispheric asymmetries in visual mental imagery. Brain Struct Funct 227: 697–708. https://doi.org/10.1007/s00429-021-02277-w, with the authorization of Springer Nature.

domain-general FIN in the LH fusiform gyrus; and (iii) the relevant domain-preferring areas in the VTC (fusiform face areas for face imagery, visual word form area for word imagery, etc.). The strong left lateralization of FIN suggests a role as a bridge between semantic and visual information, enabling the building of mental images with visual characteristics from semantic memory. Domain-preferring regions tended to activate bilaterally during visual mental imagery, except for color imagery, which mainly engaged the LH anterior color patch for six participants out of ten typical imagers, the RH patch for one, and bilateral patches for the remaining two. Color imagery activated only the anterior color patch of the three color-biased areas identified by Lafer-Sousa et al. (2016) for color perception.

Liu et al. also examined healthy individuals who claimed not to experience any visual mental images, a condition known as congenital aphantasia (Zeman et al., 2015; de Vito and Bartolomeo, 2016; Keogh et al., 2021). Interestingly, individuals with aphantasia exhibited activation patterns in brain regions that closely resembled those seen in individuals with typical imagery abilities during tasks involving visual mental imagery. This result is consistent with the preserved ability of these individuals to answer questions pertaining to the visual appearance of objects (Liu and Bartolomeo, 2023). However, in aphantasic individuals, the FIN was functionally disconnected from frontoparietal networks. This finding may explain the lack of imagery phenomenology in these individuals: conscious visual experience—whether perceived or imagined—is likely to rely on the coordinated activity of frontoparietal networks and high-level visual cortex (Dehaene et al., 2006; van Vugt et al., 2018; Mashour et al., 2020; Liu et al., 2023a). Interestingly, however, aphantasic individuals exhibited greater activity than typical imagers in the RH inferior frontal gyrus and supramarginal gyrus, which are important nodes of the right-lateralized reorienting network (Corbetta et al., 2008; Bartolomeo and Seidel Malkinson, 2019; Liu et al., 2023a). Such (presumably inappropriate) activity might play a role in disrupting the subjective experience of mental imagery, for example, by interrupting ongoing activity in more dorsal frontoparietal networks (Corbetta et al., 2008).

CONCLUSIONS

Converging evidence from neurological patients and neuroimaging studies in neurotypical individuals suggests that the processes involved in object recognition follow a partly hierarchical pattern.

RH-based circuits seem to hold a relative advantage for perceptual integration, while there is a transition toward LH-based circuits for higher conceptual and semantic levels of processing. These hemispheric asymmetries in visual object processing could, to some extent, be influenced by the general need to minimize wiring (Sterling and Laughlin, 2015) in conjunction with the existence of specialized networks within each hemisphere. Examples of such circuits include the RH inferior parietal-ventrolateral prefrontal network important for swift responses to unexpected external events (Corbetta et al., 2008; Bartolomeo and Seidel Malkinson, 2023; Liu et al., 2023a) and the perisylvian language networks in the LH, crucial for semantic and linguistic processing. The RH lateralization for object perception, however, appears to be relative in nature. It is plausible that unilateral RH lesions lead to object agnosia only when they also affect the functioning of homolog LH regions (Konen et al., 2011).

The predominant LH lateralization of the semantic system (Binder et al., 2009; Fernandino et al., 2022) might also contribute to clarifying the hemispheric asymmetry of VTC circuits during voluntary visual mental imagery, a process that involves transitioning from object meaning to simulated visualization. Patients with lesions directly impacting the imagery-related VTC regions or disconnecting them from semantic networks are likely to experience general impairments in voluntary imagery (Moro et al., 2008; Thorudottir et al., 2020). Conversely, lesions or disconnections specifically targeting domain-preferring regions may lead to domain-specific patterns of mental imagery deficits (Goldenberg, 1993; Bartolomeo et al., 2002). In contrast, more posterior lesions disconnecting an intact VTC from visual input are expected to produce recognition deficits with preserved visual mental imagery (Behrmann et al., 1992; Bartolomeo et al., 2020; Bartolomeo, 2021).

Taking a broader perspective, the evidence discussed in this chapter suggests that the appropriate level of description for visual object processing goes beyond single cortical regions or whole hemispheres, and involves instead interactions in large-scale cortical circuits within and between the hemispheres.

ABBREVIATIONS

7T, seven tesla; BA, Brodmann area; ERPs, event-related brain potentials; FIN, fusiform imagery node; fMRI, functional magnetic resonance imaging; IPS, intra-parietal sulcus; LH, left hemisphere; LOC, lateral occipital complex; MNI, Montreal Neurologic Institute; MRI, magnetic resonance imaging; PET, positron emission tomography; RH, right hemisphere; ROI, region of interest; TMS, transcranial magnetic stimulation; TR, repetition time; VOSP, visual object space perception battery; VTC, ventral temporal cortex.

References

Acres K, Taylor KI, Moss HE et al. (2009). Complementary hemispheric asymmetries in object naming and recognition: a voxel-based correlational study. Neuropsychologia 47: 1836–1843. https://doi.org/10.1016/j.neuropsychologia.2009.02.024.

Alivisatos B, Wilding J (1982). Hemispheric differences in matching Stroop-type letter stimuli. Cortex 18: 5–21. https://doi.org/10.1016/S0010-9452(82)80015-4.

Andrade K, Samri D, Sarazin M et al. (2010). Visual neglect in posterior cortical atrophy. BMC Neurol 10: 68. https://doi.org/10.1186/1471-2377-10-68.

Aviezer H, Landau AN, Robertson LC et al. (2007). Implicit integration in a case of integrative visual agnosia. Neuropsychologia 45: 2066–2077. https://doi.org/10.1016/j.neuropsychologia.2007.01.024.

Ayzenberg V, Behrmann M (2022a). Does the brain's ventral visual pathway compute object shape? Trends Cogn Sci 26: 1119–1132. https://doi.org/10.1016/j.tics.2022.09.019.

Ayzenberg V, Behrmann M (2022b). The dorsal visual pathway represents object-centered spatial relations for object recognition. J Neurosci 42: 4693–4710. https://doi.org/10.1523/JNEUROSCI.2257-21.2022.

Bachoud-Lévi A-C, Bartolomeo P (2003). Mechanisms of pure Alexia: spatially based impairment, letter identification deficit, or both? Neurocase 9: 164–176. https://doi.org/10.1076/neur.9.2.164.15069.

Bartolomeo P (2002). The relationship between visual perception and visual mental imagery: a reappraisal of the neuropsychological evidence. Cortex 38: 357–378. https://doi.org/10.1016/s0010-9452(08)70665-8.

Bartolomeo P (2008). The neural correlates of visual mental imagery: an ongoing debate. Cortex 44: 107–108. https://doi.org/10.1016/j.cortex.2006.07.001.

Bartolomeo P (2021). Visual agnosia and imagery after Lissauer. Brain 144: 2557–2559. https://doi.org/10.1093/brain/awab159.

Bartolomeo P (2022). Visual objects and their colors. In: Handbook of clinical neurology, Elsevier179–189. https://doi.org/10.1016/B978-0-12-823493-8.00022-5.

Bartolomeo P, Miceli G (2023). Definition: object color agnosia. Cortex167: 65. https://doi.org/10.1016/j.cortex.2023.07.002.

Bartolomeo P, Seidel Malkinson T (2019). Hemispheric lateralization of attention processes in the human brain. Curr Opin Psychol 29: 90–96. https://doi.org/10.1016/j.copsyc.2018.12.023.

Bartolomeo P, Seidel Malkinson T (2023). Building models, testing models: asymmetric roles of SLF III networks? Comment on "Left and right temporal-parietal junctions (TPJs) as 'match/mismatch' hedonic machines: a unifying account of TPJ function" by Doricchi et al. Phys Life Rev 44: 70–72. https://doi.org/10.1016/j.plrev.2022.12.004.

Bartolomeo P, D'Erme P, Gainotti G (1994). The relationship between visuospatial and representational neglect. Neurology 44: 1710–1714. https://doi.org/10.1212/wnl.44.9.1710.

Bartolomeo P, Bachoud-Lévi A-C, Denes G (1997). Preserved imagery for colours in a patient with cerebral Achromatopsia. Cortex 33: 369–378. https://doi.org/10.1016/S0010-9452(08)70012-1.

Bartolomeo P, Dalla Barba G, Boissé MF et al. (1998). Right-side neglect in Alzheimer's disease. Neurology 51: 1207–1209.

Bartolomeo P, Bachoud-Lévi A-C, De Gelder B et al. (1998a). Multiple-domain dissociation between impaired visual perception and preserved mental imagery in a patient with bilateral extrastriate lesions. Neuropsychologia 36: 239–249. https://doi.org/10.1016/S0028-3932(97)00103-6.

Bartolomeo P, Bachoud-Lévi A-C, Degos J-D et al. (1998b). Disruption of residual reading capacity in a pure alexic patient after a mirror-image right-hemispheric lesion. Neurology 50: 286–288. https://doi.org/10.1212/WNL.50.1.286.

Bartolomeo P, Bachoud-Levi AC, Chokron S et al. (2002). Visually- and motor-based knowledge of letters: evidence from a pure alexic patient. Neuropsychologia 40: 1363–1371. https://doi.org/10.1016/s0028-3932(01)00209-3.

Bartolomeo P, Bachoud-Lévi AC, Degos JD et al. (2003). Right hemisphere contributions to residual reading in pure alexia: evidence from a patient with subsequent bilateral strokes. In: The parallel brain: cognitive neurosciences of the Corpus callosum, MIT Press, Cambridge MA, pp. 500–506.

Bartolomeo P, Vuilleumier P, Behrmann M (2015). The whole is greater than the sum of the parts: distributed circuits in visual cognition. Cortex 72: 1–4. https://doi.org/10.1016/j.cortex.2015.09.001.

Bartolomeo P, Seidel Malkinson T, de Vito S (2017). Botallo's error, or the quandaries of the universality assumption. Cortex 86: 176–185.

Bartolomeo P, Hajhajate D, Liu J et al. (2020). Assessing the causal role of early visual areas in visual mental imagery. Nat Rev Neurosci 21: 517. https://doi.org/10.1038/s41583-020-0348-5.

Bates E, Wilson SM, Saygin AP et al. (2003). Voxel-based lesion–symptom mapping. Nat Neurosci 6: 448–450. https://doi.org/10.1038/nn1050.

Bauer RM, Rubens AB (1985). Pure word deafness. In: Clinical neuropsychology, second edn. 210–213.

Beauvois M, Broadbent DE, Weiskrantz L (1982). Optic aphasia: a process of interaction between vision and language. Philos Trans R Soc Lond B Biol Sci 298: 35–47. https://doi.org/10.1098/rstb.1982.0070.

Behrmann M, Kimchi R (2003). What does visual agnosia tell us about perceptual organization and its relationship to object perception? J Exp Psychol Hum Percept Perform 29: 19–42. https://doi.org/10.1037/0096-1523.29.1.19.

Behrmann M, Plaut DC (2013). Distributed circuits, not circumscribed centers, mediate visual recognition. Trends Cogn Sci 17: 210–219. https://doi.org/10.1016/j.tics.2013.03.007.

Behrmann M, Winocur G, Moscovitch M (1992). Dissociation between mental imagery and object recognition in a brain-damaged patient. Nature 359: 636–637. https://doi.org/10.1038/359636a0.

Behrmann M, Peterson MA, Moscovitch M et al. (2006). Independent representation of parts and the relations between them: evidence from integrative agnosia. J Exp Psychol Hum Percept Perform 32: 1169–1184. https://doi.org/10.1037/0096-1523.32.5.1169.

Beis J-M, Keller C, Morin N et al. (2004). Right spatial neglect after left hemisphere stroke: qualitative and quantitative study. Neurology 63: 1600–1605. https://doi.org/10.1212/01.wnl.0000142967.60579.32.

Benson DF, Greenberg JP (1969). Visual form agnosia: a specific defect in visual discrimination. Arch Neurol 20: 82–89. https://doi.org/10.1001/archneur.1969.00480070092010.

Biederman I (1987). Recognition-by-components: a theory of human image understanding. Psychol Rev 94: 115–147. https://doi.org/10.1037/0033-295X.94.2.115.

Binder JR, Desai RH, Graves WW et al. (2009). Where is the semantic system? A critical review and meta-analysis of 120 functional neuroimaging studies. Cereb Cortex 19: 2767–2796. https://doi.org/10.1093/cercor/bhp055.

Bisiach E, Luzzatti C (1978). Unilateral neglect of representational space. Cortex 14: 129–133. https://doi.org/10.1016/S0010-9452(78)80016-1.

Boles DB (1984). Global versus local processing: is there a hemispheric dichotomy? Neuropsychologia 22: 445–455. https://doi.org/10.1016/0028-3932(84)90039-3.

Bookheimer S (2002). Functional MRI of language: new approaches to understanding the cortical organization of semantic processing. Annu Rev Neurosci 25: 151–188. https://doi.org/10.1146/annurev.neuro.25.112701.142946.

Cohen L, Verstichel P, Pierrot-Deseilligny C (1992). Hallucinatory vision of a familiar face following right temporal hemorrhage. Neurology 42: 2052. https://doi.org/10.1212/WNL.42.10.2052.

Corbetta M, Patel G, Shulman GL (2008). The reorienting system of the human brain: from environment to theory of mind. Neuron 58: 306–324. https://doi.org/10.1016/j.neuron.2008.04.017.

De Renzi E (1982). Disorders of space exploration and cognition, J. Wiley, Chichester, UK.

De Renzi E, Lucchelli F (1993). The fuzzy boundaries of apperceptive agnosia. Cortex 29: 187–215.

De Renzi E, Saetti MC (1997). Associative agnosia and optic aphasia: qualitative or quantitative difference? Cortex 33: 115–130. https://doi.org/10.1016/S0010-9452(97)80008-1.

De Renzi E, Scotti G, Spinnler H (1969). Perceptual and associative disorders of visual recognition. Relationship to the side of the cerebral lesion. Neurology 19: 634–642.

de Vito S, Bartolomeo P (2016). Refusing to imagine? On the possibility of psychogenic aphantasia. A commentary on Zeman et al. (2015). Cortex 74: 334–335. https://doi.org/10.1016/j.cortex.2015.06.013.

Dehaene S, Changeux J-P, Naccache L et al. (2006). Conscious, preconscious, and subliminal processing: a testable taxonomy. Trends Cogn Sci 10: 204–211. https://doi.org/10.1016/j.tics.2006.03.007.

Delis DC (1988). Visuospatial functioning before and after commissurotomy: disconnection in hierarchical processing. Arch Neurol 45: 462. https://doi.org/10.1001/archneur.1988.00520280116028.

Delis DC, Robertson LC, Efron R (1986). Hemispheric specialization of memory for visual hierarchical stimuli. Neuropsychologia 24: 205–214.

Ehrlichman H, Barrett J (1983). Right hemispheric specialization for mental imagery: a review of the evidence. Brain Cogn 2: 55–76. https://doi.org/10.1016/0278-2626(83)90029-5.

Etcoff NL, Freeman R, Cave KL (1991). Can we lose memories of faces? Content specificity and awareness in a prosopagnosic. J Cogn Neurosci 3: 25–41. https://doi.org/10.1162/jocn.1991.3.1.25.

Farah MJ (1984). The neurological basis of mental imagery: a componential analysis. Cognition 18: 245–272. https://doi.org/10.1016/0010-0277(84)90026-x.

Farah MJ (1990). Visual agnosia: disorders of object recognition and what they tell us about normal vision, MIT Press.

Feinberg TE, Schindler RJ, Ochoa E et al. (1994). Associative visual agnosia and alexia without prosopagnosia. Cortex 30: 395–411.

Fernandino L, Tong J-Q, Conant LL et al. (2022). Decoding the information structure underlying the neural representation of concepts. Proc Natl Acad Sci U S A 119: e2108091119. https://doi.org/10.1073/pnas.2108091119.

Fink GR, Halligan PW, Marshall JC et al. (1996). Where in the brain does visual attention select the forest and the trees? Nature 382: 626–628. https://doi.org/10.1038/382626a0.

Fink GR, Marshall JC, Halligan PW et al. (1997). Hemispheric specialization for global and local processing: the effect of stimulus category. Proc R Soc Lond B Biol Sci 264: 487–494.

Freund CS (1889). Ueber optische Aphasie und Seelenblindheit. Arch Für Psychiatr Nervenkrankh 20: 371–416.

Gilbert CD, Li W (2013). Top-down influences on visual processing. Nat Rev Neurosci 14: 350–363. https://doi.org/10.1038/nrn3476.

Goldenberg G (1993). The neural basis of mental imagery. Baillieres Clin Neurol 2: 265–286.

Goodale MA, Milner AD, Jakobson LS et al. (1991). A neurological dissociation between perceiving objects and grasping them. Nature 349: 154–156. https://doi.org/10.1038/349154a0.

Grill-Spector K (2003). The neural basis of object perception. Curr Opin Neurobiol 13: 159–166. https://doi.org/10.1016/S0959-4388(03)00040-0.

Grill-Spector K, Malach R (2004). The human visual cortex. Annu Rev Neurosci 27: 649–677. https://doi.org/10.1146/annurev.neuro.27.070203.144220.

Grill-Spector K, Weiner KS (2014). The functional architecture of the ventral temporal cortex and its role in categorization. Nat Rev Neurosci 15: 536–548. https://doi.org/10.1038/nrn3747.

Grill-Spector K, Kushnir T, Edelman S et al. (1999). Differential processing of objects under various viewing conditions in the human lateral occipital complex. Neuron 24: 187–203. https://doi.org/10.1016/S0896-6273(00)80832-6.

Grill-Spector K, Kourtzi Z, Kanwisher N (2001). The lateral occipital complex and its role in object recognition. Vision Res 41: 1409–1422.

Hajhajate D, Kaufmann BC, Liu J et al. (2022). The connectional anatomy of visual mental imagery: evidence from a patient with left occipito-temporal damage. Brain Struct Funct 227: 3075–3083. https://doi.org/10.1007/s00429-022-02505-x.

Hahn E (1895). Pathologisch-anatomische Untersuchung des Lissauer'schen Falles von Seelenblindheit. Arbeiten aus der Psychiatrischen Klinik in Breslau II: 105–129.

Humphreys GW (1999). Integrative agnosia, in: case studies in the neuropsychology of vision, Psychology Press/Taylor & Francis (UK), Hove, England, pp. 41–58.

Humphreys GW, Riddoch MJ (1984). Routes to object constancy: implications from neurological impairments of object constancy. Q J Exp Psychol A 36: 385–415.

Iidaka T, Yamashita K, Kashikura K et al. (2004). Spatial frequency of visual image modulates neural responses in the temporo-occipital lobe. An investigation with event-related fMRI. Cogn Brain Res 18: 196–204. https://doi.org/10.1016/j.cogbrainres.2003.10.005.

James TW, Culham J, Humphrey GK et al. (2003). Ventral occipital lesions impair object recognition but not object-directed grasping: an fMRI study. Brain 126: 2463–2475. https://doi.org/10.1093/brain/awg248.

Karnath H-O, Rüter J, Mandler A et al. (2009). The anatomy of object recognition—visual form agnosia caused by medial occipitotemporal stroke. J Neurosci 29: 5854–5862. https://doi.org/10.1523/JNEUROSCI.5192-08.2009.

Keogh R, Pearson J, Zeman A (2021). Aphantasia: the science of visual imagery extremes. In: Handbook of clinical neurology, Elsevier, 277–296. https://doi.org/10.1016/B978-0-12-821377-3.00012-X.

Kimchi R (1992). Primacy of wholistic processing and global/local paradigm: a critical review. Psychol Bull 112: 24–38.

Kimchi R, Merhav I (1991). Hemispheric processing of global form, local form, and texture. Acta Psychol (Amst) 76: 133–147. https://doi.org/10.1016/0001-6918(91)90042-X.

Konen CS, Behrmann M, Nishimura M et al. (2011). The functional neuroanatomy of object agnosia: a case study. Neuron 71: 49–60. https://doi.org/10.1016/j.neuron.2011.05.030.

Kosslyn SM (1994). Image and brain: the resolution of the imagery debate, The MIT Press, Cambridge, Mass.

Kosslyn SM, Behrmann M, Jeannerod M (1995). The cognitive neuroscience of mental imagery. Neuropsychologia 33: 1335–1344. https://doi.org/10.1016/0028-3932(95)00067-d.

Kourtzi Z, Kanwisher N (2001). Representation of perceived object shape by the human lateral occipital complex. Science 293: 1506–1509. https://doi.org/10.1126/science.1061133.

Kravitz DJ, Saleem KS, Baker CI et al. (2011). A new neural framework for visuospatial processing. Nat Rev Neurosci 12: 217–230. https://doi.org/10.1038/nrn3008.

Kravitz DJ, Saleem KS, Baker CI et al. (2013). The ventral visual pathway: an expanded neural framework for the processing of object quality. Trends Cogn Sci 17: 26–49. https://doi.org/10.1016/j.tics.2012.10.011.

Kurachi M, Yamaguchi N, Inasaka T et al. (1979). Recovery from alexia without agraphia: report of an autopsy. Cortex 15: 297–312. https://doi.org/10.1016/S0010-9452(79)80033-7.

Kwon M, Lee J-H (2006). Optic aphasia: a case study. J Clin Neurol 2: 258–261. https://doi.org/10.3988/jcn.2006.2.4.258.

Lafer-Sousa R, Conway BR, Kanwisher NG (2016). Color-biased regions of the ventral visual pathway lie between face- and place-selective regions in humans, as in macaques. J Neurosci 36: 1682–1697. https://doi.org/10.1523/JNEUROSCI.3164-15.2016.

Lamb MR, Robertson LC (1989). Do response time advantage and interference reflect the order of processing of global- and local-level information? Percept Psychophys 46: 254–258.

Lamb MR, Robertson LC, Knight RT (1990). Component mechanisms underlying the processing of hierarchically organized patterns: inferences from patients with unilateral cortical lesions. J Exp Psychol Learn Mem Cogn 16: 471–483.

Lissauer H (1890). Ein Fall von Seelenblindheit nebst einem Beitrage zur Theorie derselben. Arch Für Psychiatr Nervenkrankh 21: 222–270. https://doi.org/10.1007/BF02226765.

Lissauer H (1988). A case of visual agnosia with a contribution to theory. Cogn Neuropsychol 5: 157–192. https://doi.org/10.1080/02643298808252932.

Liu J, Bartolomeo P (2023). Probing the unimaginable: the impact of aphantasia on distinct domains of visual mental imagery and visual perception. Cortex 166: 338–347. https://doi.org/10.1016/j.cortex.2023.06.003.

Liu J, Spagna A, Bartolomeo P (2022). Hemispheric asymmetries in visual mental imagery. Brain Struct Funct 227: 697–708. https://doi.org/10.1007/s00429-021-02277-w.

Liu J, Bayle DJ, Spagna A et al. (2023a). Fronto parietal networks shape human conscious report through attention gain and reorienting. Commun Biol. https://doi.org/10.1101/2022.04.10.487690.

Liu J, Zhan M, Hajhajate D et al. (2023b). Visual mental imagery in typical imagers and in aphantasia: a millimiter-scale 7-T fMRI study. BioRxiv. https://doi.org/10.1101/2023.06.14.544909.

Lorenz S, Weiner KS, Caspers J et al. (2015). Two new cytoarchitectonic areas on the human mid-fusiform gyrus. Cereb Cortex. https://doi.org/10.1093/cercor/bhv225.

Mahon BZ, Caramazza A (2011). What drives the organization of object knowledge in the brain? Trends Cogn Sci 15: 97–103. https://doi.org/10.1016/j.tics.2011.01.004.

Malach R, Reppas JB, Benson RR et al. (1995). Object-related activity revealed by functional magnetic resonance imaging in human occipital cortex. Proc Natl Acad Sci U S A 92: 8135–8139. https://doi.org/10.1073/pnas.92.18.8135.

Martin M (1979). Hemispheric specialization for local and global processing. Neuropsychologia 17: 33–40.

Martinez A, Moses P, Frank L et al. (1997). Hemispneric asymmetries in global and local processing: evidence from fMRI. Neuroreport 8: 1685.

Mashour GA, Roelfsema P, Changeux J-P et al. (2020). Conscious processing and the global neuronal workspace hypothesis. Neuron 105: 776–798. https://doi.org/10.1016/j.neuron.2020.01.026.

McIntosh RD, McClements KI, Schindler I et al. (2004). Avoidance of obstacles in the absence of visual awareness. Proc Biol Sci 271: 15–20. https://doi.org/10.1098/rspb.2003.2545.

Mishkin M, Ungerleider LG, Macko KA (1983). Object vision and spatial vision: two cortical pathways. Trends Neurosci 6: 414–417. https://doi.org/10.1016/0166-2236(83)90190-X.

Moro V, Berlucchi G, Lerch J et al. (2008). Selective deficit of mental visual imagery with intact primary visual cortex and visual perception. Cortex 44: 109–118. https://doi.org/10.1016/j.cortex.2006.06.004.

Navon D (1977). Forest before trees: the precedence of global features in visual perception. Cogn Psychol 9: 353–383.

Pascual-Marqui RD, Michel CM, Lehmann D (1994). Low resolution electromagnetic tomography: a new method for localizing electrical activity in the brain. Int J Psychophysiol 18: 49–65. https://doi.org/10.1016/0167-8760(84)90014-X.

Pearson J (2019). The human imagination: the cognitive neuroscience of visual mental imagery. Nat Rev Neurosci 20: 624–634. https://doi.org/10.1038/s41583-019-0202-9.

Peyrin C, Baciu M, Segebarth C et al. (2004). Cerebral regions and hemispheric specialization for processing spatial frequencies during natural scene recognition. An event-related fMRI study. Neuroimage 23: 698–707. https://doi.org/10.1016/j.neuroimage.2004.06.020.

Poeck K (1984). Neuropsychological demonstration of splenial interhemispheric disconnection in a case of "optic anomia". Neuropsychologia 22: 707–713. https://doi.org/10.1016/0028-3932(84)90097-6.

Price CJ, Moore CJ, Humphreys GW et al. (1996). The neural regions sustaining object recognition and naming. Proc R Soc Lond B Biol Sci 263: 1501–1507. https://doi.org/10.1098/rspb.1996.0219.

Proverbio AM, Zani A (2021). Hemispheric asymmetry in visual processing: an ERP study on spatial frequency gratings. Symmetry 13: 180. https://doi.org/10.3390/sym13020180.

Ptak R, Lazeyras F (2019). Functional connectivity and the failure to retrieve meaning from shape in visual object agnosia. Brain Cogn 131: 94–101. https://doi.org/10.1016/j.bandc.2018.12.007.

Riddoch MJ, Humphreys GW (1987). A case of intergrative visual agnosia. Brain 110: 1431–1462. https://doi.org/10.1093/brain/110.6.1431.

Robertson LC, Lamb MR, Knight RT (1988). Effects of lesions of temporal-parietal junction on perceptual and attentional processing in humans. J Neurosci Off J Soc Neurosci 8: 3757–3769.

Rubens AB, Benson DF (1971). Associative visual agnosia. Arch Neurol 24: 305–316. https://doi.org/10.1001/archneur.1971.00480340037003.

Salvato G, Sedda A, Bottini G (2014). In search of the disappeared half of it: 35 years of studies on representational neglect. Neuropsychology 28: 706–716. https://doi.org/10.1037/neu0000062.

Schintu S, Hadj-Bouziane F, Dal Monte O et al. (2014). Object and space perception—is it a matter of hemisphere? Cortex 57: 244–253. https://doi.org/10.1016/j.cortex.2014.04.009.

Schnider A, Benson DF, Scharre DW (1994). Visual agnosia and optic aphasia: are they anatomically distinct? Cortex 30: 445–457. https://doi.org/10.1016/S0010-9452(13)80340-1.

Sergent J (1982). The cerebral balance of power: confrontation or cooperation? J Exp Psychol Hum Percept Perform 8: 253.

Sergent J, Ohta S, Macdonald B (1992). Functional neuroanatomy of face and object processing a positron emission tomography study. Brain 115: 15–36.

Serino A, Cecere R, Dundon N et al. (2014). When apperceptive agnosia is explained by a deficit of primary visual processing. Cortex 52: 12–27. https://doi.org/10.1016/j.cortex.2013.12.011.

Shomstein S, Behrmann M (2006). Cortical systems mediating visual attention to both objects and spatial locations. Proc Natl Acad Sci U S A 103: 11387–11392. https://doi.org/10.1073/pnas.0601813103.

Siuda-Krzywicka K, Witzel C, Chabani E et al. (2019). Color categorization independent of color naming. Cell Rep 28: 2471–2479. https://doi.org/10.1016/j.celrep.2019.08.003.

Spagna A (2022). Chapter 8—Visual mental imagery: inside the mind's eyes. In: G Miceli, P Bartolomeo, V Navarro (Eds.), Handbook of clinical neurology, the temporal lobe. Elsevier, pp. 145–160. https://doi.org/10.1016/B978-0-12-823493-8.00010-9.

Spagna A, Hajhajate D, Liu J et al. (2021). Visual mental imagery engages the left fusiform gyrus, but not the early visual cortex: a meta-analysis of neuroimaging evidence. Neurosci Biobehav Rev 122: 201–217. https://doi.org/10.1016/j.neubiorev.2020.12.029.

Sterling P, Laughlin S (2015). Principles of neural design, MIT Press, Cambridge, MA.

Stewart L, Meyer B-U, Frith U et al. (2001). Left posterior BA37 is involved in object recognition: a TMS study. Neuropsychologia 39: 1–6. https://doi.org/10.1016/S0028-3932(00)00084-1.

Thompson-Schill SL (2003). Neuroimaging studies of semantic memory: inferring "how" from "where.". Neuropsychologia 41: 280–292. https://doi.org/10.1016/S0028-3932(02)00161-6.

Thorudottir S, Sigurdardottir HM, Rice GE et al. (2020). The architect who lost the ability to imagine: the cerebral basis of visual imagery. Brain Sci 10: 59. https://doi.org/10.3390/brainsci10020059.

Toba MN, Migliaccio R, Potet A et al. (2022). Right-side spatial neglect and white matter disconnection after left-hemisphere strokes. Brain Struct Funct 227: 2991–3000. https://doi.org/10.1007/s00429-022-02541-7.

Ungerleider LG, Haxby JV (1994). 'What' and 'where' in the human brain. Curr Opin Neurobiol 4: 157–165. https://doi.org/10.1016/0959-4388(94)90066-3.

Ungerleider LG, Mishkin M (1982). Analysis of visual behaviour. In: DJ Ingle, MA Goodale, RJW Mansfield (Eds.), Two cortical visual systems. MIT Press, Cambridge.

Van Essen DC (2004). Organization of visual areas in macaque and human cerebral cortex. In: L Chalupa, L Werner (Eds.), The visual neurosciences. MIT Press507–521.

Van Kleeck MH (1989). Hemispheric differences in global versus local processing of hierarchical visual stimuli by normal subjects: new data and a meta-analysis of previous studies. Neuropsychologia 27: 1165–1178.

Van Vleet TM, Hoang-duc AK, DeGutis J et al. (2011). Modulation of non-spatial attention and the global/local processing bias. Neuropsychologia 49: 352–359. https://doi.org/10.1016/j.neuropsychologia.2010.11.021.

van Vugt B, Dagnino B, Vartak D et al. (2018). The threshold for conscious report: signal loss and response bias in visual and

frontal cortex. Science 360: 537–542. https://doi.org/10.1126/science.aar7186.
Vandenberghe R, Price C, Wise R et al. (1996). Functional anatomy of a common semantic system for words and pictures. Nature 383: 254–256. https://doi.org/10.1038/383254a0.
Warrington EK, James M (1991). The visual object and space perception battery, Bury St Edmunds Thames Val. Test Co4.
Warrington EK, Taylor AM (1978). Two categorical stages of object recognition. Perception 7: 695–705.
Weintraub S, Daffner KR, Ahern GL et al. (1996). Right sided hemispatial neglect and bilateral cerebral lesions. J Neurol Neurosurg Psychiatry 60: 342–344. https://doi.org/10.1136/jnnp.60.3.342.
Xu J, Gannon PJ, Emmorey K et al. (2009). Symbolic gestures and spoken language are processed by a common neural system. Proc Natl Acad Sci U S A 106: 20664–20669.
Yamaguchi S, Yamagata S, Kobayashi S (2000). Cerebral asymmetry of the top-down allocation of attention to global and local features. J Neurosci 20: RC72.
Zeman A, Dewar M, Della Sala S (2015). Lives without imagery—congenital aphantasia. Cortex 73: 378–380. https://doi.org/10.1016/j.cortex.2015.05.019.

Index

NB: Page numbers in *italics* refer to figures and tables.

I

K

L

M

T

U

V

W